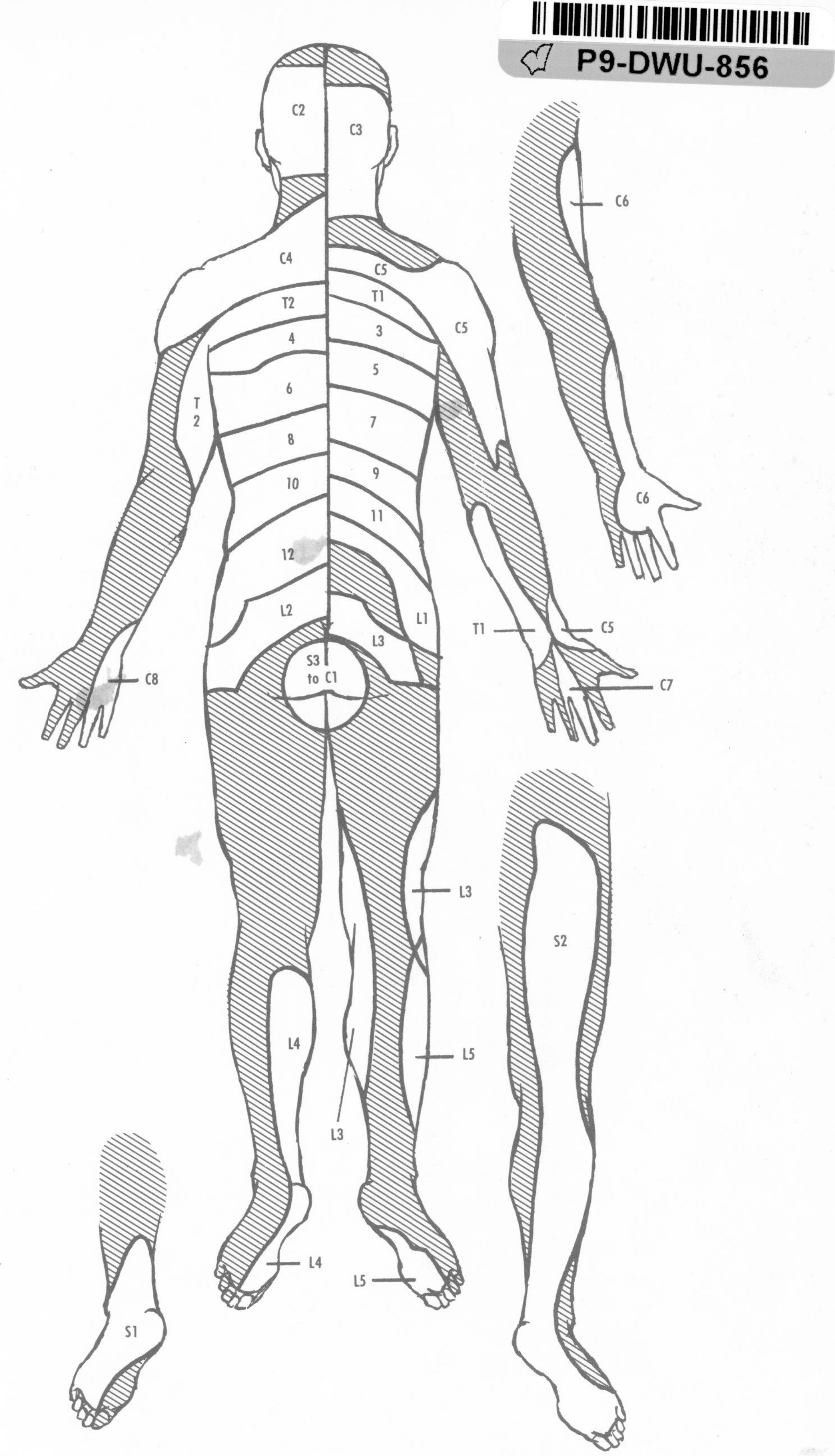
C2
C3
C6
C4
C5
T2
T1
C5
4
3
6
5
T
2
7
8
9
10
11
C6
12
L2
L1
T1
C5
L3
S3
to C1
C8
C7
L3
S2
L4
L5
L3
L4
L5
S1

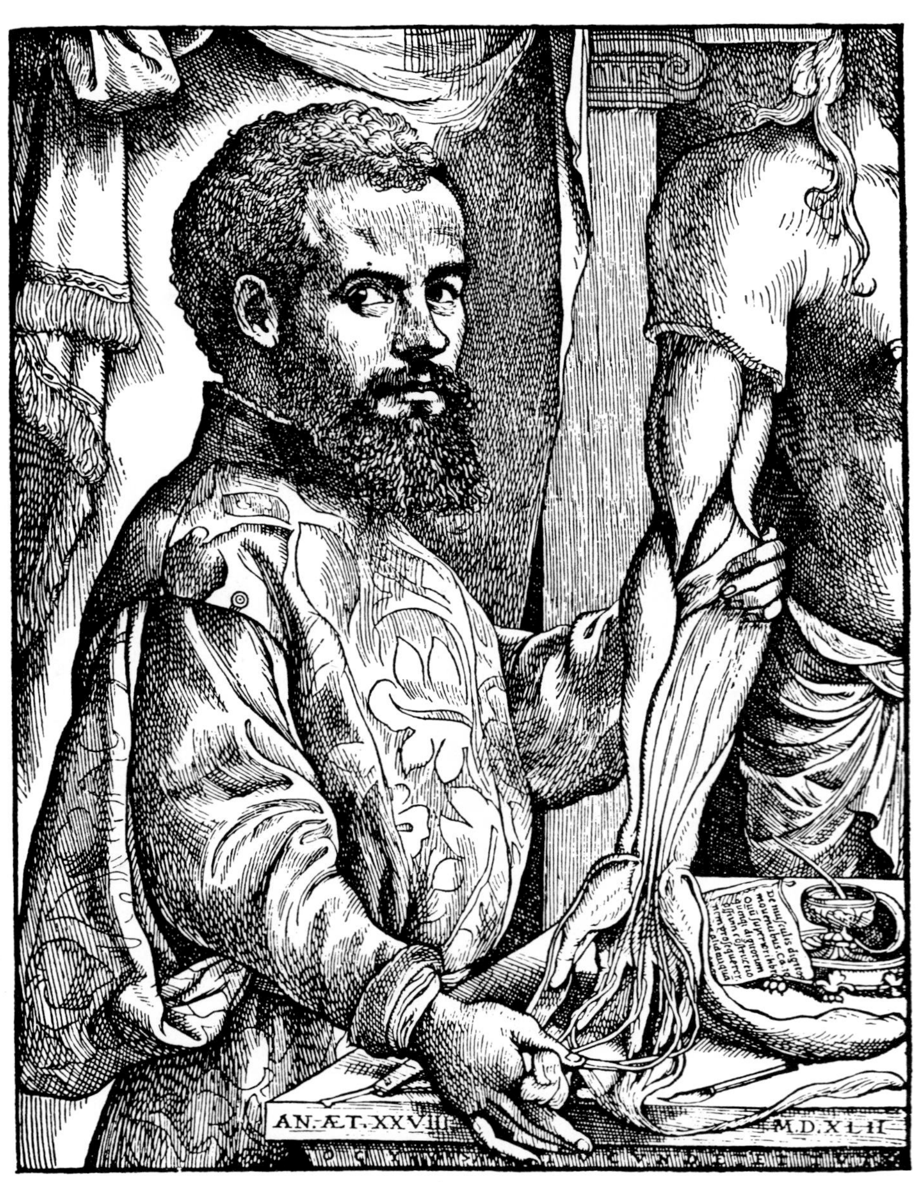

ANDREAS VESALIUS (1514–1564)

A portrait of Vesalius, whose *De humani corporis fabrica* provided the basis for modern anatomy. This portrait from the *Fabrica* is reproduced from the frontispiece of *The Illustrations from the Works of Andreas Vesalius of Brussels,* by J. B. de C. M. Saunders and C. D. O'Malley, World Publishing Co., 1950.

ERNEST GARDNER, M.D.
University of California at Davis

DONALD J. GRAY, M.S., PH.D.
Stanford University

RONAN O'RAHILLY, M.SC., M.D.
University of California at Davis

ILLUSTRATIONS BY CASPAR HENSELMANN

ANATOMY

A Regional Study of Human Structure

FOURTH EDITION

W. B. SAUNDERS COMPANY
Philadelphia, London, Toronto

W. B. Saunders Company: West Washington Square
Philadelphia, PA 19105

1 St. Anne's Road
Eastbourne, East Sussex BN21 3UN, England

1 Goldthorne Avenue
Toronto, Ontario M8Z 5T9, Canada

Library of Congress Cataloging in Publication Data

Gardner, Ernest Dean, 1915–

Anatomy: a regional study of human structure.

Includes index.

1. Anatomy, Human. I. Gray, Donald James, 1908– joint author. II. O'Rahilly, Ronan, joint author. III. Title. [DNLM: 1. Anatomy, Regional. QS4 G226a]

QM23.2.G36 1975 611 74–17753

ISBN 0–7216–4018–4

Listed here is the latest translated edition of this book together with the language of the translation and the publisher.

Portuguese (*4th Edition*) – Editora Guanabara Koogan,
Rio de Janeiro, Brazil

Spanish (*4th Edition*) – Salvat Editores, S.A.,
Barcelona, Spain

French (*4th Edition*) – Doin Editeurs S.A.
Paris, France

Anatomy ISBN 0-7216-4018-4

Last digit is the print number: 9

PREFACE TO THE FOURTH EDITION

The major aims of the present work continue to be (1) to provide a textbook for the undergraduate medical and dental student, (2) to provide information on living anatomy and to stress the importance of the relationship between structure and function, and (3), particularly by the citation of selected references, to meet the needs of the more advanced student and the postgraduate worker.

After certain preliminary matters, mostly of a systemic nature, have been considered in a series of introductory chapters, the larger portion of the book follows a regional approach. The regional plan has been adopted chiefly because most laboratory courses in human anatomy are based on regional dissection. Within any given region, however, a rigorous pursuit of the regional method, to the exclusion of systemic considerations, has not been attempted, because the present work is neither a laboratory manual nor a textbook of surgical anatomy. Thus, the position of the present book may be described as one of "moderate regionalism." The order in which Parts Two to Eight are studied can be varied to suit the needs of any given dissection schedule.

The special fields of neuroanatomy, histology, and embryology are dealt with in special textbooks, and usually in separate courses, and have therefore been omitted. It should be stressed that the brief account of the anatomy of the brain included with the Head and Neck is purely to aid the student of gross anatomy who has not yet studied neuroanatomy in detail.

The inclusion of recent references serves to emphasize that anatomy is a living discipline in which research plays an active and significant role. The abbreviations used for journals are based on the *World List of Scientific Periodicals,* third edition, 1952.

The terminology used is the *Nomina anatomica,* third edition, 1966, translated into English where applicable. The *Nomina anatomica* is based largely on the B.N.A., and, in the matter of translations, those of the Birmingham Revision (1933) have generally been followed in the present text.

In the text and in the illustrations certain abbreviations are frequently used. These include: C, cervical; Co, coccygeal; L, lumbar; N, nerve; S, sacral; T, thoracic; V, vertebra. The above abbreviations have usually been combined; for example, T.V., thoracic vertebra. Abbreviations appearing only in the illustrations include: a., artery; br., branch; g., ganglion; gld, gland; lig., ligament; m., muscle; n., nerve; plx, plexus; tr., trunk; v., vein.

In the present edition, the existing textual material and illustrations have been revised, a number of new photographs and drawings have been added, and new references have been included. Nevertheless, the total number of pages has scarcely been altered. Important concepts have been printed in bold-face type, not only to aid the beginner in learning basic information but also, by stressing items of clinical importance, to assist the student who is reviewing the subject.

A concluding section, Part Nine, has been added, in which the blood and nerve supply of certain parts of the body is summarized, thereby emphasizing the continuity between one region and another. Finally, a Table of Measurements has been appended, in which linear measurements and weights (which have been largely removed from the general textual material) of various organs and structures are listed.

The authors wish to thank many individuals, both students and teachers, for their comments and suggestions. It is again a great pleasure to acknowledge the constant encouragement and cheerful assistance of so many from the W. B. Saunders Company.

The authors will appreciate having their attention called to typographical errors as well as to errors of fact that may come to the attention of the reader. They will also appreciate having their attention directed to relevant information that may have been missed in the preparation of this book.

THE AUTHORS

CONTENTS

Part Four **THE THORAX**

Ernest Gardner, Donald J. Gray, and Ronan O'Rahilly

Part Five **THE ABDOMEN**

Ernest Gardner

Part One

GENERAL ANATOMY

INTRODUCTION 1

Ronan O'Rahilly

ANATOMY AND ITS SUBDIVISIONS

Anatomy is the science of the structure of the body. When used without qualification, the term is applied usually to human anatomy. The word is derived indirectly from the Greek *anatome,* a term built from *ana,* meaning up, and *tome*, meaning a cutting (compare the words tome, microtome, and epitome). From an etymological point of view, the term "dissection" (*dis-*, meaning asunder, and *secare*, meaning to cut) is the Latin equivalent of the Greek *anatome.*

Anatomy, wrote Vesalius in the preface to his *De Fabrica* (1543), "should rightly be regarded as the firm foundation of the whole art of medicine and its essential preliminary." Moreover, a point frequently lost to sight is that the study of anatomy introduces the medical student to the greater part of medical terminology.

Anatomy "is to physiology as geography is to history" (Fernel); that is, it provides the setting for the events. Although the primary concern of anatomy is with structure, structure and function should always be considered together. Moreover, by means of *surface* and *radiological anatomy,* emphasis should be placed on the anatomy of the living body.* As one writer expressed it, "I cannot put before you too strongly the value and interest of this rather neglected [surface] aspect of anatomy. Many a student first realizes its importance only when brought to the bedside or the operating table of his patient, when the first thing he is faced with is the last and least he has considered."[1] The classical methods of physical examination of the body and the use of some of the various "-scopes" (Gk. *skopos*, watcher), e.g., the stethoscope and the ophthalmoscope, should be included in a course in anatomy. Radiological studies facilitate achievement of "an understanding of the fluid character of anatomy and physiology of the living" (A. E. Barclay), and the importance of variation should be kept constantly in mind.

In relation to the size of the parts studied, anatomy is usually divided into (1) *macroscopic* or *gross anatomy,* and (2) *microscopic anatomy* or *histology.* The terms microscopic anatomy and histology are now used synonymously in English. The word histology is derived from the Greek words *histos*, meaning web or tissue, and *logos*, meaning a branch of knowledge. Although the present work is concerned chiefly with macroscopic anatomy, the two aspects, gross and microscopic, should be studied in close conjunction.

The body should be considered not only in its definitive form, but also developmentally. *Embryology* is the study of the embryo and the fetus, that is, the study of prenatal development. Development continues after birth, however, so that the term *developmental anatomy* may be used to include both prenatal and postnatal development. *Pediatric anatomy* is the study of the structure of the child. The study of congenital malformations is known as *teratology.*

In general, works dealing with human anatomy are arranged either (1) *systemically,* that is, according to the various systems* of the body (skeletal, muscular, digestive, etc.) or (2) *regionally,* that is, according to the natural, main subdivisions of the body (head and neck, upper limb, thorax, etc.). In this book, after the general features of the systems are dealt with in a series of introductory chapters, the remainder of the work, in general, follows a regional approach. The regional plan has been adopted chiefly because the vast majority of laboratory courses in human anatomy are based on regional dissection. Anatomy considered on a regional basis is frequently termed *topographical anatomy.*

---

*Also to be included in "living anatomy" are *endoscopic anatomy* (using various "-scopes," p. 7) and *scintigraphic anatomy* (the delineation of organs by radio-isotope scanning), together with some other specialized varieties (such as ultrasonic and thermographic delineations).

*The term *system* is used for a group of parts and organs that subserve a common function.

ANATOMICAL TERMINOLOGY

One of the following etymological works should be consulted frequently:

E. J. Field and R. J. Harrison, *Anatomical Terms: Their Origin and Derivation,* Heffer, Cambridge, 3rd ed., 1968.

H. A. Skinner, *The Origin of Medical Terms,* Williams & Wilkins, Baltimore, 2nd ed., 1961.

It has been estimated that, toward the end of the 19th century, about 50,000 anatomical names were in use for some 5000 structures in the human body. By 1895, however, a list of about 4500 terms had been prepared and was accepted at Basle. This system of nomenclature is known as the *Basle Nomina anatomica* (abbreviated B.N.A.) and is in Latin. It was foreseen that revisions would be required at a later date, and these were undertaken chiefly in Britain (*Birmingham Revision,* or B.R., 1933)[2] and in Germany (*Jena Nomina anatomica,* or J.N.A., 1935). The former terminology, prepared for and accepted by the Anatomical Society of Great Britain and Ireland, is still the best source of translations of Latin terms into English. In Paris in 1955, international agreement was reached on a Latin system of nomenclature based largely on the B.N.A. A revision of this *Nomina anatomica,*[3] translated into English where applicable, is used throughout this book, although synonyms are given frequently. Among the principles adopted in preparing the new N.A., the following are noteworthy: (1) That, with a very limited number of exceptions, each structure shall be designated by one term only. (2) That every term in the official list shall be in Latin, each country, however, to be at liberty to translate the official Latin terms into its own vernacular for teaching purposes. (3) That the terms shall be primarily memory signs, but shall preferably have some informative or descriptive value. (4) That eponyms shall not be used.[4]

In scientific parlance the word *eponym* is used for a term or phrase formed from the name of a person. Examples of eponyms are: ampere, ohm, volt(a), roentgenology, wistaria (named after C. Wistar, 1761–1818, American anatomist). Eponyms in anatomy should be avoided. They are used quite haphazardly, they give no clue as to the type of structure involved, and they are frequently misleading historically because, in many instances, the person commemorated was by no means the first to describe the structure. Poupart, for example, was not the first to note the inguinal ligament.

TERMS OF POSITION AND DIRECTION (fig. 1–1)

All descriptions in human anatomy are expressed in relation to the *anatomical position,* a convention whereby the body is erect, with the head, eyes, and toes directed forward, and with the upper limbs by the side and held so that the palms of the hands face forward. There is no implication that the anatomical position is one of rest. It is often necessary, however, to describe the position of the viscera also in the recumbent posture, because this is a posture in which patients are frequently examined clinically.

The *median plane* is an imaginary vertical plane of section that passes longitudinally through the body and divides it into right and left halves. The median plane intersects the surface of the front and the back of the body at what are called the anterior and posterior median lines. It is a common error, however, to refer to the "midline" of the body when the median plane is meant.

Any vertical plane through the body that is parallel with the median plane is

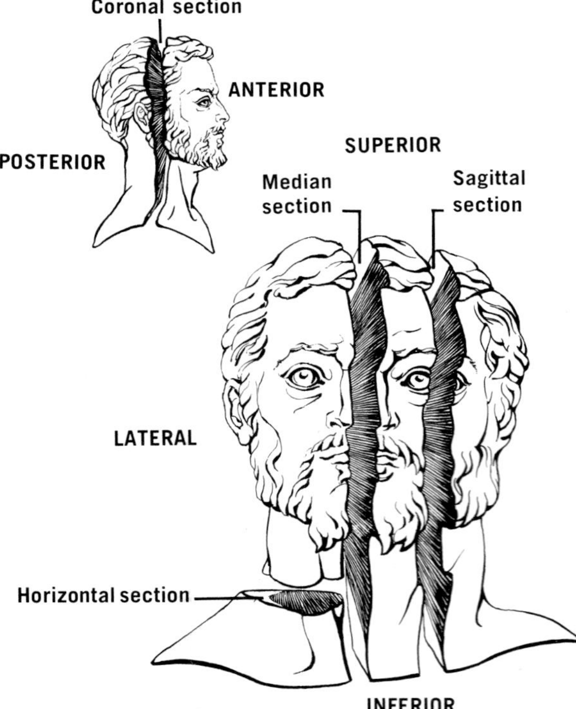

Figure 1–1 Diagrams illustrating the chief terms of position and the main planes of reference in the body.

called a *sagittal plane.* The sagittal planes are named after the sagittal suture of the skull,* to which they are parallel. In this connection it should be pointed out that the term "parasagittal" is redundant. Anything parallel with a sagittal plane is still sagittal.

Any vertical plane that intersects the median plane at a right angle and separates the body into front and back parts is termed a *frontal* or *coronal plane.*†

The term *horizontal plane* refers to a plane at a right angle to both the median and coronal planes; it separates the body into upper and lower parts. Other planes (such as transpyloric) that are employed only with reference to the trunk will be explained in the appropriate sections of the book.

It should be noted that the term "transverse" generally has the ordinary meaning of "lying or being across," that is, merely at a right angle to the longitudinal axis of a structure. Thus, a transverse section through an artery is not necessarily horizontal. A transverse section through the hand is horizontal, whereas a transverse section through the foot is coronal.

The term *medial* means nearer to the median plane, and *lateral* means farther from it. Thus, in the anatomical position, the thumb is lateral to the little finger, whereas the big toe is medial to the little toe. *Intermediate* means lying between two structures, one of which is medial and the other lateral. In the upper limb *radial* means lateral and *ulnar* means medial; in the lower limb *fibular* or *peroneal* means lateral and *tibial* means medial. The border of a limb on which either the thumb or the big toe is situated is sometimes called *pre-axial,* and the opposite border, *postaxial.* These two terms are based on the arrangement of the limbs in the embryo during the sixth postovulatory week, when the thumbs and the big toes are both on the cephalic border of the limbs.

Anterior or *ventral* means nearer the front of the body. *Posterior* or *dorsal* means nearer the back. In the case of certain parts (e.g., tongue, penis, foot), the term dorsal is given a special meaning based on comparative anatomy (posterosuperior, anterior, and superior, respectively, in the examples cited). In the upper limb the term *palmar* (known formerly as *volar*) means anterior. In the foot, *plantar* means inferior.

*L. *sagitta,* arrow.

†From L. *frons,* forehead, and from the coronal suture of the skull (L. *corona,* crown), respectively.

Superior means nearer the top or upper end of the body. *Inferior* means nearer the lower end. In the trunk, *cranial* or *cephalic* is sometimes used instead of superior, and *caudal* instead of inferior; *rostral* means nearer the "front end," which is taken to be the hypophysial area in the early embryo and the region of the nose and mouth in post-embryonic life.* In the limbs, *proximal* and *distal* are used to indicate nearer to and farther from the root or attached end of a limb, respectively.†

Internal and *external* mean respectively nearer to and farther from the center of an organ or a cavity. *Superficial* and *deep* mean nearer to and farther from the surface of the body, respectively.

The term *middle* is used for a structure lying between two others that are anterior and posterior, or superior and inferior, or internal and external.

In addition to the technical terms of position and direction, certain common expressions are also used in anatomical descriptions: front, back, in front of, behind, forward, backward, upper, lower, above, below, upward, downward, ascending, descending. These terms are free of ambiguity provided they are used only in reference to the anatomical position. A number of other common terms, such as "under," however, are generally best avoided.

HISTORY OF ANATOMY

Lack of space precludes an adequate treatment here of the history of anatomy. All that can be attempted is to present in a chronological scheme some of the chief events and personages up to the 19th century.[5] The 20th century is not included because of the increasing rapidity and ever-widening scope of advances in anatomy, and because a balanced perspective of events can be attained only after the completion of the century.

*Gk. *kranion,* skull; Gk. *kephale,* head; L. *cauda,* tail; L. *rostrum,* beak. The suffix "-ad" is sometimes added to a positional term to indicate the idea of motion. Thus, cephalad means proceeding toward the head. Such terms are useful occasionally in describing growth processes, but their application is best limited.

†These terms, however, are used in a special sense in regard to the teeth, as is explained in the chapter on the mouth.

Greek Anatomy, B.C.

Greek anatomy had its origins in Egypt.

Alcmaeon of Croton (*ca.* 500 B.C.) provided the earliest records of actual (animal) anatomical observations.

Hippocrates of Cos (*ca.* 400 B.C.) is regarded as one of the founders of the science of anatomy.

Human surface anatomy was studied for works of Greek art of the 5th century B.C.

"On Anatomy" (from Hippocratic collection, middle of 4th century B.C.) was perhaps the earliest treatise on anatomy.

"On the Heart" (from Hippocratic collection, *ca.* 340 B.C.) was the earliest complete anatomical work.

Aristotle (384–322 B.C.) was the founder of comparative anatomy.

"On Fractures and Dislocations" (from Hippocratic collection) contained the first clear description of surgical anatomy.

Herophilus of Chalcedon (*ca.* 300 B.C.) has been called the "Father of Anatomy."

Erasistratus of Chios (*ca.* 290 B.C.) has been called the "Father of Physiology," a title reserved by some, however, for Galen.

Anatomy in the Roman Empire, A.D.

"On the Naming of the Parts of the Body" by Rufus of Ephesus (*ca.* A.D. 50) was the first book on anatomical nomenclature.

The account of the anatomy of the uterus by Soranus of Ephesus (*ca.* A.D. 100) has been termed one of the best pieces of ancient descriptive anatomy.

Galen of Pergamum (*ca.* A.D. 130–200), "Prince of Physicians," demonstrated and wrote on anatomy.

14th Century

Human dissections were performed in Italy and France during the 14th century. Evidence exists that dissection began in Italy before 1240. The bull *De sepulturis* (1300) of Boniface VIII had nothing to do with anatomy.[6]

Mondino de' Luzzi (1276–1326), the "Restorer of Anatomy," performed public dissections at Bologna (1315) and wrote his *Anothomia* (1316).

15th Century

Anatomy was studied by artists, e.g., Leonardo da Vinci (1452–1519).[7]

Anatomical illustrations began to be printed during the last decade of the 15th century.

16th Century

The "Commentary" on Mondino (1521) by Berengario da Carpi (1470–1550) was the first illustrated textbook of anatomy.

Anatomical nomenclature was founded by Jacobus Sylvius (Jacques Dubois, 1478–1555).

Comparative anatomy was studied extensively, e.g., by Belon, Fabricius ab Aquapendente, and Coiter. The last two were also noted embryologists.

Anatomy was reformed by Andreas Vesalius (frontispiece) of Brussels (1514–1564) in his *De humani corporis fabrica* ("On the Workings of the Human Body") (1543).[8]

The important anatomical plates of Bartolomeo Eustachi (1524–1574) were unfortunately not published until 1714.

Other eminent anatomists at this time included Canano (a pre-Vesalian), Columbus, and Falloppio.

The compound microscope was invented about 1590 in Holland by Zacharias Jansen.

17th Century

A physiological orientation was given to anatomy by William Harvey (1578–1657) in his *Exercitatio anatomica de motu cordis et sanguinis in animalibus* (1628).

Lymphatic vessels were rediscovered during this century.

The first recorded human dissection in America took place in 1638 in Massachusetts.

Microscopic anatomy was founded by Marcello Malpighi (1628–1694).

Eminent anatomists at this time included Fabricius, Casserio, Thomas Bartholin, and Riolan the younger. Notable comparative anatomists included Swammerdam, Perrault, Duverney, and Tyson.

Alcohol was introduced as a preservative in the 1660s.

18th Century

Pathological anatomy was founded by Giovanni Battista Morgagni (1682–1771).

Eminent anatomists at this time included Albinus and Winslow.

Notable comparative anatomists included Buffon, Daubenton, Vicq d'Azyr, and John Hunter, who founded dental anatomy.

Famous anatomical museums were founded by William (1718–1783) and John (1728–1793) Hunter.

Surgical anatomical museums were founded by John (1763–1820) and Charles (1774–1842) Bell.

Modern embryology was established by Caspar Friedrich Wolff (1733–1794).

19th Century

The tissues were classified grossly in 1801 by Xavier Bichat (1771–1802).

Dissection by medical students was made compulsory (e.g., in Edinburgh, 1826; in Maryland, 1833).

"Resurrectionists" flourished in Great Britain and Ireland, 1750–1832.[9]

The last of 16 murders for the sale of bodies by William Burke and William Hare took place in Edinburgh in 1828. A murder in London three years later was followed by the passage of the Warburton Anatomy Act ("An Act for Regulating Schools of Anatomy," 1832), a provision of which permitted the use of unclaimed bodies for dissection.[10] The first anatomical act in America was passed in Massachusetts in 1831.

The "cell theory" was proposed by various workers, 1808–1831.

The theory of organic evolution was proposed as a principle in biology.

Eminent anatomists at this time included Astley Cooper.

Notable comparative anatomists included Cuvier and Meckel.

Embryology was advanced by Karl von Baer, Wilhelm His, and Wilhelm Roux.

Various "-scopes" were devised between 1819 and 1899, e.g., the stethoscope, otoscope, ophthalmoscope, laryngoscope, gastroscope, cystoscope, and bronchoscope. These instruments furthered the study of living anatomy.

Anatomical societies were founded, e.g., Anatomische Gesellschaft (1886), Anatomical Society of Great Britain and Ireland (1887), American Association of Anatomists (1888), Association des Anatomistes (1899).

Formalin was used as a fixative in the 1890s.

X-rays were discovered in 1895 by Wilhelm Conrad Röntgen (1845–1923).

Neurohistology was securely established by Ramón y Cajal (1852–1934).

ANATOMICAL LITERATURE

Books, Including Atlases

Some of the most important books on various aspects of human anatomy are listed below. (Embryology, histology, and neuroanatomy, however, are not included.) Books on individual systems (e.g., skeleton) or regions (e.g., thorax, head and neck) are cited in the appropriate chapters of this text.

Systemic Anatomy

Von Bardeleben, Benninghoff, Braus, and Rauber-Kopsch have written detailed texts in German.

Poirier, P., and Charpy, A., *Traité d' anatomie humaine*, Masson, Paris, 1903–1932, 5 volumes, some parts of which reached a 4th edition.

Quain's Elements of Anatomy, Longmans, Green, London, 11th ed., 1908–1929, 4 volumes (8 books).

Regional Anatomy

Hollinshead, W. H., *Anatomy for Surgeons*, Hoeber-Harper, New York, 2nd ed., 1968–1971, 3 volumes.

Von Lanz, T., and Wachsmuth, W., *Praktische Anatomie*, Springer, Berlin, 1935–1972, 2 volumes, some parts of which have reached a 2nd edition.

Applied Anatomy

Lachman, E., *Case Studies in Anatomy*, Oxford University Press, New York, 2nd ed., 1971.

Systemic Atlases

Kiss and Szentágothai, Sobotta, Spalteholz, Toldt, Woerdeman, and Wolf-Heidegger have produced systemic atlases.

Regional Atlases

Anson, Jamieson, and Lopez-Antunez have produced regional atlases.

Bassett, D. L., *A Stereoscopic Atlas of Human Anatomy*, Sawyer's, Portland, Oregon, 1952–1962, 8 sections.

Grant, J. C. B., *An Atlas of Anatomy*, Williams & Wilkins, Baltimore, 6th ed., 1972.

Pernkopf, E. (ed. by H. Ferner and trans. by H. Monsen), *Atlas of Topographical and Applied Human Anatomy*, Saunders, Philadelphia, 1963 and 1964, 2 volumes.

Special Atlases

Van der Schueren, G., *et al.*, *Cava vitalia*, Arscia, Brussels, 1961. Excellent illustrations of various anatomical cavities.

Yokochi, C., *Photographic Anatomy of the Human Body*, University Park Press, Baltimore, 1971. Excellent photographs arranged systemically.

Cross-Section Atlases

Eycleshymer, A. C., and Schoemaker, D. M., *A Cross-Section Anatomy*, Appleton-Century-Crofts, New York, 1911.

Morton, D. J., *Manual of Human Cross-Section Anatomy*, Williams & Wilkins, Baltimore, 2nd ed., 1944.

Roy-Camille, R., *Coupes horizontales du tronc*, Masson, Paris, 1959.

Symington, J., *An Atlas Illustrating the Topographical Anatomy of the Head, Neck, and Trunk*, Oliver and Boyd, Edinburgh, 1917. Reissued in 1956.

See also Head and Neck (p. 550).

Surface Anatomy

Hamilton, W. J., Simon, G., and Hamilton, S. G. I., *Surface and Radiological Anatomy*, Heffer, Cambridge, 5th ed., 1971.

See also Muscular System (p. 30).

Radiological Anatomy, Including Atlases

See p. 68.

Anatomical Technique

Edwards, J. J., and Edwards, M. J., *Medical Museum Technology*, Oxford University Press, London, 1959.

Montagu, M. F. A., *A Handbook of Anthropometry*, Thomas, Springfield, Illinois, 1960.

Tompsett, D. H., *Anatomical Techniques*, Livingstone, Edinburgh, 2nd ed., 1970.

Neonatal and Pediatric Anatomy

Crelin, E. S., *Anatomy of the Newborn*, Lea & Febiger, Philadelphia, 1969, and *Functional Anatomy of the Newborn*, Yale University Press, New Haven, 1973.

Peter, K., Wetzel, G., and Heiderich, F., *Handbuch der Anatomie des Kindes*, Bergmann, Munich, 1938, 2 volumes.

Scammon, R. E., *A Summary of the Anatomy of the Infant*, chapter 3 in vol. 1 of I. A. Abt (ed.), *Pediatrics*, Saunders, Philadelphia, 1923.

Symington, J., *The Topographical Anatomy of the Child*, Livingstone, Edinburgh, 1887.

Teratological Anatomy

Schwalbe, E., and Gruber, G. B., *Die Morphologie der Missbildungen des Menschen und der Tiere*, Fischer, Jena, 1906–1958, 3 parts (many subdivisions).

Periodicals

More detailed and more recent anatomical information than can be provided in textbooks, however large, should be sought in specialized monographs and in the periodical literature. The following anatomical journals are published either entirely or largely in English.

Acta Anatomica (since 1945). This is an international journal of anatomy, histology, cytology, and embryology, in which summaries of the articles are given in English, French, and German.

American Journal of Anatomy (since 1901) and Anatomical Record (since 1906). These are the official publications of the American Association of Anatomists and are published by the Wistar Institute of Anatomy and Biology.

Journal of Anatomy (since 1917), originally the Journal of Anatomy and Physiology (1866–1916). This is the official publication of the Anatomical Society of Great Britain and Ireland.

Many other journals are available in languages other than English. In German, for example, are Anatomischer Anzeiger, Archiv für Anatomie und Physiologie, Zeitschrift für Anatomie und Entwicklungsgeschichte, Zeitschrift für Zellforschung und mikroskopische Anatomie, etc.

A number of journals devoted to special anatomical fields are available, e.g., Contributions to Embryology (Carnegie Institution of Washington), Journal of Cell Science, Journal of Comparative Neurology, Journal of Embryology and Experimental Morphology, Journal of the Royal Microscopical Society, etc.

With the advent of electron microscopy, special journals devoted to ultrastructure have appeared, e.g., Journal of Cell Biology.

Moreover, the journals of related fields, such as zoology, anthropology, and genetics, frequently contain articles of anatomical interest, as do also many of the clinical journals.

REFERENCES

1. S. E. Whitnall, *The Study of Anatomy*, Arnold, London, 4th ed., 1939. A valuable scheme of anatomical clinics has been outlined by I. M. Thompson, Canad. med. Ass. J., *54*:33, 1946.
2. *Final Report of the Committee Appointed by the Anatomical Society of Great Britain and Ireland on June 22, 1928*, University Press, Glasgow, 1933.
3. *Nomina anatomica*, Excerpta Medica Foundation, Amsterdam, 3rd ed., 1966.
4. The best source-book of eponymous terms is J. Dobson, *Anatomical Eponyms*, Livingstone, Edinburgh, 2nd ed., 1962.
5. The best general introduction to the history of human anatomy is C. Singer, *A Short History of Anatomy and Physiology from the Greeks to Harvey*, Dover, New York, 1957. Another attractive work is that by J. G. de Lint, *Atlas of the History of Medicine. I. Anatomy*, Lewis, London, 1926. For the history of techniques used in gross anatomy see A. Faller, Acta anat., suppl. 7 (2 ad vol. 4), 1948.
6. M. N. Alston, Bull. Hist. Med., *16*:221, 1944.
7. For the superb drawings, see C. D. O'Malley and J. B. de C. M. Saunders, *Leonardo da Vinci on the Human Body*, Schuman, New York, 1952.
8. For the superb woodcuts, see J. B. de C. M. Saunders and C. D. O'Malley, *The Illustrations from the Works of Andreas Vesalius of Brussels*, World Publishing Co., Cleveland, 1950. See also C. D. O'Malley, *Andreas Vesalius of Brussels*, University of California Press, Berkeley, 1964.
9. See the interesting account by J. M. Ball, *The Sack-em-up Men*, Oliver and Boyd, Edinburgh, 1928.
10. N. M. Goodman, Brit. med. J., 2:807, 1944. This article should be read for its historical and legal considerations, and for its discussion of institutional and "bequest" bodies. See also A. Delmas, C. R. Ass. Anat., 52:1, 1967.

SKELETON

2

Ernest Gardner

The skeleton consists of bones and cartilages. The term *bones* refers to structures that are composed of several tissues, the predominant one being a specialized connective tissue known as *bone*. Bones provide a framework of levers, they protect organs such as the brain and heart, their marrow forms certain blood cells, and their dense substance provides for the storage and exchange of calcium and phosphate ions.

The term *osteology*, meaning the study of bones, is derived from the Greek words *osteon*, meaning bone, and *logos*, meaning a branch of knowledge. The latin term *os* is used in names of specific bones, e.g., *os coxae*, or hip bone; the adjective osseous is derived from it.

Cartilage is a tough, resilient connective tissue composed of cells and fibers embedded in a firm, gel-like, intercellular matrix (p. 15). Cartilage is an integral part of many bones, and some skeletal elements are entirely cartilaginous.

BONES

The skeleton includes the *axial skeleton* (bones of the head, neck, and trunk) and the *appendicular skeleton* (bones of the limbs). Bone may be present in locations other than in the bony skeleton. It often replaces the hyaline cartilage in parts of the laryngeal cartilages. Furthermore, it is sometimes formed in soft tissues, such as scars. Bone that forms where it is not normally present is called *heterotopic bone*.

Types of Bones

Bones may be classified according to shape: long, short, flat, and irregular.

Long Bones (fig. 2–1). Long bones are those in which the length exceeds the breadth and thickness. They include the clavicle, humerus, radius, and ulna in the upper limb, and the femur, tibia, and fibula in the lower limb. They also include the metacarpals, metatarsals, and phalanges.

Each long bone has a shaft and two ends or extremities, which are usually articular. The shaft is also known as the *diaphysis*. The ends of a long bone are usually wider than the shaft, and are known as *epiphyses*. The epiphyses of a growing bone are either entirely cartilaginous or, if

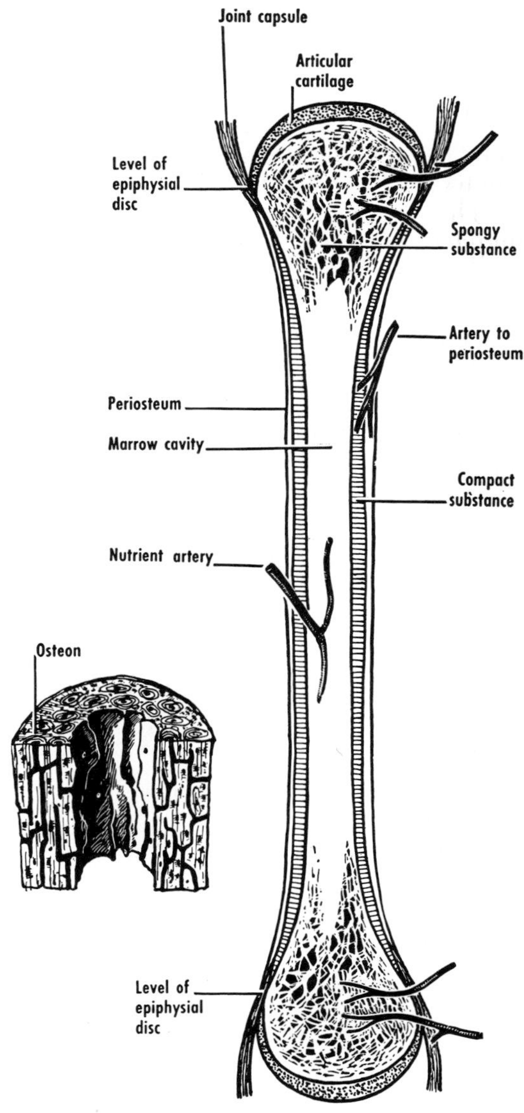

Figure 2–1 Schematic diagram of a long bone and its blood supply. The inset diagram shows the lamellae of the compacta arranged into osteons.

epiphysial ossification has begun, are separated from the shaft by cartilaginous *epiphysial discs.* Clinically, the term epiphysis usually means bony epiphysis. The part of the shaft adjacent to an epiphysial disc is usually wider than the rest of the shaft. This wider part, which contains the growth zone and newly formed bone, is called the *metaphysis.* The bony tissue of the metaphysis and of the epiphysis is continuous in the adult.

The shaft of a long bone is a tube of *compact bone* ("compacta"), the cavity of which is known as a *medullary (marrow) cavity.* The cavity contains either red or yellow marrow, or combinations of both (p. 12). The epiphysis and metaphysis consist of irregular, anastomosing bars or trabeculae, which form what is known as *spongy* or *cancellous bone.* The spaces between the trabeculae are filled with marrow. The external parts of the epiphysis and metaphysis consist of a thin layer of compact bone, and the bone on the articular surfaces of the ends is covered by cartilage, which is usually hyaline.

The shaft of a long bone is surrounded by a connective tissue sheath, the *periosteum.* Periosteum is composed of a tough, outer fibrous layer, which acts as a limiting membrane, and an inner, more cellular osteogenic layer. The inner surface of compact bone is lined by a thin, cellular layer, the *endosteum.* The periosteum is continuous at the ends of the bone with the joint capsule, but it does not cover the articular cartilage. Periosteum also serves for the attachment of muscles and tendons. The bundles of collagenous fibers of a tendon fan out in the periosteum; some continue inward and penetrate the bone. Very often, the zone of attachment of a tendon is clearly neither tendon nor bone.[1] Where tendons are in contact with bone in passing to their attachment, the underlying periosteum is sometimes fibrocartilaginous.[2]

Short Bones. Short bones are approximately equal in their main dimensions. They occur in the hands and feet, and consist of spongy bone and marrow enclosed by a thin layer of compact bone. They are surrounded by periosteum, except on their articular surfaces.

SESAMOID BONES. Sesamoid bones are a type of short bone and occur mainly in the hands and feet, embedded within tendons or joint capsules. They vary in size and number. Some clearly serve to alter the angle of pull of a tendon. Others, however, are so small that they are of scant functional importance.

ACCESSORY BONES. Accessory, or supernumerary, bones are bones that are not regularly present.[3] Such bones are usually of the short or flat type, and they occur chiefly in the hands and feet. They include some sesamoid bones and certain ununited epiphyses in the adult. They are of some medicolegal importance in that, when seen in radiograms, they may be mistaken for fractures. Callus, however, is absent, the bones are smooth, and they are often present bilaterally.

Flat Bones. Flat bones include the ribs, sternum, scapulae, and many bones of the skull. They are thin, and are usually curved or bent rather than flat. Flat bones consist of two layers of compact bone with intervening spongy bone and marrow. The intervening spongy layer in the bones of the vault of the skull is termed *diploë;* it contains many venous channels. Some bones, for example, the lacrimal bone and parts of the scapula, are so thin that they consist of only a thin layer of compact bone.

The articular surfaces of flat bones are covered with cartilage or, in the case of certain skull bones, with fibrous tissue.

Irregular Bones. Irregular bones are those that do not readily fit into other classifications. They include many of the skull bones, the vertebrae, and the hip bones. They consist mostly of spongy bone enclosed by a thin layer of compact bone. Parts of irregular bones that are quite thin may consist entirely of compact bone. *Pneumatic bones* contain air-filled cavities or sinuses.

Contours and Markings of Bones

Bones have markings and irregularities that are defined or named in various ways. Most contours and markings are seen best in dried bones, from which the periosteum and articular cartilage have been removed.

The shafts of long bones usually have three surfaces, separated from one another by three borders. Short bones frequently have six surfaces. There is much variation in the number of surfaces and borders of flat and irregular bones.

The articular surfaces are smooth, even after articular cartilage is removed. A projecting articular process is often referred to as a head, its narrowed attachment to the rest of the bone as the neck. The remainder is the body or, in a long bone, the shaft. A condyle (knuckle) is a protruding mass that carries an articular surface. A ramus is a broad arm or process that projects from the main part or body of the bone.

Other prominences, more or less in order of decreasing size, are called processes, trochanters, tuberosities, protuberances, tubercles, and spines. Linear prominences are ridges, crests, or lines, and linear depressions are grooves. Other depressions are fossae or foveae (pits). A large cavity in a bone is termed a sinus, a cell, or an antrum. A hole or opening in a bone is a foramen. If it has length, it is a canal, a hiatus, or an aqueduct. Many of these terms (e.g., canal, fossa, foramen, aqueduct, etc.) are not, however, limited to bones.

The ends of bones, except for the articular surfaces, contain many foramina for blood vessels. These foramina are most numerous near the margins of the articular surfaces. The largest ones are usually for veins. Similar vascular foramina on the shaft of a long bone are very much smaller, being barely visible, except for one or sometimes two large nutrient foramina that lead into obliquely directed canals. These canals give passage to vessels that supply the bone marrow. The direction of nutrient canals in human long bones is remarkably constant, although anomalous canals sometimes occur in the femur. The canals usually point away from the growing end of the bone, and point toward the epiphysis that unites first with the shaft. The directions of the vessels are indicated by the following jingle: To the elbow I go; from the knee I flee. In spite of the apparent association with bone growth, the direction of nutrient canals is more related to growth factors outside bone, perhaps to growth processes in limb arteries.[4]

The surfaces of bones are commonly roughened and elevated where there are powerful fibrous attachments, but are smooth where muscle fibers attach directly. Likewise, the margins to which joint capsules are attached are often marked in older bones by a thin ridge or lip of bone. Such "lipping" is apparently due to tension on the periosteum that joins the capsule at the articular margin. This pull, and also the pull where fibrous attachments are concentrated, apparently stimulates the osteogenic layer of the periosteum to form bone.

Blood and Nerve Supply

Bones are richly supplied with blood vessels, and the pattern of supply of a long bone is illustrated in figure 2–1.

Long bones are supplied by the following types of blood vessels (see fig. 2–1): (1) A *nutrient artery* (or arteries) pierces the compact bone of the shaft and divides into longitudinally directed branches that supply the marrow and compact bone as far as the metaphysis. (2) Many small branches of periosteal vessels also supply the compact bone of the shaft. (3) Metaphysial and epiphysial vessels, which arise mainly from arteries that supply the joint, pierce the compacta and supply the spongy bone and marrow of the ends of the bone. In a growing bone, the metaphysial and epiphysial vessels are separated by the cartilaginous epiphysial plate. Both groups of vessels are important for the nutrition of the growth zone, and disturbances of blood supply may result in disturbances in growth. When growth stops and the epiphysial plate disappears, the metaphysial and epiphysial vessels anastomose. The extent to which they in turn anastomose with the terminal branches of nutrient arteries is uncertain. Some blood-borne infections tend to settle in the ends of bones.

Clinical, experimental, and histological studies indicate that the blood flow through the compacta of normal adult bones is outward, that is, from the medullary arterial system first to the capillaries of the compacta, thence to the capillaries of the periosteum and of the attached muscles.[5]

Many nerve fibers accompany the blood vessels of bone. Most such fibers are vasomotor, but some are sensory and end in periosteum and in the adventitia of blood vessels. Some of the sensory fibers are pain fibers. Periosteum is especially sensitive to tearing or tension. Drilling into compact bone without anesthesia may give rise to dull pain, or an aching sensation; drilling into spongy bone may be much more painful. Fractures are painful, and an anesthetic injected between the broken ends of the bone may give relief. A tumor or infection that enlarges within a bone and causes pressure may be quite painful. Pain arising in a bone may be felt locally, that is, at the site of stimulation. However, the pain often spreads or is referred. For example, pain arising in the shaft of the femur may be felt diffusely in the lower part of the thigh, or may be felt in the knee.

Form and Architecture[6]

Bones are both rigid and elastic. They resist tensile and compressive forces equally well, and they can sustain static and dynamic loads up to many times the weight of the body. The obviously mechanical nature of the skeleton has led to many attempts to interpret the external and internal architecture of bone on mechanical grounds. Bones are admirably constructed to combine strength, elasticity, and lightness of weight, and these properties may be modified by various mechanical conditions.

The architecture of cancellous bone has been interpreted most frequently in terms of the trajectorial theory. According to this, the bony trabeculae follow the lines of maximal internal stress (trajectories) in the bone and are, therefore, adapted to resist the stresses and strains to which a bone is subjected. Some of the trabeculae are tensile-resisting, whereas others resist compressive forces.

The trajectorial theory has been severely criticized on several grounds, and is no longer accepted without reservation. The adherents of the theory believed that tensile forces are primarily responsible for bone growth, whereas compression causes atrophy. Other investigators of the subject hold the opposite view. However, under the proper conditions, either or both types of forces can stimulate bone growth. During postnatal life, function is the primary factor stimulating bone growth and determining architecture, regardless of the mechanical force involved. The postnatal influence, if any, of the biochemical, blood-forming, and other nonmechanical functions of bone on its external form and internal architecture are little known. It should be emphasized that form may depend as much on these functions as on mechanical forces.

Microscopic Structure of Bone

Bone is a specialized, constantly changing connective tissue that is composed of cells, a dense intercellular substance, and numerous blood vessels.[7] It is similar in some respects to cartilage, but differs from it in certain important ways. The structure of bone is summarized in table 2–1.

Mature bone is composed of layers and is therefore known as lamellar bone. In compact bone, these layers or lamellae are arranged into *osteons* or *haversian systems*, each of which consists of concentric *lamellae*, like tubes within tubes (fig. 2–1). The innermost tube encloses a core of tissue containing usually but a single small blood vessel. Osteons may be up to several millimeters in length, and they run longitudinally or obliquely longitudinally in the shafts of long bones. The lamellae of the trabeculae of spongy bone are arranged, not as osteons, but as flat or slightly curved sheets.

Lamellae contain spaces termed *lacunae* which are occupied by osteocytes. Canaliculi radiate from each lacuna and provide for diffusion of nutritive materials from capillaries. The life of bone is dependent upon its cells and, when a bone cell dies, adjacent bone disintegrates.

The hardness of bone results from the deposition within an organic matrix of a complex mineral substance, chiefly calcium phosphate complexes that belong to the apatite group of minerals. The mineral substance constitutes about two-thirds of bone by weight. When bone is calcined (water and organic material burnt out), it crumbles. When decalcified, bone becomes flexible. Because of their high mineral content, bones are very opaque to x-rays (p. 65).

Bone Marrow[8]

Before birth, the medullary cavities of bones, and the spaces between trabeculae, are filled with tissue called red marrow. This tissue gives rise to red blood corpuscles and to certain white blood cells (granulocytes). From infancy onward, there is a progressive diminution in the amount of blood cell-forming marrow, and a progressive increase in the amount of fat (yellow marrow).[9]

The approximate condition in the adult is considered to be as follows: Red marrow is present in the ribs, vertebrae, sternum,

TABLE 2–1 The Constituents of Cartilage and Bone

Constituents	*Hyaline Cartilage*	*Bone*
A. Cells	Chondrocytes in lacunae	Osteocytes in lacunae
B. Intercellular substance (matrix)	(a) Masked collagenous fibers (b) Ground substance, containing mucopolysaccharides (chondroitin sulfate and hyaluronic acid) and proteins	1. Organic matrix: (a) Masked collagenous fibers (ossein) (b) Ground substance, containing mucopolysaccharides (chondroitin sulfate and hyaluronic acid) and proteins 2. Inorganic matrix: Apatite crystals in ground substance
C. Additional	(Avascular and nerveless)	Blood vessels and nerves (Lymphatics in periosteum)

and hip bones. The radius, ulna, tibia, and fibula contain fatty marrow in their shafts and epiphyses. The femur and humerus usually contain a small amount of red marrow in the upper parts of the shafts, and small patches may be present in their proximal epiphyses. The tarsal and carpal bones generally contain only fatty marrow. In very advanced age, red marrow may be absent in the epiphyses and shafts of the femur and humerus as determined by ordinary inspection. However, microscopic examination of fatty marrow of any bone at any age often shows islets of blood-forming cells. Also, the relative amounts of red and yellow marrow may be altered by disease. Loss of blood, for example, may be followed by an increase in the amount of red marrow as more blood cells are formed.

Development and Growth[10]

All bones begin as mesenchymal proliferations that appear early in the embryonic period. The number of connective tissue fibers also increases. In the case of the *membrane* bones (also termed *dermal* or *investing* bones and comprising the clavicle, mandible, and certain skull bones), the cells differentiate into osteoblasts that lay down an organic matrix called osteoid. Bone salts are then deposited in this matrix. Some osteoblasts are trapped in the matrix and become osteocytes. Others continue to divide and form more osteoblasts on the surface of the bone. Bone grows only by apposition, that is, by the laying down of new bone on free surfaces.

Most bones, however, develop as cartilage bones. The mesenchymal proliferations become chondrified as the cells lay down cartilage matrix and form hyaline cartilages that have the shapes of the future bones. These cartilages are then replaced by bone, by a process that is illustrated for a long bone in figure 2–2.

Skeletal Maturation

Skeletal development involves three interrelated but dissociable components: increase in size (growth), increase in maturity, and aging. Skeletal maturation is "the metamorphosis of the cartilaginous and membranous skeleton of the foetus to the fully ossified bones of the adult."[11] Skeletal status, however, does not necessarily correspond with height, weight, or age. In fact, the maturative changes in the skeleton are intimately related to those of the reproductive system. These in turn are directly responsible for most of the externally discernible changes on which the estimation of general bodily maturity is usually based. The skeleton of a healthy child develops as a unit, and the various bones tend to keep

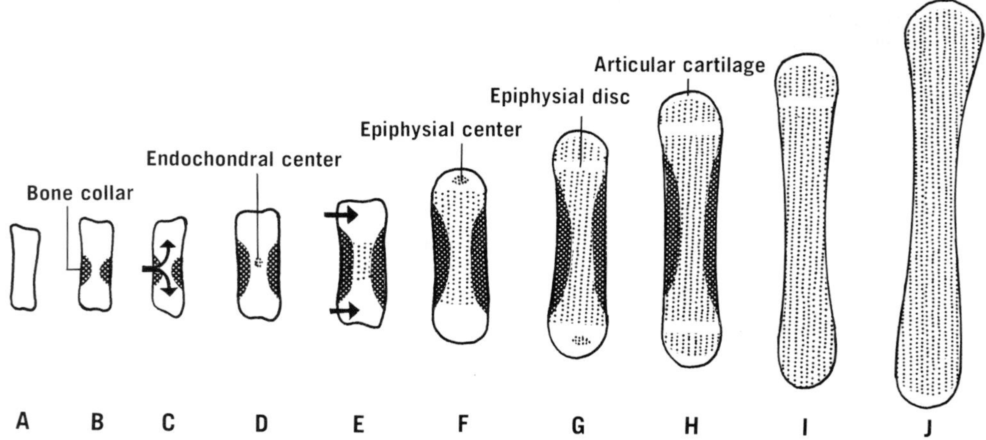

Figure 2–2 Schematic diagrams of the development of a long bone. *A*, cartilaginous model. *B*, bone collar. *C*, invasion of bone collar and cartilage. *D*, endochondral ossification begins. *E*, periosteal and endochondral ossification extend longitudinally and the cartilaginous epiphyses begin to be vascularized (indicated by arrows). *F*, an epiphysial center of ossification begins in one epiphysis. *G*, an epiphysial center begins in the other epiphysis. The first center has grown so that an epiphysial disc is present. *H*, two epiphysial discs are present. *I*, the last (second) epiphysial center to appear fuses first with the shaft. *J*, the first epiphysial center to appear (in the end from which most growth in length takes place) fuses last with the shaft.

pace with one another. Hence, radiographic examination of a limited portion of the body is believed by some workers to suffice for an estimation of the entire skeleton. The hand is the portion most frequently examined—"as the hand grows, so grows the entire skeleton," it is sometimes stated.

The assessment of skeletal maturity is important in determining whether an individual child is advanced or retarded skeletally, and, therefore, in diagnosing endocrine and nutritional disorders. Skeletal status is frequently expressed in terms of skeletal age. This involves the comparison of radiograms of certain areas with standards for those areas; the skeletal age assigned is that of the standard that corresponds most closely. Detailed standards have been published for the normal postnatal development of the hand, knee, and foot.[12] Tables showing the times of appearance of the postnatal ossific centers in the limbs are provided inside the back cover.[13]

Skeletal Maturation Periods. The following arbitrary periods are convenient in considering the progress of skeletal maturation.

1. EMBRYONIC PERIOD PROPER. This comprises the first eight postovulatory weeks of development. The clavicle, mandible, maxilla, humerus, radius, ulna, femur, and tibia commence to ossify during the last two weeks of this period.[14]

2. FETAL PERIOD. This begins at eight postovulatory weeks, when the crown-rump length has reached about 30 mm. The following elements commence to ossify early in the fetal period or sometimes late in the embryonic period: scapula, ilium, fibula, distal phalanges of hand, and certain cranial bones (e.g., the frontal).

The following begin to ossify during the first half of intrauterine life (fig. 10–4*A*, p. 66): most cranial bones and most diaphyses (ribs, metacarpals, metatarsals, phalanges), calcaneus sometimes, ischium, pubis, some segments of sternum, neural arches, and vertebral centra (C.V. 1 to S.V. 5).[15]

The following commence to ossify shortly before birth (fig. 10–4*B*, p. 66): calcaneus, talus, cuboid, usually the distal end of the femur and the proximal end of the tibia; sometimes the coracoid process, the head of the humerus, and the capitate and hamate; rarely the head of the femur and the lateral cuneiform.

3. CHILDHOOD. This is the period from birth to puberty and includes infancy (i.e., the first one or two postnatal years). Most epiphyses in the limbs, together with the carpals, tarsals, and sesamoids, begin to ossify during childhood. It is important to keep in mind that ossific centers generally appear one or two years earlier in girls than in boys. Furthermore, those epiphyses that appear first in a skeletal element usually are the last to unite with the diaphysis. They are located at the so-called "growing ends" (e.g., shoulder, wrist, knee).

4. ADOLESCENCE. This includes puberty and the period from puberty to adulthood. Puberty usually occurs at 13 ± 2 years of age in girls, and two years later in boys. Most of the secondary centers for the vertebrae, ribs, clavicle, scapula, and hip bone begin to ossify during adolescence. The fusions between epiphysial centers and diaphyses occur usually during the second and third decades. These fusions usually take place one or two years earlier in girls than in boys. The closure of epiphysial lines is under hormonal control.

5. ADULTHOOD. The humerus serves as a skeletal criterion for the transitions into adolescence and into adulthood, in that its distal epiphysis is the first of those of the long bones to unite, and its proximal epiphysis is the last (at age 19 or later). The center for the iliac crest fuses in early adulthood (age 21 to 23). The spheno-occipital junction also closes in early adulthood (age 20 to 21), and the sutures of the vault of the skull commence to close at about the same time (from age 22 onward).

Variations

Bones vary according to race, sex, and age, and from individual to individual.

Bones in females are generally lighter and smaller, because women are usually smaller and stop growing earlier. Muscular markings tend to be more pronounced in bones of males. Yet, among many bones, it may be only the "extreme" examples that can be differentiated as to sex.

Aside from size, the bones of children differ in being more resilient. When fractured, they break like a "green stick." Indeed, they sometimes bend where an adult bone would snap. Senile bones may show general atrophy and loss of compact substance, but it is very doubtful that this is an inevitable accompaniment of advancing age.

Individual variations are due to a variety of causes. Common variations are differences in size and weight of bones, and such differences are usually directly related to the height and muscular development of the individual. Bones begin to change after birth, when muscular activity is consistently present. For example, the coronoid process of the mandible is largely dependent on the muscles of mastication for its full development. Secondary markings in the form of roughened surfaces or lines begin to appear at about the time of puberty. These secondary markings are characteristic of fibrous and tendinous attachments (p. 11). Primary lines, such as the linea aspera of the femur, may become thicker and higher. Lipping of articular margins may occur with advancing age.

Bones may hypertrophy. If one bone in a limb is removed, or is congenitally absent, the fibula for example, the adjacent bone (tibia) enlarges. Conversely, bones atrophy when muscular activity is much decreased or absent, as when a limb is placed in a plaster cast or when paralytic disease is present. Organic as well as inorganic material of bone is lost; bone quality is only slightly altered.

If use-destruction (see p. 22) results in the wearing away of articular cartilage and the consequent movement of bone on bone, the surfaces in contact may become dense and polished (eburnated).

Forensic and Anthropologic Aspects[16]

When bones or fragments of bones are recovered, it is sometimes possible to determine first whether the bones are human, and then whether they are young or old, male or female. It may be difficult or impossible to determine whether a bone is human unless a relatively intact, characteristic bone is recovered. If a complete human skeleton is available, sex can be determined in about 50 per cent of children and probably 90 per cent or more of adults. Of most value in the determination of sex, in order of their importance, are (1) hip bone and sacrum, (2) skull, (3) sternum, (4) atlas, and (5) long bones. Some investigators believe that the femur is of more value than the atlas. With reasonably complete material, age can be determined to within about two years of the correct age, up to 30 years of age, and to within five to ten years of the correct age after 30.

Very great skill and experience are required to establish race.

Stature may be estimated rather accurately if a long bone is available, especially the femur. Tables have been published relating the length of long bones to stature.[17]

Because bones, and especially teeth, are resistant to decay, they are usually the only parts of the body that can be recovered an appreciable interval after burial. Hence they are a major source of information about early animal life, and it is from the study of fossilized bones that much of our knowledge of evolutionary changes has been obtained.

The study and proper interpretation of fossil bones require great experience, and a broad knowledge of variations, comparative aspects, and technical procedures. Even then, investigators may be misled.*

*"When Drs. J. S. Weiner, K. P. Oakley, and W. E. LeGros Clark . . . announced that careful study had proven the famous Piltdown skull to be compounded of both recent and fossil bones, so that it is in part a deliberate fraud, one of the greatest of all anthropological controversies came to an end."[18]

CARTILAGE

Cartilage is a tough, resilient connective tissue that is composed of cells and fibers embedded in a firm, gel-like intercellular matrix (table 2–1, p. 12). The cartilage cells *(chondrocytes)* lie in lacunae, sometimes singly, but more often in groups. A group of cells develops from a single precursor, termed a *chondroblast*. An important constituent of the matrix is a mucopolysaccharide, one of the chondroitin sulfuric acids. Adult cartilage lacks nerves, and it usually lacks blood vessels. Nutritive substances must therefore diffuse through the matrix to reach the cells. In contrast to bone, diffusion is very poor in calcified cartilage. When cartilage calcifies, chondrocytes usually die, and the cartilage is resorbed and replaced by bone. The fibers that are embedded in the matrix are collagenous or elastic. The nature and arrangement of these fibers are partly the basis for the classification of cartilage into three types—hyaline, fibrous, and elastic.

A skeletal element that is mainly or entirely cartilaginous is surrounded by a connective tissue membrane, the perichondrium, the structure of which is similar to periosteum. Cartilage grows by apposition, that is, by the laying down of new cartilage on the surface of the old. The new cartilage is formed by chondroblasts derived from the deeper cells of the perichondrium. Cartilage also grows interstitially, that is, by an increase in the size and number of existing cells and by an increase in the amount of intercellular matrix. Adult cartilage grows slowly, and repair or regeneration after a severe injury is inadequate.

Hyaline Cartilage (table 2–1). This, the most plentiful type, is so named because it has a glassy, translucent appearance owing to the character of its matrix. The matrix and the collagenous fibers that are embedded in it have about the same index of refraction. The fibers are not visible, therefore, in ordinary microscopic preparations.

The cartilaginous models of bone in the embryo consist of hyaline cartilage, as do also the epiphysial discs. Most articular cartilages, the costal cartilages, the cartilages of the trachea and bronchi, and most of the cartilages of the nose and larynx are formed of hyaline cartilage. Nonarticular hyaline cartilage has a tendency to calcify and to be replaced by bone.

Fibrocartilage. Bundles of collagenous fibers are the prominent constituent of fibrocartilage. The bundles are visible in ordinary microscopic preparations, in contrast to hyaline cartilage. The matrix is less in amount than in hyaline cartilage, and the chondrocytes are scattered. Fibrocartilage is present in certain cartilaginous joints, and it forms articular cartilage in a few joints, for example, the temporomandibular.

Elastic Cartilage. This type of cartilage resembles hyaline cartilage except that its fibers are elastic. It rarely if ever calcifies with advancing age. Elastic cartilage is present in the auricle and the auditory tube, and it forms some of the cartilages of the larynx.

REFERENCES

1. G. Mollier, Morph. Jb., *79*:161, 1937. H. Biermann, Z. Zellforsch., *46*:635, 1957.
2. D. L. Stilwell, Jr., and D. J. Gray, Anat. Rec., *120*:663, 1954.
3. R. O'Rahilly, J. Bone Jt Surg., *35A*:626, 1953; Clin. Orthopaed., *10*:9, 1957.
4. H. Hughes, Acta anat., *15*:261, 1952. V. R. Mysorekar, J. Anat., Lond., *101*:813, 1967.
5. M. Brookes *et al.*, Lancet, *1*:1078, 1961.
6. F. G. Evans, *Stress and Strain in Bones*, Thomas, Springfield, Illinois, 1957, and *Mechanical Properties of Bone*, Thomas, Springfield, Illinois, 1973.
7. A. W. Ham, J. Bone Jt Surg., *34A*:701, 1952. F. C. McLean, Science, *127*:451, 1958.
8. K. Rohr, *Das menschliche Knochenmark*, Georg Thieme, Stuttgart, 3rd ed., 1960.
9. A. Piney, Brit. med. J., *2*:792, 1922. P. Sturgeon, Pediatrics, Springfield, *7*:577, 642, 774, 1951.
10. Based on E. Gardner, Osteogenesis in the Human Embryo and Fetus, in G. H. Bourne (ed.), *The Biochemistry and Physiology of Bone*, cited below. For a discussion of the classic experiments on bone growth, see A. Keith, *Menders of the Maimed*, Lippincott, Philadelphia, 1952 (1919).
11. R. M. Acheson, J. Anat., Lond., *88*:498, 1954. F. Falkner (ed.), *Human Development*, Saunders, Philadelphia, 1966.
12. S. I. Pyle and N. L. Hoerr, *A Radiographic Standard of Reference for the Growing Knee*, Thomas, Springfield, Illinois, 1969. S. I. Pyle, A. M. Waterhouse, and W. W. Greulich (eds.), *A Radiographic Standard of Reference for the Growing Hand and Wrist*, Press of Case Western Reserve University (Year Book Medical Publishers, Chicago), 1971. N. L. Hoerr, S. I. Pyle, and C. C. Francis, *Radiographic Atlas of Skeletal Development of the Foot and Ankle*, Thomas, Springfield, Illinois, 1962.
13. S. M. Garn, C. G. Rohmann, and F. N. Silverman, Med. Radiogr. Photogr., *43*:45, 1967.
14. R. O'Rahilly and E. Gardner, Amer. J. Anat., *134*:291, 1972.
15. C. R. Noback and G. G. Robertson, Amer. J. Anat., *89*:1, 1951. R. O'Rahilly and D. B. Meyer, Amer. J. Roentgenol., *76*:455, 1956.
16. W. M. Krogman, *The Human Skeleton in Forensic Medicine*, Thomas, Springfield, Illinois, 1962. S. Smith and F. S. Fiddes, *Forensic Medicine*, Churchill, London, 10th ed., 1955. D. H. Enlow, in *Studies on the Anatomy and Function of Bone and Joints*, Springer, New York, 1966. See also J. Glaister and J. C. Brash, *Medico-legal Aspects of the Ruxton Case*, Livingstone, Edinburgh, 1937. The extent and character of mutilation of two murder victims provided a problem of anatomical reconstruction unparalleled in criminal records.
17. A. Trotter and G. C. Gleser, Amer. J. phys. Anthrop., *10*:463, 1952; *16*:79, 1958. L. H. Wells, J. forensic Med., *6*:171, 1959.
18. W. L. Straus, Jr., Science, *119*:265, 1954.

GENERAL READING

Bourne, G. H. (ed.), *The Biochemistry and Physiology of Bone*, Academic Press, New York, 2nd ed., 1971–1972, 3 volumes.

Crock, H. V., *The Blood Supply of the Lower Limb Bones in Man*, Livingstone, Edinburgh, 1967. A beautifully illustrated account of the vascular anatomy of these bones.

Enlow, D. H., *Principles of Bone Remodeling*, Thomas, Springfield, Illinois, 1963. An excellent review with original observations.

Frazer's *Anatomy of the Human Skeleton*, revised by A. S. Breathnach, Churchill, London, 6th ed., 1965. A detailed synthesis of skeletal and muscular anatomy arranged according to regions.

Gray, D. J., Organ Systems in Adaptation: The Skeleton, in Handbook of Physiology, *Adaptation to the Environment*, American Physiological Society, Washington, D.C., Sect. 4, 1964.

Hancox, N. M., *Biology of Bone*, Cambridge University Press, London, 1972. Discusses the properties of bone as a tissue, and the activities and control of cells involved in the deposition and resorption of bone.

McLean, F. C., and Urist, M. R., *Bone*, University of Chicago Press, Chicago, 3rd ed., 1968. Describes mechanisms of calcification and properties of the mineral content.

Vaughan, J. M., *The Physiology of Bone*, Clarendon Press, Oxford, 1970. An excellent, balanced account of bone as a tissue and its complex role in mineral homeostasis.

JOINTS

3

ERNEST GARDNER

In everyday language, the term *joint* means a place at which two things are joined together. In anatomical usage, a joint has been described as "the connexion subsisting in the skeleton between any of its rigid component parts, whether bones or cartilages." *Articulation,* a term that has the same Latin origin as the word *article,* is synonymous with joint. Terms such as *arthrology,* which means the study of joints, and *arthritis,* which means inflammation of joints, are of Greek origin *(arthron).*

Joints vary widely in structure and arrangement and are often specialized for particular functions. Nevertheless, joints have certain common structural and functional features. They may be classified on the basis of their most characteristic structural features into three main types: fibrous, cartilaginous, and synovial.

FIBROUS JOINTS

The bones of a fibrous joint (sometimes called a *synarthrosis*) are united by fibrous tissue. There are two types of fibrous joints: *sutures* and *syndesmoses.* With few exceptions, little if any movement occurs at either type. The joint between a tooth and the bone of its socket is termed a *gomphosis* and is sometimes classed as a third type of fibrous joint.

Sutures. In the sutures of the skull, the bones are connected by several fibrous layers. The mechanisms of growth at these joints (about which there is still dispute) are especially important in accommodating the growth of the brain.

Syndesmoses. A syndesmosis is a fibrous joint in which the intervening connective tissue is considerably greater in amount than in a suture. Examples are the tibiofibular syndesmosis and the tympanostapedial syndesmosis.

CARTILAGINOUS JOINTS

The bones of cartilaginous joints are united either by hyaline cartilage or by fibrocartilage.

Hyaline Cartilage Joints. This type of joint, which is sometimes called a *primary cartilaginous joint* and sometimes a *synchondrosis,* is a temporary union. The hyaline cartilage that joins the bones is a persistent part of the embryonic cartilaginous skeleton and as such serves as a growth zone for one or both of the bones that it joins. Most hyaline cartilage joints are obliterated, that is, replaced by bone, when growth ceases. Examples of hyaline cartilage joints include epiphysial plates, and the spheno-occipital and neurocentral synchondroses.

Fibrocartilaginous Joints. In this type of joint, which is sometimes called a *secondary cartilaginous joint* and sometimes an *amphiarthrosis,* and to which the term *symphysis* has also been applied, the skeletal elements are united by fibrocartilage during some phase of their existence. The fibrocartilage is usually separated from the bones by thin plates of hyaline cartilage. Fibrocartilaginous joints include the pubic symphysis and the joints between the bodies of the vertebrae.

SYNOVIAL JOINTS

Synovia is the fluid present in certain joints, which are consequently termed synovial. Similar fluid is present in bursae and in synovial tendon sheaths.

General Characteristics

Synovial joints, which are often termed *diarthrodial joints,* possess a cavity and are specialized to permit more or less free movement. Their chief characteristics (fig. 3–1) are as follows:

The articular surfaces of the bones are covered with cartilage, which is usually hyaline in type. The bones are united by a *joint capsule* and ligaments. The joint capsule consists mostly of a *fibrous layer* (the term joint capsule is often used to refer specifically to the fibrous layer), the inner surface of which is lined by a vascular, connective tissue, the *synovial membrane,*

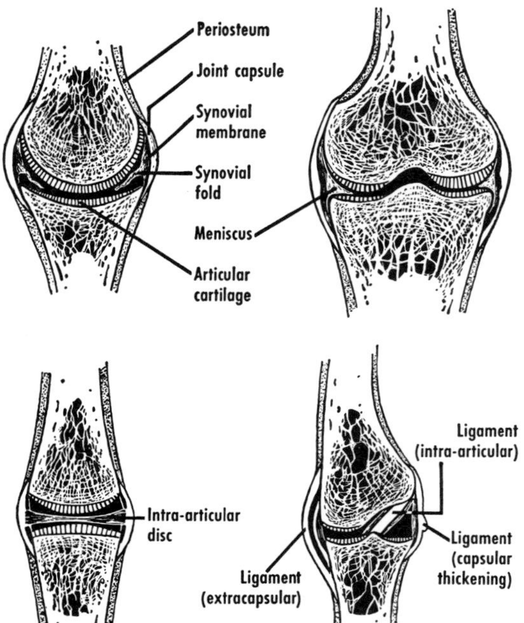

Figure 3–1 Schematic diagrams of synovial joints. The joint cavity in each is exaggerated, as is the thickness of the synovial membrane. Articular cartilage, menisci, and intra-articular discs are not covered by synovial membrane, but intra-articular ligaments are.

which produces the synovial fluid (synovia) that fills the *joint cavity* and lubricates the joint. The joint cavity is sometimes partially or completely subdivided by fibrous or fibrocartilaginous *discs* or *menisci*.

Types of Synovial Joints

Synovial joints may be classified according to axes of movement. This classification assumes the existence of three mutually perpendicular axes. A joint that has but one axis of rotation, such as a hinge joint or pivot joint, is said to have one degree of freedom. Saddle and ellipsoidal joints have two degrees of freedom. Each can be flexed or extended, abducted or adducted, but not rotated, at least independently. A ball-and-socket joint has three degrees of freedom.

Synovial joints may also be classified according to the shapes of the articular surfaces of the constituent bones. These shapes determine the type of movement and are partly responsible for determining the range of movement. The more common types of synovial joints are plane, hinge, and condylar. Ball-and-socket, ellipsoidal, pivot, and saddle joints are less frequent.

Plane Joint. The articular surfaces of a plane joint are usually slightly curved. They permit gliding or slipping in any direction, or the twisting of one bone on the other.

Hinge Joint or Ginglymus. A hinge joint is uniaxial and permits movement in but one plane. At an interphalangeal joint, for example, the movements are flexion and extension.

Condylar Joint. The articular area of each bone of a condylar joint consists of two distinct articular surfaces, each called a condyle. Although resembling a hinge joint in movement, a condylar joint permits several kinds of movements. The knee joint is a condylar joint.

Spheroidal or Ball-and-Socket Joint. In this type of joint, of which the shoulder joint is an example, a spheroidal surface of one bone moves within a "socket" of the other bone, about three axes. Flexion, extension, adduction, abduction, and rotation can occur, as well as a combination of these movements termed *circumduction*. In circumduction, the limb is swung so that it describes the side of a cone, the apex of which is the center of the "ball."

Ellipsoidal Joint. In this type of joint, which resembles a ball-and-socket joint, the articulating surfaces are much longer in one direction than in the direction at right angles. The circumference of the joint thus resembles an ellipse. It is biaxial and the radiocarpal joint is an example.

Pivot or Trochoid Joint. This type of joint, of which the proximal radioulnar joint is an example, is likewise uniaxial, but the axis is vertical, and one bone pivots within a bony or an osseoligamentous ring.

Saddle or Sellar Joint. This type of joint is shaped like a saddle; an example is the carpometacarpal joint of the thumb. It is a biaxial joint.

Movements[1]

Active Movements. Three types of active movements occur at synovial joints. Usually one speaks of movement *of* a part, movement *at* a joint; thus, flexion of the forearm, flexion at the elbow. These active movements are (1) gliding or slipping movements, (2) angular movements about a horizontal or side-to-side axis (flexion and extension) or about an anteroposterior axis (abduction and adduction), and (3) rotary movements about a longitudinal axis (medial and lateral rotation). Whether one, several, or all types of movement occur at a particular joint depends upon the shape and ligamentous arrangement of that joint.

The range of movement at joints is limited by (1) the muscles, (2) ligaments and capsule, (3) the shapes of the bones, and (4) the opposition of soft parts, such as the meeting of the front of the forearm and arm during full flexion at the elbow. The range of motion varies greatly in different individuals. In some, such as trained acrobats, the range of joint movement may be extraordinary. Such individuals, however, must train from early in life.

Passive and Accessory Movements. Passive movements are produced by an external force, such as gravity or an examiner. For example, the examiner holds the subject's wrist so as to immobilize it. He can then flex, extend, adduct, and abduct the subject's hand at the wrist, movements that the subject can normally carry out actively.

By careful manipulation, the examiner can also produce a slight degree of gliding and rotation at the wrist, movements that the subject cannot actively perform himself. These are called accessory movements (often classified with passive movements), and are defined as movements for which the muscular arrangements are not suitable, but which can be brought about by manipulation.

The production of passive and accessory movements is of value in testing and in diagnosing muscle and joint disorders.

Structure and Function

The classifications of synovial joints just listed indicate as simply as possible the shapes of the articular surfaces and the movements that can occur at the specific types of joints. They do not take into account, however, the complexity of joint mechanics and the fact that articular movements must be appreciated in terms of spherical as well as plane geometry.

The lubricating mechanisms of synovial joints are such that the effects of friction on articular cartilage are minimized.[2] The coefficient of friction during movement is less than that of ice sliding on ice. This is brought about by the nature of the lubricating fluid (viscous synovial fluid), by the nature of the cartilaginous bearing surfaces that adsorb and absorb synovial fluid, and by a variety of mechanisms that permit a replaceable fluid rather than an irreplaceable bearing to reduce friction.

Synovial Membrane and Synovial Fluid. Synovial membrane is a vascular connective tissue that lines the inner surface of the capsule, but does not cover articular cartilage. It contains cells that are indistinguishable from connective tissue cells elsewhere, at least with ordinary histologic stains. Synovial membrane differs from other connective tissues in that it produces a ground substance which is a fluid rather than a gel. The most characteristic morphologic feature of synovial membrane is a capillary network adjacent to the joint cavity. A variable number of villi, folds, and fat pads project into the joint cavity from the synovial membrane, the surface of which is otherwise relatively smooth. The tissue deep to the surface cells may be fibrous, areolar, or fatty, and it varies in thickness. Synovial membrane also contains lymphatic vessels and a few nerve fibers.

Synovial membrane is responsible for the formation of synovial fluid, which is a sticky, viscous fluid much like egg-white in consistency. The main function of synovial fluid is lubrication, but it also nourishes articular cartilage. The viscous nature of synovial fluid is due almost entirely to the presence of a non-sulfated mucopolysaccharide known as hyaluronic acid. Synovial fluid can be otherwise considered as a dialysate of blood plasma. It also normally contains a few cells (mostly mononuclear) derived from the lining tissue. Pathological processes that affect synovial membrane alter the cellular content of the fluid. Withdrawal of synovial fluid and determination of its cellular and chemical content and physical characteristics can be a valuable diagnostic aid.[3]

Cracking in joints, such as that which may occur when the fingers are suddenly pulled, is generally due to the sudden development of a partial vacuum in the joint cavity as the articular surfaces are pulled apart.[4] The partial vacuum is occupied by water vapor and blood gases under reduced pressure. Other types of cracks or noises (more often snaps) are probably due to the sudden slipping of a tendon or ligament over a bony or cartilaginous process or prominence.

Articular Cartilage. Adult articular cartilage is an avascular, nerveless, and relatively acellular tissue. The part immediately adjacent to bone is usually calcified. It is resilient and elastic, and has lost most of its power of growth and repair. However, in normal use some replacement of articular cartilage may occur.[5]

Cartilage is elastic in the sense that, when it is compressed, it becomes thinner but, on release of the pressure, it slowly regains its original thickness. Intermittent pressure causes cartilage to thicken by taking up water, and may well be an important factor in the diffusion of nutritive materials through cartilage.

Articular cartilage undoubtedly derives its main nourishment from synovial fluid, but other possible sources are shown in figure 3–2.

Articular cartilage is not visible in ordinary radiograms. Hence the so-called "radiological joint space" (p. 65) is wider than the true joint space.

Joint Capsule and Ligaments. In most joints, the capsule is composed of bundles of collagenous fibers, which are arranged somewhat irregularly in contrast to their more regular arrangements in tendons and many ligaments. These bundles, and the bundles in some ligaments, tend to spiral. Such an arrangement renders them sensitive to tension in most positions that the joint happens to occupy. Therefore the slightest movement alters the tension or torsion in the bundles and this change in tension in turn stimulates proprioceptive nerve endings in the capsule and ligaments (see below). In a few joints (middle ear) the capsule and ligaments are composed almost entirely of elastic fibers.

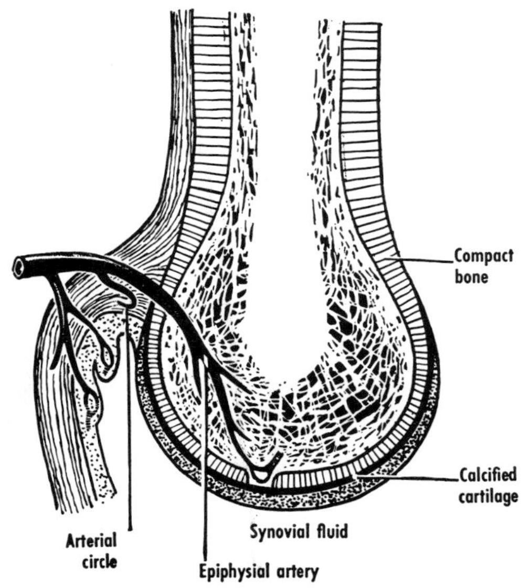

Figure 3–2 Schematic diagram of possible sources of nutrition of articular cartilage. These are (1) synovial fluid, (2) diffusion from capillaries in adjacent bone marrow, and (3) diffusion from capillaries derived from the arterial circle around the joint at the line of capsular attachment. The thickness of compact bone of the articular surface is exaggerated, as is also that of the layer of calcified cartilage.

Ligaments are classified as *capsular, extracapsular,* and *intra-articular* (see fig. 3–1). Most ligaments serve as sense organs in that nerve endings in them are important in reflex mechanisms and in the detection of movement and position. Ligaments also serve different kinds of mechanical functions. For example, the term *collateral ligament* is applied to an extracapsular ligament that remains taut throughout the range of joint movement. The *cruciate ligaments* of the knee (intra-articular) slacken in some movements and become taut in others. Strong ligaments are usually found at the sides of joints, particularly those that are hinge or ellipsoid in type.

The relationship of the epiphysial plate to the line of capsular attachment is important (fig. 2–1, p. 9). For example, the epiphysial plate is a barrier to the spread of infection between the metaphysis and the epiphysis. If the epiphysial line is intra-articular, then part of the metaphysis is also intra-articular, and a metaphysial infection may involve the joint. Likewise, in such instances, a metaphysial fracture becomes intra-articular, always a serious matter from the standpoint of damage to articular surfaces. If the capsule is attached directly to the periphery of the epiphysial plate, damage to the joint may involve the plate and thereby interfere with growth.

INTRA-ARTICULAR STRUCTURES. Menisci, intra-articular discs, fat pads, and synovial folds (see fig. 3–1) are structures that aid in spreading synovial fluid throughout the joint. Thus they assist in joint lubrication.

Intra-articular discs and menisci, which are composed mostly of fibrous tissue but may contain some fibrocartilage, have other important functions.[6] They are attached at their periphery to the joint capsule, and they are usually present in joints where flexion and extension are associated with gliding (see knee, p. 223), a combination that requires a rounded (male) surface, and a relatively flattened (female) surface. The menisci and discs, the mobility of which is under ligamentous or muscular control, help to prevent instability and yet allow a considerable range of gliding.

PERIARTICULAR TISSUES. The term periarticular tissues is a general one, and refers to fascial investments around the joint. These blend with capsule and ligaments, with musculotendinous expansions that pass over or blend with the joint capsule, and with the looser connective tissue that invests the vessels and nerves approaching the joint. The periarticular tissues contain many elastic fibers, blood vessels, and nerves.

Joints are often injured, and they are subject to many disorders, some of which involve the periarticular tissues as well as the joints themselves. Increased fibrosis (adhesions) of the periarticular tissues may limit movement almost as much as does fibrosis within a joint.

Absorption from Joint Cavity.[7] A capillary network and a lymphatic plexus lie in the synovial membrane, adjacent to the joint cavity. Diffusion takes place readily between these vessels and the joint cavity. Therefore, traumatic infection of a joint may be followed by septicemia. However, colloidal particles, if above a certain size, are usually phagocytized when placed in a joint. Most substances in the blood stream, normal or pathological, easily enter the joint cavity.

Blood and Nerve Supply. The pattern of articular blood and nerve supply is illustrated in figure 3–3. Articular and epiphysial vessels arise more or less in common. Most epiphysial vessels enter a long bone at or near the line of capsular attachment and form a prominent arterial network around the joint.

Articular vessels ultimately break up into a rich capillary network, which is especially prominent in the cellular and areolar areas of synovial membrane.[8] Lymphatic vessels accompany blood vessels and form plexuses in the synovial membrane and capsule.

It is ordinarily believed that changes in temperature, humidity, or pressure make joints more sensitive, or painful, and there is undoubtedly some truth in these beliefs. It may well be that such changes reflexly alter the blood flow. However, very little is known about the control of blood flow in joints.

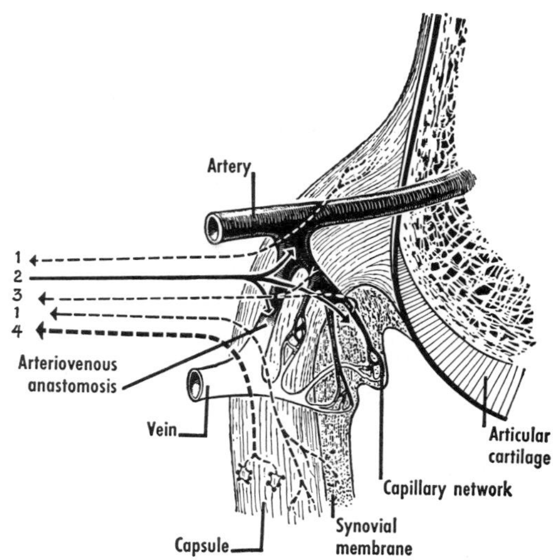

Figure 3–3 Schematic diagram of the blood and nerve supply of a synovial joint. An artery is shown supplying the epiphysis, joint capsule, and synovial membrane. Note the arteriovenous anastomosis. The nerve contains (1) sensory fibers (mostly pain) from the capsule and synovial membrane, (2) autonomic fibers (postganglionic sympathetic to blood vessels), (3) sensory fibers (pain, and others with unknown functions) from the adventitia of blood vessels, and (4) proprioceptive fibers, from Ruffini endings, and from small lamellated corpuscles (not shown). Arrows indicate direction of conduction.

The principles of distribution of nerves to joints were best expressed by Hilton: "The same trunks of nerves, whose branches supply the groups of muscles moving a joint, furnish also a distribution of nerves to the skin over the insertions of the same muscles; and—what at this moment more especially merits our attention—the interior of the joint receives its nerves from the same source."[9] Articular nerves vary in number and in course, and their areas of distribution overlap within joints. Articular nerves contain sensory and autonomic fibers, the distribution of which is summarized in figure 3–3.

Some of the sensory fibers form proprioceptive endings in the capsule and the ligaments. These endings are very sensitive to position and to movement. Their central connections are such that they are concerned in the reflex control of posture and locomotion, and in the detection of position and movement.[10]

Other sensory fibers form pain endings, which are most numerous in joint capsules and ligaments. Twisting or stretching of these structures is very painful. Studies of human joints opened under local anesthesia indicate that the fibrous capsule is highly sensitive, synovial membrane relatively insensitive.[11]

Use-Destruction (Wear and Tear, Attrition)[12]

Any moving mechanical system wears with time, and human joints are no exception. Some destruction is inevitable during normal activity. The most common result is the wearing away of articular cartilage to varying degrees, sometimes to the extent of exposing, eroding, and polishing or eburnating the underlying bone. Use-destruction may be hastened or exaggerated by many factors, of which the most important are trauma, disease, and biochemical changes in articular cartilage. These factors usually act by changing articular geometry, by decreasing synovial fluid viscosity, or both.

Development of Joints[13]

Most studies of the development of joints have been concerned with the synovial joints of the limbs. These joints begin to form during the embryonic period proper. By the end of this period, they closely resemble adult joints in form and arrangement. At about this time also, or early in the fetal period, joint cavities begin to appear, the synovial membrane begins to develop and become vascularized, and synovial fluid begins to be formed.

REFERENCES

1. For a review of the methods of measuring movement at a joint and of measuring muscle strength, see N. Salter, J. Bone Jt Surg., *37B*:474, 1955. See also *Joint Motion: Method of Measuring and Recording,* Amer. Acad. Orthop. Surg., Chicago, 1965, and W. P. Beetham, Jr., *et al., Physical Examination of the Joints,* Saunders, Philadelphia, 1965.
2. E. Radin and I. L. Paul, J. Bone Jt Surg., *54A*:607, 1972. Freeman, cited below.
3. J. G. Furey, W. S. Clark, and K. L. Brine, J. Bone Jt Surg., *41A*:167, 1959.
4. J. B. Roston and R. W. Haines, J. Anat., Lond., *81*:165, 1947.
5. F. J. Sääf, Acta orthopaed. scand., suppl. 7, 1950. R. Ekholm and B. Norbäck, Acta orthopaed. scand., *21*:81, 1951.
6. C. H. Barnett, J. Anat., Lond., *88*:363, 1954. P. Ring, Arch. d'Anat., d'Hist., d'Embryol., *53*:143, 1970.
7. E. W. O. Adkins and D. V. Davies, Quart. J. exp. Physiol., *30*:147, 1940.
8. D. V. Davies and D. A. W. Edwards, Ann. R. Coll. Surg. Engl., *2*:142, 1948.
9. John Hilton, *Rest and Pain,* edited by W. H. A. Jacobson and reprinted from the last London ed., Garfield, Cincinnati, 1891, p. 165.
10. E. Gardner, Spinal Cord and Brain Stem Pathways for Afferents from Joints, in *Ciba Foundation Symposium on Myotatic, Kinesthetic and Vestibular Mechanisms,* ed. by A. V. S. de Reuck and J. Knight, Churchill, London, 1967.
11. J. H. Kellgren and E. P. Samuel, J. Bone Jt Surg., *32B*:84, 1950.
12. A. W. Meyer, Calif. west. Med., *47*:375, 1937. C. H. Barnett, J. Bone Jt Surg., *38B*:567, 1956.
13. R. O'Rahilly, Irish J. med. Sci., p. 456, October, 1957. E. Gardner, J. Bone Jt Surg., *45*:856, 1963. See also references cited below.

GENERAL READING

Barnett, C. H., Davies, D. V., and MacConaill, M. A., *Synovial Joints,* Clowes and Sons, London, 1961. An excellent treatment of the structure and function of synovial joints, with a good bibliography.

Fick, R., in *Handbuch der Anatomie des Menschen,* ed. by K. von Bardeleben, Fischer, Jena, 1896, 1934, vol. 2 (8 volumes).

Freeman, M. A. R. (ed.), *Adult Articular Cartilage,* Pitman Medical, London, 1973. Excellent accounts, including mechanisms of lubrication and properties of synovial fluid.

Gardner, E., The Structure and Function of Joints, in *Arthritis,* ed. by J. L. Hollander and D. J. McCarty, Jr., Lea & Febiger, Philadelphia, 8th ed., 1972.

Hamerman, D., Rosenberg, L. C., and Schubert, M., Diarthrodial Joints Revisited, J. Bone Jt Surg., 52A:725, 1970.

Schubert, M., and Hamerman, D., *A Primer on Connective Tissue Biochemistry,* Lea & Febiger, Philadelphia, 1968. Includes chapters on synovial fluid and articular cartilage.

MUSCULAR SYSTEM

4

ERNEST GARDNER

No outward characteristic of animal life is more distinctive than that of movement. Movement is carried out by specialized cells called muscle fibers, the latent energy of which is, or can be, controlled by the nervous system. Muscle fibers are classified as skeletal or striated, cardiac, and smooth.

Skeletal muscle fibers are long, multinucleated cells having a characteristic cross-striated appearance under the microscope. These cells are supplied by motor fibers from cells in the central nervous system. The muscle of the heart is also composed of cross-striated fibers, but their activity is regulated by the autonomic nervous system. The walls of most organs and many blood vessels contain fusiform (spindle-shaped) muscle fibers that are arranged into sheets, layers, or bundles. These cells lack cross-striations and are therefore called smooth muscle fibers. Their activity is regulated by the autonomic nervous system and certain circulating hormones, and they supply the motive power for various aspects of digestion, circulation, secretion, and excretion.

Skeletal muscles are sometimes called *voluntary* muscles, owing to the fact that they can usually be controlled voluntarily. However, many of the actions of skeletal muscles are automatic, and the actions of some of them are reflex and only to a limited extent under voluntary control. Smooth muscle and cardiac muscle are sometimes spoken of as *involuntary* muscle.

SKELETAL MUSCLES

General Characteristics

Most muscles are discrete structures that cross one or more joints and, by contracting, can cause movements at these joints. Exceptions are certain subcutaneous muscles (e.g., facial muscles) that move or wrinkle the skin or close orifices, the muscles that move the eyes, and other muscles associated with the respiratory and digestive systems.

Each muscle fiber is surrounded by a delicate connective tissue sheath, the *endomysium.* Muscle fibers are grouped into fasciculi, each of which is enclosed by a connective tissue sheath termed *perimysium.* A muscle as a whole is composed of many fasciculi and is surrounded by *epimysium,* which is closely associated with fascia and is sometimes fused with it.

The fibers of a muscle of rectangular or quadrate shape run parallel to the long axis of the muscle. The fibers of a muscle of pennate shape are parallel to one another, but lie at an angle with respect to the tendon.[1] The fibers of a triangular or fusiform muscle are not parallel, but converge upon a tendon. The number of fibers in a muscle is dependent upon its shape, being greater in a pennate muscle than in one which is equal in size but is rectangular.

The names of muscles usually indicate some structural or functional feature. A name may indicate shape, for example, trapezius, rhomboid, or gracilis. A name may refer to location, for example, tibialis posterior. The number of heads of origin is indicated by the terms biceps, triceps, and quadriceps. Action is reflected in terms such as levator scapulae and extensor digitorum. Action and shape are combined in the term pronator quadratus, and action and location are combined in the term flexor digitorum profundus.

Muscles are quite variable in their attachments; they may be absent, and many supernumerary muscles have been described. Variations of muscles are so numerous that fairly complete accounts of them are available only in special works.[2]

Individual muscles are usually described according to their origin, insertion, nerve supply, and action. In the following discussion of nerve supply, certain features of blood supply are also included.

Origin and Insertion

Most muscles are attached either directly or by means of their tendons or aponeuroses to bones, cartilages, ligaments, or fasciae, or to some combination of these.

Other muscles are attached to organs, such as the eyeball, and still others are attached to skin. When a muscle contracts and shortens, one of its attachments usually remains fixed and the other one moves. The fixed attachment is called the *origin*, the movable one the *insertion*. In the limbs, the more distal parts are usually more mobile. Therefore the distal attachment is usually called the insertion. However, the terms origin and insertion are convenient merely for purposes of description. Very often the anatomical insertion remains fixed and the origin moves. Sometimes both ends remain fixed; the muscle then stabilizes a joint. The belly of a muscle is the part between the origin and the insertion.

Blood and Nerve Supply

Muscles are supplied by adjacent vessels. The pattern of supply varies, and several fundamental patterns have been described.[3] Some muscles are supplied by vessels that arise from a single stem and enter at one end of the muscle (gastrocnemius) or in the middle of the belly (biceps brachii). Such muscles are especially liable to necrosis from interruption of blood supply. Other muscles are supplied by a succession of anastomosing vessels (adductor magnus). The blood supply of some muscles (trapezius) exhibits features of both of the preceding types. Whatever the pattern of supply, the arteries that enter a muscle branch repeatedly and form a very extensive capillary bed.

Each muscle is supplied by one or more nerves, containing motor and sensory fibers that are usually derived from several spinal nerves. Some groups of muscles, however, are supplied mainly if not entirely by one segment of the spinal cord. For example, the motor fibers that supply the intrinsic muscles of the hand arise from the first thoracic segment of the spinal cord. Not infrequently, muscles having similar functions are supplied by the same peripheral nerve.

Nerves usually enter the deep surface of a muscle. The point of entrance is known as the "motor point" of a muscle, because electrical stimulation is more effective in producing muscular contraction than stimulation elsewhere on the muscle, since the nerve fibers are more sensitive to electrical stimulation than are muscle fibers.

Each motor nerve fiber that enters a muscle supplies many muscle fibers. The parent nerve cell and its motor fiber, together with the muscle fibers supplied, make up a *motor unit*.

Denervation of Muscle. Skeletal muscle cannot function without a nerve supply. A denervated muscle becomes flabby and atrophic. The process of atrophy consists of a decrease in size of individual muscle fibers. Each fiber shows occasional spontaneous contractions termed fibrillations. In spite of the atrophy, the muscle fibers retain their histologic characteristics for a year or more,[4] eventually being replaced by fat and connective tissue. Provided that nerve regeneration occurs, human muscles may regain fairly normal function up to a year after denervation.[5]

Actions and Functions

If a nerve to a muscle is given a brief electrical shock, the muscle responds by a brief contraction or twitch. If successive stimuli are rapidly applied, the contractions may summate, producing a prolonged contraction (tetanus). If the rate of stimulation is lessened, summation may be incomplete; the tetanus is said to be incomplete. In a muscle as a whole, gradation of activity is made possible by the number of motor units. If all the motor units are activated simultaneously, the muscle contracts once. But if motor units are activated out of phase or asynchronously (nerve impulses reaching motor units at different times), tension is maintained in the muscle.

The total force exerted by a muscle is the sum of the forces exerted by its individual fibers. Thus, of two muscles equal in size, the one with more fibers is the stronger. Muscle fibers are also said to be able to shorten to at least half their resting length, and muscles with long fibers are on that account said to be capable of producing a greater range of movement. Evidence has been presented that it is the length of fasciculi which determines the range of contraction of a muscle.[1] Whatever the mechanism, long and rectangular muscles produce a greater range of movement, whereas pennate muscles exert more force. Speed and power of movement are also related to the distance between the point of action and the axis of movement of the joint. Power is greatest when the insertion is far removed from the axis, whereas speed is usually greatest when the insertion is near the axis.

So many factors are involved, and so many variations may occur, that the interpretation of the mechanics of muscles is difficult. The actions are further complicated because some muscles cross two or more joints.

A muscle cannot contract to less than a certain minimal length (active insufficiency); attempts to contract to less than this length are generally painful. For example, the hamstring muscles that cross the hip and knee joints (p. 213) cannot shorten enough to extend the hip and flex the knee completely at the same time. A muscle cannot be stretched beyond a certain point without injury (passive insufficiency). If the hips are flexed fully, as in bending forward to touch the floor, the hamstrings may not be able to lengthen enough to allow one to touch the floor without bending the knees. This is also known as the ligamentous action; it restricts movement at a joint. It is due in part to relative inextensibility of connective tissue and tendons, and can be modified greatly by training, especially when training begins early in life.

The term contracture means a more or less permanent shortening of a muscle. It may be due to continued discharges from the central nervous system, to changes within muscle fibers leading to permanent shortening, or to a pathological increase in connective tissue in a muscle.

The pattern of muscular activity is controlled by the central nervous system. Most movements, even so-called simple ones, are complex and in many respects automatic. The overall pattern of movement may be voluntary, but the functions of individual muscles are complex, variable, and often not under voluntary control. For example, if one reaches out and picks something off a table, the use of the fingers is the chief movement. But in order to get the fingers to the object, the forearm is extended (the flexors relaxing), other muscles stabilize the shoulder, and still others stabilize the trunk and lower limbs so as to ensure maintenance of posture.

Muscles may be classified according to the functions they serve in such patterns, namely as prime movers, antagonists, fixation muscles, and synergists. A special category includes those that have a paradoxical or eccentric action, in which muscles lengthen while contracting (fig. 4–1). In so doing, they perform negative work. A mus-

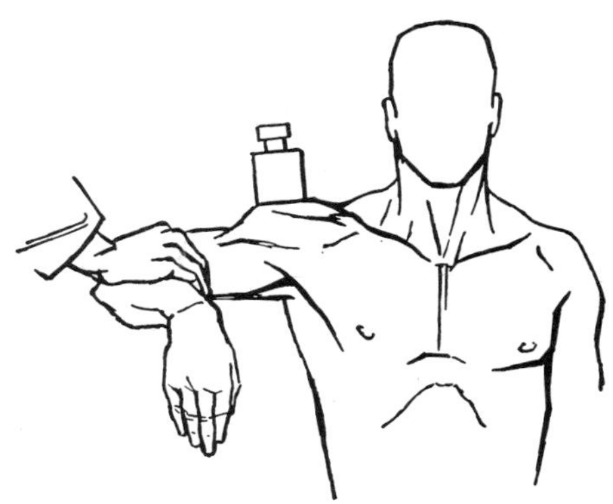

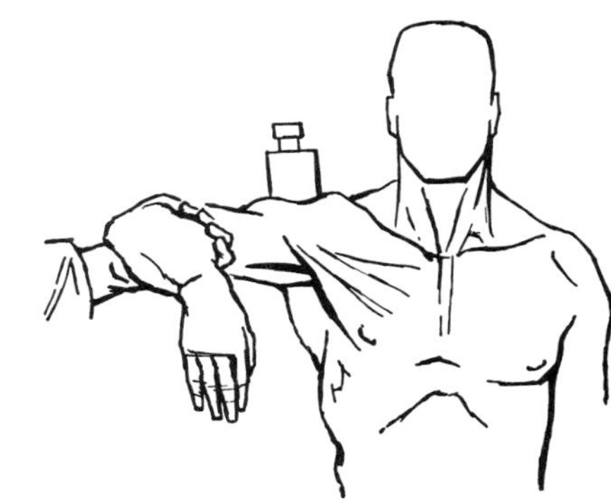

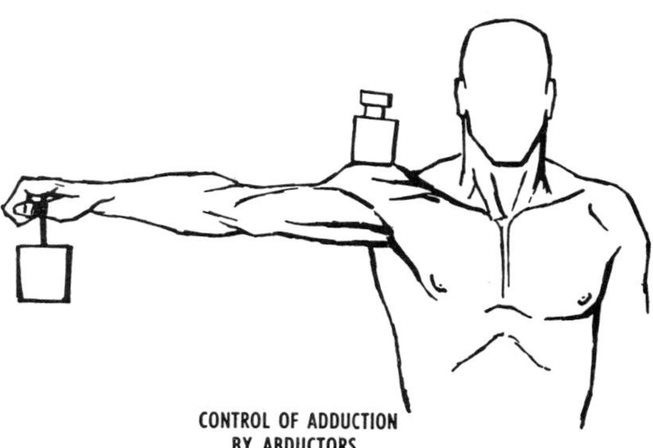

Figure 4–1 Muscle actions, drawn from life. When the subject abducts his arm against the resistance offered by the examiner, the abductor (deltoid) becomes tense. When he adducts against resistance, the deltoid relaxes and the weight sinks into it. When he adducts by lowering a pail from a horizontal position, the adductor (pectoralis major) is relaxed. The contracted deltoid controls the descent (adduction) by lengthening. In this situation, therefore, the deltoid is an antagonist to gravity, which is the prime mover, and the deltoid is doing negative work (paradoxical action). See also figure 4–2.

cle may be a prime mover in one pattern, an antagonist in another, or a synergist in another. Generally, in a description of a muscle, the action that is listed first is that which the muscle performs as a prime mover.

Prime Movers. A prime mover (fig. 4–1) is a muscle or a group of muscles that directly brings about a desired movement (flexion of the fingers in the movement cited previously). Gravity may also act as a prime mover. For example, if one holds an object and lowers it to the table, gravity brings about the lowering (fig. 4–1). The only muscular action involved is in controlling the rate of descent, an example of paradoxical action.

Antagonists. Antagonists or opponents are muscles that directly oppose the movement under consideration. Thus, the triceps brachii, which is the extensor of the forearm when acting as a prime mover, is the antagonist to the flexors of the forearm. Depending on the rate and force of movement, antagonists may be relaxed, or, by lengthening while contracting, they may control movement and make it smooth, free from jerkiness, and precise. The term antagonist is a poor one, because such muscles cooperate rather than oppose. Gravity may also act as an antagonist, as when the forearm is flexed from the anatomical position.

Fixation Muscles. Fixation muscles have a wide variety of functions. Generally they stabilize joints or parts and thereby maintain posture or position while the prime movers act.

Synergists. Synergists are a special class of fixation muscles. When a prime mover crosses two or more joints, synergists prevent undesired actions at intermediate joints. Thus, the long muscles that flex the fingers would at the same time flex the wrist if the wrist were not stabilized by the extensors of the wrist, these being synergists in this particular movement.

Testing of Muscles

Five chief methods are available to determine the action of a muscle. These are the anatomical method, palpation, electrical stimulation, electromyography, and the clinical method. No one of these methods alone is sufficient to provide full and accurate information about the actions and functions of muscles.

Anatomical Method. Actions are deduced from the origin and insertion as determined by dissection, and are checked by pulling upon the muscle, for example, during surgery. The anatomical method is usually the only way of determining the action of muscles too deep to be examined during life. Most of the actions ordinarily attributed to muscles have been determined by the anatomical method. The disadvantage of this method is that it determines all possible actions out of context. In other words, it tells what a muscle can do, but not necessarily what it actually does.

Palpation. In this method, the subject is asked to perform a certain movement and the examiner inspects and palpates the muscles taking part.

The movement may be carried out without loading or extra weight, and with gravity minimized so far as possible by support or by the recumbent position. Alternatively, the movement may be carried out against gravity, as when flexing the forearm from the anatomical position, with or without extra load. Finally it may be tested with a heavy load, most simply by fixing the limb by an opposing force. For example, the examiner requests the subject to flex his forearm and at the same time holds the subject's forearm so as to prevent flexion. Palpation of muscles that are contracting against resistance provides the best and simplest way of learning the locations and actions of muscles in the living body. Palpation is also the simplest and most direct method of testing weak or paralyzed muscles, and is widely used by those working with patients.

Palpation has certain disadvantages. Deep-seated muscles cannot be reached. When several muscles take part in a movement pattern, it may not be possible to determine the functions of each muscle by palpation alone.

Electrical Stimulation. The electrical stimulation of a muscle over its motor point causes the muscle to contract, and to remain contracted if repetitive stimulation is used. It has the advantage that it can be used in living subjects, although deep-seated muscles may be inaccessible. This is a valuable method, but one with some disadvantages. Like the anatomical method, it tells what a muscle can do, but not necessarily what its functions are. Furthermore, results may be misleading. If the deltoid muscle, which is attached to the scapula and humerus, is stimulated, both bones move and the point of the shoulder is depressed. When the deltoid acts normally, the scapula is fixed by other muscles so that the humerus can be abducted. In other words, a muscle action determined out of context may not be the natural one. The electrical method has the advantage of showing what a muscle cannot do, and this is important in evaluating actions determined by other methods.

Electromyography (fig. 4–2). The mechanical twitch of a muscle fiber is preceded by a conducted impulse that can be detected and recorded with appropriate instruments. When an entire muscle is active, the electrical activity of its fibers can be detected by electrodes placed within the muscle or on the overlying skin. The recorded response constitutes an electromyogram (EMG). Recording instruments are such that records can be obtained from several muscles simultaneously. This makes electromyography a valuable method for studying patterns of activity. Electromyography, like palpation, may therefore be classified as a natural or physiological method. Finally, the electromyographic pattern may be altered by nervous or muscular disease. Electromyography, therefore, can be used as a diagnostic aid.

The disadvantages are the same as those of palpation, namely, difficulty in assessing the precise function of a muscle that is taking part in a movement pattern. For example, electromyography consistently shows that the tensor fasciae latae muscle contracts when the leg is extended. But if care is taken to immobilize the hip so as to prevent hip flexion (one of the prime functions of the muscle), activity during leg extension disappears. Moreover, the tensor fasciae latae also contracts during abduction of the thigh, yet does not abduct the thigh when it is stimulated electrically. It may be acting as an antagonist during abduction.

Clinical Method. A study of patients who have muscles or muscle groups that are paralyzed provides valuable information about muscle function, primarily by determining which functions are lost.[6] But great caution must be exercised. In some central nervous system disorders, a muscle may be paralyzed in one movement, yet take part in another. Even in the presence of peripheral nerve injuries or direct muscle involvement, patients may learn trick movements

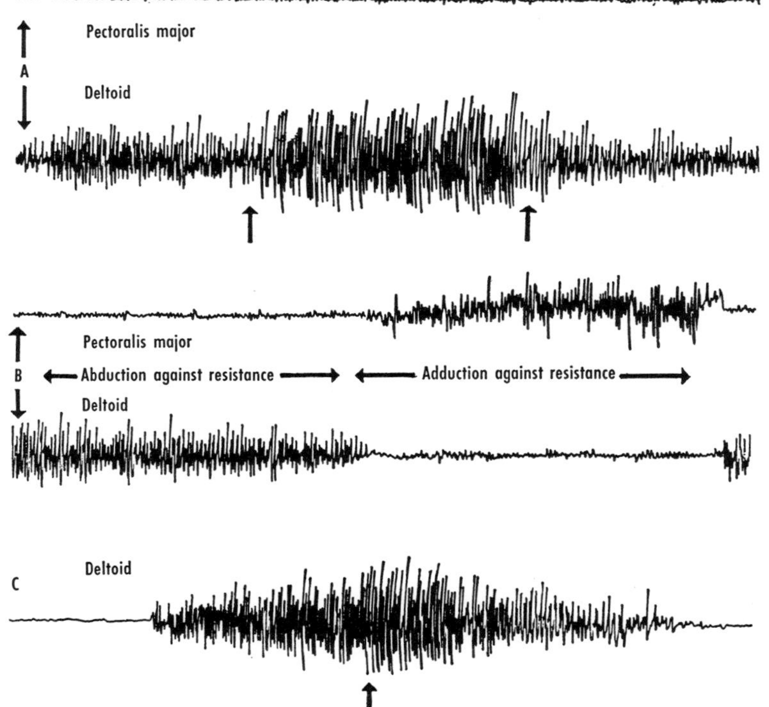

Figure 4–2 Electromyograms from a subject during muscular activity similar to that shown in figure 4–1. Electrodes on the surface of the deltoid and pectoralis major were led to amplifiers and these to an ink-writing recorder. The deflections represent the algebraic sum of the changes in electrical potential of many muscle fibers. *A*, holding the upper limb horizontally in a frontal plane. Arrow at left, below record from deltoid, attempting to elevate the upper limb further against resistance. Arrow at right, resistance released. *B*, alternating abduction and adduction against resistance. *C*, lifting a weight to a horizontal level (arrow indicates reaching this level) and lowering it.

with other muscles that compensate for or mask the weakness or paralysis.[7]

Reflexes and Muscle Tone

Many muscular actions are reflex in nature; that is, they are brought about by sensory impulses that reach the spinal cord and activate motor cells. The quick withdrawal of a burned finger and the blinking of the eyelids when something touches the cornea are examples of reflexes. It is ordinarily held that muscles that support the body against gravity possess tone, owing to the operation of stretch reflexes initiated by the action of gravity in stretching the muscles. Whether this is strictly or always true in man is open to question. There is evidence that, in an easy standing position, little if any muscular contraction or tone can be detected in the anti-gravity muscles of man.[8]

The available evidence indicates that the only "tone" possessed by a completely relaxed muscle is that due to its passive elastic tension.[9] No impulses reach a completely relaxed muscle and no conducted electrical activity can be detected.

Structure and Function

Each skeletal muscle fiber is a long, multinucleated cell that consists of a mass of myofibrils, each about 1 to 2 micrometers thick. The myofibrils lie parallel to the long axis of the fiber, which is enclosed in an apparently structureless membrane, the *sarcolemma*. The striated appearance of the muscle fiber results from the fact that different parts of the myofibrils have different indices of refraction. Corresponding parts of myofibrils are aligned (in register) with each other, giving the appearance of discs traversing the whole thickness of the fiber.

Muscle fibers vary greatly in length.[10] Most are less than 10 to 15 cm long, but some fibers may be more than 30 cm long. Some fibers are shorter than the fasciculi in which they lie, in which case muscle fibers are arranged serially, and are joined at their ends by short, fibrous bridges. Other fibers may be as long as their fasciculi.

Resting muscle is soft, freely extensible, and elastic. Active muscle is hard, develops tension, resists stretching, and lifts loads. Muscles may thus be compared with machines for converting chemically stored energy into mechanical work. Muscles are also important in the maintenance of body temperature. Resting muscle under constant conditions liberates heat, which forms a considerable fraction of the basic metabolic rate.

Of all the changes that occur after death, one of the most characteristic is stiffening of the corpse, as a result of stiffening of muscles. This change is known as rigor mortis. Its time of onset and its duration are variable. Rigor mortis is due chiefly to the loss of adenosine triphosphate from the muscles after death.

TENDONS AND APONEUROSES

The attachment of muscle to bone (or other tissue) is usually by a long, cordlike tendon or sinew, or by a broad, relatively thin aponeurosis. Tendons and aponeuroses are both composed of more or less parallel bundles of collagenous fibers. Tendons are surrounded by *epitendineum*, a thin, fibroelastic sheath of looser connective tissue that extends inward between the bundles. The surfaces of aponeuroses are covered with similar tissue. Where tendons are attached to bone, the bundles of collagenous fibers fan out in the periosteum.

Tendons are supplied by sensory fibers that reach them from muscle nerves. They also receive sensory fibers from overlying superficial nerves or from nearby deep nerves.[11] Tendons are usually supplied by arteries that anastomose within tendons to form a single longitudinal artery, accompanied by veins and lymphatic vessels.[12] The vascular needs of tendons are small. Tendons can be cut and transplanted with relative impunity. Tendons are destroyed very slowly by inflammatory processes, and infected tendons heal very slowly.

Synovial Tendon Sheaths

Where tendons run in osseofibrous tunnels, for example, in the hand and foot, they are covered by double-layered synovial sheaths (fig. 4–3). The *mesotendineum*, which is the tissue that forms the continuity between the synovial layers, carries blood vessels to the tendon. The inner layer of the synovial sheath is fused with the epitendineum. The fluid in the cavity of the sheath is similar to synovial fluid and facilitates movement by minimizing friction.

The lining of the sheath, like synovial membrane, is extremely cellular and vascular. It reacts to infection or to trauma by

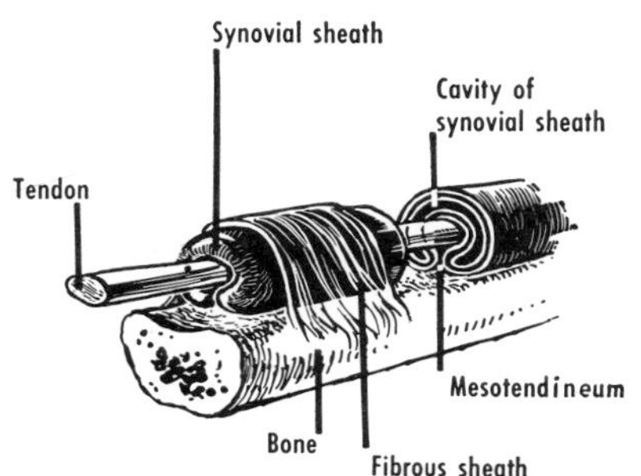

Figure 4–3 Schematic drawing of synovial and fibrous sheaths of a tendon, with a section of the synovial sheath.

forming more fluid and by cellular proliferation. Such reactions may result in adhesions between the two layers and a consequent restriction of movement of the tendon.

BURSAE

Bursae (from *L. bursa,* a purse), like synovial tendon sheaths, are connective tissue sacs with a slippery inner surface, and are filled with synovial fluid. Bursae are present where tendons rub against bone, ligaments, or other tendons, or where skin moves over a bony prominence. They may develop in response to friction. Bursae facilitate movement by minimizing friction.

Bursae are of clinical importance. Some communicate with joint cavities, and to open such a bursa is to enter the joint cavity, always a potentially dangerous procedure from the standpoint of infection. Some bursae are prone to fill with fluid when injured, for example, the bursae in front of or below the patella (housemaid's knee).

FASCIA[13]

Fascia is a packing material, a connective tissue that remains between areas of more specialized tissue, such as muscle. Fascia forms fibrous membranes that separate muscles from one another and invest them, and as such is often called deep fascia. Its several functions include providing origins and insertions for muscles, serving as an elastic sheath for muscles, and forming specialized retaining bands (retinacula) and fibrous sheaths for tendons. It provides pathways for the passage of vessels and nerves and surrounds these structures as neurovascular sheaths. It permits the gliding of one structure on another. The mobility, elasticity, and slipperiness of living fascia can never be appreciated by dissecting embalmed material.

The main fascial investment of some muscles is indistinguishable from epimysium. Other muscles are more clearly separated from fascia, and are freer to move against adjacent muscles. In either instance, muscles or groups of muscles are generally separated by intermuscular septa, which are deep prolongations of fascia.

In the lower limb, the return of blood to the heart is impeded by gravity and aided by muscular action. However, muscles would swell with blood were it not for the tough fascial investment of these muscles, serving as an elastic stocking. The investment also prevents bulging during contraction and thus makes muscular contraction more efficient in pumping blood upward.

Fascia is more or less continuous over the entire body, but it is commonly named according to region, for example, pectoral fascia. It is attached to the superficial bony prominences that it covers, blending with periosteum, and, by way of intermuscular septa, is more deeply attached to bone.

Fascia may limit or control the spread of pus. When shortened because of injury or disease, fascia may limit movement. Strips of fascia are sometimes used for the repair of tendinous or aponeurotic defects.

Most fascial layers are sparingly supplied with free nerve endings and simple, small, encapsulated endings. Proprioceptive endings are numerous in aponeuroses and retinacula.[14] This abundance suggests that these structures have a kinesthetic as well as a mechanical function.

REFERENCES

1. W. Pfuhl, Z. Anat. EntwGesch., *106*:749, 1937.
2. L. Testut, *Les anomalies musculaires chez l'homme expliquées par l'anatomie comparée,* Masson, Paris, 1884. A. F. Le Double, *Traité des variations du système musculaire de l'homme et de leur signification au point de vue de l'anthropologie zoologique,* Schleicher, Paris, 1897, 2 volumes.
3. J. Campbell and C. M. Pennefather, Lancet, *1*:294, 1919. L. B. Blomfield, Proc. R. Soc. Med., *38*:617, 1945. R. L. de C. H. Saunders *et al., The Anatomic Basis of the Peripheral Circulation in Man,* part V of W. Redisch and F. F. Tangco, *Peripheral Circulation in Health and Disease,* Grune & Stratton, New York, 1957.
4. R. E. M. Bowden and E. Gutmann, Brain, *67*:273, 1944. S. Sunderland and L. J. Ray, J. Neurol. Psychiat., *13*:159, 1950.
5. S. Sunderland, Arch. Neurol. Psychiat., Chicago, *64*:755, 1950.

6. For methods of testing as used by physical therapists, see L. Daniels and C. Worthingham, *Muscle Testing,* Saunders, Philadelphia, 3rd ed., 1972.
7. F. W. Jones, J. Anat., Lond., *54*:41, 1919. S. Sunderland, Aust. N.Z. J. Surg., *13*:160, 1944.
8. I. W. Kelton and R. D. Wright, Aust. J. exp. Biol. med. Sci., *27*:505, 1949. See also chapter on posture and locomotion (p. 249).
9. S. Clemmesen, Proc. R. Soc. Med., *44*:637, 1951.
10. B. Barrett, Acta anat., *48*:242, 1962.
11. D. L. Stilwell, Jr., Amer. J. Anat., *100*:289, 1957.
12. D. A. W. Edwards, J. Anat., Lond., *80*:147, 1946.
13. B. B. Gallaudet, *A Description of the Planes of Fascia of the Human Body,* Columbia University Press, New York, 1931. E. Singer, *Fasciae of the Human Body and Their Relations to the Organs They Envelop,* Williams & Wilkins, Baltimore, 1935.
14. D. L. Stilwell, Jr., Amer. J. Anat., *100*:289, 1957; Anat. Rec., *127*:635, 1957.

GENERAL READING

Basmajian, J. V., *Muscles Alive,* Williams & Wilkins, Baltimore, 3rd ed., 1974. An excellent and timely study of muscle functions, as revealed by electromyography.

Beevor, C., *The Croonian Lectures on Muscular Movements,* 1903, and *Remarks on Paralysis of the Movements of the Trunk in Hemiplegia,* 1909, edited and reprinted, Macmillan, London. A classic study of muscle actions based on first-hand observations of living subjects. The lower limb is not included.

Bourne, G. H. (ed.), *The Structure and Function of Muscle,* Academic Press, New York, 2nd ed., vol. 1, 1972.

Duchenne, G. B., *Physiology of Motion* (trans. and ed. by E. B. Kaplan), Saunders, Philadelphia, 1959. A classic study of muscle actions, by the use of electrical stimulation and by the study of thousands of patients with various paralyses.

Lockhart, R. D., *Living Anatomy,* Faber and Faber, London, 6th ed., 1963. Photographs of subjects showing muscles in action and methods of testing.

Royce, J., *Surface Anatomy,* Davis, Philadelphia, 1965. Photographs and key drawings of living subjects.

Wright, W. G., *Muscle Function,* Hoeber, New York, 1928. Wright examined thousands of normal and paralyzed patients. Her findings closely resemble those of Beevor and Duchenne.

NERVOUS SYSTEM 5

ERNEST GARDNER

The nervous system may be divided into the *central* nervous system, consisting of the brain and spinal cord, and the *peripheral* nervous system, consisting of the cranial, spinal, and peripheral nerves, together with their motor and sensory endings. The term *nerve* is derived from the Latin *nervus,* meaning nerve, and this in turn from the Greek *neuron,* meaning sinewlike structures, formerly including tendons and aponeuroses as well as nerves.

CENTRAL NERVOUS SYSTEM

The central nervous system is composed of billions of nerve and glial cells, together with blood vessels and a small amount of connective tissue. The nerve cells or neurons are characterized by many processes and are specialized in that they exhibit to a great degree the phenomena of irritability and conductivity. The glial cells are termed *neuroglia.* They are characterized by short processes that have special relationships to neurons, blood vessels, and connective tissue.

Brain

The brain or cerebrum is the enlarged head end of the central nervous system; it occupies the cranium or brain case. The term *cerebrum* (Latin, brain) has been used in various ways. Generally speaking it means brain; it has also been used to mean specifically the forebrain and midbrain. The adjective *cerebral* is derived from it. By contrast, *encephalon* is of Greek origin *(enkephalos).* Terms such as encephalitis, which means inflammation of the brain, are derived from it.

The human brain, like those of all other vertebrates, presents three divisions, each of which has certain relatively constant components and subdivisions. The three parts are the forebrain *(prosencephalon),* midbrain *(mesencephalon),* and hindbrain *(rhombencephalon).* The forebrain in turn has two subdivisions, *telencephalon* (endbrain) and *diencephalon* (interbrain). The hindbrain likewise has two subdivisions, the *metencephalon* (afterbrain) and the *myelencephalon* (marrowbrain). The bulk of the brain is formed by two convoluted cerebral hemispheres, which are derived from the telencephalon. The hemispheres are distinguished by the folds or convolutions of their surfaces; they form gyri, which are separated by sulci. The unpaired diencephalon lies between the hemispheres. It forms the upper part of what is generally called the *brain stem,* an unpaired stalk that descends from the base of the brain. The brain stem is formed by the diencephalon, midbrain, pons, and the myelencephalon, or medulla oblongata. The last is continuous with the spinal cord at the foramen magnum. The *cerebellum* is a fissured mass of gray matter that occupies the posterior cranial fossa and is attached to the brain stem by three pairs of peduncles. Twelve pairs of cranial nerves issue from the base of the brain and from the brain stem.

The cerebral cortex, which is the outer part of the hemispheres and is only a few millimeters in thickness, is composed of gray matter, in contrast to the interior of the brain, which is composed partly of white matter. Gray matter consists largely of the bodies of nerve and glial cells, whereas white matter consists largely of the processes or fibers of nerve and glial cells.

The interior of the cerebral hemispheres, including the diencephalon, contains not only white matter but also well-demarcated masses of gray matter known collectively as basal ganglia.* The most prominent of these are the caudate nuclei, the lentiform nuclei, and the thalami.

*In spite of this term, a collection of nerve cells within the central nervous system is usually termed a nucleus, whereas a collection of nerve cells outside the central nervous system is usually termed a ganglion.

The cerebellar cortex, like the cerebral, is composed of gray matter. The interior of the cerebellum is composed mainly of white matter, but also contains nuclei of gray matter.

The brain stem, by contrast, contains nuclei and diffuse masses of gray matter in its interior. A prominent feature is a diffuse mixture of white and gray matter, termed the *reticular formation,* which extends longitudinally throughout the brain stem.

The interior of the brain also contains cavities termed *ventricles,* which are filled with cerebrospinal fluid. These are discussed further in Part Eight, Head and Neck.

Functions. The highest mental and behavioral activities characteristic of humans are mediated by the cerebral hemispheres, in particular the cerebral cortex. Important aspects of these functions are learning and language. In addition, there are association mechanisms for the integration of motor and sensory functions. Some areas of the cerebral hemispheres control muscular activity, and their nerve cells send processes to the brain stem and spinal cord, where they are connected with motor cells, the processes of which leave by way of cranial nerves or ventral roots. Other areas are sensory and receive impulses that have reached the spinal cord by way of peripheral nerves and dorsal roots, and have ascended in the spinal cord and brain stem by pathways that consist of a succession of nerve cells and their processes. Fibers that ascend and descend in the brain and spinal cord are generally grouped into tracts. The tracts are usually named according to their origin and destination, sometimes by their position also. Thus, a corticospinal tract is a collection of fibers originating in the cerebral cortex and ending in the spinal cord; a spinothalamic tract begins in the spinal cord and ends in the thalamus.

The brain stem, which is quite similar in structure and function in all vertebrates, contains, in addition to tracts that descend and ascend through it, collections of cells that (1) comprise major integrating centers for motor and sensory functions, (2) form the nuclei of most cranial nerves (all of the cranial nerves except the first are attached to the brain stem), (3) form centers concerned with the regulation of a variety of visceral, endocrinological, behavioral, and other activities, (4) are functionally associated with most of the special senses, (5) control muscular activity in the head and part of the neck, (6) supply pharyngeal arch structures, and (7) are connected with the cerebellum.

The cerebellum is an important organ concerned with the automatic regulation of movement and posture. It functions closely with the cerebral cortex and the brain stem. Certain groups of cerebellar neurons regulate trunk muscles, others limb muscles, and still others are connected with the cerebral cortex. The anatomical arrangement of the cerebellum varies greatly among vertebrates, depending on mode of locomotion. The cerebellum is relatively more developed in primates, especially in humans.

Spinal Cord

The spinal cord is a long, almost cylindrical mass of nervous tissue, oval or rounded in transverse section, that occupies about the upper two-thirds of the vertebral canal. In contrast to the cerebral hemispheres, gray matter is found in the interior, surrounded by white matter (fig. 5–1). The spinal cord has essentially the same arrangement throughout.

The neurons of the spinal cord include (1) motor cells, the axons of which leave by way of ventral roots and supply skeletal muscles; (2) motor cells, the axons of which leave by way of ventral roots and go to autonomic ganglia (p. 35); and (3) transmission neurons and interneurons, concerned with sensory and reflex mechanisms. The white matter contains ascending and descending tracts. Some ascend to or descend from the brain, whereas others connect cells at various levels of the cord.

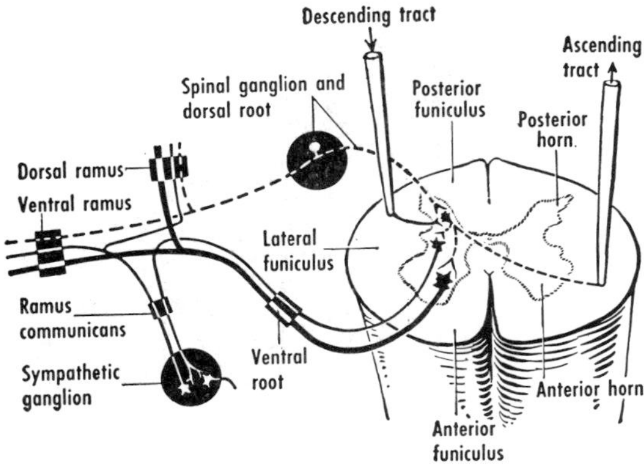

Figure 5–1 Diagram of a horizontal section of the spinal cord, with dorsal and ventral roots and a spinal nerve. The arrangement of the rami communicantes is usually much more complicated (see p. 335).

Attached to the spinal cord on each side is a series of pairs of nerves, the *spinal roots*, termed *dorsal* and *ventral* according to their position. Generally there are 31 pairs, which comprise 8 cervical, 12 thoracic, 5 lumbar, 5 sacral, and 1 coccygeal. Corresponding dorsal and ventral roots join to form a spinal nerve. Each spinal nerve divides into a dorsal and a ventral (primary) ramus, the distribution of which is described on page 33.

The spinal cord is segmental in arrangement and exhibits little of the diversification and specialization so characteristic of the brain. It carries out sensory, integrative, and motor functions, which can be categorized as reflex, reciprocal activity (as one activity starts, another stops), monitoring and modulation of sensory and motor mechanisms, and transmission of impulses to the brain.

Meninges[1]

The brain and spinal cord are surrounded and protected by layers of non-nervous tissue, collectively termed meninges. These layers, from without inward, are the *dura mater*, *arachnoid*, and *pia mater*, which are described in more detail elsewhere. The space between the arachnoid and the pia mater, the *subarachnoid space*, contains cerebrospinal fluid.

Cerebrospinal Fluid (C.S.F.)

The ventricles of the brain contain vascular choroid plexuses, from which an almost protein-free cerebrospinal fluid is formed.[2] This fluid circulates through the ventricles, enters the subarachnoid space, and eventually filters back into the venous system. Cerebrospinal fluid protects the brain and serves to minimize damage from blows to the head and neck.

The pressure of cerebrospinal fluid, which is usually between 100 and 200 mm of water, is most readily measured during lumbar puncture (p. 545). Fluid may also be withdrawn for studies of cellular and chemical content. Many neurological disorders alter the hydrodynamics of the fluid, as well as its cellular and chemical content. Consequently, studies of such changes are valuable diagnostic methods. Fluid that is withdrawn may be replaced by air or by a radio-opaque oil. Because the air or oil can be detected radiographically, the position of a mass, such as a tumor, which occupies, narrows, or blocks the subarachnoid space, may be determined. Anesthetics, such as procaine, may also be introduced for spinal anesthesia.

Blood Supply

The brain is supplied by the cerebral branches of the vertebral and the internal carotid arteries, the meninges mainly by the middle meningeal branch of the maxillary artery. The spinal cord and spinal roots are supplied by the vertebral arteries and by segmental arteries. Peripheral nerves are supplied by a number of small branches along the course of the nerves.

PERIPHERAL NERVOUS SYSTEM

A *nerve* is a collection of nerve fibers that is visible to the naked eye; the constituent fibers are bound together by connective tissue. Each fiber is microscopic in size and is surrounded by a sheath formed by a neurilemmal cell (comparable to the glial cells of the central nervous system). Hundreds or thousands of fibers are present in each nerve. Thus, according to the number of constituent fibers, a nerve may be barely visible, or it may be quite thick. A nerve as a whole is surrounded by a connective tissue sheath, the *epineurium.* Connective tissue fibers run inward from the sheath and enclose bundles of nerve fibers. Such bundles are termed *funiculi* (fasciculi); the connective tissue that encloses them is called *perineurium.*[3] The inner, smooth surface of perineurium is formed by a membrane of flattened mesothelial cells. Very small nerves may consist of only one funiculus derived from the parent nerve. Finally, each nerve fiber is enclosed by a connective tissue sheath termed *endoneurium.* The connective tissue of a nerve imparts great strength and contains the blood vessels that supply the nerve. Spinal roots lack well-defined sheaths and are much more fragile.

Nerve fibers may be classified according to the structures they supply, that is, according to function. A fiber that stimulates or activates skeletal muscle is termed a *motor* (efferent) fiber. A fiber that carries impulses from a sensory ending is termed a *sensory* (afferent) fiber. But fibers that activate glands and smooth muscle are also motor fibers, and various kinds of sensory fibers arise from endings in viscera. Consequently, a more detailed classification of functional components is sometimes used (see p. 598).

Spinal Nerves

Each lateral half of the spinal cord has attached to it the spinal roots. These consist of a dorsal root, attached to the dorsal aspect of the spinal cord, and a ventral root, attached to the ventral aspect of the spinal cord. Each dorsal root (which contains sensory fibers from skin, subcutaneous and deep tissues, and often from viscera also) is formed by neuronal processes that carry sensory impulses into the spinal cord, and which arise from neurons that are collected together to form an enlargement termed a *spinal ganglion* (fig. 5–1). Each ventral root (which contains motor fibers to skeletal muscle, and many of which contain preganglionic autonomic fibers) is formed by processes of neurons in the gray matter of the spinal cord. As mentioned previously, corresponding dorsal and ventral roots join to form a spinal nerve. Each spinal nerve then divides into a dorsal and a ventral ramus.

Distribution of Spinal and Peripheral Nerves[4]

The dorsal rami of spinal nerves supply the skin and muscles of the back. The ventral rami supply the limbs and the rest of the trunk. The ventral rami that supply the thoracic and abdominal wall remain relatively separate throughout their course. In the cervical and lumbosacral regions, however, the ventral rami intermingle to form plexuses, from which the major peripheral nerves emerge.

When the ventral ramus of a spinal nerve enters a plexus and joins other such rami, its component funiculi or bundles ultimately enter several of the nerves emerging from the plexus. Thus, as a general principle, each spinal nerve entering a plexus contributes to several peripheral nerves, and each peripheral nerve contains fibers derived from several spinal nerves. This arrangement leads to two fundamental and important types of distribution (fig. 5–2). Each spinal nerve has a pattern of distribution often referred to as *segmental* or *dermatomal.* The term *dermatome* refers to skin, and a dermatome is the area of skin supplied by the sensory fibers of a single dorsal

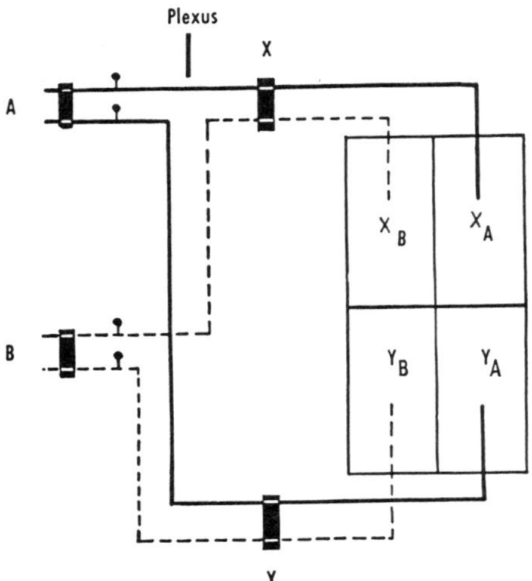

Figure 5–2 Schematic diagram of spinal and peripheral nerve distribution. Only sensory fibers to the skin are represented. Two nerve fibers of spinal nerve *A* are shown entering a plexus. One of the fibers joins peripheral nerve *X*, and the other joins peripheral nerve *Y*. Two fibers of spinal nerve *B* also join the two peripheral nerves. Thus, the areas supplied by the two spinal nerves are different from the areas supplied by the two peripheral nerves, as shown in the subdivided rectangle. Overlap is omitted. Modified from Gardner, cited in General Reading.

root through the dorsal and ventral rami of its spinal nerve.

The mixture of nerve fibers in plexuses is such that it is difficult if not impossible to trace their course by dissection, and dermatomal distribution has thus been determined by physiological experimentation and by studies of disorders of spinal nerves. Methods that have been used include stimulation of spinal roots, study of residual sensation when a root is left intact after section of the roots above and below it,[5] study of the diminution of sensation after section of a single root, and study of the distribution of the blisters that follow inflammation of roots and spinal ganglia in herpes zoster (shingles).[6] The results of such studies have yielded complex maps, chiefly because of variation, overlap, and differences in method. Variation results from intersegmental rootlet anastomoses adjacent to the cervical and lumbosacral spinal cord, and from individual differences in plexus formation and peripheral nerve distribution. Overlap is such that section of a single root does not produce complete anesthesia in the area supplied by that root; at the most, some degree of hypalgesia may result. By contrast, when a peripheral nerve is cut, the result is a central area of total loss of sensation surrounded by an area of diminished sensation. Approximately accurate dermatomes are shown in the figures inside the front cover; they are based on Foerster.[5]

There is little specific correspondence between dermatomes and underlying muscles. The general arrangement is that the more rostral segments of the cervical and lumbosacral enlargements of the spinal cord supply the more proximal muscles of the limbs, and that the more caudal segments supply the more distal muscles. A muscle usually receives fibers from each of the spinal nerves that contribute to the peripheral nerve supplying it (although one spinal nerve may be its chief supply). Section of a single spinal nerve weakens several muscles but usually does not paralyze them. Section of a peripheral nerve results in severe weakness or total paralysis of the muscles it supplies. Moreover, autonomic dysfunction occurs in the area of its distribution.

Cranial Nerves

The 12 pairs of cranial nerves are special nerves associated with the brain. The fibers in cranial nerves are of diverse functional types. Some cranial nerves are composed of only one type, others of several.

Cranial nerves differ significantly from spinal nerves, especially in their mode of embryologic development and in their relation to the special senses, and because some cranial nerves supply pharyngeal arch structures. They are attached to the brain at irregular rather than regular intervals; they are not formed of dorsal and ventral roots; some have more than one ganglion, whereas others have none; and the optic nerve is a tract of the central nervous system and not a true peripheral nerve.

Characteristic Features of Peripheral Nerves

The branches of major peripheral nerves are usually muscular, cutaneous (or mucosal), articular, vascular (to adjacent blood vessels), and terminal (one, several, or all of the foregoing types). Muscular branches are the most important; section of even a small muscular branch results in complete paralysis of all muscle fibers supplied by that branch, and may be seriously disabling. The importance of sensory loss varies according to the region. The loss is most disabling in the case of the hand and certain parts of the head and face.

Peripheral nerves vary in their course and distribution, but not as much as blood vessels do. Adjacent nerves may communicate with each other. Such communications sometimes account for residual sensation or movement after section of a nerve above the level of a communication (p. 137). Peripheral nerves have an excellent blood supply, with longitudinal channels which anastomose to such an extent that up to 15 cm of a nerve may be stripped of epineurium without depriving that stretch of nerve of its blood supply.[7]

AUTONOMIC NERVOUS SYSTEM

The term *autonomic nervous system* refers to the parts of the nervous system that regulate the activity of cardiac muscle, smooth muscle, and glands. The term, however, implies an autonomy that does not always exist. For example, skin exposed to cold air becomes blanched or pale because of a reflex constriction of blood vessels in the skin. The cold air stimulates temperature receptors in the skin, and the spinal reflex, by way of autonomic fibers to blood vessels, acts to conserve heat. Impulses also reach the brain, and a sensation of cold results. This is an example of coordination of somatic and automatic activities.

The autonomic nervous system can be considered as a series of levels which differ in function in that the higher the level, the more widespread and general its functions; the lower the level, the more restricted and specific the functions. The highest level is the cerebral cortex, certain areas of which control or regulate visceral functions. These areas send fibers to the next lower level, the hypothalamus, located at the base of the brain. The hypothalamus is a coordinating center for the motor control of visceral activity. One of its many functions, for example, is the regulation of body temperature. The hypothalamus has nervous and vascular connections with the hypophysis, by virtue of which it influences the hypophysis and, through the latter, the other endocrine glands. The hypothalamus also sends nerve fibers to lower centers in the brain stem that are concerned with still more specific functions, for example, the reflex regulation of respiration, heart rate, and circulation. These centers function by virtue of their connections with still lower centers, which are collections of nerve cells in the brain stem and spinal cord that send their axons into certain cranial and spinal nerves. It is characteristic of these axons that, unlike motor fibers to skeletal muscle, they synapse with multipolar cells outside the central nervous system before they reach the structure to be supplied. These multipolar cells are collected into ganglia; the ganglionic is the lowest level. The axons that pass from the central nervous system to these ganglion cells are termed *preganglionic fibers.* The axons of ganglion cells are called *postganglionic fibers;* all such fibers from a particular ganglion supply a specific organ or region of the body.

Sympathetic System

The sympathetic or thoracolumbar part of the autonomic system comprises the preganglionic fibers that issue from the thoracic and upper lumbar levels of the spinal cord. These fibers reach spinal nerves by way of ventral roots, and then leave the spinal nerves and reach adjacent ganglia by way of rami communicantes (fig. 5–1). These ganglia are contained in long nerve strands, the sympathetic trunks, one on each side of the vertebral column, extending from the base of the skull to the coccyx. Some preganglionic fibers synapse in ganglia of the trunk, others continue to ganglia of the prevertebral plexuses (p. 423), and still others synapse with cells in the medullae of the suprarenal glands. The postganglionic fibers either go directly to adjacent viscera and blood vessels, or return to spinal nerves by way of rami communicantes and, in the area of distribution of these nerves, supply the skin with (1) secretory fibers to sweat glands, (2) motor fibers to smooth muscle (arrectores pilorum), and (3) vasomotor fibers to the blood vessels of the limbs. Some sympathetic ganglion cells are present in the spinal nerves and in the rami communicantes. They may be collected into what are termed intermediate or accessory ganglia (p. 782).

Parasympathetic System

The parasympathetic or craniosacral part of the autonomic system comprises the preganglionic fibers that issue from the brain stem (cranial nerves 3, 7, 9, 10, 11) and sacral part of the spinal cord (second and third or third and fourth sacral segments). The ganglion cells with which these fibers synapse are in or near the organs innervated. The postganglionic fibers are very short; apparently none go to blood vessels, smooth muscle, or glands of the limbs and body wall. Most viscera, however, have a double motor supply, sympathetic and parasympathetic, sometimes with opposing functions.

Functions

By its role in central integrating mechanisms, the autonomic nervous system is involved in behavioral and neuroendocrinological mechanisms, and in the processes whereby the body keeps its internal environment constant, that is, maintains temperature, fluid balance, and ionic composition of the blood. The parasympathetic system is concerned with many specific functions, such as digestion, intermediary metabolism, and excretion. The sympathetic system is an important part of the mechanism by which one reacts to stress.

DEVELOPMENT OF NERVOUS SYSTEM

Early in embryonic development, a neural tube is formed from the ectoderm on the dorsal aspect of the embryo. The brain develops from the head end of the tube, the spinal cord from the rest of the tube. In spite of complex changes during growth and maturation, the central nervous system retains the cavity of the neural tube, except in the spinal cord, where the central canal is often partly closed by cellular proliferation. The ventricles of the adult brain develop from this cavity.

The neurons that develop in the ventral part (basal plate) of the neural tube become motor cells. The neurons that develop in the dorsal part (alar plate) become associated with sensory and reflex paths. Some of the cells in the neural crest, which is a longitudinal mass of ectodermal cells that develops on each side of the neural tube, give rise to the unipolar cells of the spinal ganglia. Some of the cells migrate and develop into neurons of the autonomic ganglia and into cells of the medullae of the suprarenal glands.

REFERENCES

1. J. W. Millen and D. H. M. Woollam, Brain, *84*:514, 1961; *The Anatomy of the Cerebrospinal Fluid*, Oxford University Press, London, 1962.
2. H. Davson, *Physiology of the Cerebrospinal Fluid*, Churchill, London, 1967. R. Katzman and H. Pappius, *Brain Electrolytes and Fluid Metabolism*, Williams & Wilkins, Baltimore, 1973. Millen and Woollam, cited in 1 above.
3. T. R. Shanta and G. H. Bourne, in *The Structure and Function of the Nervous System*, Academic Press, New York, vol. 1, 1968. W. E. Burkel, Anat. Rec., *158*:177, 1967.
4. E. Gardner, in *Peripheral Neuropathy*, ed. by P. J. Dyck, P. K. Thomas, and E. H. Lambert, Saunders, Philadelphia, 1975.
5. O. Foerster, Brain, *56*:1, 1933. O. Bumke and O. Foerster, *Handbuch der Neurologie*, Springer, Berlin, vol. 5, 1936.
6. H. Head and A. W. Campbell, Brain, *23*:353, 1900.
7. S. Sunderland, Arch. Neurol. Psychiat., Chicago, *54*:280, 1945. G. Causey, Ann. R. Coll. Surg. Engl., *16*:367, 1955.

GENERAL READING

Bossy, J., *Atlas du système nerveux*, Éditions Offiduc, Paris, 1971.

Brodal, A., *Neurological Anatomy*, Oxford University Press, London, 1969.

Delmas, J., and Laux, G., *Système nerveux sympathique*, Masson, Paris, 1952.

Ford, D. H., and Schadé, J. P., *Atlas of the Human Brain*, Elsevier, Amsterdam, 1966.

Gardner, E., *Fundamentals of Neurology*, Saunders, Philadelphia, 6th ed., 1975. A concise account of the nervous system with pertinent references. Useful for orientation and review.

Hovelacque, A., *Anatomie des nerfs crâniens et rachidiens et du système grand sympathique chez l'homme*, Doin, Paris, 1927, 2 volumes.

Miller, R. A., and Burack, E., *Atlas of the Central Nervous System in Man*, Williams & Wilkins, Baltimore, 1968.

Pitres, A., and Testut, L., *Les nerfs en schémas*, Doin, Paris, 1925.

de Ribet, R.–M., *Les nerfs rachidiens*, Doin, Paris, 1953. For the volume on the cranial nerves, see the readings for chapter 53.

Roberts, M., and Hanaway, J., *Atlas of the Human Brain in Section*, Lea & Febiger, Philadelphia, 1970.

Singer, M., and Yakovlev, P. I., *The Human Brain in Sagittal Section*, Lea & Febiger, Philadelphia, 1964.

Truex, R. C., and Carpenter, M. B., *Human Neuroanatomy*, Williams & Wilkins, Baltimore, 6th ed., 1969.

BLOOD VESSELS, LYMPHATIC SYSTEM 6

DONALD J. GRAY

BLOOD VESSELS

The blood vessels consist of a closed system of tubes, which transport blood from the heart to all parts of the body and back to the heart (figs. 6–1 and 6–2). The study of blood and lymphatic vessels is called *angiology.* The prefix *angi-* is of Greek origin (*angeion,* vessel). It is used in terms such as angiography, angioma, and angiopathy.

The heart is a muscular pump, the primary function of which is to propel blood through this system to a network of simple endothelial tubes, where exchanges can occur. The blood vessels carry blood to the lungs, where carbon dioxide is exchanged for oxygen. They also carry blood to the intestine, where nutritive materials in fluid form are absorbed, and to the ductless glands, where hormones pass through their walls to reach the blood. The products of digestion, hormones, enzymes, and oxygen contained in the blood as it passes through the blood vessels are responsible for the quality and quantity of the fluid in the tissues of the body. The functions of the tissues, such as contraction of muscles and secretion of glands, are at least partly dependent upon the composition of the tissue fluid. The waste products of tissue fluid are transported by the blood vessels to the kidneys, intestines, lungs, and skin, where they are excreted. The stability of the internal environment (*milieu intérieur*) is thus dependent upon a proper functioning of the blood vessels and upon the composition of the blood contained in them. The maintenance of this internal environment is mentioned in the discussion of the autonomic nervous system (p. 36).

The circulation of the blood was discovered by William Harvey (p. 6), who in 1628 published the results of his studies, upon which much of physiology is based.

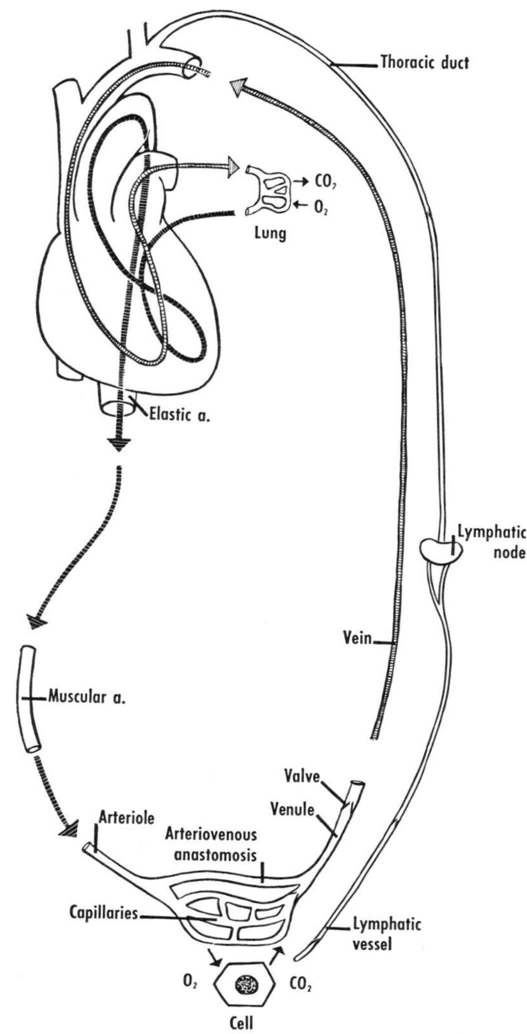

Figure 6–1 Schematic diagram of the circulatory system.

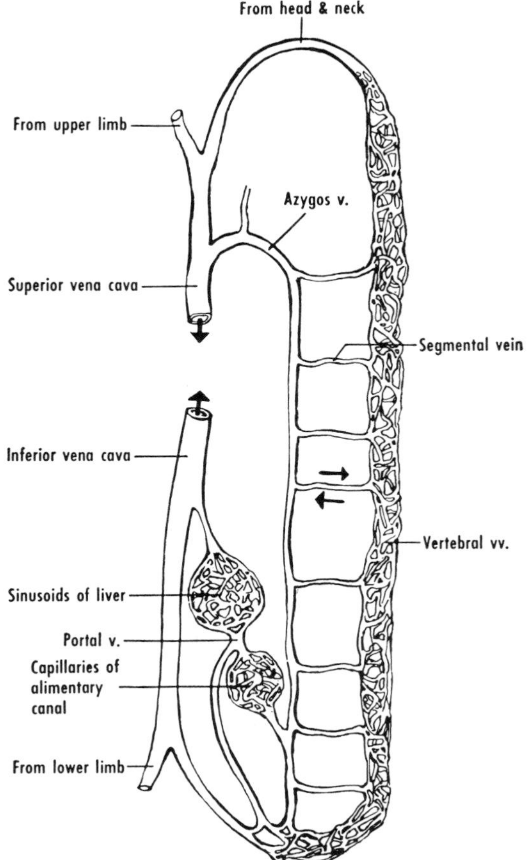

Figure 6–2 Schematic representation of venous circulation. Note the portal circulation, and also that two major systems, the azygos and the vertebral, bypass the caval system. Arrows indicate direction of flow of blood. Based on Herlihy.[1]

Circulations

Blood leaving and returning to the heart passes in two different circuits, the pulmonary circulation and the systemic circulation. In the former, the blood passes through the pulmonary arteries to the lungs and returns through the pulmonary veins to the heart. In the latter, the blood passes through the aorta to all parts of the body and returns to the heart through the superior and inferior venae cavae and the cardiac veins.

Types of Blood Vessels

Arteries. Arteries are yellowish or bluish gray in the living individual and are characterized by their pulsation. When cut, an artery bleeds in spurts. It also shortens, and, if the artery is not of large size, its ends retract in such a way that bleeding stops.

Blood pressure is usually considered to mean the pressure in the arteries, especially that in the brachial artery. Because the pressure falls so little in passing from the largest to the smallest arteries, it is essentially the same in the brachial artery as it is in the arteries proximal and distal thereto. The systolic pressure is that in the arteries at the end of the contraction (systole) of the left ventricle and in young adult males is normally from about 120 to 130 mm of mercury (the range in 95 per cent of individuals is from about 95 to 150 mm of mercury). The diastolic pressure is that in the arteries at the end of the resting phase (diastole) of the left ventricle and in young adult males is normally from about 75 to 80 mm of mercury (the range in 95 per cent of individuals is from about 55 to 95 mm of mercury). The difference between the systolic and diastolic pressures is called the pulse pressure. Normal arterial pressures are maintained by (1) the pumping action of the left ventricle, (2) the peripheral resistance, (3) the quantity of blood in the arteries, (4) the viscosity of the blood, and (5) the elasticity of the arterial walls.

The pulse is a wave of expansion and contraction of an artery and is a reflection of the pressure created by the ejection of blood from the heart. This wave is propagated through the blood column and arterial wall to the periphery. The pulse wave, which progresses about 5 to 8 meters per second in the brachial artery, travels 10 to 15 times faster than the blood.

On the basis of their structure, arteries are classified as (1) large or elastic, (2) distributing, medium-sized, or muscular, and (3) arterioles. The changes in structure between an elastic and a muscular artery may be quite gradual.

ELASTIC ARTERIES. The elastic arteries include the aorta, the brachiocephalic trunk, and the common carotid and subclavian arteries. The elasticity of the wall of the aorta permits considerable expansion. The aorta thus acts as a reservoir, and converts the intermittent flow of blood from the heart into a continuous, but pulsatile, stream. Its elastic recoil is responsible for the diastolic pressure, which propels the blood during diastole. This recoil also closes the aortic valve and drives blood into the coronary arteries.

The pulmonary trunk and pulmonary arteries are also of the elastic type. They carry blood for a relatively short distance, however, and their walls offer much less resistance to its flow. The arterial pressure in the pulmonary circuit is about one-sixth that in the aorta, or about 20 mm of mercury.

MUSCULAR ARTERIES. The muscular arteries are the branches and the continuations of the elastic arteries. Their walls contain relatively less elastic tissue and relatively more smooth muscle, which, with appropriate stimulation, contracts and decreases the caliber of the vessel. Most of the arteries of the body are of this type.

Muscular arteries differ in structure in various parts of the body. Those in the cranial cavity, for example, contain relatively less smooth muscle in their walls.

Branches arise from main trunks at an acute angle (e.g., superior mesenteric artery), at a right angle (e.g., renal arteries), or at an obtuse angle (recurrent arteries). The rate of flow in branches arising at a right or obtuse angle is often less than that in branches arising at an acute angle. Pathological changes, such as arteriosclerosis, are more apt to occur in the part of a branch near its origin.

ARTERIOLES. The arterioles are the smallest divisions of the arteries. They are said to be arteries that are less than 100 micrometers in diameter, but the relatively thick wall in relation to the small lumen is a more reliable criterion than is the total diameter. Their walls consist mostly of smooth muscle.

The arterioles provide the greatest resistance to the flow of blood, and their constriction serves to reduce the pressure of the blood before it enters the capillaries. The pressure of the blood as it flows through the arterioles decreases by about 50 to 60 mm of mercury.

Capillaries.[2] The capillaries comprise an anastomosing network into which arterioles empty. Their walls act as a semipermeable membrane, which permits the passage through it of water, crystalloids, and some plasma protein, but which is impermeable to large molecules. Oxygen and nutritive materials pass through the wall of the arteriolar end of the capillary into the tissues. Waste products and carbon dioxide return from the tissues through the wall of the venous end. Capillaries are present in greater numbers in active tissues, such as muscles, glands, liver, kidneys, and lungs. Many of them are closed, however, when these tissues are inactive. They are fewer in number in less active tissues, such as tendons and ligaments. The cornea, the epidermis, and hyaline cartilage are devoid of capillaries.

The sum of the cross-sectional area of all capillaries is from 600 to 800 times the cross-sectional area of the aorta. This total area is termed the capillary bed or capillary lake. Conditions for the interchange of substances through the capillary wall are favorable because of the low pressure and the slow rate of flow in these vessels. The pressure in the capillaries is only one-fourth (or less) as great as that in the aorta, and the rate of flow is 600 to 800 times slower. In passing from the arteriolar to the venous end of the capillary, the blood pressure decreases by about 15 to 20 mm of mercury.

In flowing through a capillary bed, blood may pass through only some of the capillaries, which are then called preferred channels.

Sinusoids. Sinusoids are wider than capillaries and are also more tortuous. They take the place of capillaries in the liver, the spleen, the bone marrow, the carotid and coccygeal bodies, the adenohypophysis, the suprarenal cortex, and the parathyroid glands. Sinusoids are also present in the heart. Unlike the capillaries, their lining cells, many of which are phagocytic, are supported by reticular tissue.

Cavernous Tissue. Cavernous tissue is the name given to the numerous blood-filled spaces and their walls in the corpora cavernosa and corpus spongiosum of the penis and in the corpora cavernosa of the clitoris. The endothelium of these spaces is similar to that of capillaries, but the partitions between them contain smooth muscle. Similar tissue is located in the lining of the nasal cavity.

Venules. Venules collect blood from the capillary plexus and join similar vessels to form veins.

Veins. Veins are dark blue in the living individual. They do not normally pulsate, and therefore bleeding from them does not occur in spurts. The veins are more numerous than arteries. Furthermore, their walls are thinner, and their diameters are usually larger than those of the corresponding arteries.

The pressure in the veins gradually decreases from that in the venous end of the capillaries to that in the veins emptying into the right atrium, in which it is slightly above zero. Both the pressure and the rate of flow are subject to variations, however, and are affected by the following factors: (1) the contraction of the left ventricle, (2) the quantity of blood permitted by the arterioles to enter the capillaries and then the veins, (3) the action of the right atrium and ventricle, (4) the pressure within the thorax (which is normally below that of the atmosphere), (5) the massaging effect of the skeletal muscles, and (6) the effect of gravity (which accounts for marked differences in hydrostatic pressure above and below the heart, especially when the body is erect).

With a few exceptions, the deep veins accompany arteries and have the same names. Most of those that accompany arteries below the elbow, below the knee, and in a few other locations are paired. These pairs, often called *venae comitantes*, communicate with one another along their course. Superficial veins run independently of arteries.

Although most of the blood from the body returns to the heart by way of the venae cavae, alternate pathways are provided. The main alternate routes, which do not accompany arteries, are the azygos system (p. 327), the vertebral system (p. 327), and the portal system (p. 417). These are represented schematically in figure 6–2. All three of these communicate with one another, and any one of them can form the main pathway for venous return when the others are partially or completely blocked.

A portal system is one in which blood, after being collected from one set of capillaries, passes through a second set of capillary-like vessels before it returns to the systemic circulation. For example, the blood collected from the capillaries of the stomach, most of the intestine, the pancreas, the spleen, and the gall bladder is conducted by way of the portal vein to the liver, where it passes through sinusoids before entering the inferior vena cava by way of the hepatic veins (fig. 36–4, p. 398).

VALVES. Valves are present in many veins. When closed, they prevent the reverse flow of blood. They consist of folds of the inner layer and have from one to three cusps. The free edges of the cusps are directed toward the heart. The circumference of a vein is sometimes greater at the site of a valve.

The valves in a tributary are often located just distal to its opening into another vein. Valves are most numerous in the veins of the limbs. They are absent in most veins of the trunk, including those of the portal and vertebral systems. They are usually not present in veins closer to the heart than the internal jugular, subclavian, and femoral veins, although they are sometimes present in the common and external iliac veins.

The valves serve to prevent the return of blood into the veins of the head and limbs when the pressure in the abdomen is increased, as during defecation, or when the pressure in the thorax is increased, as during expiration.

The valves in the superficial veins of the forearm can be readily demonstrated in the living individual (p. 99).

Anastomoses. In certain locations, arteries anastomose with each other. Such communications exist in the palm of the hand, in the sole of the foot, at the base of the brain, near the intestines, around joints, and in the heart. In the event that one of the arteries entering into such an anastomosis becomes occluded or is ligated, a collateral circulation can be established through the other.

Collateral circulation can sometimes be established through capillaries, especially in young individuals. By the addition of various tissues to its wall, a capillary can be converted into an artery or a vein.

Blood does not always pass through a capillary network in being transported from an arteriole to a venule. Certain locations are provided with *arteriovenous anastomoses*,[3] which bypass the capillaries. The walls of these shunts are thicker than those of capillaries and do not permit the exchange of materials. Arteriovenous anastomoses are very widely distributed, and are present in locations such as the skin of the nose, lips, eyelids, and palm of the hand, at the tip of the tongue, and in the intestine. Many arteriovenous anastomoses are very complex in arrangement and it has been suggested that they might be neurovascular end organs. Others, when open, shunt blood away from the capillaries; the regions supplied by these capillaries are thus deprived of blood. In areas subjected to cooling, they help to prevent loss of heat. For example, cold air coming into contact with the skin reflexly causes an opening of the arteriovenous anastomoses. Because of the decreased amount of blood in the capillaries, the skin becomes pale and less heat is lost from it. The presence of arteriovenous anastomoses in the intestine makes it possible for blood to bypass the capillaries except during periods of need, such as during digestion. An increased rate of flow through arteriovenous anastomoses and preferred channels results in an increased venous pressure, which, in turn, aids the return of blood to the heart.

End Arteries. Some arteries supply limited areas of tissues or organs without anastomosing with arteries supplying adjacent areas. They are called anatomical end arteries. The artery that supplies the retina of the eye is an example, and its occlusion results in blindness. An artery anastomosing so poorly with one supplying an adjacent territory that an adequate blood supply is not maintained after its occlusion is termed a *functional end artery*. Such arteries supply segments of the brain, kidneys, spleen, and intestines. The type of arterial distribution varies from one organ to another.[4]

Structure of Blood Vessels

Arteries consist of three coats: (1) a *tunica intima* lined by epithelium that is supported by a small amount of loose connective tissue; (2) a *tunica media*, which is the thickest coat and consists of varying proportions of smooth muscle and elastic tissue; and (3) a *tunica externa (tunica adventitia)*, which is the strongest of the three coats and consists of both collagenous and elastic fibers. The tunica externa contains small blood vessels, called *vasa vasorum*. It also contains autonomic and sensory nerve fibers, some of which are sensitive to painful stimuli. The puncture of an artery can be very painful, and an artery handled roughly may go into spasm.

The **capillaries** have only one coat, and this is formed by endothelium. The capillary walls of different organs have been classified on the basis of their fine structure.[5]

The **venules** consist of endothelium supported by a small amount of collagenous tissue and, in the larger venules, by a few smooth muscle fibers also.

Veins vary considerably in structure. Their walls are thinner and their caliber is greater than those of corresponding arteries. The tunica media is much thinner than that of arteries, and the smooth muscle in this coat may be arranged circularly, longitudinally, or spirally. The tunica externa is frequently the thickest coat, and vasa vasorum are present in greater numbers than in the similar coat of arteries.

LYMPHATIC SYSTEM

The lymphatic system[6] comprises the lymphatic vessels and the lymphatic or lymphoid tissue. Lymphatic tissue is present in certain organs, such as the intestine, and it forms other organs, such as lymph nodes.

Lymphatic Vessels

Nutritive and other materials are continuously passing through the walls of the blood capillaries into the tissue fluid. Most of these materials readily re-enter these capillaries, but the large protein molecules do not. It is therefore necessary that another system of capillaries be available, the function of which is to absorb these molecules and return them to the blood stream. The lymphatic vessels act as a specialized mechanism to serve this need. Because they are closely associated with the lymphatic tissues, they also carry lymphocytes from these tissues to the blood stream. The fluid transported by the lymphatic vessels is called lymph.

Although lymphatic vessels had been seen several hundred years B.C., little attention was paid to them until Aselli, in 1627, described these vessels in the mesentery of the dog. The term lymphatics was introduced by Bartholin in 1653.

The lymphatic vessels consist of (1) capillaries, which are simple, endothelial tubes; (2) collecting vessels, to whose endothelium some smooth muscle and fibrous connective tissue have been added; and (3) trunks (fig. 31–4, p. 330), whose adventitia contains greater amounts of connective tissues and smooth muscle.

The lymphatic capillaries are thin and transparent, and are held open by the attachment of their walls to surrounding tissues. Larger and more irregular than blood capillaries, they exist in the form of closed networks, which communicate with one another so freely that it is almost impossible to stop completely the flow of lymph in a given area. The capillaries contain few valves, but valves are numerous in the collecting vessels. The valves usually have two cusps, and they ensure that the lymph flows in one direction only, toward the heart. The vessels are constricted at the sites of the valves, and therefore present a knotted or beaded appearance. They tend to run in groups, rather than as single vessels, and often accompany veins.

Lymphatic capillaries are present in most locations in which blood capillaries are situated. They are abundant in the skin and mucous membranes, and are especially numerous around orifices, such as the mouth and anus.

The lymphatic capillaries in the mucous membrane of the small intestine have projections that end blindly at the tips of the villi. These projections are called *lacteals,* and they carry the chyle, or emulsified fat, that is produced during the process of digestion.

Lymphatic vessels are located in the endocardium, epicardium, and pericardium, and in the pleurae covering the lungs. They begin near the alveoli of the lungs, and leave through the hilus with the pulmonary artery and veins. They are present in synovial membranes and periosteum, and in the capsules and trabeculae of glands.

Lymphatic vessels are absent in the central nervous system, in skeletal muscles (but not in the connective tissue covering them), in bone marrow, in the splenic pulp, and in avascular structures, such as hyaline cartilage, nails, and hair.

Lymph

Lymph is the fluid absorbed by the lymphatic capillaries. It is clear and colorless, except in the vessels from the intestine, in which it is creamy white after digestion. It shares with the other extracellular fluids the responsibility for maintaining the constancy of the internal environment of the body, the *milieu intérieur* of Claude Bernard.

The term lymph is derived from the Latin word *lympha,* which means water, especially clear river or spring water.

Flow of Lymph. During periods of inactivity of an area or part, the flow of lymph is relatively slow. Muscular activity results in a more rapid and regular flow. The flow of lymph increases during peristalsis and also with increases in respiratory movements and cardiac activity. It increases with elevations of venous pressure, but is little affected by elevated arterial pressures. It can be increased by massage, by passive motion, and, to a small degree, by the pulsation of adjacent arteries. The obstruction of the flow of lymph from a given area results in the accumulation of abnormally large amounts of tissue fluid in that area. Such an accumulation is called lymphedema.

Lymphatic Tissue and Organs

Lymphatic or **lymphoid tissue** consists of lymphocytes of different sizes, supported by reticular cells and fibers, and by collagenous, elastic, and smooth muscle fibers. It is present in (1) lymph nodes, (2) mucous membranes, (3) the thymus, (4) the spleen, and (5) bone marrow.

The lymphatic tissue in mucous membranes has efferent vessels only. Localized aggregations of such tissue form the palatine, pharyngeal, and lingual tonsils, and the solitary and aggregated lymph nodules in the intestine.

The thymus is characterized by an arrangement of the lymphatic tissue into lobules, rather than nodules.

The spleen consists largely of lymphatic tissue, but it is designed to filter blood, rather than lymph. It is supplied with neither afferent nor efferent lymphatic vessels.

Lymph Nodes. The lymph nodes are variable in size, shape, and color. Their greatest diameter ranges from 1 to 20 mm or more. The nodes draining the liver are often brown, those draining the lung are often black, and those draining the small intestine are often creamy white. Nodes usually occur in groups, but sometimes they occur singly. They are frequently located along the course of blood vessels, and many are situated along the digestive canal. The nodes of the inguinal region can usually be palpated; those of some other regions can be felt if they are enlarged.

Lymph nodes, as well as lymphatic tissue generally, are relatively prominent at birth. They grow rapidly until late childhood, after which they decrease in both absolute and relative weight. They may enlarge again in response to inflammation (lymphadenitis).[7] Lymph nodes decrease in size during malnutrition and after irradiation.

STRUCTURE. Each lymph node is surrounded by a fibrous capsule, from which trabeculae extend inward. Afferent lymphatic vessels enter through the capsule, and efferent vessels leave through the hilus, an indentation through which blood vessels also enter and leave.

Functions. In the normal, healthy body, the production of lymphocytes is the main function of lymphatic tissues and organs, and it is the only function that is conclusively established. Lymphocytes have important roles in the development of antibodies and immune reactions.

The action of the lymphatic tissues in serving as filters under certain pathological conditions has given rise to the barrier theory, according to which these tissues play an important role in the defense mechanisms of the body. Inanimate particles, such as carbon, are held up to a large extent in the lymphatic tissues. Bacteria, viruses, cancerous cells, and red corpuscles are retained in varying numbers. The lymphatic tissues are barriers only up to a degree, however, and their efferent vessels actually facilitate the spread of infectious and malignant growths to other organs and tissues.

Hemal Nodes.[8] The hemal nodes are located mainly in the lumbar and cervical regions, in front of the vertebral column. They are smaller and much fewer in number than lymph nodes. They are connected with blood vessels, but have no connections with lymphatic vessels. Their structure is similar to that of lymph nodes. However, their sinuses are filled with blood, rather than lymph. Their function is unknown.

REFERENCES

1. W. F. Herlihy, Med. J. Aust., *1*:661, 1947.
2. A. Krogh, *The Anatomy and Physiology of Capillaries*, Steckert-Hafner, Stuttgart, 2nd ed., 1929, reprint, 1959.
3. M. Clara, *Die arterio-venösen Anastomosen*, Springer, Wien, Zweite Auflage, 1956. J. D. Boyd, Lond. Hosp. Gaz., *42*(8): Clin. Suppl., 1939. A. R. Hale and G. E. Burch, Medicine, Baltimore, *39*:191, 1960.
4. G. Lazorthes and G. Bastide, C. R. Ass. Anat., *42*:879, 1956.
5. H. S. Bennett, J. H. Luft, and J. C. Hampton, Amer. J. Physiol., *196*:381, 1959.
6. L. Allen, Annual Rev. Physiol., *29*:197, 1967. J. M. Yoffey, Discovery, *27*:24, 1966.
7. F. A. Denz, J. Path. Bact., *59*:575, 1947.
8. A. W. Meyer, Amer. J. Anat., *21*:375, 1917. H. E. Jordan, Anat. Rec., *59*:297, 1934.

GENERAL READING

Abramson, D. I. (ed.), *Blood Vessels and Lymphatics*, Academic Press, New York, 1962.

Abramson, D. I., *Circulation in the Extremities*, Academic Press, New York, 1967.

Benninghoff, A., Blutgefässe und Herz, in *Handbuch der mikroskopischen Anatomie des Meschen*, ed. by W. von Möllendorff, Springer, Berlin, vol. 6, part 1, 1930.

Franklin, K. J., *A Monograph on Veins*, Thomas, Springfield, Illinois, 1937.

Handbook of Physiology: Section 2, Circulation, ed. by W. F. Hamilton, American Physiological Society, Washington, D.C., 1962.

Hellman, T., Lymphgefässe, Lymphknötchen und Lymphknoten, in *Handbuch der mikroskopischen Anatomie des Menschen*, ed. by W. von Möllendorff, Springer, Berlin, vol. 6, part 1, 1930.

Kinmonth, J. B., *The Lymphatics*, Arnold, London, 1972.

McDonald, D. A., *Blood Flow in Arteries*, Physiol. Soc. Monogr., 7, Williams & Wilkins, Baltimore, 1960.

Orbison, J. L., and Smith, D. E. (eds.), *The Peripheral Blood Vessels*, Internat. Acad. Pathol. Monogr. no. 4, Williams & Wilkins, Baltimore, 1963.

Poirier, P., and Cunéo, B., Les Lymphatiques, in *Traité d'anatomie humaine*, ed. by P. Poirier and A. Charpy, Masson, Paris, vol. 2, part 4, 1902.

Rouvière, H., *Anatomie des lymphatiques de l'homme*, Masson, Paris, 1932.

Yoffey, J. M., and Courtice, F. C., *Lymphatics, Lymph and the Lymphomyeloid Complex*, Academic Press, New York, 1970.

VISCERA 7

DONALD J. GRAY

The viscera are the internal organs of the body; most of them are located in the thoracic, abdominal, and pelvic cavities. The internal organs are conventionally divided into four groups or systems, each of which has a common function or functions. These four groups are the digestive system, the respiratory system, the urogenital system, and the ductless glands or endocrine system.

Some of the viscera are glands, which are collections of specialized, secreting cells. These glands are of different sizes and shapes, and they are either exocrine or endocrine. The exocrine glands are provided with ducts, through which their secretions pass to the cavities of other organs. The endocrine or ductless glands have no ducts, and their secretions enter the blood stream through the walls of capillaries.

Viscera is the plural of the Latin word *viscus*, which means internal organ. The Greek work *splanchnos* also means internal organ, and the adjective splanchnic is derived from it.

Structure of Hollow Organs

Most of the viscera are in the form of hollow, tubular organs, which vary in size and shape. The walls of these organs have several layers (fig. 35–8, p. 384). From within outward these layers are called: (1) mucous membrane or mucosa, (2) submucosa, (3) muscular coat, and (4) adventitia, which may be either a fibrous or a serous coat.

Mucous Membrane. The mucous membrane is lined by epithelium, which differs in type in different organs and sometimes in different parts of the same organ. It often contains glands. The loose connective tissue adjacent to the epithelium is termed the *tunica,* or *lamina, propria.* One or two thin layers of smooth muscle, the *lamina muscularis mucosae,* are present in the outermost part of the mucous membrane of many organs.

Submucosa. The submucosa consists of loose connective tissue containing blood vessels, lymphatic vessels, nerve fibers (submucous plexus), and sometimes glands. It is the layer that gives strength to the wall. Many histologists do not recognize the presence of a submucosa in organs that do not have a lamina muscularis mucosae.

Muscular Coat. The muscular coat consists of one, two, or three layers of smooth muscle. In some organs, however, bundles of smooth muscle are so intermingled that definite layers cannot be recognized. Nerve cells and plexuses of nerve fibers (myenteric plexus) are located in this coat. They are often situated between layers, and sometimes within them.

Fibrous Coat. The fibrous coat is the outer, connective tissue layer of many of the tubular organs. Other organs have an outermost *serous coat,* which is smooth and glistening, and which is covered by a layer of mesothelial cells. Still other organs are covered partly by a fibrous and partly by a serous coat.

Special names are given to the serous coat or membranes associated with certain organs or in certain locations. That covering the lungs is called pleura, that covering the heart is called pericardium, and that covering many of the abdominal and pelvic organs is called peritoneum.

DIGESTIVE SYSTEM

The digestive system is adapted for (1) the ingestion and mastication of food, (2) the secretion of chemical substances that produce chemical changes in the food, (3) the absorption and assimilation of the nutritive materials, and (4) the elimination of waste products.

The digestive system consists of a hollow tube, which extends from the lips to the anus, and various glands whose secretions empty into the cavity of this tube and aid in the process of digestion.

The hollow tube has three successive divisions: the mouth, the pharynx, and the alimentary canal. The alimentary canal consists successively of the esophagus, the stomach, the small intestine, and the large intestine.

The glands belonging to the digestive system are the salivary glands, the liver, and the pancreas. The salivary glands comprise the paired parotid, submandibular, and sublingual glands, and many small glands, such as the labial, buccal, and lingual glands, which are located in the mucous membrane of the mouth. The secretions of all of these empty into the cavity of the mouth and form the saliva. Secretions from the liver and pancreas, which are located in the abdomen, empty into the first part of the small intestine.

Each organ of the digestive system performs a specific function or functions. In the mouth, the food is moistened and is masticated by the teeth. It is propelled by the pharynx and esophagus into the stomach, where it is mixed with the gastric juice and coverted into chyme. It is digested in the small intestine by secretions from glands in the walls of the intestine and from the liver and pancreas. Water is absorbed through the walls of the large intestine, which propels the waste products to the anus, where they are eliminated as feces.

RESPIRATORY SYSTEM

The respiratory system consists of a conducting portion and a respiratory portion, and, from a functional standpoint, also the thoracic cage and the diaphragm. Air is transported to the lungs in the conducting portion, which comprises the external nose, the nasal cavity and paranasal sinuses, the pharynx, the larynx, and the trachea. As the air passes through these organs, it is filtered, washed, humidified, and warmed or cooled by their mucous membranes.

The respiratory portion consists of the lungs, each of which is covered by a double layer of pleura. The smaller passages within the lungs are closely associated with capillaries, and here the carbon dioxide in the blood is exchanged for oxygen in the air. The respiratory portion is unable to function without the aid of the diaphragm and thoracic cage.

In addition to its respiratory function, this system is concerned with speech, in which the larynx plays an especially important role. Part of the pharynx belongs to the digestive, as well as to the respiratory, system.

UROGENITAL SYSTEM

The urogenital system consists of the urinary and genital organs, which are often included in one system because they are closely associated with each other developmentally. Furthermore, in the male, one of the organs, the urethra, serves for the transmission of both urine and seminal fluid.

Urinary Organs

The urinary organs comprise the left and right kidneys, the left and right ureters, the urinary bladder, and the urethra. The kidneys are the most important of all excretory organs. They are responsible for the maintenance of the ionic balance of the blood and for the elimination from the blood of waste products that would be harmful if allowed to accumulate in the body. They also function in helping to maintain a proper volume of blood and of tissue fluid. Urine, which is excreted by the kidneys, passes downward through the ureters into the urinary bladder, where it is stored until it is discharged through the urethra. Except for the urethra, which is much longer in the male, the urinary organs of the two sexes are similar.

Male Genital Organs

The male genital organs consist of (1) two testes, which produce spermatozoa; (2) a series of ducts, through which spermatozoa pass from the testis to reach the exterior; (3) various glands, which contribute secretions to the seminal fluid; and (4) an external genital organ, the penis.

The ducts leading from each testis are successively the epididymis, the ductus deferens, and the ejaculatory duct. The last opens into the urethra.

The glands belonging to this group of organs are the prostate, the seminal vesicles, and the bulbourethral glands. The prostate is a single organ, whereas the other organs are paired. The secretions from all of these glands empty into the urethra.

Female Genital Organs

The female genital organs consist of two ovaries, two uterine tubes, a uterus, a vagina, and the external genital organs.

The oöcytes are produced in the ovaries, and are conveyed from these organs through the uterine tubes to the cavity of the uterus. The uterine tubes transmit spermatozoa in the opposite direction, and fertilization of an oöcyte usually occurs within a tube.

If an oöcyte is fertilized, it normally becomes embedded in the wall of the uterus, where it develops and grows into a fetus. At the end of prenatal development, the fetus passes through the uterus and vagina, which together are called the birth canal.

DUCTLESS GLANDS*

The ductless or endocrine glands have two main characteristics. (1) Their secretions do not leave the glands by means of ducts, but pass directly into the blood stream, in which they are carried to all parts of the body. (2) Their secretions contain chemical substances, called hormones, which play a very important role in reproduction, growth, and metabolism.

The ductless glands are not connected with one another structurally, and they are located in widely separated regions of the body. Some of them comprise entire organs, such as the hypophysis and thyroid gland. Others, such as the pancreatic islets and the interstitial cells of the testis, are not distinct anatomical entities, but consist of groups of cells located in other organs of the body. The ductless glands show marked differences in structure, function, and mode of development.

---

*C. D. Turner and J. T. Bagnara, *General Endocrinology*, Saunders, Philadelphia, 5th ed., 1971.

Other organs are sometimes classified as ductless glands because of the endocrine-like substances that they secrete. These are the kidney, liver, and stomach. Still other organs, such as the pineal body and the thymus, are often included.

Hormones

A hormone is a chemical substance that is secreted by specialized cells in a restricted part of the body and when transported, usually by the circulation, to another part of the body may exhibit a high degree of specificity in the regulation of the rate of activity of the cells and in the integration of their activity with other parts of the organism. The word hormone is derived from the Greek word *hormanein,* meaning to set in motion, or to urge on.

Some hormones, such as those secreted by the pars distalis of the hypophysis, are called *tropic* or *trophic* (e.g., thyrotropic and gonadotropic), because they affect other organs. The tropic hormones regulate the functional states of other endocrine glands, and control, either directly or indirectly, a wide variety of physiological responses. The organs primarily affected by these are often called target organs. The qualitative and quantitative response of the target and other organs to hormones is influenced by genetic factors and by other factors, such as race, age, season, temperature, diet, and the presence of disease.

Under some conditions, hormones may have certain reciprocal interactions. For example, the adrenocorticotropic hormone, which is secreted by the pars distalis of the hypophysis, stimulates the cortex of the suprarenal gland to secrete certain steroids. When such stimulation is carried to excess, however, hormones from the suprarenal cortex inhibit the pars distalis, and the cortex of the suprarenal glands shrinks. Also, gonadotropins from the pars distalis are necessary for the normal functioning of the ovary and testis. The removal of the gonads reciprocally affects the hypophysis and the suprarenal glands.

Hormones are usually effective in very small quantities. However, the responses to them are relatively slow, compared with responses of organs to nervous stimuli. The chemical structures of a number of hormones are known, but the exact mechanism by means of which hormones exert their effects remains to be established.

SKIN, HAIR, AND NAILS 8

RONAN O'RAHILLY

LAYERS OF SKIN

The term *common integument* is used to include skin and subcutaneous tissue, hair, nails, and breast. The last-named will be described with the upper limb. The *skin (cutis)* provides a waterproof and protective covering for the body, contains sensory nerve endings, and aids in the regulation of temperature. The skin is important, not only in general medical diagnosis and in surgery, but also as the seat of many diseases of its own. The study of these is called *dermatology* (Gk. *derma,* skin).

The temperature of the skin in general is normally about 32 to 36° C (90 to 96° F), but, in the region of the toes, it is several degrees lower.

The area of the body surface (about 2 square meters) is important in measurements of basal metabolism, that is, the energy output under standardized, resting conditions. The area of an individual can be estimated from the weight and height by special formulae. Children have a relatively larger area (about 1/5 square meter in the newborn) in relation to volume and weight than do adults. In the treatment of extensive burns it is important to be able to assess the percentage of the body surface involved.

The skin (fig. 8–1) varies in thickness from about 1/2 to 3 mm. It is thicker on the dorsal and extensor than on the ventral and flexor aspects of the body. It is thinner in infancy and in old age. The stretching of the abdominal skin during pregnancy may result in red streaks (*striae gravidarum*) that remain as permanent white lines (*lineae albicantes*).

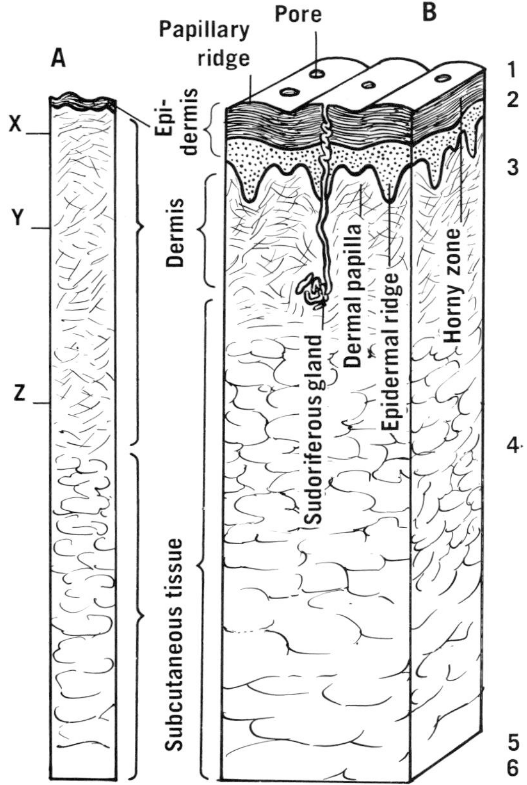

Figure 8–1 General view of skin and subcutaneous tissue. *A*, "thin" skin from abdomen; *B*, "thick" skin from palm of hand. *X*, *Y*, and *Z* represent the levels of a superficial (Thiersch) graft, a split thickness graft (including one-third to one-quarter of the dermis), and a full thickness (Wolfe) graft, respectively. *1* to *6* represent the levels of degrees of burns, according to Dupuytren's classification; other classifications of burns are also used.

The skin consists of two quite different layers: (1) the epidermis, a superficial layer of stratified epithelium that develops from ectoderm, and (2) the dermis or corium, an underlying layer of connective tissue that is largely mesodermal in origin. The dermis makes up the bulk of the skin and will be discussed first.

Corium (Dermis)

The corium or dermis, 1/2 to 2 1/2 mm in thickness,[1] contains downgrowths from the epidermis, such as hair follicles and glands. The dermis presents a superficial papillary layer of loose, delicate, collagenous, and elastic fibers, together with fibroblasts, mast cells, and macrophages. Elevations (papillae) project toward the epidermis. The thicker, deep reticular layer of the dermis consists of dense, coarse bundles of collagenous fibers. Some of the fibers enter the subcutaneous tissue, where they form bundles between lobules of fat. Smooth muscle is found in some regions (areola and nipple, scrotum and penis, and perineum). In some areas, muscle fibers of skeletal type (e.g., platysma) may be inserted into the skin. Tattooing of the skin is produced by the introduction of foreign particles, such as carbon, into the dermis.

The skin is distensible and elastic, but the elasticity seems to lessen with age. The skin lies on the *subcutaneous tissue* ("superficial fascia"), a layer of fatty areolar tissue that overlies the more densely fibrous fascia. It should be remembered that fat is liquid, or nearly so, at body temperature. The subcutaneous tissue serves as a depot for fat storage and aids in preventing loss of heat. When a pinch of skin is picked up, subcutaneous tissue is included. A hypodermic injection is one given into the subcutaneous tissue.

When the skin is punctured by a sharp, circular instrument, a linear wound results. The slit so produced is known as a *cleavage line*. Surgical incisions, however, are generally best placed parallel to flexure, crease, or wrinkle lines.[2] Flexure lines are not necessarily guides to the sites of joints. Thus well-marked flexure creases are to be seen opposite the middle of the front of the proximal phalanx. The main creases of the hand appear very early in fetal life and are not caused by movements of the digits.

Epidermis

The skin is covered by a film of emulsified material produced by glands and by cornification. The epidermis, 0.04 to 0.4 mm in thickness,[3] is an avascular layer of stratified squamous epithelium that is thickest on the palms and the soles. A biopsy specimen of the epithelium can be used for chromosomal sex detection. The epidermis, where it is thick (e.g., on the palms and the soles), presents five layers, as listed in table 8–1. In the outer layers, which may conveniently be grouped as the horny zone, the cells become converted into soft-keratin flakes that are worn away from the surface continuously. The *stratum corneum* is a tough, resilient, semitransparent cellular membrane that acts as a barrier to water transfer. Under normal conditions, mitotic figures are practically confined to the deepest layer, the *stratum basale,* which is, therefore, the normal germinative layer of the epidermis. The various layers show the stages through which the basal cells pass prior to their keratinization and shedding. The cells of the epidermis are replaced approximately once per month. Keratin is a protein that is present throughout the epidermis, perhaps in a modified form. It is readily hydrated, hence the swelling of skin on immersion in water, and dryness of the skin is due chiefly to a lack of water. Keratin may be considered as a holocrine (p. 50) secretion of the epidermis.

Human epidermis displays a rhythmic mitotic cycle. Mitosis is more active at night,[4] and it is stimulated by a loss of the superficial or horny zone. A basement membrane is probably not found, but the dermis is limited from the epidermis by a submicroscopic membrane.[5] The cells of the stratum spinosum present cytoplasmic striations *(tonofibrils),* many of which end in dense plaques *(desmosomes)* in so-called intercellular bridges (no cytoplasmic continuity is found between the spinous cells). A part, or the whole thickness, of the epidermis may be raised up in the form of blisters by plasma when the skin is damaged (e.g., by a second-degree burn), and prolonged pressure and friction result in callosities and corns.

TABLE 8–1 Arrangement of Layers of Body Surface

Skin	Epidermis	Stratum corneum Stratum lucidum Stratum granulosum Stratum spinosum Stratum basale (or cylindricum)
	Corium or dermis	Papillary layer Reticular layer
Subcutaneous tissue		
Fascia		

Several pigments, including melanin, melanoid, carotene, reduced hemoglobin, and oxyhemoglobin, are found in the skin. Melanin, which is situated chiefly in the stratum basale of the epidermis, protects the organism from ultraviolet light.

When an area of epidermis, together with the superficial part of the underlying dermis, is destroyed, new epidermis is formed from hair follicles, and also from sudoriferous and sebaceous glands, where these are present. If the injury involves the whole thickness of the dermis (e.g., in a deep burn), however, epithelization can take place only by a growing over of the surrounding edge of the epidermis or, alternatively, by the use of an autograft. Free skin grafts of the epidermis and a part or all of the thickness of the dermis can be applied, and vascularization takes place through connections between the subcutaneous vessels and those in the graft. A defect of the skin that extends into the dermis is termed an ulcer.

Lines of thickened epidermis known as *papillary ridges* form a characteristic pattern on the palmar aspect of the hand and the plantar aspect of the foot. They are concerned with tactile sensation.[6] They contain the openings of the sweat glands and overlie grooves in the dermis; these grooves are situated typically between rows of double ridges known as *dermal ridges* (fig. 8–1).[7] The papillary ridges appear in fetal life in a pattern that remains permanently. They are especially well developed in the pads of the digits, and finger prints in adults and foot prints in infants are used as a means of identification of an individual.[8]

SPECIALIZED STRUCTURES OF SKIN

Sweat (Sudoriferous) Glands

The sweat or sudoriferous glands (L. *sudor,* sweat, and *ferre,* to bear) regulate body temperature because perspiration withdraws heat from the body by the vaporization of water. The sweat glands develop in the fetus as epidermal downgrowths that become canalized. They are simple tubular glands, each having a coiled secretory unit in the dermis or in the subcutaneous tissue, and a long, winding duct that extends through the epidermis and opens by a pore on the surface of the skin (fig. 8–1). Sweat glands are particularly numerous in the palms and the soles, where they open on the summits of the papillary ridges. The chief stimuli to sweating are heat and emotion. Emotional perspiration occurs characteristically on the forehead, axillae, palms, and soles.

Large sweat glands in certain locations, such as the axilla, areola, external acoustic meatus, and eyelid, develop from hair follicles and differ from the more common (eccrine) glands in being apocrine;[9] that is to say, portions of the secreting cells disintegrate in the process of secretion. The perspiration from the apocrine glands is rich in organic material that is susceptible to bacterial action, resulting in an odor.

Water passes through the epidermis also by diffusion. This is termed insensible perspiration because it cannot be seen or felt.

Hairs

Hairs (or *pili; pilus* in the singular) are a feature characteristic of mammals. The functions of hair include protection, regulation of body temperature, and facilitation of evaporation of perspiration; hairs also act as sense organs. Hairs develop in the fetus as epidermal downgrowths that invade the underlying dermis. Each downgrowth terminates in an expanded end that becomes invaginated by a mesodermal *papilla*. The central cells of the downgrowth become keratinized to form a hair, which then grows outward to reach the surface. The hairs first developed constitute the *lanugo* or down, which is shed shortly before birth; the fine hairs that develop later constitute the *vellus*. Although hairs on many portions of the human body are inconspicuous, their actual number per unit area (40 to 880 per square cm, depending on the region) is large. In a few places (such as the palms and the soles, and the dorsal aspect of the distal phalanges) the skin is *glabrous*, that is, devoid of hair. Human scalp hairs have an average diameter of 65 micrometers.[10]

The shaft of a hair consists of a *cuticle* and a *cortex* of hard-keratin surrounding, in many hairs, a soft-keratin *medulla* (fig. 8–2). Pigmented hairs contain melanin in the cor-

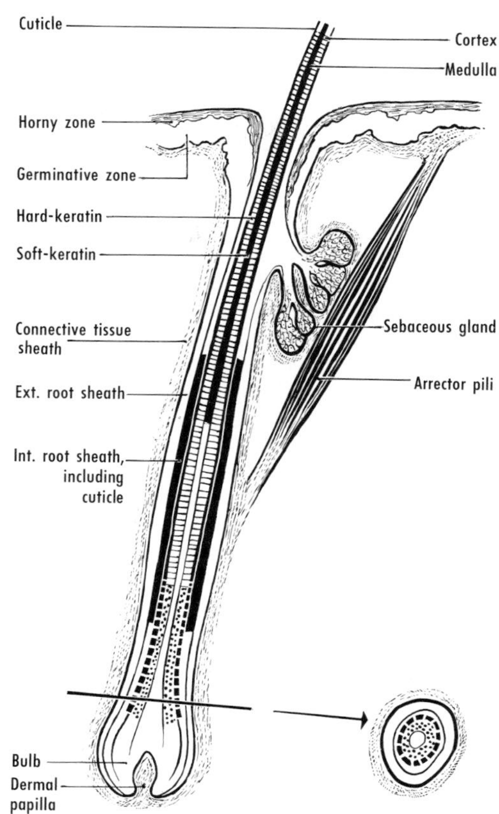

Figure 8–2 Diagram of a hair follicle. The follicle consists of an *external root sheath*, mainly the basal-cell layer of the epidermis, and an *internal root sheath* of soft-keratin, which includes a *cuticle* firmly anchored to that of the shaft of the hair.

tex and the medulla, but pigment is absent from the surrounding sheaths. The color of hair depends mainly on the shade and the amount of pigment in the cortex and, to a lesser extent, on air spaces in the hair. In white hairs pigment is absent from the cortex and the contained air is responsible for the whiteness (as it is also in the case of foamy water); "gray hair" (*canities*) is generally a mixture of white and colored hairs. Sudden blanching of the hair has been reported by so many competent observers that the existence of the phenomenon is probable.[11] Oxidation of melanin results in a colorless compound; hence dark hair can be bleached with hydrogen peroxide.

The *root* of a hair is situated in an epidermal tube known as the hair *follicle*, sunken into either the dermis or the subcutaneous tissue. The follicle is dilated at its base to form the *bulb (matrix)*.[12]

The growth of hair takes place by cell proliferation and by increase in cell volume. A fine nerve plexus surrounds the bulb, and hairs act as organs of touch.

In the obtuse angle between the root of a hair and the surface of the skin, a bundle of smooth muscle fibers, known as an *arrector pili muscle,* is usually found. It extends from the deep part of the hair follicle to the dermis. On contraction it makes the hair erect. The arrectores pilorum are innervated by sympathetic fibers and contract in response to emotion or cold. This results in an unevenness of the surface called "goose pimples," or "goose skin."

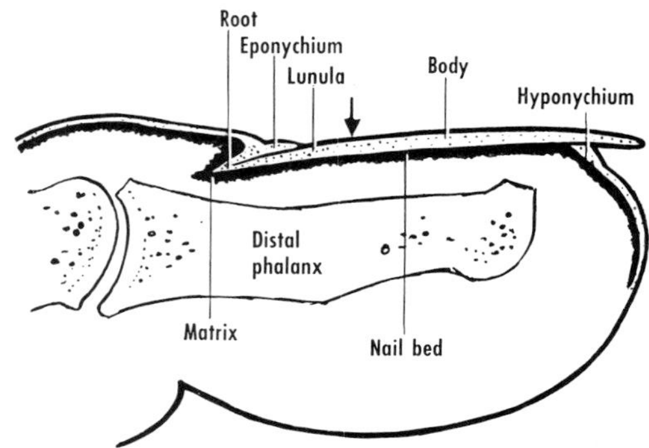

Figure 8–3 Diagram of sagittal section of a fingernail. Arrow indicates junction between root and body of nail.

Sebaceous Glands

Sebaceous glands (L. *sebum,* tallow) develop from the epidermis in the fetus, usually from the walls of hair follicles (most sebaceous glands are appendages of the external root sheaths of hairs). Sebaceous glands are absent from the palms and the soles. They are simple alveolar glands that form lobes in the dermis, generally in the acute angle between an arrector pili and its hair follicle. The basal cells of the gland proliferate, accumulate fat droplets, and are excreted as sebum through a short, wide duct into the lumen of the hair follicle. This type of gland, in which the secretory cells disintegrate, is termed *holocrine.* Sebaceous glands are under hormonal control. Contraction of the arrector pili may perhaps aid in expelling the sebum. Sebum keeps the stratum corneum pliable and, in cold weather, conserves body heat by hindering evaporation. Fat-soluble substances may penetrate the skin through hair follicles and sebaceous glands. Hence ointment vehicles are used when penetration is desired. Medicaments should be rubbed into the skin.

Sebaceous glands that are not related to hairs are found in the eyelids as tarsal glands; these are said to be apocrine in type (p. 49), as are also the ceruminous glands of the external acoustic meatus. Seborrhea involves an excessive secretion of sebum; the sebum may collect on the surface as scales known as dandruff. Acne is a chronic inflammatory condition of the sebaceous glands. When the exit from a sebaceous gland becomes plugged, a blackhead (comedo) forms; complete blockage may result in a wen (sebaceous cyst). At birth an infant is covered with *vernix caseosa,* a mixture of sebum and desquamated epithelial cells.

Nails

The nails (or *ungues; unguis* in the singular) are hardenings of the horny zone of the epidermis. They overlie the dorsal aspect of the distal phalanges (fig. 8–3). They protect the sensitive tips of the digits and, in the case of the fingers, serve in scratching. Nails develop in the fetus as epidermal thickenings that undercut the skin to form folds from which the horny substance of the nail grows distally.

The horny zone of the nail is composed of hard-keratin and has a distal, exposed part or *body,* and a proximal, hidden portion or *root.* The root is covered by a distalward prolongation of the stratum corneum of the skin. This narrow fold is composed of soft-keratin and is termed the *eponychium* (Gk. *onyx,* nail). Distal to the eponychium is the "half-moon," or *lunula,* a part of the horny zone that is opaque to the underlying capillaries.

Deep to the distal or free border of the nail, the horny zone of the fingertip is thickened and is frequently termed the *hyponychium.* The horny zone of the nail is attached to the underlying *nail bed.* The *matrix,* or proximal part of the bed, produces hard-keratin. Further distally, however, the bed may also generate nail substance.[13] Moreover, the most superficial layer of the nail may be produced by the epithelium immediately dorsal to the root and proximal to the eponychium.[14] The growth of the nail (normally about 2 to 4½ mm per month[15]) is affected by nutrition, hormones, and disease. Nail growth involves considerable protein synthesis, as a result of which nonspecific changes occur in the nails in response to various local and systemic disturbances. White spots indicate incomplete keratinization.

Hard-keratin is found in the nails and in the cortex of hairs. Soft-keratin is located in the epidermis, and in the medulla and internal root sheath of hairs.

BLOOD SUPPLY AND INNERVATION OF SKIN

The skin has a profuse blood supply, which is important in temperature regulation. The subcutaneous arteries form a network in the subcutaneous tissue and from this is derived a subpapillary plexus in the dermis. Capillary loops in the dermal papillae arise from the subpapillary plexus, and from these loops the avascular epidermis is bathed in tissue fluid. When the skin near a nail is cleared by oil, the dermal loops can be seen with a microscope in the intact, living body. Moreover, the capillaries and the subpapillary plexus can conveniently be studied *in vivo* in dekeratinized skin and with the aid of a low-power microscope.[16] A subpapillary plexus of venules gives the skin its pink color; the vessels become dilated when the skin is heated, and thereby make it look red. Most birthmarks consist of dilated capillaries (hemangioma). The dermis contains a lymphatic plexus that drains into the collecting vessels in the subcutaneous tissue. The cutaneous lymphatics can be shown *in vivo* by injecting vital dyes; "every intradermal injection is an intralymphatic one."[17]

The subpapillary plexus in the dermis, particularly in the digits, nose, and lips, has a number of arteriovenous anastomoses (A.V.A.). These connect arterioles to venules and thereby bypass the capillary bed. They have thick muscular walls supplied by sympathetic fibers and, on contraction, can act as sphincters to direct blood through the capillary bed. Thus they can control the amount of blood entering the capillaries, a useful asset in structures that have an intermittent metabolism. They dilate when the skin becomes cold, thereby shunting blood from capillary beds and reducing heat loss. A number of alternative pathways through capillary beds are available. The shorter and more direct routes are termed preferred channels.

The skin has a rich sensory innervation (fig. 8–4). The cutaneous nerves pierce the fascia and ramify in the subcutaneous tissue to form plexuses both there and in the

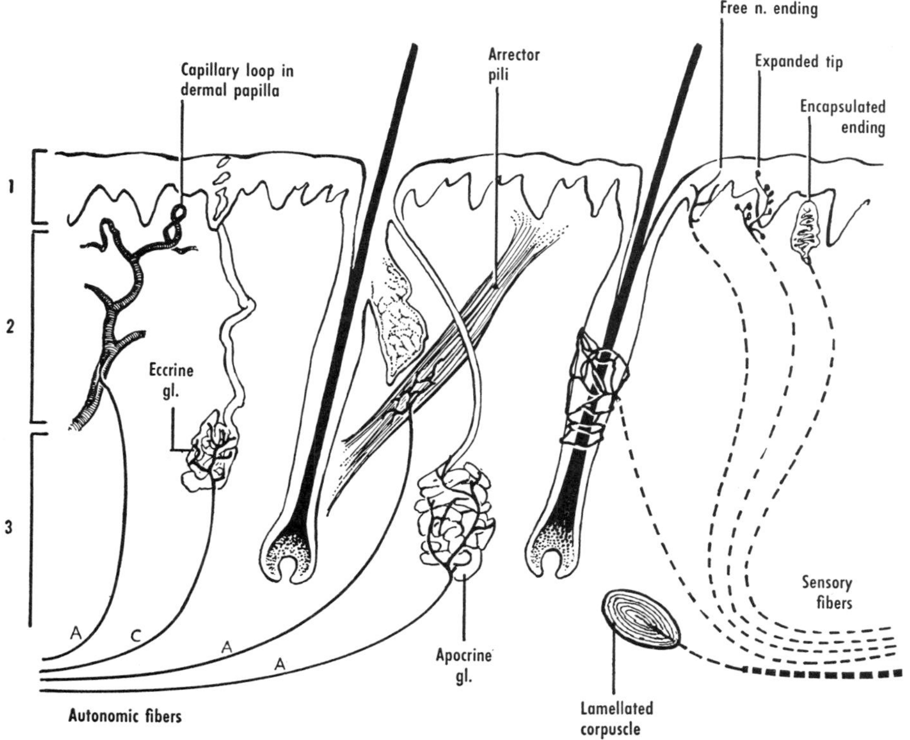

Figure 8–4 Diagram of innervation of skin. The numerals *1*, *2*, and *3* indicate the epidermis, dermis, and subcutaneous tissue respectively. The letters *A* and *C* stand for adrenergic and cholinergic nerve fibers respectively.

dermis. Finer axonal ramifications may run between the deeper cells of the epidermis. The cutaneous nerves supply both the skin and the subcutaneous tissue. The area of distribution of a given nerve, however, varies, and considerable overlapping of adjacent nerve territories takes place (p. 34).

Nerves supplying the skin may form several different types of nerve endings, and these endings have been related in a general way to the basic types of sensations that can be appreciated in the skin and the subcutaneous tissue, namely, pain, touch, temperature changes, and pressure or deep touch. Complex sensations, such as a sense of vibration and the ability to discriminate between stimuli applied simultaneously at two different points, seem to depend on combinations of these basic modalities. Itching may be regarded as a mild type of pain.

Hairy skin contains simple, free endings and plexuses around the hair follicles. Skin without hair, that of the palm, for example, presents the three types of sensory endings that are characteristic of the somatic nervous system:[18] (1) free nerve endings arising from small myelinated fibers, (2) expanded tips,* and (3) encapsulated endings.†

The basic types of sensation, however, can be elicited from both hairy and glabrous skin. Hence correlations between the type of sensation and a specific type of nerve ending are not justified. "There is not and never really has been convincing histological evidence for the commonly accepted statement that morphologically specific nerve endings subserve each of the primary modalities of cutaneous sensibility," that is, touch, warmth, cold, and pain.[19]

Lamellated corpuscles‡ are particularly large, encapsulated endings that are found chiefly in the subcutaneous and deeper tissues.

The various endings in the skin tend to be arranged in groups, each containing a number of endings. If, for example, the sense of touch is tested with a fine camel's hair brush, the sensation is elicited from multiple small "touch spots," whereas the intervening skin is devoid of sensation.

*E.g., Merkel and Ruffini endings.

†E.g., Meissner, Krause, and Pacinian endings.

‡Frequently named after Vater and/or Pacini.

Cutaneous nerves also contain motor fibers to the walls of arteries and arterioles (vasoconstrictor) and to the arrectores pilorum. These fibers belong to the sympathetic system and are adrenergic. The eccrine sweat glands, however, are supplied mainly by cholinergic fibers, whereas the apocrine glands are innervated chiefly by adrenergic fibers. The cutting of appropriate sympathetic fibers (sympathectomy) results in a deficiency of sweat (anhidrosis) and in an increased electrical resistance in the skin.

REFERENCES

1. W. F. W. Southwood, Plast. reconstr. Surg., *15*:423, 1955.
2. E. H. Courtiss *et al.*, Plast. reconstr. Surg., *31*:31, 1963.
3. J. T. Whitton, Hlth Phys., *24*:1, 1973.
4. L. E. Scheving, Anat. Rec., *135*:7, 1959.
5. A. S. Breathnach, *An Atlas of the Ultrastructure of the Human Skin,* Churchill, London, 1971.
6. N. Cauna, Anat. Rec., *119*:449, 1954.
7. L. W. Chacko and M. C. Vaidya, Acta anat., *70*:99, 1968.
8. H. Cummins and C. Midlo, *Finger Prints, Palms and Soles,* Dover, New York, 1961 (1943).
9. H. J. Hurley and W. B. Shelley, *The Human Apocrine Sweat Gland in Health and Disease,* Thomas, Springfield, Illinois, 1960.
10. S. C. Atkinson, F. E. Cormia, and S. A. Unrau, Brit. J. Derm., *71*:309, 1959.
11. A. J. Ephraim, Arch. Derm., *79*:228, 1959.
12. A. M. Kligman, J. invest. Derm., *33*:307, 1959.
13. A. Jarrett and R. I. C. Spearman, Arch. Derm., *94*:652, 1966.
14. B. L. Lewis, Arch. Derm., *70*:732, 1954.
15. M. S. Sibinga, Pediatrics, *24*:225, 1959.
16. M. J. Davis and J. C. Lawler, Arch. Derm., *77*:690, 1958.
17. S. S. Hudack and P. D. McMaster, J. exp. Med., *57*:751, 1933.
18. M. R. Miller, H. J. Ralston, and M. Kasahara, Amer. J. Anat., *102*:183, 1958. M. R. Miller and M. Kasahara, Amer. J. Anat., *105*:233, 1959. R. K. Winkelmann, *Nerve Endings in Normal and Pathologic Skin,* Thomas, Springfield, Illinois, 1960. D. Sinclair in Jarrett, cited below.
19. G. Weddell, E. Palmer, and W. Pallie, Biol. Rev., *30*:159, 1955.

GENERAL READING

Advances in Biology of Skin. A volume on a different topic appears annually.

Champion, R. H., *et al., An Introduction to the Biology of the Skin,* Blackwell, Oxford, 1970.

Horstman, E., and Dabelow, A., Die Haut. Die Milchdrüse, in *Handbuch der mikroskopischen Anatomie des Menschen,* ed. by W. von Möllendorff and W. Bargmann, Springer, Berlin, vol. 3, part 3, 1957.

Jarrett, A. (ed.), *The Physiology and Pathophysiology of the Skin,* Academic Press, New York, 1973. Volume 2 is devoted to nerves and blood vessels.

Montagna, W., and Parakkal, P. F., *The Structure and Function of Skin,* Academic Press, New York, 3rd ed., 1974. A good introduction, including histochemistry and electron microscopy.

Pinkus, H., Die makroskopische Anatomie der Haut, in *Normale und pathologische Anatomie der Haut,* ed. by O. Gans and G. K. Steigleder, Springer, Berlin, vol. 2, 1964.

Szabó, G., The Regional Anatomy of the Human Integument. . ., Phil. Trans. B., *252*:447, 1967.

DEVELOPMENT AND GROWTH

9

DONALD J. GRAY

The term *development* means the series of changes that the organism undergoes until it reaches maturity. It is generally used to include all prenatal and postnatal changes except *growth*, which means an increase in size. The term development is sometimes used synonymously with *differentiation*, which denotes an increase in complexity of the body and its parts. Differentiation results in a modification of these parts for the performance of particular functions. It may not be apparent as soon as it occurs, and sometimes requires physiological methods for evaluation. Later, however, it becomes manifest as histogenesis and organogenesis.

Growth may result from an increase in the number of cells, an increase in the size of the cells, or an increase in the amount of nonliving material. It can be measured by physical methods, and is therefore easier to assess than is differentiation.

Cell division, resulting in an increase in number of cells (hyperplasia), occurs most rapidly early in prenatal life, but the enlargement of cells (hypertrophy) occurs later. Each cell is characterized by a life cycle peculiar to itself. Some cells of the nervous system remain throughout the life of the organism, whereas the life span of certain blood cells is only a month or so. Similarly, organs have their own life cycles. Although most of them survive throughout life, some of them, such as the thymus, persist through earlier periods and then decline.

Normal development and growth do not proceed uniformly. Neither do they proceed haphazardly, and various changes occur at predictable periods. The changes that take place in the body as a whole, in its parts, and in the different organs and tissues do not take place at the same time, and they do not proceed at equal rates. Development and growth are usually considered to cease at maturity, or at the end of almost one-third of the average span of life.

Although a knowledge of development and growth is primarily of interest to the pediatrician, it also provides others with a basis for an understanding of the functional activity of the individual, both before and after maturity. It offers a desirable background for the study of physiology, pathology, physical diagnosis, and surgery. A proper assessment of these processes is often useful and is sometimes necessary in the diagnosis of disease or disability.

PERIODS OF LIFE BEFORE MATURITY

The union of the oöcyte and spermatozoon results in a unicellular zygote, which is often called a "fertilized ovum." As soon as the zygote undergoes cleavage, it is frequently termed an embryo. Inasmuch as the zygote gives rise to the extraembryonic membranes as well as the embryo, some embryologists prefer not to call the organism an embryo until the embryonic disc is formed. After all of the main systems and organs of the body have become differentiated, and after the major features of the external form of the body have been established, the organism is called a fetus. It is known as an infant as soon as it is outside of the body of the mother, even before the umbilical cord is cut.

The period of time required for an individual to reach maturity has been subdivided into a number of intervals, which are summarized in table 9–1. The subdivisions are somewhat arbitrary, and, in most cases, no sharp line can be drawn between two adjacent intervals. The length of many of the various periods varies with heredity, race, sex, and environment.

The period of development before birth is called the prenatal period, and it is often spoken of as divided into first, second, and third trimesters, three periods of equal length. It extends over approximately 10

Table 9–1 Summary of Periods of Life Before Maturity

Prenatal period	
Embryonic period	First 8 postovulatory weeks
Fetal period	From 8 postovulatory weeks to term
Postnatal period	
Infancy	First year
Childhood	
Early	1 to 6 years
Late	6 years to puberty*
Adolescence	Puberty and period from puberty to maturity†

*In girls, puberty begins with the onset of menstruation (menarche), which occurs at 13 ± 2 years. No such clear-cut criterion appears in boys, but puberty is considered to begin at 15 ± 2 years, because the skeletal age of boys at this time corresponds to that of girls at the menarche.

†Girls are considered to have reached maturity at about 18 years, boys at about 20 years.

lunar months (about 280 days from the onset of the last menstrual period, or 266 days after the presumed date of ovulation).[1] It is subdivided into (1) an embryonic period, and (2) a fetal period.

Embryonic Period. The embryonic period comprises the first eight weeks after ovulation, and the age of the embryo during this time can be referred to in terms of postovulatory weeks. In obstetric practice, however, the term menstrual weeks is usually used. It is often assumed that menstruation occurred two weeks before ovulation, and the so-called menstrual age is commonly about two weeks greater than the postovulatory age.

After fertilization has occurred, the precise time of ovulation cannot always be learned, and the time of onset of the last menstruation is often impossible to establish. However, a rough approximation of the age of an embryo can be obtained by relating its measurements to standard tables or curves (fig. 9–1). Before the flexures become apparent, the greatest length of the embryo is used. After the flexures appear, the crown-rump (C.R.) length is usually employed; this is the distance from vertex to breech and corresponds to the sitting height postnatally.

The external form and the structural organization of an embryo give more reliable information as to age than do the dimensions.

Fetal Period. The fetal period extends from eight postovulatory weeks, when the embryo has reached approximately 30 mm C.R. length, to term. Although other criteria are sometimes employed, the C.R. length is the most useful in estimating the age of the fetus (fig. 9–1). The crown-heel (C.H.) length is sometimes used, particularly in later stages, because it can be compared with the postnatal height of the individual.

Postnatal Period. The postnatal period extends from birth to maturity. It is commonly subdivided into infancy, early childhood, late childhood, and adolescence (see table 9–1).

BODY AS A WHOLE

Weight. During the entire period of development and growth, weight is the best single indicator. The weight of the newborn baby is several billion times that of the oöcyte, but the weight of the adult is only about 20 times that of the newborn. The two periods of most rapid growth occur during fetal life, particularly in the last trimester, and during adolescence.

The weight of newborn babies at term ranges from 6 to 10 pounds (2¾ to 4½ kilograms), and averages 7½ pounds (3½ kilograms). It is affected by the length of gestation, by the sex of the baby, and by the age and parity of the mother. The birth weight is less when babies are premature, when they are female, when they are in multiples, and when their mothers are younger. During the first few days after birth, babies usually lose 5 to 6 per cent of the birth weight. This loss

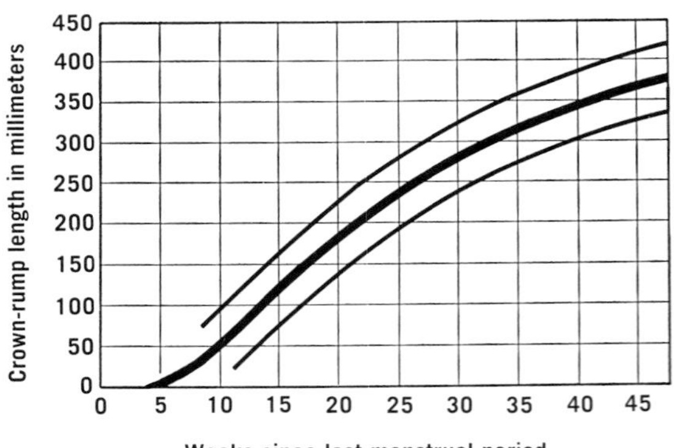

Figure 9–1 Correlation of prenatal age with crown-rump length. Based on E. Boyd, *Outline of Physical Growth and Development*, Burgess, Minneapolis, 1941. The heavy line indicates the 50th percentile. The lines on each side of this enclose between them 82 per cent of the specimens.

is the result of a diminished fluid intake and is restored in 7 to 10 days. Premature babies lose more weight and require a longer time to regain it. The weight increases to 12½ pounds (5⅔ kilograms) at three months. The birth weight is tripled by the end of the first year, and is quadrupled by the end of the second.

Girls are relatively lighter than boys of the same age during infancy and childhood, but are usually heavier than boys around puberty. After the first two or three years of the adolescent period, boys are again heavier than girls.

Length. Throughout the period of development and growth, the relative increase in length is much smaller than that in weight. The embryo increases from about 5 mm at the end of four postovulatory weeks to about 30 mm at the end of eight. On the basis of C.R. measurements, growth in length is quite gradual until term, when the C.R. length is from 310 to 350 mm. When total length is considered, the most rapid absolute increase occurs during the fourth lunar month, during which the length of the fetus increases by about 80 to 100 mm. After the fourth month, the increments in absolute and relative growth in length gradually decline. The crown-heel length is usually between 480 and 520 mm at term.

The increase in length between birth and maturity is about 3½ times. A baby adds almost one-third to its birth length in the first six months, and about one-half by the end of the first year. The birth length is doubled by about four years. The increase in length is relatively slow throughout childhood. The rapid spurt that begins just before puberty starts earlier and is completed sooner in girls than in boys. Growth in length usually ceases at about 18 years in girls and 20 years in boys.

Surface Area.[2] The relation of the surface area to the mass of the body exerts a great influence upon the metabolism of the individual. The relatively large area in the newborn baby results in much greater heat loss.

The surface area at birth is 2000 to 2500 square cm. It is doubled during the first year and tripled by the middle of childhood. At maturity, it is about seven times greater than at birth. It undergoes a relative decrease during postnatal life, however, from 800 square cm per kilogram at birth to 300 square cm per kilogram in the adult.

PARTS OF THE BODY

Each part of the body has its own pattern of development and growth. In general, the changes that occur proceed in two directions, from the cranial end caudalward, and from the mid-dorsal region ventrolaterally.

Head.[3] The head forms about one-fourth of the body at birth, and about one-twelfth at maturity. The diameter of the skull becomes decreased during a normal parturition by a narrowing of the sutures and fontanelles, and the head appears distorted for several days after birth. The anterior fontanelle is about 2½ cm wide at birth, and its width may increase for two or three months thereafter. It then gradually decreases in width and becomes obliterated at about the end of infancy.

In the embryonic period, the cranial part of the head is much larger than the facial. During late fetal and early postnatal life, however, the increased growth of the facial part tends to reduce this relationship, although the cranial part always remains larger. The eruption of the teeth and the concomitant growth of the maxilla and mandible during childhood make important contributions to the facial part. The growth of the external acoustic meatus and mastoid process during infancy and childhood contributes to the cranial part, which has almost reached its adult dimensions by late childhood. The growth in length of the head exceeds the growth in width through the entire period of development and growth.

The circumference of the head is a valuable measurement because it is related to the intracranial volume and, therefore, to the growth of the brain. An unusually large circumference may indicate hydrocephaly, and an unusually small circumference may indicate microcephaly. The average circumference of the head at birth is 35 cm (about 14 in), and is about the same as the circumference of the thorax. It is about 46½ cm at age one year, 49 cm at two years, and 50 cm at three years. The increase between three years and adulthood is only about 5 cm.

Trunk. The trunk forms about 45 to 50 per cent of the length of the entire body at all stages. The individual parts of the trunk, however, reach their maximal proportions at different periods. The thoracic part attains its greatest relative length earlier than the pelvic part, which does not reach its maximal proportion until adolescence.

The combined length of the head and trunk, termed the sitting height, is a useful indicator of development and growth. The sitting height is about 70 per cent of the total length of the body at birth, 57 per cent at three years, 52 per cent at puberty, and 53 to 54 per cent in the adult. The increase in length of the trunk after growth of the lower limbs has ceased accounts for the increase in proportion in the adult over that at puberty.

During the fetal period, the trunk is ovoid in shape, and its circumference is largest at the umbilicus. The circumferences of the thorax and head are about the same during infancy. After about two years, the circumference of the thorax becomes increasingly greater than that of the abdomen.

The transverse and anteroposterior diameters of the thorax are about the same in the infant, but the transverse is about three times greater than the anteroposterior in the adult.

The cervical flexure, which is present prenatally, becomes accentuated when the infant lifts its head. After the erect posture is assumed, it becomes even more pronounced, and the lumbar flexure begins to develop. The cavity of the pelvis enlarges, and many abdominal organs move downward.

Limbs. The relative increases in length and weight of the upper limb differ markedly from increases in length and weight of the lower limb. The upper limb accounts for about 3 per cent of the total body weight by the beginning of the fetal period. At birth, it accounts for 8 to 9 per cent, and it maintains this relationship thereafter. The lower limb also accounts for about 3 per cent of the weight of the body at the beginning of the fetal period, but its proportion has increased to about 15 per cent at birth and to about 30 per cent in the adult.

The lower limb is approximately equal in length to the upper at two postnatal years. As a result of an increase in both relative and absolute length, it is about one-sixth longer than the upper limb in the adult.

The relative proportions of the arm, forearm, and hand, and of the thigh, leg, and foot are reached quite early in the prenatal period.

The relative increase in length of the lower limb, accompanied by a lesser increase in the trunk, results in a downward shift of the midpoint of the body, which is the level above and below which the lengths of the two halves of the body are equal. These increases also result in a downward shift of the center of gravity, which is the level above and below which the weights of the two halves of the body are equal. At the beginning of the fetal period, the midpoint is at the junction of the neck and thorax. It moves downward to the level of the umbilicus by term and to the level of the pubic crest by adulthood. The center of gravity moves from the cervical region in the embryo to the opening for the inferior vena cava in the diaphragm at term, and to a point at the level of the sacral promontory in the adult (p. 520).

Systems and Organs

Skin and Subcutaneous Tissue. The skin is thin and easily injured at birth. It is covered by primary hairs, called *lanugo*, which are lost after a few weeks, and by a cheesy coating, termed *vernix caseosa*, which is a mixture of sloughed epidermal cells and the secretion of sebaceous glands. Both sweat and sebaceous glands are present in the fetal period, but they are relatively inactive in the fetus and in the first few weeks after birth. Both types of glands undergo considerable development and growth during adolescence.

Hair associated with the secondary sexual characteristics appears in late childhood and early adolescence. In the male, hair makes it appearance in the different regions of the body in the following sequence: in the pubic region, in the axilla, on the face, and on the chest and limbs. The hair extending upward from the pubis to the umbilicus is terminal, rather than pubic, hair. It is more characteristic of males, but may be present in females. Pubic hair usually appears in females just before the menarche, and axillary hair is first present about six months later.

Pigmentation of the skin of the external genital organs, the axillae, and the areolae occurs during the development of the other secondary sexual characteristics.

The increase in subcutaneous tissue and its contained fat occurs at unequal rates, and therefore the amount of this tissue cannot be regarded as a criterion of nutritional status at all stages. A spurt in the accumulation of subcutaneous fat occurs during the last third of the prenatal period and during the first nine months after birth. During the second year, the amount of fat begins to decrease, and by five years the thickness of the subcutaneous tissue is about half that at one year. The amount of fat increases again early in adolescence, especially in girls, in whom localized accumulations appear around the hips (at the same time the pelvis is increasing in width) and in the breasts.

The weight of the skin is about 1/25 of the total weight of the body at term, and it is approximately 1/16 of the total weight in the adult. The subcutaneous tissue is present in very small amounts up to five prenatal months, and it then increases rapidly until it is about one-fourth of the total weight at birth. At maturity, it represents approximately one-tenth of the total weight in males, and about one-fifth in females. The weight of the skin and subcutaneous tissue together forms about one-fourth of the total weight at term, and slightly less than one-fourth at maturity.

Skeleton. In the prenatal period, the growth of the skeleton is relatively slow until the last two months, during which it rapidly accelerates. At term, the weight of the skeleton is from one-sixth to one-fifth the weight of the body. During the period from birth to maturity, the increase in weight is similar to that in the entire body and the weight of the skeleton at maturity is about 20 times that at birth. The most rapid postnatal increase occurs during adolescence.

The first appearance and later fusion of the various centers of ossification follow a quite definite pattern and time schedule between birth and maturity (p. 13). The interpretation of the radiographic appearance of bones offers valuable information for evaluating normal and abnormal growth. The status of the skeleton at any one time is termed the *skeletal* (or *bone*) *age*, and it is a much better criterion of the stage of development and growth than is the chronological age. Skeletal age is subject to racial and sexual variations, and is influenced by nutritional and other environmental factors.[4] It has a good correlation with height, weight, and sexual development.

Muscles. The skeletal musculature is characterized by a slow growth in the embryo, but, beginning in the middle of prenatal life, it gains more in relative weight than any other system. Its increment during childhood and adolescence is as great as that of all other tissues, organs, and systems combined. Its postnatal growth results from an increase in size, rather than an increase in number, of fibers.

The skeletal muscles form about one-fourth of the weight of the body at birth, one-third in early adolescence, and two-fifths at maturity. The most rapid gain is in adolescence, during which the strength of the muscles is doubled.

Nervous System. The central nervous system forms about one-seventh of the total body weight at birth, and 1/50 at maturity.

The relative size of the head in the embryo is a reflection of the disproportionate size of the brain at this stage of development. The birth weight of the brain is doubled during the first year, and trebled during the third. Nine-tenths or more of its adult weight is reached by six years, and its adult weight may be reached by ten years.

The postnatal growth of the spinal cord is relatively less than that of the vertebral column. The caudal end of the cord, which reaches the upper margin of the third lumbar vertebra in the newborn, is usually one or two vertebral levels higher in the adult (p. 540).

Heart. The heart weighs about 20 gm at birth. This weight is doubled in the first year, and tripled at three to four years. Its most rapid postnatal growth takes place in late childhood and early adolescence. Its weight at maturity is about 12 times its birth weight.

Lymphoid Tissues. The lymphoid tissues are relatively large at birth. They grow rapidly until late childhood, after which they undergo a decrease in both absolute and relative weight. During infancy and childhood, they respond to infection by a rapid swelling and hyperplasia. It may be significant that the lymphoid tissues are in their maximal stage of development during the periods when infections of the respiratory and digestive systems are most frequent.

The spleen varies in weight at birth and throughout postnatal life. It does not undergo a significant atrophy in the adult.

Viscera.[5] The weight of the viscera as a whole is about 9 per cent of that of the body in the newborn, and about 5 to 7 per cent in the adult.

The increase in weight of the internal organs, both individually and collectively, is slow during the first five prenatal months.

It is much more rapid during the remainder of the fetal period. Until term, the rates of increase in size of the various organs are similar, but the postnatal growth of the genital organs differs greatly from that of the organs comprising the digestive, respiratory, and urinary systems. In the latter group, the weight increases rapidly during infancy and the first part of early childhood, but much more slowly later in childhood. Just before puberty, these organs again grow rapidly, and then revert to a slower rate of growth during adolescence.

DIGESTIVE SYSTEM. Many organs in the thorax and abdomen move caudalward in the postnatal period. The postnatal growth of the stomach is most rapid during the first three months. Its long axis in the infant is horizontal. The length of the small intestine increases by approximately one-half during the first year, and is doubled by puberty. The large intestine appears to be disproportionately large during the first four or five years. The weight of the liver, which is about 4 per cent of that of the body at birth, has increased by about 10 times at puberty.

Saliva has been found in the mouth of an infant before food has been ingested. Gastric juice and some of the digestive enzymes are present before birth. A dark greenish brown or black substance, called *meconium*, is present in the intestines at birth. Some meconium is found in the stool within the first 12 hours, and all of it usually disappears after a few days.

RESPIRATORY SYSTEM. The caudal migration of the trachea, bronchi, and lungs is relatively rapid until the third prenatal month, after which it becomes slower. The bifurcation of the trachea, which is located at the level of the third or fourth thoracic vertebra at birth, is about two vertebral levels lower at 12 years. During this time, the upper end of the epiglottis moves from the level of the first cervical vertebra to that of the third.

In males, the size of the larynx at puberty is almost double its size at birth. During the same period in females, the larynx increases in size by about one-third. The small lumen present in this organ during infancy and early childhood contributes to the incidence of croup during respiratory infections at this time.

The birth weight of the lungs is doubled in the first six months and tripled in the first year. Its increase is about 20 times by maturity. A rapid increase in the content of elastic tissue occurs during the first few postnatal months.

URINARY SYSTEM. The kidneys grow slowly in the earlier part of the prenatal period and rapidly in the later part. At six postnatal months, they are twice their weight at birth, at one year three times, at five years five times, and at puberty ten times.

The last renal tubules are not formed until infancy. All of the glomeruli become larger after birth, but the size of the peripheral ones increases more rapidly. The lobation that is characteristic of the fetal kidney is lost soon after infancy.

The urinary bladder, which lies partly in the abdomen during infancy, becomes pelvic in position during childhood (p. 460).

GENITAL SYSTEM. The organs of the genital system in both males and females have a pattern of development and growth that departs from the general pattern of other viscera. Under the influence of maternal hormones, the breasts of both sexes are enlarged at birth, and secretions from their glandular tissues are found in the majority of newborn babies.

The weight of the testes in the adult is about 40 times the weight at birth. The rate of increase is most rapid during infancy, and is also especially rapid during adolescence. In the prenatal period, the testis is located at the site of the deep inguinal ring at the end of the fourth month, and passes through the inguinal canal during the seventh month. It usually arrives in the scrotum during the eighth month.

The seminiferous tubules are filled with cells at birth and acquire lumina during childhood.

The birth weight of the ovary is doubled during the first six months and increases by about 30 times by adulthood. The cortex is relatively thicker in the newborn than at maturity.

The stimulation of the fetal uterus by materal hormones is responsible for the relatively large size of this organ in the newborn. In the absence of this stimulation after birth, the weight of the uterus decreases by about one-half during the first few weeks. The birth weight is not regained for 10 to 11 years. The cervix is much larger than the body of the uterus until adolescence, during which the body of the organ grows at a much more rapid rate until the adult proportions are reached.

The female external genital organs, like the uterus, are relatively large in the newborn. They decrease in size during the first few weeks after birth.

ENDOCRINE ORGANS. The pattern of development and growth of the hypophysis, the suprarenal glands, and the thyroid gland also differs from that of the general visceral group.

The relative weight of the suprarenal glands at birth is about 20 times that in the adult. The absolute birth weight decreases by about one-half during infancy, and is not regained until puberty. The suprarenal glands then grow slowly until maturity, when they are almost twice the birth weight.

The weight of the thyroid gland at maturity is about 12 times its weight at birth; that of the hypophysis is about five to six times.

NORMAL STANDARDS OF DEVELOPMENT AND GROWTH

The physical status and the progress of development and growth can be assessed by reference to standard tables and curves derived from up-to-date studies of children of the same sex, race, and environment (fig. 9–2). The normal standards have wide

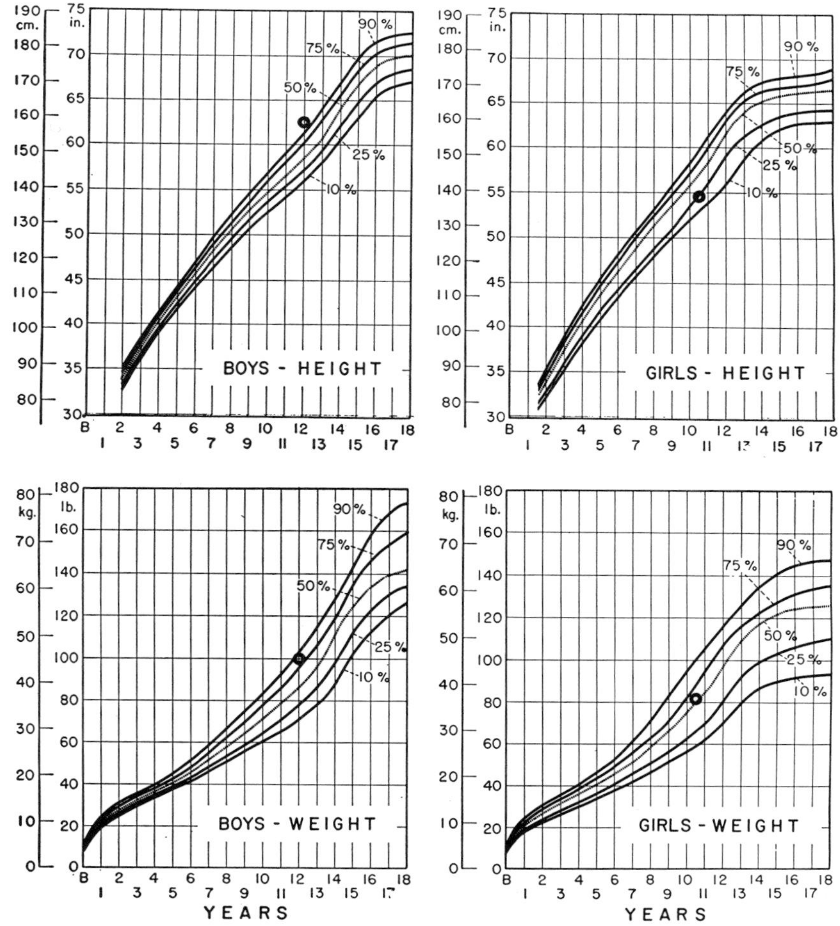

Figure 9–2 Percentile charts for height and weight, boys and girls. Based on E. Boyd, *An Introduction to Human Biology and Anatomy for First Year Medical Students*, Child-Research Council, Denver, Colorado, 1952. The circles on the two charts on the left mark the height and weight of a 12 year old boy who is taller than 90 per cent and heavier than 75 per cent of boys of his age. The circles on the two charts on the right mark the height and weight of a 10½ year old girl who is as tall as 25 per cent and is as heavy as 50 per cent of girls of her age.

ranges of variability, which are a reflection of the different patterns of development and growth of different children. Tables and curves derived from measurements of the same children at various times are termed longitudinal; those derived from mean values of different children at similar times are called cross-sectional. The former are more meaningful, the latter being used primarily as points of reference. A knowledge of statistics is desirable for the evaluation of either type.

Comparison with normal standards should not be used as a substitute for clinical examination. Marked deviations of any measurements from those of a particular age or sex group deserve careful study, and attempts to determine the cause or causes of such deviations should be made.

Certain measurements are more significant than others, and, therefore, are more frequently used. They include the following:

1. Percentage values by sex and age for weights and lengths.
2. Sitting height (may indicate disproportionate growth).
3. Circumference of head (at the level of the occiput behind, and just above the supraorbital ridges in front).
4. Circumference of chest (at the level of the nipples, subject recumbent).
5. Circumference of leg (greatest circumference; reflects nutritional status).
6. Bicristal diameter (the distance between the tubercles of the crests of the ilium; reflects general stockiness).

The following articles contain tables and curves that are useful in the assessment of physical status with reference to development and growth:

Bayer, L. M., and Bayley, N., *Growth Diagnosis*, University of Chicago Press, Chicago, 1959.

Jackson, R. L., and Kelly, H. G., Growth Charts for Use in Pediatric Practice, J. Pediat., *27*:215, 1945.

Lucas, W. P., and Pryor, H. B., Range and Standard Deviations of Certain Physical Measurements in Healthy Children, J. Pediat., *6*:533, 1935.

Meredith, H. V., Stature and Weight of Children of United States, with Reference to Influence of Racial, Regional, Socioeconomic and Secular Factors, Amer. J. Dis. Child., *62*:909, 1941.

Meredith, H. V., A "Physical Growth Record" for Use in Elementary and High Schools, Amer. J. publ. Hlth., *39*:878, 1949.

National Center for Health Statistics, Height and Weight of Children, United States, in *Vital and Health Statistics*, PHS Pub. No. 1000, series 11, no. 4, Public Health Service, U.S. Government Printing Office, Washington, D.C., September, 1970.

National Center for Health Statistics, Height and Weight of Youths 12–17 Years, United States, in *Vital and Health Statistics*, Series 11, no. 124, Public Health Service, U.S. Government Printing Office, Washington, D.C., January, 1973.

Olson, W. C., *Child Development*, Heath, Boston, 1949.

Olson, W. C., and Hughes, B. O., Growth of the Child as a Whole, chap. 12 in Barker, R. G., Kounin, J. S., and Wright, H. F. (eds.), *Child Behavior and Development*, McGraw-Hill, New York, 1943.

Pryor, H. B., Charts of Normal Body Measurements and Revised Width-Weight Tables in Graphic Form, J. Pediat., *68*:615, 1966.

Shuttleworth, F. K., *The Adolescent Periods: A Graphic and Pictorial Atlas*, Monogr., Soc. Res. Child Develm., vol. 3, no. 3, National Research Council, Washington, D.C., 1938.

Simmons, K., *Growth and Development*, Monogr., Soc. Res. Child Develm., vol. 9, no. 1, National Research Council, Washington, D.C., 1944.

Sontag, L. W., and Reynolds, E. L., The Fels Composite Sheet: I. A Practical Method for Analyzing Growth Progress; II. Variations in Growth Patterns in Health and Disease, J. Pediat., *26*:327, 336, 1945.

Stuart, H. C., Standards of Physical Development for Reference in Clinical Appraisement, J. Pediat., *5*:194, 1934.

Vickers, V. S., and Stuart, H. C., Anthropometry in the Pediatrician's Office: Norms for Selected Body Measurements Based on Studies of Children of North European Stock, J. Pediat., *22*:155, 1943.

FACTORS RESPONSIBLE FOR VARIATIONS

Variations in development and growth may depend upon a single factor, but more frequently they depend upon several factors, which may be interrelated. Normal, as well as abnormal, development and growth depend upon and are affected by the following: heredity, race, sex, congenital factors, hormones, nutrition, secular trends, climate, seasons, activity, illnesses, and the normal and abnormal functions of the various tissues and organs.

Each of the tissues and organs of the body originates from a definite and specific primordium. Furthermore, each differentiates at a different time and at a different rate. Normal differentiation depends upon the intrinsic properties of the tissues and organs and upon the irreversible changes that they undergo. The period of accelerated activity characterizing differentiation shifts from one structure to another, and such an interval in the differentiation of a structure is called its critical period. During such a period, the differentiation of a particular structure is dominant, and appears to have a depressing effect upon the differentiation of other tissues and organs. If the structure fails to take advantage of this favorable time, it is never again able to express itself, and cannot compete with other organs reaching their critical periods. It may completely fail to develop, or it may develop abnormally.

Deleterious influences exerted at a certain stage of development generally produce defects of all tissues and organs differentiating at that time. Similar influences at other stages may have little effect upon these tissues and organs, but they may affect others having different critical periods.

REFERENCES

1. H. Gray, Stanford med. Bull., *20*:24, 1962.
2. E. Boyd, *The Growth of the Surface Area of the Human Body*, University of Minnesota Press, Minneapolis, 1935.
3. A. G. Brodie, Amer. J. Anat., *68*:209, 1941. J. D. Boyd, Amer. J. Dis. Child., *76*:53, 1948.
4. C. C. Francis, Amer. J. Dis. Child., *57*:817, 1939.
5. C. M. Jackson, Anat. Rec., *3*:361, 1909.

GENERAL READING

Baldwin, B. T., *The Physical Growth of Children from Birth to Maturity*, State University of Iowa Press, Iowa City, 1922.

Boyd, E., *An Introduction to Human Biology and Anatomy for First Year Medical Students*, Child-Research Council, Denver, Colorado, 1952.

Boyd, E., *Outline of Physical Growth and Development*, Burgess, Minneapolis, 1941.

Crelin, E. S., *Anatomy of the Newborn: An Atlas*, Lea & Febiger, Philadelphia, 1969.

Crelin, E. S., *Functional Anatomy of the Newborn*, Yale University Press, New Haven and London, 1973.

Falkner, F. (ed.), *Human Development*, Saunders, Philadelphia, 1966.

Harris, J. A., Jackson, C. M., Paterson, D. G., and Scammon, R. E., *The Measurement of Man*, University of Minnesota Press, Minneapolis, 1930.

Huxley, J. S., *Problems in Relative Growth*, Methuen, London, 1932.

Jackson, C. M., On the Prenatal Growth of the Human Body and the Relative Growth of the Various Organs and Parts, Amer. J. Anat., 9:119, 1909.

Needham, J., *Biochemistry and Morphogenesis*, Cambridge University Press, London, 1942.

Peter, K., Wetzel, G., and Heiderich, F., *Handbuch der Anatomie des Kindes*, Bergmann, München, 1938, 2 vols.

Scammon, R. E., A Summary of the Anatomy of the Infant and Child, in *Pediatrics*, ed. by I. A. Abt, Saunders, Philadelphia, vol. I, chap. 3, 1923.

Scammon, R. E., and Calkins, L. A., *The Development and Growth of the External Dimensions of the Human Body in the Fetal Period*, University of Minnesota Press, Minneapolis, 1929.

Simmons, K., *Growth and Development*, Monogr., Soc. Res. Child Develm., vol. 9, no. 1, National Research Council, Washington, D.C., 1944.

Sinclair, D., *Human Growth After Birth*, Oxford University Press, London, 2nd ed., 1973.

Symposia of the Society for the Study of Human Biology, volume III, Human Growth, ed. by J. M. Tanner, Pergamon, New York, 1960.

Tanner, J. M., *Growth at Adolescence*, Blackwell, Oxford, 2nd ed., 1962.

Thompson, D. W., *On Growth and Form*, Cambridge University Press, London, 1942.

Watson, E. H., and Lowrey, G. H., *Growth and Development of Children*, Year Book Publishers, Chicago, 5th ed., 1967.

White House Conference on Child Health and Protection, *Growth and Development of the Child*, pt. II, Anatomy and Physiology, Century, New York, 1933.

Wilmer, H. A., Changes in Structural Components of Human Body from Six Lunar Months to Maturity, Proc. Soc. exp. Biol., N.Y., *43*:545, 1940.

Zuckerman, S., et al., A Discussion on the Measurement of Growth and Form, Proc. R. Soc. B., *137*:433, 1950.

RADIOLOGICAL ANATOMY

10

RONAN O'RAHILLY

GENERAL ASPECTS

The technical aspects of the nature and production of x-rays are the province of physicists and electrical engineers, and the details of radiographic processes concern radiographers. The radiologist, on the other hand, is concerned chiefly with the interpretation of radiograms and fluoroscopic images. This presupposes a knowledge of anatomy. Radiography has proved particularly valuable in the detection of the early stages of deep-seated disease, when the possibility of cure is greatest. There is little departure from the normal, however, during these early stages; hence knowledge of the earliest detectable variations, that is, of "the borderlands of the normal and early pathological in the skiagram" (Köhler), is of great medical importance. Radiodiagnosis is the most important method of non-destructive testing of the living body.

X-RAYS

An x-ray tube is shown in figure 10–1.

X-rays were discovered in 1895 by the German physicist, Wilhelm Conrad Röntgen, who later issued a report "On a New Kind of Rays." The story of the discovery of x-rays is one of the most fascinating in the history of science.

X-rays are believed to be of the same nature as the rays of visible light. These, together with ultraviolet, infrared, and radio waves, are all termed *electromagnetic waves*, that is, wavelike packages of energy that arise in association with an acceleration of electrons. All these waves travel at the same speed *(c)*, 186,000 miles per second, but they differ in wavelength (λ) and in frequency *(v)* (cycles per second), according to the equation $c = v\lambda$. The wavelength of x-rays is extremely short; those used in medical radiography are about 0.1 to 0.5 Å.* Gamma rays are physically identical with short x-rays but they are emitted by certain radioactive elements.

Properties of X-rays

The following properties of x-rays are of particular relevance here.

Penetrating Effect. X-rays penetrate solid matter; those of shorter wavelength have a greater power of penetration and are known as "hard rays." In their passage through matter, x-rays are absorbed; the amount of absorption depends on the atomic number and on the density of absorbing substance. Thus bone, owing to its calcium content, absorbs x-rays far more readily than do the soft tissues of the body. Radiography is based on the differential absorption of x-rays. Structures readily penetrated by x-rays are described as radiolucent; substances penetrated with difficulty, or not at all, as radio-opaque.

Photographic Effect. When x-rays strike a photographic emulsion, they produce an effect similar to that which is produced by light. The image can be examined after development and fixation of the film. The resulting film is termed a radiogram (or radiograph), or skiagram.

Fluorescent Effect. When x-rays strike certain metallic salts (phosphors), they cause them to fluoresce, that is, light waves are produced. Fluoroscopy depends on this characteristic, as does also the use of intensifying screens, which give off light and enhance the photographic effect.

Quality and Quantity of X-radiation

The term *quality* refers to the penetrating ability of x-rays, which varies inversely with the wavelength because rays of short wavelength (hard rays) have more energy. This energy depends on that of the electrons, which in turn depends on the speed of the electrons, and this is determined by the kilovoltage applied to the tube. The higher the kilovoltage, the greater the amount of energy in the x-rays produced and the greater the penetrating power of the rays. An ordinary x-ray beam consists of rays of differing wavelengths. As the kilovoltage is increased, there is a general increase in the number of waves of the various lengths, but there is also a relative increase in the number of waves of shorter wavelength.

*An Ångström unit is 0.0001 micrometer; a micrometer (or "micron") is one millionth of a meter, or one thousandth of a millimeter.

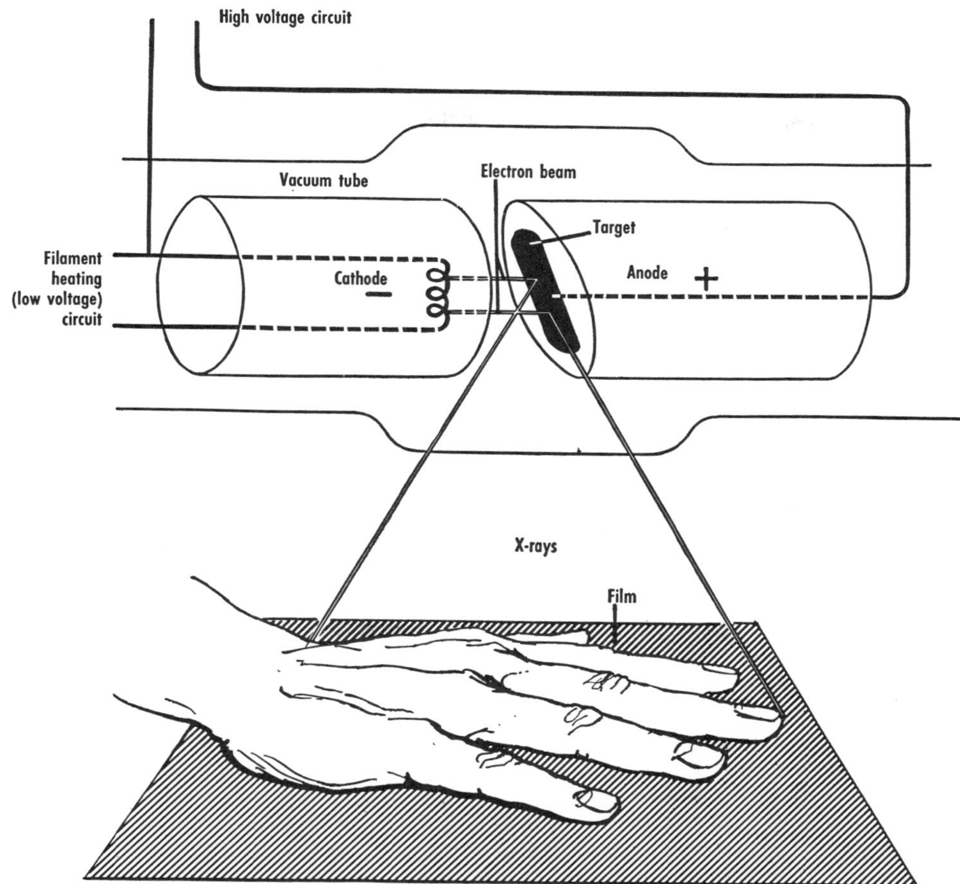

Figure 10–1 Diagram of an x-ray tube. The tube is a glass container that has a high vacuum. A hot filament cathode (heated by a separate low voltage circuit) acts as the source of electrons. The anode contains a tungsten target. When a kilovoltage is applied between the cathode and the anode (by means of the high voltage circuit), the electrons are driven over to the anode at a high speed. When the electrons strike the target, x-rays are produced.

Quantity or *dosage* refers to the amount of ionizing energy at a given place in an x-ray beam. It is measured in units called *roentgens.* Intensity is quantity per unit time. As the milliamperage in the filament heating current is increased, more electrons are liberated and strike the target. This results in an increase in the number of x-rays, but the quality is not affected.

The quality of a radiogram depends on distortion (figs. 10–2 and 10–3), detail (or definition or sharpness), density (the blackness of an x-ray film), and contrast.

RADIO-OPACITY

The following structures produce the usual radiographic image, and they are arranged in order of increasing radio-opacity (i.e., whiteness on negative film, blackness on fluoroscopic screen or on positive print) for a constant thickness:

1. Air, as found, for example, in the trachea and lungs, the stomach and intestine, and the paranasal sinuses. Also oxygen when injected into the ventricles of the brain.
2. Fat.
3. Soft tissues, e.g., heart, kidney, muscles.
4. Lime (calcium and phosphorus), for example, in the skeleton.
5. Enamel of the teeth.
6. Dense foreign bodies, for example, metallic fillings in the teeth. Also radio-opaque contrast media, such as a barium meal in the stomach.

When the density of a structure is too

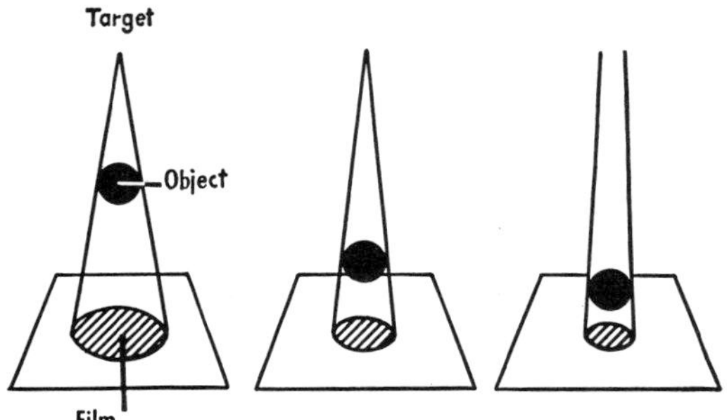

Figure 10–2 Image magnification. X-ray images are shadows, and the geometry of their formation is similar to that relating to shadows formed by ordinary light. Thus the image becomes more enlarged, the nearer the object is placed to the target, or source of radiation. In the second diagram the image of the object is smaller than in the first diagram, because the object is farther from the target. In the third diagram the target is more than 2 meters from the film, and the magnification is negligible (teleradiography).

similar to that of adjacent structures, it is possible to use contrast media in some sites. Contrast media are classified as radiolucent (e.g., oxygen) and radio-opaque (e.g., barium).

POSITIONING

The views used in radiography are named from the part of the body that is nearest the film, for example, anterior, right lateral, left anterior oblique. Alternatively, the terms anteroposterior and postero-anterior are used when the x-rays have passed through the object from front to back (tube in front of object, film behind) or from back to front (tube behind object, film in front), respectively. Details of the views commonly employed are given in special books on radiographic positioning.

FLUOROSCOPY

A fluorescent screen consists of cardboard coated with a thin layer of fluorescent material (phosphor), such as zinc cadmium sulphide. When the screen is activated by x-rays, light is emitted. The fluorescent layer is covered with a sheet of lead glass that absorbs the x-rays but through which the fluoroscopist can see the pattern of brightness produced by the x-rays. The fluoroscopic image is best observed in a darkened room and when the eyes have become dark-adapted.

The brightness, contrast, and sharpness of a fluoroscopic image are generally inferior to those of a good radiogram. Increase in brightness, however, may be obtained by intensifying the image electronically. Moreover, the fluoroscopic image can be photographed. The chief advantages of fluoroscopy are the ability to observe the motion of parts of the subject and the ability to change the position of the subject during the examination.

SPECIAL PROCEDURES

Fluoroscopy and teleradiography (see fig. 10–2) have been referred to already, and the use of contrast media has been mentioned.

The ordinary radiogram gives no impression of depth other than that dependent on anatomical expectation. In the case of many areas, it is common, therefore, to take at least two views, one at a right angle to the other (biplane radiography).

Stereoradiography. Stereoradiography is a procedure whereby dual radiograms are made, corresponding to the

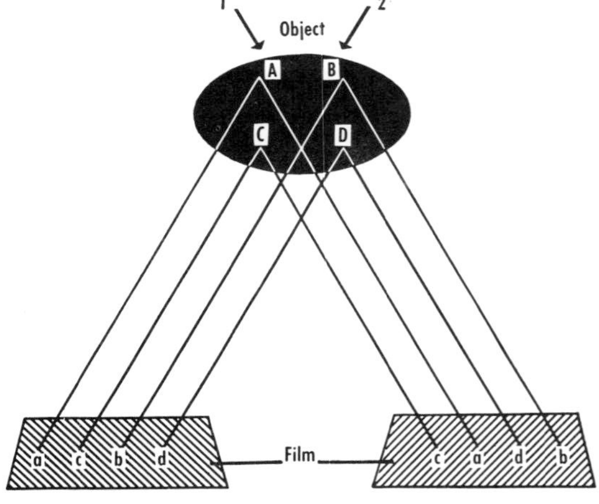

Figure 10–3 Dissociation of planes in oblique views (after Tillier). When the incident radiation is in the direction of arrow *1*, the images of the parts *A*, *B*, *C*, *D* of the object are projected in the order *c*, *a*, *d*, *b* on the film. With incident radiation in the direction of arrow 2, however, the order is *a*, *c*, *b*, *d*. Thus when the tube is displaced to the right, the upper plane (*AB*) becomes displaced to the left in relation to the lower plane (*CD*).

points of view of the two eyes. Two views of the same object are taken, each at a slightly different angle. This is accomplished by shifting the x-ray tube about 6 cm (the interpupillary distance) between the two exposures. The resultant films are placed in a special viewer (stereoscope) in order to examine them stereoscopically (in "3D").

Cineradiography. Cineradiography is the technique of recording a moving radiographic image (e.g., that of a joint in motion) upon cinematographic film. This is performed generally by photographing the image on a fluorescent screen with a cine-camera.

Tomography. In tomography (Gk. *tomos*, a section, as in microtome), a radiogram is made of a selected layer of the body (see fig. 26–3*B*, p. 261). Both tube and film are rotated during the exposure, but in opposite directions, resulting in blurring of the tissue planes other than that where the amount of motion is virtually nil (fig. 63–16).

Xeroradiography. Xeroradiography involves the principle of xerography. A selenium surface is given a positive electrostatic charge and is exposed as in film radiography. Where x-rays pass through soft tissues and reach the selenium, positive charges are leaked, whereas little charge is lost deep to bone. The latent electrostatic image is made visible by spraying with a negatively charged powder, and is then transferred to plastic-coated paper (fig. 18–24*C*).

SKELETAL RADIOLOGY

The skeleton, owing to its high radio-opacity, is generally the most striking feature of a radiogram. It is important to appreciate, however, that many of the organs and soft tissues of the body can be investigated radiographically.

The organs will be considered in the appropriate regions—the heart and lungs, for example, with the thorax. In many cases the contrast between an organ and its surroundings can be accentuated by the introduction of either a radio-opaque or a radiolucent contrast medium. Thus, a barium meal shows up the stomach and intestine, iodized oil can be injected into the bronchi or into the ducts of the salivary glands, and an organic iodine compound can be injected intravenously to outline the gallbladder. Iodinated organic compounds can be injected into blood vessels. Air can be injected into the subarachnoid space, the ventricles of the brain, or the cavity of the knee joint. These examples will be elaborated further in discussion of the appropriate regions, and the present section is confined to a general discussion of normal skeletal radiology.

GENERAL FEATURES OF A LONG BONE

Radiographically, the compact substance is seen peripherally as a homogeneous band of lime density. A nutrient canal may be visible as a radiolucent line traversing the compacta obliquely. In some areas the compacta is thinned to form a cortex. The cancellous or spongy substance is seen particularly toward the ends of the shaft as a network of lime density presenting interstices of soft-tissue density. Islands of compacta are visible occasionally in the spongiosa. The bone marrow and the periosteum present a soft-tissue density and are not distinguishable as such.

In many young bones the uncalcified portion of an epiphysial disc or plate can be seen radiographically as an irregular, radiolucent band termed an epiphysial line. When an epiphysial line is no longer seen, it is said to be closed, and the epiphysis and diaphysis are said to be united or fused. The radiographic appearance of fusion, however, precedes the disappearance of the visible epiphysial disc as seen on the dried bone.[1]

The term metaphysis is defined in various ways. Radiologically it comprises the calcified cartilage of an epiphysial disc and the newly formed bone beneath it.[2]

Transverse lines of increased radio-opacity are frequently seen in children in the shafts of long bones near the site of the epiphysial lines. The transverse lines on the film are produced by transverse strata in the bone. These strata result when, perhaps owing to an acute illness, cartilaginous growth ceases temporarily but osteoblasts continue to form bone in a horizontal layer rather than in the usual manner.[3] They are not lines of arrested growth.[4]

GENERAL FEATURES OF A JOINT

The articular cartilage presents a soft-tissue density and is not distinguishable as such. The so-called "radiological joint space," that is, the interval between the radio-opaque epiphysial regions of two bones, is

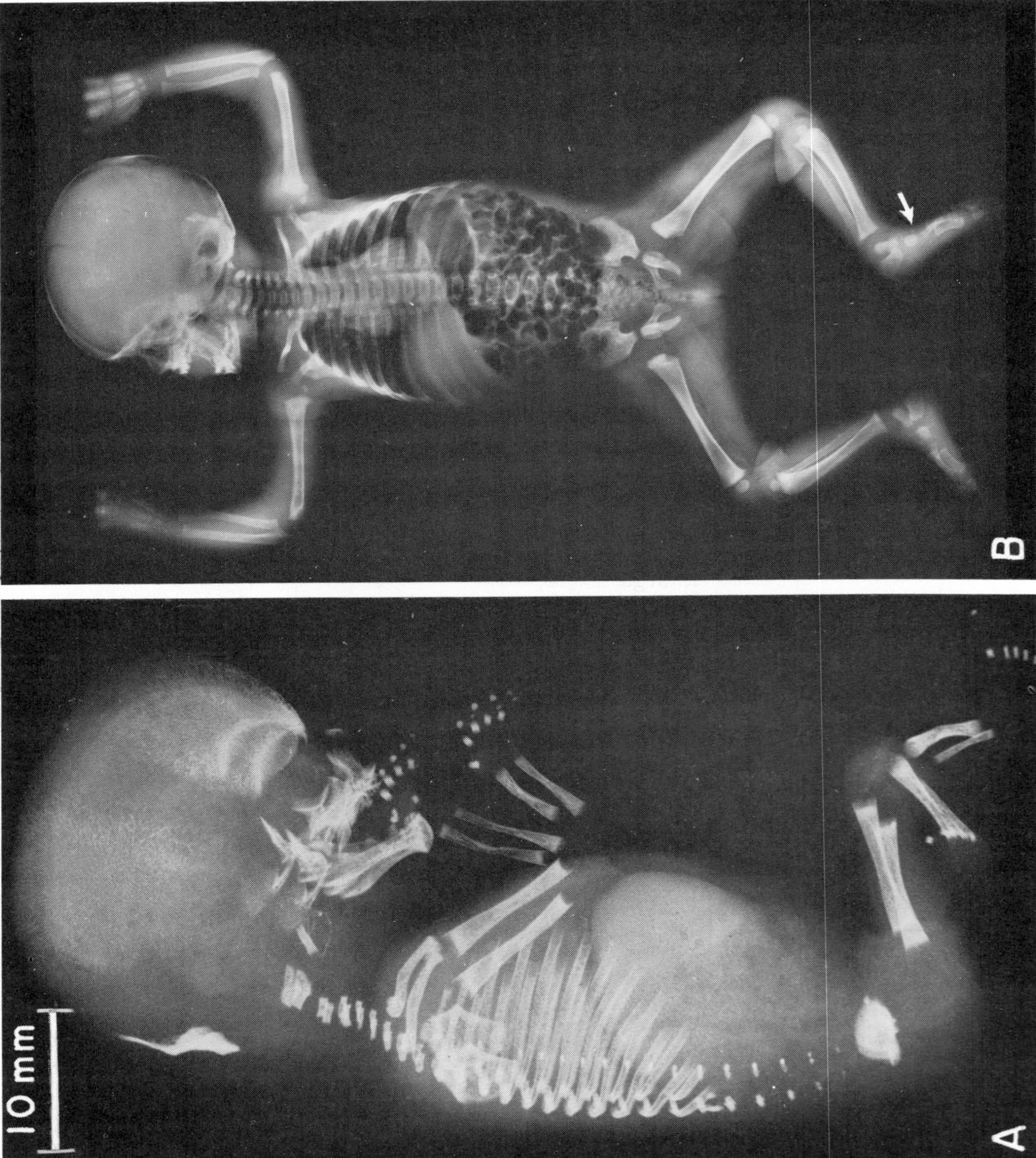

Figure 10–4 See facing page for legend.

Figure 10–4 Radiograms of fetus and infant. *A*, lateral radiogram of fetus aged 11 postovulatory weeks (69 mm C.R.[6]). The radio-opacity has been increased by the use of silver. In the head and neck, portions of the occipital, parietal, frontal, nasal, maxillary, zygomatic, sphenoid, temporal, and mandibular bones, together with the arches of the cervical vertebrae, can be identified. The trabecular network of the parietal, frontal, and squamous part of the temporal bone is apparent. The exoccipitals overlap each other and are located immediately above the cervical vertebrae; the basi-occipital lies obliquely between the exoccipitals and the mandible, overlapped partly by the tympanic ring of each side. In each upper limb, note the scapula and clavicle, and the shafts of the humerus, radius, ulna, metacarpals, and most phalanges. In each lower limb, note the ilium and the shafts of the femur, tibia, fibula, and metatarsals. The distal phalanx of one big toe can be seen in the lower right-hand corner. No carpals or tarsals are visible. In this instance, 13 ribs are present on each side, but the cervical ribs are difficult to make out in this view. The neural arches extend from C. V. 1 to L. V. 5, the vertebral centra from C. V. 7 to S. V. 2.

B, radiogram of living female infant aged two months. In the head, note the anterior fontanelle and the upper part of the coronal suture, the auricle and tympanic cavity, and the teeth (calcified areas in the crowns of the lower deciduous molars show well). In each upper limb, in addition to the shafts of the long bones, note the epiphysis for the head of the humerus, and note the capitate and hamate. In each lower limb, note the epiphyses for the lower end of the femur and the upper end of the tibia, and note the talus, calcaneus, and cuboid. The arrow points to the center for the lateral cuneiform. The head of the femur is not yet visible. In the thorax, note the ribs, vertebrae, globular heart, and the position of the diaphragm. In the abdomen, note the radio-opacity of the liver and the radiolucent, loculated appearance of gas in the intestine. In the pelvis, note the ilium, ischium, and pubis on each side, and the sacral vertebrae in the median position. The pelvis is relatively small in childhood. Note the wide gap between the parts of the pubic bones that are ossified.

A is twice, and *B* is approximately one-fourth, natural size.

occupied almost entirely by the two layers of articular cartilage, one on each of the adjacent ends of the two bones. On a radiogram the "space" is usually 2 to 5 mm in width in the adult. The joint cavity is rarely visible. (It can be seen under certain conditions, for example, distraction in the shoulder and knee joints.) The "radiological joint line," that is, the junction between the radio-opaque end of a bone and the radiolucent articular cartilage, is actually the junction between a zone of calcified cartilage over the end of the bone and the uncalcified articular cartilage.

SKELETAL MATURATION

The development of bones and skeletal maturation have been discussed in Chapter 2. Radiograms of a fetus and an infant are shown in figure 10–4. Tables showing the times of appearance of the postnatal ossific centers in the limbs are provided inside the back cover.[5] The time of appearance that is given is the age at which 50 per cent of normal children show a certain center radiographically while the remaining 50 per cent do not yet show it (see fig. 10–5). It is important to realize that a considerable range of variation occurs on each side of the median figure. Thus, in the case of the center for the medial epicondyle of the humerus, 5 per cent of girls may be expected to show the center by about two years, 50 per cent by three and one-half years, and 95 per cent by five years. The corresponding figures for boys are 5 per cent by four and one-half years, 50 per cent by six and one-half years, and 95 per cent by eight and one-half years.

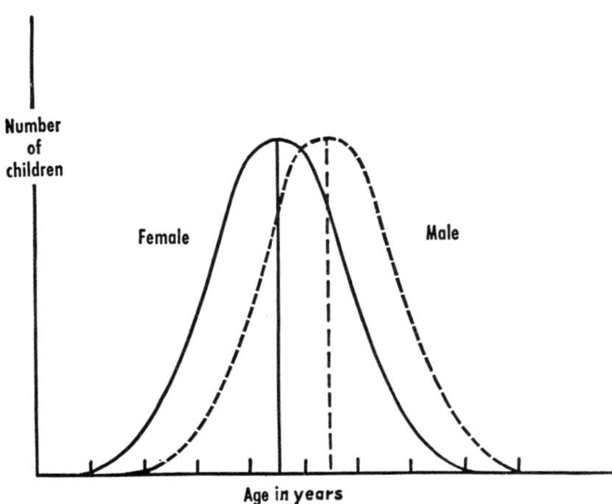

Figure 10–5 Hypothetical graph showing the time of initial appearance of a given ossific center, as detected radiographically in a large number of children. The solid line represents girls, the interrupted line boys. The corresponding vertical lines indicate the median time of appearance, that is, the age at which 50 per cent of the girls or boys, respectively, showed the center in question while the remaining 50 per cent did not yet show it.

REFERENCES

1. J. A. Keen, S. Afr. med. J., *24*:1086, 1950.
2. S. I. Pyle and N. L. Hoerr, *Radiographic Atlas of Skeletal Development of the Knee*, Thomas, Springfield, Illinois, 1955.
3. R. H. Follis and E. A. Park, Amer. J. Roentgenol., *68*:709, 1952.
4. P. S. Gindhart, Amer. J. phys. Anthrop., *31*:17, 1969.
5. The figures are based largely on S. M. Garn, C. G. Rohmann, and F. N. Silverman, Med. Radiogr. Photogr., *43*:45, 1967.
6. For further views see R. O'Rahilly and D. B. Meyer, Amer. J. Roentgenol., *76*:455, 1956. For a discussion of the radiological estimation of fetal maturity *in utero*, see J. B. Hartley, Brit. J. Radiol., *30*:561, 1957.

GENERAL READING

X-rays

The physical aspects are discussed by O. Glasser *et al.*, *Physical Foundations of Radiology;* T. A. Longmore, *Medical Photography, Radiographic and Clinical*; and J. Selman, *The Fundamentals of X-Ray and Radium Physics.*

Handbooks of Radiology

Diethelm, L., *et al.*, *Handbuch der medizinischen Radiologie*, Springer, Berlin, 1965–, many volumes.

Schinz, H. R., *et al.*, *Roentgen-Diagnostics*, Grune and Stratton, New York, 1951. A 6th edition in German began to appear in 1965.

Skeletal Radiology

Brailsford, J. F., *The Radiology of Bones and Joints*, Churchill, London, 5th ed., 1953.

Köhler, A., and Zimmer, E. A., *Borderlands of the Normal and Early Pathologic in Skeletal Roentgenology*, Grune and Stratton, New York, 3rd ed., 1968. Based on the 11th edition in German.

Radiological Atlases

Grashey, R., and Birkner, R., *Atlas typischer Röntgenbilder vom normalen Menschen*, Urban and Schwarzenberg, Munich, 10th ed., 1964.

Lusted, L. B., and Keats, T. E., *Atlas of Roentgenographic Measurement*, Year Book Publishers, Chicago, 3rd ed., 1972.

Schmidt, H., *Radiotomographic Anatomic Atlas*, Hafner, Darien, Connecticut, 1970.

Takahashi, S., *An Atlas of Axial Transverse Tomography and its Clinical Application*, Springer, New York, 1969.

See also Head and Neck (pp. 550 and 580).

Radiological Anatomy

Tillier, H., *Normal Radiological Anatomy*, trans. by R. O'Rahilly, Thomas, Springfield, Illinois, 1968.

Part Two

THE UPPER LIMB

Ernest Gardner
Donald J. Gray
Ronan O'Rahilly

Introduction

The upper limb, like the lower, is connected by a girdle to the trunk, and presents three segments, the arm, forearm, and hand. The shoulder girdle, which is formed by the scapulae and clavicles, is completed in front by the manubrium of the sternum, with which the medial ends of the two clavicles articulate. It is deficient behind. The upper limb is characterized by considerable mobility. Many of the movements depend upon the support and stability provided by muscles that have an extensive origin from ribs and vertebrae. Hence the muscles of the pectoral region and the superficial muscles of the back are included in the description of the upper limb.

The N. A. recommends the term "limb" rather than "extremity," which should be restricted in its use to the ends of structures, such as bones.

The upper limbs first appear as minute buds in embryos of about 5 mm in length, that is, at about four postovulatory weeks of age. Each limb bud elongates and develops in proximodistal sequence (for example, the forearm appears before the hand). A few days after the limbs can first be seen, nerves grow into them, and the skeleton and muscles become differentiated. Shortly afterward, fingers can be recognized.

Latin Names and English Equivalents for Parts of the Upper Limb

Latin Name	*English Equivalent*
Humerus	Shoulder
Axilla	Armpit
Brachium	Arm (upper arm)
Cubitus	Elbow
Antebrachium	Forearm
Carpus	Wrist
Manus	Hand
Palma	Palm
Digiti manus	Fingers
Pollex	Thumb

The common name for most parts of the upper limb are listed here. Their Latin equivalents are included,* and it will be noted that many other terms are derived from these. For example, the word "manual," meaning pertaining to the hand, is used in everyday speech.

*Apart from the word *humerus,* which refers specifically to one bone, there is no real equivalent in Latin for the English word denoting the shoulder region.

GENERAL READING

In addition to the following, which deal chiefly with the limbs, see also the references cited in the introductory chapters.

Castaing, J., and Soutoul, J. H., *Atlas de coupes anatomiques. I. Membre supérieur,* Maloine, Paris, 1967.

Henry, A. K., *Extensile Exposure,* Livingstone, Edinburgh, 2nd ed., 1957.

Lockhart, R. D., *Living Anatomy,* Faber & Faber, London, 6th ed., 1963.

Royce, J., *Surface Anatomy,* Davis, Philadelphia, 1965.

BONES OF UPPER LIMB

11

The shoulder girdle consists of the scapulae, or shoulder blades, and the clavicles, or collar bones. Each clavicle articulates with the scapula laterally and with the manubrium of the sternum medially. The humerus is the bone of the arm. It articulates with the scapula above, and with the bones of the forearm, the radius and ulna, below. The radius joins the bones of the carpus or wrist.

CLAVICLE

The clavicle, or collar bone (figs. 11–1 to 11–5), extends from the upper border of the manubrium of the sternum to the acromion of the scapula and thereby joins the trunk with the upper limb. It is a long bone, and has a rounded medial end, a flattened lateral end, and a shaft that has a double curve in a horizontal plane. The clavicle of a living subject can be palpated throughout its entire extent. Occasionally the clavicle is pierced by one of the supraclavicular nerves.

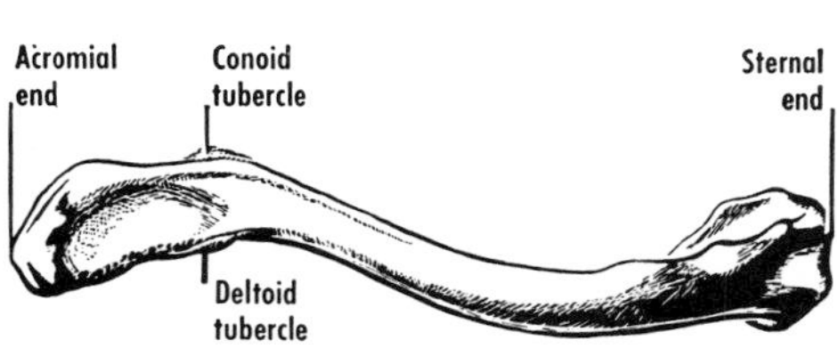

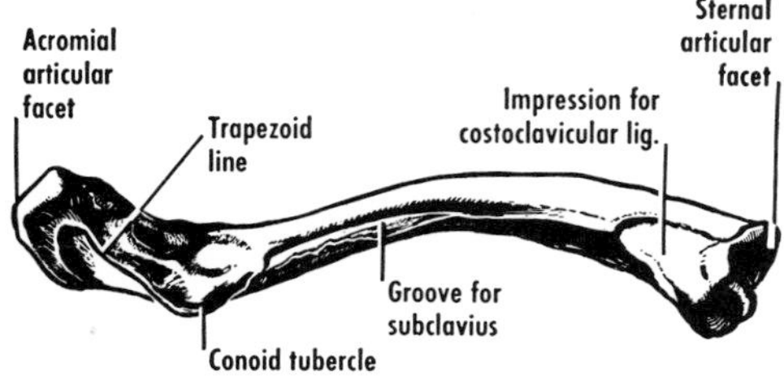

Figure 11–1 The right clavicle, viewed from in front and above, and from below.

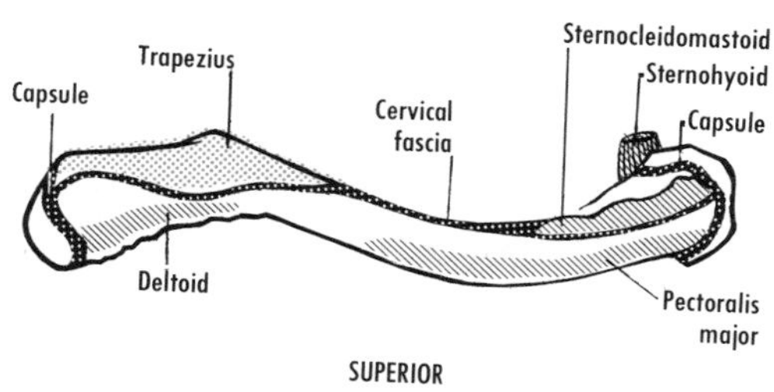

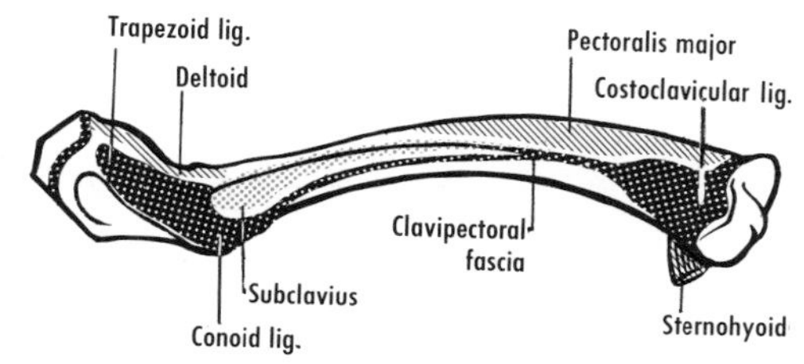

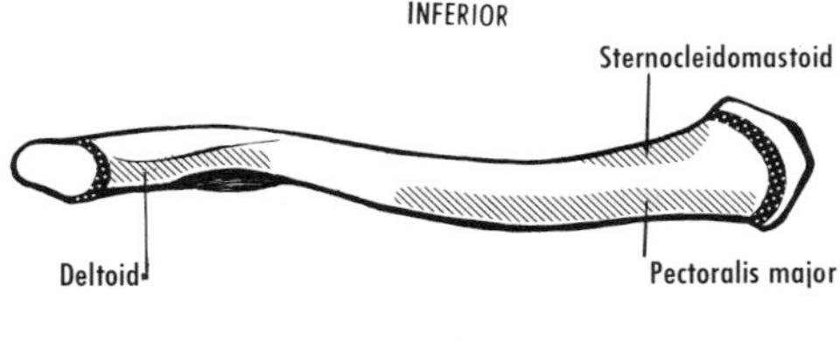

Figure 11–2 Muscular, ligamentous, and fascial attachments of the right clavicle.

The medial two-thirds of the bone is convex forward, whereas the lateral third is concave forward. The side to which a clavicle belongs can be determined by placing the rounded end medially, the concavity of the adjacent curve posteriorly, and the smooth surface of the shaft upward.

The medial or *sternal end* has an *articular surface* for the sternoclavicular fibrocartilage, which intervenes between the clavicle and the clavicular notch of the manubrium. The articular surface is commonly prolonged below, for articulation with the first costal cartilage. The slightly larger size of the medial end of one clavicle, when compared with that of the contralateral clavicle, is associated with handedness.

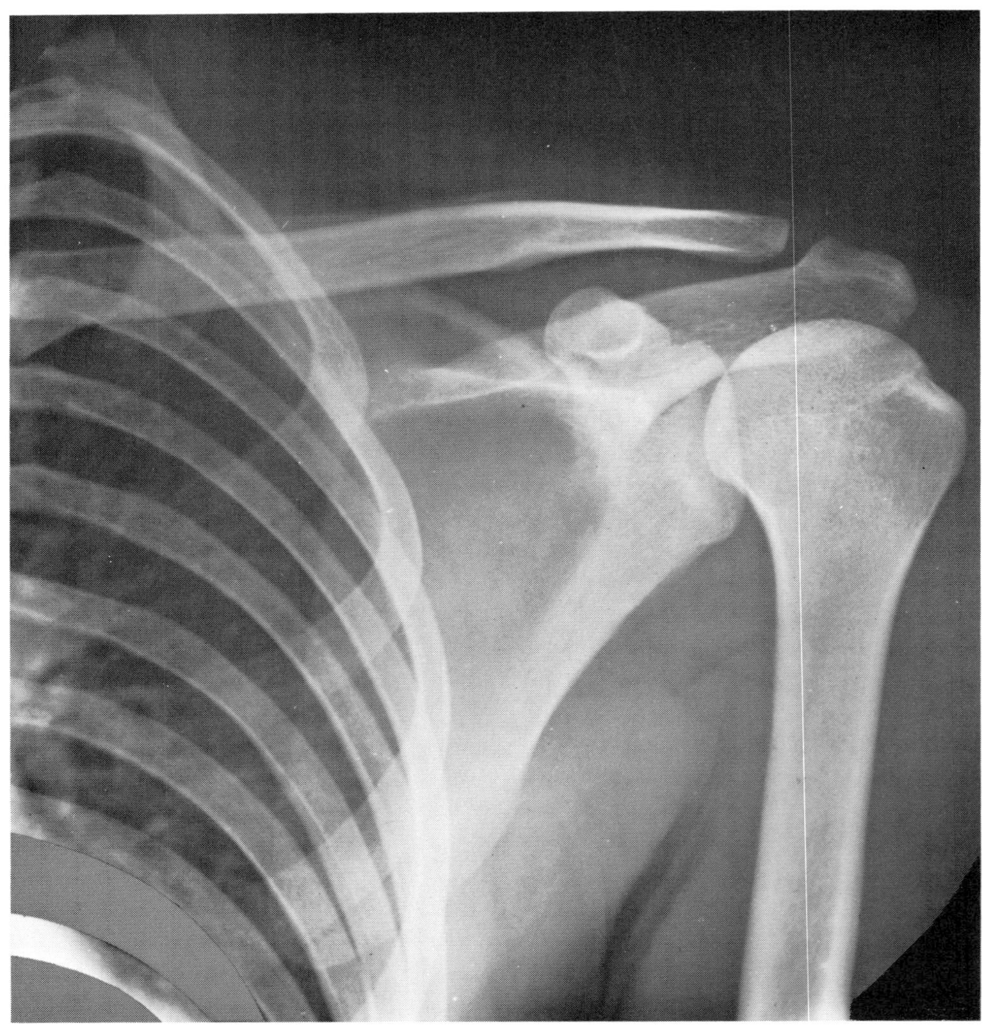

Figure 11–3 Shoulder of adult. Note acromioclavicular joint, glenoid cavity, coracoid process, and inferior angle of scapula.

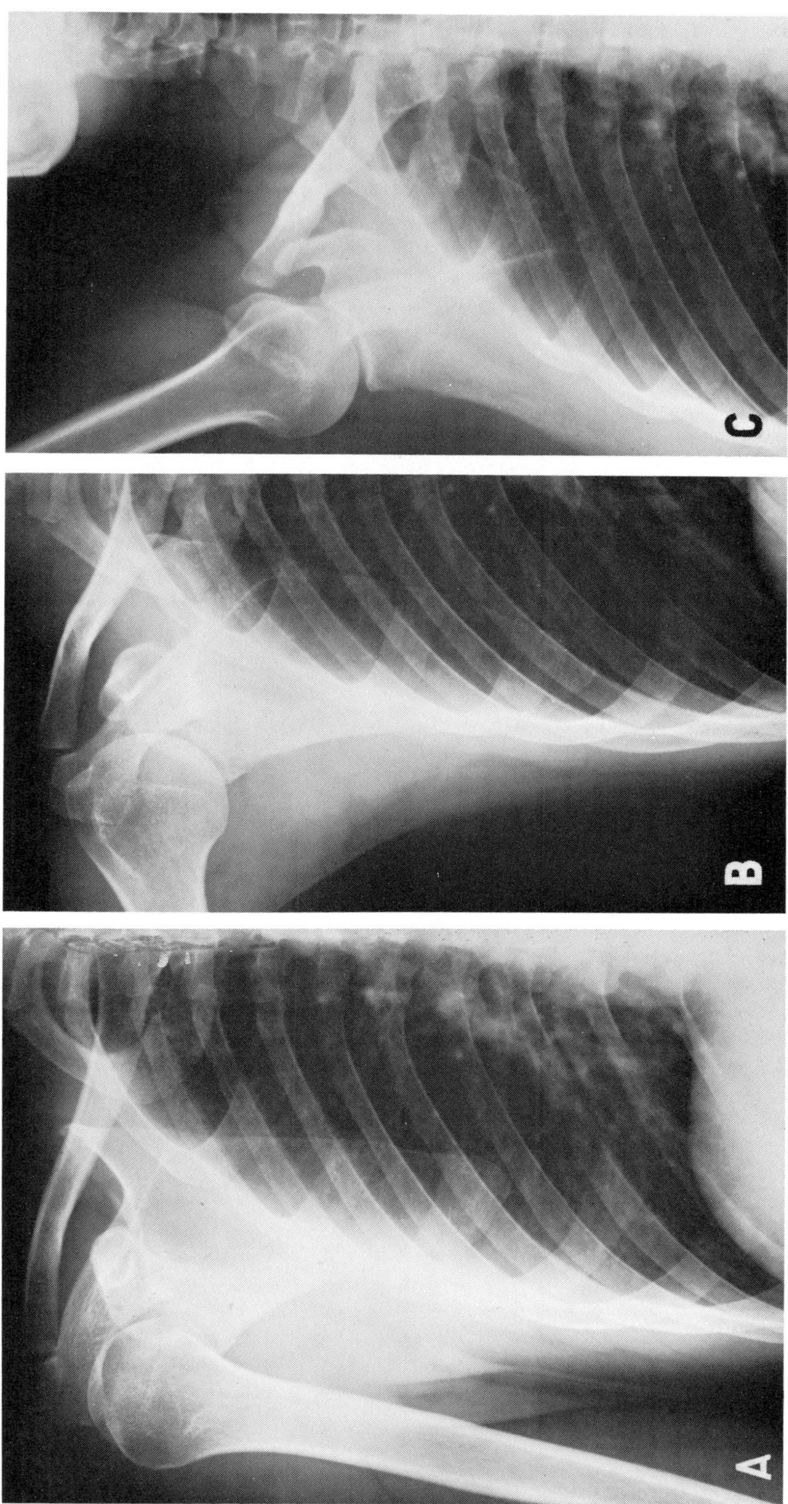

Figure 11–4 Shoulder during abduction in a coronal plane (compare figure 13–12). *A*, resting position. *B*, elevation of the arm to a right angle. *C*, elevation of the limb above the head.

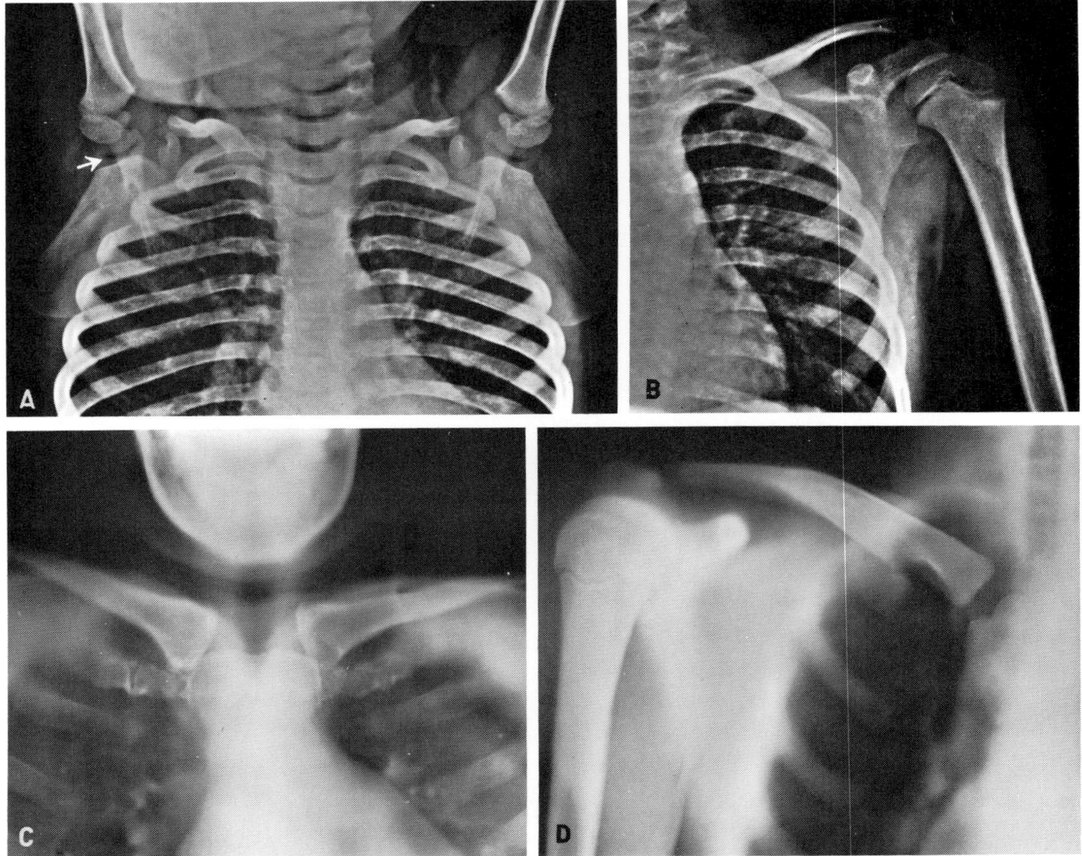

Figure 11–5 Shoulder. *A*, pneumoarthrogram produced bilaterally in a child by distraction at the shoulder joint (*arrow*). Note epiphyses for head and greater tubercle of humerus, and for coracoid process of scapula. *B*, shoulder of child. Note epiphysis formed of head and greater tubercle of humerus. *C*, tomogram of manubrium sterni. Note jugular notch, medial ends of clavicles, and first costal cartilages (calcified). *D*, tomogram of sternoclavicular joint. Note also epiphysial line of humerus. *A* and *B*, courtesy of S. F. Thomas, M.D., Palo Alto Medical Clinic, California. *C* and *D*, courtesy of Bernard S. Epstein, M.D., New Hyde Park, New York.

The medial two-thirds of the shaft has three surfaces, *anterior, posterior,* and *inferior,* separated by rounded, often indistinct borders. The inferior surface, which is often reduced to a ridge, presents a rough *impression for the costoclavicular ligament.* Just lateral to this is a shallow groove for the subclavius muscle, bounded by ridges or lips for the clavipectoral fascia. The anterior surface (its upper part is sometimes designated *superior* surface) is smooth, as is the concave posterior surface, which arches over the brachial plexus and subclavian vessels.

The lateral third of the shaft is flattened and has superior and inferior surfaces, separated by anterior and posterior margins. The anterior margin may present a small deltoid tubercle. The inferior surface presents, near its back part, a *conoid tubercle* for the conoid ligament. The *trapezoid line,* which is a triangular roughened area for attachment of the trapezoid ligament, runs forward and laterally from this tubercle. The conoid tubercle may form a synovial joint with the coracoid process of the scapula.

The lateral or *acromial end* presents an *articular surface* for the medial aspect of the acromion. This surface generally faces somewhat downward as well as laterally, and the clavicle tends to override the acromion.

Ossification[1]

The clavicle is unusual in that it is the first bone to begin ossification, and it does so in membrane, without a preceding cartilaginous phase. Two adjacent centers appear during the sixth postovulatory week and fuse almost immediately afterward. Shortly thereafter, cartilage develops at the ends of the clavicle and forms growth zones similar to those of other long bones. In cleidocranial dysostosis, a rare disease characterized by defects in the bones that ordinarily ossify intramembranously, the clavicle is one of those that may be defective or absent..

Usually, only one epiphysial center develops. This appears in the sternal end during adolescence and fuses with the shaft by the third decade. An epiphysial center occasionally appears in the acromial end during adolescence and quickly fuses with the shaft.

SCAPULA[2]

The scapula, or shoulder blade (figs. 11–3 to 11–11), is a large, flattened, triangular bone that is connected to the sternum by the clavicle, articulates with the humerus, and is applied to the posterolateral aspect of the upper part of the thorax. It consists of a body, a spine that ends laterally at the acromion, and a coracoid process.

The side to which a scapula belongs can be determined from the following information: the concave surface is anterior; a large process, the spine, projects from the posterior surface and extends laterally as the acromion; the acromion and the glenoid cavity (articular surface for the humerus) belong to the upper and lateral part of the bone.

The scapula is very mobile, and thus the shoulder girdle has a wide range of movement, but in the anatomical position the scapula is related to the posterolateral aspects of the second to the seventh ribs. In this position, the glenoid cavity looks forward as well as laterally, and abduction of the arm in the plane of the scapula takes the arm forward as well as laterally.

The body of the scapula is triangular in shape and has, therefore, two surfaces, costal and dorsal; three borders, superior, medial, and lateral; and three angles, superior, lateral and inferior.

The concave, *costal surface* is applied against the thorax, from which it is separated by the serratus anterior. The concavity of the costal surface is known as the *subscapular fossa,* and presents several low

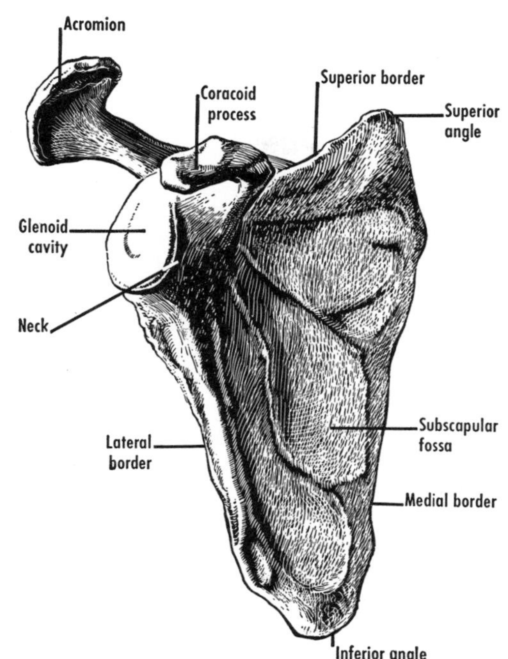

Figure 11–6 Right scapula, costal aspect, anatomical position.

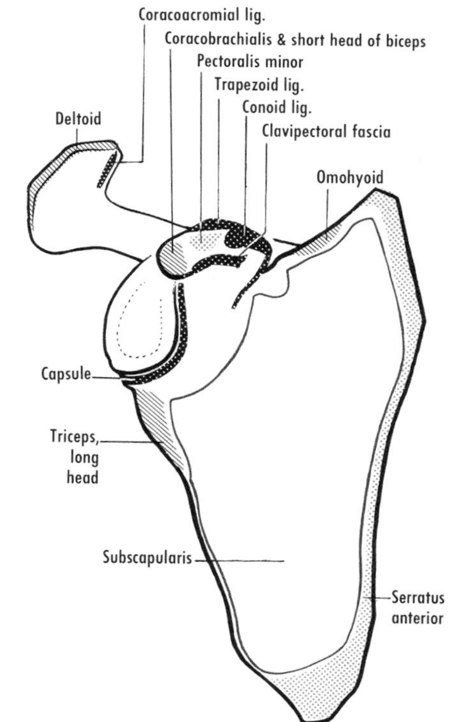

Figure 11–7 Right scapula, muscular and ligamentous attachments, costal aspect. The line of separation of the serratus anterior and subscapularis indicates the epiphysial line of the medial border.

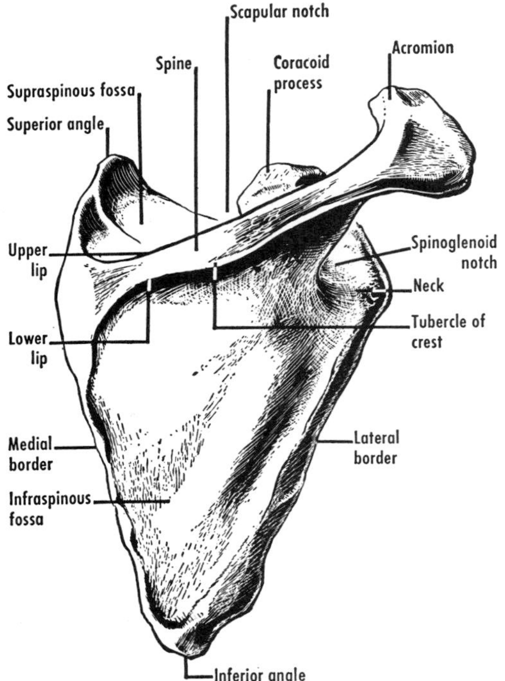

Figure 11–8 Right scapula, dorsal aspect, anatomical position.

ridges. These ridges mark the attachment of intramuscular tendinous slips.

The *dorsal surface* is divided into two unequal parts by the spine. The smaller, upper part and the upper surface of the spine form the *supraspinous fossa*. Most of the lower part of the dorsal surface is concave and together with the lower surface of the spine forms the *infraspinous fossa*. The two fossae communicate laterally by means of the *spinoglenoid notch*. This wide notch separates the lateral border of the spine from the neck and head of the scapula. The bone of the fossae is often quite thin.

The *superior border* is thin and sharp, and is interrupted at its junction with the coracoid process by the *scapular notch*. This notch varies in depth and width, and is often partially or completely bridged by an ossified superior transverse scapular ligament. The suprascapular nerve traverses the notch.

The *medial border* is usually convex, is often straight, but is sometimes concave.[3]

The *lateral border* is a thin but rough ridge that is marked in its uppermost part by the *infraglenoid tubercle*. The lateral border usually presents a groove for the circumflex scapular artery. The medial and lateral borders of the scapula are indistinctly palpable in the living.

The *superior angle* marks the junction of the superior and medial borders. The *inferior angle*, at the junction of the medial and lateral borders, can be palpated readily *in vivo*. The inferior angle moves extensively as the arm is abducted, and is an important landmark in studying the movements of the scapula. In the anatomical position, it is approximately at the level of the spine of the seventh thoracic vertebra; it lies over the seventh rib or the seventh intercostal space.

The *lateral angle* is at the junction of the lateral and superior borders. It is thickened to form the *head* of the scapula, which is joined to the rest of the scapula by the *neck*. The lateral surface of the head forms the concave *glenoid cavity*, slightly raised at its periphery. The glenoid cavity, which is directed forward and laterally, is usually piriform in shape, somewhat narrower

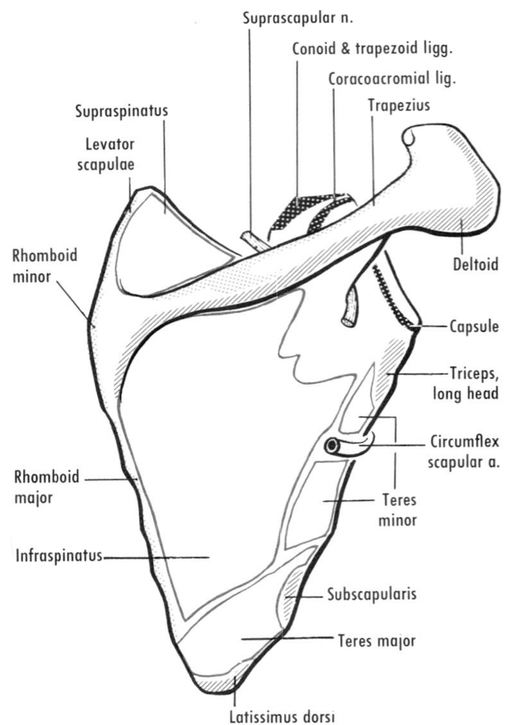

Figure 11–9 Right scapula, muscular and ligamentous attachments, dorsal aspect. The extension of the subscapularis origin to the dorsal aspect is inconstant.

above than below, and presents a slight notch at its anterior margin. The *supraglenoid tubercle* is a small roughened area immediately above the uppermost part of the margin of the glenoid cavity.

The *spine* of the scapula is a triangular plate, the front of which is attached to the body of the bone, and which continues laterally as the *acromion*. It is approximately opposite the third thoracic spine in the anatomical position. Its lateral border forms a part of the spinoglenoid notch. The posterior border or crest is subcutaneous and readily palpable. It has prominent superior and inferior lips for the insertion of the trapezius and the origin of the deltoid, respectively. The lower lip commonly shows a tubercle near its medial end.

The *acromion*, which ossifies independently, is ordinarily continuous with the spine, but may be separated from it by cartilage and fibrous tissue. It is variable in shape and may present an easily palpable *acromial angle* between its lateral and posterior margins. The lateral border, the tip, and the adjacent part of the upper surface give origin to the deltoid; the rest of the surface is subcutaneous. **The arm is measured**

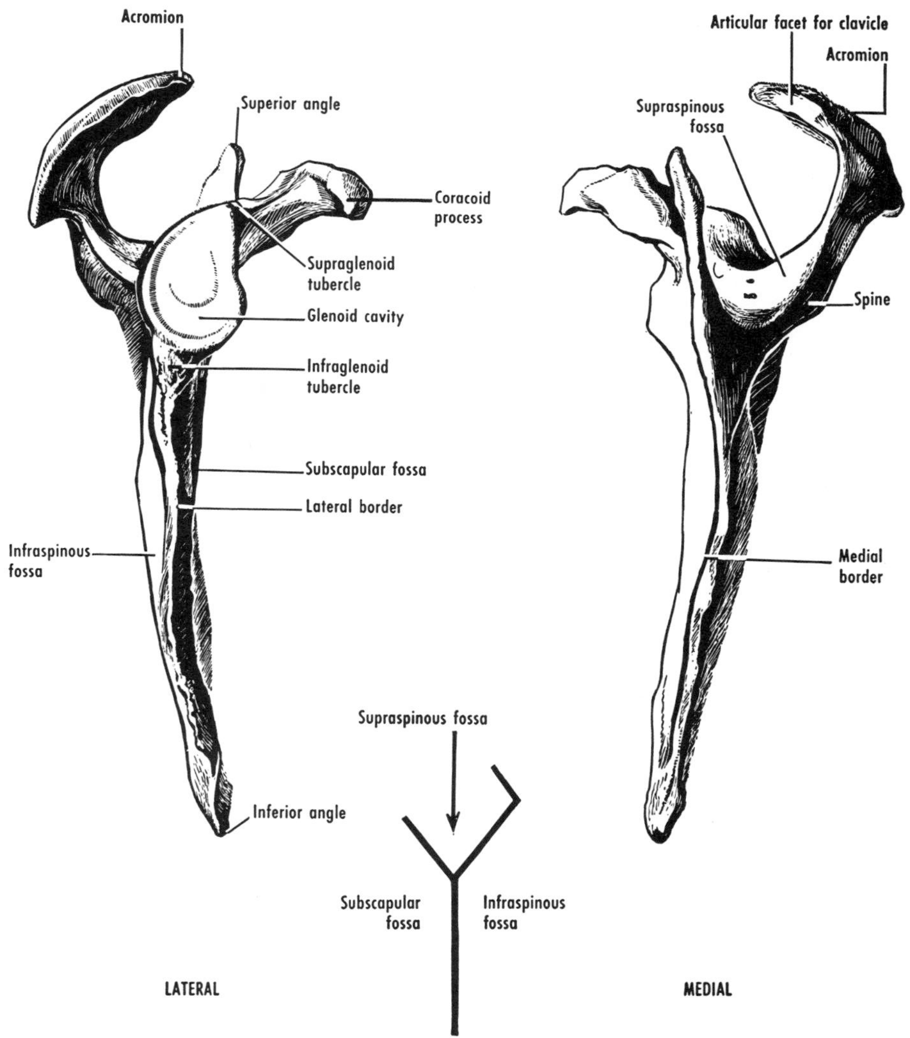

Figure 11–10 Right scapula from lateral and medial aspects. The slight notch in the anterior border of the glenoid cavity, said to be for the subscapularis tendon, marks the junction of the separate ossific centers for the glenoid cavity. The inset diagram illustrates that the upper and lower parts of the body form an angle, at the level of the spine, which contributes to the depth of the subscapular fossa.

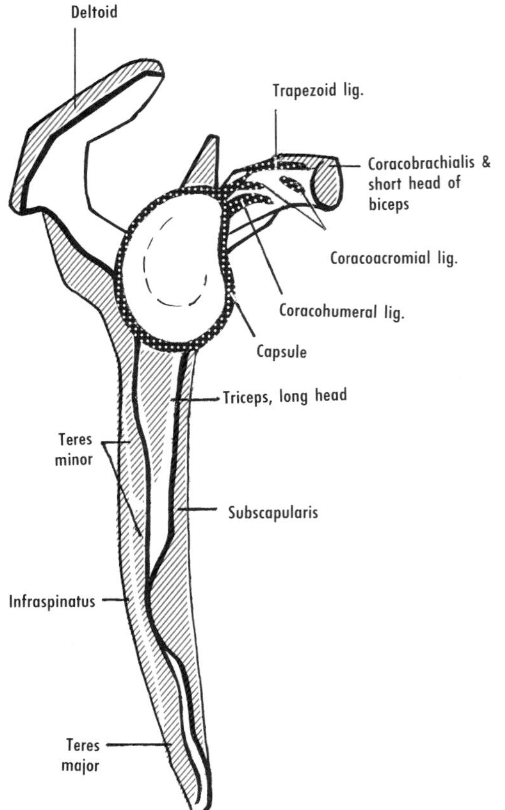

Figure 11–11 Right scapula, muscular and ligamentous attachments, lateral aspect. The origin of the tendon of the long head of the biceps from the supraglenoid tubercle and the glenoid lip is not shown.

clinically from the tip of the acromion to the lateral epicondyle of the humerus. The inferior surface of the acromion is ordinarily smooth and concave, but not infrequently presents a facet. Medially the acromion presents an articular facet for the clavicle. This facet faces upward and medially and varies in shape. The coracoacromial ligament is attached to the most anterior part of the acromion, which may be prolonged by partial ossification of this ligament.

The *coracoid process*, located above the neck and glenoid cavity, projects forward and a little laterally. Its tip can be palpated under cover of the anterior margin of the deltoid, below the junction of the lateral and intermediate thirds of the clavicle. The coracoid process consists of a vertical part that is flattened superoinferiorly, and a horizontal part that may form a synovial joint with the conoid tubercle of the clavicle.

Ossification

The scapula begins to ossify by the eighth postovulatory week. A center appears in the body, near the neck, and then spreads throughout most of the body. Two other centers appear later in the coracoid process. One, the chief coracoid center, appears during the first year, sometimes at birth. Another, for the base of the coracoid and glenoid rim, appears by puberty. Both fuse with the body during adolescence.

Epiphysial centers make their appearance in the acromion (usually two centers), medial border, inferior angle, and sometimes the tip of the coracoid during puberty or adolescence. Fusion with the body is completed by adolescence, or later in the case of the inferior angle and medial border.

HUMERUS

The humerus (figs. 11–3 to 11–5, 11–12 to 11–19) is the bone of the arm and shoulder. It articulates with the scapula at the shoulder and with the radius and ulna at the elbow. It consists of a shaft and two ends, proximal and distal. The side to which a humerus belongs can be determined by placing the bone with its rounded end upward and the enlargements or tubercles, separated by a groove, facing anteriorly. The head is now directed medially.

The proximal end consists of the head, anatomical neck, and two tubercles, greater and lesser, which are separated from each other by the intertubercular groove. The head, somewhat less than half a sphere, faces medially, upward, and backward. It can sometimes be felt in the uppermost part of the axilla. The *anatomical neck* is a slight constriction immediately adjacent to the head. The *greater tubercle* projects laterally, beyond the acromion, so that, unless the shoulder is dislocated, a ruler will not simultaneously make contact with the lateral epicondyle and the acromion. It is covered by the deltoid muscle, which is responsible for the normal rounded contour of the shoulder, and shows impressions for muscular attachments. The *lesser tubercle* forms the most anterior part of the proximal end of the humerus, and may be palpable through the deltoid muscle in the living subject. The *intertubercular groove* separates the greater and lesser tubercles and passes distally onto the shaft. The greater tubercle is continued downward as the *crest of the greater tubercle*, which forms the lateral lip of the intertubercular groove. The medial lip of the groove *(crest of the lesser*

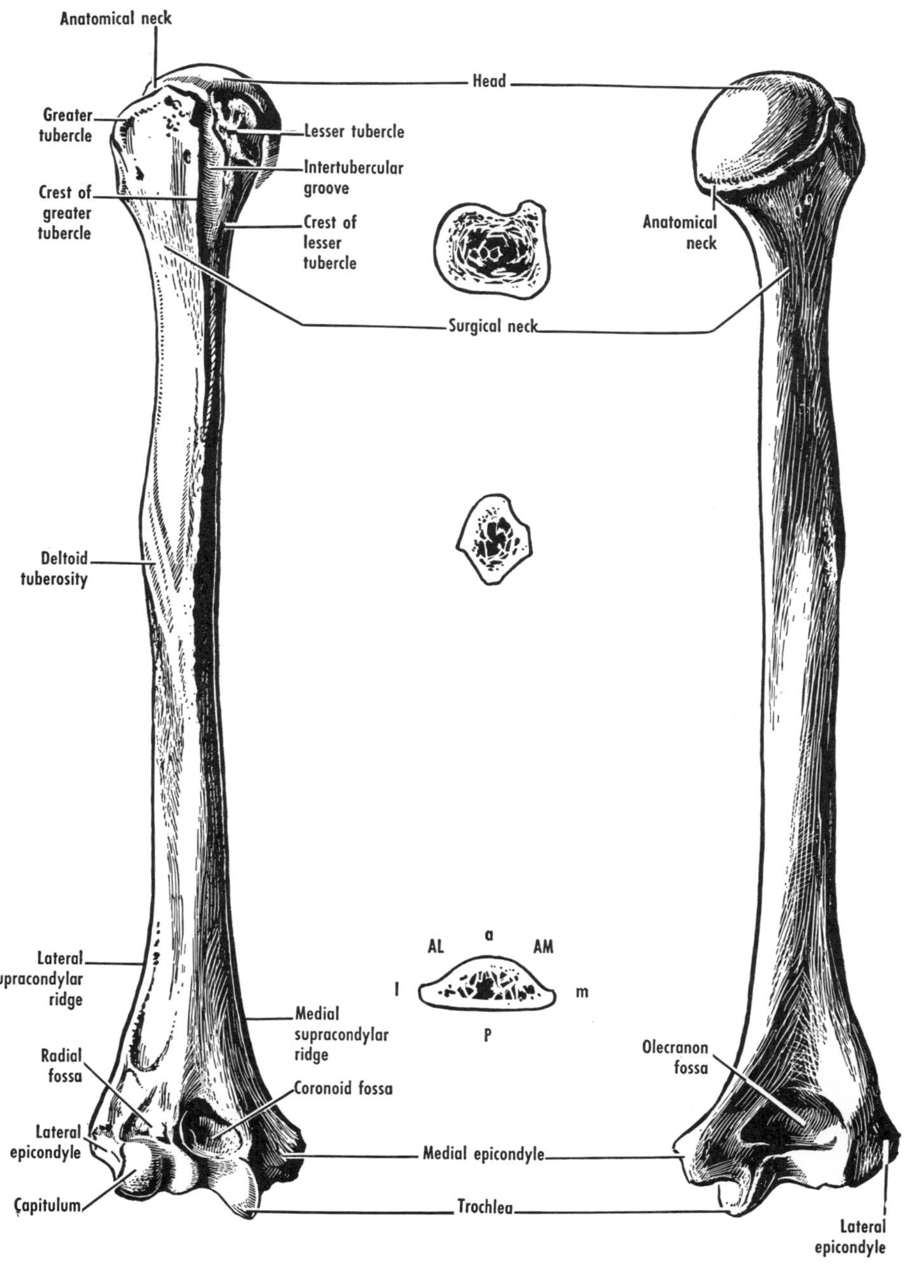

Figure 11–12 Right humerus. In the lowermost cross-section, the capital letters indicate surfaces, and the small letters indicate borders.

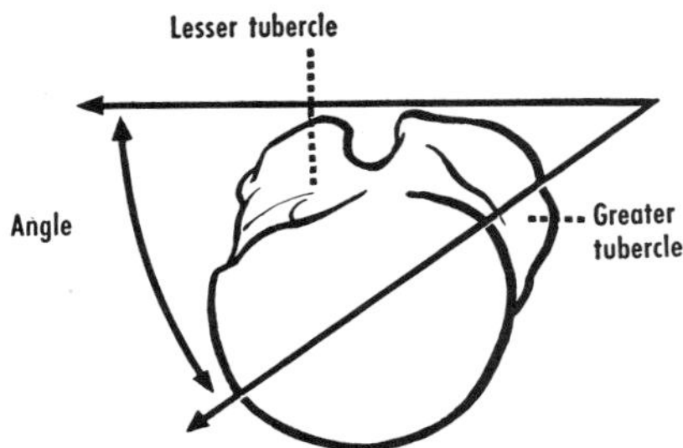

Figure 11–13 The humerus from above, showing the angle of humeral torsion. The upper arrow indicates the direction in which the medial epicondyle points and indicates an approximately horizontal axis through the epicondyles. The lower arrow indicates the long axis of the head. The angle between the arrows indicates the amount of torsion. Methods of measuring the angle of humeral torsion have varied widely.[4]

tubercle) is formed by the sharp border of the lesser tubercle and its downward continuation. The proximal end of the humerus is joined to the shaft by the *surgical neck,* a common site of fracture. The axillary nerve and the posterior circumflex artery lie in contact with the surgical neck.

The *shaft* has three surfaces, anterolateral, anteromedial, and posterior, separated by three borders, anterior, medial, and lateral.

The *medial border* is continuous above with the crest of the lesser tubercle and below with the medial supracondylar ridge. One or two nutrient foramina located on or near this border lead into canals that are directed distally. The *anterior border* is continuous with the crest of the greater tubercle above and the ridge separating the trochlea from the capitulum below. The *lateral border,* indistinct above, is continuous below with the lateral supracondylar ridge.

The *posterior surface* is marked by a wide, ill-defined *groove for the radial nerve,* which runs downward and laterally, and separates the origin of the lateral head of the triceps above from the origin of the medial head below.

The *anterolateral surface* presents near the middle a large rough area, the *deltoid tuberosity,* for the insertion of the deltoid muscle.

The uppermost part of the *anteromedial surface* forms the rough floor of the intertubercular groove; the rest of the surface is smooth. On rare occasions, a *supracondylar process*[5] of variable size projects from the anteromedial surface. A foramen completed by a fibrous band that connects the medial epicondyle and the process may transmit the brachial artery and median nerve.

The distal end consists of the condyle, the medial epicondyle, and the lateral epicondyle. The *medial epicondyle* points in approximately the same direction as the head of the humerus. However, the main axis of the distal end lies at an angle to that of the head (fig. 11–13). The medial epicondyle is rough in front and gives origin to the flexor muscles of the forearm. The ulnar nerve lies behind the epicondyle, in the *groove for the ulnar nerve,* and can be palpated there. The *lateral epicondyle* gives origin to the supinator and extensor muscles of the forearm, and to the anconeus. The *condyle* includes the trochlea and capitulum, and the coronoid, olecranon, and radial fossae. The *trochlea* runs in a spiral direction from anterior to posterior aspect. Its medial margin projects more than its lateral, so that its long axis is set obliquely to that of the shaft. Therefore, in the anatomical position the forearm forms a "carrying angle" of about 170 degrees with the arm. This angle disappears, however, during either flexion or pronation of the forearm. No significant sexual difference exists.[6] The *coronoid fossa* is located above the trochlea in front, the larger and deeper *olecranon fossa* above it behind. The bone separating these fossae may be extremely thin or even deficient. The *capitulum* is limited to the anterior and inferior aspects of the distal end and it articulates with the head of the radius, the margin of which fits into a depression that limits the capitulum medially. The very shallow *radial fossa* is located above the capitulum in front.

Because of their contact with the humerus, the axillary, radial, and ulnar nerves may be injured in fractures of the surgical neck, shaft, and medial epicondyle, respectively.

Ossification

A periosteal collar appears during the seventh postovulatory week. A center is usually present in the head at birth. The centers for the greater and lesser tubercles appear during infancy or very early childhood (fig. 11–5*A* and *B*, p. 74); the lesser center is usually not seen radiographically because it is projected on to that for the greater tubercle. The three centers of the upper end unite in early childhood to form a single epiphysis that fuses with the shaft in late adolescence. Most

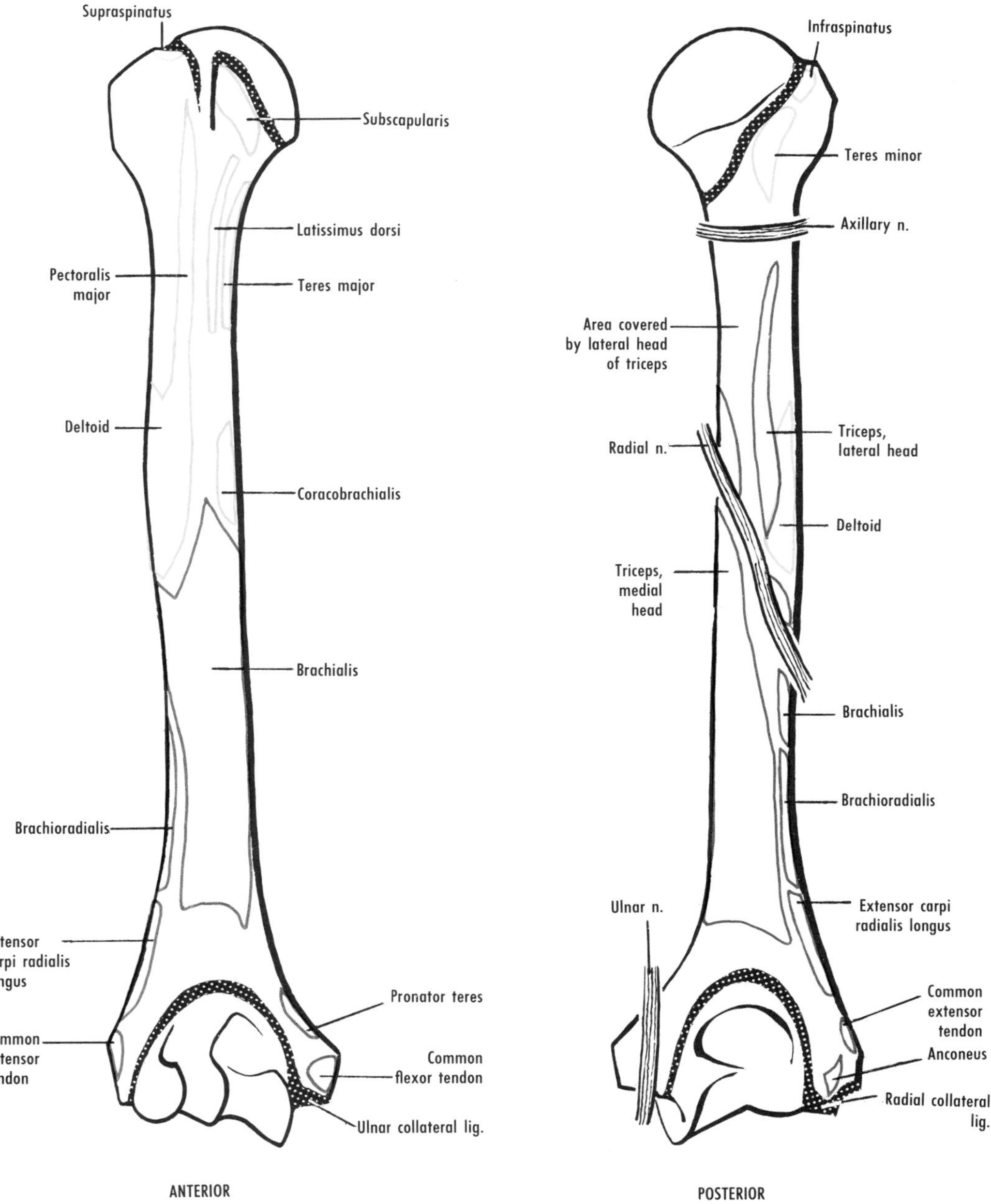

Figure 11–14 Anterior and posterior views of right humerus, showing muscular and ligamentous attachments. Note that the insertion of the deltoid is fused with the pectoralis major in front, and with the lateral head of the triceps behind.

growth in length occurs at the upper end of the humerus.

The lower end has four centers which, in order of appearance, are: capitulum and lateral part of trochlea (fig. 11–19*A* and *C*, p. 84), medial epicondyle (fig. 11–19*B* and *C*), medial part of trochlea, and finally, the lateral epicondyle. The pattern of fusion with the shaft is variable. The three lateralmost centers may unite, and the single center thereby formed fuses with the shaft during puberty. The medial epicondyle fuses shortly thereafter.

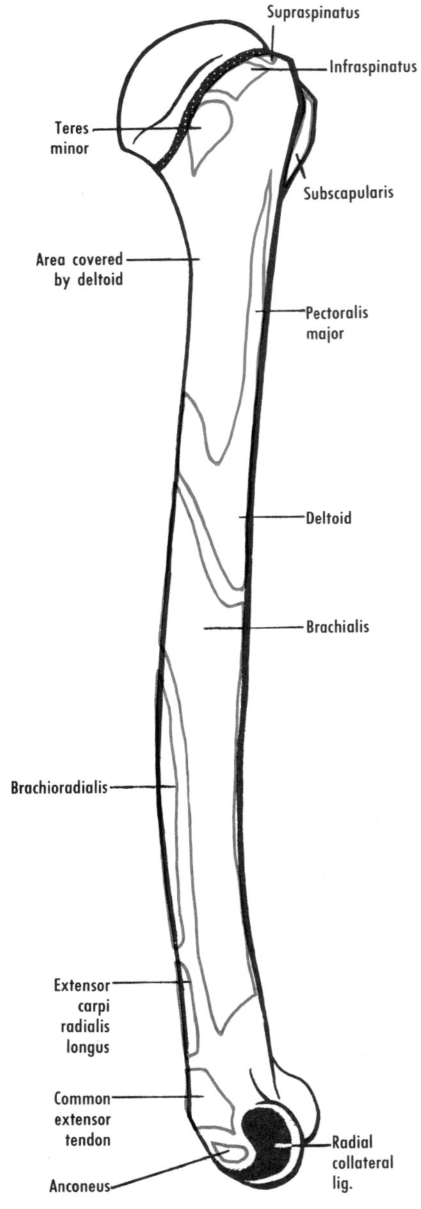

Figure 11–15 Lateral view of right humerus, showing muscular and ligamentous attachments.

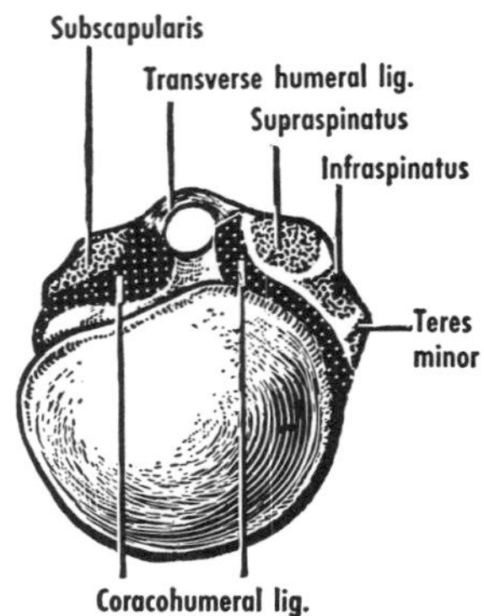

Figure 11–16 The right humerus from above, showing muscular and ligamentous attachments.

The epiphysial lines of the humerus are shown in figure 11–17.

RADIUS

The radius (figs. 11–18 to 11–26) is the shorter and more lateral of the two bones of the forearm. It articulates with the humerus proximally, the carpus distally, and the ulna medially, and it consists of a shaft and two ends, proximal and distal. To identify the

(*Text continued on page 88.*)

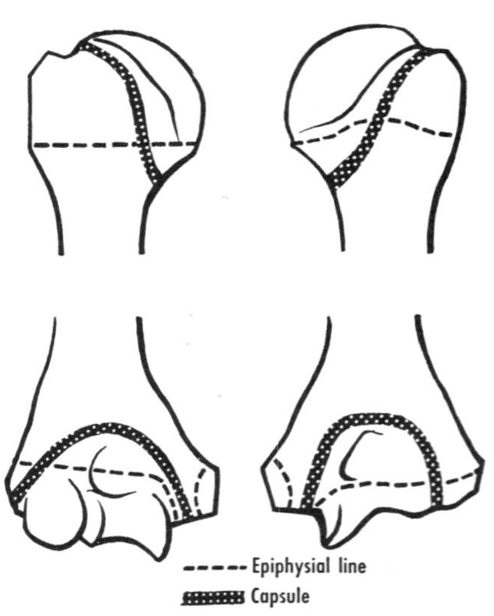

Figure 11–17 The upper and lower parts of the right humerus, showing the usual position of the epiphysial lines and the usual line of attachment of the joint capsule. The epiphysial lines at both ends are partly extracapsular. Modified from Mainland.[7]

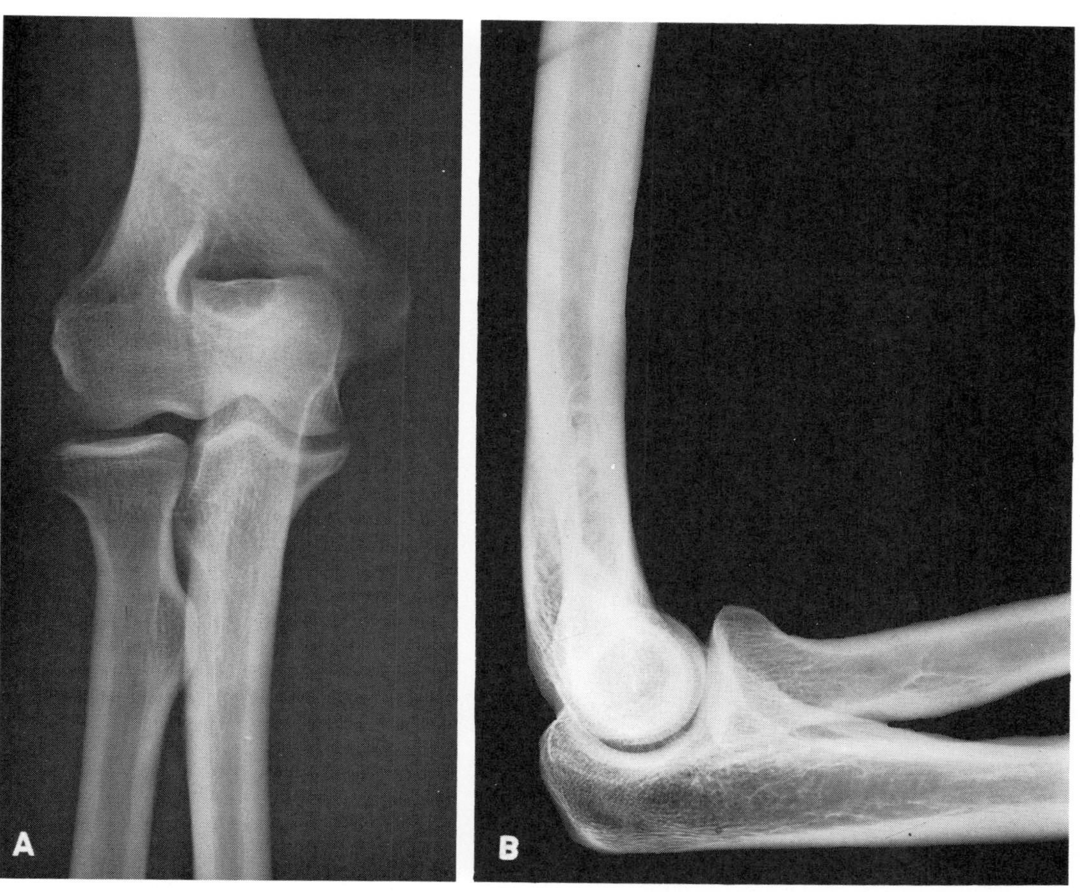

Figure 11–18 Elbows of adults. *A*, anteroposterior view. Note olecranon fossa, trochlea, medial epicondyle of humerus; head and tuberosity of radius; olecranon and coronoid process of ulna. *B*, lateral view. Note olecranon and coronoid process of ulna. Courtesy of Sir Thomas Lodge, The Royal Hospital, Sheffield, England.

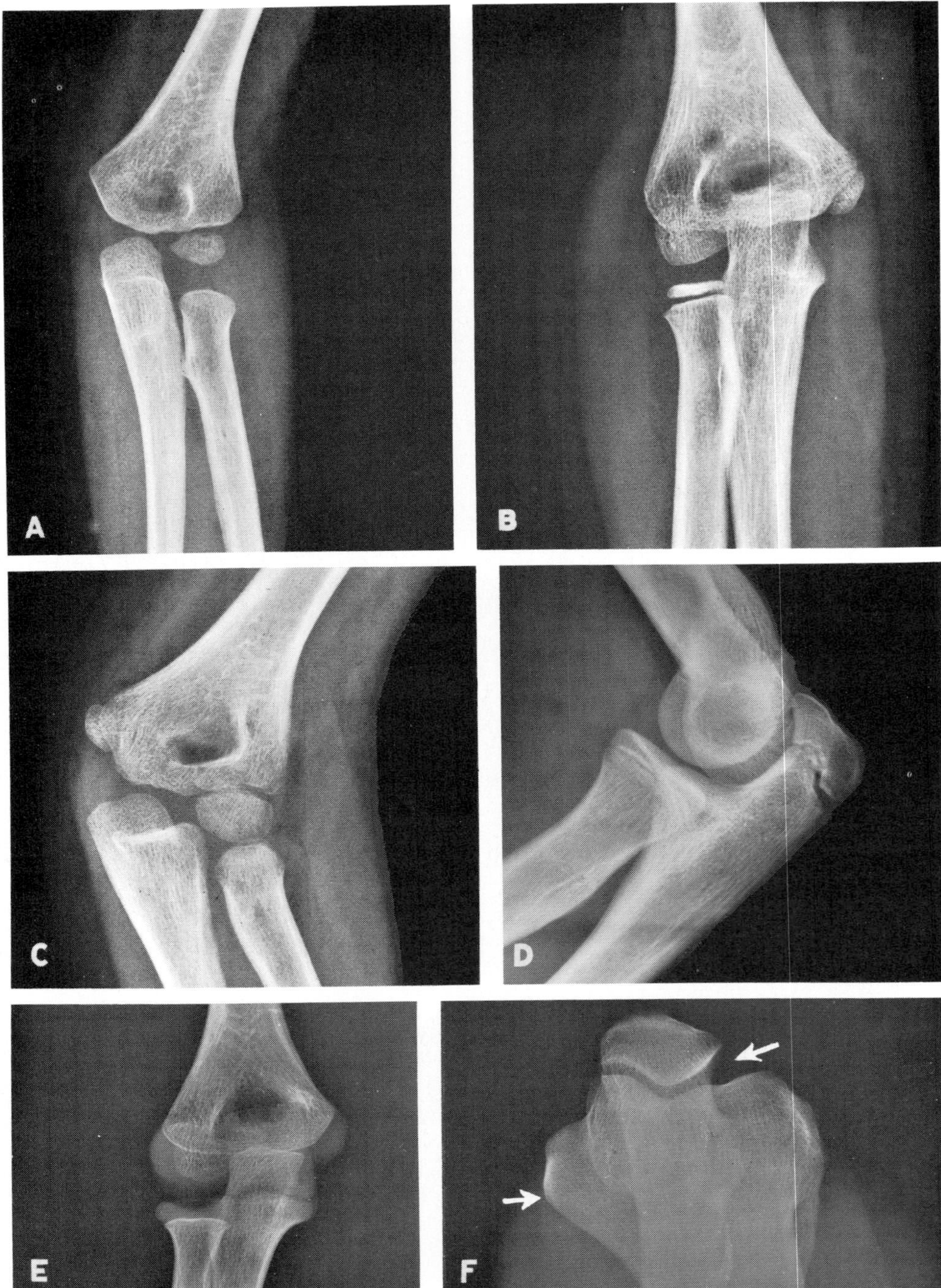

Figure 11–19 Elbow. *A*, elbow of child. Note epiphysis for capitulum and lateral part of trochlea of humerus. Ulna is at left. *B*, elbow of child. Note additional epiphyses for medial epicondyle of humerus and head of radius. *C*, elbow of child, oblique view, showing epiphyses for capitulum and lateral part of trochlea, and for medial epicondyle. *D*, epiphysis for upper end of ulna. Note also epiphysis for head of radius. *E*, radiogram of dried bones of boy aged five. Note outline of cartilage. *F*, flexed elbow of adult. Note medial epicondyle (*arrow on left*), and joint line between olecranon and trochlea (*arrow on right*). *A*, *B*, and *C*, courtesy of S. F. Thomas, M.D., Palo Alto Medical Clinic, California. *D*, courtesy of G. L. Sackett, M.D., Painesville, Ohio. *F*, courtesy of V. C. Johnson, M.D., Detroit, Michigan.

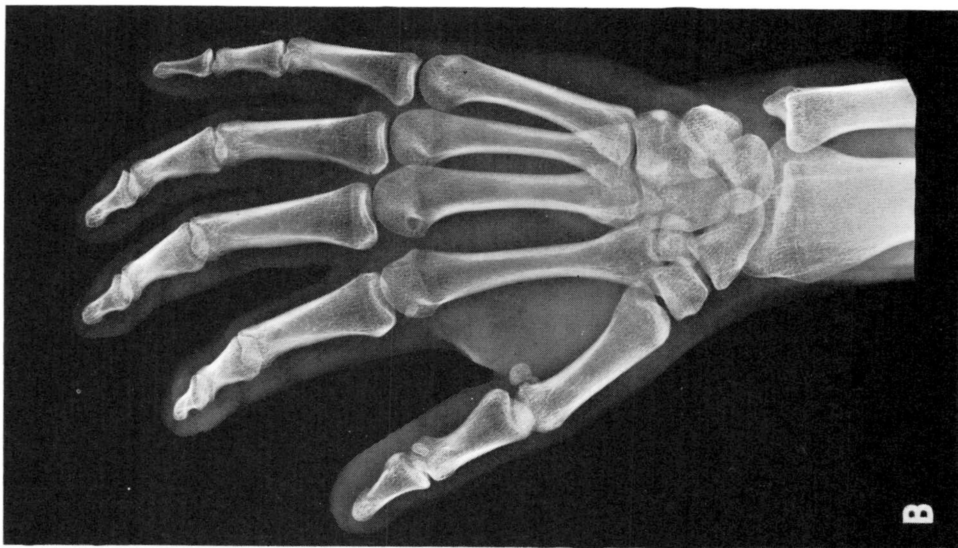

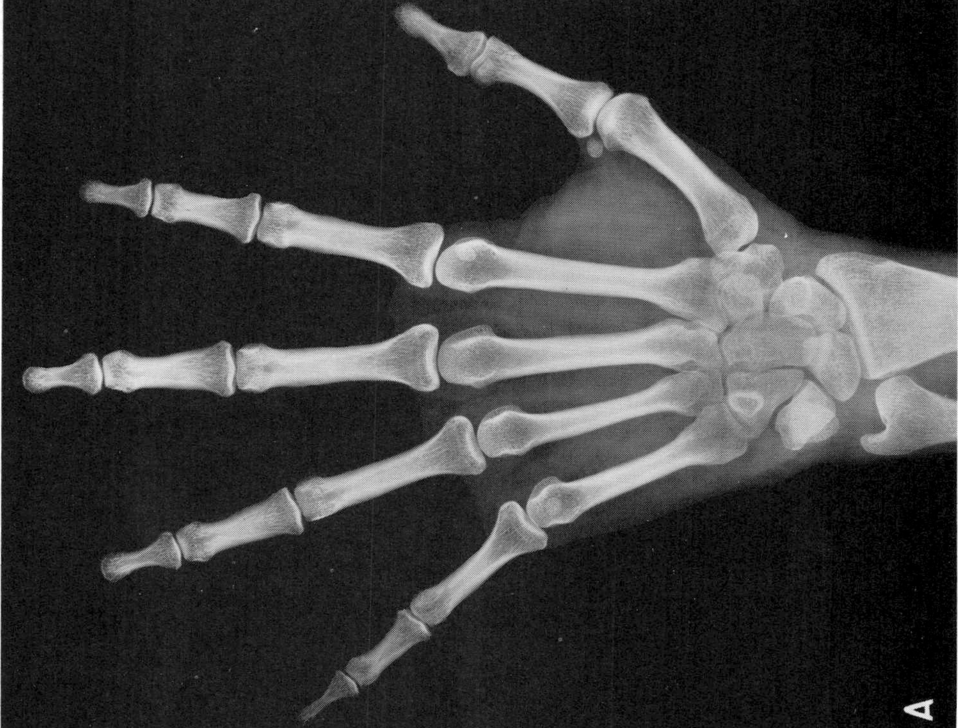

Figure 11–20 Hands of adults. *A*, posteroanterior view. Note hook of hamate, and sesamoid bones of first, second, and fifth fingers. *B*, oblique view. *A* and *B*, courtesy of S. F. Thomas, M.D., Palo Alto Medical Clinic, California.

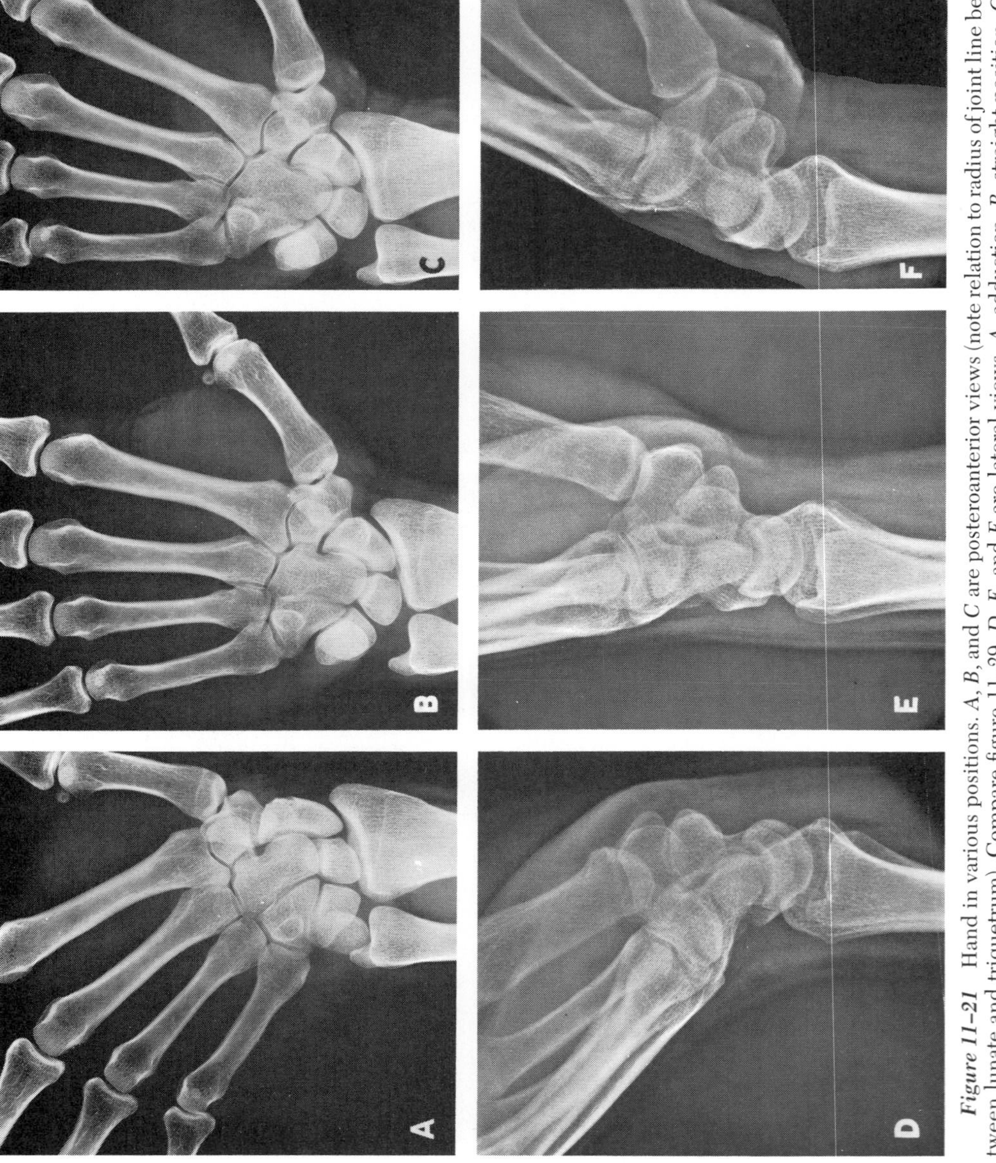

Figure 11–21 Hand in various positions. *A*, *B*, and *C* are posteroanterior views (note relation to radius of joint line between lunate and triquetrum). Compare figure 11–29. *D*, *E*, and *F* are lateral views. *A*, adduction. *B*, straight position. *C*, abduction. *D*, extension. *E*, straight position. Note lunate, capitate, scaphoid, and trapezium. *F*, flexion.

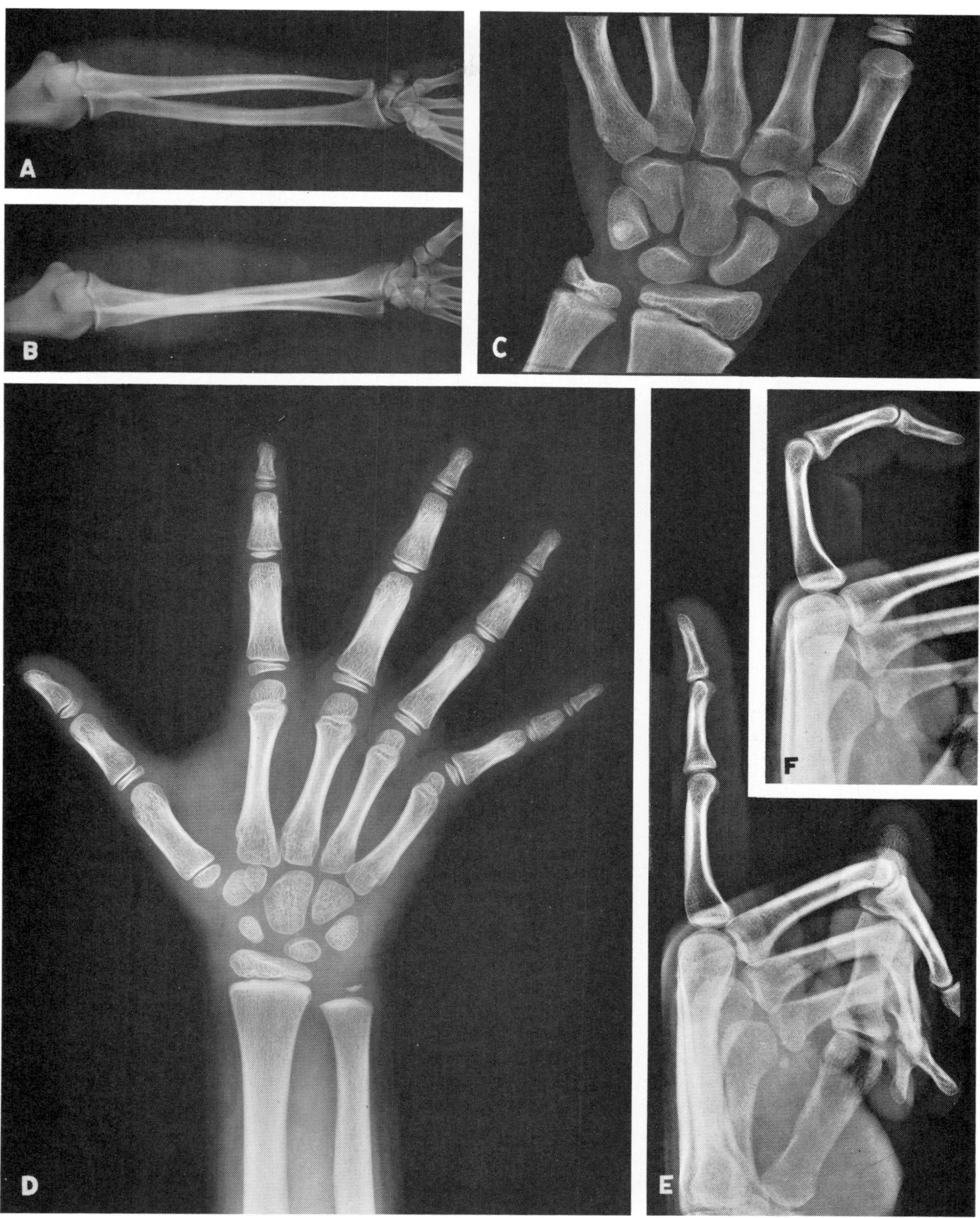

Figure 11–22 Various views of hand. *A* and *B*, forearm and hand in supination (*A*) and pronation (*B*). *C*, hand of child. Note epiphyses for lower ends of radius and ulna, for base of first metacarpal, and an accessory epiphysis for base of second metacarpal. *D*, hand of child. The pisiform does not yet show. Note epiphyses for metacarpals and phalanges. *E* and *F*, index finger in extension (*E*) and flexion (*F*). Note shift in position (relative to heads of proximal and middle phalanges) of bases of middle and distal phalanges. *A*, *B*, *E*, and *F*, courtesy of S. F. Thomas, M.D., Palo Alto Medical Clinic, California. *C*, courtesy of J. Lofstrom, M.D., Detroit Memorial Hospital.

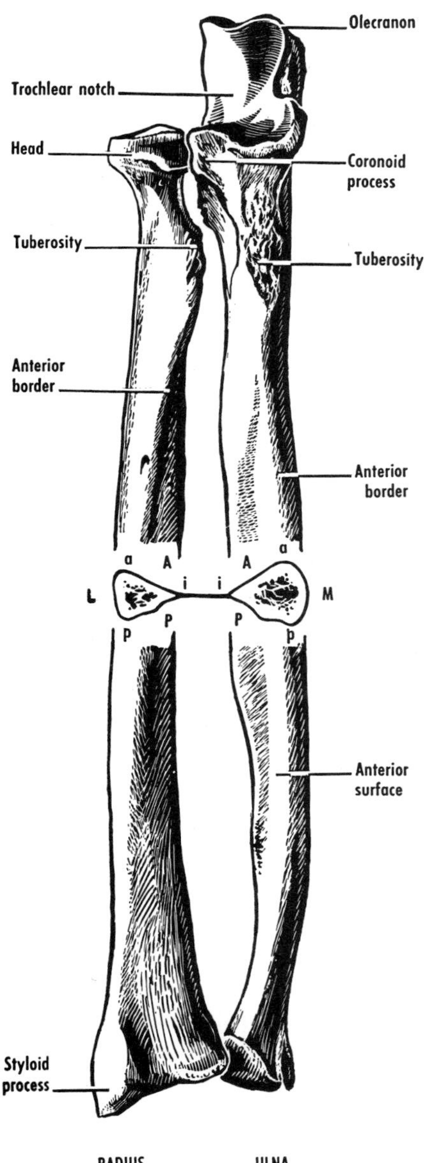

Figure 11–23 Right radius and ulna, anterior view. Cross-sections of midportions of shafts show arrangement of surfaces and borders. Capital letters indicate surfaces, and small letters indicate borders.

bone as to side, the larger end is placed distally, the smooth concave surface of this end anteriorly, and the pointed process laterally and distally. The sharp border ending above at a tuberosity then faces the ulna and hence the median plane of the body.

The upper end consists of a head, neck, and tuberosity. Proximally the *head* is concave and articulates with the capitulum of the humerus. Its *articular circumference* is widest medially, where it articulates with the ulna. It is enclosed in the rest of its extent by the annular ligament that covers the neck distally. The head of the radius can be palpated in the living just below the lateral epicondyle, especially during rotation of the forearm. The *tuberosity of the radius* is located on its front and medial side, just distal to the *neck*, which separates the upper end from the shaft.

The *shaft* has anterior, posterior, and lateral surfaces, and anterior, posterior, and interosseous borders. The lower part of the anterior surface may be marked by a rough, obliquely placed line, the pronator ridge. The nutrient artery of the shaft, directed toward the elbow, usually enters the anterior surface near the junction of its proximal and middle thirds.[8]

The *interosseous border* gives attachment to the interosseous membrane except for its uppermost part, to which the oblique cord is attached. Distally this border divides into anterior and posterior parts (the pronator quadratus muscle is inserted between them); the membrane is attached to the posterior part. The *posterior border* is indistinct above, and also below, where it ends at the dorsal tubercle on the back of the lower end. The upper part of the *anterior border* is sometimes called the oblique line. This border is continuous below with the anterior border of the styloid process.

The expanded distal end of the radius presents on its medial surface the concave *ulnar notch*, the lower margins of which give attachment to the articular disc. The lateral surface is marked by the *styloid process*, to the apex of which the radial collateral ligament is attached. The styloid process is palpable in an area located between the extensor pollicis longus and brevis tendons known as the "anatomical snuff-box." **The styloid process of the radius is about 1 cm distal to that of the ulna. This relationship is important in the diagnosis of fractures of the lower end of the radius, and in determining whether such fractures have been properly reduced.**

The anterior aspect of the distal end of the radius is smooth. The convex dorsal aspect is marked near its middle by the easily palpable *dorsal tubercle*. The *carpal articular surface* of the radius, concave from front to back and from side to side, presents a medial quadrilateral facet for articulation

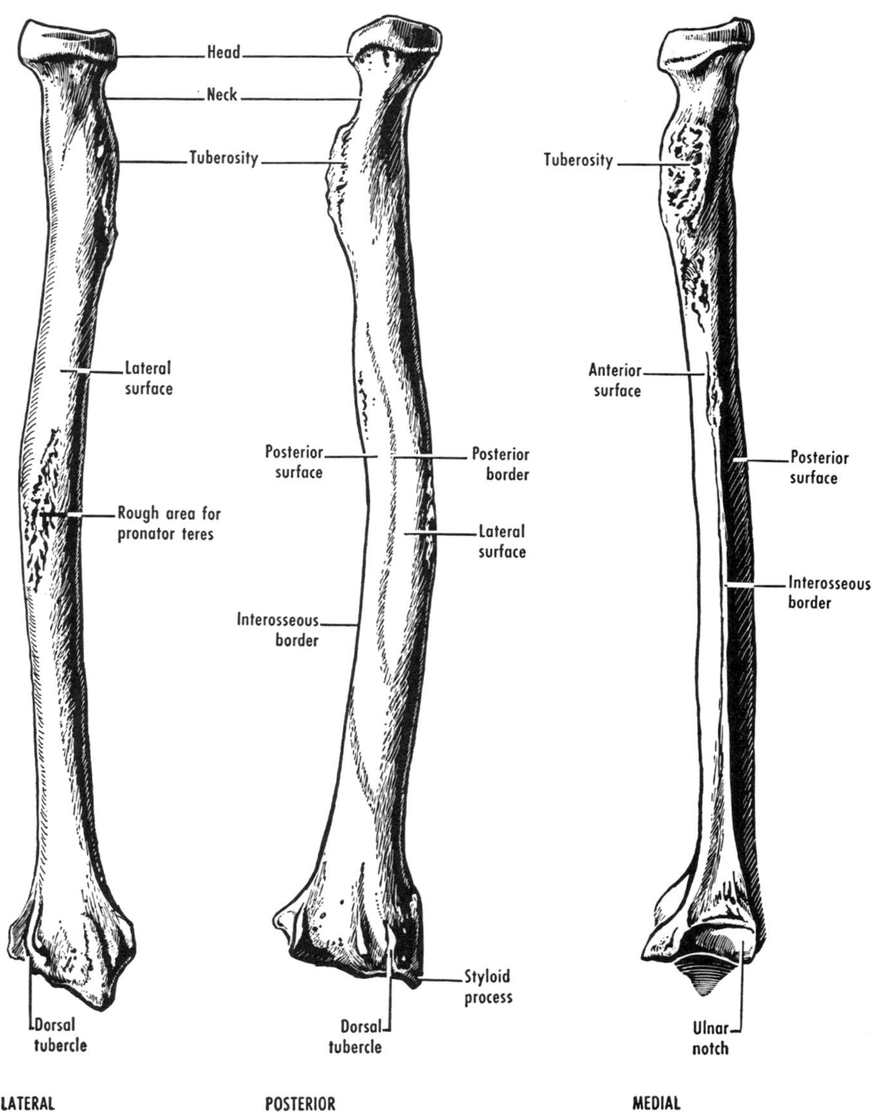

Figure 11–24 Right radius. In the lateral view, note the shallow groove immediately to the right of the styloid process; this is occupied by the tendons of the abductor pollicis longus and extensor pollicis brevis. In the posterior view, note that the dorsal tubercle is grooved; the groove is occupied by the tendon of the extensor pollicis longus. The tendons of the extensor carpi radialis longus and brevis lie to the radial side of the tubercle; the tendons of the extensor indicis and extensor digitorum lie to the ulnar side.

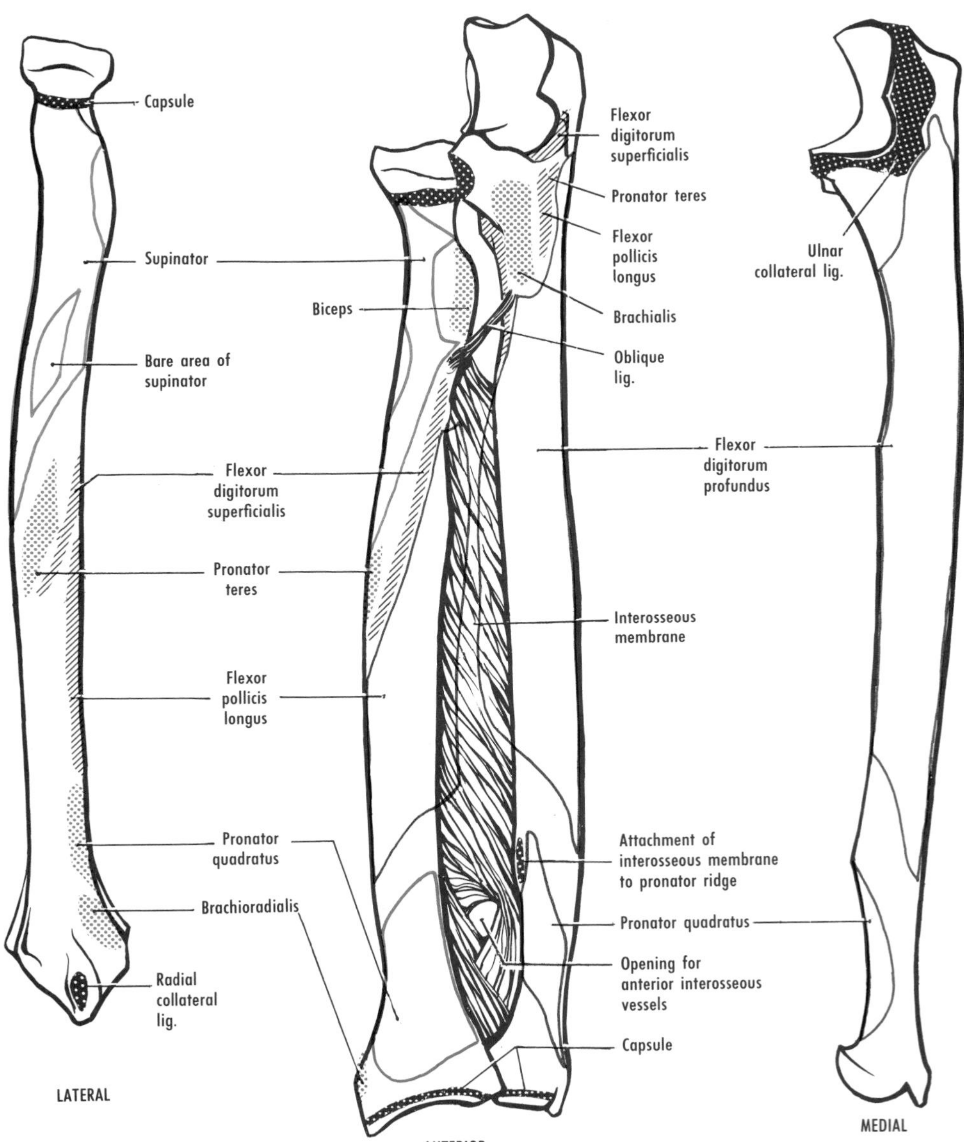

Figure 11–25 Muscular and ligamentous attachments of right radius and ulna. About midway on the shaft of the radius is a rough area for the insertion of the pronator teres, below which the shaft is covered by the tendons of the brachioradialis and the extensor carpi radialis longus and brevis. The interosseous membrane gives origin in part to the flexor pollicis longus and flexor digitorum profundus.

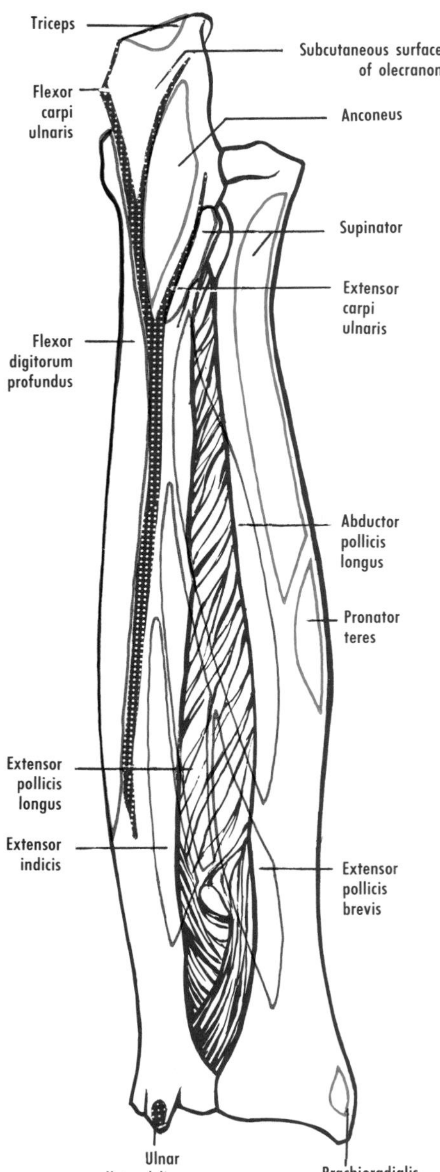

Figure 11–26 The right radius and ulna, showing muscular and ligamentous attachments, posterior aspect.

with the lunate bone, and a lateral triangular facet for articulation with the scaphoid. These articulations form a part of the radiocarpal or wrist joint.

Grooves for the extensor tendons are found on the lateral and posterior aspects of the lower end of the radius (fig. 11–24). The tendons are enveloped in synovial sheaths and are covered by the extensor retinaculum, which is attached to the intervening ridges, including the dorsal tubercle.

Ossification

A periosteal collar appears during the seventh postovulatory week. An epiphysial center appears in the distal end during infancy and another in the head during childhood (figs. 11–19*D* and 11–22*C*, pp. 84, 87). That for the proximal end fuses during puberty, and that for the distal end shortly thereafter. A center for the tuberosity, when present, appears at about the time of puberty and fuses shortly thereafter. Growth in length of the radius occurs mainly at the distal end. The epiphysial lines of the radius are shown in figure 11–28.

ULNA

The ulna (figs. 11–18 to 11–23, 11–25 to 11–28) is the longer and more medial bone of the forearm. It articulates with the humerus above, the articular disc below, and the radius laterally. It has a shaft and two ends, and it is readily palpable in its entire extent. The side to which an ulna belongs can be determined by placing its larger end proximally, the large notch at this end anteriorly, and the sharp border on the shaft of the bone laterally.

The proximal end includes the olecranon and the coronoid process, the trochlear notch, and the radial notch.

The *olecranon* is the projection on the back of the elbow, which is especially prominent when the forearm is flexed. It is the part that rests on a table, when leaning on the elbow. Its superior surface is rough behind for the insertion of the triceps muscle. Ossification in the tendon accounts for the occasional proximal prolongation of the posterior part of this surface. The articular capsule is attached along the anterior margin. The posterior or subcutaneous surface, ovoid or triangular, is relatively smooth and is covered by a bursa. The anterior surface, which forms the upper part of the *trochlear notch,* is separated by a transverse ridge or groove from the lower part of the notch. A longitudinal ridge divides the trochlear notch into medial and lateral parts, which articulate with corresponding parts of the trochlea of the humerus.

The *coronoid process* projects forward and fits into the coronoid fossa of the humerus when the forearm is flexed. Its superior surface forms the lower part of the trochlear notch. The coronoid process is prolonged

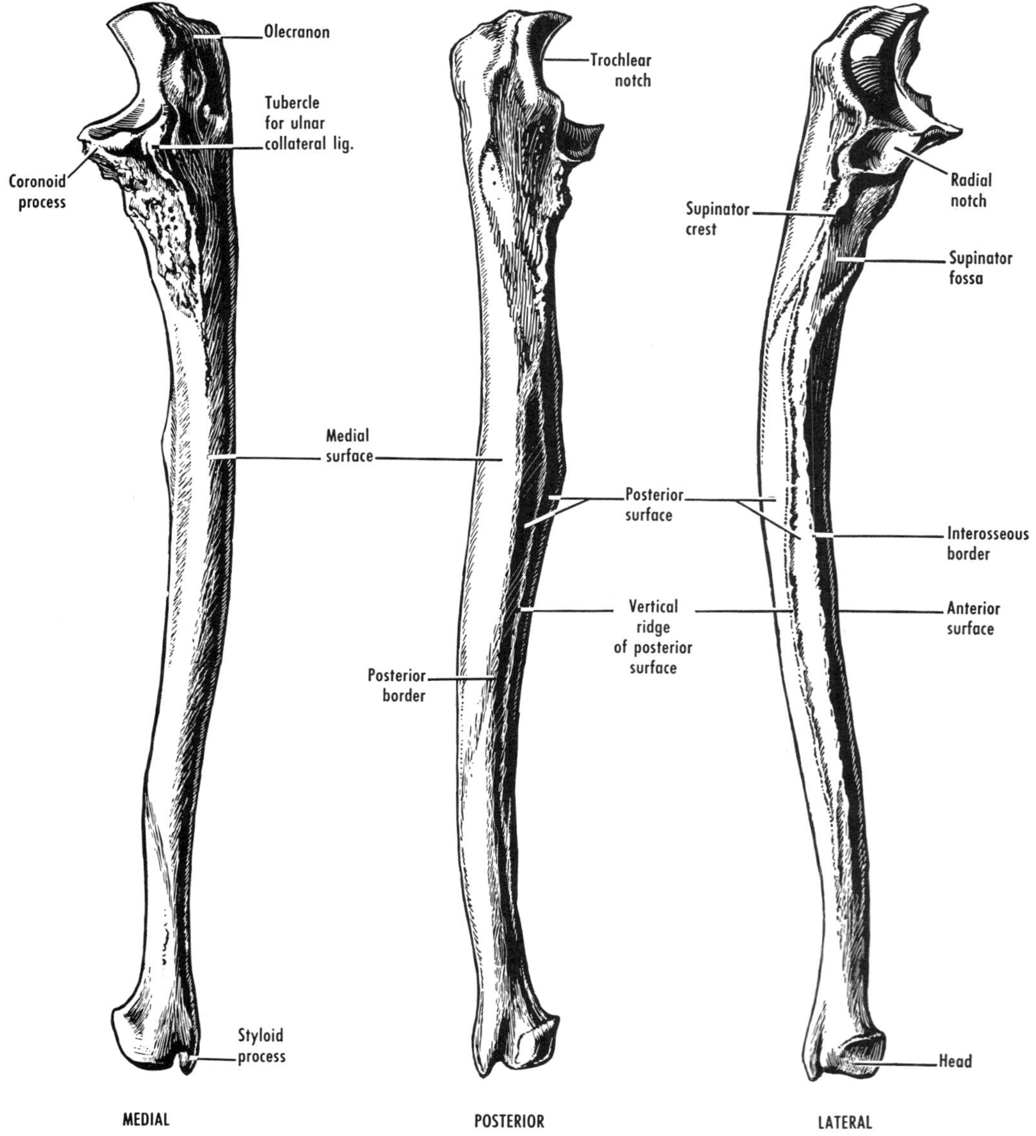

Figure 11–27 Right ulna.

downward, having medial and lateral margins that enclose an area for insertion of the brachialis; the roughened lower part of the area is the *tuberosity of the ulna.* The upper part of the medial margin has a tubercle for the ulnar collateral ligament of the elbow joint and the flexor digitorum superficialis. The ulnar nerve can be palpated against this tubercle. The lateral margin of the coronoid process continues into the front margin of the *radial notch,* to the margins of which the annular ligament is attached.

The *shaft* has anterior, posterior, and medial surfaces and anterior, posterior, and interosseous borders.

The *anterior surface* is grooved longitudinally in its upper two-thirds or three-fourths. The remainder, convex in cross-section, is often separated from the upper part by a rough, obliquely placed line, the pronator ridge. A nutrient artery, directed toward the elbow, enters the anterior surface in the middle third of the shaft. The *medial surface* is smooth and rounded, and is subcutaneous in its lower part. The *posterior surface* is marked proximally by a ridge, the oblique line, which joins the posterior border and supinator tubercle at the lower end of the *supinator crest* (fig. 15–4, p. 134). A considerable extent of the posterior surface is divided by a longitudinal ridge into a medial smooth part (covered by the extensor carpi ulnaris), and a lateral part roughened by muscular origins.

The *interosseous border* is sharp proximally but indistinct below. Above, it is continued by two lines that pass to the margins of the radial notch and enclose the supinator fossa. This fossa lodges the tuberosity of the radius in front during pronation, whereas behind, it gives origin in part to the supinator. The posterior of the two lines ascending from the interosseous border joins the oblique line at the supinator tubercle. The latter is the lowest point of the supinator crest, which forms the posterior margin of the radial notch (see fig. 15–4). The oblique line is often joined by the vertical line that subdivides the posterior surface. The *anterior border* is rounded, and extends from the coronoid process above, where it is continuous with the medial margin of the process, to the styloid process below. The *posterior* or subcutaneous *border* reaches the back of the olecranon above, and the styloid process below. It is readily palpable throughout its whole length and separates the flexor from the extensor muscles of the forearm.

The distal end includes the *head of the ulna* with its *styloid process*. The styloid process is small and conical, and is medial and posterior to the rest of the head, from which it is separated by a groove to which the articular disc is attached. This disc separates the ulna from the carpal bones, and the lower surface of the head articulates with it. The *articular circumference* of the head articulates with the ulnar notch of the radius. The head can readily be seen and felt in the pronated forearm but is obscured by the radius during supination.

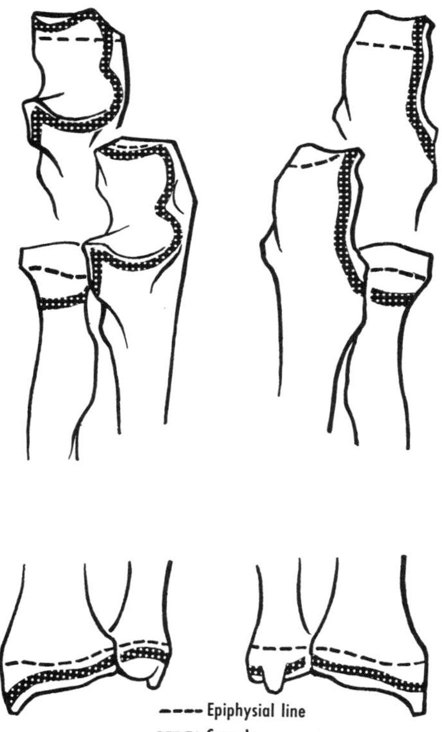

Figure 11–28 The upper and lower parts of the right radius and ulna, showing the usual position of the epiphysial lines and the usual line of attachment of the joint capsule. The epiphysial line of the head of the radius is intracapsular, that of the upper end of the ulna partly or entirely extracapsular, and those of the lower end extracapsular. The additional views of the ulna (upper two figures) show a variation in the position of the epiphysial line. Modified from Mainland.[7]

Ossification

A periosteal collar appears during the seventh postovulatory week. Epiphysial centers for the proximal and distal ends develop during childhood (figs. 11–19*D* and 11–22*C*, pp. 84, 87). The center for the distal end appears first. The proximal epiphysis fuses during puberty; that for the distal end fuses shortly thereafter. The proximal epiphysis, which is located in the upper part of the olecranon, is extremely variable in size, and it may or may not be included in the elbow joint (fig. 11–28). Growth in length of the ulna takes place mainly at the distal end.

CARPUS

There are ordinarily eight carpal bones, arranged in two rows of four (figs. 11–20 to 11–22, 11–29 to 11–31). From lateral to medial the bones of the proximal row are the scaphoid, lunate, triquetrum, and the pisiform, which lies in front of the triquetrum. Those in the distal row are the trapezium, trapezoid, capitate, and hamate. Each carpal, with the exception of the pisiform, has several facets for articulation with

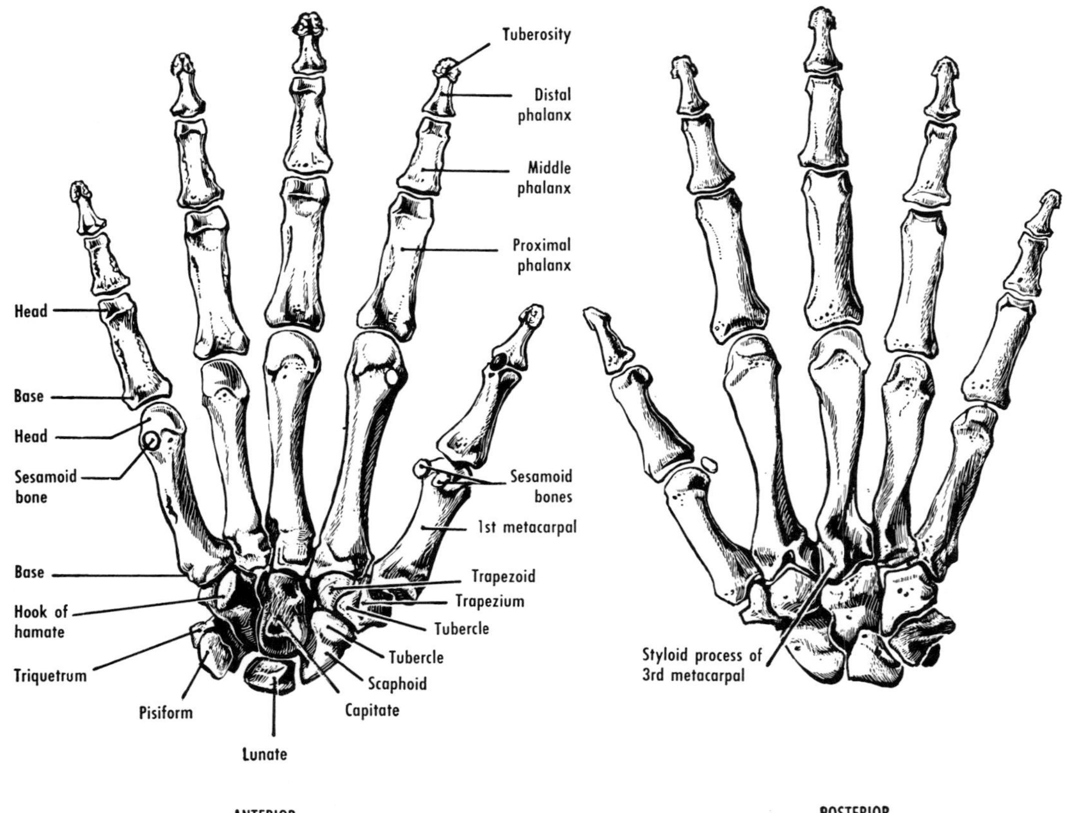

Figure 11–29 Bones of right hand, anterior and posterior aspects. The sesamoids shown are those commonly present.

neighboring bones. More attention should be paid to studying the carpus as a whole, rather than to mastering the minutiae of individual and isolated carpal bones.

The intact carpus is markedly convex from side to side behind, and concave in front. As might be expected, therefore, the posterior surfaces of the carpal bones are larger than the anterior; in the lunate, however, the converse holds, as a result of which dislocation of this bone is forward. The concavity of the carpus is maintained by the flexor retinaculum, which roofs it to form the *carpal canal* or tunnel for the flexor tendons. The flexor retinaculum is attached to bony pillars on each side, that is, to the triquetrum (and pisiform) and the hook of the hamate medially, and to the tubercles of the scaphoid and trapezium laterally. These bony prominences are palpable in the living hand.

The pisiform can be identified on the front of the medial border of the wrist, and can be moved from side to side when the hand is relaxed. The tendon of the flexor carpi ulnaris descends to it and can be felt along the medial border of the forearm. The hook of the hamate can be felt on deep pressure over the medial side of the palm, about 2 cm distal and lateral to the pisiform.

The tubercles of the scaphoid and trapezium are felt usually as a continuous bony mass at the proximal limit of the ball of the thumb. The guide to these bones is the tendon of the flexor carpi radialis, the most lateral of the tendons that descend on the front of the forearm. The tubercle of the scaphoid lies partly under cover, and partly on the lateral side, of the tendon. **The scaphoid is important because it is the carpal bone most frequently fractured, generally across its "waist."** The scaphoid and trapezium form the floor of the "anatomical snuff-box."

Interosseous ligaments bind the scaphoid, lunate, and triquetrum together, and separate the wrist joint from the midcarpal joint. The proximal aspect of these three bones is convex from side to side and from before backward. They articulate with the radius and articular disc to form the wrist joint. That the convexity extends farther posteriorly than anteriorly is associated with the greater range of extension than flexion at this joint. In adduction of the hand, the lunate articulates with the radius only, whereas in the straight position, or in abduction, it articulates with both the radius and the articular disc (see figs. 11–21, p. 86, and 16–16, p. 153). The lunate lies approximately between the two main skin creases on the front of the wrist. The pisiform articulates with the front of the triquetrum only, and is set farther forward than the other carpal bones. It does not take part in the wrist joint.

The carpal bones of the distal row are bound together by interosseous ligaments, which often separate the midcarpal from the carpometacarpal joints. The hamate and the head of the capitate occupy a socket formed by the scaphoid, the lunate, and the triquetrum. The front of the capitate is anchored to adjacent bones by the radiate ligament, and its medial surface to the hamate by a strong interosseous ligament. The capitate is in line with the third metacarpal. The hook of the hamate and the tubercle of the trapezium lie farther forward, and form the medial and lateral walls of the carpal canal.

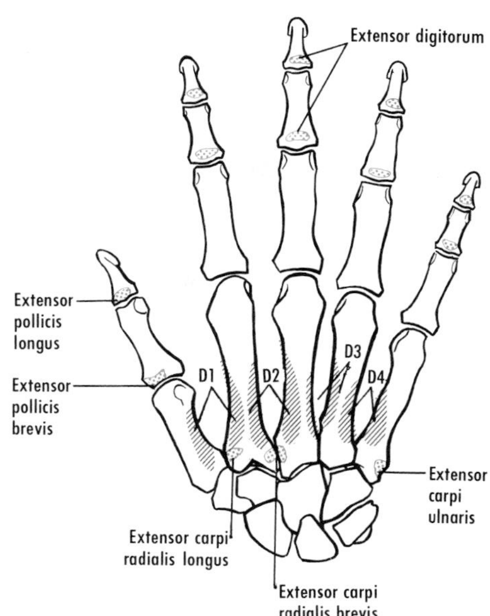

Figure 11–31 Bones of right hand, showing muscular and tendinous attachments, posterior view. Each dorsal interosseous muscle (*D*) arises from shafts of adjacent metacarpals.

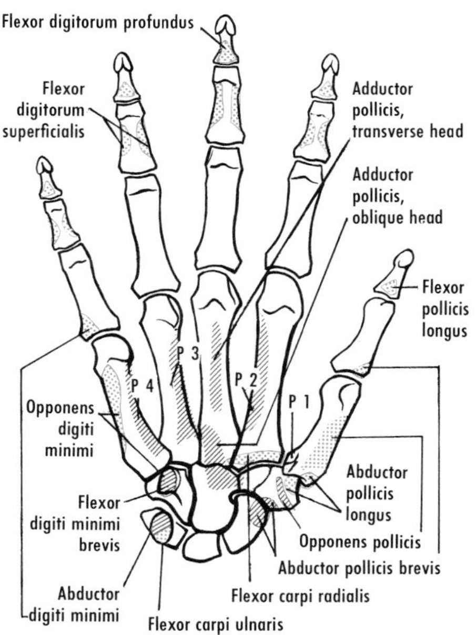

Figure 11–30 Bones of right hand, showing muscular attachments, anterior view. The flexor pollicis brevis is not shown. Of the interossei, only the palmar (*P*) are shown.

Scaphoid (Navicular). This bone, the largest in the proximal row, presents anteriorly a palpable *tubercle* to which the flexor retinaculum is attached. The pattern of blood supply is disputed, depending upon the method of study used.[9] However, there appear to be instances in which fracture through the center ("waist") of the bone deprives the proximal fragment of its blood supply, and it then dies.

Lunate. This bone is supplied by blood vessels that enter it through the capsular attachments in front and behind. Hence, dislocations are liable to be attended by interruption of the blood supply.

Triquetrum. This bone is pyramidal in shape. Proximally, it lies in contact with the articular disc laterally, and it gives attachment to ligaments medially, including the flexor retinaculum.

Pisiform. This is the smallest of the carpal bones and the last to ossify. It is easily palpated, and is movable from side to side when the muscles attached to it are relaxed.

Trapezium. Anteriorly, this bone presents a *tubercle* that can be palpated by deep pressure, more easily when the hand is extended. The tendon of the flexor carpi radialis runs in a groove on its medial side. Distally, a saddle-shaped facet articulates with the first metacarpal.

Trapezoid. Wider behind than in front, this bone gives attachment to ligaments posteriorly and anteriorly.

Capitate. This is the largest of the carpal bones

and the first to ossify. It has a rounded head above, which fits into a concavity of the lunate and scaphoid.

Hamate. This bone is easily recognized by the palpable and prominent *hook* on its palmar aspect. Laterally, the hook bounds the carpal canal and is related to the flexor tendons of the little finger. The front of the hook gives attachment to the flexor retinaculum and the pisohamate ligament.

Accessory Ossicles. Rarely, these small carpals may be found in the interstices between the carpal bones.[10] More than 20 have been described and named, and the possibility of their presence must be kept in mind in interpreting radiograms of the hand. The best known is the *centrale,* which is a rare ossicle lying dorsally between the scaphoid, capitate, and trapezoid. Carpal fusion may also occur. The most common is a fusion of the lunate and triquetrum *(os lunatotriquetrum).*

Ossification

Radiography of the hand is frequently used in an assessment of skeletal maturation. For purposes of comparison, a series of standards is available.[11] Each carpal bone ordinarily ossifies from one center, which appears postnatally (fig. 11–22). Ossification begins first in the capitate and hamate, where it may start before birth.[12]

METACARPUS

The metacarpals, or bones of the metacarpus (figs. 11–29 to 11–32), connect the carpus above with the phalanges below and are numbered from one to five, from the thumb to the little finger. The first is the shortest, the second the longest, and they decrease in length from the second to the fifth. From the palmar aspect, each is slightly concave lengthwise, and contributes to the concavity of the palm. Posteriorly, although covered by the long extensor tendons of the fingers, they can be palpated in their entire extent. Each metacarpal bone consists of a *shaft* and two ends. The distal ends or *heads* articulate with the proximal phalanges and form the knuckles of the fist. The anterior aspect of the articular surface of the head extends farther proximally than does the posterior and is notched in its middle so as to form two facets, especially prominent in the first metacarpal, where they articulate with sesamoids. A tubercle is located on each side near the back of the head, and immediately in front of this a small fossa gives attachment to the capsule and the collateral ligament. The four medial metacarpals are indirectly connected with one another by the deep transverse metacarpal ligament. The proximal end or *base* is broader on its posterior aspect than on its anterior. The adjacent sides of the bases have facets for the neighboring metacarpals; the first metacarpal, however, does not articulate with the second.

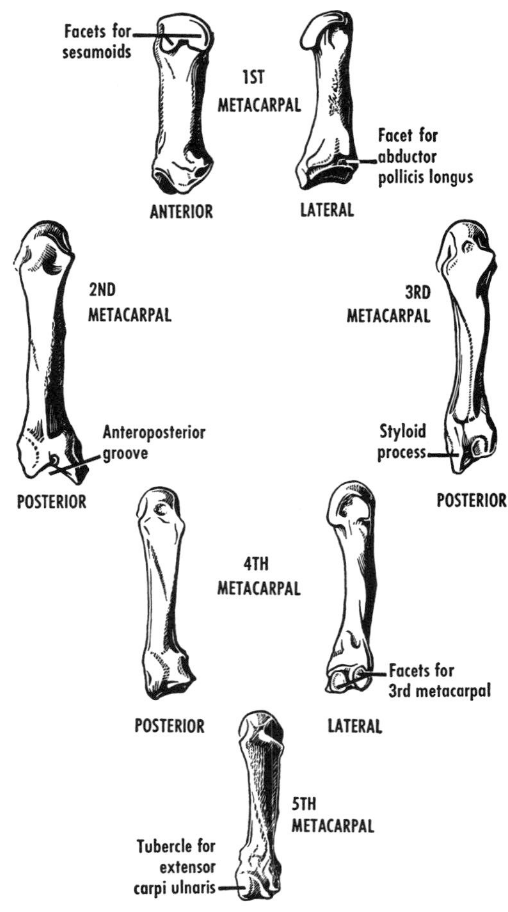

Figure 11–32 Chief identifying characteristics of the metacarpals. First metacarpal: short and broad; two facets for sesamoids; saddle-shaped carpal articular surface. The head is occasionally quite flat.[14] Second metacarpal: deep anteroposterior groove in base; four articular facets. Third metacarpal: styloid process (sometimes a separate bone, the os styloideum). Fourth metacarpal: quadrangular base with two facets on lateral side for third metacarpal. Fifth metacarpal: tubercle for insertion of extensor carpi ulnaris; articular facet on lateral side of base.

In addition to characteristics common to all five metacarpal bones, each has characteristics peculiar to it alone (fig. 11–32).[13]

Ossification

A periosteal collar and an endochondral center appear at the beginning of the third prenatal month.

Epiphysial centers appear distally (heads) in the four medial bones and proximally (base) in the first during infancy (fig. 11–22*D*). (Multiple centers, which later fuse, are normally present in the first metacarpal bone and in the first proximal phalanx.)[15] These centers unite at the end of puberty. The first occasionally has a center also in its head, and the second may have one in its base (fig. 11–22*C*); these accessory centers (called "pseudoepiphyses") are usually seen to be partly united with the shaft in a radiogram, and there is disagreement as to their mode of origin.[16]

PHALANGES

Each finger has three phalanges (figs. 11–29 to 11–31, 11–33), except the thumb, which has only two. The proximal phalanx joins the metacarpal, the distal phalanx is free at its distal end, and the middle phalanx is placed between these two. Each phalanx has a *base*, directly proximally, a *head* at its distal end, and an intervening *shaft*. The knuckles of the fingers are formed by the heads of the proximal and middle phalanges.

Proximal Phalanges. The base presents a concave, rounded facet, directed proximally, for the head of the corresponding metacarpal bone. The head articulates with the base of the middle phalanx by means of a trochlear surface, comprising two slight condyles. The posterior surface of the shaft is rounded, but the anterior surface is flat. The fibrous flexor sheath is attached to the margins of the anterior surface.

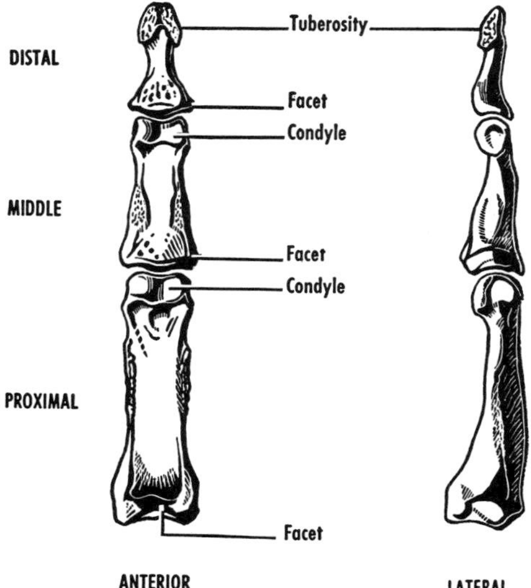

Figure 11–33 Phalanges of right middle finger.

Middle Phalanges. The articular surface of the base presents two facets for the corresponding condyles of the proximal phalanx. The head bears a trochlear surface that is similar to that of the proximal phalanx.

Distal Phalanges. The articular surface of the base presents two facets for the corresponding condyles of the middle phalanx. The distal phalanges are characterized by *tuberosities*, which are rough expansions at their distal ends, and which cover more of the anterior than of the posterior aspect.

Ossification

Bone appears in each proximal and distal phalanx during the third prenatal month, and in the middle phalanx during the fourth month. Epiphysial centers appear in all phalanges at their bases and are present by very early childhood (fig. 11–22*C* and *D*). These centers have united by the end of puberty. The distal phalanges are unusual in that ossification begins at the ends of the phalanges rather than in the form of a collar around the middle of the shaft.[17] This gives rise to the characteristic distal enlargement of the phalanx.

SESAMOID BONES

These small, rounded bones are related to the anterior aspects of some of the metacarpophalangeal and interphalangeal joints (fig. 11–29; see also figs. 11–20, p. 85, and 16–17, p. 153). They are generally embedded in the palmar ligaments of these joints. Two sesamoid bones are almost constantly present in front of the head of the first metacarpal.[18] Others frequently appear in relation to the interphalangeal joint of the thumb, the lateral side of the metacarpophalangeal joint of the second finger, and the medial side of the metacarpophalangeal joint of the fifth finger. Sesamoids in relation to other joints of the hand are uncommon.

REFERENCES

1. C. Zawisch, Z. mikr.-anat. Forsch., *59*:187, 1953. A. R. Koch, Acta anat., *42*:177, 1960. R. O'Rahilly and E. Gardner, Amer. J. Anat., *134*:291, 1972.
2. Variations in human scapulae are described by D. J. Gray, Amer. J. phys. Anthrop., *29*:57, 1942, and by A. Hrdlička, Amer. J. phys. Anthrop., *29*:73, 363, 1942.
3. W. W. Graves, Arch. intern. Med., *34*:1, 1924.
4. F. G. Evans and V. E. Krahl, Amer. J. Anat., *76*:303, 1945. V. E. Krahl and F. G. Evans, Amer. J. phys. Anthrop., *3*:229, 1945.
5. R. J. Terry, Amer. J. phys. Anthrop., *4*:129, 1921; *ibid.*, *14*:459, 1930.
6. W. B. Atkinson and H. Elftman, Anat. Rec., *91*:49, 1945. F. L. D. Steel and J. D. W. Tomlinson, J. Anat., Lond., *92*:315, 1958.
7. D. Mainland, *Anatomy*, Hoeber, New York, 1945.
8. S. Shulman, Anat. Rec., *134*:685, 1959.
9. B. E. Obletz and B. M. Halbstein, J. Bone Jt Surg., *20*:424, 1938. J. Taleisnik and P. J. Kelly, J. Bone Jt Surg., *48A*:1125, 1966.

10. R. O'Rahilly, J. Bone Jt Surg., *35A*:626, 1953; Clin. Orthopaed, *10*:9, 1957.
11. S. I. Pyle, A. M. Waterhouse, and W. W. Greulich (eds.), *A Radiographic Standard of Reference for the Growing Hand and Wrist,* Press of Case Western Reserve University (Year Book Medical Publishers, Chicago), 1971.
12. A. Christie, Amer. J. Dis. Child., *77*:355, 1949.
13. Common variations in the metacarpal bones are described by I. Singh, J. Anat., Lond., *93*:262, 1959.
14. H. Harris and J. Joseph, J. Bone Jt Surg., *31B*:547, 1949. J. Joseph, J. Anat., Lond., *85*:221, 1951.
15. A. F. Roche and S. Sunderland, J. Bone Jt Surg., *41B*:375, 1959.
16. R. W. Haines, J. Anat., Lond., *117*:145, 1974.
17. F. A. Dixey, Proc. R. Soc., *31*:63, 1881. O. Schuscik, Anat. Anz., *51*:118, 1918.
18. The incidence of sesamoids is given by J. Joseph, J. Anat., Lond., *85*:230, 1951, and their prenatal development and incidence are discussed by D. J. Gray, E. Gardner, and R. O'Rahilly, Amer. J. Anat., *101*:169, 1957. See also J. Sokolowska-Pituchowa and C. Miaśkiewicz, Folia morphol., Warsaw, *24*:136, 1965, and *26*:24, 1967.

VEINS, LYMPHATIC DRAINAGE, AND BREAST 12

VEINS

The blood from the upper limb is returned to the heart by two sets of veins, superficial and deep. The deep veins, except for the axillary, are usually arranged in pairs that have cross-connections between them. The paired veins accompany most of the arteries that are the size of the brachial or smaller, and are termed *venae comitantes.* Both sets, superficial and deep, are provided with valves, and both drain ultimately into the axillary vein.

Superficial Veins (fig. 12–1)

The anatomical disposition of superficial veins and their tributaries varies widely. The superficial veins lie in the subcutaneous tissue for most of their course and return nearly all of the blood. Blood from the hand drains chiefly into the *dorsal venous network* on the back of the hand. This network receives dorsal digital veins and communicates with deep veins.[1] Two or more veins ascend from the network; the two more prominent ones are known as the cephalic and the basilic, each of which has several valves.

The *cephalic vein,* which begins as a continuation of the radial side of the dorsal network, winds anteriorly around the lateral border of the forearm and reaches the front of the elbow region. Next it ascends along the lateral margin of the biceps, pierces the fascia, and enters the groove between the deltoid and pectoralis major muscles, where it is accompanied by the deltoid branch of the thoracoacromial artery. It then pierces the clavipectoral fascia and ends in the axillary vein. Occasionally it communicates with the external jugular vein by a tributary that passes in front of the clavicle. Near its termination it receives a few tributaries that accompany branches of the thoracoacromial artery. The cephalic vein may be accompanied by an *accessory cephalic vein.* The brachial part of the cephalic vein may be small or absent, and its antebrachial part then drains chiefly into the basilic vein (fig. 12–1).

The *basilic vein,* which begins as a continuation of the ulnar side of the dorsal network, ascends on the medial side of the forearm. As it approaches the elbow, it winds anteriorly around the medial border of the forearm to reach the front of the medial epicondyle of the humerus. Next, it ascends along the medial margin of the biceps, pierces the fascia at the middle of the arm, and accompanies the brachial artery to the axilla, where it joins the brachial veins and becomes the axillary vein.

The *median antebrachial vein,* which begins on the front of the hand, ascends on the front of the forearm and joins either the basilic or the median cubital vein.

In front of the elbow, the cephalic and basilic veins are often connected by the *median cubital vein,* an anastomosing channel that runs upward and medially from the

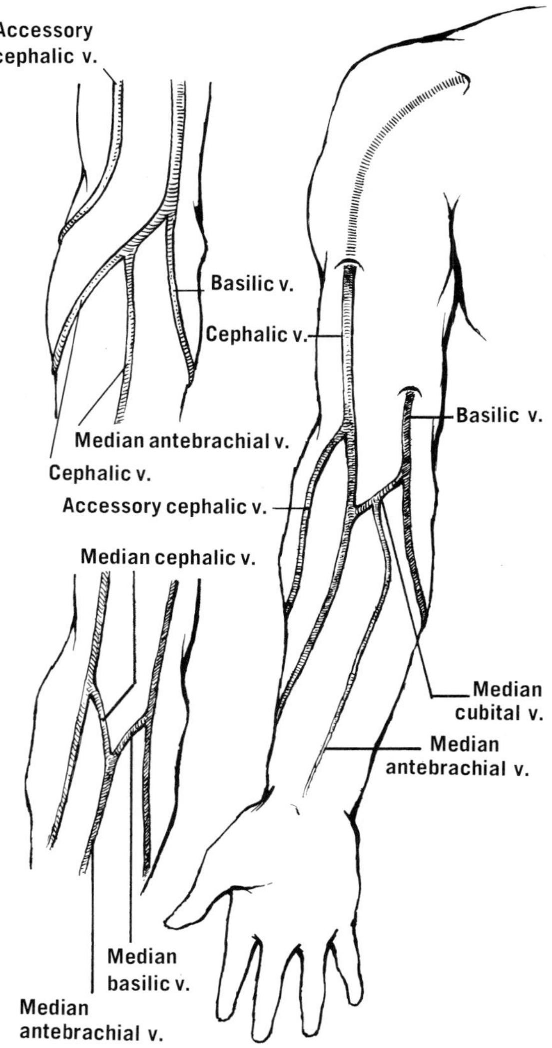

Figure 12–1 Diagrams of some common patterns of superficial veins of the upper limb. Only the larger channels at the elbow are shown; these are the ones most likely to be visible through the skin.

cephalic to the basilic (fig. 12–1; fig. 14–4, p. 127). It crosses superficially the bicipital aponeurosis that separates the vein from the underlying brachial artery and median nerve, and it may run between branches of the medial and lateral antebrachial cutaneous nerves. It usually communicates here with the deep veins of the forearm[2] and often receives a superficial tributary or tributaries from the front of the forearm.

The arrangement of superficial veins in front of the elbow is extremely variable; but certain patterns can be discerned,[3] some of which are shown in figure 12–1. **In the living, certain veins are more prominent, notably the median cubital or one of its tributaries, and these are used for withdrawing blood, for intravenous injections, for blood transfusions, and for the introduction of catheters in cardiac catheterization.**

The valves in the superficial veins of the forearm can be readily demonstrated by Harvey's classical experiment (1628).[4]

"Let an arm be ligated above the elbow in a living human subject as if for a blood-letting. At intervals there will appear . . . certain so to speak nodes and swellings . . . and those nodes are produced by valves, which show up in this way in the outer part of the hand or of the elbow. If by milking the vein downwards with the thumb or a finger, you try to draw blood away from the node or valve, you will see that none can follow your lead because of the complete obstacles provided by the valve; you will also see that the portion of the vein between the swelling and the drawn-back finger has been blotted out, though the portion above the swelling or valve is fairly distended. If you keep the blood thus withdrawn and the vein thus emptied, and with your other hand exert a pressure downwards towards the distended upper part of the valves, you will see the blood completely resistant to being forcibly driven beyond the valve.

". . . the function of the venous valves appears to be . . . to close accurately and thus prevent backflow of the passing blood.

"When, however, you take your finger away, you will see the stretch of vein fill up again from the parts below, . . . Whence it is clearly established that the blood moves in the veins from the parts below to those above and to the heart, and not in the opposite way."

Deep Veins

The deep veins accompany the arteries and have similar names. Like the superficial, they drain ultimately into the axillary vein. The axillary vein (p. 116) begins at the lower border of the teres major as a continuation of the brachial vein. At the outer border of the first rib, it continues as the subclavian vein.

LYMPHATIC DRAINAGE[5]

A rich cutaneous lymphatic plexus on the fingers drains into plexuses on the dorsum and palm of the hand. These in turn drain upward in a medial group of lymphatic vessels that accompanies the basilic vein, and a lateral group that follows the course of the cephalic vein (fig. 12–2). The medial channels may pass through the *cubital*, or *supratrochlear, nodes*, which are one or two superficially placed nodes above the

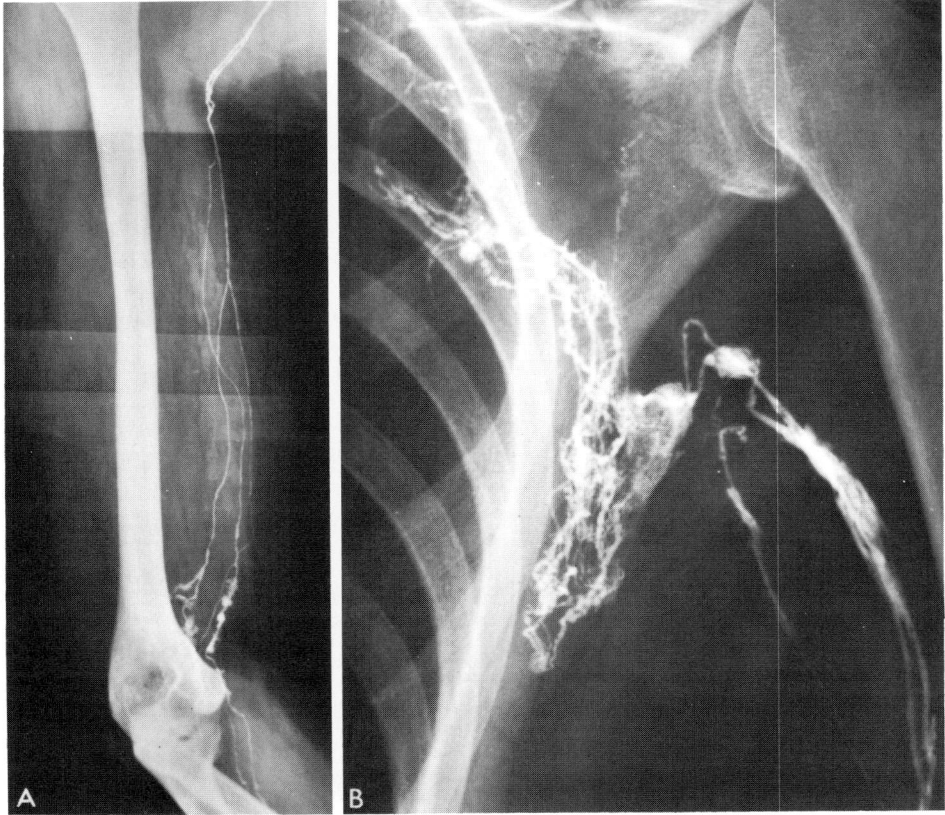

Figure 12–2 A, normal lymphangiogram of the right arm, showing the medial group of collecting vessels. B, normal axillary lymphangiogram (left side), showing the collecting trunks passing first to a small node and then to a large collecting node, from which a network of vessels spreads throughout the axilla. Note the close relationship to the lateral aspect of the thorax. Courtesy of Prof. J. B. Kinmonth, F.R.C.S.,[5] and Edward Arnold.

medial epicondyle, medial to the basilic vein. The channels then ascend in the arm, join the deep lymphatics, and enter the lateral axillary nodes (fig. 12–3). The lateral channels mostly cross the arm and join the medial group. Some continue with the cephalic vein and enter the *deltopectoral nodes*, a group sometimes termed *infraclavicular*, situated below the clavicle on the cephalic vein and draining the "vaccination area." There are also deep lymphatic vessels in the upper limb. These accompany the radial, ulnar, interosseous, and brachial arteries and end in the lateral axillary nodes.

Ultimately (1) the lymphatic vessels of the upper limb, (2) most of those from the breast, and (3) the cutaneous vessels from the trunk above the level of the umbilicus drain into the axillary nodes.

Axillary Nodes

These nodes, important and numerous and lying mostly against the lateral aspect of the thoracic wall (fig. 12–1), are arbitrarily divided into five groups (fig. 12–4).

1. The *lateral nodes* lie behind the axillary vein. They drain the upper limb.
2. The *pectoral nodes* lie along the lateral thoracic veins at the lower border of the pectoralis minor. They drain most of the breast (see fig. 12–5), but any group of axillary nodes may receive direct connections from the mammary gland.
3. The *posterior*, or *subscapular, nodes* lie along the subscapular vein at the lateral border of the scapula. They drain the posterior part of the shoulder region.
4. The *central nodes* lie near the base of the

axilla and receive the lymph from the preceding three groups. They form the largest group, and are the group that is most often palpable.

5. The *apical nodes* lie medial to the axillary vein and above the upper border of the pectoralis minor, in contrast to the preceding groups, which lie below that muscle. They are situated behind the clavipectoral fascia. **The apical nodes receive the lymph from all the other groups and sometimes directly from the breast, and they are drained by two or three subclavian trunks, which enter the jugular-subclavian venous confluence, or join a common lymphatic duct (fig. 31–3, p. 330), or empty into lower deep cervical nodes.**

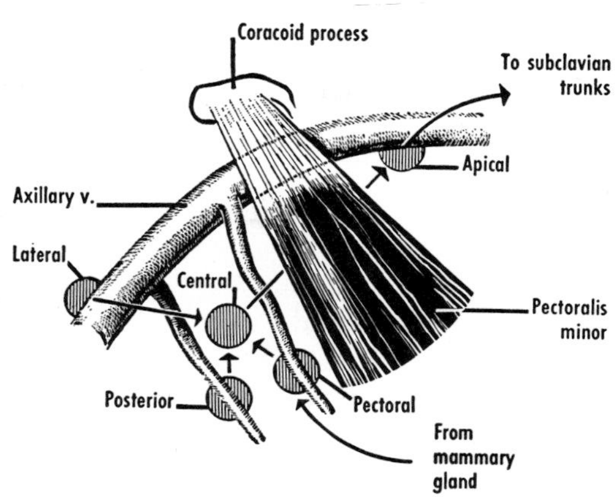

Figure 12–4 Diagram of the axillary nodes. Each circle represents a group of nodes, and the arrows represent the direction of flow in lymphatic vessels.

The lateral and central groups usually consist of 10 to 14 nodes each. The other groups usually consist of only one to seven nodes each.

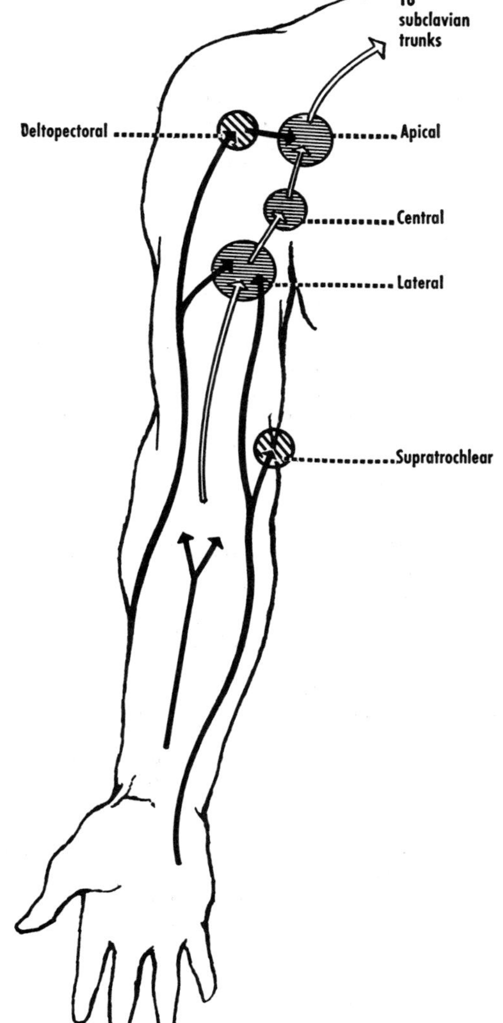

Figure 12–3 Schematic representation of lymphatic drainage of the upper limb. The supratrochlear and deltopectoral nodes receive many superficial lymphatic vessels, and are sometimes called superficial nodes. The lateral nodes are sometimes termed collecting nodes.

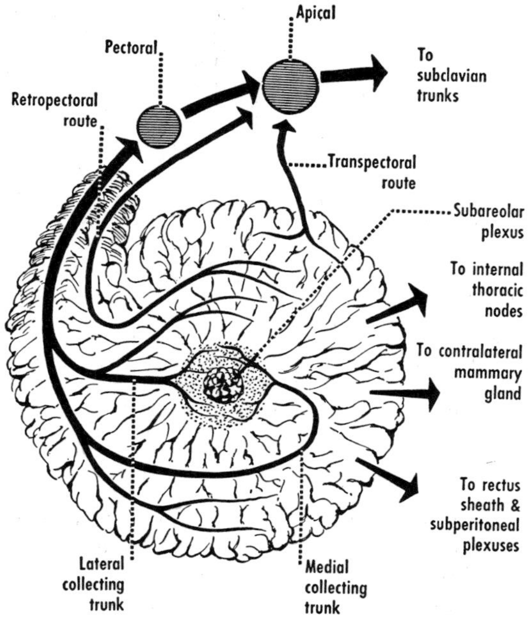

Figure 12–5 Schematic representation of the lymphatic drainage of the right breast.

THE BREAST

The breast overlies the pectoralis major, serratus anterior, and external oblique muscles. It usually extends from the second to the sixth ribs, and from the sternum to the midaxillary line. The breasts more often than not are equal in size. If one is larger and lower, it is usually the right.[6] The extent of glandular tissue (mammary gland) is greater than that of the breast; the former usually extends into the axilla to a varying degree ("axillary tail"), and it may reach the clavicle above, the epigastric fossa below, the median plane medially, and the edge of the latissimus dorsi laterally.[7] **The superolateral quadrant of the breast contains a greater bulk of glandular tissue and is the most frequent site of mammary tumors.**

The *mammary gland* is located between the superficial and deep layers of the subcutaneous tissue. The superficial layer, seldom identified in the cadaver, can be detected surgically. Posterior to the breast are the deep layer of the subcutaneous tissue, the retromammary space (areolar tissue), and the fascia covering the pectoralis major and serratus anterior. Deep projections of mammary parenchyma sometimes penetrate the superficial part of the pectoralis major.

The parenchyma consists of about 15 to 20 compound alveolar glands or *lobes*, each with a separate *lactiferous duct* opening on the nipple. The ducts may have dilatations (*lactiferous sinuses*) near their termination. The ducts can be injected with a radio-opaque medium and then visualized radiographically.[8]

The stroma of the gland consists of adipose and fibrous tissue, intermingled with the epithelial parenchyma to such an extent that blunt dissection is impossible. Anteriorly, the superficial layer of the subcutaneous tissue sends, in the words of Sir Astley Cooper, "large, strong, and numerous fibrous... processes, to the posterior surface of the skin which covers the breast... It is by these processes that the breast is suspended in its situation, and I shall therefore call them ligamenta suspensoria"[9], or *suspensory ligaments.* Their pathological contraction (as in carcinoma) results in characteristic retraction or dimpling of the skin.

The glands develop in the embryo from two vertical ectodermal thickenings, the mammary crests, on the ventrolateral aspects of the trunk. Accessory glands (polymastia) are present on rare occasions and are generally, but not always, situated on the line of the embryonic mammary crest.[10]

The *nipple* is a prominence, often at the level of the fourth intercostal space, which contains the minute openings of the lactiferous ducts of the gland. It is composed mostly of smooth muscle fibers, arranged circularly so that by contraction they compress ducts and cause the nipple to become erect. Longitudinal fibers may depress or retract the nipple. The nipple is surrounded by pigmented skin, the *areola,* which becomes brown during pregnancy and retains that color thereafter. The areola contains sweat glands, sebaceous glands which form tubercles that enlarge during pregnancy, and accessory mammary glands with miniature ducts opening through the areolar epithelium. The nipple is richly supplied with nerve fibers and contains end-organs of varying types, which are located chiefly in the dermis.[11]

Growth

At puberty, in the female, the breasts grow and the areolae enlarge and become more pigmented. The ducts bud and form lobules (gland fields). True secretory alveoli, however, do not develop until pregnancy. The mammary gland involutes after the menopause. The glandular elements decrease or disappear and are replaced by fibrous tissue, and often by fat as well, the amount of the latter varying greatly.

Male Mammary Gland

The male gland sometimes remains as only a group of epithelial cords, but more often a system of ducts develops in it. Little or no fat or fibrous tissue forms, however, and the gland remains small and flat. Some increase in size is common during puberty.

Blood and Nerve Supply

The mammary gland is extremely vascular and is supplied mainly by the perforating branches of the internal thoracic artery (p. 270) and by several branches of the axillary (chiefly the lateral thoracic).[12] The venous drainage is important, not only because the veins indicate the lymphatic pathways, but because carcinoma may metastasize by way of veins. Superficial veins drain through the perforating branches of the internal thoracic or the superficial veins of the lower part of the neck, and can be photographed in infrared light.[13] Deep veins from the breast drain into the perforating tributaries of the internal thoracic, into the axillary, and into the intercostal veins. The connections of the last mentioned with the vertebral plexus of veins (p. 327) provide a route for carcinomatous metastasis to bones and to the nervous system.

Intercostal nerves convey sensory fibers to the skin of the breast and autonomic fibers to the smooth muscle and blood vessels.

Lymphatic Drainage (fig. 12–5)

The lymphatic drainage of the breast is of clinical importance, owing to its role in the spread of malignant tumors. The lymphatic vessels of the skin of the breast, except for that of the areola and nipple, drain into the axillary, deep cervical, and deltopectoral nodes, and also into the parasternal (internal thoracic) nodes of both sides. The lymphatic vessels of the areola and nipple drain with those of the parenchyma of the gland.

The gland is drained by perilobular and subareolar

plexuses. The perilobular plexus drains toward the subareolar plexus, from which collecting trunks arise, among which are lateral and medial trunks that pass around the edge of the pectoralis major, penetrate the axillary fascia to enter the base of the axilla, and end in axillary nodes. Direct routes to nodes at the apex of the axilla are sometimes present (e.g., through or between the pectoral muscles).

The axillary nodes act as a series of filters between the breast and the venous circulation. Carcinoma cells that enter a lymphatic vessel usually have to pass through two or three groups of nodes before reaching the venous circulation.

Collecting vessels from the central and medial parts of the breast follow the perforating blood vessels through the pectoralis major, and end in the parasternal (internal thoracic) nodes behind the internal intercostal muscles and in front of the endothoracic fascia. These nodes, commonly only 1 or 2 mm in diameter, are about three to five in number on each side. Lymphatic routes across the median plane, either in the skin or in the pectoral fascia, are sometimes present. These may account for metastasis of breast carcinoma to the opposite axilla. Still other lymphatic vessels may reach the plexus on the sheath of the rectus abdominis and the subperitoneal and subphrenic plexuses.

REFERENCES

1. G. Piolino, Arch. Anat., Strasbourg, *40*:55, 1957.
2. G. Winckler, Arch. Anat., Strasbourg, *36*:177, 1953.
3. J. F. Doyle, Irish J. med. Sci., *1*:131, 1968. A. Bouchet *et al.*, Bull. Ass. Anat., *56*:971, 1972. A. Halim and S. H. M. Abdi, Anat. Rec., *178*:631, 1974.
4. William Harvey, *Movement of the Heart and Blood in Animals: An Anatomical Eassy,* trans. by K. J. Franklin, Blackwell, Oxford, 1957.
5. J. B. Kinmonth, *The Lymphatics,* Arnold, London, 1972.
6. B. Skerlj, Anthrop. Anz., *12*:304, 1935.
7. N. F. Hicken, Arch. Surg., Chicago, *40*:6, 1940.
8. N. F. Hicken *et al.*, Amer. J. Roentgenol., *39*:321, 1938. J. N. Wolfe, *Mammography,* Thomas, Springfield, Illinois, 1967. A. Willemin, *Mammographic Appearances,* Karger, Basel, 1972.
9. A. Cooper, *The Anatomy of the Breast,* Longman, Orme, Green, Brown, and Longmans, London, 1840.
10. H. Speert, Quart. Rev. Biol., *17*:59, 1942. R. Purves and J. A. Hadley, Brit. J. Surg., *15*:279, 1927.
11. E. P. Cathcart, F. W. Gairns, and H. S. D. Garven, Trans, R. Soc. Edinb., *61*:699, 1948. M. R. Miller and M. Kasahara, Anat. Rec., *135*:153, 1959.
12. B. J. Anson, R. R. Wright, and J. A. Wolfer, Surg. Gynec. Obstet., *69*:468, 1939. J. W. Maliniac, Arch. Surg., Chicago, *47*:329, 1943.
13. L. C. Massopust and W. D. Gardner, Surg. Gynec. Obstet., *91*:717, 1950. K. Bowes, Ann. R. Coll. Surg. Engl., *6*:187, 1950.

GENERAL READING

Dabelow, A., Die Milchdrüse, in *Handbuch der mikroskopischen Anatomie des Menschen,* ed. by W. von Möllendorff and W. Bargmann, Springer, Berlin, Vol. 3, Part 3, 1927.

Haagensen, C. D., *Diseases of the Breast,* Saunders, Philadelphia, 2nd ed., 1971. Includes structure and function of the breast.

SHOULDER AND AXILLA 13

MUSCLES OF PECTORAL REGION

The muscles of the pectoral region are the pectoralis major, pectoralis minor, subclavius, and serratus anterior. They form a ventrally situated group that connects the upper limb with the thoracic skeleton. The pectoralis major is the only one of this group that is inserted into the humerus. The others are inserted into the shoulder girdle. All four muscles are supplied by branches of the brachial plexus.

The subcutaneous tissue of the chest is continuous with that of the neck, the upper limb, and the abdomen. Its upper part may have a slightly reddish color owing to fibers of the platysma deep to it. Many fibers of the platysma take origin from the skin. The subcutaneous tissue usually contains fat, and it encloses the mammary gland.

The fascia of the pectoral region is attached to the clavicle and sternum. It invests the pectoralis major *(pectoral fascia)* and, at the inferolateral margin of that muscle, continues to the latissimus dorsi. Here it divides into two layers that ensheathe the latissimus dorsi and are attached behind to the spines of the thoracic vertebrae. In the interval between the pectoralis major and the latissimus dorsi, the fascia is thicker and, as the *axillary fascia,* forms the floor of the axilla. A layer that ascends from the axillary fascia and ensheathes the pectoralis minor is sometimes called the suspensory ligament of the axilla, because traction by it produces the hollow of the armpit when the arm is abducted. This layer is continued upward as the *clavipectoral fascia;* it ensheathes the subclavius and becomes attached to the clavicle (fig. 13–1).

The clavipectoral fascia blends medially with the fascia covering the first two intercostal spaces, and is attached to the first rib. Laterally the fascia extends to the coracoid process and joins the fascia of the biceps and coracobrachialis. The part between the first rib and the coracoid process is often thickened to form the costocoracoid ligament. Behind, the clavipectoral fascia blends with the sheath of the axillary vessels. It is pierced by the cephalic vein, thoracoacromial artery, and lateral pectoral nerves.

Pectoralis Major. The pectoralis major is a large, fan-shaped, multilaminar muscle, which arises from the anterior surface of the medial half of the clavicle, from the anterior surface of the sternum and the first six costal cartilages, and from the aponeurosis of the external oblique muscle of the abdomen. The bilaminar tendon of insertion is attached to the crest of the greater tubercle of the humerus, the two laminae being arranged in a **U**-shape.[1]

The cephalic vein ascends in the interval between the pectoralis major and the deltoid. These muscles are separated at their clavicular attachments by the deltopectoral triangle, which varies from little more than a gap for the cephalic vein to an area several centimeters in width. The rounded, lower border of the pectoralis major forms the anterior axillary fold.

NERVE SUPPLY AND ACTION. It is supplied by the lateral and medial pectoral nerves. It adducts the arm; the clavicular

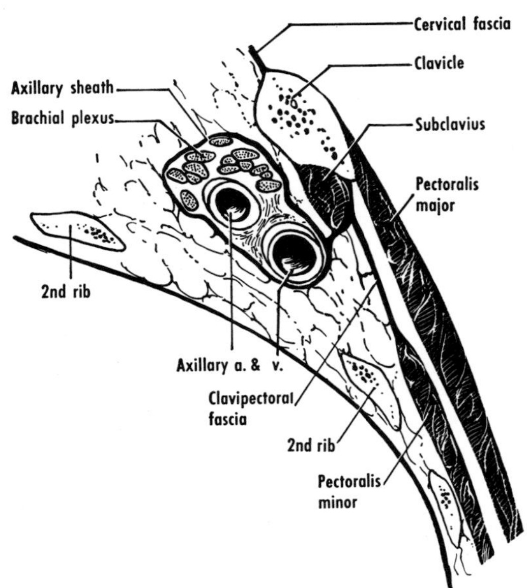

Figure 13–1 Diagram of the clavipectoral fascia and its relation to the axillary sheath. Sagittal section.

part also rotates the arm medially[2] and flexes it, whereas the sternocostal part depresses the arm and shoulder. In climbing, when the arms are fixed, this muscle draws the body upward. It aids also in throwing, pushing, and shoveling. Owing to its attachment to costal cartilages, the pectoralis major is potentially capable of elevating the upper ribs during forced inspiration, but it apparently does not do so during ordinary breathing. This potential action, however, is made use of when the arms are elevated during artificial respiration (p. 301).

Pectoralis Minor. The pectoralis minor lies behind the pectoralis major and in front of the second part of the axillary artery. It arises from the front of the external surfaces of the second to the fifth ribs. It runs upward and laterally and is inserted into the coracoid process.

NERVE SUPPLY AND ACTION. It is supplied by the medial and lateral pectoral nerves. The muscle probably depresses the point of the shoulder.[3]

Subclavius. The subclavius arises by a tendon from the junction of the first rib with its costal cartilage, and it is inserted by muscle fibers into a groove on the lower surface of the clavicle.

NERVE SUPPLY AND ACTION. The nerve to the subclavius arises from the upper trunk of the brachial plexus. It may constitute an accessory phrenic nerve (p. 333). The muscle may assist in depressing the lateral part of the clavicle.

Serratus Anterior. The serratus anterior forms the medial wall of the axilla (see fig. 13–6). It is a large muscle that arises by a series of slips from the external surfaces of the upper eight ribs. It is inserted into the costal surface of the scapula (1) at the superior angle, (2) along the medial border, and (3) at the inferior angle. About half the muscle is inserted at the inferior angle. The lower slips of origin interdigitate with those of the external oblique.

NERVE SUPPLY AND ACTION. It is supplied by the long thoracic nerve, which runs on its superficial aspect. It is a powerful muscle, which rotates the scapula so that the inferior angle moves laterally. It thus plays an important role in abduction of the arm and elevation of the arm above the horizontal. It pulls the scapula forward in throwing and pushing. It is not a muscle of forced respiration.[4] Paralysis of the serratus anterior is characterized by winging of the scapula, that is, the medial border of the bone stands away from the chest wall. Elevation of the arm above the horizontal is then virtually impossible.

Variations and Occasional Muscles[5]

The *sternalis* is a muscle that is occasionally present on the origin of the sternocostal head of the pectoralis major. Variable muscular slips are sometimes found in the region of the axilla and may form bands, termed *axillary arches,* which extend between the latissimus dorsi and pectoralis major. Muscular bundles may extend from the ribs to the medial epincondyle *(chondroepitrochlearis),* or from the latissimus dorsi to the medial epicondyle *(dorsoepitrochlearis).* Other variations have also been described. For example, the pectoral muscles are absent in rare instances.[6]

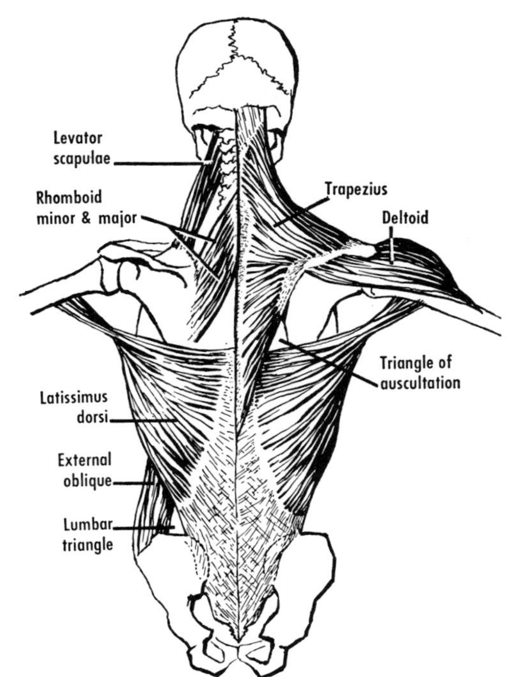

Figure 13–2 The superficial muscles of the back. Note how the latissimus dorsi covers the inferior angle of the scapula. The left trapezius is removed to show the rhomboids and levator scapulae. Based on Mollier.[7]

SUPERFICIAL MUSCLES OF BACK

This dorsal group of muscles connects the upper limb to the vertebral column. It includes the latissimus dorsi and trapezius, located superficially, and the underlying levator scapulae, rhomboid minor, and rhomboid major (fig. 13–2). The latissimus dorsi is inserted into the humerus, the others into the shoulder girdle. These muscles, although topographically related to the back, receive their nerve supply from the ventral rami of cervical nerves. The trapezius also receives fibers from the accessory nerve.

Trapezius. The trapezius is large, triangular, and superficially placed on the back of the neck and thorax. It is responsible for the sloping ridge of the neck. The trapezius muscles of the two sides together form a trapezoid. Each usually arises from the spine of the seventh cervical vertebra, from the spines and supraspinous ligaments of all the thoracic vertebrae, and from the ligamentum nuchae, and often from the superior nuchal line and external protuberance of the occipital bone.[8] The inferior limit of its origin is variable and the two muscles are often asymmetrical in this respect. In the region of the seventh cervical vertebra, the aponeurosis of origin is wider than it is elsewhere, and the absence of the more bulky muscle fibers here is made evident superficially by a depressed area.

The trapezius has a continous insertion on two bones. Its uppermost part is attached to the lateral third of the clavicle, its middle part to the acromion and the crest of the spine of the scapula, and its lowermost part to the tubercle of the crest. This last part is usually separated from the spine by a bursa.

NERVE SUPPLY AND ACTION. It is innervated by the accessory nerve and, by way of the cervical plexus, by the third and fourth cervical nerves. Clinical evidence indicates that in man the cervical components, usually regarded as wholly sensory, may contain some motor fibers. The uppermost fibers of the trapezius, together with the levator scapulae, elevate the shoulder and, acting with those of the opposite side, keep the shoulders braced by pulling the scapulae backward.[9] Their weakness results in drooping shoulders. The middle and lower parts of the muscle act with the rhomboids in retracting and steadying the scapula (squaring the shoulders). The trapezius also has an important function in rotating the scapula during abduction and elevation of the arm (p. 120).

Latissimus Dorsi. The latissimus dorsi is large, triangular, and superficially placed, except in its uppermost part, where it is covered by the trapezius. Near its lower attachment it forms the posterior boundary of the lumbar triangle (fig. 13–2; see also p. 352). The latissimus dorsi and teres major form the posterior axillary fold and contribute to the posterior wall of the axilla.

The latissimus dorsi arises (1) from the spines of the lower six thoracic vertebrae, (2) indirectly from the spines of the lumbar and sacral vertebrae through its attachment to the posterior layer of the thoracolumbar fascia, and (3) from the iliac crest. As the muscle passes toward its insertion, it receives slips from the outer surfaces of the lower three of four ribs (these slips interdigitate with the lower ones of the external oblique) and, usually, from the inferior angle of the scapula. The muscle then spirals around the lower edge of the teres major and is inserted into the floor of the intertubercular groove. The tendons of the latissimus dorsi and teres major are commonly fused.

NERVE SUPPLY AND ACTION. It is supplied by the thoracodorsal nerve. It is a powerful adductor and extensor. It plays a considerable role in the downstroke of the arm in swimming, and it is also used in rowing, climbing, hammering, and in supporting the weight of the body on the hands. Its scapular attachment may help to keep the inferior angle of the scapula against the chest wall. Through its attachment to ribs, it is an accessory muscle of respiration. It can be felt to contract when coughing.

Levator Scapulae, Rhomboid Minor, and Rhomboid Major. The levator scapulae, thin and straplike, arises from the posterior tubercles of the transverse processes of the first four cervical vertebrae. It is inserted into the medial border of the scapula at the level of and above the spine. The insertions of the levator scapulae and the two rhomboids are usually continous along the medial border of the scapula.[10]

The rhomboid major and minor are often fused. The minor is the uppermost slip; it arises from the spines of the seventh cervical and first thoracic vertebrae and from the lower part of the ligamentum nuchae, and is inserted into the medial border of the scapula at the level of the root of the spine. The major arises from the spines and supraspinous ligaments of the second to fifth thoracic vertebrae and is usually inserted into the medial border of the scapula below the spine, occasionally by way of a tendinous arch.

NERVE SUPPLY AND ACTION. The levator scapulae is supplied by small branches from the third and fourth cervical nerves (C3, 4) that enter its superficial aspect above. The dorsal scapular nerve may give branches to its deep surface as it descends to supply both rhomboids.

The levator scapulae elevates the scapula and may act in concert with the trapezius in shrugging the shoulders. However, it may also act with the rhomboids, which retract and fix the scapula.

Triangle of Auscultation

The upper border of the latissimus dorsi is overlapped by the lateral border of the trapezius. The angle thus formed is converted into a triangle by the medial border of the underlying scapula. This interval, the floor of which is formed by the rhomboid major, is called the triangle of auscultation (fig. 13–2).

MUSCLES OF SHOULDER

This group of muscles consists of the deltoid, supraspinatus, infraspinatus, teres minor, teres major, and subscapularis. They arise from the shoulder girdle and are inserted into the humerus. All are supplied by ventral rami of the fifth and sixth cervical spinal nerves through branches of the brachial plexus. The fascia of the shoulder region is tough, and is characteristically firmly bound to the muscles it invests.

Deltoid. The deltoid is a thick, coarse-textured muscle, superficially placed and responsible for the characteristic roundess of the shoulder.

This muscle, the origin of which embraces the insertion of the trapezius (fig. 13–3), arises from the front of the superior surface of the lateral third of the clavicle, from the lateral margin and adjoining upper surface of the acromion, and from the inferior lip of the crest of the spine. Its clavicular and scapular parts converge to be inserted, together with the acromial part, into the deltoid tuberosity of the humerus.

NERVE SUPPLY AND ACTION. It is supplied by the axillary nerve. The acromial or middle part is a powerful abductor of the arm, but its line of pull is such that when acting alone it abducts the arm in the plane of the scapula. The spinous or posterior part of the deltoid extends the arm and rotates it laterally. When the posterior and middle parts act together, they abduct the arm in a coronal plane. The clavicular or anterior

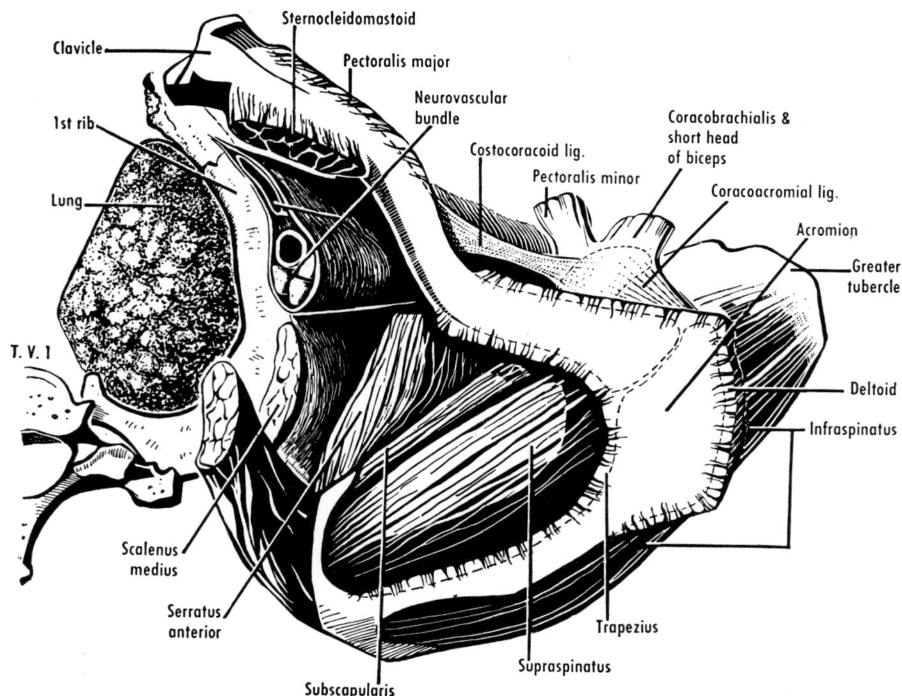

Figure 13–3 The shoulder from above. Note the relation of the neurovascular bundle (subclavian vessels and brachial plexus) to the clavicle. Note also that the insertion of the trapezius in the concavity formed by the spine, acromion, and clavicle is embraced by the origin of the deltoid from the convexity.

part of the deltoid flexes the arm and rotates it medially. When all three parts of the deltoid contract simultaneously, the arm is abducted or elevated in the plane of the scapula, and the clavicular and spinous parts counterbalance each other. The question of planes of movement at the shoulder joint is considered on page 118, where the important roles of other muscles are also considered.

The deltoid has an important function as a stabilizer in many movements, particularly those involving horizontal movements, for example, drawing a line across a blackboard or sliding a book along a shelf. These horizontal movements are sometimes termed horizontal abduction or adduction. The deltoid may directly control adduction, or may modify the actions of the adductors (see figs. 4–1 and 4–2, pp. 25, 27).

Supraspinatus. The supraspinatus arises from the medial two-thirds of the supraspinous fossa and the overlying fascia. Its tendon of insertion blends with the capsule of the shoulder joint (fig. 13–4) and is attached to the highest of three facets on the greater tubercle of the humerus. The tendon forms the floor of the subdeltoid bursa.

NERVE SUPPLY AND ACTION. It is supplied by the suprascapular nerve. It aids the deltoid in abduction of the arm. Both muscles contract simultaneously when abduction begins. The supraspinatus, the infraspinatus, the teres minor, and the subscapularis keep the head of the humerus in place and prevent it from being pulled up against the acromion by the deltoid. When the deltoid is paralyzed, the supraspinatus usually cannot fully abduct the arm, and if the supraspinatus is paralyzed, normal abduction may be difficult or impossible.

Infraspinatus. The infraspinatus is covered in its upper part by the deltoid laterally and the trapezius medially. The muscle arises from the medial two-thirds of the infraspinous fossa and from the lower surface of the spine of the scapula. Its tendon blends with the capsule of the shoulder joint and is inserted into the middle facet on the greater tubercle of the humerus. A bursa

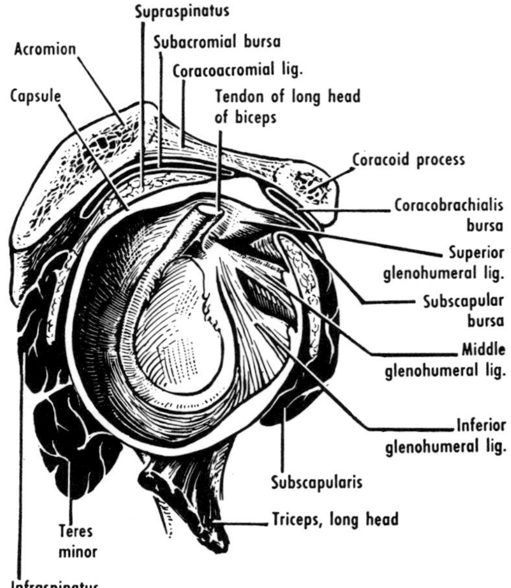

Figure 13–4 The musculotendinous cuff and the capsule of the shoulder joint, shown after cutting the cuff and removing the humerus. The tendons of the supraspinatus, infraspinatus, teres minor, and subscapularis blend with the capsule and form the cuff (also called the rotator cuff). This is incomplete below. The figure also shows three glenohumeral ligaments; these are variable in size and position. The subscapular bursa communicates with the joint cavity between the superior and middle glenohumeral ligaments.

is usually present between this muscle and the spine of the scapula near the spinoglenoid notch, and another may be present between the tendon and the joint capsule.

Teres Minor. The teres minor, which may be inseparable from the infraspinatus, arises from the lateral margin of the infraspinous fossa. Its tendon of insertion is first adherent to the capsule of the shoulder joint and is then attached to the lowest facet on the greater tubercle of the humerus and the area immediately below this.

NERVE SUPPLY AND ACTION. The infraspinatus is supplied by the suprascapular nerve, the teres minor by the axillary nerve. Both muscles rotate the arm laterally and both aid in keeping the head of the humerus in place during abduction.

Teres Major. The teres major arises from the dorsal surface near the inferior angle, and is inserted into the crest of the lesser tubercle below the insertion of the subscapularis. The tendons of the latissimus dorsi and teres major are commonly fused. The teres major, together with the latissimus dorsi and subscapularis, forms the posterior wall of the axilla.

NERVE SUPPLY AND ACTION. It is supplied by the lower subscapular nerve. It acts with the latissimus dorsi in adducting the arm. It is probable that the action of the teres major is largely static; that is, when the scapula is fixed, it helps to maintain the arm in adduction against resistance.[11]

Subscapularis. The subscapularis forms a part of the posterior wall of the axilla. It arises from almost the whole of the subscapular fossa. Its tendon of insertion passes in front of the capsule of the shoulder joint, to which it is adherent, and is attached to the lesser tubercle of the humerus and its crest.

NERVE SUPPLY AND ACTION. It is supplied by the subscapular nerves from the posterior cord. It is a strong medial rotator of the arm and helps to hold the head of the humerus in the glenoid cavity.

Triangular and Quadrangular Spaces

The three-sided interval bounded by the teres minor and subscapularis above, the teres major below and the surgical neck of the humerus laterally, is divided longitudinally by the long head of the triceps into a triangular space medially, and a quadrangular space laterally (fig. 13–5). The circumflex scapular vessels pass through the triangular space, and the axillary nerve and posterior circumflex humeral vessels traverse the quadrangular space.

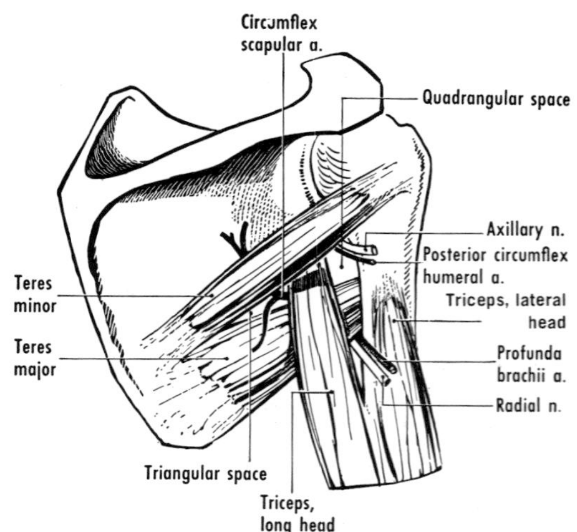

Figure 13–5 Schematic drawing of the triangular and quadrangular spaces.

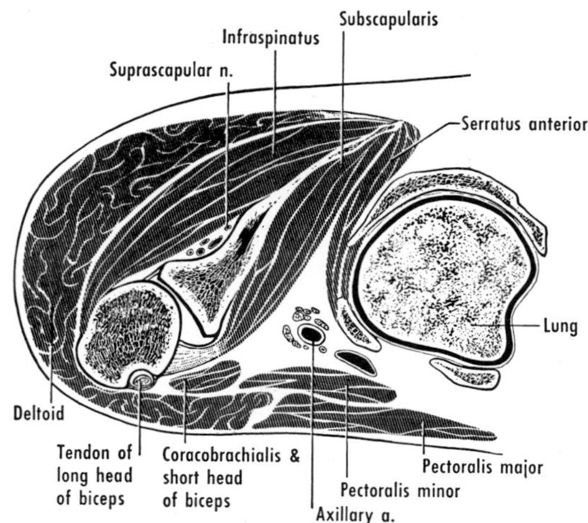

Figure 13–6 Horizontal section through the lower part of the right shoulder. Note the divergence of the subscapularis and serratus anterior to form the boundaries of the axilla. Fasciae are omitted. Based on Symington.[12]

AXILLA

The axilla is a pyramidal interval between the arm and the chest wall (fig. 13–6). Its base, formed by the axillary fascia, extends between the inferolateral margins of the pectoralis major and latissimus dorsi muscles. These margins form prominent anterior and posterior axillary folds in the living. Its apex is the interval between the posterior border of the clavicle, the superior border of the scapula, and the external border of the first rib. Through it the axillary vessels and their accompanying nerves pass from the neck toward the arm. The anterior wall of the axilla is formed by the pectoralis major and minor. The posterior wall is formed by the subscapularis, teres major, and latissimus dorsi. The axilla is bounded medially by the upper ribs and their intercostal muscles, and by the serratus anterior, and is limited laterally by the intertubercular groove of the humerus. The biceps and coracobrachialis descend between the anterior and posterior walls of the axilla.

The axilla contains the axillary artery and vein, a part of the brachial plexus and its branches, the lateral cutaneous branches of some of the intercostal nerves, the long thoracic nerve, the intercostobrachial nerve, a part of the cephalic vein, and the axillary lymph nodes. A downward prolongation of the prevertebral fascia forms the axillary sheath, which encloses the axillary vessels and adjacent nerves (fig. 13–1).

NERVES OF UPPER LIMB

Brachial Plexus

The nerves to the upper limb arise from the brachial plexus,[13] a large and very important structure situated partly in the neck and partly in the axilla. The brachial plexus is formed by the union of the ventral rami of the lower four cervical nerves (C5, 6, 7, 8) and the greater part of the ventral ramus of the first thoracic nerve (T1), but frequently receives contributions from the fourth cervical or the second thoracic nerve also, or from both. When the fourth cervical contribution is large and the first thoracic contribution is small, the plexus is described as being *prefixed* in relation to the vertebral column. By contrast, when the contributions of the first and second thoracic nerves are large, the plexus is termed *postfixed.* When the first ribs are rudimentary, the second thoracic nerve gives a large contribution to the brachial plexus.[14]

The brachial plexus next descends in the lower part of an area of the neck known as the posterior triangle (p. 680). Here it is situated above the clavicle, and posterior and lateral to the sternocleidomastoid muscle. It lies above and behind the third part of the subclavian artery and is crossed by the inferior belly of the omohyoid muscle. In this situation, the plexus may be injected with a local anesthetic. **The brachial plexus can be palpated in the living both above and below the omohyoid, in the angle between the clavicle and sternocleidomastoid. From the point of view of surface anatomy, the brachial plexus in the neck lies below a line from the posterior margin of the sternocleidomastoid at the level of the cricoid cartilage to the midpoint of the clavicle.**

The chief structures superficial to the brachial plexus in the neck are the platysma, the supraclavicular nerves, the external jugular vein, the inferior belly of the omohyoid, and the transverse cervical and descending scapular arteries (p. 705), which usually cross or pass between the trunks of the brachial plexus.

The brachial plexus descends behind the concavity of the medial two-thirds of the clavicle (see fig. 13–3), and accompanies the axillary artery under cover of the pectoralis major. Its cords are arranged around the second part of that vessel behind the pectoralis minor (fig. 13–7). The plexus is enclosed with the axillary vessels in the axillary sheath. At the inferolateral border of the pectoralis minor, in front of the subscapularis, it gives off its terminal branches.

Although variations occur in the manner in which the parts of the plexus are formed, a common arrangement is shown in figures 13–8 and 13–9. The ventral rami of the fifth and sixth cervical nerves unite to form the *upper trunk,* the seventh remains single as the *middle trunk,* and the eighth cervical and first thoracic unite to form the *lower trunk* (fig. 27–6, p. 274). Each trunk then divides into an anterior and a posterior division, and these divisions provide a general indication of which fibers are going to the front of the limb and which to the back. The anterior divisions of the upper and middle trunks unite to form the *lateral cord.* The anterior division of the lower trunk remains single as the *medial cord.* The three posterior divisions unite to form the *posterior cord.* The three cords, largely behind the first part of the axillary artery, tend to wind below the second portion of the vessel, and each now comes to occupy that side of the artery after which it is named—lateral, medial, posterior. Finally, at the inferolateral border of the pectoralis minor, the cords divide into terminal branches. Each terminal branch, owing to its complex formation in the plexus, contains fibers derived from several spinal nerves.

The brachial plexus is thus seen to be composed successively of (1) ventral rami and trunks situated in the neck in relation to the subclavian artery, (2) divisions that are usually described as being located approximately behind the clavicle, and (3) cords and branches situated in the axilla in relation to the axillary artery. The various nerve bundles have descended from the neck and come to meet, and thereafter accompany, the more superficially placed arterial tube that has ascended from the thorax. The lower trunk lies on the first rib behind the subclavian artery. Injuries to the brachial plexus are of great importance and are discussed on page 766.

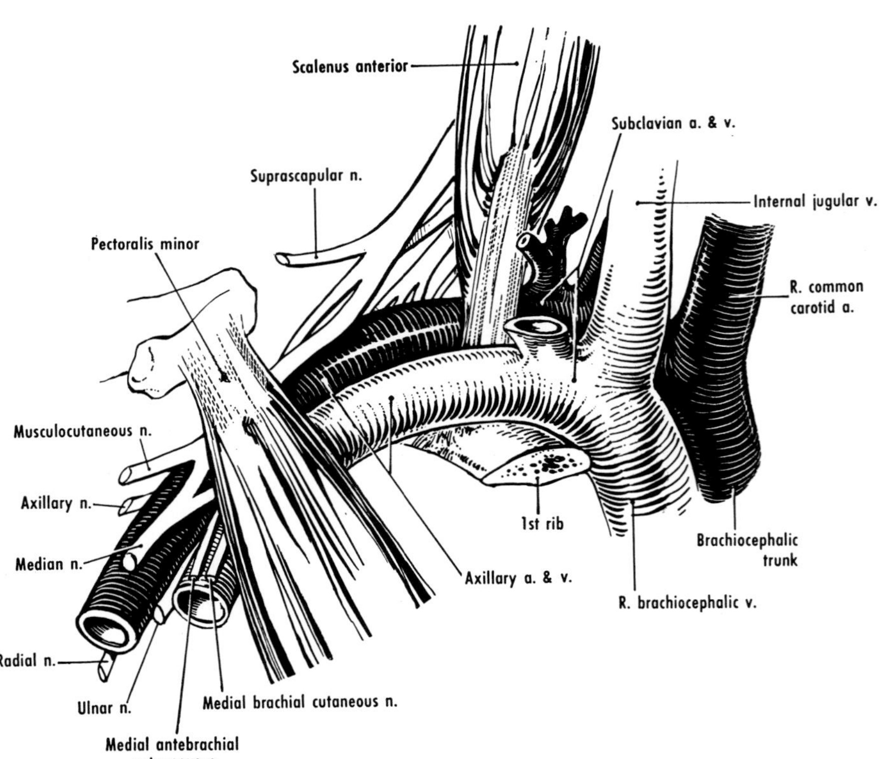

Figure 13–7 The relationships of the brachial plexus and the axillary vessels. Note that the subclavian artery and vein are separated by the scalenus anterior.

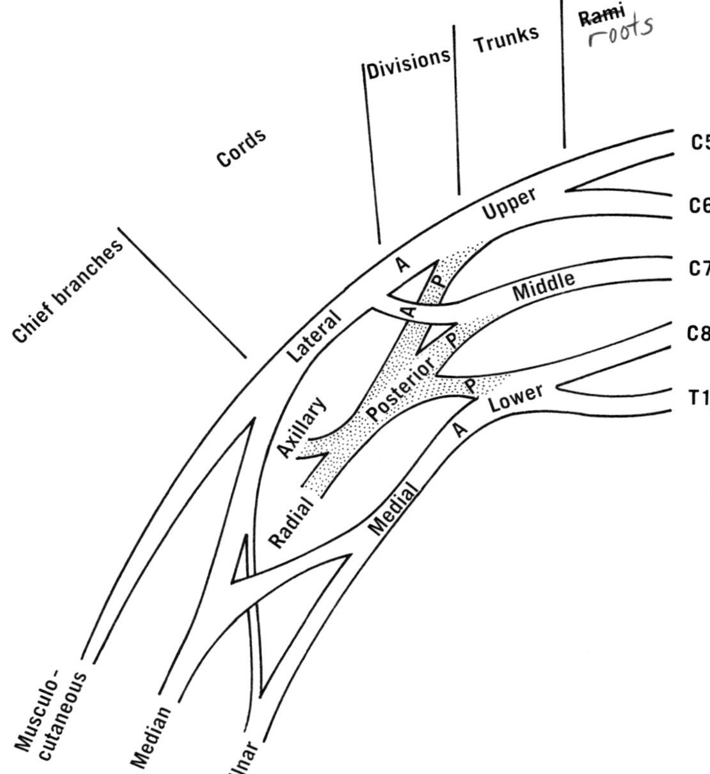

Figure 13–8 Schematic diagram of the anterior aspect of the right brachial plexus. *A*, anterior divisions of trunks. *P*, posterior divisions of trunks. The muscles to which the rami are ultimately distributed are listed in Table 13–1, p. 113.

Branches of the Ventral Rami. The branches of the ventral rami of nerves participating in the brachial plexus include the dorsal scapular nerve, the long thoracic nerve, and small twigs to the scalene and longus colli muscles.

The *dorsal scapular nerve* (chiefly C5), pierces the scalenus medius, runs deep to the levator scapulae (which it sometimes supplies), and enters the deep surface of the rhomboids.

The *long thoracic nerve* usually arises by three roots (C5, 6, 7). The upper two roots pierce, and the lowest passes in front of, the scalenus medius. The nerve descends behind the brachial plexus and the first part of the axillary artery, and runs on the external surface of the serratus anterior, to which it gives numerous branches.

Branches of the Trunks. The upper trunk gives off two muscular branches, the nerve to the subclavius and the suprascapular nerve. The lower trunk may give origin to the medial pectoral nerves, and the anterior divisions of the upper and middle trunks may give origin to the lateral pectoral nerves.

The *nerve to the subclavius* (chiefly C5) descends behind the clavicle and in front of the brachial plexus and third part of the subclavian artery, and ends in the subclavius. It also supplies the sternoclavicular joint. Frequently it contributes fibers to the phrenic nerve by means of a communicating branch (accessory phrenic nerve, p. 333). Rarely, the entire phrenic nerve may arise from this branch.

The *suprascapular nerve* (C5, 6) passes laterally and backward to the scapular notch, through which it passes, below the superior transverse scapular ligament. It supplies the acromioclavicular and shoulder joints and the supraspinatus muscle, and, accompanied by the suprascapular artery, it passes through the spinoglenoid notch to end in the infraspinatus.

Branches of the Cords. These have

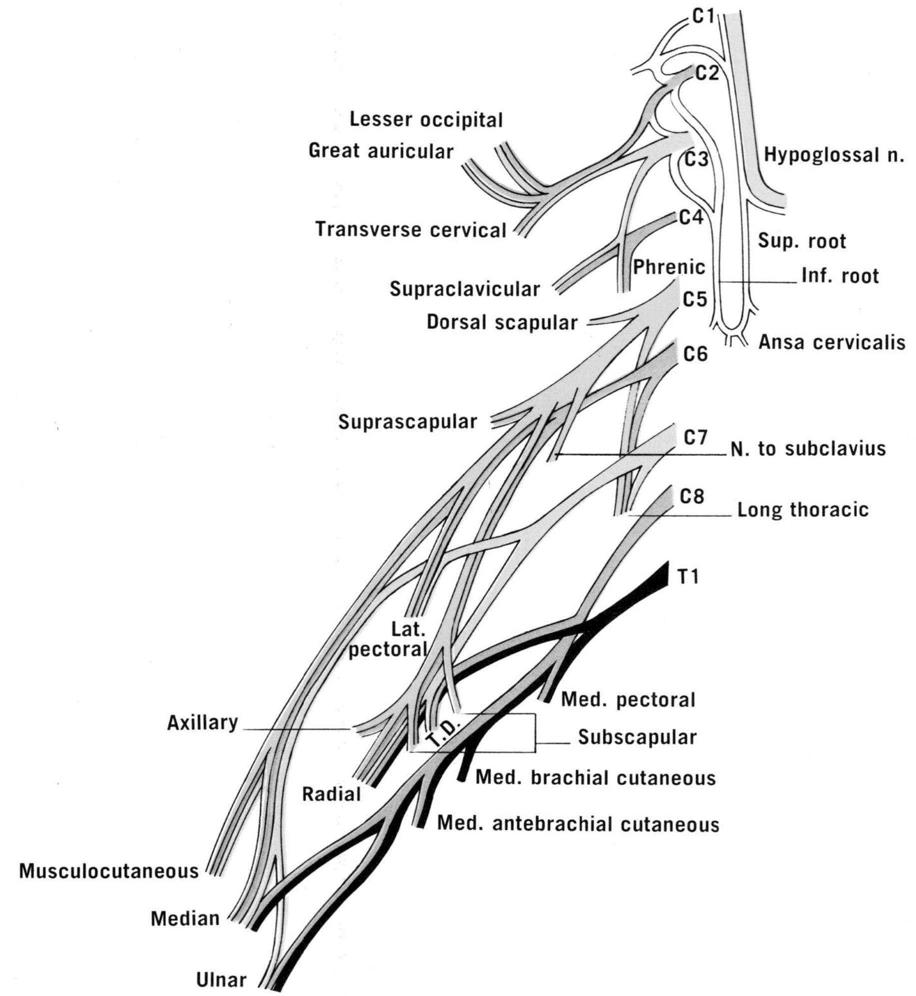

Figure 13–9 Simplified scheme of cervical and brachial plexuses, showing the distribution of nerve fibers in the roots of origin. (Brachial plexus based partly on Seddon.[15]) T.D., thoracodorsal.

the same relationship to the third part of the axillary artery as the cords from which they are derived have to the second portion of that vessel. The branches of the cords are as follows:

From the lateral cord: lateral pectoral, musculocutaneous, lateral head of median. The ulnar nerve usually has a lateral head also.

From the posterior cord: upper subscapular, thoracodorsal, lower subscapular, radial, axillary, articular.

From the medial cord: medial pectoral, medial brachial cutaneous, medial antebrachial cutaneous, ulnar, medial head of median.

The most important branches of the brachial plexus are the median, ulnar, radial, musculocutaneous, and axillary nerves. The first four of these continue distally into the limb, and the details of their course and distribution are given later.

The *median nerve*[16] ([C5], C6, 7, 8, T1) (see p. 768) arises by lateral and medial roots from the lateral and medial cords, respectively. The heads unite either lateral or anterior to the axillary artery, but occasionally the lateral head may be carried as far as the arm by the musculocutaneous nerve. The level at which the heads unite varies, and either head may be double. **The median nerve is ultimately distributed to the skin on the**

TABLE 13–1 Segmental Innervation of Muscles of Limbs*

Upper Limb		*C3*	*4*	*5*	*6*	*7*	*8*	*T1*
Levator scapulae, trapezius		■	■					
Rhomboids				■				
Deltoid, supraspinatus, infraspinatus, teres major, subscapularis, biceps, brachialis, brachioradialis, supinator				■	■			
Serratus anterior, pectoralis major (lat.)				■	■	■		
Pronators					■			
Triceps					■	■	■	
Most extensors of hand and fingers						■	?	
Latissimus dorsi						■	■	
Flexores digitorum superficiales et profundi						■	■	■
Pectoralis major (med.), palmaris longus, flexor pollicis longus							■	■
Muscles of hand							?	■

Lower Limb	*L2*	*3*	*4*	*5*	*S1*	*2*		
Psoas major, sartorius, pectineus, adductor longus	■	■						
Iliacus, quadriceps femoris, adductor brevis, gracilis	■	■	■					
Obturator externus		?	■					
Adductor magnus		■	■	■	■			
Tensor fasciae latae			■	■	?			
Glutei medii et minimi, plantaris, popliteus, muscles of front of leg			■	■	■			
Muscles of lateral side of leg			?	■	■			
Quadratus femoris, semimembranosus, tibialis posterior			?	■	■	?		
Semitendinosus			?	■	■	■		
Gluteus maximus, obturator internus, biceps (long head), flexor digitorum longus, flexor hallucis longus				■	■	■		
Muscles of foot: mostly one or more of these nerves:				■	■	■		
Piriformis				?	■	■		
Gastrocnemius, soleus					■	■		

*Adapted from various sources. It should be emphasized that, in the case of many muscles, some of these figures are uncertain.

front of the lateral part of the hand, to most of the flexor muscles on the front of the forearm, to most of the short muscles of the thumb, and to the elbow joint and many of the joints of the hand.

The *ulnar nerve* (C7, 8, T1) (pp. 124 and 769) arises from the medial cord and usually also has a lateral root (containing C7 fibers) that springs from the lateral root of the median nerve, or from the lateral cord. The ulnar nerve at its origin lies between the axillary artery and vein, and in front of the teres major. **The ulnar nerve is ultimately distributed to the skin on the front and back of the medial part of the hand, to some of the flexor muscles on the front of the forearm, to**

many of the short muscles of the hand, and to the elbow joint and many of the joints of the hand.

The *musculocutaneous nerve* (C5, 6, 7) (see p. 768) usually arises from the lateral cord and pierces the coracobrachialis, but it is extremely variable.[17] It may carry a part or all of the lateral head of the median nerve and give these fibers to the medial head of the median nerve by a communication in the arm. In other words, the lateral cord divides lower than usual. In other cases the converse may occur; that is, part or all of the musculocutaneous nerve may travel with the lateral head of the median and subsequently be given back as a communication to the musculocutaneous or, in the absence of the musculocutaneous nerve as an entity, the muscles and skin that it ordinarily supplies receive branches from the median nerve. **The musculocutaneous nerve is ultimately distributed to the flexor muscles on the front of the arm, to the skin on the lateral side of the forearm, and to the elbow joint.** The nerve to the coracobrachialis often arises separately from the lateral cord of the brachial plexus rather than as a branch of the musculocutaneous nerve.

The *radial (musculospiral) nerve*[18] ([C5], C6, 7, 8 [T1]) (see p. 767) may be regarded as the continuation of the posterior cord. It is the largest branch of the brachial plexus. At its origin it lies behind the axillary artery and in front of the subscapularis. It supplies chiefly the back of the limb, but as its name indicates, it occupies the lateral aspect for a part of its course. **The radial nerve is ultimately distributed to the skin on the back of the arm, forearm, and hand, to the extensor muscles on the back of the arm and forearm, and to the elbow joint and many of the joints of the hand. The radial nerve may be injured in the axilla by pressure of a crutch or by hanging the arm over the back of a chair.**

The *axillary (circumflex) nerve* (C5, 6) is a branch of the posterior cord. **The axillary nerve supplies the deltoid and teres minor, shoulder joint, and skin on the back of the arm.** It lies in front of the subscapularis, behind the axillary artery, and lateral to the radial nerve. At the lower border of the subscapularis it turns posteriorly and passes through the quadrangular space with the posterior circumflex humeral artery, between the long and lateral heads of the triceps. It lies below the capsule of the shoulder joint to which it sends a twig. It passes medial to the surgical neck of the humerus and divides into two branches under cover of the deltoid. The anterior branch winds around the humerus deep to the deltoid, which it supplies. A few filaments are said to pierce the muscle and become cutaneous. The posterior branch supplies the teres minor and the deltoid. The branch to the teres minor frequently has an enlargement on it[19] that is due to an increase in the amount of connective tissue. The posterior branch then turns around the posterior border of the deltoid and supplies an area of skin on the back of the arm as the *upper lateral brachial cutaneous nerve* (previously known as the *lateral brachial cutaneous nerve*). Above this level, the skin of the shoulder is supplied by the supraclavicular nerves. The level of the axillary nerve is indicated by a horizontal plane through the middle of the deltoid.

Several *lateral pectoral nerves* (C5, 6, 7)[20] arise from the lateral cord or from the anterior divisions of the upper and middle trunks. They cross in front of the axillary vessels, pierce the clavipectoral fascia, and end in the pectoralis major. They send a loop across the first part of the axillary artery to join the medial pectoral nerves and, by this means, contribute fibers to the pectoralis minor. They supply the acromioclavicular joint and often the shoulder joint also.

Several *medial pectoral nerves* (C8, T1) arise from the medial cord or from the lower trunk, come forward between the axillary artery and vein, pierce and supply the pectoralis minor, and end in the overlying pectoralis major. Not infrequently some branches turn around the lower border of the pectoralis minor to reach the pectoralis major.

A variable number of subscapular branches arise from the posterior cord, and a twig is often given to the shoulder joint as well. The *upper subscapular nerve* (or *nerves*) (C5) supplies the subscapularis. Next to arise is the *thoracodorsal nerve* (C7, 8), which descends first with the subscapular artery and then with the thoracodorsal artery, and supplies the latissimus dorsi. The *lower subscapular nerve* (or *nerves*) (C5, 6) supplies the subscapularis and the teres major.

The *medial antebrachial cutaneous nerve* (C8, T1) arises from the medial cord. It lies between the axillary artery and vein, and descends medial to the brachial artery. It may give a branch to the arm,[21] and then, below the middle of the arm, it pierces the fascia, becomes subcutaneous, and divides into *anterior* and *ulnar branches.* The anterior branch passes superficial or deep to the median cubital vein and supplies the front and medial side of the forearm as far as the wrist. The ulnar branch supplies the skin on the medial and posteromedial aspects of the forearm. Some of its twigs anastomose with the posterior antebrachial cutaneous nerve and with the dorsal branch of the ulnar nerve.

The *medial brachial cutaneous nerve* (T1), a branch of the medial cord, is the smallest branch of the brachial plexus. It crosses in front of or behind the axillary vein to gain the medial side of the vein, and supplies the skin on the medial and posterior aspects of the arm. It communicates with the intercostobrachial nerve (p. 274).

BLOOD VESSELS

Axillary Artery. **The main artery carrying blood to the upper limb has as its most im-**

portant function the supply of the vital centers of the medulla oblongata. It is called by different names (subclavian, axillary, brachial) in different parts of its course (fig. 64–1, p. 766). On the left side of the body the subclavian artery arises directly from the arch of the aorta, whereas on the right side it springs from the brachiocephalic trunk, which in turn is a branch of the arch of the aorta. From behind the sternoclavicular joint, each subclavian artery ascends into the neck to form an arch, which lies on the first rib and extends above the clavicle. **It is a matter of practical importance in first aid for severe arterial bleeding in the upper limb that the main artery to the upper limb can be compressed downward against the first rib in the angle between the clavicle and the posterior margin of the sternocleidomastoid.**

At the apex of the axilla, where the subclavian artery reaches the outer border of the first rib (fig. 13–7, p. 110), its name is changed to axillary artery. **For descriptive purposes the axillary artery is commonly divided into three portions by the pectoralis minor. The first part is above the muscle, the second behind it, and the third below it.** The first part of the axillary artery, together with the axillary vein and the brachial plexus, is enclosed in the axillary sheath, a prolongation of the prevertebral layer of the cervical fascia where it covers the scalene muscles (p. 714).

The lower half of the second part and the entire third part of the axillary artery are fairly superficial and can be compressed against the humerus when the arm is elevated.

The arterial tube to the upper limb, in ascending from the thorax, meets the more deeply placed brachial plexus in the neck. The cords of the plexus descend behind and below the first part of the axillary artery, and the three cords (lateral, posterior, and medial) come to occupy the aspects of the second part of the artery indicated by their names. At the lateral and lower border of the pectoralis minor, the cords divide into their branches, and each bears the same relationship to the third part of the axillary artery as the cord from which it arose bears to the second portion. Thus the musculocutaneous and median nerves are on the lateral side, and the medial head of the median nerve crosses in front of the vessel to unite with the lateral head. The axillary and radial nerves are posterior to the artery, whereas the ulnar and medial brachial and antebrachial cutaneous nerves are on the medial side.

The axillary artery is related posteriorly, first to the external intercostal muscle of the first intercostal space, to the first digitation of the serratus anterior and to the long thoracic nerve, and then to the posterior wall of the axilla (subscapularis, latissimus dorsi, and teres major). The axillary vein lies on its medial side, and distally the coracobrachialis is lateral to the artery. The axillary artery is covered in front by the pectoralis minor (in its second part), by the clavipectoral fascia above that muscle, and by muscular bands (axillary arches) below, when these last are present. Except at its distal end, where it is superficial, the axillary artery is covered throughout by the pectoralis major.

At the base of the axilla, where the axillary artery leaves the inferior border of the teres major and comes to lie against the triceps in the arm, its name is changed to brachial.

BRANCHES (fig. 64–1, p. 766). The axillary artery gives branches to adjacent muscles, especially to the subscapularis, and has about six named branches. These branches vary considerably in their level of origin and in their pattern of branching,[22] but their pattern of distribution is relatively constant. Each intercostal space except the first is supplied from two or more axillary sources.

1. The *highest* or *superior thoracic artery* is a small, extremely variable branch of the first part of the axillary artery. It supplies adjacent muscles.

2. The *thoracoacromial artery*, a branch of the first or second part of the axillary artery, is a short trunk, the branches of which pass forward to pierce the clavipectoral fascia. The *acromial branch* ramifies on the acromion, the *clavicular branch* supplies the subclavius muscle, the *pectoral branch* supplies the pectoralis major and minor, and the *deltoid branch* descends in company with the cephalic vein.

3. The *lateral thoracic artery*, also from the second part of the axillary artery, is an extremely variable branch that descends along the lateral border of the pectoralis minor and gives off *lateral mammary branches*.

4. The *subscapular artery* is the largest branch of the third part (sometimes of the second part) of the axillary. It arises opposite the lower border of the subscapularis, along which it descends in relation to the lateral border of the scapula. It gives off a large branch, the *circumflex scapular artery*, and then continues as the *thoracodorsal artery*, which accompanies the thoracodorsal nerve and supplies the wall of the thorax. The circumflex scapular artery is often larger than the thoracodorsal. It passes backward through the triangular space, where it may groove the lateral border of the scapula, and then ramifies in the infraspinous fossa.

5. The *anterior circumflex humeral artery* (from

the third part) is an inconstant branch that winds around the front of the surgical neck of the humerus.

6. The *posterior circumflex humeral artery* (from the third part) is a large branch that passes backward through the quadrangular space in company with the axillary nerve. A descending branch anastomoses with the profunda brachii artery. The posterior circumflex humeral artery may spring from the subscapular artery, or may arise in common with the anterior circumflex humeral artery.

Collateral Circulation. An extensive arterial anastomosis is present around the scapula; the chief vessels concerned are: (1) the subscapular and circumflex scapular arteries, along the lateral border of the scapula; (2) the descending scapular artery (p. 705), along the medial border of the bone; (3) the suprascapular artery (p. 705), near the upper margin of the scapula, and in the supraspinous and infraspinous fossae; (4) other and smaller contributions (e.g., branches of the intercostal arteries). All these vessels form extensive networks on both the costal and the dorsal surface of the scapula. The anastomosis usually enables a collateral circulation to become established after ligature of the third part of the subclavian, or the first part of the axillary artery.

Axillary Vein. The axillary vein (fig. 13–7) begins at the lower border of the teres major, where the basilic vein joins the brachial veins. The brachial veins may first unite to form a single vein, which then joins the basilic vein. The axillary vein ascends through the axilla, along the medial side of the axillary artery. It is provided with one or more valves.[23] Its anterior and posterior relations are those of the axillary artery. The lateral and apical groups of axillary lymph nodes are closely related to the axillary vein. It receives tributaries that correspond to the branches of the axillary artery, and generally have similar names. The veins corresponding to the branches of the thoracoacromial artery, however, do not unite and form a common stem. Some empty into the axillary vein, but others empty into the cephalic vein. Proximally, the axillary vein receives the cephalic vein.

The axillary vein commonly receives, directly or indirectly, the *thoracoepigastric veins*[24] and thereby provides a collateral route for venous return if the inferior vena cava becomes obstructed (p. 420).

At the outer border of the first rib, the axillary vein continues as the subclavian vein. The subclavian vein lies in front of the subclavian artery and is separated from it by the scalenus anterior. It is also at a lower level than the artery, so that it does not rise above the clavicle. Behind the medial end of the clavicle, it unites with the internal jugular vein to form the brachiocephalic vein. The right and left brachiocephalic veins then unite in the thorax to form the superior vena cava, which enters the right atrium.

JOINTS OF SHOULDER

Shoulder Joint. **The shoulder (glenohumeral) joint is a large, freely movable, ball-and-socket joint between the glenoid cavity of the scapula and the head of the humerus** (figs. 13–4, 13–6, and 13–10; see also figs. 11–3 to 11–5, pp. 72–74). The shallow glenoid cavity is deepened somewhat by the fibrous or fibrocartilaginous *glenoid lip* attached to its margins. The articular surface of the glenoid cavity is small in area compared with that of the humerus, and the glenoid cavity can hardly be regarded as a true socket, comparable to the acetabulum of the hip joint. The loose mechanical fit of the shoulder joint permits great freedom of movement. Strength and stability are imparted by adjacent muscles and tendons, in particular by the musculotendinous cuff (fig. 13–4).

The joint capsule is attached to the margin of the glenoid cavity, where it often fuses to some extent with the external aspect of the glenoid lip. Distally it is attached to the anatomical neck of the humerus, except medially, where it is reflected downward for a centimeter or more onto the shaft of the humerus. Deep to the tendons of the musculotendinous cuff, the capsule is fused with the tendons. Between the supraspinatus and subscapularis tendons, it is thickened by the *coracohumeral ligament*. Anteriorly, it is usually thickened to form several variable bands, the *glenohumeral ligaments*, which extend from the glenoid lip to the anatomical neck of the humerus (fig. 13–4). The *transverse humeral ligament* bridges the intertubercular groove (fig. 13–6), and holds the tendon of the biceps in the groove.

The upper epiphysial line of the humerus is extracapsular, except medially, where the capsule is reflected onto the shaft of the humerus. The laxity of the capsule here permits abduction. But the same laxity accounts for the fact that dislocations of the shoulder joint, which are not infrequent, are often subglenoid. The head of the humerus dislocates in the area of the lower reflection of the capsule.

SYNOVIAL MEMBRANE AND BICEPS TENDON. The synovial membrane that lines the capsule is continuous with the lining of the subscapular bursa; a subscapular recess is thus formed. The subcoracoid bursa, which lies between the coracoid process and the joint capsule, may also communicate with the joint cavity.

As the biceps tendon traverses the joint cavity it is enclosed in a tubular sheath of synovial membrane,

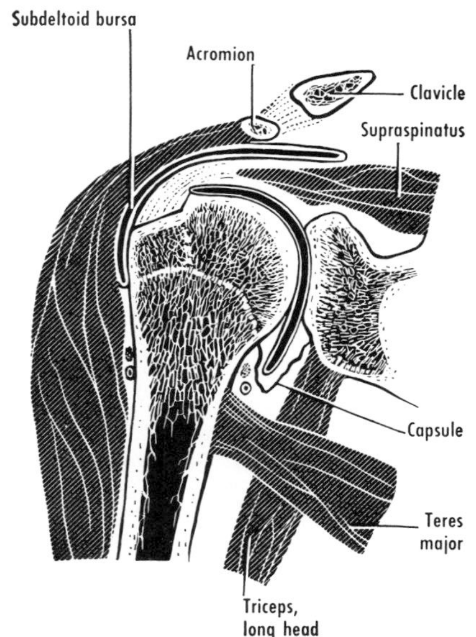

Figure 13–10 Diagram of a coronal section through the shoulder joint. The width of the joint cavity is exaggerated.

which accompanies the tendon down the intertubercular groove.

The acromion, the coracoid process, and the coracoacromial ligament constitute the coracoacromial arch. The *subdeltoid bursa* lies between the tendon of the supraspinatus below and the deltoid and coracoacromial arch above (fig. 13–10). Part of the bursa lies deep to the acromion *(subacromial bursa)*; this part is generally not separate from the subdeltoid. The bursa constitutes a lubricating mechanism between the cuff and coracoacromial arch during movement of the shoulder joint. The tendons related to the bursa are especially susceptible to wear and tear during movement.[25]

NERVE SUPPLY.[26] The shoulder joint is supplied by the axillary, suprascapular, and lateral pectoral nerves, by the posterior cord of the brachial plexus, and by sympathetic fibers from the stellate and adjacent ganglia or from the sympathetic trunk.

Sternoclavicular Joint. **This joint is formed by the medial end of the clavicle, the sternum, and the first costal cartilage (fig. 11–5C and *D*, p. 74). On the basis of movements that occur, it can be classified as a ball-and-socket joint.**

The bones are united by a fibrous capsule that surrounds the joint, encloses the medial epiphysis of the *clavicle*, and is said to be important in maintaining the clavicle when the trapezius is inactive.[27] The capsule is strengthened in front and behind by the *anterior* and *posterior sternoclavicular ligaments*. An additional band, the *interclavicular ligament*, extends across the suprasternal notch and reinforces the capsule superiorly. The capsule is reinforced below by the more laterally placed, strong *costoclavicular ligament*,[28] which ascends from the first costal cartilage to an impression on the inferior aspect of the medial end of the clavicle.

The articular surfaces, especially that on the clavicle, are chiefly fibrocartilaginous. They are somewhat curved but are separated by a densely fibrous or fibrocartilaginous *articular disc*,[29] the periphery of which blends with the capsule. The disc is attached to the sternum below and the clavicle above, and thus helps to prevent the clavicle from being driven medially (see below).

Connections Between Clavicle and Scapula. The subclavius muscle sometimes extends to the upper border of the scapula, and the costocoracoid ligament constitutes an additional connection. The chief connections are the following:

ACROMIOCLAVICULAR JOINT. This is a plane joint between an ovoid facet on the medial border of the acromion and a similar facet on the lateral end of the clavicle (fig. 11–3, p. 72). The articular surfaces are mainly fibrocartilaginous. The joint capsule is short and taut. A pad of fibrocartilage frequently projects into the joint from above, and may partition the joint completely.

CORACOCLAVICULAR LIGAMENT. This is a strong band, consisting of two ligaments, the *conoid* and *trapezoid*, which are often separated by a bursa. The conoid ligament extends upward and slightly backward from the coracoid process to the conoid tubercle on the inferior surface of the clavicle. The trapezoid ligament extends from the upper surface of the coracoid process to the trapezoid line of the clavicle. The coracoclavicular ligament reinforces the acromioclavicular joint. Thus, in a fall on the hand and outstretched arm, the acromion tends to be driven under the clavicle (owing to the upward tilt of the articular surface). This is resisted mainly by the horizontally directed trapezoid ligament. The tendency of the transmitted force to drive the clavicle medially is resisted by the articular disc at the sternoclavicular joint. If the coracoclavicular ligament ruptures, the acromioclavicular joint is dislocated.

A synovial joint is sometimes present between the coracoid process and the clavicle.[30]

Scapular Ligaments. Several ligaments connect one part of the scapula with another, and thus do not bridge a joint.

Coracoacromial Ligament (see fig. 13–4). This strong band, which is functionally related to the shoulder joint and is discussed further with that joint, extends from the lateral side of the coracoid process to the acromion. It sometimes diverges as two bands.

The coracoacromial ligament, together with the acromion and the coracoid process, forms a protective arch or roof above the supraspinatus tendon and the head of the humerus.

Superior Transverse Scapular Ligament. This bridges the scapular notch and converts it into a foramen that transmits the suprascapular nerve. It may be partially or completely ossified.

Inferior Transverse Scapular Ligament. As the supraspinatus and infraspinatus muscles extend to the greater tubercle, their fasciae fuse and form a fibrous arch for the suprascapular vessels and nerve as they pass from the supraspinous to the infraspinous fossa. This arch, which may be thin or even absent, constitutes the inferior transverse scapular ligament. It extends from the lateral border of the spine of the scapula to the neck of this bone.

MOVEMENTS OF SHOULDER

Movements at the Shoulder Joint (figs. 13–11 and 13–12).[31] Movements at the shoulder joint are abduction and adduction, flexion and extension, and circumduction and rotation. **The shoulder joint has greater freedom and range of movement than any other joint, owing in large part to the scapular movement that generally accompanies movement at the shoulder joint.**[32] Abduction and lateral rotation are controlled mainly by fibers from the fifth cervical segment of the spinal cord, adduction and medial rotation

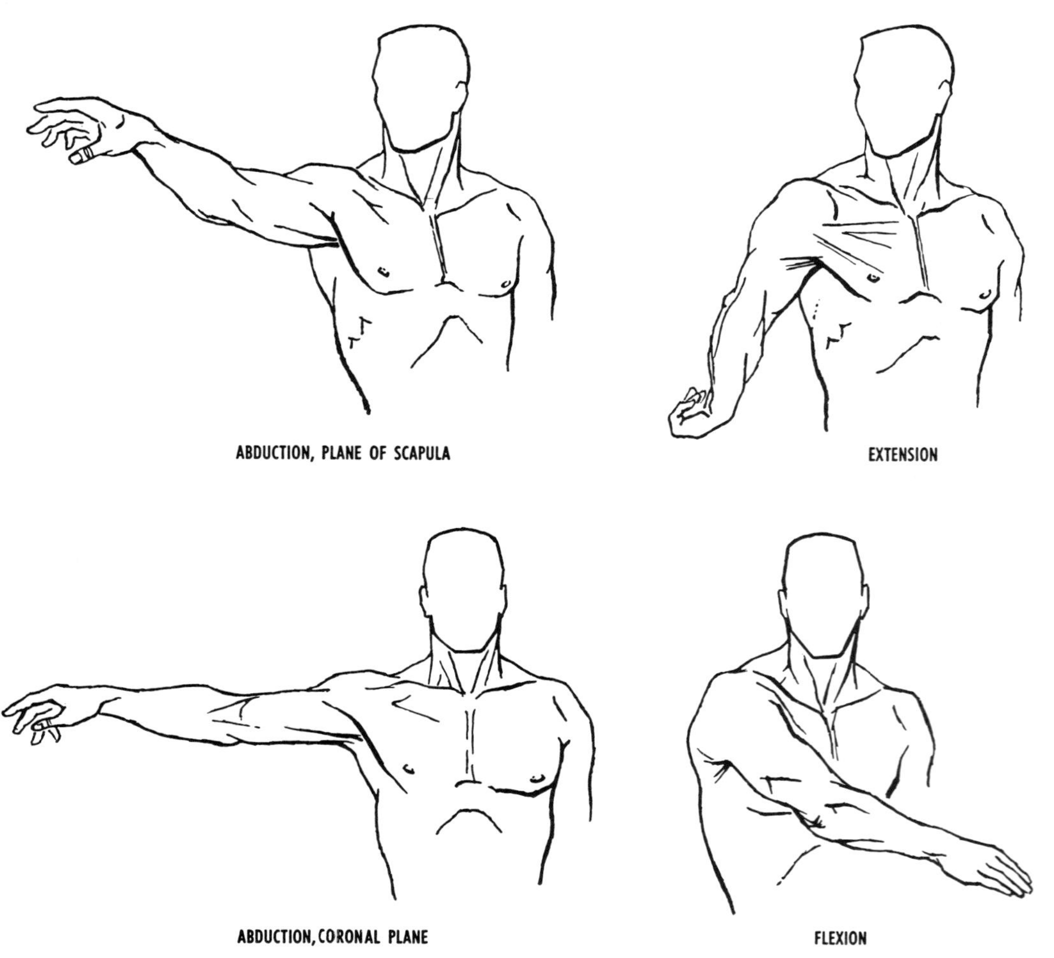

Figure 13–11 Movements at the shoulder.

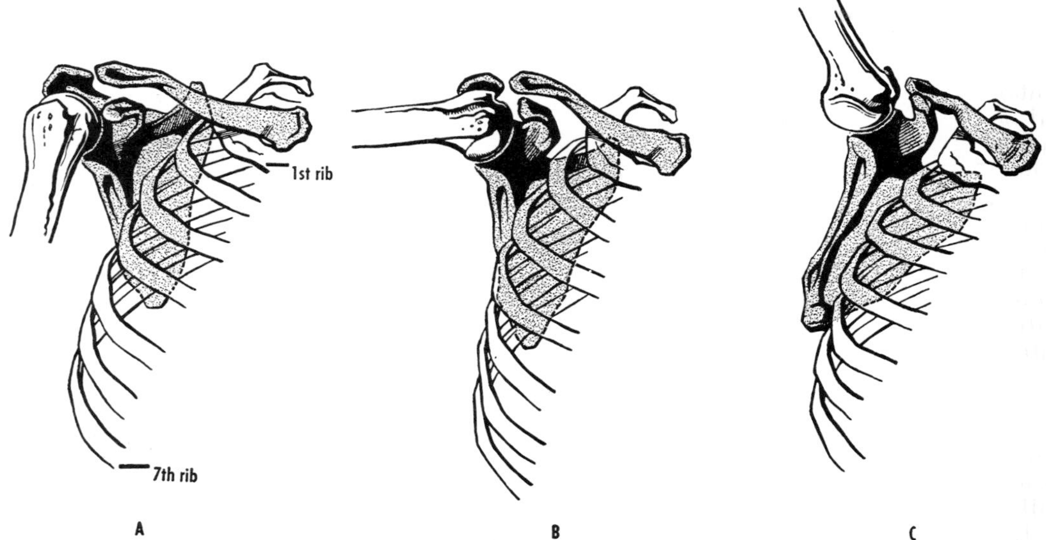

Figure 13–12 Drawings from tracings of radiograms of a right shoulder taken during abduction in a coronal plane (fig. 11–4, p. 73). *A*, arm is at the side. *B*, arm has been abducted to a right angle. *C*, arm is almost fully elevated. Note the degree to which the scapula rotates. Note also the lateral rotation of the humerus as evidenced by the change in position of the tubercles and the intertubercular groove.

by fibers from the sixth, seventh, and eighth segments.

The planes of movement are defined as follows. When the supraspinatus and the deltoid abduct the arm, they abduct it in the plane of the body of the scapula, that is, about midway between a sagittal and a coronal plane, upward and forward.[33] Abduction in a coronal plane is much more complex; it involves extension and lateral rotation of the humerus as well as abduction. In this discussion, *movements are defined in terms of the plane of the body of the scapula.* Abduction is movement forward and laterally, away from the trunk, and adduction is the converse. In flexion, the arm is carried forward and medially, across the front of the chest, and in extension it is carried backward and laterally, away from the chest. In circumduction all these movements are combined, whereas during rotation the humerus turns around its long axis. If the arm if fixed, as in climbing, the converse movements of the scapula on the humerus occur. Certain movements are difficult to define. Thus, movement in a horizontal plane (moving a book from one part of a shelf to another, or drawing a line on a blackboard) is often called horizontal abduction or adduction.

Because the glenoid cavity and the head of the humerus do not fit snugly, there is always a tendency for dislocation during movement. The muscles of the musculotendinous cuff serve to prevent this by holding the head of the humerus in place.

ABDUCTORS AND ADDUCTORS. The deltoid is the chief abductor, assisted by the supraspinatus, which contracts simultaneously with the deltoid.

One of the functions of the supraspinatus is to hold the head of the humerus in place and thereby to prevent the deltoid from pulling the head of the humerus upward, under the acromion and coracoacromial ligament.[34] The other muscles of the cuff are probably just as important in this regard. It is ordinarily held that, if the supraspinatus is paralyzed or its tendon ruptured, abduction is severely impaired. However, there is evidence that, with a complete rupture of the supraspinatus, abduction may be virtually normal. The other muscles of the cuff hold the head of the humerus in place.)[35]

When the deltoid is paralyzed, normal abduction is impaired, because the supraspinatus by itself usually cannot fully abduct the arm. Nevertheless, in some patients, abduction may be virtually normal when the deltoid is paralyzed.[36]

The chief adductors against resistance are the pectoralis major, latissimus dorsi, and teres major, aided perhaps by the posterior part of the deltoid. If the arm is lowered (adducted) from an elevated position, the descent is controlled by the middle part of the deltoid and by the supraspinatus, so long as no resistance is met.

The biceps may sometimes help to

maintain the arm in abduction. The long head of the triceps is said to aid in adduction, but it is more likely that it helps to hold the head of the humerus in place.

FLEXORS AND EXTENSORS. The chief flexors are the pectoralis major (clavicular part) and the anterior part of the deltoid, aided by the coracobrachialis and biceps. When flexion begins from a fully extended position, the sternocostal part of the pectoralis major is active initially. Both heads of the biceps seem to be active during flexion. The chief extensors are the latissimus dorsi and the posterior part of the deltoid. The latissimus dorsi is especially powerful when extension is begun against resistance, from a fully flexed position.

ROTATORS. The chief medial rotator is the subscapularis, aided by the pectoralis major, the anterior part of the deltoid, and the latissimus dorsi. The relative importance of these muscles varies according to the position of the arm; the subscapularis is the most powerful when the arm is hanging at the side. The chief lateral rotators are the infraspinatus and teres minor, aided by the posterior part of the deltoid.

Movements of the Shoulder Girdle. **The important movements of the shoulder girdle are displacements of the scapula. These include (1) elevation and depression of the scapula, (2) rotation, (3) lateral or forward movement, and (4) medial or backward movement. During movement the acromion is kept from the chest wall by a strut, the clavicle.** The lateral end of the clavicle travels in an arc, the radius of which is the clavicle and the pivot the sternoclavicular joint. The medial border of the scapula, however, is held against the chest wall and travels in a different arc, that of the chest wall. Hence the angle between the scapula and clavicle (at the acromioclavicular joint) must be altered continually. The clavicle rotates around its long axis during movement of the scapula, and impairment of this rotation, as well as fixation at either the acromioclavicular or the sternoclavicular. joint, interferes with scapular movement. The clavicle imparts both stability and precision to movements of the scapula.[37]

Muscles attached to the scapula are so arranged that none can bring about simple linear displacements. Thus, in order to elevate or depress the scapula, combinations of muscles are necessary.

In elevation of the scapula, as in shrugging the shoulders, the trapezius (upper fibers) and the levator scapulae are the prime movers; their opposite rotatory effects counterbalance each other. Depression of the scapula is due to gravity, controlled by relaxation of the aforementioned muscles, but, if depression is carried out against resistance, the lower fibers of the trapezius and the serratus anterior contract.

Forward movement of the scapula on the chest wall (as in pushing, thrusting, or punching) causes the glenoid cavity to look more anteriorly. The serratus anterior is the prime mover; the rhomboids are the antagonists. Backward movement (as in bracing the shoulders) causes the glenoid cavity to face more laterally. The prime movers are the trapezius muscles and the rhomboids.

When the scapula is rotated so that the inferior angle moves laterally and the glenoid cavity upward, the serratus anterior is the prime mover, aided by the trapezius. Such rotation generally accompanies elevation of the arm. The opposite movement, if carried out against resistance, is due to the levator scapulae and rhomboids.

Although the serratus anterior is an antagonist of the trapezius in backward movement of the scapula, and *vice versa,* both coordinate to rotate the scapula. Thus the functions of muscles vary according to the movement needed.

Combined Scapular and Glenohumeral Movements. **Little if any motion occurs at the shoulder joint without accompanying movement or displacement of the rest of the shoulder girdle.** This can be illustrated by a brief account of elevation of the arm to a vertical position, beginning by abducting the arm in a coronal plane. The deltoid and supraspinatus begin the movement. The accompanying scapular motion is at first variable and irregular, but, shortly after abduction has started, the scapula begins to rotate, although to a lesser extent than the humerus. Thus, the humerus can be elevated on the scapula to about 120 degrees, and the scapula can be rotated on the chest wall about 60 degrees. The combined movements permit the elevation of the arm to a fully vertical position. If the shoulder joint is surgically fused, the arm can be elevated about 60 degrees by virtue of scapular rotation.

In scapular rotation, the inferior angle moves laterally, and the lateral angle upward and medially (serratus anterior and trapezius). The clavicle likewise rotates on its long axis. Fixation of the clavicle at the acromioclavicular joint or at the coracoclavicular ligaments limits or prevents full elevation. During the early phases of elevation, clavicular angulation is greatest at the sternoclavicular joint; during the later phases, it is greatest at the acromioclavicular joint.

During elevation in a coronal plane, the humerus is rotated laterally (fig. 13–12); otherwise the greater tubercle would abut against the acromion.[38] If lateral rotation is prevented, elevation is limited, and paralysis of lateral rotators may simulate weakness of elevation. True abduction, in the plane of the scapula, does not involve humeral rotation. The importance of lateral rotation in elevation can be tested as follows: with the arms hanging at the sides, flex the forearms in a sagittal plane until the fingers point forward (about 90 degrees of flexion). Then abduct both arms in a coronal plane. It will be difficult or impossible to elevate them much above the horizontal; the greater tubercle now abuts against the acromion and coracoacromial ligament. Next rotate the arms so that the fingers point upward. Elevation to the vertical can now be carried out.

REFERENCES

1. G. T. Ashley, Anat. Rec., *113*:301, 1952.
2. O. Machado de Sousa, F. Berzin, and A. C. Berardi, Electromyography, 9:407, 1969.
3. G. H. Koepke *et al.*, Arch. phys. Med., *36*:271, 1933.
4. W. T. Catton and J. E. Gray, J. Anat., Lond., *85*:412, 1951.
5. R. N. Barlow, Anat. Rec., *61*:413, 1934. F. Tischendorf, Z. Anat. EntwGesch., *114*:216, 1948.
6. H. W. Jones, Brit. med. J., 2:59, 1926. J. B. Brown and F. McDowell, Surgery, 7:599, 1940. F. Parenti, Chir. Org. Mov., *45*:34, 1957.
7. S. Mollier, *Plastische Anatomie,* Bergmann, Munich, 2nd ed., 1938 (reprinted in 1967).
8. L. E. Beaton and B. J. Anson, Anat. Rec., *83*:41, 1942.
9. M. M. Wiedenbauer and O. Mortensen, Amer. J. phys. Med., *31*:363, 1952. But see J. G. Bearn, J. Anat., Lond., *101*:159, 1967, who points out that the trapezius may be inactive when the shoulder is depressed, and that the ligaments of the sternoclavicular joint help to maintain clavicular poise.
10. R. A. Macbeth and C. P. Martin, Anat. Rec., *115*:691, 1953.
11. V. T. Inman and J. B. de C. M. Saunders, J. Bone Jt Surg., *26*:1, 1944. H. L. Broome and J. V. Basmajian, Anat. Rec., *170*:309, 1971.
12. J. Symington, *The Topographical Anatomy of the Child,* Livingstone, Edinburgh, 1887.
13. A. T. Kerr, Amer. J. Anat., *23*:285, 1918. R. Fenart, Acta anat., *32*:322, 1958.
14. D. R. Dow, J. Anat., Lond., *59*:166, 1925.
15. H. Seddon, *Surgical Disorders of the Peripheral Nerves,* Churchill Livingstone, Edinburgh, 1972.
16. For variations, see M. Borchardt and Dr. Wjasmenski, Beitr. klin. Chir., *107*:553, 1917; K. Buch-Hansen, Anat. Anz., *102*:187, 1955.
17. H. V. Vallois, Arch. Anat., Strasbourg, *1*:183, 1922. H. Ferner, Z. Anat. EntwGesch., *108*:567, 1938. K. Buch-Hansen, Anat. Anz., *102*:187, 1955. J. P. Neidhardt *et al.,* Lyon Chir., *64*:268, 1968.
18. For variations, see M. Borchardt and Dr. Wjasmenski, Beitr. klin. Chir., *117*:475, 1919.
19. G. Gitlin, J. Anat., Lond., *91*:466, 1957.
20. A. S. Tavares, Acta anat., *21*:132, 1954.
21. S. Aiyama, Acta anat. Nippon, *47*:1, 1972 (Ex. med., *27*:169, 1973).
22. C. F. DeGaris and W. B. Swartley, Amer. J. Anat., *41*:353, 1928. M. Trotter *et al.,* Anat. Rec., *46*:133, 1930. D. F. Huelke, Anat. Rec., *135*:33, 1961. J. A. Keen, Amer. J. Anat., *108*:245, 1961.
23. The normal axillary venogram is described by C. J. Rominger, Amer. J. Roentgenol., *80*:217, 1958.
24. F. T. Lewis, Amer. J. Anat., *9*:33, 1909. L. C. Massopust and W. D. Gardner, Surg. Gynec. Obstet., *91*:717, 1950.
25. A. W. Meyer, J. Bone Jt Surg., *20*:491, 1922 and *29*:341, 1931; Calif. West. Med., *47*:375, 1937; Arch. Surg., Chicago, *35*:646, 1937.
26. E. Gardner, Anat. Rec., *102*:1, 1948. M. Wrete, Acta anat., *7*:173, 1949.
27. J. G. Bearn, J. Anat., Lond., *101*:159, 1967.
28. A. J. E. Cave, J. Anat., Lond., *95*:170, 1961.
29. Different forms of the disc are described by A. Beau, P. Quéreux, and P. Vassal, C. R. Ass. Anat., *42*:287, 1955.
30. R. D. Moore and R. R. Renner, Amer. J. Roentgenol., *78*:86, 1957. O. J. Lewis, J. Anat., Lond., *93*:296, 1959.
31. K.-H. Knese, Z. Anat. EntwGesch., *115*:115, 1950.
32. V. T. Inman, J. B. de C. M. Saunders, and L. C. Abbott, J. Bone Jt Surg., *26*:1, 1944. See also R. D. Lockhart, J. Anat., Lond., *64*:288, 1930; G. H. Fisk and G. Colwell, Arch. phys. Med., *35*:149, 1954; E. N. Duvall, Arch. phys. Med., *36*:149, 1955.
33. T. B. Johnston, Brit. J. Surg., *25*:252, 1937. S. D. Doody, L. Freedman, and J. C. Waterland, Arch. phys. Med., *51*:595, 1970.
34. E. A. Codman, *The Shoulder,* Todd, Boston, 1934. H. F. Moseley, Brit. J. Surg., *38*:340, 1951. M. Renard *et al.,* C. R. Ass. Anat., *51*:878, 1967.
35. O. Olsson, Acta chir. scand., Suppl. 181, 1953. B. von Linge and J. D. Mulder, J. Bone Jt Surg., *45B*:750, 1963.
36. L. J. Pollock, J. Amer. med. Ass., *79*:526, 1922. O. S. Staples and A. L. Watkins, J. Bone Jt Surg., *25*:85, 1943. E. Dehne and R. M. Hall, J. Bone Jt Surg., *41A*:745, 1959.
37. V. T. Inman and J. B. de C. M. Saunders, Calif. West. Med., *65*:158, 1946. See also M. A. MacConaill, Proc. R. Irish Acad. B, *50*:159, 1944.
38. L. McGregor, Brit. J. Surg., *24*:425, 1937. C. P. Martin, Amer. J. Anat., *66*:213, 1940.

ARM AND ELBOW 14

MUSCLES OF ARM

The muscles of the front of the arm are the biceps brachii, coracobrachialis, and brachialis. All are supplied by the musculocutaneous nerve. The triceps brachii is the muscle of the back of the arm. The basic arrangement of the muscles and nerves in the arm and forearm is shown in figure 14–1.

The *brachial fascia* is a thin, loose sheath that receives expansions from the tendons of the deltoid and pectoralis major muscles. Below, it thickens over the triceps. On each side it gives off an intermuscular septum to the corresponding supracondylar ridge and epicondyle of the humerus. The septa delimit an anterior fascial compartment, which contains the biceps, coracobrachialis, and brachialis, together with the brachioradialis and extensor carpi radialis longus, and a posterior compartment that contains the triceps.

The subcutaneous tissue over the olecranon and tendon of the triceps contains a *subcutaneous olecranon bursa,* the walls of which may be thickened and trabeculated ("miner's elbow").

Biceps Brachii. The biceps brachii arises from the scapula by two heads. It is inserted into the tuberosity of the radius, the antebrachial fascia, and, by way of this fascia, into the ulna.

The *short* or *medial* head of the biceps arises in common with the coracobrachialis from the tip of the coracoid process. The *long* or *lateral head* arises by a tendon from the supraglenoid tubercle and from the adjacent fibrocartilaginous glenoid lip (fig. 13–4, p. 108). As it descends in the intertubercular groove, the tendon is held in the groove by the transverse humeral ligament and by part of the tendon of insertion of the pectoralis major. A third head of origin is sometimes present and arises from the shaft of the humerus.[1] The tendons of origin give way to two bellies that unite and continue into a tendon that is readily palpable and is inserted into the posterior part of the tuberosity of the radius. A bursa is present between the tendon and the anterior part of the tuberosity and may surround the tendon. Part of the tendon is continued by means of an aponeurotic expansion, the *bicipital aponeurosis,* into the fascia of the forearm and thence to the ulna.[2]

Coracobrachialis. The coracobrachialis arises in common with the short head of the biceps from the tip of the coracoid process (fig. 13–3, p. 107). Its belly is usually pierced by the musculocutaneous nerve. It is inserted into the middle third of the medial border of the humerus. A bursa is usually present between its tendon of origin and the subscapularis.

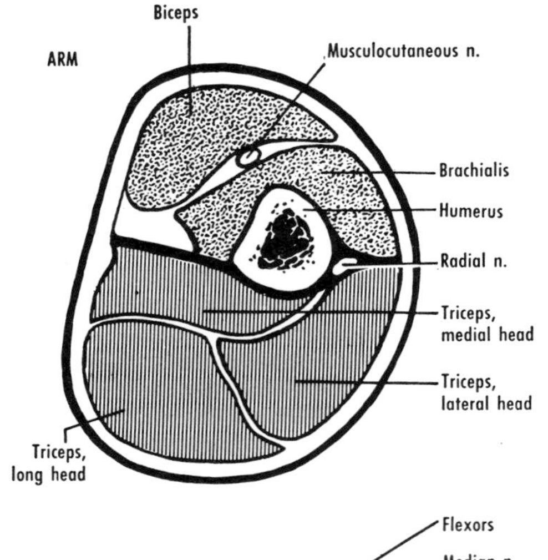

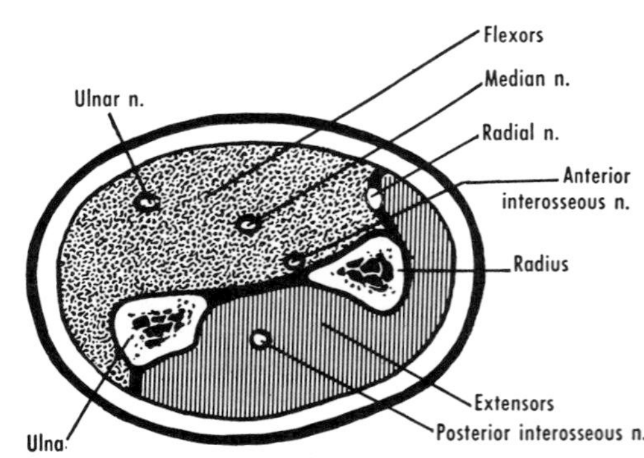

Figure 14–1 The basic arrangement of muscles and nerves in the arm and forearm. In the arm the anterior compartment is occupied by forearm flexors, supplied by the musculocutaneous nerve, the posterior compartment by forearm extensors, supplied by the radial nerve. In the forearm, the anterior muscles are supplied by the median and ulnar nerves, the posterior by the radial nerve (mostly by its posterior interosseous branch).

Brachialis. The brachialis arises from the distal two-thirds of the anteromedial and anterolateral surfaces of the humerus. Its origin embraces the insertion of the deltoid. It is inserted into the capsule of the elbow joint and the roughened anterior surface of the coronoid process and the tuberosity of the ulna.

NERVE SUPPLY AND ACTION. All three muscles are supplied by the musculocutaneous nerve. The branch to the coracobrachialis not uncommonly arises separately from the brachial plexus (p. 113). The brachialis is also supplied by the radial nerve.[3]

The flexion of the forearm (or, alternatively, the twitch of the muscle without movement) that follows tapping of the tendon of insertion of the biceps is known as the biceps jerk. The center for this reflex is in the fifth and sixth cervical segments of the spinal cord.

The coracobrachialis and biceps assist in flexing the arm. The brachialis and biceps are the chief flexors of the forearm. In addition, the biceps is a supinator of the forearm, and is active in flexion at the shoulder joint.

Triceps Brachii. The triceps brachii forms the bulk of the back of the arm. It has three heads of origin, arranged in two planes (fig. 14–2). The long and lateral heads occupy a superficial plane, whereas the medial head is on a deeper plane.

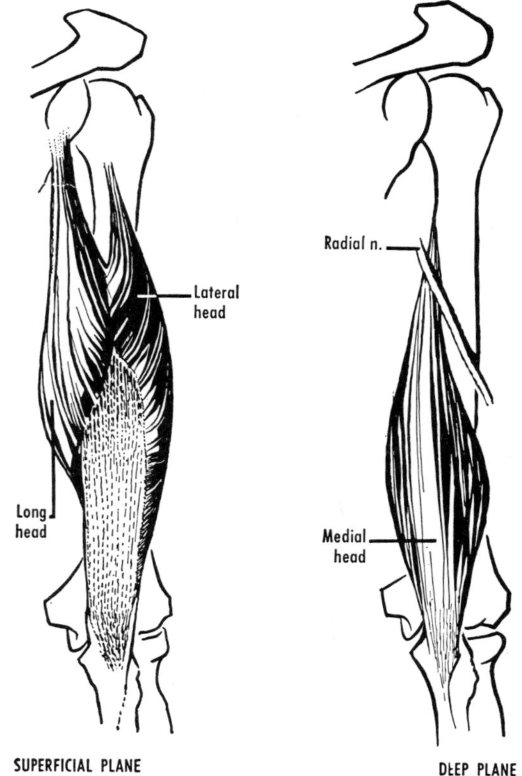

Figure 14–2 The superficial and deep planes of the triceps.

The *long head* arises from the infraglenoid tubercle of the scapula. In descending it separates the triangular from the quadrangular space (p. 108). The origins of the lateral and medial heads are separated by the groove for the radial nerve. The *lateral head* arises from the posterior surface of the humerus above this groove, the *medial head* below it. The triceps is inserted into the posterior part of the upper surface of the olecranon and, by means of what is sometimes termed the "tricipital aponeurosis," into the fascia of the forearm. A few fibers (subanconeus) may be inserted into the capsule of the elbow joint. A subcutaneous olecranon bursa is almost constantly found between the skin and the olecranon.

NERVE SUPPLY AND ACTION. The heads are supplied separately by branches of the radial nerve. The triceps, in particular the medial head, is the extensor of the forearm and takes part in pushing, throwing, hammering, and shoveling. The lateral and long heads are recruited when the movement meets resistance. The long head may also help to hold the head of the humerus in place. **The extension of the forearm (or, alternatively, a twitch without movement) that follows tapping of the tendon of insertion of the triceps is known as the triceps jerk. The center for this reflex is in the sixth and seventh cervical segments of the spinal cord.**

NERVES OF ARM

The muscles on the front of the arm are supplied by the musculocutaneous nerve, the triceps by the radial nerve. The three chief nerves to the forearm and hand (median, ulnar, and radial) arise in the axilla from the cords of the brachial plexus, accompany the axillary artery and the uppermost part of the brachial artery, and descend in the arm. The median and ulnar nerves pass directly to the forearm, but the radial nerve gives off muscular and cutaneous branches in the arm.

Musculocutaneous Nerve. The musculocutaneous nerve (p. 113, and fig. 64–4, p. 768) is a branch of the lateral cord. It often communicates with the median nerve in the arm and may arise from that nerve. If it arises in the axilla, it usually pierces the coracobrachialis. **The musculocutaneous nerve descends between the biceps superficially and the brachialis deeply, and reaches the lateral side of the arm. It supplies the coracobrachialis, biceps, and brachialis, the elbow joint, and, on rare occasions, the brachioradialis.**[4] Finally, it becomes the *lateral antebrachial cutaneous nerve,* which pierces the fascia just lateral to the biceps tendon, just at the elbow crease.[5] It divides into anterior and posterior branches, one or both of which pass posterior to the cephalic vein (fig. 14–4), and which supply the skin of the lateral half of the forearm as far as the wrist. Either branch may supply a variable area of skin on the dorsum of the hand.

Median Nerve. The median nerve (p. 112; fig. 64–4, p. 768) is formed on the lateral aspect of the axillary artery by heads derived from the lateral and medial cords of the brachial plexus. It continues on the lateral side of the brachial artery. **At about the middle of the arm the median nerve gradually crosses in front of, but occasionally behind, the brachial artery, and then descends on the medial side of that vessel. In the cubital fossa it lies behind the median cubital vein and under cover of the bicipital aponeurosis, and gives a branch to the elbow joint. It then enters the forearm between the two heads of the pronator teres.** It gives off its branches in the forearm and hand. Its surface anatomy is similar to that of the brachial artery.

Ulnar Nerve. The ulnar nerve (p. 113 and fig. 64–5, p. 769), a branch of the medial cord of the brachial plexus, descends medial to the axillary artery and continues on the medial side of the brachial artery. **At the middle of the arm the ulnar nerve pierces the medial intermuscular septum and descends with the superior ulnar collateral artery and the ulnar collateral nerve to the back of the medial epicondyle,**[6] **where it often gives a small twig to the elbow joint and where it may have a connective tissue enlargement.**[7] **It then enters the forearm between the two heads of the flexor carpi ulnaris.**

Radial Nerve. The radial nerve (p. 114 and fig. 64–3, p. 767), a branch of the posterior cord, descends behind the axillary artery. It continues behind the brachial artery but very soon dips backward with the profunda brachii artery. **The radial nerve winds around the humerus, under cover of the lateral head of the triceps. It first lies on the medial head of the triceps;**[8] **more distally it occupies the groove for the radial nerve. A short distance below the insertion of the deltoid, it pierces the lateral intermuscular septum and comes forward to the cubital fossa, where it lies deeply in the groove between the brachialis medially and brachioradialis laterally. At or below the level of the lateral epicondyle, it divides into superficial and deep branches.**

BRANCHES. The *posterior brachial cutaneous nerve* arises in the axilla, crosses the tendon of the latissimus dorsi, and supplies the skin on the back of the arm nearly as far as the olecranon.

Several *muscular branches* are given to the three heads of the triceps. One of the branches to the medial head accompanies the ulnar nerve in a part of its course and is called the ulnar collateral nerve. Branches are also given to the anconeus and elbow joint.

The *lower lateral brachial cutaneous nerve* usually arises directly from the radial nerve.[9] It supplies the lateral surface of the lower part of the arm.

The *posterior antebrachial cutaneous nerve* arises in the groove, pierces the lateral head of the triceps, and supplies the skin of the back of the forearm as far as the wrist.

Muscular branches are given to the brachialis, brachioradialis, extensor carpi radialis longus, and commonly to the extensor carpi radialis brevis also. One or more branches are given to the elbow joint.

The *deep branch of the radial nerve* arises above or below the lateral epicondyle. It winds laterally around the radius, between the superficial and deep layers of the supinator, and continues as the posterior interosseous nerve that supplies the muscles on the back of the forearm (p. 137).

The *superficial branch of the radial nerve* is the direct continuation of the radial nerve into the forearm; it is described on page 137.

ARTERIES OF ARM

Brachial Artery. The brachial artery (fig. 64–1, p. 766) is the continuation of the axillary from the lower border of the teres major, that is, from the distal limit of the posterior wall of the axilla. Its uppermost portion has the same neural relationships as the terminal part of the axillary artery. The median nerve lies lateral, the radial nerve posterior, and the ulnar and the medial antebrachial cutaneous nerves medial. The medial brachial cutaneous nerve is separated from it by the basilic or the axillary vein. The axillary nerve leaves the axilla through the quadrangular space and has no relationship to the brachial artery.

The brachial artery lies superficially on the medial side of the arm. Its upper part lies medial to the humerus, but its lower portion is in front of that bone. The artery therefore can be compressed laterally against the humerus above, and posteriorly below. Behind, it lies successively on the triceps and the brachialis. The biceps and coracobrachialis are lateral to the artery, and partly overlap it above, where its pulsations can be felt. This vessel is used in sphygmomanometry. About the middle of the arm the median nerve gradually crosses in front of (occasionally behind) the artery from the lateral to the medial side.

At the elbow the brachial artery lies in the center of the cubital fossa (see below). Here the biceps tendon is on its lateral side and the median nerve on its medial side, and it is crossed by the bicipital aponeurosis, which separates it from the median cubital vein. Just below the elbow joint, opposite the neck of the radius, the brachial artery divides into the radial and ulnar arteries for the supply of the forearm and hand. The surface anatomy of the brachial artery is described on page 157.

Apart from the brachial veins that accompany it, the brachial artery is accompanied in the middle of the arm by the basilic vein, which pierces the fascia at this level and continues upward to join the brachial veins and become the axillary vein.

Variations of the brachial artery and its branches are common.[10] In order of frequency they are: (1) high bifurcation of the brachial artery, (2) persistent median artery (p. 138), and (3) superficial ulnar artery. High origin of a forearm artery is often associated with a more superficial course than usual for the vessel concerned. Such a vessel is susceptible to damage during venipuncture in the cubital fossa. Aberrant vessels may connect the axillary or brachial artery with one of the forearm arteries, usually the radial. Sometimes the distal portion of the brachial artery lies more medially and, together with the median nerve, may pass behind a supracondylar process of the humerus (p. 80).

BRANCHES. In addition to *muscular branches*, and a *nutrient branch* for the humerus,[11] the brachial artery has the following named branches:

1. The *profunda brachii* artery arises from the back of the brachial near its origin, sometimes from the subscapular artery or from the third part of the axillary, and not infrequently in common with the superior ulnar collateral or posterior circumflex humeral artery.[12] It descends across the back of the humerus with the radial nerve in the spiral groove. It sends a *deltoid branch* upward. At the lateral side of the arm, the profunda brachii divides into a *radial collateral artery*, which comes forward with the radial nerve, and a *middle collateral artery*, which reaches the back of the lateral epicondyle.

2. The *superior ulnar collateral artery* arises near the middle of the arm and accompanies the ulnar nerve to the back of the medial epicondyle.

3. The *inferior ulnar collateral artery* arises a short distance above the elbow, and, passing behind the median nerve, divides into branches that reach the front and back of the medial epicondyle.

Collateral Circulation. The anastomosis around the elbow joint is formed by the following arteries. In front of the lateral epicondyle, the radial collateral artery joins the radial recurrent artery, which is a branch of the radial. Behind the lateral epicondyle, the middle collateral artery joins the interosseous recurrent artery (derived ultimately from the ulnar). In front of the medial epicondyle, an anterior branch of the inferior ulnar collateral artery joins the anterior ulnar recurrent artery (from the ulnar). Behind the medial epicondyle, the superior ulnar collateral artery, together with a posterior branch of the inferior ulnar collateral, joins the posterior ulnar recurrent artery (from the ulnar). There are also transverse connections, for example, between the posterior branches of the profunda brachii and the inferior ulnar collateral artery. Any one of these vessels may be small or missing, and in these cases is compensated for by other branches contributing to this network.

RELATIONSHIPS IN ARM

Some relationships in the middle of the arm (that is, slightly below the upper level in figure 14–3) may now be summarized. At this level, the deltoid is inserted laterally and the coracobrachialis medially. In front of the coracobrachialis, the medial antebrachial cutaneous nerve emerges through the fascia. The basilic vein usually pierces the fascia here to ascend with the brachial artery, and the brachial artery is crossed by the median nerve. More deeply, the ulnar nerve and superior ulnar collateral artery pass backward through the medial intermuscular septum. A short distance below the middle of the arm, the radial nerve and the radial collateral artery come forward through the lateral intermuscular septum to reach the front of the arm.

Cubital Fossa. The cubital fossa is a

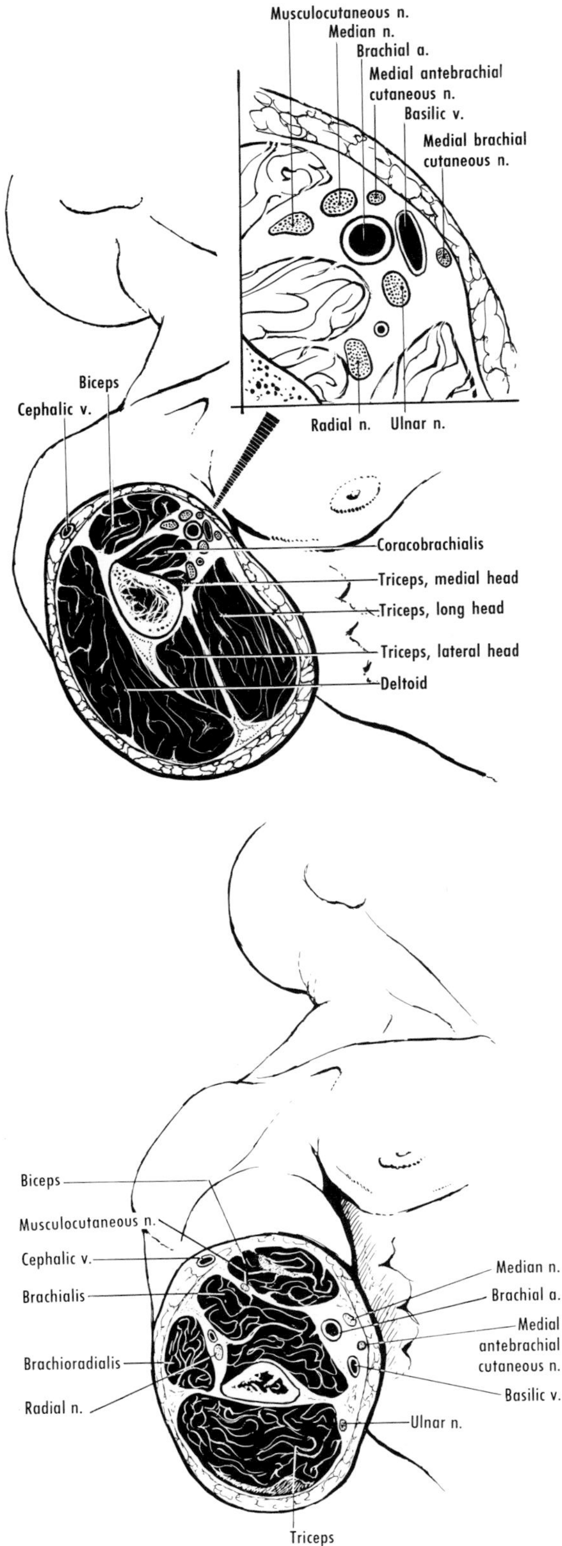

Figure 14–3 Horizontal sections through the upper and lower parts of the arm.

V-shaped interval in front of the elbow (fig. 14–4). The limbs of the **V** are formed by two forearm muscles, the brachioradialis laterally and the pronator teres medially. These muscles arise from the lateral and medial supracondylar ridges, respectively, and they approach each other as they descend to be inserted into the radius. The upper limit of the fossa is an imaginary horizontal line between the epicondyles of the humerus. The floor of the fossa is formed by the brachialis and by the supinator.

The contents of the cubital fossa are the biceps tendon, brachial artery, and median nerve, from lateral to medial side. The brachial artery usually divides at the apex of the fossa into its terminal branches, the radial and ulnar arteries. The median nerve lies about halfway between the biceps tendon and the medial epicondyle. The fossa also contains the radial nerve, deeply placed in the groove between the brachioradialis and brachialis. In front of, and either above or below, the lateral epicondyle, it divides into its deep branch, which pierces the supinator, and its superficial branch, which proceeds into the forearm under cover of the brachioradialis.

The fascia that roofs the whole cubital fossa is related superficially to the cephalic vein and lateral antebrachial cutaneous nerve and to the basilic vein and medial antebrachial cutaneous nerve. This fascial roof is strengthened by the bicipital aponeurosis, which extends from the biceps downward and medially to the antebrachial fascia and thence to the ulna. The aponeurosis roofs over the brachial artery and median nerve and is crossed almost at a right angle by the median cubital vein, which connects the cephalic to the basilic vein. **The median cubital vein is frequently used for intravenous injections and blood transfusions. Its close relationship to the underlying brachial artery and median nerve should be kept in mind.**

JOINTS OF ARM

Elbow Joint. **This hinge joint is formed between the humerus and the bones of the forearm, and is indicated approximately by a horizontal line about 2 to 3 cm below the epicondyles** (figs. 14–5 to 14–7; see also figs. 11–18 and 11–19, pp. 83, 84). It may be subdivided into *humeroradial* and *humeroulnar joints.*

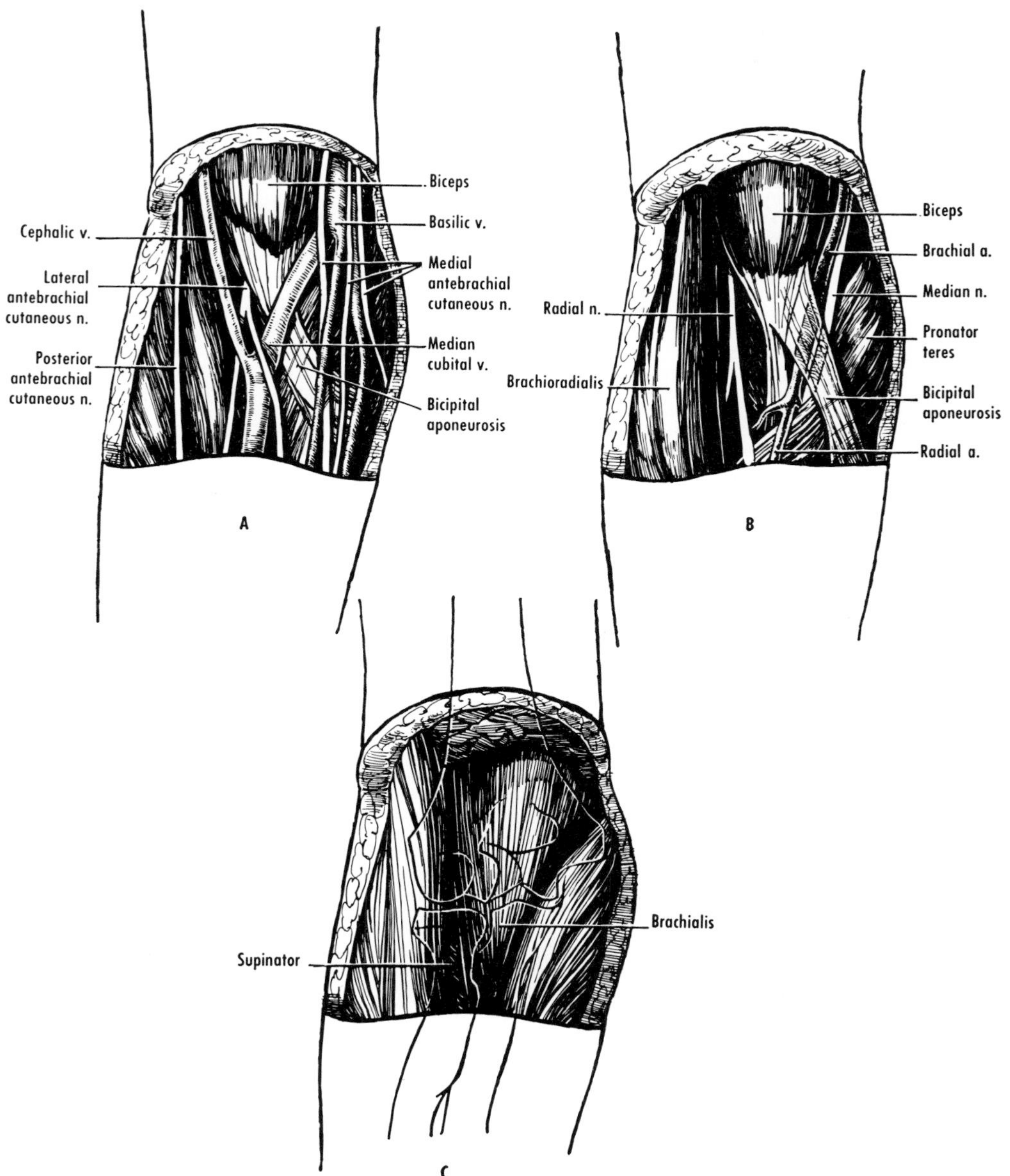

Figure 14–4 The cubital fossa. *A*, the superficial nerves and veins. *B*, the contents of the cubital fossa. *C*, the floor of the fossa.

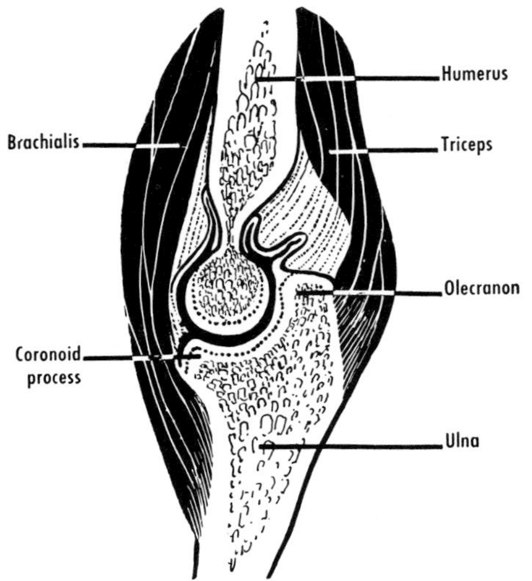

Figure 14–5 Diagram of a sagittal section through the humeroulnar part of the elbow joint. The width of the joint cavity is exaggerated.

The elbow and proximal radioulnar joints share a common cavity and certain ligaments, but they will be described separately.

The articular surface of the humerus is covered by hyaline cartilage, except on the medial surface of the trochlea. The cartilage of the trochlear notch of the ulna is interrupted by transversely placed fibrous tissue across the depth of the notch. The head of the radius is covered by hyaline cartilage that extends onto the circumference of the head and thereby into the proximal radioulnar joint. The posterior part of the capsule (posterior ligament) is thin and may be deficient medially. The anterior part (anterior ligament) is irregularly arranged; some of the deeper fibers of the brachialis are inserted into it.

The *radial collateral ligament* is a strong band that extends fanwise from the lateral epicondyle to blend with the annular ligament. It sends fibers anteriorly and posteriorly, to the coronoid process and the olecranon, respectively. The origins of the superficial extensors of the forearm and the supinator are fused with the radial collateral ligament.

The *ulnar collateral ligament* is attached above to the medial epicondyle, mainly to its lower aspect. Some fibers form a strong band directed anteriorly to the tubercle on the medial side of the coronoid process. Short posterior fibers extend from the medial epicondyle to the olecranon, forming a concavity for the ulnar nerve.

The synovial membrane of the elbow is continuous with that of the proximal radioulnar joint. Fat pads are commonly present.

NERVE SUPPLY. The joint is supplied by the musculocutaneous, median, radial, and ulnar nerves.[13]

Proximal Radioulnar Joint. **The head of the radius fits into the radial notch of the ulna and forms a pivot joint.** It is surrounded by the strong, densely fibrous *annular ligament*, which is attached to the anterior and posterior margins of the notch. This ligament is fused above with the capsule of the elbow joint and the radial collateral ligament, but is loosely attached to the neck of the radius below.[14]

The synovial membrane is reflected below, between the radius and the ulna, to form a pouch supported by a band termed the *quadrate ligament*, which helps to stabilize the proximal radioulnar joint.[15]

NERVE SUPPLY. Chiefly the radial, musculocutaneous, and median nerves.

Movements at Elbow and Proximal Radioulnar Joints.[16] **The elbow joint is a hinge joint, and voluntary movement is limited to flexion and extension.** It does not, however, act precisely as a hinge. The curvatures of the articular surfaces, particularly the curvature of the medial part of the trochlea, are such that during extension the angle that the supinated forearm makes with the arm (the "carrying angle") becomes apparent. The degree of curvature of the trochlea changes from front to back. This alters the angle that the ulna makes with the humerus and, as a consequence, the lower end of the ulna moves laterally during extension, medially during flexion. The carrying angle disappears when the forearm is flexed.

The movements at the elbow joint are flexion and extension and, between the radius and ulna, supination and pronation.

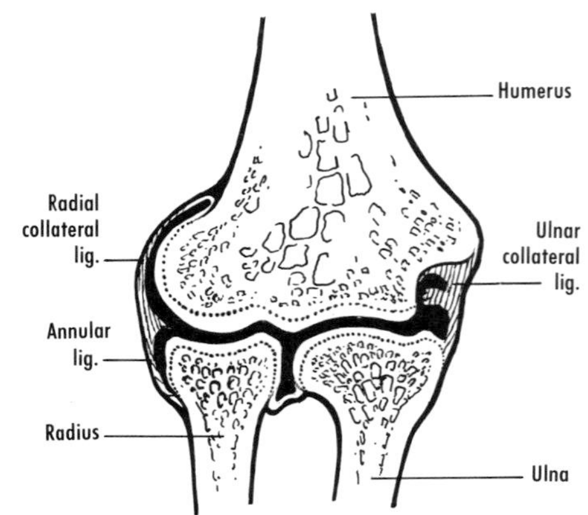

Figure 14–6 Coronal section through the elbow joint. The width of the joint cavity is exaggerated.

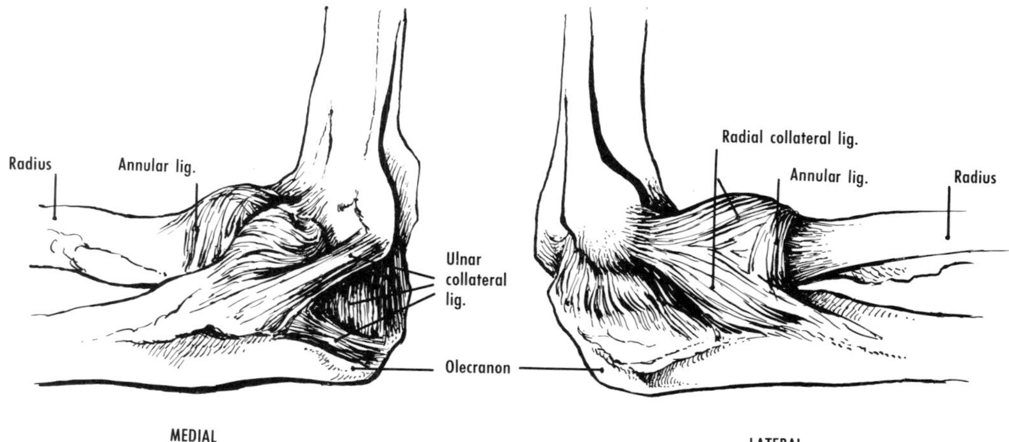

Figure 14–7 Ligaments of right elbow joint.

Flexion is controlled by fibers from the fifth and sixth cervical segments of the spinal cord, pronation and supination by fibers from the sixth, and extension by fibers from the seventh and eight.

The term supination is used for the position of the forearm and hand when the palm faces forward, as in the anatomical position. The term pronation is used when the palm faces backward. These terms are also used for the movements that bring about these positions. Pronation may be approximately defined as medial rotation about a longitudinal axis, and supination as lateral rotation, but both movements are considerably more complex. The axis of movement is represented by a line drawn from the center of the head of the radius to the lower end of the ulna. Hence, during either pronation or supination, the upper end of the radius merely rotates within the annular ligament. But its lower end describes an arc, so that in pronation it moves forward and medially around the lower end of the ulna, carrying the hand with it. The shafts of the radius and ulna therefore cross each other. Unless the elbow is fixed, rotation of the humerus accompanies rotation of the forearm. The ulna does not remain fixed. Its lower end moves backward and laterally during supination. Supination is said to be the stronger of the two movements (the threads of screws are arranged to take advantage of this), but such is not consistently the case.[17]

MUSCLES. The flexors of the forearm,[18] in order of decreasing strength, are the brachialis, biceps, and brachioradialis. The pronator teres is also a flexor when flexion is resisted. The extensor is the triceps, particularly the medial head (p. 123). The chief pronator is the pronator quadratus, aided by the pronator teres during fast or resisted movement.[19] The chief supinator is the supinator, aided by the biceps during fast or resisted movement.[20]

The biceps is a supinator and a flexor of the supinated forearm. The pronator teres is a flexor and pronator. If the biceps and pronator teres are active during flexion, their rotating effects counterbalance each other. The various muscles can be so combined that the forearm can be flexed from any position without being rotated. The biceps and triceps are usually antagonists but can act together, as, for example, when supination and extension are combined. When supination or pronation is carried out against resistance, the triceps acts to prevent flexion at the elbow joint.

REFERENCES

1. H. W. Greig, B. J. Anson, and J. M. Budinger, Quart. Bull. Northw. Univ. med. Sch., *26*:241, 1952. See also H. Ferner, Z. Anat. EntwGesch., *108*:567, 1938.
2. E. D. Congdon and H. S. Fish, Anat. Rec., *116*:395, 1953. S. Kader, Arch. Anat., Strasbourg, *40*:157, 1957.
3. M. C. Ip and K. S. F. Chang, Anat. Rec., *162*:363, 1968.
4. R. Bauer, Anat. Anz., *128*:108, 1971.
5. I. A. Olson, J. Anat., Lond., *105*:381, 1969.
6. D. B. Apfelberg and S. J. Larson, Plast. reconstr. Surg. (Balt.), *51*:76, 1973.
7. K. S. F. Chang *et al.*, Anat. Rec., *145*:149, 1963.
8. R. O. Whitson, J. Bone Jt Surg., *36A*:85, 1964.
9. E. A. Linell, J. Anat., Lond., *55*:79, 1921. T. Kasai, Amer. J. Anat., *112*:305, 1963.
10. C. F. De Garis and W. B. Swartley, Amer. J. Anat., *41*:353, 1928. C. M. Charles *et al.*, Anat. Rec., *50*:299, 1931. L. J. McCormack, E. W. Cauldwell, and B. J. Anson, Surg.

Gynec. Obstet., *96*:43, 1953. H. T. Weathersby, Sth. med. J., Nashville, *49*:46, 1956. J. H. Keen, Amer. J. Anat., *108*:245, 1961.
11. P. G. Laing, J. Bone Jt Surg., *38A*:1105, 1956.
12. C. M. Charles *et al.*, Anat. Rec., *50*:299, 1931.
13. E. Gardner, Anat. Rec., *102*:161, 1948. E. B. Kaplan, J. Bone Jt Surg., *41A*:147, 1959.
14. B. F. Martin, J. Anat., Lond., *91*:584, 1957.
15. M. Spinner and E. B. Kaplan, Acta orthopaed. scand., *41*:632, 1970.
16. K.-H. Knese, Z. Anat. EntwGesch., *115*:162, 1950.
17. H. D. Darcus, J. Anat., Lond., *85*:55, 1951. N. Salter and H. D. Darcus, J. Anat., Lond., *86*:197, 1952.
18. J. V. Basmajian and A. Latif, J. Bone Jt Surg., *39A*:1106, 1957. O. DeSousa *et al.*, Anat. Rec., *139*:125, 1961. J. E. Pauly, J. L. Rushing, and L. E. Scheving, Anat. Rec., *159*:47, 1967. See also J. V. Basmajian, cited on p. 30.
19. J. V. Basmajian and A. Travill, Anat. Rec., *139*:45, 1961.
20. A. Travill and J. V. Basmajian, Anat. Rec., *139*:557, 1961.

THE FOREARM 15

The muscles of the forearm consist of an anterior and a posterior group. Those of the anterior group are the flexors of the wrist and fingers, and the pronators. Those of the posterior group are the extensors of the wrist and fingers, and the supinator.

The arrangement of muscles and nerves of the forearm is shown in figure 15–1. The synovial sheaths of the individual muscles are described and illustrated on pages 136, 141, and 144.

The fascia of the forearm, termed the *antebrachial fascia,* forms a sheath for these muscles and contributes to their origins. It is attached behind to the olecranon and the posterior border of the ulna. It receives tendinous expansions from the triceps and biceps, and it forms the extensor retinaculum on the back of the lower end of the forearm (p. 135).

MUSCLES OF FRONT OF FOREARM

The muscles on the front of the forearm may be divided into five superficial muscles and three deep. One of the superficial muscles (flexor digitorum superficialis) is more deeply placed and is sometimes regarded as forming a middle layer. **The superficial group arises mostly from the front of the medial epicondyle of the humerus by a common tendon, and from adjacent fascia, and is supplied chiefly by the median nerve. The deep group is supplied mostly by the anterior interosseous nerve, a branch of the median. Those muscles in both groups that are not supplied by the median nerve are supplied by the ulnar nerve.** Figure 15–2 shows the arrangement of the tendons of these muscles in the lower part of the forearm.

Superficial Group

Pronator Teres. The pronator teres arises from the medial supracondylar ridge

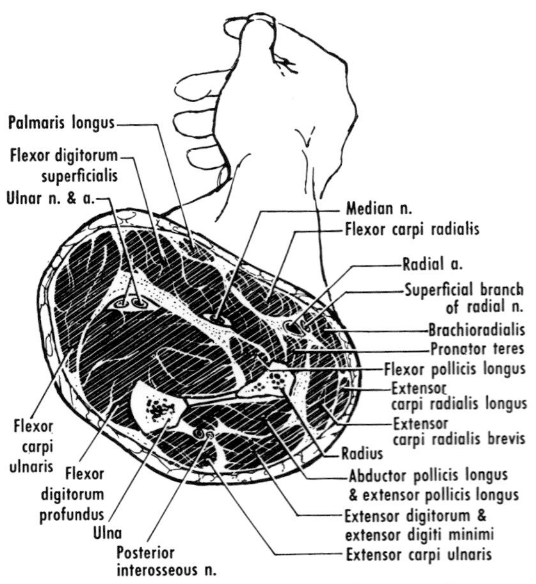

Figure 15–1 Diagram of a horizontal section through the middle of the forearm. Note that the ulnar artery and nerve are bound with the profundus, whereas the median nerve descends on the deep surface of the superficialis.

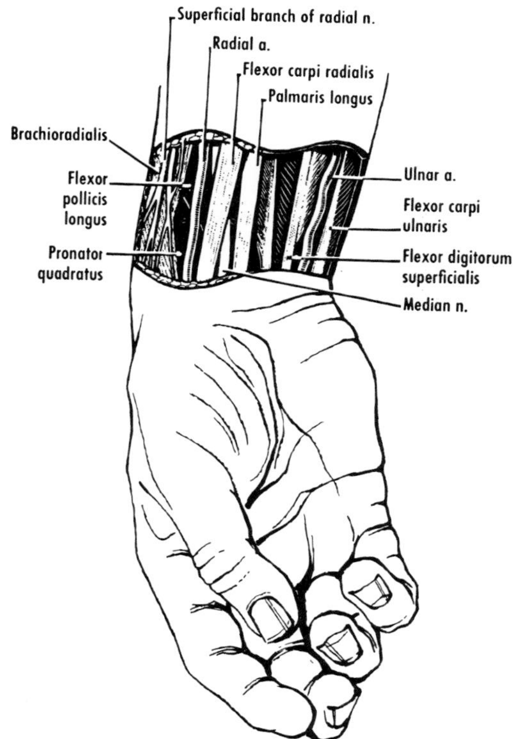

Figure 15–2 The structures in the lower part of the front of the forearm. The flexor pollicis longus is still muscular at this level. The radial artery passes from its anterior surface to that of the pronator quadratus. The flexor carpi radialis is tendinous, whereas the flexor carpi ulnaris is partly muscular. The latter covers the ulnar nerve. The median nerve is quite superficial in the interval between the palmaris longus and flexor carpi radialis. If the palmaris longus is absent, the nerve is even more exposed.

and from the medial epicondyle of the humerus. Usually a second and deeper head (often just a thin slip) takes origin from the coronoid process of the ulna. The muscle passes downward and laterally, and is inserted into a roughened area on the middle of the lateral surface of the radius.

NERVE SUPPLY AND ACTION. It is supplied by the median nerve, which usually passes between the two heads.[1] It pronates and flexes the forearm.

Flexor Carpi Radialis. The flexor carpi radialis arises by the common tendon from the medial epicondyle. It is inserted chiefly into the front of the bases of the second and third metacarpal bones.

NERVE SUPPLY AND ACTION. It is supplied by the median nerve. It flexes the hand and acts with the radial extensors in abducting the hand. It is a synergist together with the flexor carpi ulnaris and steadies the wrist during extension of the fingers.

Palmaris Longus. The palmaris longus arises by the common tendon from the medial epicondyle. Its long tendon of insertion is attached to the front of the flexor retinaculum and to the apex of the palmar aponeurosis. The palmaris longus is often absent. Its incidence is related to the side of the body, to sex, and to racial group.[2]

NERVE SUPPLY AND ACTION. It is supplied by the median nerve. It is thought to tense the palmar aponeurosis in movements of the hand, particularly the thumb.

Flexor Carpi Ulnaris. The flexor carpi ulnaris arises by the common tendon from the medial epicondyle, but also has a second head from the olecranon and from the posterior border of the shaft of the ulna. It is inserted into the pisiform and, by means of the pisohamate and pisometacarpal ligaments, into the hook of the hamate and base of the fifth metacarpal. Its tendon is a guide to the ulnar nerve and artery on its lateral side.

NERVE SUPPLY AND ACTION. It is supplied by the ulnar nerve, which passes between the two heads of the muscle. It flexes the hand and combines with the extensor carpi ulnaris to adduct the hand. It acts as a synergist by steadying the pisiform during abduction of the little finger by the abductor digiti minimi. The fibers of the two are frequently continuous. It also acts synergistically with the flexor carpi radialis in steadying the wrist during extension of the fingers and, together with the extensor carpi ulnaris, it steadies the hand during extension and abduction of the thumb.

Flexor Digitorum Superficialis. The flexor digitorum superficialis is more deeply placed than the flexor carpi ulnaris. One head, thick and strong, arises by the common tendon from the medial epicondyle of the humerus. A second head of origin, often thin and weak, arises from the upper part of the anterior border of the radius. The two heads are connected by a fibrous bridge across the median nerve and ulnar artery. The combined muscle mass then separates into a superficial and a deep part. The superficial part forms two tendons, one each for the third and fourth fingers. The deep part also forms two tendons, one each for the second and fifth fingers. The deep muscle mass often has a rounded intermediate tendon. The four tendons for the digits pass through the carpal canal, enclosed in a common synovial sheath with the flexor profundus tendons. Under cover of the palmar aponeurosis, the tendons diverge and each passes deep to the fibrous sheath of the appropriate finger. Opposite the proximal phalanx, each tendon splits into two slips that embrace a flexor profundus tendon. After reversing their surfaces, they reunite behind it and then diverge to be inserted into the margins of the anterior surface of the middle phalanx.

In a finger, a superficialis tendon is enveloped in the same digital synovial sheath as its accompanying profundus tendon, which lies deep to it. Both the superficialis and profundus tendons are anchored to the phalanges and interphalangeal joints by vascular fibrous bands called *vincula*, which, covered by folds of the synovial sheath, act as mesotendons and carry the blood supply to the tendons.[3]

NERVE SUPPLY AND ACTION. It is supplied by the median nerve. It flexes the middle phalanges on the proximal.

Deep Group

Flexor Digitorum Profundus.[4] The flexor digitorum profundus has an extensive origin from most of the anterior surface of the ulna and the adjacent portion of the medial surface and coronoid process. It also originates from the posterior border of the ulna and from the front of the interosseous membrane. It passes through the carpal canal and is enclosed in a common synovial sheath with the flexor superficials. It divides into four tendons, one for each of the medial four fingers. Each passes deep to the fibrous sheath of its digit, behind the corresponding superficialis tendon, and is enveloped in the same digital synovial sheath. Opposite the proximal phalanx, each profundus tendon is embraced by the divisions of the superficialis tendon and is then inserted into the front of the base of the distal phalanx. Each profundus tendon has vincula similar to those of the flexor superficialis.

In the palm, each flexor profundus tendon gives origin to a lumbrical muscle.

NERVE SUPPLY AND ACTION. The lateral part of the muscle is supplied by the anterior interosseous nerve (a branch of the median), the medial part by the ulnar nerve. It flexes the distal phalanges on the middle, but this movement is generally accompanied by flexion of the middle phalanges by the flexor digitorum superficialis.

Flexor Pollicis Longus. The flexor pollicis longus arises from most of the anterior surface of the radius and from the adjacent portion of the interosseous membrane. It usually has an origin from the medial epicondyle also, and often from the coronoid process.[5] The tendon of insertion passes through the carpal canal, behind the flexor retinaculum, enclosed in a special synovial sheath. It extends deeply along the medial side of the thenar eminence between the two sesamoids of the thumb, under cover of the fibrous sheath, and is inserted into the palmar aspect of the base of the distal phalanx of the thumb.

NERVE SUPPLY AND ACTION. It is supplied by the anterior interosseous nerve. It flexes the distal phalanx of the thumb.

Pronator Quadratus. The pronator quadratus arises from the anterior surface and border of the distal part of the ulna, along the pronator ridge. It passes laterally and is inserted into the anterior surface and border of the lower third of the radius.

NERVE SUPPLY AND ACTION. It is supplied by the anterior interosseous nerve. It pronates the forearm.

MUSCLES OF BACK OF FOREARM

The muscles on the back of the forearm are chiefly the extensors of the wrist and fingers. They may be divided into seven superficial muscles and five deep. Most of the superficial group arises from the back of the lateral epicondyle of the humerus by a common tendon. The muscles are supplied by the radial nerve or by its deep branch (or by the posterior interosseous nerve).

Superficial Group

Brachioradialis. The brachioradialis arises from the upper part of the lateral supracondylar ridge of the humerus and is inserted into the lateral surface of the radius, just above the styloid process.

NERVE SUPPLY AND ACTION. It is supplied by the radial nerve. It flexes the forearm.

Extensor Carpi Radialis Longus, Extensor Carpi Radialis Brevis. These muscles have similar actions. The longus arises from the lower part of the lateral supracondylar ridge and is inserted into the back of the base of the second metacarpal. The brevis arises from the lateral epicondyle by the common tendon. It is inserted into the back of the bases of the second and third metacarpals. The tendons of both muscles, with their synovial sheath, pass deep to the extensor retinaculum.

NERVE SUPPLY AND ACTION. The radial nerve supplies the longus. The brevis is supplied by either the radial nerve or by its deep branch.[6] The two muscles extend the hand, generally in conjunction with the extensor carpi ulnaris. (This extension is a normal accompaniment of finger flexion when a fist is made.) Otherwise the two radial extensors abduct the hand as they extend. Pure abduction is produced when they act with the flexor carpi radialis. The radial extensors and the extensor carpi ulnaris act synergistically to steady the wrist during flexion of the fingers.

Extensor Digitorum. The extensor digitorum arises from the lateral epicondyle by the common tendon. Above the wrist it divides into four tendons that pass deep to the extensor retinaculum, enclosed in a synovial sheath with the extensor indicis. On the back of the hand, the tendons diverge but are connected by bands. Usually the tendons for the second and third fingers are connected by a transverse band. The tendon for the fourth finger sends a slip to that for the third. The tendon for the fifth finger divides into two slips, one of which joins the tendon for the fourth finger, and the other unites with the lateral part of the extensor digiti minimi.

On the dorsum of each finger is found a fibrous sheet known as the extensor expansion or dorsal aponeurosis, which contains transverse fibers that form what is called a hood. The fibrous sheet is penetrated by the extensor tendon, which then divides into three parts, a central slip and two collateral bands. The central slip, which may be attached to the base of the proximal phalanx, is inserted into the dorsum of the base of the middle phalanx. The collateral bands fuse with expansions derived from the insertions of the interosseous and lumbrical muscles. The reinforced collateral bands converge and unite to be inserted on the dorsum of the base of the distal phalanx. The extensor expansion is described further on page 147.

NERVE SUPPLY AND ACTION. It is supplied by the deep branch of the radial nerve. It extends the proximal phalanges on the metacarpals. Its tendency to hyperextend these joints is counterbalanced by the flexors of these joints, namely the lumbricals and interossei. The antagonistic action of these flexors enables the extensor digitorum to act as a weak extensor of the middle and distal phalanges. The extensor digitorum tends to make the fingers diverge and thus can simulate the abduction of the digits produced by the dorsal interosseous muscles.

Extensor Digiti Mimimi. The extensor digiti minimi arises by the common tendon from the lateral epicondyle. The muscle gives way to a tendon that passes deep to the extensor retinaculum, enclosed in a synovial sheath. The tendon then divides into two slips, the more lateral of which joins the tendon of the extensor digitorum. Both are inserted into the extensor aponeurosis of the little finger.

NERVE SUPPLY AND ACTION. It is supplied by the deep branch of the radial nerve. It extends the proximal phalanx of the little finger.

Extensor Carpi Ulnaris. The extensor carpi ulnaris arises by the common tendon from the lateral epicondyle and from the oblique line and posterior border of the ulna. It is inserted into a tubercle on the medial side of the base of the fifth metacarpal.

NERVE SUPPLY AND ACTION. It is supplied by the deep branch of the radial nerve. It extends the hand and acts with the radial extensors. Pure adduction is produced when the muscle acts with the flexor carpi ulnaris.

Anconeus. The anconeus arises from the back of the lateral epicondyle of the humerus and is inserted into the lateral surface of the olecranon and the adjacent part of the posterior surface of the ulna.

NERVE SUPPLY AND ACTION. It is supplied by the radial nerve. It is active during supination and pronation, probably as a joint stabilizer, and it assists the triceps.[7]

Deep Group

Supinator (fig. 15–3). The supinator is largely concealed by the superficial mus-

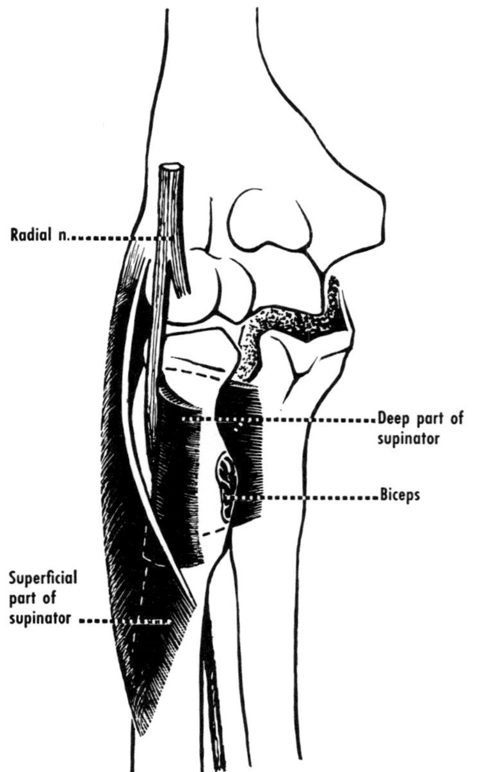

Figure 15–3 Schematic diagram of the two parts of the supinator. The superficial part descends obliquely; the deep part runs horizontally.

cles. **An understanding of the arrangement and relationships of the supinator is the key to an understanding of the elbow region.**[8] It arises chiefly from the lateral epicondyle of the humerus. Additional origins are shown in figure 15–4. Its fibers are arranged in two layers, separated by the deep branch of the radial nerve. The more vertical superficial fibers are inserted into an oblique line on the radius extending between the tuberosity and the insertion of the pronator teres.

The deep layer arises largely from the supinator fossa and crest, and from the oblique line of the ulna. The fibers almost completely encircle the radius and are inserted into the upper third of the shaft. The radius is often bare between the insertions of the superficial and the deep part. The deep branch of the radial nerve passes between the superficial and deep layers of the supinator, sometimes comes in contact with the bone, and meets the posterior interosseous vessels at the lower edge of the muscle.

NERVE SUPPLY AND ACTION. It is supplied by the deep branch of the radial nerve. It supinates the forearm.

Abductor Pollicis Longus. The abductor pollicis longus arises from the upper part of the posterior surface of the interosseous membrane and the adjacent parts of the radius and ulna. It is inserted into the lateral side of the base of the first metacarpal and usually into the trapezium. Frequently it sends a slip to the abductor pollicis brevis. Accessory slips or tendons are generally present.[9] Its tendon and that of the extensor pollicis brevis cross the extensor carpi radialis brevis and longus obliquely and pass deep to the extensor retinaculum, enclosed in a synovial sheath.

NERVE SUPPLY AND ACTION. It is supplied by the posterior interosseous nerve. It abducts the first metacarpal at the carpometacarpal joint, and stabilizes that bone during movements of the phalanges.

Extensor Pollicis Brevis. The extensor pollicis brevis arises from the distal part of the posterior surface of the radius below the origin of the abductor policis longus, with which it remains closely related, and from the adjacent part of the interosseous membrane. It is inserted into the back of the

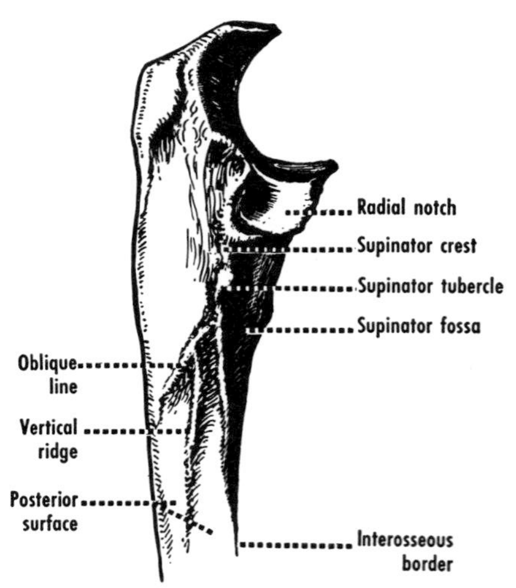

Figure 15–4 The upper end of the ulna. The edge of the radial notch, the supinator fossa and crest, and the oblique line give attachment to the supinator, and the oblique line separates it from the insertion of the anconeus. Aponeurotic fibers of the extensor carpi ulnaris also arise from the oblique line.

proximal phalanx of the thumb and continues toward the distal phalanx, which it commonly reaches. It blends with the extensor pollicis longus and forms a strong dorsal aponeurosis over the proximal phalanx.

NERVE SUPPLY AND ACTION. It is supplied by the posterior interosseous nerve. It extends the thumb.

Extensor Pollicis Longus. The extensor pollicis longus arises from the middle of the posterior surface of the ulna and from the adjacent part of the interosseous membrane. It is inserted into the dorsal aspect of the base of the distal phalanx of the thumb,. Its tendon lies in a groove medial to, or on the medial aspect of, the dorsal tubercle of the radius. It passes deep to the extensor retinaculum, enclosed in a synovial sheath, and crosses the extensor carpi radialis longus and brevis obliquely.

NERVE SUPPLY AND ACTION. It is supplied by the posterior interosseous nerve. It extends the distal phalanx of the thumb. It is said that, when the thumb is in full extension, it can adduct the thumb, owing to the oblique course of its tendon around the dorsal tubercle.

Extensor Indicis. The extensor indicis arises from a small area on the distal part of the posterior surface of the ulna and from the interosseous membrane. It is inserted into the extensor expansion of the index finger.

Anatomical Snuff-box. When the thumb is extended, a hollow, known as the "anatomical snuff-box," is readily visible between the tendon of the extensor pollicis longus medially and the tendons of the extensor pollicis brevis and abductor pollicis longus laterally (figs. 15–5 and 15–6). The floor of the hollow is formed by the scaphoid and the trapezium. It is limited proximally by the styloid process of the radius. The radial artery crosses the floor. The terminal digital branches of the superficial branch of the radial nerve can be felt crossing the tendon of the extensor pollicis longus.

Extensor Retinaculum. This is a thickening of the fascia on the back of the distal end of the forearm. It extends from the anterior border of the radius to the styloid process of the ulna and to the back of the triquetrum. It is crossed by the superficial branch of the radial nerve and by the dorsal branch of ulnar nerve.

From its deep aspect, septa attached to ridges on the radius and ulna form six compartments. Each compartment has a synovial sheath. The compartments contain the following tendons from lateral to medial side (fig. 15–7): (1) Abductor pollicis longus and extensor pollicis brevis; these may occupy separate compartments; (2) extensor carpi radialis longus and brevis; (3) extensor pollicus longus; (4) extensor digitorum and extensor indicis, (5) extensor digiti minimi, and (6) extensor carpi ulnaris.

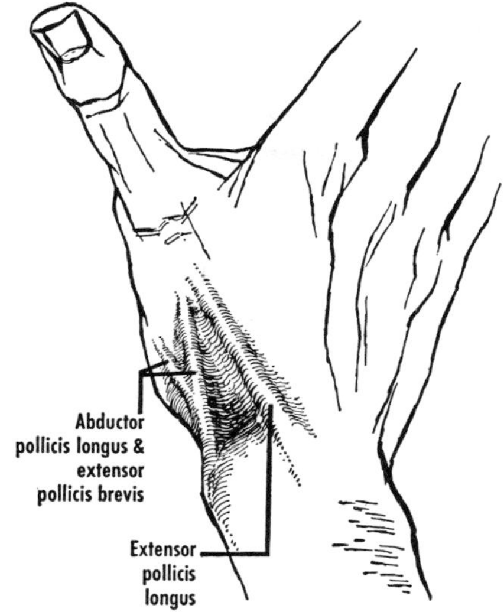

Figure 15–5 The "anatomical snuff-box." Note how the tendons of the abductor pollicis longus and extensor pollicis brevis diverge in proceeding distally.

The first group lies on the lateral side of the distal end of the radius, the second group lies on the posterolateral aspect, the third and fourth groups lie on the posterior aspect, the fifth lies posteriorly between the radius and ulna, and the sixth lies between the head of the ulna and its styloid process.

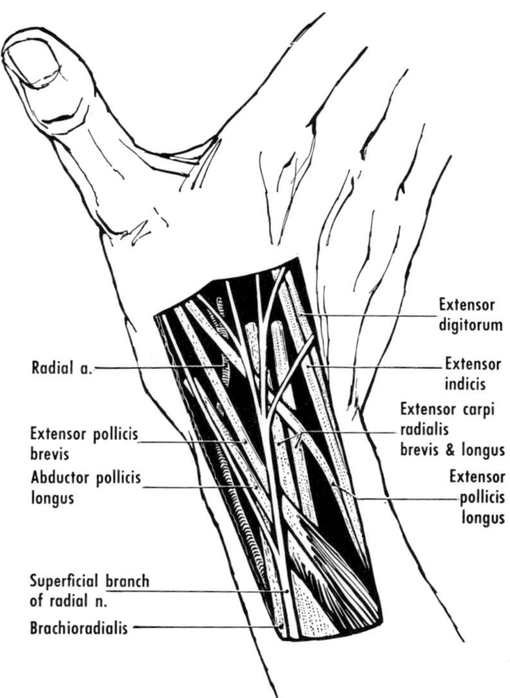

Figure 15–6 The "anatomical snuff-box."

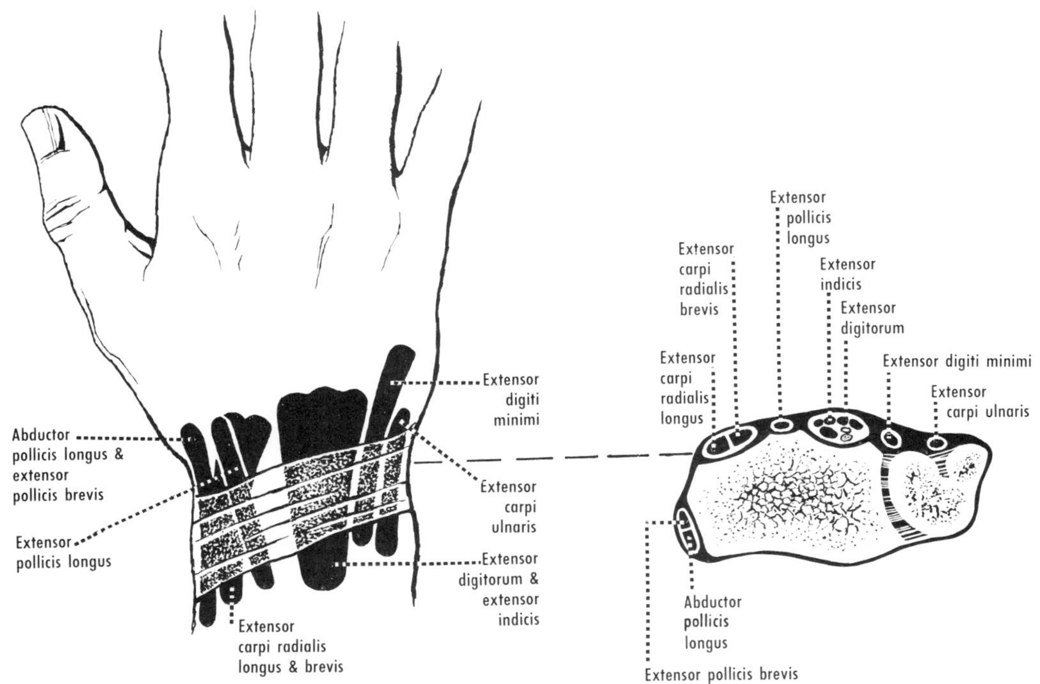

Figure 15–7 Schematic representation of the extensor tendons and their osteofibrous canals deep to the extensor retinaculum. The horizontal section at the right is through the lower ends of the radius and ulna.

NERVES OF FOREARM

Median Nerve (p. 112 and fig. 64–4, p. 768). The median nerve leaves the cubital fossa, usually by passing between the two heads of the pronator teres, and is separated from the ulnar artery by the deep or ulnar head of that muscle. Occasionally the nerve passes deep to both heads, or it may pierce the superficial or humeral head. In the forearm and hand it is accompanied by the median artery, a branch of the anterior interosseus. **The median nerve passes behind the tendinous arch connecting the two heads of the flexor digitorum superficialis, and it remains under cover of this muscle, adherent to its deep aspect and lying on the flexor digitorum profundus, until it approaches the wrist.** Here it becomes more superficial by emerging between the flexor digitorum superficialis and the flexor carpi radialis, and is partly covered by the palmaris longus when present. The median nerve enters the hand by passing through the carpal canal, behind the flexor retinaculum and in front of the flexor tendons. The median and ulnar nerves may communicate in the forearm, across the flexor digitorum profundus. The surface anatomy of the median nerve is described on page 158.

BRANCHES. In the cubital fossa, a bundle of *muscular* branches leaves the median nerve from its medial side. These supply the pronator teres, flexor carpi radialis, palmaris longus, and flexor digitorum superficialis.

The *anterior interosseous nerve* arises from the back of the median nerve in the cubital fossa. In company with the anterior interosseous artery, it descends on the front of the interosseous membrane between the adjacent margins of the flexor pollicis longus and flexor digitorum profundus, both of which it supplies. It then passes behind the pronator quadratus, supplies it, and ends in twigs to the wrist and intercarpal joints. It may communicate with the ulnar nerve across the flexor profundus.

In the lower part of the forearm, the median nerve gives off an inconstant *palmar branch,* which supplies a small area of the skin of the palm.

Ulnar Nerve (fig. 64–5, p. 769). The ulnar nerve lies in the groove on the back of the medial epicondyle of the humerus, where it can be felt and rolled against the bone, and it can also be palpated against the medial surface of the coronoid process. The origin of the name "funny bone" or "crazy bone" has been attributed to the sensation elicited by a light blow to the ulnar nerve in this region. **The ulnar nerve enters the forearm between the heads of origin of the flexor carpi ulnaris.** It then lies on the flexor digitorum profundus, covered by the flexor carpi ulnaris. At the junction of the upper and middle thirds of the forearm, the ulnar artery meets it and accompanies it on its lateral side from there downward. It gives branches to this artery. **In the distal part of the forearm, the ulnar nerve becomes superficial and lies between the flexor carpi ulnaris and the flexor digitorum superficialis. The ulnar nerve and artery enter the hand by passing in front of the flexor retinaculum, lateral to the pisiform, between this bone and the hook of the hamate.** They are covered by a slip of the flexor retinaculum.

The ulnar nerve may communicate with the median nerve in the forearm, across the flexor digitorum profundus. Such a communication may sometimes transmit important fibers from one nerve to the other. In rare instances, for example, the median nerve may carry all the fibers to the short muscles of the hand and give most of these to the ulnar by way of a communication in the forearm. In such cases a severance of the ulnar nerve above the elbow is not followed by paralysis of the short muscles of the hand.

BRANCHES. *Muscular branches* are given to the two muscles between which the ulnar nerve lies, the flexor digitorum profundus and the flexor carpi ulnaris.

In the middle of the forearm, the ulnar nerve gives off a large, cutaneous *dorsal branch,* which descends between the ulna and the flexor carpi ulnaris, turns posteriorly at the level of the wrist,[10] and is distributed to the hand (p. 149).

In the lower part of the forerm the ulnar nerve gives off a variable *palmar branch,* which crosses the flexor retinaculum and supplies the skin on the medial side of the palm.

Radial Nerve (fig. 64–3, p. 767). After piercing the lateral intermuscular septum in the arm, the radial nerve lies deeply between the brachioradialis and the brachialis. Here it divides into superficial and deep branches, either at the level of or below the lateral epicondyle.

BRANCHES. The radial nerve gives numerous branches in the arm, which have been described with that region. Of relevance here are the *posterior antebrachial cutaneous nerve* (p. 124) and *muscular branches* to the brachioradialis and extensor carpi radialis longus, and often the brevis. The terminal branches are superficial and deep.

The superficial branch, or continuation of the radial nerve, is cutaneous and articular in its distribution. It descends into the forearm under cover of the brachioradialis and lies successively on the insertions of the supinator and of the pronator teres. In this portion of its course, it accompanies the radial artery, lying lateral to and supplying it. In the distal part of the forearm, it winds dorsalward and becomes subcutaneous. It supplies the lateral side of the dorsum of the hand and communicates with the lateral antebrachial cutaneous nerve. Either this latter or the posterior antebrachial cutaneous nerve (a branch of the radial in the arm) may encroach on the distribution of the radial nerve in the hand. The radial nerve next divides into a number of *dorsal digital nerves* that supply the thumb, index finger, and part of the middle finger.[11] These branches reach as far distally as the thumbnail, but usually no further than the proximal phalanx of the other digits. The dorsal innervation here is completed distally by the digital nerves of the median. **The rare opportunity of identifying cutaneous nerves in the intact living body can be had by stroking a fingernail down the taut tendon of the extensor pollicis longus, where several branches of the radial nerve can be felt.**

The deep branch of the radial nerve is muscular and articular in its distribution. It arises under cover of the brachioradialis and winds laterally around the radius, between the superficial and deep layers of the supinator, which it supplies. **The deep branch often makes contact with a bare area of the radius and is vulnerable in fractures in this region.** On reaching the back of the forearm, it lies between the superficial and deep extensors, giving branches to the superficial group, and is accompanied by the posterior interosseous artery. For the rest of its course, it is termed the *posterior interosseous nerve,* a name that has also been applied to the deep branch in its entirety. In the distal part of the forearm, it passes onto the interosseous membrane by going deep to the extensor pollicis longus. It then lies, together with the anterior interosseous artery, in the groove for the extensor digitorum on the back of the radius. It ends on the back of the carpus in an enlargement from which twigs are distributed to the wrist and intercarpal joints.

In its course through the forearm, the deep branch supplies the supinator (and often the extensor carpi radialis brevis), the extensor digitorum, the extensor digiti minimi, and the extensor carpi ulnaris. The posterior interosseous nerve supplies the abductor pollicis longus, the extensor pollicis brevis, the extensor pollicis longus, and the extensor indicis.

ARTERIES OF FOREARM

Radial Artery (fig. 64–1, p. 766). The radial artery is the smaller terminal division of the brachial artery. It begins in the cubital fossa, opposite the neck of the radius. **In the distal part of the forearm, the radial artery is superficial and lies on the lateral side of the flexor carpi radialis tendon, which serves as a guide to it. Its pulsations can readily be felt here, supplying information of clinical importance, such as the rate, rhythm, compressibility, and condition of the arterial wall.** The radial artery lies successively on the biceps tendon, supinator, pronator teres, flexor digitorum superficialis, flexor pollicis longus, pronator quadratus, and the lower end of the radius. In the middle third of the forearm, the superficial branch of the radial nerve lies lateral to it. The radial artery leaves the forearm by winding dorsalward across the carpus, and its further course is described with the hand.

The radial artery varies, and it may be absent. It may arise in the arm (see p. 125), or even from the axillary artery. Occasionally, it runs a very superficial course throughout the forearm.

BRANCHES. The *radial recurrent artery*[12] passes between the radial nerve and its deep (interosseous) branch, and takes part in the anastomosis around the elbow joint.

Superficial palmar and *palmar carpal branches* arise in the lowermost part of the forearm.

Ulnar Artery (fig. 64–1, p. 766). The ulnar artery is the larger terminal division of the brachial. It begins in the cubital fossa, opposite the neck of the radius. It runs downward and medially in the upper third of the forearm, then directly downward, and it lies on the flexor digitorum profundus. In the oblique part of its course, it is covered by the muscles arising from the medial epicondyle. The median nerve crosses this portion of the ulnar artery, but the two structures are separated by the deep, or ulnar, head of the pronator teres when that head is present. In the distal two-thirds of the forearm, the ulnar nerve lies on the medial side of the artery. In the middle third, both artery and nerve are covered by the flexor carpi ulnaris; in the distal third, they lie lateral to its tendon and are superficial. The pulsations of the artery can be felt near the wrist. The ulnar artery leaves the forearm by passing in front of the flexor retinaculum on the lateral side of the pisiform, gives off the deep palmar branch, and continues as the superficial palmar arch. These are described with the hand. Like the radial, the ulnar artery has venae comitantes, is apt to vary, and may be absent.

BRANCHES. In addition to branches to adjacent muscles, the ulnar artery has the following named branches:

1. The *ulnar recurrent artery* is a small trunk from which *anterior and posterior branches* arise. These branches, which may arise separately, run to the front and back of the medial epicondyle, respectively, where they take part in the anastomosis around the elbow joint.

2. The *common interosseous artery* is a short trunk that arises in the lower end of the cubital fossa, passes backward, and divides into anterior and posterior interosseous arteries.

The *anterior interosseous artery* descends on the front of the interosseous membrane, accompanied by the anterior interosseous nerve. It pierces the interosseous membrane and descends to join the dorsal carpal network. It supplies nutrient branches to the radius and ulna, sends a branch behind the pronator quadratus to the palmar carpal network, and gives off the *median artery*, a long branch that accompanies the median nerve in the forearm and hand.

The *posterior interosseous artery* passes backward, above the upper border of the interosseous membrane, and descends on the back of the forearm, between the superficial and deep groups of muscles, accompanied by the posterior interosseous nerve. It ends by anastomosing with the anterior interosseous artery and with the dorsal carpal network. Near its origin it gives an *interosseous recurrent artery* that ascends under cover of the anconeus to the back of the lateral epicondyle.

3. *Palmar* and *dorsal carpal branches* arise in the lowermost part of the forearm. They are described with the hand.

INTEROSSEOUS MEMBRANE

The interosseous membrane is a thin, but strong, fibrous sheet connecting the shafts of the radius and ulna and giving attachment to muscles (see figs. 11–25 and

11–26, pp. 90, 91). Its fibers are directed mainly downward and medially, from radius to ulna. Distally, it is pierced by the anterior interosseous vessels and is continued into the fascia on the posterior surface of the pronator quadratus. The membrane is supplied by the interosseous nerves, and lamellated corpuscles are numerous in its substance.

The *oblique cord* is a thin ligamentous structure that extends from the lateral side of the tuberosity of the ulna down to the radius, just below its tuberosity.[13] There is a gap between the oblique cord and the interosseous membrane through which the posterior interosseous artery passes to the back of the forearm.

REFERENCES

1. L. E. Beaton and B. J. Anson, Anat. Rec., *75*:23, 1939. R. W. Jamieson and B. J. Anson, Quart. Bull. Northw. Univ. med. Sch., *26*:34, 1952. See also H. Ferner, Anat. Anz., *84*:151, 1937.
2. J. W. Thompson, J. McBatts, and C. H. Danforth, Amer. J. phys. Anthrop., *4*:205, 1921. See also A. F. Reimann *et al.*, Anat. Rec., *89*:495, 1944; T. S. King and R. O'Rahilly, Acta anat., *10*:327. 1950.
3. H. Winter and H.-H. Loetzke, Anat. Anz., *119*:337, 1966. E.-M. Ziegler, Anat. Anz., *130*:404, 1972.
4. J. L. Wilkinson, J. Anat., Lond., *87*:75, 1953. B. F. Martin, J. Anat., Lond., *92*:602, 1958.
5. J. Dykes and B. J. Anson, Anat. Rec., *90*:83, 1944. V. Mangini, J. Bone Jt Surg., *42A*:467, 1960.
6. C. R. Salsbury, Brit. J. Surg., *26*:95, 1938.
7. A. A. Travill, Anat. Rec., *144*:373, 1962. J. E. Pauly, J. L. Rushing, and L. E. Scheving, Anat. Rec., *159*:47, 1967. J. V. Basmajian and W. R. Griffin, J. Bone Jt Surg., *54A*:1712, 1972.
8. J. L. Shellshear and N. W. G. Macintosh, *Surveys of Anatomical Fields,* Grahame, Sydney, 1949. See also F. Davies and M. Laird, Anat. Rec., *101*:243, 1948.
9. T. Lacey, L. A. Goldstein, and C. E. Tobin, J. Bone Jt Surg., *33A*:347, 1951. S. S. Coleman, D. K. McAfee, and B. J. Anson, Quart. Bull. Northw. Univ. med. Sch., *27*:117, 1963. M. A. Baba, Anat. Rec., *119*:541, 1954.
10. L. Fischer *et al.*, C. R. Ass. Anat. *55*:266, 1970.
11. A distribution to all five digits has been recorded. See J. B. Learmonth, J. Anat., Lond., *53*:371, 1919
12. C. R. Salsbury, Brit. J. Surg., *26*:95, 1938.
13. B. F. Martin, J. Anat, Lond., *92*:609, 1958.

THE HAND 16

The hand[1] is the part of the upper limb distal to the forearm. Its skeletal framework includes the carpus or wrist. In lay parlance, however, the term "wrist" is used for the distal end of the forearm, and a wrist watch is worn over the lower ends of the radius and ulna.

The functional importance of the hand can be gauged from the fact that many injuries to the hand result in permanent disabilities. Much of the efficiency of the hand depends on the thumb. Objects can be readily grasped between the thumb and the index finger, owing to the specialized movement of opposition (p. 154).

The activities of the hand are free motion, power grip, precision handling, and pinch.

Power grip may be defined as the forcible motions of the fingers and thumb acting against the palm. Power grip transmits force to an object and, in contrast to free motion, is static and isometric. The hand conforms to the size and shape of an object, and phalangeal rotation, abduction, and adduction may be required. Examples of power grip are spherical grip (fig. 16–21, p. 155), hook grip, disc grip, and various kinds of squeezes.

Precision handling involves a change in position of a handled object, either in space or about its own axes, and requires exact control of finger and thumb positions. Precision handling is dynamic and isotonic, in contrast to the isometric nature of power grip. Most precision handling involves two basic modes of handling objects. One mode is precision rotation and consists of the thumb and finger activity that rotates an object about one of its internal axes. An example is winding a watch. The other mode is precision translation, the moving of an object away from or toward the palm (threading a needle, and extracting the thread after threading the needle).

Pinch is compression between the thumb

and index finger (fig. 16–23, p. 156) or between the thumb and first two fingers. It is primarily a static or isometric activity.

The *position of rest* (fig. 16–1) is that assumed by the inactive hand, a position due only to the intrinsic tone of the muscles. This is the position in which the hand should be placed if immobilization becomes necessary. At rest the palm of the hand is hollowed and the fingers are flexed, the fifth most so and the index least. The thumb is opposed slightly, and its surfaces are approximately at right angles to those of the other fingers. The wrist is dorsiflexed a little. Some authors distinguish between a position of rest and a *position of function,* in which the wrist is even more dorsiflexed.

The fingers are numbered from one to five, beginning with the thumb. (The thumb should be included as a finger.) The thumb is known as the *pollex,* the second finger as the *index,* the third as the *middle finger,* the fourth as the *ring finger,* and the fifth as the *little finger.* In clinical notes it is safer to identify a given finger by name rather than by number.[2] The term *thenar* is used as an adjective with reference to the ball of the thumb, and *hypothenar* is employed with reference to the ball of the little finger.

Some general features of the hand should be noted. The third finger projects farther distally than either the fourth or second. (These two may be equal in length.) All three project farther than the little finger and thumb. These arrangements are expressed as 3>4>2>5>1, or 3>2>4>5>1, a notation known as the *digital formula.*

The skin of the dorsum of the hand is thin and mobile. The thicker, hairless skin of the palm of the hand is firmly bound to underlying subcutaneous tissue by thick, fibrous bands, which enclose or form pockets for fat. Hairs are commonly present on the backs of the proximal phalanges, and are invariably absent on the distal ones. The incidence of hairs on the middle phalanges varies widely, with apparent familial and racial tendencies.[3]

Papillary ridges form a characteristic pattern on the palmar aspect of the hand. The ridges are especially well developed in the pads of the digits, and fingerprints are used extensively as a means of identification of an individual (p. 48). The hand presents a number of flexure lines at sites of skin movement. Here the dermis of the skin is anchored to the subcutaneous tissue. It should be noticed that these lines do not necessarily indicate the sites of joints. Thus, well-marked creases are seen opposite the middle of the front of a proximal phalanx. Two well-marked flexure lines run across the palm from the medial side of the hand to either side of the base of the index finger. From the lateral margin of the hand a curved crease can be traced around the base of the thenar eminence. These lines reflect the independent movements of the index finger and thumb. The main creases of the hand appear very early in fetal life and are not caused by movements of the digits. The nails are described on page 50.

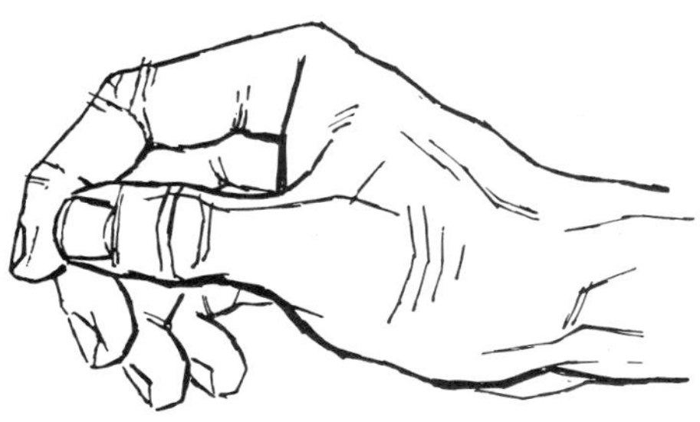

Figure 16–1 The position of the hand at rest.

FASCIA AND SYNOVIAL SHEATHS

The subcutaneous tissue on the back of the hand is loose and thin and is arranged in two layers, which join at the margins of the hand and at the webs of the fingers. The subcutaneous tissue in the palm is developed into fibrofatty pads that aid in grasping.[4]

The fascia of the hand forms a more or less continuous sheet that retains tendons in position, prevents them from ridging the skin, and provides pulleys over which they act. The fascia of the dorsum of the hand is arranged in layers,[5] which invest the tendons and the interosseous muscles.

The fascia of the front of the forearm continues distally into the hand. In front of the carpus it forms the palmar aponeurosis. Over the digits it forms strong, fibrous sheaths around the tendons.

The webs of the fingers contain transverse fibrous bands (interdigital, or natatory, ligaments), which limit flexion of a finger when the adjacent fingers are extended. Some of the fibers are continued distally along the sides of the fingers as the so-called skin ligaments.[6]

Flexor Retinaculum. **The flexor retinaculum is a transverse, fibrous band that binds the flexor tendons of the five fingers, together with their synovial sheaths and the median nerve, in the arch of the carpus. It thus converts the arch into the carpal canal** (fig. 16–2). It has four chief attachments. The proximal border

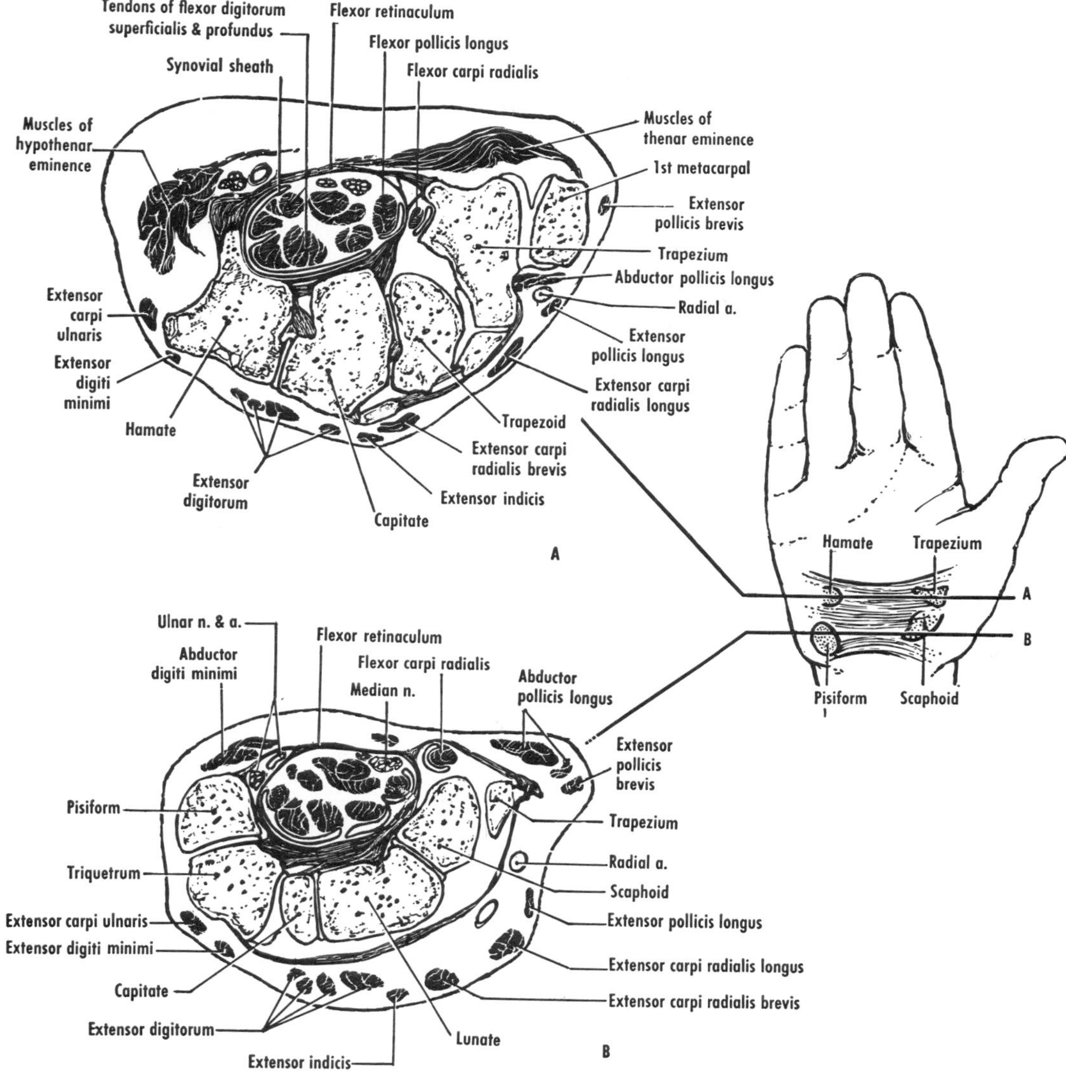

Figure 16–2 Drawings of horizontal sections through the wrist at the levels indicated on the sketch of the hand.

extends from (1) the tubercle of the scaphoid to (2) the triquetrum and pisiform,[7] and the distal border extends from (3) the tubercle of the trapezium to (4) the hook of the hamate. At least the proximal points of attachment can be felt *in vivo* and, as the proximodistal width of the retinaculum is approximately 3 cm, the quadrilateral outline of the retinaculum can be mapped on the living hand. It lies distal to the flexion creases in the skin of the wrist.

Retinacular fibers bridge the groove on the trapezium, and thus form a tunnel for the flexor carpi radialis tendon. The anterior surface of the retinaculum gives attachment to the palmaris longus and to the thenar and hypothenar muscles. This surface is crossed by the palmar branches of the median and ulnar nerves and artery. The last two may be crossed by a fascial thickening, which is attached to the pisiform and is termed the superficial part of the flexor retinaculum.

Palmar Aponeurosis. **The palmar aponeurosis is a strong, triangular membrane overlying the tendons in the palm.** Its apex is continuous with the palmaris longus (when present) and the fascia around that tendon, and is anchored to the front of the flexor retinaculum. It can be readily distinguished from the retinaculum only by the longitudinal direction of its fibers and its distal continuation into pretendinous bands. At its lateral and medial borders, it is continuous with the fascia over the thenar and hypothenar muscles, which extends deeply to reach the first and fifth metacarpal bones, respectively. These two fascial extensions isolate the central portion of the palm from the thenar and hypothenar eminences.

Distally, the palmar aponeurosis continues as four slips or pretendinous bands, overlying the flexor tendons of the medial four fingers. The bands are connected by transverse fibers that are situated a short distance proximal to the webs of the fingers, in the plane of the natatory fibers. They constitute the *superficial transverse metacarpal ligament.* Each pretendinous slip is attached to the fibrous flexor sheath of the corresponding finger. The fibrous flexor sheaths are transverse fibers that arch over the flexor tendons and their synovial sheaths. They are attached to the margins of the phalanges.

Fascial "Spaces" of Palm.[8] The flexor tendons in their synovial sheaths, together with the median nerve (closely applied to the front of the common flexor sheath), leave the carpal canal and enter the central compartment of the palm. **The central compartment is bounded in front by the deep aspect of the palmar aponeurosis, on the sides by the fascia covering thenar and hypothenar muscles, and behind by a deep pad of fat lying on the interosseous fascia and by the fascia covering the front of the adductor pollicis muscle. Uncontrolled infection (from a synovial sheath, for example) may rupture into this compartment and may spread proximally into the forearm, in front of the pronator quadratus muscle and its fascia.**

The central compartment of the palm is subdivided into fascial spaces by fibrous septa that extend from the deep aspect of the palmar aponeurosis to the adductor fascia or to the anterior interosseous fascia (fig. 16–3). These septa lie between the flexor tendons and the lumbricals and form well-defined canals. Proximally, they begin in the angle between the lumbricals and the deep flexor tendons. Distally, they merge with the deep transverse metacarpal ligament.

The fat pad that lies between the anterior interosseous fascia and the synovial sheaths is generally several centimeters long. It provides a soft and elastic cushion for the synovial sheaths of the flexors and covers the deep branch of the ulnar nerve from its entrance into the medial space to its exit into the adductor pollicis.

Synovial Flexor Sheaths (fig. 16–4). Where tendons are retained in place by fascial retinacula, they are invested by *synovial sheaths,* which facilitate gliding. **There are three synovial sheaths on the front of the wrist. That for the flexor carpi radialis tendon is short. The other two, (1) a common synovial flexor sheath enveloping all the superficialis and profundus tendons, and (2) a sheath for the flexor pollicis longus, are important for the smooth functioning of the tendons and are of clinical significance, owing to the impairment of hand function that may result from untreated infection of these sheaths.** In clinical writing, these sheaths are usually referred to as the *ulnar* and the *radial bursa,* respectively. The two sheaths extend proximally for a short distance above the flexor retinaculum, and they usually communicate with each other in the carpal canal.[9] The sheath of the flexor pollicis longus extends distally almost to the insertion of that tendon into the distal phalanx of the thumb. The common sheath of the superficialis and profundus tendons usually extends almost to the insertion of the flexor profundus into the distal phalanx of the fifth finger. In about one-fifth of instances, however, a gap separates the main sheath from the portion extending into the fifth digit. The second, third, and fourth fingers have digital sheaths also, but generally these extend proximally only as far as the necks of their metacarpal bones. There is a gap of 1 to 3 cm between their terminations and the common sheath. This interval, free of sheaths, corresponds very approximately to that between the two transverse skin creases of the palm. It can thus be seen that infection of the synovial sheath of the thumb or little finger can spread readily into the palm, and even into the forearm, whereas an infection of the sheaths of the intermediate three fingers is more likely to be localized.

The synovial sheaths in the digits have mesotendons, called vincula (p. 132), that carry the blood supply to the tendons and anchor them to the phalanges.[10]

MUSCLES OF HAND

The muscles of the hand (intrinsic) are those of the thumb (thenar muscles), those of the little finger (hypothenar muscles), the palmar and dorsal interossei, and the lumbricals. Their motor supply is derived from the first thoracic segment of the spinal cord; the fibers

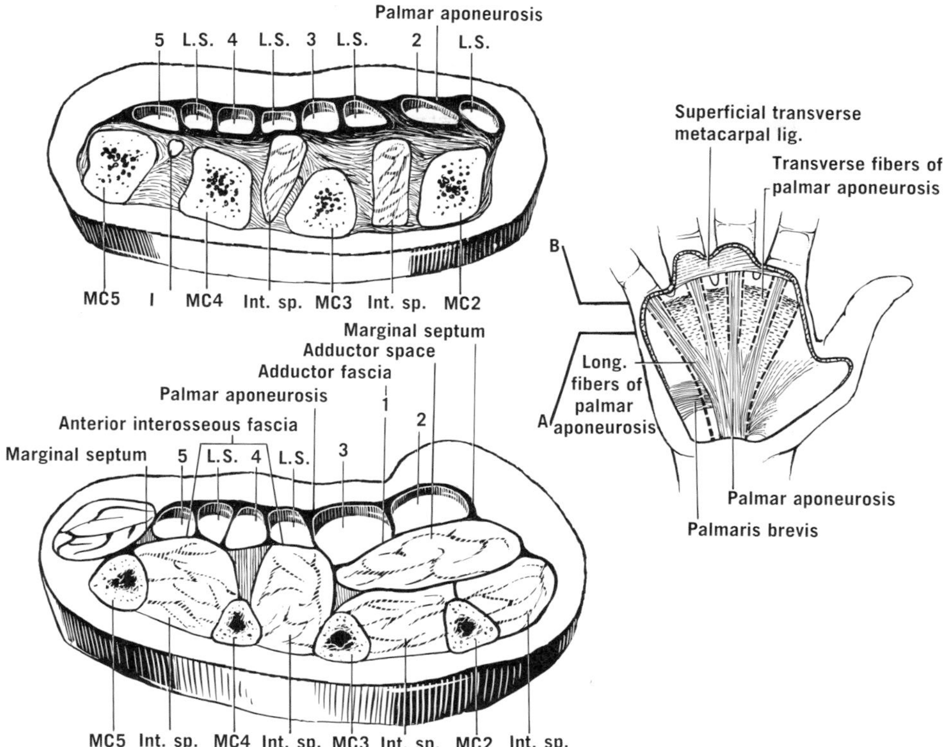

Figure 16–3 The right hand, showing the course of fibers in the palmar aponeurosis. The positions of the two marginal septa and the seven intermediate septa are marked by broken lines. *A*, transverse section through the hand at the indicated level, just distal to the thumb, showing the compartments of the central space. Note that the two shortest of the paratendinous septa are not included in the section. *B*, transverse section at the level of the heads of the metacarpals, showing the subdivision of the central space into eight compartments. L.S., lumbrical space. Int. sp., and I., interosseous space. MC, metacarpal. Based on Bojsen-Møller and Schmidt,[8] by permission of the authors and publisher.

reach the muscles by way of the median and ulnar nerves.[11]

Muscles of Thumb

The pincerlike grip between the thumb and index finger is such a useful asset to all, and such an indispensable feature to most workers, that study of the thumb is of great importance. The short muscles of the thumb (fig. 16–5) are the abductor pollicis brevis, the flexor pollicis brevis, the opponens pollicis, the adductor pollicis, and the first palmar and first dorsal interossei. The last two are discussed on page 146.

Abductor Pollicis Brevis, Flexor Pollicis Brevis, Opponens Pollicis. These muscles form the thenar eminence and arise from the front of the flexor retinaculum and from the tubercle of the trapzeium. The abductor also arises from the tubercle of the scaphoid, and usually receives a slip from the tendon of the abductor pollicis longus.[12] The abductor pollicis brevis (usually two bellies) is the most superficial and forms the lateral side of the thenar eminence. The flexor pollicis brevis forms the medial part. The abductor and the flexor, which are often fused, insert by a common tendon into the

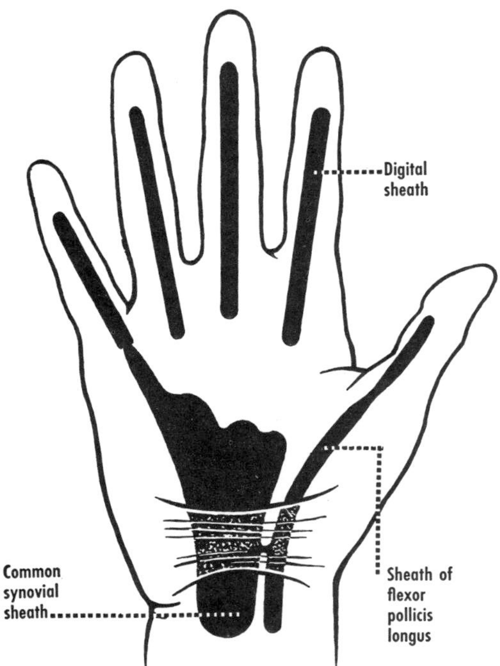

Figure 16–4 Diagram of a common arrangement of the synovial sheaths of the flexor tendons.

lateral sesamoid bone and the lateral side of the base of the proximal phalanx of the thumb, and also into the extensor expansion.

The term "deep head" of the flexor pollicis brevis has unfortunately been applied to several different muscular slips. Here it refers to a slip that arises in common with the oblique head of the adductor pollicis, and which is supplied by the ulnar nerve, but which joins the superficial head to be inserted on the lateral sesamoid.[13]

The opponens pollicis is more deeply placed and is covered by the abductor pollicis brevis. It extends from the flexor retinaculum and trapezium to the lateral border and lateral part of the front of the shaft of the first metacarpal. It is sometimes fused with the flexor pollicis brevis.

NERVE SUPPLY AND ACTION. **The three muscles are supplied by a special recurrent branch of the median nerve. This important branch lies at the distal border of the flexor retinaculum and is very superficial.** It often anastomoses with the deep branch of the ulnar nerve by a loop termed the *thenar ansa.*[14] This anastomosis accounts for the double-motor innervation of thenar muscles, especially the deep head of the flexor pollicis brevis. An important anomaly is the occasional presence of an additional motor branch of the median that approaches the thenar muscles through the flexor pollicis brevis or palmaris brevis.[15]

The actions of the three muscles are to some extent indicated by their names. Thus, the abductor abducts the thumb. But this muscle crosses two joints (carpometacarpal and metacarpophalangeal of the thumb), and it has actions at both. It acts on the carpometacarpal joint during opposition (p. 154), and rotates the proximal phalanx medially during the same movement. It is aided in these movements by the flexor pollicis brevis. The opponens acts only on the carpometacarpal joint; it rotates the first metacarpal medially during opposition. During extension of the thumb, the opponens and abductor act together to stabilize the joint.

Adductor Pollicis. The adductor pollicis is deeply placed in the palm and arises by two heads. The *oblique head* arises from the front of the base of the second metacarpal, from the capitate and trapezoid, and from the palmar ligaments. The *transverse head* arises from the longitudinal ridge on the front of the third metacarpal. Both heads, together with the first palmar interosseous muscle, are inserted by a common

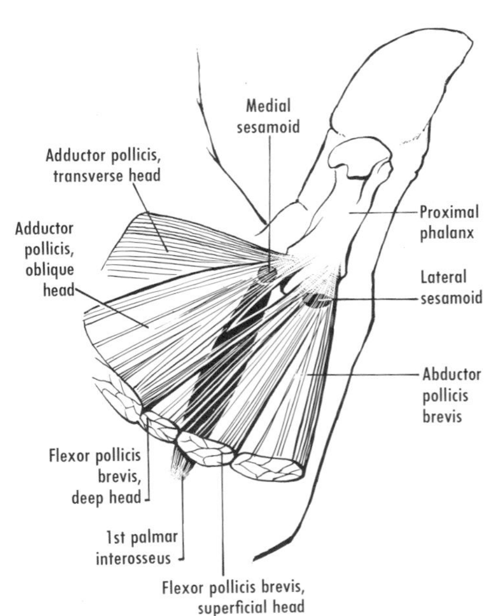

Figure 16–5 The instrinsic muscles of the thumb (opponens not shown).

tendon into the medial sesamoid bone and the medial side of the base of the proximal phalanx of the thumb, and also into the extensor expansion. A slip from the oblique head usually passes deep to the flexor pollicis longus tendon to join the flexor pollicis brevis as the deep head of the latter.

NERVE SUPPLY AND ACTION. The adductor pollicis is supplied by the deep branch of the ulnar nerve; infrequently it is innervated by the median. It adducts the thumb and, like the short flexor, aids in opposition.

In summary, most of the adductor pollicis, together with the first palmar interosseus, is inserted on the medial side of the base of the proximal phalanx and is usually supplied by the ulnar nerve. The flexor pollicis brevis and the abductor pollicis brevis are inserted into the lateral side of the base of the proximal phalanx and, together with the opponens pollicis (which is inserted into the first metacarpal), are usually supplied by the median nerve; the flexor, however, commonly receives an ulnar innervation in addition.

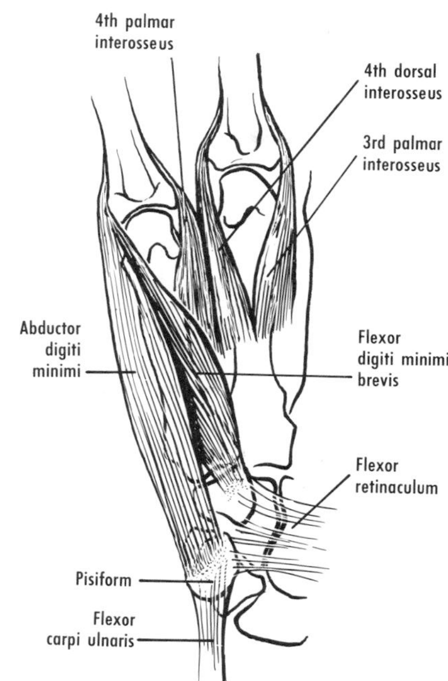

Figure 16–6 The intrinsic muscles of the little finger.

Muscles of Little Finger

Abductor Digiti Minimi, Opponens Digiti Minimi, Flexor Digiti Minimi Brevis (figs. 16–6 and 16–7). These are the short muscles of the fifth or little finger, which form the hypothenar eminence. They are supplied by the deep branch of the ulnar nerve, and their actions are indicated by their names.[16]

The abductor digiti minimi arises from the pisiform and is frequently continuous with the flexor carpi ulnaris. It is inserted into the medial side of the base of the proximal phalanx of the fifth finger. It is frequently divisible into two parts. The flexor digiti minimi brevis, which is not always present, lies lateral to the abductor. It arises from the hook of the hamate, and is inserted in common with the abductor. There is frequently a sesamoid bone in the capsule of the joint to which the tendons are related. The opponens digiti minimi is covered by the flexor and abductor, and arises from the hook of the hamate. A fibrous arch at the proximal end of the opponens transmits the deep branches of the ulnar nerve and artery. The muscle is inserted into the medial part of the front of the shaft of the fifth metacarpal bone.

NERVE SUPPLY AND ACTION. The abductor and flexor act as their names indicate. The opponens draws the fifth metacarpal bone forward, thereby deepening the hollow of the palm. The fifth finger, however, cannot oppose, although it can be opposed by the thumb.

The hypothenar muscles are covered proximally by the *palmaris brevis*, a band of subcutaneous muscle fibers that runs medially from the medial border of the palmar aponeurosis to the skin. It is innervated by the superficial branch of the ulnar nerve, assists in deepening the hollow of the palm, and protects the ulnar artery and nerve from overlying pressure.[17]

Lumbricals and Interossei

The lumbricals and interossei (figs. 16–6 to 16–8) are small but important muscles that are inserted mainly into the extensor expansion (fig. 16–9). The former are four small muscles that are associated with the tendons of the flexor digitorum profundus in the palm. The interosseous muscles lie largely between the metacarpal bones and are arranged in two groups, a palmar and a dorsal.

Lumbricals. These are numbered one to four, from lateral to medial side. The first and second arise, each by a single head, from the lateral sides of the lateral two tendons of the deep flexor, and they are innervated by digital branches of the median nerve. The third and fourth lumbricals arise, each by two heads, from the adjacent sides of the medial three tendons of the deep flexor, and they are supplied by the deep branch of the ulnar nerve. The third lumbrical may receive a twig from the median nerve also. Each lumbrical usually has the same nerve supply as its deep tendon.[18] The four lumbricals pass in front of the deep

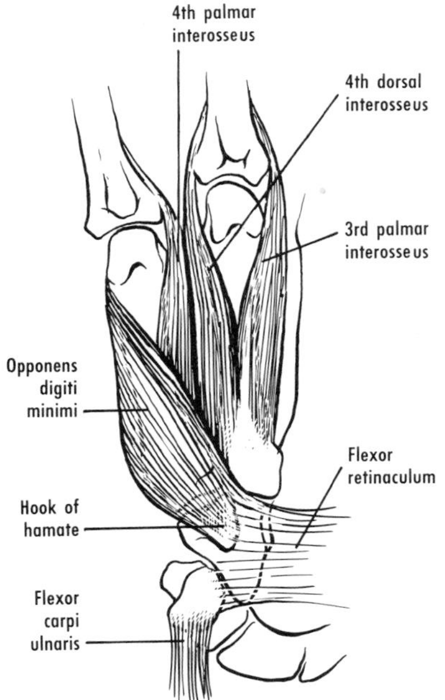

Figure 16–7 The opponens digiti minimi after removal of the overlying abductor and flexor (fig. 16–6).

transverse metacarpal ligament. They are inserted into the lateral (radial) sides of the extensor expansions of the second to fifth fingers, respectively, immediately distal to the interossei, at approximately the level of the metacarpophalangeal joints. They are chiefly extensors at the interphalangeal joints (see p. 155). Variations in origin and insertion are common.[19]

Palmar Interossei (figs. 16–8 and 16–9). These are four in number and adduct the fingers toward a line through the middle finger. Each arises by a single head from the metacarpal shaft of the finger (first, second, fourth, and fifth) that it adducts. This arrangement can best be appreciated from the diagram in figure 16–8. The palmar interossei are inserted into the extensor expansions, each on the side that will enable it to adduct the digit.

Dorsal Interossei (figs. 16–8 and 16–9). These are also four in number, and abduct the fingers away from a line through the third finger; this movement can be performed only when the fingers are extended. Each arises by two heads from the adjacent sides of two metacarpal bones. These interossei pass behind the deep transverse metacarpal ligament. The dorsal interossei are inserted into the extensor expansions of the second, third, and fourth fingers, in each case on the side that enables it to abduct the digit, and in certain fingers into the base of the proximal phalanx also (fig. 16–8). The

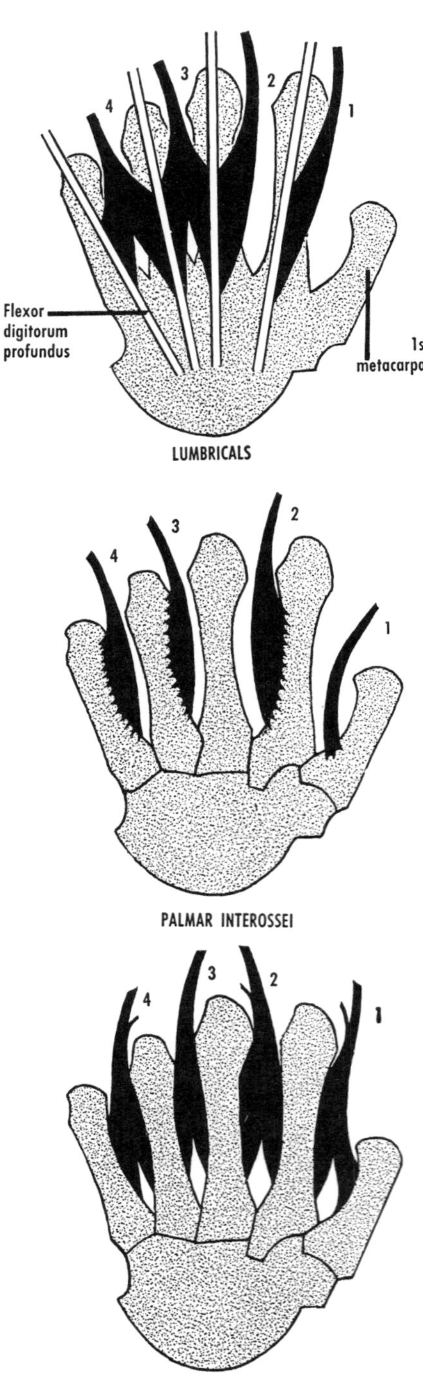

Figure 16–8 Diagrams showing arrangement of the lumbricals and interossei (see also fig. 16–9).

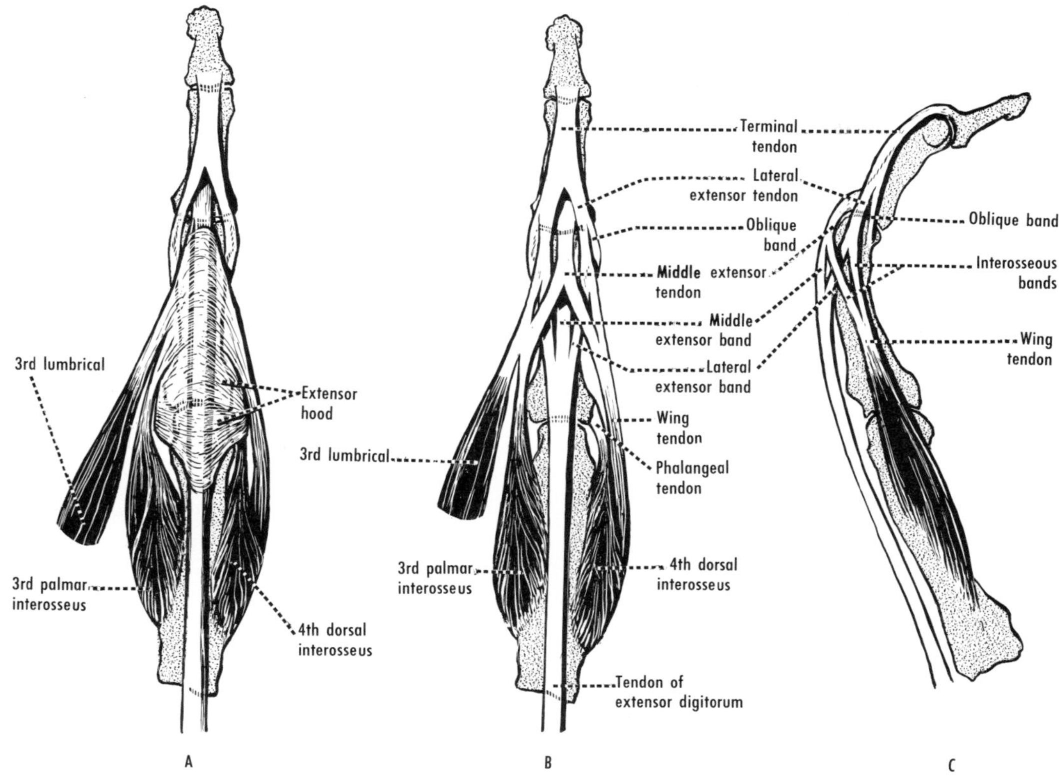

Figure 16–9 The extensor aponeurosis of the fourth finger. Based on Landsmeer.[5]

third finger has a dorsal interosseus on each side, and these two muscles are the only ones capable of producing side-to-side movement of this digit. The arrangement and abducting action of the dorsal interossei can best be understood from figure 16–8. The first dorsal interosseus can readily be seen and felt *in vivo* during abduction of the index finger against resistance.

EXTENSOR APONEUROSIS.[20] This is illustrated in figure 16–9. Note in *A* that the dorsal interosseous muscle, which is on the right side, has two tendons, one of which is inserted into the phalanx, deep to the transverse fibers that form the extensor hood, whereas the other tendon joins the hood. The lumbrical and the palmar interosseus insert only into the hood. Removal of the hood, as shown in *B*, reveals slips of the wing tendon that join the divisions of the extensor tendon. The oblique band, which is shown in *B* and *C*, extends from the front of the proximal phalanx and the fibrous sheath to the terminal tendon. When the distal interphalangeal joint is flexed, the oblique bands become taut and pull the proximal joint into flexion. When the proximal joint is extended, the oblique bands pull the distal joint into extension.

NERVE SUPPLY AND ACTION. The interossei, both palmar and dorsal, are supplied by the deep branch of the ulnar nerve. Rarely the first dorsal interosseus is innervated by the median.[21] It has been mentioned that the palmar interossei adduct, and the dorsal interossei abduct, the fingers in relation to a line through the third finger. In addition they provide rotatory components in certain movements of the hand and they flex the metacarpophalangeal joints (p. 155). For example, in combination with dorsiflexion of the hand, the interossei and lumbricals produce what has been called the **Z**-position (fig. 16–10). Their role in precision handling is discussed on page 155.

The precise details of interosseous and lumbrical movement *in vivo*, the relation between interosseous and lumbrical movements, and their correlation with the actions

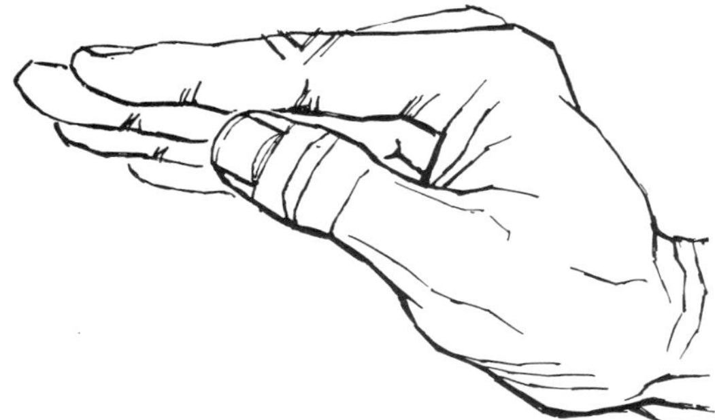

Figure 16–10 The **Z**-position of the hand; the fingers are flexed at the metacarpophalangeal joints and extended at the interphalangeal joints, and the hand is extended at the wrist.

of the long flexors and extensors of the digits are still obscure. Electromyographic and stimulatory studies of the lumbricals *in vivo* indicate that these muscles are primarily extensors at the interphalangeal joints and only weak flexors at the metacarpophalangeal joints, whereas the interossei are primarily flexors at the metacarpophalangeal joints and weak extensors at the interphalangeal joints.[22]

NERVES OF HAND

The median, ulnar, and radial nerves supply the hand (figs. 16–11 and 16–12). Branches of the lateral antebrachial cutaneous nerve, and sometimes of the posterior antebrachial cutaneous also,[23] may reach the dorsum. The territories that these nerves supply vary. **The motor fibers to the intrinsic muscles of the hand are derived from the first thoracic segment of the spinal cord; the fibers reach the muscles by way of the median and ulnar nerves.**

The superficial branch of the radial nerve, which gives cutaneous and articular branches to the hand, and the deep branch, which gives only articular branches, are described with the forearm (p. 137).

Median Nerve. In the lower part of the forearm, the median nerve gives off an inconstant *palmar branch* that crosses the flexor retinaculum and supplies a small area of the skin of the palm. The median nerve then enters the hand by passing through the carpal canal, behind the flexor retinaculum, and in front of the flexor tendons to the index finger. The median artery, when large, accompanies the median nerve in the canal and contributes to the superficial palmar arch. The nerve can be marked on the surface in the middle of the front of the wrist, medial to the easily identifiable flexor carpi radialis tendon. It lies just lateral to the palmaris longus tendon when this is present. At the distal border of the flexor retinaculum, the median nerve spreads out in an enlargement and divides into its terminal branches under cover of the palmar aponeurosis and the superficial palmar arch. Usually it divides first into lateral and medial portions.

The lateral division immediately gives an important muscular branch (sometimes termed the recurrent branch) into the base of the thenar eminence (see p. 144 for variations); it supplies the abductor pollicis brevis, flexor pollicis brevis, and opponens pollicis. The lateral division then splits into three *palmar digital nerves* for both sides of the thumb and the lateral aspect of the index finger. The digital nerve to the index finger initially gives a twig to the first lumbrical. Rarely the median nerve supplies the first dorsal interosseus or the adductor pollicis. These muscles are usually supplied by the ulnar nerve instead. There is a general tendency for the median nerve to supply the thenar muscles and for the ulnar nerve to supply most of the other short muscles of the hand, those situated medial to the flexor pollicis longus tendon. As with cutaneous innervation, however, the dividing line between the two

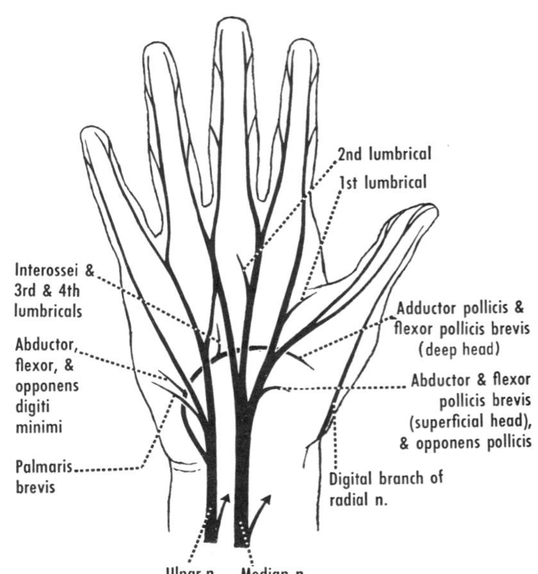

Figure 16–11 Scheme of innervation of front of hand.

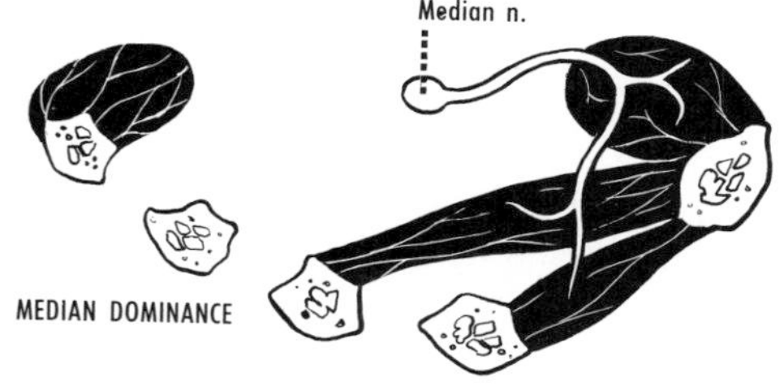

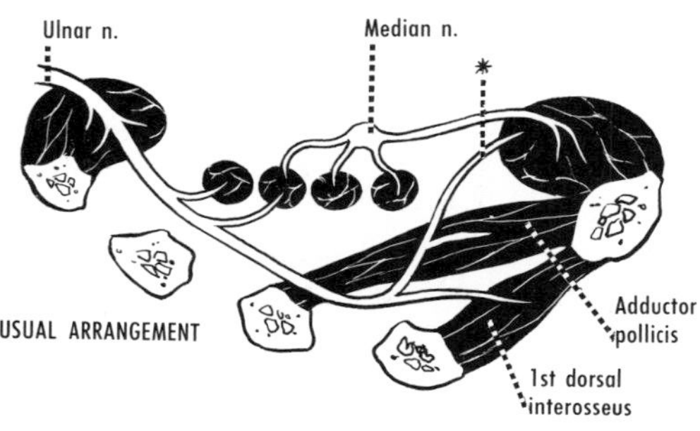

Figure 16–12 Schematic diagrams of innervation of intrinsic muscles of the hand. The asterisk indicates the branch of the ulnar to the deep head of the flexor pollicis brevis. The usual arrangement is shown in the central figure. In median dominance, all of the thenar muscles are supplied by the median nerve. In ulnar dominance, all of the thenar muscles are supplied by the ulnar nerve. Most variations, however, do not reach the extremes shown here. Based in part upon Brooks.[24]

distributions is a variable one, and either nerve may invade the territory of the other (fig. 16–12). Apparently, in very rare instances, the short muscles of the hand may be supplied by the median nerve only, in others by the ulnar nerve alone.

The medial division of the median nerve splits into two branches, each of which later divides into two palmar digital nerves, for the adjacent sides of the index and middle, and the middle and ring, fingers. The nerve destined for the index and middle fingers first gives a twig to the second lumbrical. That for the middle and ring communicates with the adjoining branch of the ulnar nerve (which sometimes takes over its distribution, or vice versa), and it usually innervates the third lumbrical in part or in whole. The digital branches of the median nerve, like those of the ulnar, lie in front of the digital arteries, on the lumbrical muscles, and are distributed to the sides and fronts of the fingers.

The digital nerves, near their terminations, send branches dorsally to supply the backs of the distal portions of the fingers (see below).

The median nerve is characterized by the twigs it gives to the bones, joints, ligaments, interosseous membrane, and blood vessels. It supplies the metacarpophalangeal and interphalangeal articulations with fibers for the appreciation of position and of movement. **Injuries to the median nerve may be followed by severe, chronic sensory and trophic disturbance, owing in part to the fact that regeneration is hindered by the complex arrangement of fibers and fascicles within the nerve.**

Ulnar Nerve. In the middle of the forearm, the ulnar nerve gives off a *dorsal branch,* which descends dorsally, between the ulna and the flexor carpi ulnaris, to the medial side of the back of the hand. After giving twigs to the skin of the back of the hand, it divides into three *dorsal digital nerves* that supply the medial side of the little finger, the adjacent sides of the little and ring fingers, and the ring and middle fingers, on their dorsal aspects. The most lateral of the dorsal digital nerves communicates with the adjoining dorsal digital nerve from the superficial branch of the radial nerve. A fourth dorsal digital nerve from the ulnar may be present and reach the index finger.

The dorsal digital branches of the radial and ulnar nerves do not reach the tips of the fingers. In the first and fifth fingers they extend as far as the nail, but in the intermediate three fingers usually only as far as the proximal or middle phalanx.[25] The dorsal innervation is then completed distally by the palmar digital branches of the median and ulnar nerves.

In the lower part of the forearm, the ulnar nerve gives off a variable *palmar branch,* which crosses the flexor retinaculum and supplies the skin on the medial side of the palm.

The ulnar nerve enters the hand external (anterior) to the carpal canal, lateral to the pisiform (between this bone and the hook of the hamate), and in front of the flexor retinaculum. The ulnar artery lies on its lateral side, and both are sometimes covered by the superficial part of the retinaculum, and by the pisohamate ligament and the palmaris brevis muscle. The nerve then divides into its two terminal branches, superficial and deep.

The *superficial branch* of the ulnar nerve gives a twig to the palmaris brevis and divides into *palmar digital nerves* for the medial side of the little finger and the adjacent sides of the little and ring fingers. These branches give cutaneous and articular twigs (to the metacarpophalangeal and interphalangeal joints), as do those of the median, and they communicate with the latter nerve. Occasionally, the ulnar nerve also supplies the adjacent sides of the ring and middle fingers. Thus, on the front of the hand, the ulnar distribution varies inversely with that of the median; on the back, the ulnar territory varies inversely with that of the radial.

The *deep branch* of the ulnar nerve passes deeply between the abductor and flexor digiti minimi muscles, both of which it supplies. It passes through a fibrous arch in the proximal end of the opponens digiti minimi and innervates that muscle. Then it curves around the hook of the hamate and extends laterally with the deep palmar arch, under cover of a pad of fat lying behind the flexor tendons. In this part of its course, it supplies all the interossei, the third and fourth lumbricals, the adductor pollicis, and usually the flexor pollicis brevis, in which it ends. **Owing to the important distribution of the ulnar nerve to the muscles responsible for the finer movements of the fingers, injuries to it cause severe disabilities (p. 769).**

To summarize, the ulnar nerve usually supplies the palmar aspects of the medial one and one-half fingers, and the dorsal aspects of the medial two and one-half fingers. The other fingers are supplied by the median on their palmar aspects, and by the radial nerve dorsally; the median nerve extends dorsally on the distal phalanx of the thumb and distal two phalanges of the first two and one-half fingers.

ARTERIES OF HAND

Radial Artery. The radial artery leaves the forearm by curving dorsalward around the radial collateral ligament and the scaphoid and trapezium, in the floor of the anatomical snuffbox (fig. 16–13). It is crossed by the superficial branch of the radial nerve and by the tendons of the abductor pollicis longus, extensor pollicis brevis, and extensor pollicis longus. It then enters the palm by passing between the heads of the first dorsal interosseous muscle. It turns medially and goes between the heads of the adductor pollicis. It anastomoses with the deep branch of the ulnar artery and forms the deep palmar arch.

BRANCHES. The radial artery gives off the following branches:

1. An inconstant *superficial palmar branch* arises in the lower part of the forearm, descends to the thenar muscles, and anastomoses with the ulnar artery to complete the superficial palmar arch (fig. 16–14).

2. A *palmar carpal branch* passes medially behind the flexor tendons and forms a network or arch with a corresponding branch from the ulnar artery.

3. A *dorsal carpal branch* runs medially deep to the extensor tendons and forms the *dorsal carpal network* with a corresponding branch from the ulnar artery. Three or more *dorsal metacarpal arteries* arise from the network, descend, and divide into *dorsal digital arteries* for the adjacent sides of the medial four fingers (fig. 16–13). The dorsal metacarpal and digital arteries anastomose with the palmar arches by means of small perforating branches.

4. *Dorsal digital arteries,* two for the thumb and one for the radial side of the index finger.

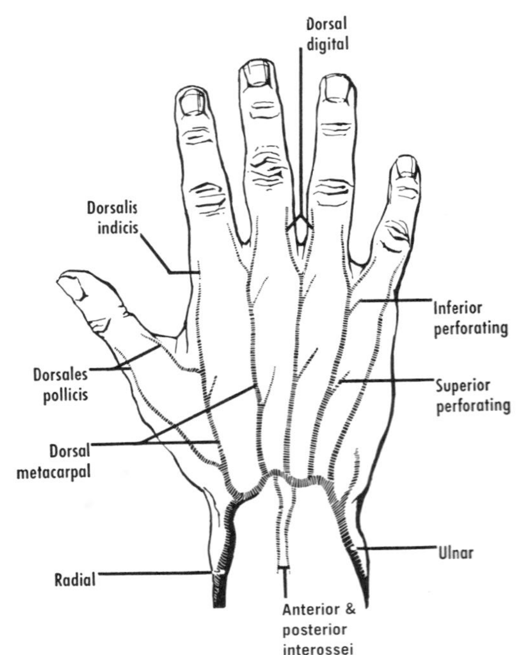

Figure 16–13 Scheme of arteries of the back of the hand.

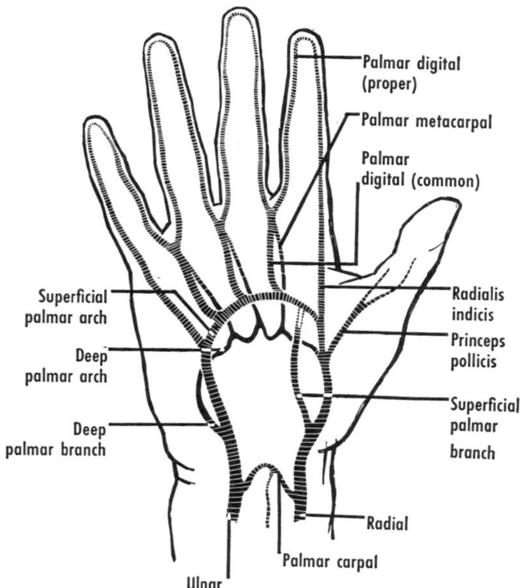

Figure 16–14 Scheme of arteries of the front of the hand. The superficial palmar arch is sometimes completed on the lateral side by the superficial palmar branch of the radial artery. (The anastomosis is shown by dotted lines.)

5. The *princeps pollicis artery* arises as the radial artery enters the palm. It descends on the first metacarpal and divides into two palmar digital arteries for the thumb. It may arise in common with the radialis indicis as a first palmar metacarpal artery, or it may give off the radialis indicis.

6. A *radialis indicis artery* runs down the radial side of the index finger. It commonly supplies both sides of the index, and the radial side of the middle finger. It is often a branch of the princeps pollicis, or it may arise from either the superficial or the deep arch. When it originates from the superficial arch, it is then distributed to the radial side of the index finger.[26]

7. The *deep palmar arch* lies on the interossei, deep to the flexor tendons, and about 1 cm proximal to the level of the superficial arch. Its deep location protects it from damage, and hence it does not have the clinical importance of the superficial arch. It gives off a variable number of deep palmar arteries, some of which are termed *palmar metacarpal arteries*.[27] These descend on the interossei and join the palmar digital arteries of the superficial arch for the adjacent sides of the medial four fingers.

Ulnar Artery. The ulnar artery enters the hand in front of the flexor retinaculum, on the lateral side of the pisiform, between this bone and the hook of the hamate. The ulnar nerve lies on its medial side, and both may be covered by the superficial fibers of the retinaculum and by the pisohamate ligament. It then divides into its two terminal branches, the superficial palmar arch and the deep palmar branch (figs. 16–14 and 64–1, p. 766).

BRANCHES. The branches of the ulnar artery include the following:

1. A *palmar carpal branch* passes laterally behind the flexor tendons and forms a network with the corresponding branch from the radial artery.

2. A *variable dorsal carpal branch* runs laterally, deep to the tendons of the flexor and extensor carpi ulnaris, and helps to form the dorsal carpal network.

3. The *superficial palmar arch* is the main termination of the ulnar artery. The completion of the arch on the radial side is extremely variable. Commonly it is completed by the radialis indicis, by the superficial palmar branch of the radial, or by the princeps pollicis. A large median artery may contribute to it. The arch lies on the flexor tendons, the lumbrical muscles, and the branches of the median nerve, under cover of the palmar aponeurosis and the palmaris brevis. In addition to a branch to the medial side of the fifth finger, the arch gives off three *common palmar digital arteries* that divide to supply the adjacent sides of the fingers.

4. The *deep palmar branch* of the ulnar artery accompanies the deep branch of the ulnar nerve between the abductor and flexor digiti minimi brevis. It joins the radial artery to form the deep palmar arch.

JOINTS OF WRIST AND HAND

Distal Radioulnar Joint. **This pivot joint is formed between the head of the ulna and the ulnar notch on the distal end of the radius.**

The *articular disc* (fig. 16–15) is a strong, triangular plate of densely fibrous tissue.[28] Its base, which is attached to the radius, may be formed of fibrocartilage. Its apex is attached to the lateral side of the base of the styloid process of the ulna. The joint cavity lies between the disc and the head of the ulna, and extends upward between the radius and ulna. It is therefore L-shaped. The disc excludes the ulna from the wrist joint. The distal surface of the disc forms a part of the wrist or radiocarpal joint, adjacent to a portion of the lunate.

The capsule is indefinite. Fibrous strands anteriorly and posteriorly fuse with the disc below, but are separated above by a proximal extension of the cavity, the *recessus sacciformis*.

MOVEMENTS. Movements are supination and pronation. These movements and the muscles producing them are described with the proximal radioulnar joint (p. 128). The radius is the bone that moves. Hence, the articular disc, which is fixed by its apex to the ulna, slides against the inferior articular surface of the head of the ulna during supination and pronation.

Radiocarpal and Carpal Joints

Radiocarpal Joints. **The radiocarpal or wrist joint is an ellipsoid joint that is formed by the radius, the articular disc, and the proximal row of carpal bones, exclusive of the pisiform (fig. 16–15). Its position is approximately indicated by a line connecting the styloid processes of the radius and ulna.**

The capsule consists mainly of localized strong bands. The broad *palmar radiocarpal ligament* extends fanwise from the lower end of the radius to the proximal row of carpals. Capsular thickenings may also be found between the styloid process of the ulna and the carpus. The thin dorsal capsule is covered, or reinforced, by the overlying *dorsal radiocarpal ligament,* a strong ligament that extends fanwise from the radius to the proximal row of carpals. The *ulnar collateral ligament* extends from the styloid process of the ulna to the triquetrum and the pisiform. The *radial collateral ligament* connects the styloid process of the radius with the scaphoid. It is crossed by the radial artery.

The synovial membrane of the wrist joint does not cover the articular disc. The cavity commonly communicates with the cavities of the intercarpal and pisotriquetral joints, but communicates with that of the distal radioulnar joint only when the disc is perforated as a result of attrition.[29]

The radiocarpal, intercarpal, and carpometacarpal joints are all supplied by the median (anterior interosseous), posterior interosseous, and ulnar (dorsal and deep branches) nerves.[30] A small branch of the superficial radial nerve enters the first interosseous space dorsally and supplies adjacent joints.

Midcarpal Joint. **The carpal bones, exclusive of the pisiform, form an important intercarpal joint, the midcarpal joint, between the proximal and distal rows** (fig. 16–15). The hamate and the head of the capitate form an ellipsoid joint by fitting into the concave socket formed by the scaphoid, lunate, and triquetrum, whereas the trapezium and trapezoid form a plane joint with the scaphoid. The bones are held together by dense, fibrous tissue continuous proximally with the capsule of the wrist joint. The union on the palmar aspect is strengthened by fibers that radiate from the capitate (*radiate carpal ligament*). The bones are also held together by strong *interosseous ligaments.*

The complicated cavity may communicate with that of the wrist, but more often it communicates with that of the carpometacarpal joint, especially on one side or the other of the trapezoid.

A variable interval,[31] occupied by synovial folds, occurs between the lunate, triquetrum, capitate, and hamate. This space is evident radiographically, and its appearance changes during movement. The lunate approaches the hamate during adduction and tends to obliterate the space. The converse occurs in abduction.

Pisotriquetral Joint. A short capsule unites the pisiform and the palmar side of the triquetrum and encloses a small synovial cavity that often communicates with the radiocarpal joint.[32] The strong *pisometacarpal ligament* extends from the pisiform to the base of the fifth metacarpal, as a direct continuation of the tendon of the flexor carpi ulnaris. The *pisohamate ligament* diverges to the hook of the hamate.

Movements at Radiocarpal and Midcarpal Joints (fig. 16–16; fig. 11–21, p. 86). These are ellipsoid joints and the movements are flexion and extension, abduction and adduction, and combinations of these. There is still considerable disagreement, however, concerning the precise details of these movements.[33]

The hand can be flexed on the forearm more that it can be extended. Flexion and extension occur at both the radiocarpal and midcarpal joints.

The hand can be adducted (ulnar deviation) more than it can be abducted (radial deviation). The latter movement is hindered by the more distally placed styloid process of the radius. Adduction occurs at both joints, but especially at the radiocarpal. During adduction, the lunate articulates completely with the radius, and the triquetrum with the disc. Abduction occurs almost entirely at the midcarpal joint.

MUSCLES. Movements at the radiocarpal and midcarpal joints are produced by the extrinsic muscles of the hand. These are

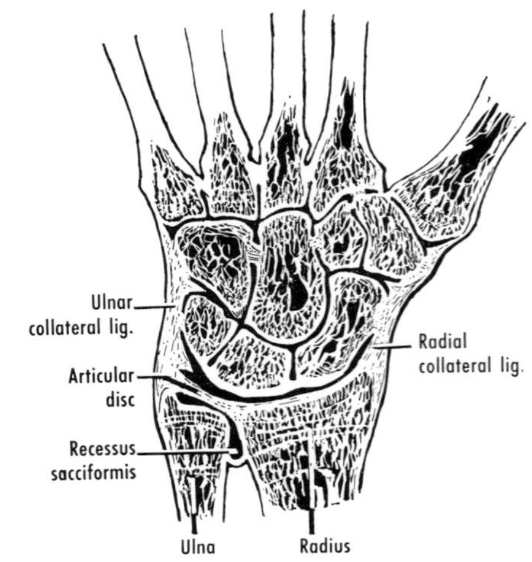

Figure 16–15 Schematic coronal section of the carpus, as if the carpus were flat.

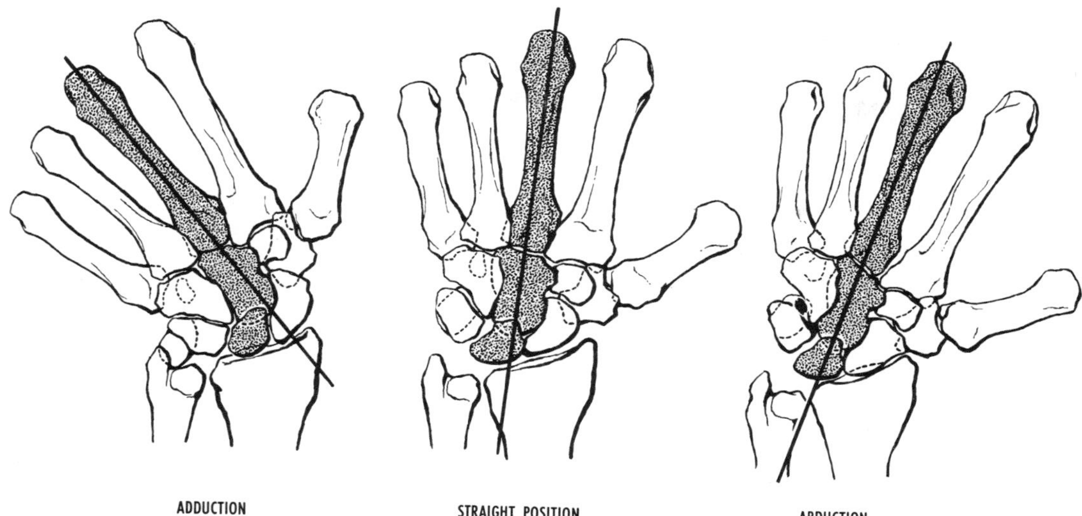

Figure 16–16 Abduction and adduction of the hand. The tracings are from posteroanterior radiograms of a left hand (fig. 11–21, p. 86). In the abducted and in the straight position, the lunate articulates superiorly partly with the radius and partly with the articular disc. In adduction, it articulates superiorly with only the radius.

the flexors (chiefly the flexor carpi ulnaris and flexor carpi radialis), the extensors (extensor carpi radialis longus and brevis, and extensor carpi ulnaris), the abductors (chiefly the flexor carpi radialis and extensor carpi radialis longus and brevis), and the adductors (extensor and flexor carpi ulnaris).

The functions of the extrinsic muscles in power grip are discussed on page 155.

Joints of the Carpus and Digits

Carpometacarpal Joints. **The carpometacarpal joint of the thumb (figs. 16–15 and 16–17) is unique among the joints of the hand. It is a saddle joint with reciprocally shaped surfaces on the trapezium and first metacarpal that give it a special freedom of movement.** It has a loosely arranged capsule that is reinforced by special ligaments.[34] Most of the muscles that are inserted distal to the joint can rotate or otherwise move the first metacarpal. It may therefore be difficult to analyze paralyses of individual muscles, because one muscle may compensate for the inadequacy of another.

The medial four carpometacarpal joints form a series of plane joints having a common irregularly shaped cavity, enclosed by a capsule on the palmar and dorsal surfaces. The cavity extends distally between the bases of the metacarpals, and proximally between the carpals. An interosseous ligament unites the capitate and hamate, and sometimes the adjacent metacarpals also. A strong ligament extends medially from the trapezium to the second and third metacarpals.[35]

Some gliding movement is possible at the carpometacarpal and intermetacarpal joints of the second to fifth fingers, more in these joints of the little finger. The fifth metacarpal, for example, can be passively moved forward and backward to a fair degree, the second metacarpal practically not at all.

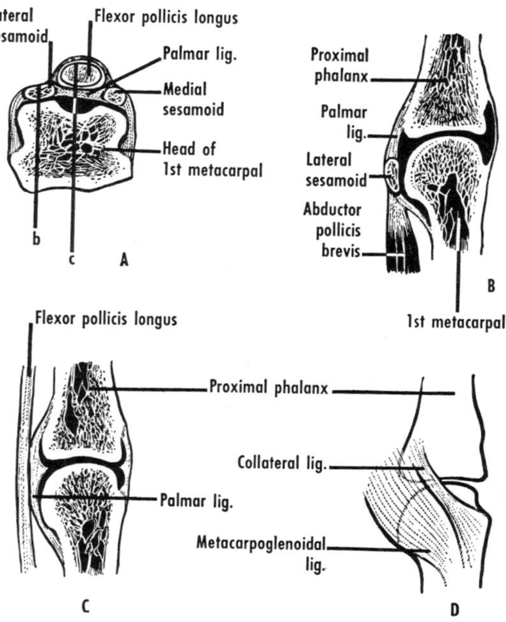

Figure 16–17 Metacarpophalangeal joint of thumb. *A*, horizontal section; *b* and *c* indicate the planes of sections *B* and *C*, which are sagittal sections. *D*, the two components of the collateral ligament.

Metacarpophalangeal and Interphalangeal Joints. **The metacarpophalangeal joints are ellipsoid, and the interphalangeal are hinge, but their ligamentous arrangements are similar** (fig. 16–17).

Each fibrous capsule is strengthened by two *collateral ligaments* that extend distally, across the joint, and are attached to the base of the phalanx. They fuse to form the anterior part of the capsule, which is a thick, densely fibrous or fibrocartilaginous pad, termed the *palmar ligament* (fig. 16–17). The palmar ligaments of the medial four metacarpophalangeal joints are connected by strong, transverse fibers of the *deep transverse metacarpal ligament.* This ligament helps to hold the metacarpal heads together. The interosseous tendons pass behind this ligament, those of the lumbricals in front. Behind, the joint capsules are formed mainly by the extensor aponeurosis.

NERVE SUPPLY. The joints are supplied by twigs from adjacent nerves, radial, median, or ulnar, depending on the location of the joint.[36]

Movements of the Hand. An understanding of the nerve supply and actions of individual muscles is important in evaluating peripheral nerve injuries. Otherwise, it is more important that the mechanisms of power grip, precision handling, and pinch be understood (p. 139). Of particular importance are movements of the thumb (figs. 16–18 and 16–23).

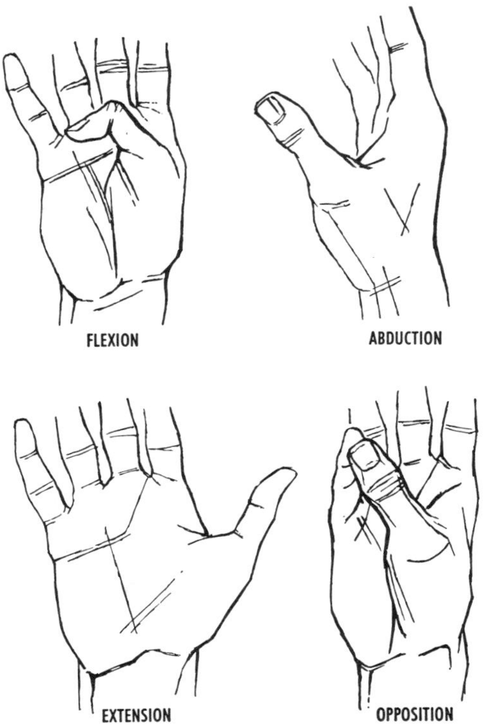

Figure 16–18 Movements of the thumb.

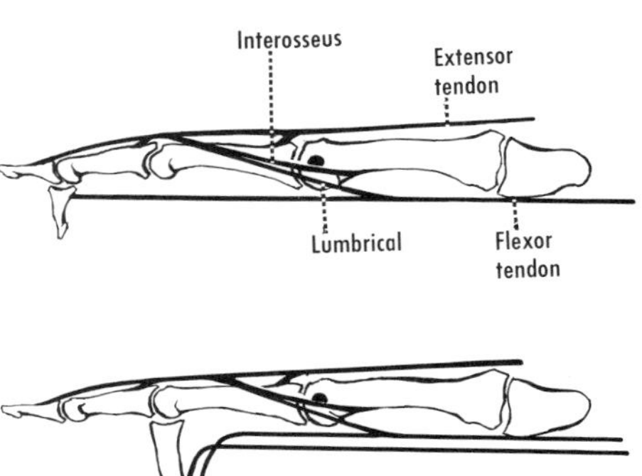

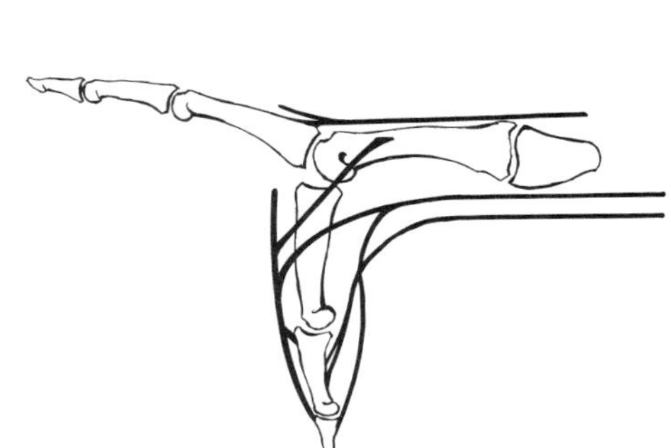

Figure 16–19 Diagrammatic representation of, from above downward, muscles controlling movements at the distal interphalangeal, proximal interphalangeal, and metacarpophalangeal joints. Based on von Lanz and Wachsmuth.[37]

The thumb is set at an angle to the plane of the palm. A forward movement of this digit as a whole, away from the palm, is called abduction, the converse movement adduction. When the hand rests on a table, palm upward, abduction moves the thumb so that it points toward the ceiling. A medial movement of the thumb in the plane of the palm is termed flexion, the converse extension. By opposition is meant the movement whereby the palmar aspect of the thumb touches the palmar aspects of the tip or front of another finger of the same hand. When the contact is not by their palmar aspect, the movement is merely one of apposition. The converse of opposition is reposition. These movements occur at all the joints of the thumb, with opposition taking place mainly at the carpometacarpal joint.

In the other fingers, the main movements occur at the metacarpophalangeal and interphalangeal joints. Flexion and extension occur at both joints (figs. 16–19 and

16–22), and an important rotatory component (figs. 16–20 and 16–21) occurs at the metacarpophalangeal joints (produced by the interossei through abduction and adduction).

In considering the hand as a whole, activities are brought about as follows:[38]

In power grip, the extrinsic muscles provide the major gripping force, and they are used in proportion to the force needed. In addition, the interossei are used as flexors and rotators at the metacarpophalangeal joints. The thenar muscles are active in most forms of the power grip, but most of the lumbricals are not significantly used.

In precision handling, gross motion and compressive forces are provided by specific extrinsic muscles. The interossei are important in imposing the necessary rotational forces, as well as flexion at the metacarpophalangeal joints. The lumbricals are chiefly extensors at the interphalangeal joints. In translation movements toward the palm, the interossei are responsible for the compression and rotation forces for the most effective positions of the fingers. In translation movements away from the palm, the handled object is driven by the interossei and the lumbricals.

The thenar muscles in precision handling act as a triad (flexor pollicis brevis, opponens pollicis, and abductor pollicis brevis). The triad provides adduction and flexion of the

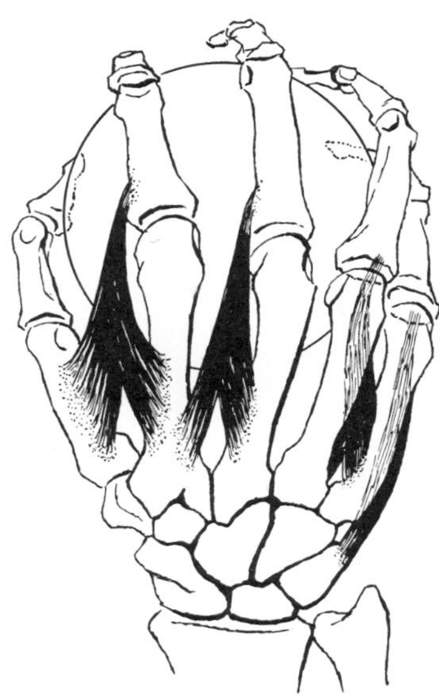

Figure 16–21 Diagram to show a power grip, with rotation at metacarpophalangeal joints while squeezing a ball. The phalangeal insertions of the dorsal interossei and of the abductor digiti minimi are shown. Based on Landsmeer.[5]

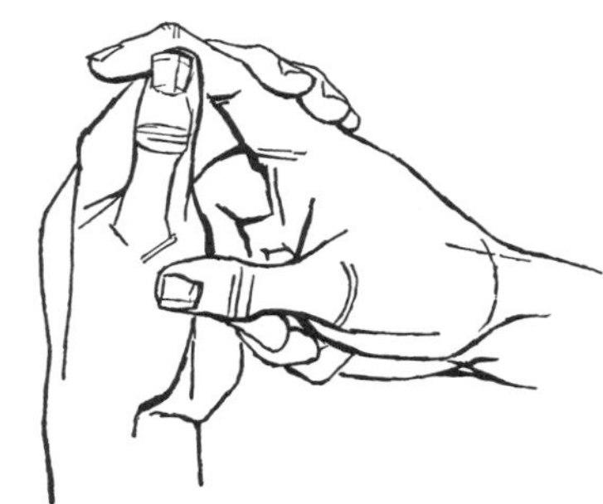

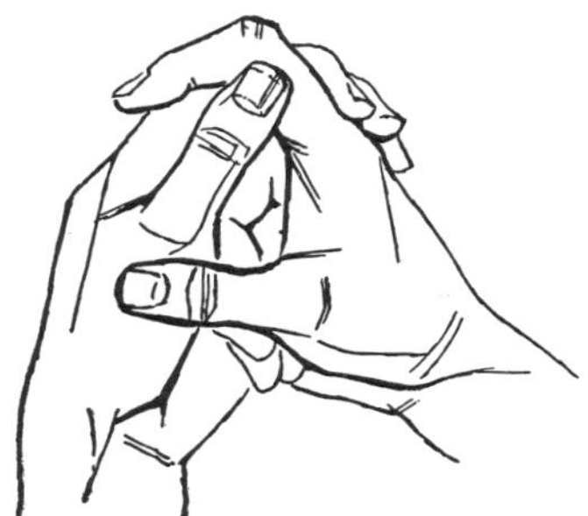

Figure 16–22 Testing of finger movements. *Upper,* if the metacarpophalangeal and proximal interphalangeal joints are stabilized as shown, flexion at the distal interphalangeal joint can be tested. *Lower,* if the metacarpophalangeal joint is stabilized as shown, flexion at the proximal interphalangeal joint can be tested.

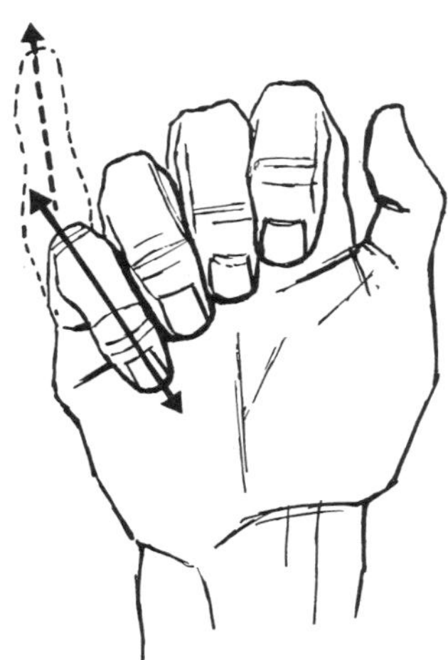

Figure 16–20 Rotation of the fingers. When the fingers are flexed they rotate slightly and are thus brought closer together in the fully flexed position. The extended little finger is shown in dotted outline. Note the difference in direction of the longitudinal axes of the flexed and of the extended little finger.

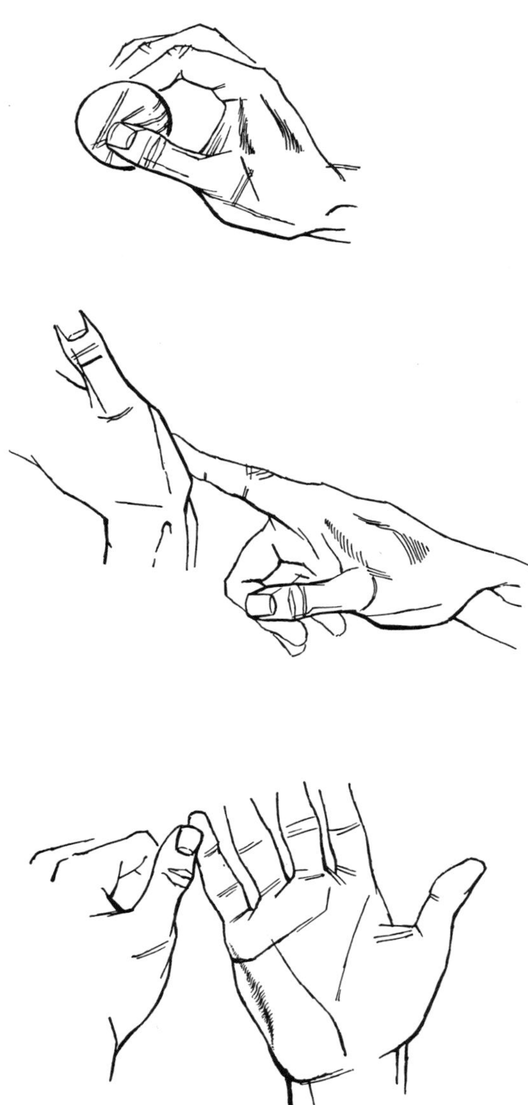

Figure 16–23 Testing of finger movements. *Upper,* the use of a pinch to test the first dorsal interosseus and adductor policis. Adduction by the other interossei can be tested by having the patient hold something, such as a piece of paper, between his fingers as the examiner tries to pull it away. *Middle,* the muscles of the thumb can be checked by palpating them as movements are made against resistance. Note the tenseness of the first dorsal interosseus of the examiner. *Lower,* test the hypothenar muscles by abducting (or flexing) against resistance.

thumb and medial rotation of the first metacarpal. The adductor pollicis comes into play when force is required to adduct the thumb toward the index finger.

In pinch, compression is provided chiefly by the extrinsic muscles, assisted by the thenar muscles and the interossei (fig. 16–23).

REFERENCES

1. F. Wood Jones, *The Principles of Anatomy as Seen in the Hand,* Baillière, Tindall, and Cox, London, 2nd ed., 1942. A classic work on the structure and function of the hand. E. B. Kaplan, *Functional and Surgical Anatomy of the Hand,* Lippincott, Philadelphia, 2nd ed., 1965. *Bunnell's Surgery of the Hand,* 5th ed. revised by J. H. Boyes, Lippincott, Philadelphia, 1970. The account of movements of the hand is based on C. Long *et al.*, J. Bone Jt Surg., *52A*:853, 1970.
2. R. Forbes, Brit. med. J., 2:851, 1955. P. H. Addison, Brit. med. J., 2:806, 1960.
3. C. H. Danforth, Amer. J. phys. Anthrop., *4*:189, 1921. *Hair, with Special Reference to Hypertrichosis,* American Medical Association, Chicago, 1925.
4. T. S. Kirk, J. Anat., Lond., *58*:228, 1924.
5. B. J. Anson *et al.*, Surg. Gynec. Obstet., *81*:327, 1945. E. B. Kaplan, cited above. J. M. F. Landsmeer, Acta anat., Suppl. 24, 1955.
6. L. W. Milford, *Retaining Ligaments of the Digits of the Hand,* Saunders, Philadelphia, 1968.
7. The pisiform is very mobile. Its removal hardly affects the attachments of the retinaculum. See A. Young and R. J. Harrison, J. Anat., Lond., *81*:397, 1947; A. K. Henry, cited on page 70.
8. M. A. Ritter, J. Brit. Soc. Surg. Hand, 5:263, 1973. F. Bojsen-Møller and L. Schmidt, J. Anat., Lond., *117*:55, 1974. For details of the web spaces, see H. G. Stack, *The Palmar Fascia,* Churchill Livingstone, Edinburgh, 1973.
9. E. W. Scheldrup, Surg. Gynec. Obstet., *93*:16, 1951.
10. The blood supply of tendons in the forearm and hand is described by J. G. Brockis, J. Bone Jt Surg., *35B*:131, 1953.
11. W. Harris, J. Anat., Lond., *38*:399, 1904, R. J. Last, J. Bone Jt Surg., *31B*:452, 1949; Brain, *74*:481, 1951.
12. J. R. Napier, J. Anat., Lond., *86*:335, 1952.
13. Based on the description of F. Wood Jones, cited above.
14. D. Harness and E. Sekeles, J. Anat., Lond., *109*:461, 1971.
15. L. Mannerfelt and C.-H. Hybbinette, Bull. Hosp. Jt Dis., N.Y., *33*:15, 1972.
16. The role of hypothenar muscles in movements of opposition of the thumb is described by S. Sunderland, Aust. N.Z. J. Surg., *13*:155, 1944.
17. M. A. Shrewsbury, R. K. Johnson, and D. R. Ousterhold, J. Bone Jt Surg., *54A*:344, 1972.
18. S. Sunderland, Anat. Rec., *93*:317, 1945.
19. E. Reinhardt, Anat. Anz., *20*:129, 1902. F. Wagenseil., Z. Morph. Anthr., *36*:39, 1937. K. F. Russell and S. Sunderland, J. Anat., Lond., *72*:306, 1938. H. J. Mehta and W. U. Gardner, Amer. J. Anat., *109*:227, 1961.
20. E. B. Kaplan, Anat. Rec., *92*:293, 1945. J. M. F. Landsmeer, Anat. Rec., *104*:31, 1949. R. W. Haines, J. Anat., Lond., *85*:251, 1951. J. M. F. Landsmeer, Acta anat., Suppl. 24, 1955.
21. S. Sunderland, Anat. Rec., *95*:7, 1946. F. Murphey, J. W. Kirklin, and A. I. Finlayson, Surg. Gynec. Obstet., *83*:15, 1946.
22. C. Long *et al.*, J. Bone Jt Surg., *52A*:853, 1970. W. H. Hollinshead, Amer. J. Anat., *134*:1, 1972.
23. J. S. B. Stopford, J. Anat., Lond., *53*:14, 1918.
24. H. St. John Brooks, J. Anat., Lond., *21*:575, 1887.
25. J. Dankmeijer and J. M. Waltman, Acta anat., *10*:377, 1950.
26. H. T. Weathersby, Anat. Rec., *122*:57, 1955. E. A. Edwards, Amer. J. Surg., *99*:837, 1960.
27. T. Murakami, Okajimas Folia anat. Jap., *46*:177, 1969.
28. K. Weigl and E. Spira, Reconstr. Surg. Traumatol., *11*:139, 1969.
29. M. Kutsuna, Jap. J. med. Sci., Trans. Abstr. I., Anat., *2*:187, 1930. B. N. Kropp, Anat. Rec., *92*:91, 1945.

30. D. J. Gray and E. Gardner, Anat. Rec., *151*:261, 1965.
31. R. O'Rahilly, Acta radiol., Stockh., *39*:401, 1953.
32. B. N. Kropp, Anat. Rec., *92*:91, 1945.
33. K. C. Bradley and S. Sunderland, Anat. Rec., *116*:139, 1953. R. D. Wright, J. Anat., Lond., *70*:137, 1935. See also M. A. MacConaill, J. Anat., Lond., *75*:166, 1941; W. W. Gilford, R. H. Bolton, and C. Lambrinudi, Guy's Hosp. Rep., *92*:52, 1943.
34. R. W. Haines, J. Anat., Lond., *78*:44, 1944. J. R. Napier, J. Anat., Lond., *89*:362, 1955. K.-O. Gedda, Acta chir. scand., Suppl. 193, 1954. A. P. Pieron, Acta orthopaed. scand., Suppl. 148, 1973.
35. H.-J. Welti, Arch. Anat., Strasbourg, *49*:481, 1966.
36. J. S. B. Stopford, J. Anat., Lond., *56*:1, 1921. See also D. L. Stilwell, Jr., Amer. J. Anat., *101*:75, 1957; D. J. Gray and E. Gardner, Anat. Rec., *151*:261, 1965.
37. T. von Lanz and W. Wachsmuth, *Praktische Anatomie. Ein Lehr- und Hilfsbuch der anatomischen Grundlagen ärtzlichen Handelns,* Springer, Berlin, 1935–1958.
38. C. Long *et al.*, J. Bone Jt Surg., *52A*:853, 1970.

SURFACE ANATOMY OF UPPER LIMB 17

Some of the muscles and tendons of the upper limb that can usually be identified by inspection are shown in figures 17–1 and 17–2.

SHOULDER AND ARM

The scapula is opposite the second to seventh ribs in the anatomical position; its spine is opposite the spine of the third thoracic vertebra. The coracoid process can be palpated below the lateral point of trisection of the clavicle, deep to the anterior edge of the deltoid. The acromion is outlined when the deltoid contracts against resistance. The acromioclavicular joint is medial to the lateral part of the acromion. **The arm is measured clinically from the tip of the acromion to the lateral epicondyle of the humerus.**

The head of the humerus may be felt deeply in the upper part of the axilla, especially in thin subjects. With the arm at the side, the head points backward as well as medially, as does the medial epicondyle. The greater tubercle of the humerus is covered by the deltoid, which contributes to the roundness of the shoulder. The greater tubercle is the most lateral bony point of the shoulder region and prevents a straight edge from touching simultaneously the tip of the acromion and the lateral epicondyle. The medial epicondyle is approximately in the same vertical plane as the head of the humerus. With the arm at the side, the medial epicondyle is approximately at the same level as the transpyloric plane. **The elbow joint is about 2 to 3 cm below the epicondyles.**

The subclavian artery extends from behind the sternoclavicular joint to about the midpoint of the clavicle, forming an arch, convex upward, 1 to 3 cm above that bone. The axillary artery extends from the midpoint of the clavicle to the medial margin of the biceps, opposite the posterior axillary fold. The brachial artery extends along the medial margin of the coracobrachialis and biceps to the cubital fossa at the level of the neck of the radius. **The brachial pulse is palpable throughout most of the course of the artery.**

The brachial plexus is palpable in the neck, and the cords may also be palpable around the axillary artery. The axillary nerve is indicated by a horizontal line through the middle of the deltoid. The median and ulnar nerves accompany the brachial artery. **The ulnar nerve passes behind the medial epicondyle, where it can be palpated, to the medial side of the coronoid process.** The radial nerve extends from the medial margin of the biceps, opposite the posterior axillary fold, obliquely across the back of the arm. It pierces the lateral intermuscular septum at the superior point of trisection of a line between the deltoid insertion and the lateral epicondyle and then descends to the front of the lateral epicondyle. **The radial nerve**

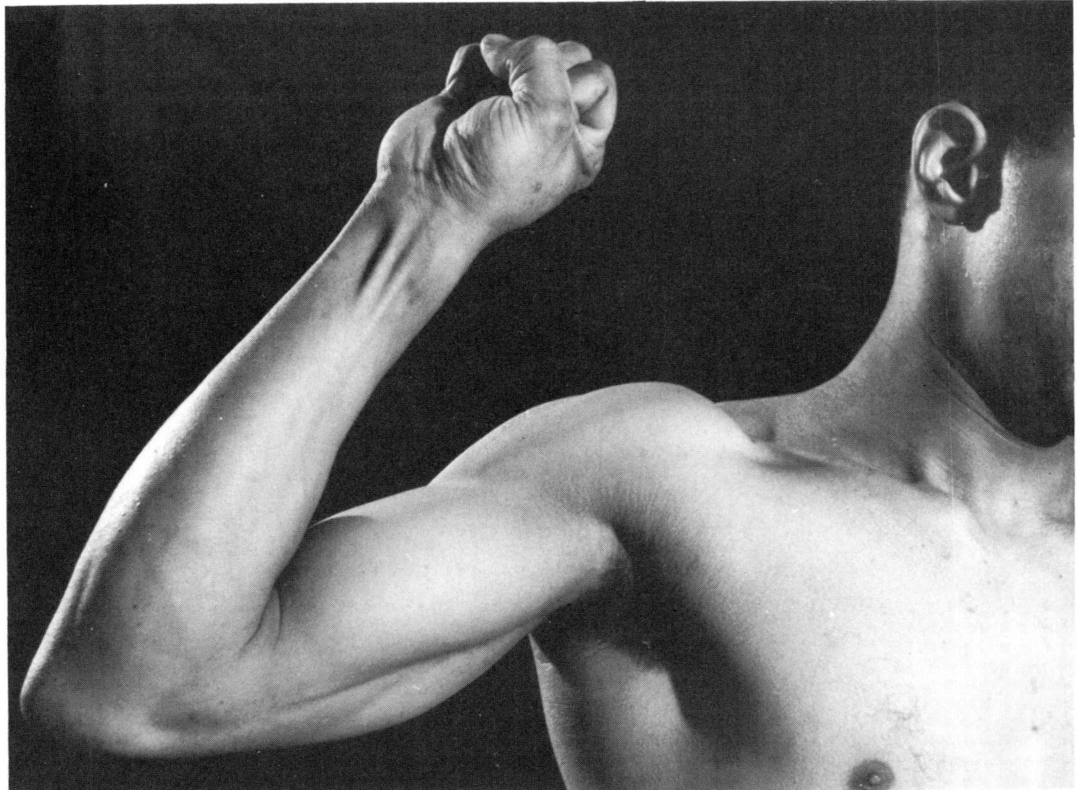

Figure 17–1 Some surface landmarks of the upper limb. Courtesy of J. Royce, Ph.D. (cited on p. 70), and Davis, Philadelphia.

(*Figure 17–1 continued on opposite page.*)

may be palpated in thin persons as it winds around the humerus, especially about 1 to 2 cm below the deltoid insertion, and also in the interval between the brachialis and brachioradialis.

FOREARM

The head of the radius can be felt just below the lateral epicondyle during pronation and supination of the forearm. The dorsal tubercle can be felt as a vertical ridge on the back of the lower end of the radius; it is grooved on its medial aspect by the tendon of the extensor pollicis longus. **The styloid process of the radius is in the proximal end of the anatomical snuff-box, about 1 cm distal in level to the styloid process of the ulna. The relative positions of the two styloid processes are disturbed in fractures at the wrist and are a clue to the proper realignment of the broken bones.**

The olecranon of the ulna forms a straight, horizontal line with the epicondyles of the humerus when the elbow is extended, and an equilateral triangle when the elbow is flexed to a right angle. **The coronoid process is not directly palpable, but the ulnar nerve can be rolled against it, just distal to the medial epicondyle.** The posterior border of the ulna is subcutaneous, is palpable throughout its entire length, and separates the flexors from the extensors. The head of the ulna forms the prominent dorsal elevation in the pronated forearm. It is covered by the radius in the supinated forearm. The styloid process is the medial and distal part of the head. The groove between them contains the tendon of the extensor carpi ulnaris.

The wrist joint is approximately indicated by a line connecting the styloid processes.

The median nerve extends down the middle of the forearm to the midpoint between the styloid processes. Just before passing behind the flexor retinaculum, it is quite superficial

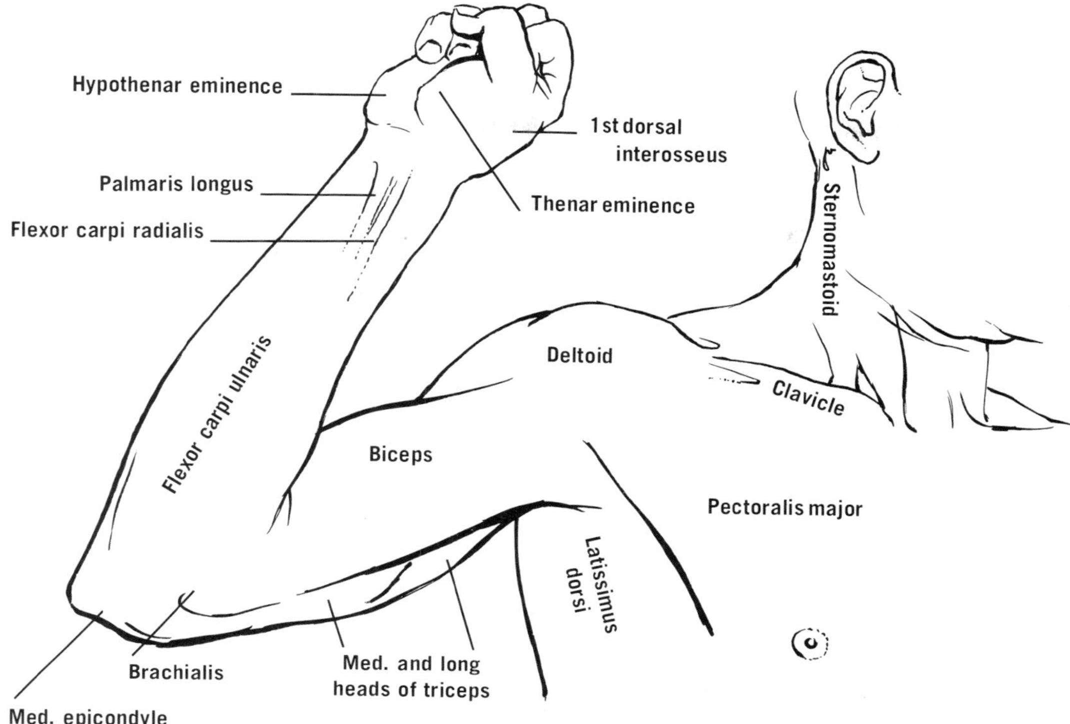

Figure 17–1 Continued.

in the interval between the tendons of the palmaris longus and the flexor carpi radialis. The tendons of the flexor carpi ulnaris, palmaris longus (when present), and flexor carpi radialis are palpable in this region. The ulnar nerve passes behind the medial epicondyle to the medial side of the coronoid process. Its course is indicated on the surface by a line from the front of the medial epicondyle to the lateral margin of the pisiform (and medial to the hook of the hamate). It is deep to the tendon of the flexor carpi ulnaris. **The superficial branch of the radial nerve continues from the cubital fossa to the inferior point of trisection of the forearm, turns deep to the brachioradialis, and enters the snuff-box.** It may be palpable for a few centimeters proximal to the snuff-box, and its terminal digital branches can be felt crossing the tendon of the extensor pollicis longus (by using a fingernail on the taut tendon).

The radial artery extends from the brachial artery to the tubercle of the scaphoid, then deep to the tendons of the snuff-box. **The radial pulse is palpable lateral to the tendon of the flexor carpi radialis (use three fingers), and also in the snuff-box.** The ulnar artery extends from the brachial artery distally and medially to the medial side of the forearm, and then to the lateral side of the pisiform.

The pattern of the superficial veins may be visible by inspection, especially over the cubital fossa. If not, they may be distended by placing a tourniquet around the arm so as to obstruct the venous return, and by clenching and opening the fist rapidly so as to increase the venous return. An appropriate vein for intravenous injection or withdrawal of blood may thereby be selected. The sites of valves are frequently visible when the veins are full. Veins often become distended when the arm is hanging by the side, especially those on the dorsum of the hand.

HAND

The scaphoid can be palpated in the snuff-box just distal to the styloid process of the radius. It comes against the palpating finger when the subject adducts his hand. The tubercle of the scaphoid can be pal-

Figure 17–2 Some surface landmarks of the back and upper limbs. Courtesy of J. Royce, Ph.D. (cited on p. 70), and Davis, Philadelphia.

(*Figure 17–2 continued on opposite page.*)

pated on the radial side of the tendon of the flexor carpi radialis. (It may be easier if the subject adducts and extends his hand slightly.) The lateral end of the distal crease of the wrist approximately indicates the tubercle. Thus, fingers palpating the tubercle and snuff-box simultaneously have the scaphoid between them (a procedure that may be used in examining for fracture of the scaphoid).

The trapezium is indistinctly palpable in the snuff-box, distal to the scaphoid. The edge of the carpometacarpal joint of the thumb may also be distinguished here. The tubercle or crest of the trapezium generally seems continuous with the tubercle of the scaphoid on palpation, although it can sometimes be felt separately.

The pisiform can be located by tracing the tendon of the flexor carpi ulnaris to it. When the hand is in a position such as to relax the muscle (frequently, when the hand is flexed and adducted), the pisiform can be felt between the thumb and index finger, and can be moved back and forth. The position of the pisiform is approximately indicated by the medial end of the distal crease of the wrist.

The hook of the hamate is approximately 2 cm distal and lateral to the pisiform, is more deeply placed, and is indistinctly palpable.

The dorsal tubercle of the radius is in line with the styloid process of the third metacarpal. The capitate and lunate occupy the interval between them.

The rounded medial side of the triquetrum can be palpated just distal to the styloid process of the ulna, when the forearm is midway between full pronation and full supination, and especially when the hand is abducted. The hamate is deep and indistinct distal to the triquetrum, although the base of the fifth metacarpal is prominent.

The heads of the metacarpals form the knuckles of the fist. The heads of the proximal and middle phalanges form the knuckles of the flexed fingers. The web

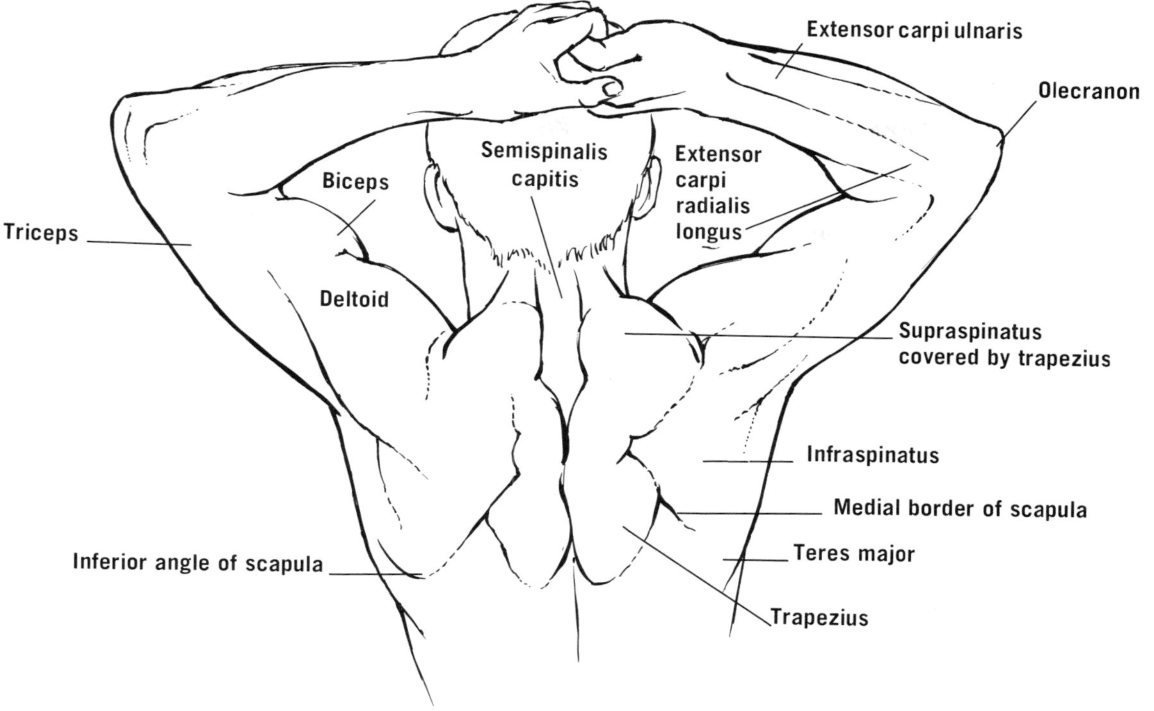

Figure 17–2 Continued.

margins and flexion creases at the junctions of the fingers and palm are distal to the metacarpophalangeal joints and opposite the middle of the proximal phalanges. The middle flexion creases indicate the proximal interphalangeal joints, but the distal flexion creases are slightly proximal to the distal interphalangeal joints.

The tendons forming the margins of the snuff-box are generally readily identified when the muscles are tensed. In some persons the several tendons of the abductor pollicis longus become visible. The tendons of the extensor carpi radialis longus and brevis may be visible and are generally palpable at their insertion when making a fist. The tendons of the interossei may be felt by placing the thumb and index finger on each side of a proximal phalanx. On flexing and extending the middle and distal phalanges, the interosseous tendons roll under the palpating fingers.

The flexor retinaculum is about 3 cm square. Its proximal margin is indicated by a line between the pisiform and the tubercle of the scaphoid, the distal margin by a line between the hook of the hamate and the tubercle of the trapezium.

The digital synovial sheaths reach the distal interphalangeal joints. Those for the intermediate three fingers begin opposite the necks of the metacarpals.

The most distal part of the superficial palmar arch is at about the level of the palmar surface of the extended thumb. Alternatively, it can be located by the two horizontal creases of the palm, between which it lies. The most distal part of the deep palmar arch is about 1 cm proximal to the superficial arch.

Part Three

THE LOWER LIMB

ERNEST GARDNER
DONALD J. GRAY
RONAN O'RAHILLY

Introduction

The lower limb, like the upper, is connected by a girdle to the trunk and presents three segments, the thigh, leg, and foot. The pelvic girdle is formed by the two hip bones, which are joined in front but are separated behind by the upper part of the sacrum. The pelvic girdle and the sacrum together form a heavy, rigid ring, termed the bony pelvis.

The lower limb is specialized for the support of weight and the management of gravity, and for locomotion. Some of the muscles that act upon it originate from the pelvic girdle, sacrum, and vertebral column. Hence it is customary in describing the lower limb to include the regions that are transitional between the trunk and the lower limb (e.g., the gluteal region, or buttock, and the inguinal region, or groin). Most persons do not use the two limbs equally, and there is no clear-cut correlation with handedness.*

*I. Singh, Acta anat., 77:131, 1970.

The lower limbs first appear as minute buds in embryos of about 5 mm in length, that is, at about four postovulatory weeks of age. The lower limb buds are slightly behind the upper in development. Each limb bud elongates and develops in a proximodistal sequence (e.g., the leg appears before the foot). A few days after the limbs can first be seen, nerves grow into them, and the skeleton and muscles become differentiated. Shortly afterward, toes can be identified.

The common names for many parts of the lower limb are listed here. Their Latin equivalents are included, and it will be noted that many other terms are derived from these. For example, the word "pedal," meaning pertaining to the foot, is used in everyday speech.

Latin Names and English Equivalents for Parts of the Lower Limb

Latin Name	*English Equivalent*
Coxa	Hip
Natis or Clunis	Buttock
Femur	Thigh
Genu	Knee
Crus	Leg
Sura	Calf
Talus	Ankle
Pes	Foot
Calx	Heel
Planta	Sole
Digiti pedis	Toes
Hallux	Big toe

GENERAL READING

In addition to the following, which deal chiefly with the limbs, see also the references cited in the introductory chapters.

Castaing, J., and Soutoul, J. H., *Atlas de coupes anatomiques. II. Membre inférieur,* Maloine, Paris, 1968.
Henry, A. K., *Extensile Exposure,* Livingstone, Edinburg, 2nd ed., 1957.
Lockhart, R. D., *Living Anatomy,* Faber and Faber, London, 6th ed., 1963.
Royce, J., *Surface Anatomy,* Davis, Philadelphia, 1965.

BONES OF LOWER LIMB

18

The hip bones, which form the pelvic girdle, meet in front at the pubic symphysis. Each articulates with the upper part of the sacrum behind. The femur is the bone of the thigh. It articulates with the hip bone above, and with the tibia below. The tibia and fibula are the bones of the leg; they join the skeleton of the foot at the ankle.

HIP BONE

The hip bone (figs. 18–1 to 18–8; see also fig. 40–2, p. 440) joins the sacrum to the femur and therefore forms the bony connection between the trunk and the lower limb. The side to which a hip bone belongs can be determined by placing the bone so that its large, cup-shaped cavity faces laterally and slightly anteriorly, and the large obturator foramen is located below and in front of this.

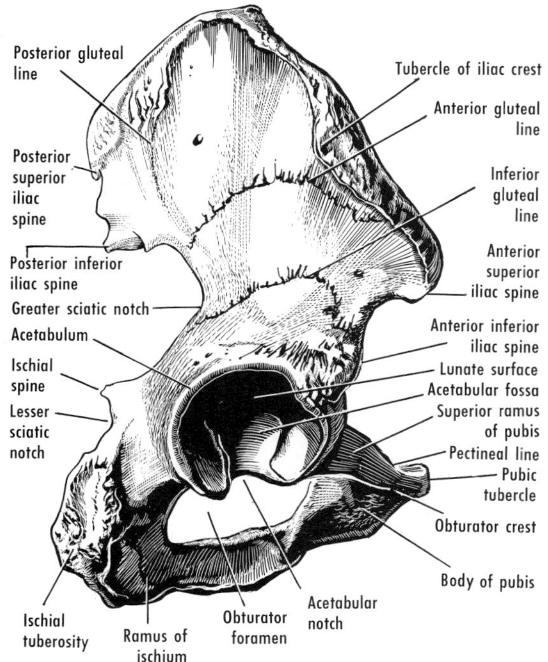

Figure 18–1 Right hip bone, lateral view, anatomical position.

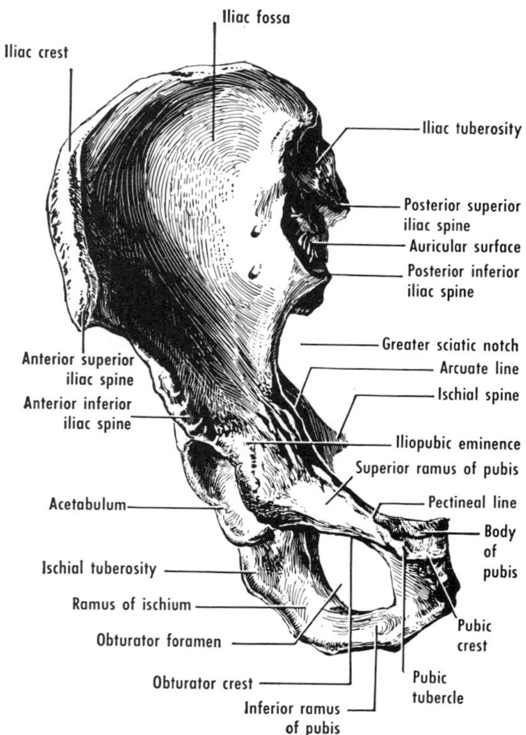

Figure 18–2 Right hip bone, anterior view, anatomical position.

Most of the borders and surfaces of the hip bone are named according to the anatomical position. It is important to realize that in this position the internal aspect of the body of the pubis faces almost directly upward and the urinary bladder rests on it. This position is readily apparent in the isolated hip bone when the bone is oriented so that the articular surface of the pubic symphysis is in a sagittal plane, and the acetabular notch points downward. The pubic tubercle and the anterior superior iliac spine are then in about the same coronal plane. This is the correct orientation of the bone in the anatomical position.

The hip bone forms the anterior and lateral walls of the bony pelvis. It joins its fellow of the opposite side to form the pubic

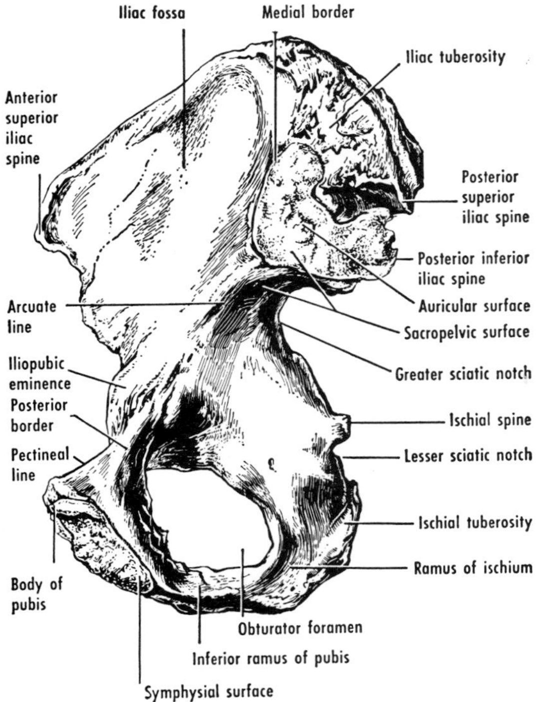

Figure 18–3 Right hip bone, medial view, anatomical position.

symphysis in front, and joins the sacrum to form the sacroiliac joint behind. Each hip bone consists of the ilium, ischium, and pubis, all three of which in the adult are fused at the acetabulum to form a single bone.

Component Elements

Ilium. The ilium consists of a *body,* which forms about two-fifths of the acetabulum, and a *wing* or *ala,* which forms the upper, expanded part of the hip bone. The body and wing are demarcated from each other on the internal aspect of the bone by the lower half of the medial border. The lower half of this border is a smooth rounded ridge that constitutes the *arcuate line* (iliac part of linea terminalis, p. 439). No demarcation is visible on the external aspect of the bone.

The body of the ilium joins the ischium and the pubis. A faint line that extends posteriorly from the margin of the acetabulum marks the fusion of the ilium with the ischium. The iliopubic eminence marks the fusion of the ilium with the pubis.

The expanded upper end of the ilium is the *iliac crest,* which can be palpated in its entire extent in the living subject. The crest is somewhat arched. It is also curved from side to side, being convex outward in front, and concave inward behind. Its anterior limit is the *anterior superior iliac spine,* to which the inguinal ligament is attached, and its posterior limit is the *posterior superior iliac spine.* Most of the crest presents *inner* and *outer lips,* which enclose a rough *intermediate line* or area. The tubercle of the crest is a thickening or projection on the outer lip about 5 cm behind the anterior superior iliac spine (at the level of L. V. 5). **The highest point of the iliac crest is somewhat behind its midpoint, at the level of L. V. 4. The supracristal plane is a horizontal plane that connects the highest points of the right and left crests.**

The wing or ala of the ilium presents

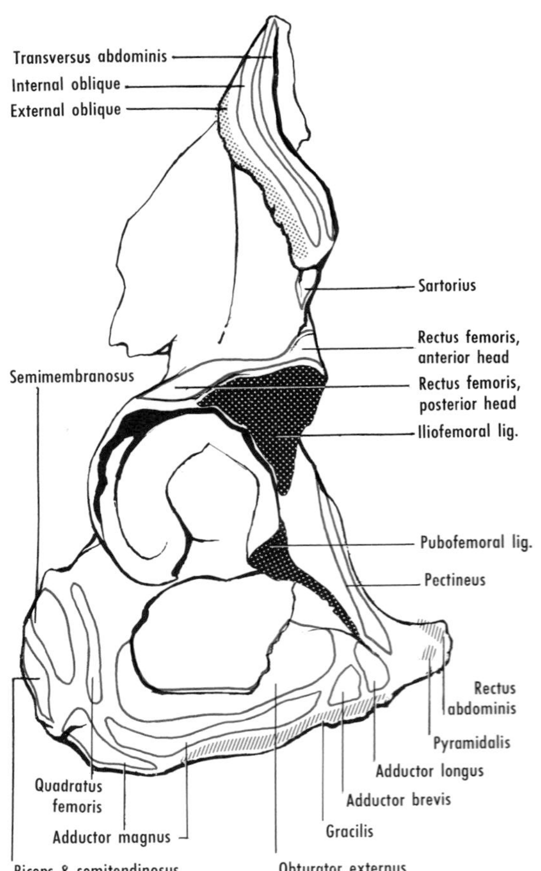

Figure 18–4 Right hip bone, inferolateral view, muscular and ligamentous attachments.

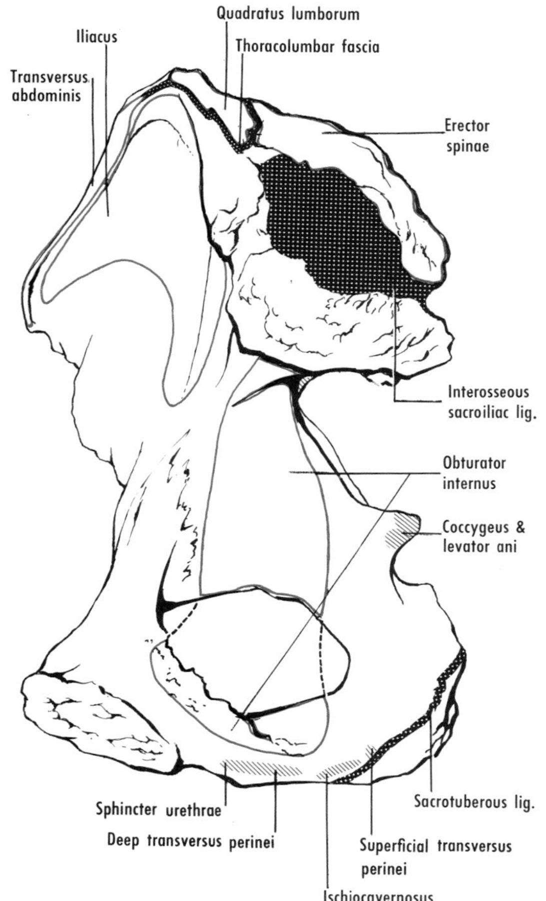

Figure 18–5 Right hip bone, medial view, muscular origins and ligamentous attachments.

the gluteal and the sacropelvic surfaces, the iliac fossa, and three borders, anterior, posterior, and medial.

The *gluteal surface* of the ilium, which is curved like the iliac crest, is a wide surface between the anterior and posterior borders. It is crossed by three curved ridges, which vary in prominence according to the muscularity of the subject. The *posterior gluteal line* begins at the crest about 5 cm in front of the posterior superior iliac spine and curves downward to the greater sciatic notch. The *anterior gluteal line* begins near the tubercle of the crest and arches across the gluteal surface toward the greater sciatic notch. The *inferior gluteal line* begins immediately above the anterior inferior iliac spine and curves backward and downward, about 2 to 3 cm above the acetabulum, toward the greater sciatic notch. It is the least distinct of the three lines.

The *iliac fossa,* situated between the anterior and medial borders, is limited above by the iliac crest and below by the lower parts of the medial border (fig. 18–3). The fossa is smooth and concave and usually contains a large nutrient foramen. The bone of the upper part of the fossa may be thin and translucent.

The *sacropelvic surface,* between the posterior and medial borders, is limited above and behind by the iliac crest. It includes the auricular surface and the iliac tuberosity. The ear-shaped *auricular surface,* which is articular, is located directly behind the iliac fossa. This surface is often flat. The *preauricular sulcus,* as a rule found only in the female, is located in front of and below the auricular surface. The *iliac tuberosity* is the rough area above and behind the auricular surface. Accessory facets for articulation with the sacrum may be present in this region (see p. 521).

The smooth, lower part of the sacropelvic surface (also known as the pelvic surface) lies between the lower part of the medial border and the greater sciatic notch. It forms a part of the lateral wall of the true pelvis.

The anterior border of the wing of the ilium extends from the anterior superior spine to the acetabulum. Its lower part presents the *anterior inferior iliac spine,* a roughened projection a little above the acetabulum. A shallow groove, below and medial to this spine, lodges the iliopsoas. The *iliopubic eminence,* just medial to this groove, is a low prominence that marks the line of fusion of the ilium with the pubis. **The anterior superior spine is an important landmark. It can usually be palpated by tracing the inguinal ligament upward, or the iliac crest downward.** In the erect position it is in about the same coronal plane as the pubic tubercle and is at about the level of S. V. 1. **The lower limb is measured clinically from the anterior superior spine to the tip of the medial malleolus of the tibia.** The measuring tape should be kept along the medial side of the patella.

The posterior border begins at the posterior superior iliac spine. It extends downward to the *posterior inferior iliac spine* and then makes an abrupt turn forward to become continuous with the posterior border of the ischium. It forms with this border the *greater sciatic notch.* A skin dimple usually

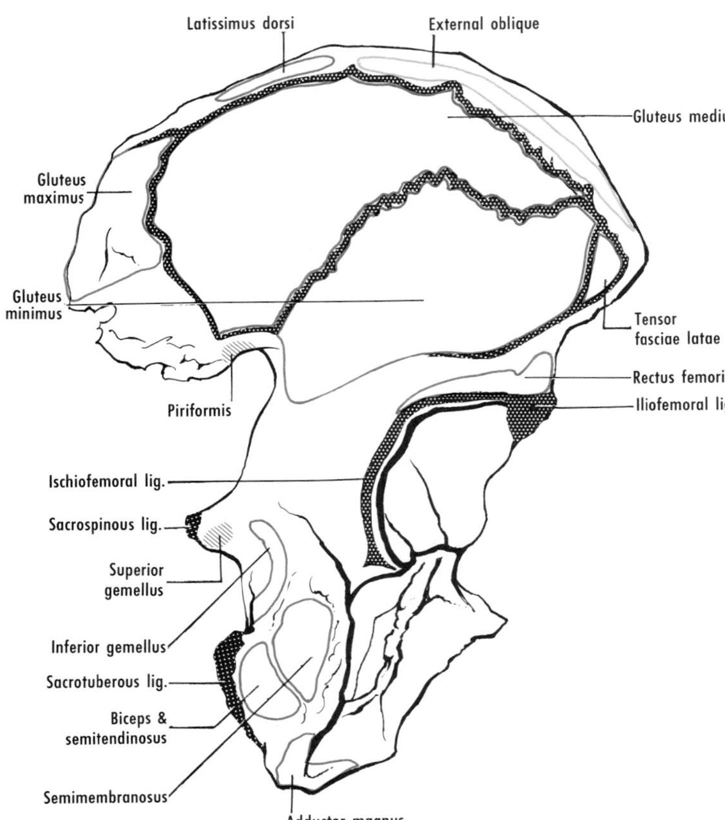

Figure 18–6 Right hip bone, posterolateral view, muscular and ligamentous attachments. The origin of the tensor fasciae latae often extends more posteriorly.

marks the position of the posterior superior spine, about 5 cm lateral to the median plane (fig. 51–1, p. 546). **A line connecting the two skin dimples is at the level of S. V. 2 and indicates the level of the sacroiliac joints.**

The medial border of the wing of the ilium begins at the iliac crest and extends downward, first as a roughened line, and then as a sharp edge bounding the auricular surface in front. It then turns abruptly forward and continues as a rounded line to the iliopubic eminence. It is this rounded lower half that forms the arcuate line.

Ischium. The ischium (from which the adjective sciatic is derived) forms the posteroinferior part of the hip bone and consists of a body and a ramus.

The *body of the ischium* has upper and lower ends. The upper fuses with the pubis and ilium, and forms part of the acetabulum. The free lower end and the roughened lower part of the dorsal surface form the *tuber of the ischium (ischial tuberosity).* The ramus projects from the lower end and fuses with the inferior ramus of the pubis below the obturator foramen.

The body has femoral, pelvic, and dorsal surfaces. The femoral surface is below the acetabulum and faces the thigh. It is limited in front by the margin of the obturator foramen; this margin is sometimes called the anterior border. The smooth pelvic surface helps to form the bony wall of the ischiorectal fossa. It is continuous above with the pelvic surfaces of the ilium and the pubis.

The dorsal surface of the ischium is continuous above with the gluteal surface of the ilium. Below, it continues into the free lower end of the ischium and with it forms the ischial tuberosity. The lateral edge of this surface may be termed the lateral border of the ischium. It separates the femoral and dorsal surfaces. The other edge, which may be termed the posterior border, separates the pelvic and dorsal surfaces; it is continuous above with the posterior border of the ilium. It helps to complete the *greater*

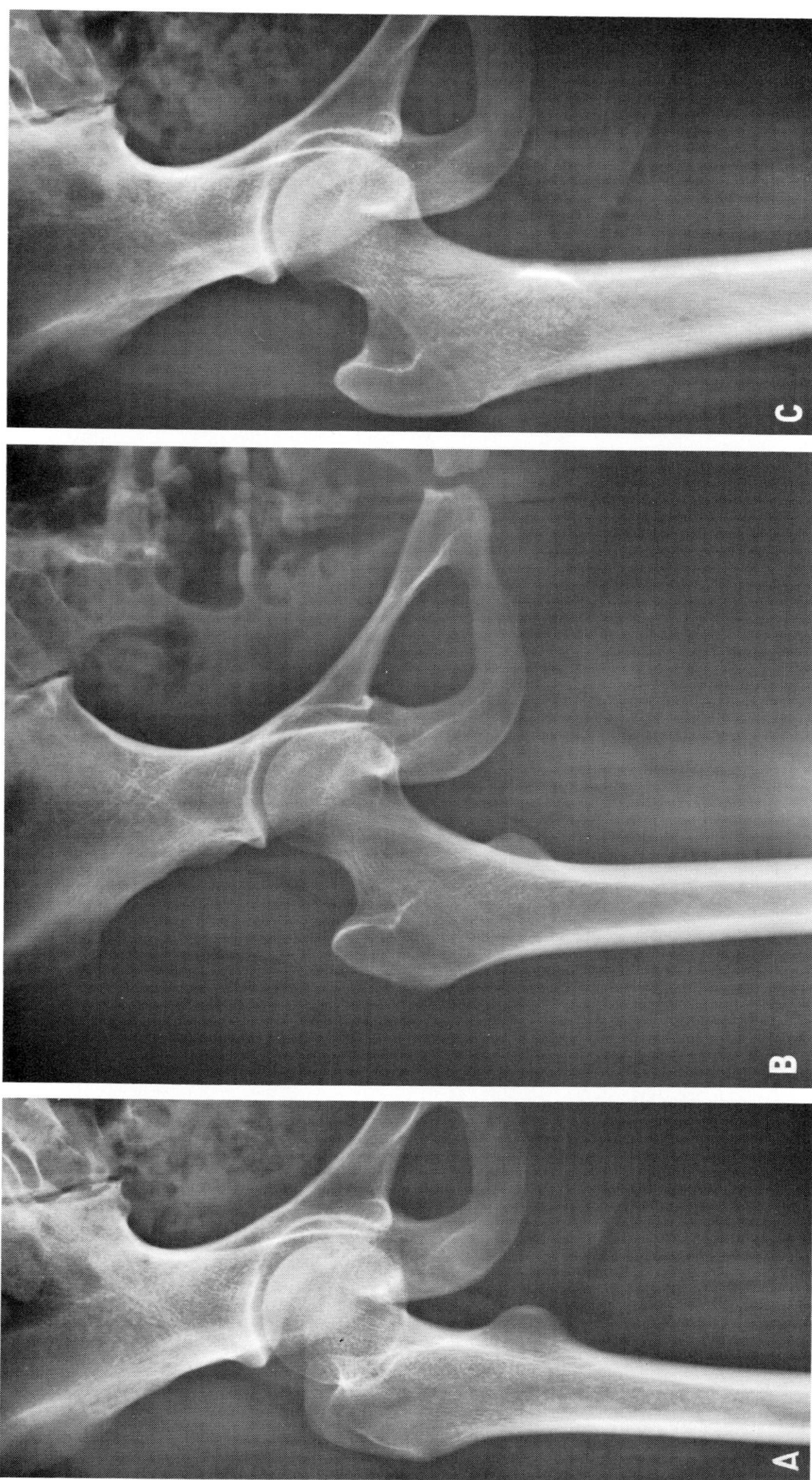

Figure 18–7 Hip in various positions. *A*, maximum lateral rotation. Note the foreshortening of the neck of the femur. The lesser trochanter is clearly visible. *B*, anatomical position. Although the neck shows well, it is still slightly foreshortened, and a slight degree of medial rotation is necessary to show the neck correctly. *C*, maximum medial rotation. Note that the lesser trochanter is now overlapped completely by the shaft of the femur.

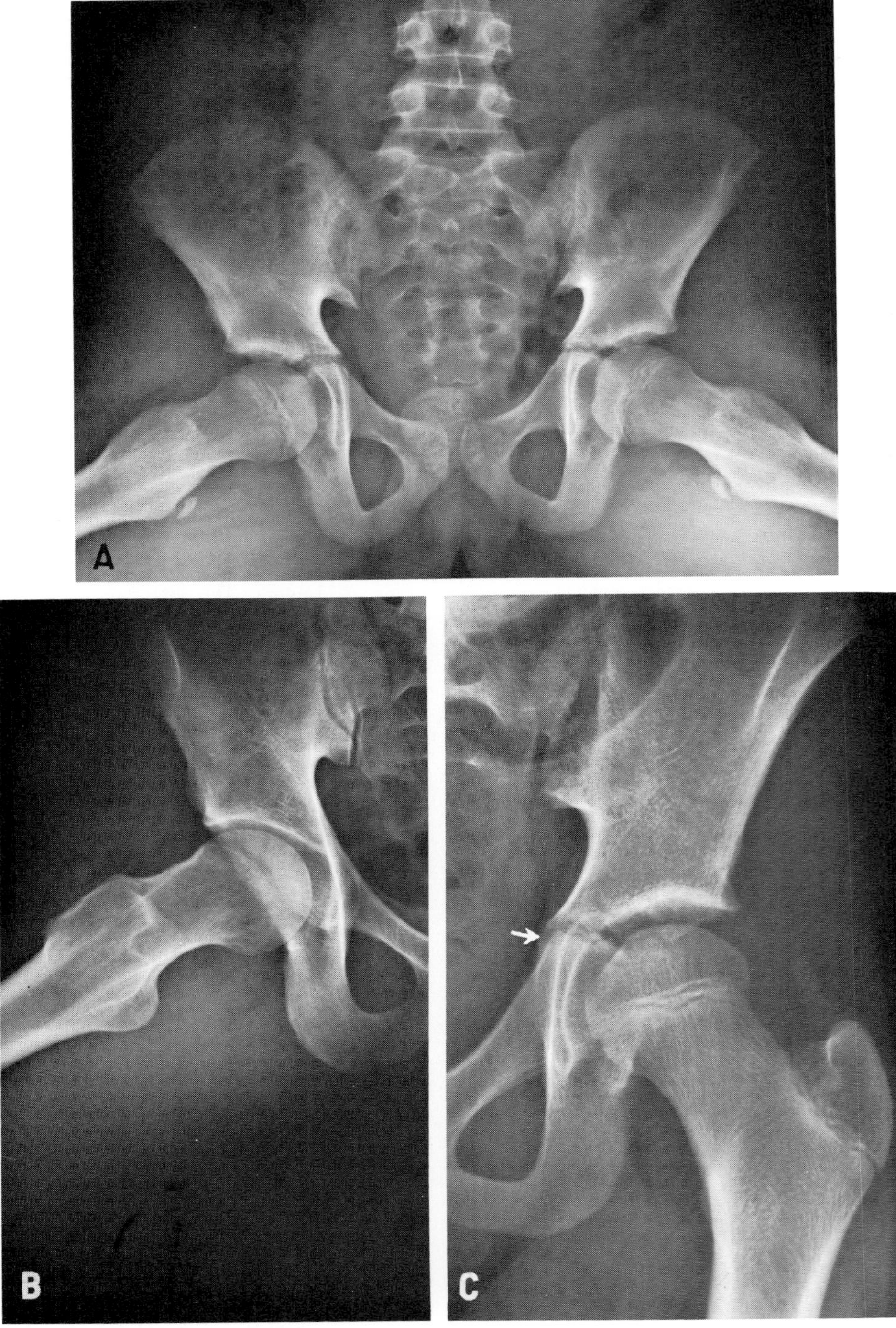

Figure 18–8 *A*, hips of child, abducted. Note epiphysis for lesser trochanter on each side. *B*, hip of adult, abducted. Note greater trochanter (*above*) and lesser (*below*). *C*, hip of child. Note site of triradiate cartilage (*arrow*), and epiphyses for head and greater trochanter of femur.

sciatic notch, below which the border is marked by the sharp, triangular *ischial spine.* The *lesser sciatic notch* is a rounded notch in the posterior border, between the spine and the tuberosity.

The sacrospinous and sacrotuberous ligaments convert the sciatic notches into the greater and lesser sciatic foramina.

The ischial tuberosity consists of a smooth upper portion and a rough lower portion. The upper part is subdivided for the attachments of the hamstrings. The tuberosity is palpable when the thigh is flexed.

The *ramus of the ischium** extends upward and medially and joins the inferior ramus of the pubis. The conjoined rami of the ischium and pubis form a bar of bone with two surfaces, external and internal, and two borders, upper and lower. A ridge usually divides the inner surface lengthwise into upper and lower parts. The sharp upper border faces the obturator foramen. The lower border is rough and somewhat everted.

Pubis. The pubis is divided into a body and two rami, superior and inferior.

The *body* is the wide, compressed part of the bone just medial to the rami.† Its symphysial or medial surface is ovoid and rough.[1] The body joins the body of the opposite pubis in the median plane to form the *pubic symphysis.* The other surfaces of the body are pelvic and femoral. The smooth pelvic surface looks upward; the urinary bladder rests on it. The femoral surface looks downward; it is roughened for the attachment of muscles. The rough anterior border is the *pubic crest,* and the lower lateral part of it is the prominent *pubic tubercle.* **The pubic tubercle is a landmark in the lower part of the abdominal wall, about 3 cm from the median plane** (figs. 33–6, p. 356, and 33–8, p. 358). **It is a guide to the inguinal ring, femoral ring, and saphenous opening. When the hip bone is placed in the anatomical position, the pubic tubercle and the anterior superior iliac spine are in the same coronal plane.** In this position, also, the pubic crest, the coccyx, the middle of the acetabulum, the head of the femur, and the tip of the greater trochanter are all in about the same horizontal plane. The sacrum is entirely, or almost entirely, above the level of the pubic symphysis.

The *superior ramus* extends upward, backward, and laterally to the acetabulum, where it fuses with the ilium and ischium. Internally it is delimited from the body of the ilium by the iliopubic eminence and from the ischium by a rough line between the iliopubic eminence and the margin of the obturator foramen.

The superior ramus has pectineal, pelvic, and obturator surfaces, and anterior, posterior, and inferior borders. The anterior border is the pecten pubis *(pectineal line),* a sharp edge that begins at the pubic tubercle and continues to the iliopubic eminence. The pubic crest and the pectineal line constitute the pubic part of the linea terminalis. The inferior border is the *obturator crest;* it runs from the pubic tubercle to the acetabular notch. The triangular pectineal surface lies between the pectineal line and the obturator crest. The posterior border is the margin of the obturator foramen as viewed from the pelvic aspect. The pelvic surface, between the pectineal line and the posterior border, is continuous with the pelvic surface of the body. The obturator surface presents the *obturator groove,* an oblique groove that lodges the obturator vessels and nerve. The obturator membrane (see below) converts the groove into the *obturator canal* (fig. 40–10, p. 445).

The *inferior ramus* is a short bar that extends from the body of the pubis backward, downward, and laterally. It meets and fuses with the ramus of the ischium.

Acetabulum

The acetabulum (figs. 18–1 and 21–9) looks downward, forward, and laterally. It is a large, cup-shaped cavity on the outer side of the hip bone, which articulates with the head of the femur to form the hip joint. The acetabulum is deficient below at the *acetabular notch.* The rough depression in the floor of the acetabulum above the notch is the *acetabular fossa.* The fossa is sometimes thin and translucent in its upper part. The remainder of the acetabulum, the *lunate surface,* is smooth and articulates

---

*Some authors describe a superior ramus of the ischium. The superior ramus of such descriptions is incorporated with the body of the ischium in the present description.

†The term "body of the pubis" has also been used to refer to the part of the pubis that forms a portion of the acetabulum.

with the head of the femur. The *acetabular lip* is attached to its peripheral margin.

The pubis forms about one-fifth of the acetabulum, the ilium almost two-fifths, and the ischium somewhat more than two-fifths (fig. 18–9*C*).

Obturator Foramen

The obturator foramen is bounded by the pubis and the ischium and their rami. The foramen is closed, except at the obturator groove, by the thin, strong *obturator membrane* (fig. 40–10, p. 445), which is attached to the margins of the foramen.

Ossification (fig. 18–9)

The three parts of the hip bone begin to ossify during the fetal period. At birth, each of these centers has formed a part of the acetabulum. The centers in the ramus of the ischium and the inferior ramus of the pubis meet by late childhood, and by this time the three primary centers are separated in the acetabulum by the **Y**-shaped *triradiate cartilage* (fig. 18–8). One or more secondary centers appear in this cartilage during late childhood. Other secondary centers (iliac crest, anterior inferior iliac spine, ischial tuberosity, and pubic symphysis) appear during puberty. The center for the tuberosity appears during adolescence.

Acetabular centers generally begin to unite during adolescence; union is completed by late adolescence or early adult life. The other centers unite during the third decade. Other secondary centers are occasionally present (e.g., pubic crest, ischial spine). The os acetabuli is an occasional ossicle present at the edge of the acetabulum, at the line of junction of the pubis with the ischium.

FEMUR

The femur, or thigh bone (figs. 18–7, 18–8, and 18–10 to 18–22), is the longest and heaviest bone of the body, its length ranging from one-fourth to one-third of that of the body. Stature can be estimated from the length of the femur.[2] In the standing position, the femur transmits weight from the hip bone to the tibia. In the living subject, the femur is so covered with muscles that it is palpable only near its upper and lower ends.

The femur consists of a shaft and two ends, upper and lower. The upper end comprises a head, a neck, and two trochanters, the greater and lesser. The lower end consists of two spirally curved condyles, the medial and lateral. The side to which a femur belongs can be determined when the rounded end or head faces upward and medially, and the convexity of the shaft faces forward.

When the femur is in the anatomical position, the two condyles are in the same horizontal plane; that is, their lower surfaces will both touch a horizontal surface, such as a table top. The shaft makes an angle of about 10 degrees with a vertical line dropped from the head of the femur (fig. 18–13). This vertical line is the axis about which medial and lateral rotation of the femur occur.

The plane of the neck of the femur, followed medially, usually lies a little in front of the plane of the condyles; the head of the femur is said to be anteverted. The

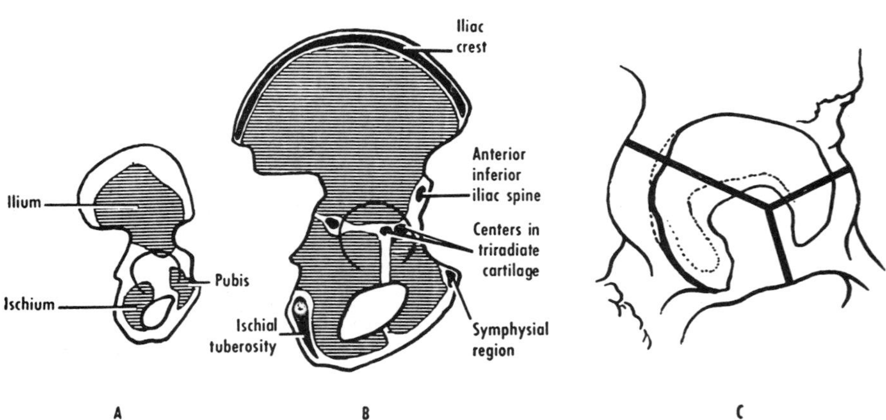

Figure 18–9 *A*, hip bone at birth. Lined regions indicate bony areas in ilium, ischium, and pubis. The rest is cartilage. *B*, hip bone at puberty, showing the increase in ossification and secondary centers (black). *C*, adult acetabulum. The lines indicate the lines of fusion of ilium, ischium, and pubis.

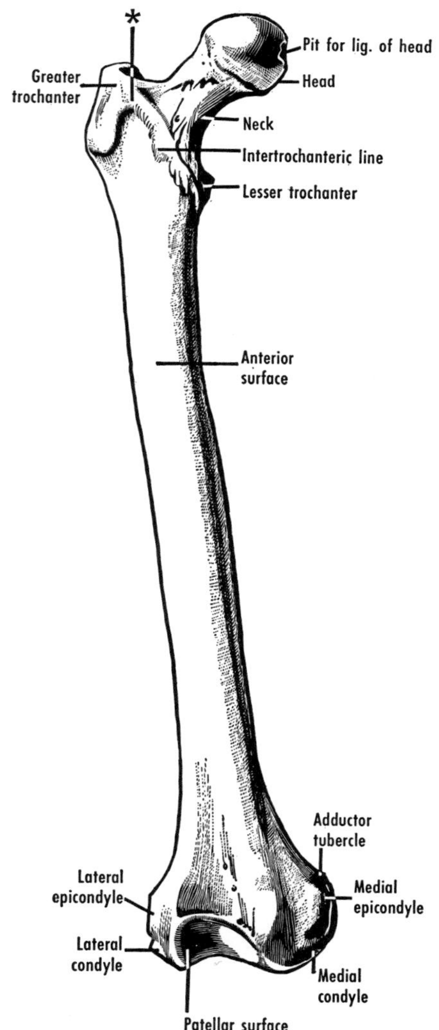

Figure 18–10 Right femur, anterior view, anatomical position. The asterisk indicates the cervical tubercle.

small, acute angle between the two planes is termed the *angle of femoral torsion,* because the shaft of the femur appears to be twisted so that the head comes to point a little forward (figs. 18–14 and 18–16). **In adults, the angle of torsion averages about 15 degrees; it is generally much greater in infancy (average 31 degrees).[3] The degree of anteversion may be altered in pathological conditions. Hence its determination may be important in diagnosis and treatment.**

The angle that the long axis of the neck makes with the long axis of the shaft is called the *angle of inclination* (fig. 18–15). It varies with age, sex, and development of the bony skeleton. It may be changed by any pathological process that weakens the neck of the femur. **When the angle of inclination is diminished, the condition is known as coxa vara; when it is increased, as coxa valga.**

The shaft of the femur is curved, with the convexity forward (fig. 18–16). The curvature is most marked in its upper part. The upper and lower parts of the shaft are compressed from before backward. *Platymeria* is an exaggerated degree of compression.

The *head* of the femur forms about two-thirds of a sphere and faces upward, medially, and slightly forward. A pit or *fovea* to

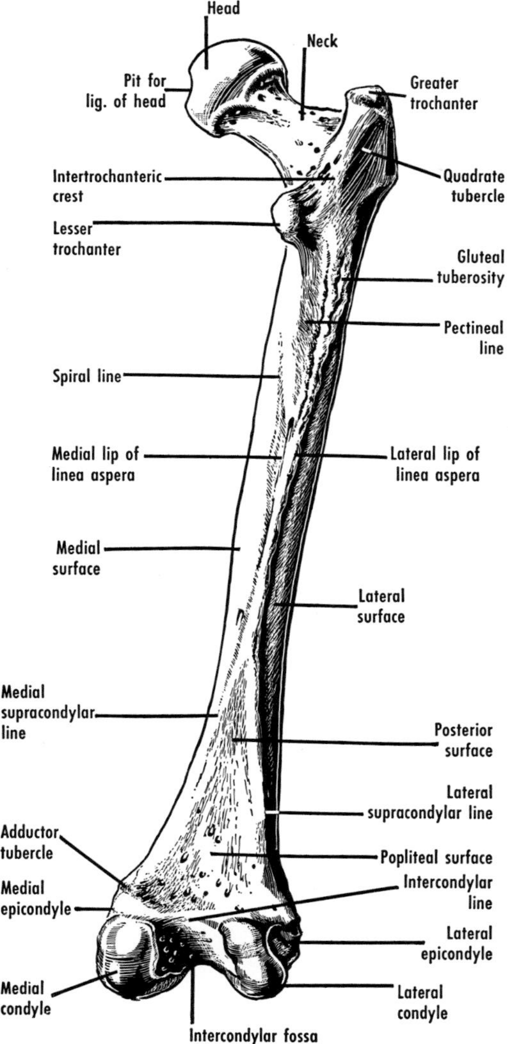

Figure 18–11 Right femur, posterior view, anatomical position.

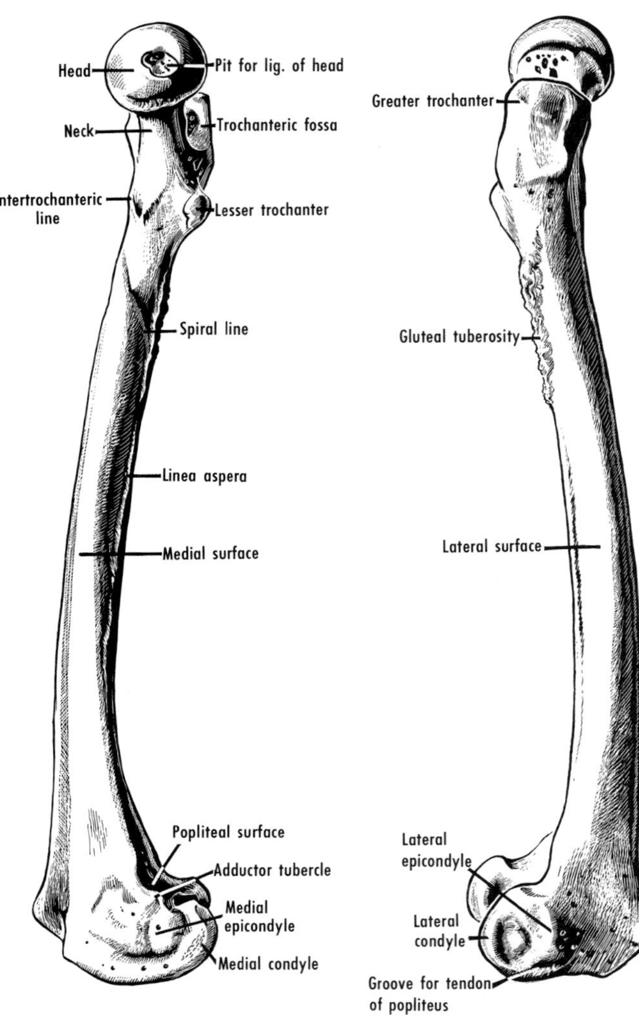

Figure 18–12 Right femur, medial and lateral views.

which the ligament of the head of the femur is attached is located slightly below and behind its center. The articular surface of the head is prolonged upon the anterosuperior region of the neck in many femora.

The *neck* is a thick bar of bone, somewhat rectangular in cross-section, that connects the head to the shaft in the region of the trochanters. The neck, except in front, is separated from the head by a sharp margin. A sagittal groove, frequently present on the anterior aspect of the neck, corresponds to the acetabular lip when the thigh is rotated medially.[4] In front, the neck and shaft are separated by the *intertrochanteric line*, which begins above at a tubercle (*cervical tubercle*) and runs downward and medially. It becomes continuous with a fainter spiral line, which winds to the back of the bone below the lesser trochanter and joins the medial lip of the linea aspera. Behind, the neck is smooth; about two-thirds of it is intracapsular. A groove for the obturator externus muscle is usually evident on the extracapsular aspect and passes obliquely upward to the trochanteric fossa. Above, the neck is short and has many nutrient foramina. Below, it runs downward and laterally to end at the lesser trochanter.

The laterally placed *greater trochanter*, which projects above the junction of the shaft with the neck, can be palpated on the lateral side of the thigh, about 10 cm below the iliac crest. **In the erect position the greater trochanter is in the same horizontal plane as the pubic tubercle, the head of the femur, and**

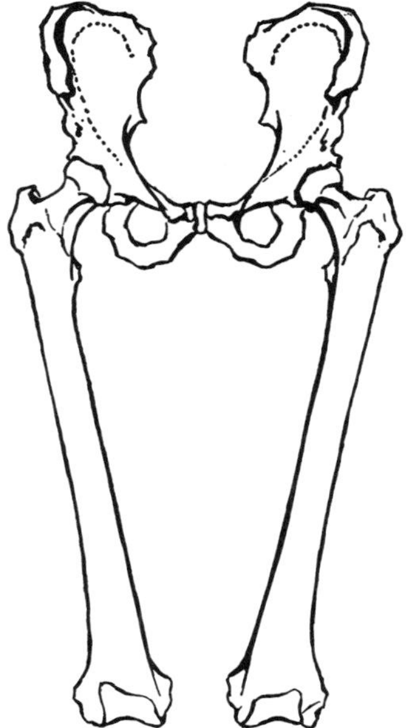

Figure 18–13 Orientation of hip bones and femora in the erect position.

the coccyx. It has medial and lateral aspects, and superior, anterior, and posterior margins. The lateral aspect is rectangular; a ridge for the insertion of the gluteus medius crosses it obliquely.

The anterior and superior margins are relatively wide. The posterior margin continues below into the intertrochanteric crest, which ends at the lesser trochanter. The medial surface presents a roughened depression, the *trochanteric fossa.*

The *intertrochanteric crest* is a ridge that connects the back of the greater trochanter with the lesser trochanter. The *quadrate tubercle* is a rounded elevation on the crest.

The rounded, conical, *lesser trochanter* extends medially from the posteromedial part of the junction of the neck with the shaft. The lesser trochanter is indistinctly palpable above the lateral end of the gluteal fold when the thigh is rotated medially.

The *shaft* of the femur presents anterior, medial, and lateral surfaces in its middle part and, in addition, a posterior surface in its upper and lower parts. It shows poorly defined medial and lateral borders but, in its middle third, a prominent posterior border, the *linea aspera* (fig. 18–20). The linea aspera has *medial* and *lateral lips,* and an intermediate area that broadens to form posterior surfaces as the lips diverge in the upper and lower thirds of the shaft. The lateral lip becomes continuous above with the *gluteal tuberosity.* The medial lip is continuous above with the spiral line. The *pectineal line* extends from the back of the lesser trochanter to the linea aspera. The lateral lip of the linea aspera is prolonged below as the lateral supracondylar line, which ends at the lateral epicondyle. The medial lip extends distally as the medial supracondylar line. It is interrupted by a smooth area related to the femoral artery and ends at the adductor tubercle. The flat posterior or popliteal surface lies between the distal prolongations of the lips of the linea aspera. One or two nutrient arteries enter the shaft at or near the linea aspera and are directed upward.

The distal end consists of two spirally curved condyles, which are continuous in front but are separated below and behind by the *intercondylar fossa.* In front, the condyles form the *patellar surface,* which is divided by a vertical groove into two unequal parts. The lateral part is wider and extends farther proximally. It articulates with

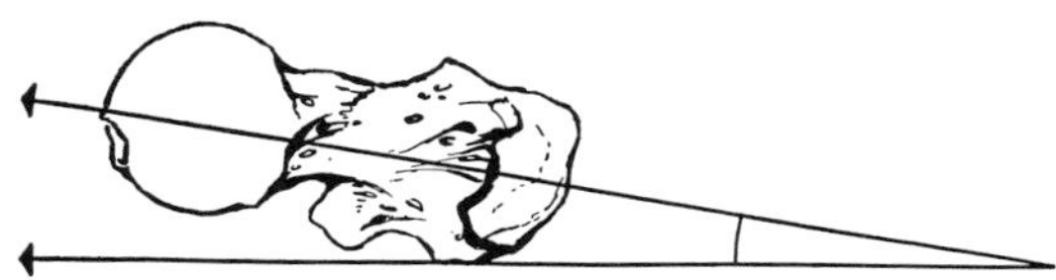

Figure 18–14 Anteversion of the head of the femur as viewed from above. The angle of torsion is the angle between the long axis of the head (*upper arrow*) and the horizontal axis of the condyles (*lower arrow*).

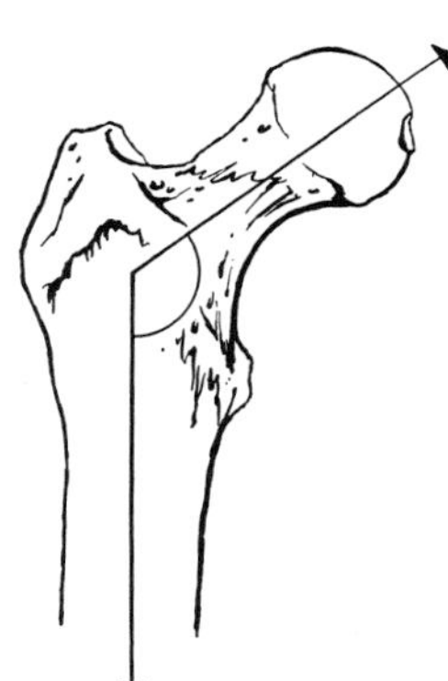

Figure 18–15 The angle of inclination. It averages about 125 degrees in adults.

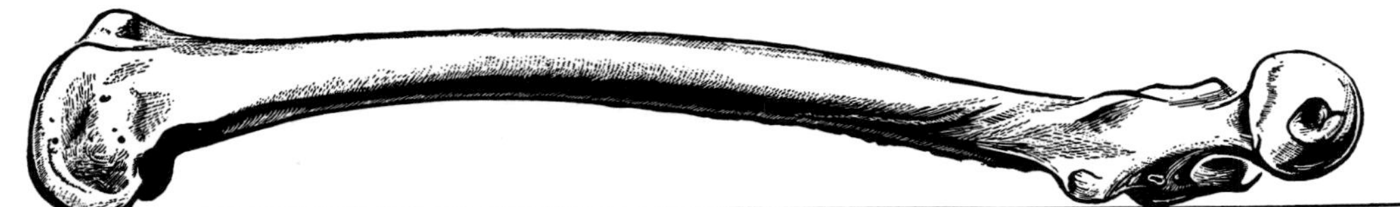

Figure 18–16 Right femur, medial view, the posterior aspect resting on a horizontal surface. Note the curvature of the shaft. Note also the anteversion (the head does not touch the surface).

the lateral articular facet of the patella. The narrower medial part articulates with the medial facet of the patella. The lower aspect of the lateral condyle is relatively broad and straight. That of the medial condyle is curved and narrow. A narrow, crescentic area adjacent to the intercondylar fossa articulates with the patella in extreme flexion. The posterior part of the medial condyle is wider and straighter than that of the lateral. The posterior parts of both articulate with the tibia only during flexion.

The medial surface of the medial condyle is rough and convex. Its most prominent part is the *medial epicondyle.* The *adductor tubercle,* a small prominence on the uppermost part of the condyle, is palpable and may be located by following the tendon of the adductor magnus to it.

The *lateral epicondyle,* smaller than the medial, is a prominence on the lateral surface of the lateral condyle. The lateral head of the gastrocnemius arises from an impression or pit immediately above the lateral epicondyle, and the popliteus from a pit immediately below it. There is often a groove that runs upward and backward from the pit, along the articular margin. This groove lodges the tendon of the popliteus when the leg is flexed. A notch in the articular margin below the pit lodges the tendon when the leg is extended.

The intercondylar fossa is limited below and at the sides by the margins of the condyles, and it is marked off from the popliteal surface of the shaft by the intercondylar line.

Structure

The femur is the classical example for the study of bony architecture. Two masses of compact bone are associated with the upper end. One is the calcar femorale (fig. 18–22), a bar of bone, that extends into the neck from the region of the lesser trochanter.[5] The other is the cervical torus, a thickened band or ridge of compacta found inconstantly[6] on the upper aspect of the neck between the head and the greater trochanter.

Blood Supply

The blood supply[7] to the head is important because it may be interrupted when the neck is fractured. Metaphysial and lateral epiphysial arteries derived from the medial circumflex artery are carried in the re-

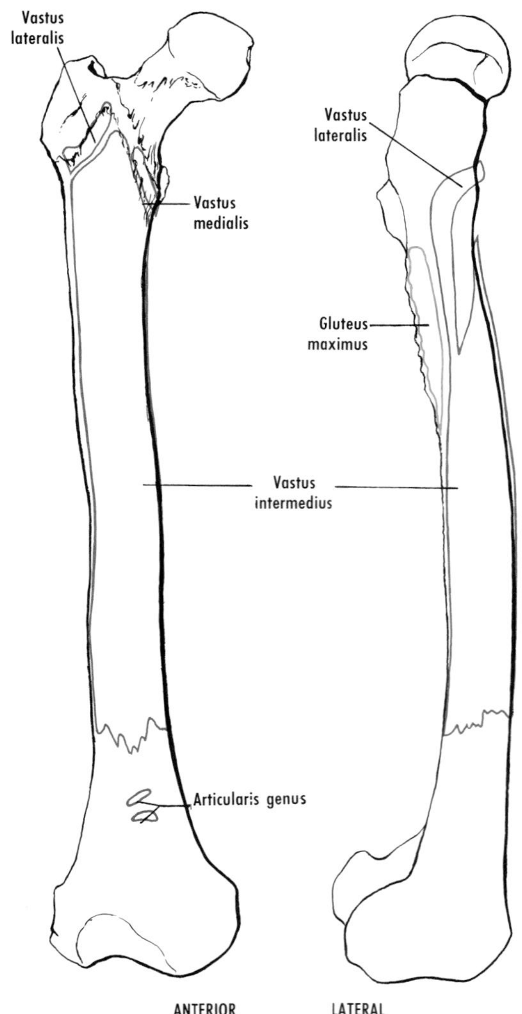

Figure 18–17 Right femur, showing muscle attachments. The lower part of the origin of the vastus intermedius fuses with that of the vastus lateralis.

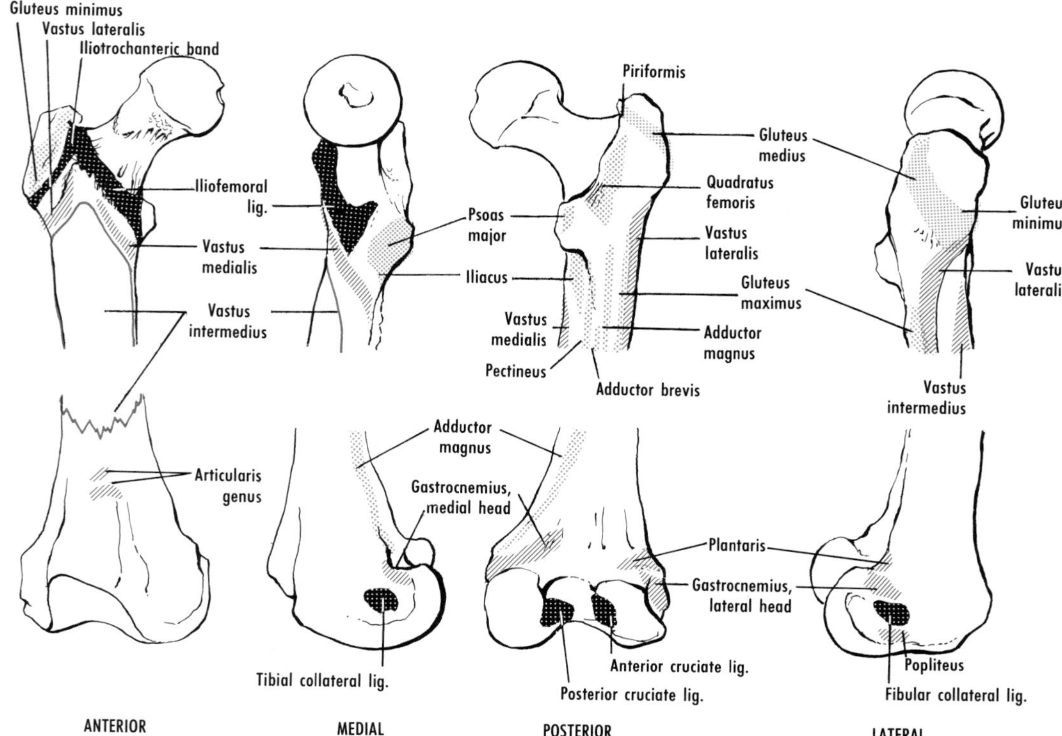

Figure 18–18 Muscular and ligamentous attachments to the upper and lower ends of the right femur. *Anterior*, the fascia that encloses the tensor fasciae latae meets at the front of that muscle, turns around the front edge of the gluteus minimus, and fuses with the rectus femoris and iliofemoral ligament at the ilium and with the tendon of the gluteus minimus below at the greater trochanter. This fascial band constitutes the iliotrochanteric band. *Medial*, the attachment of the iliofemoral ligament turns upward above the lesser trochanter, and constitutes the femoral attachment of the pubofemoral ligament. *Posterior*, see also figure 18–20 for details. *Lateral*, the gluteus medius is inserted along an oblique line on the lateral aspect of the greater trochanter, continuous in front and below with the gluteus medius (a bursa intervenes), and above and behind with the piriformis.

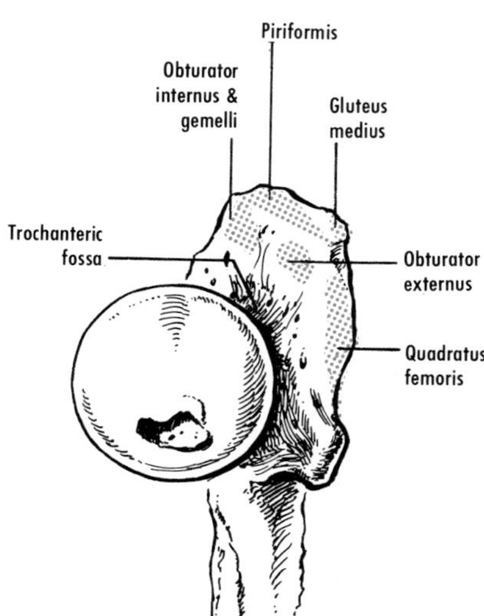

Figure 18–19 An approximately posteromedial view of the upper end of the right femur, showing muscle insertions. Note the almost continuous tendinous insertion of the obturator internus, piriformis, and gluteus medius.

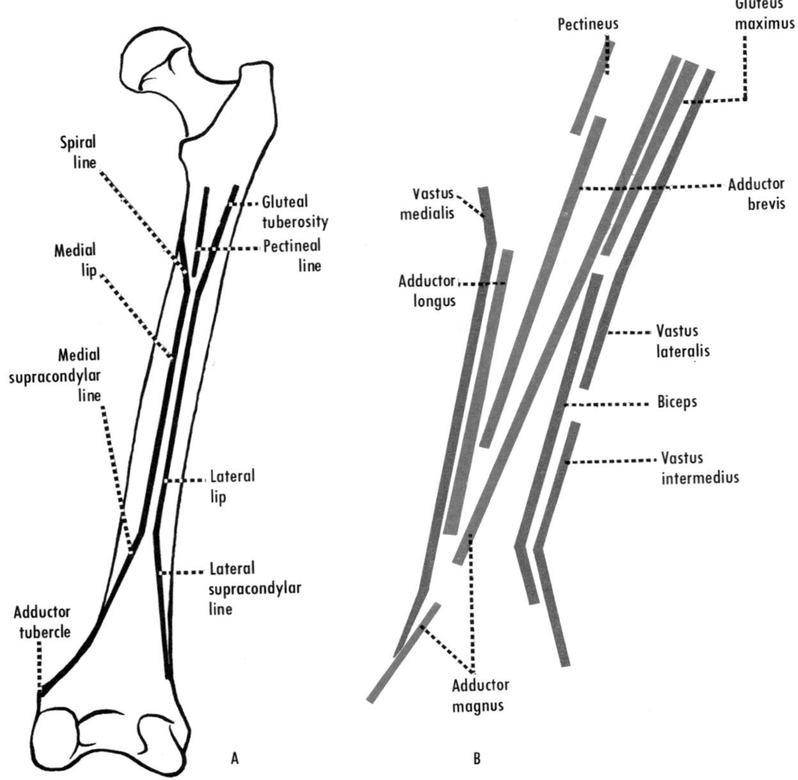

Figure 18–20 Schematic representation of the back of the femur. *A*, the linea aspera and its extensions above and below. *B*, muscle attachments.

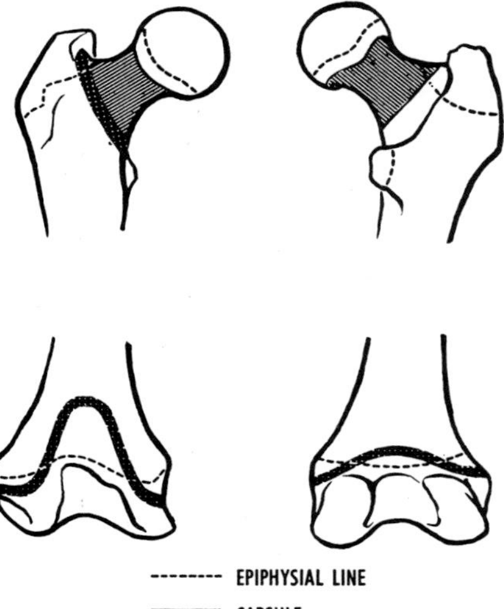

Figure 18–21 The upper and lower ends of the femur, showing the usual position of the epiphysial lines and the usual line of attachment of the joint capsule. The posterior part of the neck is covered by a reflection of synovial membrane, but has little if any capsular attachment. Based on Mainland.[8]

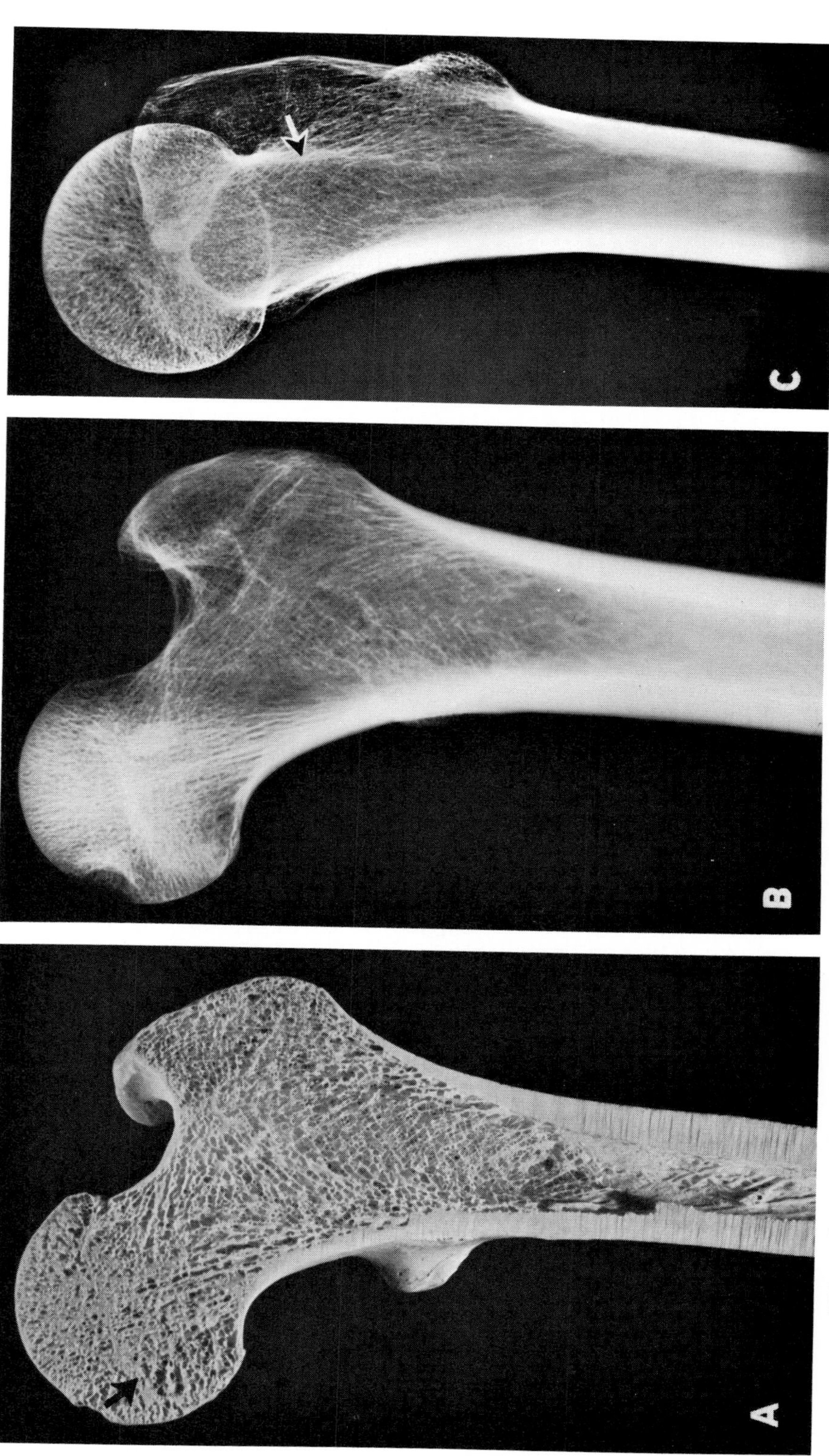

Figure 18–22 *A*, photograph of posterior half of left femur cut in a coronal plane. Note the straight and arching trabeculae. The arrow indicates the site of the former epiphysial disc. The head and greater trochanter are covered by a very thin shell of compact bone. *B*, radiogram of intact femur. Note the straight and arching trabeculae. The pit for the ligament of the head is clearly visible. *C*, lateral radiogram of intact femur. Note the calcar femorale (*arrow*).

tinacula of the neck to the head and neck of the femur. Medial epiphysial branches, which arise chiefly from the obturator artery (p. 214), enter the head by way of the ligament of the head. One, and sometimes two, nutrient arteries enter the shaft through or near the linea aspera and then course upward.

Ossification[9]

A periosteal collar is present by the seventh postovulatory week. An epiphysial center is usually present in the distal end at birth. That for the head appears during infancy, that for the greater trochanter during childhood, and that for the lesser trochanter during late childhood. Fusion of the lesser trochanter with the shaft occurs during adolescence, followed by fusion of the greater trochanter, the head, and the lower end (late adolescence or early adulthood). Growth in length of the femur occurs mainly at the lower end. Epiphyses of the femur are shown in figures 18–8 and 18–25, and the usual positions of the epiphysial lines of the femur are shown in figure 18–21.

PATELLA

The patella, or knee cap (figs. 18–23 to 18–25), is a triangular sesamoid bone, about 5 cm in diameter, that is embedded in the tendon of insertion of the quadriceps femoris muscle. When the quadriceps is relaxed, the patella can be moved from side to side and, to a lesser extent, upward and downward. It articulates behind with the patellar surface of the condyles of the femur. When the patella is placed on a table with its articular surface downward and its apex pointing away from the observer, the larger lateral facet indicates the side to which the patella belongs. The patella will tilt to this side because the lateral part, being larger, is heavier.

The patella has two surfaces, anterior and articular; three borders, superior, medial, and lateral; and an apex.

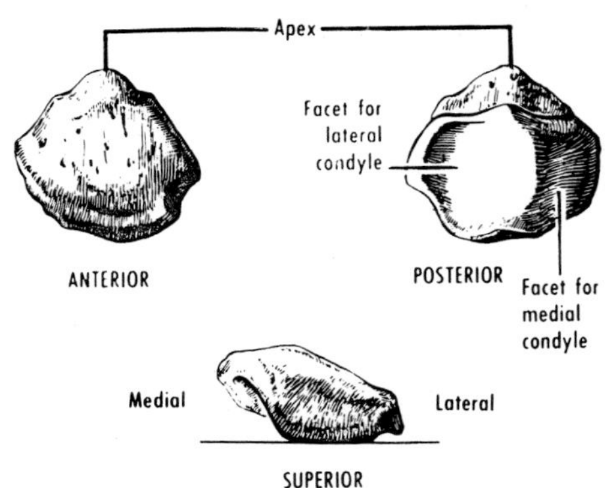

Figure 18–23 The right patella.

The *anterior surface* is convex from side to side and from above downward. It has numerous vertical ridges and many small openings for nutrient vessels. It is covered by a part of the tendon of insertion of the quadriceps femoris. This part of the tendon is continued from the *apex* to the tuberosity of the tibia and is termed the *ligamentum patellae.*

A rounded vertical ridge on the articular surface separates it into a larger, lateral articular, and a smaller, medial articular, facet. The nonarticular part of the posterior aspect is related to the infrapatellar pad of fat and gives attachment to the ligamentum patellae.

The superior border, or base, slopes downward and forward. The lateral and medial borders converge at the apex.

Ossification

The patella ossifies from several centers, which appear during early childhood and later fuse.[10] Ossification is usually completed by puberty or adolescence.

TIBIA

The tibia, or shin bone (figs. 18–24 to 18–32), is, next to the femur, the longest and the heaviest of the bones of the body. It measures about one-fourth to one-fifth of the length of the body. It is located on the anterior and medial side of the leg, where it can be palpated in its entire length. In the standing position, it transmits the weight from the femur to the bones of the ankle and foot. It can be identified as to side by placing its larger expanded end proximally, its most prominent border anteriorly, and the distal prolongation of its lower end medially.

The tibia has a shaft and two ends, upper and lower. When viewed from above, the shaft appears to be twisted, as if the upper end were rotated more medially than the lower end. **The angle between a horizontal line through the condyles and one through the malleoli indicates the degree of tibial torsion** (average, 15 to 20 degrees; range, 0 to 40 degrees).[11]

The upper end is large and is expanded for articulation with the lower end of the femur. It is bent slightly backward, and it consists of *medial* and *lateral condyles*, and a *tuberosity* (sometimes described with the shaft). The upper surface of each condyle is

(*Text continued on page 184.*)

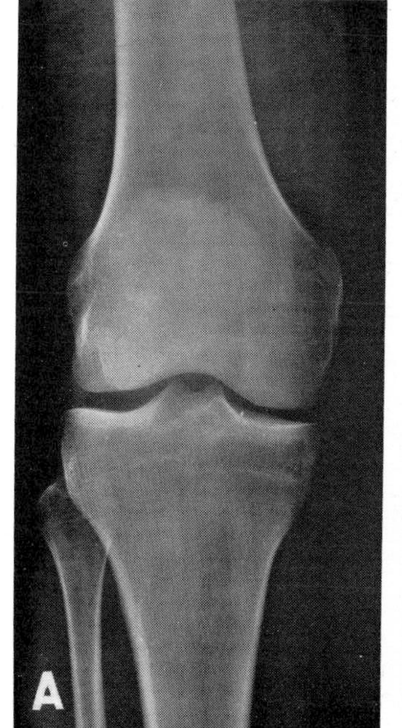

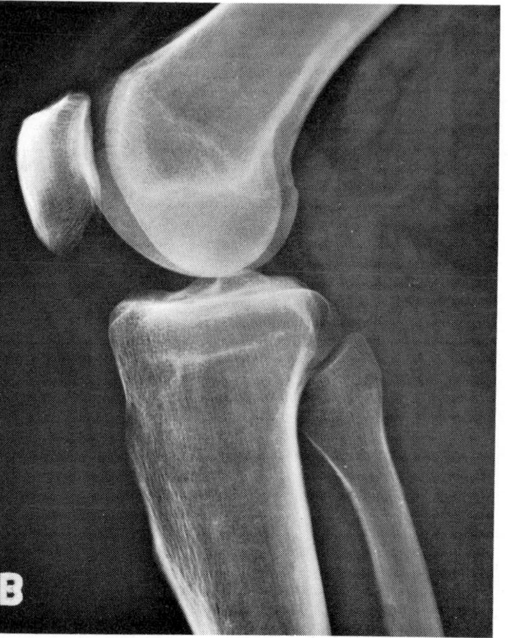

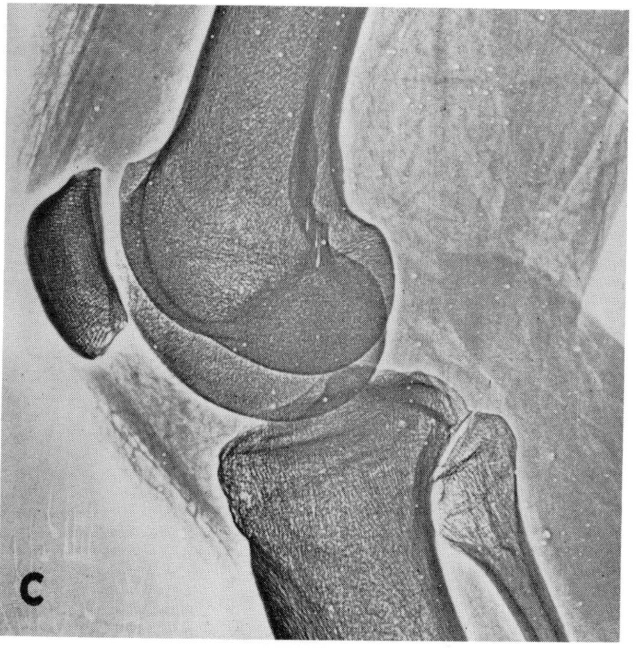

Figure 18–24 *A*, anteroposterior view of knee. Note obliquity of femur, outline of patella, radiolucent interval occupied by menisci and articular cartilage, and intercondylar eminence of tibia (showing lateral and medial intercondylar tubercles). *B*, lateral view of flexed knee. Note patella, condyles of femur, head of fibula and (superior) tibiofibular joint, and tuberosity of tibia. *C*, xeroradiogram of knee. Powder image on selenium-coated metal plate. *A* and *B*, courtesy of V. C. Johnson, M.D., Detroit, Michigan. *C*, courtesy of J. F. Roach, M.D., Albany, New York.

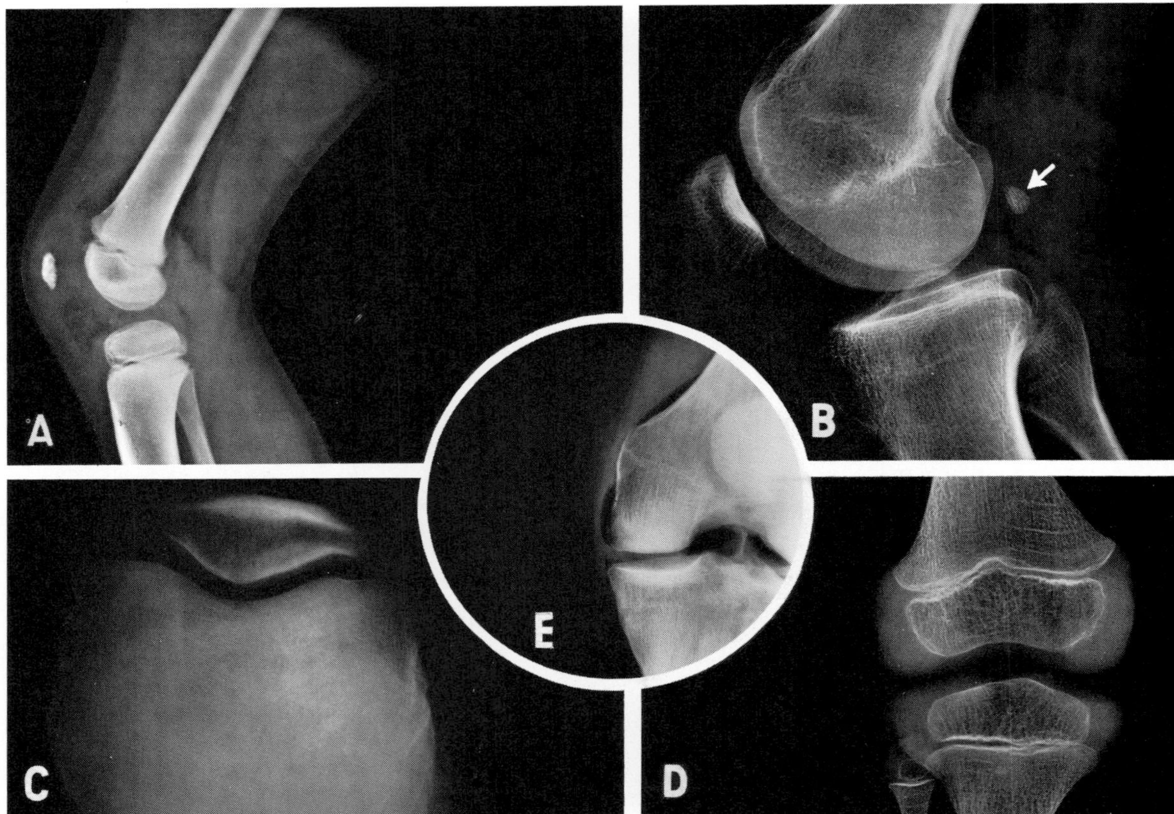

Figure 18–25 A, knee of child, lateral view. Note epiphyses for lower end of femur and upper end of tibia. The patella has begun to ossify, and the fat deep to the ligamentum patellae is visible as a radiolucent area. *B*, lateral radiogram of knee, showing fabella (*arrow*). *C*, radiogram of flexed knee. Note radiological joint space between femur and patella. The lateral condyle of the femur is that on the right-hand side of the illustration. *D*, radiogram of dried bones of boy aged five. Note outline of cartilage. *E*, pneumoarthrogram of knee produced by injecting air into the joint cavity. Note the medial meniscus and the cruciate ligaments. *A*, courtesy of V. C. Johnson, M.D., Detroit, Michigan. *E*, courtesy of Sir Thomas Lodge, The Royal Hospital, Sheffield, England.

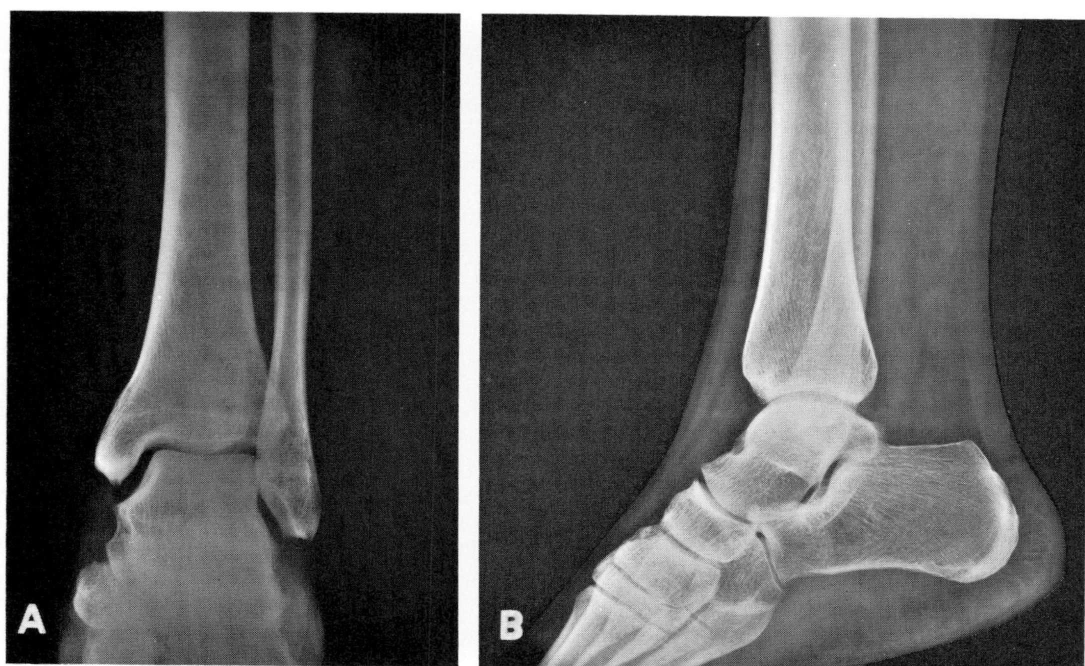

Figure 18–26 *A*, anteroposterior view of left ankle. Note the medial and lateral malleoli (and their different levels), and the trochlea of the talus. *B*, lateral view of ankle. Note line of talotibial part of joint, and outlines of talus, navicular, and calcaneus. *B*, courtesy of V. C. Johnson, M.D., Detroit, Michigan.

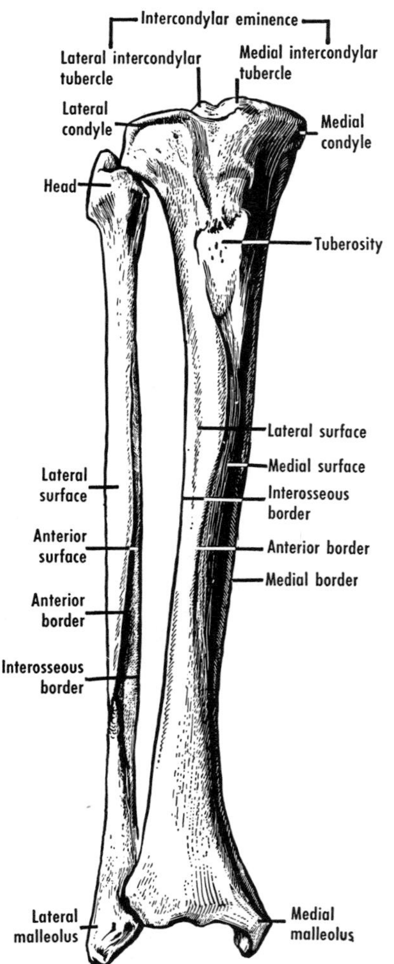

Figure 18–27 Right tibia and fibula, anterior view.

large, ovoid, and smooth, and articulates with the corresponding femoral condyle. The surfaces are separated from front to back by the *anterior intercondylar area,* the *intercondylar eminence,* and the *posterior intercondylar area.* The anterior area is the larger of the two and is depressed below the articular surface. The sides of the intercondylar eminence are prolonged proximally by the *medial* and *lateral intercondylar tubercles,* on which the articular surfaces extend. The intercondylar eminence varies greatly in shape and height, and may be absent.[12]

The lateral condyle is somewhat more prominent than the medial. The lower aspect of its posterior protrusion presents a flat, circular *articular facet* for the head of the fibula.

The medial and lateral condyles share an anterior surface of triangular shape. The apex of the triangle is formed by the *tuberosity of the tibia.* A groove limits the tuberosity above and laterally.[13] The upper part of the tuberosity is smooth and rounded. **In the position of kneeling, the body rests on the rough lower part of the tuberosity, the ligamentum patellae, the front of the tibial condyles, and the patella.**

The *shaft* of the tibia is thinnest at the junction of its middle and distal thirds, and gradually expands above and below. It presents three surfaces, medial, lateral, and posterior, and three borders, anterior, medial, and interosseous.

The *medial surface* is smooth and slightly convex. It can be felt through the skin. The *lateral surface* is slightly concave. Its lower part becomes convex and turns forward to become continuous with the front of the lower end. The *posterior surface* lies between the medial and interosseous borders. Its upper third is crossed by a rough ridge, the *soleal line,* which extends obliquely downward from the articular facet for the fibula to the medial border. An indistinct vertical line extends distally from the soleal line for a short distance and subdivides the posterior surface into medial and lateral parts. A large nutrient foramen, directed downward, is usually present on the upper third of the posterior surface.

The *anterior border* or *crest* is the most prominent of the three borders and forms the "shin." Above, it begins along the lateral margin of the tuberosity. It passes obliquely medially and becomes continuous with the anterior margin of the medial malleolus. It can be felt in its entire extent in the living. The *medial border* is poorly defined. It extends from the back of the medial condyle to the back of the medial malleolus. The *interosseous* (or lateral) *border,* better defined than the medial, gives attachment to the interosseous membrane. Above, it begins on the lateral condyle, about halfway between the fibular facet and the tuberosity. Below, it extends to the apex of the fibular notch, where it bifurcates.

The lower end of the tibia has a distal projection, the *medial malleolus,* and it has five surfaces, anterior, posterior, medial, lateral, and inferior.

The posterior surface is marked by the

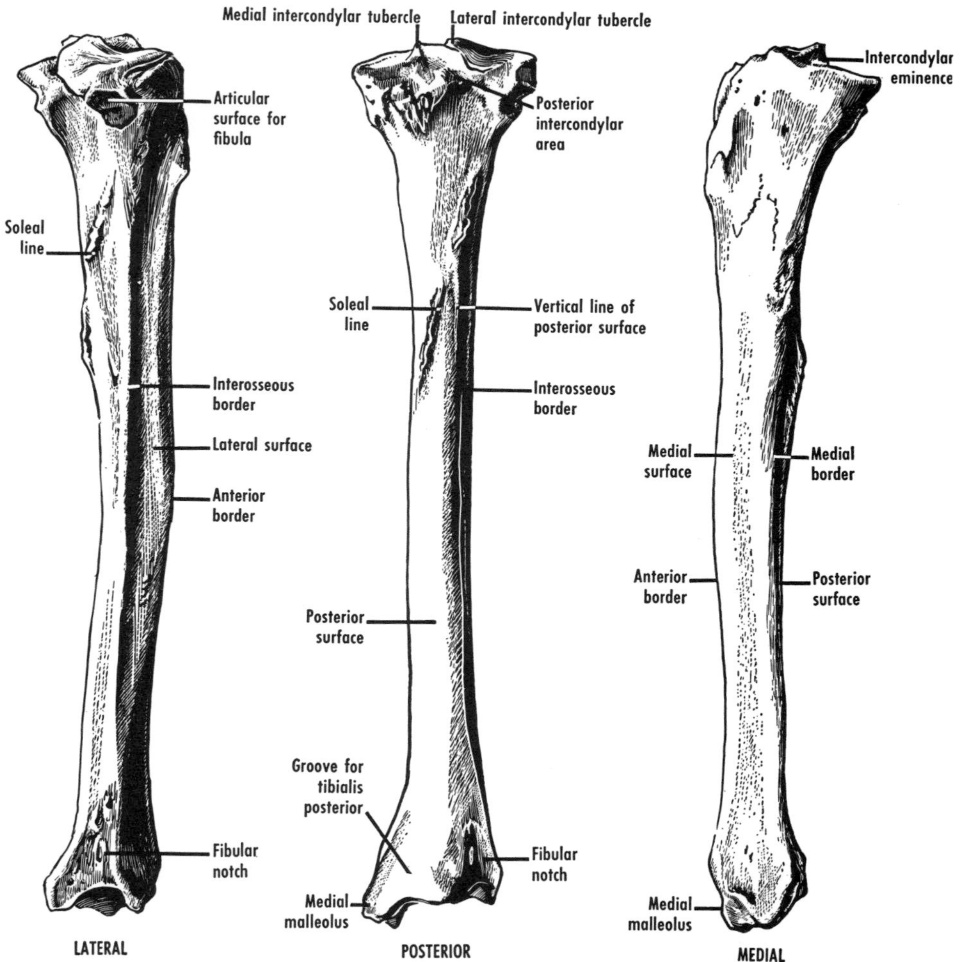

Figure 18–28 Right tibia. Note that the soleal line (posterior view) is interrupted (see figs. 22–5 and 22–6, pp. 230 and 231). The vertical line that subdivides the posterior surface separates the origins of the flexor digitorum longus and the tibialis posterior.

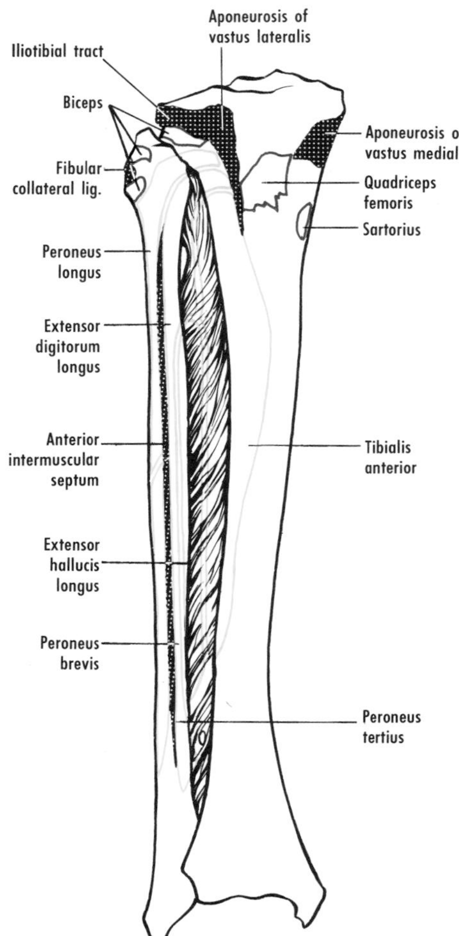

Figure 18–29 Right tibia and fibula, anterior aspect, muscular and ligamentous attachments. Note that the peroneus longus and extensor digitorum longus arise from the tibia as well as from the fibula.

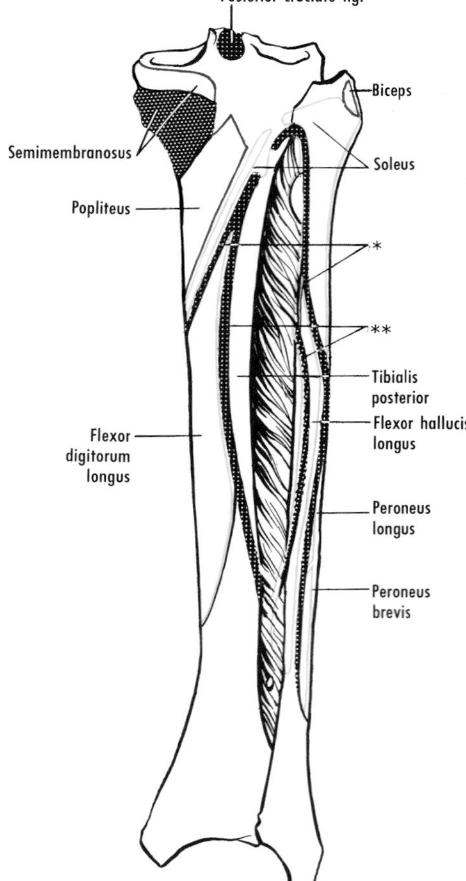

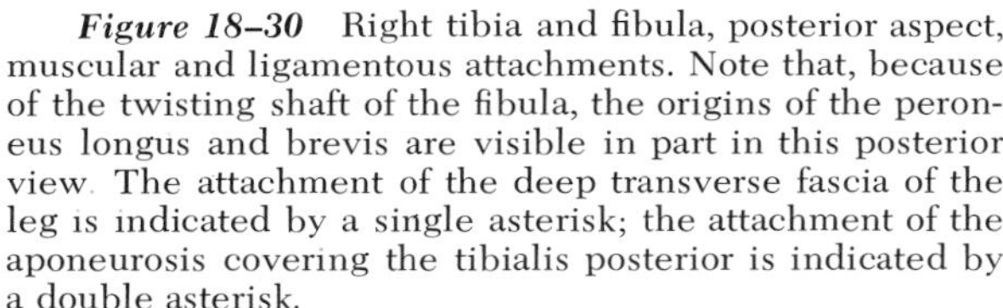

Figure 18–30 Right tibia and fibula, posterior aspect, muscular and ligamentous attachments. Note that, because of the twisting shaft of the fibula, the origins of the peroneus longus and brevis are visible in part in this posterior view. The attachment of the deep transverse fascia of the leg is indicated by a single asterisk; the attachment of the aponeurosis covering the tibialis posterior is indicated by a double asterisk.

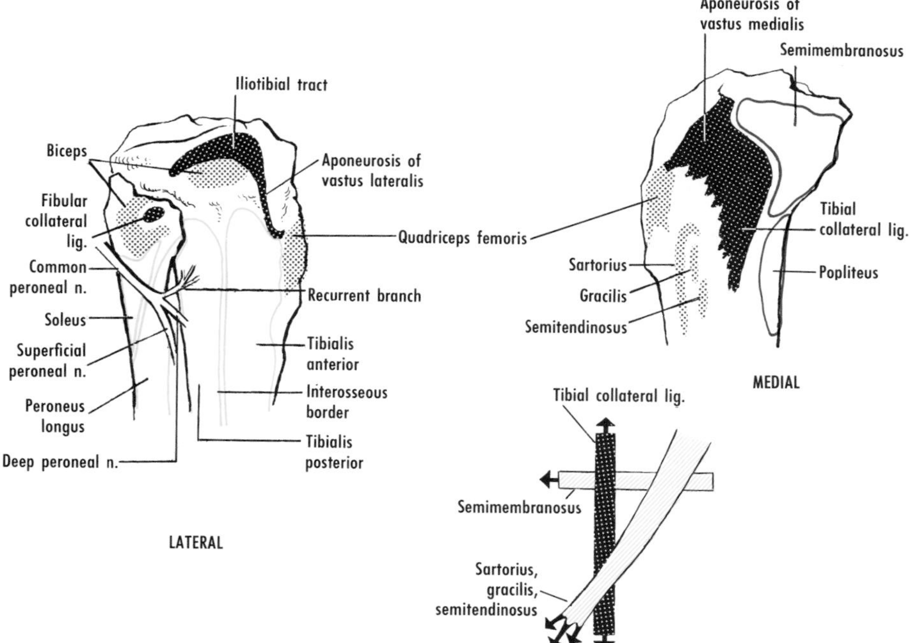

Figure 18–31 Muscular and ligamentous attachments of upper ends of the right tibia and fibula. In the lateral view, note that the interosseous border of the tibia separates the attachments of the tibialis anterior and posterior. The origin of the peroneus longus and extensor digitorum brevis from the tibia is not shown (see fig. 18–29). *Lower right,* schematic representation of relations of the tibial collateral ligament, which crosses the semimembranosus tendon, but which is deep to the tendons of the gracilis, semitendinosus, and sartorius.

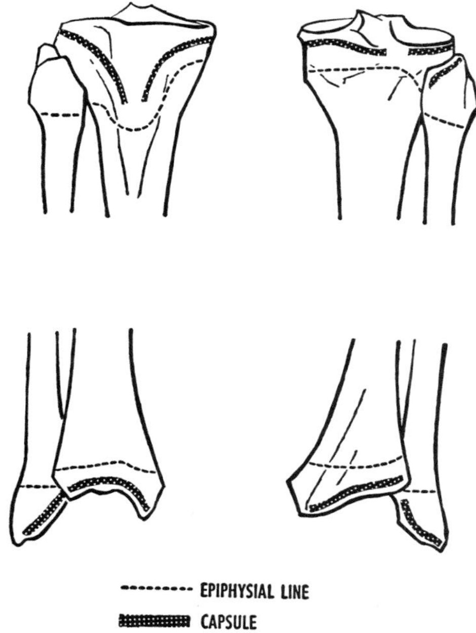

Figure 18–32 The upper and lower ends of the right tibia and fibula, showing the usual position of the epiphysial lines, and the usual line of attachment of the joint capsule. Based on Mainland.[8]

malleolar groove for the tendons of the tibialis posterior and flexor digitorum longus. The groove lies near the medial margin and continues distally onto the posterior margin of the medial malleolus. An indistinct, laterally placed groove for the flexor hallucis longus tendon is sometimes present on this surface. The lateral surface of the lower end is a wide, triangular depression, the lower part of which is smoother and deeper, and forms the *fibular notch.* The lower part of the fibula lies in this notch.

The *inferior articular surface* of the lower end of the tibia is rectangular and articulates with the anterior surface of the body of the talus. The inferior articular surface is prolonged onto the *articular surface of the malleolus,* which articulates with the medial surface of the talus.

Ossification

A periosteal collar is present by the seventh postovulatory week. The epiphysial center for the upper end is usually present at birth. That for the lower end appears during infancy. The tuberosity is formed in part by downward growth from the upper epiphysial center. An additional center in the tuberosity is commonly present; it appears during late childhood. The medial malleolus ossifies by an extension of the lower epiphysial center; a separate center sometimes appears for the tip.[14] The center for the lower end begins to fuse during puberty, that for the upper end somewhat later. Growth in length of the tibia occurs mainly at its upper end. Epiphyses of the tibia are shown in figures 18–25 and 18–34, pages 182 and 191, and the usual positions of the epiphysial lines of the tibia are shown in figure 18–32.

FIBULA

The fibula* (figs. 18–24 to 18–27, 18–29 to 18–31, and 18–33) is located on the lateral side of the leg, approximately parallel to the tibia. It is nearly as long as the tibia and is very slender. Its ends are slightly expanded. It forms synovial joints with the tibia above and the talus below. Its intermediate part is joined to the tibia by the interosseous membrane. It does not bear weight and, because muscles cover its middle part, it can be palpated only at its ends. The fibula can be identified as to side by placing its flattened end inferiorly so that the smooth, triangular facet on this end faces medially and the adjacent notch is located behind the facet.

The fibula has a shaft and two ends, upper and lower.

The upper end, or *head,* which articulates with the back part of the tibia, can be palpated immediately below the posterior part of the lateral condyle of the tibia. **The head of the fibula is on the same level as the tuberosity of the tibia, and can be located by tracing the biceps tendon downward.** The medial part of the upper aspect of the head bears a circular *articular surface* for the lateral condyle of the tibia. This surface faces forward, upward, and medially. A rough area lateral to this surface gives attachment to the tendon of the biceps femoris and to the fibular collateral ligament. The *apex* (or *styloid process*) is prolonged upward from the lateral and posterior aspects. **The common peroneal nerve winds from its position behind the head and becomes lateral to the fibula at the "neck," where it can be rolled between the finger and the bone.**

The *shaft* arches forward as it descends to the lateral malleolus. Hence the plane of the interosseous membrane is sagittal; that is, it extends from front to back. Only in the lower part of the leg is it in a coronal plane, that is, directed from medial to lateral. **The torsion of the fibula and the changing relationships of the fibula and tibia must be kept in mind in order to understand the topography of the leg.** The surfaces and borders of the shaft* vary considerably, especially in the lower fourth, which tends to be flattened from side to side. A well-developed shaft presents three surfaces, anterior, posterior, and lateral, and three borders, anterior, interosseous, and posterior. There is also a medial crest on the posterior surface. A nutrient foramen is present in the middle third of the shaft. It is occasionally directed upward.[15]

The *anterior surface,* which lies between the anterior and interosseous borders, is narrow above, but wider below. The *posterior surface* is subdivided by a prominent *medial crest,* which begins at the neck. Below, it passes obliquely anteriorly to join the interosseous border. The *lateral surface,* lateral to the anterior border, turns in its lower part to face posteriorly and become continuous with the posterior aspect of the lateral malleolus. A groove lodging the tendons of the peroneus longus and brevis is often present on the lower part of this surface.

The *anterior border* is a sharp, even line that extends from the neck to the lower end, where it passes laterally and bifurcates to enclose a triangular subcutaneous area just above the lateral malleolus. The *interosseous border,* which is often poorly marked, gives attachment to the interosseous membrane. This border begins above at the neck, near the anterior border. It runs parallel to the anterior border until it reaches the lower part of the shaft. Here it bifurcates to enclose a triangular area that is located above the articular facet on the lateral malleolus, and which serves for the attachment of the interosseous tibiofibular ligament. The *posterior border* is a distinct edge that begins above at the neck. Near the middle of the shaft, it winds posteriorly and medially, and ends as the medial margin of the posterior surface of the malleolus.

*L., *fibula.* "Peroneus" and "peroneal" are derived from the corresponding Greek name, *perone.*

*The surfaces and borders of the fibula are named here in accordance with the B.R., because the names in the N.A. are unsatisfactory.

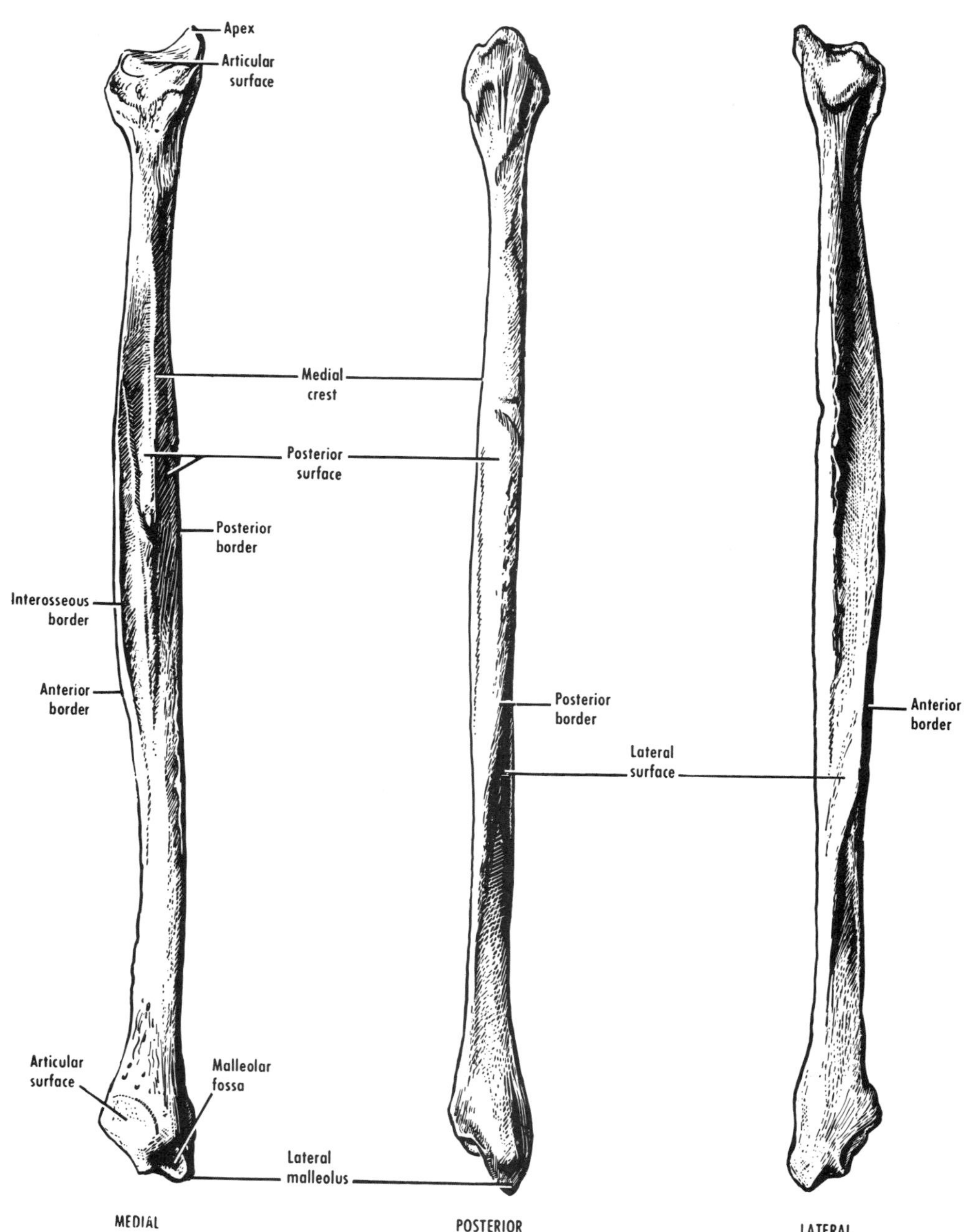

Figure 18–33 Right fibula. Note that (1) in the medial view the medial crest subdivides the posterior surface (see fig. 22–10); (2) in the posterior view the lateral surface turns posteriorly in passing downward; and (3) in the lateral view the anterior surface turns laterally in passing downward.

The shaft of the fibula presents an appearance of being twisted laterally through about a quarter of a right angle. This is correlated with the disposition of the overlying muscles.

The lower end of the fibula is the lateral malleolus, which is more prominent than the medial, is more posterior, and extends about 1 cm more distally. It articulates with the lateral surface of the talus, which fits between the two malleoli. The convex, lateral aspect of the lateral malleolus is continuous above with the triangular, lateral extension of the anterior surface of the shaft. The medial aspect of the malleolus presents in front the triangular *articular surface* or facet for the talus; the upper margin of this surface articulates with the tibia. The *malleolar fossa* lies behind the articular surface on the medial aspect. The wide posterior margin of the malleolus presents a groove, continued distally from the lateral surface of the shaft, for the peroneal tendons.[16]

Ossification

A periosteal collar is present by the eighth postovulatory week, sometimes earlier. The fibula is exceptional in that, although growth in length occurs mainly at its upper end (as in the tibia), the epiphysis for its lower end appears first,[17] during infancy; however, the lower epiphysis fuses first, during adolescence. The center for the upper end appears during early childhood and fuses during late adolescence. The nutrient artery is directed away from the growing end, that is, away from the knee. The epiphysis for the lower end of the fibula is shown in figure 18–34*A*, and the usual positions of the epiphysial lines of the fibula are shown in figure 18–32.

TARSUS

Although a knowledge of the chief characteristics of the individual tarsal bones is necessary for a detailed understanding of the structure of the foot, it is more important to study the skeleton of the foot as a whole and to identify the chief bony landmarks in the living foot.

The tarsus (figs. 18–26, 18–34 to 18–40) usually comprises seven bones, of which one, the talus, articulates with the bones of the leg. The seven bones, in order of approximately decreasing size, are: calcaneus, talus, cuboid, navicular, medial cuneiform, lateral cuneiform, and intermediate cuneiform. Tarsal anomalies are not infrequent, and they may be involved in disorders of the foot.

The talus ends in front in a rounded head, which is directed forward and medially, and which rests on a projection of the calcaneus, termed the sustentaculum tali. The front ends of the talus and calcaneus are more or less flush and form the transverse tarsal joint by articulating with the navicular and cuboid, respectively. The three cuneiforms lie between the navicular and the first three metatarsals, whereas the cuboid articulates directly with the fourth and fifth metatarsals.

The superior surface of the tarsus is convex, being particularly so from side to side. The prominent body of the talus projects upward and presents the articular surfaces (collectively termed the trochlea) for the leg bones. The body and head of the talus are separated from each other by a short neck, under the lateral side of which is a depression, the *sinus tarsi*, between the talus and calcaneus (fig. 18–35*B*). The sinus narrows medially and extends as the tarsal canal toward the back of the sustentaculum tali (p. 245).

In front of the navicular the intermediate cuneiform is seen to have a characteristically square superior surface associated with the backward projection of the second metatarsal between the medial and lateral cuneiforms. The line of the tarsometatarsal joints is thus irregular. The lateral part of this line curves backward in association with the narrow lateral side of the cuboid, which is partly overlapped by the prominent tuberosity of the fifth metatarsal.

The inferior aspect of the tarsus is concave and is limited behind by the prominence of the heel, and the tuber calcanei. The tuber presents medial and lateral processes or tubercles (the medial is much the larger), which give attachment to the plantar aponeurosis and some of the muscles of the sole. The heel rests on these two tubercles. The lower aspect of the sustentaculum tali has a groove that is continuous with one on the back of the talus, the continuous furrow lodging the tendon of the flexor hallucis longus. The head of the talus can be seen in the gap between the sustentaculum and the navicular. This interval is bridged by the strong plantar calcaneonavicular ligament, which completes the socket of the important talocalcaneonavicular joint.

The medial side of the navicular presents a prominent tuberosity. In front of the navicular, the three wedge-shaped cun-

(*Text continued on page 194.*)

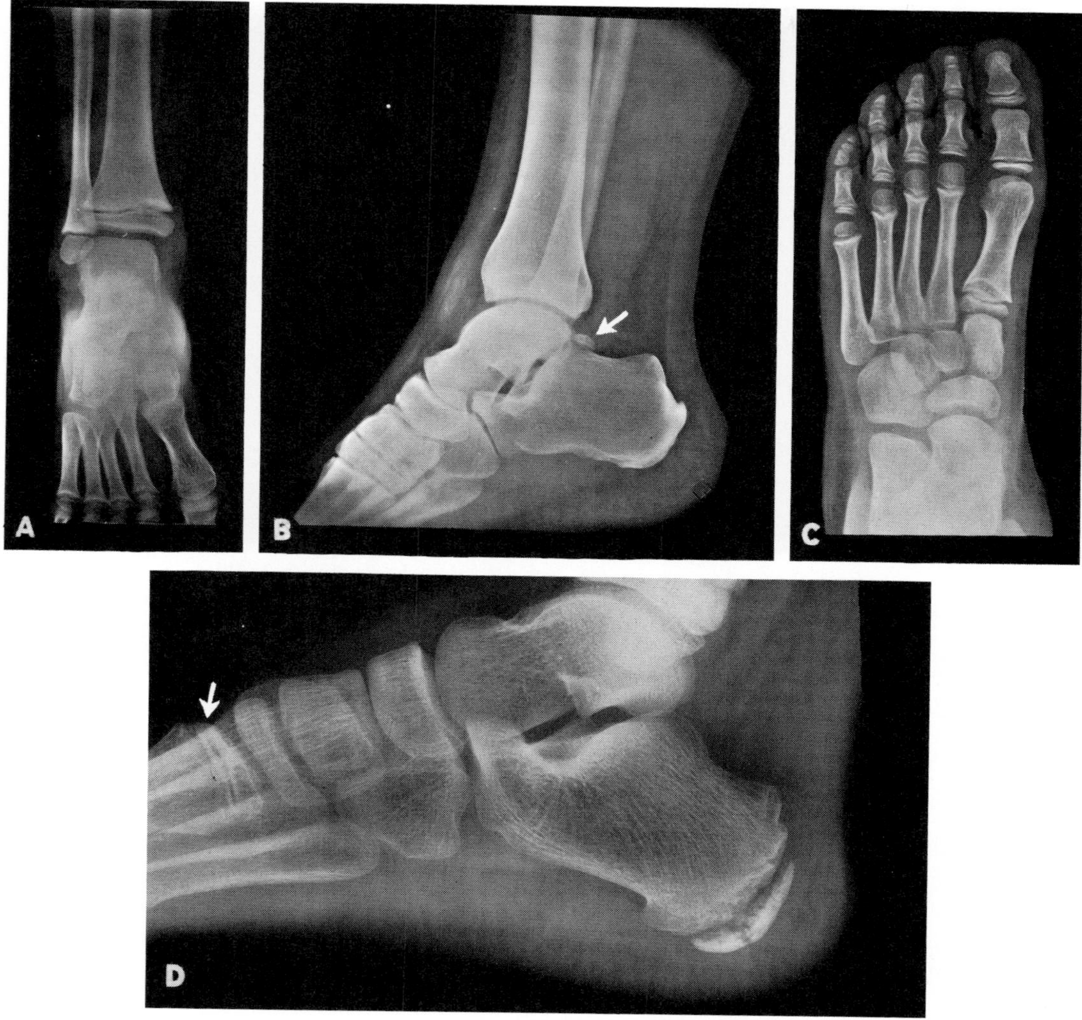

Figure 18–34 Ankle and foot. *A*, ankle of child. Note epiphyses for lower ends of fibula and tibia. The epiphysial line of the fibula is in line with the ankle joint. *B*, lateral view of adult ankle. Note os trigonum (*arrow*) at back of talus. *C*, foot of child. Note epiphyses for metatarsals and phalanges. Note also the irregularity in the ossification of the phalanges of the fifth toe. *D*, lateral view of foot of child. Note epiphyses for base of first metatarsal (*arrow*) and for calcaneus. *A*, *B*, and *C*, courtesy of V. C. Johnson, M.D., Detroit, Michigan. *D*, courtesy of George L. Sackett, M.D., Painesville, Ohio.

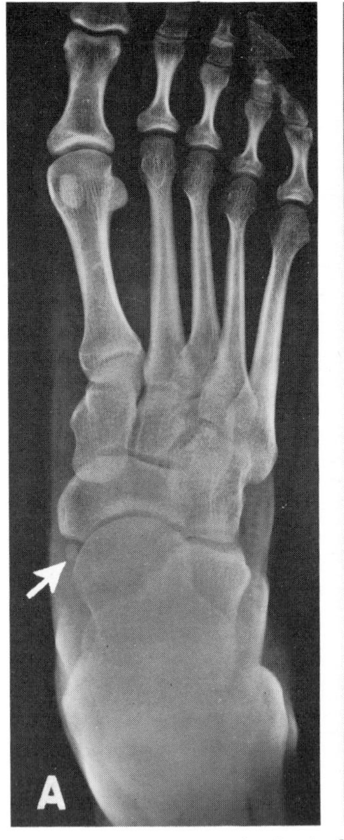

Figure 18–35 A, dorsoplantar view of foot. At least a portion of all seven tarsal bones can be identified. Note the os tibiale externum (*arrow*) near the tuberosity of the navicular; it is present bilaterally in this subject. Note also the sesamoid bones below the head of the first metatarsal. *B*, lateral view of foot. Note the navicular and its tuberosity (overlapped by head of talus), the cuboid and a peroneal sesamoid, and the tuberosity of the fifth metatarsal. The numerals *1*, *2*, and *3* indicate the lines of the first, second, and third cuneometatarsal joints, respectively. *C*, oblique view of foot. Note the region where the calcaneus may meet the navicular. A peroneal sesamoid (*arrow*) is visible near the tuberosity of the cuboid. *A*, *B*, and *C*, courtesy of V. C. Johnson, M.D., Detroit, Michigan.

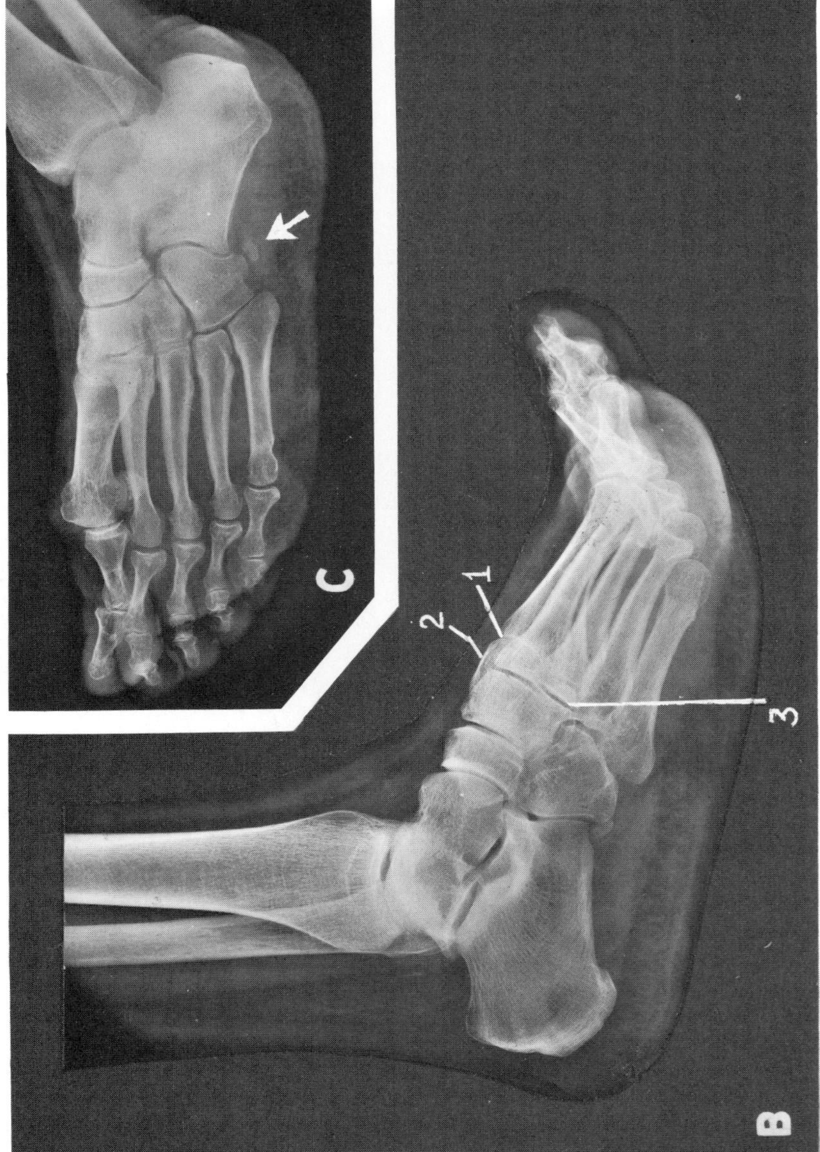

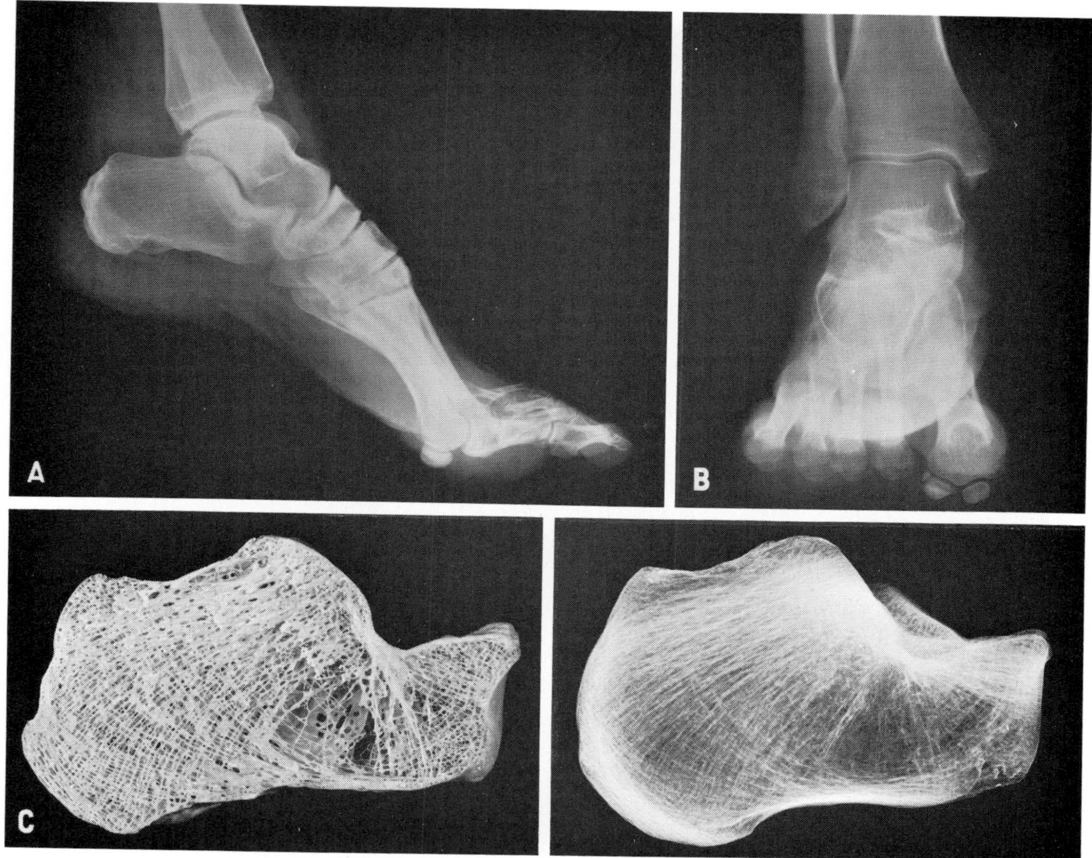

Figure 18–36 A, lateral view of foot, subject not bearing weight. *B*, posterior view of left foot to show sesamoid bones at first metatarsophalangeal joint. The toes are markedly dorsiflexed. *C*, photograph of half of calcaneus cut in a sagittal plane. Note the various curved trabeculae. *D*, radiogram of an intact, dried calcaneus. Note the various curved trabeculae. *A*, from *Medical Radiography and Photography;* courtesy of Felton O. Gamble, D.C.S., Tucson, Arizona.

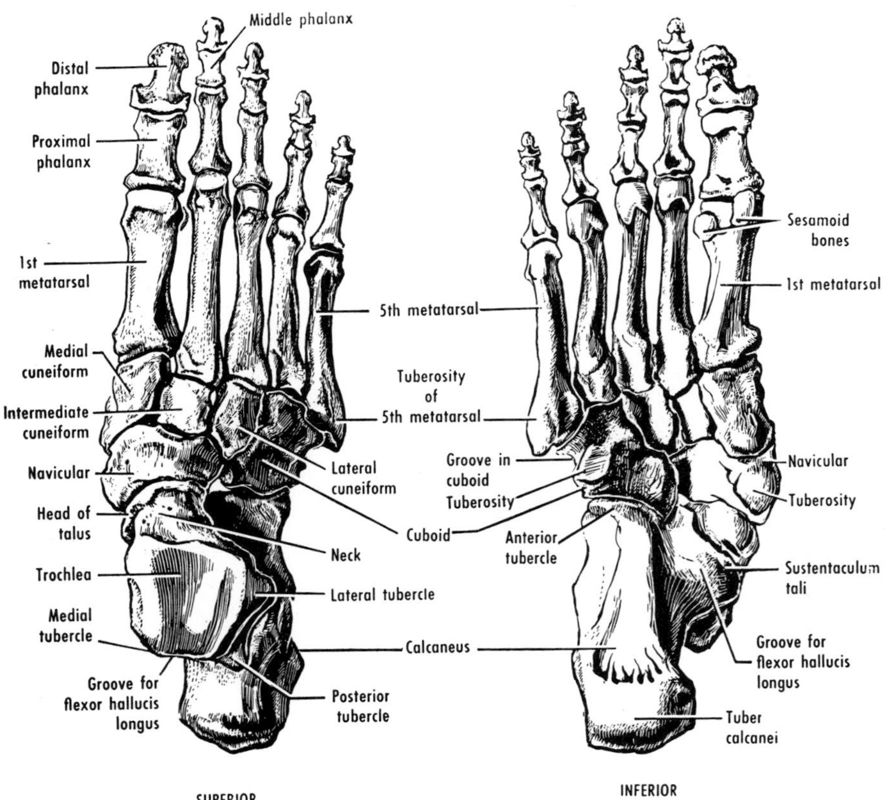

Figure 18–37 Bones of the right foot.

eiforms can be seen, the base of the medial cuneiform and the edges of the other two. The cuboid lies lateral to the navicular and the cuneiforms. **These bones (navicular, cuneiforms, and cuboid), together with the five metatarsals, form the transverse arch of the foot.** The inferior surface of the cuboid is grooved obliquely near the bases of the lateral two metatarsals. The groove may be occupied partly by the tendon of the peroneus longus, and it is directed toward the junction of the medial cuneiform and the first metatarsal, into both of which bones the tendon is inserted. The groove is bounded behind by the tuberosity of the cuboid. Behind this a shallow concavity on the cuboid and the calcaneus lodges the short plantar ligament. The more superficially placed long plantar ligament extends from the greater part of the lower surface of the calcaneus (in front of the tubercles) to the tuberosity of the cuboid.

On the medial side of the foot, a longitudinal arch is formed by the calcaneus, talus, navicular, cuneiforms, and the first three metatarsals.

In the living foot the sustentaculum tali can be felt approximately 1 to 2 cm below the medial malleolus, and the head of the talus can be recognized in front of the malleolus. The tuberosity of the navicular can be felt in front of the sustentaculum, and the back of the tuberosity may be taken as the medial end of the transverse tarsal joint. The medial cuneiform and the first metatarsal can be identified in front of the tuberosity.

On the lateral side of the foot, a longitudinal arch is formed by the calcaneus, cuboid, and the lateral two metatarsals.

The only readily identifiable points on the lateral side of the living ankle and foot are the lateral malleolus and the tuberosity of the fifth metatarsal. The lateral end of the transverse tarsal joint is about halfway between these bony landmarks.

Talus. The talus or ankle bone is the only tarsal bone without muscle or tendon attachments. It comprises a head, a neck, and a body.

The *body* is formed by the larger, posterior part of the talus and presents the *trochlea*, which includes three surfaces that articulate with the bones of the leg. The superior surface articulates with the tibia. The *lateral malleolar facet* is large and triangular. Its base faces upward, and the *lateral process of the talus** is at its apex. The *medial malleolar facet* is comma-shaped.

The posterior aspect of the body of the talus presents the *posterior process of the talus*, which is marked by the vertical *groove for the tendon of the flexor hallucis longus muscle*. This groove continues into one on the lower surface of the sustentaculum tali. The posterior process presents an elevation on each side of the groove on the back of the talus. One is the *medial tubercle*, the other the *lateral tubercle* (sometimes called posterior tubercle). The *os trigonum*, when present (fig. 18–34*B*), is generally considered to be a separate lateral tubercle.

The lower aspect of the body of the talus presents, from behind forward, the oblong *posterior calcanean facet* of the subtalar joint (fig. 18–41), the *middle calcanean facet*, which articulates with the upper surface of the sustentaculum tali, and the *anterior calcanean facet* on the head. This last facet and the head form the ball for the socket of the talocalcaneonavicular joint.

The *neck of the talus* is directed medially as well as forward; it forms an angle with the body (fig. 18–42) of about 15 degrees.[18] This angle is slightly greater in the newborn.

In certain peoples two small facets are frequently found on the neck of the talus, in a position to articulate with the tibia during the extreme dorsiflexion of squatting.[19] These "squatting" facets are present in fetuses,[20] but disappear after birth in people who do not habitually squat.[21]

The *head of the talus* presents a convex articular surface for the navicular, and a medial facet for the plantar calcaneonavicular ligament. The main blood supply of the talus is by way of vessels that enter the tarsal canal and the sinus tarsi.[22]

Calcaneus. Also known as the os calcis, the calcaneus transmits much of the weight of the body from the talus to the ground. Its structure, which is shown in figure 18–36*C* and *D*, has received considerable attention from the standpoint of its function in bearing and transmitting weight. Its trabeculae, the arrangement of which varies with sex, are readily demonstrated radiographically.[23]

The front half of the upper aspect of the calcaneus is prolonged medially to form the *sustentaculum tali*,

*In the B.R., the lateral process is termed lateral tubercle, a term which has since been applied in N.A. to one of the tubercles of the posterior process.

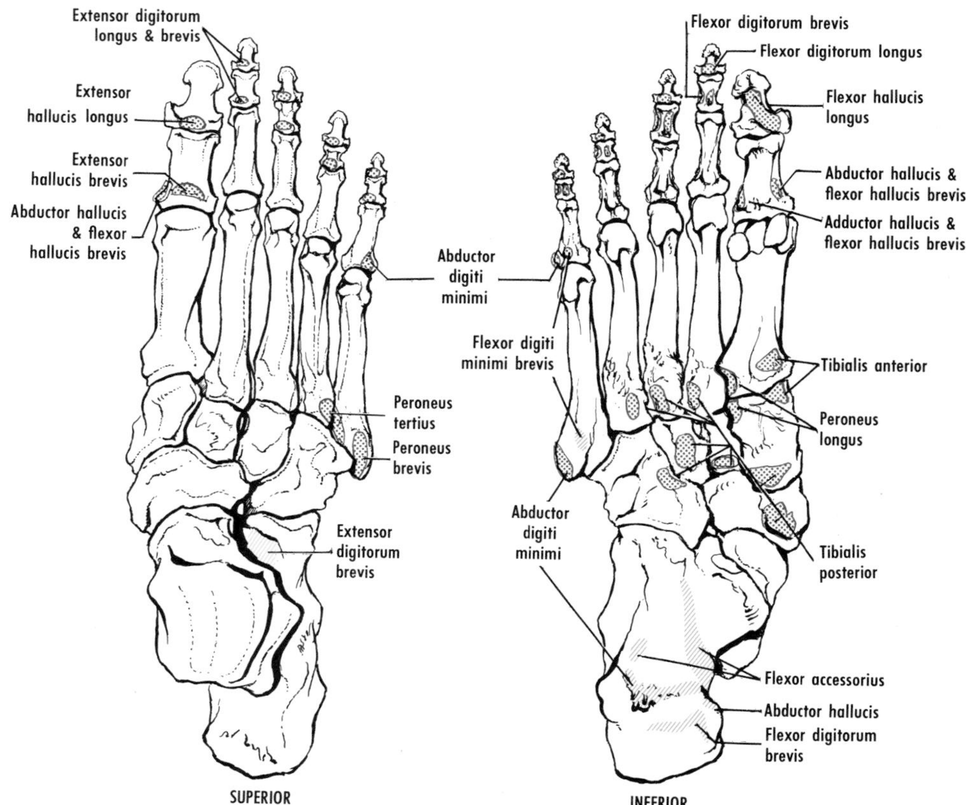

Figure 18–38 Bones of the right foot, muscular and ligamentous attachments. The attachments of the interossei are omitted. See figures 22–3 and 22–4, pp. 228 and 229, for further details of the calcaneus.

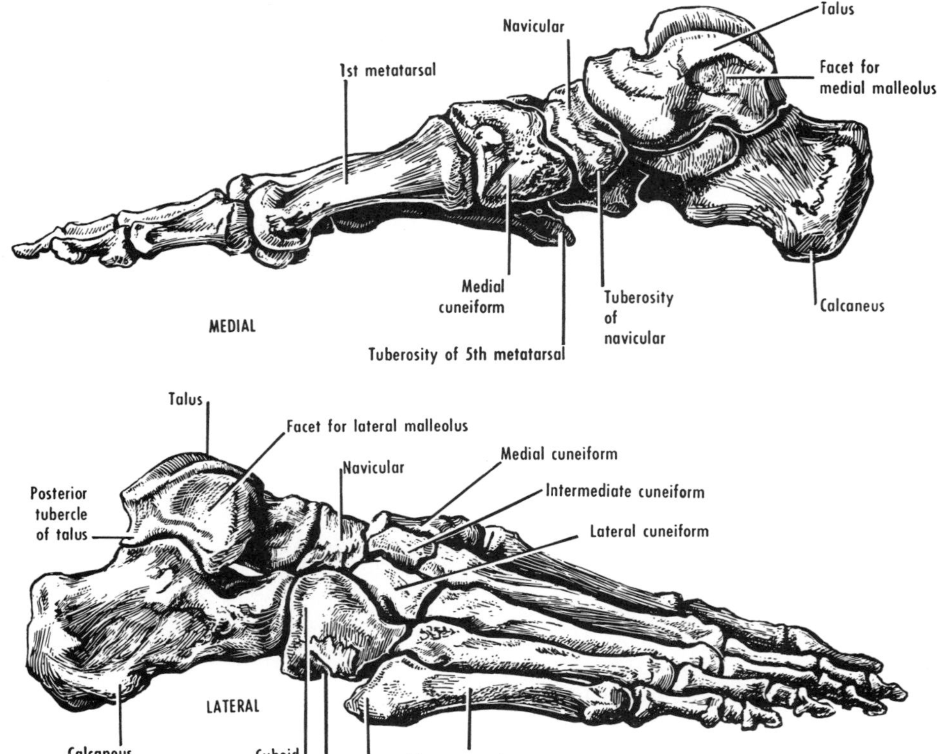

Figure 18–39 Bones of the right foot.

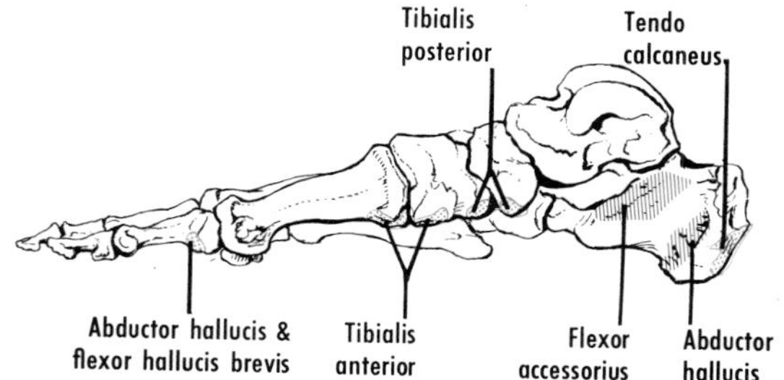

Figure 18–40 Bones of the right foot, medial aspect, muscular and tendinous attachments.

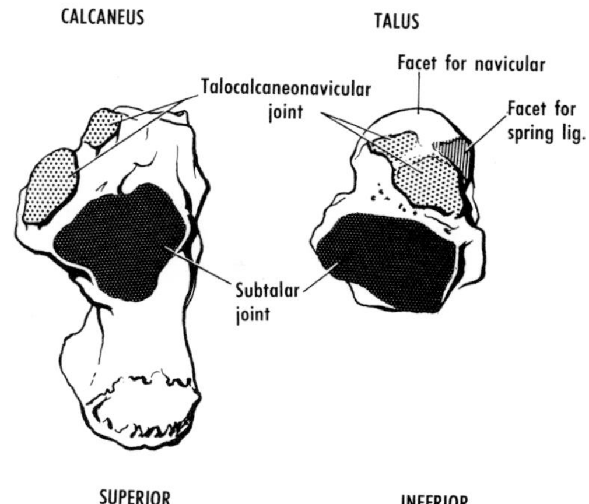

Figure 18–41 The right calcaneus and talus, showing corresponding articular facets.

which can be felt *in vivo* just below the medial malleolus. The upper surface of the sustentaculum forms the middle facet for the talus. This facet is continued forward as the *anterior facet* for the head of the talus (fig. 18–41). The *posterior facet* for the talus articulates with the posterior facet of the body of the talus to form the *talocalcanean joint.* The convex, expanded, posterior aspect of the calcaneus forms a part of the *tuber calcanei.*

The lower aspect of the calcaneus presents an eminence near its front end, the *anterior tubercle.* Two processes or tubercles, on which the bone rests, extend onto the lower surface from the tuber calcanei. These are the *medial* and *lateral processes.* The medial is the longer.

The concave medial aspect of the calcaneus is bridged by the flexor retinaculum. The lower surface of the sustentaculum tali presents the *groove for the flexor hallucis longus tendon.*

The lateral aspect of the calcaneus is marked by a projection, the *retrotrochlear eminence,* to which the calcaneofibular ligament is attached. A projection of variable size, the *peroneal trochlea* (trochlear process, peroneal tubercle) may sometimes be found in front of the eminence (fig. 18–43). The peroneal trochlea and retrotrochlear eminence are collectively designated the *peroneal process.*

The anterior aspect of the calcaneus is formed by the saddle-shaped *cuboid articular surface.*

Navicular. The navicular is situated between the talus behind and the three cuneiform bones in front. The narrow, rough lateral surface sometimes presents a small facet that articulates with the cuboid. The medial surface projects downward to form the *tuberosity of the navicular.* The tuberosity can be palpated about 3 cm below and in front of the medial malleolus.

Cuneiforms. The cuneiform bones, so called because they are wedge-shaped, lie between the navicular behind and the first three metatarsals in front; they are medial to the cuboid. Behind, they are almost flush, for they all articulate with the front of the navicular. The medial and lateral cuneiforms, however, project farther forward than does the intermediate. The resulting gap is occupied by the base of the second metatarsal. The stability of the foot is increased because of this arrangement.

Cuboid. The cuboid can be palpated on the lateral aspect of the foot. Its posterior surface has a saddle-shaped facet for the calcaneus. Its inferomedial angle is prolonged to form a "calcanean process." The anterior surface presents medial and lateral facets for the fourth and fifth metatarsals.

The lower surface is characterized in front by a groove that runs obliquely forward and medially from the lateral aspect. The tendon of the peroneus longus occasionally occupies the groove.[24] The long plantar ligament is attached to both lips of the groove. A ridge that limits the groove behind becomes flattened laterally to form the *tuberosity of the cuboid,* a promontory on the smooth, anterior slope of which the peroneus longus tendon plays.

The medial surface articulates with the lateral cuneiform, and sometimes with the navicular. The small lateral aspect is deeply notched by a continuation of the groove from the lower surface; it is largely overlapped by the tuberosity of the fifth metatarsal.

Accessory Ossicles. These may be found in the interstices between the tarsal bones. About 28 have received distinctive names.[25] Some are called *sesamoids,* for example, the bone in the peroneus longus tendon (os sesamoideum peroneum) (fig. 18–35*C*). The os tibiale externum is an ossicle near the tuberosity of the navicular. The os trigonum is at the

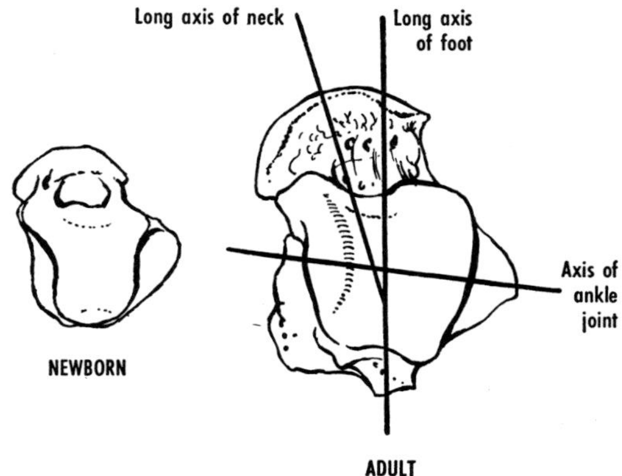

Figure 18–42 The talus of a newborn and of an adult, drawn to scale, and showing the angle that the neck makes with the body.

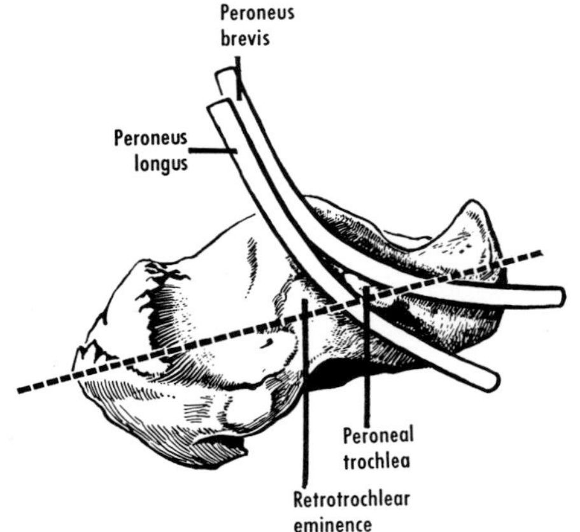

Figure 18–43 Right calcaneus, lateral aspect, modified from David B. Meyer. The dotted line is an arbitrary line used to locate structures on the lateral calcanean surface.

posterior tubercle of the talus (fig. 18–34*B*). These three ossicles are the most common.

Ossification (fig. 18–34)

Periosteal bone sometimes forms on part of the calcaneus at about the fourth prenatal month.[26] Otherwise, ossification begins in the calcaneus before birth as an endochondral center. Endochondral ossification begins in the talus shortly before birth. Epiphyses appear for the tuber calcanei[27] and posterior tubercle of the talus[28] during childhood. Fusion of the tuber calcanei is often radiographically complete at about puberty, and that of the lateral tubercle of the posterior process of the talus a few years earlier.

METATARSUS

The *metatarsals,* or bones of the metatarsus, connect the tarsus behind with the phalanges in front and are numbered from one to five, from the big to the little toe. Each metatarsal has a base, a shaft, and a head.

The *base* is the larger, proximal end. It is wedge-shaped, and the base of the wedge faces upward. The base articulates behind with the tarsus and on one or both sides with adjacent metatarsals.

The *shaft,* concave below and convex above, decreases in size as it passes forward. The upper surface is covered by the extensor tendons and is wider toward the ends than in the middle. The lower border begins at the end of the wedge of the base. It passes distally and bifurcates near the head to form two forks, each of which ends in an enlargement on the lower aspect of the head. The medial and lateral borders end distally in tubercles on the medial and lateral aspects of the head, respectively.

The *head* articulates with the base of the proximal phalanx by a convex articular surface that extends farther proximally on the lower aspect than on the upper. The lower aspect is notched, however, and the middle portion, which is nonarticular, lodges the flexor tendon. Tubercles at the junction of the head and shaft above give attachment to collateral ligaments.

The metatarsals are longer and thinner than the metacarpals. In addition to characteristics common to all five metatarsal bones, each has characteristics peculiar to it alone.[29]

First Metatarsal. This bone, which is relatively short and thick, carries much more weight than do the other metatarsals. It can be palpated in its entire length. The articular surface of the head has two grooves, which are separated by a ridge and which articulate with the sesamoid bones. The surface of the head is sometimes flattened.[30] Occasionally the first metatarsal is congenitally very short and thick. This, however, is seldom a cause of disability of the foot.[31]

Second Metatarsal. The second is the longest of the metatarsals. It is prolonged behind, between the medial and lateral cuneiforms.

Third Metatarsal. This is shorter than the second.

Fourth Metatarsal. This is shorter than the third. The heavy base has an oblique plane and the appearance of a notch and a twist.

Fifth Metatarsal. The fifth is usually longer than the third and fourth. Its base is characterized by a process on its lateral side, the *tuberosity,* which projects backward as well as laterally.

Ossification (fig. 18–34)

Periosteal collars appear around the metatarsals during early fetal life. Epiphysial centers appear distally (heads) in the lateral four metatarsals and in the base of the first during early childhood. (Multiple epiphysial centers are frequently present, especially in the first metatarsal and first phalanx.)[32] They fuse during adolescence.[33] The first metatarsal, like the first metacarpal, may have an epiphysis for its head as well as its base. Not infrequently the tuberosity of the fifth metatarsal has an epiphysis that appears during late childhood and fuses during puberty.[34]

PHALANGES

Each toe has three phalanges, except the first, which has only two, and the fifth, which often has only two. Compared with those in the hand, the phalanges of the foot have thin, rounded shafts and large ends.

Those of the big toe, however, are short, broad, and strong.

Proximal Phalanges. Each of these has a base, a shaft, and a head. The *base* articulates with the corresponding metatarsal by an ovoid, concave facet. The *shaft* is narrow and is slightly concave below. The *head* has a groove in the center and elevations on each side. The proximal phalanx of the big toe has an obliquity of its base that is correlated with the lateral divergence of the big toe (*valgus deviation*).[35] It is stated, however, that in peoples other than Europeans, this divergence may be almost nonexistent.[36]

Middle Phalanges. These are short and become smaller, proceeding laterally from the second toe. The middle phalanx of the fifth, when present as a separate bone, is often only an irregular nodule. The base of each articulates with the proximal phalanx by two depressions that are separated by a ridge. The shaft is flat, and the head presents a trochlear surface for the distal phalanx.

Distal Phalanges. The distal phalanges of the medial toes are larger. Each distal phalanx consists of a wide base and a distal end. The distal end is expanded to form the (ungual) *tuberosity*, which covers more of the lower than of the upper aspect of the phalanx.

The middle and distal phalanges of the little toe are often fused.[37] The fusion occurs in cartilage before birth.[38] Similar fusions may occur in other toes (the fourth, particularly), but they are less frequent.

Ossification (fig. 18–34)

The phalanges usually begin to ossify during fetal life. The epiphysial centers for the bases appear during early childhood and fuse during puberty. The middle phalanges of the third to fifth toes may not have epiphyses.

SESAMOID BONES

These small, rounded bones are related to the lower aspects of some of the metatarsophalangeal and interphalangeal joints.[39] They are generally embedded in, or covered by, the plantar ligaments of these joints. Two sesamoid bones are almost constantly present below the head of the first metatarsal (p. 247; fig. 18–36*B*). Rarely, there may be four. An interphalangeal sesamoid is sometimes present in the big toe, occasionally a metatarsophalangeal sesamoid in the little toe, and, very occasionally, an interphalangeal sesamoid in the little toe. These sesamoids develop before birth and begin to ossify during late childhood.

REFERENCES

1. T. W. Todd, Amer. J. phys. Anthrop., *4*:1, 1921.
2. M. Trotter and G. C. Gleser, Amer. J. phys. Anthrop., *10*:463, 1952.
3. K. Pearson and J. Bell, *Long Bones of the English Skeleton*, Draper's Company Research Memoirs, Biometric Series 10 and 11, Cambridge University Press, London, 1919 (dried bones). A. R. Shands and M. K. Steel, J. Bone Jt Surg., *40A*:803, 1958; G. Fabry, G. D. MacEwen, and A. R. Shands, J. Bone Jt Surg., *55A*:1726, 1973 (radiographic methods in living subjects).
4. A. W. Meyer, Amer. J. phys. Anthrop., *7*:257, 1924; Amer. J. Anat., *55*:469, 1934. For other details of the neck, see also T. Walmsley, J. Anat., Lond., *49*:305, 1915.
5. M. Harty, J. Bone Jt Surg., *39A*:625, 1957.
6. A. W. Meyer, Amer. J. Anat., *55*:469, 1934.
7. F. G. St. Clair Strange, *The Hip*, Heinemann, London, 1965.
8. D. Mainland, *Anatomy*, Hoeber, New York, 1945.
9. W. J. L. Felts, Amer. J. Anat., *94*:1, 1954 (prenatal development). S. I. Pyle and N. L. Hoerr, *Radiographic Standard of Reference for the Growing Knee*, Thomas, Springfield, Illinois, 1969. S. Scheller, Acta radiol., Stockh., Suppl. 195, 1960.
10. H. Hellmer, Acta radiol., Stockh., Suppl. 27, 1935. J. McKenzie and E. Naylor, J. Anat., Lond., *91*:583, 1957. S. I. Pyle and N. L. Hoerr, cited above.
11. C. G. Hutter and W. Scott, J. Bone Jt Surg., *31A*:511, 1949. H. Rosen and H. Sandick, J. Bone Jt Surg., *37A*:847, 1955. Tibial torsion is less during infancy, and it varies independently of femoral torsion. See H. Elftman, Amer. J. phys. Anthrop., *3*:255, 1945.
12. B. Giorgi, Clin. Orthopaed., *8*:209, 1956.
13. E. S. R. Hughes and S. Sunderland, Anat. Rec., *96*:439, 1946.
14. H. D. W. Powell, J. Bone Jt Surg., *43B*:107, 1961.
15. V. R. Mysorekar, J. Anat., Lond., *101*:813, 1967.
16. M. E. Edwards, Amer. J. Anat., *42*:213, 1928.
17. F. G. Ellis and J. Joseph, J. Anat., Lond., *88*:533, 1954. N. L. Hoerr, S. I. Pyle, and C. C. Francis, *Radiographic Atlas of Skeletal Development of the Foot and Ankle*, Thomas, Springfield, Illinois, 1962.
18. C. H. Barnett, J. Anat., Lond., *89*:225, 1955.
19. M. I. Satinoff, J. hum. Evol., *1*:209, 1972.
20. R. H. Charles, J. Anat., Lond., *28*:1, 271, 1893–94. I. Singh, J. Anat., Lond., *93*:540, 1959.
21. C. H. Barnett, J. Anat., Lond., *88*:509, 1954.
22. E. Wildenauer, Z. Anat. EntwGesch., *115*:32, 1950. R. A. Haliburton *et al.*, J. Bone Jt Surg., *40A*:1115, 1958. G. L. Mulfinger and J. Trueta, J. Bone Jt Surg., *52B*:160, 1970.
23. G. Sassu, Arch. ital. Anat. Embriol., *62*:330, 1957. F. Morin, Arch. ital. Anat. Embriol., *49*:92, 1943.
24. M. E. Edwards, Amer. J. Anat., *42*:213, 1928. E. Wildenauer and W. Müller, Z. Anat. EntwGesch., *115*:443, 1951.
25. Two general accounts, with a review of the literature, are provided by R. O'Rahilly, J. Bone Jt Surg., *35A*:626, 1953, and Clin. Orthopaed., *10*:9, 1957.
26. E. Hintzsche, Z. mikr.-anat. Forsch., *21*:531, 1930. A. Hasselwander, in Peter, Wetzel, and Heiderich, *Handbuch der Anatomie des Kindes*, Bergmann, Munich, 1938, 2 volumes.
27. E. Ruckensteiner, *Die normale Entwicklung des Knochensystems im Röntgenbild*, Thieme, Leipzig, 1931. See also V. V. Harding, Child Develpm., *23*:181, 1952, and Hoerr, Pyle, and Francis, cited above.
28. A. McDougall, J. Bone Jt Surg., *37B*:257, 1955.
29. Common variations in the metatarsal bones are described by I. Singh, J. Anat., Lond., *94*:345, 1960.
30. J. Joseph, J. Anat., Lond., *85*:221, 1951.
31. R. I. Harris and T. Beath, J. Bone Jt Surg., *31A*:553, 1949.

32. A. F. Roche and S. Sunderland, J. Bone Jt Surg., *41B*:375, 1959.
33. H. Flecker, Amer. J. Roentgenol., *68*:37, 1952.
34. C. T. Holland, J. Anat., Lond., *55*:235, 1921.
35. J. L. Wilkinson, J. Anat., Lond., *88*:537, 1954. C. H. Barnett, J. Anat., Lond., *96*:171, 1962.
36. N. A. Barnicott and R. H. Hardy, J. Anat., Lond., *89*:355, 1955.
37. P. Venning, Amer. J. phys. Anthrop., *14*:1, 1956.
38. E. Gardner, D. J. Gray, and R. O'Rahilly, J. Bone Jt Surg., *41A*:847, 1959. See also D. Trolle, *Accessory Bones of the Human Foot*, trans. by E. Aagensen, Munksgaard, Copenhagen, 1948.
39. A. H. Bizarro, J. Anat., Lond., *55*:256, 1921. M. S. Burman and P. W. Lapidus, Arch. Surg., Chicago, *22*:936, 1931. S. N. Kassatkin, Z. Anat. EntwGesch., *102*:635, 1934. C. A. Hubay, Amer. J. Roentgenol., *61*:493, 1949.

VEINS AND LYMPHATIC DRAINAGE OF LOWER LIMB

19

VEINS[1]

SUPERFICIAL VEINS

Dorsal digital veins that run along the two dorsal margins of each toe unite at the webs of the toes to form *dorsal metatarsal veins*. These then empty into the *dorsal venous arch*, which overlies the metatarsal bones and is situated in the subcutaneous tissue. This arch receives communications from the plantar venous arch (p. 240). Proximally it is connected with the irregular *dorsal venous network of the foot*.

Great (Long) Saphenous Vein* (fig. 19–1). This vein, which is also called *large* or *greater*, begins at the junction of the dorsal digital vein of the medial side of the big toe with the dorsal venous arch. It passes in front of the medial malleolus and crosses the medial surface of the tibia obliquely, in company with the saphenous nerve. It ascends along the medial border of the tibia, comes to lie behind the medial condyles of the tibia and femur, and then courses upward along the medial side of the thigh. It comes to overlie the femoral triangle and pierces the cribriform fascia, which occupies the saphenous opening in the fascia lata. It then pierces the femoral sheath and ends in the femoral vein. The great saphenous vein is often enlarged and tortuous, and its valves, which are distributed more or less irregularly over its whole length,[2] may be defective. Such defective vessels are known as *varicose veins*.

In addition to many unnamed tributaries, the great saphenous vein receives the posterior arch vein and anterior vein of the leg, which join it near the knee, and an *accessory saphenous vein* (usually lateral, sometimes medial), which joins it at or near the saphenous opening. Other tributaries form a variable pattern[3] at the saphenous opening (fig. 19–1). **Communications**

*The term saphenous in Greek means visible, but as applied to the vein is thought to be Arabic in origin (*al-safin*, meaning hidden). In most of its course, the vein lies on and is closely applied to the fascia, and hence is generally invisible. Near the knee it is often more superficial and may be immediately deep to the skin.

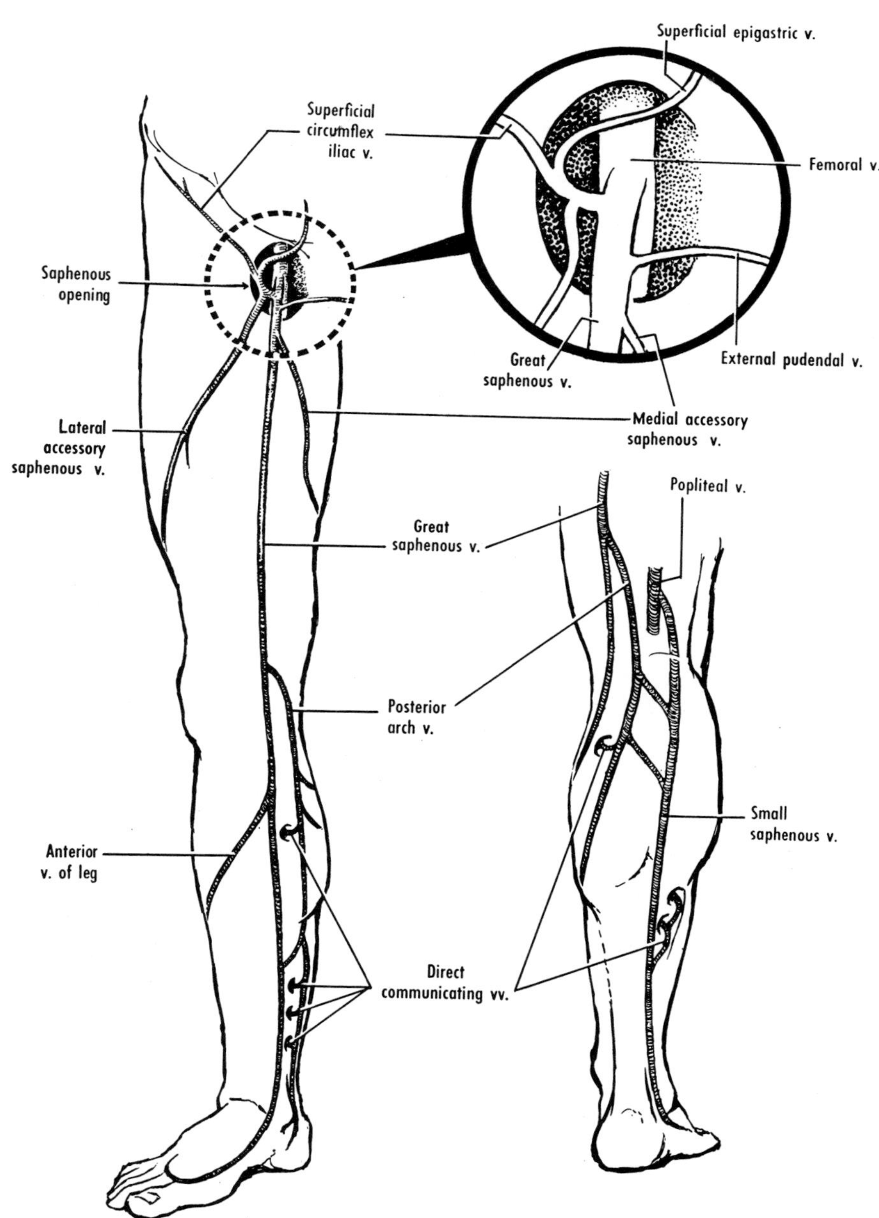

Figure 19–1 A simplified representation of the superficial veins of the lower limb. The major tributaries are shown and also the major communicating veins above the ankle, according to Dodd and Cockett.[1] Details of the veins of the foot[3] have been omitted.

occur between the superficial epigastric vein and the lateral thoracic vein, by way of the thoracoepigastric veins. In the event of obstruction to the superior or inferior vena cava, such communications may become enlarged.

Small (Short) Saphenous Vein. This vein, which is also called *lesser*, begins at the junction of the dorsal digital vein of the lateral side of the little toe with the dorsal venous arch. It ascends along the lateral border of the tendo calcaneus, behind the lateral malleolus (fig. 19–1). It then courses up the back of the leg, first between the subcutaneous tissue and fascia, and then in a tunnel formed by two layers of fascia,[4] in company with the sural nerve. It next passes between the heads of the gastrocnemius and pierces the fascia of the popliteal fossa. It ends in a variable fashion, often in the popliteal or the great saphenous, sometimes in deep veins or muscular veins of the lower part of the thigh, and occasionally in the veins of the calf muscles. It has several irregularly placed valves.

DEEP VEINS
(figs. 19–2 and 19–4)

The deep veins begin in the foot as plantar digital veins on the plantar surfaces of the toes (p. 240). The principal deep veins are the *femoral* and the *popliteal*, and the veins accompanying the anterior tibial, posterior tibial, and peroneal arteries and their branches. These veins have many valves, those in the veins of the leg being placed every few centimeters or so. The femoral vein is most frequently valved in its proximal part; one valve just above the saphenofemoral junction is constantly present.[2] The soleus muscle contains many large, long veins, which often show widenings or sinuses.[5] **Most blood returns by way of**

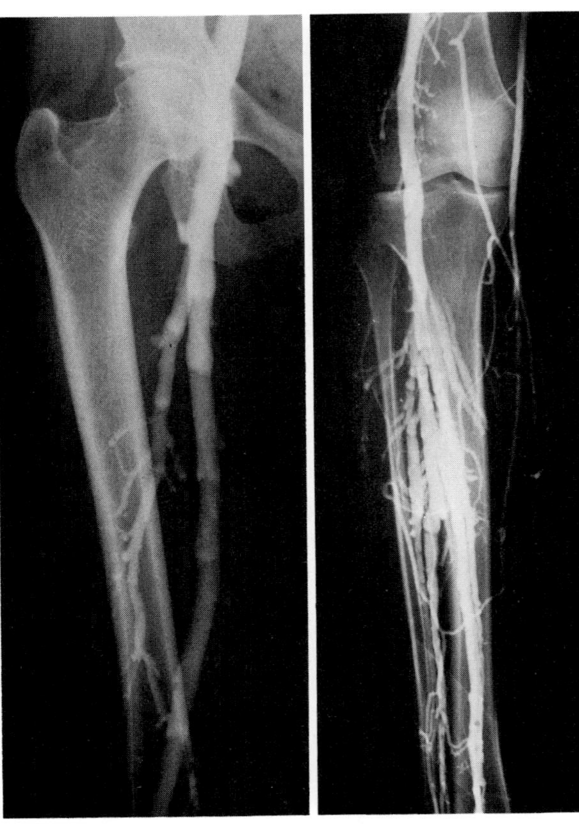

Figure 19–2 Normal venograms showing, *right*, deep veins of the leg, the popliteal vein, and the great saphenous vein, and *left*, the femoral and deep femoral veins. Note the slight bulges at the valves, in some of which the cusps can be distinguished. Courtesy of G. M. Stevens, M.D., Palo Alto Medical Clinic, California.

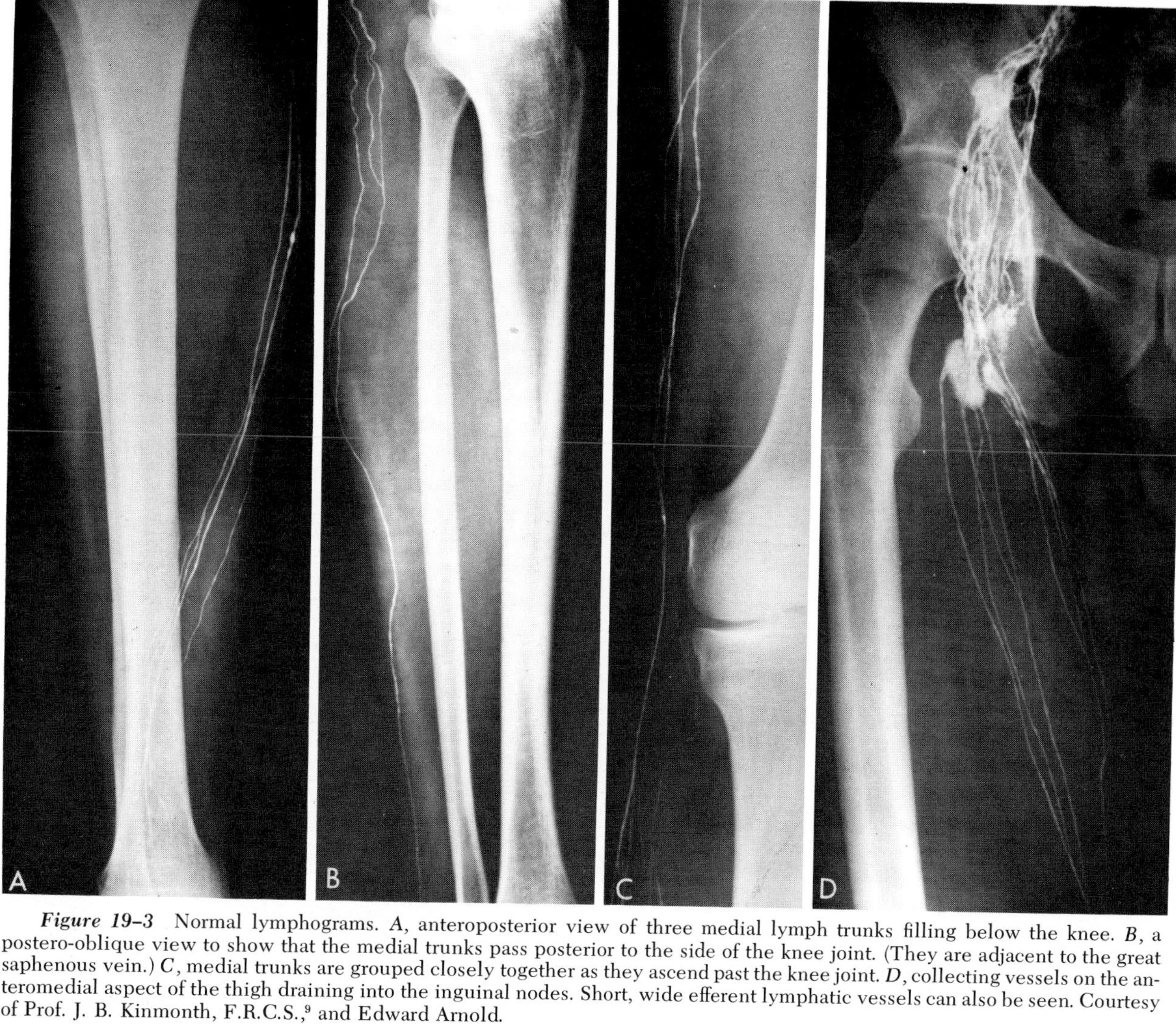

Figure 19–3 Normal lymphograms. *A*, anteroposterior view of three medial lymph trunks filling below the knee. *B*, a postero-oblique view to show that the medial trunks pass posterior to the side of the knee joint. (They are adjacent to the great saphenous vein.) *C*, medial trunks are grouped closely together as they ascend past the knee joint. *D*, collecting vessels on the anteromedial aspect of the thigh draining into the inguinal nodes. Short, wide efferent lymphatic vessels can also be seen. Courtesy of Prof. J. B. Kinmonth, F.R.C.S.,[9] and Edward Arnold.

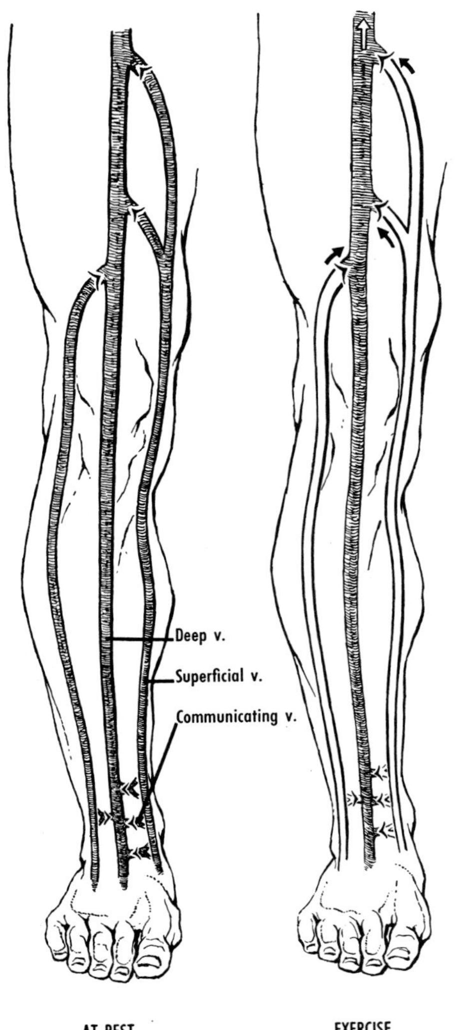

Figure 19–4 Schematic representation of venous circulation. The changes in venous pressure during exercise are such that nearly all blood returns by the deep veins, which receive blood from superficial veins by way of the communicating veins above the ankle.

deep veins, and the numbers and connections of venous channels are such that there are many alternate routes, even when the femoral vein is ligated.

PERFORATING, OR COMMUNICATING, VEINS

The perforating veins connect the superficial and deep veins (figs. 19–1 and 19–4). There are four types of perforating veins, direct, indirect, mixed, and atypical.[6] Direct perforating veins connect a superficial vein and a main deep vein. One perforating vein is usually present in the thigh and another in the upper part of the leg. An important series is found in the lower part of the leg, and still others are present in the foot. In those of the foot, the valves are positioned so that blood flows from deep to superficial.[7] Above the ankle, each perforating vein has a valve near each end, and these are positioned so that blood flows from superficial to deep.

The indirect perforating veins are connections between superficial and muscular veins. The latter in turn empty into the main deep veins. Indirect perforating veins are small, numerous, and variable. A significant number usually perforate the gluteus maximus and join the underlying gluteal veins.[8] Some perforating veins consist of both direct and indirect channels, and a few have atypical courses or valvular arrangements.

VENOUS RETURN

Muscular action, combined with the arrangement of valves, is an important factor in returning blood from the lower limb, and the flow of blood is markedly reduced when one stands quietly. During exercise, the pressure changes are such that above the ankle blood from superficial veins flows inward to the deep veins, which thus carry all of the load (fig. 19–4). The superficial veins can be obliterated without seriously affecting the circulation, provided that the deep veins are intact.

LYMPHATIC DRAINAGE[9]
(figs. 19–3 and 19–5)

The main collecting lymphatic vessels are deep and superficial. The deep ones ascend along blood vessels and usually empty into the popliteal nodes. The superficial lymphatic vessels—together with those from the gluteal region, the anterior and lateral abdominal wall, the external genitalia (except the glans of the penis or clitoris), the

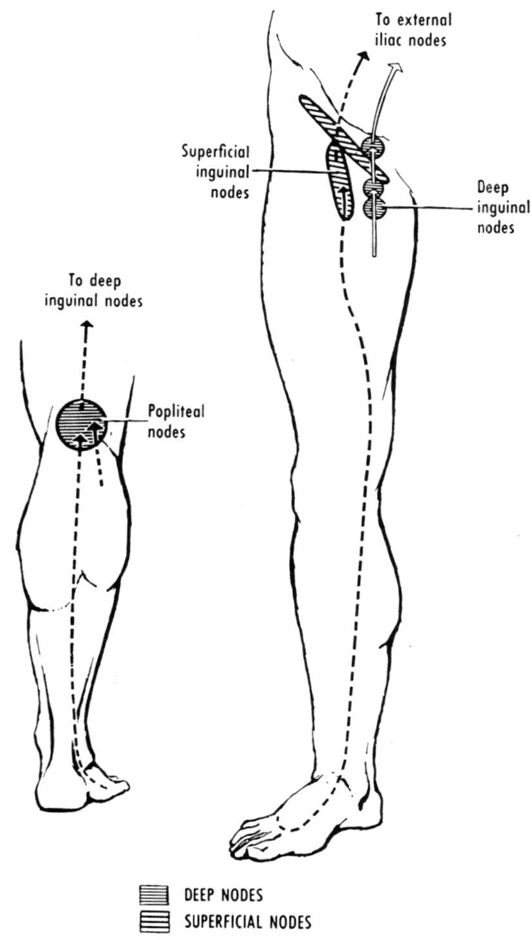

Figure 19–5 Schematic representation of lymphatics of lower limb. The two lower arrows in the figure on the left indicate that the popliteal nodes receive both superficial and deep lymphatic vessels.

uterus (only in part, by way of vessels along the round ligament), and the anus—converge toward the groin and drain into the inguinal nodes (fig. 19–5).

The superficial vessels of the lower limb comprise two sets, a medial one of three to seven main trunks, and a lateral one of one to two main trunks. The medial trunks end in the inguinal nodes. The lateral trunks either join the medial trunks above the knee or end in the popliteal nodes.

The *popliteal nodes* are small, one to five in number, and lie deep to the fascia of the popliteal fossa. Their afferents are the deep trunks along the tibial vessels, and the lateral group of superficial trunks. Their efferents accompany the femoral vessels to the deep inguinal nodes.

The *inguinal nodes* are three to fourteen in number, are mostly subcutaneous in position, beginning just below the saphenofemoral junction, and are often palpable in the living subject. They are usually described as superficial and deep, but this division has no physiological or clinical significance. The few deep nodes (one to three) lie deep to the fascia lata, along the medial side of the femoral vein. One may be present in the femoral ring.

The efferents of the inguinal nodes enter the external iliac nodes and ultimately drain into the lumbar (aortic) nodes and vessels (fig. 38–4, p. 421).

REFERENCES

1. H. Dodd and F. B. Cockett, *The Pathology and Surgery of the Veins of the Lower Limb,* Livingstone, Edinburgh, 1956. I. F. K. Muir, E. H. Mucklow, and A. J. H. Rains, Brit. J. Surg., *42*:276, 1954. J. Ludbrook, *Aspects of Venous Function in the Lower Limbs,* Thomas, Springfield, Illinois, 1966. M. C. Conrad, *Functional Anatomy of the Circulation to the Lower Extremities,* Year Book Medical Publishers, Chicago, 1971.
2. L. B. Kwakye, Acta morphol. neerl.-scand., 9:41, 1971.
3. Accounts of the venous patterns and anomalies in the region of the saphenous opening are given by E. H. Daseler *et al.,* Surg. Gynec. Obstet., *82*:53, 1946, by A. R. Mansberger *et al.,* Surg. Gynec. Obstet., *91*:533, 1950, and by A. Morin *et al.,* C. R. Ass. Anat., *55*:459, 1970.
4. J. F. Doyle, Irish J. med. Sci., 6th series, 317, 1967. E. Stolic, C. R. Ass. Anat., *55*:1016, 1970.
5. L. B. Kwakye, Acta morphol. neerl.-scand., 9:281, 1972.
6. E. Stolic, Bull. Ass. Anat., *56*:1164, 1972.
7. E. P. Lofgren *et al.,* Surg. Gynec. Obstet., *127*:289, 1968.
8. J. F. Doyle, Irish J. med. Sci., *3*:285, 1970.
9. E. H. Daseler, B. J. Anson, and A. F. Reimann, Surg. Gynec. Obstet., *87*:679, 1948. Y. Tezuka, Kumamoto med. J., *6*:1, 1954. J. J. Pflug and J. S. Calnan, Brit. J. Surg., *58*:925, 1971. J. B. Kinmonth, *The Lymphatics,* Arnold, London, 1972.

GLUTEAL REGION 20

The skin of the buttock is supplied by the superior, middle, and inferior clunial nerves (pp. 532 and 209), by the lateral cutaneous branches of the subcostal and iliohypogastric nerves (p. 426), and by the perforating cutaneous nerve (p. 455).

FASCIA OF GLUTEAL REGION

The subcutaneous tissue of the buttock is usually thick and fatty. A bursa is found in this layer over the greater trochanter. The fascia of the gluteal region encloses the gluteus maximus, continues forward as a strong aponeurotic sheet ("gluteal aponeurosis") over the gluteus medius, and splits around the tensor fasciae latae. Here its deep layer fuses with the capsule of the hip joint and the posterior or reflected head of the rectus femoris. Below, the gluteal aponeurosis continues distally as the iliotibial tract of the fascia lata; above, it is attached to the iliac crest and, behind, to the sacrotuberous ligament; elsewhere it is continuous with the fascia lata. It is bound to the skin along the gluteal fold, below the lower border of the gluteus maximus.

The arrangement of vessels and nerves is such that the upper and lateral quadrant of the buttock and also the anterior part of the gluteal region (the part containing the tensor fasciae latae) are relatively avascular and free of major nerves. Hence these regions are commonly used for intramuscular injections.

MUSCLES OF GLUTEAL REGION

Gluteal Muscles and Tensor Fasciae Latae

The gluteus maximus, gluteus medius, and gluteus minimus, from superficial to deep in that order, form the bulk of the buttock. These muscles are supplied by the gluteal vessels and nerves, which reach them through the greater sciatic foramen. The tensor fasciae latae, which is functionally associated with the gluteal muscles as well as with the flexors of the thigh, is supplied by the superior gluteal nerve and is therefore described with the gluteal muscles.

Gluteus Maximus. This coarsely fasciculated muscle, the thick cranial part of which is uniquely human,[1] arises from the ilium behind the posterior gluteal line, from the dorsal surfaces of the sacrum, coccyx, and sacrotuberous ligament, from the aponeurosis of the erector spinae, and from the gluteal aponeurosis. It is inserted partly into the gluteal tuberosity of the femur, but mainly into the iliotibial tract of the fascia lata and thereby into the lateral lip of the linea aspera and the lateral condyle of the tibia.

The gluteus maximus leaves the tuberosity of the ischium uncovered when the thigh is flexed, as in the sitting position. There is generally a large bursa (or several small ones) between the muscle and the greater trochanter, another between it and the upper part of the vastus lateralis, and often one over the ischial tuberosity.

NERVE SUPPLY AND ACTION. It is supplied by the inferior gluteal nerve. It is a powerful extensor of the thigh, and of the pelvis or trunk upon the fixed lower limbs. It nevertheless has no major postural functions,[2] is relaxed when an individual stands upright, and is used but little in ordinary walking. The gluteus maximus acts when force is necessary and is thus important in running, climbing, and similar activities, including rising from a sitting position. By paradoxical action, it regulates flexion of the hip in the process of sitting down. Furthermore, by acting from a fixed insertion, it can extend the trunk, and it is important in extension from a stooped position (p. 538). Because its line of pull is below and behind the hip joint, the gluteus maximus is said to rotate the thigh laterally, thereby opposing the gluteus medius. Its upper part may abduct.

Owing to its intimate relation to the ischiorectal fossa, the gluteus maximus can compress the fossa, and thus indirectly has an effect on the anal canal.

Gluteus Medius. This muscle arises from the ilium between the anterior and posterior gluteal lines and from the overlying gluteal aponeurosis. The muscle fibers converge on a short, strong tendon, which has an oblique insertion on the lateral surface of the greater trochanter. A bursa lies deep to the tendon at its insertion.

Gluteus Minimus. This arises from the ilium between the anterior and inferior gluteal lines. It often blends with the gluteus medius in front and the piriformis behind. It is inserted on the anterior border of the greater trochanter.

NERVE SUPPLY AND ACTION. Both gluteus muscles are supplied by the superior gluteal nerve. They abduct the thigh and rotate it medially. They have powerful actions on the pelvis when the thigh is fixed and are especially important in walking. During walking the gluteus medius and minimus of the limb on the ground abduct the pelvis, that is, tilt or hold it so that the pelvis on the side of the free or swinging limb is prevented from sagging (p. 252). The foot of the free limb can thereby clear the ground.

Paralysis of the gluteus medius leads to a characteristic lurching or waddling gait. The side of the pelvis opposite the paralyzed muscle sags, and, in order that the limb of that side can clear the ground, the trunk is inclined toward the paralyzed side.

Tensor Fasciae Latae. This arises from the outer lip of the iliac crest and from the anterior superior iliac spine. It is inserted into the iliotibial tract.

NERVE SUPPLY AND ACTION. It is supplied by the superior gluteal nerve. It flexes the thigh and rotates it medially. In flexing, it acts in conjunction with the iliopsoas. If the iliopsoas is paralyzed, the tensor fasciae latae hypertrophies. As a medial rotator, the tensor acts in conjunction with the gluteus medius and minimus. It also contracts during other hip movements, particularly during abduction, in which movement it is probably a synergist or fixator. The muscle has no direct action on the leg.

Lateral Rotators of the Thigh

These are six relatively small muscles, which for the most part are covered by the gluteus maximus. They are the piriformis, obturator internus, superior gemellus, inferior gemellus, quadratus femoris, and obturator externus. The obturator externus is supplied by the obturator nerve, the others by branches of the sacral plexus.

Piriformis. This arises mainly from the pelvic surface of the sacrum (S. V. 2 to 4) and the sacrotuberous ligament, and from the ilium just below the posterior inferior spine. The piriformis leaves the pelvis through the greater sciatic foramen (fig. 20–1; fig. 40–12, p. 446), and is inserted into the upper border of the greater trochanter.

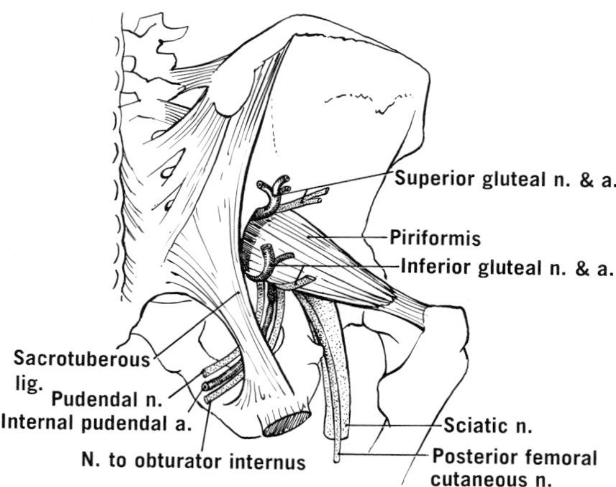

Figure 20–1 The arrangement of the structures emerging from the greater sciatic foramen. The foramen gives exit to the piriformis, to seven nerves (sciatic, posterior femoral cutaneous, superior gluteal, inferior gluteal, pudendal, nerve to obturator internus, and nerve to quadratus femoris), and to three groups of vessels (internal pudendal, superior gluteal, and inferior gluteal). The nerve to the quadratus femoris and the veins that accompany the arteries are not shown.

NERVE SUPPLY. It is supplied by branches of the ventral rami of the first and second sacral nerves. These branches usually enter the pelvic surface of the muscle directly.

Obturator Internus. This arises from the pelvic surface of the obturator membrane and the hip bone (fig. 40–12, p. 446). Above, the bony origin extends from the greater sciatic notch to the pectineal line, and below, it extends along the pubic and ischial margins of the obturator foramen. The muscle fibers converge upon a fasciculated tendon that leaves the pelvis through the lesser sciatic foramen, turns acutely forward, and is inserted into the medial surface of the greater trochanter, just in front of the insertion of the piriformis.

NERVE SUPPLY. It is supplied by a nerve that arises from the sacral plexus (p. 455), and which ends in the perineal surface of the obturator internus. The extrapelvic part of the muscle may receive an additional branch, either directly from the sacral plexus or by way of the nerve to the quadratus femoris.

Superior Gemellus, Inferior Gemellus. These are two small muscles that arise from the ischial spine and tuberosity, respectively, and insert into the upper and lower margins of the obturator internus tendon, respectively.

NERVE SUPPLY. The superior gemellus is usually supplied by the nerve to the obturator internus, and the inferior by the nerve to the quadratus femoris.

Quadratus Femoris. This short, thick muscle extends from the ischial tuberosity to the intertrochanteric crest. A small offshoot is inserted into the quadrate tubercle.[3]

NERVE SUPPLY. It is supplied by a nerve that arises from the sacral plexus (p. 455).

Obturator Externus. This muscle rotates the thigh laterally, and, because its insertion is topographically related to the preceding muscles, it is described here. It arises from the external surface of the pubis and ischium along the margin of the obturator foramen and from the obturator membrane. The muscle fibers converge upon a thick tendon that passes across the back of the hip joint and is inserted into the trochanteric fossa of the femur (fig. 21–10, p. 222).

NERVE SUPPLY. It is supplied by the posterior division of the obturator nerve.

Actions. The six muscles just described rotate the thigh laterally and stabilize the hip joint. The piriformis and obturator internus are inserted above the level of the head of the femur and they can therefore abduct the thigh also. This action, the importance of which is questionable, is said to be most prominent when the thigh is extended. The quadratus femoris and obturator externus are inserted somewhat lower than the level of the head of the femur and it may be possible for them to adduct the thigh.

VESSELS AND NERVES OF GLUTEAL REGION

Vessels

The gluteal arteries arise, directly or indirectly, from the internal iliac artery, but the patterns of origin are extremely variable (p. 449).

Superior Gluteal Artery. This artery, the largest branch of the internal iliac, lies between the lumbosacral trunk and the first sacral nerve (fig. 41–2, p. 451). It then leaves the pelvis through the greater sciatic foramen, above the piriformis. In the gluteal region, under cover of the gluteus maximus, it divides into superficial and deep branches.

BRANCHES. In the pelvis, it supplies adjacent muscles and the hip bone. In the gluteal region, the *superficial branch* divides immediately into twigs that enter the gluteus maximus. Some of them pierce it to reach the overlying skin. The *deep branch* runs forward between the gluteus medius and minimus and divides into *superior* and *inferior branches,* which accompany the branches of the superior gluteal nerve and supply adjacent muscles.

Inferior Gluteal Artery. Another branch of the internal iliac, this passes backward, between the first and second, or second and third, sacral nerves, and leaves the pelvis through the greater sciatic foramen, below the piriformis (fig. 41–2, p. 451). In the gluteal region, it lies under cover of the gluteus maximus and descends medial to the sciatic nerve, in company with the posterior femoral cutaneous nerve. It lies behind the obturator internus, gemelli, quadratus femoris, and adductor magnus.

BRANCHES. Inside the pelvis it gives muscular branches, twigs to the bladder, seminal vesicles, and prostate, and occasionally a branch that takes the place of the middle rectal artery. Outside the pelvis, it supplied the gluteal and hamstring muscles, the hip joint, and the overlying skin, and takes part in the cruciate anastomosis (p. 217). Coccygeal branches pierce the sacrotuberous ligament and supply the soft tissue on the back of the coccyx. The *companion artery of the sciatic nerve* is a slender branch that descends on or in the sciatic nerve and supplies it.

Veins. The *superior* and *inferior gluteal veins* are usually double. They accompany the arteries and empty into the internal iliac vein. They communicate with tributaries of the femoral vein and provide an important route for return of blood from the lower limb, because they can return the blood if the femoral vein is ligated.

Nerves

Several important nerves from the sacral plexus (p. 454) either supply or traverse the gluteal region (fig. 20–2). They are the superior and inferior gluteal nerves, the pudendal nerve, special muscular branches, the posterior femoral cutaneous nerve, and the sciatic nerve. The pudendal nerve enters the gluteal region through the greater sciatic foramen and re-enters the pelvis through the lesser sciatic foramen. It is described with the pelvis (p. 455).

Superior Gluteal Nerve (L4, 5, S1) (fig. 41–4, p. 454; fig. 64–8, p. 772). This nerve passes backward through the greater

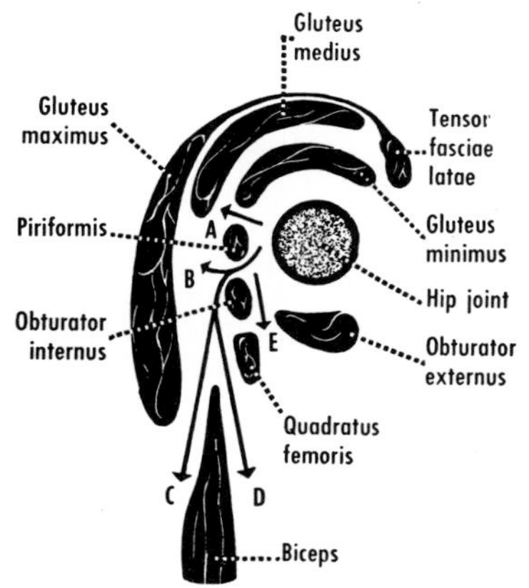

Figure 20–2 Schematic sagittal representation of the muscular planes about the hip joint. The arrows indicate structures emerging from the greater sciatic foramen as follows: *A*, superior gluteal nerve and vessels; *B*, inferior gluteal nerve and vessels; *C*, posterior femoral cutaneous nerve; *D*, sciatic nerve; *E*, nerve to the quadratus femoris.

sciatic foramen, above the piriformis. An upper branch supplies the gluteus medius, and a lower branch supplies the gluteus minimus and medius, the tensor fasciae latae, and the hip joint.

Inferior Gluteal Nerve (L5, S1, 2; sometimes from the common peroneal component of the sciatic nerve[4]) (fig. 41–4, p. 454; fig. 64–8, p. 772). This passes through the greater sciatic foramen, below the piriformis. It breaks up at once into branches that enter and supply the overlying gluteus maximus.

Posterior Femoral Cutaneous Nerve. This is a branch of the sacral plexus (S1 to S3) (fig. 20–1; fig. 41–4, p. 454) that enters the gluteal region through the greater sciatic foramen, below the piriformis. It descends deep to the gluteus maximus, in company with the inferior gluteal artery and sciatic nerve. It becomes superficial near the popliteal fossa and accompanies the small saphenous vein to the middle of the calf of the leg, where its terminal filaments anastomose with the sural nerve.

BRANCHES. While deep to the gluteus maximus, the posterior femoral cutaneous nerve gives off *inferior clunial nerves (gluteal branches)* to the skin of the buttock, *perineal branches* that cross the hamstrings and supply the skin of the external genitalia, and *femoral* and *sural branches* to the skin on the back of the thigh and calf. Some branches may reach the heel.

Sciatic Nerve. **The largest nerve in the body, this actually consists of two nerves, the tibial and peroneal, that are bound together.** The sciatic nerve is a branch of the sacral plexus (L4 to S3) (fig. 20–1; fig. 64–8, p. 772). It leaves the pelvis and enters the gluteal region by passing through the greater sciatic foramen, below the piriformis.[5] Sometimes, however, the peroneal component pierces the piriformis, or even emerges above this muscle, and it then remains separate for the rest of its course. The sciatic nerve descends under cover of the gluteus maximus, between the greater trochanter and ischial tuberosity. In front, it lies successively on the ischium, the gemelli and the obturator internus, and the quadratus femoris. The nerve then enters the thigh. Its subsequent course is described on page 213.

REFERENCES

1. J. T. Stern, Amer. J. phys. Anthrop., *36*:315, 1972.
2. E. Karlsson and B. Jonsson, Acta morph. neerl.-scand., *6*:161, 1964.
3. S. Sunderland, J. Anat., Lond., *72*:309, 1938.
4. S. Załuska and Z. Urbanowicz, Folia morphol., Warsaw, *30*:167, 1971.
5. L. E. Beaton and B. J. Anson, Anat. Rec., *70*:1, 1937. P'an Ming-Tzu, Amer. J. phys. Anthrop., *28*:375, 1941.

THIGH AND KNEE 21

The skin of the thigh is supplied by cutaneous branches of the femoral and obturator nerves, and by the lateral and posterior femoral cutaneous nerves, the ilioinguinal nerve, and the femoral branch of the genitofemoral nerve. The lateral cutaneous branches of the subcostal and iliohypogastric nerves, which supply the skin of the buttock, may also supply the upper and front parts of the thigh.

FASCIA OF THIGH

The subcutaneous tissue often contains much fat. It is quite thick in the groin, where it forms two layers that enclose the superficial inguinal lymph nodes, the great saphenous vein, and smaller vessels. The thin, membranous deep layer is best marked on the medial side of the great saphenous vein just below the inguinal ligament. It covers the saphenous opening (where it is termed the *cribriform fascia),* fuses with the femoral sheath and lacunar ligament, and, laterally, fuses with the fascia lata below and parallel to the inguinal ligament. Thus, fluid that collects deep to the subcutaneous tissue of the abdomen cannot extend into the thigh.

The fascia of the thigh *(fascia lata)* is attached to the subcutaneous portions of the hip bone (e.g., the iliac crest), sacrum, and coccyx, and to the inguinal and sacrotuberous ligaments. Inward extensions from its deep surface to the femur form the *lateral* and *medial intermuscular septa.* Thus, three compartments are formed, anterior, posterior, and medial (fig. 21–1).

The fascia lata also blends with the aponeurotic insertions of the vastus medialis and lateralis to form the medial and lateral retinacula of the patella. That part of the fascia lata overlying the vastus lateralis forms the *iliotibial tract* (fig. 21–1; see also fig. 25–2, p. 255),[1] which in turn extends inward to the lateral lip of the linea aspera and the lateral supracondylar line as the lateral intermuscular septum. Above, the tract continues to the iliac crest as the gluteal aponeurosis. Below, it blends with the lateral retinaculum of the patella. The gluteus maximus and tensor fasciae latae are inserted into the tract and, with the tract and septum, form a continous, strong, musculoligamentous apparatus that is an important mechanism in maintaining posture and in locomotion.

Deep to the sartorius, fascial fibers form a dense membrane termed the subsartorial fascia, which binds the vastus medialis to the adductor longus and magnus, and covers the femoral vessels in the adductor canal. The medial intermuscular septum, which is thinner and less distinct than the lateral, extends inward from the subsartorial fascia to the medial supracondylar line and the medial lip of the linea aspera.

Saphenous Opening (fossa ovalis) (figs. 19–1 and 21–2). This is a large, ovoid gap in the fascia lata, approximately 4 cm below and lateral to the pubic tubercle. It varies in size and shape, lies in front of the femoral vein, and transmits the great saphenous vein to the femoral vein. The fascia lata is fused with the inguinal ligament from the anterior superior iliac spine to the pubic tubercle. It is fused with the lacunar ligament and thus reaches the pectineal line. From the pubic tubercle, it is reflected downward and laterally, lateral to the great saphenous vein, forming here the *falciform margin,* which is adherent to the femoral sheath. The upper part of this margin is the *superior cornu,* and the lower the *inferior cornu.* The latter continues behind the great saphenous vein and then turns deeply and upward to cover the pectineus muscle. The fascia lata, in forming the saphenous opening, may be said to spiral, first downward and laterally, then medially and upward, and finally superficial to deep. The saphenous opening is closed by the cribriform fascia,

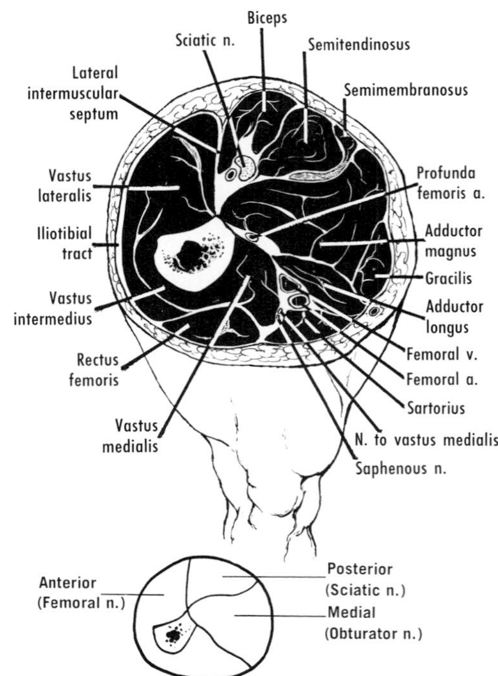

Figure 21–1 Drawing of a horizontal section through the middle of the thigh. The three major muscular compartments are shown in the inset figure: anterior (supplied by the femoral nerve), posterior (supplied by the sciatic nerve), and medial (supplied by the obturator nerve).

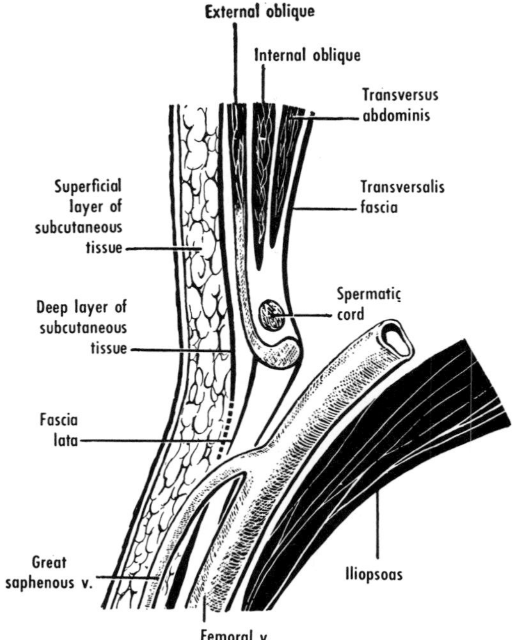

Figure 21–2 Schematic representation of the anterior abdominal wall, inguinal ligament, and saphenous opening in a sagittal plane. See also figure 33–9, page 359.

which is pierced by the great saphenous vein and some of its tributaries, and by small vessels.

The topography of the region of the saphenous opening may appear different in the living, owing to the distention of the femoral vein by blood.[2]

Femoral Sheath, Femoral Triangle, and Adductor Canal

Femoral Sheath (fig. 21–3). **The uppermost parts of the femoral artery and vein lie behind the inguinal ligament in the vascular compartment, situated in the groove between the iliopsoas and the pectineus. The iliopsoas and the femoral nerve occupy the more laterally placed muscular compartment. The two compartments are separated by thickened fascia that forms the iliopectineal arch or septum. The femoral artery and vein, and the medially placed femoral canal, are enclosed in a fascial funnel known as the femoral sheath.** The fascia transversalis forms the front of the femoral sheath, and the fascia iliaca forms its back. The anterior wall of the sheath is pierced by the femoral branch of the genitofemoral nerve and by the great saphenous vein. The femoral sheath is only a few centimeters long. Inferiorly the sheath tapers and becomes indefinite as it fuses with the adventitia of the vessels. Within the femoral sheath, from lateral to medial, are the femoral artery, femoral vein, and femoral canal, separated by two anteroposterior septa.

The femoral canal is situated in front of the pectineus, and it contains fat and a few lymph vessels. Its upper end or base, termed the *femoral ring* is closed by extraperitoneal tissue known as the *femoral septum,* and a lymph node may be found here. The septum is covered superiorly by parietal peritoneum. The lateral margin of Henle's ligament (p. 360) bounds the femoral ring medially.

The femoral canal is important surgically because of its relation to hernia. **A femoral hernia is the protrusion of extraperitoneal tissue, with or without an abdominal viscus, through the femoral ring.** It may pass down the femoral canal, and through the femoral sheath at the lower end of the canal, at about the level of the lacunar ligament. The hernial sac is formed by parietal peritoneum, but the external coverings of the hernia are frequently fused. The hernia may pass through the saphenous opening. Clinically, the neck of the hernia is found at a site on the thigh just inferior and lateral to the pubic tubercle.

Femoral Triangle (fig. 21–4). **The femoral triangle is located in the upper third of the front of the thigh. It contains the femoral vessels and nerve. It is bounded laterally by the medial border of the sartorius, medially by the medial border of the adductor longus, and superiorly by the inguinal ligament. Its roof is formed by the fascia lata and the cribriform fascia. Its floor is formed by the iliopsoas, pectineus, and adductor longus** (fig. 21–5).

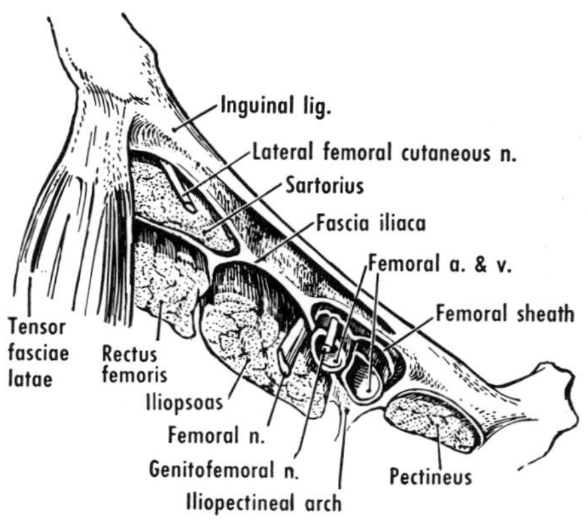

Figure 21–3 The structures that descend behind the inguinal ligament.

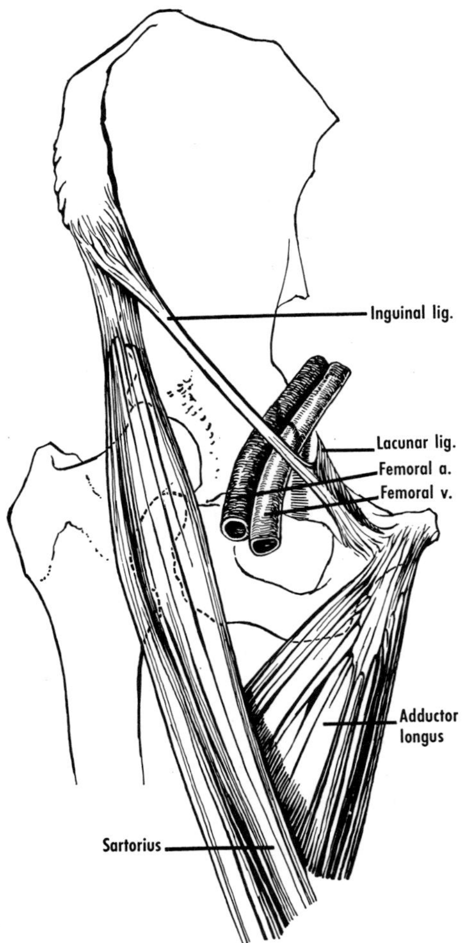

Figure 21–4 The femoral triangle (sartorius, adductor longus, inguinal ligament).

In the femoral triangle, the femoral artery is covered anteriorly by the front of the femoral sheath (fig. 21–3), and by the cribriform fascia above and the fascia lata below. Behind, the artery lies on the back of the femoral sheath and the psoas major, which separates it from the head of the femur. Below this, it is separated from the pectineus and adductor longus by the femoral vein. The vein, which lies behind the artery in the lower part of the femoral triangle, winds to its medial side above. Lateral to the femoral artery are the femoral nerve above, and the saphenous nerve and the nerve to the vastus medialis below.

Adductor Canal. **The adductor (or subsartorial) canal* is located in the middle third of the medial part of the thigh. The canal contains the femoral vessels, the saphenous nerve, and usually the nerve to the vastus medialis.** It is bounded laterally by the vastus medialis, and medially by the adductor longus and often the adductor magnus (fig. 21–1).[3] Superficially, it is covered by the sartorius and by the subsartorial fascia, which connects the lateral and medial boundaries.

In the adductor canal, the femoral artery is covered in front by the fascial roof of the canal, the subsartorial plexus of nerves, and the sartorius. Posteriorly, the artery is separated from the adductor longus and magnus by the femoral vein. The vastus medialis lies laterally. The saphenous nerve accompanies the femoral artery throughout the canal; it is lateral, anterior, and medial to the vessel, in turn.

*Described by John Hunter (1728–1793), who ligated the femoral artery in the canal in patients with popliteal aneurysm.

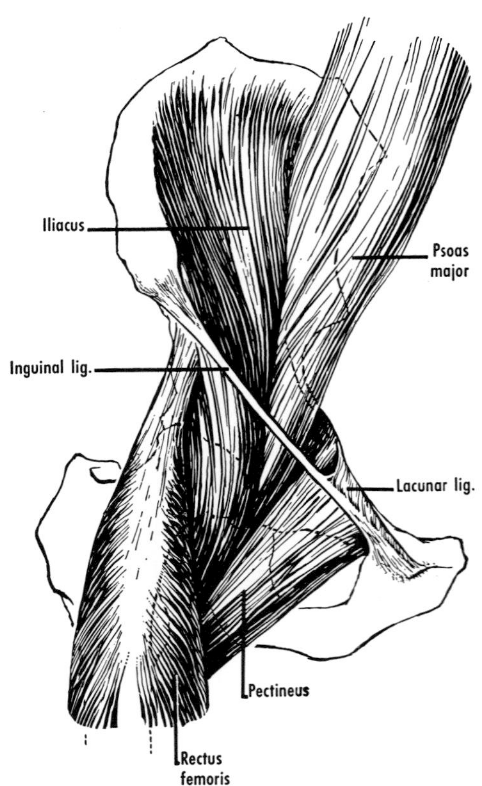

Figure 21–5 The floor of the femoral triangle. The adductor longus, which also forms a part of the floor, is not shown. See figure 21–4.

BACK OF THIGH

Muscles

The muscles of the back of the thigh are the biceps femoris, semitendinosus, and semimembranosus. Except for the short head of the biceps, which is supplied by the peroneal portion of the sciatic nerve, these muscles arise from the tuberosity of the ischium, are supplied by the tibial part of the sciatic nerve, and cross two joints. They are collectively known as the hamstrings.

A part of the adductor magnus aids the hamstrings in extending the thigh. It is described with the adductors (p. 214).

Biceps Femoris. The *long head* of this muscle arises from the medial facet on the ischial tuberosity, in common with the semitendinosus.[4] Some fibers are continuous with the sacrotuberous ligament. In the lower part of the thigh, the muscle gives way to a tendon that is joined by the short head. The *short head* arises from the lateral lip of the linea aspera, from the upper part of the lateral supracondylar line, and from the lateral intermuscular septum. The combined tendon, which is palpable and visible (fig. 25–1), forms the lateral boundary of the popliteal fossa. It descends to the head of the fibula and the fascia of the leg. Part of the tendon is prolonged to the fibular collateral ligament, to the lateral condyle of the tibia, and to adjacent fascia (fig. 18–31, p. 187).[5]

Semitendinosus. This muscle arises in common with the long head of the biceps. In the middle of the thigh its belly gives way to a long, slender tendon, which is palpable and visible as one of the medial boundaries of the popliteal fossa (fig. 25–1). This tendon crosses the semimembranosus and is inserted into the fascia of the leg and the upper part of the medial surface of the tibia, deep to the sartorius and distal to the gracilis. The *bursa anserina* (tibial intertendinous bursa) is found at these tendinous insertions; it separates the tendons from the tibial collateral ligament.

Semimembranosus. This arises by a flattened tendon, chiefly from the lateral facet on the ischial tuberosity, and also from the adjacent part of the ischial ramus. The tendon becomes muscular in the upper part of the thigh. The tendon of insertion[6] begins about the middle of the thigh. It consists of superficial and deep parts. An expansion of the superficial part turns upward and laterally as the oblique popliteal ligament of the knee joint; the remainder forms the fascia of the popliteus and is attached to the medial border and the soleal line of the tibia. The deep part of the tendon forms (1) a thick cord that extends into and is attached in a groove on the medial condyle of the tibia, deep to the tibial collateral ligament, and (2) a short slip that is attached to a tubercle below the groove. Several bursae are associated with the insertions. One between the main tendon and the medial head of the gastrocnemius often communicates with the cavity of the knee joint.

Actions. The hamstrings cross the hip and knee joints. They are the main extensors of the thigh and flexors of the leg, especially during walking. When the leg and thigh are fixed, they can extend the trunk.

The ability of the hamstrings to act on one of the two joints depends upon the position of the other joint. Thus, if the knee is fully flexed, the hamstrings are so shortened that they cannot contract further and act on the hip. Likewise, if the hip is fully extended, the hamstrings are so shortened that they cannot act upon the knee.

These muscles cross two joints and therefore have ligamentous actions (p. 25). If the hip is fully flexed, as in attempting a high kick, the hamstrings are so lengthened that it is generally difficult or impossible to extend the knee fully at the same time (except by continued practice). In fact, the taut hamstrings tend to draw the knee into flexion. Likewise, if the knee is fully extended, the hamstrings are so lengthened that it is difficult to flex the hip fully. Without practice, it is often difficult to touch the toes with the fingers without bending the knees.

Nerves

Posterior Femoral Cutaneous Nerve (p. 209).

Sciatic Nerve. The sciatic nerve (p. 209; fig. 64–8, p. 772), having entered the gluteal region through the greater sciatic foramen, descends under cover of the gluteus maximus (see fig. 20–1). It enters the thigh, where it lies anteriorly on the adductor magnus and is crossed posteriorly by the long head of the biceps. On its medial side are the inferior gluteal artery (which gives a special branch, the companion artery of the sciatic nerve) and the posterior femoral cutaneous nerve. Although separation of the sciatic into the tibial and common peroneal nerves may occur at any level in the gluteal region or the thigh, it usually occurs in the lower third of the thigh. If the division occurs in the sacral plexus, the common peroneal generally adopts a more posterior course by piercing the piriformis.

BRANCHES. The branches of the sciatic nerve arise mostly from its medial side. A series of twigs, derived from the tibial nerve, supply the semitendinosus, the semimembranosus, the long head of the biceps, and the adductor magnus. A branch to the short head of the biceps arises from the common peroneal nerve.

MEDIAL SIDE OF THIGH

The muscles of the medial side of the thigh are the pectineus, adductor longus, adductor brevis, adductor magnus, gracilis, and obturator externus. The pectineus and adductor brevis are topographically a part also of the front of the thigh. The obturator externus belongs in part to the gluteal region, with which it is described. It rotates the thigh laterally; the chief function of the others is adduction of the thigh.

Pectineus. This forms the medial part of the floor of the femoral triangle (p. 211). It arises from the pectineal line of the pubis, descends behind the lesser trochanter, and is inserted into the upper half of the pectineal line of the femur.

Adductor Longus. The medial border of this muscle forms the medial boundary of the femoral triangle (p. 211). It arises from the femoral surface of the body of the pubis, below the crest, and is inserted into the medial lip of the linea aspera.

Adductor Brevis. This muscle is largely concealed by the adductor longus and the pectineus. It extends from the body and inferior ramus of the pubis to the pectineal line and the upper part of the linea aspera. The anterior branch of the obturator nerve lies in front of it; the posterior branch lies behind it.

Adductor Magnus. This large, triangular muscle consists of: (1) an *adductor part,* which runs chiefly from the ischiopubic ramus to the linea aspera and is supplied by the obturator nerve; and (2) an *extensor part,* which runs chiefly from the ischial tuberosity to the adductor tubercle and is supplied by the sciatic nerve. The adductor part is the anterior and upper part of the muscle (adductor minimus); it arises to some extent from the body of the pubis, but mainly from the margin of the ischiopubic ramus between the gracilis and the obturator externus. The fibers run nearly horizontally to the back of the femur. The extensor part, sometimes termed the ischiocondylar portion, arises medial to the adductor part, from the ischial ramus and tuberosity. The insertion is by muscle fibers into the linea aspera and medial supracondylar ridge, and by a tendon into the adductor tubercle.

The attachment to the linea aspera is interrupted by three or four fibrous arches for perforating branches of the profunda femoris artery. Between the insertion on the medial supracondylar line and the adductor tubercle there is a longer fibrous arch for the passage of the femoral vessels into the popliteal fossa.

Gracilis. This long, thin muscle arises from the lower margin of the body and inferior ramus of the pubis and is inserted on the upper part of the medial surface of the shaft of the tibia.

Nerve Supply and Actions. Most of these muscles are supplied by the obturator nerve. The pectineus[7] is usually supplied by the femoral nerve, sometimes by the obturator nerve as well, or by the accessory obturator nerve. The extensor part of the adductor magnus is supplied by the tibial part of the sciatic nerve.

The three adductors (aided somewhat by the pectineus) are powerful muscles and are used in all movements in which the thighs are pressed together. They are important stabilizers during flexion and extension. The longus and magnus are active during medial rotation, but the importance of this action is uncertain. The extensor part of the adductor magnus aids the hamstrings in extending the thigh.

The gracilis acts at both the hip and the knee, but is chiefly a flexor, adductor, and medial rotator, especially during the swing phase of walking. It has no important postural function.

Vessels and Nerves

Obturator Artery. This artery, a branch of the internal iliac, is described with the pelvis (p. 451). Its *anterior* and *posterior branches* wind around the margin of the obturator foramen, supply adjacent muscles, and anastomose at the lower margin. The posterior gives off an *acetabular branch* that passes through the acetabular notch and is the chief source of medial epiphysial twigs to the head of the femur.[8]

Obturator Nerve (L3, 4, sometimes L2 or L5 also) (fig. 38–9, p. 427; fig. 64–8, p. 772). This nerve arises from the lumbar plexus in the substance of the psoas major. It emerges at the medial margin of the psoas, at the level of the pelvic inlet, behind the common iliac vessels. It then accompanies (lying above) the obturator vessels to the obturator groove, where it divides into anterior and posterior branches. These pass through the obturator foramen to reach the thigh, where they are separated by the adductor brevis. The trunk and one or both branches give twigs to the hip joint.

BRANCHES. The *anterior branch* of the obturator nerve lies in front of the obturator externus and the adductor brevis, and behind the pectineus and the adductor longus. It ends along the latter muscle, supplies it and the gracilis, adductor brevis, and sometimes the pectineus also, and ends as a filament to the femoral artery and subsartorial plexus. Twigs from it supply the overlying skin and occasionally the knee joint.

The *posterior branch* of the obturator nerve pierces the obturator externus. It descends in front of the adductor magnus, behind the adductor brevis. It ends by piercing the magnus (sometimes in company with the femoral artery), descending on the popliteal artery, and perforating the oblique popliteal ligament to supply the knee joint. It supplies the obturator externus, the adductor magnus, and sometimes the adductor brevis.

Accessory Obturator Nerve (L3, 4 or L2, 3). When present (p. 428), this emerges at the medial margin of the psoas major and enters the thigh in front of the pubis. It communicates with the anterior branch of the obturator nerve and supplies branches to the pectineus and the hip joint.

FRONT OF THIGH

Muscles

The muscles of the front of the thigh are the iliopsoas, the quadriceps femoris, and the sartorius. The tensor fasciae latae, pectineus, and adductor longus, topographically related in part to the front of the thigh, have already been described.

Iliopsoas (figs. 21–5; 33–13, p. 364). This is the great flexor of the thigh and trunk. Its broad lateral part, the iliacus, and its long medial part, the psoas major, arise from the iliac fossa and the lumbar vertebrae, respectively.

ILIACUS. The iliacus arises from the upper part of the iliac fossa and from the ala of the sacrum and adjacent ligaments. A slip that arises from the lower part of the anterior inferior iliac spine is sometimes called the iliacus minor. Most of the fibers of the iliacus are inserted into the lateral aspect of the tendon of the psoas major; some reach the lesser trochanter.

PSOAS MAJOR. The psoas major arises by fleshy slips from the intervertebral disc above each lumbar vertebra and from the adjacent margins of the vertebrae. It arises also from fibrous arches and from the transverse processes of the lumbar vertebrae. The medial arcuate ligament of the diaphragm arches obliquely over the upper part of the psoas major. The thick, elongated muscle thus formed descends along the brim of the pelvis and enters the thigh behind the inguinal ligament. The tendon that arises on the lateral side of the psoas muscle passes in front of the hip joint and is inserted on the lesser trochanter. A bursa, which may communicate with the cavity of the hip joint, commonly separates this tendon from the joint capsule.

The *psoas minor*, which is often absent, is described on page 365.

NERVE SUPPLY AND ACTION. The psoas major is supplied by the lumbar plexus (L2, 3, and sometimes L1 or 4), which is formed in the substance of the muscle. The iliacus is supplied by the femoral nerve by way of branches that arise in the abdomen.

The iliopsoas is the chief flexor of the thigh[9] and, when the thigh is fixed, of the trunk. It advances the limb during walking. The psoas major bends the vertebral column to one side (lateral flexion) and controls trunk deviation when sitting. It is a postural muscle and is active in standing subjects. It is a lateral rotator of the thigh, but this action is unimportant.

Quadriceps Femoris. **This is one of the largest and most powerful muscles in the body. It consists of (1) a biarticular muscle, the rectus femoris, which extends from the hip bone to the tibia, and (2) three monoarticular muscles, the vastus lateralis, vastus intermedius, and vastus medialis, which arise from the front and sides of the femur and extend to the tibia. These four muscles combine into an aponeurotic and ten-**

dinous insertion into the tibia. A large sesamoid bone, the patella, is present in the tendinous part of the insertion.

RECTUS FEMORIS. The rectus femoris arises by an anterior head from the anterior inferior iliac spine, and by a posterior or reflected head from the posterosuperior aspect of the rim of the acetabulum. The two heads which are closely related to the iliofemoral ligament (fig. 18–4, p. 166), unite and continue into the muscle belly, which in turn becomes tendinous in the lower part of the thigh. Part of the tendon is inserted into the base of the patella; the remainder continues to the tuberosity of the tibia.[10] The part of the tendon of insertion between the patella and the tuberosity is known as the *ligamentum patellae.*

VASTUS LATERALIS. The vastus lateralis is a thick muscle that has a narrow origin, from above downward, from the intertrochanteric line, the greater trochanter, the gluteal tuberosity, the upper part of the lateral lip of the linea aspera, and the lateral intermuscular septum. It has an extensive coattachment with the gluteus maximus. Its muscle fibers continue into a broad aponeurosis that blends with the lateral side of the rectus femoris tendon. Most of the aponeurosis fuses with the overlying fascia (p. 210) and continues to the lateral border of the patella and of the ligamentum patellae, and to the lateral condyle of the tibia.

VASTUS MEDIALIS. The vastus medialis is a thick, powerful muscle that usually forms a characteristic bulge in the lower medial part of the thigh. It covers but is not attached to the medial surface of the femur. It arises, from above downward, from the intertrochanteric line, the spiral line, and the medial intermuscular septum. Its lower part is fused with the adductor magnus and adductor longus. Its muscle fibers continue into an aponeurosis that blends with the medial side of the rectus femoris tendon. Most of the aponeurosis fuses with the overlying fascia (p. 210) and continues to the medial border of the patella and of the ligamentum patellae, and to the medial condyle of the tibia.

VASTUS INTERMEDIUS. The vastus intermedius has a fleshy origin from the anterior and lateral surfaces of the upper two-thirds of the shaft of the femur and from the distal half of the lateral intermuscular septum. Here it is fused with the vastus lateralis, and thus has a bony origin from the lateral lip of the linea aspera and from the lateral supracondylar line. Its muscle fibers give way to an aponeurosis that joins the deep aspect of the tendon of the rectus and the other vasti.

From their superficial aspects the vastus medialis and intermedius appear to be fused, but their bony origins are separated by a narrow interval.

NERVE SUPPLY AND ACTION. The quadriceps femoris is supplied by the femoral nerve. It extends the leg and controls its flexion. The rectus femoris arises from the hip bone and therefore flexes the thigh as well as extends the leg. The rectus aids the iliopsoas and has been termed the "kicking muscle." As a two-joint muscle, its actions have advantages and disadvantages comparable with those of the hamstrings.

The quadriceps femoris is especially important in climbing, running, jumping, rising from a sitting position, and walking up and down stairs. It must be emphasized that in many phases of such activities the leg is fixed and the thigh moves. A strong quadriceps is not essential for ordinary walking on level ground; the ability to walk up or down stairs is a more accurate index of its strength. When the quadriceps is weak or paralyzed, the patient must put his knees into extension, either with his hands or with mechanical aids. Standing may still be possible, however, and ordinary walking of an awkward type may still be accomplished on a level surface.

It is usually stated that the vastus medialis contracts strongly only during the last phase of extension, but this has not been confirmed. There are, nevertheless, some differences in timing and intensity of action of the different parts of the quadriceps.[11]

Owing to the fact that the femora are set obliquely, there is an angle at the knee, an angle that is often more acute in females, and exaggeration of which is termed genu valgus (knock-knee). Owing to this angle, the patella tends to move laterally when the leg is extended. This lateral movement is accentuated by the lateral pull of the vastus lateralis. When dislocation of the patella occurs, the dislocation is nearly always lateral, and it happens more often in women. The lateral pull, however, is counterbalanced by the medial, more horizontal pull of the vastus medialis and by the fact that the lateral condyle of the femur has a more forward projection and a deeper slope for the larger, lateral facet of the patella. Thus, there is a mechanical impediment to lateral dislocation.

KNEE JERK. **The knee jerk is elicited by tapping the ligamentum patellae. The sudden**

stretch of the muscle reflexly induces contraction. The center for this reflex is in the third lumbar segment of the spinal cord.

BURSAE. A *subtendinous prepatellar bursa* may be present between the patella and the tendinous fibers that descend in front of it, a *subfascial prepatellar bursa* between the tendinous fibers and the overlying fascia, and a *subcutaneous infrapatellar bursa* in front of the patella and ligamentum patellae. The subcutaneous bursa is almost always present. Its wall, like that of the subcutaneous olecranon bursa, may be thickened and multilocular. Branches of superficial nerves may traverse the subcutaneous bursa. A *deep infrapatellar bursa* is also present deep to the tibial attachment of the ligamentum patellae.

Articularis Genus. This is a small, unimportant muscle that consists usually of two slips that arise from the front of the lower part of the femur and are inserted on the upper part of the capsule of the knee joint.[12]

Sartorius. This muscle arises from the anterior superior iliac spine and the area immediately below it. It takes a long, spiral course over the front of the thigh and is inserted into the upper part of the medial surface of the tibia, where it covers the tendons of the gracilis and semitendinosus. The bursa anserina is associated with these tendons. The sartorius forms the lateral boundary of the femoral triangle and covers the adductor canal.

NERVE SUPPLY AND ACTION. It is supplied by the femoral nerve. It acts chiefly as a flexor of the thigh and leg, and is most active during hip flexion.[13]

Femoral Vessels

Femoral Artery. The femoral artery is the continuation of the external iliac below the level of the inguinal ligament. In the upper third of the thigh, it is relatively superficial in the femoral triangle. In the middle third, it lies deeply in the adductor canal. In the lower third, the name of the artery is changed to popliteal as it passes through the adductor opening (fig. 64–2, p. 766).

BRANCHES. In the proximal part of its course, the femoral artery gives off the superficial epigastric artery, the superficial circumflex iliac artery, the superficial external pudendal and the deep external pudendal artery. More distally, it gives off the profunda femoris and the descending genicular arteries.

1. The *superficial epigastric artery* pierces the femoral sheath and fascia lata, and courses toward the umbilicus to anastomose with the inferior epigastric artery (p. 361).

2. The *superficial circumflex iliac artery* pierces the femoral sheath and fascia lata and runs toward the anterior superior iliac spine, where it anastomoses with the deep circumflex iliac artery (p. 362).

3. The *external pudendal arteries* (sometimes described as superficial and deep) emerge through the saphenous opening, run medially and upward across the spermatic cord (or the round ligament in the female), and give *inguinal branches* to the skin and muscles of that region, and *anterior scrotal* (or *labial*) *branches.*

4. *Profunda femoris artery.* Between 1 and 5 cm below the inguinal ligament, the femoral artery, directly or indirectly, gives off circumflex arteries, which supply the greater part of the thigh. In about half of instances, these arteries arise from the femoral artery by a common trunk, the profunda femoris artery.[14] This artery, as it runs medially from behind the femoral artery, gives off its circumflex branches, and then descends on the medial side of the femur, lying on the adductor brevis and magnus. It ends by passing through a fibrous arch in the adductor magnus as the last perforating artery. The profunda femoris gives off muscular branches and several (often three) *perforating arteries.* These pass through fibrous arches in the insertions of the adductor brevis and magnus. They supply the hamstrings and anastomose with one another in the vastus lateralis. Deep to the gluteus maximus the first perforating artery anastomoses with the inferior gluteal and with the transverse branches of the lateral and medial circumflex arteries. This union is known as the cruciate anastomosis. The nutrient arteries to the femur arise from one or more of the perforating arteries, usually the first, or first and second. The continuation of the profunda femoris is often referred to as the fourth perforating artery.

The *lateral circumflex artery,* which may arise directly from the femoral, runs laterally among the branches of the femoral nerve and then behind the sartorius and rectus femoris, gives off an *ascending branch* that anastomoses with the superior gluteal artery, a *transverse branch* that winds around the femur and enters the cruciate anastomosis, and a *descending branch* that reaches the knee. The last branch may arise separately from the profunda femoris.

The *medial circumflex artery,* which may arise separately from the femoral, runs backward between the psoas major and pectineus, toward the acetabulum. It gives off an *acetabular branch,* which anastomoses with the acetabular branch of the obturator artery and which may send a medial epiphysial branch to the head of the femur, and then divides into an *ascending branch,* which anastomoses with the gluteal arteries, and a *transverse branch,* which enters the cruciate anastomosis.

Twigs from the circumflex arteries reach the head of the femur by way of the retinacula on the neck of the femur, and are termed lateral epiphysial arteries.

5. The *descending genicular artery* arises from the femoral just before the latter ends. It divides immediately into *saphenous* and *articular branches.* The saphenous accompanies the saphenous nerve to the knee

and anastomoses with the medial inferior genicular artery. The articular branches descend in the substance of the vastus medialis to the knee joint.

Femoral Vein. The femoral vein, the lower part of which may be double, is the continuation of the popliteal veins above the opening in the adductor magnus. It ascends through the adductor canal, lying posterolateral and then posterior to the femoral artery. Next, it passes through the femoral triangle, lying posterior and then medial to the femoral artery. It enters the femoral sheath lateral to the femoral canal and ends behind the inguinal ligament by becoming the external iliac vein. It usually has two or three valves; one is placed at the upper end of the vein, and another just above the opening for the profunda femoris vein. Its chief tributaries are the profunda femoris, the medial and lateral circumflex veins, and the great saphenous vein.[15] A plexiform anastomosis between the femoral vein and the profunda femoris vein is commonly present in the adductor canal.

Nerves

Femoral Nerve (chiefly L4, plus L2 and L3) (fig. 38–9, p. 427; fig. 64–8, p. 772). This is the largest branch of the lumbar plexus. It arises in the substance of the psoas major and emerges from the lateral border of that muscle a little below the iliac crest. It then descends in the groove between the iliacus and psoas major muscles and enters the thigh behind the middle of the inguinal ligament in the muscular compartment, lateral to the femoral vessels. Entering the femoral triangle, the femoral nerve breaks up into a number of terminal branches.

BRANCHES. In the abdomen, the femoral nerve may give off the lateral femoral cutaneous nerve. In the iliac fossa, it supplies the iliacus and the femoral artery.[16] The nerve to the pectineus arises here (or in the femoral triangle), and passes behind the femoral sheath to supply the pectineus and the hip joint.

The terminal branches of the femoral nerve are sometimes classified into an anterior division (anterior cutaneous and branch to the sartorius) and a posterior division (muscular and saphenous).

The *anterior cutaneous branches* of the anterior division are subdivided into the intermediate and medial cutaneous nerves. The intermediate cutaneous nerve, which is usually double, gives branches to the sartorius and supplies the skin on the front of the thigh; distally it contributes to the patellar plexus. The medial nerves cross superficial to the femoral vessels at the apex of the femoral triangle, supply the skin on the medial side of the thigh, and contribute to the subsartorial and patellar plexuses. The *muscular branch* of the anterior division goes directly to the sartorius.

The *muscular branches* of the posterior division supply the quadriceps femoris and the articularis genus. The branch to the rectus femoris also supplies the hip joint. The branches to the vasti send filaments to the knee joint.

The *saphenous nerve*[17] is regarded as the termination of the femoral nerve. It descends with the femoral vessels through the femoral triangle and subsartorial canal, crosses the femoral artery from lateral to medial side, and then becomes cutaneous. It gives a twig to the knee joint, contributes to the subsartorial and patellar plexuses, and then descends in the leg with the great saphenous vein, supplying the skin on the medial side of the leg and foot. The saphenous nerve may be joined by filaments constituting an accessory femoral nerve. This is a common variant that arises from the lumbar plexus, runs a separate course into the thigh, and usually ends by joining one of the branches of the femoral nerve. There may also be an accessory saphenous nerve (from the femoral or saphenous nerves) that descends superficial to the adductor canal to the medial side of the calf.[18]

The subsartorial plexus consists of communications, deep to the sartorius, between branches of the medial femoral cutaneous nerves and the saphenous and obturator nerves. The patellar plexus, in front of the knee, is formed by communications between branches of the intermediate, medial, and lateral femoral cutaneous nerves and the saphenous nerve.

Lateral Femoral Cutaneous Nerve (figs. 33–8, p. 358; 38–9, p. 427). This may arise from the femoral nerve; it is otherwise an independent branch of the lumbar plexus (L2, or L2 and L3, or L1 and L2). It emerges from the lateral border of the psoas major, crosses the iliacus obliquely, and enters the thigh by passing behind the inguinal ligament, near the anterior superior iliac spine. It divides into anterior and posterior branches that supply the skin on the anterior and lateral aspects of the thigh.

Another cutaneous nerve relevant to this region is the femoral branch of the genitofemoral nerve. The latter is a branch of the lumbar plexus. Its femoral branch enters the thigh behind the inguinal ligament, on the lateral side of the femoral artery, pierces the anterior wall of the femoral sheath, and supplies the skin superficial to the femoral triangle, lateral to the territory of the ilioinguinal nerve.

POPLITEAL FOSSA

The popliteal fossa (figs. 21–6 and 21–7) is a diamond-shaped area at the back of the knee. Its upper boundaries are the biceps laterally and the semitendinosus and semimembranosus medially. Its lower boundaries are the plantaris and the lateral head of the gastrocnemius laterally, and the medial head of the gastrocnemius medially. The roof is formed by the popliteal fascia, which is put on the stretch when the knee is extended. The floor is formed, from above downward, by the popliteal surface of the femur, the oblique popliteal ligament of the knee, and the fascia over the popliteus muscle. **The popliteal fossa contains the common peroneal and tibial nerves, the popliteal vessels, the posterior femoral cutaneous nerve, the genicular branch of the obturator nerve, the small saphenous vein, lymph nodes, bursae, and fat.**

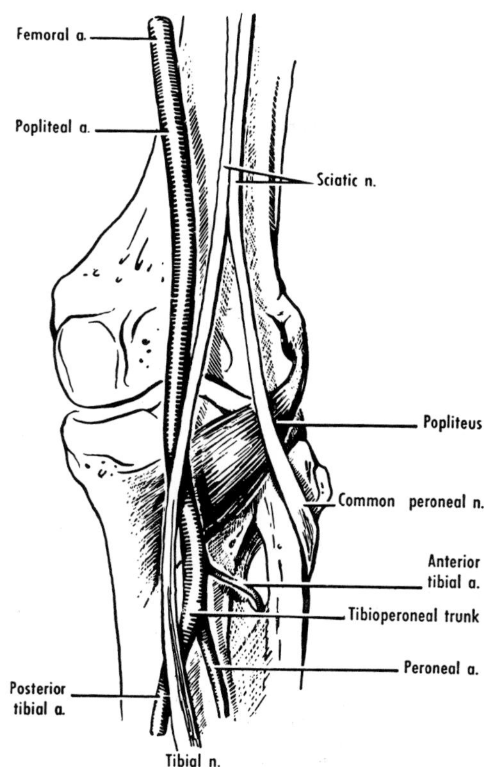

Figure 21–7 Structures in the right popliteal fossa; popliteal veins omitted. Note relationship of the tibial nerve first to the popliteal and then to the posterior tibial artery. Note also that the anterior tibial artery passes laterally, not anteriorly, through an arch in the interosseous membrane.

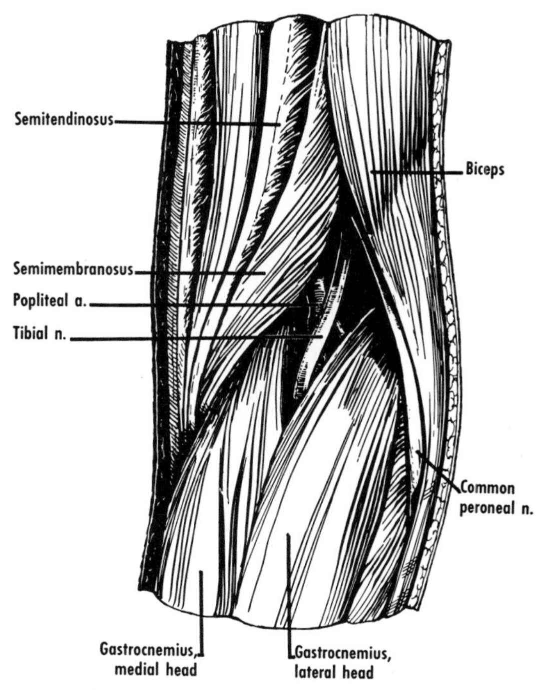

Figure 21–6 The right popliteal fossa.

Vessels

Popliteal Artery. In the popliteal fossa, the popliteal artery lies successively on the popliteal surface of the femur (from which it is separated by fat only), the oblique popliteal ligament, and the popliteus. Posteriorly, it is related successively to the lateral border of the semimembranosus, the popliteal veins and the tibial nerve, and the gastrocnemius and plantaris. The popliteal veins cross behind the artery from lateral to medial side when traced downward, and the tibial nerve crosses behind the popliteal vein, also from lateral to medial side.

BRANCHES (fig. 64–1, p. 766). One of several cutaneous branches *(superficial sural artery)* accompanies the small saphenous vein. Of the various muscular branches, the *sural arteries* constitute the sole blood supply to the gastrocnemius. Five named genicular arteries arise from the popliteal. The *medial* and *lateral superior genicular arteries* pass medially and laterally, respectively, above the corresponding condyle of the femur and head of the gastrocnemius, and deep to the hamstrings. They take part in the anastomosis around the knee joint. The *middle genicular artery* runs directly forward, pierces the oblique popliteal ligament, and enters the knee joint. The *medial* and *lateral inferior genicular arteries* pass medially and laterally, respectively, lying on the popliteus and under cover of the corresponding head of the gastrocnemius. Each then runs deep to the corresponding collateral ligament and ends by taking part in the anastomosis around the knee joint.

The terminal branches of the popliteal artery are the *anterior* and *posterior tibial arteries.* They arise at the lower border of the popliteus. The posterior tibial at this level is sometimes referred to as the tibioperoneal trunk.

The anastomosis around the knee joint is formed by the two superior and the two inferior genicular arteries of the popliteal, by the descending branch of the lateral circumflex artery and the descending genicular artery, and by the circumflex fibular and anterior recurrent arteries from the anterior tibial. Cross connections between the medial and lateral sides occur superficial and deep to the quadriceps, and also deep to the ligamentum patellae.

Popliteal Veins. Usually two in number,[19] these are formed at the knee by the venae comitantes of the anterior and posterior tibial arteries. They are bound closely to the popliteal artery and lie at first posteromedial to it and lateral to the tibial nerve. As they ascend through the popliteal fossa, the popliteal veins lie behind the popliteal artery, between this vessel and the overlying tibial nerve.

Above, the veins are posterolateral to the artery. The popliteal veins have several valves. They receive tributaries corresponding to the branches of the popliteal artery and also the small saphenous vein. They end by passing through the opening in the adductor magnus and becoming the femoral vein (or veins).

Nerves

Common Peroneal (Lateral Popliteal) Nerve (L4 to S2) (fig. 64–8, p. 772). Usually incorporated in the sciatic nerve in the gluteal region and thigh, the common peroneal nerve descends separately through the popliteal fossa. It follows closely, and is in part concealed by, the medial edge of the biceps. It crosses superficially the lateral head of the gastrocnemius to reach the back of the head of the fibula. It then winds laterally around the neck of that bone (where it is often palpable, and where it is susceptible to injury) under cover of the peroneus longus. Here it divides into its terminal branches, the deep peroneal and superficial peroneal nerves,

BRANCHES. While a part of the sciatic nerve, the common peroneal nerve supplies the short head of the biceps and sometimes the knee joint also. In the popliteal fossa, it supplies the knee joint and gives rise to a branch that divides into the *lateral sural cutaneous nerve* (for the skin on the lateral side of the leg) and the *peroneal communicating branch* (which usually joins the medial sural cutaneous nerve to form the sural nerve; see below). At the neck of the fibula, the common peroneal nerve gives off a small recurrent nerve that supplies the knee and tibiofibular joints, and the tibialis anterior. Sometimes the common peroneal nerve gives a branch or branches to the peroneus longus, and to either the tibialis anterior or the extensor digitorum longus, or to both.

Tibial (Medial Popliteal) Nerve (L4 to S3) (p. 214; fig. 64–8, p. 772). This nerve, incorporated (like the common peroneal nerve) in the sciatic nerve in the gluteal region and thigh, also descends separately through the popliteal fossa. It then lies on the popliteus muscle, under cover of the gastrocnemius, and at the lower border of the popliteus passes deep to the fibrous arch of the soleus to the back of the leg. Its further course is described on page 234.

BRANCHES. While it is incorporated in the sciatic nerve, the tibial nerve supplies the semitendinosus, semimembranosus, the long head of the biceps, and the adductor magnus. In the popliteal fossa twigs are given to the knee joint. *Muscular branches* supply the gastrocnemius, soleus, plantaris, popliteus, and tibialis posterior. A branch of the nerve to the popliteus, *the interosseous nerve of the leg,* passes distally on the interosseous membrane and reaches the level of the tibiofibular syndesmosis. The *medial sural cutaneous nerve* descends between the two heads of the gastrocnemius and joins the peroneal communicating branch of the common peroneal nerve to form the *sural nerve.*[20] The medial sural cutaneous nerve may continue as the sural nerve, and rarely the lateral sural cutaneous nerve extends to the foot as the sural nerve. The sural nerve lies on the tendo calcaneus and then, in company with the small saphenous vein, passes to the back of the lateral malleolus. It contributes *lateral calcanean branches* to the skin on the back of the leg and the lateral aspect of the foot and heel, gives twigs to the ankle joint and adjacent tarsal joints, and continues forward to the lateral side of the little toe as the *lateral dorsal cutaneous nerve.* It supplies the little toe and the adjacent digital joints, and communicates with the superficial peroneal nerve. Its cutaneous distribution is occasionally much greater.[21]

JOINTS

Hip Joint[22]

The hip joint is a very strong and stable ball-and-socket joint, formed by the acetabulum of the hip bone and the head of the femur (figs. 18–7 and 18–8, pp. 169 and 170). The bones of the hip joint are surrounded by powerful muscles and are united by a strong, dense capsule. More than half of the head of the femur is contained within the acetabulum, which is deepened by the *acetabular lip,* and is completed below by the *transverse ligament,* which bridges the acetabular notch. The acetabular lip is a densely fibrous or fibrocartilaginous structure that rims the edge of the acetabulum. It continues across the acetabular notch where, by joining more deeply placed fibers, it forms the transverse ligament. This ligament does not fill the notch completely.

The hip joint is admirably constructed to combine relatively free movement with the support and transmission of weight. In standing, for example, the entire weight of the upper part of the body is transmitted through the hip bones to the head and neck of each femur.

The capsule of the hip joint (figs. 21–8 to 21–10) is attached to the margin of the acetabulum. In front, it blends with the acetabular lip, below with the transverse ligament. The capsule extends to the femur, where it is attached chiefly to the intertrochanteric line. Some parts of the capsule are thicker than others, and are termed ligaments.

The strongest and most important ligament is the *iliofemoral ligament.* It is attached above to the anterior inferior iliac spine and to the area behind this, where it fuses with the reflected tendon of the rectus femoris and the adjacent fascia. Below, it is attached to the intertrochanteric line of the femur. The *pubofemoral ligament* extends from the pubic part of the acetabulum and the superior ramus of the pubis horizontally to the lower part of the intertrochanteric line. The part of the capsule between the iliofemoral and pubofemoral ligaments is often thin. The bursa that intervenes between it and the psoas may communicate with the hip joint.

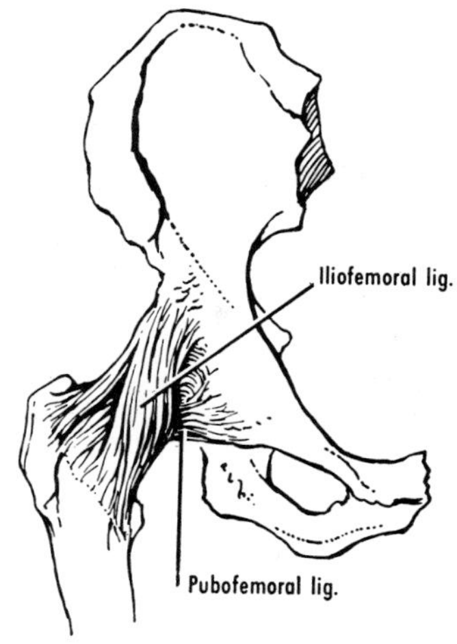

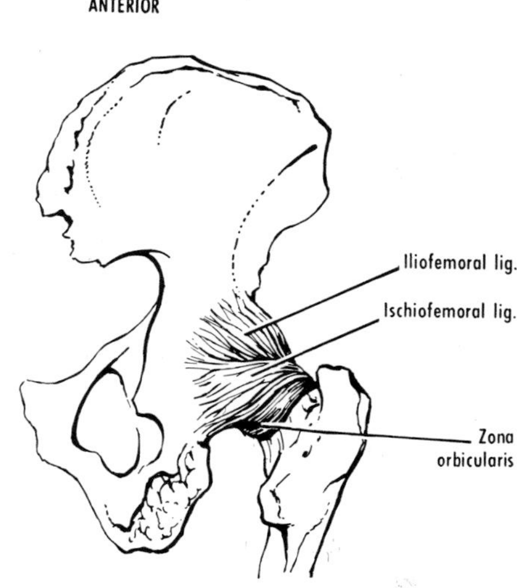

Figure 21–8 The capsule of the hip joint. Note that circular fibers form the zona orbicularis and that the neck of the femur is not completely covered.

The part of the capsule that is attached to the acetabulum behind extends horizontally, across the neck of the femur, and fuses with the iliofemoral ligament. This part is named the *ischiofemoral ligament,* but only the lower part of the ligament reaches the femur directly. It does so by spiraling upward to the junction of the neck with the greater trochanter. The more deeply placed fibers of the ischiofemoral ligament encircle the neck of the femur and form the *zona orbicularis* (fig. 21–8). **The arrangement of the posterior part of the capsule is such that the lateral one-third to one-half of the posterior aspect of the neck of the femur is uncovered,**

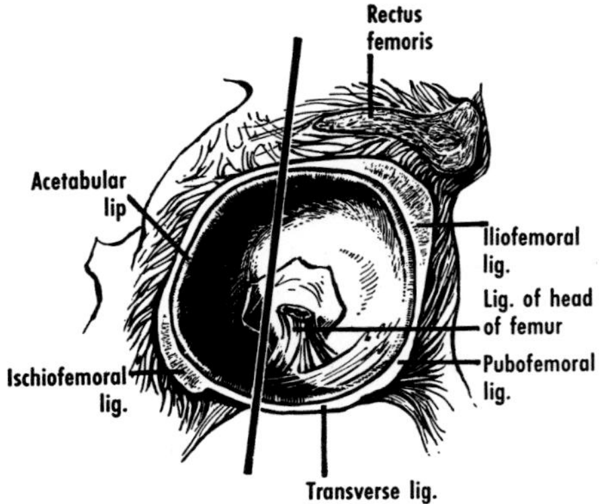

Figure 21–9 The acetabulum and hip joint capsule after removal of the femur. Note how the capsule varies in thickness. The line indicates the plane and position of the section in figure 21–10.

that is, is extracapsular. The extracapsular aspect is covered by the tendon of the obturator externus.

The iliofemoral ligament is notable for its thickness and strength, but much of the rest of the capsule is nearly as thick, so that separate ligaments are often difficult to distinguish.

Wherever capsular fibers are attached to the femur, they tend to be reflected as retinacula along the neck toward the head of the femur, carrying lateral epiphysial vessels and reflections of synovial membrane with them.

The *ligament of the head of the femur* (ligamentum teres) is a flat or triangular band that arises by ischial and pubic roots from the margins of the acetabular notch and from the transverse ligament. It is attached to the pit on the head of the femur. It transmits medial epiphysial vessels to the head of the femur. It is of doubtful mechanical significance.

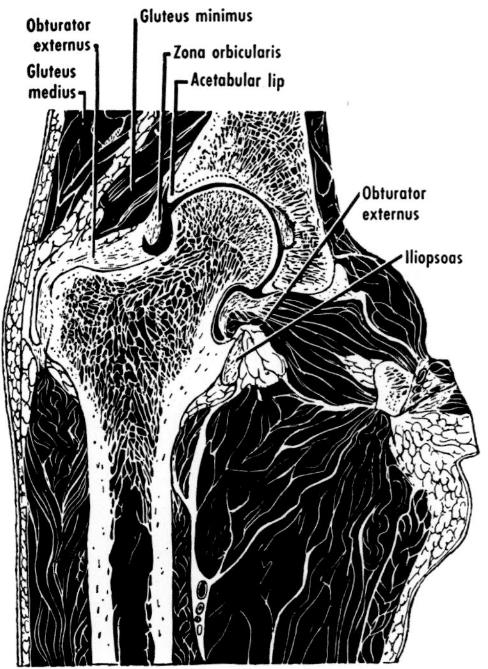

Figure 21–10 Drawing of a coronal section of the hip joint in a plane indicated in figure 21–9.

A thin layer of *synovial membrane* lines the inner aspect of the capsule and is reflected above onto the outer aspect of the acetabular lip, and below onto the neck of the femur. At the transverse ligament it covers the fat that fills the fossa, from which it is prolonged as a tubular investment for the ligament of the head. The synovial membrane forms a visible pouch or recess as it is reflected onto the neck.

NERVE SUPPLY. The hip joint is supplied by the femoral, obturator, and superior gluteal nerves, by the nerve to the quadratus femoris, and by the accessory obturator nerve when that nerve is present.[23]

Movements at the Hip Joint.[24] The movements of the thigh at the hip joint are flexion and extension, abduction and adduction, circumduction, and rotation. The movements of the trunk at the hip joint are equally important, as when a person lifts his trunk from the supine position.

Flexion and extension of the thigh occur about a horizontal axis through the head of the femur. The capsule is slack when the hip is flexed. If the knee is also flexed (to relax the hamstrings), the thigh can be brought up against the anterior abdominal wall. Not all of this movement occurs at the hip joint; some is due to flexion of the vertebral column. During extension, the capsule, especially the iliofemoral ligament, becomes taut. The hip can usually be extended only slightly beyond the vertical. Extension, combined with moderate abduction and medial rotation, locks the hip joint. The joint is mechanically most stable precisely when most weight is borne.

Abduction and adduction occur about an anteroposterior axis through the head of the femur. Abduction is usually somewhat freer than adduction.

Rotation occurs about a vertical axis extending from the head of the femur and approximately through the center of the medial femoral condyle. This axis is not the long axis of the femur. Rotation can be carried through about one-sixth of a circle when the thigh is extended, and more when it is flexed.

In circumduction, the limb swings around a cone, the apex of which is at the head of the femur.

FLEXORS AND EXTENSORS. The iliopsoas, the tensor fasciae latae, and the rectus femoris flex the thigh. They are aided by the adductors, and by the sartorius. The iliopsoas is the strongest of the flexors. The tensor fasciae latae is also a medial rotator. In pure flexion, its rotating action is neutralized by the action of the lateral rotators.

The extensors are the hamstrings and the gluteus maximus. The gluteus maximus is relatively inactive unless forceful extension is necessary.

ABDUCTORS AND ADDUCTORS. The gluteus medius and minimus abduct the thigh. These muscles, or at least their anterior parts, are also medial rotators in pure abduction. Their rotating effects are neutralized by the lateral rotators. The tensor fasciae latae also contracts during abduction, but it is probably acting as a fixator.

The adductors are the three named adductors (longus, brevis, and magnus), aided by the pectineus, and perhaps by the gracilis.

ROTATORS. The tensor fasciae latae, gluteus medius, and gluteus minimus rotate the thigh medially. The role of other muscles is controversial. For example, the adductors contract during medial rotation, but in so doing they are acting as adductors to counteract the abducting action of the true medial rotators.

The lateral rotators are the short muscles of the gluteal region, that is, the obturator internus and externus, superior and inferior gemelli, piriformis, and quadratus femoris, aided by the gluteus maximus.

Knee Joint[25]

The articular surfaces of the knee joint (figs. 21–11 to 21–14) are characterized by their large size and by their complicated and incongruent shapes, which have an important bearing on the movements at this joint.

The articular surfaces are the condyles of the femur, the condyles of the tibia, and the patella (fig. 18–24, p. 181). The femur slants medially at the knee, whereas the tibia is nearly vertical. The angle between the vertical axes of the femur and tibia is about 10 to 12 degrees. The angle is exaggerated in knock-knee (genu valgus).

The articular capsule that invests the joint is usually thin and, in some areas, deficient. It is attached to the femur above the intercondylar fossa, to the margins of the femoral condyles, to the margins of the patella and the ligamentum patellae, and to the margins of the tibial condyles. The patella and ligamentum patellae serve as a capsule in front.

As the capsule extends from the femur to the tibia, it is attached to the outer aspects of the menisci. The part of the capsule between the menisci and the tibia is sometimes termed the coronary ligament. On the medial side of the joint, the capsule is usually fused with the posterior part of the tibial collateral ligament. On the lateral side of the joint, a strong thickening of the capsule (short lateral ligament) extends from the lateral epicondyle to the head of the fibula. It lies deep to the

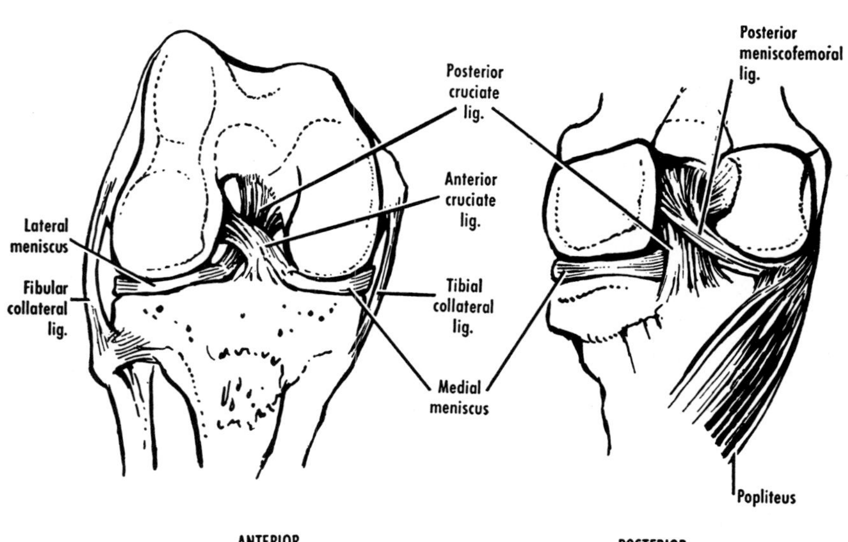

Figure 21–11 Anterior view of a flexed right knee joint in which the transverse ligament is absent. In the posterior view of the right knee joint, note that the popliteus arises in part from the lateral meniscus. The attachments of the collateral ligaments to the menisci are omitted.

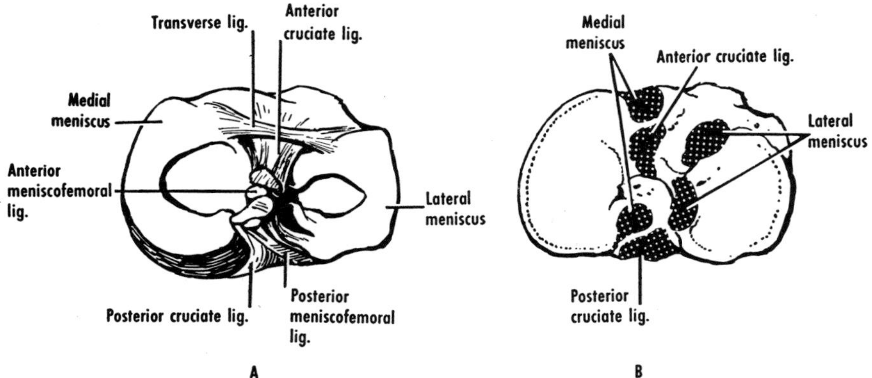

Figure 21–12 *A*, the cruciate ligaments and the menisci from above. Note the differences in size and shape of the menisci. *B*, the right tibia from above, showing attachments of menisci and cruciate ligaments.

fibular collateral ligament, and forms a part of the origin of the popliteus tendon.

The capsule is strengthened on both sides by the aponeurotic expansions of the vasti and the overlying fascia. The combined fascial-aponeurotic sheets are known as the *medial* and the *lateral retinaculum of the patella,* respectively.

At the back of the joint, the capsule forms a thin layer deep to the heads of the gastrocnemius. There is usually a bursa between each head and the capsule. The medial bursa usually communicates with the knee joint cavity and with the adjacent bursa of the semimembranosus tendon. Between the heads of the gastrocnemius, the capsule is greatly thickened by a strong expansion from the tendon of the semimembranosus. This expansion, the *oblique popliteal ligament,* runs upward and laterally, across the back of the joint, to the lateral condyle of the femur and to the lateral head of the gastrocnemius.

EXTRACAPSULAR LIGAMENTS. The *tibial collateral ligament* is a broad, flat band that extends from the medial epicondyle of the femur to the medial surface of the tibia.[26] The ligament lies immediately external to the capsule; its deep portion is attached to the capsule (medial and posterior), to the outer aspect of the medial meniscus, and to the tibia above the groove for the semimembranosus tendon. One or more bursae may be present deep to the ligament. The tibial collateral ligament and the widespread expansions of the semimembranosus tendon are important supports for the medial side of the joint.[27] Together with the fibular collateral ligament, they help to prevent hyperextension of the knee joint.

The *fibular collateral ligament,* more rounded and cordlike, extends from the lateral epicondyle of the femur to the head of the fibula. Its deep aspect is related to the short lateral ligament. Its lower end is covered by the biceps tendon (a bursa intervening), and it is separated from the lateral meniscus by the tendon of the popliteus muscle. The important stabilizing structures on the lateral side of the joint are the fibular collateral ligament, the biceps tendon, the popliteus tendon, and the iliotibial tract.[28]

INTRA-ARTICULAR LIGAMENTS. These consist of the cruciate ligaments and the menisci. The tendon of the popliteus muscle is also intra-articular in a part of its course.

The *anterior* and *posterior cruciate ligaments* extend from the bone adjacent to the intercondylar fossa of the femur to the

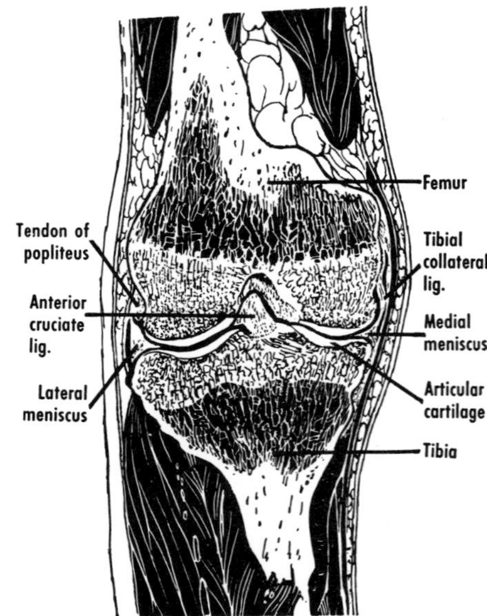

Figure 21–13 Coronal section of knee joint.

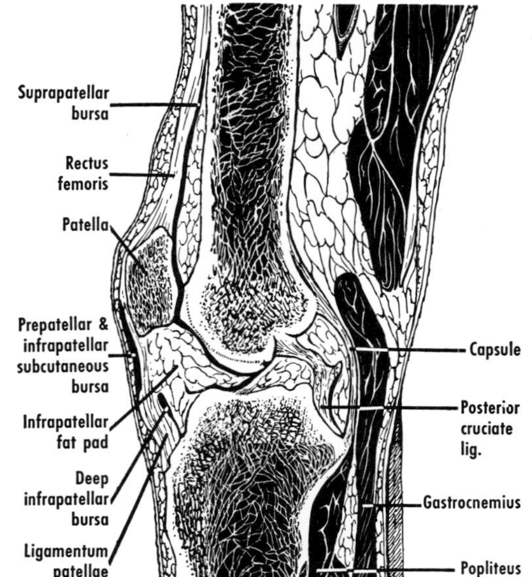

Figure 21–14 Sagittal section of knee joint. The width of the articular and bursal cavities is exaggerated.

tibia, in front of and behind the intercondylar eminence, respectively. The ligaments are named anterior and posterior according to their tibial attachments. Behind, they are continous with the capsule, from which the synovial membrane is prolonged around them. There may be a bursa between the ligaments. The two ligaments cross each other somewhat like the limbs of the letter **X**.*

The *medial* and *lateral menisci* or *semilunar cartilages* are crescent-shaped, densely fibrous structures that rest on the articulating surfaces of the upper end of the tibia. They deepen the concavity of the tibial condyles, act as cushions or shock absorbers, and facilitate lubrication. Each meniscus is wedge-shaped in section, with a thick, external aspect that is fused with the capsule, and a thin, free internal border. The ends or horns of the menisci are attached to the tibia, in front of and behind the intercondylar eminence. The *transverse ligament* is a fibrous band that connects the menisci in front. It is sometimes absent. The medial meniscus forms a semicircle, whereas the lateral forms nearly a circle. The lateral meniscus is often grooved posterolaterally by the tendon of the popliteus, which arises in part from its posterior part. Variations in shape and arrangement of the menisci are common,[29] and oblique intermeniscal ligaments may be present.[30] Rarely one of the menisci, usually the lateral, is discoid in shape.[31]

Commonly a fibrous band extends upward and medially from the posterior part of the lateral meniscus to the medial femoral condyle. It may pass behind the posterior cruciate ligaments as the *posterior meniscofemoral ligament,* or in front of it as the *anterior meniscofemoral ligament,* or it may split around it.[32] The lateral meniscus has less of a capsular attachment. It is freer to move, and it is likely that the popliteus pulls this meniscus backward during flexion, so that it follows the backward movement of the femur.

SYNOVIAL MEMBRANE. The cavity of the knee joint (fig. 18–25*E*, p. 182) is the largest in the body. Its synovial membrane lines the fibrous capsule and, like it, is attached to the edges of the menisci. It is reflected forward from the posterior wall of the joint onto the cruciate ligaments, forming a common covering for both ligaments. Between the patella and the tibia, it lines the *infrapatellar fat pad.* A median fold, or folds, of synovial tissue, the *infrapatellar synovial fold,* extends backward from the fat pad to the intercondylar fossa of the femur. An *alar fold* on each side diverges from the median fold to the lateral edges of the patella.

The joint cavity is prolonged for 5 cm or more above the patella as the *suprapatellar bursa.* This bursa lies deep to the articularis genus and vastus intermedius; at times it may extend halfway up the front of the femur. It may also be prolonged distally along the tendon of the popliteus as the *subpopliteal recess.*

NERVE SUPPLY.[33] The knee joint is supplied by twigs from the muscular branches of the femoral nerve and from the saphenous nerve. It receives a variable number of branches from the tibial, common peroneal, and obturator nerves.

Movements at the Knee Joint.[34] That the knee joint is usually considered to be a hinge joint implies that its movements are flexion and extension about a horizontal axis. **However, the knee joint is more truly a condylar joint, and its movements are complex. The shapes and curvatures of the articular surfaces are such that hinge movements are combined with gliding and rolling, and with rotation about a vertical axis.** In the extended limb, the axis of rotation extends from the head of the femur to the medial condyle of the tibia. Hence the lateral condyle of the femur swings around this vertical axis through the medial condyle.

When the thigh is flexed at the knee and the leg remains fixed (as when assuming the sitting position), the thigh rotates laterally during the first part of flexion, and the femur rolls backward on the tibia. Conversely, when the thigh is extended, it ro-

*"Represent the ligaments by your lower limbs while standing, i.e., cross your right leg (right anterior cruciate ligament) in front of your left. Rotate your trunk to right and left" (Mainland).

tates medially during the last part of extension. At the end of extension, the joint is said to be "locked" or "screwed home." This may not be truly the case, but at the end of extension the articular surfaces are more congruent, and the joint is most stable mechanically.

If the thigh is fixed and the leg is free to move (as when swinging the leg while sitting in a chair), the tibia rotates on the femur. Lateral rotation of the tibia is the equivalent of medial rotation of the femur. Hence, the first part of flexion of the leg at the knee is accompanied by medial rotation of the leg, and the terminal part of extension is accompanied by lateral rotation of the leg.

When the knee is fully extended, all ligaments are taut. The posterior cruciate ligament limits forward slipping of the femur on the tibia; the anterior cruciate ligament limits backward slipping.[35] In full flexion, the fibular collateral ligament is relaxed; it tightens with rotation in either direction.

Muscles. The quadriceps femoris extends the leg. The flexors are the hamstrings, aided by the gracilis, sartorius, and gastrocnemius. It is unlikely that the popliteus contributes toward flexion of the knee.

The biceps femoris is the principal lateral rotator of the leg on the thigh, especially when the knee is flexed; the semitendinosus is a medial rotator. The popliteus, acting from a fixed femur, is also a medial rotator.[36] Acting from a fixed tibia, it can rotate the femur laterally, and is considered to be important in "unlocking" the joint.

Tibiofibular Joint

This is a small, plane joint between the facet on the head of the fibula and that on the back of the lateral condyle of the tibia. It is frequently referred to as the superior, or proximal, tibiofibular joint. Thickenings of its capsule in front and behind form the *anterior* and *posterior ligaments of the head of the fibula,* respectively.

The joint cavity may communicate with the subpopliteal recess, and thereby with the cavity of the knee joint, or it may communicate directly with the knee joint cavity.

REFERENCES

1. E. B. Kaplan, J. Bone Jt Surg., *40A*:817, 1958.
2. W. J. Lytle, Ann. R. Coll. Surg. Engl., *21*:244, 1957.
3. L. Olivieri, Quad. Anat. prat., *10*:470, 1955.
4. B. F. Martin, J. Anat., Lond., *102*:345, 1968.
5. R. S. Sneath, J. Anat., Lond., *89*:550, 1955. J. L. Marshall, F. G. Girgis, and R. R. Zelko, J. Bone Jt Surg., *54A*:1444, 1972.
6. A. J. E. Cave and C. J. Porteous, Ann. R. Coll. Surg. Engl., *24*:251, 1959. See also A. Faller, Acta anat., *6*:92, 1948; E. B. Kaplan, Bull. Hosp. Jt Dis., *18*:51, 1957.
7. T. Sugihara, Okajimas Folía anat. jap., *28*:377, 1956.
8. H. T. Weathersby, J. Bone Jt Surg., *41A*:261, 1959.
9. J. V. Basmajian, Anat. Rec., *132*:127, 1958. R. D. Keagy, J. Brumlik, and J. J. Bergan, J. Bone Jt Surg., *48A*:1377, 1966. P. Fitzgerald, Irish J. med. Sci., *2*:31, 1969.
10. E. S. R. Hughes and S. Sunderland, Anat. Rec., *96*:439, 1946. See also O. J. Lewis, J. Anat., Lond., *92*:587, 1958.
11. D. A. Brewerton, Ann. phys. Med., *2*:164, 1955. M. Ravaglia, Chir. Org. Mov., *44*:498, 1957. J. V. Basmajian, T. P. Harden, and E. M. Regenos, Anat. Rec., *172*:15, 1972.
12. L. J. A. DiDio, A. Zappalá, and W. P. Carney, Acta anat., *67*:1, 1967.
13. C. E. Johnson, J. V. Basmajian, and W. Dasher, Anat. Rec., *173*:127, 1972.
14. H. D. Senior, Amer. J. Anat., *33*:243, 1924. G. D. Williams *et al.*, Anat. Rec., *46*:273, 1930. J. A. Keen, Amer. J. Anat., *108*:245, 1961.
15. C. M. Charles *et al.*, Anat. Rec., *46*:125, 1930. E. A. Edwards and J. D. Robuck, Surg. Gynec. Obstet., *85*:547, 1947.
16. F. R. Wilde, Brit. J. Surg., *39*:97, 1951.
17. J. Pürner, Anat. Anz., *129*:114, 1971.
18. H. Sirang, Anat. Anz., *130*:158, 1972.
19. A. F. Williams, Surg. Gynec. Obstet., *97*:769, 1953. See also H. Dodd and F. B. Cockett, *The Pathology and Surgery of the Veins of the Lower Limb*, Livingstone, Edinburgh, 1956.
20. D. F. Huelke, Amer. J. phys. Anthrop., *15*:137, 1957; Anat. Rec., *132*:81, 1958. See also C. Kosinski, J. Anat, Lond., *60*:274, 1926; D. D. Williams, Anat. Rec., *120*:533, 1954.
21. J. R. Barbour, Med. J. Aust., *1*:275, 1947. J. J. Joyce and M. Harty, Clin. Orthopaed., *98*:27,1974.
22. F. G. St. Clair Strange, *The Hip*, Heinemann, London, 1965.
23. E. Gardner, Anat. Rec., *101*:353, 1948. L. G. Wertheimer, J. Bone Jt Surg., *34A*:477, 1952.
24. W. H. Roberts, Anat. Rec., *147*:321, 1963.
25. M. Harty and J. J. Joyce, AAOS Instructional Course Lectures, *20*:206, 1971.
26. A. Jost, Arch. Anat., Strasbourg, *1*:245, 1922. O. C. Brantigan and A. F. Voshell, J. Bone Jt Surg., *25*:121, 1943. R. J. Last, J. Bone Jt Surg., *32B*:93, 1950. C. H. Barnett, J. Anat., Lond., *88*:59, 1954. E. B. Kaplan, Surg. Gynec. Obstet., *104*:346, 1957.
27. L. F. Warren, J. L. Marshall, and F. Girgis, J. Bone Jt Surg., *56A*:665, 1974.
28. E. B. Kaplan, Bull. Hosp. Jt Dis., *18*:51, 1957.
29. C. M. Charles, Anat. Rec., *63*:355, 1935.
30. A. Lahlaidi, C. R. Ass. Anat., *56*:1046, 1972.
31. E. B. Kaplan, J. Bone Jt Surg., *39A*:77, 1957. J. A. Ross, I. C. K. Tough, and T. A. English, J. Bone Jt Surg., *40B*:262, 1958.
32. E. B. Kaplan, Bull. Hosp. Jt Dis., *17*:176, 1956. L. Heller, Anat. Rec., *130*:314, 1958. L. Candiollo and G. Gautero, Acta anat., *38*:304, 1959. L. Heller and J. Langman, J. Bone Jt Surg., *46B*:307, 1964.
33. E. Gardner, Anat. Rec., *101*:109, 1948.
34. M. A. MacConaill, J. Anat., Lond., *66*:210, 1932. K.-H. Knese, Z. Anat. EntwGesch., *115*:287, 1950. C. H. Barnett, J. Anat., Lond., *87*:91, 1953. J. W. Smith, J. Anat., Lond., *90*:236, 1956.
35. O. C. Brantigan and A. F. Voshell, J. Bone Jt Surg., *23*:44, 1941. See also R. W. Haines, J. Anat., Lond., *75*:373, 1941; J. C. Kennedy, H. W. Weinberg, and A. S. Wilson, J. Bone Jt Surg., *56A*:223, 1974.
36. C. H. Barnett and A. T. Richardson, Ann. phys. Med., *1*:177, 1952. See also R. J. Last, J. Bone Jt Surg., *32B*:93, 1950.

THE LEG

22

The skin and subcutaneous tissue of the leg are supplied by the saphenous, posterior femoral cutaneous, medial sural cutaneous, lateral sural cutaneous, sural, and superficial peroneal nerves, and sometimes by the obturator also.

FASCIA OF LEG

The fascia of the leg is continuous with the fascia lata at their common attachments to the condyles of the tibia and to the head of the fibula. It receives expansions from the biceps tendon laterally, and from the tendons of the semitendinosus, gracilis, and sartorius medially. The semimembranosus contributes to the fascia over the popliteus.

The fascia is attached to the anterior border of the tibia; it encircles the leg to reach the medial border of the tibia. Inward extensions from the deep surface of the fascia to the anterior and to the posterior border of the fibula form the *anterior* and the *posterior intermuscular septum*, respectively. Thus, three compartments are formed, anterior, lateral, and posterior (fig. 22–1). The posterior compartment is subdivided by a fascial septum, the deep transverse fascia of the leg, which extends from the medial border of the tibia to the posterior border of the fibula. Above, this septum is attached to the soleal line. Below, it becomes a strong layer, which merges with the outer layer of fascia and contributes to the flexor retinaculum.

The fascia of the anterior compartment is thick and dense; it forms a tight investment for the muscles and gives origin to them in part. The fascia of the upper part of the lateral compartment is likewise dense. The tight fascial investment of these muscles helps to prevent undue swelling of the muscles during exercise and thereby facilitates venous return.

Below, at the ankle, the fascia is continuous with that of the foot. At the level of the malleoli it forms three retinacula, flexor, extensor, and peroneal, which bind down the respective tendons.

FRONT OF LEG

Muscles

The muscles of the front of the leg are the tibialis anterior, extensor hallucis longus, extensor digitorum longus, and peroneus tertius. They arise from bone, from the investing fascia, from the interosseous membrane, and from the adjacent intermuscular septum. They are supplied by the deep peroneal nerve. The tibialis anterior usually, and the extensor digitorum longus often, receive a branch from the common peroneal nerve also.

These muscles dorsiflex the ankle. In addition, some invert the foot and others evert it. **In dorsiflexion, the foot is bent at the ankle so that the dorsum or upper aspect of the foot moves away from the ground (toward the front of the leg). In plantar flexion, the opposite movement occurs. In inversion, the sole of the foot is turned medially; in eversion, it is turned laterally. Dorsiflexion and plantar flexion occur at the ankle joint; inversion and eversion at the subtalar and transverse tarsal joints.**

Tibialis Anterior. This muscle arises from the lateral condyle of the tibia, from the upper two-thirds of the lateral surface of its shaft, and from the interosseous membrane. It is inserted into the medial side of the medial cuneiform and the base of the first metatarsal.

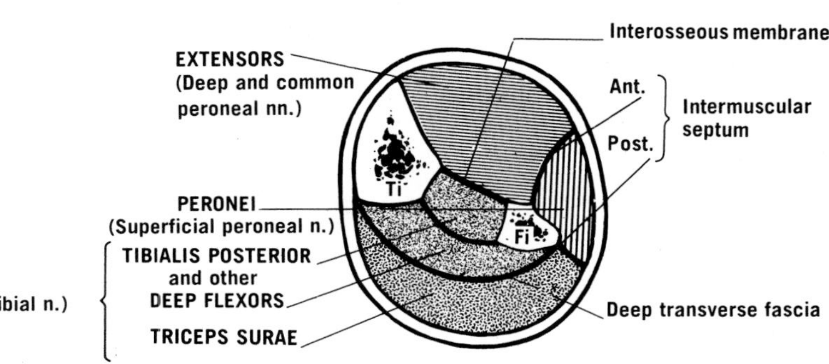

Figure 22–1 Diagram of a horizontal section approximately through middle of leg, showing arrangement of muscle groups. The posterior group is supplied by the tibial nerve, the lateral by the superficial peroneal, and the anterior by the common and the deep peroneal. Note that the fibula is on a plane posterior to that of the tibia. (Compare fig. 22–10.)

ACTION. It dorsiflexes and inverts the foot.

Extensor Digitorum Longus. This arises from the lateral condyle of the tibia, from the upper three-fourths of the anterior surface of the shaft of the fibula, and from the interosseous membrane. Its tendon divides into four tendons in front of the ankle; these are inserted into the lateral four toes. Each forms a membranous expansion over the dorsum of a metatarsophalangeal joint, where it fuses with the capsule. Near the proximal interphalangeal joint, the expansion divides into three slips. The central part of the aponeurosis continues distally to the base of the middle phalanx. The collateral slips continue to the base of the distal phalanx.

ACTION. The extensor digitorum longus extends the toes, mainly at the metatarsophalangeal joints. It also dorsiflexes and everts the foot.

Peroneus Tertius. This is the lower lateral part of the extensor digitorum longus. Its size and the degree of its separation from the long extensor are variable. It arises from the lower fourth of the anterior surface of the fibula and from the interosseous membrane. It may insert into the base of the fifth (or fourth) metatarsal, but more often it becomes continuous with the fascia.

Extensor Hallucis Longus. This has a narrow origin from the middle half of the anterior surface of the fibula and from the interosseous membrane. It is inserted into the superior aspect of the base of the distal phalanx of the big toe.

ACTION. It extends the big toe and aids in dorsiflexion of the foot.

Extensor Retinacula (figs. 22–2 to 22–4). These are thickenings of the fascia at the ankle and on the dorsum of the foot. They bind the tendons in place and prevent "bowstringing." Synovial sheaths of the extensor tendons descend behind the retinacula (fig. 23–1, p. 237).

The extensor retinacula are termed superior and inferior. The *superior extensor retinaculum* is an indistinct thickening that extends between the anterior borders of the tibia and fibula, just above the ankle. The *inferior extensor retinaculum* is ≻ shaped. Its stem or apex springs from a medial root in the tarsal canal, and from intermediate and lateral roots in the sinus tarsi (fig. 22–4).[1] The stem forms a loop for the extensor tendons and then divides into upper and lower bands. The upper band runs to the medial malleolus; the lower passes to the medial side of the foot, where it spreads out in the fascia. Either of the two bands that diverge from the stem may lie superficial or deep to the tibialis anterior tendon. Septa extend inward from the bands and stem, and form slings. These slings prevent

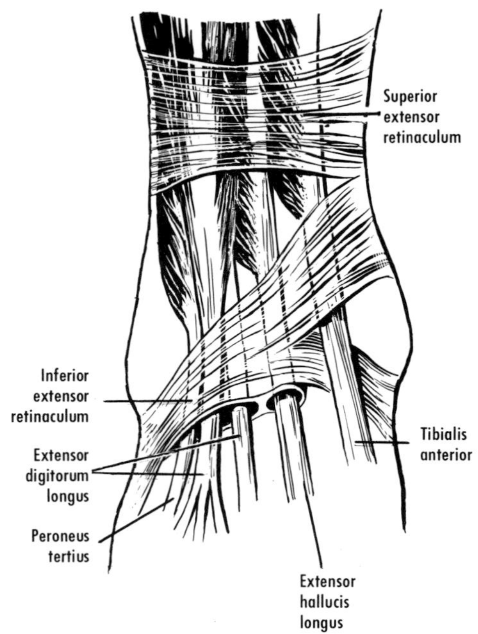

Figure 22–2 The extensor retinacula, right foot.

medial displacement of the long extensors during inversion of the foot.

Vessels and Nerves

Anterior Tibial Artery. The smaller of the terminal divisions of the popliteal, this artery is deeply placed in its upper third but is readily accessible distally. It begins at the

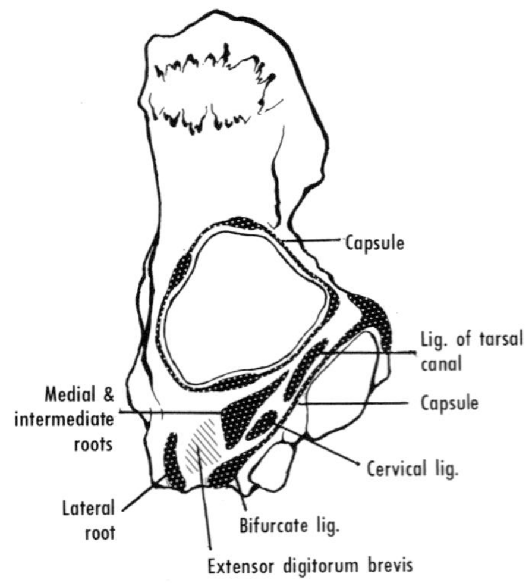

Figure 22–3 The right calcaneus, upper aspect, showing muscular and ligamentous attachments in the floor of the sinus tarsi and tarsal canal.

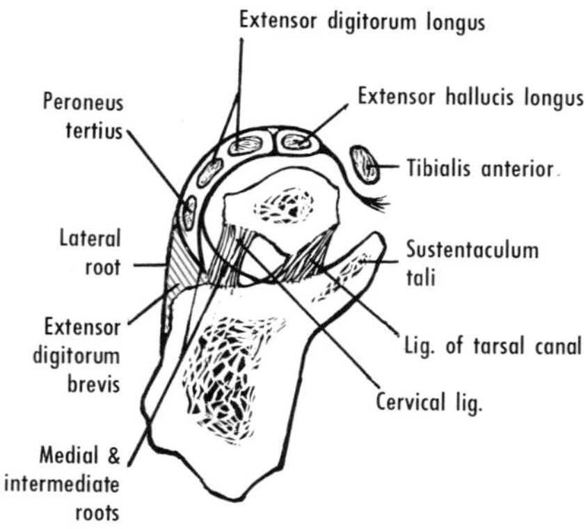

Figure 22–4 Schematic section (approximately coronal) of right calcaneus and talus. This is a composite representation, not all of the structures being present in any one section.

lower border of the popliteus, passes laterally through the fibrous arch of the tibialis posterior (between its tibial and fibular heads), and then passes through the fibrous arch of the interosseous membrane, to meet its companion nerve. It descends on the front of the interosseous membrane, but in the lower part of its course it lies directly on the tibia. The anterior tibial artery is accompanied by two veins and by the deep peroneal nerve. On the dorsum of the foot, it continues into an arterial network. It often has a well-defined direct continuation, the dorsalis pedis artery (p. 240).

BRANCHES. These supply the adjacent muscles and skin of the front of the leg. The *circumflex fibular branch* (occasional) usually arises from the posterior tibial. The *posterior tibial recurrent artery* (inconstant) ascends in front of the popliteus. The *anterior tibial recurrent artery* ascends in the tibialis anterior and contributes to the anastomosis around the knee joint. The *medial anterior malleolar artery* arises above the ankle joint. It descends to the medial malleolus, where it anastomoses with branches of the posterior tibial and medial plantar arteries. The *lateral anterior malleolar artery* passes behind the extensor digitorum longus and anastomoses with the perforating branch of the peroneal artery. Both malleolar arteries contribute to the networks around the ankle joint.

Deep Peroneal Nerve. This is one of the terminal branches of the common peroneal nerve. At the level of the neck of the fibula, the common peroneal passes through a fibrous arch in the aponeurosis of the soleus, between the soleus and the peroneus longus,[2] and divides into deep and superficial peroneal nerves. The deep continues around the neck of the fibula, pierces the anterior intermuscular septum and the extensor digitorum longus, and then descends on the interosseous membrane. It meets the anterior tibial artery, and both pass deep to the extensor retinacula. In the foot, where the nerve lies about midway between the malleoli, it divides into its terminal branches, medial and lateral.

BRANCHES. Branches are given to the tibialis anterior, extensor hallucis longus, extensor digitorum longus, peroneus tertius, and to the ankle joint. Of the terminal divisions, the medial, or digital, branch lies lateral to the dorsalis pedis artery and divides into *dorsal digital nerves* for the adjacent sides of the first and second toes.[3] It also gives some articular twigs and communicates with the superficial peroneal nerve. The lateral branch passes laterally across the tarsus, deep to the extensor digitorum brevis. It ends in an enlargement (compare the posterior interosseous nerve at the wrist) that gives several branches to the extensor digitorum brevis and adjacent joints. It may also send some twigs (perhaps afferent) to the first three dorsal interossei.

LATERAL SIDE OF LEG

Muscles

The two muscles of this group, the peroneus longus and the peroneus brevis, lie between the anterior and posterior intermuscular septa, lateral to the fibula. Like the muscles of the anterior group, they are tightly invested by fascia. They arise in part from this fascia and the adjacent septa. Both are supplied by the superficial peroneal nerve; the peroneus longus is often supplied by the common peroneal also.

Peroneus Longus. The peroneus longus arises from the lateral condyle of the tibia, from the head of the fibula, and from the upper two-thirds of the lateral surface of the shaft of the fibula.

Its tendon curves behind the lateral malleolus in a common synovial sheath with the peroneus brevis. Then, after passing through a notch on the lateral side of the cuboid, it crosses the sole of the foot obliquely (fig. 23–11, p. 245) to be inserted into the lateral side of the medial cuneiform and the adjacent aspect of the base of the first metatarsal. Below the cuboid the tendon contains either a sesamoid bone or a fibrocartilaginous thickening. In its course across the foot, the tendon has a distal synovial sheath within a fibrous sheath formed by a prolongation of the long plantar ligament.

ACTION. The peroneus longus is a plantar flexor of the foot and an evertor. It acts on the medial side of the foot by depressing the first metatarsal; it thereby enables the inverted foot to remain plantigrade.

Peroneus Brevis. This muscle lies deep to the longus; it arises from the lower two-thirds of the lateral surface of the fibula. Its tendon winds around the back of the lateral malleolus, and then turns forward, above the peroneal trochlea, to the tuberosity of the fifth metatarsal. Commonly a small slip from the tendon either joins the extensor digitorum longus tendon to the little toe or continues forward to the proximal phalanx.

ACTION. It everts the foot.

Peroneal Retinacula. These are superior and inferior. The *superior peroneal retinaculum* extends from the malleolus to the calcaneus and binds the peroneal tendons behind the malleolus. The *inferior peroneal retinaculum* binds the tendons to the lateral side of the calcaneus.

Nerves

Superficial Peroneal (Musculocutaneous) Nerve. This is one of the terminal branches of the common peroneal. It descends in front of the fibula and between the perinei and the extensor digitorum longus. In the lower part of the leg, it divides into medial and intermediate dorsal cutaneous branches.

BRANCHES. *Muscular branches* are given to the peroneus longus and brevis. The branch to the latter is often prolonged to the extensor digitorum brevis and adjacent joints and ligaments, and is termed the accessory deep peroneal nerve.[4]

The two terminal cutaneous divisions pass in front of the extensor retinacula to the dorsum of the foot. The *medial dorsal cutaneous nerve* divides into a branch for the medial side of the big toe, a branch that communicates with the deep peroneal nerve, and a branch that divides into dorsal digital branches for the adjacent sides of the second and third toes. The *intermediate dorsal cutaneous nerve* divides into two branches, each of which divides into dorsal digital nerves for the adjacent sides of the third and fourth, and the fourth and fifth toes. The distribution of the cutaneous branches, however, is subject to considerable variation. The beds of the nails and the tips of the toes are supplied by the plantar digital branches of the medial and lateral plantar nerves.

BACK OF LEG

Muscles (figs. 22–5 to 22–8)

The superficial muscles of this group are the gastrocnemius and soleus, which form the triceps surae, and the plantaris. The deep muscles are the popliteus, tibialis posterior, flexor digitorum longus, and flexor hallucis longus. The last three muscles, which are separated from the superficial group by the deep transverse fascia of the leg, have in addition to their bony origins an extensive attachment to the underlying interosseous membrane.

NERVE SUPPLY. **All the muscles of the back of the leg are supplied by the tibial nerve.**

Triceps Surae. This consists of the gastrocnemius and the soleus.

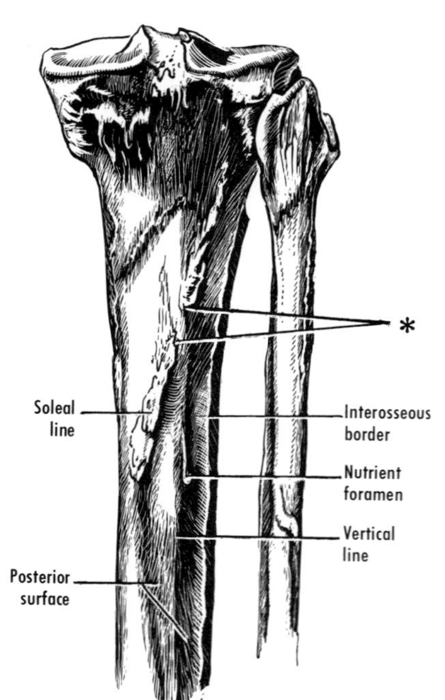

Figure 22–5 The upper parts of the tibia and the fibula, from behind. Note the interruption in the soleal line (*) for the groove for the nutrient artery to the tibia. Based on Shellshear and Macintosh.[1]

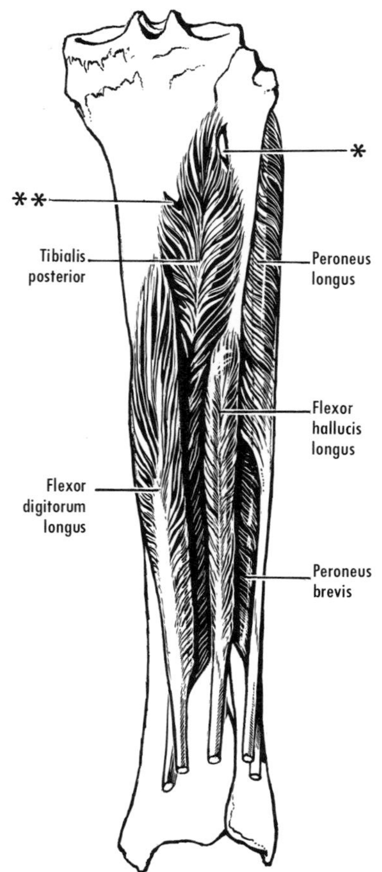

Figure 22-6 Origins of deep muscles of calf. Note the fibrous arch in the tibialis posterior for the anterior tibial vessels (*). Note also the arch in the tibialis posterior for the nutrient artery to the tibia (**). Based on Shellshear and Macintosh.[1]

GASTROCNEMIUS. The gastrocnemius has two large heads that arise from the lower end of the femur and end at about the middle of the leg in a common tendon. The *lateral head* arises from the upper part of the lateral aspect of the lateral condyle of the femur. A bursa often intervenes between the lateral head and the joint capsule. The lateral head may contain a sesamoid bone, the *fabella* (fig. 18-25*B*, p. 182), which may have a facet for articulation with the lateral condyle of the femur, and which is connected to the fibula by the fabellofibular ligament.[5] The *medial head*, which rarely contains a sesamoid, arises from the popliteal surface of the femur above the medial condyle and from the upper part of the medial condyle near the adductor tubercle. This head is commonly separated from the capsule by a bursa, which usually communicates with the joint cavity and with an adjacent bursa of the semimembranosus.

The bellies of the muscle converge upon a membranous sheet, which fuses with the subjacent tendon of the soleus to form the *tendo calcaneus.**

*The synonym *tendo Achillis* was first used in 1693.[6] Legend has it that his mother held Achilles by his heels when dipping him in the river Styx to make him invulnerable. At the siege of Troy, Achilles was mortally wounded by an arrow in his heel.

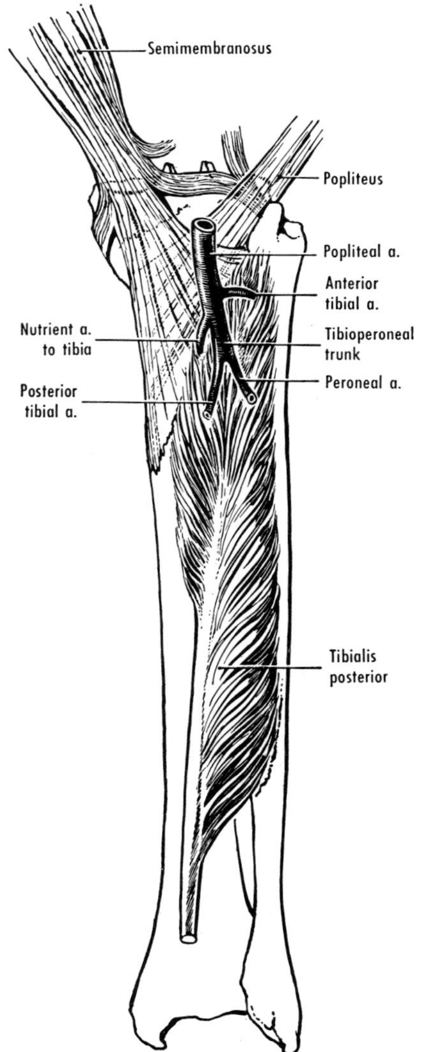

Figure 22-7 The divisions of the popliteal artery and their relations to the tibialis posterior. Based on Shellshear and Macintosh.[1]

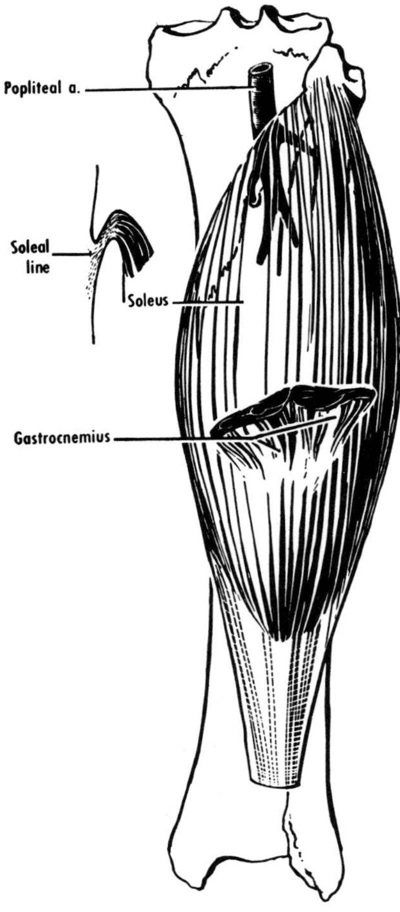

Figure 22–8 The soleus superimposed on the structures shown in figure 22–7. Note that its apex is at the fibula. Note also from the inset that the soleus ascends from the soleal line and then turns downward. Thus, the upper border of the muscle is at a higher level than the soleal line. Based on Shellshear and Macintosh.[1]

The gastrocnemius is one of the few muscles with but one source of blood supply. Each head is supplied by a branch of the popliteal artery.[7]

SOLEUS.[8] The soleus is a thick, flat muscle that arises mainly by a cone-shaped origin from the back of the head and from the upper part of the posterior surface of the fibula, and from the posterior intermuscular septum. It has an additional origin from a *tendinous arch* that extends from a tubercle on the neck of the fibula to the soleal line of the tibia, and, to a varying extent, from the medial border of the tibia. The tendinous arch lies behind the lower part of the popliteus muscle, the popliteal vessels, and the tibial nerve.

The tendon of the soleus fuses with the tendinous sheets of the gastrocnemius to form the tendo calcaneus. This thick, strong tendon, the fibers of which are arranged in a somewhat spiral fashion, is inserted into the posterior aspect of the calcaneus. A bursa lies between the tendon and the bone.

ACTION. **The triceps surae is an important postural and locomotor muscle. It is a plantar flexor of the foot, it acts mainly on the lateral side of the foot, and it tends to invert the foot.** The triceps surae is important in walking, running, jumping, and dancing. Its use depends partly upon the fact that the gastrocnemius is a two-joint muscle; it can flex the knee as well as plantar flex the foot. If, however, the knee is fully flexed, the gastrocnemius is so shortened that it can shorten no more. It cannot then plantar flex the foot. This is carried out by the deep flexor muscles. If the foot is fully plantar flexed, the gastrocnemius cannot flex the knee. Conversely, by ligamentous action, if the knee is fully extended, the lengthened gastrocnemius tends to pull the foot into plantar flexion. And if the foot is fully dorsiflexed, the lengthened gastrocnemius tends to pull the knee into flexion.

ANKLE JERK. **The ankle jerk is a reflex twitch of the triceps surae induced by tapping the tendo calcaneus. The reflex center is in the fifth lumbar or first sacral segment of the spinal cord.**

Plantaris. This muscle is variable in size and extent, and may be absent.[9] It arises from the lower part of the lateral supracondylar line and from the popliteal surface of the femur, just above the lateral head of the gastrocnemius. Its thin, membranous tendon descends between the gastrocnemius and the soleus to be attached to the medial side of the tendo calcaneus or to the posterior aspect of the calcaneus. Sometimes it spreads diffusely into the fascia at the medial side of the ankle. It is supplied by the tibial nerve and its action is unimportant.

Popliteus. The popliteus has two origins, femoral and meniscal.[10] The femoral arises by a strong rounded tendon from the pit at the end of the groove on the lateral aspect of the lateral condyle of the femur (it occupies the groove during flexion). It is joined by a band from the head of the fibula, and the two have an expansion to the capsule known as the *arcuate popliteal ligament.*[11] The meniscal origin arises by tendinous fibers from the back of the lateral meniscus. The muscle passes downward and medially to the triangular surface of the tibia above the soleal line.

ACTION.[12] It is a prime medial rotator of the tibia; it is also a lateral rotator of the femur when the tibia is fixed. It pulls the lateral meniscus backward at the beginning of flexion. Any flexing action of the popliteus is of doubtful importance. The muscle is active during crouching; presumably it prevents the femur from sliding forward on the tibia.

Flexor Digitorum Longus. This arises from the middle half of the posterior surface of the tibia, below the soleal line. Its tendon descends behind the medial malleolus,[13] and then turns forward, at or just below the medial edge of the sustentaculum tali. In the sole it lies below the flexor hallucis longus, from which it receives a tendinous slip. The tendon divides into four parts, one for each of the lateral four toes. The quadratus plantae is inserted into the tendon near its division (p. 238). The lumbricals arise at the points of division. Each tendon enters a fibrous sheath and courses forward to its insertion on the distal phalanx. Within the sheath it perforates the accompanying short flexor tendon. Both tendons are invested by a synovial sheath, and both are connected to phalanges by vincula.[14]

ACTION. It flexes the distal phalanges of the lateral four toes.

Flexor Hallucis Longus. This arises from the lower two-thirds of the posterior surface of the fibula and from the posterior intermuscular septum. Its tendon passes deep to the flexor retinaculum. It occupies first the groove on the posterior surface of the talus and then the groove on the under surface of the sustentaculum tali. It runs obliquely across the sole of the foot, above the tendon of the flexor digitorum longus, to which it gives a slip. It is inserted on the inferior aspect of the base of the distal phalanx of the big toe. In extending forward, it crosses the head of the flexor hallucis brevis, and then runs between the two sesamoids, into a fibrous sheath below the proximal phalanx. The synovial sheaths of this and the other flexor tendons are described on page 236.

ACTION. It flexes the distal phalanx of the big toe.

Tibialis Posterior. **This muscle, an understanding of which is the key to an understanding of the leg, is deeply placed beneath the two long flexor muscles, and has an extensive origin from the interosseous membrane, the fibula, and the tibia.** The posterior surface of the fibula is subdivided by the medial crest into a posterior part for the flexor hallucis longus and an anterior part for the tibialis posterior. The fibular origin of the latter extends up to and includes the capsule of the tibiofibular joint. It is to be emphasized that the plane of the interosseous membrane, from the tibia to the fibula, runs from anterior to posterior. Hence, the tibialis posterior extends forward and upward on the interosseous membrane to the tibia, between the interosseous border and the oblique line. The upper part of the muscle, between the tibia and the fibula, constitutes a fibrous arch behind which the anterior tibial vessels pass from medial to lateral.

The upper limit of the tibial origin of the tibialis posterior is the upper half or two-thirds of the soleal line. This line is often interrupted near its upper end by a smooth area with a tubercle at each end (fig. 22–5). A fibrous arch here transmits the nutrient artery of the tibia.

The tendon of the tibialis posterior descends behind the medial malleolus, where it is bound down by the flexor retinaculum. It continues forward, under cover of the abductor hallucis, and spreads out immediately below the plantar calcaneonavicular ligament. Here it contains fibrocartilage or a sesamoid. It inserts into the tuberosity of the navicular; offshoots continue to the cuneiforms, the cuboid, the sheath of the peroneus longus, and the bases of the second to fourth metatarsals. The tendon has a long synovial sheath.

ACTION. **The tibialis posterior is the principal invertor of the foot.**

Flexor Retinaculum. This is a rather imperfectly delimited thickening of the fascia. Above, it is directly continuous with the layers of fascia in front of and behind the triceps surae. Below, it is continuous with the tendinous origin of the abductor hallucis. The flexor retinaculum extends between the medial malleolus and the medial side of the calcaneus. Septa that extend inward from its deep aspect join bone and the deltoid ligament and form four compartments or tunnels.[13] These contain the following structures, from before backward (fig. 22–9): (1) the tendon of the tibialis posterior and its synovial sheath, (2) the tendon of the flexor digitorum longus and its sheath, (3) the posterior tibial vessels and the tibial nerve, and (4) the tendon of the flexor hallucis longus and its sheath.

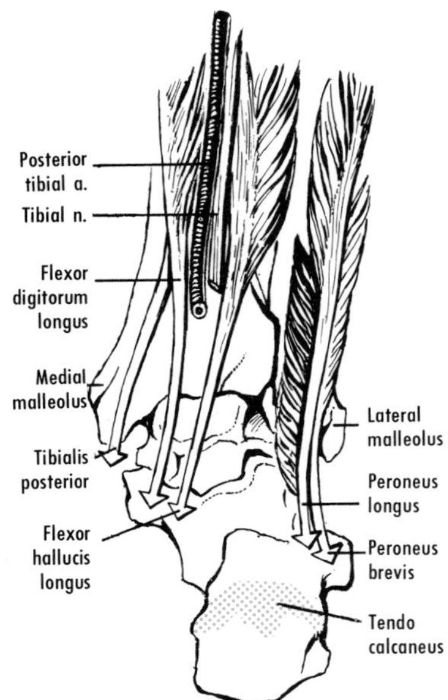

Figure 22–9 Schematic posterior view of the arrangement of the tendons behind the ankle. Note that the tibialis posterior is immediately posterior to the medial malleolus and that the flexor digitorum longus, having overlapped the tendon of the tibialis posterior above, is now lateral. Next, the tibial nerve and posterior tibial artery (veins omitted) descend for a short distance in immediate contact with the tibia. The tendon of the flexor hallucis longus is contained within a groove between the medial and posterior tubercles of the talus.

Vessels

Posterior Tibial Artery. This is the larger of the terminal divisions of the popliteal. It continues the line of the latter vessel and begins at the lower border of the popliteus. Anteriorly, it lies successively on the tibialis posterior, the flexor digitorum longus, and the posterior surface of the tibia. Posteriorly, the artery is covered by the deep transverse fascia of the leg and the soleus and gastrocnemius. Distally, it becomes more superficial but, at its division into medial and lateral plantar arteries, it lies deep to the flexor retinaculum and abductor hallucis. The tibial nerve is successively medial, posterior, and lateral to the artery.

BRANCHES (fig. 64–2, p. 766). The posterior tibial artery supplies adjacent muscles and gives a nutrient artery to the tibia. This artery, the largest nutrient artery to a long bone, passes through the fibrous arch of the tibialis posterior. In addition, there are the following named branches:

1. The *circumflex fibular branch* winds laterally around the neck of the fibula, through the soleus, and contributes to the anastomosis around the knee joint.

2. The *peroneal artery*[15] is usually as large as the posterior tibial and continues approximately the line of that vessel. It arises below the lower border of the popliteus, crosses the tibialis posterior, and descends along the medial crest of the fibula. Distally, it lies on the interosseous membrane, passes behind the lateral malleolus, and anastomoses with the dorsalis pedis and lateral plantar arteries. Its branches are (a) muscular, (b) the nutrient artery to the fibula, (c) a *communicating branch* to the posterior tibial, (d) perforating, and (e) the lateral malleolar. The *perforating branch* passes forward between the interosseous membrane and the interosseous tibiofibular ligament and descends in front of the ankle joint to anastomose with neighboring arteries. If the anterior tibial artery is small or absent, the peroneal is large and may, by means of its perforating branch, replace the dorsalis pedis artery. The *lateral malleolar branches* contribute to the network about the lateral malleolus and end as *calcanean branches*.

3. A *medial malleolar branch* (or branches) ramifies on the medial malleolus. It gives off calcanean branches to the *calcanean network*.

The terminal branches are the medial and lateral plantar arteries (p. 240).

Posterior Tibial Veins (Venae Comitantes). These are formed by the union of the medial and lateral plantar veins. They drain adjacent structures, receive the peroneal veins, and unite with the anterior tibial veins to form the popliteal veins. These deep veins return most of the blood from the leg and the foot. Blood from the superficial veins reaches them by way of communicating veins.

Nerves

Tibial Nerve. This nerve descends through the popliteal fossa. It then lies on the popliteus muscle, under cover of the gastrocnemius. At the lower border of the popliteus, it passes deep to (in front of) the tendinous arch of the soleus, descends first on the tibialis posterior and the flexor digitorum longus, and then on the tibia. Becoming more superficial and crossing the posterior tibial artery posteriorly to gain its lateral side, it ends under cover of the flexor retinaculum by dividing into the medial and lateral plantar nerves.

BRANCHES. Below the popliteal fossa, *muscular branches* are given to the soleus, tibialis posterior, flexor hallucis longus, and flexor digitorum longus. The branch or branches to the flexor hallucis descend with the peroneal artery. *Medial calcanean branches* are distributed to the skin of the heel and sole. A twig is given to the ankle joint. The terminal branches are the medial and lateral plantar nerves (p. 241).

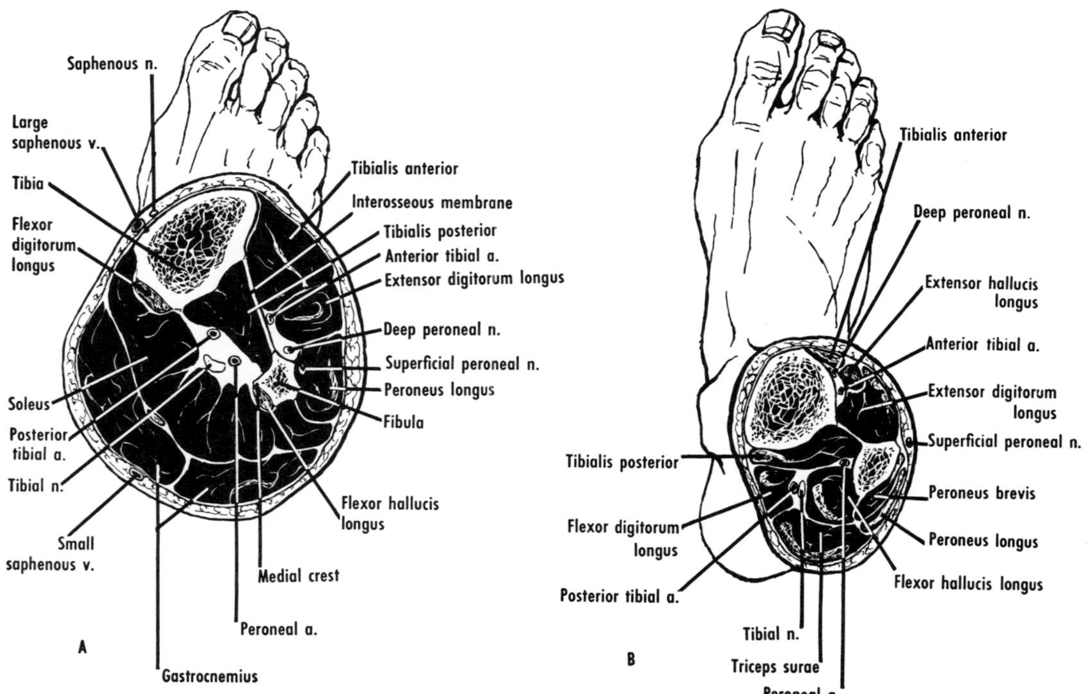

Figure 22–10 *A*, horizontal section through upper part of right leg. Note the posterior position of the fibula with respect to the tibia. *B*, horizontal section through lower part of right leg. Note the shift in the relative positions of the tibia and fibula.

INTEROSSEOUS MEMBRANE

This membrane is attached to the interosseous borders of the tibia and the fibula. In the upper part of the leg, its plane is nearly anteroposterior; in the lower part it is more nearly mediolateral (fig. 22–10). The anterior tibial vessels enter the anterior compartment of the leg through an arch in the upper and posterior part of the interosseous membrane.

REFERENCES

1. J. W. Smith, J. Anat., Lond., *92*:616, 1958. J. L. Shellshear and N. W. Macintosh, *Surveys of Anatomical Fields*, Grahame, Sydney, 1949.
2. H. Gloobe and D. Chain, Acta anat., *85*:84, 1973.
3. J. R. Barbour, Med. J. Aust., *1*:275, 1947. See also C. Kosinski, J. Anat., Lond., *60*:274, 1926.
4. G. Winckler, Arch. Anat., Strasbourg, *18*:181, 1934. E. H. Lambert, Neurology, *19*:1169, 1969.
5. E. B. Kaplan, J. Bone Jt Surg., *43A*:169, 1961.
6. J. H. Couch, Canad. med. Ass. J., *34*:688, 1936.
7. J. Campbell and C. M. Pennefather, Lancet, *1*:294, 1919. E. A. Edwards, Surg. Gynec. Obstet., *97*:87, 1953.
8. K. Trzenschik and H. H. Loetzke, Anat. Anz., *124*:297, 1969.
9. E. H. Daseler and B. J. Anson, J. Bone Jt Surg., *25*:822, 1943.
10. C. M. Fürst, *Der Musculus Popliteus und seine Sehne,* E. Malmströms, Lund, 1903. E. Weinberg, Arch. Anat., Strasbourg, *9*:254, 1929. R. J. Last, J. Bone Jt Surg., *32B*:93, 1950. K. O. Mörike, Anat. Anz., *133*:265, 1973.
11. J. F. Lovejoy and T. P. Harden, Anat. Rec., *169*:727, 1971.
12. C. H. Barnett and A. T. Richardson, Ann. phys. Med., *1*:177, 1952. J. V. Basmajian and J. F. Lovejoy, J. Bone Jt Surg., *53A*:557, 1971.
13. The retromalleolar arrangements are described by G. Leutert and R. Arnold, Anat. Anz., *122*:455, 1968.
14. E.-M. Ziegler, Anat. Anz., *130*:404, 1972.
15. A. Bouchet, L. Fischer, and R. Tiano, Bull. Ass. Anat., *58*:233, 1973.

FOOT AND ANKLE 23

The foot is the part of the lower limb distal to the leg. It is often compared with the hand, and they have many similarities, but the hand is a tactile, grasping organ, whereas the functions of the foot are support and locomotion.[1] These functions are discussed in the next chapter.

The toes are numbered from one to five, beginning with the big toe, or the hallux. The big toe usually projects farther than the other toes, and the digital formula is 1>2>3>4>5. Occasionally it is 1=2>3>4>5 or 2>1>3>4>5. Rarely the middle toe projects farther than the others.

The skin on the dorsum of the foot is thin and mobile. Hair on the dorsum of the foot is sparse, unevenly distributed, and often nearly completely absent. It is commonly present on the dorsum of the proximal phalanges, occasionally on the dorsum of the middle phalanges, but not on the dorsum of the distal phalanges.[2] The hair over the proximal phalanx of the big toe generally forms a well-marked tuft in men; it is often absent in women.

The skin of the sole of the foot is thick, and may be still thicker in the form of calluses. It is firmly bound to the subcutaneous tissue by thick fibrous strands, which enclose pockets of fat. Lines of thickened epidermis, known as *papillary ridges*, form a characteristic pattern on the sole of the foot, so that footprints, like fingerprints, may be used as a means of identification. The foot presents a number of flexure lines at the sites of skin movement, but they are not as well developed or important as those of the hand.

FASCIA OF FOOT

The subcutaneous tissue in the sole of the foot is greatly thickened by fibrofatty pads that are important weight-bearing structures.

The fascia of the foot is continuous above with the fascia of the leg. That on the dorsum of the foot is a thin membranous layer that ensheathes the tendons.[3] Above, it merges with the extensor retinacula. At the sides of the foot the fascia blends with the plantar aponeurosis.

Plantar Aponeurosis. **The fascia on the sole of the foot is specialized as the plantar aponeurosis (see fig. 23–12). This is a strong aponeurotic sheet that is divided into three parts, central, medial, and lateral.** The strong central part is attached behind to the medial process of the tuber calcanei. It extends forward, widening as it does so, and divides into five processes, one for each toe. Transverse fibers help to bind these processes together. A superficial slip from each process ends in the skin of the groove or crease between toes and sole. The rest of the process divides into two slips that fuse with the fibrous tendon sheath and attach to the plantar ligament of the metatarsophalangeal joint. The plantar aponeurosis thus forms a strong mechanical tie, especially marked for the big toe and medial arch, between the calcaneus and each proximal phalanx.[4]

The thin medial part of the fascia covers the lower part of the abductor hallucis, and at its junction with the central part, extends upward as a vertical septum that marks off a medial or big toe compartment.[5] The lateral part of the fascia extends as an aponeurotic band from the lateral process of the tuber calcanei to the tuberosity of the fifth metatarsal. It covers the abductor digiti minimi, and at its junction with the central part, extends upward and marks off a lateral or little toe compartment.

Fascial "Spaces" of Foot.[6] Two spaces are located on the dorsum of the foot, one subcutaneous, the other subaponeurotic. The important spaces in the sole of the foot are in the central compartment, which comprises four spaces that lie successively above the central part of the plantar aponeurosis. The first, or lowest, space lies between the aponeurosis and the flexor digitorum brevis, the second between the brevis and the quadratus plantae, and the third between the quadratus plantae and the tarsal bones and associated ligaments. The fourth, and uppermost, is an oblique space above the adductor hallucis.

Synovial Tendon Sheaths.[7] Three synovial sheaths are found in front of the ankle, deep to the extensor retinacula, for the (1) tibialis anterior, (2) extensor hallucis longus, and (3) extensor digitorum longus and peroneus tertius. The common arrangement of these sheaths is shown in figure 23–1.

As the tendons of the extensor digitorum longus extend over the metatarsophalangeal joints, they may be invested by synovial sheaths. The tendons of the extensor digitorum brevis are commonly so invested.

A single synovial sheath behind the lateral malleolus contains the tendons of the peroneus longus and brevis. This sheath subdivides, so that each tendon has a prolongation from the common sheath. A distal sheath invests the peroneus longus tendon from the cuboid nearly to its insertion. The proximal and distal sheaths may communicate with each other.

Three synovial sheaths are present behind the medial malleolus, for the (1) tibialis posterior, (2) flexor digitorum longus, and (3) flexor hallucis longus. These sheaths may communicate with one another, but they do not join any of the digital sheaths. In the toes, each of the long flexor tendons possesses a digital sheath, which it shares with the short flexor tendon.

MUSCLES OF FOOT

The muscles of the foot are the extensor digitorum brevis on the dorsum (which is supplied by the deep peroneal nerve), and the

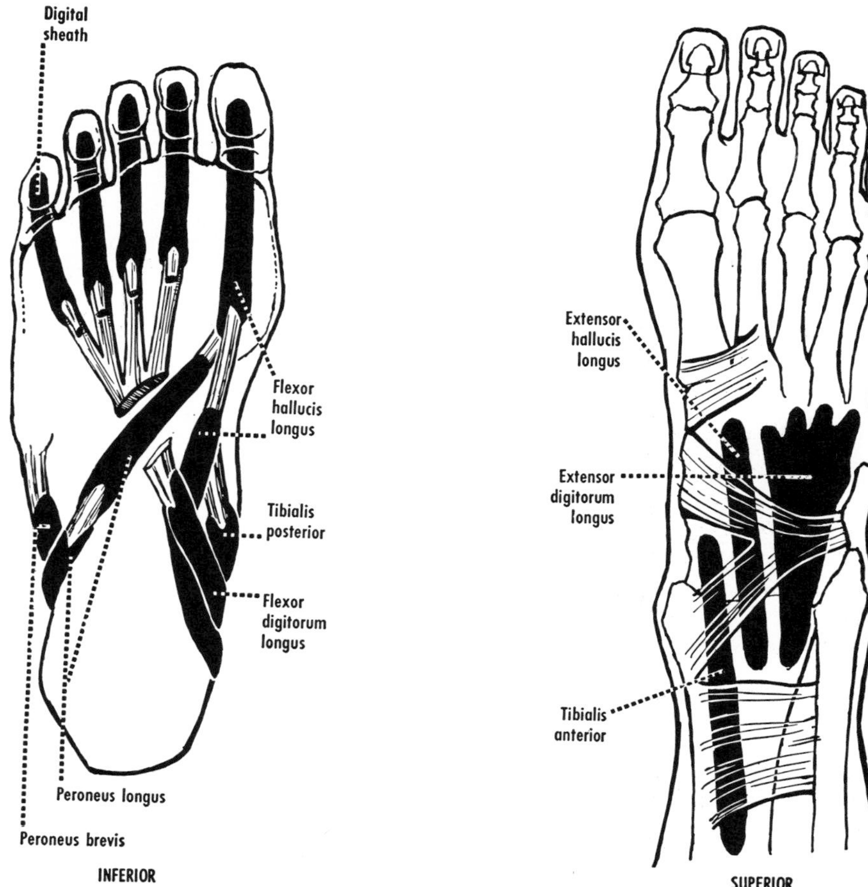

Figure 23–1 Synovial tendon sheaths of the foot and ankle. The more distal and somewhat less constant sheaths of the extensor digitorum longus and brevis have been omitted.

muscles of the big and little toes, the quadratus plantae, the flexor digitorum brevis, the lumbricals, and the interossei (which are supplied by the lateral and medial plantar nerves).

Muscles of Dorsum of Foot

Extensor Digitorum Brevis. This is the only muscle of the dorsum of the foot. It arises from the floor of the sinus tarsi, and from the limbs of the extensor retinaculum (see figs. 22–3 and 22–4, pp. 228, 229). It usually divides into four tendons, which are inserted into the four medial toes by fusing with the long extensor tendons. The most medial part of this muscle, the *extensor hallucis brevis,* ends separately at the base of the proximal phalanx of the big toe.

NERVE SUPPLY AND ACTION. It is supplied by the deep peroneal nerve, and sometimes by the accessory deep peroneal nerve also (p. 230). The muscle aids in extending the four medial toes at the metatarsophalangeal and interphalangeal joints.

Muscles of Sole of Foot

The muscles of the sole of the foot (fig. 23–2) are individually of little importance, but collectively they are significant in posture and locomotion, and they strongly support the arches of the foot during movement (p. 248). These muscles are arranged in three groups, a medial group for the big toe, a central group, and a lateral group for the little toe. From the standpoint of dissection, however, it is simpler to consider them in layers. The most superficial (i.e., lowest) layer comprises the abductor hallucis, the flexor digitorum brevis, and the abductor digiti minimi. The second layer consists of the quadratus plantae, the lumbricals, and the tendons of the flexor hallucis longus and flexor digitorum longus. The third layer comprises the flexor hallucis brevis, the ad-

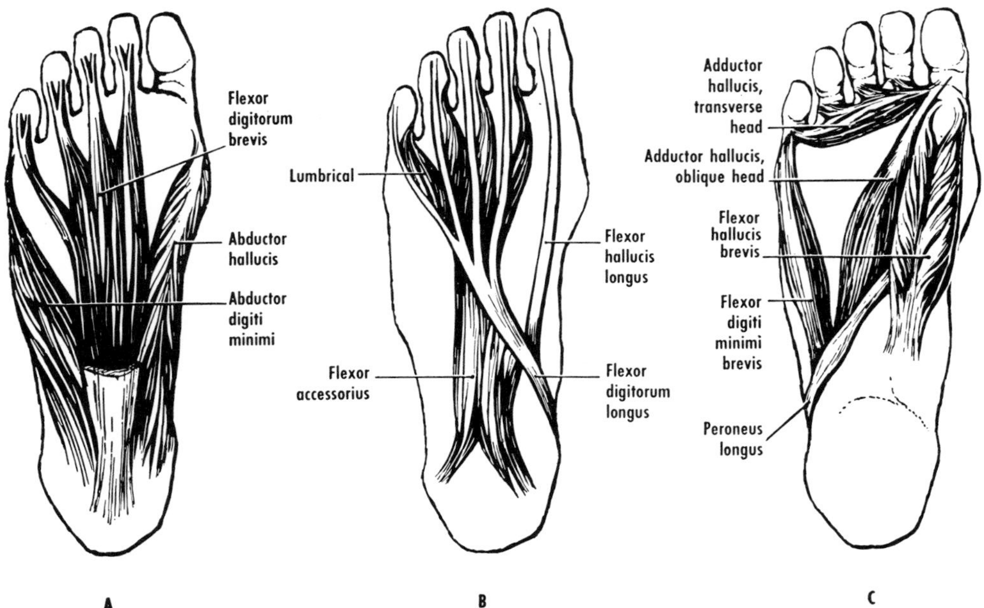

Figure 23–2 Muscles of sole of foot, shown in successive layers from below upward. *A*, the first or most superficial layer. *B*, the second layer. *C*, the third layer, including the peroneus longus, the insertion of which belongs to the fourth layer. See figures 23–4, 23–5, and 23–11 for the fourth layer.

ductor hallucis, and the flexor digiti minimi brevis. The fourth or uppermost layer consists of the interossei and the tendons of the tibialis posterior and peroneus longus.

The terms abduction and adduction of the toes are used with reference to an axis through the second toe. Thus, abduction of the big toe is a medial movement, away from the second toe.

FIRST LAYER

Abductor Hallucis. This muscle arises from the medial process of the tuber calcanei and adjacent aponeuroses. In common with the medial head of the flexor hallucis brevis, it is inserted into the medial sesamoid and the base of the proximal phalanx of the big toe. It is supplied by the medial plantar nerve and it flexes and abducts the big toe.

Flexor Digitorum Brevis. This muscle extends forward from the medial process of the tuber calcanei and divides into four tendons for the four lateral toes. Each tendon enters the fibrous flexor sheath in common with the long flexor tendon. It is perforated by the long flexor tendon and then divides to insert along the sides of the middle phalanx. It is supplied by the medial plantar nerve and it flexes the four lateral toes at the proximal interphalangeal joint.

Abductor Digiti Minimi. This muscle arises diffusely from the lateral process and adjacent parts of the tuber calcanei, and is inserted on the lateral side of the proximal phalanx of the little toe. It is supplied by the lateral plantar nerve and it abducts and flexes the little toe. The fibers of the plantar part of the muscle are usually inserted on the fifth metatarsal, and may then constitute a separate muscle (*abductor ossis metatarsi quinti*).

SECOND LAYER

Quadratus Plantae (Flexor Accessorius). This muscle, supplied by the lateral plantar nerve, arises by medial and lateral heads from the tuber calcanei and adjacent fascia and ligaments, and is inserted into the deep surface of the tendon of the flexor digitorum longus in the region of its division.[8] The quadratus plantae has a straight pull from the heel to the obliquely placed flexor tendon (see p. 252 for a suggested action).

Lumbricals. These are four in number. All arise from the long flexor tendons[9] (fig. 23–3), and each is inserted on the medial side of the base of the proximal phalanx of the respective toe (second to fifth). A few tendinous fibers reach the extensor aponeurosis. The tendon of each lumbrical runs below the deep transverse metatarsal ligament to its insertion; those of the interossei run above.

The first lumbrical is supplied by the medial plantar nerve, the others by the lateral plantar. Like the lumbricals of the hand, they increase slack in the flexor tendons. They also aid the interossei in flexion at the metatarsophalangeal joints. Whether they aid in abduction and adduction is uncertain.

THIRD LAYER

Flexor Hallucis Brevis. This muscle arises from the adjacent septum and from the metatarsal extension of the tibialis posterior. It is inserted by medial and lateral parts, the medial into the medial sesamoid and the base of the proximal phalanx of the big toe, in common with the abductor hallucis, and the lateral into the lateral sesamoid and the base of the proximal phalanx, in common with the adductor hallucis. It is supplied by the medial plantar nerve and it flexes the big toe.

Adductor Hallucis. This has *oblique* and *transverse heads*, both of which are supplied by the lateral plantar nerve. The oblique head arises chiefly from the sheath of the peroneus longus, and is inserted into the lateral sesamoid and the proximal phalanx, in common with the lateral part of the flexor hallucis brevis. Functionally, it is a part of this flexor. The transverse head arises from the deep transverse metatarsal ligament, and extends medially to the fibrous sheath of the flexor hallucis longus. The transverse head acts as a contractile tie for the heads of the metatarsals.

Flexor Digiti Minimi Brevis. This is a small muscle that is supplied by the lateral plantar nerve. It arises from the sheath of the peroneus longus and is inserted into the base of the proximal phalanx of the little toe. Presumably it flexes the little toe. Some of the muscle fibers are inserted into the fifth metatarsal and may form a separate muscle, the *opponens digiti minimi.* The term opponens, however, refers to a nonexistent function.

FOURTH LAYER

Dorsal and Plantar Interossei. These are supplied by the lateral plantar nerve and are arranged as shown in figures 23–4 and 23–5. Both groups are more prominent as seen from below, so much so that they scarcely merit the terms dorsal and plantar.

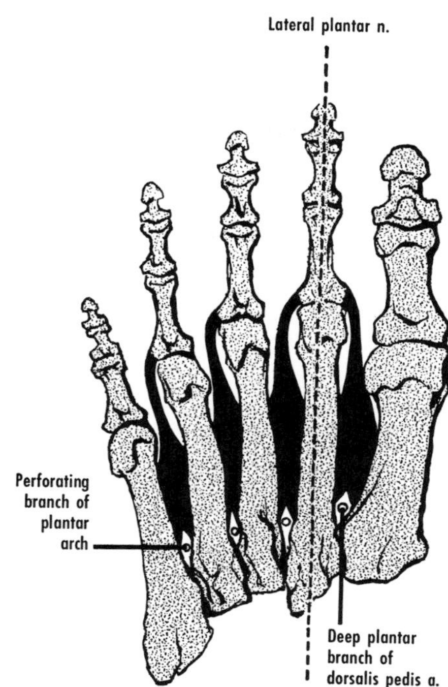

Figure 23–4 The arrangement of the dorsal interossei of the right foot, plantar aspect. They are supplied by the lateral plantar nerve.

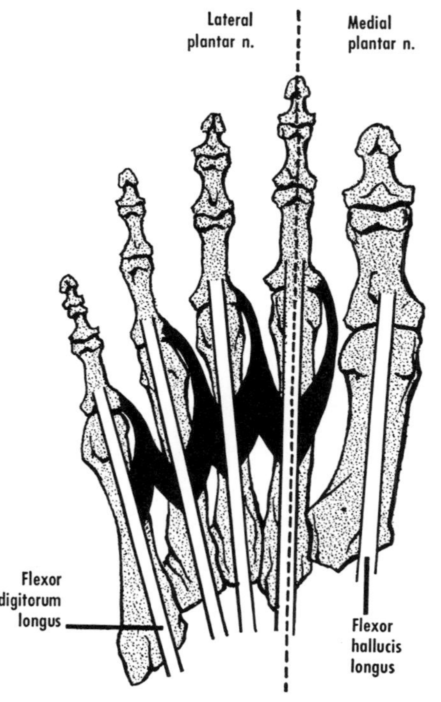

Figure 23–3 The arrangement of the lumbricals of the right foot, plantar aspect. The lumbricals on the lateral side of the longitudinal axis are supplied by the lateral plantar nerve; that on the medial side is supplied by the medial plantar nerve.

Each of the three plantar interossei arises from the medial side of the base of its respective metatarsal (third, fourth, and fifth) and from the sheath of the peroneus longus. The tendons pass forward, above the deep transverse metatarsal ligament, to the medial side of the base of the proximal phalanx.

Each of the four dorsal interossei arises from the shafts of the adjacent bones.[10] Each tendon passes forward, above the deep transverse metatarsal ligament, to the base of the proximal phalanx.

Neither group of interossei is inserted into the extensor aponeurosis.[11] The tendons of both groups may contribute to the capsules of the metatarsophalangeal joints.

ACTION. The interossei (and lumbricals) flex the metatarsophalangeal joints, and by acting as antagonists to the long extensors allow the latter to act from fixed toes and thereby to move the leg. The interossei can also adduct and abduct, with respect to an axis through the second toe (figs. 23–4 and 23–5), but these are unimportant actions. Perhaps more important is the fact that, because the interossei arise from adjacent metatarsals, they hold these bones together and thereby strengthen the metatarsal arch.

Plantar Skin Reflex. **When the skin of the sole is slowly stroked along the outer border from the heel forward, the toes flex. But in patients with certain disorders of the motor path-**

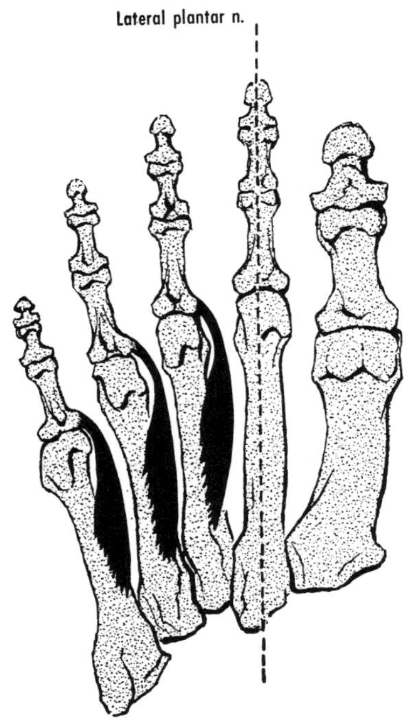

Figure 23–5 The arrangement of the plantar interossei of the right foot, plantar aspect. They are supplied by the lateral plantar nerve.

ways of the brain and spinal cord (and in infants before they walk) similar stimulation of the sole results in a slow dorsiflexion of the big toe and a slight spreading of the other toes. This response is known as the *Babinski reflex*.

VESSELS OF FOOT[12]

Medial Plantar Artery (fig. 23–6). Usually the smaller of the terminal branches of the posterior tibial, this artery arises under cover of the flexor retinaculum. At first deep to the abductor hallucis, it passes forward in the sole and comes to lie between the abductor hallucis and the flexor digitorum brevis. A fairly constant *superficial branch* continues directly forward and supplies the medial side of the big toe.

During its course, the medial plantar artery gives off cutaneous, muscular, and articular branches. Its *deep branch* gives three superficial digital branches, which anastomose with the medial three plantar metatarsal arteries.

Lateral Plantar Artery. This also arises under cover of the flexor retinaculum. It passes forward and laterally in the sole, between the flexor digitorum brevis below and the quadratus plantae above (i.e., between the first and second layers of the muscles of the sole), giving off calcanean, cutaneous, and muscular branches. From the base of the fifth metatarsal, it turns medially and helps to form the plantar arch (fig. 23–6).

Dorsalis Pedis Artery. Although variable in size and course, this vessel is nevertheless palpable and is important clinically in assessing peripheral circulation. It is the continuation of the anterior tibial from a point midway between the malleoli (fig. 23–6). It is crossed by the inferior extensor retinaculum and the extensor hallucis brevis. It lies successively on the capsule of the ankle joint, the head of the talus, the navicular, and the intermediate cuneiform. Laterally are the medial terminal branch of the deep peroneal nerve and the extensor digitorum longus and brevis. The tendon of the extensor hallucis longus crosses either the anterior tibial or the dorsalis pedis artery and comes to lie on the medial side of the latter. The dorsalis pedis terminates in a *deep plantar branch* at the proximal end of the first intermetatarsal space and passes to the sole between the heads of the first dorsal interosseus. There it forms the plantar arch.

BRANCHES. The branches of the dorsalis pedis are variable in size, pattern, and frequency, and are shown and named in figure 23–6. They form an arterial network on the dorsum of the foot, a network that has a relatively constant pattern, but the individual branches of which vary in size. The dorsalis pedis, for example, may be too small to be palpable.

Plantar Arch. The plantar arch is formed by the lateral plantar artery, is accompanied by the deep branch of the lateral plantar nerve, and lies between the third and fourth layers of the muscles of the sole. The arch gives off four *plantar metatarsal arteries*, from which *perforating branches* arise that ascend through the interosseous spaces and anastomose with the dorsal metatarsal arteries. *Plantar digital arteries* are also given off; their distribution is shown in figure 23–6.

Veins. The superficial veins are described on page 200. The deep veins begin as *plantar digital veins* on the plantar aspects of the toes. These veins drain proximally and are joined by veins from a *plantar venous network* on the sole of the foot to form four *plantar metatarsal veins.* These communicate with veins on the dorsum of the foot by veins that pass upward between the heads of the metatarsals. The plantar metatarsal veins unite to form the *plantar venous arch,* from which medial and lateral plantar veins run posteriorly and unite behind the medial malleolus to form the posterior tibial veins.

NERVES OF FOOT

The nerves of the foot include the saphenous (p. 218), the sural (p. 220), the deep and superficial peroneal (pp. 229, 230), and the medial and lateral plantar nerves.

Medial Plantar Nerve. The larger of the two terminal branches of the tibial nerve, the medial plantar nerve arises under cover of the flexor retinaculum, deep to the abductor hallucis. It then runs forward in the sole and comes to lie between the abductor hallucis and the flexor digitorum brevis, lateral to the medial plantar artery. It supplies these muscles and the skin on the medial side of the foot. Its terminal branches are four *plantar digital nerves* that supply muscles (flexor hallucis brevis and first lumbrical), skin (big toe, second and third toes, and medial side of fourth toe), and adjacent joints. The plantar digital nerves extend onto the dorsum to supply the nail beds and the tips of the toes.

Lateral Plantar Nerve. This nerve arises under cover of the flexor retinaculum

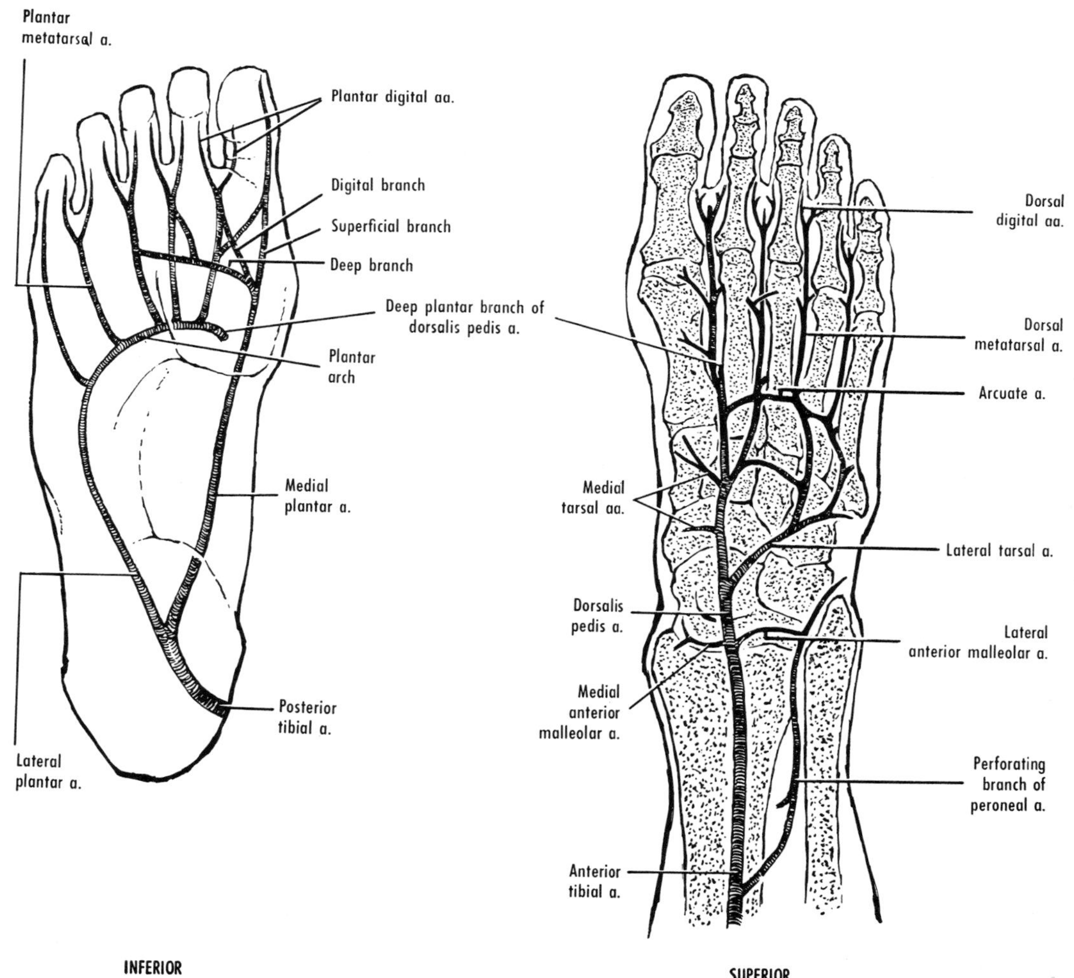

Figure 23–6 The arteries of the sole and dorsum of the foot.

and runs forward and laterally between the flexor digitorum brevis and the quadratus plantae, medial to the lateral plantar artery. It then divides into a *superficial* and a *deep branch*. During its course it supplies the quadratus plantae, the abductor digiti minimi, and the lateral side of the sole. The superficial branch supplies the flexor digiti minimi brevis, the lateral side of the sole and little toe and adjacent joints, and by *plantar digital nerves* the adjacent sides and joints of the fourth and fifth toes. The deep branch turns medially and supplies the interossei, the second, third, and fourth lumbricals, the adductor hallucis, and adjacent joints.

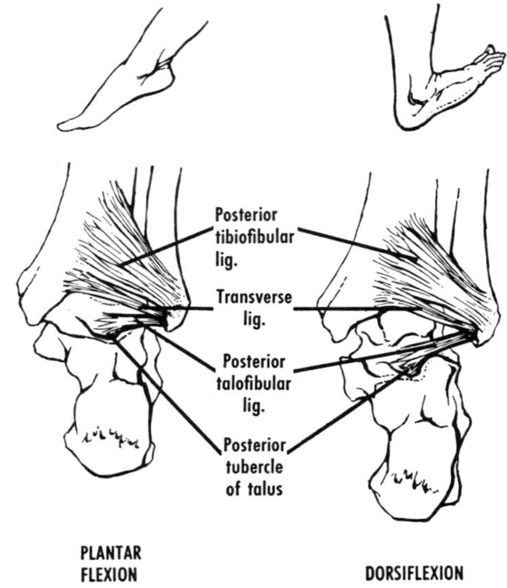

Figure 23–7 Ligaments of tibiofibular syndesmosis and of ankle joint from behind. Note the difference in position of the transverse ligament and of the posterior talofibular ligament in dorsiflexion and in plantar flexion. Based on F. Wood Jones.[1]

JOINTS

Tibiofibular Syndesmosis

The tibiofibular syndesmosis is a strong fibrous union between the lower ends of the tibia and fibula. An interosseous ligament connects the adjacent rough surfaces of the bones and helps to hold the malleoli against the talus, and is continuous above with the interosseous membrane.

The joint is strengthened in front and behind by strong bands, the *anterior* and *posterior tibiofibular ligaments*. An additional ligament, the *transverse ligament*, arises in the malleolar fossa of the fibula in common with the posterior talofibular ligament of the ankle joint (fig. 23–7). It extends to the posterior surface of the tibia, behind the talus.

Ankle Joint

The ankle (talocrural) joint is a hinge joint between the tibia and fibula for the one part and the trochlea of the talus for the other (fig. 18–26, p. 183). The tibia and the fibula form a socket, wider in front than behind, in which the talus moves.

The joint capsule is thickened on each side by several ligaments (figs. 23–8 and 23–9).

The capsular thickening on the medial side of the joint forms the *medial* or *deltoid ligament*,[13] which is attached above to the medial malleolus, and below to the talus, navicular, and calcaneus. The anterior part of the ligament reaches the neck of the talus; its most superficial part extends to the navicular. The posterior and also the deepest part of the deltoid ligament descends to the side of the talus. The intervening part extends downward to the sustentaculum tali.

Strong, fibrous septa pass inward from the deep aspect of the flexor retinaculum to the outer aspect of the deltoid ligament and form fibrous tunnels for tendons, vessels, and nerves (p. 233). The outer aspect of the deltoid ligament is therefore ridged and grooved.

Three discrete ligaments are present on the lateral side of the joint. These are often collectively termed the lateral ligament. The portions of the capsule that intervene between the ligaments are quite thin. The *anterior talofibular ligament* extends from the lateral malleolus to the neck of the talus. The *posterior talofibular ligament* is a fairly discrete band that arises in the malleolar fossa in common with the transverse ligament. It extends to the posterior tubercle of the talus (fig. 23–7). The intervening *calcaneofibular ligament* is ridged by septa extending inward from the peroneal retinacula.

The strong medial and lateral ligaments prevent anterior and posterior slipping of the talus, although they allow free dorsiflexion and plantar flexion. In dorsiflexion, the broad front part of the trochlea of the talus is forced into the narrower posterior part of the tibiofibular socket, and the tibia and fibula are pushed apart slightly. Aided by the strong ligaments of the tibiofibular syndesmosis, the malleoli grip the talus tightly.

The transverse ligament and the posterior talofibular ligament are roughly parallel with each other when the foot is plantar flexed. During dorsiflexion, the ligaments separate like the blades of a scissors (fig. 23–7). The transverse ligament slides against the posterior aspect of the trochlea. In fact, it forms one of the articular surfaces of the ankle joint, deepening the socket for the talus.

SYNOVIAL MEMBRANE. A synovial-lined recess of the joint cavity often extends upward into the lower part of the tibiofibular syndesmosis. Fat pads and synovial folds are present in the anterior and posterior part of the ankle joint.

NERVE SUPPLY. The joint is supplied by the tibial, sural, deep peroneal, and saphenous nerves, and by the accessory deep peroneal nerve when present.[14]

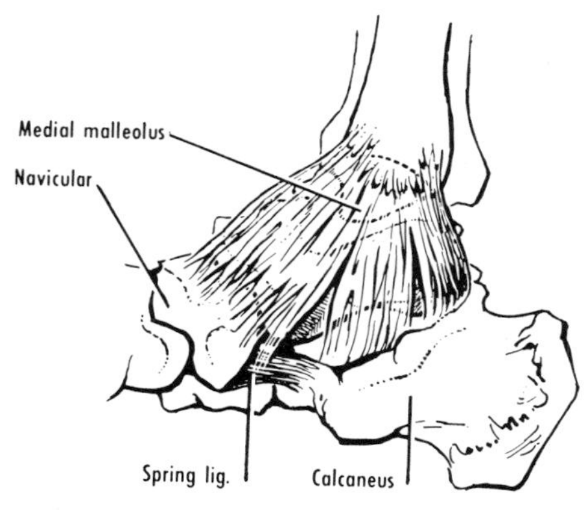

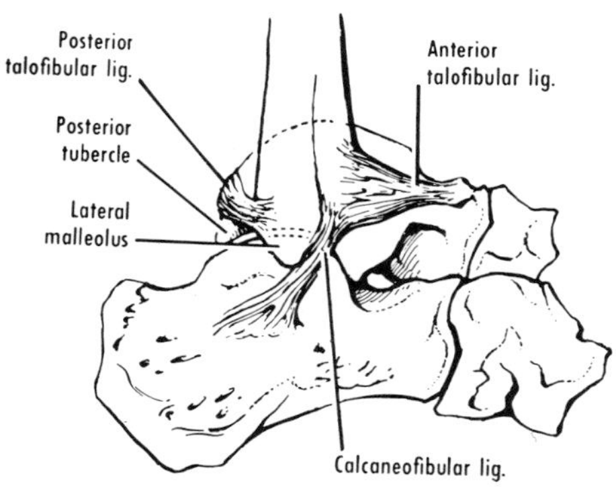

Figure 23–8 Ligaments of the ankle joint. The medial view shows the medial ligament, which forms a dense, almost continuous ligament. The ligaments on the lateral side, however, are usually separated from one another. (See figure 23–7 for another view of the posterior talofibular ligament.)

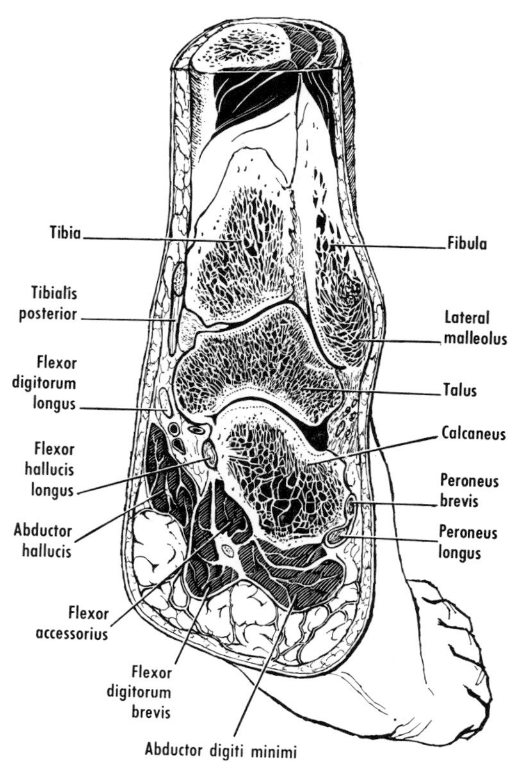

Figure 23–9 Coronal section of the ankle joint. The width of the articular cavities is exaggerated.

Movements at the Ankle Joint (figs. 23–10, 23–13). **The movements are dorsiflexion and plantar flexion. The axis of movement passes approximately through the malleoli.** It is roughly at a right angle to a line passing forward from the heel through the third toe.

Starting with the foot at a right angle to the leg, it is possible to plantar flex the foot to a greater degree than it can be dorsiflexed. But there is much individual variation, and the range of movement can be increased by training. In toe dancing, for example, the dorsum of the foot is in line with the front of the leg. Part of this movement, however, involves joints other than the ankle.

In forcible weight-bearing dorsiflexion, as in squatting, the front margin of the lower articular surface of the tibia nearly reaches the neck of the talus.

MUSCLES. The triceps surae and peroneus longus plantar flex the foot. Acting together, their individual actions of inversion and eversion are neutralized. But the gastrocnemius is a two-joint muscle, and

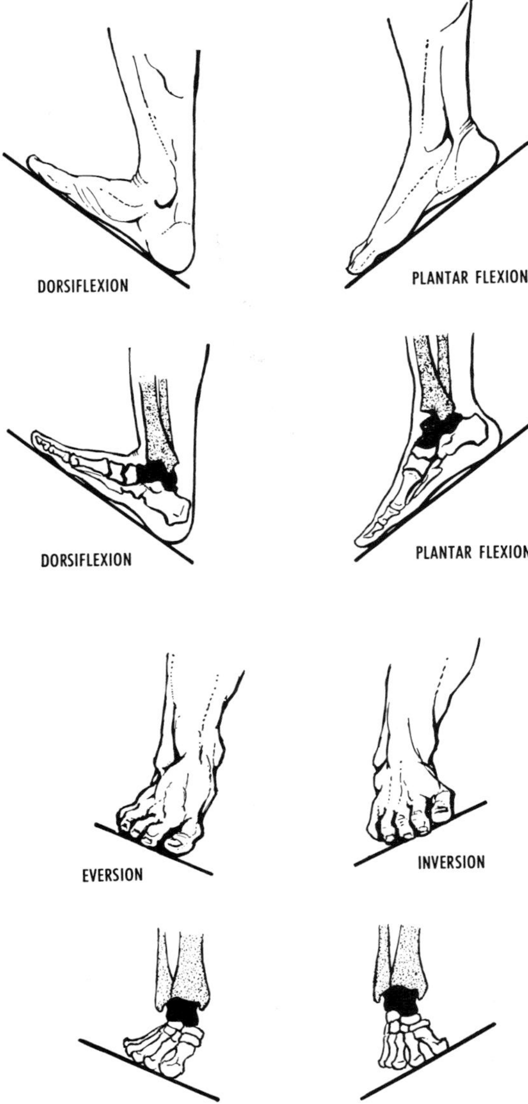

Figure 23–10 Movements of the foot and ankle. Dorsiflexion and plantar flexion are shown as in walking up and down hill. Movement occurs at the ankle joint. Eversion and inversion are shown as in standing sideways on a hill. Movement occurs at the tarsal joints, the talus remaining fixed. Based on S. Mollier.[15]

when the knee is fully flexed the triceps surae cannot shorten enough to act on the foot. Plantar flexion is then carried out by the tibialis posterior and peroneus longus, aided probably by the flexor digitorum longus and flexor hallucis longus.

The tibialis anterior and extensor digitorum longus dorsiflex the foot. Acting together, their individual actions of inversion and eversion are neutralized. The extensor hallucis longus may aid in dorsiflexion.

Intertarsal Joints

The talus moves with the foot during dorsiflexion and plantar flexion. But during inversion and eversion, which occur at intertarsal joints, the talus moves with the leg. **The most important intertarsal joints are the subtalar, the talocalcaneonavicular, and the calcaneocuboid. The last two form the transverse tarsal or midtarsal joint.** The other intertarsal joints are the cuneocuboid, intercuneiform, and cuneonavicular.

Subtalar (Talocalcanean) Joint. This is a separate joint lying behind the tarsal canal. It is formed by the facet on the lower part of the body of the talus and the posterior facet on the upper part of the calcaneus. The articular margins are connected by a short capsule that is thickened on each side. The anterior part of the capsule lies in the tarsal canal.

Talocalcaneonavicular Joint.[16] This joint, a part of the *transverse tarsal joint,* lies in front of the tarsal canal. It resembles a ball-and-socket joint, in which the head of the talus fits into a socket formed by the navicular in front and the calcaneus below.

The considerable interval between the navicular and the calcaneus is occupied by the *plantar calcaneonavicular ligament.*[17] This ligament connects the sustentaculum tali with the navicular and completes the socket. The tendon of the tibialis posterior lies immediately below it. The ligament is composed of dense fibrous tissue and is nonelastic, and a part of its articular surface resembles articular cartilage.[18]

Calcaneocuboid Joint. This is also a part of the transverse tarsal joint and resembles a saddle joint, although it lacks the freedom of motion of such a joint. The two bones are connected by a capsule and by special ligaments. The *bifurcate ligament* is a strong ligament that arises from the upper part of the calcaneus, in the floor of the sinus tarsi. It divides, one part going to the navicular and the other to the cuboid.

The tension that develops during the support of body weight is taken up by strong ligaments on the plantar aspect of the tarsus (figs. 23–11 and 23–12). The *long plantar ligament* arises from most of the plantar aspect of the calcaneus. It extends forward to the tuberosity of the cuboid. Some of its

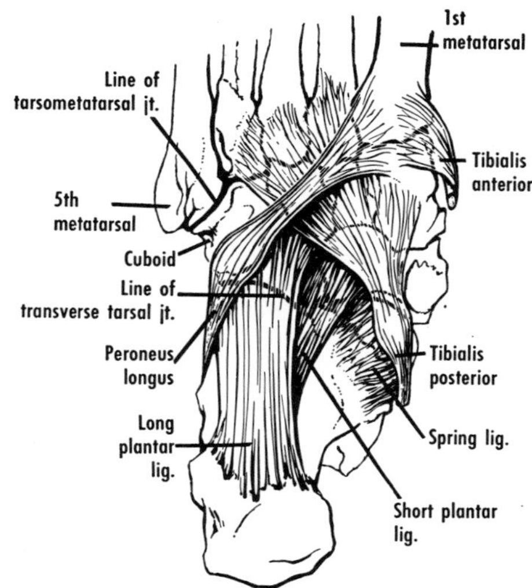

Figure 23–11 The tendons and ligaments of the foot, plantar aspect. Note the oblique course of the peroneus longus tendon and the widespread insertion of the tibialis posterior.

superficial fibers continue into the sheath of the peroneus longus and thereby reach the bases of the lateral three metatarsals. The *plantar calcaneocuboid ligament* (or short plantar ligament) arises from the front of the lower surface of the calcaneus, deep to the long plantar ligament. It extends somewhat obliquely to the ridge behind the groove in the cuboid.

The cavities of the subtalar and transverse tarsal joints do not communicate with each other. The cavity of the transverse tarsal joint forms an irregular plane across the tarsus.

Tarsal Canal and Sinus Tarsi (figs. 22–3 and 22–4, pp. 228, 229).[19] The grooves on the lower part of the talus and the upper part of the calcaneus form the tarsal canal, which separates the subtalar joint behind from the talocalcaneonavicular joint in front. This canal runs obliquely forward and laterally, and expands at its anterolateral end to form the sinus tarsi. The sinus contains blood vessels (most of the blood supply of the talus is derived from these vessels), ligaments, and fat. The bifurcate ligament springs from the floor of the sinus tarsi. Lateral to the bifurcate ligament is the origin of the extensor digitorum brevis. Behind it is a strong (cervical) ligament, which extends upward to the neck of the talus. Part of the inferior extensor retinaculum enters the sinus tarsi and continues into the ligament of the tarsal canal (fig. 22–4, p. 229). It has been postulated that the cervical ligament limits inversion, and that the ligament of the tarsal canal limits eversion.

Cuneocuboid, Intercuneiform, and Cuneonavicular Joints. The cuneocuboid joint and the two intercuneiform joints are plane joints, which are united by short plantar, dorsal, and interosseous ligaments. The cuneonavicular joint is a plane joint between the cuneiforms and the navicular, the articular cartilage of which is faceted for the cuneiforms. The cavities of these intertarsal joints often communicate; the cuneocuboid, however, may be separate. The cuboid and navicular are usually united by fibrous tissue, but a small synovial cavity may be present between them.

Nerve Supply. The plantar aspects of the intertarsal joints are supplied by the medial or lateral plantar nerves, and their dorsal aspects chiefly by the deep peroneal nerve (sometimes by the dorsal cutaneous and accessory deep peroneal nerves also).[20]

Movements at the Intertarsal Joints (figs. 23–10 and 23–13). **The chief movements of the foot distal to the ankle joint are inversion and eversion.** In inversion, the sole of the foot is directed medially (or the equivalent movement in which the foot is fixed and the leg moves). In eversion, the sole of the foot is turned so that it faces laterally (or the equivalent movement in which the foot is fixed and the leg moves). Inversion and eversion take place mainly at the subtalar and transverse tarsal joints. The axes of movement at these joints are placed obliquely with reference to the standard planes. Hence, none of the standard movements (flexion-extension, adduction-abduction, supination-pronation) can occur by itself. Each of the movements that occur is a combination of two or more of the primary movements. Inversion comprises supination, adduction, and plantar flexion. Eversion includes pronation, abduction, and dorsiflexion.

It is unfortunate that the terms used to describe the movements of the foot are employed differently by various authors. Usually, however, supination and pronation refer to the medial and lateral rotation about

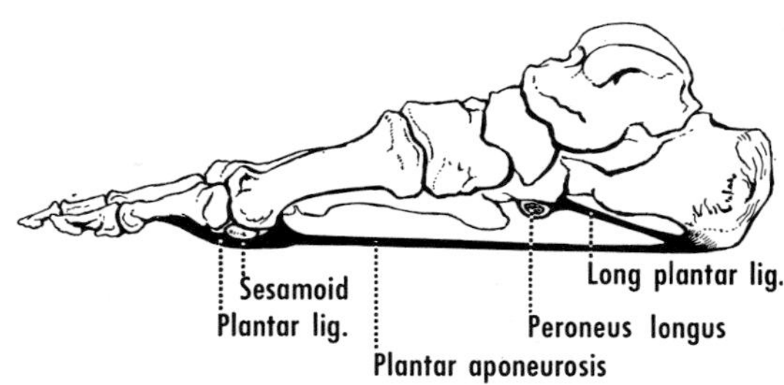

Figure 23–12 Schematic representation of the plantar aponeurosis and the long plantar ligament. Note that the main attachment of the long plantar ligament is to the cuboid behind the peroneus longus tendon; a part of it continues into the fibrous sheath of the tendon. The plantar aponeurosis, shown here extending to the big toe, continues into the plantar ligament of the metatarsophalangeal joints and thereby onto the proximal phalanges.

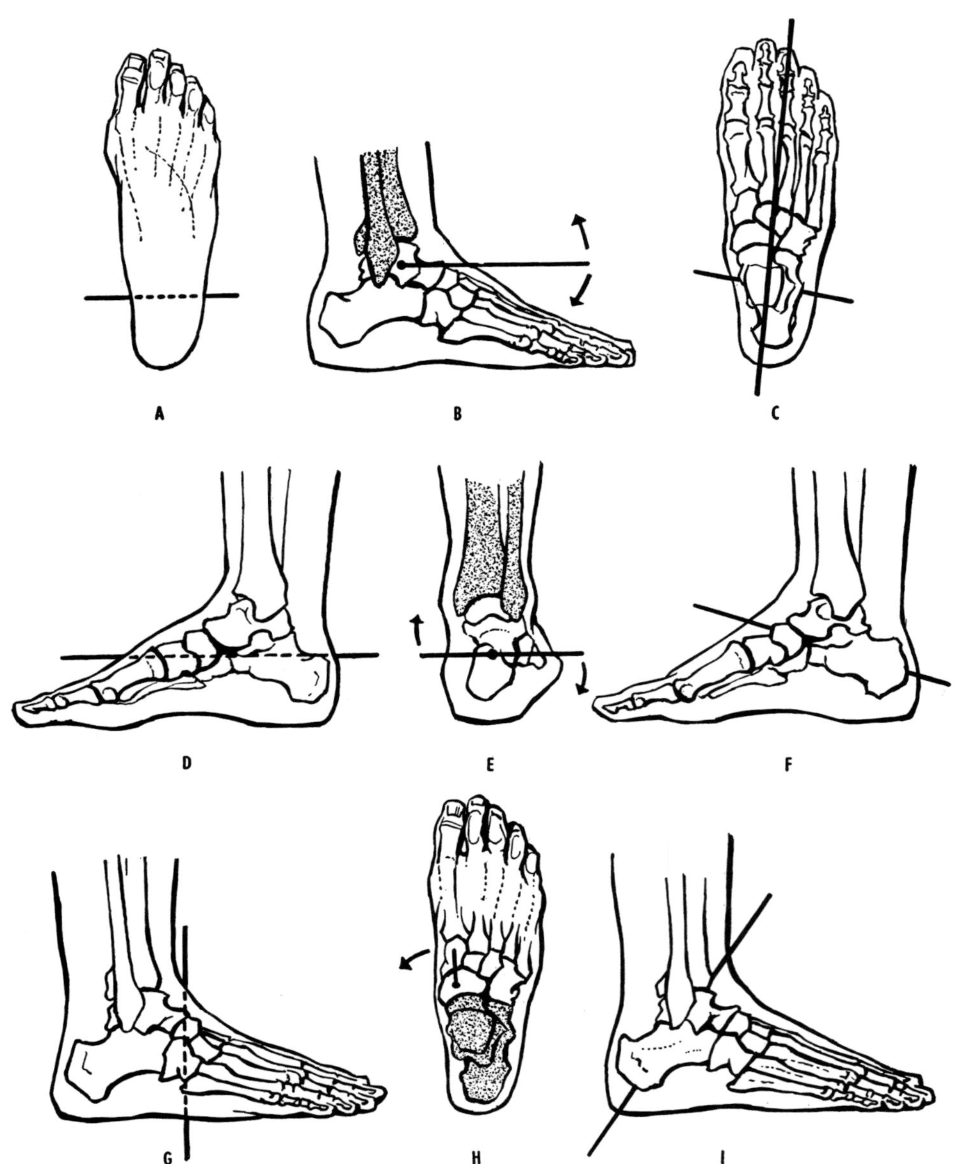

Figure 23–13 Schematic representation of the oblique axes of movements of the ankle and foot.[21]

Upper row. If the axis of movement of the ankle were horizontal, as shown in *A*, flexion and extension would occur in a sagittal plane, as shown by arrows in *B*. But, as shown in *C*, the axis of the ankle is obliquely vertical.[22] Moreover, the axis changes with movement, being inclined downward and laterally during dorsiflexion, downward and medially during plantar flexion. The obliquity of the axis is largely responsible for the inversion that automatically accompanies plantar flexion. The inversion is only slightly reduced in patients with fused subtalar and midtarsal joints.[23]

Middle row. If there were an anteroposterior axis, such as is shown in *D*, supination-pronation would occur in a coronal plane (*E*). But the axes of the intertarsal joints are oblique (one of the axes, that of the transverse tarsal joint, is shown in *F*), and true supination-pronation is consequently impossible.

Lower row. If there were a vertical axis, such as is shown in *G*, adduction-abduction of the forefoot would occur in a horizontal plane (*H*). But, again, the axes of the intertarsal joints are oblique (an example of one of these axes, that of the talocalcaneonavicular joint, is shown in *I*), and true adduction-abduction is therefore impossible.

an anteroposterior axis. Abduction and adduction refer to those movements of the front part of the foot about a vertical axis. It should be re-emphasized that none of these movements can occur by itself.

The intertarsal joints anterior to the transverse tarsal joint are individually of minor importance. Some gliding occurs at each, however, and collectively their elasticity and the multitude of small movements that they permit are of considerable significance.

MUSCLES. The tibialis posterior and the tibialis anterior, aided by the triceps surae, invert the foot. The peroneus longus and the extensor digitorum longus, aided by the peroneus tertius, evert the foot.

Tarsometatarsal and Intermetatarsal Joints

These are plane joints that allow simple gliding. The range of movement at each joint is small, but the several joints together give elasticity to the foot, allow twisting of the front part of the foot, and thereby increase the range of movements begun at more proximally situated joints. They are supplied by the deep peroneal, medial and lateral plantar, and sural nerves.

Tarsometatarsal Joints. The medial cuneiform and the first metatarsal have an independent joint cavity (fig. 18–35*B*, p. 192).* The cavities of the intermediate and lateral cuneometatarsal joints communicate (around the interosseous ligaments) with the intercuneiform and the cuneonavicular joints. The second metatarsal fits in a socket formed by the three cuneiforms. There is a single cavity for the joint between the cuboid and the fourth and fifth metatarsals.

The tarsometatarsal joints are bound together by articular capsules, and by *plantar, dorsal and interosseous ligaments.*

Intermetatarsal Joints. There is no joint between the bases of the first and second metatarsals, although a bursa is sometimes present. The other joints are united by *plantar* and *dorsal ligaments,* and by *interosseous ligaments,* which connect the nonarticular parts of the adjacent surfaces of the bases. The joint cavities are forward projections of the tarsometatarsal joints.

*The tarsometatarsal joint of the big toe is a distinctly human feature.[24] Opposition can occur at this joint in anthropoids, but not in man. Those who become skillful in using their toes, as do those born without hands, do so by movements at the metatarsophalangeal and interphalangeal joints.

Metatarsophalangeal and Interphalangeal Joints

The metatarsophalangeal joints are ellipsoid, and the interphalangeal joints are hinge, but the ligamentous arrangements of the two joints are similar.

Each joint is united by an *articular capsule* strengthened by two *collateral ligaments* that extend distally across the joint and are attached to the base of the phalanx and plantar part of the capsule. The plantar part of the capsule is a thick, densely fibrous or fibrocartilaginous pad, the *plantar ligament.* The plantar ligaments of the metatarsophalangeal joints are connected by strong, transverse fibers of the *deep transverse metatarsal ligament.*[25] This ligament helps to hold the metatarsal heads together. It forms a strong tie for this part of the foot, and it is aided by the adductor hallucis (transverse head) that arises from its plantar aspect. The tendons of the interossei pass above this ligament, those of the lumbricals below it.

Above, the joint capsules consist mainly of the extensor aponeurosis. A small fibrous pad usually extends downward for a short distance.

NERVE SUPPLY. The joints are supplied on their plantar and dorsal aspects by the neighboring digital nerves.[20]

Movements. Flexion, extension, abduction, and adduction occur at the metatarsophalangeal joints. The range of extension is greater than the range of flexion.[26]

The flexors are the lumbricals, the interossei, the flexor hallucis brevis, and the flexor digiti minimi brevis. The extensors are the extensor hallucis longus and the extensor digitorum longus and brevis. The toes are abducted (with reference to the second toe) by the abductor hallucis, abductor digiti minimi, and dorsal interossei. They are adducted by the adductor hallucis and plantar interossei.

The interphalangeal joints are hinge joints; flexion is the freer movement. The flexors are the flexor digitorum longus and brevis and the flexor hallucis longus. The extensors are the extensor hallucis longus and the extensor digitorum longus and brevis.

Sesamoid Mechanism. The metatarsophalangeal joint of the big toe is of special interest. Two grooves on the plantar aspect of the head of the first metatarsal articulate with the sesamoids that are embedded in the plantar ligament. Ligaments bind each sesamoid to the sides of the metatarsal head. The most medial slip of the plantar aponeurosis divides into two. Each division passes to a sesamoid and thereby, through the plantar ligament, is firmly connected to the phalanx (fig. 23–12).[27] The two sesamoids of the big toe take the weight of the body, especially during the latter part of the stance phase of walking, and they intervene between the head of the metatarsal and the soft tissues of the ball of the great toe.

The sesamoid mechanism is deranged in bunions and in hallux valgus. These two are associated disorders. A bunion is a swelling medial to the joint, due to a thickening of the wall of a bursa that is generally present here. In hallux valgus the big toe is displaced laterally, the angulation occurring at the metatarsophalangeal joint. The plantar ligament and the sesamoids are displaced laterally, and the ligaments of the medial side of the joint are stretched.[28]

ARCHES AND FLAT FEET

The arches of the foot (figs. 23–14 to 23–16) are the longitudinal and the transverse. On the medial side of the foot, a *longitudinal arch* is formed by the calcaneus, talus, navicular, cuneiforms, and the first three metatarsals. On the lateral side of the foot, a longitudinal arch is formed by the calcaneus, cuboid, and the lateral two metatarsals. The *transverse* or *metatarsal arch* is formed by the navicular, cuneiforms, and cuboid, together with the five metatarsals. These bony arches, which are the result of the intrinsic mechanical arrangement of the bones, are supported by ligaments.[29] During movement, they receive additional support from muscles,[30] chiefly from those that invert and evert the foot.

The arches of the foot are present nearly as soon as the skeletal elements are in their definitive form during intrauterine life. Prenatally, fat is distributed throughout the sole of the foot and forms the sole pad, which is responsible for the convexity of the sole of the fetal foot. At birth and during infancy, the sole pad still masks the arches of the skeleton, and the sole of the foot may appear flat. Later in life, the encapsulated pockets of subcutaneous fat thin out in areas that are not in contact with the ground. Hence, in most adults, the medial arch can be recognized in footprints.

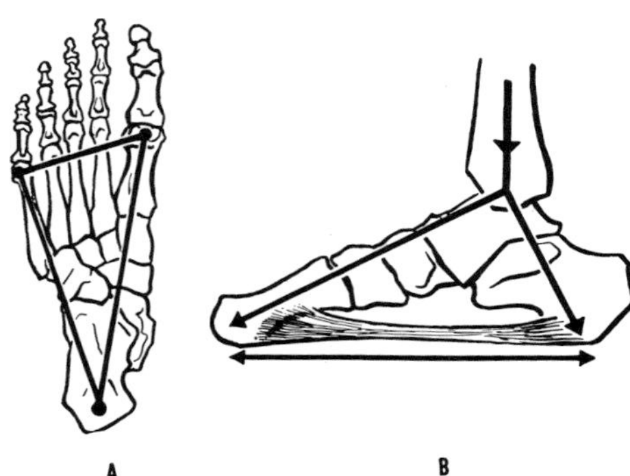

Figure 23–15 A, the three main points of weight bearing in the foot. *B*, the medial part of the longitudinal arch; the arrows indicate the distribution of weight that tends to flatten the arch. The connection between the calcaneus and the metatarsal schematically represents the ligamentous support of the arch. Based on Mollier.[15]

The extent to which the sole of the foot makes contact with the ground is not necessarily an accurate index of the height of the bony arches. The term *flatfoot (pes planus)*, refers to any one of many conditions. Strictly speaking, a flat foot is one with a simple depression of the longitudinal arch. In such instances, the foot is constructed so that it has a low arch when the foot is in the ideal posture for a particular individual (fig. 23–14*B*). This type of foot is not uncommon,[31] and it is not pathological; it represents the normal contour of a strong and stable foot.

The converse of flatfoot is *pes cavus* (fig. 23–14*C*), in which the longitudinal arch is excessively or abnormally high.

The term *clubfoot*, or *talipes*, is used for a foot that appears twisted out of shape or position. There are several varieties, and all are congenital.

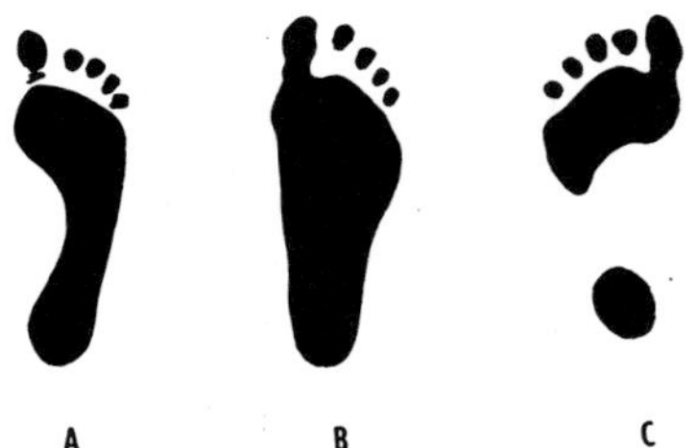

Figure 23–14 Footprints. *A*, normal; *B*, flatfoot; *C*, high longitudinal arch.

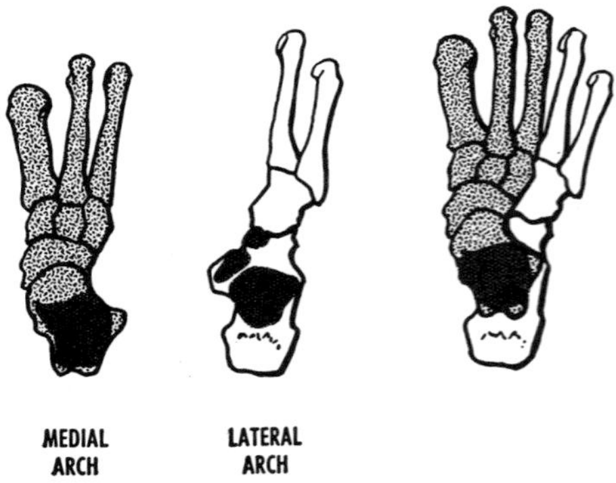

Figure 23–16 The bony components of the longitudinal arches. The third figure shows both arches.

REFERENCES

1. F. Wood Jones, *Structure and Function as Seen in the Foot*, Ballière, Tindall, & Cox, London, 2nd ed., 1949. E. de Doncker and C. Kowalski, Acta orthop. Belgica, *36*:377, 1970.
2. C. H. Danforth, Amer. J. phys. Anthrop., *4*:189, 1921; *Hair, with Special Reference to Hypertrichosis*, American Medical Association, Chicago, 1925.
3. P. Bellocq and P. Meyer, Acta anat., *30*:67, 1957.
4. J. H. Hicks, J. Anat., Lond., *88*:25, 1954.
5. B. F. Martin, J. Anat., Lond., *98*:437, 1964.
6. M. Grodinsky, Surg. Gynec. Obstet., *49*:737, 1929.
7. A. G. H. Lowell and H. H. Tanner, J. Anat., Lond., *42*:415, 1908. M. Grodinsky, Surg. Gynec. Obstet., *51*:460, 1930. F. Wood Jones, cited above.
8. T. E. Barlow, J. Anat., Lond., *87*:308, 1953. G. Winckler and G. Gianoli, Arch. Anat., Strasbourg, *38*:47, 1955.
9. Variations in the origins of the lumbricals are described by R. Schmidt and R. Schultka, Z. Anat. EntwGesch., *126*:172, 1967.
10. Additional origins of the first dorsal interosseus are described by J. C. Lamont, J. Anat., Lond., *42*:236, 1908. See also A. E. Harbeson, J. Anat., Lond., *72*:463, 1938.
11. A. Forster, Arch. Anat., Strasbourg, *7*:247, 1927. J. T. Manter, Anat. Rec., *93*:117, 1945.
12. H. Radke, Fortschr. Röntgenstr., *85*:580, 1956. E. A. Edwards, Acta anat., *41*:81, 1960. J. F. Huber, Anat. Rec., *80*:373, 1941. H. M. Vann, Anat. Rec., *85*:269, 1943. T. Murakami, Okajimas Folí anat. jap., *48*:295, 1971.
13. The deltoid ligament contains an appreciable amount of elastic tissue. See T. J. Harrison, J. Anat., Lond., *85*:432, 1951.
14. E. Gardner and D. J. Gray, Anat. Rec., *161*:141, 1968. H. Lippert, Z. Anat. EntwGesch., *123*:295, 1962. J. Champetier, Acta anat., *77*:398, 1970.
15. S. Mollier, *Plastische Anatomie*, Bergmann, Munich, 2nd ed., 1938.
16. E. Barclay Smith, J. Anat., Lond., *30*:390, 1896.
17. R. von Volkmann, Anat. Anz., *131*:425, 1972, and *134*:460, 1973.
18. R. H. Hardy, J. Anat., Lond., *85*:135, 1951.
19. J. W. Smith, J. Anat., Lond., *92*:616, 1958. D. R. Cahill, Anat. Rec., *153*:1, 1965. See also G. Winckler, Arch. Anat., Strasbourg, *18*:181, 1934.
20. E. Gardner and D. J. Gray, Anat. Rec., *161*:141, 1968.
21. J. H. Hicks, J. Anat., Lond., *87*:345, 1953. On movement at the subtalar and other tarsal joints, see also M. C. Hall, Canad. J. Surg., *2*:287, 1959, and A. Huson, An Anatomical and Functional Study of the Tarsal Joints, thesis, Leiden, 1961.
22. M. Harty, Lancet, *2*:275. 1953. C. H. Barnett and J. R. Napier, J. Anat., Lond., *86*:1, 1952. C. H. Barnett, J. Anat., Lond., *87*:499, 1953.
23. C. H. Barnett, J. Anat., Lond., *89*:225, 1955.
24. F. Wood Jones, J. Anat., Lond., *63*:408, 1929.
25. F. Wood Jones, cited in reference 1. R. W. Haines, J. Anat., Lond., *87*:460, 1953.
26. J. Joseph, J. Bone Jt Surg., *36B*:450, 1954.
27. J. H. Hicks, J. Anat., Lond., *88*:25, 1954.
28. R. W. Haines and A. McDougall, J. Bone Jt Surg., *36B*:272, 1954.
29. J. V. Basmajian and G. Stecko, J. Bone Jt Surg., *45A*:1184, 1963. R. Mann and V. T. Inman, J. Bone Jt Surg., *46A*:469, 1964.
30. N. Suzuki, Nagoya med. J., *17*:57, 1972.
31. R. I. Harris and T. Beath, J. Bone Jt Surg., *30A*:116, 1948.

POSTURE[1] AND LOCOMOTION[2] 24

POSTURE

The act of standing consists of a series of relatively immobile attitudes, separated by brief intervals of movement during which swaying occurs.[3]

When one is in the easy standing position, that is, upright, comfortably balanced, with the feet slightly apart and rotated somewhat laterally (slightly toed-out), few muscles of the back and lower limbs are active during the immobile periods. The mechanical arrangements of muscles and joints are such that a minimum of muscular activity is necessary to maintain this position. The position of the line of gravity, which is determined by the distribution of body weight, is a major factor in determining the degree of muscular activity involved in maintaining all phases of posture. **The line of gravity extends upward through the junctions of the curves of the vertebral column, and downward, behind the hip joint, but in front of the knee and ankle joints.**[4] As a rough approximation, the line of gravity may be considered to parallel the anterior border of the tibia.

In the easy standing position, the hip and knee joints are extended and are in their most stable positions. Because the line of gravity passes behind the hip joint, the weight of the body tends to extend the joint further. Hyperextension is resisted by the joint capsule, especially by the iliofemoral ligament. The line of gravity passes in front of the knee joint, which tends to be hyperextended. This is resisted by the ligamentous apparatus of the knee and by the ligamentous action of the hamstrings (sometimes by their active contraction).

If the line of gravity moves backward less than 2 to 3 cm, the supporting mechanism of the knee collapses. This is illustrated by the childhood trick of surreptitiously pushing at the backs of the knees of an unsuspecting person. A considerable amount of flexion occurs before the extensors reflexly contract.

The line of gravity passes in front of the ankle joint, and the weight of the body tends to cause forward sway (dorsiflexion) at that joint. The ankle is less stable than the hip and knee, and the forward sway is checked by contraction of the calf muscles.

The standing position has great lateral stability, dependent particularly upon the arrangement of the fascia lata, the iliotibial tract, the fibular collateral ligament of the knee, and the tibialis anterior. The last, acting from a fixed foot, checks lateral sway at the ankle. The gluteus medius and minimus are relaxed during lateral sway.

LOCOMOTION

Locomotion is an extraordinarily complex function, of which only a brief account is given here. It is laboriously learned, and is almost completely automatic. Disturbances of gait (and also of posture) are important signs in many central nervous system disorders.

The pattern of walking may be altered by many factors. These include habit or style (pigeon-toed, and toe-out, or splay-footed, gaits), poor posture, overweight, footwear (such as high heels), and individual differences in coordination. Patterns of locomotion may differ greatly, not only in sequence or type of muscular activity, but also in the amount of energy required. The gluteus maximus and quadriceps femoris, for example, are much more active and important in walking up and down hill.[5] The list of examples could be expanded almost indefinitely, were one to discuss the varied and exceedingly complex patterns of athletics and the dance.

The movements of the lower limb during walking on level ground may be divided into "swing" and "stance" phases. The swing phase occurs when the limb is off the ground, and the stance phase when it is in contact with the ground, bearing weight. A cycle of walking is the period from the heel-strike of one foot to the next heel-strike of the same foot.

The center of gravity moves upward and downward twice during each cycle. That is, the body is lifted when a limb is extended during its stance phase, and it is lifted again when the other limb is extended during its stance phase. These vertical displacements are visible as the "bobbing" up and down of the head. The amount of vertical displacement is about 5 cm. Note from figure 24–1 that, when the body is directly over the stance limb, that limb is not fully extended. In other words, the center of gravity when a person is standing upright is at a higher level than at any time during walking; a person is about 1 cm "shorter" when he walks.

During walking, the center of gravity also moves about 5 cm from side to side. This displacement is evident when the person is watched from in front or from behind.

If the lower limbs were rigid levers, the displacement of the center of gravity would be much greater, and more energy would be required in walking.

Three basic types of movements are involved: flexion and extension at the hip, knee, and ankle joints, and at the front part of the foot; abduction and adduction, chiefly at the hip joint; and rotation, chiefly at the hip and knee joints. The muscles do not always act in clear-cut patterns as prime movers and antagonists, but tend to stabilize joints during weight bearing and push-off, and to accelerate and decelerate movements imparted by gravity.

Flexion and Extension. In figure 24–1A, the subject is viewed from the side so as to illustrate flexion and extension. The right foot is beginning the swing phase; preparatory to push-off, the ankle is relatively rigid

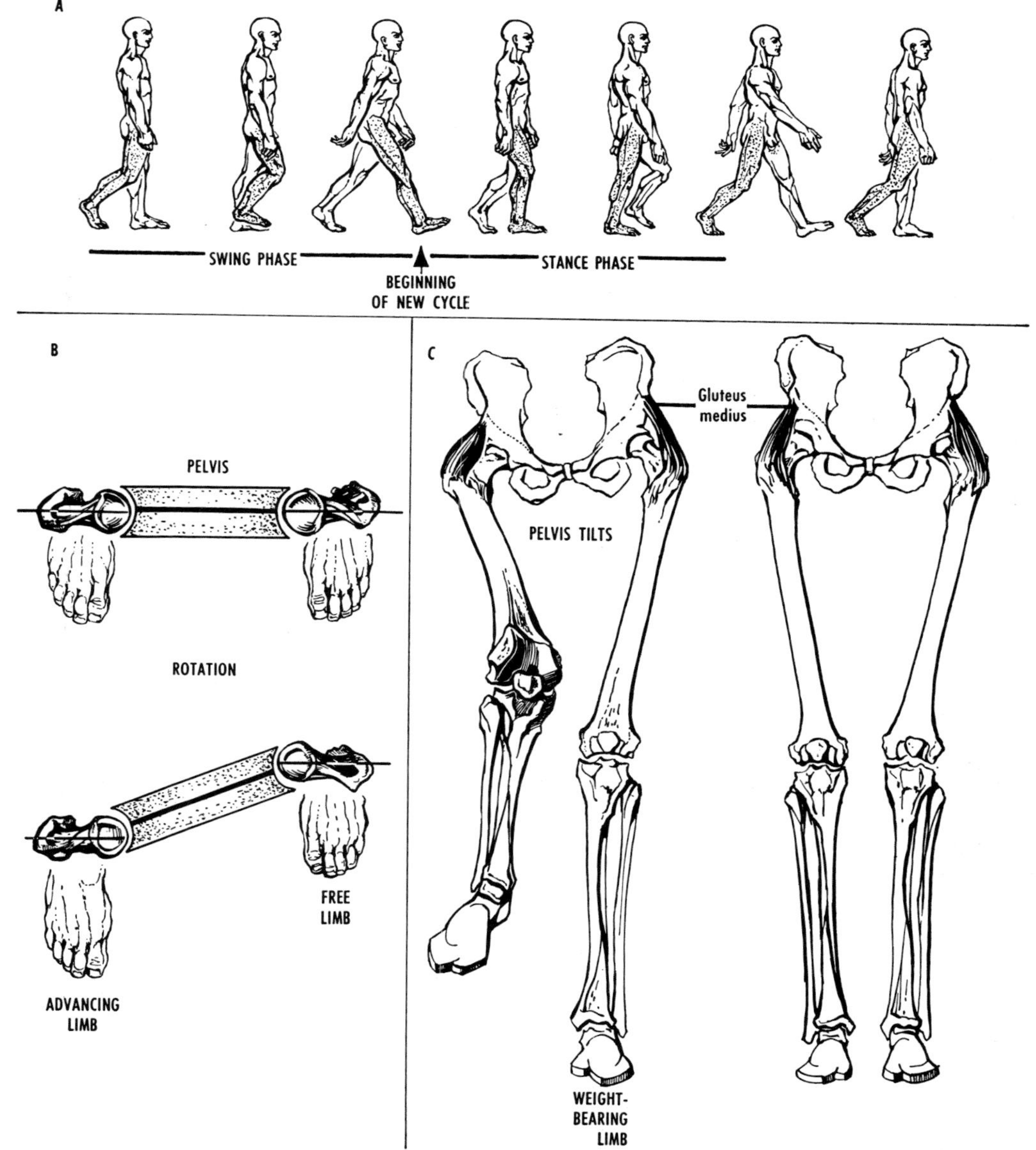

Figure 24–1 *A*, the swing and stance phases of the right lower limb. *B*, rotation between the femur and the pelvis viewed from above. *C*, abduction and adduction viewed from in front.

and the foot is bent at the metatarsophalangeal joints. The hip, knee, and ankle flex during the first part of the swing phase. The limb then begins to extend, and is fully extended when the heel strikes the ground; this is the beginning of the stance phase. The knee thereupon flexes slightly, and it is flexed slightly when weight is fully borne. It again extends fully at the end of the stance phase, preceding the push-off into the swing phase. There is therefore extension at the knee joint twice during the stance phase.

The hip flexors are most active during the early part of the swing phase, whereas the extensors are maximally active at heel-strike. Not only do they extend the hip, but they decelerate extension of the leg so as to plant the heel on the ground. The quadriceps femoris acts during two parts of the cycle, shortly after full extension at heel-strike (to support the knee) and again at take-off. The quadriceps takes less part in level walking than is ordinarily believed. The plantar flexors of the foot are most active during the latter half of stance, especially during take-off. The dorsiflexors of the foot are active after heel-strike (when they decelerate the foot and prevent it from slapping the ground, and also at the beginning of the swing phase when they facilitate the clearing of the foot from the ground).

During the latter part of the stance phase, the toes tend to flex and grip the ground or walking surface.* The long extensors and the intrinsic muscles of the foot stabilize the toes and provide fixed origins so that the long flexors and extensors can act on the leg.†

Stabilization is important during walking. An example is the control of pelvic tilt by the hip abductors. The invertors and evertors of the foot are the principal stabilizers during the stance phase. They also aid in the support of the arches of the foot during the stance phase, as do the intrinsic muscles of the foot.

Abduction and Adduction. In figure 24–1*C*, the pelvis is viewed from the front. When weight is borne upon one limb during the stance phase, the pelvis sags or tilts toward the ground on the free or swing side, owing to the effect of gravity. The sag is minimized by the abductors of the hip on the stance side. They contract strongly, acting on the pelvis from a fixed femur. During walking, this pelvic tilt alternates from side to side.

Rotation. In figure 24–1*B*, the pelvis and the hip joints are viewed above. During walking, the segments of the limb rotate about longitudinal axes (for example, medial rotation of the femur on the tibia at the end of leg extension).[7]

Note that, as the limb advances, the femur rotates laterally with respect to the hip bone. The net result is that the feet are kept pointed more or less straight ahead.

*The weight of the body flattens the arch; this is accentuated at take-off. The increased tension in the plantar aponeurosis tends to flex the joints, and thus takes part in "gripping," and also helps to provide stable toes.[6]

†F. Wood Jones suggested that in certain phases of the cycle of walking the quadratus plantae contracts and pulls on the tendon of the flexor digitorum longus. The belly of the latter would thus relax (and would therefore not hinder dorsiflexion at the ankle), but its action on the toes would be maintained.

REFERENCES

1. B. Åkelblom, *Standing and Sitting Posture,* A.-B. Nordiska Bokhandeln, Stockholm, 1948. J. Joseph, *Man's Posture: Electromyographic Studies,* Thomas, Springfield, Illinois, 1960. J. V. Basmajian, *Muscles Alive,* Williams & Wilkins, Baltimore, 3rd ed., 1974.
2. Based upon *Fundamental Studies of Human Locomotion and Other Information Relating to Design of Artificial Limbs,* University of California report to National Research Council, vols. 1 and 2, 1945–47; Basmajian, cited above; S. Carlsöö, *How Man Moves,* trans. by W. P. Michael, Heinemann, London, 1973; and upon papers by V. T. Inman, J. Bone Jt Surg., *29*:607, 1947; A. S. Levens, V. T. Inman, and J. A. Blosser, J. Bone Jt Surg., *30A*:859, 1948; J. B. de C. M. Saunders, V. T. Inman, and H. D. Eberhart, J. Bone Jt Surg., *35A*:543, 1953.
3. J. W. Smith, Acta orthopaed. scand., *23*:159, 1954.
4. M. G. Fox and O. G. Young, Res. Quart. Amer. Ass. Hlth. phys. Educ., *25*:277, 1954.
5. J. Joseph and R. Watson, J. Bone Jt Surg., *49B*:774, 1967.
6. J. H. Hicks, J. Anat., Lond., *88*:25, 1954.
7. M. Harty, Lancet, *2*:275, 1953.

SURFACE ANATOMY OF LOWER LIMB 25

HIP AND THIGH

In the erect position, the pubic crest, coccyx, middle of the acetabulum, head of the femur, and tip of the greater trochanter are all in approximately the same horizontal plane. The sacrum is entirely, or almost entirely, above the level of the pubic symphysis. The highest point of the iliac crest is behind the midpoint of the crest. The supracristal plane connecting these highest points is at the level of the spine of L.V. 4. The tubercle of the iliac crest is at the level of L.V. 5.

The anterior superior iliac spine is an im-

portant landmark. It is located by tracing the iliac crest downward or the inguinal ligament upward. In the erect position, it is in approximately the same coronal plane as the pubic tubercle and is about at the level of the first sacral vertebra. The length of the lower limb is measured clinically from the anterior superior iliac spine to the tip of the medial malleolus; the tape is kept to the medial side of the patella.

The pubic tubercle is about 3 cm from the median plane. It is a guide to the superficial inguinal ring, the femoral ring, and the saphenous opening. It can be located by tracing the tendon of the adductor longus upward to it.

A skin dimple often marks the posterior superior iliac spine, and this in turn indicates the level of the sacroiliac joint and S.V. 2.

The ischial tuberosity is palpable when the thigh is flexed.

The greater trochanter is approximately 10 cm below the iliac crest. In the erect position, it is in the same horizontal plane as the pubic tubercle, head of the femur, and coccyx. The hip joint is thus indicated by this plane, about 1 cm below the intermediate third of the inguinal ligament. A line from the ischial tuberosity to the anterior superior iliac spine crosses the tip of the greater trochanter.

The adductor tubercle is the highest point of the medial condyle of the femur. It may be located by following the tendon of the adductor magnus downward to it.

The exit of the superior gluteal artery and nerve from the pelvis is indicated by the superior point of trisection of a line from the posterior superior iliac spine to the upper end of the greater trochanter (fig. 25–1).

The exits of the inferior gluteal artery and nerve, internal pudendal artery, and pudendal nerve are indicated by the inferior point of trisection of a line from the

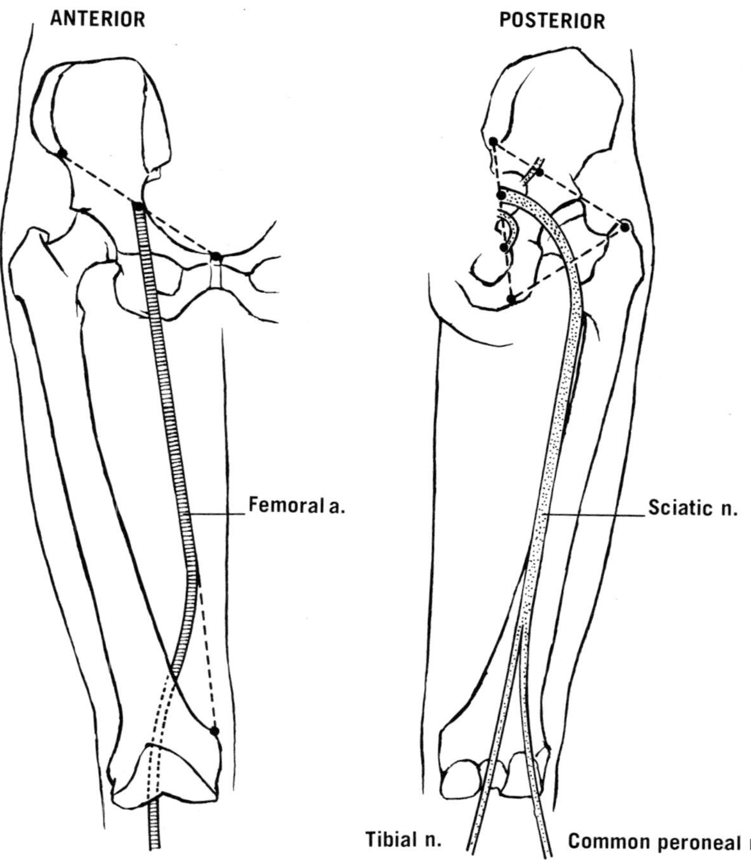

Figure 25–1 Surface anatomy of the femoral artery, gluteal nerves, and sciatic nerve.

posterior superior iliac spine to the ischial tuberosity.

The sciatic nerve emerges from the pelvis about a third of the way along a line from the posterior superior iliac spine to the ischial tuberosity (superior point of trisection) (fig. 25–1). It then descends, slightly lateral to the ischial tuberosity (midpoint of line between ischial tuberosity and greater trochanter), down the middle of the back of the thigh. It ends about halfway down the thigh or in the upper part of the popliteal fossa.

The femoral artery is indicated by the upper two-thirds of a line from the midinguinal point (midpoint between the anterior superior iliac spine and the pubic symphysis) to the adductor tubercle (fig. 25–1). Its pulsations can be felt when the thigh is flexed, abducted, and rotated laterally. **The artery can be compressed, especially by pressing directly backward at the midinguinal point.** The adductor canal is in the middle third of the thigh, deep to the sartorius. The femoral nerve descends behind the middle of the inguinal ligament (lateral to the artery) and divides after a short course.

LEG (fig. 25–2)

The lowest level of the knee joint is at the level of the margins of the tibial condyles, 1 cm or more below the apex of the patella. The head of the fibula (located by tracing the biceps tendon downward) and the tibial tuberosity are at the same level, about 1 cm below the knee joint. The tuberosity and the anterior border of the tibia are subcutaneous and constitute what is known as the "shin."

The medial malleolus is subcutaneous. Its tip is in a plane anterior to and above the tip of the lateral malleolus. The tendons of the tibialis posterior and the flexor digitorum longus are often palpable behind the malleolus. The triangular subcutaneous surface of the lower part of the fibula is palpa-

Figure 25–2 Some surface landmarks of the lower limb. Courtesy of J. Royce, Ph.D. (cited on p. 164), and Davis, Philadelphia.

(*Figure 25–2 continued on opposite page.*)

ble and is continuous with the lateral surface of the lateral malleolus. The tip of the lateral malleolus is about 1 cm distal to that of the medial malleolus, and it is more posterior. The tendons of the peronei are palpable behind it.

The highest level of the ankle joint is about 1 cm above the tip of the medial malleolus.

The popliteal artery is indicated by a line from the superior angle of the popliteal fossa to the middle of the back of the leg at the level of the tibial tuberosity. Its pulsations can sometimes be detected when the knee is flexed passively. The common peroneal nerve descends from the sciatic nerve, at the superior angle of the popliteal fossa, along the medial margin of the biceps to the back of the head of the fibula. It then winds forward around the neck. **The common peroneal nerve is palpable against the biceps and the fibula.**

The tibial nerve begins at the superior angle of the popliteal fossa and descends, first with the popliteal artery and then with the posterior tibial artery. Its course is indicated by a line from about the level of the tuberosity of the tibia, downward to the midpoint between the medial malleolus and the heel. **The posterior tibial artery is often palpable between the malleolus and the tendo calcaneus.**

The anterior tibial artery is indicated by a line from the midpoint between the tibial tuberosity and the fibula, downward to the midpoint between the malleoli anteriorly. The deep peroneal nerve extends from the lateral side of the neck of the fibula, and then turns medially to join and descend with the anterior tibial artery, to the midpoint between the malleoli. The superficial peroneal nerve becomes cutaneous in the lower third of the leg.

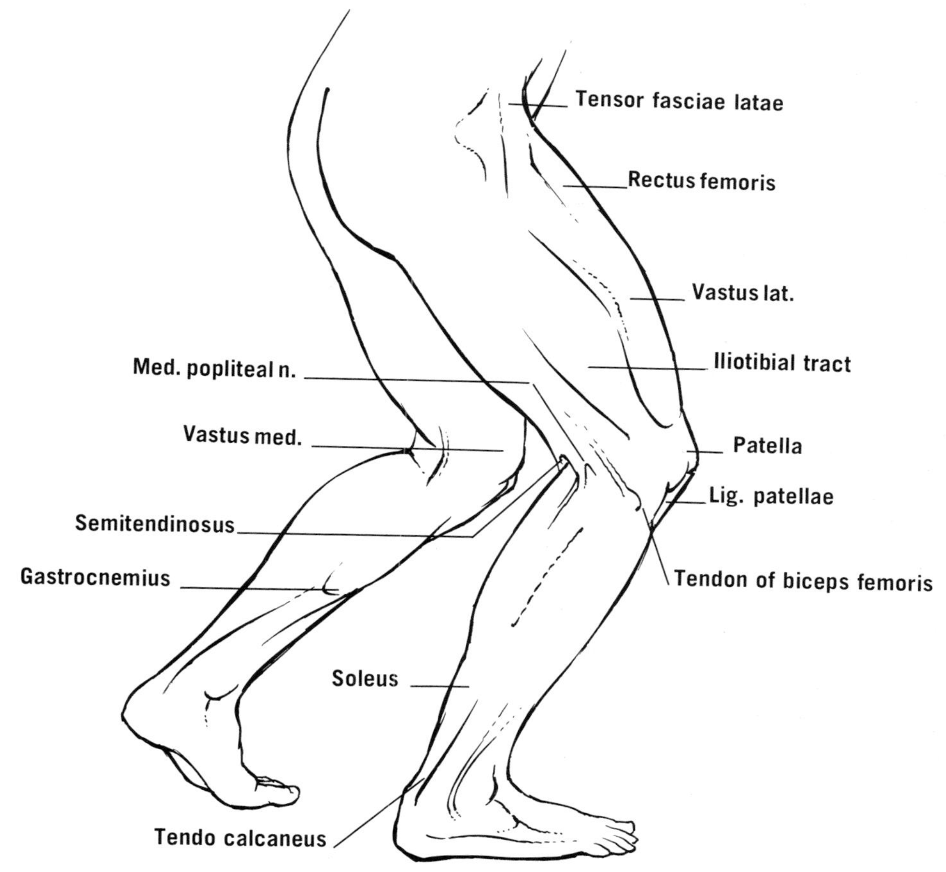

Figure 25–2 Continued.

FOOT AND ANKLE

The medial edge of the sustentaculum tali is approximately 2 to 3 cm below the tip of the medial malleolus. The peroneal trochlea, when present, is about 2 to 3 cm below the tip of the lateral malleolus.

The posterior end of the talus can sometimes be felt vaguely between the medial malleolus and the tuber calcanei when the foot is dorsiflexed. The medial part of the head can be felt between the medial malleolus and the tuberosity of the navicular in the everted foot. The lateral part of the head may be palpated on the dorsum of the inverted foot. The body may be palpated on the dorsum of the plantar-flexed foot. **The various parts of the talus are more readily palpable in children.**

The tuberosity of the navicular is palpable about 3 cm below and anterior to the tip of the medial malleolus and about 2 to 3 cm in front of the sustentaculum tali.

The medial cuneiform is located by tracing the tendon of the tibialis anterior forward to it in the dorsiflexed foot. The first metatarsal is anterior to the medial cuneiform. The medial edge of its base is approximately 4 to 5 cm anterior to the tuberosity of the navicular. The tuberosity of the fifth metatarsal is at the middle of the lateral border of the foot.

The transverse tarsal joint is indicated by a line from the back of the tuberosity of the navicular to a point halfway between the lateral malleolus and the tuberosity of the fifth metatarsal.

The tarsometatarsal joint line is indicated roughly by a line curving laterally and backward, from just behind the base of the first metatarsal to immediately behind the tuberosity of the fifth metatarsal. The base of the second metatarsal, however, projects behind this line.

The dorsalis pedis artery extends from midway between the malleoli anteriorly to the posterior end of the first intermetatarsal space. **The pulsations of the dorsalis pedis artery are often not palpable, but when they are, they may be felt on the lateral side of the extensor hallucis longus tendon.**

The lateral plantar artery and nerve are indicated by a line from between the medial malleolus and heel to the medial side of the base of the fifth metatarsal. The medial plantar artery and nerve begin at the same proximal point and extend to the posterior end of the first intermetatarsal space.

The great saphenous vein begins on the dorsum of the foot, anterior to the medial malleolus. It then ascends along the medial border of the tibia and passes behind the medial condyle of the femur. It is indicated then by a line from the adductor tubercle to the saphenous opening.

The small saphenous vein begins on the dorsum of the foot and ascends behind the lateral malleolus to the popliteal fossa.

Part Four

THE THORAX

ERNEST GARDNER

DONALD J. GRAY

RONAN O'RAHILLY

Introduction

The thorax lodges the heart and lungs, and many other important structures. Its skeletal framework, which encloses and protects these and some of the abdominal organs, consists of the thoracic vertebrae and intervertebral discs, the ribs and the costal cartilages, and the sternum (fig. 26–1). The thoracic skeleton is constructed so that by appropriate movements the volume of the thoracic cavity can be varied.

The *thoracic cavity* communicates with the front of the neck by the *superior thoracic aperture,* or *thoracic inlet.* The inlet is bounded by the upper margin of the first thoracic vertebra behind, the upper border of the manubrium in front, and the first pair of ribs and their cartilages at the sides. The inlet measures about 5 cm from front to back, and about 10 cm from side to side. Owing to the obliquity of the first ribs, the inlet slopes downward and forward. It is occupied on each side by the apices of the lung and pleura and by the neurovascular bundle for the upper limb; it is occupied in the median plane by the vessels of the head and neck and by viscera.

The thoracic cavity communicates with the abdomen by the *inferior thoracic aperture,* or *thoracic outlet,* which is closed by the diaphragm. The outlet is large and uneven in outline; it is bounded by the twelfth thoracic vertebra, the twelfth pair of ribs, the free edges of the lower six pairs of costal cartilages, and the xiphisternal joint.

The seventh to tenth costal cartilages meet on each side and their medial borders form the costal margin. The costal margins form the sides of the *infrasternal* (*subcostal*) *angle;* the xiphisternal joint forms its apex. The xiphoid process extends downward into the infrasternal angle. The slight depression in front of the process is the *epigastric fossa* ("pit of the stomach"). The infrasternal angle is commonly between 70 and 110 degrees.

At birth the thorax is nearly circular in section, but between infancy and puberty it gradually becomes more elliptical, until in the adult it is wider from side to side than from front to back. The maximum eccentricity of the thoracic diameters is reached one to two years earlier in girls than in boys. The shape of the chest also varies from person to person. The extremes are represented by those with broad chests and wide infrasternal angles (hypersthenic individuals), and those with thin, narrow chests and narrow infrasternal angles (asthenic individuals). The ratio of chest depth (anteroposterior diameter) to chest width (transverse diameter) is the thoracic index.

The right side of the thorax is often larger, the right clavicle more prominent, the right shoulder higher, the muscles of the right side bigger, and the curve of the vertebral column slightly convex to the right.

GENERAL READING

Edwards, E. A., Malone, P. D., and Collins, J. J., *Operative Anatomy of Thorax,* Lea & Febiger, Philadelphia, 1972.

Kubik, S., *Surgical Anatomy of the Thorax,* trans. by S. E. Connelly, reviewed and edited by J. E. Healey, Saunders, Philadelphia, 1970.

SKELETON OF THORAX

26

The skeleton of the thorax includes the sternum, the ribs, the costal cartilages, and the thoracic vertebrae and intervertebral discs (fig. 26–1).

STERNUM

The sternum is a flat bone that forms a portion of the bony wall of the thorax. It consists of three parts, the manubrium, body, and xiphoid process, from above downward (figs. 26–2 and 26–3). **Because of its accessibility and the thinness of its compacta, the sternum can be punctured by a needle, and marrow can be aspirated for study.** Blood may also be transfused into the marrow of the sternum.

Manubrium. The manubrium is the widest and thickest of the three parts. The concave middle part of the upper border is the *jugular notch,* which can easily be palpated and which is, on the average, at the level of T.V. 3 (at T.V. 2 in forced inspiration, and at T.V. 4 in forced expiration).[1] Laterally the upper border presents the concave *clavicular notches* for the sternal articular surfaces of the clavicles. The part of the manubrium between the clavicular notches is the thickest part of the entire sternum. The first costal cartilage is attached to the rough upper part of the lateral border, and the lower part of this border presents a facet which, together with an adjacent facet on the body, articulates with the second costal cartilage.

The lower border of the manubrium articulates with the upper border of the body at a slight angle, termed the *sternal angle.* A transverse ridge that marks the angle can be felt and often seen in the living, about 5 cm below the jugular notch. **The sternal angle is an important bony landmark at the level of T.V. 4 or T.V. 5. It marks not only the junction of the manubrium with the body of the sternum, but also the level of the second costal cartilages. Hence it is a reference point in counting ribs.** Rarely, however, the sternal angle may occur at the level of the third costal cartilages. The manubrium and body of the sternum are usually joined by fibrocartilage; the junction is sometimes ossified.

Body. The body of the sternum is about twice as long as the manubrium and is widest at the level of the fourth or fifth costal cartilage. Three more or less distinct lines usually cross the front of the body. These mark the lines of fusion of earlier separate ossific centers. In children, for example, the body is composed of several ossific centers united by cartilage. The lowest of the three transverse lines may be interrupted by a foramen, which is filled with hyaline cartilage in the living. The third, fourth, and fifth costal cartilages fit into notches in the lateral border of the body at the levels of these lines. The sixth costal

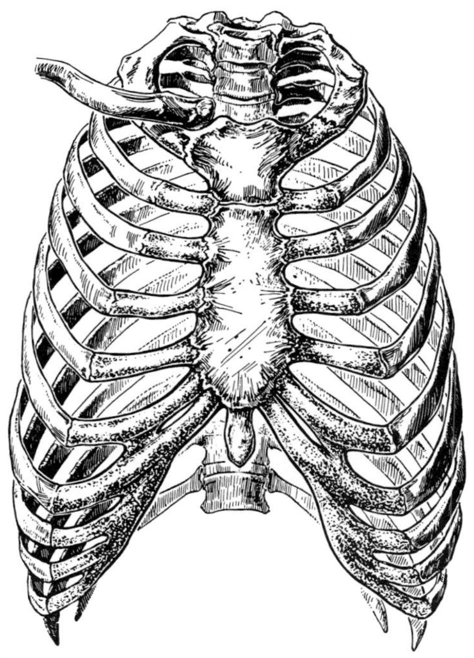

Figure 26–1 The bones of the thorax. Note that the upper two and one-half and the lower two and one-half thoracic vertebrae are visible.

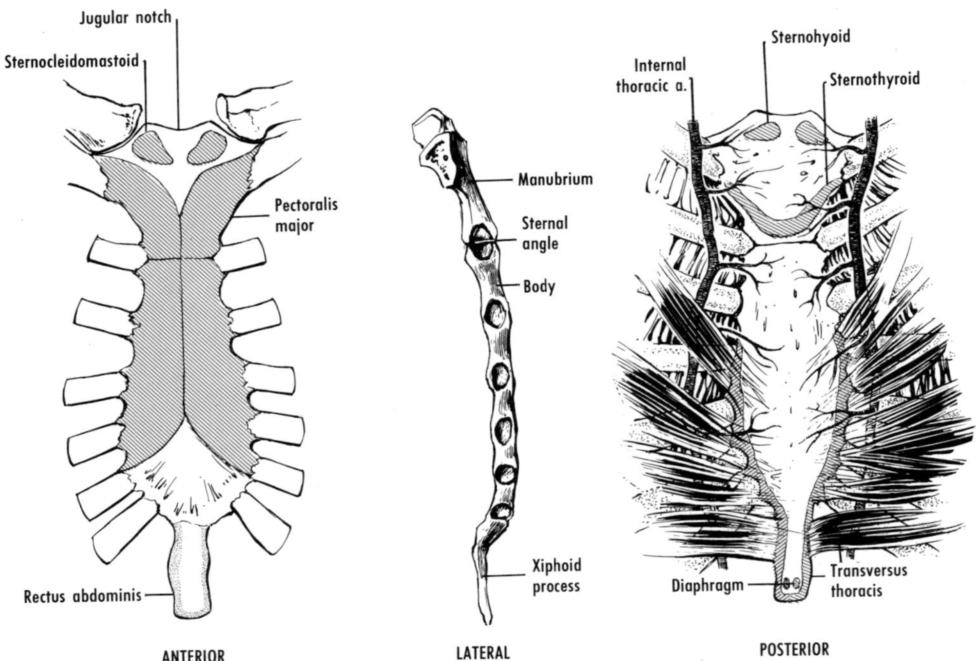

Figure 26–2 The sternum and its muscular attachments.

cartilage fits into a notch in the lateral border of the lowest segment, and the seventh fits into a notch usually shared by the body and the xiphoid process.

The posterior surface of the body, which is slightly concave from above downward, is usually smoother than the anterior, which is slightly convex. The lower margin of the body is separated from the xiphoid process by fibrocartilage until old age, when the two parts usually become fused.

Xiphoid Process. The xiphoid process, the smallest of the three parts of the sternum, is variable in size and shape. It may be bifid or perforated. It is thinner than the body. In living subjects a depression can be felt and often seen in front of the xiphoid process, the epigastric fossa or "pit of the stomach." In the adult, the xiphoid process consists of hyaline cartilage surrounding a central core of bone. This core increases with age.

The *xiphisternal joint* is at the apex of the infrasternal angle. It can be felt as a short horizontal ridge in the upper part of the epigastric fossa. It lies at about the level of T.V. 10 or T.V. 11.[2]

Ossification and Variations

Variations in shape are usually correlated with the pattern of development. Two longitudinal bands connect the anterior ends of the developing ribs early in embryonic life. These bands fuse and form a single median structure, the sternum. Complete fusion predisposes to a series of single, median ossific centers that are present in the manubrium and in each of the four body segments by birth. However, there may be no center for the fourth segment. Incomplete fusion predisposes to bilateral ossification, especially in the third and fourth segments. The degree to which there are bilateral ossific centers determines the shape of the adult bone.[3] Still less intimate fusion leads to a sternal foramen or to a complete sternal fissure.

An ossific center appears in the xiphoid process during childhood. The centers in the body fuse between late childhood and early adulthood. The xiphoid process does not ossify completely. Suprasternal ossicles are present occasionally.[4] They ossify during puberty and may articulate with the manubrium, or may fuse with it.

The sternum may be much sunken or depressed ("funnel chest"), but such a condition does not necessarily cause symptoms.[5]

RIBS

There are usually 12 ribs (fig. 26–4) on each side of the body. They are elongated,

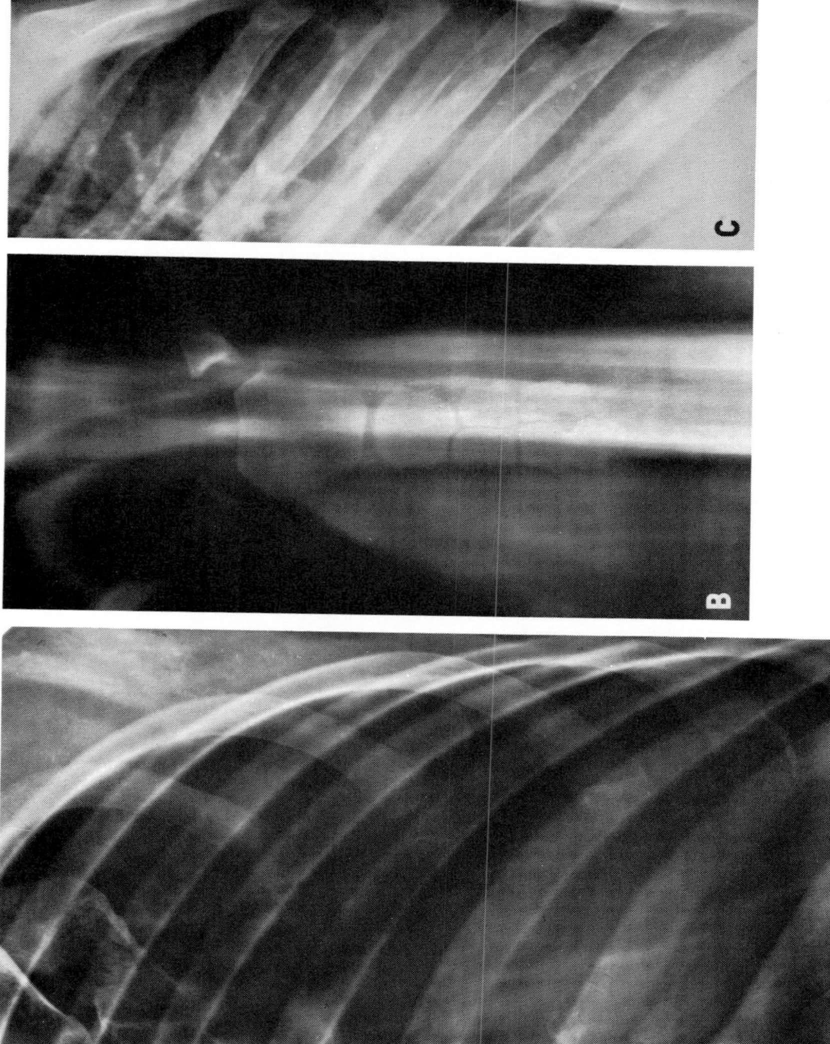

Figure 26–3 Sternum. *A*, oblique view of thorax to show the sternum. Note the manubrium, the body, and the xiphoid process. Note also the costal notches on each side for the costal cartilages of the first seven ribs. Some of the costal cartilages are calcified. The radio-opaque bars seen crossing the sternum obliquely are the back portions of ribs. *B*, tomogram of sternum of child. Note that some of the ossific centers have not yet united. With the exception of its first piece, the body of the sternum is being ossified by bilateral centers. The large shadow on the left side of the illustration is produced by the heart. *C*, lateral view of sternum of child. The various ossific centers have not yet united. The manubriosternal joint is indicated by an arrow. *B* and *C*, courtesy of Bernard S. Epstein, M.D., New Hyde Park, New York.

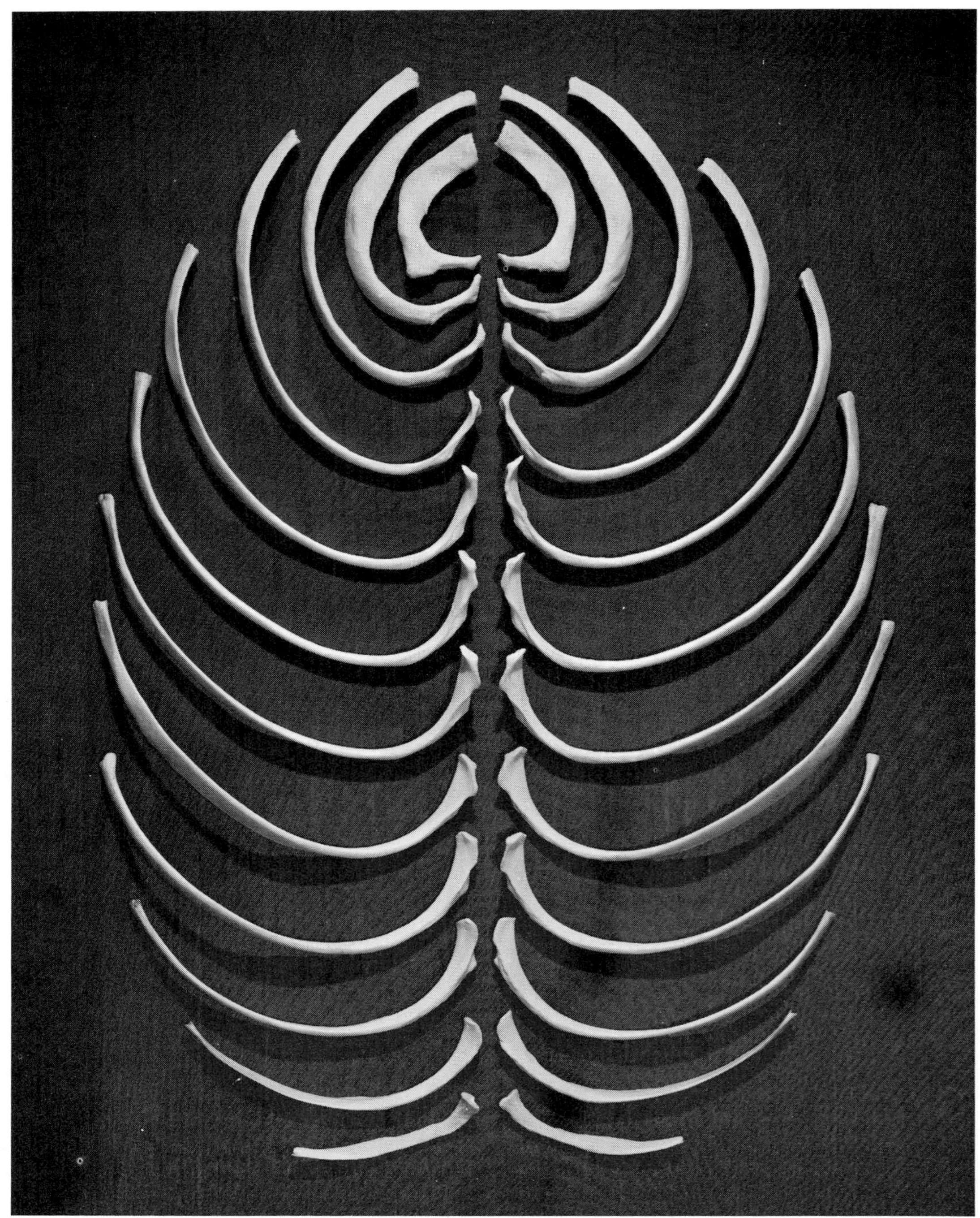

Figure 26–4 Photograph of ribs.

flattened bones, which curve downward and forward from the thoracic vertebrae. They help to protect the thoracic contents, and also the organs of the upper part of the abdomen. The ribs increase in obliquity from the first to the ninth, which is the most oblique of all. The ribs, as well as the costal cartilages, increase in length from the first to the seventh, and then decrease to the twelfth. The first seven, and sometimes eight, ribs are connected to the sternum by their costal cartilages and are called *true ribs*. Of the remaining five, which are called *false ribs*, the eighth, ninth, and usually the tenth, by means of their costal cartilages, join the costal cartilage immediately above, whereas the eleventh and twelfth are free. These last two are called *floating ribs*.

Supernumerary ribs are common. When present, they may be in the lumbar region, but they are more likely to be in the cervical region (p. 703). The anterior ends of ribs may be wide, and they are sometimes bifid. Two adjacent ribs are sometimes partially fused.

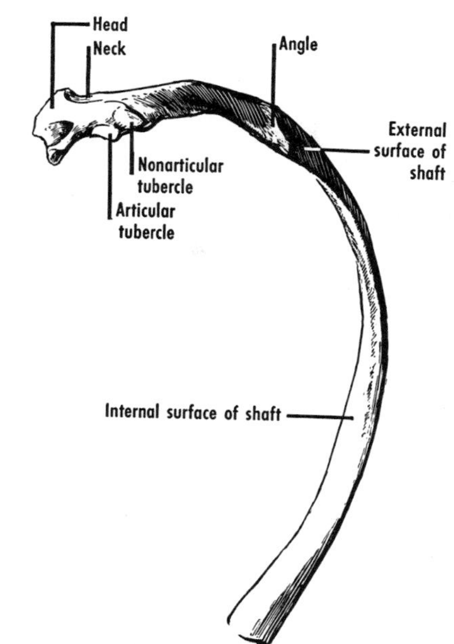

Figure 26–6 The right seventh rib from below and behind.

A Typical Rib

The third to ninth ribs have characteristics in common and are known as typical ribs (figs. 26–5 to 26–7). Each has a head, neck, and shaft. The *head* presents an *articular surface* that is divided into two facets by a *crest*. The larger, lower facet articulates with the superior costal facet of the vertebra corresponding in number to the rib, and the smaller, upper facet articulates with the inferior costal facet of the suprajacent vertebra.

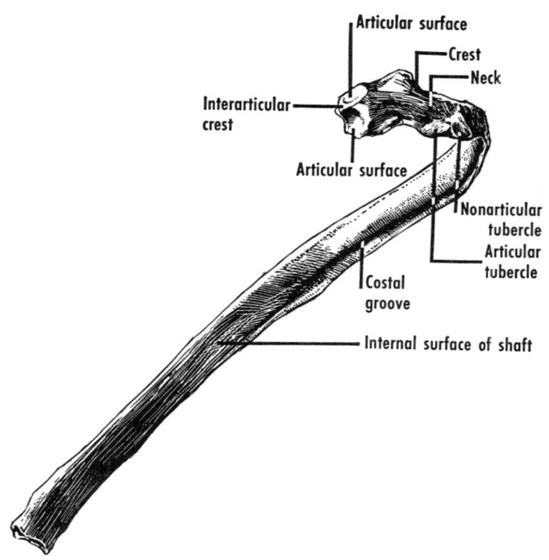

Figure 26–5 The internal aspect of the right seventh rib. Note the slope downward and forward and the twist of the shaft.

The *neck*, which is located between the head and the tubercle, presents a *crest* on its upper border. There is often an inferior crest on the neck. The junction of the neck and shaft is marked on the outer surface by the tubercle, the *articular surface* of which joins the facet on the transverse process of its vertebra.

The *shaft* of the rib passes backward and laterally a short distance and then turns forward and laterally. This turn is the *angle* of the rib (figs. 26–6 and 26–7). The shaft continues to curve, turning forward, medially, and downward. The shaft is also twisted, so that the outer surface tends to look upward as well as laterally as it is followed forward.

The backward and lateral direction of a rib from its head is such that a line connecting the angles of a pair of ribs passes through the tip of a spinous process. Thus, a

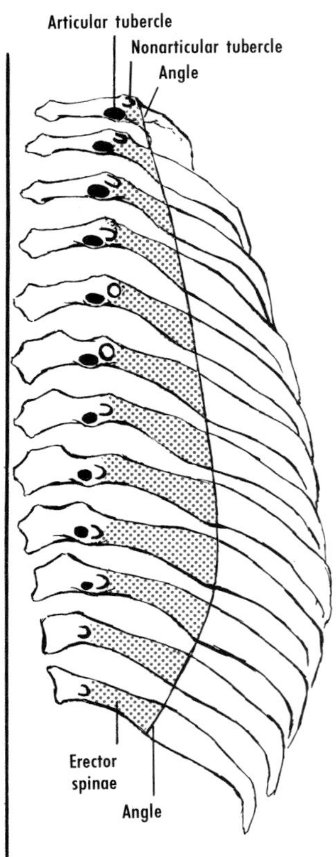

Figure 26–7 Diagrammatic representation of the ribs from behind, showing the articular and nonarticular parts of the tubercles. The line connecting the angles of the ribs marks the lateral extent of the erector spinae and of its fascia. Note the variation in the position of the angle.

person lying on his back is supported by the spinous processes and the angles of the ribs.

The convex, external surface of the shaft gives attachment to or is covered by muscles. The concave, inner surface is marked in its lower part by the *costal groove,* which is widest and deepest behind. The lower border of the rib limits the costal groove below. This border is sharp behind but rounded in front. The upper border is rounded behind and sharper in front. A cup-shaped depression in the anterior end of the shaft receives the costal cartilage.

First Rib. The first rib (figs. 26–8 and 26–9) lies at the upper limit of the thorax, where it helps to bound the thoracic inlet. It is broad and flat; its surfaces face upward and downward, and its borders inward and outward. It is wider and more curved than the succeeding ribs and is also usually the shortest. It slopes downward and forward from its vertebral to its sternal end. The small articular surface of the head usually presents only one facet, which articulates with the first thoracic vertebra. The long, slender, and rounded neck lies immediately behind the apex of the pleura and lung. The upper surface of the shaft presents near its middle the inconstant *groove for the subclavian artery,* which also lodges the lower trunk of the brachial plexus. The *tubercle for the scalenus anterior muscle* is located in front of this groove, near the inner border of the shaft. A *groove for the subclavian vein* is located in front of the tubercle.

The inferior surface of the first rib is often slightly grooved by the first intercostal nerve and by its collateral branch.[6] These grooves are variable in incidence and distinctness. The first rib may be present as only a rudimentary structure, which may be fused with the second rib near the scalene tubercle.[7] In such instances, other skeletal anomalies that distort the thoracic inlet are usually present (p. 703).

Second Rib. The second rib, which is about twice as long as the first, is strongly curved but not twisted. The articular surface of the head, like similar surfaces of a typical rib, presents two facets, which articulate with the first and second thoracic vertebrae. The second rib has an angle, a tubercle, and a poorly marked costal groove. The outer surface of the shaft near its middle presents the special characteristic of the second rib, the *tuberosity for the serratus anterior muscle,* for the attachment of the second and part of the first digitation of that muscle.

Tenth, Eleventh, and Twelfth Ribs. The tenth rib may resemble a typical rib, but the articular surface of the head usually has only one facet, for the tenth thoracic vertebra. It may be difficult to distinguish the tenth from the eleventh rib.

The eleventh rib has a single facet on the articular surface of the head. The angle and costal groove are poorly marked. The tubercle, when present, is small and has no articular facet. The anterior end of the rib is usually pointed.

The twelfth rib is small and slender.

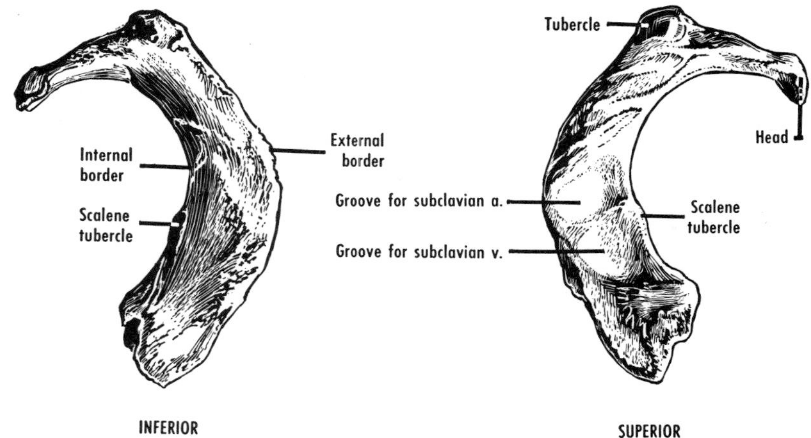

Figure 26–8 The first rib, from below and from above.

Like the eleventh, it also presents a single facet on the articular surface of the head. The tubercle, angle, neck, and costal groove are absent or, at best, poorly marked. The upper border is rounded, whereas the lower is sharp. The twelfth rib is variable in length, and may be 11 to 14 cm long.[8] **Differences in length are important in surgical approaches to the kidney.**

Costal Cartilages

The costal cartilages, more rounded than the ribs, consist of bars of hyaline cartilage, which, beginning in the fourth decade, may become ossified.[9] At one end they fit into depressions in the anterior ends of the ribs. At their opposite ends, the upper seven, sometimes eight, articulate with the sternum. The ends of the eighth, ninth, and usually the tenth connect with the costal cartilage immediately above and form the costal margin. The ends of the eleventh and twelfth lie between the muscles of the abdominal wall.

The first and second costal cartilages incline downward as well as medially, the third is approximately horizontal, and the fourth usually begins to slope upward. The fifth to the tenth continue in the direction of the corresponding rib for about 3 cm and then turn upward. The costal cartilages are elastic, can withstand considerable twisting, and impart resiliency to the chest wall.

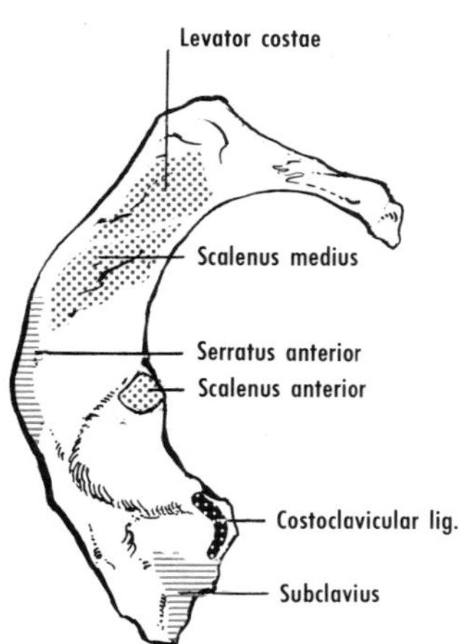

Figure 26–9 The first rib from above, showing muscular and ligamentous attachments. The scalenus minimus (not shown) is also inserted into the first rib.

Ossification

A primary center is present near the angle of the rib by the end of the embryonic period proper. Secondary centers, one for the head and two for the tubercles of typical ribs, appear at about the time of puberty and fuse with the ribs during late adolescence or early adulthood. The first, and the seventh to tenth inclusive, have one secondary center for the tubercle. The eleventh and twelfth ribs have only one secondary center, for the head.

THORACIC VERTEBRAE

The thoracic vertebrae and intervertebral discs are described on pages 516 and

533. The superior costal facet of a typical thoracic vertebra, along with the intervertebral disc and the inferior costal facet of the suprajacent vertebra, forms a socket that receives the head of the corresponding rib.

REFERENCES

1. P. Bellocq and J. G. Koritké, C. R. Ass. Anat., *42*:313, 1955.
2. A. B. Appleton, J. Anat., Lond., *72*:317, 1938.
3. G. T. Ashley, J. Anat., Lond., *90*:87, 1956; Amer. J. phys. Anthrop., *14*:449, 1956.
4. W. M. Cobb, J. Anat., Lond., *71*:245, 1937. K. Kinoshita, Kyushu J. med. Sci., *7*:63, 1956.
5. W. Evans, Brit. Heart J., *8*:162, 1946. I. D. Sutherland, J. Bone Jt Surg., *40B*:244, 1958.
6. A. J. E. Cave, J. Anat., Lond., *63*:367, 1929.
7. T. W. Todd, J. Anat., Lond., *46*:244, 1912. J. C. White, M. H. Poppel, and R. Adams, Surg. Gynec. Obstet., *81*:643, 1945.
8. J. Minet, Arch. Mal. Reins, *9*:47, 1935. F. A. Hughes, J. Urol., *61*:159, 1949.
9. Ossification is the true process, although the term calcification is commonly used. See J. B. King, Brit. J. Radiol., *12*:2, 1939. The pattern of the ossification differs according to sex. See S. Navani, J. R. Shah, and P. S. Levy, Amer. J. Roentgenol., *108*:771, 1970.

THORACIC WALL AND MEDIASTINUM 27

THORACIC WALL

MUSCLES

The muscles of the thoracic and abdominal walls are mostly arranged in external, middle, and internal layers. In the thorax, the external intercostal muscles form the external layer, and the internal intercostal muscles form the middle layer (figs. 27–1 and 27–2). The innermost intercostals, the subcostals, and the transversus thoracis form the internal layer. The diaphragm separates the thoracic and abdominal cavities. The internal layer of muscles, and the sternum, costal cartilages, and ribs, are separated from the costal pleura by a small amount of loose connective tissue, the *endothoracic fascia.*

Other muscles that contribute to the thoracic wall include certain muscles of the upper limb, the muscles of the abdominal wall, and certain muscles of the back, all of which lie external to the ribs and intercostal spaces. The levatores costarum are topographically associated with the muscles of the back, but are functionally associated with the intercostal muscles and are described with them.

External Layer

External Intercostals. The external intercostal muscles are attached to the lower margins of each of the first eleven ribs. Their fibers pass downward and forward to the upper margin of the rib below. The lower seven muscles are intimately connected with the external oblique. The external intercostals extend from the tubercles of the ribs behind to the area of the costochondral junctions in front, where they give way to the *external intercostal membranes.* Thus, the muscles are entirely interosseous in position.

The external intercostals are supplied by the corresponding intercostal or thoracoabdominal nerves. They elevate the ribs, and are therefore considered to be muscles of inspiration.

Levatores Costarum. These muscles arise from the transverse processes of the seventh cervical to eleventh thoracic vertebrae (fig. 27–5). Each is inserted into the external surface of the subjacent rib, between the tubercle and the angle. The lowermost

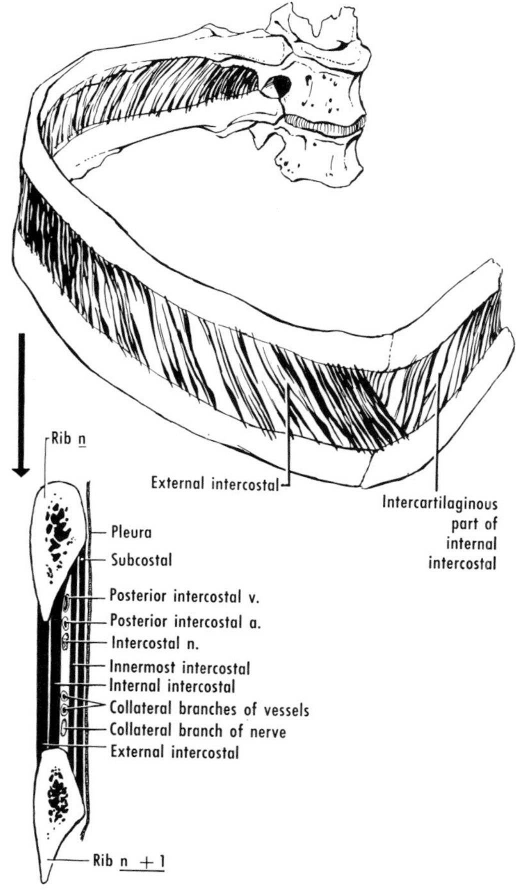

Figure 27–1 The intercostal muscles. The upper figure shows the direction of fibers of the external and internal intercostals. The arrow indicates the location of the section of the lower figure (near the angles of the ribs). The subcostal muscle, which is limited to one intercostal space in the diagram, may cross one or two ribs (see text). The endothoracic fascia, which lies between pleura and muscle, is omitted.

muscles in part extend over one rib to insert into the next rib below.

They are supplied by the dorsal rami of the eighth cervical and first to eleventh thoracic nerves.[1] They elevate the ribs and have an unimportant inspiratory function.

Middle Layer

Internal Intercostals. The internal intercostal muscles are attached to the lower margins of the ribs and costal cartilages, and to the floor of the costal groove when it is present. Their fibers pass downward and backward to the upper margins of the subjacent ribs and costal cartilages. The internal intercostals extend from the medial ends of the intercostal spaces to the angles of the ribs, where they give way to the *internal intercostal membranes.* Thus, the internal intercostal muscles consist of intercartilaginous and interosseous parts. The lowermost muscles are intimately connected with the internal oblique.

The internal intercostal muscles are supplied by the corresponding intercostal or thoracoabdominal nerves. They are muscles of expiration, except for the intercartilaginous parts in the upper four or five spaces (the "parasternal" intercostals), which are inspiratory in function (p. 277).

Internal Layer

Innermost Intercostals.[2] These may be regarded as parts of the internal intercostals, from which they are separated by the intercostal vessels and nerves. They are not well developed and may be absent in the upper spaces. They extend between the costal groove above and the upper margin of the rib below, and are supplied by intercostal or thoracoabdominal nerves. In the anterior ends of the lowermost spaces, they are fused with the diaphragm. Their action is unknown.

Subcostals. Variable in number, these are better developed in the lower part of the thorax. They arise from the lower margins of the ribs near their angles and are inserted into the upper margins of the second or third rib below. They are supplied by intercostal or thoracoabdominal nerves, and they probably elevate the ribs.

Transversus Thoracis[3] ***(Sternocostalis)*** (fig. 26–2, p. 260). This muscle arises by aponeurotic slips from the posterior surface of the xiphoid process and body of the sternum, and is inserted into the internal surface of the second to sixth costal cartilages. It is supplied by the corresponding intercostal nerves, and has an unimportant expiratory function.

Diaphragm

The diaphragm is the most important muscle of respiration, but it is not essential. It separates the thoracic and abdominal cavities (fig. 33–13, p. 364). Each half of the muscular part of the diaphragm is divided into three parts; sternal, costal, and lumbar. The three parts are inserted into the central tendon, a structure of trifoliate shape that lies just below the heart and has no bony attachments.[4] The tendon contains the foramen for the inferior vena cava.

Sternal Part (fig. 29–7, p. 290). Narrow slips arise from the back of the xiphoid

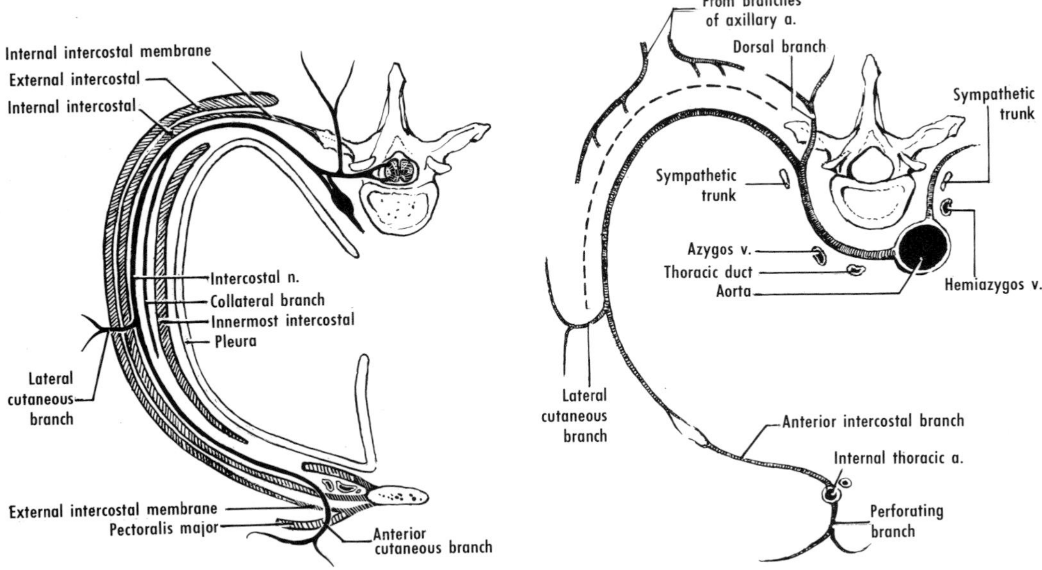

Figure 27–2 Diagrammatic representation of nerves, arteries, and muscles of the thoracic wall. See figure 27–1 for position of collateral branches. Note that the intercostal vessels pass behind the longitudinally disposed structures of the posterior mediastinum. The thickness of the intercostal muscles is exaggerated.

process and descend to the central tendon. In the cadaver, however, owing to the postmortem relaxation of the diaphragm, the sternal part appears to ascend from its origin. On each side, a small gap, known as the *sternocostal triangle,* is located between the sternal and costal parts and transmits the superior epigastric vessels and a few lymphatics. This gap may be the location of a diaphragmatic hernia.

Costal Part. The costal parts, which form the right and left "domes," arise from the inner surfaces of the lower six costal cartilages and lower four ribs. They interdigitate with the transversus abdominis at their costal attachments. The fibers are inserted into the anterolateral parts of the central tendon. Several small fibrous arches often connect adjacent costal cartilages, especially the last two or three, and some fibers arise from these arches. The portion of the diaphragm arising from the last two ribs is thin and often deficient.

Lumbar, or Vertebral, Part. Each lumbar part arises from two fibrous arches and from the bodies of the upper lumbar vertebrae (fig. 33–13, p. 364). The two fibrous arches are the *medial* and *lateral arcuate ligaments.* The medial ligament is a thickening of the fascia over the upper part of the psoas major. It extends from the body of L.V. 1 or 2 to the tip of the transverse process of the same vertebra, from which the lateral ligament extends to the eleventh or twelfth rib. The lateral ligament is a thickening of the fascia over the upper part of the quadratus lumborum. Muscle fibers extend upward from both ligaments to the central tendon.

The portion of the costal part of the diaphragm that arises from the eleventh and twelfth ribs is often separated from the lumbar part by an interval called the *vertebrocostal trigone*. When such an interval is present, it is occupied by loose connective tissue that separates the pleura above from the suprarenal gland and upper pole of the kidney below.

The part of the diaphragm arising from the lumbar vertebrae forms two muscular crura that ascend to the central tendon. The *right crus* arises from the upper three or four vertebrae, and the *left crus* from the upper two or three. The crura are united in front of the aorta by a fibrous arch, the *median arcuate ligament,* thereby forming the aortic opening (p. 414). The larger right crus splits around the esophagus (see figs. 28–2, p. 280, and 33–13, p. 364).[5] Part of the right crus continues into the suspensory ligament of the duodenum. The left crus, which ascends to the left of the esophagus, is smaller and much more variable. A slip from the left crus may enter into the formation of the esophageal opening.

Nerve Supply. The diaphragm and adjacent pleura and peritoneum are supplied by the phrenic nerves (fig. 31–7, p. 333), each of which is distributed to one half of the diaphragm. The left half of the right crus, which lies to the left of the median plane, is supplied by the left phrenic nerve.[6] The peripheral part of the diaphragm is also supplied with sensory and vasomotor fibers from the thoracoabdominal nerves.

ACTION[7]

The diaphragm descends when it contracts, and draws the central tendon downward with it. The volume of the thorax is thereby increased and intrathoracic pressure decreased. The volume of the abdominal cavity is decreased, however, and intra-abdominal pressure is raised.

The costal part of the diaphragm raises and everts the costal margin. The movements of the diaphragm are also important in the circulation of the blood. The decreased intrathoracic pressure and increased abdominal pressure that accompany the descent of the diaphragm facilitate the return of blood to the heart.

Each half of the diaphragm has a separate nerve supply, and paralysis of one half does not affect the other half. However, the two halves of the diaphragm usually contract synchronously. The diaphragm is under voluntary control only to a certain extent. No one can voluntarily hold his breath to the point of asphyxiation.

Hiccups are spasmodic, sharp contractions of the diaphragm.

SHAPE AND RELATIONS

In the median plane, the sternal part of the diaphragm descends from its origin and then curves upward over the liver to the central tendon, in conformity with the diaphragmatic surface of the heart (fig. 29–7, p. 290). The heart and pericardium rest on the central tendon. The downward inclination of the diaphragm from its sternal origin is ordinarily not seen in a cadaver. In sagittal sections, each costal part or dome can be seen to arch over the abdominal viscera.

Openings. **The diaphragm has three major openings. The *aortic opening,* which lies behind the crura, transmits the aorta, and often the thoracic duct and greater splanchnic nerves as well. The *esophageal opening,* in the right crus, transmits the esophagus and the vagus nerves. The *foramen for the inferior vena cava,*[8] in the right half of the central tendon, transmits the inferior vena cava, the right phrenic nerve, and lymphatic vessels from the liver.** Sometimes the right hepatic vein traverses this opening before joining the inferior vena cava. The splanchnic nerves, sympathetic trunk, subcostal vessels and nerves, superior epigastric vessels, musculophrenic vessels, and azygos and hemiazygos veins either pierce the diaphragm or are related to it.

Position of the Diaphragm. **In the erect position, and in the midphase of respiration, the highest parts of the domes of the diaphragm are at about the same level as the apex of the heart** (p. 308). Changing from the erect to the supine posture has little effect on the total movement of the diaphragm, but the resting level of the diaphragm rises (the excursion during inspiration is increased), and thoracic volume is decreased.

In the lateral posture (lying on one's side), the dome of the diaphragm on the "lower" side rises farther into the thorax, much as the whole diaphragm does in the supine position (fig. 32–7, p. 346). The excursion of the "lower" dome and of the corresponding lung is increased during inspiration.[9]

Development and Congenital Anomalies. The diaphragm is formed chiefly from the *septum transversum.* Outright defects of the diaphragm are usually congenital and are uncommon, as are accessory slips and duplications.[10] However, variations in the degree of development of the muscular parts are fairly common. A pleuroperitoneal canal, usually the left, may persist; the result is a gap in the costal part of the diaphragm. The diaphragm may be almost completely absent on one side; this represents a severe grade of persistent pleuroperitoneal canal.

Diaphragmatic Hernias.[11] A diaphragmatic hernia is a displacement of an abdominal organ or structure through a weak area or defect in the diaphragm into the thoracic cavity. Such a hernia may be congenital or may be acquired after birth, and it may be due to trauma. Almost every abdominal viscus has been found in the thorax in different patients.

Congenital hernias may occur through gaps in the costal part of the diaphragm (failure of closure of pleuroperitoneal opening), through the esophageal opening, or through the sternocostal triangle. Acquired hernias may occur in the same locations. Most diaphragmatic hernias, whether congenital or acquired, occur through the esophageal opening and are termed *hiatal hernias.*

Diaphragmatic hernias usually have sacs, which

consist of peritoneum and which are formed as an organ pushes peritoneum ahead of it as it enters the thoracic cavity. However, if a pleuroperitoneal canal fails to close, the pleural and peritoneal cavities communicate, and a hernia through the defect will not have a sac. The preformed sac of a paraesophageal hernia may be due to persistence of the right pneumatoenteric recess (upper part of embryonic lesser sac). The continuity is usually obliterated before birth, although the upper end of the recess may persist as the infracardiac bursa (p. 303).

BLOOD VESSELS AND LYMPHATIC DRAINAGE

The thoracic wall is supplied by the internal thoracic and highest intercostal arteries, which are branches of the subclavian, and by the posterior intercostal and subcostal arteries, which are mostly branches of the aorta. The thoracic wall is also supplied by branches of the axillary artery.

Arteries

Internal Thoracic (Internal Mammary) Artery (figs. 26–2 and 27–2). The origin of this artery is variable, but it usually arises from the lower aspect of the first part of the subclavian. Here it lies close to the medial margin of the scalenus anterior and usually opposite the thyrocervical trunk. It passes downward, forward, and medially, behind the sternocleidomastoid, the clavicle, and the subclavian and internal jugular veins. It rests on the pleura behind, and it is crossed by the phrenic nerve, which gives filaments[12] to it and which passes obliquely from its lateral to its medial side.

It passes downward through the thorax, behind the upper six costal cartilages and the intervening internal intercostal muscles, just lateral to the sternum. Behind, it lies on the pleura, except where it is separated from it by the transversus thoracis muscle. It is accompanied by two venae comitantes and by lymphatic vessels.

The internal thoracic artery ends at the sixth intercostal space by dividing into its terminal branches, the superior epigastric and musculophrenic arteries.

BRANCHES. The main distribution is to the thoracic and abdominal walls. The branches include a variable number of *mediastinal, thymic,* and *bronchial branches.*

The *pericardiacophrenic artery* arises in the upper part of the thorax, accompanies the phrenic nerve to the diaphragm, and helps to supply the pleura and pericardium.

Two *anterior intercostal branches* are present in each of the upper six intercostal spaces. Both run laterally, the upper one joining the posterior intercostal artery, and the lower one joining the collateral branch of the posterior intercostal.

A *perforating branch* is located in each of the upper six intercostal spaces. The perforating branches accompany the anterior cutaneous branches of the intercostal nerves and supply the pectoralis major muscle and the skin over it. The second, third, and fourth supply *mammary branches.*[13]

A *lateral costal branch,*[14] when present, passes downward and laterally behind the ribs and anastomoses with the posterior intercostal arteries. It is occasionally as large as the internal thoracic artery.

The *superior epigastric artery* is the medial of the two terminal branches. It passes behind the seventh costal cartilage and between the sternal and costal origins of the diaphragm (sternocostal triangle). It descends between the rectus abdominis muscle and the posterior layer of its sheath, and then enters this muscle and anastomoses with the inferior epigastric artery (fig. 33–5, p. 355). It supplies the diaphragm and anterior abdominal wall; a branch courses in the falciform ligament to the liver.

The *musculophrenic artery* is the lateral of the two terminal branches. It passes downward and laterally behind the costal attachments of the diaphragm. It pierces the diaphragm behind the eighth costal cartilage and ends at approximately the tenth intercostal space, where it anastomoses with the deep circumflex iliac artery and with the last two intercostal arteries. It supplies the diaphragm and the muscles of the abdominal wall, and gives off two anterior intercostal branches in each of the seventh, eighth, and ninth spaces.

Highest Intercostal Artery.[15] This artery arises from the costocervical trunk of the subclavian and then descends in front of the neck of the first rib. The sympathetic trunk is on its medial side, the first thoracic nerve on its lateral side. It gives off the first posterior intercostal artery at the lower border of the neck of the first rib and continues in front of the neck of the second rib to become the second posterior intercostal artery. The course and distribution of these two arteries are similar to those of the other posterior intercostal arteries. Variations are common.

Posterior Intercostal Arteries (figs. 27–2 and 27–3). The first two posterior intercostal arteries arise from the highest intercostal artery, the remaining nine from the back of the aorta. The right arteries pass across the front of the vertebral column and are therefore longer than the left. Most of them pass behind the esophagus, thoracic duct, and azygos vein. The left arteries course behind the hemiazygos or accessory

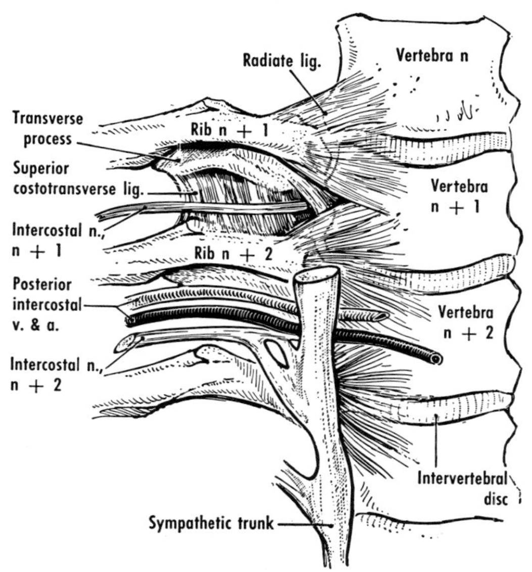

Figure 27–3 Intercostal vessels and nerve. A part of the sympathetic trunk is shown, including some rami communicantes.

hemiazygos vein. Both right and left arteries lie behind the pleura and are crossed by the sympathetic trunks; the splanchnic nerves also cross the lower ones. After passing in front of or behind an intercostal nerve, each artery reaches the angle of the rib above. It then enters the costal groove, where it lies between the corresponding nerve and vein, and passes forward between the innermost and internal intercostal muscles. At the anterior end of the intercostal space, its terminal branches anastomose with the anterior intercostal branches of the internal thoracic or musculophrenic arteries. The tenth and eleventh arteries pass into the abdominal wall and anastomose there with branches of the superior epigastric, subcostal, and lumbar arteries.

The *dorsal branch* passes backward in company with the dorsal branch of the corresponding thoracic nerve and divides into a *muscular* and a *spinal branch.* The muscular branch in turn gives rise to a *medial* and a *lateral cutaneous branch.* Each supplies the muscles and skin of the back. The spinal branch passes through the intervertebral foramen and helps to supply the spinal cord (p. 541).

The *collateral branch* arises near the angle of the rib and then runs forward to the anterior end of the space, where it anastomoses with the lower of the two anterior intercostal branches of the internal thoracic and musculophrenic arteries.

The lateral cutaneous branch accompanies the corresponding branch of the nerve through the overlying muscles. The branches of the third, fourth, and fifth posterior intercostals give small *mammary branches.*

BRANCHES. The *right bronchial artery* often arises from the aorta as a common trunk with the right third posterior intercostal artery (p. 299).

Subcostal Arteries. These two arteries are in series with the intercostal arteries. Each enters the abdomen with the corresponding nerve, courses behind the lateral lumbocostal arch, descends between the kidney and the quadratus lumborum, pierces the transversus abdominis, and anastomoses with adjacent arteries.

Collateral Circulation. The anastomoses between the internal thoracic, the posterior intercostals, and the inferior epigastric provide an important collateral route in cases of obstruction of the aorta, such as coarctation of the aorta.

Veins

Internal Thoracic Veins. The venae comitantes of the internal thoracic artery unite and form a trunk that empties into the corresponding brachiocephalic vein (sometimes, on the right side, into the superior vena cava). The right and left pairs of venae comitantes communicate by a vein that passes immediately in front of the xiphoid process.[16]

Posterior Intercostal and Subcostal Veins. The posterior intercostal and subcostal veins receive tributaries corresponding to the branches of these arteries. The first posterior intercostal vein of each side passes over the apex of the lung and its pleura and ends usually in the brachiocephalic vein (fig. 31–1, p. 327), sometimes in the vertebral vein. It may also join the *superior intercostal vein,* which is formed by the second and third, and sometimes the fourth, posterior intercostal veins. The right vein joins the azygos, and the left passes upward across the arch of the aorta and joins the left brachiocephalic vein. The remaining posterior intercostal veins on the right side join the azygos, those on the left the hemiazygos and accessory hemiazygos.

The right subcostal vein enters the thorax behind the right lateral arcuate ligament and joins the right ascending lumbar vein to form the azygos vein (p. 327). The left subcostal vein enters the thorax behind the left lateral arcuate ligament and joins the left ascending lumbar vein to form the hemiazygos vein.

Venous circulation and routes of collateral return are discussed on page 327.

Lymphatic Drainage

The lymph nodes of the thorax are visceral or parietal according to their location. The visceral nodes are discussed on page 329. The parietal nodes are parasternal, phrenic, and intercostal.

Parasternal (Internal Thoracic) Nodes. These nodes are present along the upper part of the internal thoracic artery, one or two in each of the upper four or five spaces.[17] They receive lymphatic vessels from the medial part of the breast, from the diaphragm, from the intercostal spaces, and from the costal pleura. Their efferent vessels collect into a single trunk that usually joins the bronchomediastinal trunk of the same side (fig. 31–3, p. 330).

The parasternal nodes provide a route by which cancer of the breast can spread to the lungs and mediastinum, or even downward to the liver (nodes on the diaphragm drain the liver).

Phrenic (Diaphragmatic) Nodes. Several groups of these nodes are present on the thoracic surface of the diaphragm. They receive lymphatic vessels from the lower intercostal spaces and from the pericardium, diaphragm, and liver, and send their efferent vessels to the parasternal nodes. Several nodes on the posterior part of the diaphragm send their efferent vessels to the posterior mediastinal nodes. Still other nodes, located near the left phrenic nerve and the inferior vena cava, receive lymphatic vessels from the diaphragm, liver, stomach, and esophagus, and send efferent vessels to the phrenic nodes in front of and behind them.

Intercostal Nodes. One or two small nodes are present at the vertebral end of each intercostal space. They receive lymphatic vessels from structures along the adjacent blood vessels, and from the pleura. The nodes in the upper intercostal spaces drain into the thoracic duct. Those in the lower spaces generally drain into a vessel on each side that descends to the cisterna chyli.

THORACIC NERVES

Each of the 12 thoracic spinal nerves gives off a meningeal branch (p. 532), and then, after emerging from an intervertebral foramen, divides into a dorsal and a ventral ramus. These rami (fig. 27–4) contain motor fibers to muscle, sensory fibers from skin and deep tissues, and postganglionic sympathetic fibers to blood vessels, sweat glands, and arrectores pilorum.

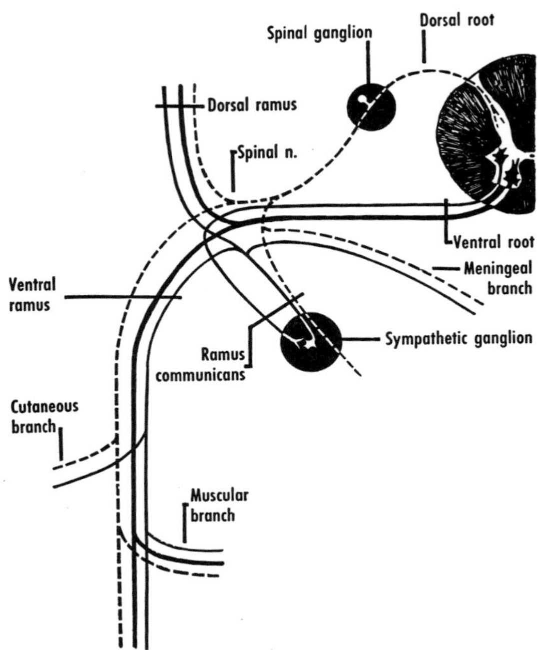

Figure 27–4 Functional components of a thoracic spinal nerve. For purposes of simplification, each component is shown as a single fiber. Motor fibers to skeletal muscle are shown as heavy lines, sympathetic fibers as light lines, and sensory fibers as interrupted lines. The branches of the ventral ramus to the pleura and peritoneum are not shown.

Dorsal Rami

The dorsal rami (figs. 27–2, 27–4, and 27–5) pass backward and supply the muscles, bones, joints, and skin of the back (p. 532).

Ventral Rami

Each ventral ramus is connected to the sympathetic trunk by a variable number of rami communicantes, and each runs a separate course forward, supplying the skin, muscles, and serous membranes of the thoracic and abdominal walls. **The distribution of the ventral rami is segmental, but overlap of adjacent nerves is so great that section of three consecutive nerves is necessary to produce complete anesthesia and paralysis within the middle one of the three intercostal spaces supplied.**

The ventral rami of the first eleven thoracic spinal nerves are called intercostal nerves. However, the first three ventral rami give branches to the upper limb as well as to the thoracic wall, the seventh to eleventh ventral rami are thoracoabdominal in their distribution, and the ventral ramus of the twelfth nerve is subcostal rather than intercostal in position.

Typical Intercostal Nerves (figs. 27–2 to 27–4). The fourth, fifth, and sixth intercostal nerves are typical intercostal nerves and supply only the thoracic wall, including the intercostal, subcostal, serratus posterior superior, and transversus thoracis muscles. Each emerges from behind a detached part of the superior costotransverse ligament. It then lies behind the pleura and in front of, first, the main part of the superior costotransverse ligament, and then the internal intercostal membrane. It passes below the neck of the rib numerically corresponding to it and enters the costal groove, where it lies below the corresponding posterior intercostal vessels. In its course forward, it lies first on the pleura and endothoracic fascia, then between the innermost and internal intercostal muscles, and finally on the transversus thoracis muscle and the internal thoracic vessels. At the anterior end of the intercostal space, medial to these vessels, it turns forward through the internal intercostal muscle, the external intercostal membrane, and the pectoralis major muscle. It is distributed as the *anterior cutaneous branch* to the skin and subcutaneous tissue of the front of the chest and gives off *medial mammary branches*.

At the angle of the rib, each intercostal nerve gives off a branch to the external intercostal muscle and then a lateral cutaneous and a collateral branch. Usually a fine communicating branch is sent upward across the deep aspect of the rib to the neighboring intercostal nerve. Each nerve gives twigs to the parietal pleura and each communicates with the sympathetic trunk by means of one to four rami communicantes (p. 335).

The *collateral branch* passes forward in the lower part of the intercostal space. This branch may rejoin the intercostal nerve but, if so, again leaves it. The collateral branch ends as a lower anterior cutaneous nerve of the corresponding intercostal space.

The *lateral cutaneous branch* becomes superficial after obliquely piercing the intercostal and serratus anterior muscles. It then divides into anterior and posterior branches, which supply the skin and subcutaneous tissue of the thorax. Some of the anterior branches give off *lateral mammary branches*.

Special Nerves. The first, second, and third intercostal nerves are special in that they supply the arm as well as the thorax. The first thoracic nerve is the largest of the thoracic spinal nerves. It divides opposite the superior costotransverse ligament into a larger upper and smaller lower part. The upper part joins the brachial plexus; the lower part becomes the first intercostal nerve (fig. 27–6), which has a distribution similar to that of a typical intercostal nerve,[18] except that its lateral cutaneous branch supplies the skin of the axilla and may communicate with the intercostobrachial nerve (and sometimes with the medial brachial cutaneous nerve also).

The second intercostal nerve, which may contribute to the brachial plexus, otherwise has an intercostal distribution similar

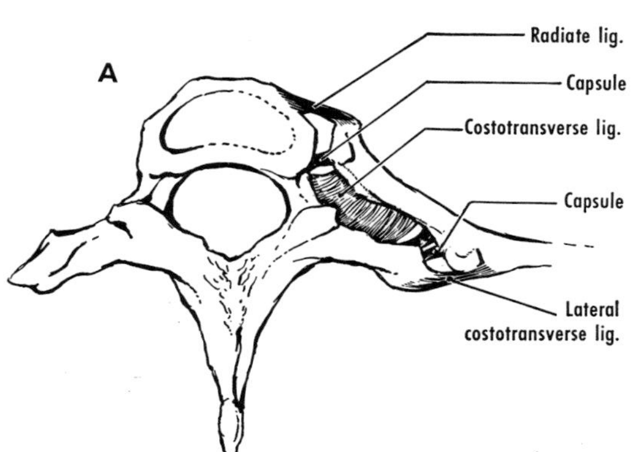

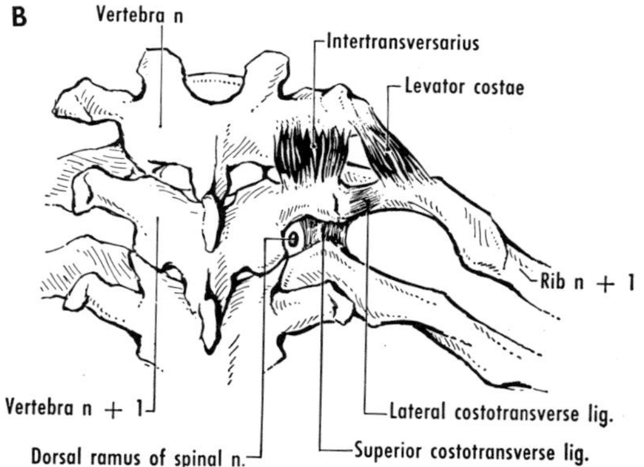

Figure 27–5 Diagrammatic representation of costovertebral joints viewed from above *(A)* and from behind *(B)*.

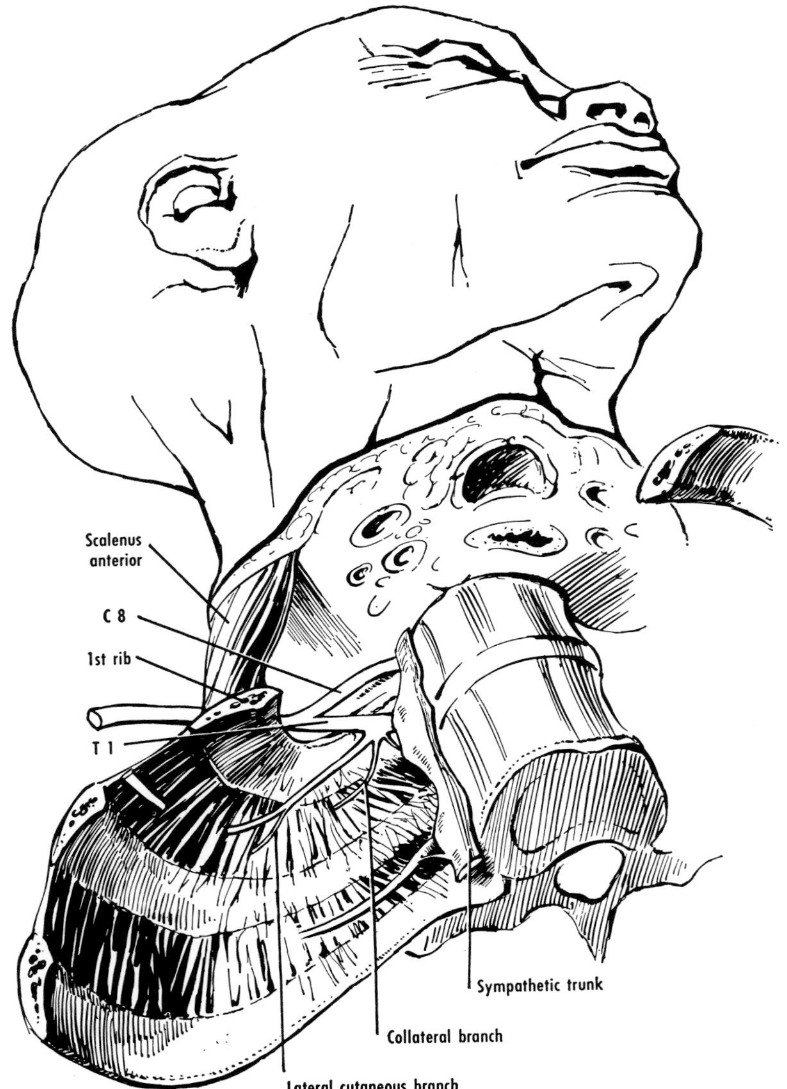

Figure 27–6 The ventral ramus of the first thoracic nerve, viewed from below. Note how the upper division of T1 joins C8 and forms the lower trunk of the brachial plexus, which rests on the first rib. Part of the sympathetic trunk is shown. The cervicothoracic ganglion is tightly bound to the first thoracic nerve by rami communicantes, but these lie posteriorly and are hidden.

to that of a typical intercostal nerve. Its lateral cutaneous branch passes into the arm as the *intercostobrachial nerve.* It pierces the intercostal and serratus anterior muscles, becomes superficial at the posterior axillary fold, and supplies the skin and subcutaneous tissue on the back and medial side of the arm as far as the elbow. It commonly anastomoses with the medial and posterior brachial cutaneous nerves, and it may communicate with the lateral cutaneous branches of the first and third intercostal nerves. It also supplies the axillary arch when that muscle is present.

The third intercostal nerve has a distribution similar to that of a typical intercostal nerve. However, its lateral cutaneous branch often gives a twig to the medial side of the arm, and another which becomes incorporated with the intercostobrachial nerve.

Thoracoabdominal Nerves (figs. 33–11 and 33–12, pp. 362, 363). The seventh to eleventh intercostal nerves are also special in that they supply the abdominal as well as the thoracic wall. They course forward and downward to the anterior ends of the intercostal spaces. Here they pass first between the chondral attachments of the diaphragm and the transversus abdominis. (The seventh, eighth, and ninth nerves pass behind the corresponding costal cartilages.) They continue between the transversus and internal oblique muscles, and then between the rectus abdominis and the posterior wall of its sheath. Here, each nerve divides into two branches. The larger one breaks up in a plexiform arrangement that supplies the rectus and from which an *anterior cutaneous branch* pierces the rectus and the anterior layer of its sheath to supply the overlying skin. The smaller branch also supplies the rectus and may pierce the muscle and become cutaneous.

The *lateral cutaneous branches* of the thoracoabdominal nerves pierce the external oblique and divide into anterior and posterior branches. These supply the skin of the back, side, and front of the abdominal wall.

The thoracoabdominal nerves supply the intercostal, subcostal, serratus posterior inferior, transversus abdominis, external and internal oblique, and rectus abdominis muscles, and give sensory twigs to the adjacent diaphragm, pleura, and peritoneum.

Subcostal Nerve. The ventral ramus of the twelfth thoracic nerve is special in that it is subcostal rather than intercostal in position, and is known as the subcostal nerve. It enters the abdomen behind the lateral arcuate ligament, courses downward and laterally behind the kidney (figs. 37–7, p. 411, and 38–9, p. 427), pierces the transversus abdominis, and passes between this muscle and the internal oblique. It then enters the sheath of the rectus, turns forward through its anterior layer, and becomes superficial about halfway between the umbilicus and the pubic symphysis.

Its lateral cutaneous branch descends through the internal and external oblique, and becomes superficial above the iliac crest. It supplies the skin and subcutaneous tissue of the gluteal region and lateral side of the thigh as far as the level of the greater trochanter of the femur.

The subcostal nerve supplies parts of the transversus, oblique, and rectus muscles, and usually the pyramidalis. Twigs are also given to the adjacent peritoneum.

JOINTS

The joints of the thorax comprise those between (1) ribs and vertebrae, (2) ribs and costal cartilages, (3) costal cartilages, (4) costal cartilages and sternum, and (5) the parts of the sternum itself. They also include the joints between vertebrae (see p. 533).

Costovertebral Joints

The costovertebral joints (figs. 27–3 and 27–5) are the joints of the heads of the ribs and the costotransverse joints. They are supplied by the thoracic dorsal rami.

Joints of Heads of Ribs. The articular surface of the head of a typical rib (second to ninth, inclusive) articulates with the inferior and superior costal facets of the bodies of two adjacent vertebrae and the intervertebral disc between these facets. A short, horizontally placed ***intra-articular ligament*** extends from the crest of the head of the rib to the intervertebral disc and separates an upper from a lower joint cavity.

The *articular capsule* completely surrounds the joint. It is thickened in front to form the *radiate ligament* (figs. 27–3 and 27–5A).

The heads of the first, tenth, eleventh, and twelfth ribs each articulate with only one vertebra, and their joints have only a single cavity.

Costotransverse Joints. The articular surface of the tubercle of a typical rib articulates with the costal facet on the transverse process of the corresponding vertebra.

The articular capsule is reinforced by accessory ligaments. These are the *costotransverse*, the *lateral costotransverse*, and the *superior costotransverse ligaments* (fig. 27–5).

The eleventh and twelfth ribs lack tubercles and therefore have no costotransverse joints. The *lumbocostal ligament* connects the twelfth rib with the tips of the transverse processes of the first and second lumbar vertebrae, and runs behind the quadratus lumborum.

Costochondral and Interchondral Joints

Costochondral Joints. These are the hyaline cartilaginous joints between the costal cartilages and the depressions in the ends of the shafts of the ribs.

Interchondral Joints. Each of the fifth to the eighth, and sometimes the ninth, costal cartilages articulates with the cartilage immediately below, just medial to where these cartilages turn upward in front. Each of these joints usually contains a synovial cavity.[19]

Sternocostal Joints

The sternocostal *(sternochondral)* joints are formed by the articulation of the medial ends of the first seven costal cartilages with the costal notches on the lateral margin of the sternum.[20] In the first joint, the hyaline cartilage of the notch is usually directly continuous with the costal cartilage. This joint sometimes contains a cavity.

The second to seventh joints are generally considered to be synovial, but they may be partially or completely filled with fibrocartilage.

The capsule of each joint is reinforced in front by the *radiate sternocostal ligament*, which interlaces with the periosteum and the tendinous origin of the pectoralis major to form a thick, dense *sternal membrane*. A horizontal *intra-articular ligament* often divides the second joint into upper and lower parts.

The *costoxiphoid ligaments*, when present, extend from the front of the xiphoid process to the seventh and sometimes also the sixth costal cartilage.

Joints of the Sternum

Manubriosternal Joint. The fibrocartilaginous joint between the manubrium and the body of the sternum may become ossified.[21]

Xiphisternal Joint. The cartilaginous union between the xiphoid process and the body of the sternum begins to ossify in adult life, and in old age is commonly completely ossified.

MOVEMENTS OF THE THORACIC CAGE

The frequency of movement of the joints of the thorax is greater than that of any other combination of joints, except possibly those between the auditory ossicles. The range of movement of any one of the thoracic joints is small, but any disorder that reduces their mobility hampers respiration.

Axes of Movement. The second to sixth ribs each move around two axes. Movement at the costovertebral joint about a side-to-side axis results in raising and lowering the sternal end of the rib, the so-called "pump-handle" movement (fig. 27–7). Since ribs slope downward, any elevation, for instance, during the inspiratory phase of breathing, results in an upward and forward movement of the sternum and an increase in the anteroposterior diameter of the thorax. Movement at the costovertebral joint about a front-to-back axis leads to depression or elevation of the middle of the rib, the so-called "bucket-handle" movement (fig. 27–7). This type of movement, which increases the transverse diameter of the thorax and which also occurs during inspiration, takes place chiefly at the seventh to the tenth costotransverse joints. The articular tubercles of these joints are flat. Thus the ribs can move up and down and thereby allow the "bucket-handle" movement.

The axes of rotation of the first rib differ somewhat from those of the other ribs.[22] There is little if any movement of the first ribs during quiet breathing, although rotation about the axis of the neck of each first rib causes a little raising or lowering of the sternum, and results in a slight change in the anteroposterior diameter of the thoracic inlet.

The movements of the seventh to tenth ribs occur about axes similar to those of the second to sixth. However, owing to the fact that their costal cartilages turn upward, an

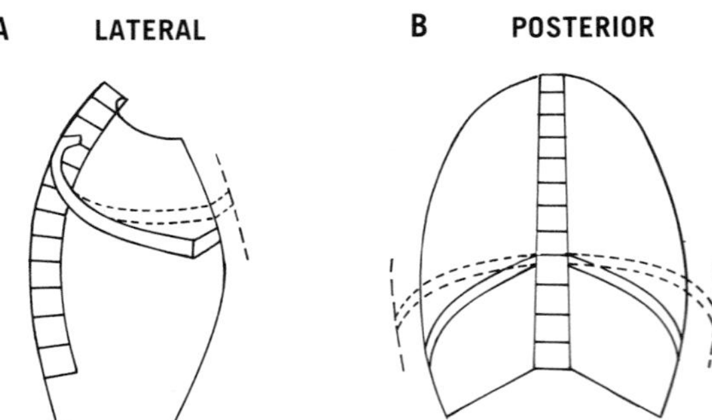

Figure 27–7 Representation of certain movements of the ribs. In *A*, when the upper ribs are elevated, the anteroposterior diameter of the thorax is increased ("pump-handle" movement). In *B*, the lower ribs move laterally when they are elevated, and the transverse diameter of the thorax is increased ("bucket-handle" movement).

elevation of the anterior ends of the ribs (the "pump-handle" movement) tends to be associated with a backward movement of the sternum, which is possible only through bending at the manubriosternal joint.

Elevation of the upper ribs leads to elevation and forward movement of the sternum. Elevation of the lower ribs leads to backward movement of the sternum. Hence in such movements there is bending at the manubriosternal joint and twisting of the costal cartilages. Ossification at the manubriosternal joint and of the costal cartilages impairs movement of the thoracic cage.

Thoracic Volume. **A movement of only a few millimeters of the bony cage forward, upward, or laterally is sufficient to increase the volume of the thoracic cage by almost one-half liter. This is the usual volume of air that enters and leaves the lungs during quiet breathing.**[23]

In deep breathing the excursions of the bony cage are greater, the back is extended, and the normal curvature of the thoracic spine is straightened. This results in a further increase of the anteroposterior diameter of the thorax.

The descent of the diaphragm, which increases the height of the thoracic cavity, is the other major factor in increasing the volume of the thorax, although in unilateral, or even bilateral, paralysis of the diaphragm, there may be no significant disability.[24] Diaphragmatic breathing, which involves abdominal muscles to a varying degree, is often called abdominal breathing. In general, both thoracic and abdominal breathing are used in the same individual to varying degrees.

Muscles of Respiration.[7] The diaphragm is the most important muscle of respiration, and increases the volume of the thoracic cavity, as described previously. The mechanical actions of the intercostal muscles are not yet completely established, but it is likely that the external intercostals and the intercartilaginous parts of the internal intercostals elevate the ribs, and thus are inspiratory in function, whereas the interosseous portions of the internal intercostals depress the ribs and are therefore expiratory in function. Acting simultaneously, as during forced expiration, the intercostal muscles probably maintain tension on the intercostal spaces and prevent them from bulging outward as the intrathoracic pressure is increased. However, the important muscles of expiration, including forced expiration, are the external abdominal muscles (see below).

In the inspiratory phase of quiet breathing, the diaphragm, the "parasternal" intercostals, and the external intercostals posteriorly are active in all subjects, and the scalene muscles in some subjects. Expiration is chiefly passive, being dependent upon such factors as the elasticity of the lungs (p. 301), but the interosseous internal intercostals of the seventh to tenth interspaces are regularly active during the last part of expiration.

The pattern is similar during moderately increased breathing, at ventilatory rates up to 50 liters a minute. Intercostal activity spreads, the outer layer in any region being inspiratory, and the deeper layer, which comes into action irregularly, being expiratory.

As breathing becomes more vigorous (between 50 and 100 liters per minute), the sternomastoids and the extensors of the vertebral colum are active toward the end of inspiration, and the anterolateral external abdominal muscles become increasingly active during expiration. These muscles, which compress the abdominal viscera and are active in coughing, straining, and vomiting, draw the ribs down and are the most important expiratory muscles.

In greatly increased breathing (above 100 liters per minute), all of the accessory muscles of respiration become active throughout inspiration, and all of the external abdominal muscles are active throughout expiration.

The scalene and sternomastoid muscles are the only accessory muscles of inspiration that have any significant activity, and the sternomastoids are more important than the scalenes. The sternomastoids, like the scalenes, can move the ribs up and down, and are usually active during high rates of ventilation. Many other muscles have potential respiratory functions because they are attached to the ribs, but aside from the erector spinae and latissimus dorsi, which contract strongly during coughing, their role in respiration is negligible.

Muscular control of expiration is important in speaking and singing. The ability of a trained singer to hold a prolonged note, for example, depends on the co-ordinated activity of several muscles. The quadratus lumborum, attached below the hip bone, is able to fix the twelfth rib and thus to provide a stable attachment for the diaphragm, which maintains tension as it slowly relaxes. The abdominal muscles contract slowly; their expiratory action is opposed by the diaphragm. The intercostal muscles control tension in the intercostal spaces. The entire mechanism makes possible an even, precisely controlled flow of air upward through the larynx.[25]

MEDIASTINUM

The thoracic cavity contains the lungs, pleurae, and, in the mediastinum, certain other structures, chiefly the heart. **The mediastinum, which is the interval between the two pleural sacs, is commonly considered as comprising a *superior mediastinum*, above the level of the pericardium, and three lower divisions, termed *anterior*, *middle*, and *posterior*** (fig. 27–8). The *middle mediastinum* contains the pericardium and heart and the immediately adjacent parts of the great vessels, together with the main bronchi and the other structures of the roots of the lungs. The *anterior mediastinum* is located in front of the pericardium and behind the sternum. Its chief component is the thymus, which also occupies the front part of the superior mediastinum. The *posterior mediastinum* is situated behind the pericardium. It contains, among other structures, the esophagus and the thoracic aorta, which have descended into it from the superior mediastinum. The *superior mediastinum* contains the esophagus and the trachea behind, the thymus (or its remains) in front, and, in between, the great vessels related to the heart and pericardium.

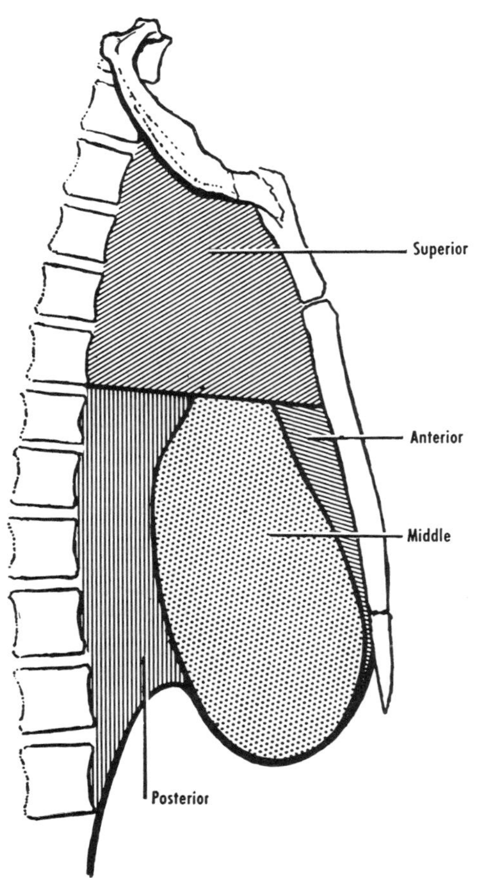

Figure 27–8 Diagram of the divisions of the mediastinum. The boundaries shown here are those in erect, living subjects (see figs. 29–7 and 29–8).

Loose connective tissue, often infiltrated with fat, pervades the mediastinum and surrounds and suspends the organs. This tissue becomes more fibrous and rigid with age, and the mediastinal structures become correspondingly less mobile. The support of mediastinal structures is due in part to the continuity of vessels and organs from the neck, but also to the arrangement of the mediastinal connective tissue.[26] This connective tissue extends to the anterior thoracic wall as sternopericardial ligaments; it connects organs, and it reaches the posterior thoracic wall.

A significant part of the mediastinum can be visualized and surgical procedures can be carried out by means of a tubular, lighted instrument termed a *mediastinoscope,* commonly inserted through a suprasternal incision. The procedure is known as *mediastinoscopy.*

REFERENCES

1. A. B. Morrison, J. Anat., Lond., 88:19, 1954. R. Steubl, Z. Anat. EntwGesch., *128*:211, 1969. W. Langenberg and S. Jüschke, Z. Anat. EntwGesch., *130*:255, 1970.
2. F. Davies, R. J. Gladstone, and E. P. Stibbe, J. Anat.,

Lond., *66*:323, 1932. See also T. Walmsley, J. Anat., Lond., *50*:165, 1916; A. J. E. Cave, J. Anat., Lond., *63*:367, 1929; M. A. H. Siddiqi and A. N. Mullick, J. Anat., Lond., *69*:350, 1935.
3. J. Satoh, Okajimas Foliá anat. jap., *48*:103, 1971.
4. D. M. Blair, J. Anat., Lond., *57*:203, 1923.
5. A. Low, J. Anat., Lond., *42*:93, 1907. J. L. Collis, T. D. Kelly, and A. M. Wiley, Thorax, *9*:175, 1954.
6. J. L. Collis, L. M. Satchwell, and L. D. Abrams, Thorax, *9*:22, 1954. G. S. Muller Botha, Thorax, *12*:50, 1957. R. Scott, Thorax, *20*:357, 1965. R. Shehata, Acta anat., *63*:49, 1966.
7. For the functional anatomy of respiration, see E. J. M. Campbell, E. Agostini, and J. N. Davis, *The Respiratory Muscles*, Lloyd-Luke, London, 2nd ed., 1970, and R. M. Peters, *The Mechanical Basis of Respiration*, Little, Brown, Boston, 1969. See also A. Taylor, J. Physiol., *151*:390, 1960.
8. W. F. Walker and H. D. Attwood, Brit. J. Surg., *48*:86, 1960.
9. R. D. Adams and H. C. Pillsbury, Arch. intern. Med., *29*:245, 1922.
10. L. Allen, J. thorac. Surg., *19*:290, 1950. T. B. Sappington and R. A. Daniel, J. thorac. Surg., *21*:212, 1951.
11. S. W. Harrington, Ann. Surg., *122*:546, 1945; Surg. Gynec. Obstet., *86*:735, 1948. P. R. Allison, Surg. Gynec. Obstet., *92*:419, 1951. N. R. Barrett, Brit. J. Surg., *42*:231, 1954. J. J. Schlegel, Ergebn. Chir. Orthop., *41*:350, 1958.
12. A. A. Pearson and R. W. Sauter, Thorax, *26*:354, 1971.
13. B. J. Anson, R. R. Wright, and J. A. Wolfer, Surg. Gynec. Obstet., *69*:468, 1939.
14. B. N. Kropp, J. thorac. Surg., *21*:421, 1951.
15. E. Ennabli, Arch. Anat. Path., *14*:98, 1966.
16. E. P. Chung, J. thorac. cardiovasc. Surg., *63*:880, 1972.
17. E. P. Stibbe, J. Anat., Lond., *52*:257, 1918.
18. A. J. E. Cave, J. Anat., Lond., *63*:367, 1929.
19. C. Briscoe, J. Anat., Lond., *59*:432, 1925.
20. D. J. Gray and E. D. Gardner, Anat. Rec., *87*:235, 1943. M. Williams, J. Morph., *101*:275, 1957.
21. M. Trotter, Amer. J. phys. Anthrop., *18*:439, 1934.
22. R. W. Haines, J. Anat., Lond., *84*:94, 1946.
23. The human thoracic diameters at rest and during activity are described by P. R. Davis and J. D. G. Troup, J. Anat., Lond., *100*:397, 1966.
24. A. L. Banyai, Arch. Surg., Chicago, *37*:288, 1938.
25. M. H. Draper, P. Ladefoged, and D. M. Whitteridge, Brit. med. J., *1*:1837, 1960.
26. P. Marchand, Thorax, *6*:359, 1951. D. L. Bassett, Anat. Rec., *133*:248, 1959.

ESOPHAGUS, TRACHEA, AND BRONCHI 28

THORACIC PART OF ESOPHAGUS[1]

The esophagus, or gullet (figs. 28–1 to 28–3), extends from the lower end of the pharynx to the cardiac opening of the stomach. It has cervical (p. 691), thoracic, and abdominal (p. 376) parts.

The esophagus begins at the level of the cricoid cartilage (C.V. 6). It pierces the diaphragm at about the level of T.V. 11 or 12. In the erect position it is 25 to 30 cm long, slightly more than twice as long as the trachea, and it is 1 to 2 cm shorter in women.

The esophagus is a median structure that lies first behind the trachea and then behind the left atrium. It begins to deviate to the left below the left main bronchus. When it pierces the diaphragm it often makes an abrupt turn to the left. In the posterior mediastinum it is related to the vertebral column as a string is related to a bow. Hence there is a (retrocardiac) space between it and the vertebral column, a space that is visible radiographically in oblique and lateral views. Here, branches from the aorta reach the esophagus between the layers of pleura that form the retroesophageal recesses.

The esophagus serves mainly for conduction of food and liquid, and has been successfully replaced by a nonmuscular tube. **The esophagus is quite distensible, and will accommodate nearly anything that can be swallowed.** An object as large as an upper denture has been known to reach the stomach without causing great discomfort.

The muscular layer of the esophagus is

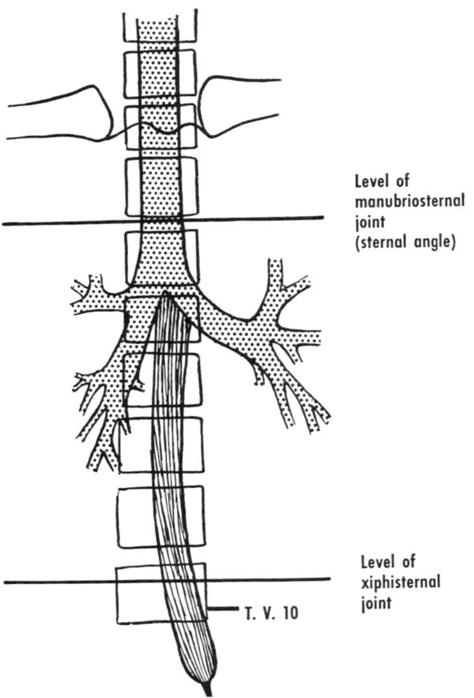

Figure 28–1 The trachea and esophagus in relation to vertebral and sternal levels in the erect position.

striated muscle above, and smooth below. Rarely, skeletal muscle is present in the lower part of the esophagus.[2]

Blood Supply and Lymphatic Drainage. The esophagus is supplied by the inferior thyroid and the bronchial arteries, by direct branches of the aorta, and by the phrenic and left gastric arteries.[3] Esophageal veins drain into adjacent veins. **An anastomosis of veins of the lower part of the esophagus with the left gastric vein is one of the important communications between the portal and systemic systems** (p. 418).

The lymphatic vessels of the thoracic part of the esophagus drain into the phrenic, posterior mediastinal, and tracheal nodes.

Nerve Supply. Special motor fibers in the vagus nerve supply the skeletal muscle.

Parasympathetic preganglionic fibers reach the esophagus by way of the vagus nerves and synapse with ganglion cells in the esophagus. The postganglionic fibers supply smooth muscle and glands, which they usually activate. Preganglionic sympathetic fibers arise in the lower part of the thoracic spinal cord and synapse in ganglia of the sympathetic trunks. Postganglionic fibers enter the esophageal plexus by visceral branches of the trunks and by branches of the greater splanchnic nerves. They probably act conversely to the parasympathetic fibers.

Pain fibers from the esophagus accompany sympathetic fibers to the sympathetic trunks, through which they pass to rami communicantes and spinal nerves. They enter the spinal cord by way of dorsal roots. A vague, deep-seated pain may be provoked from the

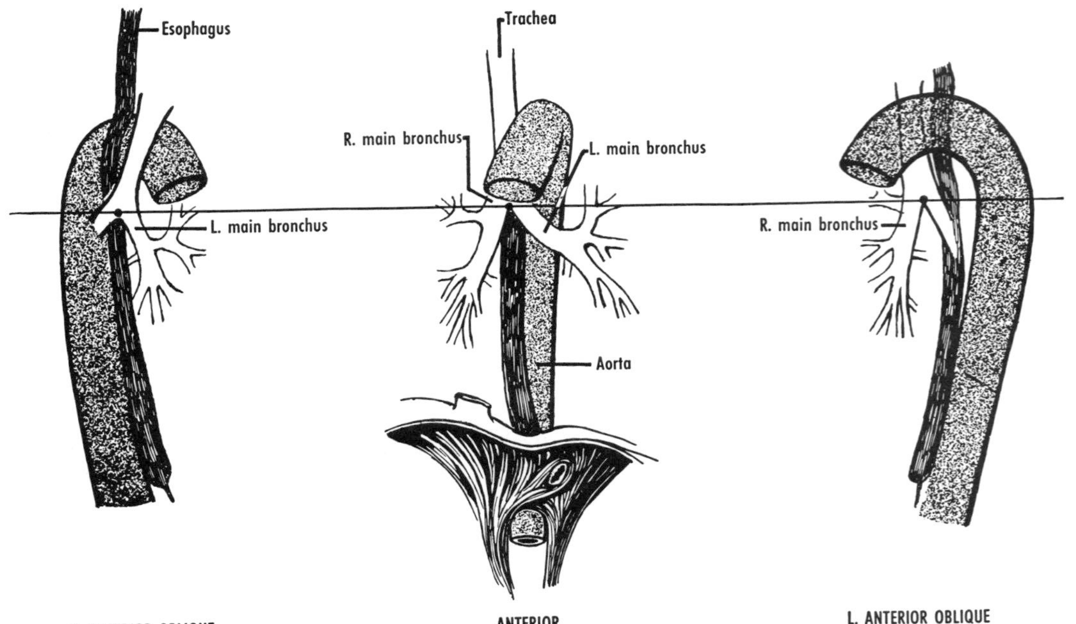

Figure 28–2 The relations of the trachea, bronchi, esophagus, and aorta to one another. In the right anterior oblique view, the right lobar and segmental bronchi are omitted because they are not clearly visible in radiograms of this view. For similar reasons, the left lobar and segmental bronchi are omitted from the left anterior oblique view.

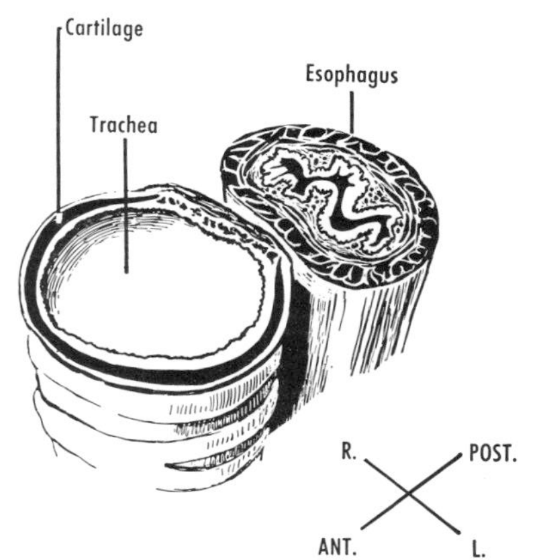

Figure 28–3 Diagrammatic horizontal sections of the trachea and esophagus.

esophagus, especially from its lower part, the stimulation of which may cause pain that is felt deep to the sternum or in the epigastrium. Such pain resembles that arising from the stomach or heart ("heartburn").

Radiological Anatomy. If a thick barium paste is swallowed, the lumen of the esophagus can be visualized radiographically (fig. 28–4). Adjacent structures produce impressions on the esophagus (p. 343).[4] Alterations in the position or shape of these structures by disease may, in turn, alter the impressions they normally make. **The esophagus presents constrictions (1) at its beginning, (2) frequently where it is crossed by the left main bronchus, and (3) commonly where it passes through the diaphragm.**

Swallowing

The process of swallowing (p. 749) may be watched fluoroscopically. A thin paste or fluid containing barium is swallowed so rapidly that the process cannot be analyzed by the naked eye. The material is "shot down" to the cardiac orifice, where its passage may be slowed before it enters the stomach. (The sphincteric arrangements at the cardioesophageal junction are discussed on page 376).

A thick barium paste or a bolus of food passes more slowly down the esophagus. It may be further slowed at the various constrictions and impressions. The skeletal muscle in the upper part of the esophagus begins to be replaced by smooth muscle at the level of the clavicles. The shift of contraction from skeletal to smooth muscle may account for a momentary slowing in the passage of food at this level.

TRACHEA

The trachea, or windpipe (figs. 28–1 to 28–3), begins in the neck, where it is continuous with the lower end of the larynx. It descends in front of the esophagus, enters the superior mediastinum, and divides into right and left main bronchi. The trachea is essentially a median structure, but near its lower end it deviates slightly to the right. On that account, **the left main bronchus crosses in front of the esophagus.** The trachea is very mobile and can be moved about readily during surgical procedures.

The trachea has 16 to 20 **C**-shaped bars of hyaline cartilage which provide the rigidity that prevents the trachea from collapsing. It has longitudinally arranged elastic fibers that render the trachea elastic enough so that it stretches and descends, with the roots of the lungs, during inspiration. Its elasticity aids in the recoil of the lungs during expiration.

In the erect subject, **the trachea divides at the level of T.V. 5 or 6** (sometimes T.V. 7).[5] The level is higher in the cadaver. The trachea moves during respiration and with movements of the larynx. The level of bifurcation varies accordingly. The trachea is about 9 to 15 cm in length, slightly less than one-half the length of the esophagus. The length of the trachea varies with the individual, with age, and with the phase of respiration.[6]

The *carina* is a ridge on the inside, at the bifurcation of the trachea. It is formed by a backward and somewhat downward projection of the last tracheal cartilage. It is a landmark during bronchoscopy and it separates the upper ends of the right and left main bronchi. Usually, the carina is to the left of the median plane. Its uppermost part is sometimes membranous rather than cartilaginous.

Relations. **The arch of the aorta is at first in front of the trachea, and then on its left side, immediately above the left main bronchus.** The brachiocephalic and left common carotid arteries are at first in front and then on its right and its left side, respectively. The

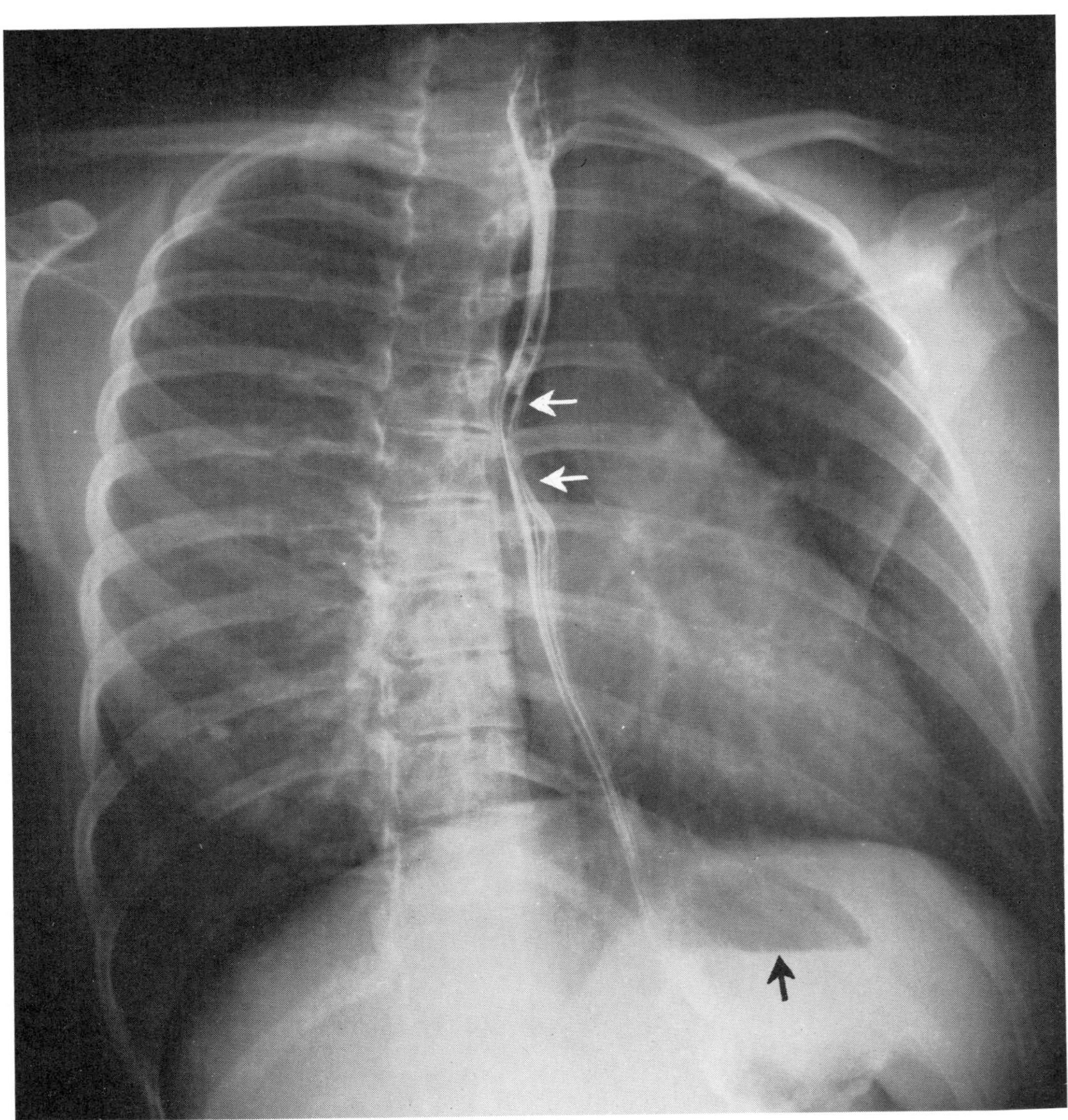

Figure 28–4 Oblique (R.A.O.) view of thorax. The esophagus is shown coated with barium paste. Note the vertical folds of the mucosa and the indentations produced by the arch of the aorta (*upper arrow*) and by the left main bronchus (*lower arrow*). Note also the right and left domes of the diaphragm (on the left and right sides of the illustration, respectively), and the fluid level in the stomach (*lowermost arrow*). Courtesy of D. L. Bassett, M.D., and Sawyer's Inc. From D. L. Bassett, *A Stereoscopic Atlas of Human Anatomy*, Sawyer's, Portland, Oregon, 1958, section 4, reel 129, view 6, Copyright 1958, Sawyer's Inc., U.S.A.

esophagus lies behind it. On account of its vascular relationships, the trachea is somewhat closer to the apex of the right lung than it is to the left. Other relationships are shown in figures 28–1 and 28–2.

Blood Supply and Lymphatic Drainage. The trachea is supplied mainly by the inferior thyroid arteries, but it also receives branches from the superior thyroid, bronchial, and sometimes the internal thoracic arteries. It is drained mainly by the inferior thyroid veins.

The lymphatic vessels of the trachea drain into adjacent lymph nodes (cervical, tracheal, and tracheobronchial).

Nerve Supply. Preganglionic parasympathetic fibers in the vagus nerves are given to the trachea by direct branches of the vagus nerves and by branches of the recurrent laryngeal nerves. The fibers synapse with ganglion cells in the wall of the trachea. The postganglionic fibers supply smooth muscle and glands. Their function appears to be the stimulation of these structures. Postganglionic sympathetic fibers that reach the trachea from the sympathetic trunks probably act in the converse manner on smooth muscle and glands.

The vagus nerves also carry pain fibers that supply the mucous membrane. Irritation of the mucous membrane generally causes pain or coughing. (Electric stimulation of the tracheobronchial mucosa in human patients causes pain that is referred to the neck or to the front of the chest on the same side.[7]) If the irritation occurs suddenly, as when an irritating gas is inhaled, for example, respiration may be reflexly stopped.

Radiological Anatomy. The trachea is usually visible above the arch of the aorta in ordinary posteroanterior radiograms of the chest. The air within it gives its lumen a greater translucency.

MAIN BRONCHI

Each main bronchus extends from the bifurcation of the trachea to the hilus of the corresponding lung. The *right main bronchus* may be considered as comprising an upper part, from which the segmental bronchi for the upper lobe arise, and a lower part, from which the segmental bronchi for the middle and lower lobes emerge (fig. 28–5). The *left main bronchus* divides into two lobar bronchi, one each for the upper and the lower lobes of the left lung. The upper lobar bronchus may be considered as having an upper division and a lower, or lingular, division.

The right main bronchus, about 2.5 cm in length, is shorter, wider, and more nearly vertical than the left. Because it is almost in a direct line with the trachea, foreign objects passing through the trachea usually enter the right main bronchus. The left main bronchus, 5 cm or more in length, crosses in front of the esophagus (fig. 28–2) and forms the second indentation of that organ as seen radiographically. Both bronchi are mobile, elastic structures, and change in length (as does also the trachea.) They possess cartilaginous rings that become plates when the bronchi become intrapulmonary at the lung roots.[8] Respiratory changes in position of the lung roots are described elsewhere (p. 301).

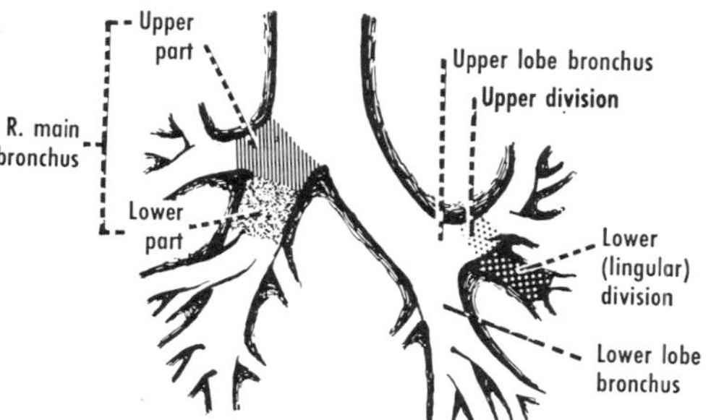

Figure 28–5 The lobar bronchi. See page 294 for the segmental bronchi.

Blood Supply, Lymphatic Drainage, and Nerve Supply. The bronchi are supplied by the bronchial arteries. Venous drainage is into the bronchial veins. Their lymphatic vessels drain into adjacent nodes (bronchopulmonary and tracheobronchial).

The nerve supply is similar to that of the treachea, and reaches the bronchi by way of the cardiac and pulmonary plexuses. The bronchi are relatively insensitive to pain, although stimulation of their mucous membrane usually causes coughing.

REFERENCES

1. A general reference, with extensive bibliography, is J. Terracol and R. H. Sweet, *Diseases of the Esophagus*, Saunders, Philadelphia, 1958.
2. L. B. Arey and M. J. Tremaine, Anat. Rec., *56*:315, 1933.
3. A. L. Shapiro and G. L. Robillard, Ann. Surg., *131*:171, 1950. L. L. Swigart *et al.*, Surg. Gynec. Obstet., *90*:234, 1950.
4. W. Evans, *The Course of the Oesophagus in Health, and in Disease of the Heart and Great Vessels*, Spec. Rep. Ser. med. Res. Coun., Lond., No. 208, 1936.
5. The bronchial and the subcarinal angles are given by R. S. Turner, Anat. Rec., *143*:189, 1962.
6. J. E. Jesseph and K. A. Merendino, Surg. Gynec. Obstet., *105*:210, 1957. H. Pineau, I. Eralp, and A. Delmas, Arch. Anat. Path., *20*:395, 1972.
7. D. R. Morton, K. P. Klassen, and G. M. Curtis, Surgery, *28*:699, 1950.
8. F. Vanpeperstraete, Adv. Anat., Embryol., Cell Biol., *48*:1973.

PLEURA AND LUNGS

29

PLEURA

The two lungs, each with its pleural sac, are contained in the thoracic cavity (figs. 29–1 and 29–2). The pleura is a thin, glistening, slippery serous membrane that lines the thoracic wall and mediastinum, where it is called *parietal pleura.* It is reflected from the mediastinum to the lung, where it is called *visceral,* or *pulmonary, pleura.* The visceral pleura covers the lungs and dips into its fissures. The opposed surfaces of the parietal and visceral pleura slide smoothly against each other during respiration. The potential space between them, the *pleural cavity,* contains a film of fluid of capillary thinness.

The complete extent of the pleura is not accessible to direct study in the intact, living individual and, except in certain regions or with special techniques, the pleura is not visible radiographically.

Both the parietal and the visceral pleura can stretch under prolonged tension. If a lung is congenitally absent, or if it is collapsed, the other lung may expand and, together with its pleura, balloon over in front of the heart and fill the area void of lung.[1]

Blood Supply and Lymphatic Drainage. The parietal pleura is supplied mainly by branches of the posterior intercostal, internal thoracic, superior intercostal, internal thoracic, and superior phrenic vessels. Lymphatic vessels drain into the adjacent lymph nodes of the thoracic wall. These nodes in turn may drain into axillary nodes.

The visceral pleura is supplied by bronchial arteries, but its venous blood is drained by pulmonary veins. Lymphatic vessels are numerous, and drain toward nodes at the hilus.[2]

Nerve Supply. The intercostal and thoracoabdominal nerves give sensory twigs to the costal part of the parietal pleura. The thoracoabdominal nerves and the subcostal nerve give sensory branches to the peripheral part of the diaphragmatic pleura. Sensory fibers from the phrenic nerve supply the mediastinal pleura and the central part of the diaphragmatic pleura. **Parietal pleura, especially its costal part, is very sensitive to pain.**[3] When costal pleura is irritated, pain is felt locally, deep to a rib or intercostal space. Irritation of the part of the diaphragmatic pleura supplied by a thoracic nerve causes a more diffuse pain, which often radiates into the abdominal wall and lumbar region. Irritation of pleura supplied by the phrenic nerve causes referred pain, sometimes to the region of the ear, but most often to the neck, and particularly to the skin over the trapezius muscle as far as the tip of the shoulder (p. 332). The visceral pleura is insensitive.

Radiological Anatomy. Parietal pleura is visible radiographically only in certain regions or in special views.

Visceral pleura is usually not visible radiographically, except where it presents an edge to the x-ray beam, as in special views that demonstrate the fissures. The lobule of the azygos vein may also be delimited by virtue of the pleural reflection between it and the right upper lobe (p. 343). Any disease process that results in thickening of the pleura may render it visible radiographically.

General Arrangement of Parietal Pleura

The parietal pleura has costal, mediastinal, and diaphragmatic parts, and a cupola (figs. 29–3 to 29–6).

Costal Pleura. **The costal pleura is separated from the sternum, costal cartilages, ribs, and muscles by a small amount of loose connective tissue, the *endothoracic fascia,* which provides a natural cleavage plane for surgical separation of the pleura from the thoracic wall.**

In front, the costal pleura turns sharply onto the mediastinum, where it is continuous with the mediastinal pleura. The edge of the reflection is called the *anterior border* of the pleura. The part of the pleural cavity between the layers of this pleural reflection is the *costomediastinal recess.* Below, the costal pleura is continuous with the diaphragmatic pleura, the *costodiaphragmatic recess* being created by its reflection. The edge of the recess, which is termed the *inferior border* of the pleura, usually does not extend as far inferiorly as the costal attachments of the diaphragm. Hence a part of the diaphragm is left uncovered. Behind, the costal pleura turns forward over the sides of the bodies of the vertebrae, and is again continuous with the mediastinal

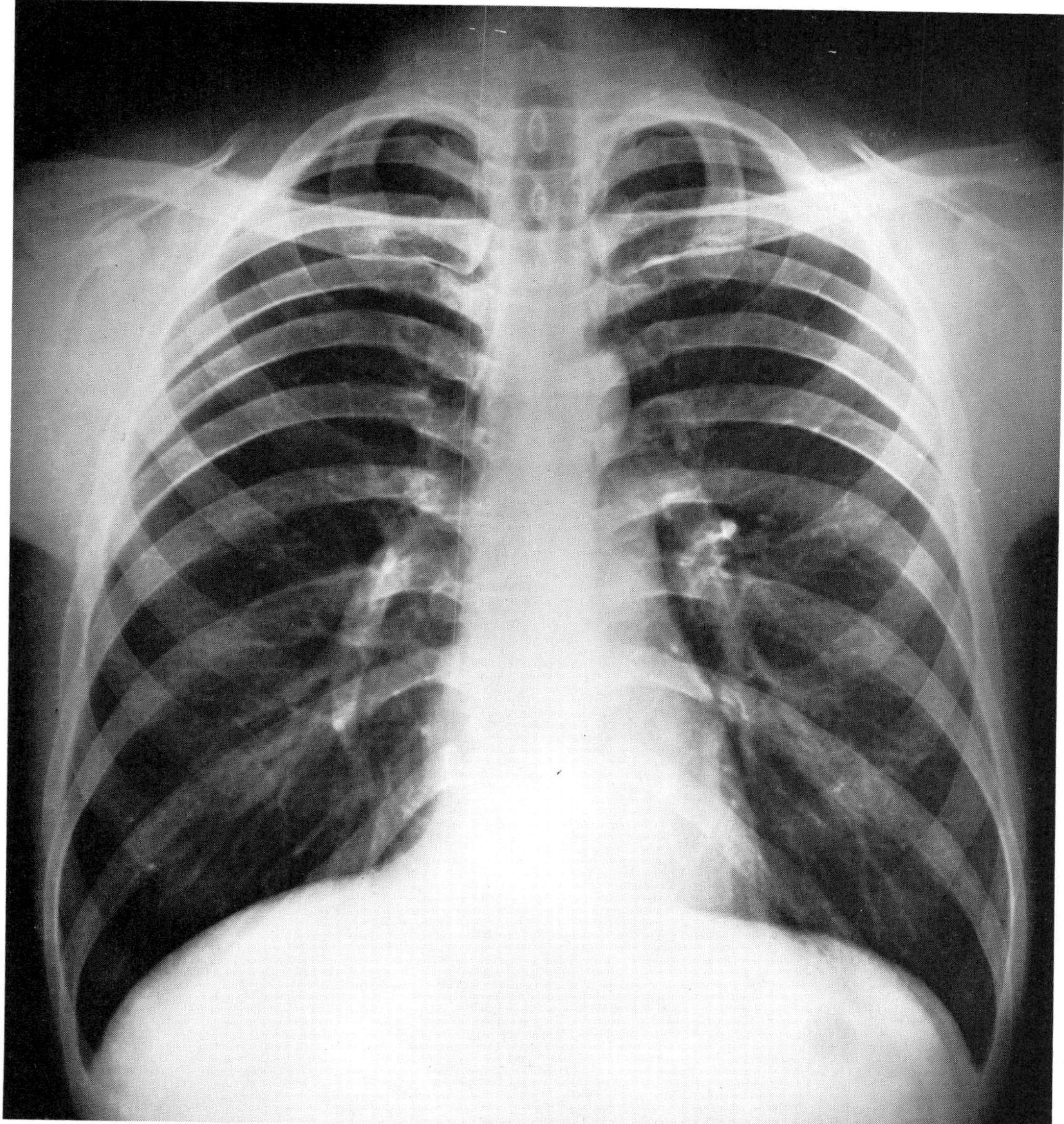

Figure 29–1 Thorax of an adult. Note clavicles, ribs, diaphragm, cardiovascular shadow (including right atrium, aortic knuckle, and left ventricle), trachea, and lungs (including vascular markings). Large descending branches of the pulmonary arteries are visible on each side of the heart.

pleura. The turn forward at the region of continuity forms an indefinite *posterior border* of the pleura.

The course of the anterior border of the pleura is subject to variation (fig. 29–3).[4] **In the living adult subject, the two anterior borders probably meet at or near the median plane during part of their course.**[5] In infants, however, they are separated by the thymus. The divergence of the left anterior border may leave a part of the pericardium uncovered (bare area of the heart or pericar-

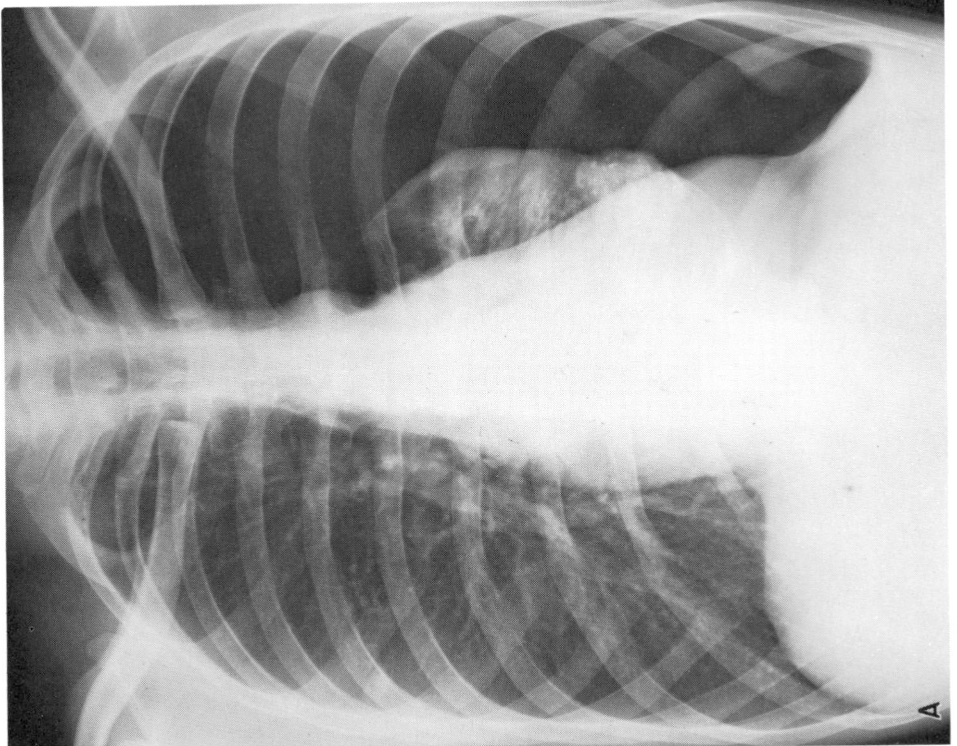

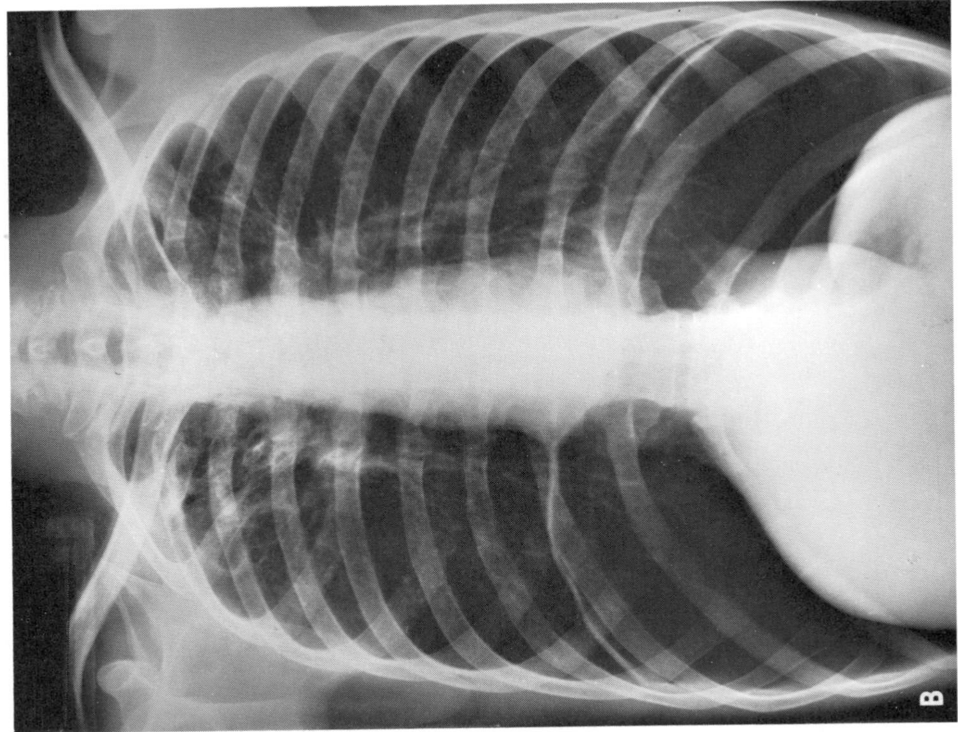

Figure 29–2 Thorax and abdomen. *A*, pneumothorax. The left lung has been collapsed (note its small size). The presence of air in the pleural cavity accounts for the radiolucent appearance of most of the left side of the chest. *B*, pneumoperitoneum. Air in the peritoneal cavity has displaced the abdominal viscera downward, and the right and left domes of the diaphragm have become outlined as radio-opaque lines.

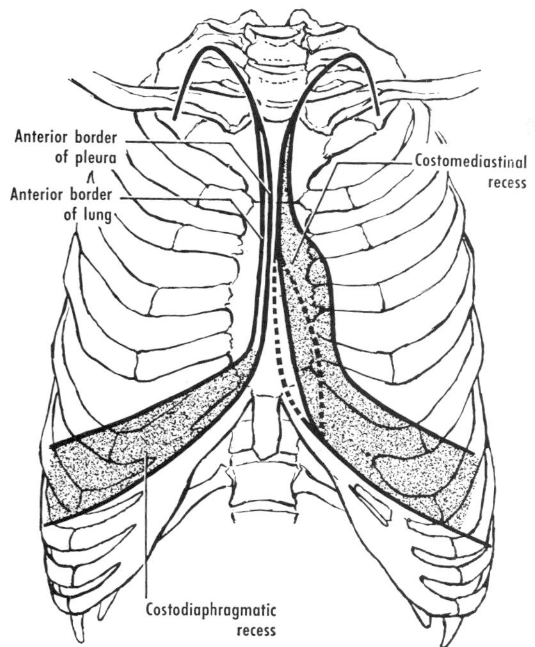

Figure 29–3 The pleurae from the front. The interrupted lines represent variants between which the left anterior border of the pleura lies in 70 per cent of instances. The costodiaphragmatic and costomediastinal recesses are indicated by stippling. Based on Woodburne.[4]

dium), but such an arrangement seems to be uncommon after childhood. The inferior border of the pleura also varies in level (fig. 29–4).[6] Behind, it is closely related to the twelfth rib and, indirectly, to the kidney (fig. 37–7*B*, p. 411).

Even during quiet breathing, the lung undoubtedly extends into the costomediastinal and costodiaphragmatic recesses, but its presence there cannot usually be detected by percussion.

Mediastinal Pleura. Above the root of the lung, the mediastinal pleura is a continuous sheet that extends from the costomediastinal recess to the posterior border, where it becomes the costal pleura. At the root of the lung, the mediastinal pleura turns laterally, enclosing the structures of the root and becoming continuous with the visceral pleura. Below the root of the lung, the mediastinal pleura turns laterally as a double layer (figs. 29–7 and 29–8), which extends between the esophagus and the lung. This double layer is termed the *pulmonary ligament.* It is continuous above with the reflection around the lung root. It tapers from above downward, and ends below in a free border.

The part of the mediastinal pleura covering the pericardium is adherent to it, except along the phrenic nerve and accompanying vessels. The mediastinal pleura tends to insinuate itself between mediastinal structures. The pleurae of the right and left sides approach each other above the level of the arch of the aorta, behind the esophagus. That of the right side often crosses the median plane. The reflection of pleura behind the esophagus forms a *retroesophageal recess* on each side (fig. 29–6). Each recess is probably occupied by a part of the lung, and contributes to the *retrocardiac space* seen radiographically.[7] A small *infrapericardial recess* of the right pleural sac sometimes extends beneath the pericardium, just behind the inferior vena cava.

Diaphragmatic Pleura. This part of

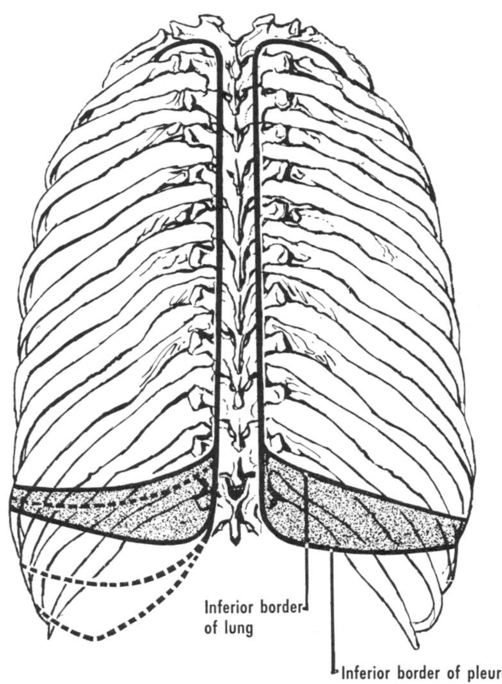

Figure 29–4 The pleurae from behind. The interrupted lines on the left represent common variations in the position of the inferior border of the pleura. The costodiaphragmatic recesses are stippled. Based on Lachman.[6]

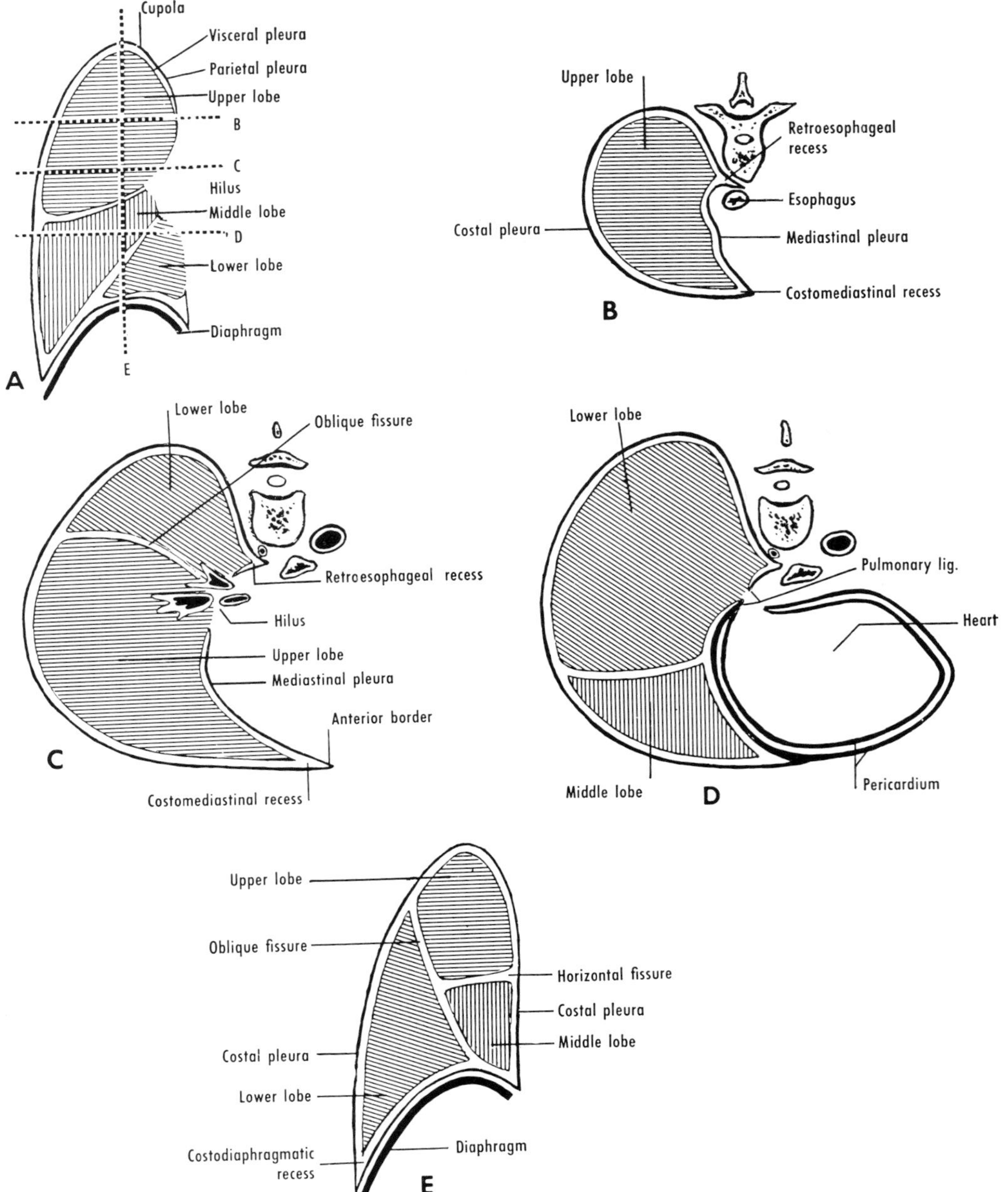

Figure 29–5 Diagrams of pleural reflections. *A*, coronal section of right lung and pleura. Lines *B* to *E* indicate the respective planes and levels of sections shown in diagrams *B* to *E*. *B*, upper horizontal section. Note costomediastinal and retroesophageal recesses. *C*, middle horizontal section. Note that the anterior border of the pleura forms the edge of the costomediastinal recess and that the oblique fissure reaches almost to the hilus. *D*, lower horizontal section, showing also relationships to the pericardium. Note that the pulmonary ligament is formed by the double reflection of the pleura below the hilus of the lung (fig. 29–7, p. 290). The mediastinal pleura is adherent to the fibrous pericardium, except where the phrenic nerve descends between them (not shown). *E*, sagittal section.

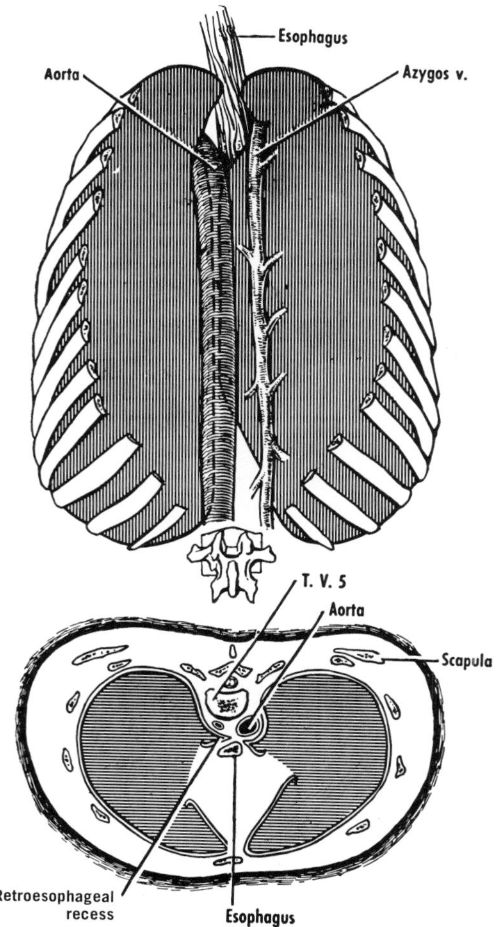

Figure 29–6 *Upper figure,* a posterior view, and *lower figure,* a horizontal section at the level of T.V. 5, showing relationships of lungs and pleura. The interrupted lines in the upper figure indicate the edges of the retroesophageal recesses. Based on Lachman.[6]

the parietal pleura covers most of the diaphragm, except the central tendon. A thin layer of endothoracic fascia, the *phrenicopleural fascia,* connects it with the diaphragm.

Cupola of the Pleura (Cervical Pleura). The costal and mediastinal parts of the parietal pleura are continuous over the apex of the lung, where they form the cupola or dome of the pleura. The cupola is strengthened by a thickening of the endothoracic fascia, the *suprapleural membrane,* which is attached to the inner margin of the first rib and to the transverse process of the seventh cervical vertebra. Usually some muscle fibers (*scalenus minimus*) and fibrous tissue are inserted into this membrane and onto the first rib.

The cupola of the pleura and the apex of the lung are at the level of the first rib behind. However, because the first rib slopes downward, the lung and pleura extend above the level of the anterior part of the rib into the root of the neck. Their highest point is usually indicated by the spine of the seventh cervical vertebra behind. They lie behind the sternocleidomastoid muscle, 2 to 3 cm or more above the level of the medial third of the clavicle.

The sympathetic trunk, first thoracic nerve, and vessels of the first intercostal space lie behind the cupola. Disorders of the lung and pleura in this region, as well as of adjacent bones, may involve these vessels and nerves. Involvement of the first thoracic nerve will result in paralysis of the intrinsic muscles of the hand; involvement of the sympathetic trunk will lead to Horner's syndrome (p. 706).

LUNGS

The lungs are the organs of respiration. The adjective pulmonary is from the Latin *pulmo,* lung. **Each lung is attached to the heart and trachea by its root and pulmonary ligament. It is otherwise free in the thoracic cavity.** The lungs are light, soft, spongy, elastic organs. Healthy lungs always contain some air; they float in water and crepitate when squeezed. A lung filled with fluid resulting from disease may not float in water. A lung from a fetus or newborn infant is light pink in color and firm to the touch. If the infant has not drawn a breath, the lung will not float. The surface of an adult lung is usually mottled, and presents dark gray or bluish patches on a bluish background. The increasing coloration with age is due to impregnation with inhaled atmospheric dust. The upper parts of the lung expand mainly in a horizontal plane during breathing, and coloration of the upper parts of the lungs tends to occur in bands, deep to and corresponding to the intercostal spaces.

The main bronchus that enters the hilus

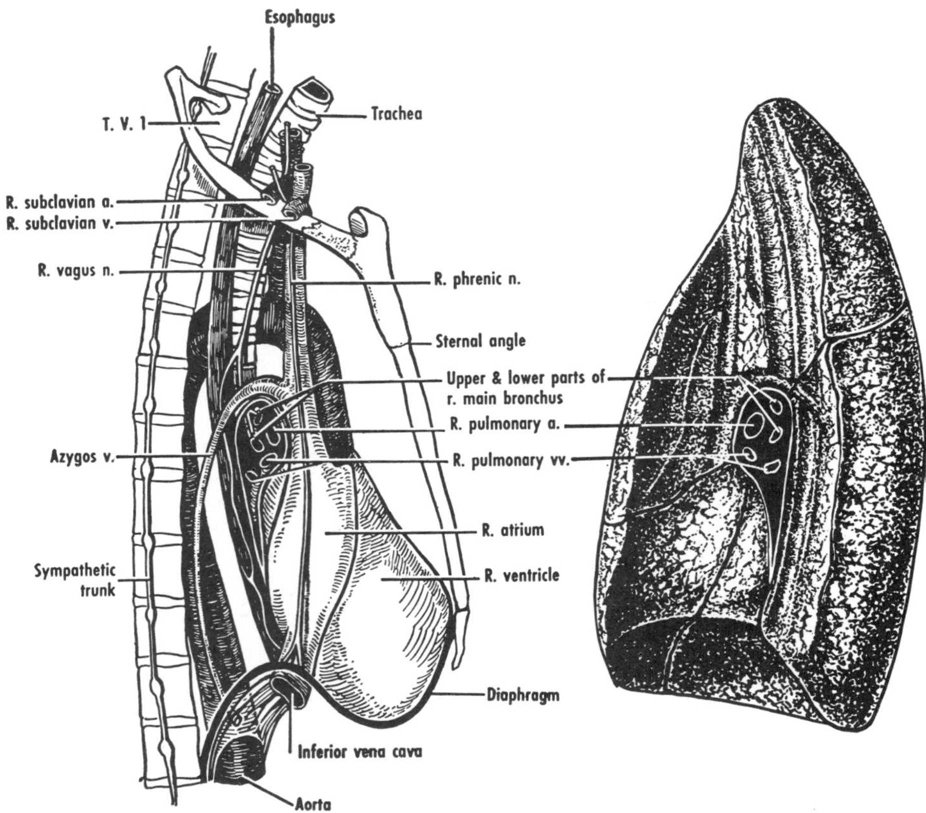

Figure 29–7 Mirror image views of right lung and mediastinal structures. The line of reflection of the parietal to the visceral pleura is shown as a white line around the hilus, prolonged below as the two-layered pulmonary ligament. Impressions produced by mediastinal structures have been indicated on the medial surface of the lung as they might appear in a hardened lung. Based on Mainland and on Mainland and Gordon.[9]

of each lung divides and subdivides within the substance of the lung and forms a system of branching air tubes termed the *bronchial tree.* The tubes carry air to the alveoli, where respiratory exchange with the blood occurs.

The right lung is heavier than the left. It is shorter because the right dome of the diaphragm is higher (the right lobe of the liver pushes it up), and it is wider because the heart and pericardium bulge more to the left.

Each lung presents an apex, a base, three surfaces (costal, medial, and diaphragmatic), and three borders (anterior, inferior, and posterior). Interlobar surfaces are also present but are hidden in the depths of the fissures. The left lung is divided into upper and lower lobes by an oblique fissure. The right lung is divided into upper, lower, and middle lobes by an oblique and a horizontal fissure (fig. 29–9).

The bronchi and pulmonary vessels extend from the trachea and heart, respectively, to each lung, and collectively form the root of the lung on each side. The hilus is the part of the medial surface where these structures enter the lung.

When a lung is hardened by embalming, it retains impressions of adjacent structures, whereas lungs freshly removed generally do not. For example, the overlying ribs and costal cartilages leave impressions on hardened lungs. The aorta and azygos vein usually form well-marked grooves on the medial surfaces of the left and right lungs, respectively. These and other cadaveric markings serve to point up the relationships of the lungs (figs. 29–7 and 29–8).

ANATOMICAL FEATURES

Surfaces and Borders

Apex. The apex is rounded. It has the same relationships as the cupola of the pleura (p. 705). The apex of the right lung is smaller than that of the left, and is closer to the trachea. This probably explains why the percussion note of the right apex may be higher and less resonant than that of the left.[8]

Costal Surface. The convex costal surface fits the part of the thoracic wall formed by the sternum, ribs, and costal cartilages. The costal surface joins the medial surface at the anterior and posterior borders, and the diaphragmatic surface at the inferior border.

Medial Surface. The medial surface has vertebral and mediastinal parts. The *vertebral part* is applied to the sides of the bodies of the vertebrae. The *mediastinal part* is related to the middle, posterior, and superior parts of the mediastinum. The *cardiac impression,* produced by the heart and pericardium, is deeper in the left than in the right lung. The *hilus* is a wedge-shaped area above and behind the cardiac impression; it contains the blood vessels, lymphatic vessels, nerves, and bronchi entering or leaving the lung. The arrangement of structures within the hilus differs in the right and left lungs.

Diaphragmatic Surface. The concave diaphragmatic surface corresponds to the dome of the diaphragm. The diaphragmatic surface of the right lung is more deeply concave than that of the left, in conformity with the higher position of the right dome of the diaphragm. The diaphragmatic surface of the right lung is related to the right lobe of the liver, that of the left to the fundus of the stomach, to the spleen, sometimes to the left

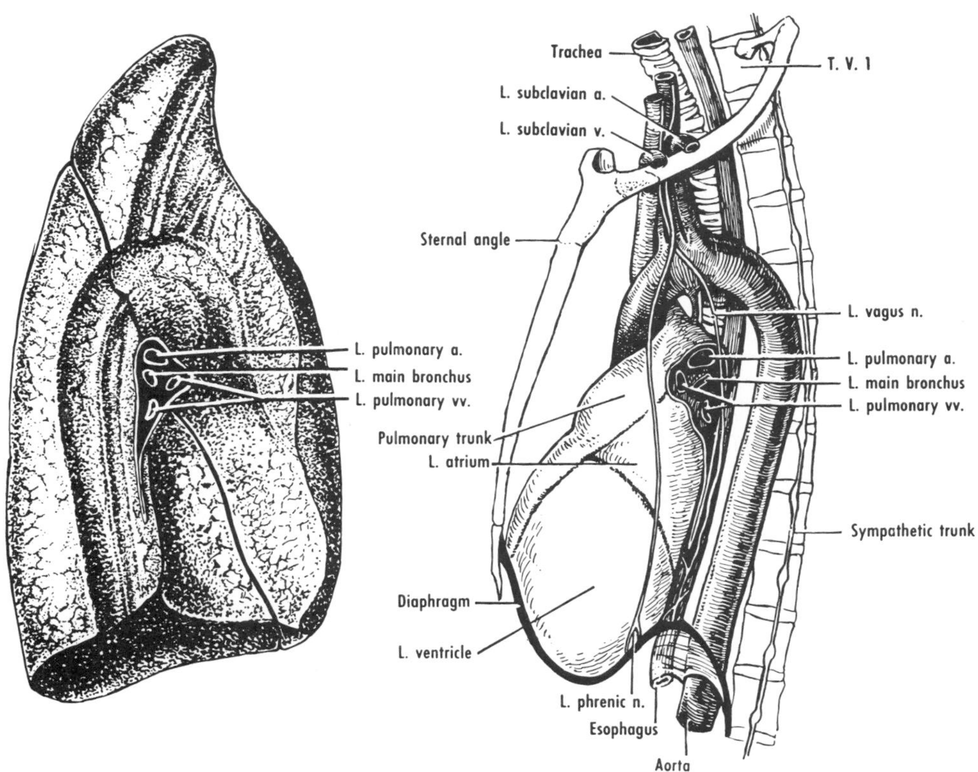

Figure 29–8 Mirror image views of left lung and mediastinal structures. The line of reflection of the parietal to the visceral pleura is shown as a white line around the hilus, prolonged below as the pulmonary ligament. Impressions produced by mediastinal structures have been indicated on the medial surface of the lung as they might appear in a hardened lung. Based on Mainland and on Mainland and Gordon.[9]

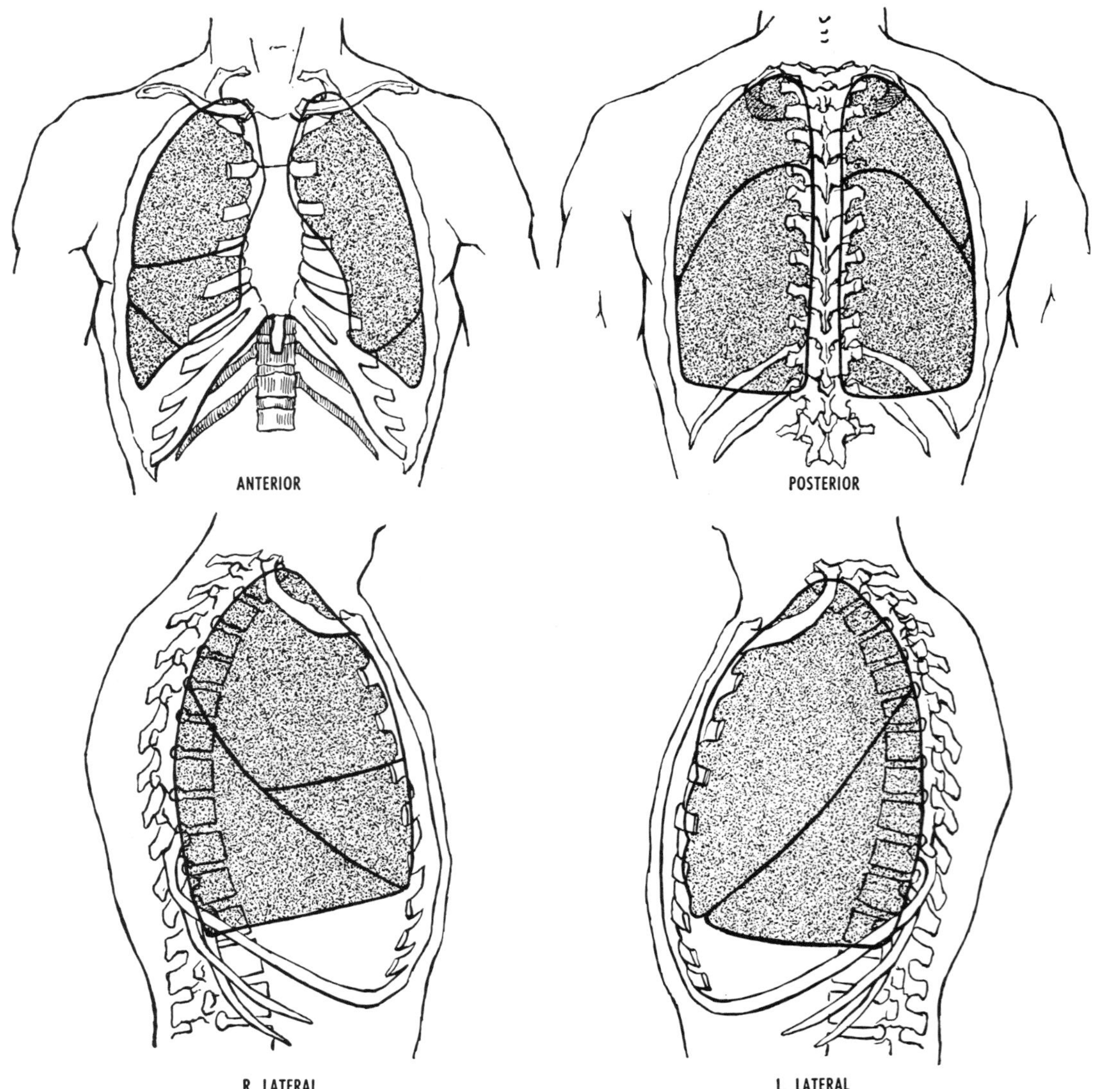

Figure 29–9 The lobes and fissures of the lungs. Based on Brock.[10]

flexure of the colon, and fairly often to the left lobe of the liver.

Anterior Border. This corresponds more or less to the anterior border of the pleura (fig. 29–3). It is uncertain whether the costomediastinal recess of the pleura is completely filled by the lung during quiet breathing, but it is during a deep inspiration. The anterior border of the left lung probably deviates more to the left (*cardiac notch*) than does that of the pleura, but the size of this notch during life is uncertain. A similar but smaller notch is occasionally found in the right lung. The *lingula* is a small, tongue-shaped portion of the upper lobe of the left lung that lies between this notch and the oblique fissure; it corresponds to the middle lobe of the right lung.

Inferior Border. The inferior border separates the diaphragmatic from the costal and medial surfaces. This edge of the lung occupies the costodiaphragmatic recess of the pleura during all phases of respiration. However, the inferior border as demonstrated by percussion is at a higher level (fig. 29–4), a level that descends during a deep inspiration. The explanation is that during quiet breathing the part of the lung in the costodiaphragmatic recess is too thin to be demonstrable by percussion. During a deep inspiration, the lung expands in the space it already occupies. **For surgical purposes, one should assume that the lung and pleura are coextensive. The liver, stomach, spleen, colon, and kidney, as well as the peritoneal cavity, extend to a higher level than the periphery of the diaphragm and the inferior border of the lung. Any perforation of the lower intercostal spaces must be considered as an abdominal as well as a thoracic wound.**

The inferior border as outlined by percussion begins at about the level of the xiphisternal joint. It then extends laterally in a somewhat straighter line than the pleura, about two intercostal spaces higher (figs. 29–3 and 29–4). It crosses the sixth rib in the midclavicular line, the eighth rib in the midaxillary line, and is then directed toward the spine of the tenth thoracic vertebra. Considerable individual variation occurs in the level of the inferior border. During a deep inspiration, the apparent level descends at least two intercostal spaces.

Posterior Border. The relationships of the posterior border are the same as those of the posterior border of the pleura.

Lobes and Fissures

The left lung is divided into upper and lower lobes by a long, deep, oblique fissure, which extends inward, almost to the hilus. The *upper lobe,* which lies above and in front of this fissure, includes the apex and anterior border of the lung. The lingula of the left lung corresponds to the middle lobe of the right lung. The larger *lower lobe* lies below and behind this fissure, and includes nearly all of the base and most of the posterior part of the lung.

The right lung is divided into *upper, middle,* and *lower lobes* by an oblique and a horizontal fissure. The oblique fissure separates the lower lobe from the middle and upper lobes. The horizontal fissure extends forward from the oblique fissure and separates the upper and middle lobes. The middle lobe is usually triangular or wedge-shaped in outline.

Oblique Fissure (fig. 29–9). On the right side, the oblique fissure usually begins at the level of the head of the fifth rib, and it may be even lower. (It may also be at the level of the head of the fourth rib.) The origin of the left oblique fissure is usually at a higher level than that of the right.

Viewed from the costal surface in the living subject, the oblique fissure curves downward, following the line of the sixth rib. It ends near the sixth costochondral junction, where it meets the inferior border of the lung. When the arm is abducted and the hand placed on the back of the head, the medial border of the scapula approximately indicates the oblique fissure.

Horizontal Fissure. This begins at the oblique fissure near the midaxillary line, at about the level of the sixth rib. It extends forward in an extremely variable fashion to the anterior border at the level of the fourth costal cartilage.

Bronchi and Root of the Lung

The root of the lung, which is formed by the structures entering and emerging at the hilus, connects the medial surface of each lung to the heart and trachea. The chief structures in the root are the bronchi and pulmonary vessels. Other structures are nerves, bronchial vessels, and lymphatic vessels and nodes, all embedded in connective tissue. The root is enveloped in pleura, which is prolonged below as the pulmonary ligament.

The trachea and main bronchi occupy a plane posterior to that occupied by the heart and great vessels.[9] This relationship is maintained in the root of the lung, where, from before backward, are situated veins, artery, and bronchus, with the artery above the veins. Figures 29–7 and 29–8 illustrate some of the important relationships.

Bronchopulmonary Segments (fig. 29–10).[10] The lung may be considered as being divided into smaller and smaller segments, each segment being the area of distribution of a specific bronchus (fig. 28–5, figs. 29–11 to 29–14). Eventually, these conducting tubes end in tiny air spaces termed *alveoli.* The wall of an alveolus is the alveolar membrane, or gas-blood barrier, through which oxygen and carbon dioxide diffuse. The area of the lung at the hilus lacks respiratory tissue and is called the nonrespiratory or nonexpansile part of the lung.

The term bronchopulmonary segment is applied to the largest segments within a lobe. The bronchus supplying a segment is a direct branch of a lobar, or "division," bronchus. It is therefore a third order branch, the first being a main bronchus, the second a lobar, or "division," bronchus. Bronchopulmonary segments are separated from one another by connective tissue septa. The septa are continuous with the visceral pleura and they send prolongations into the segments. These septa prevent diffusion of air into a segment the bronchus of which has been blocked. The bronchopulmonary segments are named (compare fig. 29–10 and table 29–1), but there is considerable variation in the segments and their bronchi and blood vessels (see below).

Pulmonary disorders may be localized in, or confined to, a bronchopulmonary segment. Further clinical importance of segments lies in the fact that surgical removal of a segment for tumor or infection is feasible. Also, the segment

(*Text continued on page 298.*)

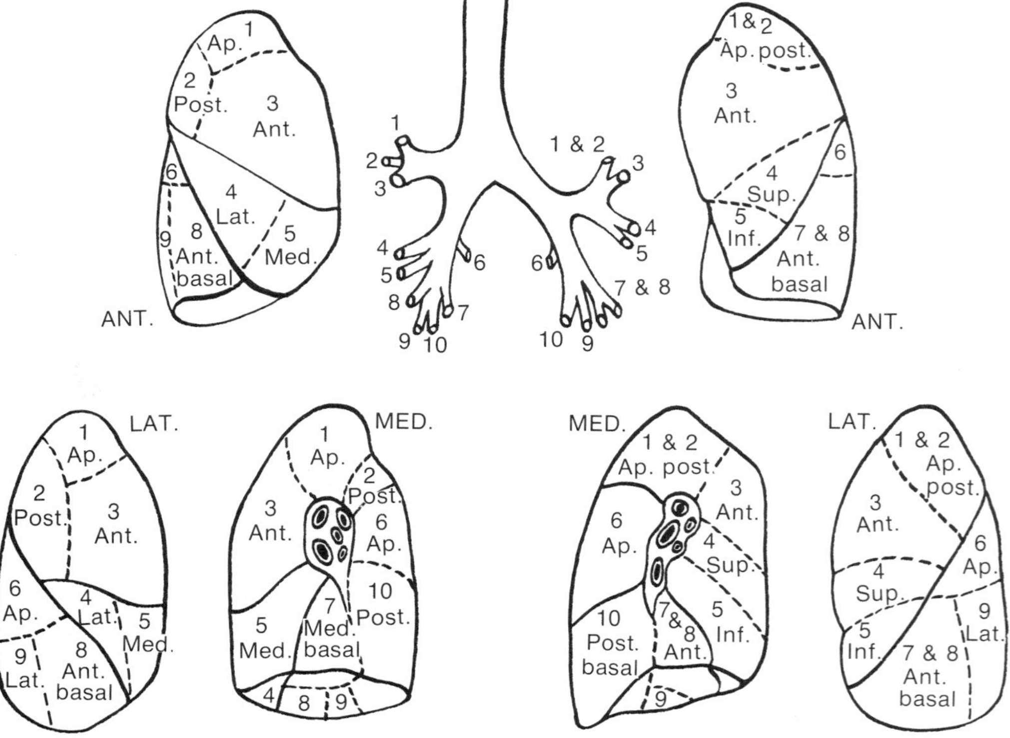

Figure 29–10 The segmental bronchi and bronchopulmonary segments. The bronchi and segments are numbered so as to facilitate comparison with figures 29–11 to 29–13. For names, see table 29–1, p. 298.

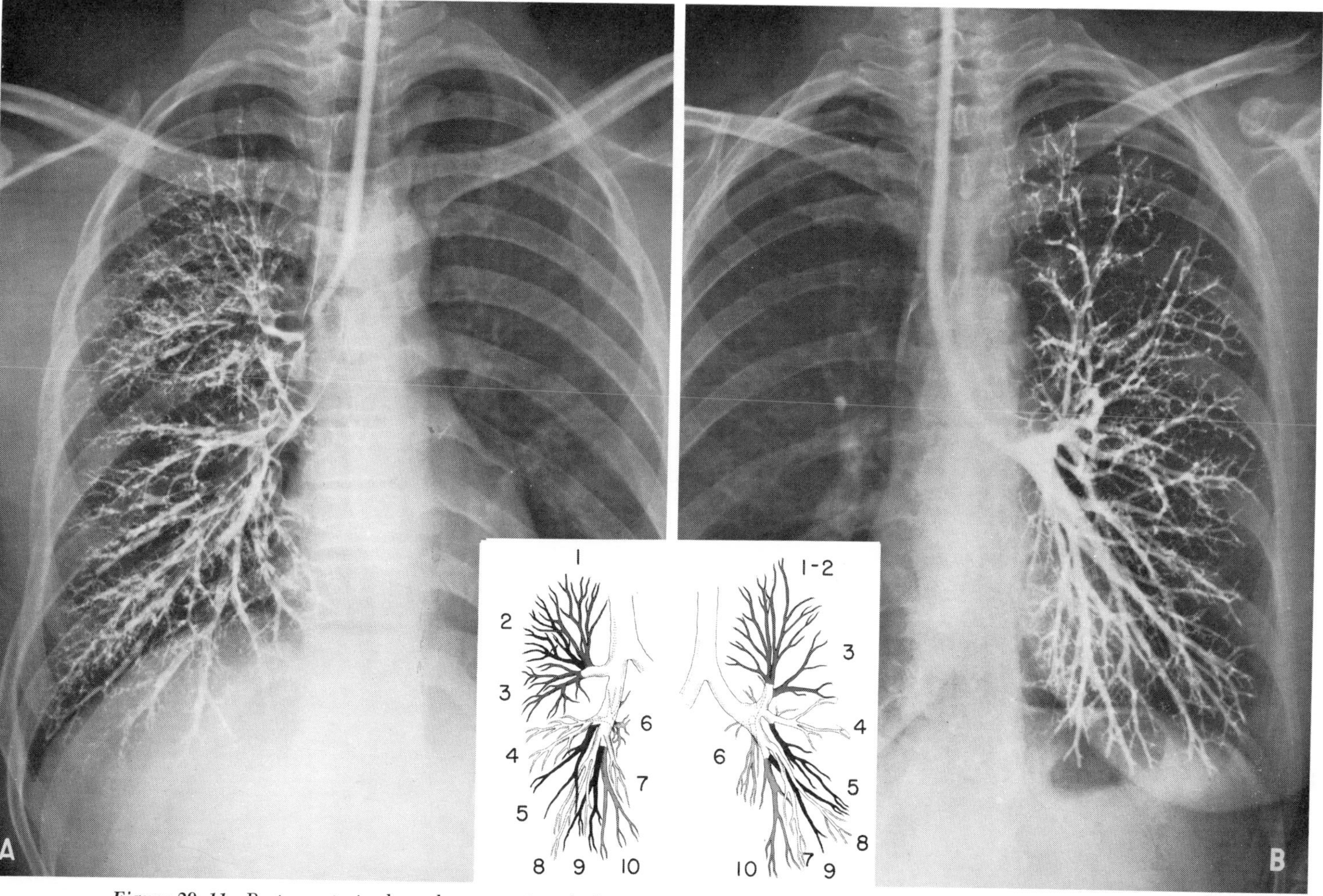

Figure 29–11 Posteroanterior bronchograms. *A*, right lung. *B*, left lung. From *Medical Radiography and Photography;* courtesy of J. Stauffer Lehman, M.D., and J. Antrim Crellin, M.D., Philadelphia, Pennsylvania. For terminology see table 29–1, p. 298.

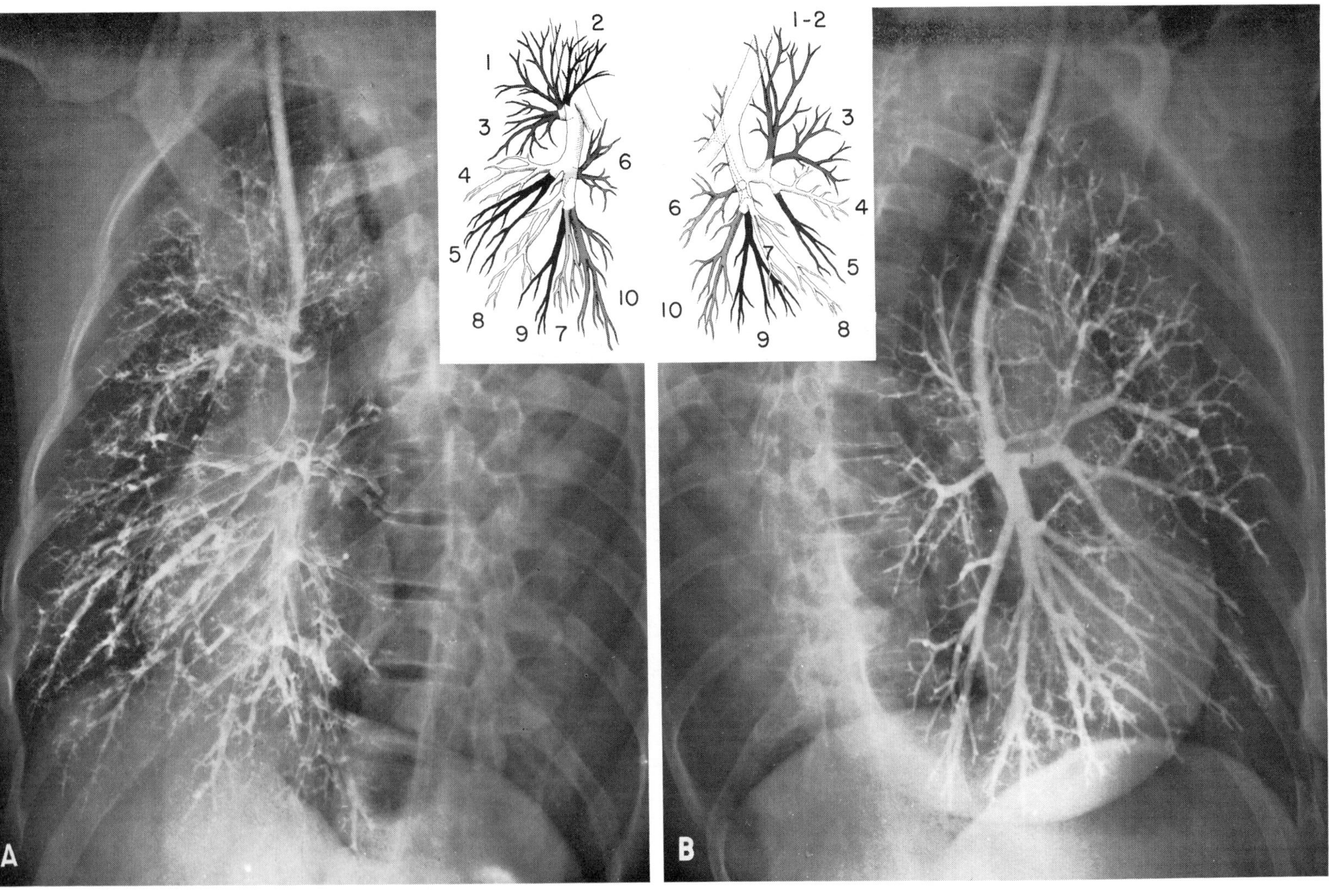

Figure 29–12 Oblique bronchograms. *A*, right lung, left anterior oblique view. *B*, left lung, right anterior oblique view. From *Medical Radiography and Photography.* Courtesy of J. Stauffer Lehman, M.D., and J. Antrim Crellin, M.D., Philadelphia, Pennsylvania. For terminology see table 29–1, p. 298.

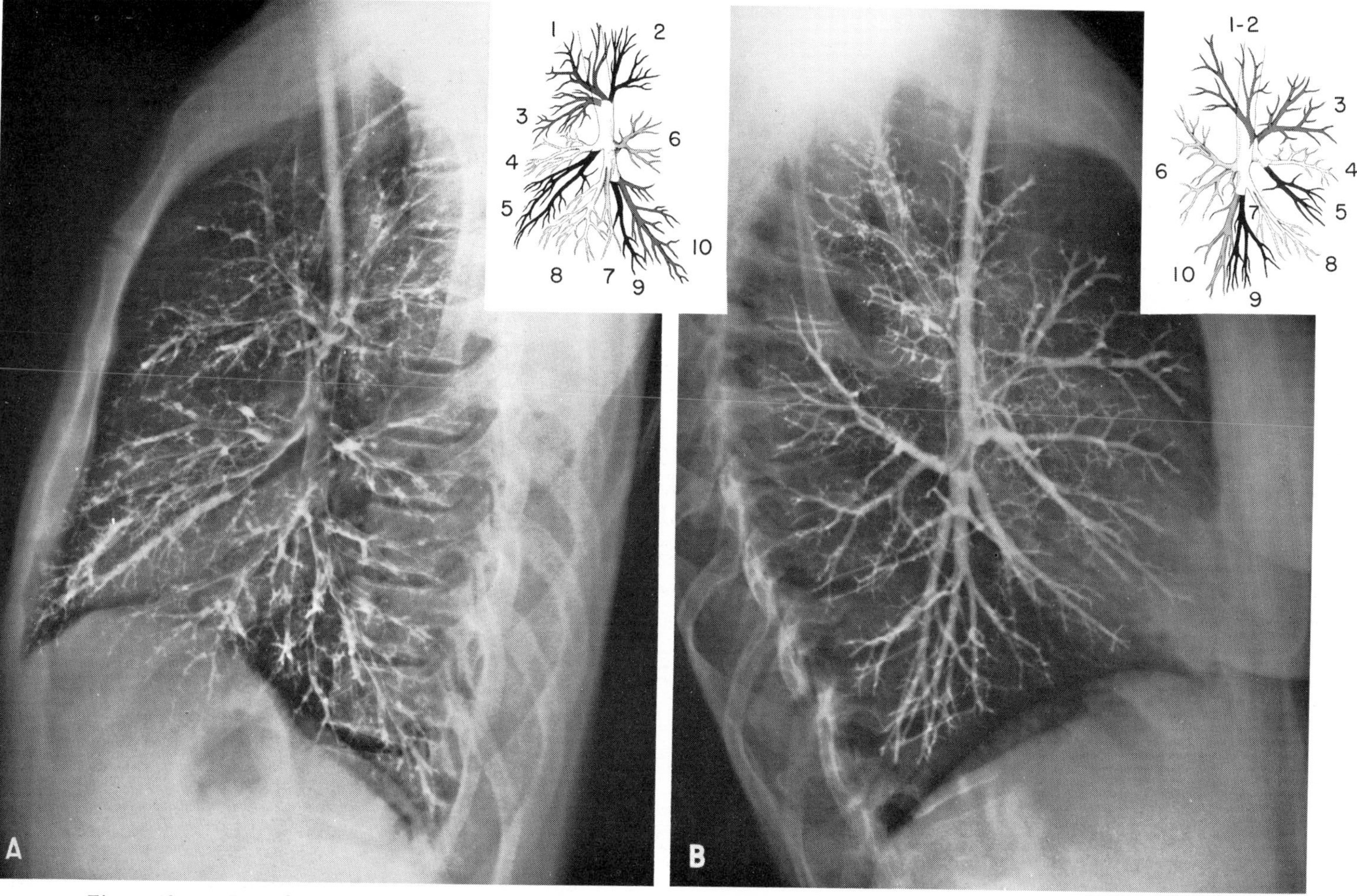

Figure 29–13 Lateral bronchograms. *A*, right lung. *B*, left lung. From *Medical Radiography and Photography;* courtesy of J. Stauffer Lehman, M.D., and J. Antrim Crellin, M.D., Philadelphia, Pennsylvania. For terminology see table 29–1, p. 298.

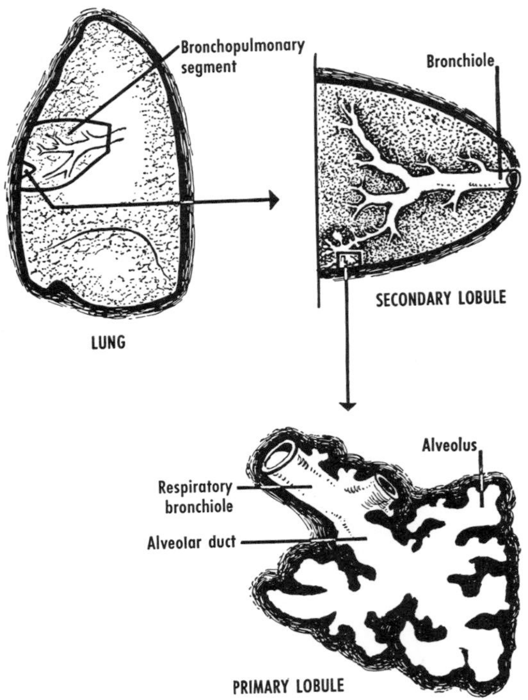

Figure 29–14 Diagrammatic representation of a bronchopulmonary segment and primary and secondary lobules. Based on Miller.

involved in a disorder may be located by radiography or bronchoscopy, and the information gained may determine the course of treatment.

The branches of the pulmonary arteries accompany the bronchi and tend to correspond to the segments. They are more variable, however, in that a branch may supply more than one segment, or, more often, the number of named branches to a lobe may exceed the number of segments. **Pulmonary veins do not accompany the bronchi. They are intersegmental in drainage and are therefore guides to intersegmental planes.**[11]

Variations

Fissures are often absent or incomplete, especially the horizontal. On the other hand, a supernumerary fissure may partition the lung to form an extra lobe. Variations in the pattern of branching of the segmental and subsegmental bronchi are common, with the result that the bronchopulmonary segments vary in number and position. For example, an additional subapical (subsuperior) segment is sometimes present in either the left or the right lower lobe, and a medial basal (cardiac) segment may be present in the left lower lobe. Rarely a segmental bronchus may arise from the trachea.[12]

The common variations in the lungs are of little functional importance, but they may be important in surgical and diagnostic procedures. Some variations may be recognized radiographically. One such is the *lobule of the azygos vein.*[13] Several kinds of these have been described, but the one referred to here is formed when the azygos vein, instead of arching over the hilus, arches over the upper part of the right lung. In so doing it appears to "cut" deeply into lung tissue and to isolate partially a medial part of the lung, called the lobule of the azygos vein. Parietal pleura is carried into the fissure with the vein. The four layers of pleura (mesoazygos) in the fissure, together with the vein at the bottom of the fissure, may be visible radiographically.

Radiological Anatomy

This is discussed in greater detail on page 343. The lungs are rendered translucent by the air they contain, and this translucency is increased during inspiration. Hence much of the lungs can be delimited radiographically (see figs. 29–1 and 29–2). However, the parts that extend below the levels of the domes of the diaphragm are obscured by the liver under the right dome, and by the fundus of the stomach, the liver, the spleen, and sometimes the left flexure of the colon,

TABLE 29–1 Bronchopulmonary Segments

Right Lung	*Left Lung*
Superior Lobe	*Superior Lobe*
1. Apical	1 and 2. Apicoposterior
2. Posterior	
3. Anterior	3. Anterior
Middle Lobe	
4. Lateral	4. Superior lingular
5. Medial	5. Inferior lingular
Inferior Lobe	*Inferior Lobe*
6. Apical (superior)	6. Apical (superior)
7. Medial basal (cardiac)	7 and 8. Anterior basal (Medial basal [cardiac] is independent in *ca.* 1/3 of instances)
8. Anterior basal	
9. Lateral basal	9. Lateral basal
10. Posterior basal	10. Posterior basal

under the left dome. The heart and great vessels also obscure portions of the lungs.

The arteries at the hilus are visible radiographically and form a pattern that extends into the lung. The bronchi are usually not visible, except occasionally when they are seen end-on, and neither are lymph nodes, unless they are calcified or fibrosed.

BLOOD SUPPLY, LYMPHATIC DRAINAGE, NERVE SUPPLY, AND DEVELOPMENT

Arteries and Veins

Blood that is to be oxygenated is carried by the pulmonary arteries. The tissues of the lungs are nourished by the bronchial arteries. In rare instances, anomalous arteries to the lower parts of the lungs may arise from the lower part of the thoracic aorta or from the upper part of the abdominal aorta.[14]

Pulmonary Arteries. The intrapulmonary branches of the pulmonary arteries accompany the bronchi and lie in their connective tissue sheaths. They end in capillary networks in the alveolar ducts and sacs, and in the alveoli.

Pulmonary Veins. The pulmonary veins, which lack valves, collect arterial blood from the respiratory part of the lung, and venous blood from the visceral pleura and from the bronchi. The first few divisions of the main bronchi, however, are drained by bronchial veins. Pulmonary veins are intersegmental in location. They course in connective tissue septa toward the hilus and sometimes cross a fissure.

Usually a single pulmonary vein leaves each lobe. The right upper and middle lobar veins typically join near the hilus to form the right upper pulmonary vein. Commonly, then, four pulmonary veins (right and left upper and right and left lower) pass to the left atrium of the heart (p. 324). Variations in size and number are common. Occasionally, one or more pulmonary veins enter the right atrium or superior vena cava.[15] Difficulty may then result if the lung tissue drained by the remaining veins (into the left atrium) becomes diseased, and oxygenation is interfered with.

Bronchial Arteries. There is usually one bronchial artery on the right, which often arises from the aorta as a common trunk with the right third posterior intercostal artery, but which may arise from the upper left bronchial artery. There are usually two bronchial arteries on the left, which arise from the aorta. Many patterns and anomalies of the bronchial arteries have been described,[16] including origins from the internal thoracic artery, from a subclavian artery, or from an inferior thyroid artery. Several longitudinal, anastomosing branches of each bronchial artery accompany the intrapulmonary bronchi as far as the respiratory bronchioles. They supply oxygenated blood to the nonrespiratory tissues of the lungs, including the nerves, the walls of the pulmonary vessels, and a part of the visceral pleura.

Bronchial Veins. The venous blood from the first few divisions of the bronchi is carried by bronchial veins to the azygos, hemiazygos, or posterior intercostal veins. All other venous blood is carried by the pulmonary veins.

Lymphatic Drainage

Deep lymphatic vessels drain the bronchial tree, the pulmonary vessels, and the connective tissue septa. The deep vessels have few valves. They intercommunicate, and also communicate with superficial vessels in the septa of the secondary lobules. The superficial vessels have numerous valves. Lymph in both groups of vessels flows toward the hilus, where the vessels end in the pulmonary and bronchopulmonary nodes. These in turn drain into the tracheobronchial nodes (p. 329).

If the parietal and visceral pleurae become fused, lymphatics in the lung and visceral pleura may drain into axillary nodes. (The presence of carbon particles in the axillary nodes is presumptive evidence of such a fusion.)

Nerve Supply[17]

The anterior and posterior pulmonary plexuses in front of and behind the root of the lung are formed by branches from the vagus nerves and the sympathetic trunks (p. 336). Groups of parasympathetic ganglion cells are present in the plexuses and along the bronchial tree.

Autonomic Fibers. Preganglionic parasympathetic fibers from the vagus nerves synapse with ganglion cells, the axons of which supply the smooth muscle and glands of the bronchial tree. They are probably excitatory to these structures.

Postganglionic sympathetic fibers arise from the upper four or five thoracic sympathetic ganglia, and reach the plexuses by direct branches. They supply blood vessels, and the smooth muscle and glands of the bronchial tree. They are probably inhibitory to the smooth muscle of the bronchi and bronchioles.

Sensory Fibers. These are of vagal origin. Some form sensory endings in the walls of the pulmonary vessels, especially the veins, in both their pulmonary and extrapulmonary course. The functions of these endings are uncertain, although mechanical stimulation of those in the vessel walls may result in a severe fall in blood pressure and heart rate. Still other vagal fibers form sensory endings in the visceral pleura and in the walls of the bronchi and bronchioles, and probably between alveoli also. These endings are concerned in the reflex control of respiration. Still other sensory endings are present in the bronchial mucous membrane; their irritation provokes coughing.

Development of Respiratory System[18]

The epithelium of the respiratory system arises as a diverticulum of the foregut. This diverticulum grows caudally, and divides into two. Each division (main bronchus) grows into the thoracic cavity, as if it were invaginating the pleural sac ahead of it, and continues to divide. The diverticulum carries with it the mesenchyme that gives rise to the connective tissue, muscle, and cartilage of the lung. Alveoli begin to be formed by the second trimester, and infants from about the seventh month of pregnancy may be viable, that is, may be able to live *ex utero.*

If the trachea and esophagus fail to separate properly during development, abnormal communications may be left between them. These communications constitute an important anomaly known as *tracheoesophageal fistula.* In the most common form, the upper part of the esophagus ends blindly, whereas the lower part arises from the trachea. The most severe anomaly, fortunately not common, is absence of a lung.[19]

RESPIRATION

The major function of the lungs is to oxygenate the mixed venous blood. This involves (1) *ventilation,* a process that includes both volume and distribution of the air reaching the alveoli; (2) *diffusion,* the process by which oxygen and carbon dioxide pass through the alveolar membrane; and (3) *blood flow in the pulmonary capillaries,* which must be adequate in volume and even in distribution.

Ventilation is brought about chiefly by the muscles of respiration. Although these are skeletal in type, the respiratory process is reflexly controlled by the respiratory centers in the brain stem. The activity of these centers can be voluntarily modified, but their reflex functions cannot be completely suppressed. The volume of respiration is indicated by certain anatomical measurements termed *primary lung volumes.* These are (1) residual volume, (2) expiratory reserve volume, (3) tidal volume, and (4) inspiratory reserve volume. Residual volume, which is indicated by the innermost circle in figure 29–15, is the volume of air that remains after as much air as possible has been voluntarily expelled from the lungs. (The volume of a collapsed lung is less than the residual volume.) Tidal (500 to 600 ml) volume (depth of breathing) is the volume of gas entering or leaving the respiratory tract with each breath. It is also the difference between quiet inspiration and quiet expiration. Pulmonary physiologists also define capacities. For example, vital capacity is the difference between total lung capacity (all four primary volumes; indicated in figure 29–15 by the outermost line) and residual volume. It is the maximal volume of air that can be forcibly expelled from the lungs after a maximal inspiration.

With regard to the distribution of air during ventilation, it is important to recognize that the flow of air through the lungs is not uniform at low ventilatory rates. For example, in normal subjects in a standing position, up to the first 500 ml of air inspired from residual volume is distributed almost exclusively to the apical regions of the lungs.[20] Among the factors responsible for this unequal distribution are regional differences in resistance to flow in the small airways (less than 2 mm in diameter). Hence, as the resistance to air flow increases (e.g., because of loss of lung elasticity with age), the chances are that parts of the lungs in elderly individuals will be poorly ventilated during normal breathing. Moreover, respiratory disease at any age may interfere with air flow. Thus, simple tests designed to measure the volume change (closing volume) during the second half of vital capacity determination are useful in detecting early small-airway disease.[21] More complex tests of airflow and distribu-

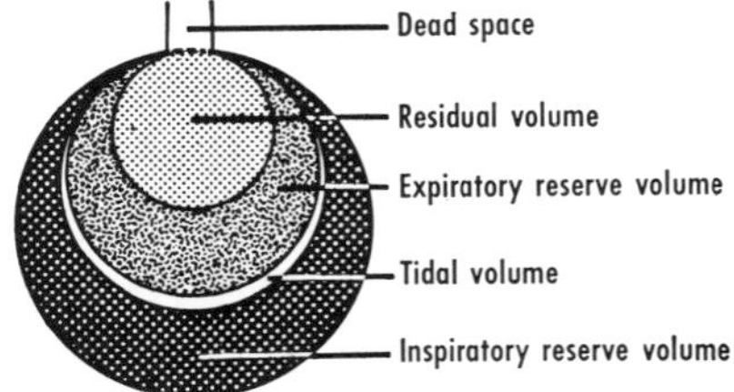

Figure 29–15 The primary lung volumes. The outermost line represents the greatest size to which the lung can expand. Based on Comroe *et al.*

tion and of respiratory efficiency are also available for the study of a broader spectrum of respiratory diseases.[22]

The average frequency of breathing is about 11 to 14 per minute in healthy adults under basal metabolic conditions. The average respiratory rate at birth is 39 per minute.[23] The average volume of air entering or leaving the respiratory passages with each breath (tidal volume, or depth of breathing, figure 29–10) is 500 to 600 ml.

The movements of the thoracic bony cage and of the diaphragm are described elsewhere (p. 276). These movements and their role in altering intrathoracic pressure must be considered in connection with the functions of the lungs and pleura.

The parietal and visceral layers of the pleura and the intervening capillary layer of fluid can be compared to two pieces of glass separated by a thin film of fluid. The pieces of glass can slide easily against each other, but considerable force is necessary to pull them apart. The layers of the pleura likewise slide easily against each other, but under normal conditions never separate. If the pleural layers did not adhere to each other, the lungs could not expand. Two forces aid in the adherence of the layers. One is the atmospheric pressure on the outside of the thoracic wall. The other is the intra-alveolar pressure which, because of its connection through the air passage with the exterior, is equal to the atmospheric pressure when breathing is suspended.

Two forces tend to pull the pleural layers apart. One is the elasticity of the thoracic wall, which is directed outward. (The ends of a cut rib spring outward.) The other is the retractive power of the lungs. During inspiration, the elastic tissues of the trachea, bronchi, lungs, and thoracic wall are stretched. The retractive power of the lungs is increased by being stretched, but it is still minimal compared with the atmospheric pressure and with the adhering power of the layers of pleura. The potential energy created by the contraction of the inspiratory muscles is stored in the elastic tissues. When the muscles relax (quiet expiration is a passive process completed within three seconds), the stretched elastic tissues recoil with a force that depends on the volume of air in the lungs at the end of inspiration and on the compliance (lack of resistance to airflow) of the airways. Even at rest, the elastic tissues tend to pull against the pleural layers, because the lungs are not completely emptied. The lungs of an infant tend to empty completely of air during expiration and, being relaxed, exert no pull against the pleura at the resting expiratory level.[24] The thorax grows faster than the lungs, however, and shortly after birth the lungs begin to be stretched so that, as in the adult, even at rest they pull against the pleural layers.

Another major factor in retractive power is surface phenomena at the gas-tissue interface, that is, surface tension of the mucous lining or mucous material of the alveoli.[25]

During inspiration the thoracic cavity enlarges and the intra-alveolar pressure is reduced. If the pressure in the great veins and the atria is measured, it is found to be negative at rest, and to decrease during inspiration. The difference in pressure between capillaries and the right atrium therefore becomes greater. One result is an increase in the return of blood to the heart. The converse occurs during expiration. The pressure of the cerebrospinal fluid is directly related to venous pressure. Accordingly, there are pressure changes in the former corresponding to the phases of respiration.

Movement of the Bronchial Tree. During inspiration, the bronchial tree elongates and the roots of the lungs move downward, outward, and forward, and the trachea and main bronchi descend.[26] These movements are not marked during quiet breathing, but they are considerable with a deep inspiration, when the bifurcation of the trachea may descend 5 cm or more. The forward movement of the roots provides space, so to speak, into which the posterior parts of the lungs can expand. Any increase in the rigidity of the mediastinum, or hindrance to movement of the root of the lung, will hamper ventilation of the posterior parts of the lung. It must be emphasized, however, that ventilation has great reserves. An individual with both lower lobes, the right middle lobe, and the lingula removed has lived without noticeable ill effects.[27]

The lungs also show differential expansion that is related to the respiratory movements. Most of the expansion of the upper parts of the lungs, for example, occurs as the thoracic cage increases in diameter, whereas the lower parts of the lungs expand mainly as the diaphragm descends. It is quite likely that these differential movements of the upper and lower parts of the lung are facilitated by the oblique fissures.

Artificial Respiration. Many methods of artificial respiration have been devised to provide adequate ventilation and to mimic natural ventilation as closely as possible.

The best is the mouth-to-mouth or mouth-to-nose method. The patient is placed on his back and his head is tilted back into a "chin-up" position, with the neck stretched. With one hand, the operator pulls the lower jaw anteriorly. With the other, he pinches the nose shut or seals the mouth to prevent air leakage. He then takes a deep breath and blows into the mouth or nose of the patient until the patient's chest rises. He then stops and allows expiration to occur. The cycle of forced-inspiration-passive-expiration is repeated 12 to 20 times per minute. For infants, the operator seals both mouth and nose with his mouth, and blows with small puffs of air.

A recommended manual technique is the arm lift-back pressure method,[28] in which the patient lies prone and the operator kneels at his head. Inspiration is accomplished by lifting and drawing the arms toward the operator. This procedure lifts the weight of the body from the chest, the pectoral muscles elevate the ribs, and the spine is extended. Expiration is produced by pressing on the patient's back.

REFERENCES

1. J. W. Pierson, J. Amer. med. Ass., *105*:399, 1935.
2. T. C. Pennell, J. thor. cardiov. Surg., 52:629, 1966.
3. J. A. Capps, Arch. intern. Med., 8:717, 1911. J. A. Capps and G. H. Coleman, *An Experimental and Clinical Study of Pain in the Pleura, Pericardium and Peritoneum,* Macmillan, New York, 1932.
4. R. T. Woodburne, Anat. Rec., 97:197, 1947.
5. G. J. Noback, Anat. Rec., 52 (Suppl.):28, 1932.
6. E. Lachman, Anat. Rec., 83:521, 1942; Amer. J. Roentgenol., *56*:419, 1946.
7. H. C. Maier, Amer. J. Roentgenol., *43*:168, 1940. E. Lachman, Anat. Rec., *83*:521, 1942.
8. A. F. Hewat, Edinb. med. J., *45*:326, 1938.
9. For a discussion of important relations and surgical anatomy, see D. Mainland, *Anatomy,* Hoeber, New York, 1945. See also D. Mainland and E. J. Gordon, Amer. J. Anat., *68*:457, 1941; R. Brock and L. L. Whytehead, Brit. J. Surg., *43*:8, 1955; E. M. Kent and B. Blades, J. thorac. Surg., *12*:18, 1942; and P. Caulouma, Gaz. Hôpit., *130*:251, 1958.
10. C. L. Jackson and J. F. Huber, Dis. Chest, 9:319, 1943. J. F. Huber, J. natn. med. Ass., *41*:49, 1949. R. C. Brock, *The Anatomy of the Bronchial Tree,* Cumberledge, London, 2nd ed., 1954. E. A. Boyden, *Segmental Anatomy of the Lungs,* Blakiston Division, McGraw-Hill, New York, 1955.
11. B. H. Ramsey, Surgery, *25*:533, 1949.
12. W. Woźniak, Folia Morphol. Praha, *14*:148, 1966.
13. L. E. Etter, Amer. J. Roentgenol., *58*:726, 1947. B. J. Anson *et al.,* Quart. Bull. Northw. Univ. med. Sch., *24*:285, 1950.
14. A. Bruwer, O. T. Clagett, and J. R. McDonald, J. thorac. Surg., *19*:957, 1950.
15. H. Brody, Arch. Path. (Lab. Med.), *33*:221, 1942. O. C. Brantigan, Surg. Gynec. Obstet., *84*:653, 1947.
16. J. F. Menke, Anat. Rec., *65*:55, 1936. R. O'Rahilly, H. Debson, and T. S. King, Anat. Rec., *108*:227, 1950.
17. O. Larsell and R. S. Dow, Amer. J. Anat., *52*:125, 1933. J. B. Gaylor, Brain, *57*:143, 1934. H. Spencer and D. Leob, J. Anat., Lond., *98*:599, 1964. F. L. Dwinnell, Jr., Amer. J. Anat., *118*:217, 1966.
18. L. J. Wells and E. A. Boyden, Amer. J. Anat., *95*:163, 1954. E. A. Boyden, Amer. J. Surg., *89*:79, 1955.
19. L. B. Thomas and E. A. Boyden, Surgery, *31*:429, 1952. A. R. Valle, Amer. J. Surg., *89*:90, 1955. R. O'Rahilly and E. A. Boyden, Z. Anat. EntwGesch., *141*:237, 1973.
20. J. Milic-Emili *et al.,* J. appl. Physiol., *21*:749, 1966.
21. D. S. McCarthy *et al.,* Am. J. Med., *52*:747, 1972.
22. J. A. Burdine *et al.,* J. nucl. Med., *13*:933, 1972.
23. H. J. Boutourline-Young and C. A. Smith, Amer. J. Dis. Child., *80*:753, 1950.
24. T. G. Heaton, Canad. med. Ass. J., *39*:275, 1938.
25. J. Mead, Physiol. Rev., *41*:281, 1961.
26. C. C. Macklin, Amer. J. Anat., *35*:303, 1925; Physiol. Rev., 9:1, 1929; Amer. Rev. Tuberc., 25:393, 1932.
27. E. A. Graham, Surgery, *8*:239, 1940.
28. A. S. Gordon *et al.,* J. Amer. med. Ass., *147*:1444, 1951. Council on Physical Medicine and Rehabilitation, J. Amer. med. Ass., *147*:1454, 1951.

GENERAL READING

Comroe, J. H., *et al., The Lung,* Year Book Medical Publishers, Chicago, 2nd ed., 1962. A valuable account of pulmonary physiology.

Fraser, R. G., and Paré, J. A. P., *Structure and Function of the Lung,* Saunders, Philadelphia, 1971. Deals chiefly with the radiology of the airways and vessels.

The following are concerned with the normal anatomy and structure of the lung:

Engel, S., *The Child's Lung,* Arnold, London, 1947.

Lauweryns, J. M., The blood and lymphatic microcirculation of the lung, in *Pathology Annual,* ed. by S. C. Sommers, Appleton-Century-Crofts, New York, vol. 6, 1971.

Miller, W. S., *The Lung,* Thomas, Springfield, Illinois, 2nd ed., 1947.

Nagaishi, C., *et al., Functional Anatomy and Histology of the Lung,* University Park Press, Baltimore, 1972.

Policard, A., *Le Poumon,* Masson, Paris, 2nd ed., 1955.

von Hayek, H., *Die menschliche Lunge,* Springer-Verlag, Berlin, 2nd ed., 1970.

HEART AND PERICARDIUM

30

The pericardium[1] is the fibroserous sac (figs. 30–1 and 30–2) that encloses the heart and with it occupies most of the middle mediastinum (fig. 27–8, p. 278). The inner part of the outer fibrous layer is lined by a serous membrane that is reflected onto the surface of the heart. It forms a closed sac that contains a film of fluid. **The pericardium, with its fluid, lubricates the moving surfaces of the heart, holds the heart in position and prevents its dilatation, and constitutes an important hydrostatic system.**

Fibrous Pericardium. The strong, outer part of the pericardium is a dense layer of interlacing collagenous bundles with a network of elastic fibers in its deepest part.[2] Below, it blends with the central tendon of the diaphragm, to which it is firmly bound in front and on the right (pericardiacophrenic ligament). The union with the diaphragm elsewhere is looser. The pericardium is pierced by the inferior vena cava and blends with its adventitia. Occasionally a tiny *infracardiac bursa* is found between the pericardium and the esophagus.[3] This bursa is formed from the upper end of the right pneumatoenteric recess of the embryo. Fringes or folds of fat are often found at the edges of the junction of the pericardium and diaphragm on the right and left sides. This extrapericardial fat may be visible radiographically, especially on the left, where it may obscure the lower left corner of the cardiac shadow.

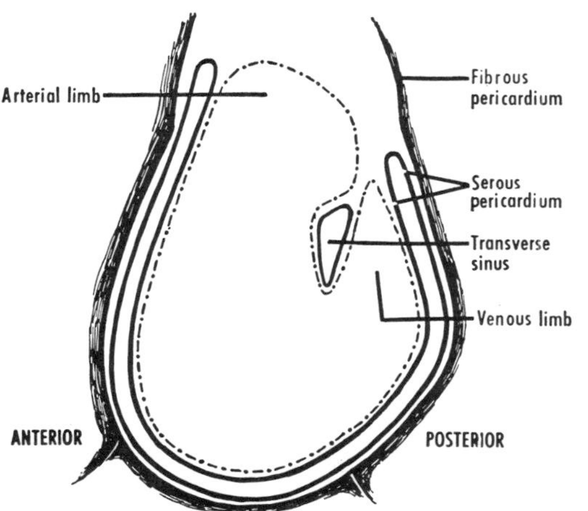

Figure 30–1 Diagrammatic representation of a sagittal section through the heart and pericardium. Note how the serous layer of pericardium is reflected onto the heart and forms a double layer.

Behind, the pericardium is bound by loose connective tissue to the structures in the posterior mediastinum. It is closely related to the esophagus and the thoracic aorta. On its lateral aspects it is adherent to the mediastinal pleura, except where separated from it by the phrenic nerves and their accompanying vessels. In front, the pericardium forms the posterior boundary of the anterior mediastinum. Two variable, fibrous strands, the *sternopericardial ligaments,* connect the pericardium above and below with the posterior surface of the sternum.

Above and behind, the fibrous pericardium gradually blends with the superior vena cava, the pulmonary trunk and arteries, the four pulmonary veins, and the ligamentum arteriosum.

The fibrous pericardium is so unyielding, and is so closely related to the great vessels, that, if fluid accumulates rapidly in the pericardial cavity, the heart may be compressed and venous return impeded. A few cubic centimeters of blood in the pericardial cavity may cause serious symptoms. On the other hand, if fluid accumulates slowly, the pericardium will gradually yield and distend, so that a considerable amount of fluid can be accommodated.

Congenital absence of the pericardium has been recorded.[4] There were no noticeable effects on the heart in such instances.

Serous Pericardium (fig. 30–1). This is a closed sac, the outer *parietal layer* of which lines the inner surface of the fibrous pericardium and is reflected onto the heart, where it is termed the *visceral layer* or *epicardium.* As the visceral layer is re-

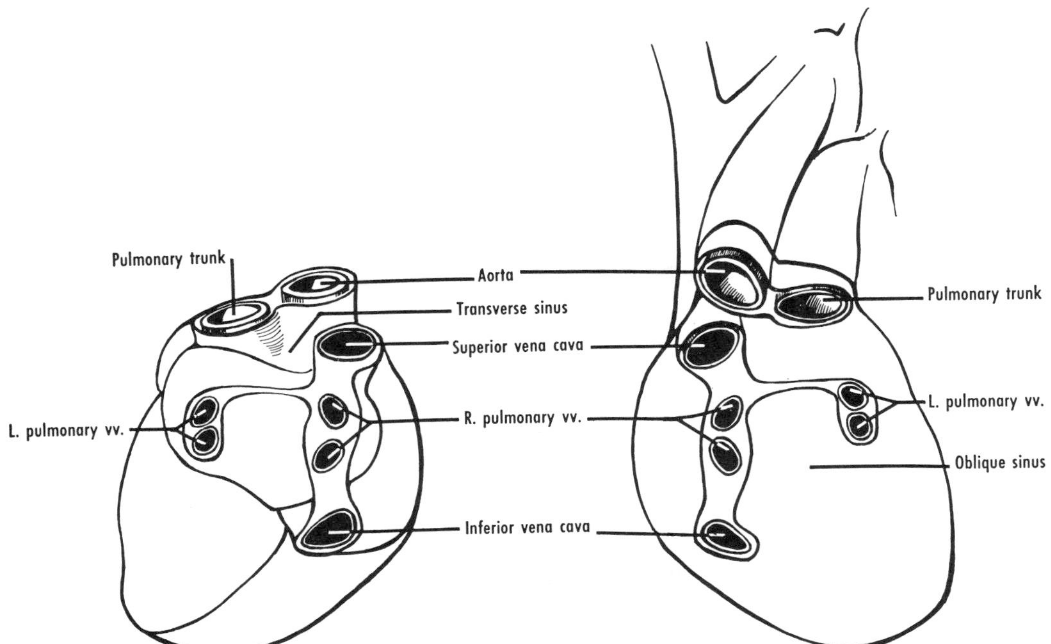

Figure 30–2 Mirror images of the pericardial reflections. *Left figure,* the reflections onto the heart, viewed from behind. *Right figure,* the heart is removed and the posterior part of the pericardium is viewed from in front. The reflection at the veins forms an irregular continuous line that begins at the inferior vena cava, extends up to the lower right pulmonary vein, and turns to the left across the left atrium to the left pulmonary veins. The irregular space thereby bounded is the oblique sinus of the pericardium.

flected onto the heart, it partially ensheathes the great vessels. **The visceral and parietal layers, the opposing surfaces of which are lined by mesothelium, are separated by a potential space, the *pericardial cavity,* and are moistened by a film of fluid.**

In the embryo the parietal layer is reflected onto the heart at the arterial and venous ends. The aorta and pulmonary trunk, which develop at the arterial end, are enclosed in the adult by a common sheath of the visceral layer. When the pericardial sac of the adult is opened from the front, it is possible to pass a finger behind the aorta and pulmonary trunk, in front of the left atrium and the superior vena cava. This passage is the *transverse sinus of the pericardium* (figs. 30–1 and 30–2).

The reflection of the pericardium at the veins is more complex. It forms an irregular line around a space termed the *oblique sinus of the pericardium,* which is bounded on each side by the serous pericardial folds at the entrances of the right and left pulmonary veins. Between the veins, the sinus is limited by the reflection of the serous layer onto the inner aspect of the fibrous layer. If the pericardial sac is opened from in front, a finger can be placed in the oblique sinus from below, where it is open. The *fold of the left vena cava* is a small fold of serous pericardium between the left pulmonary artery and the upper left pulmonary vein, behind the left end of the transverse sinus. It encloses the *ligament of the left vena cava,* which is a remnant of the left rostral cardinal vein. A small vein, *the oblique vein of the left atrium,* extends from the lower end of this ligament to the coronary sinus (p. 318).

Blood and Nerve Supply. The pericardium is supplied by the pericardiacophrenic branches of the internal thoracic arteries, and by pericardial branches of the bronchial, esophageal, and superior phrenic arteries. These vessels have extracardiac anastomoses with the coronary arteries (p. 318). The epicardium is supplied by the coronary arteries.

The pericardium is supplied by branches of the phrenic nerve, which contain vasomotor and sensory fibers.[5] Pericardial pain is felt diffusely behind the sternum, and may radiate to the thoracic wall and the abdomen,[6] but the pericardium is less sensitive than the pleura.

The epicardium is supplied with vasomotor and sensory fibers from the coronary plexuses, but no pain results from stimulation of the epicardium.

HEART

The heart (the adjective cardiac is from the Greek *kardia*, meaning heart) is situated in the middle mediastinum. It is divided into right and left halves by an obliquely placed, longitudinal septum. Each half consists of a chamber called an *atrium*, which receives blood from the veins, and a chamber called a *ventricle*, which propels the blood into and along the arteries. The heart lies more to the left side of the median plane. In the living subject, its long axis is directed from behind forward, to the left, and downward.

The superior vena cava, inferior vena cava, and intrinsic veins of the heart discharge venous blood into the right atrium (figs. 30–3 to 30–5). Blood then enters the right ventricle, from which it is ejected into the pulmonary trunk. The right and left pulmonary arteries carry blood to the lungs, and the pulmonary veins return blood to the left atrium. The blood then enters the left ventricle and is ejected into the aorta. The important valves of the heart are four in number: the *right* and *left atrioventricular*, the *pulmonary* between the right ventricle and the pulmonary trunk, and the *aortic* between the left ventricle and the aorta.

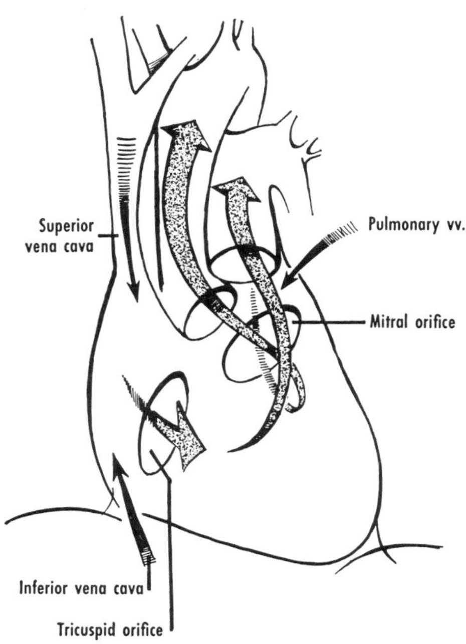

Figure 30–3 The circulation of blood through the chambers of the heart. Note that blood flows almost horizontally forward from the right atrium to the right ventricle.

The heart is composed, from without inward, of epicardium, myocardium, and endocardium. The *epicardium* is the visceral pericardium, and is often infiltrated with fat. The coronary vessels that supply the heart run in the epicardium before entering the myocardium. The *myocardium* is composed mainly of cardiac muscle fibers. It also contains a connective tissue skeleton, which supports and gives attachment to the musculature. The thickness of the myocardial layer is proportional to the amount of work that it does. The ventricles do more work than the atria, and their walls are thicker. The pressure in the aorta is higher than that in the pulmonary trunk, and the wall of the left ventricle is more than twice as thick as that of the right. The *endocardium* is the smooth endothelial lining of the interior of the heart.

SIZE AND POSITION

Size. The determination of heart size in the living subject is an important but difficult clinical problem.[7] A heart that is narrow (as seen in anterior radiograms) may have considerable anteroposterior depth (as seen in lateral radiograms). Likewise, an apparently large "transverse" heart may be no larger in volume than a narrow one, because the anteroposterior depth of a transverse heart may be less. A tumor or other mass may contribute to the cardiac silhouette and give the appearance of an increase in heart size.

Elaborate tables have been published relating heart size to age, sex, height, weight, and surface area of the body. Most of the tables use the surface area of the heart, as determined from anterior radiograms, as an index of heart size.

CARDIAC DIAMETERS. Cardiac diameters include the long diameter, that is, the length of the long axis, and the broad diameter, that is, the greatest diameter of the cardiac shadow perpendicular to the long diameter. Another diameter that is used in estimating heart size is the maximal transverse diameter, which is the maximal distance from the median plane to the right side of the cardiac silhouette, added to the maximal distance from the median plane to the left side of the silhouette. Usually, about two-thirds of the diameter is to the left of the median plane.

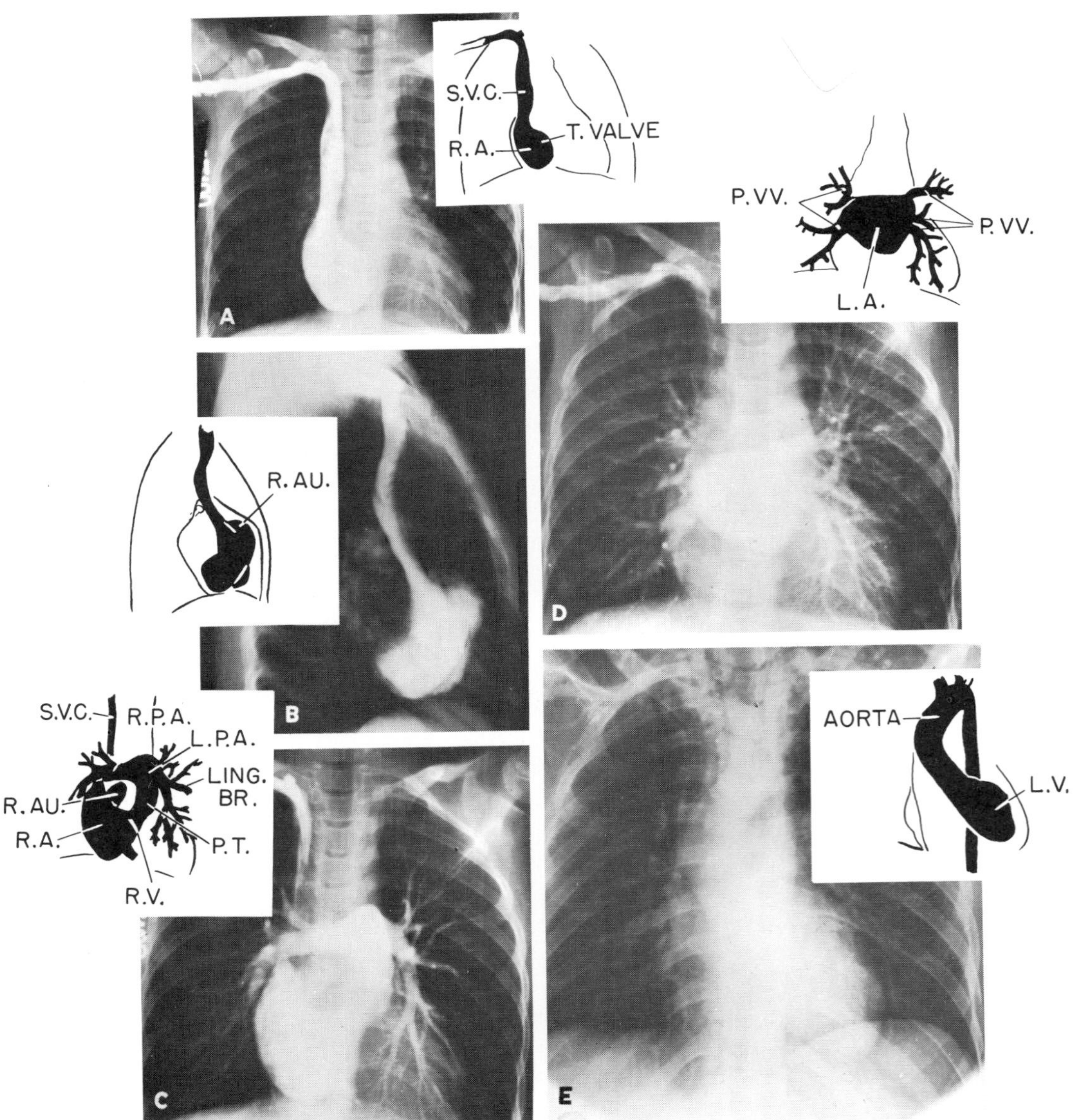

Figure 30–4 Angiocardiograms. *A* and *B*, simultaneously exposed anteroposterior and lateral angiocardiograms, showing the superior vena cava and the right atrium. *C*, right atrium, right ventricle, and pulmonary arterial system. The right ventricle is in partial systole. *D*, left atrium and pulmonary venous system. *E*, left ventricle and aorta.

Abbreviations: L.A., left atrium; LING. BR., lingular branch; L.P.A., left pulmonary artery; L.V., left ventricle; P.T., pulmonary trunk; P.VV., pulmonary veins; R.A., right atrium; R.AU., right auricle; R.P.A., right pulmonary artery; R.V., right ventricle; S.V.C., superior vena cava; T. VALVE, tricuspid valve.

From R. N. Cooley and R. D. Sloan, *Radiology of the Heart and Great Vessels*, Williams & Wilkins, Baltimore, 1956. Courtesy of R. N. Cooley, M.D., University of Texas, Galveston, and the Williams & Wilkins Co.

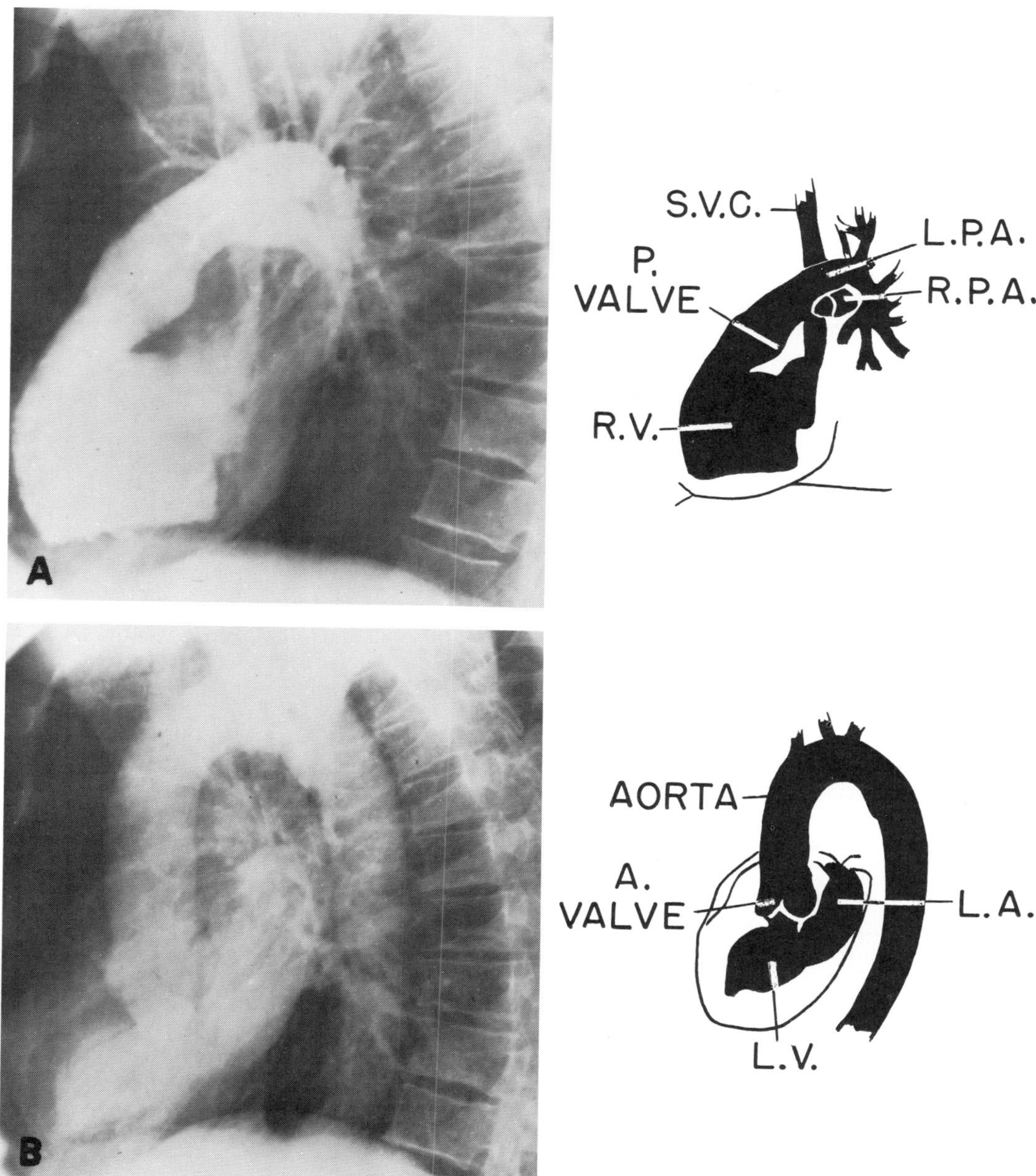

Figure 30–5 Angiocardiograms. *A*, lateral view of heart: right chambers and pulmonary trunk. *B*, lateral view of heart: left chambers and aorta.

Abbreviations: A. VALVE, aortic valve; L.A., left atrium; L.P.A., left pulmonary artery; L.V., left ventricle; P. VALVE, pulmonary valve; R.P.A., right pulmonary artery; R.V., right ventricle; S.V.C., superior vena cava.

From R. N. Cooley and R. D. Sloan, *Radiology of the Heart and Great Vessels*, Williams & Wilkins, Baltimore, 1956. Courtesy of R. N. Cooley, M.D., University of Texas, Galveston, and the Williams & Wilkins Co.

Position. In young adult males, erect and in the midphase of respiration, the average position of the heart with respect to the anterior thoracic wall is illustrated in figure 30–6. The heart is considerably higher in the cadaver, owing to the relaxation of the diaphragm and the upward, post-mortem shift of the abdominal viscera. In the living, the position of the heart, particularly its so-called left border, can be outlined approximately by percussion, but there is often considerable discrepancy between the position of this border as determined by percussion and that found by radiographic methods. **Locating the apex beat is probably a better guide to the position of the left border than is percussion.**

APEX AND APEX BEAT.[8] The so-called apex of the heart is often rounded and is usually ill-defined radiographically. When an apex can be recognized radiographically, it is usually at the level of the sixth costal cartilage, below and medial to where the apex beat can be felt. The so-called apex beat, which is an impulse imparted by the heart, can be felt on the front of the left side of the chest in most individuals. The site, also called the *point of maximal pulsation,* is usually in the fourth or fifth intercostal space, about 6 or 7 cm from the median plane (with considerable variation). The apex beat is produced by a complex movement of the left ventricle during contraction. Although this beat is a fairly reliable guide to the position of the left border, the apex beat may be felt outside the cardiac area in some subjects (see fig. 32–1, p. 341).

Orientation.[9] **The atria, which form the base of the heart, lie behind the ventricles.** The long axis of the heart extends from the base to the apex; it is directed from the center of the base, from behind forward, downward, and to the left. The right atrium lies behind the right ventricle, which occupies the right and front part of the heart. **The planes of the atrioventricular orifices are more vertical than horizontal (fig. 30–3), and the blood flows almost horizontally forward from the atria to the ventricles,** especially from the right atrium to the right ventricle. Depending upon whether the long axis of the heart is more horizontal or more vertical, a heart may be "transverse" (a condition commonly associated with infancy, obesity, or pregnancy), or "long," "narrow," or "vertical." Most hearts fall between these two categories and are termed "oblique."

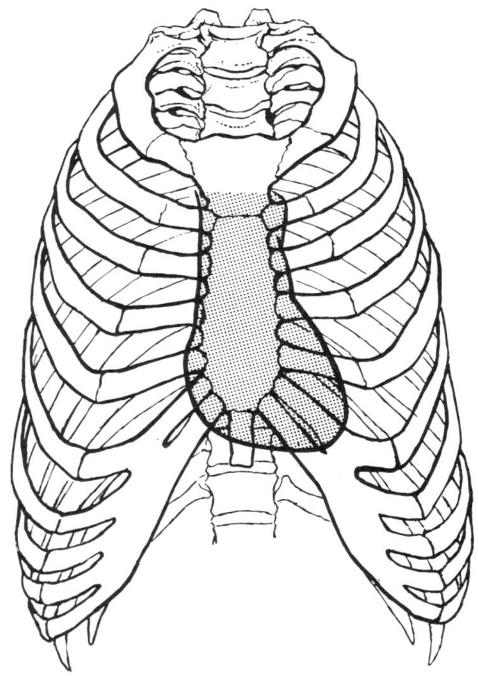

Figure 30–6 The cardiovascular shadow (of a "vertical" heart) in relation to the bony cage. Depending on posture and phase of respiration, the lower margin of the heart may be at a still lower level, as much as 5 cm below the xiphisternal joint (figs. 29–7 and 29–8, pp. 290, 291). See figure 32–3, p. 344, for the composition of the cardiovascular shadow.

Variations in Size, Shape, and Position. The size, shape, and position of the heart may vary from individual to individual, and also from time to time in the same individual (see figs. 32–6 to 32–9).

BODY TYPE. Tall, slender individuals are more likely to have "vertical" hearts (similar to the shape seen after a deep inspiration). Heavy-set, stocky people are more apt to have "transverse" hearts (similar to the shape seen after a deep expiration).

Deformities of the thoracic wall or vertebral column may markedly affect the shape and position of the heart, often without causing symptoms. A striking example is occasionally seen in individuals with retrocurvature of the sternum (funnel chest). The sternum is depressed, and the heart is pushed backward or to one side.

AGE. The heart of the newborn and of

the infant is small, but it is large in proportion to the chest, and is usually globular in shape. It appears particularly large at the end of expiration. Part of its apparent large size is due to the rotation associated with the large liver. The heart is more "transverse" than in the adult, and is at least one intercostal space higher in position. The average pulse rate of the newborn is 120 to 140 per minute.

POSTURE. During assumption of the supine position, the heart moves upward and backward. This shift is due mainly to the upward movement of the diaphragm. The heart also becomes more "transverse," so that the apex beat is displaced laterally. Likewise, when one is lying on one's side, the apex beat shifts toward that side. In the supine position, the heart moves backward, away from the anterior chest wall, and the apex beat is more difficult to feel.

RESPIRATION. **The position and movements of the diaphragm are the most important factors that determine the position of the heart.** The pericardium is firmly attached to the central tendon of the diaphragm. The position of the heart therefore varies with the position of the diaphragm. Changes in position of the heart during quiet breathing are hardly noticeable, but, with a deep inspiration, the heart descends and rotates to the right so that it becomes more "vertical" and narrower. It also moves backward, and the apex beat is lower and more medial. The converse occurs in a deep expiration. The apex beat may then be as high as the third or fourth intercostal space.

ANATOMY

Radiological Anatomy

The heart causes a dense shadow in radiograms (fig. 29–1, p. 285), but this shadow merges with those cast by the great vessels. The cardiovascular shadow is described in some detail on page 344.

External Anatomy

The fixed heart is usually described as having an apex, a base, and three surfaces: sternocostal, diaphragmatic, and pulmonary, or left. Borders are also described, but they are indefinite and usually cannot be distinguished *in vivo*. Nor can an apex often be made out; the apical region is frequently rounded.

The *base* of the heart is formed by the atria. It is directed backward, and the atria lie mainly behind the ventricles. The superior and inferior venae cavae and the pulmonary veins enter the heart at the base. The interatrial septum is sometimes indicated as a slight groove at the base, immediately to the right of the right pulmonary veins. Each atrium continues anteriorly, on each side of the aorta and pulmonary trunk, as an ear-shaped appendage, the *auricle* (fig. 30–7). The right auricle often overlaps the root of the aorta in front. In clinical usage, an atrium is sometimes called an auricle, and the appendage is called an auricular appendage.

There may be a slight groove on the right or lateral wall of the right atrium, extending from the front of the orifice of the superior vena cava to the right of the inferior vena cava. This groove is the *sulcus terminalis,* the external indication of a well-developed muscular band, the *crista terminalis,* which projects into the interior of the right atrium. The upper part of the sulcus terminalis is occupied by the sinuatrial node.

The atria and ventricles are separated by the *coronary,* or *atrioventricular, groove,* which lodges the coronary sinus, the right coronary artery, and the end of the left coro-

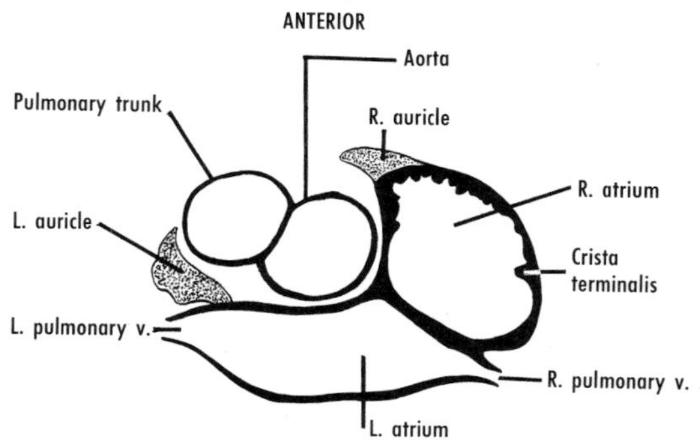

Figure 30–7 Diagram of a section of the heart to show the usual relationships of the atria, auricles, pulmonary trunk, and aorta. Note that the left atrium lies behind the pulmonary trunk and the aorta, from which it is separated by the transverse sinus of the pericardium. Modified from Walmsley.

nary artery. This groove is well marked behind, but is interrupted in front by the aorta and pulmonary trunk.

The *sternocostal surface* of the heart is formed mainly by the right ventricle. A part of the right ventricle is prolonged upward into the pulmonary trunk. This part is the *conus arteriosus*, or *infundibulum.* The interventricular branch of the left coronary artery may be lodged in a shallow *anterior interventricular groove* (actually superior) on the left part of the sternocostal surface. A *posterior interventricular groove* (actually inferior) may be present on the diaphragmatic surface. It lodges the interventricular branch of the right coronary artery. These grooves, which indicate the interventricular septum, are commonly obliterated by epicardial fat.

The *left,* or *pulmonary, surface* is formed mainly by the left ventricle, which produces the cardiac impression on the medial surface of the left lung. The *diaphragmatic surface* is formed by both ventricles. It rests chiefly on the central tendon of the diaphragm.

Internal Anatomy of the Atria
(fig. 30–8)

The inner surfaces of both auricles are ridged by muscular elevations, the *musculi pectinati.* The inner surface of the left atrium is smooth. That of the right atrium is partly ridged by musculi pectinati, which extend from the auricle to the crista terminalis.

Right Atrium. The posterior and septal walls are smooth. The musculi pectinati begin on the right side of the atrium at a vertical muscular ridge, the crista terminalis, the external indication of which is the sulcus terminalis. The *sinus venarum cavarum* is the region of the atrium into which the superior and inferior venae cavae empty. The part of the atrial wall between the two orifices of the venae cavae forms a rather variable, rounded, horizontally placed elevation, the *intervenous tubercle,* consisting mainly of a band of muscle.

On rare occasions the *opening of the superior vena cava* is guarded by a partial valve. The *valve of the inferior vena cava* is a variable semilunar fold that lies in front of and sometimes partly covers the *opening of the inferior vena cava.* This valve is often fenestrated,[10] it may be absent, and it is probably of no functional significance in the adult. The *valve of the coronary sinus* is an occasional, often fenestrated leaf of variable size. It is related to the circular opening of the sinus immediately in front and to the left of the valve of the inferior vena cava.[11] Rarely, one or more pulmonary veins empty into the right atrium.

A number of small openings are present in the walls of the atrium. These openings, which are termed *foramina venarum minimarum,* are the terminations of small venous channels, the *venae cordis minimae.* The openings occur in all chambers of the heart and several types have been described.[12]

The *right atrioventricular,* or *tricuspid, orifice* is guarded on the ventricular side by the tricuspid valve. The orifice is usually large enough to admit three fingers.

Interatrial Septum. In the right atrium, the lower part of the septum contains an ovoid, depressed area, the *fossa ovalis.* A rounded fold, the *limbus fossae ovalis (anulus ovalis),* bounds the fossa above, in front, and behind. The upper part of the fossa may be separated from the limbus by the *foramen ovale,*[13] an aperture of variable size that represents the persistence of the fetal foramen ovale by which the atria communicate with each other.

The fossa ovalis may be recognized in the left atrium as a translucent region of the interatrial septum. The upper edge of this region is the free border of the *valve of the foramen ovale.* This fold may form a sickle-shaped ridge, but it is often only an interlacing network of fibers.

Left Atrium. The cavity of the left atrium is prolonged on each side as pouches for the *openings of the pulmonary veins.* Musculi pectinati are confined mainly to the auricle. Foramina venarum minimarum are present in the wall of the atrium. The *left atrioventricular,* or *mitral, orifice* is guarded on the ventricular side by the mitral valve, and is usually large enough to admit two fingers.

Internal Anatomy of the Ventricles
(fig. 30–8)

The ventricular part of the heart has four openings, an atrioventricular and an aortic on the left, and an atrioventricular and a pulmo-

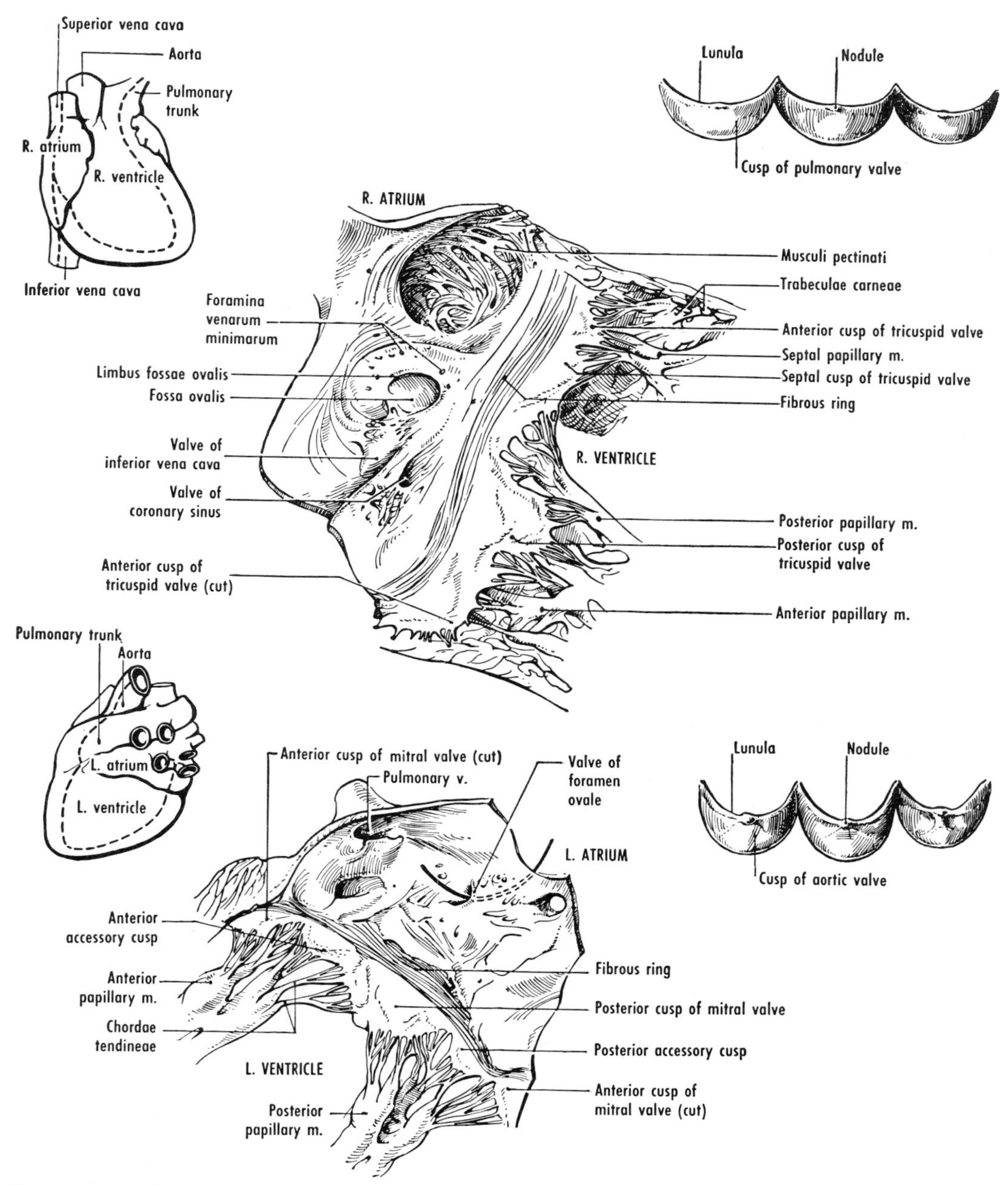

Figure 30–8 The internal anatomy of the heart. Drawn from a heart that was opened by incisions indicated by the interrupted lines in the two figures at the left. *Upper middle,* internal anatomy of right atrium and right ventricle. The walls were spread out so that the entire circumference of the right atrioventricular orifice is shown. *Lower middle,* internal anatomy of left atrium and ventricle. The walls were spread out so that the entire circumference of the left atrioventricular orifice is shown. A patent foramen ovale in this heart is indicated by the probe. *Figures at right,* the valves of the pulmonary trunk and aorta.

The atrioventricular orifices are shown in the same orientation as in the living (see fig. 30–3).

nary on the right. A mass of dense connective tissue occupies the interval between the atrioventricular and aortic openings, and is continuous with the fibrous rings around these openings and with the upper part of the interventricular septum.

The inner surfaces of the ventricles (except for the infundibulum) are irregular, owing to the projection of muscle bundles, the *trabeculae carneae.* Three kinds of trabeculae carneae occur: ridges, bridges, and pillars. (1) Ridges, or columns, are bundles that are raised in relief from the ventricular wall. (2) Bridges are rounded bundles, free in their middle and attached at each end to the ventricular wall. (3) Pillars are the *papillary muscles,* which are cone-shaped muscles, the bases of which are attached to the ventricular wall. Their apices continue into fine, tendinous cords, the *chordae tendineae,* which are attached to the apices, borders, and ventricular surfaces of the cusps of the atrioventricular valves.

The valvular apparatus in each ventricle consists of the fibrous ring around the atrioventricular opening, the valve proper, the chordae tendineae, and the papillary muscles. The chordae tendineae that are attached to the free margin of the valve prevent eversion of the valve. Those that are attached to the ventricular surface steady and strengthen the valve.

Each atrioventricular valve has cusps, the bases of which are attached to the fibrous ring that surrounds the opening. The atrial surfaces of the cusps are smooth, whereas the ventricular surfaces are roughened for attachment of the chordae tendineae. The free borders of the cusps often have small nodular thickenings. Atrial muscle fibers and a capillary network may be present in the bases of the cusps, but continue into the cusps for at most,[14] a few millimeters. The remainder of the cusps consists of rather dense, avascular connective tissue, covered on each surface by endocardium.

The *semilunar valves* of the aorta and of the pulmonary trunk are situated at the roots of these vessels. Each has three cusps that consist of avascular fibrous tissue covered on each surface by intima. The free edge of each cusp has a small central thickening of fibrous tissue, the *nodule.* Extending from each side of the nodule is a narrow, thin, crescentic area termed the *lunula,* which lacks fibrous tissue. The spaces between the cusps and the walls of the vessels are the *aortic* and *pulmonary sinuses,* respectively.

Cardiac valves are not visible radiographically unless they are calcified, but they may be delimited during angiocardiography. By this method, the aortic valve has been observed to be almost in the center of the cardiac shadow as seen from the front. The mitral valve is close by, a little lower and to the left.

Foramina venarum minimarum are present in both ventricles, but they are not as numerous as in the atria.

The right ventricle usually has (1) a tricuspid valve, (2) a trabeculated septal surface, and (3) a supraventricular crest and infundibulum. The left ventricle usually has (1) a bicuspid valve, (2) a smooth septal surface, and (3) no crest or infundibulum. These internal morphologic criteria are important because, in some anomalies, a morphologic right ventricle may be present on the left side, and a morphologic left ventricle on the right side.

Right Ventricle. The cavity of the right ventricle is **U**-shaped, but with the **U** on its side: **⊂** (fig. 30–9). The lower limb of the **U**, which receives blood from the right atrium, is the venous or inflowing part of the

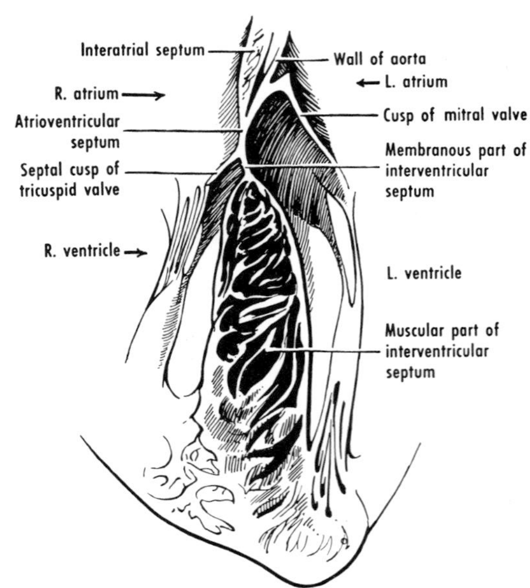

Figure 30–9 Drawing of the interventricular septum in section.

ventricle. The *conus arteriosus,* which is the arterial or outflowing part of the ventricle, is the upper and anterior limb. It ends in the pulmonary trunk. A thick muscular ridge, the *supraventricular crest,* lies in the angle between the two parts. The walls of the conus arteriosus are usually smooth. The junction of the conus arteriosus with the pulmonary trunk is a region in which the wall of the artery is composed of dense fibrous tissue that encircles the pulmonary valve and is continuous with the fibrous ring of the cardiac skeleton. This part of the wall of the vessel is sometimes termed the root of the pulmonary trunk.

The right ventricle lies in front of the right atrium, the plane of the *right atrioventricular orifice* is nearly vertical, and blood flows in a horizontal direction from the atrium to the ventricle (fig. 30–3). The opening is guarded by the *right atrioventricular,* or *tricuspid, valve,* which has three cusps, *anterior, posterior* (*inferior*), and *septal* (*medial*) (fig. 30–10). In the living subject the posterior cusp lies below the stream of blood from the atrium, whereas the anterior cusp intervenes between the opening and the conus arteriosus. The septal cusp is single, the anterior may be scalloped, and the posterior commonly has a variable number of scallops.[15] The papillary muscles corresponding to the cusps are *anterior, posterior (inferior),* and *septal,* of which the largest and most constant is the anterior. It arises from the anterolateral walls of the ventricle and from the septomarginal trabecula. The posterior papillary muscle is irregular in size and position. The septal papillary muscle is often formed of several muscles, of which one may be larger and more constant.

The *septomarginal trabecula (moderator band)* is a more or less isolated trabecula, of the bridge type, which extends from the interventricular septum to the base of the anterior papillary muscle, in the lower part of the ventricle. It contains Purkinje fibers from the right limb of the atrioventricular bundle (see p. 316).

The cusps of the *pulmonary valve* are two in front and one behind (fig. 30–10). The terminology of the pulmonary and aortic valves is important in considering and comparing congenital cardiac defects in which the great vessels are misaligned, out of position, or transposed.[16] The terminology is confused because a system of naming the cusps according to their positions in a heart that is not oriented in its usual *in situ* relationships is in common use; it places the cusps one in front and two behind, an orientation that is present during embryonic development but only occasionally in the adult. The differences in terminology are shown in figures 30–10 and 30–14.

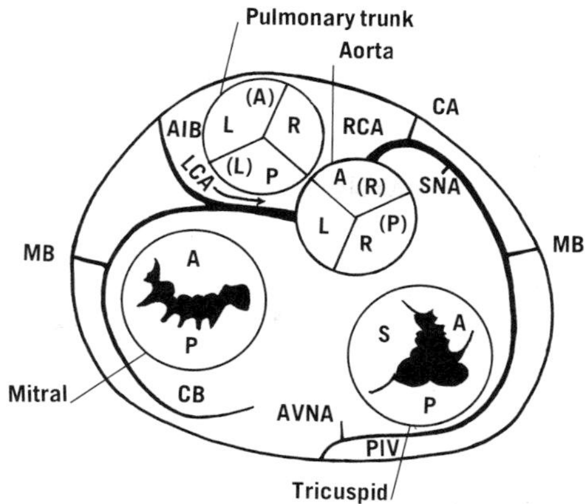

Figure 30–10 Diagram of the heart valves, their cusps, and the coronary arteries, showing their usual positions *in situ.* The pulmonary cusps are left, L, right, R, and posterior, P. The aortic cusps are anterior, A, left, L, and right, R. Note that a slight, clockwise rotation of the heart would be sufficient to make the pulmonary cusps anterior, A, left, L, and right, R, which are the names given in the *Nomina anatomica.* The aortic cusps would be left, L, right, R, and posterior, P. The tricuspid cusps are septal, S, anterior, A, and posterior, P. The mitral cusps are anterior, A, and posterior, P.

Abbreviations: RCA, right coronary artery; CA, branch to conus arteriosus; SNA, sinus node artery; MB, marginal branch; PIV, posterior interventricular branch; AVNA, atrioventricular node artery; LCA, left coronary artery; AIB, anterior interventricular branch; CB, circumflex branch.

Interventricular Septum (fig. 30–9). This septum is a strong, obliquely placed partition that consists of a *membranous* and a *muscular part.*

The membranous part of the septum is thin, smooth, and fibrous in structure. Commonly, the septal cusp of the tricuspid valve is attached to the right side of the upper portion of the membranous part in such a way that the right side of the septum is related to the right atrium above the valve, and to the right ventricle below the valve. Consequently, the part of the septum above the valve intervenes between the right atrium and the left ventricle. This part is termed the *atrioventricular septum.*

One surface of the interventricular septum looks forward and to the right and bulges into the cavity of the right ventricle. The other surface looks backward and to the left and is concave toward the left ventricle. The septum extends from the apical region of the heart to the interval that separates the pulmonary and tricuspid openings from the aortic and mitral orifices. The edges of the septum are sometimes indicated on the surface of the heart by slight interventricular grooves.

Left Ventricle. Arterial pressure is much higher in the systemic than in the pulmonary circulation, the left ventricle performs more work, and its wall is usually more than twice as thick as that of the right. The lower, or inflowing, part of the cavity of the left ventricle communicates with the left atrium. The upper and anterior part is the *aortic vestibule,* the walls of which are mainly fibrous. The vestibule leads into the aorta. The junction of the aortic vestibule and the aorta is a region in which the wall of the aorta is composed of dense fibrous tissue that encircles the valve and is continuous with the fibrous ring of the cardiac skeleton. This part of the wall of the aorta is sometimes termed the root of the aorta.

The left ventricle is mostly in front of the left atrium, the plane of the *left atrioventricular orifice* is nearly vertical, and blood flows from atrium to ventricle in an obliquely forward direction, from right to left. The opening is guarded by the left *atrioventricular,* or *mitral, valve.* A continuous ring of valvular tissue surrounds the opening and is attached to the fibrous ring. Two major cusps *(anterior,* or *aortic,* and *posterior,* or *ventricular)* project from the ring of valvular tissue.[17] The posterior cusp is notched, usually in such a way as to present three scallops.[18] The anterior cusp is much less commonly scalloped. There are usually two papillary muscles, *anterior* and *posterior.*

The cusps of the *aortic valve* (figs. 30–10 and 30–14) are generally one in front and two behind, although embryonic relations (two in front and one behind) are maintained in some instances. Hence, there are problems in terminology, just as there are for the pulmonary cusps.

Cardiac Skeleton and Myocardium

Cardiac Skeleton. The cardiac skeleton (fig. 30–11) consists of fibrous or fibrocartilaginous tissue that surrounds the atrioventricular and semilunar openings, gives attachment to the valves and the muscular layers, and is continuous with the roots of the aorta and pulmonary trunk and with the membranous part of the interventricular septum.

Myocardium. Cardiac muscle fibers are arranged into complex sheets and bands of muscle. The atrial and ventricular musculatures are separate; the conducting system is their only muscular connection.

MUSCULATURE OF THE ATRIA. The musculature of the atria consists of superficial bundles common to both atria, and deep bundles, some of which are restricted to one atrium (such as musculi pectinati), and others of which pass into the interatrial septum. A variable amount of cardiac muscle may extend upward along the superior vena cava, and across the coronary sinus, external to the walls of these vessels. Myocardial fibers are arranged circularly around the openings of the pulmonary veins,[19] and may extend along the veins as sleeves.

MUSCULATURE OF THE VENTRICLES.[20] Two main groups of ventricular myocardial muscle sheets occur, a superficial spiral and a deep constrictor. The fibers in one sheet are more or less at right angles to those in the other. The deep constrictor sheet, most pronounced in the left ventricle, compresses the ventricles like a clenched fist. The contraction of these deep fibers leads to a decrease in diameter of the mitral and tricuspid openings, the muscle fibers acting like sphincters. Because of the spiral arrangement of the superficial bundles, the heart is twisted during systole, like a cloth being wrung out.

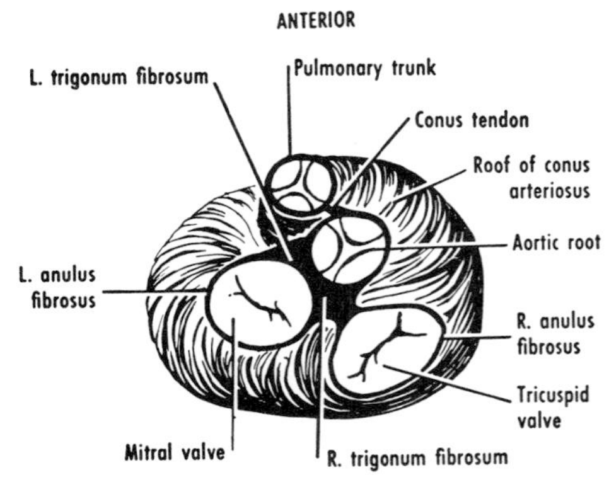

Figure 30–11 Diagrammatic representation of the cardiac skeleton, as seen from above. Names have been given to different parts of the skeleton, but it is a continuous mass of connective tissue that surrounds the openings as shown. Modified from Walmsley.

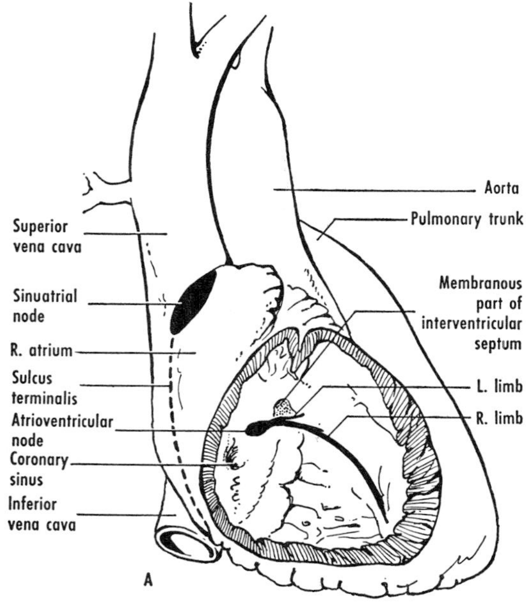

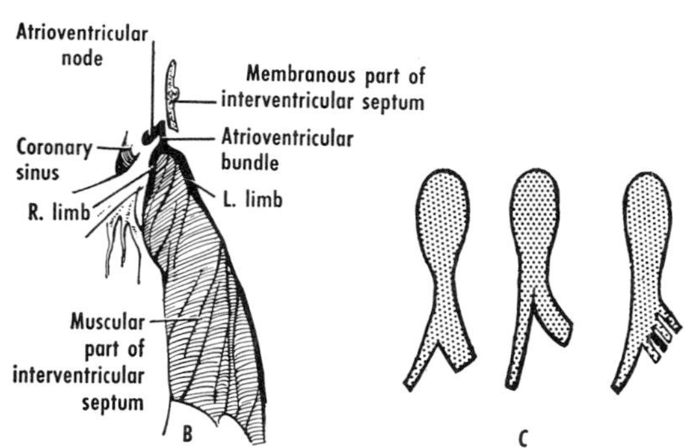

Figure 30–12 Diagrams of the conducting system. *A*, showing the positions of the sinuatrial and atrioventricular nodes. The atrium and right ventricle are opened and the interventricular septum is exposed. *B*, diagram of the atrioventricular bundle and its branches, viewed from above. *C*, diagram showing three types of left limb of the atrioventricular bundle, viewed from above. *B*, based on Bassett, *A Stereoscopic Atlas of Human Anatomy*. *C*, based on Walls.[26]

CONDUCTING SYSTEM

The conducting system (fig. 30–12) consists of specialized muscle fibers (specialized in that they conduct impulses) that connect certain "pacemaker" regions of the heart with cardiac muscle fibers. The intrinsic, rhythmic contractions of cardiac muscle fibers are regulated by pacemakers, and the intrinsic rhythmicity of the pacemakers is regulated in turn by nerve impulses from vasomotor centers in the brain stem.

In the embryo, the cardiac muscle fibers begin to contract rhythmically and synchronously before nerve fibers reach the heart. Muscle fibers appear first in the ventricular region and begin to beat. Later, muscle fibers develop in the atrial region, and begin to beat at a faster rate than do those of the ventricles. In the adult, if the conducting system between the atria and ventricles is destroyed (complete heart block), the ventricles and atria beat at different rhythms. The rate of the ventricles may be 30 per minute or less. However, even this low rate of activity may be sufficient to maintain an adequate circulation.

The conducting system of the adult heart comprises the sinuatrial node, the atrioventricular node, and the atrioventricular bundle with its two limbs and the subendocardial plexuses of Purkinje fibers. The impulse begins at the sinuatrial node, activates the atrial musculature, and is thereby conveyed to the atrioventricular node. Some of the atrial muscle fibers form bundles that pass more or less directly from the sinuatrial to the atrioventricular node.[21] These internodal tracts may be functionally specialized, but whether they are important in normal physiological activity has not yet been determined. The atrioventricular bundle, its two limbs, and the Purkinje fibers conduct the impulse from the atrioventricular node to the ventricular myocardium.

Sinuatrial Node (Sinus Node). **Ordinarily the pacemaker for the heart, this node is located at the anterolateral region of the junction of the superior vena cava and the right atrium, near the upper end of the sulcus terminalis.**[22] It may also be located by tracing the sinus node artery to it. This artery, which usually arises from the right coronary artery (fig. 30–10), but which may arise from the left, runs through the sinuatrial node.[23]

Lying just beneath the epicardium, the sinuatrial node is a pale, curved fusiform mass, about 7 mm long and not much more than a millimeter in thickness.[24] It contains a network of specialized cardiac muscle fibers that become continuous with the muscle fibers of the atrium at the periphery of the node. The fibers in the node are supplied by autonomic fibers. There are numerous ganglion cells near the node, some of which are grouped together as small ganglia.

Atrioventricular Node. This node, which is somewhat smaller than the sinuatrial,[25] is located beneath the endocardium of the right atrium, in that part of the interatrial septum that forms, or is continuous with, the right fibrous trigone, immediately above the opening of the coronary sinus. Like the sinuatrial node, it consists of a meshwork of specialized cardiac muscle fibers. These are continuous with (1) atrial muscle fibers (see discussion above) and (2) the atrioventricular bundle. The node is supplied by autonomic nerve fibers. The blood supply is usually by the posterior interventricular artery or by the right coronary itself.

Atrioventricular Bundle. This collection of specialized fibers leaves the atrioventricular node and passes upward in the right fibrous trigone to the membranous part of the interventricular septum. From there it runs forward and divides into right and left crura that straddle the muscular part of the septum.[26] The *right crus,* or *limb,* is a rounded bundle that continues forward toward the apical region, enters the septomarginal trabecula, and reaches the ventricular wall and the anterior papillary muscle. Its fibers then form a subendocardial plexus of Purkinje fibers in the papillary muscles and the wall of the right ventricle. The *left crus,* or *bundle,* which consists of one to three bands or flat strands, runs forward toward the apical region. It courses just beneath the endocardium covering the left surface of the muscular part of the septum. The fibers reach the papillary muscles and ramify subendocardially as a plexus of Purkinje fibers. The fibers of the atrioventricular bundle and its crura begin to show the characteristics of Purkinje fibers as they descend in the interventricular septum. Purkinje fibers are slightly larger and somewhat lighter-staining than ordinary cardiac muscle fibers, and their striations are less evident. They may be traced directly into ordinary cardiac muscle fibers.[27]

The bundle and its crura are surrounded in their whole length by a fibrous sheath that isolates them from the adjacent myocardium. The sheath is continued as a delicate investment for the Purkinje fibers.

BLOOD SUPPLY,[28] LYMPHATIC DRAINAGE, AND NERVE SUPPLY

The heart is supplied by the right and left coronary arteries (fig. 30–13), which usually arise from the ventral and the left aortic sinuses, respectively (figs. 30–10 and 30–14). The heart is drained by a number of veins. Some of these empty directly into the chambers of the heart, whereas others drain into the coronary sinus, which in turn empties into the right atrium. The coronary arteries and their first few orders of branches course in and supply the epicardium. Subsequent branches penetrate the myocardium. The coronary arteries are supplied by sensory and autonomic fibers of the coronary plexuses.

Coronary Arteries

There is no sharp line of demarcation between the ventricular distribution of the coronary arteries. Not infrequently, the pattern is as follows: The right coronary artery supplies the right ventricle (except the left part of its anterior wall), the right part of the posterior wall of the left ventricle, and some of the interventricular septum. The left coronary artery supplies most of the left ventricle, part of the right ventricle, and most of the interventricular septum. When the distribution of the right coronary artery extends further toward the left ventricle in front, the right coronary artery is said to be preponderant. Left coronary artery preponderance is present when the circumflex branch is longer than usual and crosses the posterior interventricular groove.[29] It is likely that in most instances the heart is equally supplied by the coronary arteries.

Most of the blood in the coronary arteries returns to the chambers of the heart by way of veins. Some may return directly to the heart by way of special sinusoids in the myocardium, or by way of tiny branches of arterioles in the endocardium that open directly into the chambers, and some blood leaves by way of extracardiac anastomoses.[30]

Right Coronary Artery. This artery arises from the anterior (right) aortic sinus. It proceeds to the right, emerges between the pulmonary trunk and the right auricle, and runs in the coronary groove to the back of the heart, where it anastomoses with the left coronary artery. During the first part of its course, it gives branches to the right ventricle. The first of these ventricular branches supplies the conus arteriosus; it often arises separately from the aorta and is

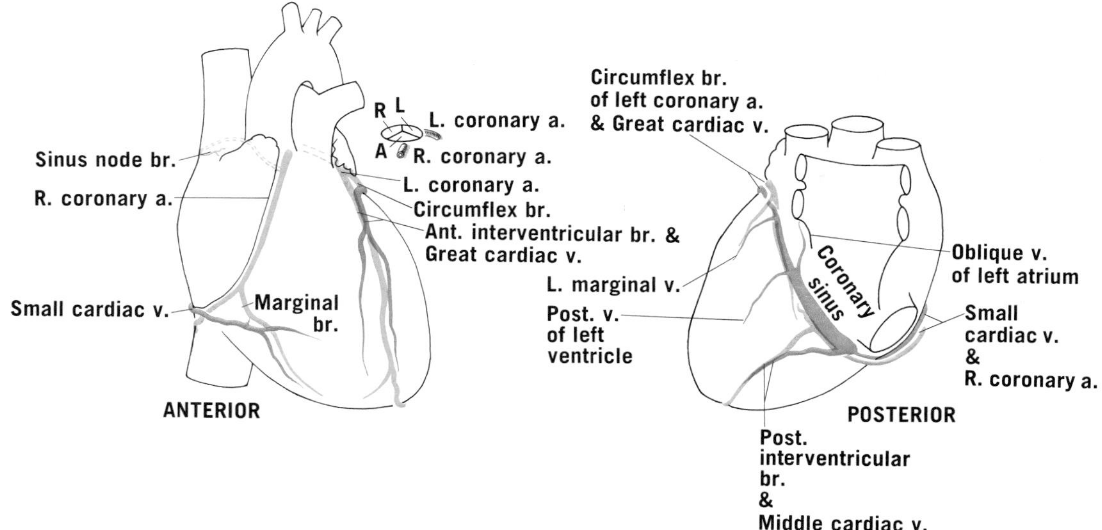

Figure 30–13 The coronary vessels and their major named branches. The veins shown are those that drain into the coronary sinus.

then named the *conus artery*. A relatively constant marginal branch descends along the right ventricle to the apex. Commonly, the first part of the right coronary artery gives off a branch termed the *sinus node artery*, which runs upward and medially, supplying the right atrium. It encircles the level of the opening of the superior vena cava and enters the sinuatrial node. The nodal artery may arise from a branch of the

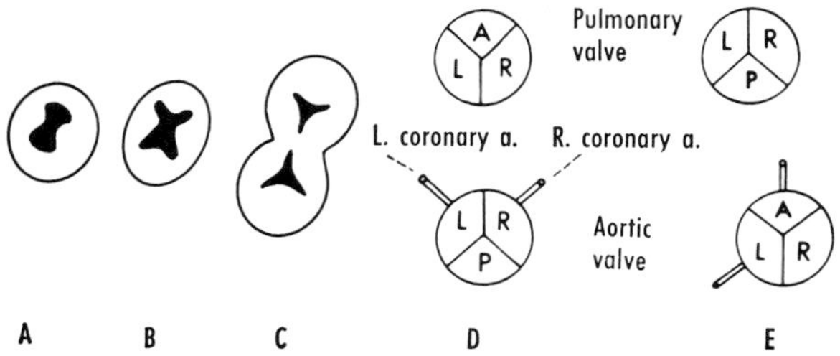

Pulmonary valve

L. coronary a. R. coronary a.

Aortic valve

A B C D E

Figure 30–14 Schematic representation of the partitioning of the truncus arteriosus and the development of the semilunar valves. *A*, truncus arteriosus. *B*, *C*, and *D*, an intermediate position during the rotation that the heart undergoes during development, the basis for naming of the valves in the *Nomina anatomica*. *E*, the usual position of the valves of the adult heart *in situ*. The terminology is different from that recommended by the NA, but is the one employed here. Based on Kramer.[31] A terminology based on embryonic relationships has been proposed by Kerr and Goss.[32]

right coronary artery, or from the left coronary artery or one of its branches.

As the right coronary artery continues, it gives additional branches to the right atrium[33] and the right ventricle. It enters the coronary groove and reaches the posterior interventricular groove, where it gives off several branches, one of which is the *posterior interventricular branch*. The right coronary artery then continues across the posterior interventricular groove and anastomoses with the circumflex branch of the left coronary artery.

The posterior interventricular artery courses in the posterior interventricular groove to the apical region. It may arise from the circumflex branch of the left coronary artery. It supplies the adjacent parts of both ventricles and a part of the interventricular septum. The artery that supplies the atrioventricular node usually arises from the first part of the posterior interventricular branch, sometimes from the right coronary itself (see circumflex branch below).

Left Coronary Artery. This artery arises from the left aortic sinus, behind the pulmonary trunk. It courses between this trunk and the left auricle, gives off an *anterior interventricular branch* (which descends to the apical region), supplies the left atrium, and, as the *circumflex branch,* continues into the left part of the coronary groove, where it anastomoses with the right coronary artery.

The anterior interventricular branch, which can be considered to be a direct continuation of the left coronary artery, descends in the anterior interventricular groove to the apex of the heart. It turns around the apex and ascends for a variable distance in the posterior interventricular groove, where it meets the terminal branches of the posterior interventricular branch of the right coronary artery. The anterior interventricular branch supplies both ventricles and furnishes the chief supply of the interventricular septum.

The circumflex branch supplies the adjacent part of the left ventricle (a marginal branch is relatively constant), the left atrium,[33] and often the septum. Posteriorly, in the coronary groove, the circumflex branch may cross the posterior interventricular groove, supplying the atrioventricular node and giving rise to the posterior interventricular branch.

Variations and Anomalies.[34] There may be but one coronary artery, and there may be more than two. Branches of the coronary arteries often vary in number, size, and distribution. A common variation is one in which both interventricular branches arise from the left coronary artery. Occasionally there are two anterior interventricular branches. Often there are several posterior interventricular branches rather than one. Anomalies may be secondary to certain malformations of the heart. They may also be of primary and major significance, such as a direct communication of a coronary artery with a chamber of the heart.

Intracardiac and Extracardiac Anastomoses. In most areas of the heart, there are numerous small arterial and precapillary anastomoses.[35] Although these anastomoses appear to be inadequate to provide good collateral circulation if a coronary artery or one of its major branches is suddenly occluded, these small anastomotic channels can enlarge considerably during the course of an occlusion that develops slowly. In such instances, an individual might later survive complete occlusion of one coronary artery, or even of both.

The coronary arteries supply the epicardium, which also receives small branches from other arteries[36] (aorta, internal thoracic, superior phrenic, posterior intercostal, tracheal, esophageal, and bronchial). The anastomoses between these vessels and the coronary arteries may enlarge and provide a collateral circulation when coronary occlusion develops slowly.

Venous Drainage

The heart is drained partly by veins that empty into the coronary sinus (fig. 30–13) and partly by small veins that empty directly into the chambers of the heart.

The direct veins include two or three small vessels, the *anterior cardiac veins,* which drain the front of the right ventricle, cross the coronary groove, and end directly in the right atrium. One of these drains the lower margin of the heart and is sometimes termed the right marginal vein. The *venae cordis minimae* are very small veins that begin in the substance of the heart and end directly in its cavities (chiefly the atria).[37]

The *coronary sinus* lies in the coronary groove, between the left atrium and left ventricle. It is a short, but relatively wide, trunk that ends in the right atrium, between the opening of the inferior vena cava on the right and the tricuspid opening in front. The right side of its opening is guarded by the valve of the coronary sinus. The coronary sinus receives the following tributaries: (1) The *great cardiac vein,* which ascends in the anterior interventricular groove and then continues as the coronary sinus. Just before becoming the coronary sinus, it receives the left marginal vein, which drains the left margin of the heart. (2) The *posterior vein of the left ventricle,* which is often double and which may end in the middle cardiac vein. (3) The *middle cardiac vein,* which ascends in the posterior interventricular groove and may empty directly into the right atrium. (4) The *small cardiac vein* from the right margin of the right ventricle. It may empty directly into the right atrium and it may receive the right marginal vein. (5) The *oblique vein of the left atrium,* which is the remains of the embryonic left common cardinal vein, and a part of the left rostral cardinal. It empties into the beginning of the coronary sinus.

Lymphatic Drainage[38]

Lymphatic capillaries drain into vessels in the epicardium, where they follow the coronary arteries and end in right and left collecting trunks. The right trunk runs to the superior (anterior) mediastinal nodes. The left reaches a node (the "caval node") of the superior tracheobronchial group, between the aorta and the superior vena cava.

Nerve Supply

The heart is supplied by autonomic nerve fibers and sensory fibers from the

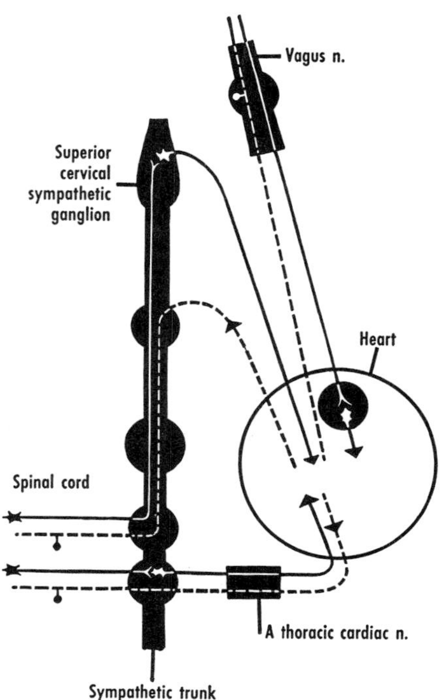

Figure 30–15 Schematic representation of the sympathetic and sensory (interrupted lines) fibers to the heart. For vagal fibers, see figure 31–8, p. 334.

vagus nerves and the sympathetic trunks (fig. 30–15). Many ganglion cells are present in the atria, particularly near the nodes, and in the vicinity of the veins. Some are also present in the ventricles. These nerve cells, which are mostly parasympathetic, occur in both the epicardium and the myocardium. Complex sensory nerve endings are found in the heart, especially in the atria near the openings of veins, and in the walls of the great veins.

Autonomic Fibers.[39] The nerve supply of the heart is of undoubted importance, although a transplanted heart can function without an extrinsic nerve supply. Preganglionic sympathetic fibers arise from the first to fourth (and sometimes the fifth and sixth) thoracic levels of the spinal cord. They synapse in cervical and thoracic ganglia, and postganglionic fibers to the heart are carried by cardiac branches of the cervical and thoracic parts of the sympathetic trunk. The preganglionic parasympathetic fibers in the vagus nerves are carried by the cervical and thoracic cardiac branches of the vagus nerves to ganglion cells in the heart. The postganglionic fibers of both systems supply the sinuatrial and atrioventricular nodes and the coronary vessels.

The various cardiac nerves are extremely variable in their topography and course, and are best grouped according to their level of origin, as follows.

1. The *cervical cardiac nerves* (often superior and middle) arise from the cervical sympathetic trunk, from the ganglia, or from both, and are usually joined by cervical cardiac branches of the vagus. The conjoined nerves then descend in front of, or behind, the arch of the aorta and enter the cardiac plexus.

2. Several *cervicothoracic nerves* (also named inferior cervical cardiac) arise from the cervicothoracic ganglion and the ansa subclavia, and are usually joined by cervicothoracic cardiac branches of the vagus. The conjoined nerves then run anterior or posterior to the arch of the aorta to the cardiac plexus.

3. *Thoracic cardiac nerves* arise from the upper four or five thoracic sympathetic ganglia and, together with thoracic cardiac branches of the vagus and left recurrent laryngeal nerves, go directly to the cardiac plexus, especially to the posterior walls of the atria.

Sensory Fibers. The parent fibers of complex sensory endings in the heart ascend in the vagus nerves. Experimental work indicates that these endings are all involved in the reflex control of blood pressure, blood flow, and heart rate.

Free endings are present in the connective tissue of the heart and in the adventitia of the blood vessels. Their parent fibers proceed to the thoracic and lower cervical parts of the sympathetic trunks. They enter the spinal cord by way of the upper four or five thoracic dorsal roots. Section of these dorsal roots, or of the rami communicantes to the corresponding spinal nerves, will generally abolish cardiac pain. Cardiac pain is usually referred to the left shoulder and medial side of the left arm, forearm, and hand (ulnar distribution), although it may also be felt in the side of the chest, and occasionally in the neck, ear, lower jaw, or diffusely, deep to the sternum.

CARDIAC CYCLE (fig. 30–16)

The contraction of the heart is termed systole, its relaxation, diastole. When the ventricles are filled, they begin to contract. The increase in intraventricular pressure causes the atrioventricular valves to shut, and the vibrations resulting from this closure are a major cause of the first "heart sound." During the phase of increasing intraventricular pressure (isometric contraction of the ventricles), the atrioventricular valves are steadied, and are prevented from being pushed into the atria by the contraction of the papillary muscles and of the superficial layers of the myocardium.

When intraventricular pressures surpass those in the aorta and pulmonary trunk, the aortic and pulmonary valves open, and blood is ejected into these arteries. The deep constrictor layer of the ven-

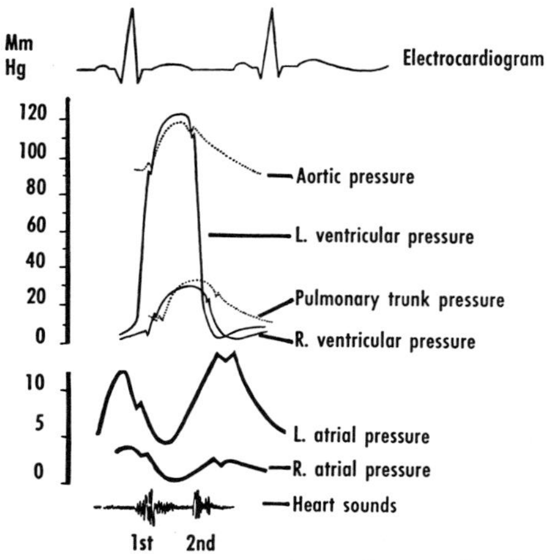

Figure 30–16 Time relationships and values of events in the cardiac cycle in man. Based on O. Bayer *et al.*[40]

tricles is responsible for this ejection of blood. The highest pressure reached during the ejection phase is the systolic blood pressure. The closure of the aortic and pulmonary valves is a major cause of the second "heart sound."

During ventricular systole, the heart twists so that more of the left ventricle comes to face forward. At the end of ventricular contraction, and after the aortic and pulmonary valves close, the ventricular musculature is relaxed. Intraventricular pressure drops until it is lower than that in the atria. The atrioventricular valves then open, and blood flows from the atria into the ventricles. The ventricles dilate as they fill with blood (ventricular diastole). During this time, pressure in the arteries drops to its lowest level (diastolic blood pressure). The atria contract during this phase (atrial systole), although to a much lesser degree than the ventricles do later. During ventricular filling, the atrioventricular valves remain in midposition; the push against the cusps by inflowing blood is balanced by eddies set up on their ventricular surfaces.

If the valves of the heart are diseased in such a way that blood leaks past them, or are stiffened so that they impede the flow of blood, abnormal vibrations occur and can be heard as murmurs (see p. 342).

Certain conditions in which the heart stops beating may require that an attempt be made to start the beat. **Heart beat may be started by electric stimulation, by massaging the heart after the chest has been opened, or by closed-chest cardiac massage.** In the last method, with the patient supine, firm pressure is applied vertically downward on the lower part of the sternum, about 60 times per minute.

As in the case of skeletal muscle, a conducted impulse precedes contraction. A record of this electric activity of the heart is termed an *electrocardiogram* (ECG or EKG). The form, direction, and amplitude of the various deflections in a normal ECG depend upon the position of the electrodes with reference to the heart, and therefore upon the position of the heart, the size of the chest, the position of the body, and the phase of respiration.

DEVELOPMENT OF HEART AND BLOOD VESSELS

The heart and blood vessels develop very early in the embryonic period proper and are the first organ system to become functional. The nutritional and respiratory needs of the embryo can be met only by a circulatory mechanism, and the circulation of blood has begun probably by the end of the third week after fertilization. Muscle cells begin to form and the heart begins to beat.

The heart is formed from two simple cellular tubes, the endothelial or endocardial tubes, which form in the cephalic end of the embryo. The cephalic end of each tube turns dorsally as the first aortic arch, and then continues caudally as one of the two dorsal aortae. The caudal end of each endocardial tube is joined by a vessel formed by the fusion of the corresponding vitelline and umbilical veins. The two endocardial tubes soon begin to fuse. The growing heart becomes **S**-shaped, and its originally caudal venous end comes to be dorsal and slightly cranial to the arterial end, which subdivides to form the aorta and pulmonary trunk (figs. 30–14 and 30–17). The external form of the heart is more or less established during the second month. The morphologic details of the development of the heart and blood vessels are complicated, and should be sought in textbooks of embryology.

Fetal Circulation (fig. 30–18)

The following account is based chiefly on studies made in sheep.[41] Angiographic studies of human fetuses indicate that the sequence of events in humans is similar.[42]

The oxygenated blood returning from the placenta in the umbilical vein(s) is largely shunted by the ductus venosus to the inferior vena cava. Mixture of oxygen-

Figure 30–17 Schematic representation of the spiral septum in the truncus arteriosus and bulbus cordis.

ated and deoxygenated blood occurs in the inferior vena cava, because this channel also receives blood from the caudal parts of the embryo and from the hepatic and vitelline veins. Most of the blood in the inferior vena cava, on entering the heart, passes to the left of the edge of the septum secundum, through the foramen ovale, and into the left atrium. Some of the blood entering from the inferior vena cava mixes with that entering from the superior vena cava.

The blood in the left atrium (together with the venous blood from the lungs) enters the left ventricle and the aorta. Most of this blood (largely oxygenated) goes to the head, neck, and upper limbs. Some descends in the aorta (mixed with the venous blood from the ductus arteriosus) and is distributed to the trunk, lower limbs, and placenta.

Venous blood from the upper limbs, from the head and neck, and from most of the body wall enters the right atrium

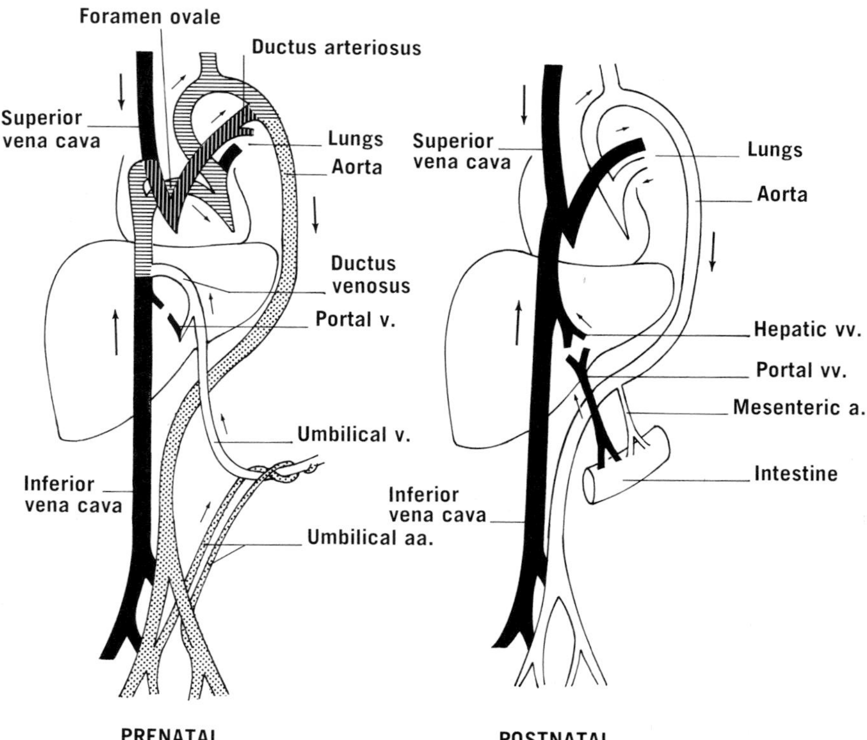

Figure 30–18 Schematic representation of the fetal and postnatal circulations. The differences in shading represent differences in oxygenation of the blood. The vessels with the darkest shading contain the least oxygenated blood.

through the superior vena cava. In the atrium it mixes with a small amount of blood from the inferior vena cava and then reaches the right ventricle and the pulmonary trunk. Some of this blood reaches the lungs and is returned (still as venous blood) to the left atrium (where it mixes with oxygenated blood from the placenta). Most of the blood in the pulmonary trunk, however, is shunted through the ductus arteriosus to the aorta, by which most of it reaches the placenta for oxygenation.

The fetal circulation is therefore arranged to take up oxygen from the maternal circulation in the placenta, to direct most of the oxygenated blood to the head and beck, and to shunt venous blood past the lungs to the placenta.

Although the fetal lungs have no respiratory function, the amount of blood circulating through them becomes considerable during the last part of the prenatal period. The lungs are not expanded, and the resistance to flow of blood is higher than it is postnatally. The right ventricle pumps blood against this resistance and against the aortic pressure. Before birth, the wall of the right ventricle is as thick as that of the left, or thicker.

Changes at Birth. Soon after the first breath, the two umbilical arteries contract and prevent blood from leaving the body of the child. Blood can still be returned from the placenta to the child, because the umbilical vein and the ductus venosus do not contract as early as the umbilical arteries. The vessels gradually become fibrous structures: the umbilical arteries become the medial umbilical ligaments (p. 453), the intra-abdominal part of the umbilical vein becomes the ligamentum teres of the liver (p. 394), and the ductus venosus becomes the ligamentum venosum (p. 394).

With the first breath, the pattern of circulation changes. Venous blood is directed to the lungs for oxygenation rather than to the placenta.

As the lungs expand, their resistance to blood flow drops, and pulmonary blood flow increases. The ductus arteriosus contracts. With the drop in pulmonary blood pressure, blood flows from the aorta to the left pulmonary artery. This reversal in flow through the ductus arteriosus persists for several hours or days. More blood reaches the lungs, and more returns to the left atrium. The pressure in the left atrium comes to equal that in the right. With approximately equal pressures in the atria, the limbus fossae ovalis and the valve of the foramen ovale come together and usually close the foramen ovale. Even if the foramen does not close, the equality of pressure prevents any significant flow of blood between the atria. The functional closure of the foramen ovale occurs soon after birth, but anatomical closure requires weeks or months. When the ductus arteriosus becomes completely closed, the adult type of circulation is attained. Although functional closure of the ductus arteriosus, like functional closure of the foramen ovale, occurs within a few hours or a few days after birth, anatomical closure again requires weeks or months. Connective tissue grows into its lumen and steadily obliterates it; the ductus gradually changes into the fibrous *ligamentum arteriosum.* At birth the right ventricle outweighs the left; however, within a month after birth, the left ventricle is heavier.[43] The difference in thickness gradually increases until the adult ratio is reached.

In summary, **the following changes occur at or shortly after birth. (1) The umbilical arteries become obliterated and form the medial umbilical ligament, (2) the umbilical vein and ductus venosus become obliterated and form the ligamentum teres and ligamentum venosum, (3) the ductus arteriosus becomes obliterated and forms the ligamentum arteriosum, and (4) the foramen ovale closes.**

Congenital Malformations of the Heart[44]

The importance of congenital heart disease in pediatric cardiology can scarcely be overemphasized. An understanding of congenital heart disease is dependent upon a knowledge of the development of the heart and blood vessels, and upon a knowledge of fetal circulation, changes in circulation at birth, and circulatory and respiratory physiology.

Congenital heart disease results from a structural abnormality of the heart, which is probably present at birth. There are three major groups of abnormalities: (1) abnormal communications between the pulmonary and systemic circulations, (2) valvular and vascular lesions, and (3) transposition of the great arteries, veins, or individual cardiac chambers.

The clinical manifestations of congenital heart disease are complex and depend upon the extent to which the circulation is altered and the oxygenation of blood compromised. Moreover, the structural abnormality may involve the conducting system and thereby interfere with the cardiac rhythm.

REFERENCES

1. H. Milhiet and P. Jager, *Anatomie et Chirurgie du Péricarde,* Masson, Paris, 1956. J. P. Holt, Amer. J. Cardiol., *26*:455, 1970.
2. G. T. Popa and E. Lucinescu, J. Anat., Lond., *67*:78, 1932.
3. J. L. Bremer, Anat. Rec., *87*:311, 1943. S. J. Viikari, Ann. Chir. Gyn. Fenn., *39* (Suppl.):3, 1950.
4. R. L. Moore, Arch. Surg., Chicago, *11*:765, 1925. H. Southworth and C. S. Stevenson, Arch. intern. Med., *61*:223, 1938. S. Sunderland and R. J. Wright-Smith, Brit. Heart J., *6*:167, 1944.
5. F. Morin and E. Bonivento, Arch. ital. Anat. Embriol., *43*:56, 1940. B. Delaloye, Arch. Anat., Strasbourg, *40*:131, 1957.
6. J. Alexander, A. G. Macleod, and P. S. Barker, Arch. Surg., Chicago, *19*:1470, 1929. See also J. A. Capps, Arch. intern. Med., *8*:717, 1911; J. A. Capps and G. H. Coleman, *An Experimental and Clinical Study of Pain in the Pleura, Pericardium and Peritoneum,* Macmillan, New York, 1932.
7. F. J. Hodges, Amer. J. Roentgenol., *42*:1, 1939. W. J. Comeau and P. D. White, Amer. J. Roentgenol., *47*:665, 1942.
8. R. O'Rahilly, Amer. Heart J., *44*:23, 1952.
9. R. Walmsley, Brit. Heart J., *20*:441, 1958.
10. E. D. U. Powell and J. M. Mullaney, Brit. Heart J., *22*:579, 1960.
11. Variations in the opening of the coronary sinus are described by H. K. Hellerstein and J. L. Orbison, Circulation, *3*:514, 1951.
12. R. T. Grant and L. E. Viko, Heart, *15*:103, 1929. K. Unger, Z. Anat. EntwGesch., *108*:356, 1938.
13. B. M. Patten, Amer. J. Anat., *48*:19, 1931. G. A. Seib, Amer. J. Anat., *55*:511, 1934. R. R. Wright, B. J. Anson, and H. C. Cleveland, Anat. Rec., *100*:331, 1948.
14. D. R. Dow and W. F. Harper, J. Anat., Lond., *66*:610, 1932. W. F. Harper, J. Anat., Lond., *73*:94, 1938; J. Anat., Lond., *75*:88, 1940.
15. M. D. Silver *et al.,* Circulation, *43*:333, 1971.
16. R. J. Merklin, Amer. J. Anat., *125*:375, 1969.
17. The functional anatomy of the mitral valve has been described by M. A. Chiechi, W. M. Lees, and R. Thompson, J. thorac. Surg., *32*:378, 1956; by J. C. Van der Spuy, Brit. Heart J., *20*:471, 1958; and by L. A. Du Plessis and P. Marchand, Thorax, *19*:221, 1964.
18. J. H. Lam *et al.,* Circulation, *41*:449, 1970. N. Ranganathan *et al.,* Circulation, *41*:459, 1970.
19. H. Nathan and M. Eliakim, Circulation, *34*:412, 1966.
20. R. L. Flett, J. Anat., Lond., *62*:439, 1928. J. S. Robb and R. C. Robb, Amer. Heart J., *23*:455, 1942. R. F. Rushmer, Physiol. Rev., *36*:400, 1956. M. Lev and C. S. Simkins, Lab. Invest., *5*:396, 1956.
21. T. N. James, Amer. Heart J., *66*:498, 1963.
22. T. N. James, Anat. Rec., *141*:141, 1961.
23. R. Ryback and N. Mizeres, Anat. Rec., *153*:23, 1965.
24. R. C. Truex, M. Q. Smythe, and M. J. Taylor, Anat. Rec., *159*:371, 1967.
25. T. N. James, Amer. Heart J., *62*:756, 1961.
26. E. W. Walls, J. Anat., Lond., *79*:45, 1945.
27. R. C. Truex and W. M. Copenhaver, Amer. J. Anat., *80*:173, 1947. J. H. Kugler and J. B. Parkin, Anat. Rec., *126*:335, 1956.
28. L. Gross, *The Blood Supply to the Heart,* Hoeber, New York, 1921. W. Spalteholz, *Die Arterien der Herzwand,* S. Hirzel, Leipzig, 1924. D. E. Gregg, Physiol. Rev., *26*:28, 1946. T. N. James, *Anatomy of the Coronary Arteries,* Hoeber, New York, 1961. G. Baroldi and G. Scomazzoni, *Coronary Circulation in the Normal and Pathologic Heart,* Office of Surg. Gen., Dept. Army, Washington, D.C., 1967.
29. S. H. Ahmed *et al.,* Acta anat., *83*:87, 1972.
30. J. T. Wearn *et al.,* Amer. Heart J., *9*:143, 1933.
31. T. C. Kramer, Amer. J. Anat., *71*:343, 1942.
32. A. Kerr and C. M. Goss, Anat. Rec., *125*:777, 1956.
33. L. J. A. DiDio and T. W. Wakefield, Acta cardiol., Brux., *27*:565, 1972. Nguyen Huu, J. P. Leroy, and A. Tiercelin, Bull. Ass. Anat., *57*:905, 1973.
34. J. E. Edwards, Circulation, *17*:1001, 1958. J. A. Ogden and J. M. Kabemba, Acta cardiol., Brux., *25*:487, 1970.
35. W. Laurie and J. D. Woods, Lancet, *2*:812, 1958. L. Reiner *et al.,* Arch. Path. (Lab. Med.), *71*:103, 1961. T. N. James, Amer. Heart J., *62*:756, 1961.
36. C. L. Hudson, A. R. Moritz, and J. T. Wearn, J. exp. Med., *56*:919, 1932.
37. R. F. Butterworth, J. Anat., Lond., *88*:131, 1954.
38. L. R. Shore, J. Anat., Lond., *63*:291, 1929. P. R. Patek, Amer. J. Anat., *64*:203, 1939.
39. N. J. Mizeres, Amer. J. Anat., *112*:141, 1963. J. P. Ellison and T. H. Williams, Amer. J. Anat., *124*:149, 1969.
40. O. Bayer *et al., Atlas intracardioler Druckkurven,* Thieme, Stuttgart, 1959.
41. A. E. Barclay, K. J. Franklin, and M. M. L. Prichard, *The Foetal Circulation,* Blackwell, Oxford, 1944. E. C. Amoroso *et al.,* J. Anat., Lond., *76*:240, 1942. J. A. Keen, J. Anat., Lond., *77*:104, 1942.
42. J. Lind and C. Wegelius, Cold Spr. Harb. Symp. quant. Biol., *19*:109, 1954.
43. J. A. Keen, J. Anat., Lond., *77*:104, 1942. E. N. Keen, J. Anat., Lond., *89*:484, 1955. J. L. Emery and A. Mithal, Brit. Heart J., *23*:313, 1961.
44. H. B. Taussig, *Congenital Malformations of the Heart,* Harvard University Press, Cambridge, 2nd ed., 1960. A. S. Nadas and D. C. Fyler, *Pediatric Cardiology,* Saunders, Philadelphia, 3rd ed., 1972.

GENERAL READING

Tandler, J., *Anatomie des Herzens,* vol. 3, pt. 1 of K. von Bardeleben, *Handbuch der Anatomie des Menschen,* Fischer, Jena, 1896–1934, 8 volumes.

Walmsley, T., *The Heart,* vol. 4, pt. 3 of Quain's *Elements of Anatomy,* Longmans, Green, London, 11th ed., 1929.

BLOOD VESSELS, LYMPHATIC DRAINAGE, AND NERVES OF THORAX

31

BLOOD VESSELS

PULMONARY CIRCULATION

Although the pulmonary trunk and arteries carry venous blood, they are arteries in the sense that they carry blood away from the heart at a relatively high pressure (systolic pressure 20 to 30 mm of mercury), and in a pulsatile fashion. Moreover, their walls, like the wall of the aorta, are elastic in structure.

Pulmonary Trunk

The pulmonary trunk[1] extends from the conus arteriosus of the right ventricle to the concavity of the arch of the aorta, at the left of the ascending aorta. It is invested with fibrous pericardium, and, together with the aorta, is included in a loose sheath of serous pericardium that sometimes forms a pouch extending as far as the ligamentum arteriosum.[2] After a course of about 5 cm, the pulmonary trunk divides into right and left pulmonary arteries.

The course of the pulmonary trunk may be indicated by a line extending from the center of the upper part of the cardiac outline to the left side of the sternal angle, behind the left second costal cartilage. The trunk forms the left border of the vascular shadow seen in anterior radiograms, below the aortic knuckle (fig. 32–3).

The pulmonary arteries and their branches are largely responsible for the normal shadows seen radiographically in the roots and hili of the lungs.

Right Pulmonary Artery. The right pulmonary artery is longer and wider than the left. It passes below the arch of the aorta, in front of the right main bronchus to the hilus of the right lung (fig. 29–7, p. 290). Before entering the hilus, it gives off a branch from which the segmental arteries to the upper lobe arise. The branches to the other lobes arise in the hilus.

Left Pulmonary Artery. The left pulmonary artery is shorter and narrower than the right. It extends laterally into the root of the left lung, in front of the left main bronchus (fig. 29–8, p. 291). At the hilus it divides into branches that accompany the bronchi to the upper and lower lobes, and from which segmental arteries arise. The left pulmonary artery is connected to the arch of the aorta by the *ligamentum arteriosum,* the fibrous remains of the ductus arteriosus.

Pulmonary Veins

There are usually five pulmonary veins, one from each lobe. The right upper and middle veins usually join, so that four veins, upper and lower on each side, enter the left atrium. The right lower vein passes behind the right atrium, between the openings of the superior and inferior venae cavae. The right upper vein crosses behind the superior vena cava. Both right veins enter a prolongation or pouch of the left atrium, as do the left veins. All the pulmonary veins receive partial reflections of the serous layer of the pericardium, and they help to bound the oblique sinus of the pericardium. None of the pulmonary veins or their tributaries has valves. Variations are discussed in connection with the lungs (p. 299), and relationships to the pericardium and atria are discussed on pages 304 and 309.

SYSTEMIC CIRCULATION

The chief systemic supply of the thorax is derived from branches of the aorta. The axillary (p. 114) and internal thoracic (p. 270) arteries and the costocervical trunk (p. 705) also contribute.

Aorta

The aorta is the main systemic artery of the body. It is divided into the ascending aorta, the arch of the aorta, and the descending aorta. The part of the descending aorta in the thorax is called the thoracic aorta. Branches arise from each of the parts. The aorta is an elastic artery, the thick tunica media of which is composed largely of plates of elastic tissue. The vessel is admirably constructed to withstand the systolic blood pressure and to provide elastic recoil. Its elasticity diminishes with age. The walls of the aorta are supplied by vasa vasorum given off by its various branches. The walls of the ascending aorta and the arch of the aorta contain pressoreceptors, the parent fibers of which ascend in the vagus nerves to the vasomotor centers in the brain stem. Stimulation of these receptors by a rise in blood pressure leads reflexly to a fall in blood pressure and heart rate. In many lower vertebrates the sensory fibers from the aorta ascend in a single branch of the vagus nerve, the "aortic depressor nerve," but a single nerve does not occur in humans.

Ascending Aorta. The ascending aorta, which lies in the middle mediastinum, extends from the root of the aorta upward and slightly to the right, to the level of the sternal angle just to the right of the median plane. It is invested with fibrous pericardium and shares a serous reflection with the pulmonary trunk. It is about 3 cm in diameter and has a course of about 5 cm. Ordinarily the ascending aorta does not extend far enough to the right to take part in the right border of the cardiovascular shadow. It may, however, impart a pulsation (as observed fluoroscopically) to the superior vena cava, which does take part in this border.

The root of the aorta is dilated, owing to three bulgings in its wall, the *sinuses of the aorta*. Each sinus is related to a cusp of the aortic valve (p. 312), and is named accordingly. At its origin, the ascending aorta has the pulmonary trunk and conus arteriosus in front, the left atrium and transverse sinus of the pericardium behind. Higher up, it is overlapped by the right pleura and lung and lies in front of the right pulmonary artery and right main bronchus. A ridge of fatty tissue partly encircles the middle part of the ascending aorta. Below, this expands into a cushion opposite the edge of the right auricle.[3]

The branches of the ascending aorta are the right and left coronary arteries (p. 316).

Arch of the Aorta. The ascending aorta continues into the arch of the aorta,[4] which runs to the left, in front of the trachea, and then turns backward and downward, above the left bronchus, to the left of the trachea and esophagus (see figs. 28–2, 32–5, and 32–6, pp. 280, 345, 346). **The arch of the aorta occupies an almost sagittal plane in the superior mediastinum, behind the lower part of the manubrium of the sternum.** It forms a prominence that is visible radiographically as the "aortic knuckle" (fig. 32–3, p. 344).

The arch is related on its left to the left phrenic nerve, the left vagus nerve, the left superior intercostal vein, and cardiac branches of the left vagus and the sympathetic trunk. Above are the three branches of the arch; these are crossed in front by the left brachiocephalic vein. Below are the bifurcation of the pulmonary trunk and the root of the left lung. The ligamentum arteriosum connects the arch to the left pulmonary artery. The left recurrent laryngeal nerve hooks below it, and then ascends. The arch of the aorta usually causes an impression on the left side of the esophagus, an impression that often merges with and is indistinguishable from that caused by the left main bronchus.

The branches of the arch of the aorta are the brachiocephalic trunk, the left common carotid artery, and the left subclavian artery. After giving off these branches the aorta is diminished in diameter.

A slight constriction (isthmus aortae) immediately distal to the left subclavian artery indicates a critical area. A severe constriction here may occur during development (coarctation of the aorta). If the coarctation occurs distal to the opening of the ductus arteriosus into the aorta, an adequate collateral circulation develops before birth. If, however, the coarctation occurs proximal to the ductus arteriosus, an adequate collateral circulation does not develop and the infant may die when the ductus closes.[5] The important collateral channels are (1) anterior intercostal branches of the internal thoracic artery, which anastomose with posterior intercostal arteries, and the superior epigastric artery, which anastomoses with the inferior epigastric; (2) branches of the subclavian artery (p. 703), which, by anastomosing with arteries about the scapula, indirectly anastomose with posterior intercostal arteries.[6] An additional collateral channel may be furnished by the anterior spinal arteries from the vertebral arteries (p. 541). The anterior spinal artery is reinforced at various levels by medullary branches of segmental arteries (fig. 50–3, p. 542).

BRACHIOCEPHALIC TRUNK. The brachiocephalic trunk is the first branch of the arch. It extends from behind the lower part of the manubrium sterni to the level of the right sternoclavicular joint. Behind this joint it divides into right subclavian and right common carotid arteries. It is at first in front of, and then to the right of, the trachea. The right brachiocephalic vein is to its right; it is crossed in front by the left brachiocephalic vein.

LEFT COMMON CAROTID ARTERY. This arises slightly to the left of the brachiocephalic trunk. It extends upward, at first in front and then to the left of the trachea, and enters the neck behind the left sternoclavicular joint.

LEFT SUBCLAVIAN ARTERY. This arises behind the left common carotid artery, ascends lateral to the trachea, and leaves the thorax behind the left sternoclavicular joint.

VARIATIONS. Minor variations in the branches of the arch of the aorta are fairly frequent,[7] a common origin of the brachiocephalic trunk and the left common carotid artery, for example. Few of the variations have any functional significance, but they provide interesting exercises in analyzing the development of the embryonic aortic arches. Some variations may cause symptoms because of their position, or they may be of surgical significance.[8] For instance, an aortic ring that encircles the trachea and esophagus may press on these structures. The brachiocephalic trunk, when it arises in common with the left common carotid artery, may cross in front of the trachea. The right subclavian artery may have an anomalous origin from the thoracic aorta (see vas aberrans, below); in such instances it may be associated with a right thoracic duct. These variations are important in surgical operations on the trachea and esophagus. The *thyroidea ima artery* is an inconstant branch of the arch of the aorta or of the brachiocephalic trunk; it ascends in front of the trachea to the thyroid gland and should be kept in mind in median approaches to the trachea (p. 690).

Thoracic Aorta. The thoracic aorta descends in the posterior mediastinum, from the arch to about the level of T.V. 12. There it traverses the aortic opening in the diaphragm and becomes the abdominal aorta. In the upper part of its course, it is to the left of the vertebral column. It gradually reaches the front of the vertebral column, where it lies behind the esophagus, and it enters the abdomen in the median plane. The thoracic duct is to its right. The azygos vein may be to its right or left. In front, from above downward, are the root of the left lung, the pericardium, the esophagus, and the diaphragm.

The branches of the thoracic aorta may be classified as parietal and visceral.

The *parietal* branches include the third to the eleventh posterior intercostal arteries and the subcostal arteries (p. 270). They also include several small *superior phrenic arteries* to the posterior part of the diaphragm, and the *vas aberrans*. The last named is an inconstant branch that represents the embryonic right dorsal aorta. It passes upward and to the right, behind the esophagus. Occasionally the right dorsal aorta persists and the right fourth aortic arch disappears. The right subclavian artery then arises from the thoracic aorta, and its first part is formed by the vas aberrans.

The *visceral* branches of the thoracic aorta are *bronchial* (usually two left bronchial arteries, and occasionally one right, p. 299), *esophageal* (usually two), *pericardial*, and *mediastinal*. The bronchial arteries also supply adjacent lymph nodes, pulmonary vessels, pericardium, and esophagus.

Veins

The veins of the thoracic cavity are thin-walled and valveless, and have longitudinally disposed smooth muscle in their adventitia.

Brachiocephalic Veins. Each brachiocephalic vein is formed by the union of the internal jugular and subclavian veins, behind the corresponding sternoclavicular joint. The right brachiocephalic vein descends more or less vertically. The left crosses the superior mediastinum obliquely, in front of the branches of the arch of the aorta. The two veins join and form the superior vena cava, at about the level of the sternal angle, behind the second costal cartilage on the right side. Both veins receive several tributaries (fig. 31–1).

Superior Vena Cava. The superior vena cava descends on the right side of the ascending aorta, receives the azygos vein, and ends in the right atrium. The superior vena cava and the right brachiocephalic vein form the upper right margin of the cardiovascular shadow (fig. 32–3, p. 344).

Inferior Vena Cava. After the inferior vena cava traverses its opening in the central tendon of the diaphragm, it has an intrathoracic course of 2 to 3 cm and then enters the right atrium. Its opening in the atrium is guarded by an imperfect valve. The inferior vena cava may be visible radiographically between the diaphragm and the lower right part of the cardiovascular shadow (see fig. 32–3). The right hepatic vein sometimes traverses the caval opening in the diaphragm before joining the inferior vena cava.

Azygos System of Veins. Most of the blood from the back and from the walls of the thorax and abdomen drains into veins that lie alongside of the bodies of the vertebrae. These veins form what is termed the azygos system.[9] The terminal veins of this system are the azygos, hemiazygos, and accessory hemiazygos veins (fig. 31–1). The azygos system is so variable that no short account of it can include the variations.

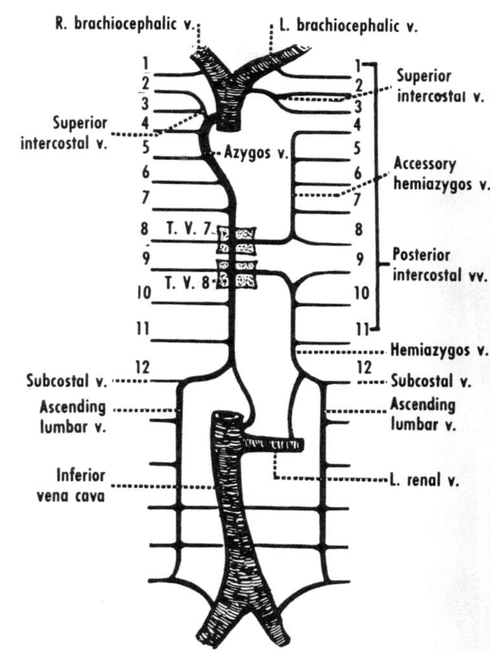

Figure 31–1 Diagrammatic representation of veins of the thorax.

The *azygos vein* (fig. 31–2) is commonly formed by the junction of the right subcostal and right ascending lumbar veins. It ascends through the posterior and superior mediastina and in front of the posterior intercostal arteries, adjacent to the aorta. It arches over the root of the right lung, and ends in the superior vena cava. In its upward course, the azygos vein on rare occasions deviates and demarcates a lobule of the right lung (the lobule of the azygos vein, p. 298). During its ascent, the azygos vein often runs a part of its course on the left side of the vertebral column.[10] The lower part of the azygos vein is usually connected to the back of the inferior vena cava by a small vein or fibrous cord. The tributaries of the azygos vein are the right superior intercostal vein, the fourth to eleventh posterior intercostal veins of the right side, and the hemiazygos or accessory hemiazygos veins, or both.

The *hemiazygos* and the *accessory hemiazygos veins* form an extremely variable arrangement on the left side.

The hemiazygos vein usually begins by the union of the left subcostal and ascending lumbar veins, and is commonly connected with the left renal vein. Its tributaries are the lower posterior intercostal veins, and some mediastinal and esophageal veins. It ends in the azygos vein.

The accessory hemiazygos vein usually begins at the fourth intercostal space as a continuation of the vein of that space. It descends, receiving tributaries from spaces above and below, as well as from the bronchial and mediastinal veins, and joins either the hemiazygos or the azygos vein.

Anastomoses and Routes of Venous Return. **Extensive anastomoses among the caval, the azygos, and the vertebral systems provide multiple routes for the return of blood to the heart** (fig. 6–2). In effect, the azygos and vertebral systems bypass the caval system.

The *vertebral venous system*[11] consists of plexuses of thin-walled veins that drain the back, the vertebrae, and the structures in the vertebral canal (see p. 531). They may be demonstrated in the living by intraosseous venography.[12] These plexuses communicate above with the intracranial veins and below with the portal system, and empty into the vertebral, posterior intercos-

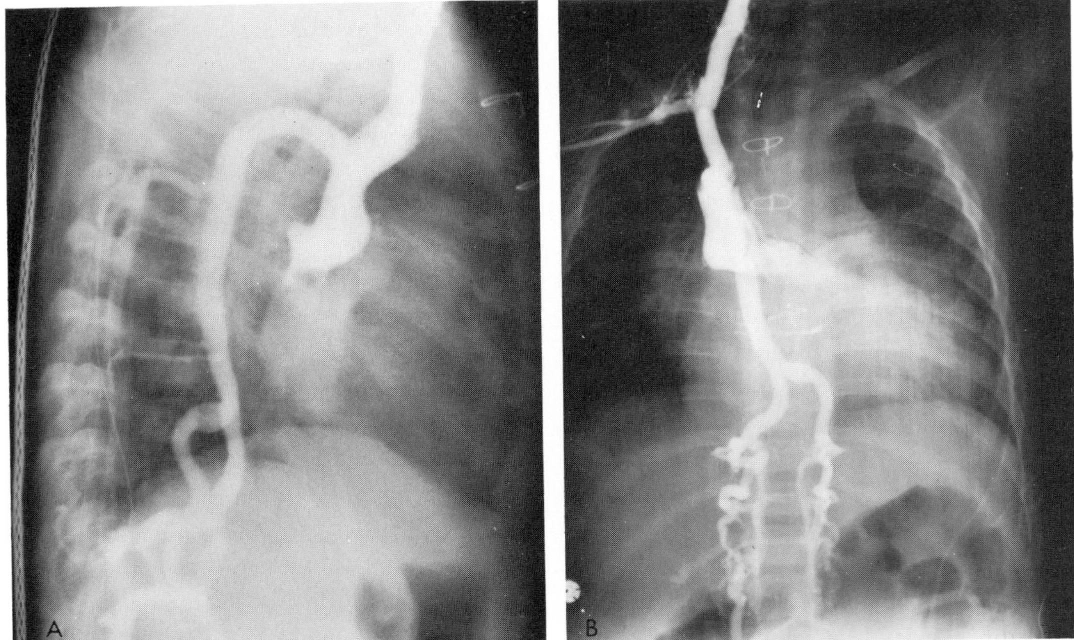

Figure 31–2 Angiograms of the azygos system. *A*, lateral; *B*, anteroposterior. Note reflux into the hemiazygos veins, especially in *B*. Courtesy of R. A. Castellino, M.D., Stanford University, California.

tal, lumbar, and lateral sacral veins. The veins in the vertebral plexus are valveless, and blood may flow in either direction in them. An increase in intrathoracic or intra-abdominal pressure, such as occurs during coughing or straining, or during expiration, may cause blood in the vertebral plexuses to flow away from the heart, either upward or downward. The increased pressure in these venous plexuses is reflected as an increase in the pressure of cerebrospinal fluid. Cyclic changes in venous pressure, or changes due to coughing or straining, are clinically important because they may facilitate the spread of tumor cells and infectious organisms. For example, cells from pelvic, abdominal, thoracic, or breast tumors may enter the venous system and may be carried into the vertebral plexuses during reversal of blood flow. These cells may ultimately lodge in the vertebrae, the spinal cord, or the brain.

Other collateral channels are furnished by veins of the thoracic wall that drain upward into the axillary and internal thoracic veins, and anastomose below with veins that empty into the tributaries of the inferior vena cava and of the portal system. Some of the more important anastomoses of this group are made by the superficially placed *thoracoepigastric veins*, which connect the lateral thoracic with the superficial epigastric vein on each side.

If the superior vena cava is obstructed above the azygos vein, blood can drain downward, into the veins of the body wall, then into the internal thoracic veins and iliac veins, and return to the heart by the inferior vena cava and azygos systems. Blood from the head and neck can also enter the vertebral plexuses and thereby return to the heart. Obstruction of the superior vena cava between the azygos vein and the right atrium is more serious because this leaves only the inferior vena cava as a route of venous return. However, a case of complete closure of the opening of the superior vena cava has been described.[13]

If the inferior vena cava is obstructed, blood can flow upward in veins of the body wall and in the vertebral plexuses, and so reach the superior vena cava.

Patients can therefore survive occlusion of nearly any vein in the body, provided that, in the case of major veins such as the superior or inferior vena cava, the occlusion is not too rapid. Even when occlusion of these veins has been sudden, as by surgical ligation, instances of survival have been recorded.

LYMPHATIC DRAINAGE

LYMPH NODES

The nodes of the thorax are classified as parietal and visceral, according to their location. Their lymphatic vessels communicate rather freely. The parietal nodes, which are parasternal, phrenic, and intercostal, are described on page 271.

Visceral Nodes

The visceral lymph nodes drain the lungs, pleurae, and mediastinum. They are irregularly arranged in the roots of the lungs, along the trachea and bronchi, and in the superior and the posterior mediastinum (fig. 31–3).

Nodes in Roots and Hili of Lungs. The nodes in the roots and the hili of lungs are subdivided into several groups. A few small *pulmonary nodes* are present along the larger bronchi near the hilus, within the substance of the lung. The *bronchopulmonary nodes* are embedded in the root of the lung, mainly at the hilus. The bronchopulmonary and pulmonary nodes drain upward, into the tracheobronchial nodes.[14] The *tracheobronchial nodes* form an *inferior* group, in the angle of bifurcation of the trachea, and *superior* groups, in the angle between the trachea and bronchus on each side.[15] Those tracheobronchial nodes that are related to the lower part of the right main bronchus are sometimes termed the "lymphatic sump"; they are invaded by tumors from both upper and lower lobes.[16]

The tracheobronchial nodes receive, by way of the pulmonary and bronchopulmonary nodes, the lymphatic vessels of the lung and visceral pleura, the bronchi, the lower part of the trachea, and the heart. The lymphatic drainage of the right lung to the tracheobronchial nodes is ipsilateral, that of the left is often bilateral. (The left lower lobe drains to the right superior tracheobronchial nodes.) The efferents of the tracheobronchial nodes ascend on the trachea.

Lymphatic nodes in the roots of the lungs commonly are secondarily involved by infections, such as tuberculosis, and by tumors of the lungs and mediastinum. Their density may increase so that they become visible radiographically, particularly if they become calcified.

Tracheal (Paratracheal) Nodes. These are scattered along each side of the trachea, extending upward into the neck. They receive lymphatic vessels from the trachea and esophagus, and from the tracheobronchial nodes. Their efferents join those of the mediastinal nodes to form a *bronchomediastinal (mediastinal) trunk* on each side of the trachea. This trunk also receives the efferent vessels from the parasternal nodes.

Mediastinal Nodes. These include a few small nodes scattered in the superior mediastinum, the so-called innominate nodes (also called *anterior mediastinal nodes,* in spite of their location in the superior mediastinum). They receive vessels from the thymus, the walls of the great veins (one of the nodes is called the "caval node"), the pericardium, and the heart. Their efferents join those from the trachea, bronchi and lungs to form the bronchomediastinal trunk. The *posterior mediastinal nodes* are a few nodes around the lower part of the thoracic portion of the esophagus. The lowest ones of this group rest on the diaphragm, near or indistinguishable from the more posterior of the phrenic nodes. The posterior mediastinal nodes receive lymphatic vessels from the esophagus and pericardium, and a few from the lower lobes of the lungs. Their efferents pass directly to the thoracic duct and to the descending intercostal lymphatic trunks. A few ascend to the tracheal group.

LYMPHATIC VESSELS

All of the lymphatic drainage of the thorax is directed toward the bronchomediastinal trunks, the thoracic duct, and the descending intercostal lymphatic trunks (fig. 31–3).

The lymphatic trunks are extremely variable. Additional trunks, such as paratracheal and posterior mediastinal, are often present. The left bronchomediastinal trunk may end in the thoracic duct, or it may empty into one of the veins near the end of

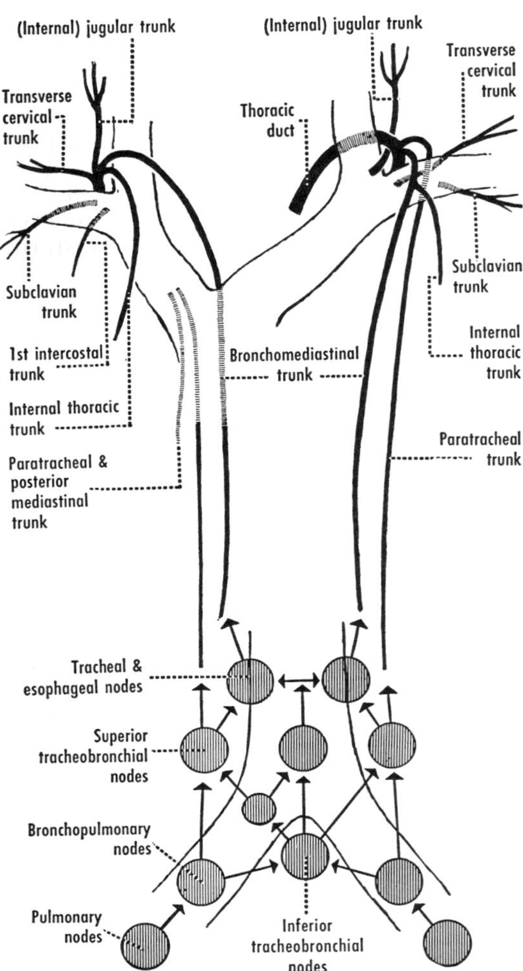

Figure 31–3 Schematic representation of the visceral lymph nodes and collecting trunks of the thorax.

the thoracic duct. The right bronchomediastinal trunk forms various combinations with the right subclavian and jugular trunks. Rarely all three unite to form a *right lymphatic duct (right thoracic duct),* which then empties directly into the junction of the internal jugular and subclavian veins.

Thoracic Duct. The thoracic duct[17] extends from the upper part of the abdomen to the neck, where it ends in one of the large veins (figs. 31–4 and 31–5). It begins in the abdomen at the junction of the intestinal, lumbar, and descending intercostal trunks (p. 422). The junction either consists of a dilatation called the *cisterna chyli,* or else is a plexus-like arrangement.[18] The thoracic duct usually passes through or near the aortic opening of the diaphragm (there may be two or more ducts at the level of the diaphragm) and ascends in the posterior mediastinum on the right side of the aorta, between the aorta and the azygos vein, if the azygos vein is on the right side. At about the level of T.V. 5 or 6, it begins to cross obliquely to the left, behind the esophagus. This crossing is generally completed in the upper thoracic region. It ascends on the left side of the esophagus, passes behind the left subclavian artery, and enters the neck, where it forms an arch that may reach the level of C.V. 7, several cm above the level of the clavicle. It then turns forward and downward, usually forming multiple channels that unite into a common trunk that ends in the left internal jugular vein (fig. 31–6).[19]

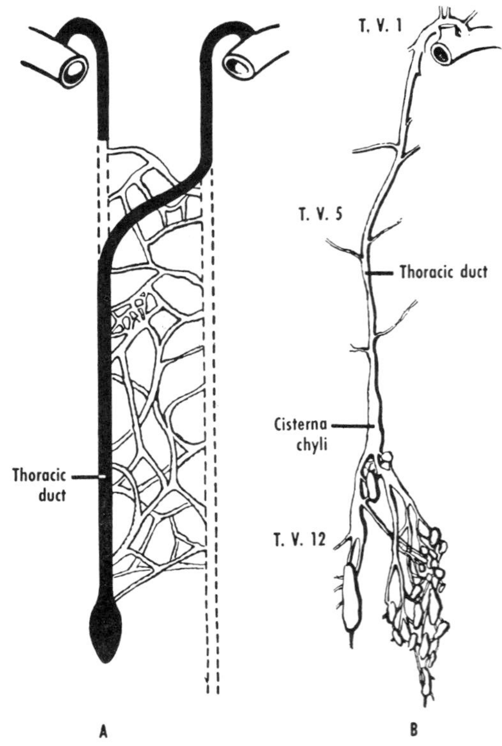

Figure 31–4 Thoracic duct. *A*, schematic representation of development (postulated but not proved). *B*, common arrangement of thoracic duct, crossing vertebral column at T.V. 5 or 6. Based on Davis.[17]

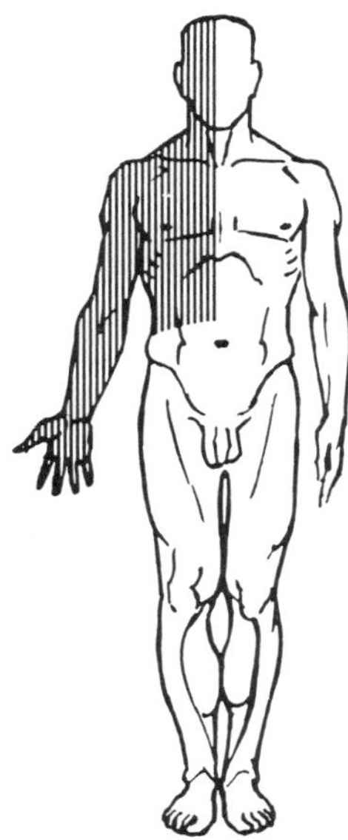

Figure 31–5 All except the shaded area is drained by the thoracic duct.

Variations in the thoracic duct are common. Its thoracic course varies, it may occasionally be multiple, it may end as multiple channels or as a single trunk, it may end in the left subclavian or left brachiocephalic vein, and occasionally it is completely left-sided. Rarely, it may end on the right side.

In addition to the lymphatic trunks that it receives at its origin, the thoracic duct receives efferents from the posterior mediastinal and upper intercostal nodes. In the neck or in the upper part of the thorax it generally receives the left jugular and subclavian trunks, and often the left bronchomediastinal trunk also.

Most of the lymph in the body reaches the venous system by way of the thoracic duct (fig. 31–5), but the anastomoses of collecting channels in the thorax and neck are so extensive that no serious effects result if the thoracic duct is ligated.

THYMUS

The thymus of the adult is an irregularly shaped mass located partly in the neck and partly in the thorax.[20] It is composed of one to three irregular lobes, most commonly two. Each lobe consists of numerous lobules surrounded, at least partially, by thin, connective tissue capsules. The lympho-

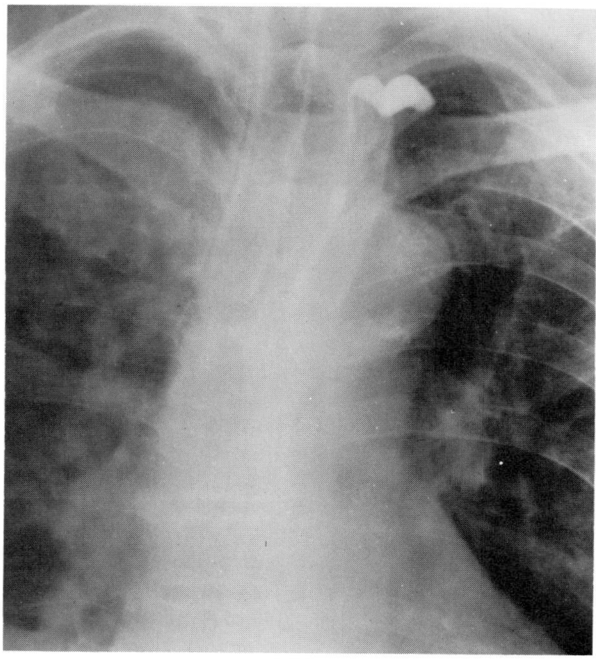

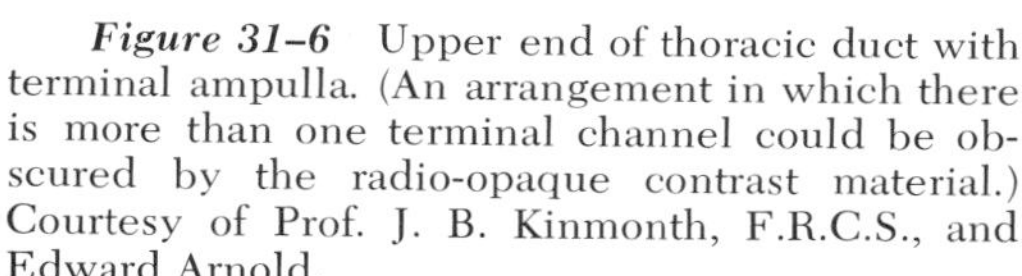

Figure 31–6 Upper end of thoracic duct with terminal ampulla. (An arrangement in which there is more than one terminal channel could be obscured by the radio-opaque contrast material.) Courtesy of Prof. J. B. Kinmonth, F.R.C.S., and Edward Arnold.

cytes formed in the thymus play a major role in the development and maintenance of the immune system.

The thoracic part of the thymus usually lies behind the manubrium or the upper part of the body of the sternum, but it has been found as low as the level of the xiphoid process. The cervical part of the thymus, which often consists only of a fibrous band containing thymic tissue, lies on the front and sides of the trachea, behind the sternohyoid and sternothyroid. It is connected by fibrous strands with the tissues around the thyroid gland.

The thymus weighs only a few grams at birth, but grows steadily and reaches its greatest size at puberty. It then begins to regress. Much of its substance is replaced by fat and fibrous connective tissue, but thymic tissue never disappears completely. At birth the thymus is a broad, irregularly lobulated gland, which is often bilobed. Its upper extensions reach as far as the thyroid gland, but the bulk of the thymus lies in the superior mediastinum. There, for the first few years of life, it causes a distinctive radiographic shadow.[21]

The thymus has a plentiful blood supply from nearby vessels (inferior thyroid, internal thoracic, anterior intercostals). This supply persists even after involution of the organ. Its lymphatic drainage is profuse. Most of its efferent vessels go to the innominate nodes, but some empty directly into adjacent veins.

NERVES

The nerves of the thorax are the thoracic, phrenic, and vagus nerves, the sympathetic trunks, and the autonomic plexuses. The thoracic nerves are described on page 272.

Phrenic Nerves

The phrenic nerves supply the diaphragm. Each nerve generally arises from the fourth and fifth cervical nerves (p. 711), and may have a contribution from the third and sometimes from the second or sixth. Each nerve enters the thorax after passing in front of the scalenus anterior. Each is accompanied in its course through the thorax by the pericardiacophrenic branches of the internal thoracic vessels, each gives branches to the pericardium, the mediastinal pleura, and the central part of the diaphragmatic pleura, and each divides into three to five branches at the pericardiodiaphragmatic angle.[22] The most posterior branch on each side pierces the diaphragm and supplies the crus (the left part of the right crus is supplied by the right phrenic); the right posterior branch lies immediately lateral to the inferior vena cava, to which it gives a twig. All other branches pierce the diaphragm and are distributed to it from below. Both phrenic nerves give filaments to the plexuses along the inferior phrenic arteries and to the central part of the diaphragmatic peritoneum.

Right Phrenic Nerve. The right phrenic nerve descends at the right side of the superior vena cava and right atrium, in front of the root of the right lung, between the pericardium and mediastinal pleura.

Left Phrenic Nerve. The left phrenic nerve descends between the left subclavian and left common carotid arteries, lateral to the vagus nerve and the arch of the aorta. It passes in front of the root of the left lung between the mediastinal pleura and the pericardium, and its branches pierce the diaphragm immediately to the left of the pericardium.

Functional Components. The phrenic nerves contain motor, sensory, and sympathetic fibers (fig. 31–7). The motor fibers supply the diaphragm. Most of the sensory fibers are pain fibers from the diaphragmatic peritoneum, from the diaphragmatic and mediastinal pleurae, and from the pericardium. Pain from the area of supply of a phrenic nerve is usually referred to the skin over the trapezius, that is, from the lower part of the neck to the tip of the shoulder.[23] Pain is sometimes referred to the region of the ear also. In such instances, the phrenic nerve probably has a considerable contribution from the second or third cervical nerve, nerves which supply this region by way of the great auricular nerve. Referred pain due to irritation of the part of the pleura supplied by the phrenic nerve is indistinguishable from that due to irritation of the part of the peritoneum supplied by the

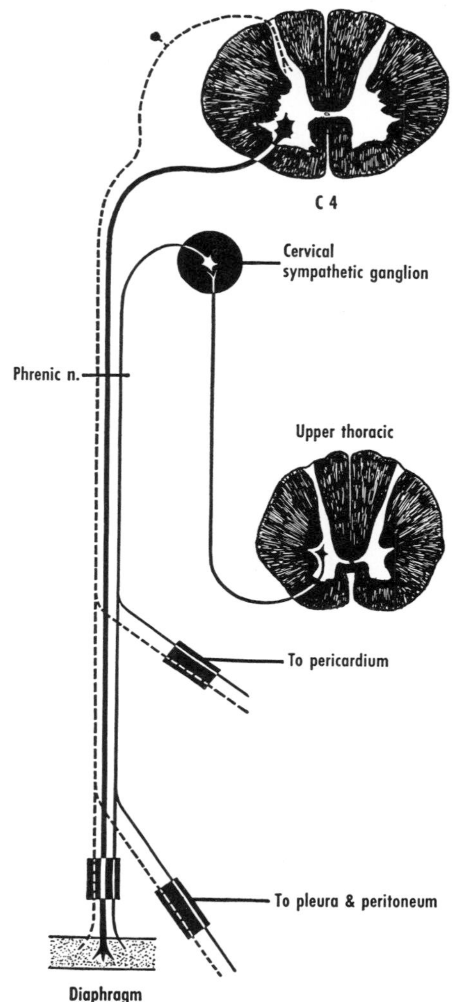

Figure 31–7 Functional components of the phrenic nerve. For purposes of simplification, each component is shown as a single fiber.

phrenic nerve. The sympathetic fibers are vasomotor.

Accessory Phrenic Nerves. The contribution of the fifth cervical nerve to the phrenic often arises from, or is a continuation of, the nerve supplying the subclavius muscle. In some instances this contribution may run a separate course into the thorax before joining the phrenic nerve and is then termed the accessory phrenic nerve. Occasionally the entire phrenic nerve arises in this way. There may also be accessory contributions that are not connected with the nerve to the subclavius. The accessory phrenic nerve usually descends in front of the subclavian vein, whereas the phrenic nerve passes behind it.

If an accessory phrenic nerve is present, section of or injury to the phrenic nerve in the neck will not paralyze the corresponding half of the diaphragm completely; the accessory nerve still provides some motor supply.

Vagus Nerves

After the vagus nerves leave the posterior cranial fossa and descend through the neck, they enter the thorax, where they contribute to the pulmonary plexuses and then continue to the esophagus, where they form the esophageal plexus (p. 336). At the lower part of the esophagus, the plexus collects into an *anterior* and a *posterior vagal trunk*, which descend through the esophageal opening of the diaphragm (p. 268). The anterior trunk descends on the anterior aspect of the esophagus, and the posterior on the posterior aspect. Each trunk contains fibers from both the right and left vagus nerves,[24] and each gives branches to the stomach (p. 422).

Each vagus has a *recurrent laryngeal* branch, which supplies the trachea, esophagus, and larynx. Each has a variable number of *cervical cardiac* branches, which arise in the neck. These usually join the cardiac branches of the cervical sympathetic ganglia. *Cervicothoracic cardiac* branches arise near the thoracic inlet and join cervicothoracic branches of the sympathetic trunks. Branches are also given to the bronchi and the esophagus, directly and by way of associated plexuses.

Right Vagus Nerve. The right vagus nerve crosses in front of the first part of the right subclavian artery, behind the superior vena cava, and descends in the superior mediastinum at the right side of the trachea. Thoracic cardiac branches are given off below the origin of the recurrent laryngeal nerve. At the root of the lung, the right vagus breaks up into trunks that contribute to the pulmonary plexuses and then continue into the esophageal plexus.

The *right recurrent laryngeal nerve* arises from the right vagus as the latter passes in front of the subclavian artery. It hooks below the artery, then behind it, and ascends between the trachea and esophagus, both of which it supplies. Its further course is described with the neck (p. 699).

Left Vagus Nerve. The left vagus nerve enters the thorax between the left common carotid and left subclavian arteries, behind the left brachiocephalic vein. It descends in the superior mediastinum, crosses the left side of the arch of the aorta, gives some branches to the heart (thoracic cardiac nerves), contributes to the pulmonary plexuses, and then enters the esophageal plexus. Just above the arch of the aorta, the nerve is crossed superficially by the left phrenic nerve.

The *left recurrent laryngeal nerve* leaves the left vagus at the arch of the aorta, hooks below the arch to the left of the ligamentum arteriosum, and then ascends at the right side of the arch, between the trachea and esophagus. It gives branches to the aorta, trachea, and esophagus, and to the heart (thoracic cardiac nerves). **The left recurrent laryngeal nerve is liable to be damaged by disorders of the aorta (such as aneurysms) or of the mediastinum (tumors).** Such damage may be irritative at first, leading to coughing, as if the mucous membrane of the trachea and larynx were being irritated. Destruction of the nerve is followed by hoarseness and by paralysis of the ipsilateral vocal cord.

It is believed that both recurrent laryngeal nerves owe their adult relationships to their relationships in the embryo to the sixth aortic arches, namely, that they pass caudal to these vessels on their way to the larynx. On the right side, the dorsal part of the sixth arch (homologue of the ductus arteriosus) disappears, the fifth arch regresses, and the right recurrent laryngeal nerve comes to lie below the fourth arch (beginning of the subclavian artery). On the left side, the fifth arch also regresses, but the sixth persists, first as the ductus arteriosus, then as the ligamentum arteriosum. Hence, in the adult, the left recurrent laryngeal nerve passes caudal to the ligamentum arteriosum before ascending to the larynx.

Functional Components. Figure 31–8 illustrates the components of the entire nerve. The vagus nerves contain motor fibers to the muscles of the pharynx and larynx. These motor fibers, however, leave the medulla oblongata by way of the cranial part of the accessory nerve (p. 700). The thoracic branches of the vagus nerves contain parasympathetic and sensory fibers.

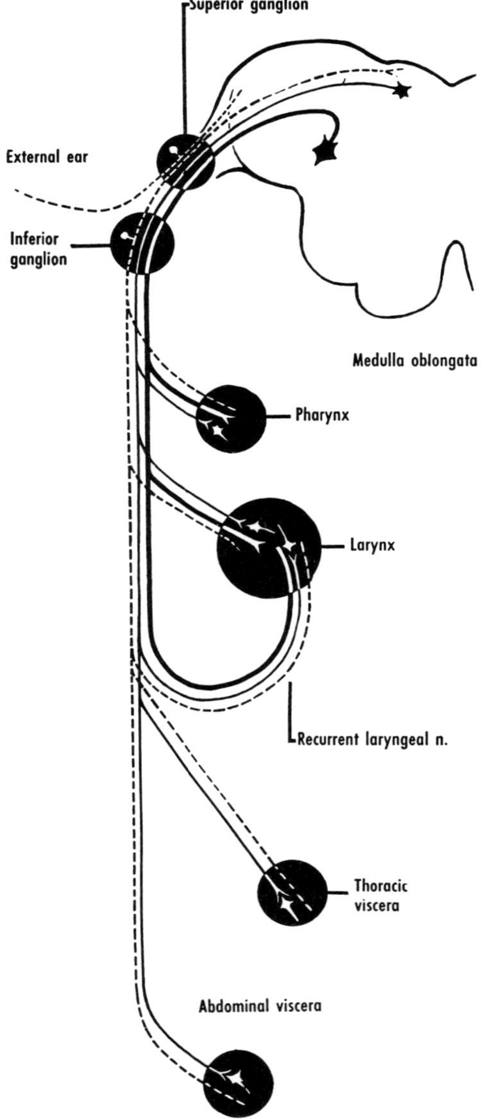

Figure 31–8 Functional components of the vagus nerve. For purposes of simplification, each component is shown as a single fiber. The distinction between accessory and vagal components is not shown (see fig. 60–22, p. 700).

The important parasympathetic fibers are those to the heart, which are involved in the regulation of the heart beat. Other parasympathetic fibers supply the smooth muscle and glands of the trachea, bronchi, bronchioles, esophagus, and abdominal viscera.

Many sensory fibers are concerned with pulmonary and cardiovascular reflexes. Other sensory endings occur in the mucous membrane of the trachea, bronchi, and bronchioles. Irritation of these endings leads to coughing and reflex inhibition of respiration. Coughing from tracheal irritation may be accompanied by a deep-seated sensation that can be painful. Still other sensory fibers in the vagus nerves come from the esophagus and abdominal viscera.

Sympathetic Trunks and Ganglia

The sympathetic trunks (fig. 64–15, p. 783) enter the thorax from the neck, and descend in front of the heads of the ribs and in front of the posterior intercostal vessels and accompanying nerves. The thoracic parts of each of the trunks commonly have 11 or 12 separate ganglia of varying size; occasionally there are 10 or 13. The first thoracic ganglion is often fused with the inferior cervical sympathetic ganglion to form the *cervicothoracic*, or *stellate, ganglion*. The second thoracic ganglion is occasionally fused with the first. The remaining thoracic ganglia generally lie at the levels of the corresponding intervertebral discs. Sometimes the portion of the sympathetic trunk between two adjacent ganglia is double; at other times it is very slender. The sympathetic trunks enter the abdomen by piercing the crura of the diaphragm or by passing behind the medial arcuate ligaments.

The trunks and ganglia are connected with the ventral rami of the thoracic nerves by rami communicantes; they give branches to the adjacent viscera and blood vessels, and they send splanchnic nerves into the abdomen.

Preganglionic fibers for the heart and coronary vessels arise from the upper four to five or six segments of the thoracic part of the spinal cord, and synapse in the corresponding ganglia, as well as in cervical ganglia. Postganglionic fibers reach the heart by way of cervical, cervicothoracic, and thoracic cardiac nerves.

Preganglionic fibers for the bronchial tree and pulmonary vessels arise from the upper four to five or six segments of the thoracic part of the spinal cord. The postganglionic fibers leave the corresponding ganglia in direct branches to the pulmonary plexuses and in thoracic cardiac nerves.

Preganglionic fibers for the aorta arise from the upper segments of the thoracic part of the spinal cord. Postganglionic fibers reach the aorta and the proximal parts of its branches as direct branches of the upper thoracic ganglia.

Preganglionic fibers for the esophagus arise mainly from the lower segments of the thoracic part of the spinal cord. Postganglionic fibers reach the thoracic part of the esophagus by direct branches and by way of the greater splanchnic nerves.

Preganglionic fibers for the blood vessels, sweat glands, and arrectores pilorum of the thoracic and abdominal walls and back arise from all levels of the thoracic part of the spinal cord. Postganglionic fibers enter the spinal nerves by way of rami communicantes, and reach these structures by coursing in the dorsal and ventral rami, and in the meningeal branches of the spinal nerves.

Rami Communicantes. Each ganglion has one to four rami communicantes, which connect it with the corresponding nerve, and often with the nerves above and below. The ramus or rami containing the most postganglionic fibers tend to connect with the corresponding nerve. The rami containing the most preganglionic fibers are more oblique in direction (coming from the spinal nerve above or below), and they are more lateral in position (farther from the spinal cord).[25] Accessory sympathetic ganglia occur along the ventral rami and rami communicantes of the lower thoracic nerves (p. 782).

Preganglionic sympathetic fibers in the spinal nerves reach the sympathetic trunk by way of rami communicantes. Postganglionic fibers from the trunk and ganglia return to these nerves by way of rami communicantes. Sensory (pain) fibers from the thoracic and abdominal viscera pass through the sympathetic trunk into the rami communicantes, and thus reach the spinal nerves and the dorsal roots.

Visceral Branches. Cardiac branches arise from the ansa subclavia, the cervicothoracic ganglion, and the upper four or five thoracic ganglia. Branches of the upper four or five thoracic ganglia are also given to the pulmonary plexuses. Some filaments reach the aortic plexus, and a few enter the esophageal plexus. The major visceral branches are the three splanchnic nerves. The term splanchnic is also applied to certain visceral branches of the lumbar sympathetic trunks and of the sacral nerves.

GREATER SPLANCHNIC NERVE. This nerve is formed by three or four large roots and an inconstant number of smaller ones, which arise from the sympathetic trunk and ganglia, usually between the fifth and tenth ganglia. The roots run obliquely downward and forward toward the aorta, unite, pierce the crus,[26] and end in the celiac ganglia and plexuses. A rather large and relatively constant *splanchnic ganglion* and a variable number of smaller ganglia occur along this nerve near the diaphragm.

The greater splanchnic nerves give filaments to the aorta, esophagus, and pleura. One or both of the nerves is frequently closely applied to the aorta.

LESSER SPLANCHNIC NERVE. This nerve, which may be absent, is usually formed by one to three rootlets that arise from the lower thoracic ganglia. It descends slightly lateral to the greater splanchnic nerve, pierces the crus, and joins the aorticorenal ganglion and the celiac plexus. It often gives filaments to the splanchnic ganglion, and communicates with the renal and superior mesenteric plexuses. The greater and lesser splanchnic nerves are sometimes fused.

LOWEST SPLANCHNIC NERVE. This small nerve, which may be absent, usually arises from the last thoracic ganglion. It enters the abdomen at the medial side of the sympathetic trunk and joins the aorticorenal ganglion and adjacent plexus.

Composition of Sympathetic Trunk and Branches. The autonomic composition is considered in more detail later (p. 782). It should be emphasized here, however, that the sympathetic trunk and its branches also contain sensory fibers. Most of these are pain fibers from the thoracic and abdominal viscera and from blood vessels. These sensory fibers traverse the sympathetic trunks and rami communicantes to reach the spinal nerves and dorsal roots and thereby enter the spinal cord. Their levels of entry from a specific organ often correspond to the levels of outflow of sympathetic fibers to that organ.

Autonomic Plexuses

Many of the branches of the vagus nerves and sympathetic trunks that supply the viscera and blood vessels in the thorax intermingle and form plexuses in which the individual vagal fibers, or bundles of such fibers, lie side by side with individual sympathetic and sensory fibers, or bundles of such fibers, derived from the sympathetic trunks. The plexiform nature of the nerve supply to the viscera is accentuated by the connective tissue in which the nerves are embedded. Indeed, if the connective tissue is removed, the plexuses are much less evident and the various branches can often be traced directly to the organs they supply.

Cardiac Plexus. The nerves that supply the heart tend to converge in front of the lower part of the trachea, behind the arch of the aorta, but may be followed separately to the heart. A variable number of *cardiac ganglia,* some of which are fairly large, occur along the cervicothoracic nerves. On their way to the heart, the various nerves give fine twigs to the adventitia of the ascending aorta, the arch of the aorta, and the pulmonary trunk and arteries.

The cardiac plexus consists of the plexiform arrangement that is related to the trachea, arch of the aorta, and pulmonary trunk, and which, upon reaching the heart, forms right and left coronary and right and left atrial plexuses in the epicardium. The plexiform arrangement is also connected with the pulmonary plexuses.

Pulmonary Plexuses. Although these are often described as anterior and posterior, they are interconnected, and they are also connected with a portion of the cardiac plexus. As the vagus nerves descend to the roots of the lungs, they break up into several strong trunks, which are joined by branches from the sympathetic trunks and cardiac plexus. The larger trunks thus formed lie behind the root of the lung and constitute the posterior pulmonary plexus. A few lie in front of the root and constitute the anterior pulmonary plexus. Branches from these plexuses accompany the blood vessels and bronchi into the lungs.

Below the roots of the lungs, the plexuses unite into trunks which then enter the esophageal plexus.

Esophageal Plexus. This is a plexus of variable arrangement that is formed by the vagus nerves after they leave the pulmonary plexuses. Each vagus enters the fibrous layer of the esophagus and breaks up into a variable number of branches that intercommunicate and also join similar branches of the contralateral nerve. The vagus nerves lose their identity in this plexus. At the lower end of the esophagus, the part of the plexus in front of the esophagus collects into an *anterior vagal trunk* (or trunks). That behind the esophagus collects into a *posterior vagal trunk* (or trunks). Each trunk has fibers from both vagus nerves. Both continue through the esophageal opening in the diaphragm, the anterior trunk to the anterior surface of the stomach, the posterior to the posterior surface.

The esophageal plexus receives filaments from the sympathetic trunks and from the greater splanchnic nerves. Probably all of the fibers in these filaments supply the thoracic part of the esophagus; it is uncertain whether any sympathetic fibers in them accompany the vagal trunks into the abdomen.

The portion of the esophageal plexus in the upper part of the thorax and in the neck is much less distinct than that in the lower part of the thorax, where it consists largely of a plexiform arrangement of the vagus nerves on their way to the abdomen. The esophagus in the neck and upper part of the thorax receives parasympathetic fibers mainly by way of the recurrent laryngeal nerves.

Thoracic Aortic Plexus. The thoracic aorta receives filaments from the sympathetic trunks and vagus nerves. These filaments ramify in the adventitia and form a delicate plexus from which fine twigs can be traced a short way along the branches of the aorta. The plexus is continuous through the aortic opening of the diaphragm with the plexus of the abdominal aorta and also with the celiac plexus.

REFERENCES

1. Anomalies of the pulmonary vessels and their surgical significance are discussed by C. W. Findlay and H. C. Maier, Surgery, *29*:604, 1951.
2. H. W. Greig *et al.*, Quart. Bull. Northw. Univ. med. Sch., *28*:66, 1954.
3. W. W. Parke and N. A. Michels, Anat. Rec., *154*:185, 1966.
4. Various measurements of the arch of the aorta and its branches are provided by N. L. Wright, J. Anat., Lond., *104*:377, 1969.
5. R. C. Bahn, J. E. Edwards, and J. W. DuShane, Pediatrics, *8*:192, 1951.
6. J. E. Edwards *et al.*, Proc. Mayo Clin., *23*:333, 1948.
7. J. J. McDonald and B. J. Anson, Amer. J. phys. Anthrop., *27*:91, 1940. J. D. Liechty, T. W. Shields, and B. J. Anson, Quart. Bull. Northw. Univ. med. Sch., *31*:136, 1957. A. Pontes, *Artérias supra-aórticas*, thesis, University of Brazil, Rio de Janeiro, 1963.
8. R. E. Gross and P. F. Ware, Surg. Gynec. Obstet., *83*:435, 1946.
9. G. A. Seib, Amer. J. phys. Anthrop., *19*:39, 1934. D. Bowsher, J. Anat., Lond., *88*:400, 1954. C. H. Barnett, R. J. Harrison, and J. D. W. Tomlinson, Biol. Rev., *33*:442, 1958.
10. H. Nathan, Thorax, *15*:229, 1960.
11. O. V. Batson, Amer. J. Roentgenol., *78*:195, 1957. H. J. Clemens, *Die Venensysteme der menschlichen Wirbelsaüle*, Walter de Gruyter, Berlin, 1961.
12. R. Schobinger, Angiology, *11*:283, 1960.
13. O. F. Kampmeier, Anat. Rec., *19*:361, 1920.
14. H. C. Nohl, Thorax, *11*:172, 1956.
15. H. P. Nelson, J. Anat., Lond., *66*:228, 1932.
16. H. C. Nohl-Oser, Ann. R. Coll. Surg. Engl., *51*:157, 1972.
17. F. R. Sabin, Amer. J. Anat., *9*:43, 1909. H. K. Davis, Amer. J. Anat., *17*:211, 1915. S.-I. Jacobsson, *Clinical Anattomy and Pathology of the Thoracic Duct*, Almqvist & Wiksell, Stockholm, 1972.
18. A. Rosenberger and H. L. Abrams, Amer. J. Roentgenol., *111*:807, 1971.
19. P. Kinnaert, J. Anat., Lond., *115*:45, 1973.
20. R. H. Bell *et al.*, Quart. Bull. Northw. Univ. med. Sch., *28*:156, 1954.
21. H. A. Judson, Radiology, *30*:636, 1938.
22. M. W. Thornton and M. R. Schweisthal, Anat. Rec., *164*:283, 1969.
23. J. A. Capps, Arch. intern. Med., *8*:717, 1911. J. A. Capps and G. H. Coleman, *An Experimental and Clinical Study of Pain in the Pleura, Pericardium, and Peritoneum*, Macmillan, New York, 1932. Z. Cope, Brit. J. Surg., *10*:192, 1922. J. C. Hinsey and R. A. Phillips, J. Neurophysiol., *3*:175, 1940. F. S. A. Doran and A. H. Ratcliffe, Brain, *77*:427, 1954.
24. H. A. Teitelbaum, Anat. Rec., *55*:297, 1933. J. van Geertruyden, Arch. Anat., Strasbourg, *32*:221, 1949. R. G. Jackson, Anat. Rec., *103*:1, 1949.
25. J. Pick and D. Sheehan, J. Anat., Lond., *80*:12, 1946.
26. H. Loeweneck, H. J. Stork, and P. Loeweneck, Anat. Anz., *126*:531, 1970.

SURFACE ANATOMY, PHYSICAL EXAMINATION, AND RADIOLOGICAL ANATOMY

32

SURFACE ANATOMY

The sternal angle is an important landmark, which serves as a reference point in counting ribs. It is usually palpable and is sometimes visible about 5 cm below the jugular notch of the sternum. The second costal cartilage extends laterally on each side from the angle, and the second intercostal spaces can thereby be located below the cartilage and rib. Occasionally, however, the sternal angle is at the level of the third costal cartilages. In practice, it is usually simpler to count interspaces than ribs. The first rib is difficult to palpate, but the first intercostal space can be located just below the clavicle and can be used as a starting point in counting ribs if the sternal angle is absent. It is unreliable to count ribs from the twelfth upward, because the twelfth rib may be absent, or it may be too short to project lateral to the erector spinae muscle.

In young adult males, the jugular notch is at the level of about T.V. 3, and the xiphisternal joint at about T.V. 10 or 11. The manubrium is about 5 cm in length, and the sternal angle is at the level of T.V. 4 or 5. The levels are about one vertebra higher in women.

The infraclavicular fossa is a slight depression just below the lateral part of the clavicle.

In women, the breast overlies the pectoralis major. The base of the breast is more or less circular and covers an area from the second to sixth ribs, with the "axillary tail" extending into the axilla. The position of the mass of the breast is extremely variable. **The nipple is more constant in position in men; it lies in front of the fourth intercostal space or adjacent ribs, just lateral to the midclavicular line.**

In the anatomical position, the scapula is generally opposite the second to seventh ribs. The medial end of the spine of the scapula is usually opposite the spine of T.V. 3. The triangle of auscultation is described on page 106.

The surface anatomy of the thoracic vertebrae is described elsewhere (p. 546).

Lines and Planes of Reference

The median plane is the most important plane. The relative positions of deeper structures, such as the apical region of the heart, should be given in terms of distance from the median plane.

The midclavicular line[1] is commonly used as a reference. It extends downward from the midpoint of the clavicle. It is roughly equivalent to a vertical line from a point midway between the median plane and the lateral border of the acromion.

Other vertical lines include: (1) lateral sternal line, along the lateral margin of the sternum; (2) parasternal line, midway between the lateral sternal and midclavicular; (3) nipple line, lateral to the midclavicular line; (4) anterior axillary line, downward from the anterior axillary fold, which is formed by the lower margin of the pectoralis major; (5) midaxillary line, midway between anterior and posterior axillary lines, downward to the tubercle of the iliac crest; (6) posterior axillary line, downward from the posterior axillary fold, which is formed by the latissimus dorsi; (7) scapular line, through the inferior angle of the scapula in the anatomical position; (8) paravertebral line, through the tips of the transverse processes.

Internal Organs

Diaphragm. **In the erect position and midphase of respiration, the highest parts of the domes of the diaphragm are at about the same level as the apex of the heart, that is, the fifth intercostal space or sixth rib in the midclavicular line, and T.V. 10 or 11.** The right dome is commonly about 1 cm higher. During quiet breathing, the diaphragm undergoes an excursion of about 1/2 cm. During deep breathing, the excursion may be as much as 10 cm.

Trachea. **In the erect subject, the trachea divides at the level of T.V. 5 to 7.** The level is higher in the cadaver.

Pleura. The anterior border extends downward from the cupola, behind the sternoclavicular joint, to the middle of the sternal angle, and then to about the level of the xiphisternal joint. From here it turns laterally, as the inferior border. Its course is indicated by a line that crosses the eighth rib in the midclavicular line, the tenth rib in the midaxillary line, and is then directed toward the spine of T.V. 12 (for variations, see p. 285). The posterior border extends upward toward the cupola, about 2 to 3 cm from the median plane (but see retroesophageal recesses, p. 287).

Lungs. The anterior and posterior borders correspond approximately to the anterior and posterior borders of the pleura. **The inferior border, as determined by percussion, begins at about the level of the xiphisternal joint. It extends laterally, about two intercostal spaces higher than the inferior border of the pleura. It crosses the sixth rib in the midclavicular line, the eighth rib in the midaxillary line, and is then directed toward the spine of T.V. 10. During a deep inspiration, the apparent level descends at least two intercostal spaces.**

The right oblique fissure usually begins at the level of the head of the fifth rib; the left usually begins at a somewhat higher level. Each oblique fissure curves downward, following the line of the sixth rib and ending at the sixth costochondral junction, where it meets the inferior border. When the arm is abducted and the hand placed on the back of the head, the medial border of the scapula approximately indicates the oblique fissure.

The horizontal fissure begins at the oblique fissure near the midaxillary line, at about the level of the sixth rib. It extends forward in a highly variable fashion to the anterior border at the level of the fourth costal cartilage.

Heart. **The apex beat or point of maximal pulsation is usually felt in the fourth or fifth intercostal space, about 6 or 7 cm from the median plane (with considerable variation).** In the erect position, the lower border of the heart is below the level of the xiphisternal joint, and may be as much as 5 cm below.

Blood Vessels. The course of the pulmonary trunk may be indicated by a line extending from the center of the cardiac shadow to the left side of the sternal angle, behind the left second costal cartilage.

The ascending aorta extends upward and slightly to the right to the level of the sternal angle, just to the right of the median plane. The arch of the aorta lies behind the lower part of the manubrium of the sternum.

The brachiocephalic trunk is indicated by a line from the lower part of the manubrium to the right sternoclavicular joint. The left common carotid and left subclavian arteries are indicated by a line from the manubrium to the left sternoclavicular joint.

Each brachiocephalic vein is formed behind the sternoclavicular joint. The two veins unite to form the superior vena cava at the level of the sternal angle, behind the right second costal cartilage.

PHYSICAL EXAMINATION

The classical methods of physical examination of the living body, in the order in which they are employed, are inspection, palpation, percussion, and auscultation. It is important to remember that, in using these methods, similar points on the two sides of the chest should be compared systematically.

Respiratory System

Inspection. The chest should be inspected from the front, back, and sides, and also viewed from above downward by looking over the shoulders of the patient. The trachea and the apex beat should be sought, and their positions confirmed by palpation. The scapulae, clavicles, ribs, and sternal angle should be observed. The intercostal spaces are broader in front than they are behind. The subcostal angle is commonly between 70 and 110 degrees.

The expansion of the chest may be measured by placing a tape measure just below the nipples and instructing the patient to breathe in and out deeply. The chest should expand equally on the two sides. The degree of movement of the abdominal wall in breathing, as compared with movement of the chest, should be observed.

The rate of respiration is normally 11 to 14 per minute. It is faster in children (the neonatal rate is two to three times as fast) and slower in elderly people. It is about one-fourth of the pulse rate. When difficulty in breathing (dyspnea) is present, the sternocleidomastoids and the muscles of the external nose come into play.

Deformities of the chest, including depressions in the sternum, although they predispose to respiratory disease, are frequently present without disease of the thoracic contents. Thus nasal obstruction due to adenoids results in defective filling of the lungs that may cause alteration in the configuration of the chest.

Palpation. The hands should be warm and palpation should be gentle.

The position of the trachea should be confirmed by placing a finger above the jugular notch of the manubrium.

VOCAL FREMITUS. By placing the flat of the hand on the chest while the patient says "ninety-nine" or "one, one, one," the examiner can detect vibrations transmitted from the larynx through the trachea, bronchi, lungs, and chest wall.

Percussion. Vibrations can be set up in the tissues and organs of the body by means of a sharp tap. The middle finger of the left hand (pleximeter) is placed in firm contact with the chest wall and the other fingers of this hand are raised. A sharp tap is then made on the middle phalanx of this finger with either the index or middle finger of the right hand (plexor, or percussor). The percussor is kept perpendicular to the pleximeter. The movement takes place at the wrist, which should be released immediately after striking. Deeper structures and a larger mass of tissues can be set in vibration by heavier percussion.

Air-containing organs, especially the lungs, yield a note termed resonance. Extreme resonance, called tympany, may be encountered over the stomach and intestines. The note produced by percussion of solid organs, such as the liver and the blood-filled heart, is termed dullness.

The thorax is resonant, except over the heart and great vessels. The apices of the lungs may be examined by percussion of the clavicles.

Auscultation. That the breath and heart sounds can be heard more clearly and conveniently through some conductor was appreciated by René Laennec at the beginning of the 19th century. The stethoscope has undergone considerable modification since that time.

During quiet breathing, the character of the breath sounds should be listened to over many different areas, similar points on the two sides of the chest being selected for comparison. The upper lobes are examined below the clavicles and over the supraspinous fossae; the lower lobes over the back of the lower (seventh to tenth) ribs.

The breathing heard over most of the chest is termed "vesicular"; sometimes it is too faint to be heard except during deep respiration. The inspiratory sound is longer than the expiratory, the latter frequently being inaudible. Vesicular breathing is characterized by a soft, rustling sound, caused by the passage of air into the numerous alveoli.

The breathing over the region of the trachea and main bronchi is termed "bronchovesicular"; that is, the vesicular breathing is modified by a bronchial (a better term is laryngeal) or tubular element, a loud, harsh sound caused by vibrations of the vocal folds and adjacent pharynx, larynx, and trachea. The inspiratory and expiratory sounds are equal in length, and there is a pause between them. The laryngeal or tubular element may be dominant over solidified portions of diseased lungs or over cavities.

The breath sounds are harsher in children (puerile breathing) and after exercise.

VOCAL RESONANCE. Vibrations transmitted from the larynx to the chest wall while the subject says "ninety-nine" or "one, one, one" can be heard through a stethoscope. High-pitched sounds are appreciated through the stethoscope, whereas only low-pitched ones are palpable as vocal fremitus. The resonance is loudest over the region of the trachea and main bronchi. Vocal resonance is increased over solidified portions of diseased lungs and may sometimes be related to underlying cavities. In some instances even syllables of the whispered voice may be heard.

Narrowing or partial obstruction in the tracheobronchial tree may result in various types of adventitious sounds, the details of which are described in works on physical diagnosis.

Circulatory System

Inspection. The position of the apex beat may be determined in some people by inspection.

Palpation. **The heart rate is commonly 70 per minute, but considerable variation (50 to 90) occurs. The neonatal rate is twice as fast.**

The flat of the hand should be placed over the area of cardiac pulsation. The apex beat, if it can be detected, should then be localized by one or two fingers. The farthest inferolateral point at which the palpating fingers are lifted is generally used in measuring the distance of the apex beat from the median plane. **The apex beat is commonly 6 to 7 cm or more from the median plane and is generally in the fourth or fifth intercostal space.** It is not necessarily, however, within the cardiac area (fig. 32–1). The apex beat is often both visible and palpable, but it may be diffuse.

Vibrations communicated to the palpating hand from the heart or great vessels are termed "thrills," and are indicative of organic disease.

Percussion. The technique of percussion has already been described. Percussion has limited value in assessing the outline of the heart; accurate determination of cardiac size necessitates radiography. By careful percussion and after much practice, however, the position of the right and left borders of the heart may be ascertained roughly. Percussion is begun below the axilla and is carried out along an intercostal space. The pleximeter finger is usually kept parallel with the border of the heart, and it is moved medially a centimeter or so at a time. Before the right margin of the heart is percussed, the upper limit of liver dullness should be sought by percussing downward in each intercostal space until a change of note is found (usually in the fifth space). The pleximeter finger is then placed above this level and parallel with the sternum, and moved medially in an effort to locate the right margin of the heart.

Some authors distinguish (1) relative or deep cardiac dullness obtained by fairly heavy percussion and delimiting approximately the cardiac outline, and (2) absolute or superficial dullness obtained by light percussion and supposedly corresponding to an area of heart not covered by the left lung.

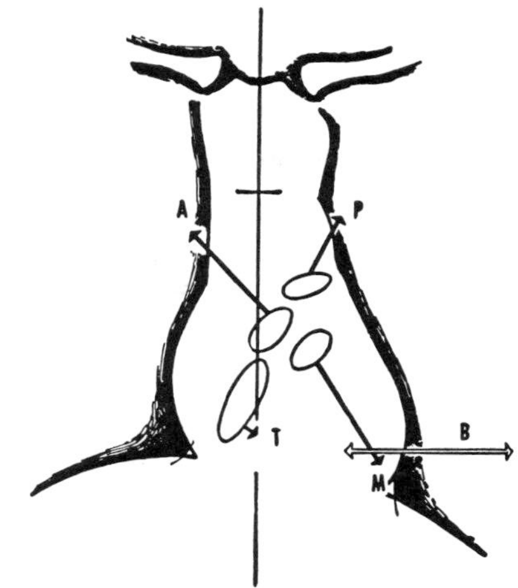

Figure 32–1 Areas of auscultation. Arrows indicate direction of conduction of heart sounds from the valves; the sounds are heard best at the indicated points. *A*, aortic; *P*, pulmonary; *T*, tricuspid; *M*, mitral. The arrow *B* indicates the common limits of the area in which the apex beat may be felt.

Auscultation. The rate and rhythm of the heart beat may be confirmed and compared with those of the pulse.

HEART SOUNDS. It is important to appreciate that considerable differences in the intensity and character of the heart sounds are found among normal individuals. Obesity and well-developed musculature are associated with decreased intensity, whereas the heart sounds are usually clearly audible in children.

The first heart sound is attributed to ventricular contraction and the closure of the atrioventricular valves, the second sound to closure of the aortic and pulmonary valves. A third sound, heard early after the second, is detected occasionally near the apical region. The sounds have been likened very approximately to the syllables "lubb-dupp," followed by a pause. They have been approximated in musical notation as follows:

The area of maximal intensity of the heart sounds for each valve corresponds not to the anatomical location of the respective valve but to the area where the cavity in which the valve lies is nearest the body surface and as far as possible from other valves, and which is "distal" to the valve with reference to the direction of flow in the blood stream. **The areas of maximal audibility (fig. 32–1) are:**

Pulmonary: over the left second intercostal space;

Aortic: over the right second intercostal space;

Mitral: over the apical region;

Tricuspid: over the lower part of the body of the sternum.

The first heart sound is usually loudest at the apical region, the second at the pulmonary and aortic auscultatory areas. In young persons, the second sound is louder at the pulmonary area than at the aortic, and the converse is true in elderly people. The heart sounds tend to be louder and sharper after excitement or exercise. Asynchronous closure of either the two atrioventricular valves or of the aortic and pulmonary valves may result in a reduplication of the first or second sound, respectively; reduplication is usually, but not always, pathological.

Apart from pericardial or pleural friction rubs, most additional sounds fall into the category of "murmurs," which arise at the valves or the great vessels. There are probably several factors involved in their production, but an important one is a disproportion between an orifice and the cavity into which the blood is pouring. Some murmurs are considered physiological, for example, systolic bruits over the pulmonary auscultatory area, which are frequent in healthy individuals and are most marked when the patient is lying down. Murmurs are classified according to the phase of the cardiac cycle during which they occur (diastolic murmur or systolic murmur), and the valve at which they occur (e. g., mitral diastolic murmur).

RADIOLOGICAL ANATOMY

LUNGS AND ESOPHAGUS

The chief x-ray methods used in the examination of the chest are fluoroscopy, radiography, tomography, and bronchography.

Fluoroscopy

Fluoroscopy permits a study of the pulsations of the heart and aorta and examination of the respiratory movements of the diaphragm and ribs, and it allows rotation of the patient during the procedure. The usual projections are those shown in figure 32–2. A definite order of procedure is recommended in fluoroscopy: first, a general survey of the chest; then a small-field study of the apices; inspection of the middle and lower lung fields in various projections; and finally a study of the heart and great vessels. The order of procedure, however, varies from one radiologist to another.

Radiography

The most frequently employed view in radiography of the chest is anterior (posteroanterior projection). The x-ray tube is located behind the erect patient and the film is placed vertically in front of the chest. The scapulae are excluded from the lung fields by abduction and medial rotation of the arms and by placing the arms around the cassette. The tube is centered opposite T.V. 4 and is preferably 2 meters from the film. The exposure (less than 1/10 second) is made while the patient holds his breath suspended after a deep inspiration. Stereoscopic anterior films are sometimes made. High-voltage techniques are used occasionally in order to render the clavicles, ribs, and heart less noticeable.

The following criteria have been proposed for assessing the quality of a general radiogram of the chest: (a) The outline of the vertebral column (the body of T.V. 3 in particular), but not the intervertebral spaces, should be barely visible through the cardiac shadow; (b) the clarity of the lung vessel pattern should be dis-

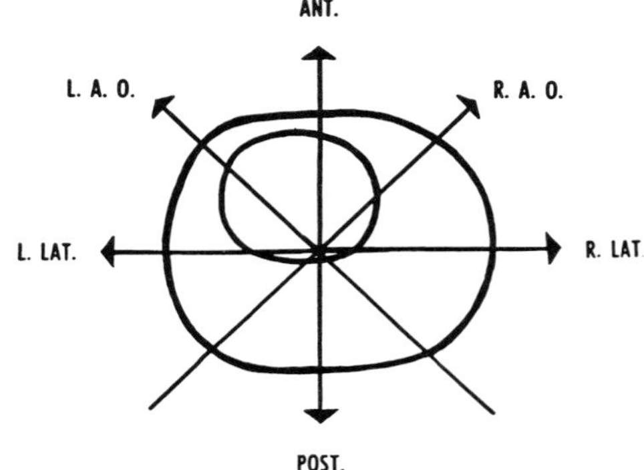

Figure 32–2 The usual views used in fluoroscopy and radiography are anterior, posterior, right anterior oblique (R.A.O.; fencing position), and left anterior oblique (L.A.O.; boxing position). Other views sometimes used are right and left lateral. Based on Zdansky.[2]

cernible clearly, even at the bases; (c) the ribs should be outlined clearly in the scapular region.

A definite order of inspection is recommended in the examination of a radiogram of the chest. For example:

Chest Wall (clavicles, sternum, scapulae, spine, ribs, breasts, soft tissues, such as the sternocleidomastoids). Variations and anomalies of the ribs may be present for example, cervical ribs, forked ribs.

Diaphragm. The diaphragm is not directly visible radiographically but, by virtue of the density of the liver below and the translucency of the lung above, the position of each dome is delimited. Hence diaphragmatic movements are readily followed fluoroscopically, and the shape and position of each dome can be determined in the living.

It is usual to refer to the arcuate density at the lower portion of the lungs as "the diaphragm." The edge of the shadow is the upper surface of the diaphragm, but the thickness of the shadow is composed of the diaphragm and of abdominal organs such as the liver and the spleen. The domes of the diaphragm are usually opposite T.V. 10 or 11, the right almost invariably being higher than the left. The diaphragm appears relatively lower on sitting (because the abdominal muscles are relaxed and the abdominal viscera sink somewhat), and higher on lying down. It is relatively higher also during youth, in women, and in the presence of obesity. It is also higher on the recumbent (fixed) side in the lateral decubitus position.

Cardiovascular Shadow and Trachea (see fig. 29–1, p. 285). The heart and the great vessels are discussed later. The trachea appears as a radiolucent band, usually in the median plane above, but a little to the right in its lower part. It extends from the level of C.V. 6 to about T.V. 5 to 7. The bifurcation is approximately one vertebra higher in the supine position, two vertebrae higher in an infant. Its level varies slightly with the phases of respiration.

Lungs, Pleurae, Hili. The main constituent of the hilar shadows and of the lung markings is the pulmonary vessels (see fig. 29–1). The branches of the pulmonary arteries are chiefly responsible, and they follow the distribution of the bronchial tree. The vascular pattern varies in prominence from one individual to another. The bronchi and lung tissue are radiolucent, but the main bronchi may frequently be recognized within the dense hilar shadows. The hilar shadows are produced by vascular, bronchial, lymphatic, and connective tissue components. Normal hilar lymph nodes, however, are not visible as such. The interlobar fissure between the upper and middle lobes is occasionally discernible on the right side of the body. Right and left lung fields should be equally clear.

The normal pleura is visible radiographically only when an extensive area of the plane of a layer of pleura is in the plane of the x-ray beam.

The lobule of the azygos vein is a medial portion of the apex of the right lung which, when present, is partly separated from the remainder of the upper lobe by a pleural septum (mesoazygos). This septum varies in position among different people and may be horizontal, oblique, or vertical. It may be detected radiographically close to the right border of the shadow of the superior mediastinum, and typically its shadow has the shape of an inverted tear-drop or a comma.[3]

For diagnostic convenience the lungs, as seen in an anterior radiogram, may be considered to present three or four arbitrary fields each: apical, upper, middle, and lower. The lines of division used are the clavicles and horizontal lines through the front portions of the second and fourth ribs. These topographic areas are useful in recording the presence of lesions when it is not possible to commit oneself to a lobar or a segmental localization. It should be kept in mind that about one-quarter of the total lung fields is not visible on conventional films because it is obscured by the cardiovascular shadow and by subdiaphragmatic structures.

Certain differences are found between the two sides of the body. On the right side, the diaphragm and the liver are higher. On the left side, the cardiac outline is prominent and may present a so-called "apex"; the aortic knuckle is visible, and there may be air in the fundus of the stomach or in the colon. The rare conditions of dextrocardia and situs inversus viscerum, however, should be kept in mind.

Esophagus (see fig. 28–4, p. 282). The esophagus presents a series of impressions, all of which are concave to the left in both anterior and right anterior oblique views. Some of these impressions may be detected radiographically when the esophagus is visualized during the swallowing of a barium paste. The impressions are due to the arch of the aorta, the left bronchus and right pulmonary artery (this impression is seldom visible), and the descending aorta just above the diaphragm. Because of the intimate relation of the esophagus to the aorta and to the left atrium, contrast filling of the esophagus is used for detecting enlargement of these structures.

Tomography

Tomography may be employed to show the carina free of surrounding structures, the bronchi in the hili, or the azygos vein. In pathological conditions, tomography may sometimes be of considerable use in the detection of pulmonary lesions.

Bronchography

The bronchi are generally visible only with the aid of special techniques. Bronchography is the procedure whereby the bronchial tree is demonstrated radiographically after being outlined by a contrast medium. In this procedure, an iodine compound is injected by way of an intratracheal catheter. The catheter may be introduced either through the mouth or through the nose. The most frequently employed views are the lateral (for either lung), the left anterior oblique (for the right lung), and the right anterior oblique (for the left lung). Examples of bronchograms are given in figures 29–11 to 29–13 (pp. 295–297), and the arrangement of the bronchi with reference to the individual bronchopulmonary segments is shown in figure 29–10 (p. 294).

HEART

The chief radiographic methods used in the examination of the heart and great vessels are fluoroscopy, radiography and teleradiography, and special methods such as orthodiagraphy, kymography and electrokymography, cardiac catheterization, angiocardiography, and aortography.

Fluoroscopy

Fluoroscopy permits a plastic conception of the size and shape of the heart to be attained, and allows the recognition of pulsations and movements.

Radiography and Teleradiography

The most frequently employed view in radiography of the heart is anterior (posteroanterior projection); the x-ray tube is located behind the erect subject and the film is placed vertically in front of his chest. The exposure is made at the end of a full but not forced inspiration. The phase of the cardiac cycle during which the exposure is made is generally disregarded. Short exposure times are necessary to obtain a sharp image of the heart and great vessels.

In teleradiography (Gk. *tele-*, far, as in telescope and telephone), the x-ray tube is placed at least 2 meters from the film. Magnification is reduced considerably by this procedure, generally to about 5 per cent.

Special Methods

Cardiac Catheterization. A radio-opaque catheter is introduced into a peripheral vein and passed under fluoroscopic control into the right side of the heart (right atrium, right ventricle, pulmonary trunk, and a pulmonary artery, successively). Intracardiac pressures can be recorded, and samples of blood removed. As an alternative route, the catheter may be passed directly into the left atrium through a bronchoscope or an esophagoscope.

Angiocardiography (Gk. *angeion*, vessel, as in angiology). The passage of a radio-opaque medium (an iodine compound) injected into a peripheral vein is followed through the heart and great vessels by means of serially exposed radiograms (see figs. 30–4 and 30–5, pp. 306, 307). Alternatively, cineradiography may be employed. The medium first fills the right side of the heart and later the left. Angiocardiography permits a study of the circulation through the intact, functioning heart, and is particularly useful in the investigation of congenital cardiac anomalies.

Aortography. The passage of a radio-opaque medium injected into the aorta (ascending, arch, descending thoracic, or abdominal) is followed through its branches by means of serially exposed radiograms (fig. 41–1, p. 450). The renal arteries and their branches may be demonstrated by aortography.

The Heart and Cardiovascular Shadow

In an anterior view, the borders of the cardiovascular shadow are generally produced as follows (figs. 32–3 and 32–4). The

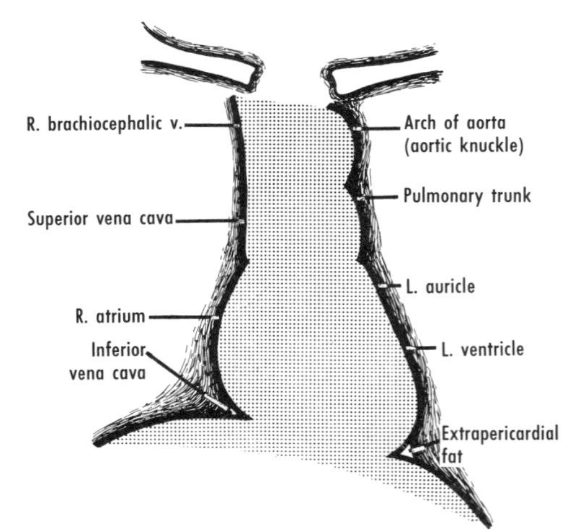

Figure 32–3 Composition of margins of cardiovascular shadow. Based on Zdansky.[2]

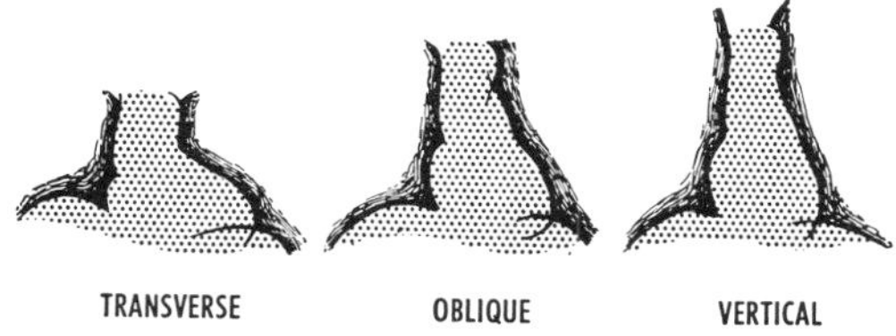

Figure 32–4 Types of cardiovascular shadows.

right border: right brachiocephalic vein, superior vena cava (occasionally the ascending aorta), right atrium (right ventricle sometimes), and inferior vena cava or right hepatic vein. The left border: arch of the aorta (forming a prominence known as the aortic knuckle, or knob), pulmonary trunk (conus arteriosus rarely), left auricle and left ventricle, and an extrapericardial pad of fat.

The lower left part of the cardiac silhouette may present what is best termed "the region of the apex," because a number of hearts do not possess an anatomical apex.[4] Of those that do, the radiographic "apex" is usually below the level of the shadow of the diaphragm.[5] The relationship of the apex beat to the cardiac silhouette[6] is discussed on page 308.

Three chief types of cardiovascular shadows are commonly described (fig. 32–4): the transverse type, characteristic of the obese, the pregnant, and infants; the oblique type, found in most persons; and the vertical type, present in people with a narrow chest.

The heart appears relatively larger in infancy and childhood (partly owing to rotation associated with the large size of the liver). It appears very large at the end of expiration. The upper part of the cardiovascular shadow is superimposed on the shadow of the thymus, the diaphragm is high, and an aortic knob is not visible. During the first month of infancy, the heart displays right ventricular dominance, as a result of which the cardiac outline resembles a "sheep's nose," or a wooden shoe (*coeur en sabot*).

In a right anterior oblique (R.A.O., fig. 32–5) projection, the back of the heart lies in

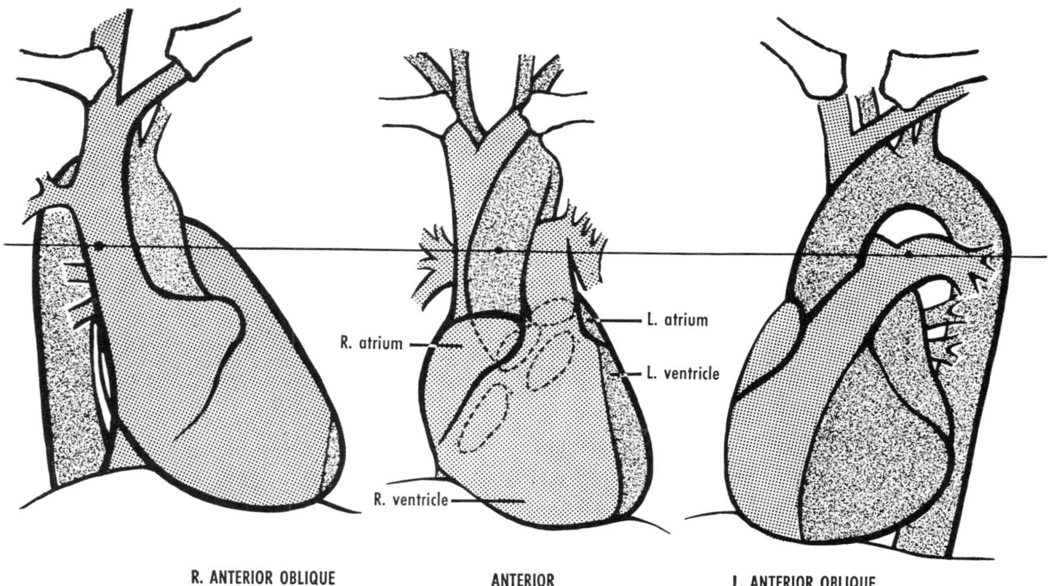

Figure 32–5 The heart and great vessels. The horizontal line indicates the level of bifurcation of the trachea. Based on Zdansky.[2]

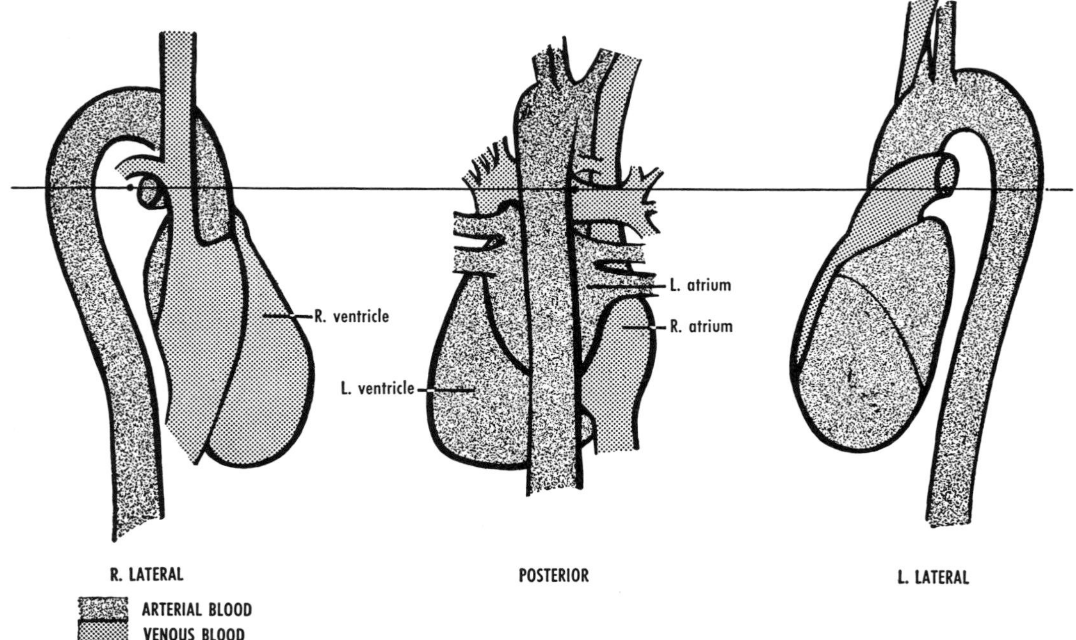

Figure 32–6 The heart and great vessels. The horizontal line indicates the level of bifurcation of the trachea. Based on Zdansky.[2]

front of the spine. Below the carina and pulmonary vessels, the posterior border of the cardiovascular shadow is formed by the left atrium and the inferior vena cava. A clear area behind the left atrium, the retrocardiac space, contains the esophagus. Another clear area, the prevertebral window or retrovascular space, is visible behind the trachea. The anterior border of the cardiovascular shadow is formed by the ascending aorta, the pulmonary trunk and conus arteriosus, the right ventricle, and the apical region.

In a left anterior oblique (L.A.O., fig. 32–5) projection, the back of the heart shadow lies clear of the spine. The anterior

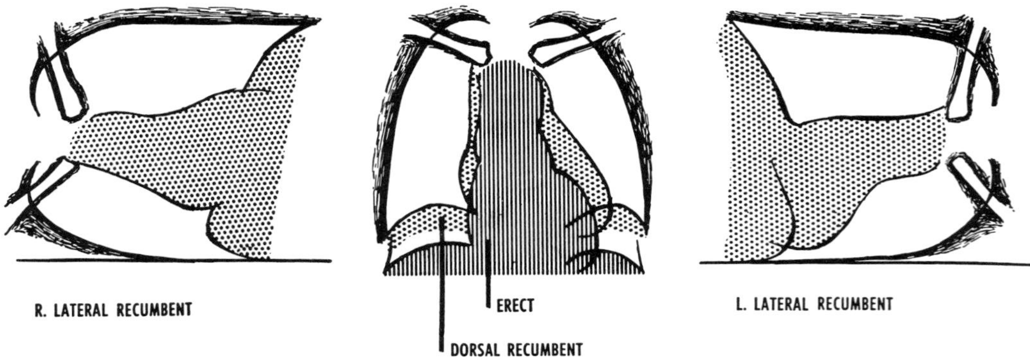

Figure 32–7 Variations in shape and position of the heart according to posture. In the middle view, note that, on assuming the dorsal recumbent position, the heart appears to increase in size. In the lateral recumbent position, the mediastinum sinks to one side, owing to the effect of gravity. The diaphragm on the lower side ascends, whereas on the upper side it descends. Based on Zdansky.[2]

border of the cardiovascular shadow is formed by the superior vena cava, the ascending aorta, the right atrium, and right ventricle. Frequently the arch of the aorta is well seen and a clear area, the "aortic window," is present below it. The posterior border of the cardiovascular shadow is formed slightly by the left auricle but chiefly by the left ventricle. In an L.A.O. view, the rays are frequently in the plane of the interventricular septum.

In lateral views (fig. 32–6), the right ventricle is closely related to the back of the sternum. Above this, the front of the silhouette is formed by the conus arteriosus and pulmonary trunk, and the ascending aorta. In front of these structures the retrosternal space indicates the location of the anterior mediastinum. The entire course of the thoracic aorta can sometimes be seen in a lateral view. The posterior part of the cardiovascular shadow is formed chiefly by the left atrium, also by the left ventricle, and sometimes by the inferior vena cava.

Position of Heart. Cardiac position and configuration depend chiefly on the diaphragm, and the position of the diaphragm depends mainly on posture[7] and respiration. In the erect position (fig. 32–7), the diaphragm descends and the heart rotates. The heart then presents a decrease in its transverse diameter, in its frontal area, and in its volume. In the erect position, the heart lies at the level of T.V. 7 to 10. In recumbency it rises about one vertebra. In the erect position, the lower border of the heart may be 5 cm or more below the xiphisternal joint in the median plane.[8]

During inspiration the heart appears more vertical (fig. 32–8). As a result, the hili of the lungs are more readily seen. The radiological appearance of the heart and lungs varies widely in the same individual during different phases of respiration (fig. 32–9).

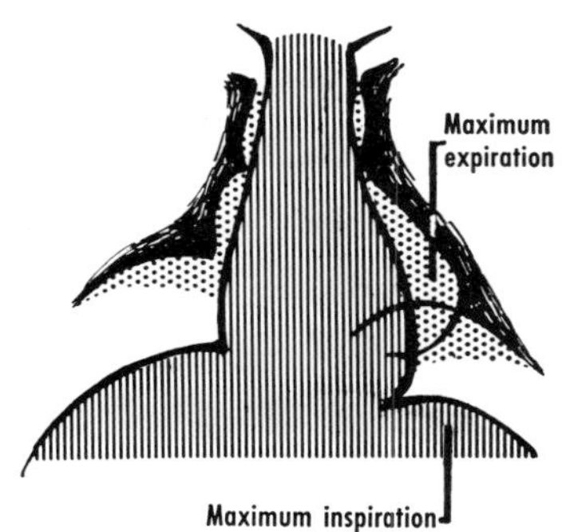

Figure 32–8 The shape of the heart at maximum inspiration and maximum expiration.

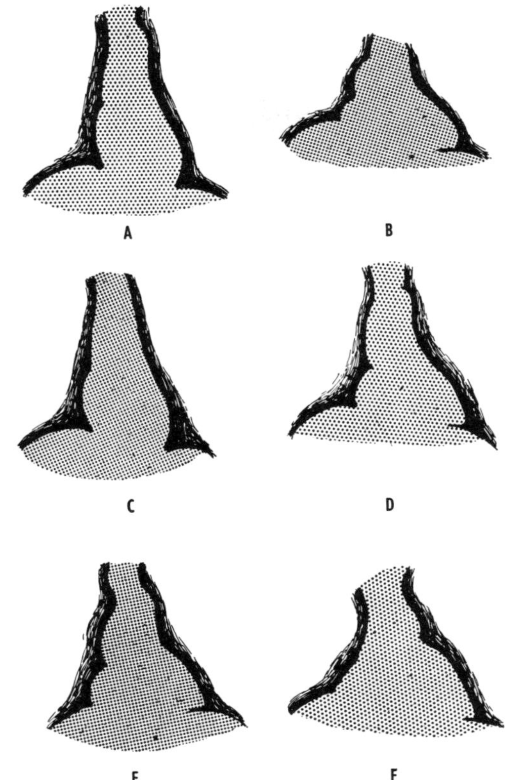

Figure 32–9 Changes in shape of the cardiovascular shadow of one person during various respiratory procedures. *A*, end of a normal inspiration, breath held without forcing. *B*, end of a normal expiration, breath held without forcing. *C*, great positive pressure developed by closing the glottis after inspiration, and forcibly attempting expiration (Valsalva test). The intrathoracic pressure is thereby increased, causing a decrease in blood flow into, and a reduction in size of, the heart. *D*, the glottis closed at the end of normal expiration, and expiration forced. *E*, the glottis closed at the end of normal inspiration, and extreme effort made to inspire. *F*, the glottis closed at the end of normal expiration, and extreme effort made to inspire (Müller's test). The intrathoracic pressure is thereby decreased, causing an increase of blood flow into, and an increase in size of, the heart. Based on Crowden and Harris.[9]

Heart Size. In clinical practice the size of the heart is estimated by assessing an anterior teleradiogram in terms of the individual and his habitus (p. 305).

REFERENCES

1. D. A. Rytand, Ann. int. Med., *69*:329, 1968.
2. E. Zdansky, *Roentgen Diagnosis of the Heart and Great Vessels*, trans. by L. J. Boyd, Grune & Stratton, New York, 1953.
3. L. E. Etter, Amer. J. Roentgenol., *58*:726, 1947. B. J. Anson *et al.*, Quart. Bull. Northw. Univ. med. Sch., *24*:285, 1950.
4. T. S. Keith, Lancet, *1*:1466, 1936.
5. Otten, cited by F. M. Groedel, *Lehrbuch und Atlas der Rontgendiagnostik in der inneren Medizin und ihren Grenzgebieten*, Flehmann, Munich, part 1, 1936.
6. R. O'Rahilly, Amer. Heart J., *44*:23, 1952.
7. J. E. Habbe, Amer. J. Roentgenol., *76*:706, 1956.
8. D. Mainland and E. J. Gordon, Amer. J. Anat., *68*:457, 1941.
9. G. P. Crowden and H. A. Harris, Brit. med. J., *1*:439, 1929.

GENERAL READING

Bailey, H., *Demonstrations of Physical Signs in Clinical Surgery*, Wright, Bristol, 14th ed., 1967.

Chamberlain, E. N., *Symptoms and Signs in Clinical Medicine*, Wright, Bristol, 7th ed., 1961.

Delp, M. H., and Manning, R. T., *Major's Physical Diagnosis*, Saunders, Philadelphia, 8th ed., 1975.

Part Five

THE ABDOMEN

ERNEST GARDNER

Introduction

The trunk comprises the thorax, the abdomen, the pelvis, and the back. The abdomen (proper) lies between the thorax and the pelvis. **The abdominal cavity (proper) is separated from the thoracic cavity above by the diaphragm, and from the pelvic cavity below and behind by an arbitrary plane passing through the terminal lines on the bony pelvis (fig. 40–1, p. 439). A considerable part of the abdominal cavity lies under cover of the thoracic bony cage.**

The abdomen is considered by some to include the pelvis, and the abdominal cavity to include the pelvic cavity. The pelvic cavity projects backward from the abdominal cavity at nearly a right angle.

The abdominal cavity contains most of the organs of the digestive system (stomach, intestine, liver, pancreas), part of the urogenital system (kidneys, ureters), the spleen, the suprarenal glands, and parts of the autonomic plexuses. It also contains the peritoneum, which is the great serous membrane of the digestive system. Many abdominal organs may lie partly or temporarily in the pelvis, and pelvic organs may at times be abdominal in position.

The abdominal wall consists, in front, of the rectus abdominis and pyramidalis muscles and the aponeuroses of three muscles (external oblique, internal oblique, transversus). The sides are formed by these three muscles, and, in part, by the iliacus muscles, and by the hip bones. Behind, the abdominal wall is formed by the bodies of, and the discs between, the five lumbar vertebrae, the crura of the diaphragm, the psoas major and minor muscles laterally, and, still more laterally, the quadratus lumborum, and, in part, the iliacus muscles and the iliac bones (fig. 39–2, p. 433). Most of the abdominal wall is arranged in layers.* These layers, which have surgical importance,† are, from without inward: (1) skin, (2) subcutaneous tissue, (3) muscles and fasciae, or bone, (4) extraperitoneal tissue, and (5) peritoneum. The abdominal wall may be the site of certain congenital defects, such as ventral hernias (p. 360).

The abdominal wall accommodates well to the expansion imposed by pregnancy, or by the continued deposition of fat, and has been known to expand enormously from slowly growing abdominal or pelvic tumors or from excessive obesity. The muscles thin out, but the skin grows, and nerves and blood vessels lengthen. Reddish lines known as striae gravidarum are sometimes seen in the skin of the abdomen during pregnancy. After parturition, these striae gradually change into thin, silvery, scarlike lines, the lineae albicantes. Lineae albicantes may also occur in men, however, and in the skin of the thighs of both men and women.

---

*C. E. Tobin, J. A. Benjamin, and J. C. Wells, Surg. Gynec. Obstet., *83*:575, 1946. M. A. Hayes, Amer. J. Anat., *87*:119, 1950.

†V. L. Rees and F. A. Coller, Arch. Surg., Chicago, *47*:136, 1943. E. W. Lampe, Surg. Clin. N. Amer., *32*:545, 1952.

ABDOMINAL WALLS

33

ANTEROLATERAL ABDOMINAL WALL

MUSCLES

The muscles on each side are two in front, the rectus abdominis and pyramidalis, and three anterolaterally, the external oblique, internal oblique, and transversus. Each of the three anterolateral muscles is sandwiched between thin layers of fascia. A variable amount of subcutaneous tissue lies between the skin and the external oblique, and a variable amount of extraperitoneal tissue between the transversalis fascia and the peritoneum.

The subcutaneous tissue contains fat, especially in its superficial part. Its deeper part tends to be more collagenous, and therefore more membranous, especially in the inguinal region (fig. 33–9, p. 359). The superficial part is continuous with similar tissue in the thigh and with the superficial fatty layer of the superficial perineal fascia (p. 495). The deeper membranous part is fastened along the inguinal ligament, and to the fascia lata for about a finger-breadth below the inguinal ligament, and is fused with the linea alba in the median plane. It is prolonged from the pubic symphysis onto the dorsum of the penis as the *fundiform ligament.* It is continuous with the dartos of the scrotum and with the membranous layer of the superficial perineal fascia. When the urethra is ruptured below the urogenital diaphragm, urine may spread throughout the anterior part of the perineum, and may then infiltrate the subcutaneous tissue and spread over the anterior abdominal wall. Its spread into the thigh is limited by the attachment of the membranous part of the subcutaneous tissue to the fascia lata.

The fascia of the abdominal wall is the thin, investing layer of the external oblique. It is continued medially over the aponeurosis of the external oblique to the linea alba. Below, at the superficial inguinal ring it fuses with the fascia on the deep surface of the external oblique and is prolonged onto the spermatic cord as the external spermatic fascia (p. 360).

External Oblique (figs. 33–1 and 33–2). The most superficial of the three anterolateral muscles, this arises by a series of fleshy slips from the external surfaces of

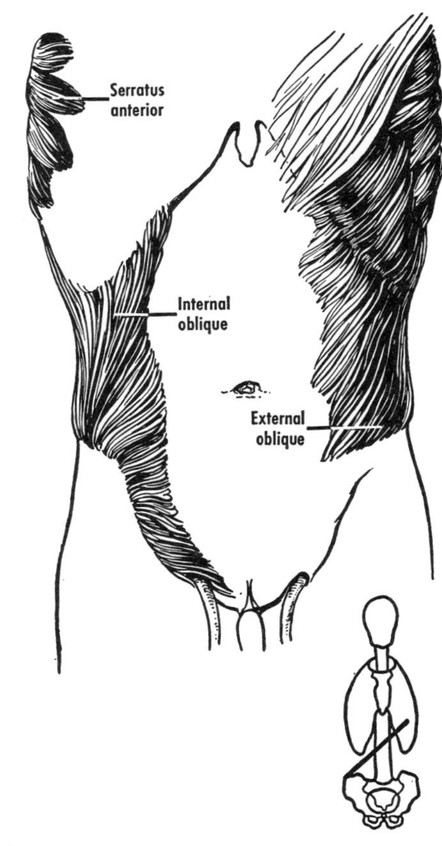

Figure 33–1 External and internal obliques. The fibers of one external oblique are roughly parallel to those of the opposite internal oblique. The lower diagram indicates the line of pull of the left external oblique and the right internal oblique. Acting together, these muscles flex and rotate the trunk.

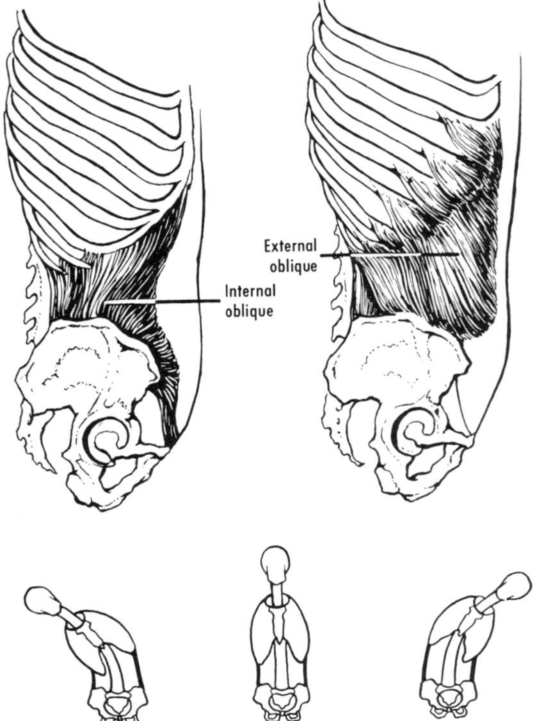

Figure 33–2 *Upper left*, internal oblique; *upper right*, external oblique. Lower figures show that the external oblique and internal oblique of one side act together in bending the trunk toward that side.

the lower eight ribs. The slips interdigitate with those of the serratus anterior and latissimus dorsi, and are often fused with the external intercostal muscles. The slips blend, their fibers running downward and medially. The fibers of the lower and posterior part of the muscle descend vertically to the external lip of the iliac crest. At or above the spinoumbilical line (anterior superior iliac spine to umbilicus) the fibers of the rest of the muscle give way to a thin, strong aponeurosis, which becomes continuous with the aponeurosis of the opposite muscle at the linea alba. **The lower edge of the aponeurosis extends, as the inguinal ligament, from the anterior superior iliac spine to the pubic tubercle** (p. 360). Medially, the aponeurosis fuses with that of the underlying internal oblique. In the upper part of the abdomen, this fusion takes place near the lateral border of the rectus. Near the pubis, however, the line of fusion is near the linea alba. At this level, therefore, the aponeurosis of the external oblique contributes little to the anterior layer of the sheath of the rectus.

The posterior border of the external oblique muscle is usually free and forms an angle with the lateral border of the latissimus dorsi. This angle is converted into the *lumbar triangle* by the iliac crest, which gives attachment to these two muscles in front and behind, respectively (fig. 13–2, p. 105). The floor of the lumbar triangle is formed by the internal oblique.[1]

NERVE SUPPLY. The thoracoabdominal nerves and the subcostal nerve.

Internal Oblique (figs. 33–1 and 33–2). This muscle arises from the thoracolumbar fascia (thus indirectly from the spinous and transverse processes of the lumbar vertebrae), by muscular fibers from the intermediate line of the iliac crest, and from the fascia iliaca.

The muscle fibers radiate (1) upward and medially to the lower three ribs, where they are continuous with the lower three internal intercostal muscles; (2) upward and medially to an aponeurosis that is attached above to the costal margin, and below is directed horizontally to the sheath of the rectus; and (3) downward toward the pubis. The relationship of the internal oblique to the inguinal canal is described later (p. 360).

The part of the internal oblique that arises from the iliac crest is often thicker than the rest of the muscle, and may be split into anterior and posterior parts.[2] The posterior part has been termed an accessory internal oblique.

NERVE SUPPLY. The lower two or three thoracoabdominal nerves and the subcostal nerve. Branches from the iliohypogastric and ilioinguinal nerves may be present, but whether they contain motor fibers is uncertain.

Transversus Abdominis (fig. 33–3). This muscle arises from the fascia iliaca, the internal lip of the iliac crest, and the thoracolumbar fascia, and from the inner surfaces of the lower six costal cartilages, where the fibers interdigitate and often fuse with the diaphragm. The muscle fibers run more or less horizontally, although the lowermost ones incline downward and run parallel to those of the internal oblique. The muscle fibers end in an aponeurosis that contributes to the sheath of the rectus. The uppermost muscle fibers course behind the rectus before giving way to the aponeurosis, which then extends to the xiphoid process. The arrangement of the lowermost part of the aponeurosis is described later (p. 358).

The fascia on the inner surface of the transversus serves as the epimysium and forms a fairly distinct layer, which is termed *transversalis fascia*.[3] The transversalis fascia is a part of the inner investing fascia of the abdominal wall and as such is continuous with the transversalis fascia of the other side behind the rectus abdominis and rectus sheath. It is continuous with the fascia iliaca, the diaphragmatic fascia, the parietal pelvic fascia, the thoracolumbar fascia at the edge of the quadratus lumborum, and the anterior layer of the femoral sheath. Some investigators, however, consider the transversalis fascia to be a part of the extraperitoneal connective tissue.[4]

NERVE SUPPLY. The thoracoabdominal nerves and the subcostal nerve. As in the case of the internal oblique, the question of a possible motor supply by the iliohypogastric and ilioinguinal nerves remains unsettled.

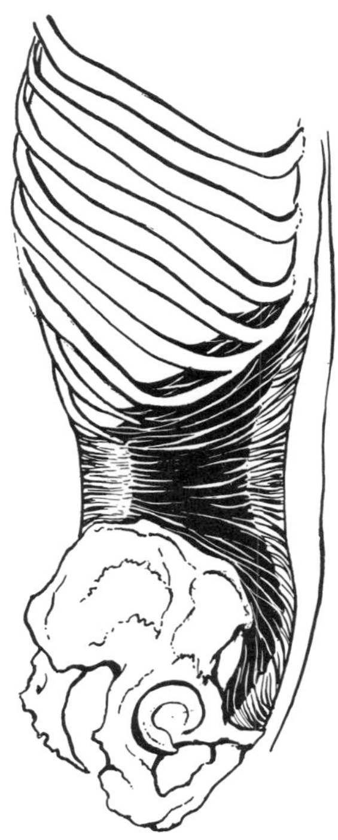

Figure 33–3 The transversus abdominis.

Rectus Abdominis (figs. 33–4 and 33–5). This is a long, thin, relatively wide muscle that is attached above to the front of the xiphoid process and to the fifth to seventh costal cartilages, and below to the pubic crest and symphysis. Three or more *tendinous intersections* traverse the muscle in front and fuse with the anterior layer of the sheath. The medial edge of the upper part of each rectus is attached to the linea alba.

NERVE SUPPLY. The thoracoabdominal nerves and the subcostal nerve.

Linea Alba and Sheath of Rectus (figs. 33–4 and 33–5). The aponeurosis of the external oblique passes in front of the rectus abdominis. The rectus sheath is formed chiefly by the aponeuroses of the internal oblique and the transversus. The two aponeuroses meet at the lateral edge of the rectus along a curved line termed the *linea semilunaris*. From the level of the xiphoid process downward for a variable distance, the transversus abdominis passes behind the rectus. Below, however, the transversus aponeurosis lies in front of the rectus. The lower limit of the part that lies behind the rectus forms a crescentic border, the *arcuate line* (or *linea semicircularis*). The level at which the arcuate line occurs is variable, and the change in course of the transversus aponeurosis may be abrupt or gradual. In the latter instance, secondary lines may be present. Below the level of the arcuate line, the transversalis fascia separates the rectus from the extraperitoneal connective tissue.

Beginning just below the level of the xiphoid process, the aponeurosis of the internal oblique divides into an anterior and a posterior layer. The anterior layer passes in front of the rectus to the *linea alba*,[5] which is a tough, tendinous raphe of interlacing fibers, and which extends from the xiphoid process to the pubic symphysis. The anterior layer is fused with the overlying external oblique aponeurosis. The posterior layer passes behind the rectus to the linea alba and is fused with the transversus aponeurosis. The division of the internal oblique aponeurosis into anterior and posterior layers is absent in the

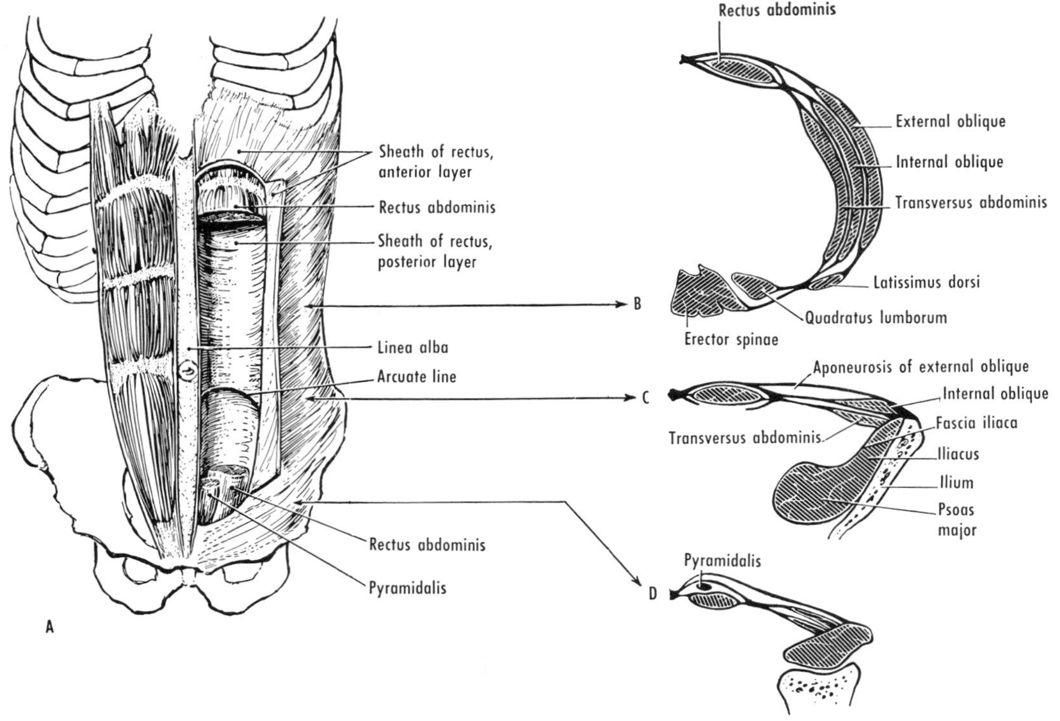

Figure 33–4 Muscles of the abdominal wall. *A*, the rectus abdominis. Arrows indicate levels of diagrams *B* to *D*, which show a common arrangement of the sheath of the rectus as seen in horizontal section. The transversalis fascia is not shown separately from the transversus aponeurosis.

lowermost part of the abdomen, where the aponeuroses of all three muscles pass in front of the rectus to the linea alba.

The formation of the rectus sheath is subject to considerable variation with respect to the levels of divisions and their arrangements,[6] and also to the role of the accessory internal oblique, which may determine the contribution of the internal oblique aponeurosis to the posterior layer of the rectus sheath.

In summary, according to the classical description, the sheath of the rectus consists of an *anterior* and a *posterior layer.* Above the level of the arcuate line, the anterior layer is formed by the aponeurosis of the internal oblique, together with the aponeurosis of the external oblique medial to its fusion with the internal. Below the arcuate line, the anterior layer is formed by the aponeuroses of the internal oblique and the transversus, together with the external oblique near the linea alba. At the level of the xiphoid process, the posterior layer is formed by the transversus abdominis and its aponeurosis. Below this level, to the arcuate line, it is formed by the aponeuroses of the internal oblique and the transversus. The anterior and posterior layers fuse in the median plane with those of the opposite side to form the linea alba.

Pyramidalis. The lower part of the anterior wall of the sheath splits and encloses the pyramidalis muscle, which extends upward from the body of the pubis to the linea alba. Unimportant and often absent, it is supplied by the subcostal nerve.

Actions. **The muscles of the abdominal wall protect the viscera and help to maintain and to increase intra-abdominal pressure. They move the trunk and help to maintain posture.**

The chief function of the rectus abdominis is flexion of the trunk against resistance and, in supine positions, lifting of the chest and, indirectly, of the head.[7] The muscle may be tested by having a supine subject flex his trunk without using his arms. The rectus also plays a role in breathing (p. 277) and in straining. Its sheath serves as a retinaculum and prevents the muscle from bowstringing.

The contraction of the external and internal obliques, of the transversus, and of the muscles of the pelvic diaphragm provides a tense abdominal wall and pelvic floor that resist the downward pressure of the thoracoabdominal diaphragm during straining and coughing. The combined actions of these muscles can result in a considerable increase in intra-abdominal pressure. The muscles are therefore important in respiration, defecation, micturition, parturition, and vomiting. The lower parts of the internal oblique and the transversus, those that are attached to the iliac crest and the fascia iliaca, are attached wholly to the pelvis and cannot act on the trunk. Their functions, aside from their effects on the inguinal canal, are therefore more or less restricted to bringing about changes in intra-abdominal pressure.

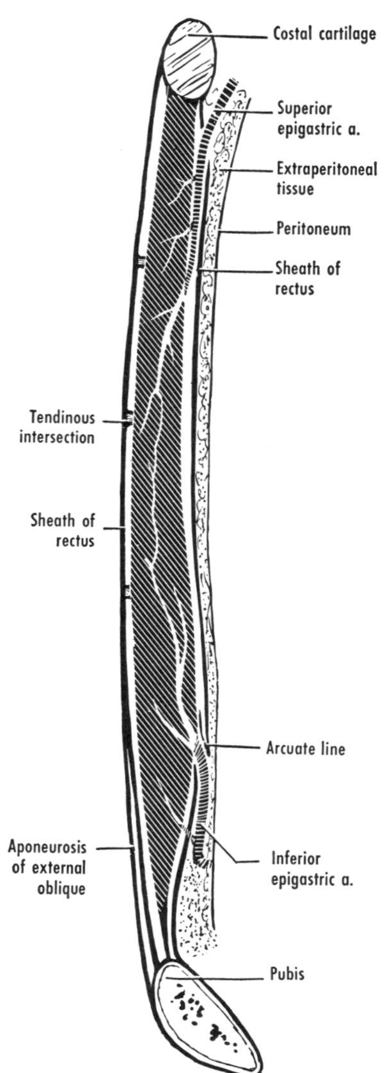

Figure 33–5 Diagram of a sagittal section of the rectus abdominis and its sheath. The fascial layer behind the rectus abdominis, below the inferior epigastric artery, is the transversalis fascia. It has not been indicated as a separate layer above the artery.

The muscles of the abdominal wall, although inactive during quiet breathing, are the most important muscles of forced expiration. They contract at the end of a maximal voluntary inspiration and thus limit it, they are active during the expiratory phases of breathing if the ventilation is great, and they are active during forced expiration and coughing (p. 277).

When the external obliques contract simultaneously, they aid the rectus abdominis muscles in flexing the trunk. This flexing action would occur during activities such as straining or weight lifting except for the counterbalancing contraction of the erectores spinae, which extend the trunk (p. 529). The oblique muscles aid the muscles of the back in rotation of the trunk (p. 539), the external oblique of one side working with the opposite internal oblique. The muscles on one side aid in bending the trunk to that side (lateral flexion). They also help to maintain balance when leaning to the opposite side or when standing on one leg. In the erect position, there is a moderate amount of activity in the obliques.

All of the abdominal muscles control hyperextension of the trunk and help to fix the thoracic bony cage during movements of the upper limbs.

Inguinal Canal

The inguinal canal is an oblique passage, 3 to 5 cm long, through the abdominal wall (figs. 33–6 to 33–9). It is occupied in males by the spermatic cord and in females by the round ligament of the uterus, and it contains the ilioinguinal nerve. The canal is a potentially weak area in males, and inguinal hernias are common. Probably more has been written, less has been agreed upon, and more confusion has existed with regard to the inguinal canal and its surgery than is the

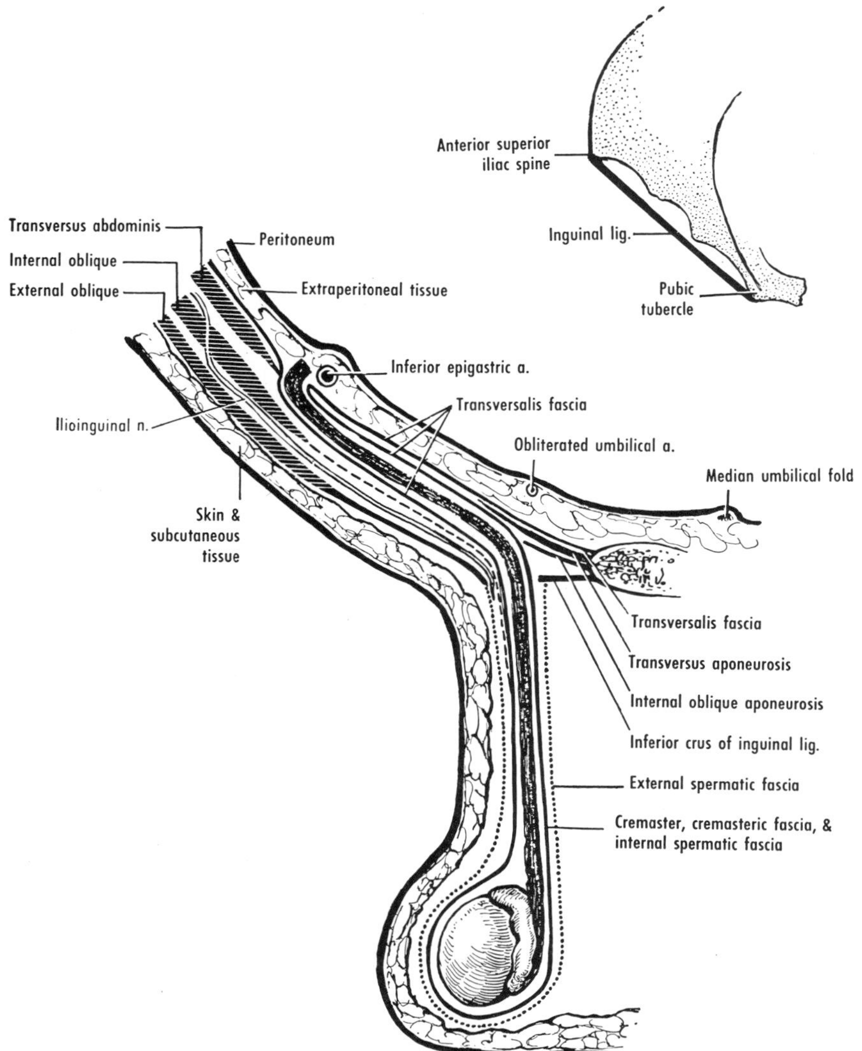

Figure 33–6 Inguinal canal and layers of the scrotum (see also fig. 43–2, p. 468). The diagram combines two planes. One is the plane of the inguinal ligament (*upper right figure*); the other is a sagittal plane through the scrotum. The separation of the layers of the canal and of the cord is exaggerated. The cremaster muscle and cremasteric fascia are shown as a direct continuation of the internal oblique and are also shown fusing with the transversalis fascia. Note that the transversalis fascia is prolonged along the ductus deferens as the internal spermatic fascia, and that it forms most of the posterior wall of the inguinal canal.

case with any other region of the body. Three important reasons are: (1) variations have not been adequately considered, (2) terms have not been defined precisely and have been used differently by different investigators, and (3) eponyms are abundant.[8]

The ductus deferens hooks around the lateral side of the inferior epigastric artery and is joined by vessels and nerves to form the spermatic cord. The ductus deferens, vessels, and nerves are embedded in a continuation of extraperitoneal connective tissue. Just above the midinguinal point, the spermatic cord traverses the *deep inguinal ring,* which is a slitlike opening in the transversalis fascia. The lower part of the deep ring is strengthened by a loop of fibers in the transversalis fascia.[9] The cord then runs obliquely downward and medially in the inguinal canal and emerges through the *superficial inguinal ring,* which is a triangular opening of variable size in the aponeurosis of the external oblique. In its course through the canal, the spermatic cord acquires sheaths from each of the layers of the abdominal wall, and these sheaths continue with the cord into the scrotum.

The anterior wall of the inguinal canal is formed by the aponeurosis of the external oblique and, laterally, by muscle fibers of the internal oblique. The posterior wall is formed by the transversus aponeurosis and transversalis fascia, being usually more aponeurotic medially and more fascial near the deep ring. Sometimes the aponeurosis of the internal oblique contributes to the posterior wall at the medial end of the canal. Above, the canal is bounded by the arching fibers of the internal oblique and the transversus abdominis. The floor is formed by the inguinal ligament and the lacunar ligament.

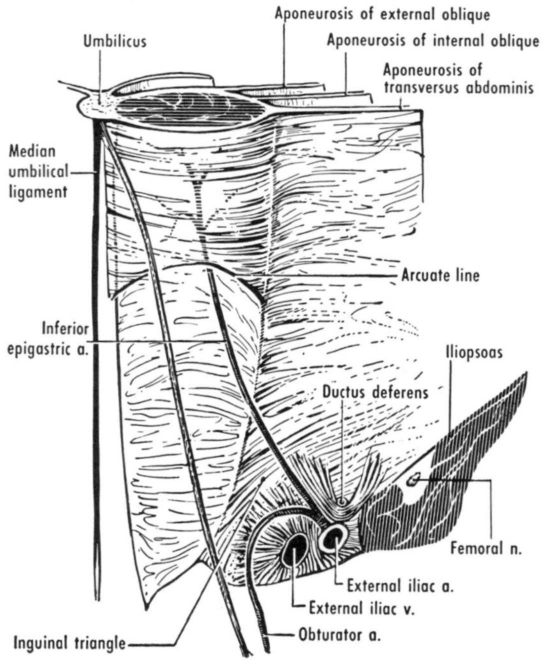

Figure 33–7 The posterior aspect of the anterior abdominal wall, with its fasciae and the peritoneum removed. The sheath of the rectus does not correspond to the classical description. The obturator artery is shown arising from the inferior epigastric. The obliterated umbilical artery is shown without a label. A direct inguinal hernia would enter the inguinal canal through its posterior wall, medial to the inferior epigastric artery.

The inferior epigastric vessels lie behind the canal, just medial to the deep ring. These vessels and their ensheathing extraperitoneal connective tissue form the lateral umbilical fold, which is visible on the inner aspect of the peritoneum. The vessels also form the lateral border of the *inguinal triangle,* the medial border of which is formed by the lateral edge of the rectus abdominis, and the inferior border or base by the pectineal ligament and the pubis. The triangle overlies the medial inguinal fossa and the lateral part of the supravesical fossa. Sometimes the urinary bladder is quite close to the medial end of the inguinal canal. The inguinal canal is present before birth but it is shorter and much less oblique than in the adult,[10] and the superficial ring lies almost directly in front of the deep ring.

The female inguinal canal is narrower than that of the male, and hernias are much less frequent. The canal transmits the round ligament of the uterus, its accompanying vessels, and the ilioinguinal nerve. The round ligament ends as fibrous strands in the subcutaneous tissue of the labium majus.

The chief protection of the inguinal canal is muscular. The muscles that increase intra-abdominal pressure and tend to force abdominal contents into the canal at the same time tend to narrow the canal and to close the rings. For example, during contraction of the abdominal muscles, fibers of the transversus move laterally and upward, and those of the internal oblique at the lateral margin of the deep ring move medially. The direction of movement produced by muscular contraction has been demonstrated by electrical stimulation of the muscles at operation.[11] As a result, the deep ring moves upward and laterally and is closed like a shutter, and the canal is lengthened and made more oblique.

The layers of the abdominal wall and their relations to the inguinal canal and spermatic cord (figs. 33–8 and 33–9) are considered in more detail on the following page.

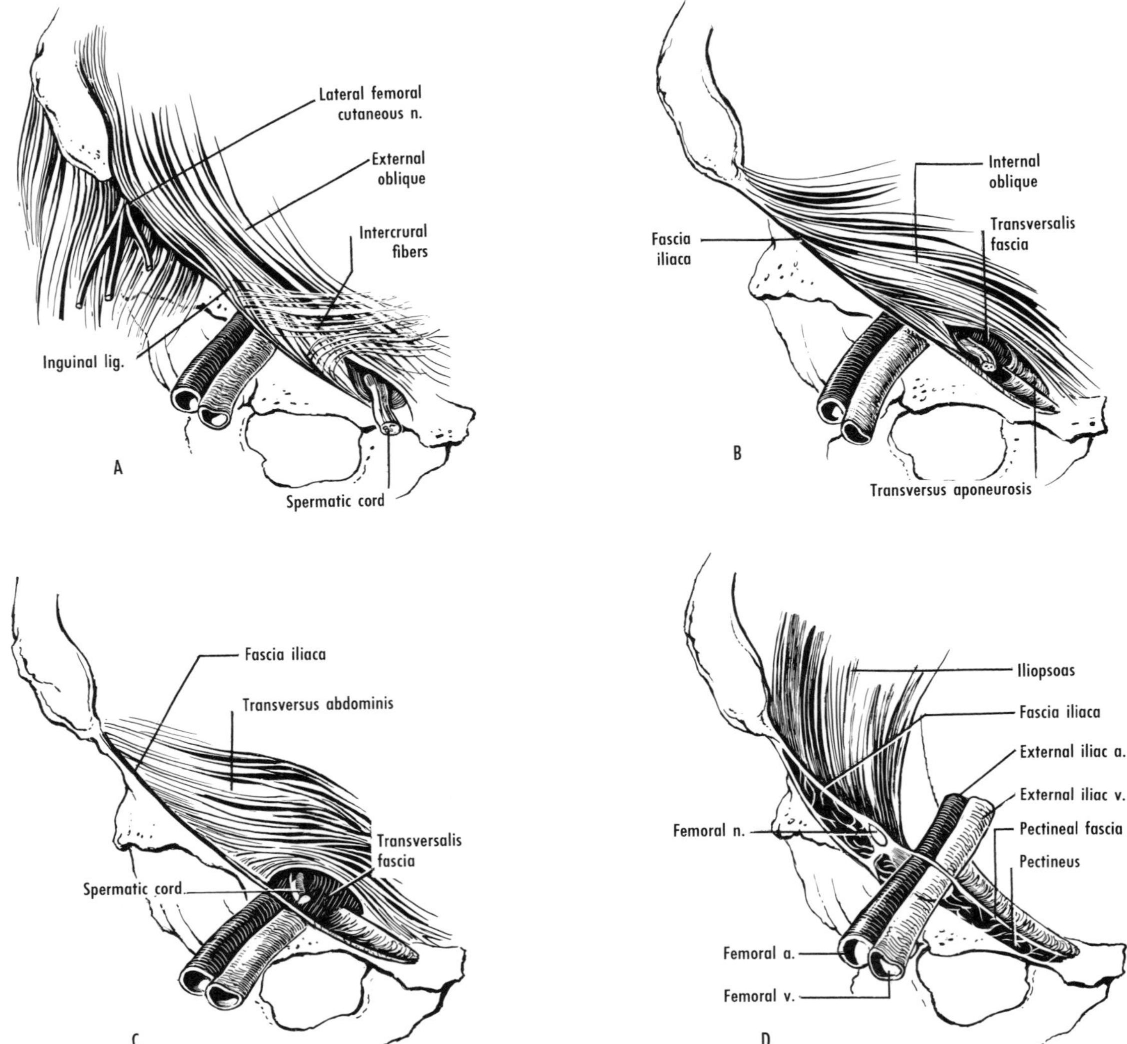

Figure 33–8 Diagrams of inguinal canal and inguinal ligament from in front, and progressing from outside inward, layer by layer. No distinction is made between muscle and aponeurosis. *A*, the external oblique, with the superficial inguinal ring and the spermatic cord on the lacunar ligament and inferior crus. *B*, the internal oblique, arising from the fascia iliaca, extends medially to the pubis. It presents an oblique opening for the spermatic cord. *C*, part of the transversus abdominis is shown arising from the fascia iliaca, extending medially, and arching over the spermatic cord to the pubis. The transversalis fascia behind the transversus extends downward to the fascia iliaca. *D*, the cut edges of the iliopsoas and the pectineus. The fascia iliaca divides around the vessels.

Transversus Abdominis and Transversalis Fascia. The transversus aponeurosis, which may pass above the spermatic cord or may be pierced by it, continues medially to the anterior layer of the rectus sheath. Below, the aponeurosis is fused with the aponeurosis of the internal oblique. The united aponeuroses form a deep arch over the vessels descending into the thigh and continue medially, to be attached to the pectineal fascia and *pectineal ligament.* The ligament consists of fibers that extend horizontally from the iliopectineal arch (p. 211) along the pectineal line, to the pubic tubercle. It covers the superior ramus, the pectineal line, and the upper part of the pectineal fascia. The transversalis fascia extends to the pubis, where it is continuous with the parietal pelvic fascia. Above the level of the pubis it invests the rectus, and between the pubis and the arcuate line it separates the rectus from the extraperitoneal connective tissue. It also con-

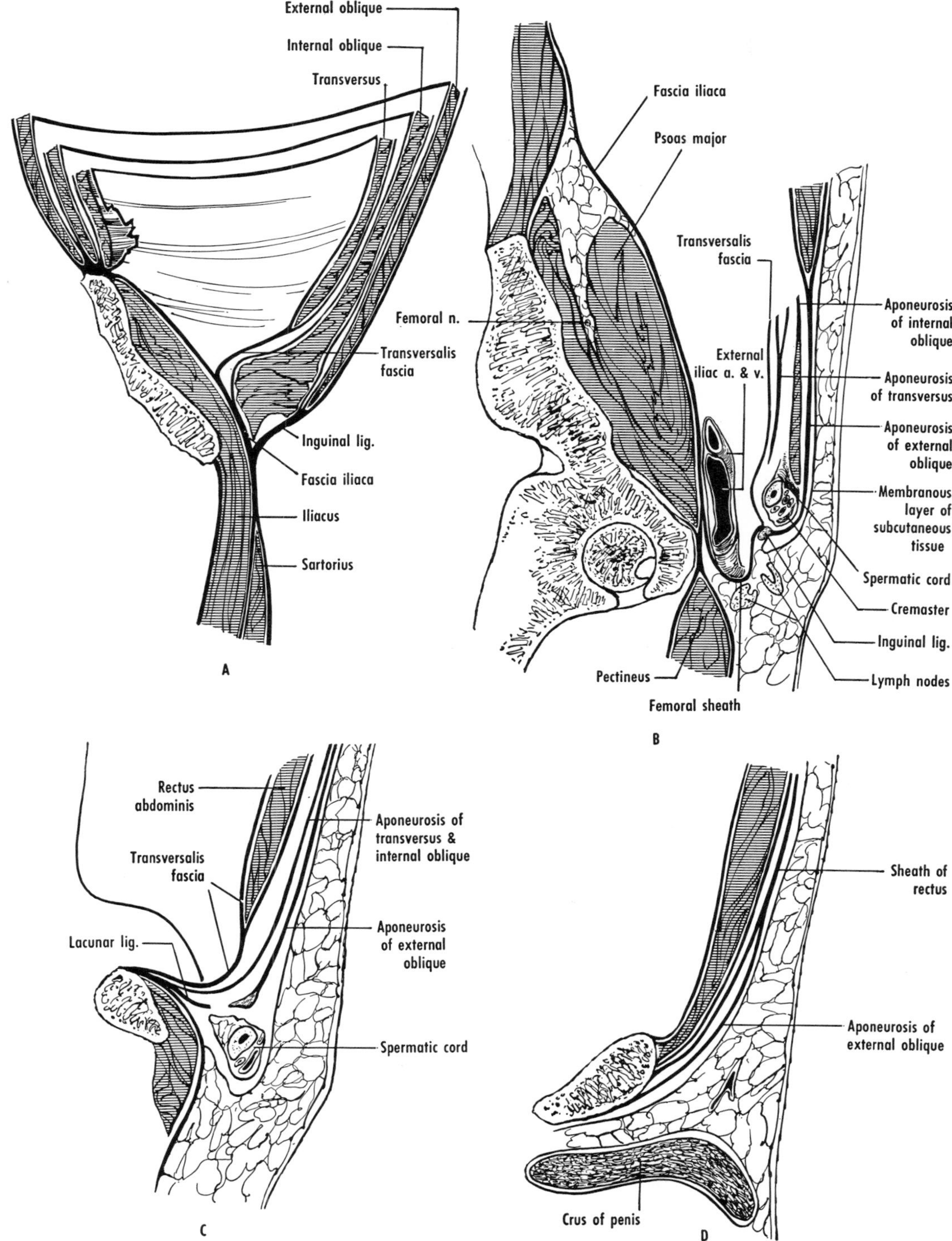

Figure 33–9 Drawings of sagittal sections through the inguinal canal. *A*, just medial to the anterior superior iliac spine. *B*, an offshoot of the fascia iliaca contributes to the femoral sheath and is continuous with the inguinal ligament, the transversalis fascia, and the membranous layer of the subcutaneous tissue. *C*, through the superficial inguinal ring. Note that the lacunar ligament extends backward and upward, *D*, through the body of the pubis. Note that the external oblique aponeurosis is fused with the sheath of the rectus above, but is separate below.

tinues below the level of the transversus aponeurosis and fuses with the fascia iliaca. It is prolonged in front of the femoral vessels into the thigh as the anterior part of the femoral sheath (p. 211), and it forms the *internal spermatic fascia*, which invests the spermatic cord.

Internal Oblique and Cremaster. The muscular fibers of the internal oblique that arise from the fascia iliaca extend as far medially as the deep inguinal ring. The spermatic cord traverses the muscular part of this muscle very obliquely, and in so doing acquires a layer of muscle fibers and investing fascia, the *cremaster muscle* and *cremasteric fascia* (p. 472).

The aponeurosis of the internal oblique continues medially to the anterior layer of the sheath of the rectus. Below, as previously mentioned, it is fused with the transversus aponeurosis. The medial portion of the fused aponeuroses is termed the *conjoined tendon* or aponeurosis (*falx inguinalis*). It continues medially in front of the rectus sheath. The conjoined tendon varies in the degree to which the internal oblique aponeurosis contributes,[12] so that the latter is usually not present behind the superficial ring. The posterior wall of the inguinal canal here is formed chiefly by a layer composed of transversus aponeurosis and transversalis fascia in a varying combination, with varying strength. It is this layer that usually resists direct inguinal hernias.

The lateral edge of the tendon of the rectus, at its attachment to the pubis, often extends laterally as a thin aponeurosis that is fused with the transversalis fascia and the transversus aponeurosis behind the superficial inguinal ring. This fusion forms what has been termed Henle's ligament; it is the medial edge of the femoral ring (p. 211).

External Oblique and Inguinal Ligament.[13] The lower edge of that part of the external oblique aponeurosis which extends from the anterior superior iliac spine to the pubic tubercle is termed the *inguinal ligament*. It is fused with the *iliopubic ligament*, which is a tough fibrous strip that is formed by the fascia iliaca and fascia lata between the ilium and the pubis. This fused strip forms a superficial arch over the vessels descending into the thigh. At its medial end, the strip extends backward to the pectineal fascia and ligament. This portion, which is arranged somewhat horizontally, is known as the *lacunar ligament*, or *pectineal part of the inguinal ligament*.

Lateral to the pubic tubercle, the external oblique aponeurosis divides into two crura, the *medial* (or superior) and the *lateral* (or inferior). The divergence of the crura forms the superficial inguinal ring. **The superficial inguinal ring normally admits the tip of the little finger, and may be found by pushing the loose scrotal skin upward along the spermatic cord to a point immediately above the pubic tubercle, and then passing backward. The tip of the finger then feels the margins of the ring.**

The medial crus continues medially and downward to the body of the pubis. The lateral crus, on which the spermatic cord rests, is the medial end of the inguinal ligament. A small band of fibers *(reflected inguinal ligament)*, sometimes well defined, extends from the pubic attachment of the lateral crus, upward and medially, behind the superior crus, and blends with the opposite external oblique aponeurosis.

The fasciae on both surfaces of the external oblique fuse at the superficial ring and form a sheath for the spermatic cord, the *external spermatic fascia*. Between the crura, the aponeurosis often forms bands of variable size and arrangement, the intercrural fibers. Similar bands may be scattered on the outer surface of the lower part of the external oblique aponeurosis (fig. 33–8*A*).

Hernias

Most hernias[14] occur in the umbilical and inguinal regions. Those in the umbilical region are classified with the ventral hernias.[15] They are usually congenital and result from an incomplete closure of the abdominal wall. Ventral hernias may also occur through defects in the linea alba, where they are also termed median hernias, or along the linea semilunaris, where they are also termed lateral hernias. Hernias may occur through the femoral ring (p. 211), and occasionally in the lumbar triangle, or through the obturator foramen.

Inguinal hernias are of two kinds, indirect or oblique, and direct.

Indirect Inguinal Hernia (fig. 33–10). **In indirect hernia, abdominal contents enter the inguinal canal at the deep ring.** Indirect hernia is more common than direct, more common in men, and more common on the right side. It is generally believed that indirect hernias are due to congenital factors, the most important of these being a partly or wholly patent processus vaginalis.

The layers of an indirect hernia are those of the spermatic cord. In a hernia of long standing, the layers may become thickened and much more readily separated or distinguished than in the normal state. Indeed, extra layers may often be found.

Direct Inguinal Hernia. **A direct hernia enters the inguinal canal through its posterior wall, medial to the inferior epigastric artery.** It therefore involves the posterior wall in the region of the inguinal triangle, above the inguinal ligament, that is, either the medial inguinal fossa or the supravesical fossa, or both. It protrudes forward to the superficial ring, but rarely through it. The primary cause is some type of weakness of the posterior wall. The sac of a direct hernia is formed by the peritoneum behind the anterior abdominal wall.

BLOOD VESSELS AND LYMPHATIC DRAINAGE

Blood Vessels

The cutaneous arteries of the abdominal wall arise from the deeper arteries (see below), and also from the superficial epigastric and superficial circumflex iliac

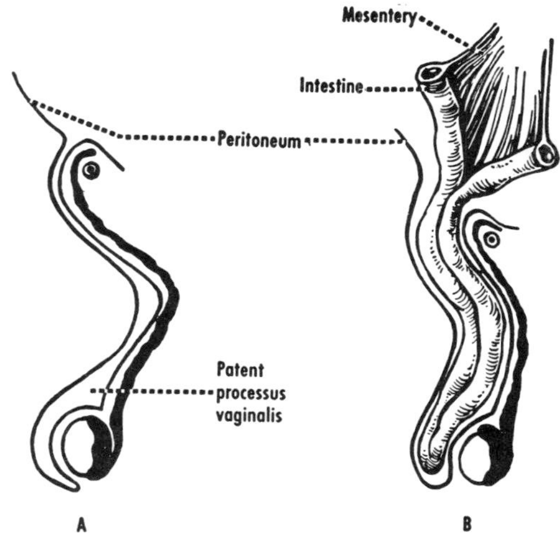

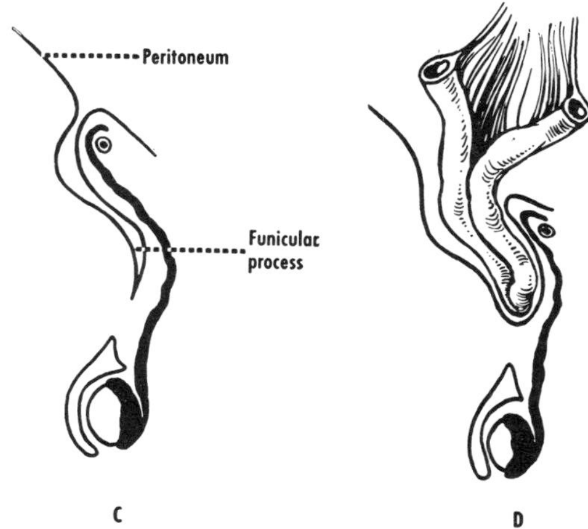

Figure 33–10 Diagrams of congenital indirect inguinal hernias. *A*, patent processus vaginalis. *B*, hernia into the process. *C*, funicular process. *D*, hernia into funicular process.

branches of the femoral artery (p. 217). The cutaneous veins that drain the area from about the level of the umbilicus downward end in the great saphenous vein (p. 200). Above, they anastomose with vessels that converge into the thoracoepigastric veins. These in turn drain into the lateral thoracic vein and thus provide a route of collateral circulation in case of caval obstruction (p. 420). Subcutaneous veins near the umbilicus anastomose with the portal vein by way of branches along the ligamentum teres of the liver.

The chief arteries of each half of the abdominal wall are two from above, the superior epigastric and musculophrenic branches of the internal thoracic artery (p. 270), and two from below, the inferior epigastric and deep circumflex iliac branches of the external iliac artery (p. 415). There are also contributions from the lumbar and subcostal arteries and from the lower posterior intercostals.

Superior Epigastric Artery (fig. 33–5). The superior epigastric artery enters the sheath of the rectus from behind the seventh costal cartilage, through the sternocostal triangle, and descends behind the rectus, supplying it and the overlying skin. One or more branches from the right artery may reach the liver along the falciform ligament. The anastomoses between superior and inferior epigastric arteries provide collateral circulation between the subclavian and external iliac arteries.

Musculophrenic Artery. This artery courses along the costal margin, behind the cartilages, supplying the intercostal spaces, diaphragm, and abdominal wall.

Inferior Epigastric Artery (figs. 33–5 and 33-7; fig. 41–1, p. 450). This arises from the external iliac artery near the midinguinal point. It ascends past the medial margin of the deep ring, where the ductus deferens hooks around its lateral side. As it continues toward the lateral edge of the rectus abdominis, it forms the lateral border of the inguinal triangle. The artery then pierces the transversalis fascia, ascends between the rectus and the posterior wall of its sheath, and ascends behind the rectus in a compartment formed by the posterior wall of the rectus sheath.[16] It supplies the rectus, adjacent muscles, and skin, and anastomoses with branches of the superior epigastric artery. **The anastomoses between superior and inferior epigastric arteries provide collateral circulation between the subclavian and external iliac arteries.** Two branches arise near the deep ring. One, the *artery to the cremaster,* enters the inguinal canal, supplies the cremaster, and anastomoses with the testicular artery. The other, the *pubic branch,* descends to the back of the pubis and anastomoses with the pubic branch of the obturator (p. 451).

Deep Circumflex Iliac Artery (fig. 41–1, p. 450). This arises from the external iliac at about the same level as the inferior epigastric. It runs laterally, first behind the inguinal ligament and then along the iliac crest, finally piercing the transversus and ramifying between that muscle and the internal oblique. Before the vessel reaches the anterior superior iliac spine, an *ascending branch* pierces the transversus, supplies that muscle and the internal oblique, and anastomoses with the musculophrenic artery.

Lymphatic Drainage

The lymphatic vessels of the skin, like the superficial veins, drain in two directions, from approximately the level of the umbilicus downward to superficial inguinal nodes, and upward from this level to axillary nodes. Some uterine lymphatics accompany the round ligament and drain into the inguinal nodes.

NERVES

The abdominal wall is supplied by the thoracoabdominal nerves and by the iliohypogastric and ilioinguinal nerves (figs. 33–11 and 33–12).

Thoracoabdominal Nerves. These nerves, the seventh to eleventh intercostals (p. 275), leave the intercostal spaces and course downward and forward between the transversus and internal oblique, supplying these muscles and the external oblique. They enter the sheath of the rectus, where branches turn forward to supply the rectus and the overlying skin. A vertical incision along the linea semilunaris will denervate the rectus, and one through the middle of the rectus will denervate its medial half. The subcostal nerve, which has a similar course, also supplies the pyramidalis. The lateral cutaneous branches of these nerves have been described (p. 275).

Iliohypogastric and Ilioinguinal Nerves. Derived chiefly from the first lumbar nerve (p. 426), these are mainly cutaneous in distribution. The ilioinguinal nerve enters the inguinal canal and accompanies the spermatic cord (or round ligament of the uterus) to the scrotum (or labium majus).

CUTANEOUS DISTRIBUTION. Each thoracoabdominal nerve supplies a band of skin by means of its lateral and anterior cutaneous branches. The overlap of adjacent nerves is such that section of a single nerve results only in diminished sensation in its area of supply.

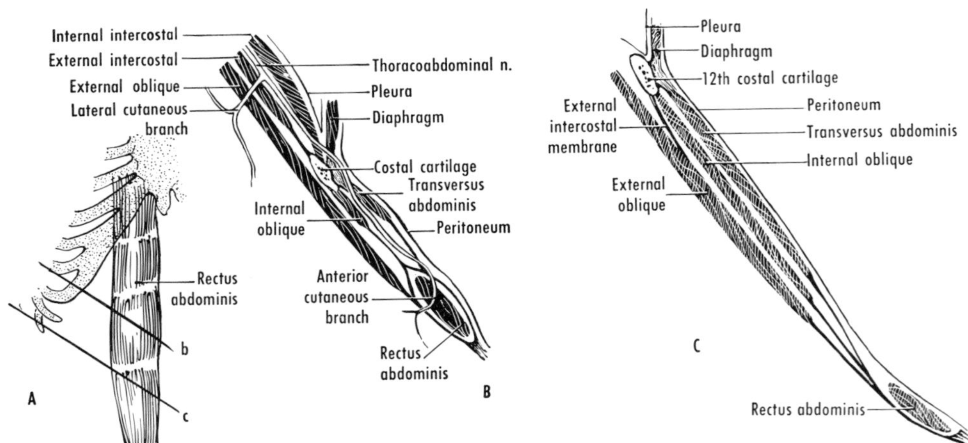

Figure 33–11 Diagrams of the muscular layers of the abdominal wall and the course of the thoracoabdominal nerves. Line *b* indicates the plane of diagram *B*, and line *c* indicates the plane of diagram *C*. Note that in *B* the diaphragm and transversus have a tendinous junction. Muscular branches of the nerve are omitted. In *C*, the part of the internal oblique immediately adjacent to, and attached to, the costal cartilage is continuous with the internal intercostal of that space. Nerves are omitted.

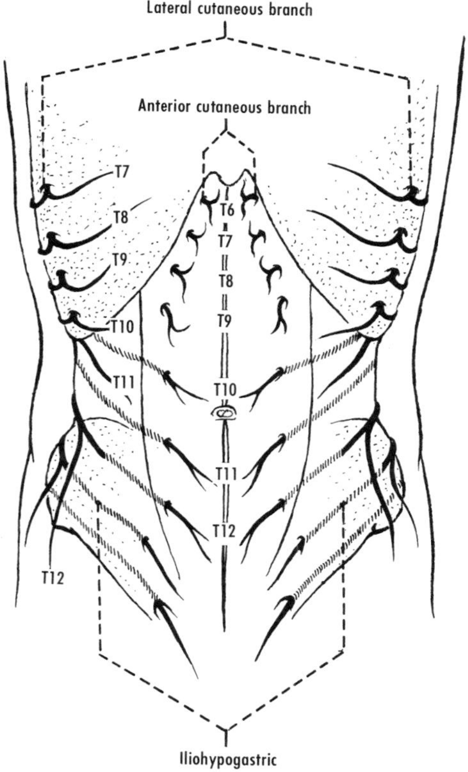

Figure 33–12 The cutaneous distribution of the thoracoabdominal nerves.

UMBILICUS

The umbilicus (L.), *omphalos* (Gk.), or navel, is a depressed or pitted scar in the median plane, somewhat nearer to the pubis than to the xiphoid process (fig. 33–7). Before birth the abdominal wall is open at the attachment of the umbilical cord, with its two arteries and vein, and the urachus. After the umbilical cord is severed at birth, a scar forms at the umbilicus. In the adult, some of the constituents of the cord can be recognized on the inner aspect of the abdominal wall, where they converge at the umbilicus.

All the layers of the abdominal wall are fused at the umbilicus. When subcutaneous fat accumulates, the skin cannot be lifted away from the fused area. Hence the skin becomes raised around the margins of the scar; the umbilicus thereby becomes depressed or pitted at a variable time after birth.

A variety of congenital anomalies may occur in the region of the umbilicus.[17] These may be broadly classified as (1) alimentary (e.g., persistence of the omphalomesenteric or vitellointestinal duct); (2) urachal (e.g., partial or complete patency of the urachus); (3) vascular (e.g., a persistent omphalomesenteric vein); and (4) somatic (e.g., faulty development of the abdominal wall, including ventral hernias. An example is *omphalocele,* which is a protrusion of the bowel through a large defect at the umbilicus).

POSTERIOR ABDOMINAL WALL

The posterior abdominal wall is composed of the bodies of, and the discs between, the five lumbar vertebrae, the psoas major and psoas minor muscles laterally, and, still more laterally, on each side, the quadratus lumborum, ilium, and iliacus (fig. 33–13). The diaphragm also contributes to the upper part of the posterior wall. The erector spinae lies behind the quadratus lumborum, as does the latissimus dorsi more superficially. The origins of the internal oblique and transversus abdominis from the thoracolumbar fascia lie at the lateral edge of the quadratus lumborum.

The aorta and the inferior vena cava lie on the front of the bodies of the vertebrae, the psoas major muscles at their sides. The upper lumbar vertebrae are partly covered by the crura of the diaphragm. The medial and lateral arcuate ligaments of the diaphragm bridge the psoas and quadratus lumborum, respectively. The kidneys and suprarenal glands lie against the wall below these arches. Still lower the cecum and ascending colon are related to the abdominal wall on the right side. The descending colon is related to the wall on the left side. Elsewhere the wall is lined by parietal peritoneum.

Iliopsoas. This is the great flexor of the thigh and trunk. Its broad lateral part, the iliacus, and its long medial part, the psoas major, arise from the iliac fossa and lumbar vertebrae, respectively. The muscle is described in detail elsewhere (p. 215).

The transversalis fascia is continuous with the anterior layer of thoracolumbar fascia in front of the quadratus lumborum

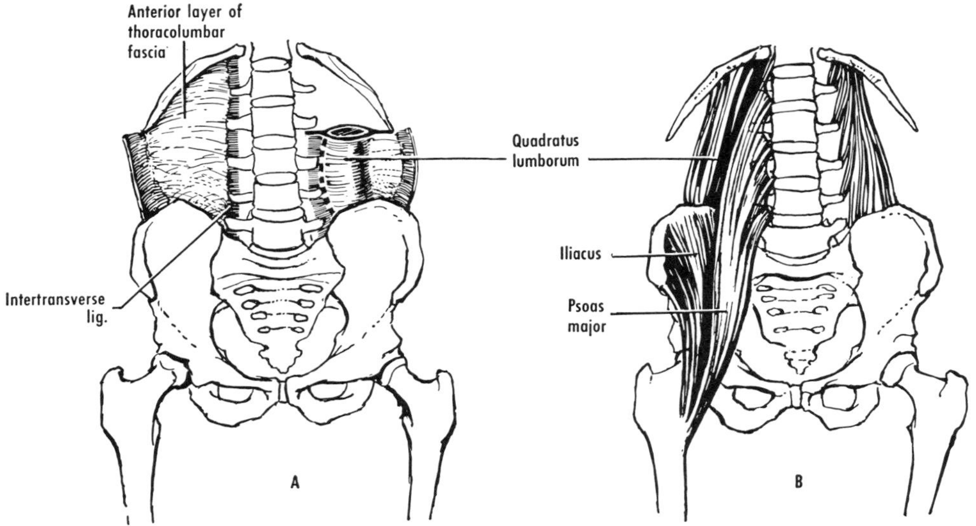

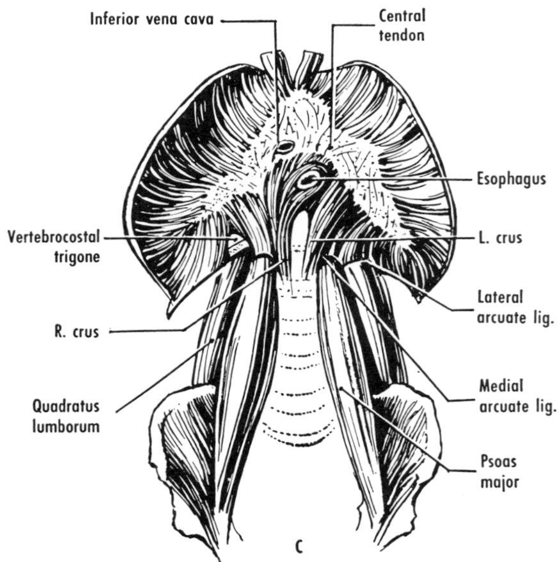

Figure 33–13 Muscles of posterior abdominal wall. *A*, the thoracolumbar fascia, shown on the right part of the figure splitting to enclose the quadratus lumborum (see figure 49–10, p. 537). *B*, psoas major, iliacus, and quadratus lumborum. *C*, relationships of diaphragm to muscles of the posterior abdominal wall. A vertebrocostal trigone is shown on one side.

(fig. 49–10, p. 537). Through this, it is continuous medially with the psoas fascia, and below with the fascia iliaca. The psoas fascia, also termed the psoas sheath, is attached laterally to the transverse processes of the lumbar vertebrae, and medially to the bodies of the lumbar vertebrae. Below, in the iliac fossa, it is continuous with the fascia iliaca. The psoas sheath forms a loose investment, and infections in the muscle (e.g., a tuberculous abscess from an infected vertebral body) may descend within the muscle into the thigh.

The *fascia iliaca* covers the iliacus. Above, it is attached to the iliac crest, together with the transversalis fascia. Below, it contributes to the femoral sheath (fig. 33–9*B*, p. 359) and to the inguinal ligament (p. 360), and continues into the thigh. The transversus and internal oblique arise in part from it (fig. 33–4, p. 354).

Psoas Minor. This is a small muscle, which, when present, arises from the bodies of T.V. 12 and L.V. 1. It is inserted into the arcuate line, reaching the iliopectineal eminence, and has an inconstant additional attachment to the fascia iliaca and the pectineal ligament. It lies on the front surface of the psoas major and is supplied by the lumbar plexus. It probably assists the psoas major in actions on the vertebral column.

Quadratus Lumborum. This is a roughly quadrilateral muscle that is attached below to the posterior part of the internal lip of the iliac crest, above to the last rib, and medially to the tips of the transverse processes of the lumbar vertebrae. It is enclosed between the anterior and middle layers of the thoracolumbar fascia (fig. 49–10, p. 537).

Nerve Supply and Action. The subcostal nerve and lumbar plexus. It probably flexes the trunk laterally, and is thought to fix the last rib and thereby aid the diaphragm (p. 277). When both muscles contract, they probably steady the trunk.

REFERENCES

1. G. W. Cooper, Anat. Rec., *114*:1, 1952.
2. H. B. Howell, Surgery, *6*:653, 1939. K. S. Chouke, Anat. Rec., *61*:341, 1935.
3. C. B. McVay and B. J. Anson, Anat. Rec., *77*:213, 1940. J. D. Rives and D. D. Baker, Ann. Surg., *115*:745, 1942.
4. D. Browne, Lancet, *1*:460, 1933. Y. Appajee, Ind. J. Surg., *7*:113, 1945.
5. H. Hadžiselimović and V. Tomić, Anat. Anz., *129*:421, 1971.
6. C. B. McVay and B. J. Anson, Anat. Rec., *77*:213, 1940. A. Ruiz Liard, M. Latarjet, and F. Crestanello, C. R. Ass. Anat., *55*:532, 1970.
7. W. F. Floyd and P. H. S. Silver, J. Anat., Lond., *84*:132, 1950. O. Machado de Sousa and J. Furlani, Acta anat., *88*:281, 1974.
8. H. F. Lunn, Ann. R. Coll. Surg. Engl., *2*:285, 1948.
9. W. J. Lytle, Brit. J. Surg., *32*:441, 1945; Ann. R. Coll. Surg. Engl., *9*:245, 1951.
10. H. Curl and R. G. Tromly, J. Anat., Lond., *78*:148, 1944. S. B. Chandler, Anat. Rec., *107*:93, 1950.
11. D. H. Patey, Brit. J. Surg., *36*:264, 1949.
12. J. H. Clark and E. I. Hashimoto, Surg. Gynec. Obstet., *82*:480, 1946.
13. R. E. Condon, Ann. Surg., *173*:1, 1971. J. F. Doyle, J. Anat., Lond., *108*:297, 1971.
14. L. M. Zimmerman and B. J. Anson, *Anatomy and Surgery of Hernia*, Williams & Wilkins, Baltimore, 2nd ed., 1967.
15. G. M. Wyburn, J. Anat., Lond., *71*:201, 1937, *72*:365, 1938, *73*:289, 1939; Brit. J. Surg., *40*:553, 1953.
16. R. Orda, Acta anat., *83*:382, 1972.
17. T. S. Cullen, *The Umbilicus and Its Diseases*, Saunders, Philadelphia, 1916. H. L. Trimingham and J. A. McDonald, Surg. Gynec. Obstet., *80*:152, 1945. See also Wyburn, cited in reference 15.

ABDOMINAL VISCERA AND PERITONEUM

34

ABDOMINAL VISCERA

The principal viscera of the abdomen proper are the stomach, the intestine, the liver and biliary system, the pancreas, the spleen, the suprarenal glands, and the kidneys and ureters. The stomach and intestine are mostly attached to the body wall by a mesentery formed by the peritoneum, whereas the three paired glands, kidneys, suprarenals, and gonads (in the abdomen proper before birth) lie retroperitoneally.

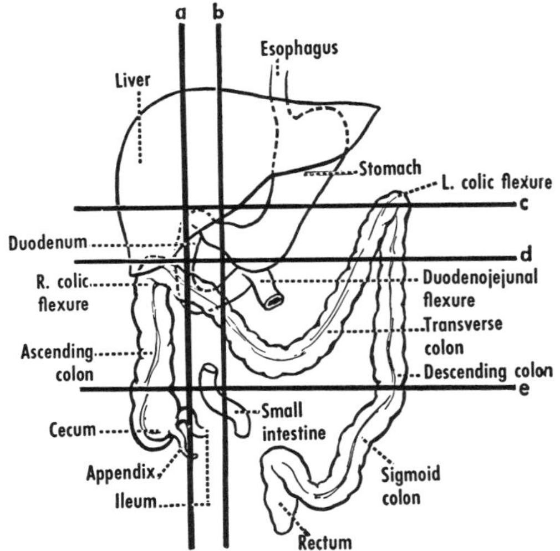

Figure 34–1 The liver and the parts of the alimentary canal. The lines *a* and *b* indicate the planes of section in figure 34–2, the lines *c*, *d*, and *e* the planes of section in figure 34–3.

The liver and pancreas are associated with the alimentary canal. The general relationships of the abdominal viscera are shown in figures 34–1 to 34–3.

The positions of the abdominal viscera vary from individual to individual and depend upon gravity, posture, respiration, and state of filling (if hollow). After death the thoracic and pelvic diaphragms relax, the abdominal viscera move upward (thereby shifting the positions of the thoracic organs), and the pelvic viscera sink. These changes are mimicked in life during deep anesthesia. Embalmed viscera are hard, immobile, and unnatural in color. Impressions produced on an organ by adjacent organs may be fixed in place by embalming, and may result in surfaces, borders, and other markings that are not present during life.

Radiological studies have given the most accurate and valuable information about the position and mobility of the viscera in living individuals. Such studies have shown that most viscera are very mobile, so much so that many have no fixed position. **"The normal abdominal viscera have no fixed shapes and no fixed positions, and every description of them must be qualified by a statement of the conditions existing at the time of observation. Moreover, profound change may be caused not only by mechanical forces but also by mental influences."**[1]

In radiological studies of more than

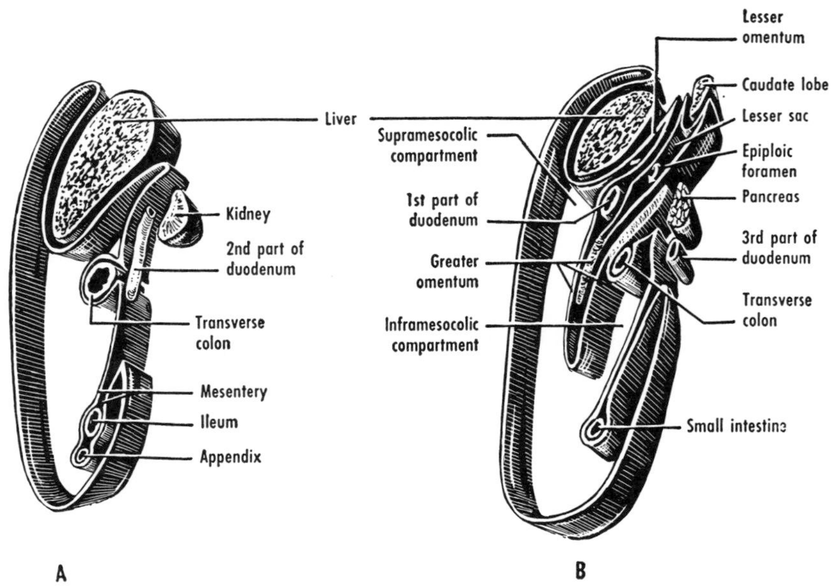

Figure 34–2 A and *B*, diagrams of sagittal sections in the planes *a* and *b* of figure 34–1, respectively, shown from the left. The separation of the various peritoneal layers and organs is exaggerated. The transverse mesocolon and greater omentum are shown in greater detail in figure 34–6.

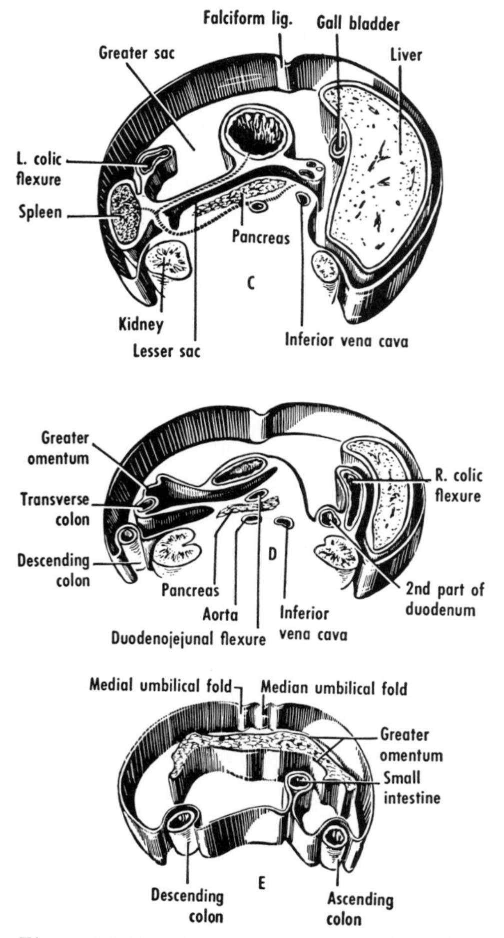

Figure 34–3 Diagrams of horizontal sections in planes *c, d,* and *e* of figure 34–1, shown from above.

1000 students in the erect position,[2] with the use of the supracristal plane as a landmark, the following positions were encountered and were normal: lowest part of the greater curvature of the stomach as much as 15 cm below the supracristal plane; pylorus usually below the transpyloric plane; transverse colon dipping into the true pelvis; lower border of the liver below the supracristal plane. Organs tend to sink in the erect position, to rise in the recumbent position, and to rise or fall according to the movements of the diaphragm. Some organs may move up or down as much as 17 cm in the shift between the standing and supine positions. Others may move hardly at all. The most mobile organs are those attached by mesenteries, whereas retroperitoneal organs are relatively less mobile. Further details are given in the chapters devoted to the specific organs.

During physical examination of the abdomen (p. 432), a few viscera may be accessible to palpation. The following structures can sometimes be palpated in normal subjects: lumbar vertebrae, lower pole of right kidney, sometimes the liver, occasionally the spleen, and the pulsations of the abdominal aorta. The body of the uterus can be palpated bimanually. Examination of abdominal viscera otherwise depends mostly on special techniques, such as those involving radiography. The chief methods of radiographic study of abdominal viscera are described on page 432.

PERITONEUM

The peritoneum[3] is a smooth, glistening, serous membrane that lines the abdominal wall, where it is known as *parietal peritoneum,* and is reflected from the wall to the various organs, the surfaces of which it covers to a variable extent. The peritoneal covering of organs is termed *visceral peritoneum.* It forms an integral part of the outermost or serosal layer of many organs. The extraperitoneal connective tissue external to the parietal peritoneum is carried with the peritoneal reflections to the organs and becomes a part of the serosal layer.

Some of the abdominal viscera, the kidneys, for example, lie on the posterior abdominal wall and are covered by peritoneum only on their anterior aspects. Such organs are said to be retroperitoneal in position. Other organs, such as most of the intestine, are almost completely invested by peritoneum. They are connected to the body wall by a mesentery.

The arrangement of the peritoneum is such that a double-layered sac is formed, comparable in this respect to the pleura and the pericardium. The peritoneal cavity is normally empty except for a thin film of fluid that keeps surfaces moist. The organs are packed so closely that the cavity is normally only a potential space of capillary thinness. **The peritoneal cavity in the male is a completely closed sac. In the female, the**

uterine tubes open into it. These tubes also open into the interior of the uterus, and the peritoneal cavity is therefore indirectly in communication with the exterior of the body. Air injected into the cavity of the uterus normally enters the uterine tubes and reaches the peritoneal cavity (p. 477). This procedure is used as a test for patency of the uterine tubes.

The two most important functions of the peritoneum are to minimize friction and to resist infection. A less important function is the storage of fat, especially in the greater omentum. The peritoneum provides a very slippery surface that permits free movement of the abdominal viscera. The peritoneum exudes fluid and cells in response to injury or infection, and tends to wall off or localize infection. The greater omentum tends to move to a site of irritation (the mechanism of movement is obscure), to become adherent to it, and thereby to increase the local blood supply. It may thus aid in preventing the spread of infection.

The surface area of the peritoneum is very great, probably equal to that of the skin, and fluid injected into the peritoneal cavity is absorbed very rapidly. Certain anesthetics, such as solutions of barbiturate compounds, may be given by intraperitoneal injection. This method is often used for anesthetizing animals.

Nerve Supply. The parietal peritoneum is supplied by the nerves to the adjacent body wall: the subdiaphragmatic part by the phrenic nerves, the remainder by the thoracoabdominal and subcostal nerves and by branches of the lumbosacral plexus.

The fibers in the nerves to the peritoneum are sensory and vasomotor. Most of the parietal peritoneum is very sensitive to pain.[4] Painful stimuli to the anterior and lateral regions are roughly localized to the point stimulated. By contrast, painful stimuli to the central part of the diaphragmatic peritoneum are referred to the shoulder (p. 332). Painful stimuli to the peripheral part of the diaphragmatic peritoneum are felt in an intercostal space. The roots of mesenteries contain pain fibers that are sensitive to stretch.

The visceral peritoneum, like the visceral pleura and pericardium, is insensitive.

Terminology. Certain terms, often arbitrary, are commonly used in connection with the peritoneum. A peritoneal reflection that connects the intestine and the body wall is usually named according to the part of the canal to which it is attached. For example, although the reflection to the jejunum and ileum is termed *the* mesentery, that to the transverse colon is the transverse mesocolon. Some peritoneal reflections between organs or between the body wall and organs, are termed ligaments (e.g., gastrohepatic ligament, falciform ligament) or folds (e.g., rectouterine fold, lateral umbilical fold). The term fold is usually applied to a peritoneal reflection with a free edge. Most ligaments contain blood vessels, and most folds are raised by underlying blood vessels, but neither provides much strength. Finally, a broad sheet or reflection of peritoneum is termed an omentum. The Greek term for omentum is *epiploon*, and the adjective epiploic is derived from it.

General Arrangement of Peritoneum

The continuity of the visceral and the parietal peritoneum in the adult is shown in the horizontal and sagittal sections in figures 34–2 and 34–3, which also show the arrangement of certain mesenteries and ligaments, and the subdivisions of the peritoneal cavity. The attachments of the peritoneum are shown in figures 34–4 and 34–5.

An incision carried through the anterior abdominal wall and the parietal peritoneum enters the peritoneal cavity. The part of the cavity thereby entered is termed the greater sac, in contrast to the lesser sac, which is a complicated recess that communicates with the greater sac by way of the epiploic foramen. The arrangement of the peritoneum, the boundaries of the greater and lesser sacs, and the attachments and contents of the mesenteries, ligaments, folds, and omenta are best learned by repeated study and by exploration with the fingers.

Greater Sac. The greater sac extends from the diaphragm to the pelvic floor. Its anterior wall contains four folds that converge at the umbilicus, and two more laterally placed folds. Of the six folds, one lies above the umbilicus, the others below. The upper fold is the *falciform ligament*, which contains in its free margin the ligamentum teres of the liver (obliterated umbilical vein). It also contains a part of what is termed the anterior abdominal fat body.[5] The lower folds are (1) the *median umbilical fold* (containing the urachus), which extends from the urinary bladder to the umbilicus; (2) the two *medial umbilical folds*, each containing an obliterated umbilical artery, which extend from the sides of the bladder to the umbilicus; and (3) the two *lateral umbilical folds* (fold of epigastric artery), which extend from the deep inguinal

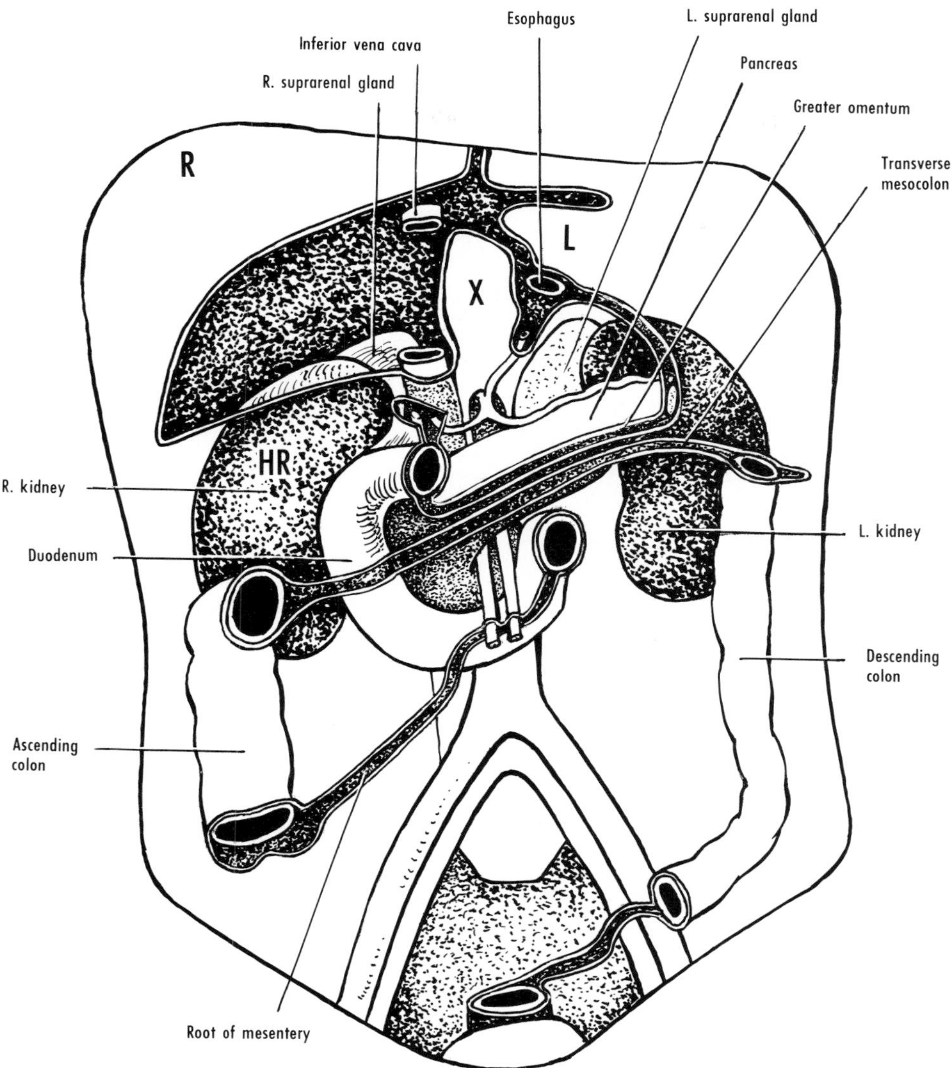

Figure 34–4 The attachments of the peritoneum to the posterior abdominal wall, viewed from in front. R, location of right subphrenic space; L, left subphrenic space; HR, hepatorenal pouch or right subhepatic space. The site of the epiploic foramen can be identified between the inferior vena cava and the cut portion of the lesser omentum. Note the superior recess (X) of the lesser sac, and the branches of the celiac trunk. A mirror image view is shown in figure 34–5. Based on O'Rahilly.[7]

ring on either side to the arcuate line (fig. 33–7, p. 357).

Three depressions on each side of the median plane are produced by the umbilical folds: (1) the *supravesical fossa,* between the median and medial umbilical folds, the lateral part of which is related to the inguinal triangle (p. 357); (2) the *medial umbilical fossa,* between the medial and lateral umbilical folds, and which is related to the inguinal triangle; and (3) the *lateral inguinal fossa,* which is lateral to the lateral umbilical fold and which overlies the site of the deep inguinal ring.

Below the umbilicus and between the two medial umbilical ligaments, the extraperitoneal connective tissue presents two additional layers of fascia, the umbilical

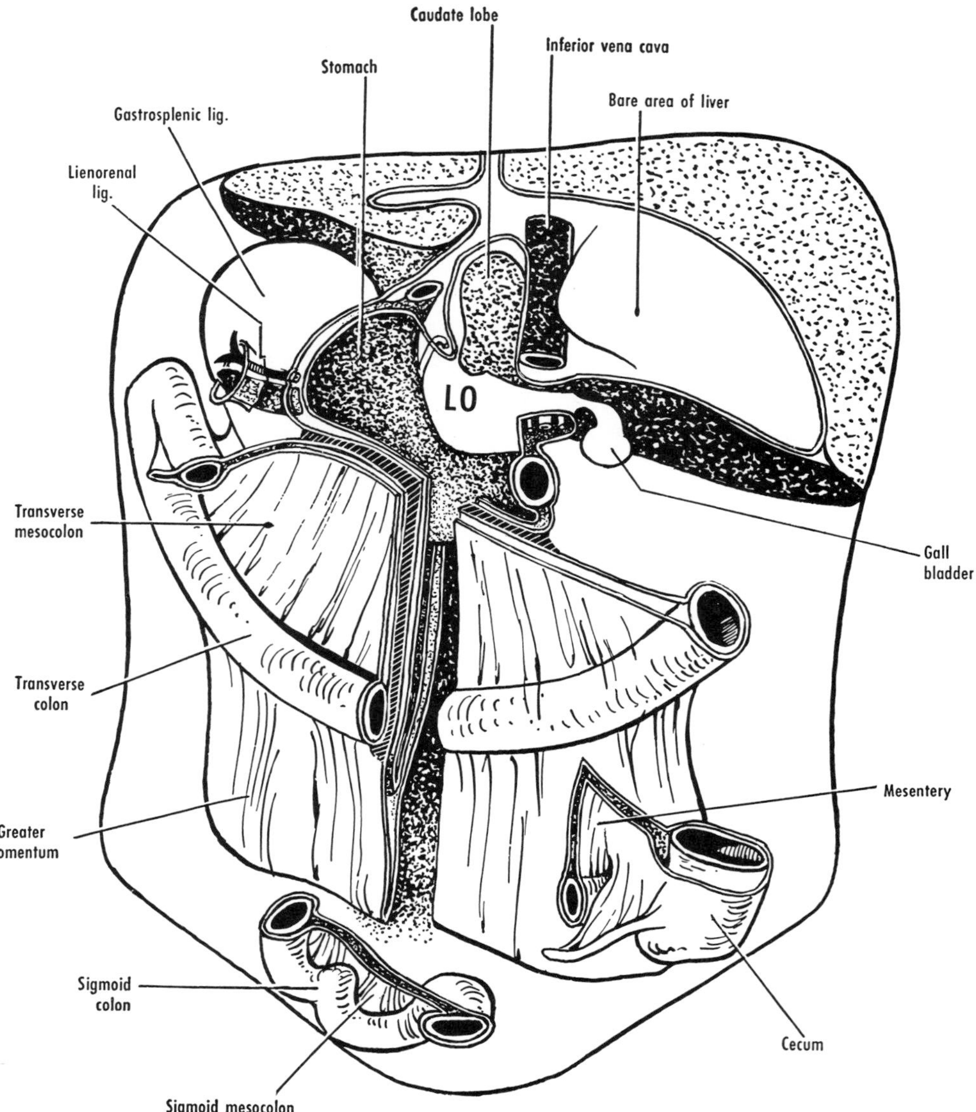

Figure 34–5 The attachments of the peritoneum to the abdominal viscera, viewed from behind. This is a mirror image of the view shown in figure 34–4. A vertical segment of the transverse mesocolon and the greater omentum has been removed to show the layers. The cross-hatching indicates fusion. For the liver, compare figure 36–2*D*, p. 396). The letters LO stand for lesser omentum. Based on O'Rahilly.[7]

vesical (or umbilicovesical) and the umbilical prevesical fasciae.[6] The umbilical vesical fascia is the deeper of the two layers. It extends downward from the umbilicus and ensheathes the median and medial umbilical ligaments. On reaching the urinary bladder, it becomes continuous with the connective tissue covering that organ. The umbilical prevesical fascia lies between the transversalis fascia in front and the umbilical vesical fascia behind. It extends between the medial umbilical ligaments, and it is fused with the umbilical vesical fascia and with the transversalis fascia along these ligaments. It is also fused with them at the umbilicus. Below, it is attached to the inferolateral surfaces of the bladder. The potential space between the transversalis fascia and the umbilical prevesical fascia, between the bladder and the pubis, is the retropubic space (p. 461).

The *greater omentum* is a prominent peritoneal fold that hangs down from the stomach, in front of the transverse colon, to which it is attached. The greater omentum is a double fold, and in the embryo is composed of four layers derived from the dorsal mesogastrium (fig. 34–6). The cavity between the middle two layers is usually obliterated, at least in the lower part of the omentum. The greater omentum usually contains lobulated fat. It tends to adhere to areas of inflammation, and is often present in hernias.

If the greater omentum is turned upward, the coils of the small intestine can be examined, and the *mesentery* followed to its *root*. **Traced upward and to the left, the root of the mesentery leads to the duodenojejunal flexure. Traced downward and to the right, it leads to the ileocolic junction.**

The greater omentum traced upward leads to the transverse colon (behind the omentum) and the stomach. A hand placed below the transverse colon meets the *transverse mesocolon,* which is attached in front to the transverse colon, and behind to the posterior abdominal wall. The attachments of the transverse mesocolon can now be

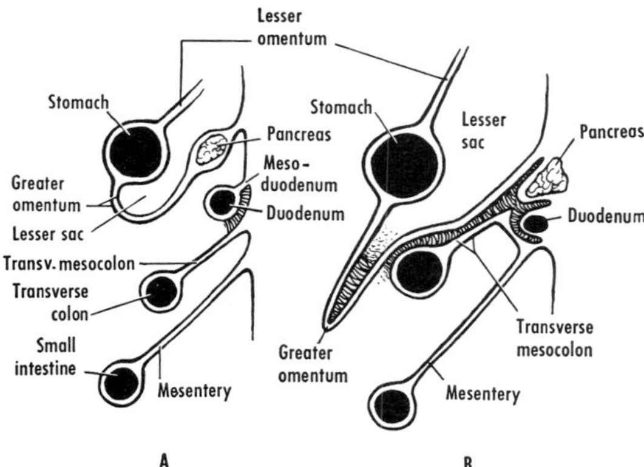

Figure 34–6 Diagrammatic representation in sagittal section of the formation of the transverse mesocolon and greater omentum. *A*, embryo; *B*, adult. The numerous short lines in the greater omentum and between (1) the posterior wall of the lesser sac and (2) the transverse mesocolon and transverse colon represent lines of fusion.

traced horizontally. A finger penetrating the transverse mesocolon enters the lesser sac. Likewise, a finger put through the greater omentum between the stomach and transverse colon enters the lesser sac.

The various parts of the large intestine can be located by following the intestine upward from the ileocolic junction, or in either direction from the transverse colon. The ascending and the descending colon usually lack a mesocolon, and in such instances are retroperitoneal. The sigmoid colon commonly has a mesocolon, which may have the shape of an inverted **V**, the apex being at the pelvic brim, in front of the left ureter.

If the falciform ligament is followed upward, it is found to be reflected onto the diaphragmatic surface of the liver. A hand placed between the liver and the diaphragm on the right is prevented from passing to the left by this reflection. Likewise, a hand placed between the liver and the diaphragm on the left cannot pass to the right. More posteriorly, the reflection of peritoneum from the diaphragm to the liver diverges and forms the upper layer of the coronary ligament on the right, and the upper layer of the left triangular ligament on the left.

The fundus of the gall bladder is usually visible at the inferior border of the liver. Traced upward, the gall bladder leads to the *lesser omentum,* which is derived from the ventral mesogastrium, and which extends between the liver and the stomach and duodenum. The free edge of the lesser omentum is at the right. **The epiploic foramen (aditus, or opening into the lesser sac) is immediately behind the free edge of the lesser omentum. A finger in the opening, and a thumb in front of the ligament, will catch the bile duct, hepatic artery, and portal vein between them (bile duct at the right, hepatic artery at the left, and portal vein behind). The inferior vena cava lies behind the epiploic foramen.** The part of the lesser omentum that extends between the liver and the duodenum is termed the *hepatoduodenal ligament.* The part that connects the liver with the stomach is termed the *gastrohepatic ligament.* The two parts are continuous.

SUBDIVISIONS OF GREATER SAC. The greater sac is subdivided by the greater omentum, transverse colon, and transverse mesocolon into an upper, anterior part, the supramesocolic compartment, and a lower, posterior part, the inframesocolic compartment. These compartments form channels or recesses that determine how or where peritoneal fluid gravitates or spreads. The inframesocolic compartment is further divided by the mesentery of the small intestine into right (upper) and left (lower) parts. The latter drains into the pelvis. The *paracolic grooves* are longitudinal depressions lateral to the ascending and descending colon. The supramesocolic compartment is subdivided by the liver into subphrenic and subhepatic spaces (table 34–1).

Lesser Sac. The lesser sac is a large, irregular space that lies mostly behind the stomach and lesser omentum (figs. 34–2 and 34–3). Its anterior wall is formed by the peritoneum that (1) forms the posterior layer of the lesser omentum, (2) covers the posterior surface of the stomach and 1 or 2 cm of the first part of the duodenum, and (3) forms the posterior of the anterior two layers of the greater omentum. The posterior wall of the lesser sac is formed by the peritoneum that covers the diaphragm, pancreas, left kidney and left suprarenal, and duodenum, and continues downward as the anterior of the two posterior layers of the greater omentum. The posterior of these two layers is fused with, but can be separated from, the upper layer of the transverse mesocolon and its continuation onto the transverse colon. The borders of the lesser sac, which are extremely variable, are illustrated in figure 34–7.

The epiploic foramen, which leads from the greater sac into the lesser sac, is a short canal that can be located by running a finger along the gall bladder to the free edge of the lesser omentum. **Two fingers can usually be inserted into the opening (the anterior and posterior walls of the opening are generally in contact). In this position, the right or free edge of the lesser omentum and its contents (bile duct, hepatic artery, and portal vein) are in front of the fingers, the liver is above, the first part of the duodenum is below, and the peritoneum covering the posterior abdominal wall and inferior vena cava lies behind.**

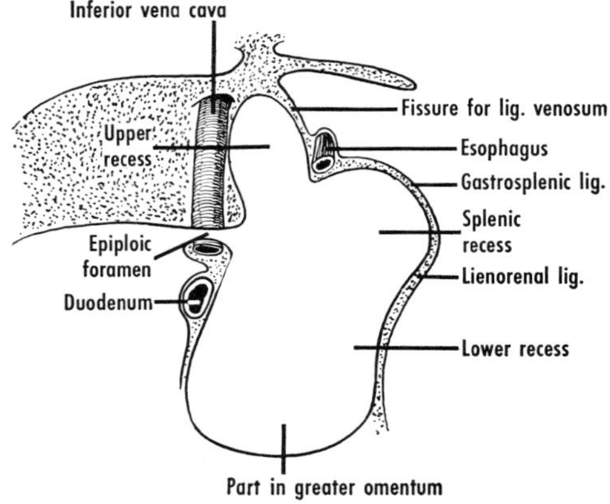

Figure 34–7 Schematic diagram of the posterior wall of the lesser sac in a coronal plane.

The lesser sac presents three recesses, upper, lower, and splenic. The *upper recess* lies behind the liver. The caudate lobe projects into it from above (fig. 34–2). The *lower recess* (omental bursa of the embryo) lies behind the stomach and in the greater omentum; a part of it extends to the left as the *splenic recess.* The posterior wall of the lesser sac is marked by two forward-projecting folds, the right and left *gastropancreatic folds,* which are formed respectively by the common hepatic artery (running downward and to the right) and the left gastric artery (running upward and to the left). The term omental bursa is sometimes used synonymously with the term lesser sac, but the omental bursa is the recess that develops in the embryo in the mesoderm lateral to the endodermal anlage of the foregut.[8] The cavity thus formed corresponds to the lower recess of the adult lesser sac. It soon joins the general coelomic cavity and also joins the more cephalically placed right pneumatoenteric recess. The upper part of the latter recess usually disappears, but may persist above the diaphragm as the infracardiac bursa (p. 303).

Minor Peritoneal Folds, Fossae, and Recesses (fig. 34–8). The minor fossae and recesses are usually formed by small peritoneal folds. Although it has been thought that a part of the intestine might become caught in one of these fossae or recesses (intra-abdominal or retroperitoneal hernia), they are usually so shallow that herniation into them is unlikely unless there is a maldevelopment of the peritoneum or the viscera.[9]

The most common of these minor and usually unimportant structures occur in association with the duodenum, the cecum, and the sigmoid colon. Nine different types of fossae and a number of folds have been described in relation to the duodenum.[10] Two small folds, the superior and inferior duodenal folds, are often present and extend to the left from the third part of the duodenum. The superior fold is formed by the inferior mesenteric vein and the ascending branch of the left colic artery. The *superior duodenal recess* lies behind the superior fold, and the *inferior duodenal recess* behind the inferior. The inferior mesenteric vein and a branch of the left colic artery may form a paraduodenal fold, behind which is a paraduodenal recess (an extension of the superior duodenal recess). The *retroduodenal recess* is occasionally present behind the fourth part of the duodenum. Rarely, a *mesentericoparietal recess* lies below the third part of the duodenum. The fold in front of it contains the superior mesenteric artery.

Three recesses may occur about the cecum. The

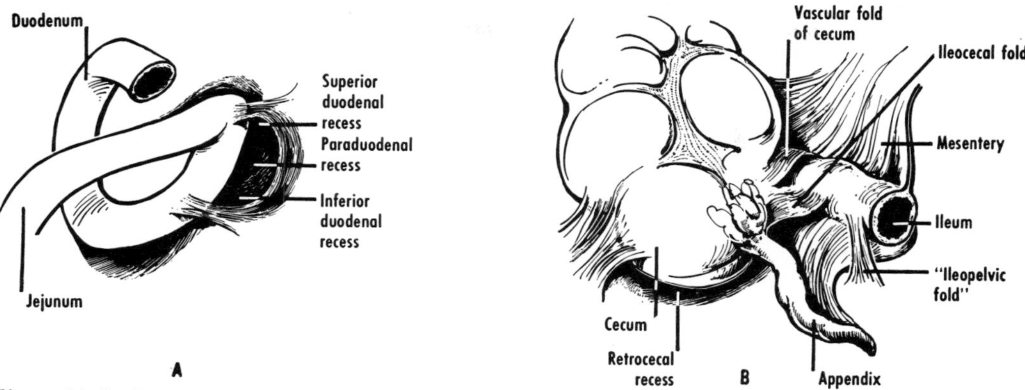

Figure 34–8 Diagrams of some of the folds, fossae, and recesses around *(A)* the duodenum and *(B)* the cecum. Based on a dissection.

superior ileocecal recess is formed by the *vascular fold of the cecum,* which contains the anterior cecal vessels, and extends from the mesentery to the ileocecal junction. The *inferior ileocecal recess* is formed by the *ileocecal fold,* which extends from the terminal part of the ileum to the base of the appendix. The *retrocecal recess* lies behind the cecum and may reach up behind the ascending colon. *Cecal folds* may connect the cecum to the abdominal wall on each side of the recess.

The *intersigmoid recess,* or *recess of the pelvic mesocolon,* is present in the fetus and in most children, but may be absent in the adult. It lies behind the apex of the mesocolon, in front of the left ureter, at the birfurcation of the left common iliac artery.

Subphrenic (Subdiaphragmatic) and Subhepatic Spaces.[11] These spaces are of clinical importance, particularly in infections, which may produce abscesses in one of them. The most important spaces in this regard are on the right side.

Table 34–1 lists the clinically important spaces and their boundaries. The two subphrenic spaces are separated by the falciform ligament. The hepatorenal pouch or recess, also known as the right subhepatic space, is marked off by a peritoneal reflection, the hepatorenal ligament, between the right kidney and the visceral surface of the liver.[12]

Development of Alimentary Canal and Peritoneum

Early in the embryonic period, the intraembryonic endoderm with its associated mesoderm consists of three parts, the foregut, the hindgut, and the intervening midgut, which communicates ventrally with the definitive yolk sac. Figure 34–9 illustrates certain features of subsequent development which help to explain the adult arrangements.

The stomach is attached to the posterior abdominal wall by a portion of the *dorsal common mesentery* known as the *dorsal mesogastrium,* and to the anterior abdominal wall by the *ventral mesogastrium.* The liver and pancreas develop at the junction of the foregut and midgut. The liver grows into the ventral mesogastrium. The pancreas develops from a ventral bud that grows into the ventral mesogastrium with the liver, and a dorsal bud that grows into the dorsal mesogastrium with the spleen. The part of the duodenum cephalic to the entrance of the bile duct is at-

TABLE 34–1 Relationships of Clinically Important Subphrenic and Subhepatic Spaces

	Important Relationships				
Space	*Anterior*	*Posterior*	*Superior*	*Right*	*Left*
Right subphrenic	Anterior abdominal wall	Upper layer of coronary ligament	Diaphragm	Diaphragm	Falciform ligament
Left subphrenic	Anterior abdominal wall	Left triangular ligament	Diaphragm	Falciform ligament	Spleen
Right subhepatic (also termed hepatorenal pouch)	Visceral surface of right lobe of liver	Right kidney	Lower layer of coronary ligament	Diaphragm	Epiploic foramen

tached to the anterior abdominal wall by the ventral mesogastrium.

The midgut undergoes rapid elongation during development and forms the caudal part of the duodenum (distal to the entrance of the bile duct), the jejunum and ileum, and the large intestine as far as the left third of the transverse colon. The remainder of the large intestine and the proximal portion of the anal canal are derived from the hindgut. The cecum and appendix develop as projections from the first portion of the large intestine. The large intestine at first ends blindly, but during the third month of intrauterine life the anal membrane between the endodermal hindgut and the ectodermal anal pit (proctodaeum) disintegrates.

During the course of development, the spleen grows to the left in such a way that the gastrolienal and lienorenal ligaments are formed (fig. 34–9*G*). The part of the ventral mesogastrium between the stomach and the liver becomes the lesser omentum, whereas that between the liver and the anterior body wall becomes the falciform ligament, the free edge of

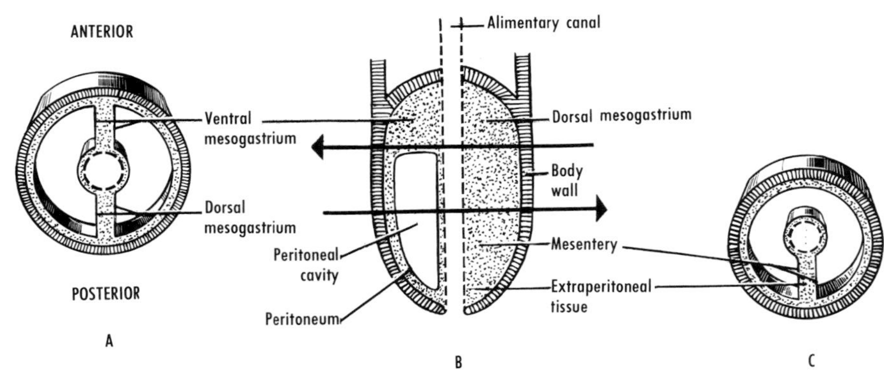

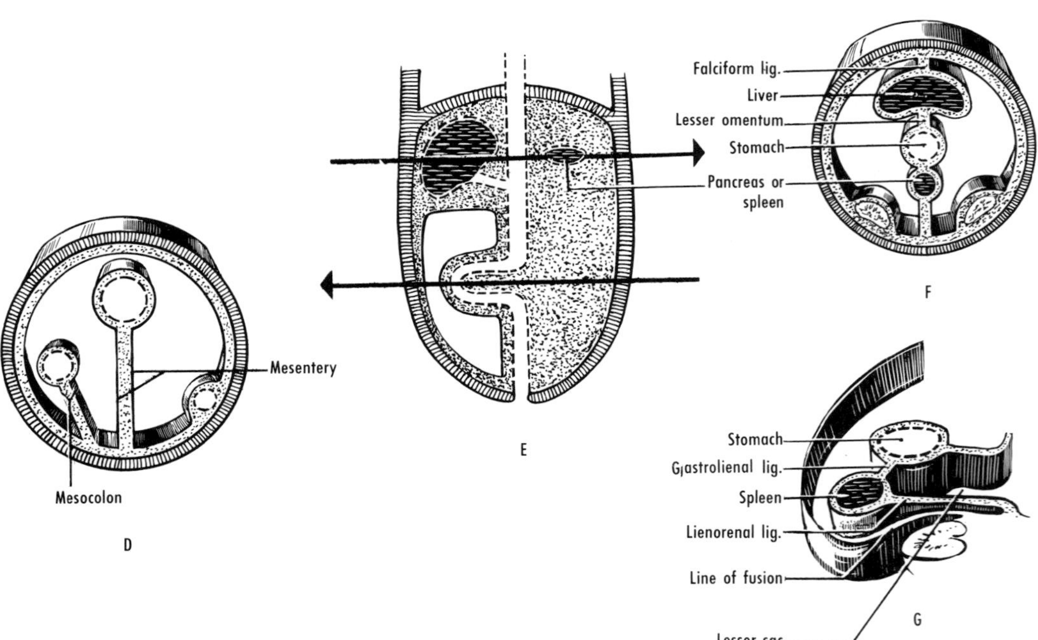

Figure 34–9 Principles of the alimentary canal, based on development. *B* and *E* are sections in the median plane. The arrows indicate the levels of sections shown in *A*, *C*, *D*, and *F*. *D*, however, represents a stage somewhat more advanced than the level indicated in *E*. The change in position of the stomach and spleen is shown in *G*. The left layer (now posterior) of the dorsal mesogastrium will fuse with the peritoneum on the posterior body wall. The two fused peritoneal layers will disappear, leaving a short lienorenal ligament.

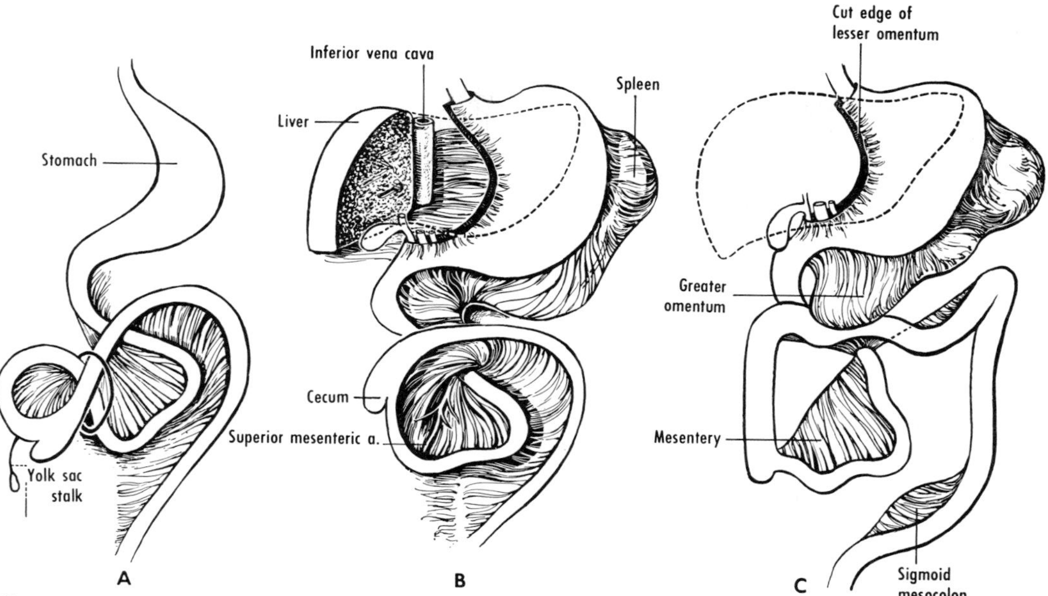

Figure 34–10 Schematic representation of mesenteric attachments and of the rotation of the gut. *A*, early fetal period. Note that the colon crosses ventral to the duodenum. The small intestine is being reduced, that is, withdrawn into the abdominal cavity through the umbilical ring. *B*, slightly later in the fetal period; the intestine is completely reduced. The axis of rotation of the gut is the superior mesenteric artery. *C*, the fixation of mesenteries in the adult. Note that every second part of the gastrointestinal canal retains a mesentery.

which contains the ligamentum teres (obliterated umbilical vein).

The part of the dorsal mesogastrium between the stomach and the pancreas and spleen grows in such a way that the greater omentum is formed. The greater omentum later becomes attached to the transverse colon; its two posterior layers fuse with the transverse mesocolon of the embryo to form the transverse mesocolon of the adult. As a result of this fusion, the pancreas and duodenum become retroperitoneal.

Rotation of Gut (fig. 34–10). The wide communication between the midgut and the definitive yolk sac becomes reduced, by the end of the first intrauterine month, to a narrow passage (vitellointestinal duct) which becomes obliterated. Occasionally a portion of the duct may persist postnatally as a diverticulum ilei. During development, the mesentery elongates and the midgut forms a loop which, early in the second month, herniates through the umbilicus and into the umbilical cord. The herniated intestine returns to the abdomen during the third month. During this withdrawal, it undergoes a rotation that results in the establishment of arrangements and relationships characteristic of the intestine in the adult. Faulty early development may result in the retention of a midgut loop in the umbilical stalk. Interference with later stages may result in nonrotation, malrotation, or internal hernia. If the cecum fails to descend, it retains its original subhepatic position.

REFERENCES

1. A. E. Barclay, *The Digestive Tract*, Cambridge University Press, London, 2nd ed., 1936.
2. R. O. Moody, J. Anat., Lond., *61*:223, 1927. R. O. Moody, R. G. Van Nuys, and W. E. Chamberlain, J. Amer. med. Ass., *81*:1924, 1923.
3. J. Brizon, J. Castaing, and F. G. Hourtoulle, *Le Péritoine*, Libraire Maloine S. A., Paris, 1956.
4. V. J. Kinsella, Brit. J. Surg., *27*:449, 1940.
5. N. G. Nordenson, T. Petrén, and P. J. Wising, Z. anat. EntwGesch., *93*:223, 1930. R. J. Merklin, Amer. J. Anat., *132*:33, 1971.
6. G. Hammond, L. Yglesias, and J. E. Davis, Anat. Rec., *80*:271, 1941.
7. R. O'Rahilly, Irish J. med. Sci., p. 663, October, 1947.
8. R. Kanagasuntheram, J. Anat., Lond., *91*:188, 1957.
9. M. Laslie, C. Durden, and L. Allen, Anat. Rec., *155*:145, 1966.
10. R. C. Bryan, Amer. J. Surg., *28*:703, 1935.
11. D. P. Boyd, New Engl. J. Med., *275*:911, 1966.
12. A. Gisel, Acta anat., *27*:149, 1956.

ESOPHAGUS, STOMACH, AND INTESTINE

35

The esophagus, stomach, and intestine constitute the alimentary canal and are derived from the foregut, midgut, and hindgut. The portion of the foregut below the diaphragm is supplied by the celiac trunk. The midgut is supplied by the superior mesenteric artery, and the hindgut by the inferior mesenteric. The entrance of the bile duct into the duodenum marks the junction of the foregut and midgut. The junction of the midgut and hindgut occurs in the left part of the transverse colon.

The esophagus or gullet is a conducting tube for food, whereas the stomach, intestine, and associated glands are concerned with the digestion of food and the excretion of undigested material.

The products of digestion pass through the epithelium of the gastric and intestinal mucosa to the blood and lymphatic capillaries. The capillaries of the gastrointestinal canal drain into veins that ultimately form the portal vein. The portal vein then breaks up into a second set of capillaries (sinusoids) in the liver. These capillaries in turn drain into veins that form hepatic veins.

The submucosa imparts strength to the alimentary canal. The chief function of the muscularis is to move the contents. The muscularis may sometimes confine the contents to one region of the canal. Most of the alimentary canal has an outermost serosal layer that is slippery and permits mobility. Some parts of the canal, however, have an outermost fibrous layer that tends to fix the organ to the abdominal wall and thereby tends to limit mobility, and some parts have a serosal layer on one aspect and a fibrous layer on another.

The alimentary canal is characterized by a sphincter mechanism at each junctional area, e.g., pharyngoesophageal, gastroesophageal, pyloric, and ileocolic. It is likely that the major function of sphincters, which are under neural and hormonal control, is to prevent regurgitation of luminal contents from one portion of the canal to another.

ABDOMINAL PART OF ESOPHAGUS

The cervical and thoracic parts of the esophagus are described elsewhere (pp. 279, 691). The lower part of the esophagus, having made a turn to the left and traversed the esophageal opening in the diaphragm, joins the stomach at its lesser curvature. This junction is known as the gastroesophageal, or cardioesophageal, junction.

The gastroesophageal junction is a region at which the passage of food into the stomach may slow down. More important, it constitutes a barrier to the reflux of contents from the stomach to the esophagus. The circular layer of smooth muscle of the lower end of the esophagus is continuous with that layer above and also continues directly into the stomach.

Above the gastroesophageal junction is a sphincteric segment[1] 1 to 4 cm long, which is located partly in the thorax, partly in the opening in the diaphragm, and partly in the abdomen. The segment is joined to the diaphragm by the phrenicoesophageal ligament. The segment is further characterized by a subdivision into an upper tubular portion, and a lower expanded portion, the vestibule. The circular muscle fibers around the junction between these two portions comprise the lower esophageal sphincter. An additional sphincter may be present around the vestibule. The normal sphincteric mechanism is represented by a zone of intraluminal resting pressure that is higher than that in the fundus of the stomach. The pressure decreases just prior to the arrival of a bolus. The rapid relaxation and subsequent after-contraction of the lower esophageal sphincter in response to swallowing are chiefly under neural control. Resting tone of the sphincter seems to be under hormonal control.[2] The closing mechanism between the esophagus and the stomach includes mucosal folds that come together when the sphincter contracts.[3]

An additional mark of the gastroesophageal junction is provided by the "sling" fibers of the stomach. These muscle fibers hook around the left side of the junction and descend toward the lesser curvature of the stomach. They can be recognized as a notch in living persons during a barium meal radiological examination.

Blood and Nerve Supply. The blood supply of the esophagus is described on page 280. The anastomoses between the esophageal and the left gastric veins, which are important portal-systemic anastomoses (p. 418), take place through venous plexuses in the lamina propria, in the submucosa, and on the outer surface of the junctional region.[4]

The gastroesophageal junction is supplied by the splanchnic nerves (from the celiac plexus by way of the left gastric plexus and other adjacent plexuses) and by the vagal trunks.[5] The autonomic fibers are distributed mainly to smooth muscle. Some sensory fibers are also present and are probably concerned with pain. Painful stimulation of the lower esophagus (e.g., by reflux of acid gastric contents) may cause pain ("heartburn") that is felt deep to the sternum or in the epigastrium. Similar pain may arise from the stomach.

Esophageal (Hiatal) Hernia. This term is applied to several types of diaphragmatic hernias through the esophageal opening in the diaphragm (p. 269).

STOMACH

The stomach (Gk. *gaster*, belly; the adjective *gastric* is from the Latin *gastricus*) presents a cardiac part, a fundus, a body, and a pyloric part (fig. 35–1A); two curvatures, greater and lesser; two walls, anterior and posterior; and two openings, cardiac and pyloric.

The cavity of the esophagus joins that of the stomach at the *cardiac opening* at the junction of the greater and lesser curvatures. The immediately adjacent portion of the stomach is the *cardiac part*. It is distinguished only by the cardiac glands in its mucosa. There is no external line of demarcation between the cardiac part and the fundus or body.

The *fundus* is the part of the stomach above the level of the entrance of the esophagus. It usually contains swallowed air (on the average, about 50 ml) and is therefore visible in ordinary radiograms of this region. The mucosa of the fundus is similar in structure to that of the body. Both contain gastric glands proper. The *body* of the stomach is the portion between the fundus and the pyloric part. There is no external line of demarcation between the body and the fundus above, or between the body and the pyloric part below. The line of demarcation between the body and the pyloric part can be accurately located only by special methods that distinguish their

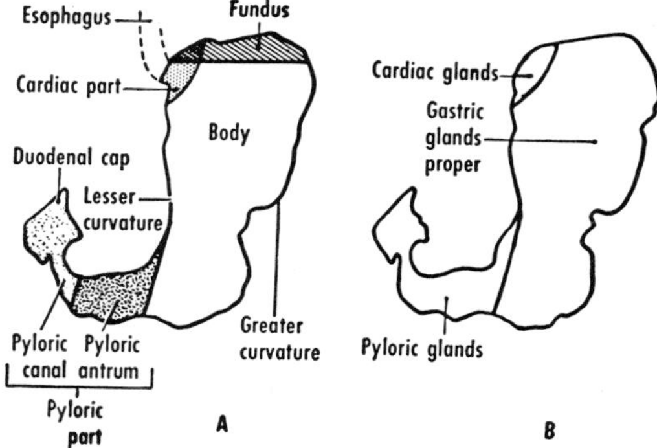

Figure 35–1 A, the parts of the stomach, anterior view. The outline of the stomach is based on a radiogram. *B*, the location of the glands.

mucosae.[6] The approximate location of the line is shown in figure 35–1.

The *pyloric part* of the stomach is the portion that is lined by mucosa containing pyloric glands. The proximal part is called the *pyloric antrum,* the distal part the *pyloric canal.* The *pyloric opening* (also termed *pylorus*) between the first part of the duodenum and the stomach is surrounded by the *pyloric sphincter.* A prominent vein, the prepyloric vein, is often present in front of the pyloroduodenal junction, and the transition from the thick, muscular pyloric part to the thin-walled duodenum is usually quite abrupt. The pyloric sphincter may be congenitally thickened, a condition known as congenital hypertrophic pyloric stenosis, and one that requires surgical intervention very early in infancy.[7]

The *greater* and *lesser curvatures* extend from the cardiac to the pyloric opening. The greater curvature is on the left, tends to be convex, and is much the longer. The lesser curvature is on the right, is much shorter, and tends to be concave. The lesser curvature usually exhibits a notch, the *incisura angularis* or angular notch, at its most dependent point. The notch is nearly always evident in radiograms of the barium-filled stomach, taken in the erect position. The stomach has *anterior* and *posterior walls,* but the directions in which they face vary according to the position and the degree of filling of the stomach.

The stomach is a very distensible organ, having a capacity of 1 to 2 liters, or more, and it does not have a fixed shape. When it is empty, it commonly resembles the letter J. The stomach may, however, be cylindrical or roughly crescentic, and its shape is readily altered by changes in posture (fig. 35–2).

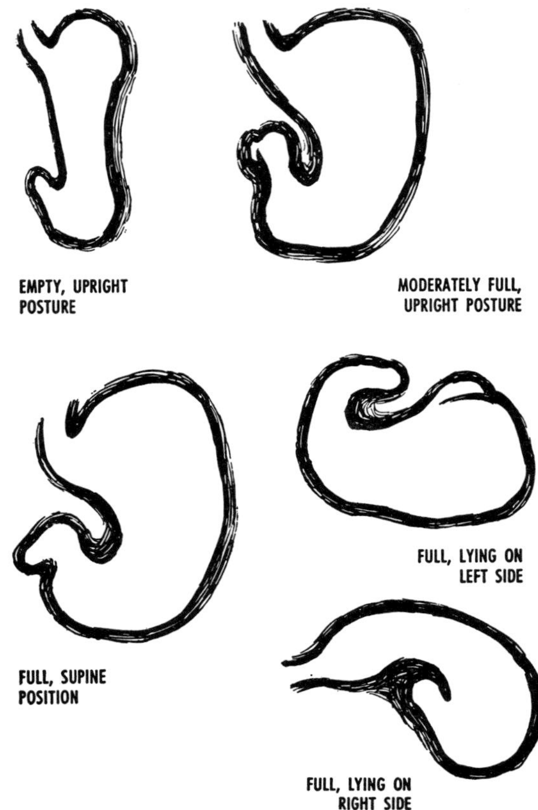

Figure 35–2 Outline drawings of the shape of the stomach under different conditions, viewed from in front. Based on Barclay.

Relations. The stomach is a mobile, easily displaced organ, and has no fixed position. When the subject is recumbent and the stomach is empty, the following are the common and important relationships. In front are the diaphragm, the liver, the anterior abdominal wall, and sometimes the transverse colon. The whole of the anterior surface is covered with peritoneum, and a part of the greater sac of the peritoneal cavity intervenes between the stomach and the structures mentioned. Behind, from above downward, are the diaphragm, the left suprarenal gland, the pancreas and a part of the left kidney, and the transverse mesocolon. The posterior surface is covered with peritoneum, except for a small "bare" area near the cardiac opening, where the stomach is in direct contact with the left crus of the diaphragm. Except in the "bare" area, the posterior surface of the stomach is separated by the lesser sac from the structures mentioned. The spleen is also related to the stomach, usually to the upper part of the greater curvature or to the adjacent part of either of the two surfaces (usually the posterior).

The relationships just given may not

apply in other positions or physiological conditions of the stomach. The following general points apply under most conditions.

The cardiac orifice is the most fixed part of the stomach. It can be marked on the surface at the left costal margin, about midway between the xiphisternal joint and the transpyloric plane, a few centimeters to the left of the median plane. The fundus fits into the curve of the left dome of the diaphragm, and moves with it. Air in the fundus causes a tympanitic note on percussion. The pyloric part is very mobile. In the recumbent position, with the stomach empty, it is at or near the transpyloric plane, a few centimeters to the right of the median plane. In the erect position or with the stomach full, it may be anywhere between the transpyloric and the supracristal plane, on either side of the median plane. The greater curvature dips even lower; the stomach may enter the true pelvis.

PERITONEAL RELATIONS. The part of the lesser omentum between the liver and the stomach is known as the *gastrohepatic ligament*. The two layers of the lesser omentum separate at the lesser curvature of the stomach, cover the anterior and posterior surfaces, with the exception of the "bare" area, as noted above, and meet at the greater curvature. The two layers continue to the left from the upper part of the greater curvature as the gastrophrenic and gastrosplenic ligaments, and they continue downward from the lower part of the greater curvature as the anterior two layers of the greater omentum (fig. 34–6*B*). The peritoneal covering of the anterior surface of the stomach is continued upward for a short distance on the anterior aspect of the esophagus.

Most of the knowledge of the shape, position, and movements of the living stomach has been obtained by radiological studies, usually after a barium meal (figs. 35–3 and 35–4). Further details of radiological anatomy are given on page 434.

Structure and Function. The stomach has a mucosa that contains pyloric, gastric, and cardiac glands. Its strong muscular layer is arranged chiefly into an inner circular and an outer longitudinal layer. The mucosa of the empty stomach may be thrown into folds termed rugae or *gastric folds*. These folds, which are not permanent, have a core of submucosa. The mucosa of the living stomach can be examined by means of an electrically lighted tubular instrument passed down the esophagus. The procedure is termed gastroscopy.

Enzymatic digestion is the chief function of the stomach. Food entering it has no definite route, except that liquids tend to take the path nearest the straight line of gravity, that is, along the lesser curvature (gastric canal). The upper half of the stomach is little more than a receptive bag that relaxes as it fills with food.

In spite of the outpouring of gastric juice, solid or semisolid food remains solid or semisolid in the stomach for a long time, sometimes several hours or more. The gastric juice converts food on the surface of the mass into a smooth, liquid mixture termed *chyme*. Once it is formed, chyme is emptied fairly quickly into the duodenum, that is to say, there is some mechanism whereby liquids are evacuated from the stomach and solids are retained.

Peristaltic movements are responsible for the emptying of the contents of the stomach into the duodenum. These movements are ringlike contraction waves that begin about the middle of the stomach and travel slowly but smoothly toward the pylorus. About three waves occur per minute, each of about 20 seconds duration. The peristaltic movements are seldom forceful, unless obstruction is present. The pyloric sphincter is an integral part of the mechanism. It contracts when a peristaltic wave reaches it, and relaxes during the intervals. Like the lower esophageal sphincter, it is under neural and hormonal control. Its chief function seems to be to prevent reflux of material from the duodenum into the stomach. Little if any peristaltic activity occurs in the fundus and body. The stomach, therefore, does not act like a mill or churn, except perhaps to a slight extent in the pyloric canal, where chyme tends to be moved back and forth before being emptied into the duodenum.

Blood Supply (fig. 35–5).[8] The arteries that supply the stomach arise, directly or indirectly, from the celiac trunk (p. 416). The arteries are the right and left gastric, the right and left gastroepiploic, the short gastric, and often the left inferior phrenic. The right and left gastric arteries approach the lesser curvature of the stomach, whereas the short gastric arteries and the branches of the gastroepiploic arteries approach the greater curvature. There are usually five or six short gastric arteries. They arise from a division of the splenic artery, from the main trunk of the splenic, or from an accessory splenic branch.

Anastomoses of the arteries that supply the stomach with those that supply the other abdominal organs provide routes of collateral circulation.[9] Within the walls of the stomach, anastomoses between gastric and esophageal arteries occur constantly, but anastomoses between gastric and duodenal arteries are scanty, at least in the walls of the pyloroduodenal junction ("bloodless line").

The veins of the stomach accompany the arteries but do not end in a common stem before entering the portal vein. They empty either into the portal vein or into one of its tributaries, in an extremely variable manner. The prepyloric vein, which indicates the junction between the pylorus and the duodenum, joins the right gastric vein. The anastomoses between the left gastric vein and the esophageal veins are important portal-systemic anastomoses (p. 418).

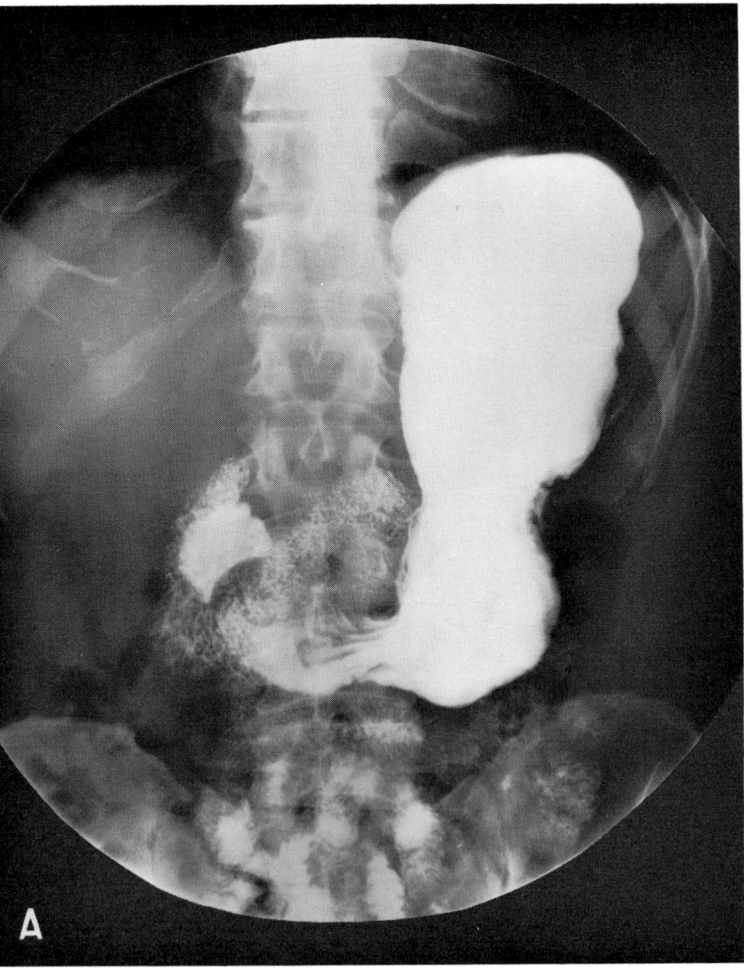

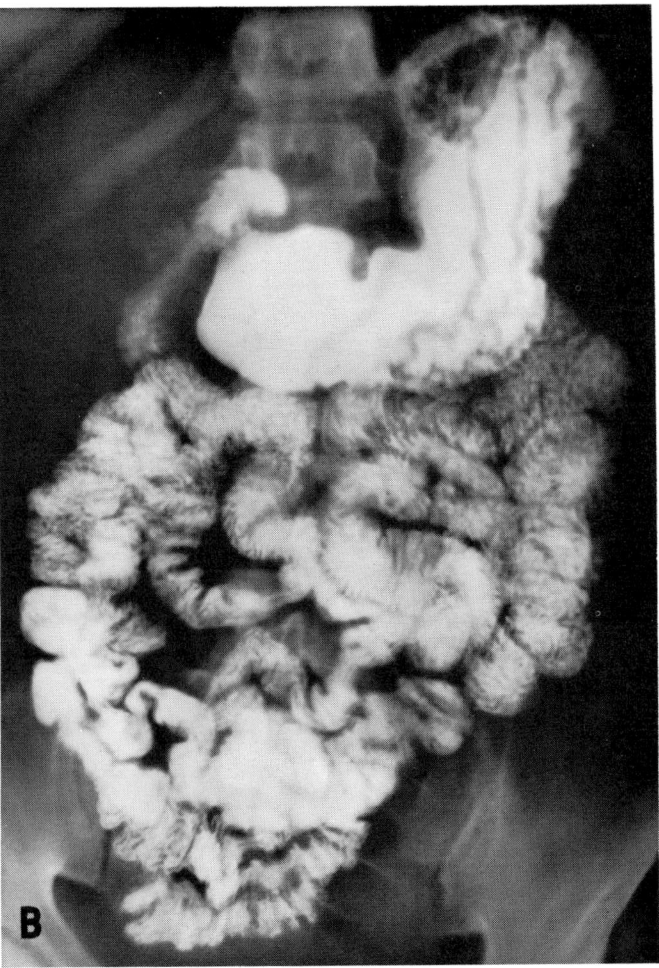

Figure 35–3 Stomach and small intestine. *A*, stomach after barium meal; prone position. Note duodenal cap, feathery pattern of barium in small intestine, twelfth rib on right side of body, and pyloric part of stomach at level of L.V. 4. *B*, small intestine 25 minutes after ingestion of barium meal. Note duodenal cap, feathery pattern of barium in jejunum, and ileum in lower part of photograph.

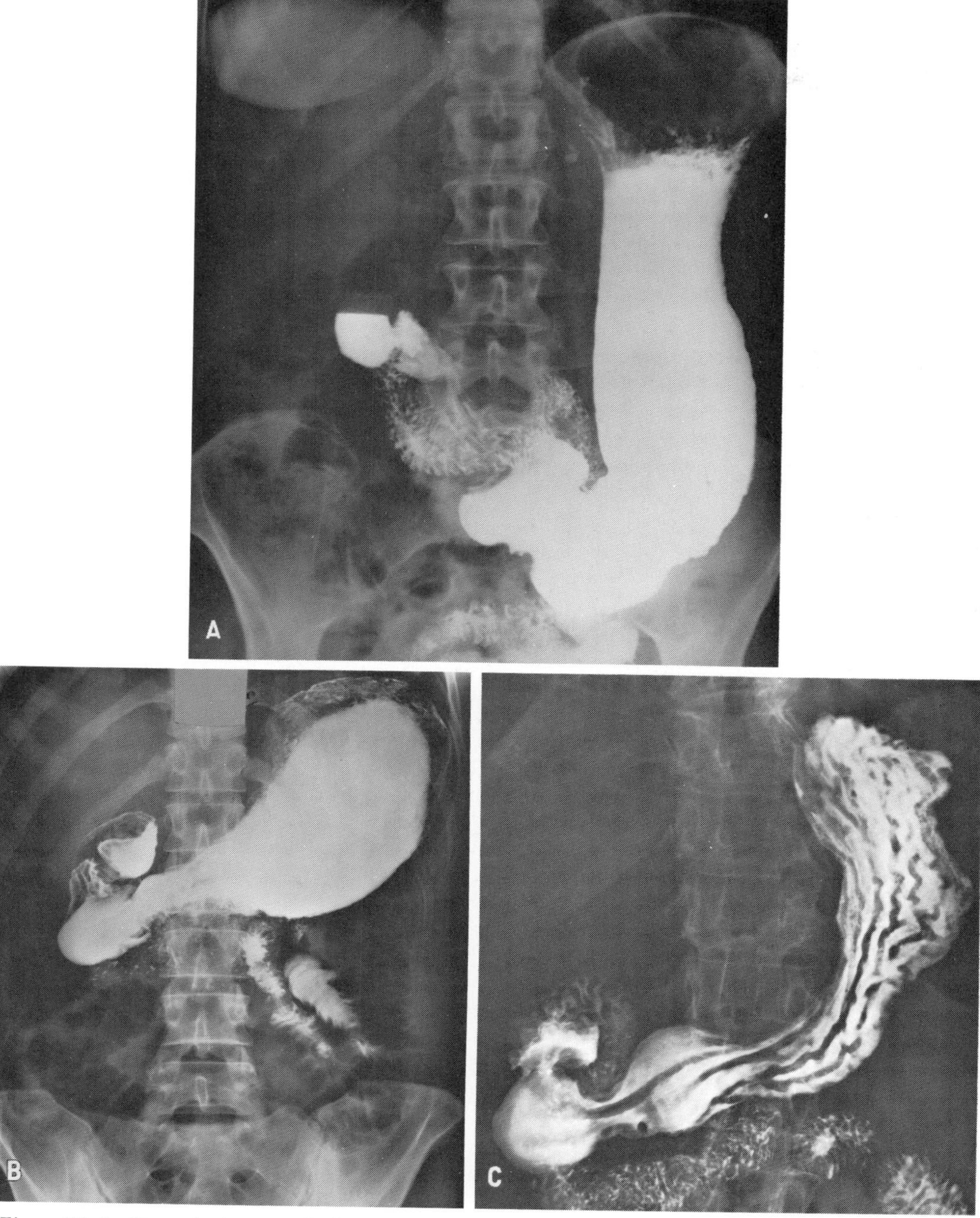

Figure 35–4 Stomach. *A*, stomach after barium meal; upright position. Some of the barium has passed through the bile and cystic ducts and is seen in the lower part of the gall bladder. (Note fluid level in both stomach and gall bladder.) The lower part of the stomach of this woman extends considerably below the supracristal plane, and the fundus of the gall bladder is at the level of L.V. 4. The shadow in the upper left corner is produced by the right breast. *B*, stomach after barium meal. Note horizontal position of stomach ("steerhorn" stomach). Duodenal cap and portions of jejunum are visible. *C*, stomach coated with barium. Note folds of mucosa of stomach, and feathery pattern of small intestine. *A*, courtesy of A. J. Chilko, M.D., New York, New York.

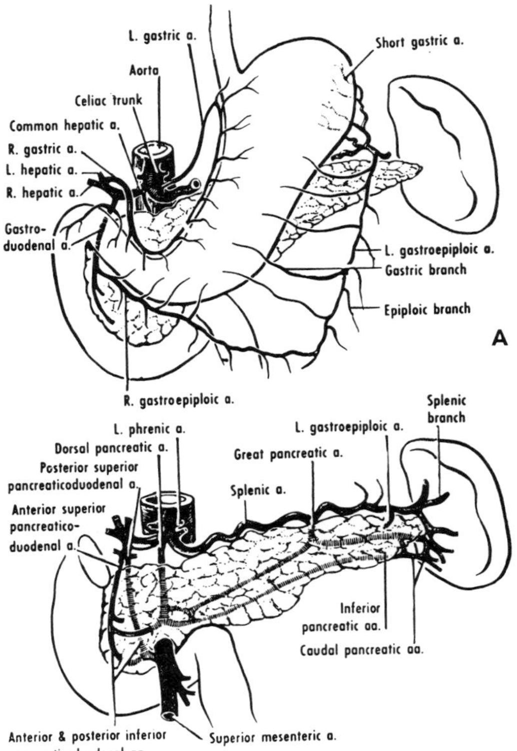

Figure 35–5 Diagrams of the arterial supply of stomach (*A*), duodenum (*B*), pancreas, and spleen. The stomach and first part of the duodenum have been removed in *B*. The pancreatic supply is based on Woodburne and Olsen.

Lymphatic Drainage.[10] The lymphatic plexuses that drain the stomach communicate with similar plexuses in the esophagus and duodenum and empty into lymphatic vessels that ultimately drain into the thoracic duct. The regional nodes lie along the adjacent arteries and are named accordingly. However, alternative and additional names are often used (fig. 35–6).

The blood and lymphatic drainage of the stomach are such that cancer can easily spread: (1) to the liver, by the portal vein or by reversal of flow in lymphatic vessels from the liver; (2) to the pelvis, by retroperitoneal lymphatic vessels; (3) to any other part of the body, both directly by veins and indirectly by way of the thoracic duct. The thoracic duct may be composed of several trunks in the neck, and one or more of these trunks may enter a cervical lymph node. There have been instances in which cancer of the stomach metastasized to a node above the left clavicle.

Nerve Supply (fig. 35–7). The stomach is supplied by the celiac plexus by way of plexuses along the arteries to the stomach, by sympathetic fibers in the left phrenic nerve, and by gastric branches of the vagal trunks (p. 422).[11]

Preganglionic sympathetic fibers reach the celiac and other ganglia by way of splanchnic nerves. Postganglionic fibers are then distributed to blood vessels and gastric musculature. Preganglionic parasympathetic fibers arise in the medulla oblongata, descend in the vagus nerves, and reach the stomach by direct gastric branches of the vagal trunks, or indirectly through the celiac plexus.

The sensory fibers in the nerves supplying the stomach are of several types. The most important are concerned with reflexes, such as reflex activation of gastric secretion. These sensory fibers ascend in the vagi. Whether pain fibers are present is open to question. The sense of hunger is said to be associated with contractions of the stomach, but this association is by no means clear.

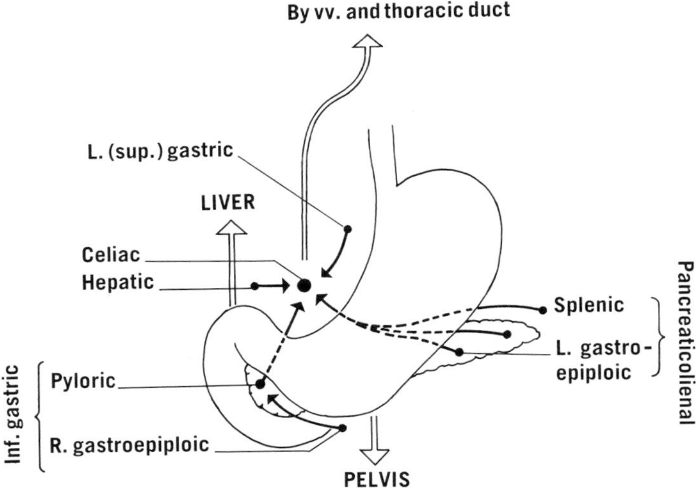

Figure 35–6 Diagram of lymphatic drainage of stomach, pancreas, and spleen. Each dot (circle) represents one or more nodes. The white arrows indicate three major directions of spread.

INTESTINE

SMALL INTESTINE

The major part of digestion occurs in the small intestine, which extends from the pylorus to the ileocolic junction, where it joins the large intestine. The small intestine consists of the duodenum, which is a short, curved portion that for the most part lacks a mesentery, and the long, coiled jejunum and ileum, which are attached to the posterior abdominal wall by a mesentery. The Greek word *enteron,* meaning gut, is used to refer to the intestine (enteritis is inflammation of the intestine) and to its peritoneal attachments (mesentery). The small intestine is an indispensable organ. Food is completely digested in it, its lining is adapted for absorption, it has a large area of surface for absorption, and it has a plentiful blood supply.

The small intestine just removed at autopsy is about 7 meters long, but varies between 5 and 8.[12] (The length of the small and large intestines after death together is 9 meters.) The length of the small intestine is correlated with stature, and is therefore slightly less (by about a meter) in women.

The small intestine has a characteristic radiographic appearance after a barium meal, owing to the circular folds and villi that impart a feathery appearance to the outline of the barium material (fig. 35–4). The duodenal cap, however, is relatively smooth (p. 434).

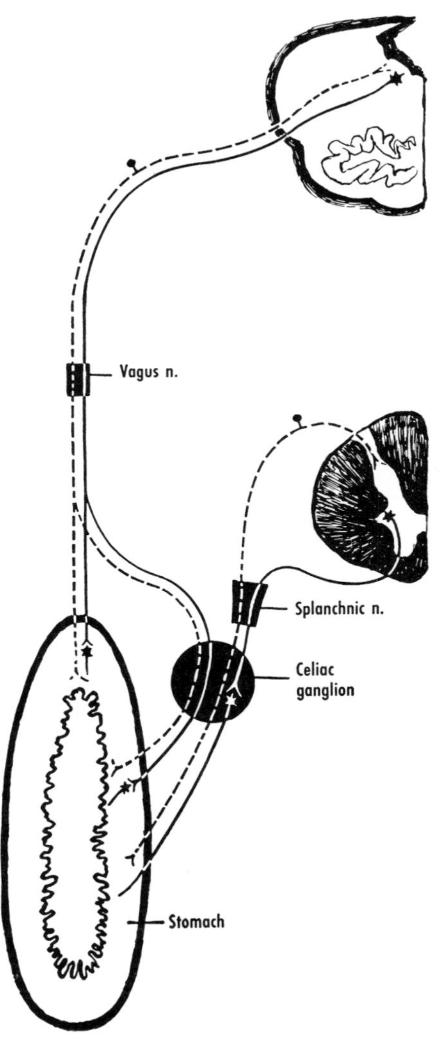

Figure 35–7 Functional components of the nerve supply of the stomach. For purposes of simplification, each component is shown as a single fiber. Autonomic fibers are shown as continuous lines, sensory fibers as interrupted lines.

Structure and Muscular Activity (fig. 35–8). The surface area of the mucosa is greatly increased by myriads of microscopic *villi* and by *circular folds.* Circular folds are permanent folds of mucosa and submucosa, which are present throughout the small intestine, except for the first few centimeters of the duodenum. Lymphatic follicles occur throughout the small intestine, and large patches of lymphatic tissue termed aggregated lymphatic follicles occur in the ileum, mainly in its lower part. The muscular coat is arranged in inner circular and outer longitudinal layers.

Chyme that enters the duodenum is soon moved on to the jejunum. Relatively little peristalsis occurs in the duodenum unless it is overloaded. The first part of the duodenum is fairly quiet, but the rest shows an irregular, quick activity that suggests a churning, milling action produced by contraction of the muscularis mucosae.

Peristalsis occurs in the jejunum and ileum, but it is not forceful, except in the case of obstruction. It is often local, that is, restricted to a loop, and it may well be of less importance than the movements of the muscularis mucosae in propelling and turning over the intestinal contents. Intestinal movements are difficult to study and analyze in man, those seen at operations are not necessarily normal, and the findings in experimental animals are not necessarily applicable to man.

The terminal part of the ileum is inert compared with the rest of the small intestine.[13] The entrance of food into the stomach tends to cause the ileum to empty into the cecum (gastroileal reflex).[14] A sphincteric mechanism at the ileocecal junction, like other

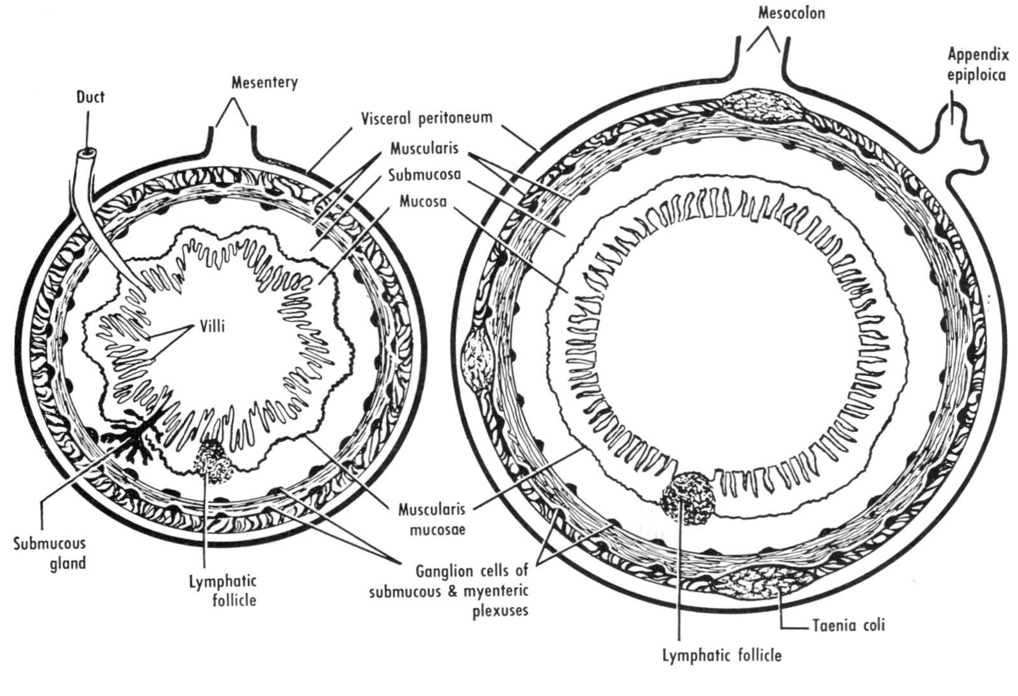

Figure 35–8 Schematic diagrams of the layers of the small intestine and large intestine. The diagram of the small intestine shows submucous glands and the opening of a duct (bile or pancreatic), although these appear only in the duodenum. Ganglion cells of the submucous and myenteric plexuses are scattered rather than in groups as shown here.

gastrointestinal sphincters, is under neural and hormonal control. The alimentary canal is active at birth. The stomach of an infant usually contains relatively more swallowed air than does that of an adult, and the infant must therefore be made to burp. The stomach of an infant also takes longer to empty than that of an adult.

Duodenum

The duodenum, which is derived from the foregut and midgut, is so called because it was estimated to be 12 fingerbreadths in length. It is variable in shape, but is usually shaped somewhat like the letter **C**, the concavity of which encloses the head of the pancreas. It has *upper* (first), *descending* (second), *horizontal* or *inferior* (third), and *ascending* (fourth) parts. The duodenum extends from the pylorus to the duodenojejunal flexure, and is 25 to 30 cm long.

1. The first part of the duodenum, which receives acid chyme from the stomach, runs to the right and backward, from the front of the vertebral column to the right side of the vertebral column and the inferior vena cava. **The beginning of the first part of the duodenum is termed the free part**[15] **because it is not attached to the posterior abdominal wall. It is very mobile and follows all the shifts in position of the pyloric part of the stomach. The free part is termed the duodenal cap in radiology; it lacks circular folds.**

In front are the liver and gall bladder; behind are the bile duct, portal vein, and pancreas. The gall bladder is so closely related to the duodenum that, after death, the duodenum is usually stained by bile that has leaked through the wall of the gall bladder.

2. The second part descends in front of the right renal vessels and a variable amount of the right kidney. In front are the liver and gall bladder, transverse colon, and small intestine.

3. The third part runs to the left, across

the right psoas major, inferior vena cava, aorta, and left psoas major. Other posterior relations include the right ureter, right testicular or ovarian vessels, and inferior mesenteric vessels. It is crossed in front by the superior mesenteric vessels and the root of the mesentery.

4. The fourth part ascends on the left side of the aorta and then turns anteriorly as the duodenojejunal flexure. The level of the flexure is almost as high as that of the first part of the duodenum. Often there is no distinction between the third and fourth parts; the two form an inferior part that ascends obliquely to the flexure.

Smooth muscle and elastic fibers, which form a triangular sheet that ascends from the back of the duodenum (from any or all parts except the first) to the right crus of the diaphragm, constitute the *suspensory muscle of the duodenum.*[16] If the muscle is short and is attached only to the duodenojejunal flexure, it may angulate the flexure and present a barrier to successful intubation.[17]

Position and Peritoneal Relations (fig. 35–9). Except for its free part, the duodenum is relatively fixed, but it nevertheless shows some variation in position. **The mobile first part is often at the level of L.V. 2 (range, T.V. 12 to L.V. 3) *in vivo*, but the level descends with age.**[18]

The first part of the duodenum is attached to the liver by the hepatoduodenal part of the lesser omentum. The beginning of the first part of the duodenum, the mobile or free part, is covered by peritoneum in front and behind. The rest of the duodenum is retroperitoneal; its posterior aspect is fixed to the posterior abdominal wall and the adjacent organs.

Bile Duct and Pancreatic Duct. **The descending part of the duodenum receives the bile duct, pancreatic duct, and accessory pancreatic duct.** The bile duct and pancreatic duct usually empty together into the *greater duodenal papilla,* a small, nipple-like projection on the interior of the posterior and medial aspect (concavity) of the second part of the duodenum, about 7 cm from the pylorus.[19] The bile duct often intertwines with the pancreatic duct, and in more than one-half of instances the two unite and form a short *hepatopancreatic ampulla,* which opens into the duodenum on the surface of the greater duodenal papilla (fig. 36–6, p. 401). The mucous membrane of the ampulla becomes continuous with that of the duodenum at the mouth of the papilla. In other instances, the two ducts meet, but open separately into the duodenum at the mouth of the papilla. Each duct usually acquires a sphincteric muscle coat; that surrounding the lower end of the bile duct is known as the *sphincter of the bile duct* or the *choledochal sphincter* (p. 401).[20] The sphincter around the pancreatic duct is sometimes poorly developed. When the two ducts form an ampulla, a *sphincter of the hepatopancreatic ampulla* may also be present. Variations in extent and thickness of the various sphincters are common.[21] The presence of a hepatopancreatic ampulla may, under certain conditions, predispose toward the reflux of bile into the pancreatic duct.

The accessory pancreatic duct empties into the *lesser duodenal papilla,* which is

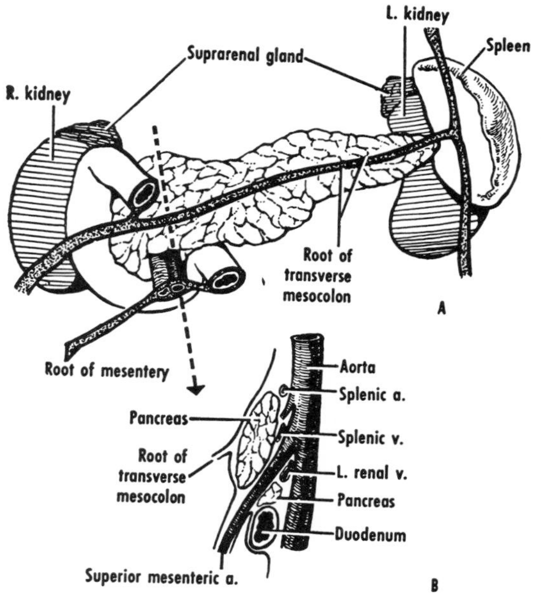

Figure 35–9 *A*, schematic diagram of the peritoneal relations of the duodenum and pancreas, from in front. Note the continuity of the mesoduodenum (of the first part of the duodenum) with the transverse mesocolon. *B*, sagittal section in the plane indicated by the arrow in *A*. Note the left renal vein, the uncinate process of the pancreas, and the duodenum in the angle between the superior mesenteric artery and the aorta.

situated on the anteromedial aspect of the descending part of the duodenum, about 2 cm higher than the greater. It is not infrequently absent.

Anomalies. Anomalies are relatively uncommon, and their origin is uncertain.[22] They include atresia (discontinuity of lumen) and stenosis (complete or incomplete).

Blood Supply, Lymphatic Drainage, and Nerve Supply. The entrance of the bile duct into the duodenum is a rough indication of the fact that the celiac trunk supplies the infradiaphragmatic portion of the foregut, and the superior mesenteric artery supplies the midgut. The branches from the celiac trunk arise from the right gastric and gastroduodenal arteries, and from the superior mesenteric artery by way of the inferior pancreaticoduodenal arteries (fig. 35–5). The principal blood supply is from the two arcades formed by the superior and inferior pancreaticoduodenal arteries (fig. 35–5). A lesser supply, principally to the first part of the duodenum, is derived from the supraduodenal and retroduodenal branches of the gastroduodenal artery (p. 416). Veins from the duodenum tend to follow the arteries, but are more variable.

The arteries approach the duodenum through its concavity. An incision along the right edge of the second part of the duodenum will therefore free or mobilize the duodenum and the head of the pancreas without endangering their blood supply.

The lymphatic vessels of the duodenum drain into anterior and posterior collecting trunks and nodes that lie in front of and behind the pancreas. Ultimately they drain into the thoracic duct.

The duodenum is supplied by autonomic and sensory fibers derived from the celiac and superior mesenteric plexuses. The nerve supply is similar to that of the rest of the small intestine (fig. 35–12).

Jejunum and Ileum

Of the coiled part of the small intestine, the jejunum constitutes about the proximal two-fifths and the ileum about the distal three-fifths.

In contrast to the ileum, the jejunum "typically" is more often empty, more vascular (redder in the living), and thicker walled, and its mesentery shows translucent areas between the vessels, owing to the absence of fat. But these "typical" differences are between the upper part of the jejunum and the lower part of the ileum. It is often difficult to distinguish between the jejunum and the ileum during an operation.

Very occasionally a remnant of the omphalomesenteric or vitellointestinal duct of the embryo persists, and in the adult may form a *diverticulum ilei* several centimeters in length.[23] In the embryo the duct is attached to the apex of the midgut loop. Hence, in the adult, a diverticulum ilei occurs at a variable distance from the cecum.

Relations. The jejunum and ileum are suspended from the posterior abdominal wall by the mesentery. They are much longer than the root of the mesentery and are arranged in coils or loops. These loops occupy much of the abdominal cavity and some of the pelvic cavity. The jejunum and ileum are much more mobile than any other part of the alimentary canal, and any one loop can occupy almost any position in the abdominal cavity. The root of the mesentery runs downward and to the right (fig. 34–4, p. 369), and the loops of the upper part of the jejunum tend to be in the left upper quadrant, whereas the end of the ileum is in the lower right quadrant. The upper left end of any loop of intestine is the gastric or proximal end.

Mesentery. The fan-shaped mesentery connects the jejunum and ileum to the posterior abdominal wall. **The border attached to the abdominal wall is the *root of the mesentery*. The root is about 15 cm long and is directed obliquely downward and to the right, from the duodenojejunal flexure to the level of the right sacroiliac joint.** Measured from the root to the intestine, the mesentery may be 20 cm or more in width in its central part. The mesentery consists of two layers of peritoneum, right and left, which pass forward to the intestine (see fig. 34–3, p. 367). The two layers also contain between them the branches of the superior mesenteric vessels, nerves, lymph nodes and vessels, and a variable amount of fat.

At the posterior abdominal wall, the lower part of the right layer of the mesentery passes to the right over the ascending colon; the upper part becomes continous with the inferior layer of the transverse mesocolon. The left layer of the mesentery passes to the left, over the descending colon.

Blood Supply (fig. 35–10).[24] The superior mesenteric artery supplies the midgut, that is, the small intestine from the entrance of the bile duct into the duodenum, and the large intestine to near the left colic flexure.

A variable number of *jejunal* and *ileal arteries* (no more than 6 or 7 large ones) arise from the convexity (left side) of the superior mesenteric artery. These branches descend in the mesentery, branching in such a way as to form a series of arcades (fig. 35–11).[25] Straight arteries from the most peripheral arcade run directly to the gut without anastomosing. They divide at the gut and give branches to one or both sides. The arcades are probably related to mobility in the sense

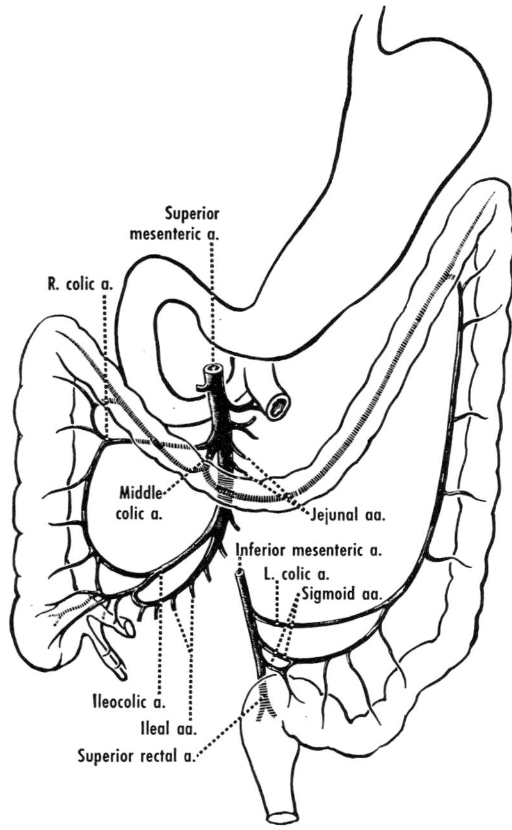

Figure 35–10 Diagram of the arterial supply of the jejunum, ileum, and colon.

that very free movement may occur without interfering with the circulation. Arcades are thus fewer in number in the less mobile upper part of the jejunum.[26] Within the wall of the intestine, the vessels form an anastomosing network of small vessels. The anastomoses, although relatively poor on the antimesenteric border,[27] are sufficient to sustain several centimeters of intestine.

Veins accompany the arteries (as do lymphatic vessels and nerve plexuses) and drain by way of the superior mesenteric vein into the portal vein.

Lymphatic Drainage. Drainage from the mucosa is into lymphatic vessels that accompany the mesenteric blood vessels. After a fatty meal, the mesenteric lymphatics contain considerable amounts of emulsified fat; the creamy-appearing lymph is termed chyle.

Lymph nodes are abundant in the mesentery and lie along the blood vessels. Lymph from the intestine courses toward the root of the mesentery and ultimately enters the thoracic duct.

Cancer of the intestine can spread to the liver by the portal vein and by lymphatic vessels (abundant communications occur between the lymphatic vessels of all the abdominal viscera), and to anywhere else in the body by veins and by the thoracic duct. The liver is frequently the organ first involved by spread of cancer from the intestine or the stomach.

The lymphatic vessels in the wall of the small intestine (and colon) run at right angles to the long axis, and cancer or tuberculosis spreading along them tends to encircle the gut and constrict it. By contrast, the vessels of the rectum run longitudinally, and obstruction, if any, is late in occurrence.

Nerve Supply (fig. 35–12). The small intestine is supplied by autonomic and sensory fibers from the celiac and superior mesenteric plexuses. The fibers accompany the arteries to the intestine. The sensory fibers include pain fibers and fibers concerned with reflex regulation of movement and secretion. The intestine is quite insensitive to most painful stimuli, including cutting and burning, but it is quite sensitive to distention. Distention results in a sensation of "cramps."

LARGE INTESTINE

The large intestine consists of the cecum and appendix, the colon,[28] which has ascending, transverse, descending, and sigmoid parts, and the rectum and anal canal.

Air and gas bubbles are often seen in the large intestine in ordinary radiograms, especially in the cecum, ascending colon, and distal part of the transverse colon.[29] Barium given by mouth or enema can be used to delimit the large intestine (fig. 35–13).

Structure and Function. The large intestine, except for the rectum and anal canal, is characterized by a mucosa with goblet cells, glands, and absorptive cells. The *appendices epiploicae* (singular, appendix epiploica) are small masses of fat, enclosed in peritoneum, which extend from the surface of the colon.

The outer longitudinal layer of muscle is thickened by the three bands, the *taeniae coli* (fig. 35–19).[30]

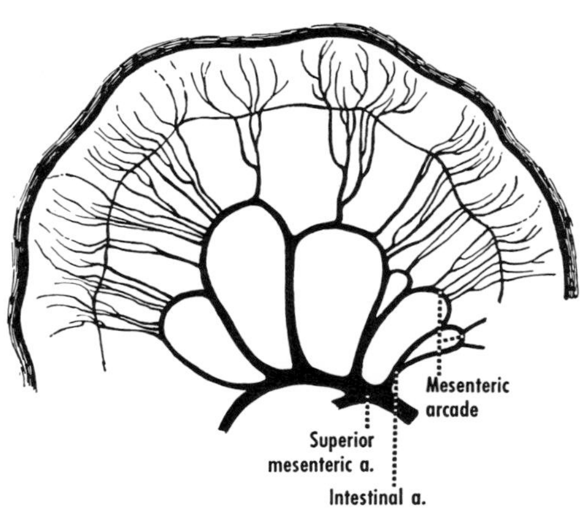

Figure 35–11 Mesenteric vascular pattern.

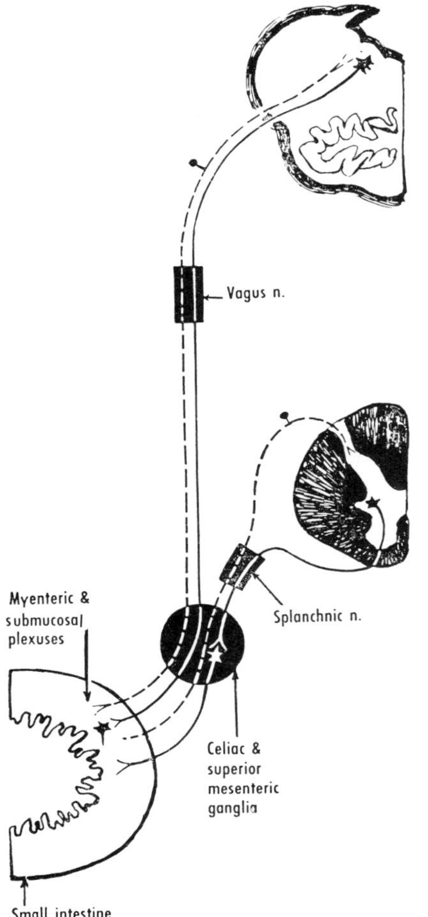

Figure 35–12 Functional components of the nerve supply of the small intestine. For purposes of simplification, each component is shown as a single fiber. Autonomic fibers are shown as continuous lines, sensory fibers as interrupted lines.

The taeniae (singular, taenia), which are about 1 cm in width, are variable in position. They are best marked on the cecum and ascending colon. Their positions on the transverse colon are variable; they are less conspicuous and tend to become diffuse on the descending colon, and they are quite indistinct on the sigmoid colon. On the rectum (p. 488), the longitudinal muscle coat is in the form of broad, diffuse bands, chiefly on the anterior and posterior aspects.

Most of the large intestine is characterized by puckerings or sacculations, termed *haustra* (singular, haustrum). It is uncertain whether the haustra are due to shortness of the taeniae as compared with the length of the colon, or whether they are due to a special arrangement of circular muscle. Haustra in living individuals shift in position, and disappear from time to time.

The large intestine is characterized by its capacity, its distensibility, the length of time it retains its contents, and the special arrangement of its musculature. It possess considerable mobility, especially the transverse and the sigmoid colon. These properties are directly related to the chief functions of the large intestine, which are the formation, transport, and evacuation of feces.[31] These functions require mobility, the absorption of water, and the secretion of mucus.

MOBILITY.[32] Movements of the large intestine are quite different from those of the small intestine. Moreover, they are so infrequent that they are not visible when studied by ordinary radiographic methods. When food enters the stomach, activity increases in the colon. The terminal part of the ileum begins to fill about 1/2 to 5 hours after barium is given orally. The ileum empties into the cecum by a sort of stripping movement, which occurs at 3 to 15 minute intervals and is 10 to 20 seconds in duration. As material enters the cecum, the latter relaxes and descends; its haustra disappear. The cecum and the ascending and transverse colon fill slowly, largely by a passive process that may take several hours. As the transverse colon fills, it becomes lower in position.

Muscular activity of the colon consists of rapid, powerful "mass movements," associated with the formation of haustra. These mass movements occur mainly in the transverse colon, but also in the descending and the sigmoid colon at intervals of two or three times a day. They move the colonic contents to the sigmoid colon. The contents of the bowel are usually held in the sigmoid colon until just before defecation begins.

ABSORPTION OF WATER. The material in the terminal part of the ileum is about 90 per cent water, most of which is absorbed from the cecum and ascending colon.

SECRETION OF MUCUS. Mucus, which is secreted profusely in response to injury or irritation, is an extremely important protective substance. It protects the mucous membrane from direct injury, dilutes irritants, and interferes with or prevents the absorption of many substances.

Blood Supply (figs. 35–10 and 35–14).[33] The large intestine is supplied by the superior and inferior mesenteric arteries, with the inferior mesenteric being distributed to the hindgut portion (from the left part of the transverse colon distally). The branches of the mesenteric arteries form a long marginal artery that in most instances extends from the cecum to the sigmoid colon. It is comparable to a series of primary arcades. Variable secondary arcades arise from it, especially where the colon is more mobile. Straight vessels that arise from the marginal artery or arcades supply the walls of the gut (fig. 35–15). The arteries that feed the marginal artery are the ileocolic, the right, middle, and left colic, and the sigmoid (p. 417).

The veins that accompany the arteries drain into the portal vein by way of the inferior and superior mesenteric veins. There are also many small, retroperitoneal veins that leave the retroperitoneal parts of the intestine and connect with the veins of the body wall (p. 418).

Lymphatic Drainage. The lymphatic drainage is similar to that of the small intestine. The vessels in the wall run mostly at right angles to the long axis of the gut, and form plexuses that drain first into nodes near the gut. These regional nodes are named according to their position: ileocolic, right, middle, and left

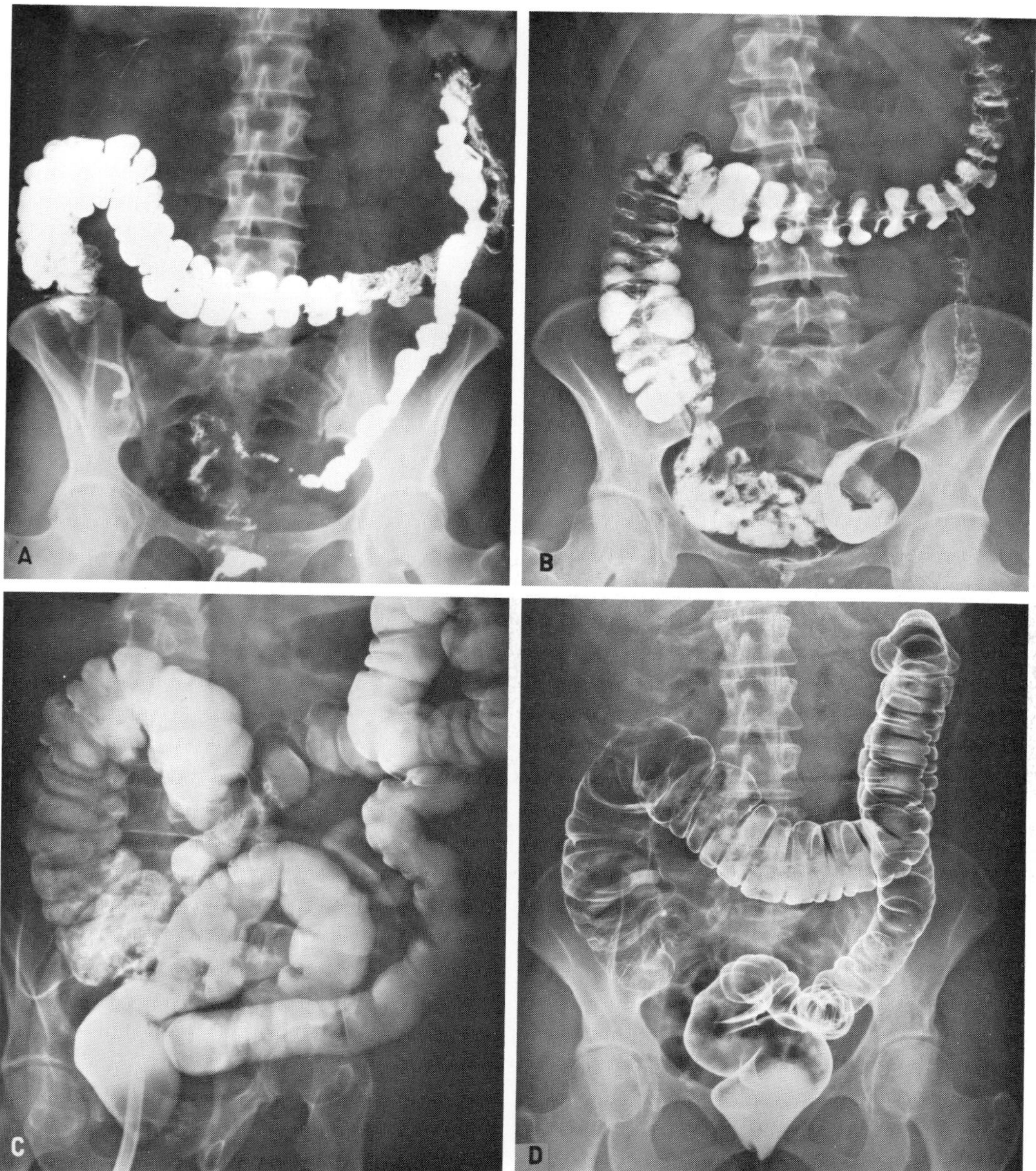

Figure 35–13 Large intestine shown by barium enema. Note different levels of transverse colon. *A*, note vermiform appendix. *B*, note ileum and pattern of colon. *C*, oblique view of colon and rectum. *D*, colon and rectum shown by double contrast enema. *A*, courtesy of Maurice C. Howard, M.D., Omaha, Nebraska. *C*, from *Medical Radiography and Photography;* courtesy of Robert A. Powers, M.D., Palo Alto, California. *D*, from *Medical Radiography and Photography;* courtesy of Eugene E. Ahern, M.D., Minneapolis, Minnesota.

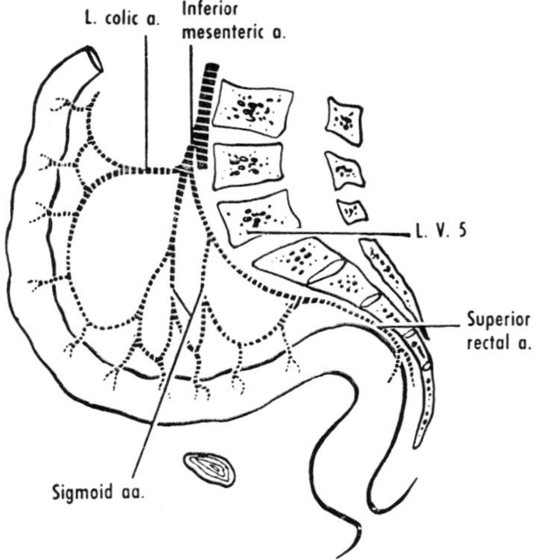

Figure 35–14 A common pattern of the inferior mesenteric artery as viewed in lateral perspective. The descending and sigmoid colon have been pulled forward.

colic, and inferior mesenteric. They drain into vessels that accompany the blood vessels. Cancer of the colon can spread to the liver by the portal vein and by lymphatics, and to other parts of the body by veins and lymphatic vessels.

Nerve Supply (fig. 35–16). Autonomic and sensory fibers reach the large intestine by way of continuations of the celiac, superior mesenteric, and inferior mesenteric plexuses that accompany the colic arteries. The parasympathetic supply to the distal part of the colon, however, is derived from the pelvic splanchnic nerves by way of the hypogastric nerves and the inferior hypogastric plexuses (p. 456),[34] through branches that supply the sigmoid colon and extend up to the middle of the descending colon, sometimes as far as the left flexure. The pain fibers in the colon are activated by distention, and enter the spinal cord by way of the splanchnic nerves. The reflex control of defecation is described elsewhere (p. 490).

Cecum and Appendix

Cecum. The cecum is the part of the large intestine that lies at and below the level at which the ileum joins the large intestine. The cecum lies in the right iliac fossa, and may reach the pelvic brim, especially when the subject is standing. It is usually surrounded by peritoneum, but it lacks appendices epiploicae. There is usually no mesocecum, but a fold on each side of the posterior aspect of the cecum may extend between the cecum and the body wall. In the embryo, the cecum is cone-shaped at first, the taeniae meet at the apex of the cone, and the appendix develops from the apex. Subsequently, the growth is such that the cecum becomes **U**-shaped and its lateral wall forms its most dependent part, as it does in the adult cecum. Also, the attachment of the appendix is on the posteromedial aspect of the cecum.

The terminal part of the ileum usually enters the large intestine posteromedially, sometimes medially, occasionally posteriorly. The last few centimeters of the ileum sometimes lack a mesentery. This part of the ileum is then related directly to the posterior body wall. The *ileocecal* opening has two folds or lips, one above and one below, that form the so-called *ileocecal (ileocolic) valve.* These folds meet at their ends and may continue as the *frenulum of the ileocecal valve,* around the circumference of the intestine, thereby marking the cecocolic junction. The folds and the frenulum vary considerably in size, thickness, and arrangement.[35] In the living individual, the ileum forms a conoid projection into the cecum, and the term ileal papilla has been proposed for this projection.[36] The ileocecal valve has little mechanical importance. After a barium enema, when the colon is filled, barium is often found in the terminal part of the ileum.

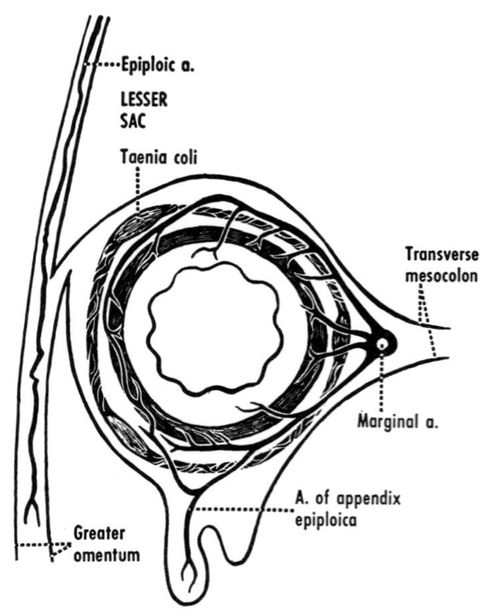

Figure 35–15 Diagram of the intrinsic arterial supply of the colon.

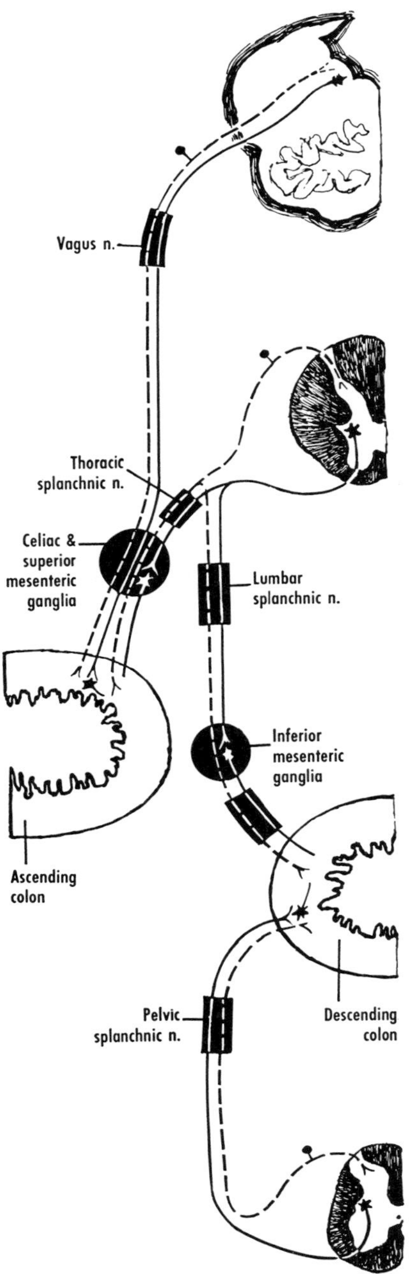

Figure 35–16 Functional components of the nerve supply of the colon. For purposes of simplification, each component is shown as a single fiber. Autonomic fibers are shown as continuous lines, sensory fibers as interrupted lines.

The important posterior relations of the cecum are with the muscle, vessels, and nerves in the iliac fossa (fig. 35–17). In front are the anterior abdominal muscles, with variable parts of the greater omentum, transverse colon, and small intestine intervening.

BLOOD SUPPLY.[37] The ileocolic artery, a branch of the superior mesenteric, gives off the *anterior* and

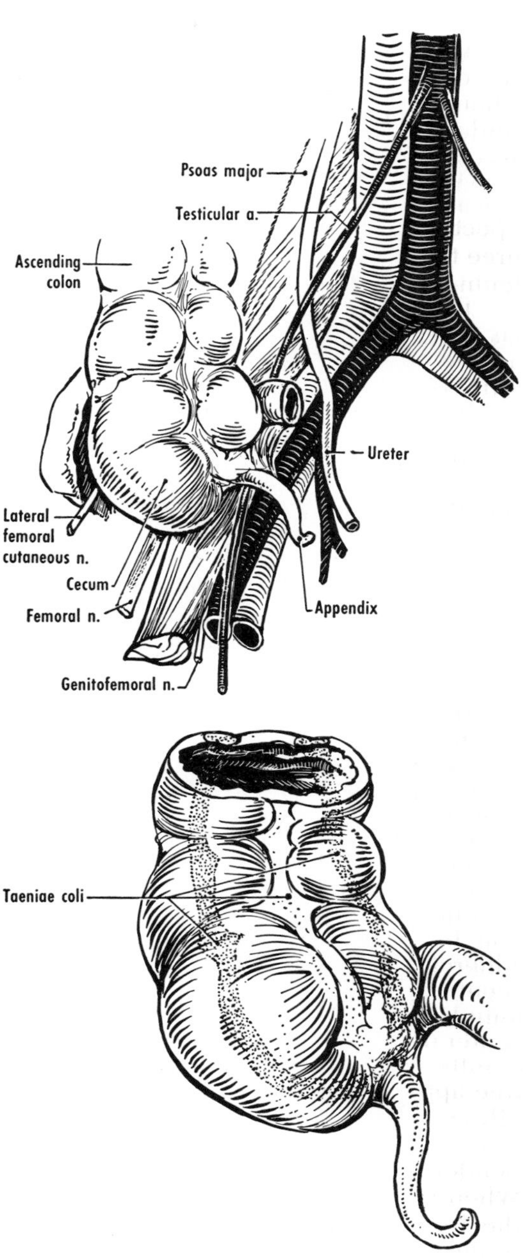

Figure 35–17 *Upper,* common relations of cecum and appendix. A free appendix may hang over the pelvic brim or may be retrocecal. *Lower,* positions of the taeniae, which meet at the appendix.

the *posterior cecal artery*. The posterior supplies most of the cecum.

Appendix. A vermiform appendix occurs only in man and the anthropoid apes, although some other mammals may have a similar organ or lymphoid structure at the apex of the cecum. It is congenitally absent in man on rare occasions.[38] The appendix typically arises from the posteromedial aspect of the cecum at the junction of the three taeniae coli, about 1 to 2 cm below the ileum. It is about 9 to 10 cm in length.

The appendix lacks sacculations and has a longitudinal muscle coat that lacks taeniae. Its mucosa is heavily infiltrated with lymphoid tissue. The part of the appendix immediately adjacent to the cecum has a narrower lumen and a thicker muscularis than does the more distal part.

The appendix lacks a true mesentery, but there is usually a peritoneal fold termed the mesentery of the vermiform appendix (mesoappendix), which contains the *appendicular artery*, a branch of the ileocolic artery. The appendix often receives additional branches from the cecal arteries.[39] The fold is often short, so that the appendix is folded or kinked on itself.

The appendix is variable in position, and may be classified as anterior, with ileal or pelvic positions, or posterior, with subcecal, retrocecal, or retrocolic positions.[40] The most common position is pelvic. Appendices may also be fixed or free. Fixed appendices, which are present in a little more than a third of instances, are usually either retrocecal or retrocolic in position. They are held in position by a short peritoneal fold or by an adherence to the posterior aspect of the cecum or colon. A free appendix may be found anywhere within a sphere, the center of which is the attachment of the appendix to the cecum.[41] The position of a free appendix is partly determined by the filling of the cecum. When the cecum is full and descends, the free appendix is dependent and usually reaches the pelvis. When the cecum is empty and contracted, the free appendix may become retrocecal in position.

The appendix is a narrow, hollow, muscular organ. If it is inflamed it tends to go into spasm. Spasm, as well as distention, causes pain which is referred to the epigastrium. If the adjacent parietal peritoneum becomes inflamed, pain is also felt in the right lower quadrant of the abdomen, and the overlying muscles often show reflex spasm. **The point of maximal tenderness to pressure may be anywhere in the right lower quadrant. McBurney originally described this point as about 2 inches from the anterior superior iliac spine on a line to the umbilicus in cases of inflamed appendix. This point, however, should not be used as an anatomical landmark—the appendix and the umbilicus are too variable in position.[42]**

Colon

Ascending Colon. The ascending colon extends upward in the right iliac fossa and on the posterior abdominal wall to the *right colic flexure*, which lies in front of a part of the right kidney. Except for its lowermost part, the ascending colon is covered with peritoneum only in front and on its sides, being related behind to the abdominal wall. An ascending mesocolon is sometimes observed. Not infrequently, in the living, the laxity of tissues is such that the ascending colon lies well away from the posterior abdominal wall and may, in fact, touch the anterior abdominal wall. Sometimes peritoneum spreads directly across the front of the ascending colon without dipping back at the sides. The pericolic "bands," or "membranes," thereby formed markedly reduce the normal mobility of the colon.

Transverse Colon. The transverse colon, which is derived from both midgut and hindgut, extends to the left from the right colic flexure. The right part is related behind to the duodenum and pancreas, but the remainder has extremely variable relations. The *left colic flexure* is usually higher, more acute, and less mobile than the right, and both flexures usually have a considerable anteroposterior component.[43] A fold of peritoneum, the *phrenicocolic ligament*, may connect the flexure to the diaphragm. **The transverse mesocolon is of a length such that most of the transverse colon loops down, often below the level of the iliac crests, and even into the true pelvis.**

TRANSVERSE MESOCOLON. This is a broad fold of peritoneum that passes forward from the pancreas and encloses most of the transverse colon. Its upper layer is adherent to, or fused with, the greater omentum (see figs. 34–5 and 34–6, pp. 370, 371). Its lower layer covers the lower part of the pancreas and the third and fourth parts of the duodenum, and is continuous with the right layer of the mesentery. The blood vessels, nerves, and lymphatic vessels of the transverse colon are contained between the layers of the mesocolon.

Descending Colon. The descending colon, which usually lacks a mesocolon, descends to approximately the pelvic brim, where the sigmoid colon begins.

Sigmoid (Pelvic) Colon. The sigmoid colon, which becomes the rectum in front of the sacrum, is distinguished by the *sigmoid mesocolon,* the line of attachment of which is variable. The sigmoid colon itself forms a loop, the shape and position of which depend much upon the degree of filling. When the sigmoid colon is full, it is relaxed, and therefore much longer, and may reach the epigastrium[44] or it may lie in the pelvis. When empty and shorter, it usually runs first forward and to the right, then backward and to the right. In Ethiopians, the sigmoid colon usually has an ascending suprapelvic loop, regardless of the state of filling.[45] A volvulus of this loop is the commonest cause of acute intestinal obstruction in this population.

The rectosigmoid junction receives branches from the superior rectal artery as well as from the sigmoid arteries.

SIGMOID (PELVIC) MESOCOLON. This is a fold of peritoneum that attaches the sigmoid colon to the pelvic wall. Its line of attachment may form an inverted **V**, the apex of which lies in front of the left ureter and the division of the left common iliac artery. An intersigmoid recess (p. 373) may be present at the apex.

REFERENCES

1. G. S. Muller Botha, *The Gastro-oesophageal Junction,* Churchill, London, 1962. G. W. Friedland *et al.,* Thorax, *21*:487, 1966. B. S. Wolf, P. Heitmann, and B. R. Cohen, Amer. J. Roentgenol., *103*:251, 1968. F. F. Zboralske and G. W. Friedland, Calif. west. Med., *112*:33, 1970. The last contains a list of the many anatomical terms and synonyms applied to the distal esophagus and gastroesophageal junction.
2. R. H. Salter, Lancet, *1*:347, 1974.
3. A. H. James, *The Physiology of Gastric Digestion,* Arnold, London, 1957. G. S. Muller Botha, Brit. J. Surg., *45*:569, 1958.
4. C. A. F. de Carvalho, Acta anat., *64*:125, 1966.
5. G. A. G. Mitchell, Brit. J. Surg., *26*:333, 1938.
6. E. Landboe-Christensen, Acta path. microbiol. scand., *54* (Suppl.):671, 1944.
7. C. A. Nafe, Arch. Surg., Chicago, *54*:555, 1947.
8. H. I. El-Eishi, S. F. Ayoub, and M. Abd-el-Khalek, Acta anat., *86*:565, 1974.
9. T. E. Barlow, F. H. Bentley, and D. N. Walder, Surg. Gynec. Obstet., *93*:657, 1951.
10. J. H. Gray, J. Anat., Lond., *71*:492, 1937. I. Donini, Acta anat., *23*:289, 1955.
11. G. A. G. Mitchell, J. Anat., Lond., *75*:50, 1940.
12. B. M. L. Underhill, Brit. med. J., *2*:1243, 1955.
13. A. E. Barclay, Radiology, *33*:170, 1939.
14. A. F. Hertz, J. Physiol., *47*:54, 1913.
15. H. Ogilvie, Lancet, *1*:1077, 1952.
16. J. C. Haley and J. H. Perry, Amer. J. Surg., *77*:590, 1949. M. Argème *et al.,* C. R. Ass. Anat., *55*:76, 1970. L. Costacurta, Acta anat., *82*:34, 1972.
17. M. O. Cantor, Surgery, *26*:673, 1949.
18. S. M. Friedman, Amer. J. Anat., *79*:147, 1946.
19. V. J. Dardinski, J. Anat., Lond., *69*:469, 1935.
20. E. A. Boyden, Surg. Gynec. Obstet., *104*:641, 1957. I. Singh, J. anat. Soc. India, *6*:1, 1957.
21. J. A. Sterling, Surg. Gynec. Obstet., *98*:420, 1954. I. Singh, J. anat. Soc. India, *5*:54, 1956.
22. H. B. Lynn and E. E. Espinas, Arch. Surg., Chicago, *79*:357, 1959. T. V. Santulli and W. A. Blanc, Ann. Surg., *154*:939, 1961. E. A. Boyden, J. G. Cope, and A. H. Bill, Anat. Rec., *157*:218, 1967.
23. L. M. Howell, Amer. J. Dis. Child., *71*:365, 1946. See also H. H. Curd, Arch. Surg., Chicago, *32*:506, 1936.
24. J. V. Basmajian, Surg. Gynec. Obstet., *101*:585, 1955. J. Sonneland, B. J. Anson, and L. E. Beaton, Surg. Gynec. Obstet., *106*:385, 1958.
25. R. J. Noer, Amer. J. Anat., *73*:293, 1943. R. Sarrazin and J. B. Levy, C. R. Ass. Anat., *53*:1503, 1969.
26. J. A. Ross, Edinb. med. J., *57*:572, 1950; Brit. J. Surg., *39*:330, 1952. T. E. Barlow, Brit. J. Surg., *43*:473, 1956.
27. F. S. A. Doran, J. Anat., Lond., *84*:283, 1950.
28. See Brit. med. J., *1*:708, 1946, for a brief discussion of the history of the terminology of the parts of the colon.
29. C. F. DeGaris, Ann. Surg., *113*:540, 1941.
30. P. E. Lineback, Amer. J. Anat., *36*:357, 1925. G. F. Hamilton, J. Anat., Lond., *80*:230, 1946.
31. T. L. Hardy, Lancet, *1*:519, 1945.
32. A. F. Hertz and A. Newton, J. Physiol., *47*:57, 1913. A. F. Hurst, Lancet, *1*:1483, 1935. A. Oppenheimer, Amer. J. Roentgenol., *45*:177, 1941. A. E. Barclay, *The Digestive Tract,* Cambridge University Press, London, 2nd ed., 1936.
33. J. C. Goligher, Brit. J. Surg., *37*:157, 1949. J. A. Steward and F. W. Rankin, Arch. Surg., Chicago, *26*:843, 1933. A. F. Castro and R. S. Smith, Surg. Gynec. Obstet., *94*:223, 1952. J. A. Ross, Edinb. med. J., *57*:572, 1950. J. G. Brockis and D. B. Moffat, J. Anat., Lond., *92*:52, 1958. J. D. Griffiths, Brit. med. J., *1*:323, 1961.
34. R. T. Woodburne, Anat. Rec., *124*:67, 1956.
35. R. E. Buirge, Anat. Rec., *86*:373, 1943.
36. L. J. A. DiDio, J. R. Marques, and E. P. Pinto, Acta anat., *44*:346, 1961.
37. W. C. O. Hall, J. Anat., Lond., *83*:65, 1949.
38. J. J. Saave, Acta anat., *23*:327, 1955.
39. M. A. Shah and M. Shah, Anat. Rec., *95*:457, 1946.
40. K. Buschard and A. Kjaeldgaard, Acta chir. scand., *139*:293, 1973.
41. C. F. DeGaris, Ann. Surg., *113*:540, 1941. H. Maisel, Anat. Rec., *136*:385, 1960.
42. R. O'Rahilly, Irish J. med Sci., p. 738, November, 1948.
43. J. P. Whalen and P. A. Riemenschneider, Amer. J. Roentgenol., *99*:55, 1967.
44. A. Oppenheimer and G. W. Saleeby, Surg. Gynec. Obstet., *69*:83, 1939.
45. F. B. Lisowski, Ethiop. med. J., *7*:105, 1969.

LIVER, BILE PASSAGES, PANCREAS, AND SPLEEN 36

The duct systems of the liver and the pancreas are derivatives of the alimentary canal. They arise in the embryo as outgrowths at about the junction of the foregut and the midgut, that is, from the region that becomes the duodenum. All vertebrates have a liver and pancreas, the exocrine secretions of which are emptied into the intestine.

LIVER

The liver (Gk. *hepar,* liver; hence the adjective *hepatic*) is a large, soft, reddish organ. **The liver is the largest gland in the body, and is an important exocrine gland, the secretion of which is termed bile. Many of the products of the hepatic cells are discharged directly into the blood stream, and are often considered to be its endocrine secretion.**

For the most part, the liver lies under cover of the thoracic bony cage, covered by the diaphragm. Its position in a living person can be demonstrated by the uptake of certain radioactive isotopes (fig. 36–1).[1] At birth it is relatively large and occupies nearly two-fifths of the abdomen. In adults the range in weight is from 1000 to 3000 gm. The liver presents two surfaces, diaphragmatic and visceral.

The *diaphragmatic surface* of the liver is smooth and shaped so that it fits the diaphragm. This surface has anterior, posterior, superior, and right aspects or parts. which are sometimes designated separately as surfaces. The diaphragmatic surface is separated in front from the visceral surface by the sharp *inferior border,* which is interrupted by the shallow *notch for the ligamentum teres.* This notch is about in the median plane.

The *visceral,* or *inferior, surface* of the liver, which is somewhat flattened, and which faces downward, backward, and to the left, presents the *quadrate* and *caudate lobes* (fig. 36–2). These are marked off by an **H**-shaped group of fissures and grooves. The cross-bar of the **H** is the *porta hepatis,* which is the hilus of the liver, and which contains the hepatic ducts and the branches of the portal vein and the hepatic artery proper. The *fissure for the ligamentum teres* extends from the notch at the inferior border to the left branch of the portal vein at the porta hepatis. The fissure contains the *ligamentum teres,* which is the obliterated remains of the left umbilical vein. The *fissure for the ligamentum venosum* extends from the porta hepatis to the inferior vena cava. This deep fissure contains the *ligamentum venosum,* the fibrous remains of the ductus venosus.

The liver can be divided into *right* and *left lobes,* which are demarcated on the diaphragmatic surface by the attachment of the falciform ligament, and on the visceral surface by the fissure for the ligamentum venosum behind and the fissure for the ligamentum teres in front. However, the liver is actually a bilateral organ, consisting of right and left halves that are sharply divided from each other. The plane of division is sometimes called the "principal plane" or "main boundary fissure"; it extends forward from the gall bladder and inferior vena cava, somewhat to the right of the falciform ligament and the median plane. The right and left halves receive, respectively, the right and left branches of the portal vein and of the hepatic artery proper, and give rise to the

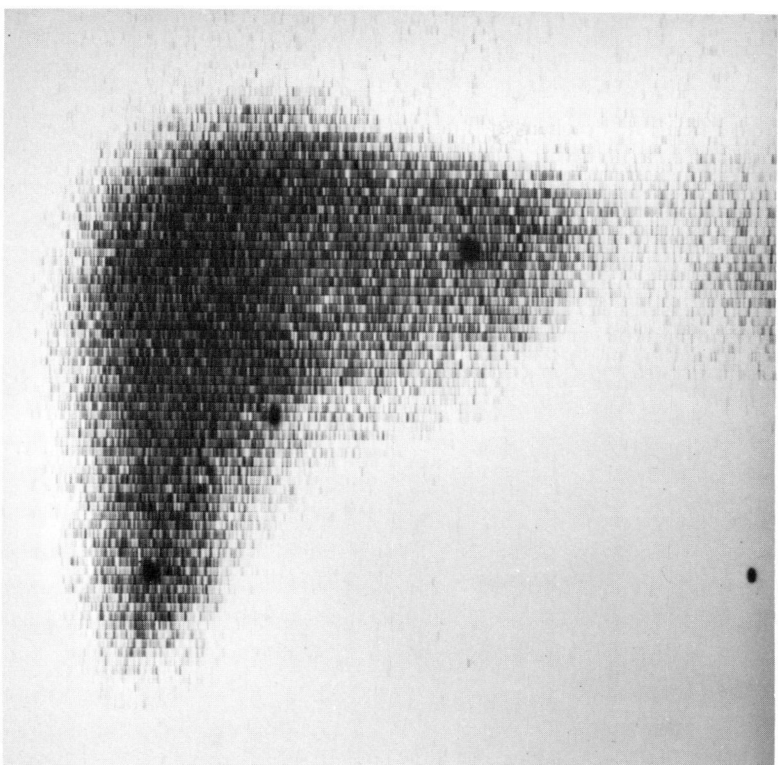

Figure 36–1 Scintigram of liver. The radioisotopic scan of the liver depends on (1) phagocytosis of particles by the stellate cells of the reticuloendothelial system, and (2) excretion by the hepatic cells into the biliary system. This anterior view shows the normal appearance of the liver. Areas of lesser density are frequently seen near the hepatic veins, the porta hepatis, the fossa for the gall bladder, and the notch for the ligamentum teres. The fainter area at the right of the photograph is produced by the spleen. From F. H. DeLand, M.D., and H. N. Wagner, M.D., *Atlas of Nuclear Medicine,* Volume 3, *Reticuloendothelial System, Liver, Spleen and Thyroid,* Saunders, Philadelphia, 1972; courtesy of the authors.

right and left hepatic ducts. The halves are nearly equal in weight, and there is little overlap in the intrahepatic distribution of vessels and ducts.

Within each half, the primary branches are sufficiently consistent so that four portal segments can be described (fig. 36–3).[2] The pattern of secondary branching is such that additional segments can also be delineated. For example, a horizontal plane in each half divides the segments into upper and lower parts, making eight liver (portal) segments. If the caudate lobe is designated separately, there are nine segments. However, the variations in secondary branching are such that general agreement has not yet been reached on terminology. Moreover, the hepatic veins are intersegmental in position, draining adjacent segments, and segmental surgery must take this into account.[3]

Unlike the bronchopulmonary segments, the hepatic segments are not separated by connective tissue septa. Furthermore, the liver is uniform in structure and function. Hence, the division into segments is significant diagnostically and surgically, but has little functional importance.

Relations. The liver moves with respiration, and it shifts in position with any postural change that affects the diaphragm. Moreover, it varies in position according to the body type of the individual. The following important relations may apply only when the body is in the recumbent position.

The convex diaphragmatic surface is in contact with the diaphragm and anterior abdominal wall, and is mostly separated from them by peritoneum. A large part of the posterior part of the diaphragmatic surface is devoid of peritoneum and is in direct contact with the diaphragm. This part constitutes the *bare area* of the liver; it is limited above and below by the layers of the coronary ligament. The bare area is triangular in outline. Its apex is at the right

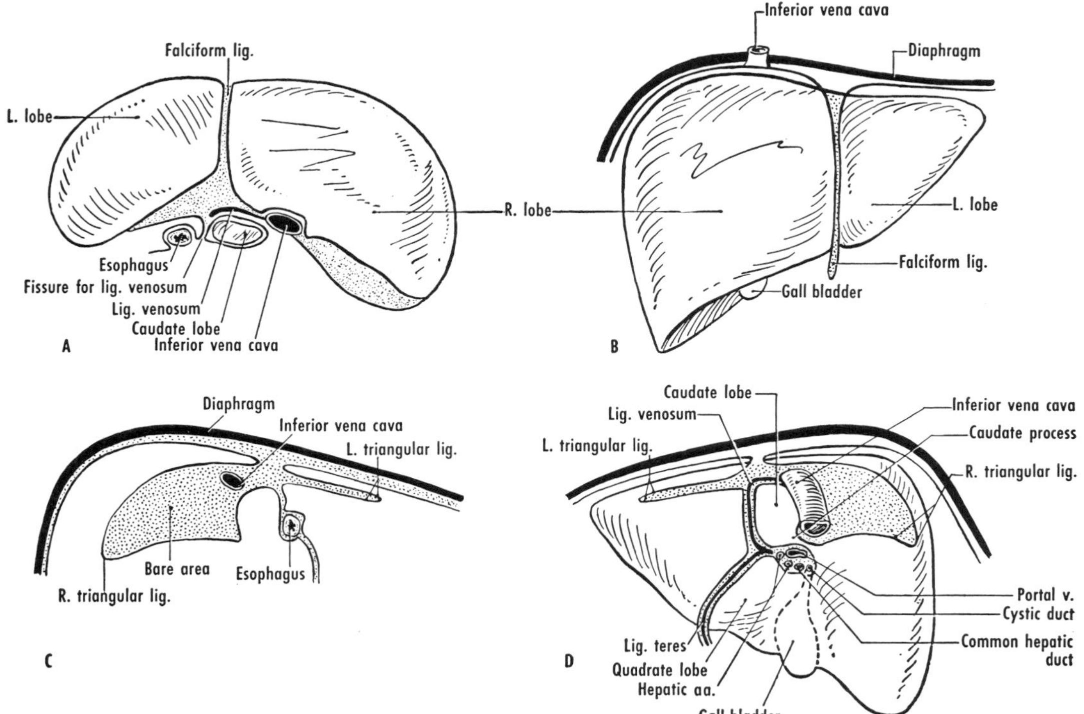

Figure 36–2 Diagrams of the liver and its peritoneal relations. The stippled areas represent surfaces not covered with peritoneum. *A*, superior view. *B*, anterior view. *C*, the diaphragm, viewed from in front, showing the position of the bare area of the liver. *D*, the visceral surface of the liver, viewed from behind.

triangular ligament. Its base, which is directed to the left, is formed by the deep *groove for the inferior vena cava.* This groove may be bridged over by a band of connective tissue or by liver tissue. A part of the bare area is related to the right suprarenal gland.

The visceral surface is related, from left to right, to: (1) the upper part of the stomach, lower end of the esophagus, and lesser omentum; (2) the pyloric part of the stomach and first part of the duodenum, at the right of the *fossa for the gall bladder;* and (3) the right colic flexure and right kidney.

PERITONEAL RELATIONS. In the embryo, the developing liver bud grows into the ventral mesogastrium and the septum

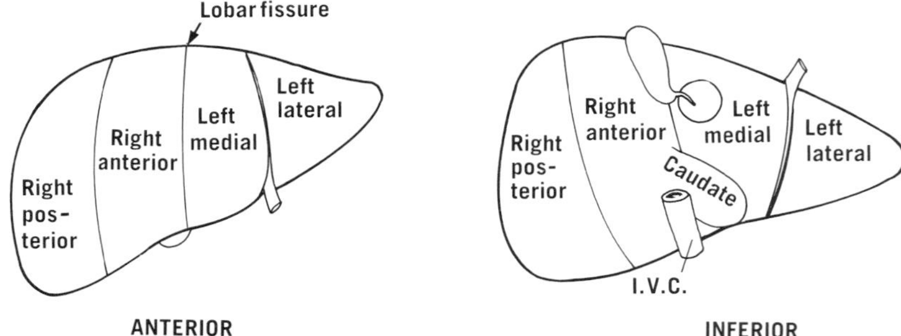

Figure 36–3 Anterior and inferior surfaces of the liver, showing simplified representation of portal segments.

transversum. The portion of the ventral mesogastrium between the liver and the body wall becomes the falciform ligament of the adult, whereas the part between the liver and the foregut becomes the lesser omentum (fig. 34–9, p. 374). The growth into the ventral mesogastrium and the septum transversum accounts for the fact that the adult liver is nearly surrounded by peritoneum, but maintains contact with the diaphragm at the bare area.

The liver is connected to the diaphragm and the anterior abdominal wall, and to the stomach and duodenum, by a number of peritoneal folds that are the reflections of the investing peritoneum of the liver (fig. 36–2). These folds are the lesser omentum, the coronary ligament, the right and left triangular ligaments, and the falciform ligament.

The *lesser omentum* extends from the liver to the lesser curvature of the stomach and the beginning of the duodenum. The attachment of the lesser omentum to the liver is L-shaped, as viewed from behind. The horizontal limb of the L corresponds to the margins of the porta hepatis, the vertical limb to the floor of the fissure for the ligamentum venosum. At the upper end of the latter fissure, the left or anterior layer of the lesser omentum continues into the posterior layer of the left triangular ligament. The right or posterior layer of the lesser omentum is indirectly continuous with the lower layer of the coronary ligament. The right part of the lesser omentum, that is, the hepatoduodenal ligament, contains the bile duct at its free margin. The hepatic artery proper lies at the left of the duct, the portal vein behind and usually somewhat to the left. There are, however, many variations in this arrangement, some of which are of surgical importance.

The reflection of peritoneum from the diaphragm to the upper and posterior parts of the diaphragmatic surface of the liver forms the *coronary ligament*. The ligament consists of an upper or anterior and a lower or posterior layer. These layers meet at the right; their junction constitutes the *right triangular ligament* (fig. 36–2). The layers of the coronary ligament diverge toward the left and enclose the triangular bare area. The upper layer of the coronary ligament is continuous at the left with the right layer of the falciform ligament; the lower layer is continuous with the right layer of the lesser omentum. The left layers of the falciform ligament and the lesser omentum meet and form the *left triangular ligament*.

The *falciform ligament* connects the liver to the diaphragm and the anterior abdominal wall. The two layers of this sickle-shaped ligament enclose the ligamentum teres, some paraumbilical veins, and a portion of a fat pad. The free edge of the falciform ligament meets the inferior border of the liver at the notch for the ligamentum teres, from which the fissure for the ligamentum teres continues onto the visceral surface. The two layers of the falciform ligament are somewhat separated on the diaphragmatic surface, as they are reflected onto the diaphragm, leaving a narrow strip devoid of peritoneum. The left layer is continuous with the left triangular ligament, the right with the upper layer of the coronary ligament.

Surface Anatomy. **The liver, for the most part, lies under cover of the thoracic bony cage, covered by the diaphragm. On the right side it extends above the level of the inferior border of the lung, and dullness is therefore encountered as one percusses downward on the lung (p. 340). In infancy and childhood, the liver extends slightly below the costal margin.[4]**

In thin subjects with narrow chests, the liver lies mainly or entirely to the right of the median plane.[5] Its inferior margin slopes sharply downward and to the right, and its lower right corner may reach the iliac crest, or even lie below it. In plump subjects with broad chests, the liver extends much more to the left of the median plane. The slope of the inferior border is much less, and the gall bladder is at the costal margin. In both types of subjects, the liver changes in position with any postural change that affects the diaphragm.

The study of liver tissue obtained for biopsy by needle puncture is of considerable value in the diagnosis and treatment of liver disease. Transthoracic puncture into the liver is made through the seventh, eighth, or ninth intercostal space, between the anterior and midaxillary lines, one interspace below the upper limit of liver dullness, while the patient is holding his breath in full expiration.

The liver is opaque to x-rays and is largely responsible for the outline of the diaphragm as seen in radiograms. Its borders and surfaces usually cannot be made out distinctly.

Structure. The liver has a uniform structure, consisting of anastomosing sheets of cells with intervening sinusoids. The liver cells secrete into minute bile canaliculi, which unite to form ductules, which in turn merge to form intrahepatic ducts. Ultimately, right and left hepatic ducts leave the liver. The liver cells are supplied by branches of the portal vein and hepatic artery that accompany the duct system and ultimately empty into the sinusoids. Thus, the hepatic cells are bathed in a mixture of venous blood from the alimentary canal, and arterial blood from the hepatic artery. The sinusoids are drained by veins that empty into the inferior vena cava by way of the hepatic veins.

The liver is covered by a thin, fibrous capsule that lies deep to the peritoneum. Thin, incomplete septa project inward. Furthermore, connective tissue sheaths penetrate the liver with blood vessels and ducts, and constitute the *perivascular fibrous capsule.* Owing to the arrangement of the connective tissue and liver cells, the surface of the liver is mottled, and its cut or broken surface granular. The liver bleeds profusely when ruptured, especially because its veins lack valves and remain open when torn.

Function. The bile that is secreted by the liver is stored in the gall bladder and is expelled into the duodenum when food arrives there. Bile salts in the bile aid in the digestion and absorption of fat, and are returned to the liver by the portal system for re-excretion. The liver also has important roles in the synthesis of proteins, in intermediary metabolism, clotting mechanisms, detoxification, and iron, copper, vitamin, and glycogen storage, and it is an important blood-forming organ in the fetus. Many of its metabolic functions depend upon receiving venous blood from the alimentary canal by the portal vein.

Liver cells can divide and replace those destroyed by disease. It has been estimated that one-third of the liver is sufficient to maintain normal liver function. Total removal of the liver is quickly fatal.

Blood Supply. The liver has a double blood supply, from the hepatic artery proper and from the portal vein. The portal vein carries venous blood from the alimentary canal to the sinusoids of the liver (fig. 36–4.)

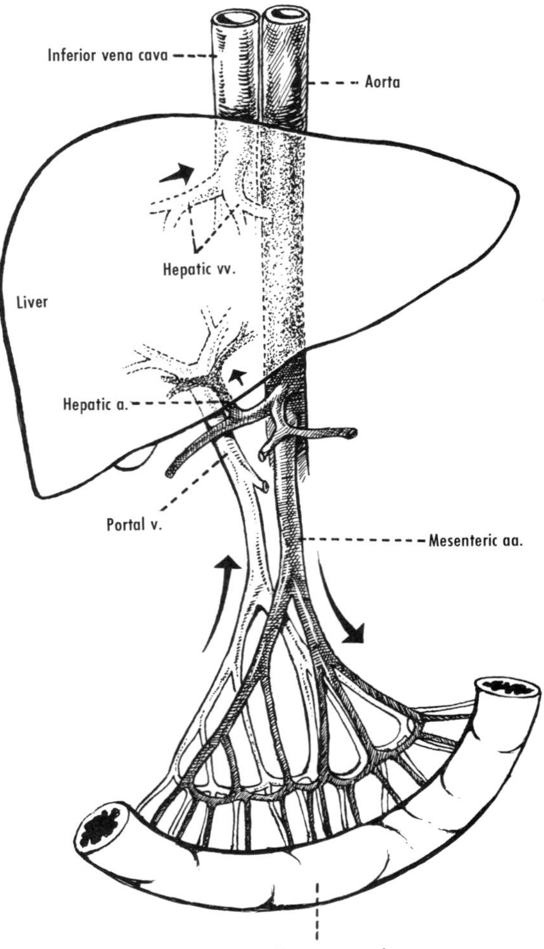

Figure 36–4 Schematic diagram of the portal circulation. Blood from the aorta supplies the alimentary canal. Venous blood from the intestine reaches the sinusoids of the liver by way of the portal vein. Venous blood from the liver reaches the inferior vena cava by way of the hepatic veins.

The common hepatic artery is classically described as ascending in the lesser omentum, as the hepatic artery proper, to the left of the bile duct, in front of the portal vein, where it divides into right and left branches (also called right and left hepatic arteries). The right branch crosses in front of the portal vein, behind the bile duct, gives off the cystic artery, and enters the liver. The left branch continues to the left half of the liver. Variations are common (p. 401); the pattern described above, although the most frequent one, is present in a minority of cases. Often one of the segmental arteries arises outside the liver. For example, in nearly half of instances the artery to the medial segment arises outside the liver and is then termed the middle branch, or middle hepatic artery. The origin of the common hepatic artery itself is subject to considerable variation.

The portal vein ascends behind the bile duct and hepatic artery proper. At the porta hepatis it divides into right and left branches, and commonly gives an additional branch to the quadrate lobe.

The central veins of the liver drain into a system of hepatic veins that eventually form three main vessels, the left, middle, and right hepatic veins.[6] The left is usually joined by the middle, which lies in the "principal plane," and empties into the inferior vena cava. The right likewise empties into the inferior vena cava. In addition, a number of small veins (short hepatic veins) pass backward from the liver into the vena cava.

Lymphatic Drainage. Deep lymphatic vessels form subperitoneal networks from which vessels reach the internal thoracic nodes. Some accompany the ligamentum teres to the umbilicus. The other major path for lymphatic vessels is along the blood vessels in the lesser omentum to the celiac nodes and thence to the thoracic duct.

The liver is a common site of metastatic cancer,

which can reach it from the thorax, from the breast (p. 272), or from any region drained by the portal vein.

Nerve Supply. Large numbers of nerve fibers reach the liver and bile passages by way of a very extensive hepatic plexus. As this plexus extends from the celiac plexus it receives additional branches from the anterior vagal trunk.

The plexus includes vasomotor fibers, fibers to smooth muscle, and pain fibers, particularly from the bile passages (p. 401).

BILE PASSAGES

The extrahepatic bile passages are the gall bladder and various ducts (fig. 36–5). The Greek word *cholos* means bile, and the combining form *chole-* is used in many terms relating to the bile passages, such as cholecystectomy and cholelithiasis.

The right and left hepatic ducts emerge from the corresponding halves of the liver and unite to form the common hepatic duct. This receives the cystic duct from the gall bladder and becomes the bile duct, which empties into the second part of the duodenum in common with or alongside the pancreatic duct. The bile passages are surgically important because of the frequency of inflammation and of gallstones. A striking feature of the bile passages is the relative absence of muscle, compared with the intestine,[7] except for the gall bladder and the lower 2 cm of the bile duct. The bile passages are therefore capable of distention, but not much contraction. The mucosa of the bile duct contains glands.

Gall Bladder

The gall bladder lies in its fossa on the visceral surface of the liver, where it is covered inferiorly and at its sides by peritoneum. The main part of the gall bladder is termed the *body*. The blind lower end of the body, at or below the inferior border of the liver, is termed the *fundus*. Above, the *neck* of the gall bladder joins the body with the *cystic duct*. The upper part of the body, the neck, and the first part of the cystic

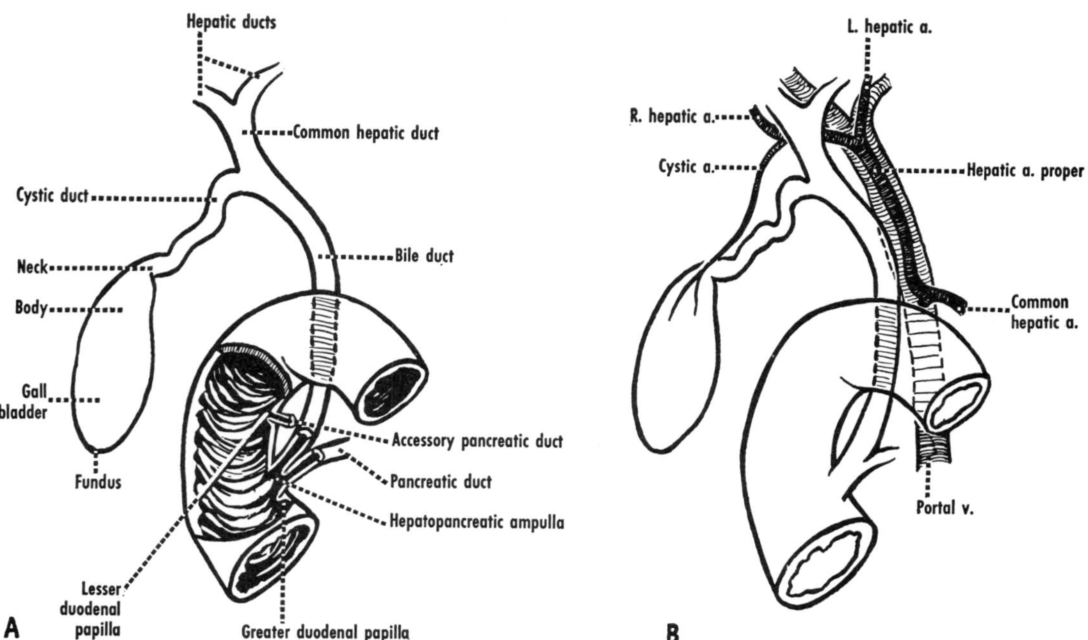

Figure 36–5 *A*, the bile passages. *B*, one of the patterns of structures in the hepatic pedicle. The common hepatic duct, the cystic duct, and the liver form the cystohepatic triangle, in the neighborhood of which are most of the structures of importance in cholecystectomy.

duct are usually **S**-shaped,[8] an arrangement that results in their being termed the siphon. The gall bladder varies greatly in size and shape. On the average it holds about 30 ml. A dilatation termed the cervical pouch is sometimes present at the junction of the body and neck, but is pathological.[9] The mucosa of the cystic duct and the neck of the gall bladder is thrown into *spiral folds*. Those of the duct are so regular that they have been termed spiral valves.

Double gall bladder and congenital absence are rare. Some animals normally lack a gall bladder (e.g., horse, deer, rat).

Relations and Surface Anatomy. When the subject is in the recumbent position, the relations of the gall bladder are the liver above, the first or second part of the duodenum, or both, behind, the transverse colon below, and the anterior abdominal wall in front.

The gall bladder varies in position with the position of the liver. When the subject is in the erect position, the gall bladder may be anywhere between the right costal margin and the linea semilunaris, and between the transpyloric and supracristal planes, depending on body type.[10] In slender women, the gall bladder may hang down below the iliac crest.

Cystic Duct. The cystic duct runs backward, downward (in living individuals, the duct usually runs upward), and to the left, and joins the common hepatic duct to form the bile duct. The cystic duct may be quite long and may descend to the duodenum before joining the common hepatic duct. On the other hand, the cystic duct may be very short. The first part of the cystic duct is usually a part of the siphon.

Hepatic Ducts and Bile Duct

Hepatic Ducts. The *right* and *left hepatic ducts* leave the corresponding halves of the liver and unite to form the *common hepatic duct*. This runs downward and to the right and joins the cystic duct to form the *bile duct*. The left hepatic duct is usually wider and joins the right at an acute angle. The most frequent variation is a junction of the two hepatic ducts at a lower level. The cystic duct usually runs parallel to the common hepatic duct before joining it.

Bile Duct. The bile duct* runs in the free edge of the lesser omentum, behind the first part of the duodenum, traverses the head of the pancreas (or is enfolded by the posterior part of the head), and enters the duodenum. In the living, it runs an angled or curved course for about 4 to 8 cm, with the concavity to the right.[11] When it reaches the concavity of the second part of the duodenum, it lies behind and slightly above the pancreatic duct. The two then run obliquely through the wall of the duodenum for almost 2 cm.

The important relations of the bile duct are as follows. The portal vein, formed behind the neck of the pancreas, ascends behind and to the left of the bile duct. The gastroduodenal branch of the common hepatic artery descends with the duct, and the hepatic artery proper ascends at the left of the duct, in front of the portal vein. The posterior superior pancreaticoduodenal branch of the gastroduodenal artery tends to spiral around the bile duct as it descends, and usually crosses in front of the retroduodenal part. Lymphatic vessels and nodes accompany the blood vessels and duct. The inferior vena cava lies behind the structures in the lesser omentum, separated from the ducts by the portal vein behind the first part of the duodenum, and by the epiploic foramen higher up. The bile duct is most accessible to surgical exposure during its course in the lesser omentum.

INTRADUODENAL PART OF BILE DUCT.[12] As the bile duct traverses the duodenal wall, it becomes constricted, its wall thickens, and its lumen narrows. During its course through the duodenal wall, the bile duct is closely associated with the pancreatic duct. They are usually united by connective tissue, especially in the latter part of their course, and they often empty into a common channel, usually termed the hepatopancreatic ampulla, which in turn empties into the duodenum at the apex of the greater duodenal papilla. These relationships are shown in figure 36–6.

*The frequently used term "common bile duct" is not found in any of the official terminologies. At the level under consideration, there is only one named bile duct; the passages within the liver, which are of a different order of magnitude, are best referred to as biliary ductules and canaliculi.

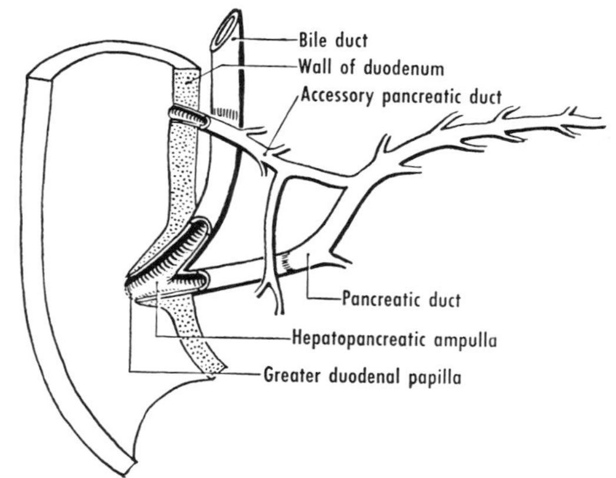

Figure 36–6 Schematic diagram of the arrangement of pancreatic and accessory pancreatic ducts.

A circular layer of smooth muscle is present around the intraduodenal part of the bile duct. It is thickest at the end of the bile duct and it is termed the sphincter of the bile duct. It is continuous with the sphincter of the ampulla when that is present. The sphincter of the bile duct apparently is not continuous with the musculature of the duodenum. It is under neural and hormonal control, and it is the chief mechanism controlling the flow of bile into the duodenum.

Functions of Bile Passages

The gall bladder receives bile, stores it, concentrates it by absorbing water and salts, and sends it to the duodenum when food arrives there. Removal of a gall bladder does not affect hepatic function. Bile that reaches the duodenum in the absence of a gall bladder is more diluted. Certain radio-opaque substances, usually organic iodine compounds, if given by mouth or injected intravenously, are excreted in the bile. Because bile is concentrated in the gall bladder, the bile passages can thereby be visualized radiographically (fig. 36–7).

Constituents of bile may be thrown out of solution and form gallstones (cholelithiasis). The common varieties contain varying amounts of cholesterol. In others, bile pigments or calcium is the predominant constituent. Gallstones are common, and are frequently squeezed into the duct. They may be passed into the duodenum, or they may obstruct the duct, especially at the point of narrowing. Pain from distention and spasm of the biliary tract is severe, especially when an obstruction is present.

Blood Supply, Lymphatic Drainage, and Nerve Supply

The cystic artery supplies the gall bladder. It arises from the right hepatic artery, and may be double or aberrant. The veins of the gall bladder for the most part enter the liver and break up into capillaries. Occasionally one or two veins from the serosal surface join the portal vein.

The hepatic and bile ducts are supplied mainly by multiple small branches from the cystic, supraduodenal, and posterior superior pancreaticoduodenal arteries.[13] One or two small veins, which communicate with those of the pancreas and duodenum, ascend near the bile and hepatic ducts and empty into the liver. In addition, a venous plexus is present on the supraduodenal part of the bile duct.[14] This plexus drains the duct and ascends to the liver.

The lymphatic vessels of the gall bladder and bile passages anastomose above with those of the liver, and below with those of the pancreas.

The nerve supply is by way of the hepatic plexus. Pain fibers that arise from the bile passages reach the spinal cord by way of the splanchnic nerves. Pain from distention or spasm can be excruciating. It is usually felt in the upper right quadrant or in the epigastrium, is often referred posteriorly, to the region of the right scapula, and is sometimes cardiac in type and distribution.

Hepatic Pedicle

The structures entering the liver at the porta hepatis constitute the hepatic pedicle (fig. 36–5). Here, and within the rest of the lesser omentum, they show many variations, some of surgical importance, to the extent that a "typical" pattern seldom exists.[15]

Arteries. The common hepatic artery may arise from the left gastric artery, from the aorta, or occasionally from the superior mesenteric artery. The right hepatic artery may arise separately from the superior mesenteric, and the left from the left gastric. The right hepatic artery may pass in front of rather than behind the bile or hepatic duct, and it is usually closely related to the cystic duct. An accessory hepatic artery is often present, and represents an extrahepatic origin of a segmental artery.[16] The right hepatic artery frequently runs parallel to the cystic duct for an appreciable distance before giving off the cystic artery, and may be inadvertently ligated if the duct is ligated.[17] The cystic artery, which usually lies in the cystohepatic triangle (fig. 36–5), is sometimes double.

Ducts. The lengths of the hepatic and cystic ducts, and the angle at which they join, vary greatly; in some the cystic and common hepatic ducts join so low that the bile duct is formed behind the duodenum. Accessory hepatic ducts occur generally from the right half of the liver, but their incidence is uncertain. These accessory ducts are segmental ducts with an extrahepatic course. Accessory hepatic ducts may also join the gall bladder directly. On the whole, major congenital variations are rare. The more important are congenital stenosis or atresia, duplication of the bile duct, and abnormal opening (into the pylorus).

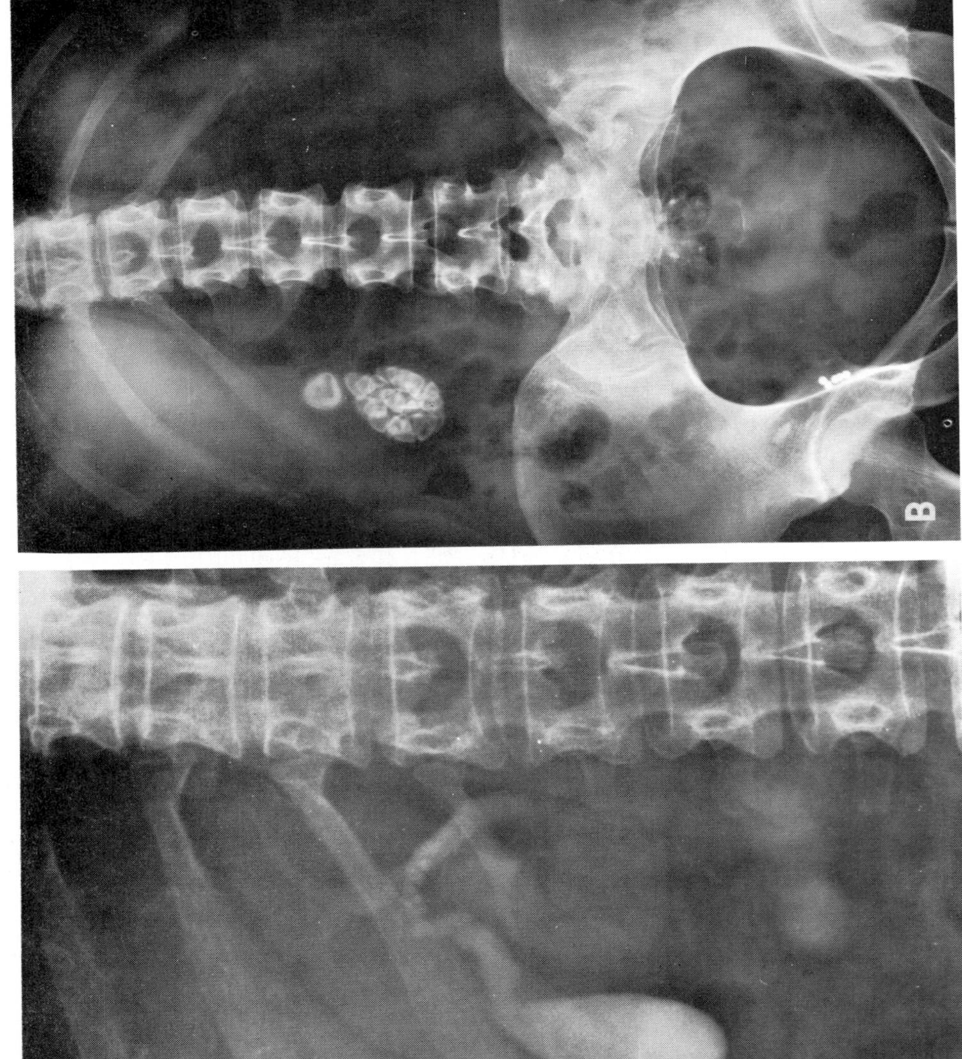

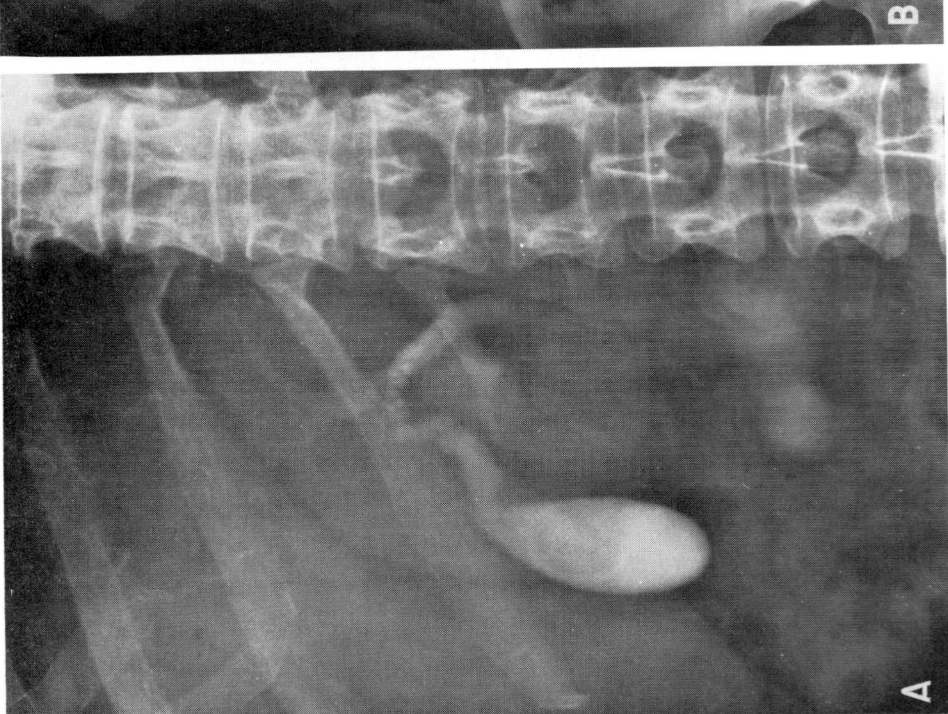

Figure 36–7 Gall bladder. A, cholecystogram showing right, left, and common hepatic ducts, cystic duct, and gall bladder (fundus at level of L.V. 2 in this person), and bile duct. Compare figure 36–5. *B*, position of gall bladder shown by radio-opaque gallstones. Note multifaceted appearance of stones. Fundus at level of upper portion of L.V. 4 in this person. A, courtesy of John Pepe, M.D., Brooklyn, New York.

PANCREAS

The pancreas,* an exocrine and endocrine gland, is a soft, fleshy organ with very little connective tissue. The pancreas consists of a head, a body, and a tail. The junction of the head and body is known as the neck. The superior mesenteric vein ends by joining the splenic vein to form the portal vein behind the neck.

The *head* lies within the curve of the duodenum and is overlapped in front by the pyloric part of the stomach and the first part of the duodenum. The bile duct, descending behind the first part of the duodenum, is at first behind the head of the pancreas but, before entering the duodenum, is usually embedded in the pancreas. The pancreaticoduodenal arcades lie in front of and also behind the head and are partly embedded in it. The *uncinate process* is a prolongation of the lower and left part of the head and projects upward and to the left behind the superior mesenteric vessels (fig. 35–9, p. 385). The superior mesenteric vein, which is at the right of the artery, passes through the *pancreatic notch* formed by this process.

The *body* and the *tail* of the pancreas extend to the left, crossing the vertebral column. The tail projects into the lienorenal ligament, where it lies in contact with the spleen. The body, which lies immediately below the celiac trunk and above the duodenojejunal flexure, is somewhat prismatic in shape; it presents three surfaces, anterior, posterior, and inferior, and three borders, superior, anterior, and inferior. The *tuber omentale* is a small projection from the superior border that is in contact with the posterior surface of the lesser omentum.

Relations. The principal structures in front of the pancreas are the stomach and sometimes the transverse colon. The important posterior relations are: (1) the inferior vena cava and aorta, and the renal and gonadal vessels behind the head; (2) the superior mesenteric and portal veins behind the neck; (3) the diaphragm, left suprarenal gland, and left kidney and renal vessels behind the body; (4) the splenic vein regularly lies behind the body and tail of the pancreas; it is sometimes partly embedded in them. The tortuous, splenic artery is located above the vein, in the region of the superior border. The tail is more mobile than the rest of the pancreas, and is therefore more variable in its relations.

PERITONEAL RELATIONS. The tail of the pancreas is surrounded by peritoneum; the pancreas is otherwise retroperitoneal. The two layers of the transverse mesocolon project forward from the pancreas (fig. 35–9, p. 385). The upper layer is continuous with the posterior layer of the greater omentum, to which it is adherent or fused. Above the line of attachment of the transverse mesocolon, the pancreas is covered in front by the peritoneum that forms the posterior wall of the lesser sac (figs. 34–2 and 34–6, pp. 366, 371). The lower layer of the transverse mesocolon covers the inferior surface of the body of the pancreas and the anterior surface of the head, from which it passes in front of the third and fourth parts of the duodenum and continues as the right layer of the mesentery.

Pancreatic Ducts[18] (fig. 36–6). The *pancreatic duct*, which is usually the main exit for pancreatic secretion, begins in the tail of the pancreas and runs to the right near the posterior surface of the gland. Near the neck it turns downward and to the right and comes into relation with the bile duct, with which it empties into the second part of the duodenum at the greater duodenal papilla. Frequently an *accessory pancreatic duct* is present. It drains a part of the head, runs upward in front of the pancreatic duct, to which it is usually connected, and empties into the duodenum at the lesser duodenal papilla (p. 385). It is patent more often than not.[19]

Anomalies. The most common anomaly is accessory or aberrant pancreatic tissue, which may appear in the stomach or duodenum (or jejunum, ileum, gall bladder, or spleen).[20] The accessory pancreatic tissue may include islet cells. The pancreas may also be divided or annular; an annular pancreas may constrict or obstruct the duodenum.

Structure and Function. The pancreas is both an exocrine and an endocrine gland. The exocrine portion consists of secretory units, the pancreatic acini. These comprise glandular cells the enzymatic secre-

*The pancreas and thymus of animals are known as "sweetbreads."

tions of which are discharged into the duct system and thereby into the duodenum.

The endocrine portion of the gland is composed of small clumps of cells, the pancreatic islets, which are scattered throughout the pancreas. Each islet is richly supplied by capillaries into which the cells discharge the hormone insulin.

Blood Supply[21] ***and Lymphatic Drainage.*** The pancreas is supplied by the pancreaticoduodenal arteries and by branches of the splenic artery (fig. 35–5, p. 382). The anterior superior and anterior inferior pancreaticoduodenal arteries form an arcade in front of the head of the pancreas, and the posterior superior and posterior inferior pancreaticoduodenal arteries form an arcade behind the head of the pancreas. Both arcades supply the pancreas and the duodenum. The pancreas also receives a number of branches from the splenic artery (p. 416). These include the dorsal pancreatic artery, the inferior pancreatic artery, the pancreatica magna artery, and the caudal pancreatic arteries. Occasionally there is an arterial shunt across the front of the head of the pancreas between the gastroduodenal and superior mesenteric arteries.[22] The veins, somewhat more variable, accompany the arteries.

The lymphatic vessels that drain the pancreas extend to all the adjacent nodes: splenic, mesenteric, gastric, hepatic, and celiac.

Nerve Supply. The pancreas is supplied by nerve fibers from the celiac and superior mesenteric plexuses. These fibers are autonomic and sensory. The sensory fibers include some concerned with reflexes and others concerned with pain. Pain fibers from the pancreas enter the spinal cord by way of the splanchnic nerves.

SPLEEN

The spleen[23] (L. *lien,* Gk. *splen;* hence the adjectives *lienal* and *splenic*) is a soft, vascular organ that lies against the diaphragm and the ninth to eleventh ribs on the left side. Although the spleen is not a digestive organ, its venous drainage is into the portal system. It is a lymphatic organ that filters blood, removes iron from hemoglobin, produces lymphocytes and antibodies, and stores and releases blood with a high concentration of corpuscles. Removal of the spleen causes no apparent disability. The spleen has *diaphragmatic* and *visceral surfaces, superior* and *inferior borders,* and *anterior* and *posterior ends* (often called medial and lateral ends). The superior border is notched; the notches represent the remains of fetal lobation. The surfaces and borders of the spleen are most notable in embalmed cadavers. In the living the spleen is easily molded by adjacent structures.

Accessory splenic tissue may occur in any portion of the abdominal cavity, but chiefly in the tail of the pancreas.[24]

Relations. The diaphragmatic surface is related to the costal part of the diaphragm. The visceral surface has gastric, renal, and colic surfaces. The *gastric surface* is related to the stomach. A long fissure, the *hilus,* which is present at the lower part of the spleen, is pierced by vessels and nerves. The *renal surface* on the lower part of the visceral surface is related to the left kidney and sometimes to the left suprarenal gland. The *colic surface,* at the anterior end, is related to the left colic flexure. The tail of the pancreas may reach the spleen between the colic surface and the hilus.

PERITONEAL RELATIONS. The spleen is surrounded mainly by peritoneum, except at the hilus. A peritoneal reflection may run from the lower pole to the greater omentum; it may be torn from the spleen during surgical retraction of the stomach to the right, resulting in bleeding.[25] The spleen develops in the dorsal mesogastrium and remains connected to the stomach by the *gastrolienal (gastrosplenic) ligament,* and to the body wall and kidney by the *phrenicolienal ligament,* the lower part of which is termed the *lienorenal ligament* (fig. 34–9, p. 374). The lienorenal ligament transmits the splenic vessels and contains the tail of the pancreas.

Surface Anatomy. The spleen is a very mobile organ. In the recumbent position its long axis is roughly parallel to the long axis of the tenth rib, and it lies against the ninth to eleventh ribs. It is usually not palpable unless it is enlarged or is markedly dislocated. The part of an enlarged, hardened spleen that can be palpated is usually the notched superior border.

Structure. The fibrous capsule gives off trabeculae to the interior of the spleen, which is composed of splenic pulp. The pulp consists of lymphatic follicles or "splenic corpuscles" (white pulp), surrounded by red pulp. The red pulp is composed of venous sinuses that are held together by a network of reticular fibers and are lined by cells that include those of the reticuloendothelial system. The blood in the arteries that enter the spleen eventually reaches the sinuses, which are drained by splenic veins.

Blood Supply*[26] *and Lymphatic Drainage. The spleen is supplied by the splenic artery, which usually arises from the celiac trunk and runs a tortuous course to the left, near the upper border of the body of the pancreas (see fig. 35–5). During its course it gives off pancreatic branches, short gastric branches, and the left gastroepiploic artery. It then divides into two or three terminal branches that undergo further division before entering the spleen through much of its visceral surface. The end of the inferior pancreatic artery may also supply the spleen.

A number of venous trunks leave the spleen, especially from its hilus, and join to form the splenic vein. The splenic vein runs to the right behind the body of the pancreas and joins the superior mesenteric vein, behind the neck of the pancreas, to form the portal vein. It often receives the inferior mesenteric vein, and sometimes the left gastric vein. The splenic and portal veins may be demonstrated by the percutaneous injection of radio-opaque material into the spleen.[27] The material enters the blood and thereby the portal system.

The lymphocytes formed in the spleen enter the blood stream. Lymphatic vessels are present only in the capsule and the larger trabeculae. They drain into adjacent nodes.

Nerve Supply. A dense network of fibers continues from the celiac plexus along the splenic artery. Most are postganglionic sympathetic fibers to the smooth muscle of the capsule, trabeculae, and the splenic vessels in the pulp.

REFERENCES

1. F. H. DeLand and H. N. Wagner, *Atlas of Nuclear Medicine,* vol. 3, *Reticuloendothelial System, Liver, Spleen and Thyroid,* Saunders, Philadelphia, 1972.
2. G. A. Kune, Aust. N. Z. J. Surg., *39*:117, 1969.
3. C.-H. Hjortsjö pioneered the concept of liver segments: Kungl. Fysiogr. Sallskapets Handl., N. F. *59*:1, 1948; Acta anat., *11*:599, 1951; *Leverns Segmentering,* Pharmacia, Uppsala, 2nd ed., 1964. Subsequently, various studies have been published and systems of nomenclature proposed: J. E. Healey, P. C. Schray, and R. J. Sorensen, J. int. Coll. Surg., *20*:133, 1953. H. Elias, Surgery, *36*:950, 1954; Recent Results in Cancer Research, *26*:116, 1970. W. Platzer and H. Maurer, Acta anat., *63*:8, 1966.
4. D. Deligeorgis *et al.,* Arch. Dis. Childh., *45*:702, 1970.
5. F. G. Fleischner and V. Sayegh, New Engl. J. Med., *259*:271, 1958.
6. K. J. Hardy, Aust. N. Z. J. Surg., *42*:11, 1972.
7. J. Kirk, Ann. R. Coll. Surg. Engl., *3*:132, 1948.
8. G. Simon, Ch. Debray, and J. A. Baumann, Ann. Rech. méd., *5*:1125, 1957.
9. F. Davies and H. E. Harding, Lancet, *1*:193, 1942.
10. N. F. Hicken, Q. B. Coray, and B. Franz, Surg. Gynec. Obstet., *88*:577, 1949. F. G. Fleischner and V. Sayegh, New Engl. J. Med., *259*:271, 1958.
11. H. Wapshaw, Brit. J. Surg., *43*:132, 1955. E. Samuel, Ann. R. Coll. Surg. Engl., *20*:157, 1957.
12. W. H. Hollinshead, Surg. Clin. N. Amer., *37*:939, 1957.
13. A. L. Shapiro and G. L. Robillard, Surgery, *23*:1, 1948. F. A. Henley, Brit. J. Surg., *43*:75, 1955.
14. J. H. Saint, Brit. J. Surg., *48*:489, 1961.
15. N. F. Hicken, Q. B. Coray, and B. Franz, Surg. Gynec. Obstet., *88*:577, 1949. E. V. Johnston and B. J. Anson, Surg. Gynec. Obstet., *94*:669, 1952. B. J. Anson, Quart. Bull. Northw. Univ. med. Sch., *30*:250, 1956. M. A. Hayes, I. S. Goldenberg, and C. C. Bishop, Surg. Gynec. Obstet., *107*:447, 1958. A. dos Santos Ferreira and A. Caria Mendes, C. R. Ass. Anat., *53*:1487, 1968.
16. J. P. J. Vandamme, J. Bonte, and G. van der Schueren, Acta anat., *73*:192, 1969. W. Feigl, W. Firbas, and H. Sinzinger, Anat. Anz., *134*:139, 1973.
17. H. K. Gray and F. B. Whitesell, Surg. Clin. N. Amer., *30*:1001, 1950.
18. T. Smanio, Int. Surg., Chicago, *52*:125, 1969.
19. W. F. Rienhoff and K. L. Pickrell, Arch. Surg., Chicago, *51*:205, 1945. W. Dawson and J. Langman, Anat. Rec., *139*:59, 1961.
20. M. Feldman and T. Weinberg, J. Amer. med. Ass., *148*:893, 1952.
21. R. T. Woodburne and L. L. Olsen, Anat. Rec., *111*:255, 1951.
22. L. L. Olsen and R. T. Woodburne, Surg. Gynec. Obstet., *99*:713, 1954.
23. L. Arvy, *Splénologie,* Gauthier-Villars, Paris, 1965.
24. G. M. Curtis and D. Movitz, Ann. Surg., *123*:276, 1946. B. Halpert and F. Györkey, Anat. Rec., *133*:389, 1959.
25. A. Gourevitch and M. D. Lord, Brit. J. Surg., *52*:202, 1965.
26. J.-L. Cayotte *et al.,* C. R. Ass. Anat., *55*:591, 1970.
27. J. A. Evans and W. D. O'Sullivan, Med. Radiogr. Photogr., *31*:98, 1955.

KIDNEYS, URETERS, AND SUPRARENAL GLANDS

37

The kidneys (L. *ren*, Gk. *nephros;* hence the adjectives *renal* and *nephric*) and ureters are a part of the urogenital system, and the suprarenal glands are a part of the endocrine system, but these three structures are related topographically. They lie on the posterior abdominal wall, on each side of the vertebral column. The kidneys maintain the ionic balance of the blood, and the loss of both of them is fatal. The suprarenals are endocrine glands, and their secretions are necessary for the continuation of life.

KIDNEYS

The kidneys are a pair of bean-shaped, reddish-brown organs covered by a thin, glistening, fibromuscular capsule that strips easily from a normal but not from a diseased kidney. Each has *anterior* and *posterior surfaces, medial* and *lateral borders,* and *upper* and *lower poles.* The lateral border is convex. The medial border is indented at the *hilus,* which leads into the *renal sinus.* The major renal vessels enter and leave the hilus, and the ureter leaves from the hilus.

The kidneys lie alongside the vertebral column, against the psoas major muscle, and in an oblique plane, between coronal and sagittal. Their long axes slope forward and laterally, as well as downward, in conformity with the long axis of the psoas major. The kidneys are about 11 to 13 cm long, the left being somewhat longer and larger than the right.[1] The outlines of the kidneys may be visible in ordinary radiograms (fig. 39–2*B*, p. 433). The calices, pelvis, and ureter may be outlined by intravenous pyelography (fig. 37–1).

Relations. The important relations of the kidney are as follows:

Above, the upper pole is covered by the suprarenal gland, which is enclosed within the renal fascia in common with the kidney.

In front, the right kidney is related to the liver, the second part of the duodenum, the ascending colon or right colic flexure, and the small intestine. The left is related to the stomach, pancreas, descending colon or left colic flexure, spleen, and small intestine. Sometimes the spleen is so closely related to the left kidney that its weight causes a bulging on the lateral border.[2] This may be evident in radiograms.

The important posterior relations are: the diaphragm, the psoas major and quadratus lumborum, and branches of the lumbar plexus, together with the twelfth rib and the lateral edge of the erector spinae. The lower pole of the kidney may be near the lumbar triangle (a kidney abscess may emerge here). **The diaphragm generally separates the lung and pleura from the upper part of the kidney. In the vertebrocostal trigone (p. 268), the upper pole of the kidney may be separated from the pleura by only a layer of connective tissue.**

Peritoneal Relations. Both kidneys are retroperitoneal. The part of the anterior surface of the right kidney related to the liver and the small intestine is covered with peritoneum. The rest of the anterior surface is devoid of peritoneum. The part of the anterior surface of the left kidney in contact with the stomach, spleen, and small intestine is covered with peritoneum.

The important relations of the kidneys

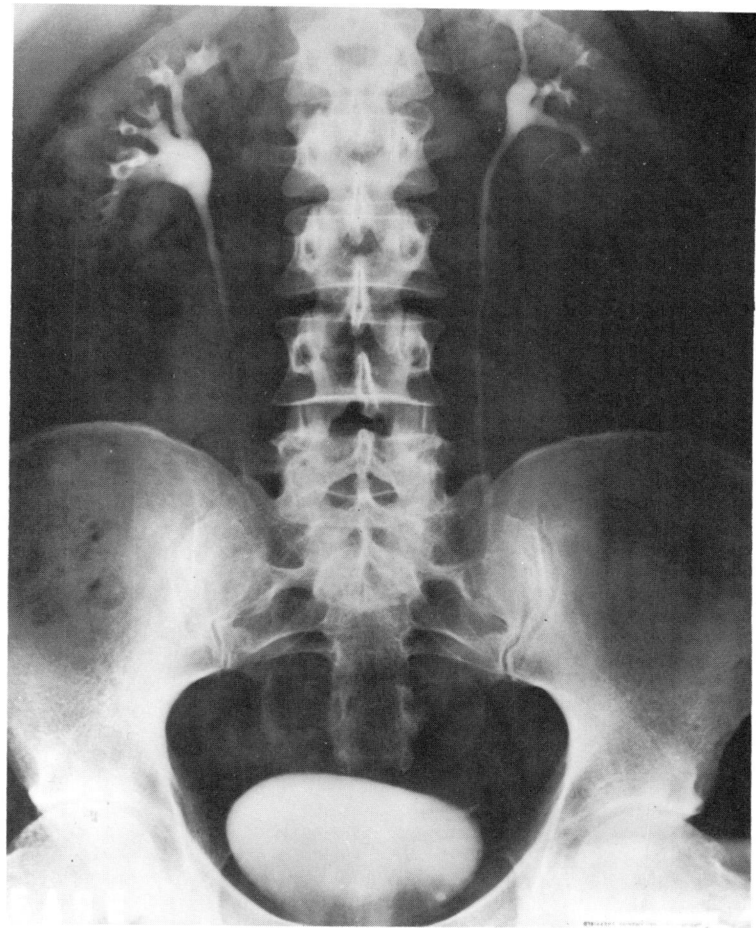

Figure 37–1 Intravenous pyelogram. Note the calices, some of which are seen from the side and others end-on, and the renal pelves, which differ in shape and level. Courtesy of Sir Thomas Lodge, The Royal Hospital, Sheffield, England.

with the great vessels are shown in figure 37–2.

Surface Anatomy.[3] **In the erect position, the kidney extends from the level of L.V. 1 to L.V. 4. The right kidney may be a little lower than the left, probably owing to the liver.** Its lower pole is sometimes palpable. The levels of both kidneys change during respiration and with changes in posture. In the recumbent position the kidneys are at the level of about T.V. 12 to L.V. 3. The amount of movement with respiration is variable. During deep breathing, the kidneys may move up and down less than 1 cm or as much as 7 cm.

A very movable (floating) kidney is sometimes found near the pelvic cavity or the anterior abdominal wall, and has been known to turn upside down and twist on its blood vessels.

Renal Sinus. The medial border of the kidney contains a vertical fissure, the hilus, which transmits the renal vessels and nerves and the upper end of the ureter. The hilus leads into a recess, the renal sinus, which is lined by a continuation of the capsule and contains the renal vessels and the *renal pelvis*.[4] The latter is the upper expanded end of the ureter. The Greek word *pyelos*, meaning a tub or basin, refers to the pelvis of the ureter, and the term pyelonephritis refers to an inflammation of the kidney and the ureter.

Within the sinus the renal pelvis divides into two or three short wide tubes, the *major calices*. Each of these subdivides

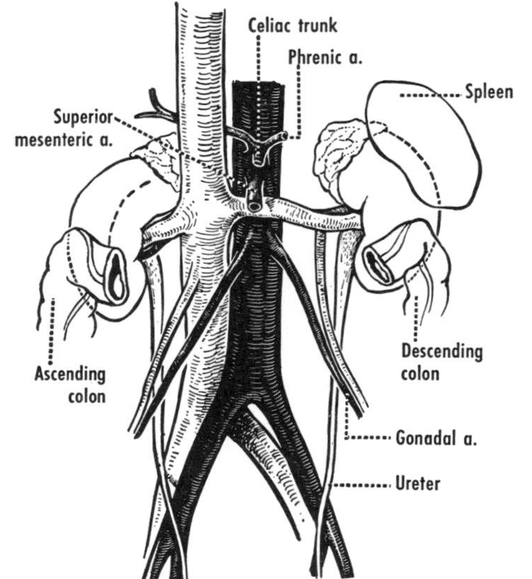

Figure 37–2 Relations of the kidney. Several branches of the aorta, such as the inferior mesenteric artery, are omitted. Based on Stirling.[5]

into seven to fourteen *minor calices.* Each minor calyx is indented in a cup-shaped fashion and is perforated by collecting tubules. Occasionally there are no major calices.[6]

Structure. Each kidney contains a million or more epithelial *renal tubules* (or *nephrons*), the functional units of the kidney, developed from the metanephros. One end of a nephron terminates blindly, the other empties into a collecting tubule, an excretory duct that conducts urine to a minor calyx. There are about 500 collecting tubules, which are developed from the ureteric bud in the embryo. The blind end of each nephron is invaginated by capillaries to form a double-layered *glomerular capsule.* The convoluted tuft of capillaries is termed a *glomerulus,* whereas the capsule and glomerulus together are termed a *renal corpuscle.*

The kidney is composed of an outer, paler cortex, and an inner, darker medulla (fig. 37–3). The *cortex* consists of renal corpuscles, parts of the secretory tubules, and the beginnings of collecting tubules. The medulla consists of *renal pyramids,* each of which contains collecting tubules and parts of the secretory tubules. The *papilla,* or apex of each pyramid, fits into

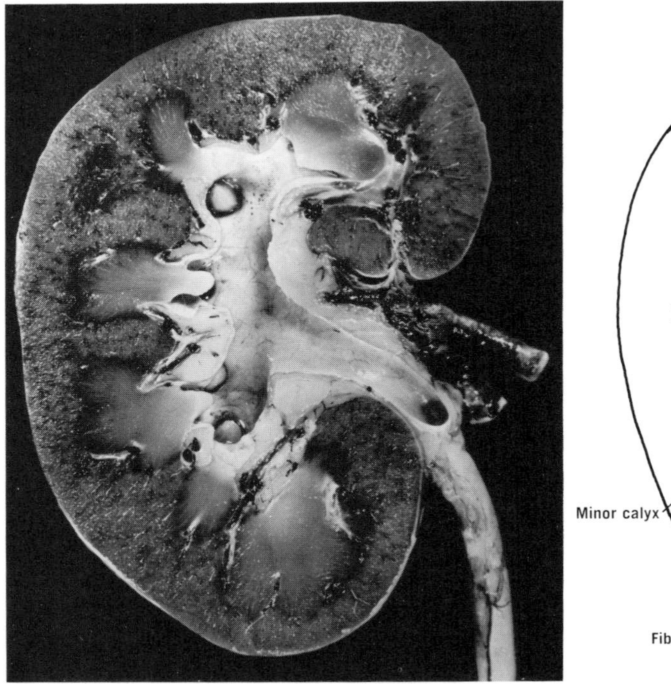

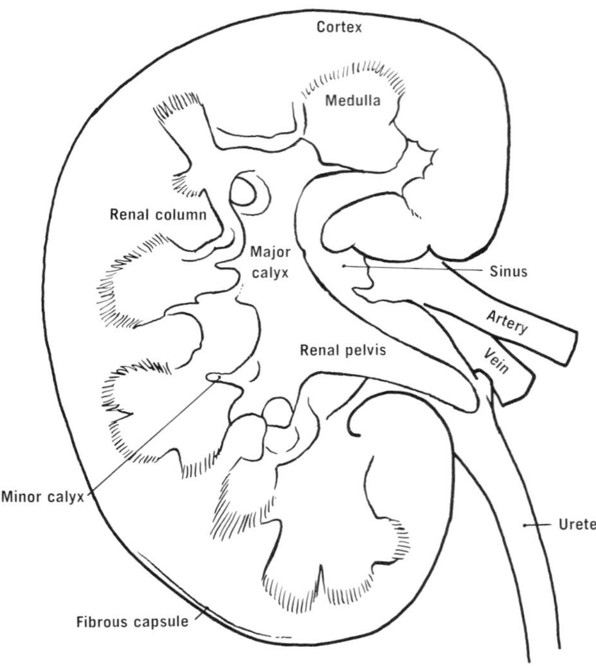

Figure 37–3 Coronal section of kidney. Courtesy of Professor C. Yokochi and Igaku Shoin, Ltd., Tokyo.

the cup-shaped indentation of a minor calyx, which is perforated by the collecting tubules. The prolongation of the tubules of a pyramid into the cortex gives the cortex a striated appearance. The striated cortex immediately external to each pyramid is termed a medullary ray. The cortical tissue that lies between two adjacent pyramids is a *renal septum* or *column*.

The lobe of a kidney is a pyramid and its associated cortex.[7] In the fetus, there are usually five to six primary branches of the ureteric bud, and hence five to six lobes. This number increases by division of the branches, and nineteen or more lobes have been reported.[8] Evidence of lobation may persist for some time after birth, even in adult kidneys.

The kidneys maintain the ionic balance of the blood. In so doing, they excrete waste products in the form of urine. Concentrations of urinary components, such as urates or other crystalline compounds, may form in a calyx or in the pelvis of the ureter and constitute what are known as kidney stones (renal calculi; nephrolithiasis). These vary in size and may be small enough to pass down the ureter. They may also become fixed in a calyx, in the pelvis, or in the ureter. If a stone is small enough to enter the ureter and large enough to obstruct it, renal colic may develop.

Blood Supply[9] and Lymphatic Drainage. The renal arteries arise from the aorta just below the origin of the superior mesenteric artery near the level of the disc between the first and second lumbar vertebrae. The right renal artery passes behind the inferior vena cava. Each renal artery descends slightly as it runs to the renal pelvis (fig. 37–4), supplying the suprarenal gland and the ureter, and then dividing into upper, lower, and posterior primary branches.[10] Two secondary branches (intermediate and middle) are often present. In addition, a suprahilar (apical) artery may occur, and may be primary or secondary in origin. Based on the arterial distribution, segments of the kidney are described, each consisting of several lobes and each supplied solely by a segmental artery. Depending upon definition, a segmental artery may be a primary branch, or the definition may be extended to include the secondary branches. Although segments do exist and are of surgical importance, there is such variation in distribution of branches from one kidney to another that no consistent pattern, common to most kidneys, exists.[11] The segmental arteries divide into interlobar arteries, the branches and distribution of which are shown in figure 37–5.

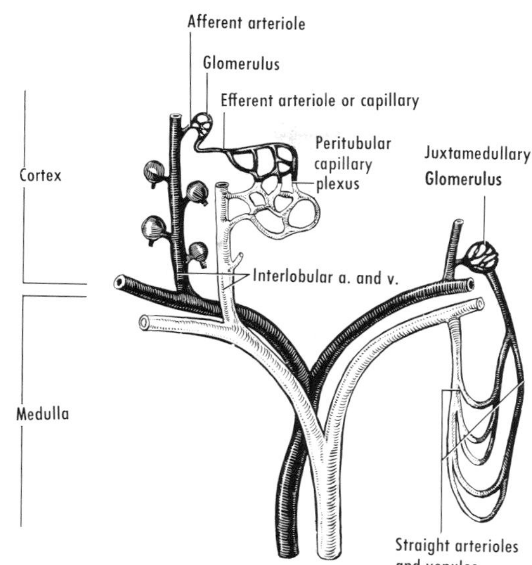

Figure 37–5 Schematic diagram of blood supply of kidney. An interlobar artery divides into arcuate arteries, from which interlobular arteries and straight arterioles arise.

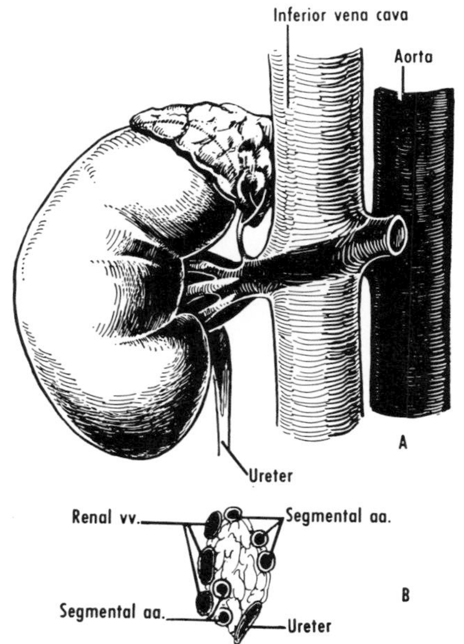

Figure 37–4 *A*, the right suprarenal gland, kidney, and renal pedicle. *B*, a cross-section of the pedicle.

Several veins drain the kidney and unite in a variable fashion to form the renal vein.[12] There is sometimes more than one renal vein on the right side, but the left kidney is usually drained by a single vein. Furthermore, the left renal vein drains an extensive area of the body, receiving blood not only from the kidney, but also from the suprarenal gland, gonad, diaphragm, and body wall (fig. 37–6).[13]

The lymphatic vessels of the kidney drain into adjacent nodes and thence into lumbar nodes.

Nerve Supply. The kidney has an extensive nerve supply from the extensions of the celiac (aorticorenal) and intermesenteric plexuses that accompany the renal artery, as well as from direct branches of the thoracic and lumbar splanchnic nerves.[14] Pain fibers, principally from the renal pelvis and the upper part of the ureter, enter the spinal cord by way of splanchnic nerves.

Malformations and Variations. Disturbances in development are responsible for a variety of renal anomalies and abnormalities. Among these are polycystic kidney; horseshoe kidney; lobation of adult

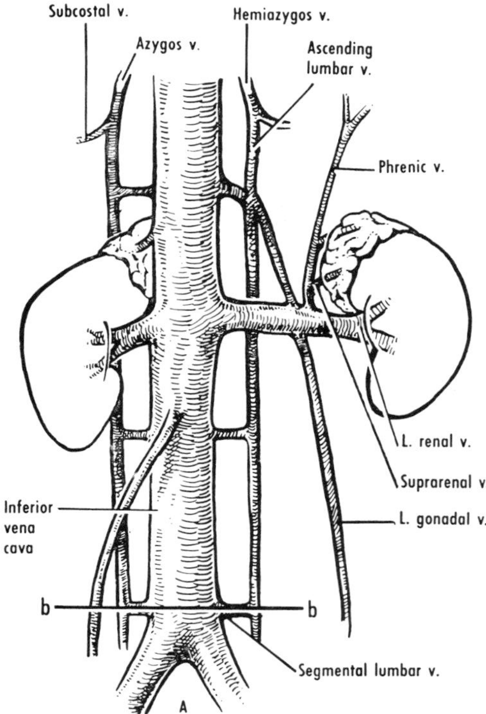

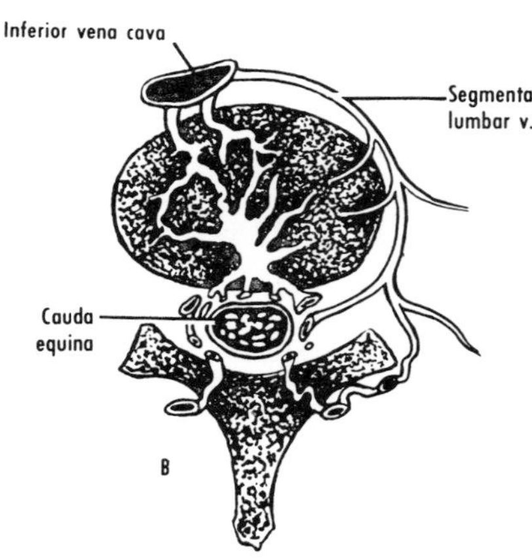

Figure 37–6 A, diagram of the renal veins. Note the extensive area drained by the left renal vein, which has tributaries from the back, abdominal wall, diaphragm, suprarenal gland, and gonad. *B*, a horizontal section in the plane *bb* indicated in *A* to show tributaries of the inferior vena cava from the vertebral veins (right segmental vein omitted).

kidneys; abnormally low kidneys, otherwise known as ectopic kidney, for example, pelvic kidney (due to failure of upward migration); duplication of kidney; and variations in branching of ureters. It has been shown that individuals with renal variations are much more susceptible to renal disease.[15] Congenital absence of one kidney is uncommon. Absence of both is incompatible with life.

Renal Pedicle

The ureter and the vessels that enter the hilus of the kidney constitute the renal pedicle. Variations in this region are common and are often important.[16]

The main relations within the pedicle are as follows: the renal vein is in front, the ureter behind, and the arteries arranged as shown in figure 37–4. A kidney removed from the body can usually be identified as to side by placing it so that the ureter is behind, pointing downward and medially. A part of the renal pelvis lies outside of the renal sinus and is therefore a part of the pedicle.

The urogenital ridge in which the kidney develops is supplied by multiple, paired vessels, of which only a few usually remain. The uppermost one, the inferior phrenic artery, supplies the diaphragm and the suprarenal gland. Another supplies the suprarenal gland, and a third, the lowermost, supplies the kidney. In addition, another artery supplies the gonad. Variations in persistence of the mesonephric vessels as regards both level and number account for many of the arterial variations.

The segmental arteries to the kidney usually arise from the renal artery near or at the hilus, but frequently one or more of the segmental arteries may arise before the renal reaches the hilus, or they may arise from the aorta or from the inferior suprarenal artery. These are called accessory arteries. The suprarenal and renal blood supplies are intimately related, so much so that the inferior suprarenal artery commonly arises from the renal, and in turn usually gives a capsular branch to the kidney, a branch from which an accessory artery may arise.The gonadal arteries are also closely related to the renal pedicle, and vary in their relations to it. One of the gonadal veins, usually the left, may hook around the renal vein, that is, pass upward and behind the renal vein, then forward over it, and finally downward in front of it.

Both renal veins tend to lie in front of the arteries; the left is necessarily the longer. Anomalous renal veins are not com-

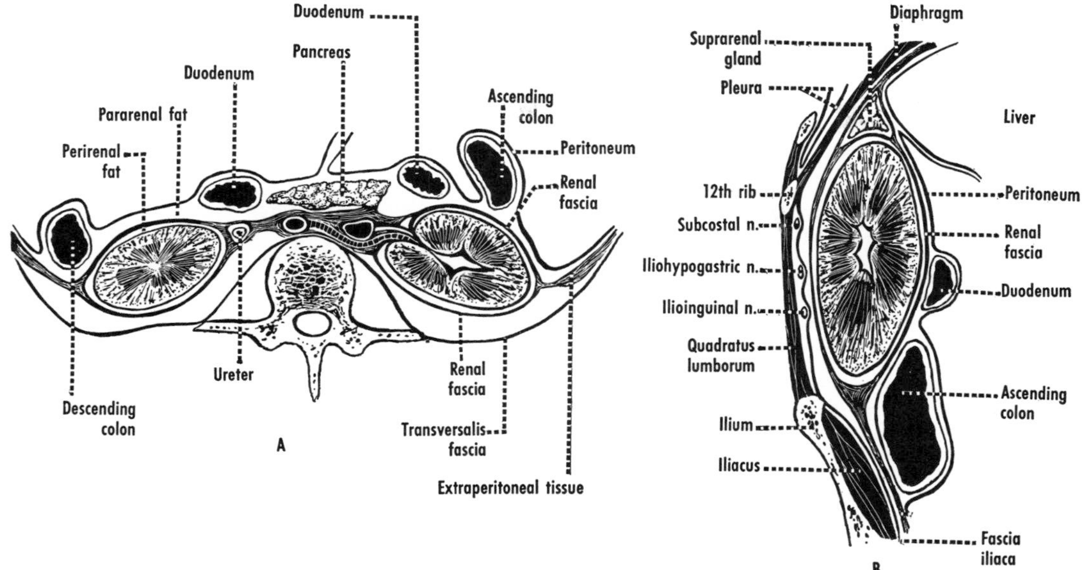

Figure 37–7 The kidneys and renal fasciae: *A*, horizontal section; *B*, sagittal section.

mon. Multiple renal veins occasionally occur; when they do, they are usually on the right side. The left renal vein is seldom multiple, but it may divide and surround the aorta (circumaortic venous ring).

The topographical arrangement of structures in the renal pedicle may be further complicated by variations in the ureter (p. 412). Thus, the most that can be said about a "typical" arrangement is that veins tend to be in front, and that arteries and veins lie mostly in front of the pelvis and ureter.

Renal Fascia

The extraperitoneal tissue lateral to the kidney condenses to form a membranous layer called the *renal fascia* which, as it passes medially, splits to enclose the kidney (fig. 37–7).[17] The anterior layer continues across the median plane, whereas the posterior layer merges with the connective tissue in front of the vertebral column. Both layers also merge with the connective tissue around the renal vessels. The two layers are fused with each other above the suprarenal gland. Below, they fuse weakly and merge with the extraperitoneal tissue that surrounds the ureter.

The renal fascia is separated from the capsule of the kidney by the perinephric space. This space is occupied by the perinephric (perirenal) fat. The fat that lies external to the renal fascia, both anteriorly and posteriorly, is called pararenal fat.

It seems well established that the renal fascia is a condensation of extraperitoneal tissue and is not derived from transversalis fascia. Renal fascia is absent if the kidneys fail to develop.[18] Furthermore, if the kidneys develop in an iliolumbar location, they have a renal fascia that is distinct from transversalis fascia or fascia iliaca.[19]

The two perinephric spaces do not communicate across the median plane. They are, however, potentially open below, to the extent that, although a perinephric abscess of one kidney will not spread across the median plane, it may emerge below, especially along the ureter. Studies in which air or fluids are injected into the perinephric spaces give slightly differing results. Air, being more diffusible, may cross the median plane.[20]

URETERS

The ureter is a muscular tube, 25 to 30 cm long, that connects the kidney with the urinary bladder. The ureter is retroperitoneal, and the upper half is abdominal, the lower half pelvic, in position. The ureter leaves the renal pelvis at or near the hilus, behind the renal vessels, and descends on the psoas major, embedded in extraperitoneal connective tissue. It crosses the common iliac artery or the first part of the external iliac, runs along the lateral wall of the pelvis, and turns medially toward the bladder. Its pelvic course in the female is of special importance (p. 464). In the region of the ischial spine, it turns downward, forward, and medially, below the uterine vessels, about 1½ to 2 cm from the cervix. The ureter is in its greatest danger here in hysterectomy.[21] Because of the asymmetry of the uterus and vagina, the left ureter is more closely related to the vagina, sometimes crossing the median plane in front of it.

The right ureter lies behind the second part of the duodenum at its origin. During its course it is crossed by the root of the mesentery and by the gonadal vessels. The left ureter likewise is crossed by gonadal vessels. At the pelvic brim it passes behind the sigmoid colon, at the apex of the sigmoid mesocolon.

The ureter may be constricted to a variable degree (1) at the junction of the ureter and renal pelvis, (2) where it crosses the pelvic inlet, and (3) during its course through the wall of the bladder. These are potential sites of obstruction.

The ureters are very distensible. They become much dilated and thick-walled when there is a chronic obstruction in their lower parts or in the urethra. An acute obstruction, such as commonly results from a kidney stone, usually causes pain.

The mucous membrane, which is lined by transitional epithelium, is thrown into folds when the ureter is empty. The thick muscular coat contains circular and longitudinal smooth muscle fibers. Urine is passed down the ureter by waves of contraction,[22] and enters the bladder in spurts at frequent intervals (one to six times per minute).

Blood Supply, Lymphatic Drainage, and Nerve Supply. The ureter is supplied by a variable number of "long arteries" from the renal, gonadal, and inferior vesical arteries, as well as by a number of other arteries.[23] The branches that reach the ureter divide into ascending and descending branches that anastomose. Veins accompany the arteries. Lymphatic vessels drain into adjacent nodes.

Nerve fibers reach the ureter from adjacent plexuses (renal and hypogastric). The plexuses contain pain fibers. Renal colic, which results from an acute distention and is usually due to an obstruction by a kidney stone, is characterized by sudden, severe pain. **Depending upon the level of obstruction, the pain in renal colic may be referred to the lumbar or hypogastric regions, to the external genitalia, or to the testis.**

Variations. The most common variation is doubling of the upper end of the ureter.[24] Less commonly, the entire ureter may be duplicated.

SUPRARENAL GLANDS

The paired suprarenal glands are small endocrine glands that weight about 3 to 6 gm each. Some of the hormones produced by the suprarenals are essential to life. Each suprarenal gland lies on the superomedial aspect of the front of the corresponding kidney (figs. 37–2 and 37–4). The right suprarenal projects somewhat behind the inferior vena cava. Each is surrounded by renal fascia, to which it is firmly attached. The glands can be demonstrated radiographically by the injection of oxygen or air into the perirenal fat. The layers of renal fascia are fused above the suprarenal glands, which are attached to the fascia. This fascia in turn is attached to the diaphragm.

In most animals the glands are termed adrenal glands, owing to the fact that they are near the kidneys, but not necessarily above them. Hence adrenaline (epinephrine) and adrenalectomy.

The right gland is somewhat pyramidal. Its base rests on the kidney. Behind, it lies against the diaphragm; in front, it is in contact with the bare area of the liver, the inferior vena cava, and with peritoneum. The left suprarenal is more flattened, and somewhat semilunar in shape. It is related behind to the diaphragm. In front, it is covered above by the peritoneum of the lesser sac and below by the pancreas. The splenic artery is an important anterior relation. Each gland possesses a hilus, from which the suprarenal vein emerges.

Structure and Function. Each suprarenal gland comprises two different endocrine components; the *cortex* and the *medulla*. The entire gland is surrounded by a connective tissue capsule. The suprarenal cortex produces steroid hormones that are important in maintaining electrolyte balance and in protein and carbohydrate metabolism. The medulla produces epinephrine and norepinephrine, the effects of which are generally similar to those resulting from stimulation of the sympathetic nervous system.

Blood Supply and Lymphatic Drainage. Multiple suprarenal arteries arise from the inferior phrenic artery, one or more inferior suprarenal arteries often arise from the renal, and a middle suprarenal artery from the aorta may reach the gland.[25] The number and patterns of arrangement of suprarenal arteries are so variable that they are not alike in any two bodies, or on the two sides of one body.

The venous drainage[26] is by the suprarenal vein, which leaves the hilus, and by many small veins that accompany the arteries. The right vein (sometimes double) enters the inferior vena cava, and the shorter left vein enters the renal vein, usually as a common trunk with the left inferior phrenic vein.

Only a few lymphatic vessels are present in the cortex, but there are many in the medulla. They accompany veins to adjacent lymph nodes.

Nerve Supply. The suprarenal glands are supplied by the celiac plexus and thoracic and lumbar splanchnic nerves. The fibers are mostly preganglionic sympathetic fibers that go directly to the cells of the medulla.

Development and Variations. Because of the development of the fetal or provisional cortex, the suprarenals are extremely large at birth, being about one third the size of the kidneys. After birth, the fetal cortex degenerates, and the absolute weight of the glands decreases. The glands do not regain their birth weight until puberty.

Accessory suprarenal tissue is commonly present in the abdomen and pelvis. The cortex develops in association with the urogenital ridge. Accessory cortical tissue may therefore be present in the vicinity of the kidney, or anywhere along the path of the descending gonads, and has been found in the pelvis and scrotum. Accessory medullary or chromaffin tissue is also a common finding.

Chromaffin System

Cells that develop from neuroectoderm and that stain with chrome salts may occur anywhere that sympathetic ganglion cells occur. They are most common in the abdominal cavity, usually near sympathetic ganglia along the aorta. These *paraganglia* or *para-aortic bodies*, together with the adrenal medullae, constitute the chromaffin system. Most of the paraganglia secrete norepinephrine.

Many of the para-aortic bodies reach their maximal size during fetal life.[27] Two of them are fairly constant in position (near the origin of the inferior mesenteric artery) and are about 1 cm long.[28] They continue to enlarge after birth, but shortly thereafter decrease considerably in size.

REFERENCES

1. H. Moël, Acta radiol., Stockh., *46*:640, 1956.
2. J. Frimann-Dahl, Acta radiol., Stockh., *55*:207, 1961.
3. R. O. Moody and R. G. Van Nuys, Anat. Rec., *76*:111, 1940.
4. P. A. Narath, *Renal Pelvis and Ureter,* Grune & Stratton, New York, 1951.
5. W. B. Stirling, *Aortography,* Livingstone, Edinburgh, 1957.
6. H. Fine and E. N. Keen, J. Anat., Lond., *100*:881, 1966.
7. J. Hodson, Brit. J. Urol., *44*:246, 1972.
8. G. Inke, M. Schneider, and W. Schneider, Anat. Anz., *118*:241, 1966. G. Inke, M. Schneider, W. Schneider, and G. Trautmann, Anat. Anz., *129*:471, 1971.
9. J. Fourman and D. B. Moffat, *The Blood Vessels of the Kidney,* Blackwell, Oxford, 1971. F. T. Graves, *The Arterial Anatomy of the Kidney,* John Wright, Bristol, 1971.
10. Ref. 6. See also H. E. Engelbrecht *et al.,* S. Afr. med. J., *43*:826, 1969. S. Poisel and H. P. Spängler, Acta anat., *76*:516, 1970.
11. Ref. 10. See also W. Woźniak, A. Kiersz, and S. Wawrzniak, Anat. Anz., *132*:332, 1972. G. Arvis, C. R. Ass. Anat., *53*:432, 1968.
12. S. Poisel and H. Sirang, Acta anat., *83*:149, 1972.
13. R. A. Davis, F. J. Milloy, and B. J. Anson, Surg. Gynec. Obstet., *107*:1, 1958.
14. G. A. G. Mitchell, J. Anat., Lond., *70*:10, 1935.
15. O. S. Culp and P. E. Hiebert, J. Urol., *51*:397, 1944. J. E. Dees, J. Urol., *46*:659, 1941. E. C. Smith and L. A. Orkin, J. Urol., *53*:11, 1945.
16. R. J. Merklin and N. A. Michels, J. int. Coll. Surg., *29*:41, 1958.
17. C. P. Martin, J. Anat., Lond., *77*:101, 1942. C. E. Tobin, Anat. Rec., *89*:295, 1944. F. Morin and P. L. Bruzzone, Arch. Anat. Anthrop., Lisboa, *26*:673, 1949. G. A. G. Mitchell, Brit. J. Surg., *37*:257, 1950.
18. J. A. Benjamin and C. E. Tobin, J. Urol., *65*:715, 1951.
19. E. H. Daseler and B. J. Anson, J. Urol., *49*:789, 1943.
20. J. Grossman, J. Anat., Lond., *88*:407, 1954.
21. J. Howkins, Ann. R. Coll. Surg. Engl., *15*:326, 1954.
22. F. Kiil, *The Function of the Ureter and Renal Pelvis,* Saunders, Philadelphia, 1957.
23. L. J. McCormack and B. J. Anson, Quart. Bull. Northw. Univ. med. Sch., *24*:291, 1950. O. Daniel and R. Shackman, Brit. J. Urol., *24*:334, 1952. E. Douville and W. H. Hollinshead, J. Urol., *73*:906, 1955.
24. C. D. Read, Ann. R. Coll. Surg. Engl., *10*:228, 1952.
25. R. Gagnon, Rev. canad. Biol., *16*:421, 1957, *25*:135, 1966. R. J. Merklin, Anat. Rec., *144*:359, 1962. J. W. Dobbie and T. Symington, J. Endocr., *34*:479, 1966.
26. R. Gagnon, Rev. canad. Biol., *14*:350, 1956. F. R. C. Johnstone, Amer. J. Surg., *94*:615, 1957.
27. W. H. Hollinshead, Quart. Rev. Biol., *15*:156, 1949.
28. G. Iwanow, Z. Anat. EntwGesch., *91*:404, 1930. R. E. Coupland, J. Anat., Lond., *86*:357, 1952.

BLOOD VESSELS, LYMPHATIC DRAINAGE, AND NERVES

38

BLOOD VESSELS

The arteries that supply the anterolateral abdominal wall are described elsewhere (pp. 217, 361). Other arteries to the abdomen arise from the abdominal aorta.

ABDOMINAL AORTA

The abdominal aorta begins at the aortic opening in the diaphragm, at about the level of T. V. 12. It descends in front of the vertebral bodies, at the left of the inferior vena cava. It deviates slightly to the left as it descends and ends at about the level of L. V. 4 by dividing into the right and left common iliac arteries. The important anterior relations are, from above downward, the pancreas and the splenic and left renal veins, the third part of the duodenum, and coils of small intestine. The celiac plexus and ganglia lie in front of the upper part of the aorta. At a somewhat lower level the intermesenteric part of the aortic plexus lies in front.

The abdominal aorta may be compressed by backward pressure on the anterior abdominal wall at the level of L. V. 4, especially in children and thin adults.

The parietal and visceral branches of the abdominal aorta may be classified as paired and unpaired (fig. 38–1).[1]

Parietal Branches

The inferior phrenic, lumbar, and common iliac arteries are paired; the median sacral is unpaired.

Inferior Phrenic Arteries. The right and left inferior phrenic arteries arise from the celiac trunk almost as often as they do from the aorta, and they often arise by a common stem. Each artery crosses the corresponding crus of the diaphragm and divides into branches that supply the diaphragm and anastomose with the pericardiacophrenic and musculophrenic arteries. Many superior suprarenal arteries arise from each inferior phrenic artery or its posterior branch. The left inferior phrenic may give a branch to the stomach, and both may give an accessory artery to the kidney.

Lumbar Arteries. These are small segmental arteries that arise from the back of the aorta. There are usually four or five pairs,[2] and any pair may arise as a common trunk, especially in the case of the lower lumbar arteries. The median sacral artery may arise from one or the other of the fifth lumbar arteries.

The lumbar arteries run between the psoas major muscle and the vertebral bodies, and divide into smaller ventral and larger dorsal branches. The *ventral branches* supply adjacent muscles and nerves, in particular the lumbar plexus, and anastomose with segmental arteries above and below. Each *dorsal branch* passes backward in company with the dorsal

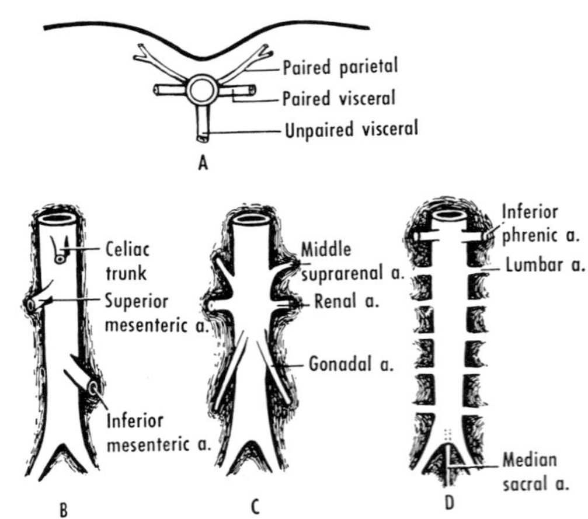

Figure 38–1 *A*, horizontal schematic representation of types of branches of aorta. *B*, unpaired visceral branches. *C*, paired visceral branches. *D*, paired and unpaired parietal branches.

branch of the corresponding lumbar nerve, and supplies the structures of the back. It gives off a *spinal branch* that enters the vertebral canal (p. 542). The dorsal branch of the fifth lumbar artery may be supplanted by the lumbar branch of the iliolumbar artery.

Common Iliac Arteries. The right and left common iliac arteries are the terminal branches of the aorta. Each runs downward and laterally and ends at the level of the lumbosacral disc by dividing into external and internal iliac arteries. The right common iliac is usually a little longer, owing to the fact that the aorta is to the left of the median plane at its bifurcation.

The ureter crosses in front of either the bifurcation of the common iliac, or the upper part of the external iliac. The left artery has in front of it the apex of the sigmoid mesocolon and the superior rectal vessels, and behind it the bodies of L. V. 4 and 5, and the psoas major. The right common iliac artery is separated from L. V. 4 and 5 and the right psoas major by the upper ends of the common iliac veins and the beginning of the inferior vena cava.

The aortic plexus is continued along the iliac vessels as the iliac plexus.

EXTERNAL ILIAC ARTERIES. The right and left external iliac arteries are the continuation of the common iliac arteries. Each descends in the iliac fossa to a point behind the inguinal ligament, where it becomes the femoral artery. The iliac plexus is continued along it to the femoral artery. Behind, it lies on the psoas major muscle. The cecum, appendix, and small intestine may be in front of the right artery, and the sigmoid colon and small intestine in front of the left. In the male, the testicular artery and ductus deferens lie in front of the lower part of the artery, and in the female the round ligament. The ureter may cross the upper part of the artery, and in the female may cross the ovarian vessels also. The external iliac arteries give small branches to adjacent structures, and each has two named branches, the inferior epigastric and the deep circumflex iliac artery (p. 361).

INTERNAL ILIAC ARTERIES. These are described on page 449.

Median Sacral Artery. This is an unpaired parietal branch that arises from the back of the aorta, a little above its bifurcation, or from one or both of the lowest lumbar arteries. It descends in front of L. V. 4 and 5, and then in front of the sacrum and coccyx, and ends in the coccygeal body (p. 447).

Visceral Branches[3]

The suprarenal, renal, and gonadal arteries are paired; the celiac trunk and superior and inferior mesenteric arteries are unpaired.

Middle Suprarenal Arteries. These are small paired vessels that arise slightly above the level of the renal arteries. They may be absent or multiple, may supply a considerable part of the suprarenal glands, or may supply mainly the perirenal fat.

Renal Arteries. These arise at about the level of L. V. 2. The right renal artery, usually lower than the left, passes behind the inferior vena cava. Each gives off one or more inferior suprarenal arteries, a branch to the ureter, twigs to adjacent fat and body wall, and then divides into its primary branches (p. 409).

Gonadal Arteries (Testicular or Ovarian). The gonads develop in the urogenital ridge, near the kidney, and receive their blood supply from the abdominal aorta (fig. 38–2).[4] The arteries that supply the gonads sometimes number three or four and have a variable level of origin with respect to the renal artery and to one another.[5] Occasionally they arise as a common trunk. The *testicular arteries* are long slender vessels that arise from the front of the aorta, or from adjacent branches of the aorta. Each one passes downward and laterally on the psoas major, and across the ureter, to which it gives a branch or branches. Reaching the deep inguinal ring, it accompanies the ductus deferens into the scrotum, where it supplies the spermatic cord and the testis. Each *ovarian artery* similarly passes downward and laterally on the psoas major and gives a branch or branches to the ureter. Each crosses the external iliac artery, enters the suspensory ligament of the ovary, and runs medially in the mesovarium. It supplies the ovary and anastomoses with the ovarian branch of the uterine artery.

Celiac Trunk. **The celiac trunk is the artery of the caudal part of the foregut.** It is a wide, short vessel that arises immediately below the aortic opening of the diaphragm,

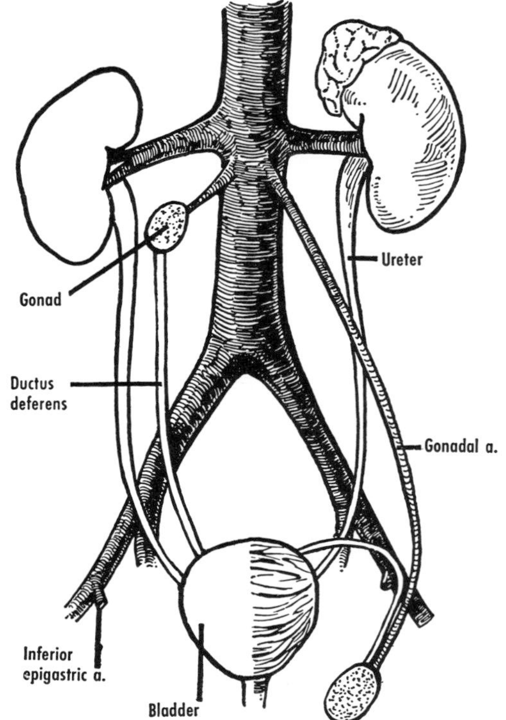

Figure 38–2 Schematic representation of blood supply of gonads and kidney. The gonad develops near the kidney (*left half of figure*). Later, the gonad descends, carrying its blood vessels and duct with it. The right half of the figure shows the adult relationship in the male.

between the crura, occasionally as a common trunk with the superior mesenteric artery. It is embedded in the dense, tough mat of the celiac ganglia and plexus. In at least one-half of instances, after a course of 1 to 3 cm it divides into left gastric, common hepatic, and splenic arteries.[6] It may also give off inferior phrenic arteries and a direct branch to the pancreas (the dorsal pancreatic artery). Any one of its branches may arise separately from the aorta or from the superior mesenteric artery.

LEFT GASTRIC ARTERY (fig. 35–5, p. 382).[7] This, the smallest branch, runs upward and to the left in the left gastropancreatic fold. On reaching the stomach, it turns and runs along the lesser curvature between the layers of the lesser omentum and ends by anastomosing with the right gastric artery. It gives *esophageal branches*, and branches to both surfaces of the stomach that anastomose with gastric branches of the splenic and gastroepiploic arteries. It may also give off the left hepatic artery.

SPLENIC ARTERY (fig. 35–5, p. 382). This is the largest branch of the celiac trunk. It has a tortuous course along the upper border of the body of the pancreas, a course so variable that no two splenic arteries are alike. During its course it gives many *pancreatic branches*, and it ends in a number of *splenic branches*. The *left gastroepiploic artery* and a number of short gastric arteries arise from one of the splenic branches or from the terminal part of the splenic artery. The left gastroepiploic artery runs from left to right, between the layers of the greater omentum. It gives branches to the stomach, and long, slender omental or *epiploic branches*. It does not anastomose directly with the right gastroepiploic artery but communicates in a variable fashion in the omentum.[8]

Some of the pancreatic branches are fairly constant and are named separately. The *dorsal pancreatic artery* (fig. 35–5, p. 382) is usually a branch of the splenic artery, but may arise from the superior mesenteric, from the hepatic, or from the celiac trunk. The continuation of the left branch of the dorsal pancreatic artery is sometimes called the *inferior*, or *transverse, pancreatic artery*. The *pancreatica magna artery*, which enters the body of the gland, is usually a large superior pancreatic branch of the splenic artery. Branches to the tail of the pancreas arise from the splenic artery or from one of its divisions, such as the left gastroepiploic artery, and are sometimes called *caudal pancreatic arteries*.

COMMON HEPATIC ARTERY (fig. 35–5, p. 382). This artery runs along the upper border of the body of the pancreas in the right gastropancreatic fold to the upper aspect of the first part of the duodenum where, in a variable fashion, it divides into the hepatic artery proper, the right gastric artery, and the gastroduodenal artery.

The *hepatic artery proper* continues upward in the edge of the lesser omentum to the liver, where it divides into *right* and *left branches*. The *cystic artery* arises from the right branch.

The *right gastric artery* is a small branch that runs along the lesser curvature of the stomach between the layers of the lesser omentum. It supplies the duodenum and stomach and anastomoses with the left gastric artery.

The *gastroduodenal* artery is a short, thick trunk that descends behind the first part of the duodenum, with the bile duct at its right. It often gives a *supraduodenal artery* to the upper aspect of the first part of the duodenum, and a number of small *retroduodenal branches* to the inferior aspect. The *posterior superior pancreaticoduodenal artery* arises behind the first part of the duodenum, tends to spiral around the bile duct, and enters into the posterior arcade (p. 404). On reaching the pancreas, the gastroduodenal artery divides into right gastroepiploic and anterior superior pancreaticoduodenal arteries. The *right gastroepiploic artery* runs to the left along the greater curvature of the stomach, between the layers of the greater omentum. It gives off gastric branches, and *epiploic branches* to the greater omentum. The *anterior superior pancreaticoduodenal artery* enters into the anterior arcade.

Superior Mesenteric Artery.[9] **The superior mesenteric artery (fig. 35–10, p. 387) is the artery of the midgut.** It arises from the front of the aorta below the origin of the celiac trunk. It supplies a part of the pancreas, all of the small intestine except a part of the duodenum, and the large intestine from the cecum to near the left colic flexure. At its origin it lies behind the pancreas and the splenic vein. From above downward, it descends in front of the left renal vein, the uncinate process of the pancreas, and the third part of the duodenum. It then enters the root of the mesentery, and runs in the root to the right iliac fossa. The superior mesenteric vein is usually on its right side.

Its first branch, the *inferior pancreaticoduodenal artery,* may arise from the first jejunal branch. It runs to the right and divides into *anterior inferior pancreaticoduodenal* and *posterior inferior pancreaticoduodenal arteries.* These arteries enter into the anterior and posterior arcades (p. 404). Both arteries may arise separately from the superior mesenteric.

Several branches that arise in a variable manner from the concavity (right side) of the superior mesenteric artery supply the large intestine. These branches are the *ileocolic,* the *right colic,* and the *middle colic arteries* (p. 388). Their anastomoses contribute to the formation of the marginal artery. The ileocolic artery has two or more branches with a variety of anastomotic communications. These branches supply the terminal part of the ileum, the cecum, and the appendix.

A varying number of *jejunal* and *ileal arteries* arise from the convexity (left side) of the superior mesenteric artery (p. 386). The first jejunal branch may give rise to the inferior pancreaticoduodenal artery. The superior mesenteric artery and its branches are accompanied by veins, and by a large number of nerve fibers and lymphatic vessels.

Inferior Mesenteric Artery. **The inferior mesenteric artery (fig. 35–10, p. 387) is the artery of the hindgut.** It arises from the aorta several centimeters above its bifurcation. It supplies the distal part of the colon, that is, from near the left colic flexure to the ampulla of the rectum. From its origin, it runs downward and to the left, on the psoas major. It crosses the inlet of the pelvis and becomes the *superior rectal artery.* The superior rectal artery crosses the left common iliac artery, where the ureter is lateral to it, at the apex of the sigmoid mesocolon. It then continues between the layers of the sigmoid mesocolon to the rectum, where it divides into two branches that continue downward in the wall of the rectum (p. 489). The inferior mesenteric artery and its branches are accompanied by nerve fibers (inferior mesenteric plexus), and by veins and lymphatics. The artery is accompanied by the inferior mesenteric vein in the lower part of its course.

Before crossing the inlet of the pelvis, the inferior mesenteric artery gives off *left colic* and *sigmoid arteries.* These form arcades that contribute to the marginal artery, and from which straight arteries reach the gut.

The anastomosis between the left and middle colic arteries at the left colic flexure is good (through the marginal artery); occasionally an intermesenteric communication connects the left colic with the middle colic or superior mesenteric arteries,[10] and rarely the middle colic arises from the inferior mesenteric artery.[11] The extramural anastomosis between the last sigmoid branch and the superior rectal artery is usually poor,[12] but within the walls of the gut may be adequate.[13]

Collateral Circulation

The collateral circulation that develops after obstruction of the abdominal aorta (provided the openings for the renal arteries are intact) is complex in detail, although simple in principle.[14] Anastomoses that bypass the obstruction form three groups, of which the first two are the most important:

(1) Longitudinal anastomoses between parietal vessels, in particular the lower intercostal, subcostal, and epigastric arteries. (2) Anastomoses between visceral branches, especially the intestinal and colic. (3) Anastomoses across the median plane, especially in the pelvis, between branches of the internal iliac arteries.

VEINS

Most of the veins of the abdomen accompany the corresponding arteries and require no separate description. Certain features of the venous system, however, merit emphasis. These are the portal system, the inferior vena cava and its tributaries, and the vertebral plexus, and their interconnections.

Portal System

Venous blood from the gastrointestinal canal is collected by the portal vein and its tributaries and is carried to the sinusoids of the

liver, from which it is ultimately drained into the inferior vena cava by the hepatic veins (fig. 36–4, p. 398).

The portal vein is formed by the junction of the superior mesenteric and splenic veins behind the neck of the pancreas (fig. 38–3); these veins may be demonstrated radiologically by the percutaneous injection of radio-opaque material into the spleen.[15] The superior mesenteric vein is extremely variable, having 10 to 25 tributaries. Yet its area of drainage and general course are fairly constant.[16]

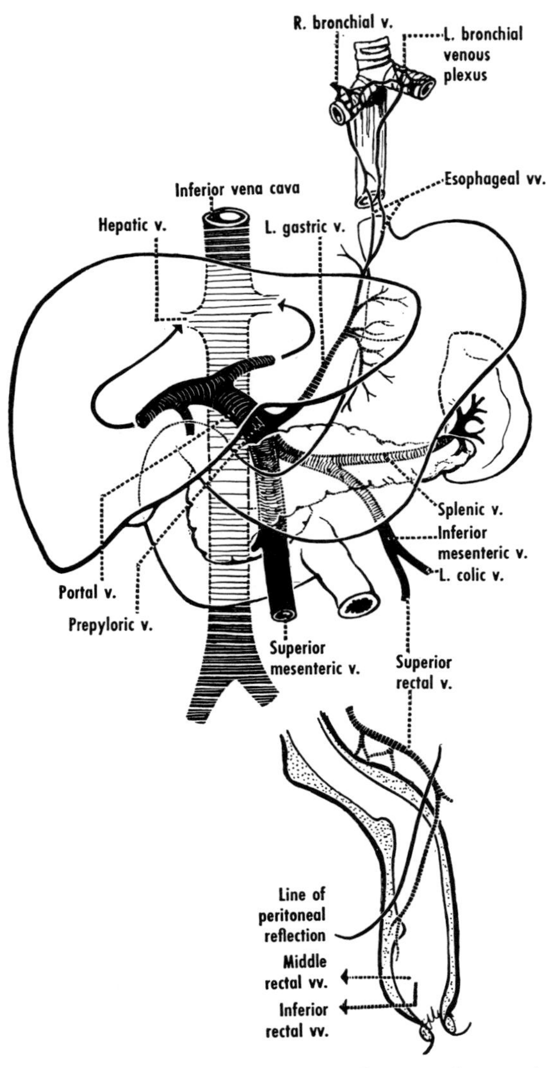

Figure 38–3 Portal vein and some of its major tributaries.

The inferior mesenteric vein may join the junction of the superior mesenteric and splenic veins, so that all three form the portal vein. In other instances, the inferior mesenteric vein joins the splenic or superior mesenteric vein in a highly variable fashion.[17] The portal vein often receives the left gastric vein.

The portal vein enters the hepatoduodenal ligament, ascends behind the bile duct and the hepatic artery, receives a variable number of smaller veins, and divides at the porta hepatis into right and left branches. Anomalies of the portal vein are rare.

Portal-systemic Anastomoses. Valves in the portal system are insignificant or absent, although valves may be present in the smaller tributaries at birth. **Therefore, the portal hypertension that results from obstruction of the portal vein readily causes enlargement of connections between the portal tributaries and the systemic veins, and reverse flow of blood into systemic veins.** The portal-systemic anastomoses are normally small, but enlarge greatly in the presence of obstruction. The important anstomoses are as follows:

1. BETWEEN INFERIOR MESENTERIC VEIN AND INFERIOR VENA CAVA AND ITS TRIBUTARIES. The superior rectal tributaries of the inferior mesenteric veins anastomose with the middle and inferior rectal veins, and blood from the portal system can reach the inferior vena cava by way of the internal and common iliac veins. The anastomoses of tributaries of the internal iliac veins also allow blood to reach the vertebral plexus and the external iliac veins, and thereby both the inferior and superior caval system (fig. 6–2, p. 38).

2. BETWEEN GASTRIC VEINS AND SUPERIOR VENA CAVA AND ITS TRIBUTARIES. The veins of the lower part of the esophagus, which anastomose above with bronchial veins and below with the left gastric vein,[18] also anastomose with the azygos system. Portal blood can therefore reach the azygos system and vertebral plexus, and thereby the superior vena cava. The gastroesophageal anastomoses are prone to become large, thin-walled varicosities in the presence of portal obstruction, and to rupture and cause serious bleeding.

3. BETWEEN RETROPERITONEAL

VEINS AND CAVAL AND AZYGOS SYSTEMS. Retroperitoneal veins are numerous, very small veins that drain the nonperitonealized surfaces of the organs (ascending and descending colon, duodenum, pancreas, liver). These veins and the retroperitoneal parts of the portal vein tributaries have small anastomoses with the segmental and phrenic veins. The connections of the segmental and phrenic veins allow blood to reach the heart by way of the caval system, both directly and by way of the azygos system and vertebral veins.

4. BETWEEN PARAUMBILICAL VEINS AND SUBCUTANEOUS VEINS. The paraumbilical veins in the falciform ligament connect the left branch of the portal vein with subcutaneous veins in the region of the umbilicus. The drainage of subcutaneous veins into the epigastric veins allows blood to reach the superior and inferior venae cavae. The paraumbilical veins also include some very small veins that drain the structures at the porta hepatis. These portal-systemic connections are unimportant. They are normally closed, or so small as to be virtually closed, and open and enlarge only with portal obstruction.[19]

During prenatal development, the two umbilical veins from the placenta enter the liver and break up into sinusoids. The right vein atrophies, and a shunt, the *ductus venosus*,[20] connects the left vein to the heart through the liver. The left umbilical vein and the ductus venosus become obliterated after birth, and form the ligamentum teres and the ligamentum venosum, respectively.

Inferior Vena Cava

The inferior vena cava[21] is a large, valveless, venous trunk that receives the blood from the lower limbs, and much of the blood from the back and from the walls and contents of the abdomen and pelvis.

It is formed by the junction of the two common iliac veins, slightly below and to the right of the bifurcation of the aorta. It ascends at the right of the aorta, through the central tendon of the diaphragm, and empties into the right atrium. From below upward it lies behind the peritoneum (crossed by the root of the mesentery and right gonadal vessels), duodenum and pancreas, portal vein, epiploic foramen, and the liver. The right renal artery crosses behind it.

The tributaries of the inferior vena cava are the common iliac, gonadal, renal, suprarenal, inferior phrenic, lumbar, and hepatic veins.

Variations. The part of the vena cava below the kidneys may be doubled or may be located on the left rather than the right. Sometimes the part of the inferior vena cava below the kidneys develops from the subcardinal vein rather than from the supracardinal (lateral sympathetic). In such instances the inferior vena cava lies in front of the ureter (preureteric vena cava).* The left inferior vena cava of the fetus persists more often than might be expected. When present in full length it connects the left renal and left common iliac veins. A segment of the left inferior vena cava sometimes persists and complicates the structure of the left renal pedicle.

Common Iliac Veins. The right and left common iliac veins are formed by the junction of the respective *external* and *internal iliac veins*, and through them the inferior vena cava drains the lower limbs and most of the pelvis. On account of the relative positions of the inferior vena cava and the aorta, the left common iliac vein lies directly below the bifurcation of the aorta. The iliac veins usually lack valves.

Gonadal and Suprarenal Veins. The right gonadal vein *(testicular* or *ovarian)* and the right suprarenal usually empty into the inferior vena cava, the left into the left renal vein.

Renal Veins. Each of the renal veins tends to lie in front of the corresponding renal artery; the left is the longer. The right renal vein, which may be multiple, receives few if any tributaries other than those from the kidney. The left renal vein, however, drains an extensive part of the body in a complex fashion (see fig. 37–6).

Inferior Phrenic Veins. The right inferior phrenic vein generally empties into the inferior vena cava. The left joins the left suprarenal (and thereby the renal), the left renal, or the inferior vena cava.

Hepatic Veins. These are short trunks (two or three major hepatic veins, several minor ones) that empty into the inferior vena cava just as it passes through the diaphragm. The right hepatic vein sometimes traverses the caval opening before joining the vena cava.

Lumbar Veins. These consist of four or five segmental pairs that accompany in part the corresponding arteries. Their dorsal branches drain the structures of the back and have free connections with the vertebral

*Synonyms are postcaval, circumcaval, deflected, and retrocaval. A ureter in this position is often obstructed, and should be suspected in obscure right-sided hydronephrosis.[22]

plexuses. The lumbar veins may empty separately into the inferior vena cava or common iliac, but are generally united on each side by a vertical connecting vein, the *ascending lumbar vein.* Each ascending lumbar vein enters the thorax behind the psoas major and medial arcuate ligament of the corresponding side. The right vein joins the right subcostal vein and forms the azygos vein (p. 327). The left joins the left subcostal and forms the hemiazygos vein.

The left upper lumbar veins and ascending lumbar vein are usually connected with the left renal vein.

Collateral Circulation. The collateral channels available in the event of an obstruction of the inferior vena cava are very numerous and complex, but can be classified in two groups, both of which are longitudinal in nature. (1) A variety of anastomoses in the pelvis and abdomen allows blood to reach the superficial and inferior epigastric veins, to ascend in these veins to the thoracoepigastric and superior epigastric veins, and thereby to reach the superior vena cava. (2) Anastomoses of tributaries of the inferior vena cava with the vertebral system of veins allow blood to ascend by way of these plexuses to the superior caval system. Blood may also descend, enter the pelvic veins, and thereby reach the epigastric veins and eventually the superior vena cava.

Vertebral Plexus

The vertebral system of veins is discussed, and its functions and clinical importance emphasized, on page 327. In the abdomen and pelvis, as well as in the thorax and in the head and neck, the main systemic channels have widespread valveless connections with the valveless venous plexus of the vertebral system. In the abdomen and pelvis, as elsewhere, the clinical importance of the vertebral plexus is due to its role in the spread of tumor cells and infections.

The same mechanisms that produce an ebb and flow of blood between the vertebral plexus and azygos system operate in the abdomen and pelvis, but with different timing. Blood returning from the lower limbs, pelvis, and abdomen depends for its flow upon pressure differences between capillaries and the venous side of the heart, and upon the very important squeezing action of muscles, combined with the arrangement of valves. The flow of blood is further aided by respiration. During inspiration, the intrathoracic pressure decreases, and the pressure difference between capillaries and heart becomes greater. During this phase, blood pours into the azygos system from the vertebral plexus. The same inspiratory excursion, by virtue of a diaphragmatic movement that carries viscera downward and compresses them, increases intra-abdominal pressure. This increase in pressure forces blood upward (valves in pelvis and lower limbs prevent downward flow), thereby aiding venous return. At the same time, blood tends to flow into the vertebral plexus. Thus, during inspiration, venous return increases, blood flows into and up the vertebral plexus from the abdomen, and out of the vertebral plexus into the thorax.

The converse changes occur during expiration. **It is clear that, unlike the direction of flow in the main systemic venous channels, which remains constant, the direction of flow in the vertebral plexus can vary according to the phase of respiration. The flow of blood from the abdomen and pelvis into the vertebral plexus is accentuated by any increase in intra-abdominal pressure due to coughing or straining.**

LYMPHATIC DRAINAGE[23]

The lymphatic vessels of the anterolateral abdominal wall are discussed on page 362. The lumbar lymphatic vessels ascend from the iliac nodes (fig. 41–6, p. 458) as two or three chains that are grouped about the aorta. Each chain consists of several vessels that are more or less parallel, and each overlies the right and left margins of the vertebral bodies. (If three chains are present, the middle one is near the median plane.) There is often a looping vessel at the right of L. V. 3 or 4 (fig. 38–4). This is the *right lower lumbar bypass vessel;* it corresponds to a gap in the chain of nodes. The ascending lymphatic chains join the thoracic duct.

There are two or three chains of *lumbar (aortic) nodes* (fig. 38–5), which are either in columns or scattered. The right and left chains overlie the transverse processes. (If there are three chains, the middle one lies in front of the aorta; nodes also lie behind the aorta in the upper lumbar and lower thoracic regions.) The nodes are vari-

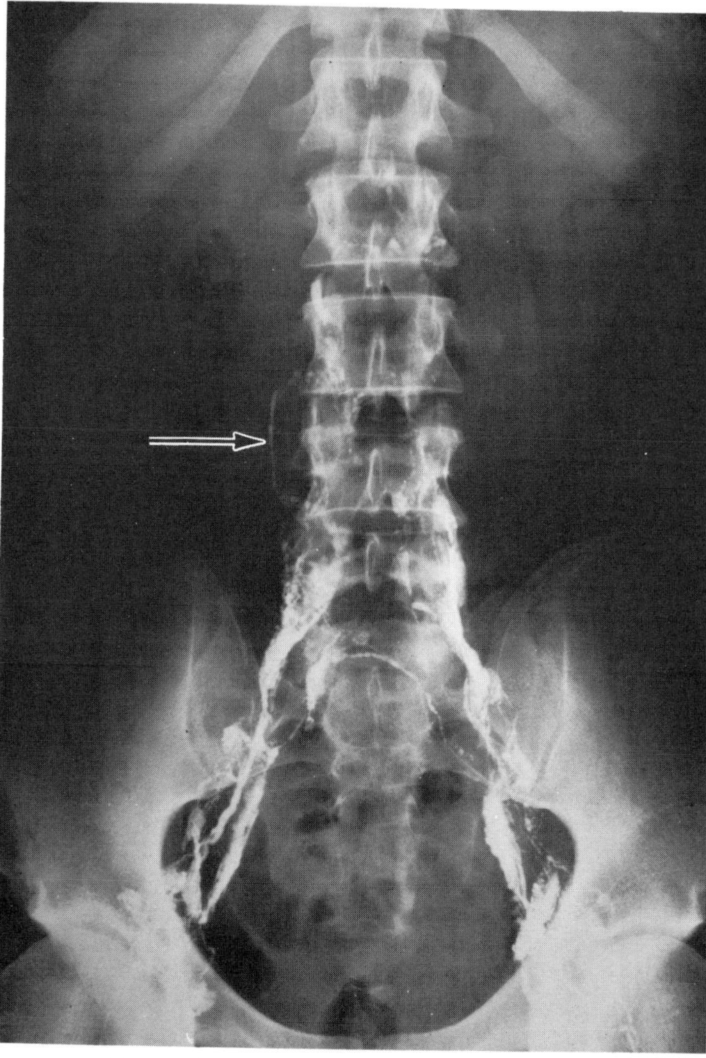

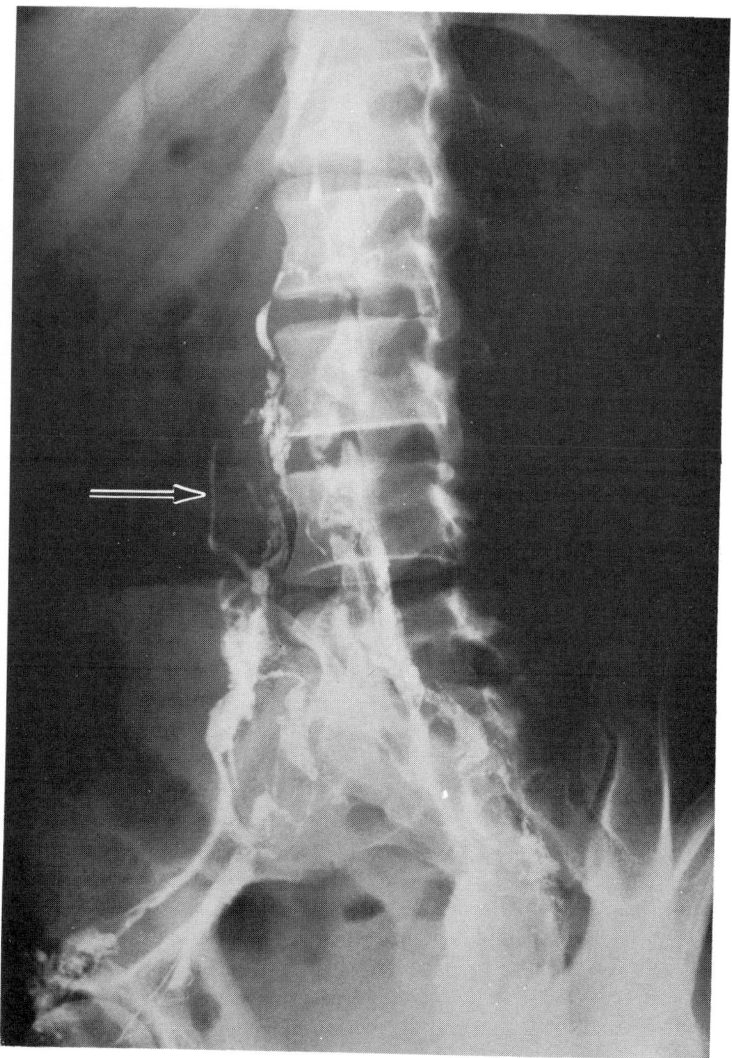

Figure 38–4 Two normal lymphangiograms (anteroposterior and oblique), showing lower lumbar nodes and vessels. Arrows indicate right lower lumbar bypass vessel. Courtesy of G. M. Stevens, M.D., Palo Alto Medical Clinic, Palo Alto, California.

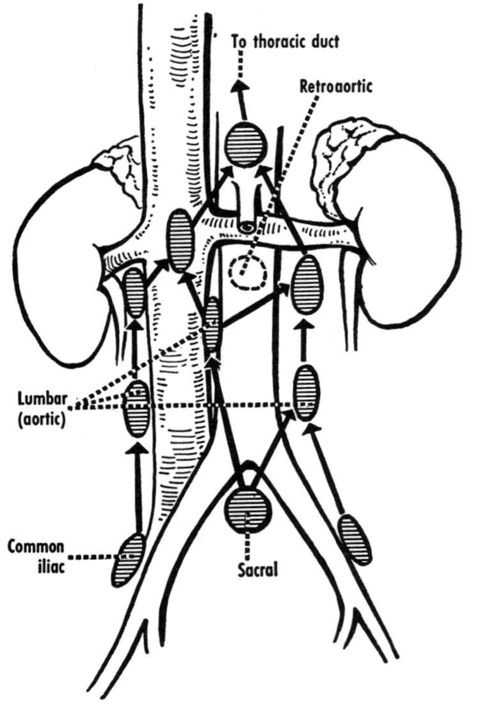

Figure 38–5 Diagram of the lymphatic vessels and nodes of the posterior abdominal wall.

able in size and in number (10 to 54). Connections between right and left sides are common (lumbar lymphatic crossover).

Other abdominal lymph nodes are regional rather than central, and are scattered along the vessels supplying the various organs.

The lower end of the thoracic duct lies behind and on the right side of the aorta, adjacent to the vertebral column and the right crus of the diaphragm. The thoracic duct (p. 330) begins in a highly variable fashion, either as an elongated, ovoid dilatation, the *cisterna chyli,* or as an irregular dilatation, or as a plexus of vessels. Several collecting ducts converge toward the cisterna chyli or plexus. These include the intestinal trunk from the nodes in front of the aorta, a pair of lumbar trunks from the more lateral nodes, and a pair of descending intercostal trunks from nodes in the lower intercostal spaces. The number of trunks varies, and trunks other than those named may be present. The thoracic duct passes upward through the aortic opening in the diaphragm and ascends in the thorax to the root of the neck, where it empties into the venous system.

NERVES

The nerves of the abdomen are the thoracoabdominal nerves, the phrenic and vagus nerves, the thoracic splanchnic nerves, the sympathetic trunk and branches, the autonomic plexuses and the lumbar plexus. The thoracoabdominal nerves have been described (p. 275).

Phrenic Nerves

The phrenic nerves, which contain motor, sensory, and sympathetic fibers (fig. 31–7, p. 333), supply the pericardium, mediastinal pleura, the central part of the diaphragmatic pleura, the diaphragm, and peritoneum. The right phrenic nerve pierces the diaphragm near the inferior vena cava, or traverses the opening for that vein. The left phrenic nerve pierces the diaphragm immediately to the left of the pericardium. Most of the motor fibers in the nerves are distributed to the diaphragm from below.

Some of the sympathetic fibers in the left nerve reach the stomach, whereas others reach the suprarenal gland.

Vagus Nerves

When the vagus nerves (p. 333) enter the esophageal plexus, they intermingle and form anterior and posterior vagal trunks. These trunks descend on the esophagus to the anterior and posterior surfaces of the stomach, respectively. Each trunk contains fibers from both right and left vagus nerves.

The anterior vagal trunk gives off several (sometimes only one) hepatic branches that course in the lesser omentum to the hepatic plexus; some fibers descend along the hepatic artery to reach the organs supplied by branches of this artery. After giving off the hepatic branches, the anterior vagal trunk gives off several gastric and celiac branches. The posterior vagal trunk likewise has a number of gastric and celiac branches.

The vagal fibers that enter the celiac plexus course in the branches of the celiac and superior mesenteric plexuses to reach the stomach, pancreas, liver, small intestine, and the large intestine as far as the left colic flexure (i.e., the derivatives of the foregut and midgut). The remainder of the large intestine receives parasympathetic fibers from the pelvic splanchnic nerves (p. 456).

Functional components are discussed on page 334.

Thoracic Splanchnic Nerves

These sympathetic nerves, which are the greater, lesser, and lowest splanchnic nerves, arise from the thoracic part of the sympathetic trunk (p. 335). They carry a major part of the sympathetic and sensory supply of the abdominal viscera.

Greater Splanchnic Nerve. This pierces the muscular part of the crus of the diaphragm, then turns medially, and enters the celiac ganglion. A fairly large *splanchnic ganglion* and several smaller ganglia are located along the nerve. After piercing the diaphragm, the right nerve lies behind the inferior vena cava, the left one behind the left suprarenal gland.

Lesser Splanchnic Nerve. This pierces the diaphragm slightly lateral to the greater splanchnic nerve. It joins the aorticorenal ganglion and gives filaments to the celiac, superior mesenteric, and renal plexuses, and often to the splanchnic ganglion.

Lowest Splanchnic Nerve. This enters the abdomen at the medial side of the sympathetic trunk and joins the aorticorenal ganglion and renal plexus.

Sympathetic Trunks and Ganglia[24]

The sympathetic trunks, the functional components of which are described on page 782, enter the abdomen by piercing the diaphragm or by passing behind the medial arcuate ligaments. They descend on the vertebral column, adjacent to the psoas major muscles. The right trunk lies behind the inferior vena cava, the left one beside the aorta. The trunks continue into the pelvis in front of the sacrum.

The sympathetic trunks are seldom symmetrical, and the lumbar ganglia are irregular in size, position, and number (usually 3 to 5).[25] There may be from two to six ganglia and sometimes a sympathetic trunk is merely an elongated, ganglionated mass. Variability in the number of lumbar ganglia appears to be due to the fact that, as ganglia develop, each one separates into two parts. These two parts later recombine and form the segmented trunk. Irregularity in recombination leads to the variations commonly present in the adult.

The identification of the proper level of a specific ganglion is very difficult. Counting from the highest lumbar ganglion found cannot be depended upon. For example, the first lumbar ganglion, when present, lies between the crus of the diaphragm and the vertebral column.[26] It is difficult to reach and is often overlooked. Ganglia are best identified by means of their rami communicantes.

Rami Communicantes. Each lumbar ganglion has two or more rami communicantes, and is connected to the ventral rami of two or more spinal nerves. The lowest ramus, which contains the most preganglionic fibers, is usually the key to the identification of a ganglion. For example, the first lumbar ganglion has rami that connect it to the twelfth thoracic and first lumbar nerves. The second lumbar ganglion, which is usually the largest and most constant of the lumbar ganglia, has rami that connect it with the first and second lumbar nerves. Identification of ganglia during surgical procedures is very uncertain, owing to the difficulty of locating and dissecting rami communicantes and ventral rami that are medial to the psoas major.

The second lumbar nerve is the lowest to contain preganglionic fibers. Consequently, if the sympathetic trunk is cut below the rami that connect it with this nerve, the preganglionic fibers to the lower limb will be severed.

Visceral Branches. These consist of four or more *lumbar splanchnic nerves*[27] of variable size that arise from the lumbar ganglia or the sympathetic trunk. The upper ones join the celiac and adjacent plexuses, the middle ones go to the intermesenteric and adjacent plexuses, and the lower ones descend to the superior hypogastric plexus.

Autonomic Plexuses

The great prevertebral plexus of the abdomen is formed by splanchnic nerves, branches from both vagus nerves, and masses

of ganglion cells, all so embedded in connective tissue that they form a very tough, dense network.

The prevertebral plexus lies in front of the upper part of the aorta; it extends along the aorta and its branches (fig. 38–6). The plexus and its peripheral extensions contain pre- and postganglionic sympathetic fibers, preganglionic parasympathetic fibers, and sensory fibers. The plexus and its extensions are continuous, but the following various parts of it are named according to the arteries with which they are associated.

Celiac and Superior Mesenteric Plexuses. The *celiac plexus* lies on the front and sides of the aorta at the origins of the celiac trunk and superior mesenteric and renal arteries. It contains the paired celiac ganglia, the superior mesenteric ganglion (or ganglia), and many small, unnamed ganglionic masses; its functional components are shown in figure 38–7. The irregularly shaped *celiac ganglia* lie at the level of origin of the celiac trunk, each on the corresponding crus of the diaphragm. The right ganglion lies behind the inferior vena cava and the head of the pancreas, and the left lies above the body of the pancreas, behind the lesser sac. The *aorticorenal ganglia,* which are sometimes partly fused with the celiac ganglia, lie near the origin of the renal arteries.[28]

Branches of the celiac plexus extend along arteries and form plexuses that are named accordingly—*hepatic, gastric, phrenic, splenic, suprarenal,* and *renal.* Small ganglia are located in these plexuses, and some are named, for example, *phrenic* and *renal.* The phrenic ganglion is at the junction of the celiac plexus and the phrenic nerve. Branches of the celiac plexus from the region of the aorticorenal ganglia also descend and, together with branches of the intermesenteric plexus, form *ureteric* and *testicular* or *ovarian* plexuses. The fibers in the testicular plexus accompany the vessels to the spermatic cord, testes, and epididymis, and those in the ovarian plexus accompany the vessels to the ovary, broad ligament, and uterine tube.

The superior *mesenteric ganglion* or *ganglia* lie immediately below, or at the sides of, the superior mesenteric artery, and are commonly fused with the celiac ganglia. The branches that accompany the artery form the *superior mesenteric plexus.*

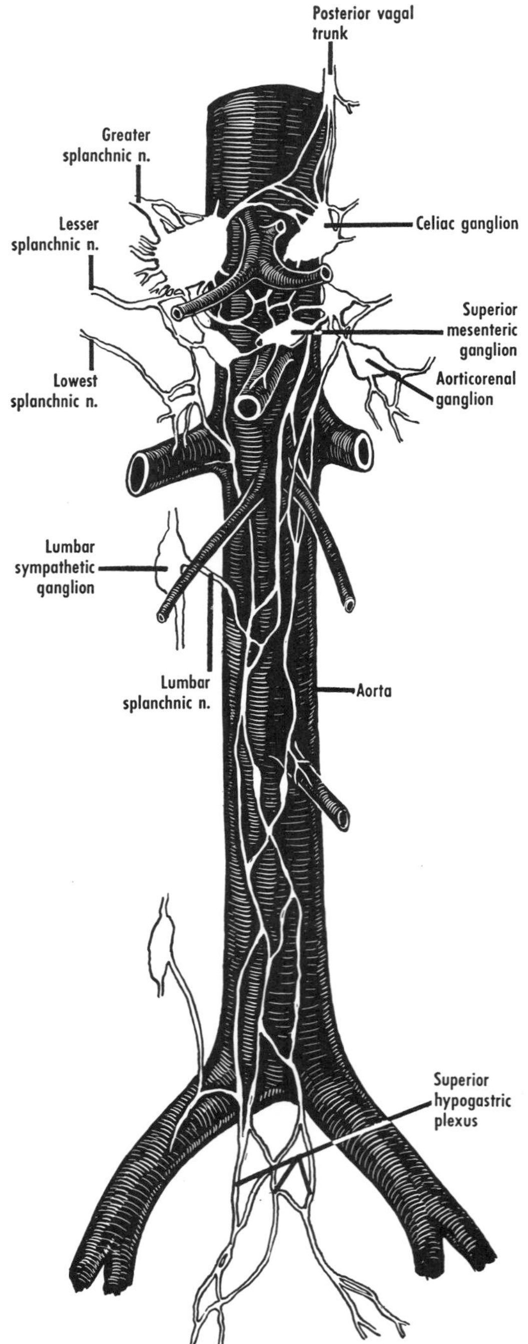

Figure 38–6 Diagram of the prevertebral ganglia and plexus. See also figure 64–15 (p. 783) for the sympathetic trunks and ganglia.

Aortic Plexus. Fibers that continue downward along the aorta constitute the aortic plexus, which, as it descends, receives branches of lumbar splanchnic nerves. The part of the plexus between the origins of the superior and inferior mesenteric arteries is also known as the *intermesenteric plexus.* It receives branches from all the lumbar splanchnic nerves.[29] Below the bifurcation of the aorta, the aortic plexus becomes the superior hypogastric plexus (p. 456), the fibers of which are derived mainly from the lumbar splanchnic nerves.

Some filaments from the aortic plexus, reinforced by branches of lumbar splanchnic nerves, form a plexus along the common and external iliac arteries.[30] This plexus is joined by a large branch of the genitofemoral nerve and continues into the thigh on the femoral artery. It contains many sensory fibers and may provide a pathway for pain fibers from the upper part of the lower limb.

Inferior Mesenteric Plexus. This plexus is an extension of the aortic plexus along the inferior mesenteric artery. One or more inferior mesenteric ganglia are present near the beginning of the artery. The plexus continues along the branches of the artery, and it forms the *superior rectal plexus,* which carries sympathetic fibers to, and afferent fibers from, the rectum.

Lumbar Plexus

The dorsal rami of the lumbar spinal nerves, which provide a part of the nerve supply of the back, are described later (p. 532). The ventral rami enter the psoas major muscle, where they combine in a variable fashion to form the lumbar plexus (figs. 38–8 and 38–9). (A division into anterior and posterior segments that then combine, as occurs in the trunks of the brachial plexus, has been described but is difficult to demonstrate.) Within the muscle the rami are connected to the lumbar sympathetic trunk by rami communicantes. **The second to fourth nerves are usually (in approximately three-fourths of instances) described as forming the lumbar plexus proper. However, the lower part of the fourth lumbar nerve and all of the fifth enter the sacral plexus (the combined trunk is known as the lumbosacral trunk), and the two plexuses are commonly known as the lumbosacral plexus.** The fourth lumbar is then the one ventral ramus that is common to both plexuses. Finally, the branches of the first lumbar nerve also are usually described with the lumbar plexus.

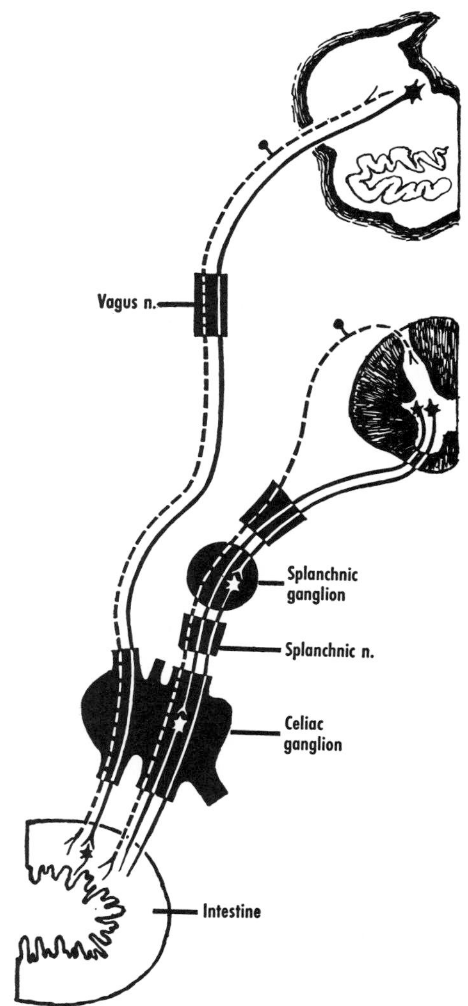

Figure 38–7 Functional components of the celiac ganglia and plexus. For purposes of simplification, each component is shown as a single fiber. Autonomic fibers, continuous lines; sensory fibers, interrupted lines. Part of one celiac ganglion and the types of fibers coursing through it and its branches are indicated (pre- and postganglionic sympathetic, preganglionic parasympathetic, and sensory).

As in the case of the brachial plexus, prefixation and postfixation of the lumbosacral plexus in the sense of complete shifts upward or downward are uncommon. Nevertheless, the plexus is often spoken of as *prefixed* when the upper level is the eleventh or twelfth thoracic nerve, and *postfixed* when the lower border is the fifth sacral or first coccygeal nerve. The total range may therefore be from the eleventh thoracic to the first coccygeal. The rami that supply the limbs, exclusive of the cu-

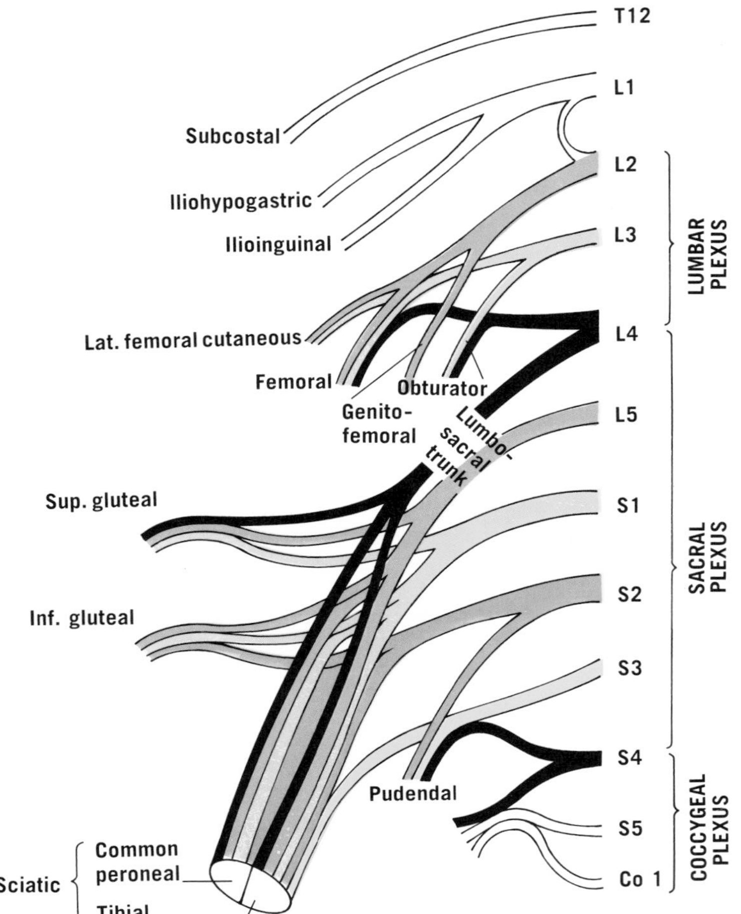

Figure 38–8 Simplified scheme of lumbosacral and coccygeal plexuses. (Lumbosacral plexus based partly on Seddon.[31]) See table 13–1 (p. 113) for ultimate distribution of rami to muscles.

taneous branches of T12 and L1, range from L1 to S3. Moreover, minor variations in pattern are common, and the right and left plexuses are seldom symmetrical.[32]

The lumbar plexus gives direct branches (L1 to L4) to the quadratus lumborum, psoas major, and psoas minor muscles. The following account of the remaining branches is based on the patterns that are most frequently encountered.[33]

First Lumbar Nerve. The first lumbar nerve, which has variable connections with the subcostal nerve and with L2, gives off muscular twigs and then divides into the iliohypogastric and ilioinguinal nerves (fig. 38–10), which emerge from the lateral side of the psoas. Strictly speaking, these are not part of the lumbar plexus, but they are usually described with it.

The first lumbar nerve resembles an intercostal nerve in giving off a collateral branch, the ilioinguinal nerve, and then continuing as the iliohypogastric nerve, which has a lateral cutaneous branch.[34] The point of origin, however, is extremely variable.

The *iliohypogastric nerve,* which may arise from T12, runs behind the lower part of the kidney and in front of the quadratus lumborum, pierces the posterior part of the transversus abdominis above the iliac crest, and divides into a lateral and an anterior cutaneous branch. The lateral cutaneous branch pierces the external and internal obliques and supplies the skin over the side of the buttock. The anterior cutaneous branch runs forward between the obliques, pierces the aponeurosis of the external oblique, and supplies the skin above the pubis. Muscular branches, if any, are probably sensory.

The *ilioinguinal nerve,* which sometimes arises from T12, and occasionally from L2, runs a similar course to the iliac crest where, having pierced the transversus and internal oblique, it continues forward to accompany the spermatic cord or round ligament through the inguinal canal. It emerges from the superficial ring and gives cutaneous branches to the thigh, and anterior scrotal or anterior labial branches.

Lateral Femoral Cutaneous Nerve. This nerve, arising from L2, from L2 and 3, or from L1 and 2, is often bound to the femoral nerve by connective tissue and may appear to arise from that nerve in the iliac fossa. It runs obliquely across the iliacus toward the anterior superior iliac spine, behind the inguinal ligament, through or in front of the sartorius, and into the thigh. Its anterior and posterior branches supply the skin of the front and side of the thigh (p. 218).

Femoral Nerve. This nerve, arising principally from L4, plus L2 and L3, is the largest branch of the lumbar plexus. Occasionally it receives a contribution from L5. It is the lowest of the branches and emerges from the lateral side of the psoas major. It descends between the psoas and iliacus and enters the thigh behind the inguinal ligament. In the iliac fossa it gives twigs to the iliacus. The nerve to the pectineus and that to the femoral artery may also arise in the iliac fossa, or behind the inguinal ligament.

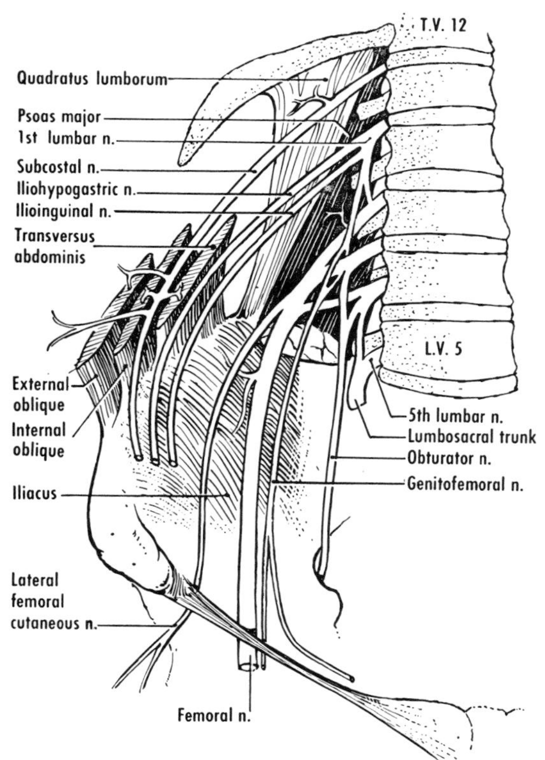

Figure 38–9 Schematic diagram of the lumbar plexus in relation to the muscle layers of the abdominal wall. The lateral cutaneous branch of the iliohypogastric nerve is not shown (see fig. 38–10). Based on Pitres and Testut.[35]

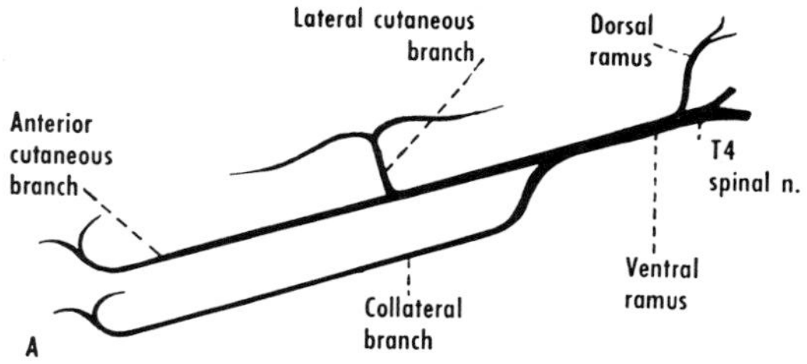

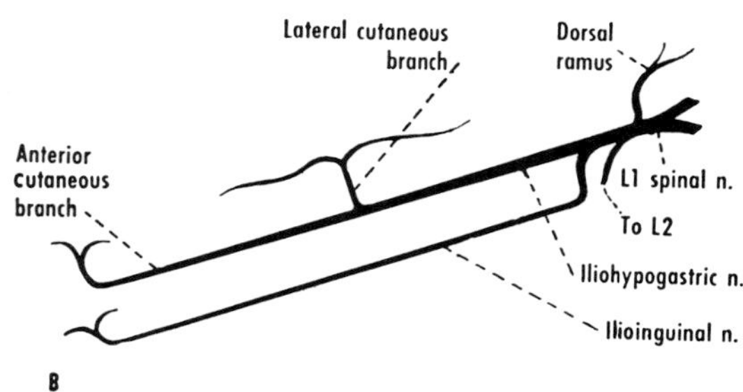

Figure 38–10 Comparison of *A*, an intercostal, and *B*, the first lumbar nerve. Based on Davies.[34]

The femoral nerve supplies the quadriceps femoris, pectineus, and sartorius muscles, the skin of the front and medial side of the thigh and medial side of the leg, and the hip and knee joints. Its course and distribution are described on page 218.

Genitofemoral Nerve. This usually arises from L2 or from L1 and 2, occasionally from L3. It descends on the front of the psoas and divides into genital and femoral branches. The *genital branch* enters the inguinal canal through the deep ring, supplies the cremaster, and continues on to supply the scrotum (or labium majus) and the adjacent part of the thigh. The *femoral branch* enters the femoral sheath, lateral to the artery, turns forward, and supplies the skin of the femoral triangle.

Obturator Nerve. This arises from L3 and 4, sometimes from L2 also, and occasionally from L5. It emerges from the medial aspect of the psoas at the inlet of the pelvis. It runs downward and forward on the lateral wall of the pelvis and enters the thigh through the obturator foramen. It supplies the adductor muscles and gracilis, the skin of the medial side of the thigh, and the hip and knee joints. Its course and distribution in the thigh are considered on page 215.

Accessory Obturator Nerve. When present,[36] this small nerve, arising chiefly from L3 and 4, descends medial to the psoas and enters the thigh by crossing above the superior ramus of the pubis deep to the pectineus. It supplies the hip joint and the pectineus (p. 215).

REFERENCES

1. Details of vertebral levels of the aorta and its branches are given by R. George, J. Anat., Lond., *69*:196, 1935; by E. W. Cauldwell and B. J. Anson, Amer. J. Anat., *73*:27, 1943; and by D. Obounou-Akong, R. M. Ouiminga, and R. Louis, C. R. Ass. Anat., *56*:1089, 1972.
2. A. H. Young, J. Anat., Lond., *39*:295, 1905. A. Rigaud, J. H. Soutoul, and C. Isabellon, C. R. Ass. Anat., *104*:699, 1959.
3. N. A. Michels, *Blood Supply and Anatomy of the Upper Abdominal Organs,* Lippincott, Philadelphia, 1955. R. A. Nebesar *et al., Celiac and Superior Mesenteric Arteries,* Little, Brown, Boston, 1969 (a correlation of angiograms and dissections).
4. J. C. B. Grant, Canad. med. Ass. J., *15*:1195, 1925.
5. G. Gerard, C. R. Soc. Biol., Paris, *74*:778, 1913.
6. H. M. Helm, Anat. Rec., *9*:637, 1915. See also reference 3.
7. H. I. El-Eishi, S. F. Ayoub, and M. Abd-el-Khalek, Acta anat., *86*:565, 1974.
8. R. E. Horton, Guy's Hosp. Rep., *101*:108, 1952. H. Ogilvie, Lancet, *1*:1077, 1952.
9. R. Sarrazin and J.-B. Levy, C. R. Ass. Anat., *53*:1503, 1969. See also reference 3.
10. G. H. Williams and E. J. Klop, Univ. Mich. med. Bull., *23*:53, 1957.
11. A. M. Vare and U. V. Karandikar, J. anat. Soc. India, *21*:74, 1972.
12. S. Sunderland, Aust. N. Z. J. Surg., *11*:253, 1942. J. C. Goligher, Brit. J. Surg., *41*:351, 1954. J. V. Basmajian, Surg. Gynec. Obstet., *99*:614, 1954.
13. J. D. Griffiths, Ann. R. Coll. Surg. Engl., *19*:241, 1956.
14. R. F. Muller and M. M. Figley, Amer. J. Roentgenol., *77*:296, 1957. See also G. J. Baylin, Anat. Rec., *75*:405, 1939, for a report of a case of complete obstruction of the aorta below the renal arteries.
15. J. A. Evans and W. D. O'Sullivan, Med. Radiogr. Photogr., *31*:98, 1955.
16. C. Gillot *et al.,* J. int. Coll. Surg., *41*:339, 1964.
17. B. E. Douglass, A. H. Baggenstoss, and W. H. Hollinshead, Surg. Gynec. Obstet., *91*:562, 1950. C. W. A. Falconer and E. Griffiths, Brit. J. Surg., *37*:334, 1950. R. S. Gilfillan, Arch. Surg., Chicago, *61*:449, 1950. L. J. A. DiDio, Anat. Rec., *141*:141, 1961. P. Barry, A. Repolt, and J.-M. Autissier, C. R. Ass. Anat., *53*:510, 1968.
18. H. Butler, Thorax, *6*:276, 1951.
19. E. A. Edwards, Arch. intern. Med., *88*:137, 1951.
20. A. D. Dickson, J. Anat., Lond., *91*:358, 1957.
21. R. A. Davis, F. J. Milloy, and B. J. Anson, Surg. Gynec. Obstet., *107*:1, 1958. E. J. Ferris *et al., Venography of the Inferior Vena Cava and Its Branches,* Williams & Wilkins, Baltimore, 1969.
22. O. S. Lowsley, Surg. Gynec. Obstet., *82*:549, 1946. J. E. Heslin and C. Mamonas, J. Urol., *65*:212, 1951. W. E. Goodwin, D. E. Burke, and W. H. Muller, Surg. Gynec. Obstet., *104*:337, 1957.
23. A. dos Santos Ferreira, *Les grandes lymphatiques: abdomino-thoraco-cervicales,* Université de Lisbonne, 1973. B. T. Jackson, Ann. R. Coll. Surg. Engl., *54*:3, 1974. See also Kinmonth, cited on page 43.
24. J. Pick and D. Sheehan, J. Anat., Lond., *80*:12, 1946.
25. R. H. Webber, Anat. Rec., *130*:581, 1958.
26. K. C. Bradley, Aust. N. Z. J. Surg., *20*:272, 1951.
27. A. Kuntz, J. comp. Neurol., *105*:251, 1956.
28. J. E. Norvell, J. comp. Neurol., *133*:101, 1968.
29. W. Wozniak, Folia Morphol., Warsaw, *24*:37, 1965.
30. F. R. Wilde, Brit. J. Surg., *39*:514, 1952.
31. H. Seddon, *Surgical Disorders of the Peripheral Nerves,* Churchill Livingstone, Edinburgh, 1972.
32. A. Rigaud *et al.,* C. R. Ass. Anat., *42*:1206, 1955.
33. R. H. Webber, Acta anat., *44*:336, 1961.
34. F. Davies, J. Anat., Lond., *70*:177, 1935.
35. A. Pitres and L. Testut, *Les Nerfs en Schémas,* Doin, Paris, 1925.
36. R. T. Woodburne, Anat. Rec., *136*:367, 1960.

SURFACE ANATOMY, PHYSICAL EXAMINATION, AND RADIOLOGICAL ANATOMY

39

SURFACE ANATOMY

A brief summary is given here of points already discussed throughout the section on the abdomen.

The xiphisternal joint is at the apex of the *infrasternal angle,* the sides of which are the seventh pair of costal cartilages. The xiphoid process extends into the angle. The slight depression in front of the process is the *epigastric fossa* ("pit of the stomach") of the anterior abdominal wall. The seventh to tenth costal cartilages meet on each side and form the costal margin. The costal margins form the sides of the infrasternal angle.

The whole of the iliac crest is generally palpable. Its highest part is situated posteriorly. The midaxillary line, continued downward, meets the tubercle of the crest. The anterior superior iliac spine often forms a visible prominence. The posterior superior iliac spine is usually indicated by a dimple. The pubic tubercle is located about 2 to 3 cm lateral to the median plane.

The linea alba forms a median furrow, broader above, and is especially evident in lean, muscular individuals when the recti are contracting. In such individuals, the linea semilunaris is evident as a shallow, curved groove on the lateral side of each rectus, and the segmentation of the recti by the tendinous intersections may also be seen. The external oblique is also prominent, often appearing as a bulge above the iliac crest.

The umbilicus is a prominent but highly variable landmark in the median plane, usually between the levels of L. V. 3 and L. V. 5. It is lower in childhood and in old age, and the variation in level may be extreme in obese individuals.

Planes and Points of Reference

The following planes and markings are commonly used as landmarks in the examination of the abdomen:

Supracristal Plane. A horizontal plane between the highest points of the iliac crests. It is at the level of the spine of L. V. 4.

Transtubercular Plane. A horizontal plane through the tubercles of the iliac crests, at the level of L. V. 5.

Transpyloric Plane. A horizontal plane approximately halfway between the jugular notch of the sternum and the pubic symphysis, at the level of L. V. 1. When the arm is at the side, the medial epicondyle of the humerus is approximately at the transpyloric plane.

Right and Left Lateral Planes. Sagittal planes midway between the median plane (the pubic symphysis) and the anterior superior iliac spine on each side, that is, sagittal planes through the midinguinal points.

Midinguinal Point. A point midway between the anterior superior iliac spine and the median plane (therefore medial to the midpoint of the inguinal ligament). It marks the ductus deferens at the deep inguinal ring, and the origins of the inferior epigastric and deep circumflex iliac arteries.

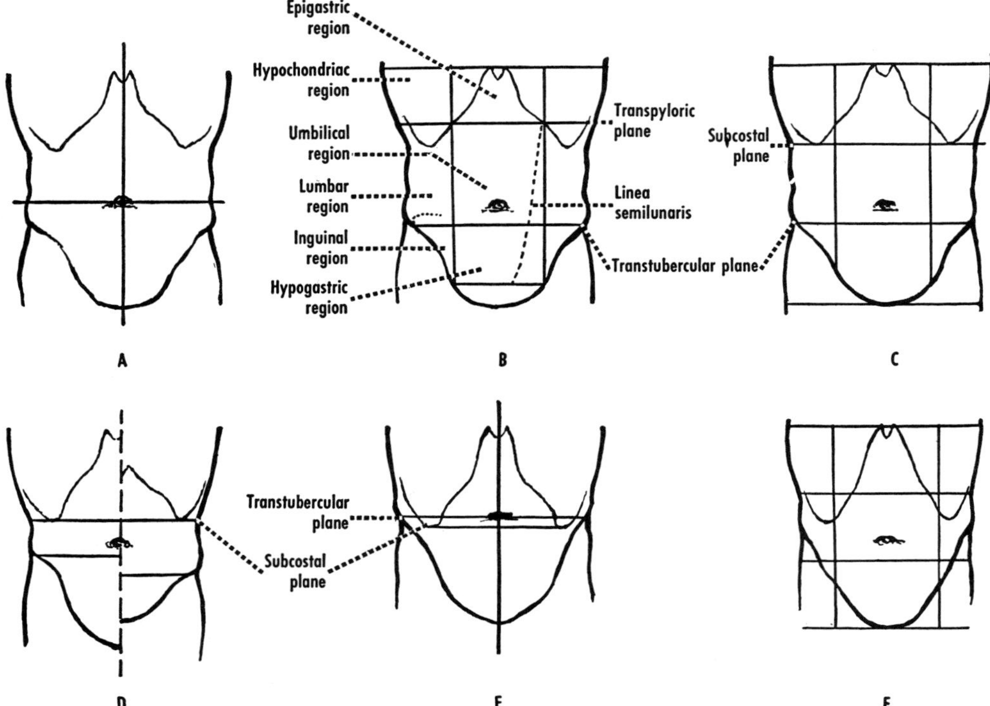

Figure 39–1 Abdominal topography. *A*, division into quadrants. *B*, division into nine regions by two vertical and two horizontal planes. *C*, division into nine regions using subcostal instead of transpyloric plane. *D*, narrow umbilical zone in a 75 year old man on the left; wide zone in a 3 year old child on the right. *E*, a 58 year old man with a subcostal plane at a lower level than the transtubercular; no umbilical or lumbar regions are present. *F*, a system of regions based on division of the linea alba into thirds. *B* to *F*, from Brown and Smith.[1]

Regions of the Abdominal Wall. In examining and describing the abdomen, it is customary to relate pain or swelling or the position of an organ to one of the regions shown in figure 39–1. **The simplest and most commonly used regional subdivisions are right and left upper and lower quadrants, formed by the median plane and a horizontal plane through the umbilicus.**

Alternatively, the abdomen may be divided into nine regions, bounded by two horizontal planes (transpyloric and transtubercular) and two sagittal planes (right and left lateral); these are also designated in clinical descriptions. The practical value of such detailed subdivision is very doubtful. That value is still further diminished by the common use of different planes, such as subcostal plane instead of transpyloric, and sagittal planes through the midpoints of the inguinal ligaments instead of through the midinguinal points. However, the names of the various regions are in common use; for example, epigastric pain is that felt in the epigastric region.

Structures in the Abdominal Wall

Inguinal Ligament. It occupies the groin, extending from the anterior superior iliac spine to the pubic tubercle. The skin crease at the junction of the thigh and abdomen lies just below and parallel to the inguinal ligament.

Deep Inguinal Ring. It is immediately above the midinguinal point, lateral to the origin of the inferior epigastric artery.

Inguinal Canal. About 3 to 5 cm long, it extends between deep and superficial rings, above the medial half of the inguinal ligament.

Superficial Inguinal Ring. About 1 cm above and lateral to the pubic tubercle.

Abdominal Viscera

Most normal viscera have no fixed shape and no fixed position. Some organs are more fixed than others, and the positions of organs in a patient under deep general anesthesia closely resemble the positions they occupy in a cadaver.

Stomach. The cardiac opening is relatively fixed, and can be marked at the left costal margin, about midway between the plane of the xiphisternal joint and the transpyloric plane, about 3 cm to the left of the median plane. The fundus corresponds to the left dome of the diaphragm. The greater curvature, extremely variable, is usually between the transpyloric and the transtubercular planes. The lesser curvature may be above or partly below the supracristal plane. **The stomach may enter the true pelvis. The pyloric part, quite mobile, is usually between the transpyloric and supracristal planes, to the right or left of the median plane—about 2 to 3 cm to the right of the median plane on the transpyloric plane when the stomach is empty and the subject recumbent.**

Duodenum. About 25 cm long. The first part is mobile, like the pylorus. The duodenojejunal flexure is relatively fixed, about 2 to 3 cm below the transpyloric plane, and to the left of the median plane.

Jejunum and Ileum. No fixed position.

Cecum and Appendix. In the right iliac fossa, sometimes at the brim of the pelvis. The position of the appendix is highly variable. A circle 18 cm in diameter, centered on the lateral point of trisection of the right spinoumbilical line, was necessary to cover the positions of the base of the appendix in 30 unselected cases.[2]

Colon. The ascending and descending colons are lateral to the right and left lateral planes, respectively. The right colic flexure is usually below the transpyloric plane, the left colic flexure above it, but both may be below the supracristal plane. The transverse colon is extremely variable, and may enter the true pelvis. The sigmoid colon, likewise highly variable in length and position, is commonly in the true pelvis.

Liver and Gall Bladder. **The liver is related above to the domes of the diaphragm, and occupies an extensive area on the right side. Its lower part on the right side may reach below the supracristal plane.** The fundus of the gall bladder is commonly in the angle between the right costal margin and the linea semilunaris, but the gall bladder may occupy almost any position on the right side. The positions of both the liver and the gall bladder vary according to body type.

Spleen. Variable in size, its long axis roughly corresponds to the long axis of the tenth rib.

Kidney. **In the erect position, it extends from about L. V. 1 to L. V. 4, the hilus being about 5 cm from the median plane.** The right kidney is usually slightly lower than the left. Both descend with inspiration and when the body assumes the erect position.

Urinary Bladder. It extends into the abdomen in children and may reach the umbilicus. In adults also, a full bladder may reach the umbilicus.

Uterus. The fundus of the pregnant uterus rises above the pubic symphysis in the third month; it reaches the supracristal plane in the sixth, and the xiphisternal joint in the eighth.

Peritoneum. The root of the mesentery extends for about 15 cm from the duodenojejunal flexure downward and to the right, to the level of the right sacroiliac joint. The root of the transverse mesocolon extends for a similar distance between the right and left colic flexures. The root of the sigmoid mesocolon is often an inverted **V**, the apex of which lies at the division of the left common iliac artery, in front of the left ureter.

Blood Vessels. The abdominal aorta begins at the aortic opening between the xiphisternal joint and the transpyloric plane. It divides at the supracristal plane (level of L. V. 4). The inferior vena cava begins at the transtubercular plane, slightly below the bifurcation of the aorta, ascends at the right side of the aorta, and pierces the diaphragm above the level of the xiphisternal joint. The inferior epigastric artery extends upward and medially, from the midinguinal point; the deep circumflex iliac artery extends laterally from the same point.

PHYSICAL EXAMINATION

The classical methods of physical examination comprise inspection, palpation, percussion, and auscultation.

Inspection. Respiratory movements can be observed best with tangential light. Muscles, especially the rectus abdominis with its tendinous intersections, can be identified in lean, muscular subjects. The level of the testes relative to each other (p. 468) should be noted.

Palpation. Both examiner and subject should be comfortable, the subject preferably on a couch or examining table. The examiner's hands should be warm, and the flat of the hand should be used. Tenseness of abdominal muscles prevents proper examination. They may be relaxed by drawing the knees up, by deep breathing, or by diversion. **With deep palpation, the following structures can sometimes be identified in normal subjects: abdominal aorta, lumbar vertebrae, lower pole of right kidney, sometimes the liver, occasionally the spleen.** To palpate deeply, let the hand follow the abdominal wall inward in deep expiration, the hand pushing deeper with successive expirations. The liver and spleen may be more readily palpable in infants and children.[3]

Stroking the skin with a sharp point may induce reflex contraction of the abdominal muscles. Any movement of the umbilicus should be noted. The superficial abdominal reflex normally varies in intensity of response, and may be diminished or absent, especially in obese individuals. The cremasteric reflex is likewise a superficial reflex. It consists of a reflex elevation of the testis following stroking or scratching of the inner side of the upper part of the thigh. The reflex is best developed in children.

The body of the uterus can be palpated bimanually (p. 485) when the index finger is placed in the vagina and the opposite hand is placed on the anterior abdominal wall.

The superficial inguinal ring can be identified and examined with the subject erect. The scrotum is invaginated with the little finger, the finger pushing upward along the spermatic cord until the pubic tubercle is reached, and then pushing backward. The ring normally admits the tip of the little finger, or even the index finger. When the subject coughs, an impulse from a hernia may be felt by the finger.

Percussion (see also p. 340). A tympanitic note is obtained over the alimentary canal, especially over the fundus of the stomach, where it is noted during percussion downward over the left lung. Dullness is obtained over the liver, the spleen, and a full bladder. Dullness over the liver is encountered in percussing downward over the right lung.

Auscultation. It is used to listen to bowel sounds. It is also employed during pregnancy to hear the fetal heart sounds through the abdominal and uterine walls. The fetal heart rate is about twice the maternal pulse rate.

RADIOLOGICAL ANATOMY

The chief methods of study include fluoroscopy and radiography, supplemented by the use of contrast media. The contrast media include gas (air or oxygen), barium suspension, and various organic iodides.

Survey of Abdomen

An anterior radiogram ("scout" or survey film) of the entire abdomen shows the lower ribs, the lumbar vertebrae, and the ilia, and should show the sacroiliac joints (fig. 39–2).

In the upper part of the abdomen, the shadow of the liver is usually well defined and demarcates the domes of the diaphragm. The spleen can often be recognized on the left side. The psoas major muscles usually present a clearly defined shadow on each side of the vertebral column; the shadow widens as it descends. The kidney shadows can often be identified lateral to the upper part of the psoas

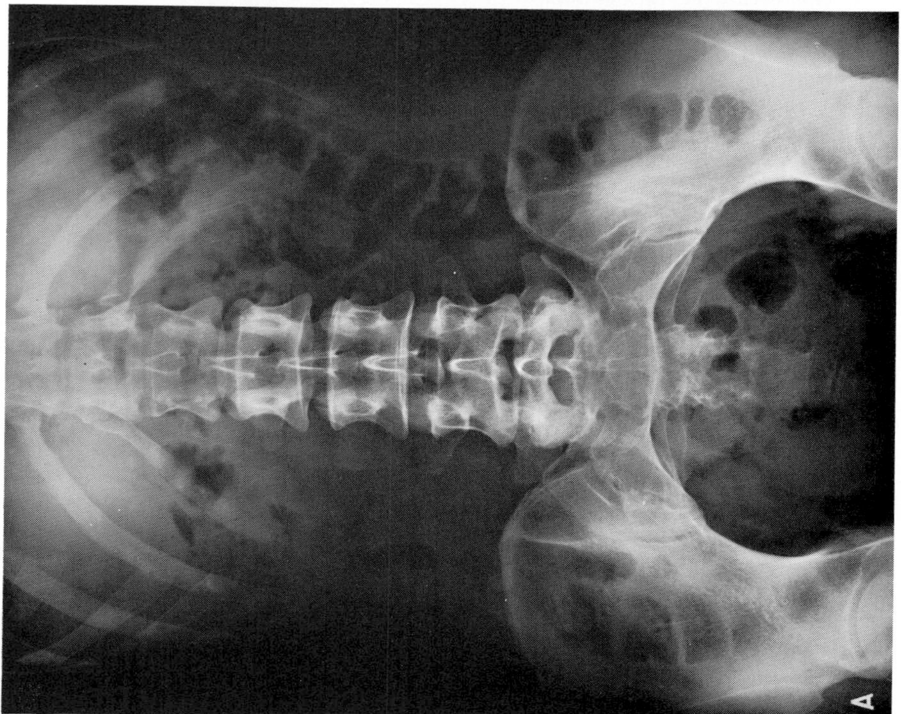

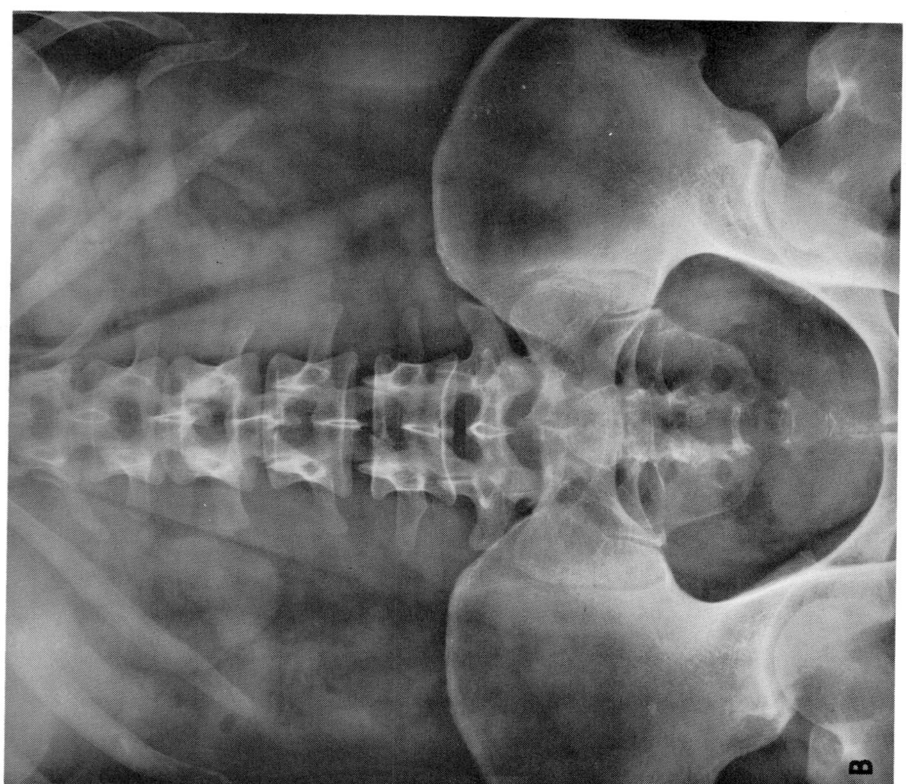

Figure 39–2 Abdomen. *A*, survey or "scout" film of abdomen, with emphasis on lumbar vertebrae. Note twelfth ribs, bodies and transverse and spinous processes of vertebrae (L.V. 4 spinous process is on supracristal plane), and gas in large intestine, particularly in the descending colon. *B*, "scout" film of abdomen, with emphasis on soft tissues. Note kidneys, psoas major muscles, and urinary bladder. The twelfth ribs are much shorter than in subject shown in A. Note transverse processes of lumbar vertebrae, and the sacroiliac joints, sacrum, and coccyx.

shadows. (The perirenal fat acts as a natural contrast medium.)

Air that has been swallowed is present in the fundus of the stomach and is visible as a translucent area. Gas may be present in the small intestine of the infant but is not normally present in that of the adult. Air and gas bubbles are usually visible in the large intestine of the adult and appear as small translucent areas, often intermixed with shadows produced by fecal material.

Alimentary Canal

Barium suspension is the contrast medium used for the alimentary canal and is given orally or by enema. If a small amount is swallowed, the process of swallowing and the entrance of the material into the cardiac part of the stomach may be watched fluoroscopically. The stomach and the duodenum may then be examined, radiograms taken as necessary, and more barium swallowed so as to fill the stomach. The passage of barium can be studied in successive radiograms, which show that, after the material enters the duodenum, it reaches the terminal part of the ileum in two hours or less, and for the most part has entered the large intestine after six hours. It may reach the rectum within 24 hours. Some barium may still be present in the large intestine after several days. The time of passage and of evacuation varies greatly.

Stomach and Duodenum (see figs. 35–3 and 35–4, pp. 380, 381). When barium enters the cardiac part of the stomach, it tends to form a triangular mass below the air in the fundus. The material then descends in a narrow stream (canalization) into the pyloric part of the stomach. By pressure on the anterior abdominal wall, the barium may be spread as a thin film over the mucosa of the stomach. The gastric folds are thereby outlined or accentuated. The position and shape of the stomach may also be readily altered by such pressure, and may be observed to change with changes in posture or with varying emotions. Peristalsis may also be observed during fluoroscopy. Peristaltic waves may be evident in radiograms as notches in the lower part of the body of the stomach or in the pyloric part.

Folds, similar to gastric folds, may be demonstrated in the beginning of the first part of the duodenum. This part of the duodenum, however, has poorly developed circular folds or lacks them. It is known as the duodenal cap, because when it is filled with barium it presents a smooth outline, much like that of the stomach.

Except for the beginning of the first part of the duodenum, the small intestine has a characteristic appearance after a barium meal, owing to the circular folds and villi, which impart a feathery or floccular appearance to the outline of the barium. Material begins to enter the duodenum within a few minutes after it has entered the stomach; the stomach is usually emptied within six hours or less. The duodenal cap generally fills and empties fairly quickly.

Jejunum and Ileum. The duodenojejunal junction or the first part of the jejunum may be often identified. Otherwise no clear-cut distinction between the jejunum and the ileum is possible, except that the terminal part of the ileum usually has a homogeneous rather than a feathery appearance. This part of the ileum may also be readily identified if some of the material has entered the cecum and ascending colon.

Large Intestine. The large intestine may be demonstrated by barium taken orally or given by enema (see fig. 35–13, p. 389). Better filling of the large intestine is obtained by a barium enema. In either event the large intestine presents a smooth outline, identified by the characteristic haustra. The outline and the haustra may be accentuated by the double contrast method (fig. 35–13*D*, p. 389), in which, after evacuation of a barium enema, air is injected through the anal canal so as to distend the intestine, the mucosa of which still retains a thin layer of barium.

When a barium enema is given, some barium often enters the terminal part of the ileum. An effective valve does not appear to be present at the ileocolic junction.

Liver and Bile Passages

Liver. The liver is opaque to x-rays and is largely responsible for the outline of the diaphragm as seen in plain radiograms. Its borders and surfaces usually cannot be made out distinctly. However, certain radioactive isotopes are concentrated by the liver and can be readily detected by scanning (fig. 36–1, p. 395). Another radiological method is portal venography, in which the contrast medium is injected into the spleen, either directly during surgery, or percutaneously. The medium enters the splenic vein and then the portal vein. The liver becomes outlined by virtue of the delineation of the intrahepatic branches of the portal vein. The shape and position of the liver are readily determined, and the positions of the hepatic segments may also be observed.

Bile Passages. Some organic iodides, when given by mouth or intravenously, reach the liver. They enter the bile and, when the gall bladder is functioning normally, render the gall bladder increasingly radio-opaque as the bile becomes more concentrated. This procedure, which is known as oral or intravenous cholecystography (see fig. 36–7, p. 402), is a test of gall bladder function. Contrast medium may also be injected into the bile duct, either directly during surgery, or by way of a tube previously put into the duct during surgery (operative and postoperative cholangiography). These procedures constitute tests of patency of the bile passages.

Particularly noteworthy are the variability in position of the gall bladder, the S-shape of the neck and cystic duct, the number and arrangement of ducts, and the curved course of the bile duct. The emptying of the gall bladder may be visualized during fluoroscopy if a fatty meal is given.

Kidneys, Ureters, and Urinary Bladder

Contrast media given during angiography may be momentarily concentrated in the kidney and may accentuate its shadows. Other organic iodides are employed which, when given intravenously, are concentrated in and excreted by the kidney (intravenous or excretion pyelography or urography; see figure 37–1, p. 407). The calices, pelvis, and ureter are clearly shown in successive radiograms taken within a short time after the administration of the compound.

A contrast medium may be injected into the bladder (cystography). Catheters may be introduced into the ureters and a contrast medium injected (retrograde, or instrumental, pyelography). The calices, pelvis, and ureters may be clearly shown by this method. Excretion pyelography, however, in addition to providing information about the structure of the kidneys, also shows the excretory power of these organs.

Peritoneal Cavity

If air or oxygen is injected into the peritoneal cavity, so as to produce pneumoperitoneum (fig. 29–2*B*, p. 286), the outlines of the diaphragm, liver, and spleen are very clearly demarcated. A small degree of pneumoperitoneum is also produced when testing for patency of the uterine tubes (p. 477). Gas may also be injected into the perirenal space in order to outline the kidneys.

REFERENCES

1. F. R. Brown and G. Smith, Lancet, *1*:10, 1945.
2. A. E. Barclay, *The Digestive Tract,* Cambridge University Press, London, 2nd ed., 1936.
3. B. McNicholl, Arch. Dis. Childh., *32*:438, 1957.

Part Six

THE PELVIS

DONALD J. GRAY

Introduction

The pelvis is the part of the trunk below and behind the abdomen. In many textbooks, the abdomen is considered as the part of the trunk below the thorax. In this event, the abdomen is subdivided into the abdomen proper and the pelvis, and the abdominal cavity is subdivided into the abdominal cavity proper and the pelvic cavity.

The pelvic cavity, as defined here, is sometimes called the lesser pelvis or true pelvic cavity. The part of the cavity that lies between the iliac fossae and above the terminal lines is then termed the greater pelvis or false pelvic cavity. The latter is best regarded as a part of the abdominal cavity. The Latin word *pelvis* means basin.

The skeletal framework of the pelvis consists of a bony ring, to which the lower limbs are attached, and which is largely covered internally and externally by muscles.

A funnel-shaped space within the pelvis is called the *pelvic cavity.* This cavity is separated from the abdominal cavity by an oblique plane passing through the terminal lines, which are located on the sacrum, ilium, and pubis (fig. 40–1). It contains the lower part of the alimentary canal, the urinary bladder, parts of the ureters, and parts of the genital system. It is of special importance in the female, because the fetus normally passes through it during childbirth.

Although the outer limits of the pelvic cavity are curved, it is convenient to describe its boundaries as two lateral walls, a posterior wall, and a floor, all of which gradually merge one with another. These boundaries are covered in part by peritoneum and by extraperitoneal tissue, which contains variable amounts of fat. Some structures can be seen through the peritoneum, and others can be palpated through it.

GENERAL READING

Francis, C. C., *The Human Pelvis,* C. V. Mosby Co., St. Louis, 1952. A small book, devoted exclusively to the anatomy of the male and female pelvis.

Smouth, C. F. V., and Jacoby, F., *Gynaecological and Obstetrical Anatomy and Functional Histology,* Arnold, London, 3rd ed., 1953. A good description of the anatomy of the female pelvis and a pertinent account of the function of the female genital organs. Contains many references.

Waldeyer, W., *Das Becken. Topographisch-anatomisch mit besonderer Berüchsichtigung der Chirurgie und Gynäkologie,* Friederick Cohen, Bonn, 1899. A comprehensive account of the anatomy of the pelvis. Many references to classical and other descriptions up to the turn of the century.

BONES, JOINTS, AND WALLS OF PELVIS

40

BONES OF PELVIS

The skeleton of the pelvis is formed by the two hip bones at the front and sides, and by the sacrum and coccyx behind (figs. 40–1 and 40–2). In the anatomical position, the anterior superior iliac spines and the pubic tubercles are approximately in the same frontal plane. The tip of the coccyx and the upper margin of the pubic symphysis are in the same horizontal plane, which is much lower than the level of the sacral promontory. **The internal aspect of the body of the pubis faces more upward than backward, and the urinary bladder rests upon it; the pelvic surface of the sacrum faces more downward than forward.**

The lesser pelvis (true pelvis) has an upper pelvic aperture, a cavity, and a lower aperture. Each has three main diameters, anteroposterior or conjugate, oblique, and transverse. Some of the more important diameters are shown in figure 40–3.

Apertures and Cavity. UPPER PELVIC APERTURE. The upper pelvic aperture (pelvic inlet, or pelvic brim) is in the plane of the terminal lines (fig. 40–1). This plane slopes downward and forward in passing from the sacral promontory to the pubic symphysis and forms an angle of about 48 degrees with the horizontal.[1] The *anteroposterior* or

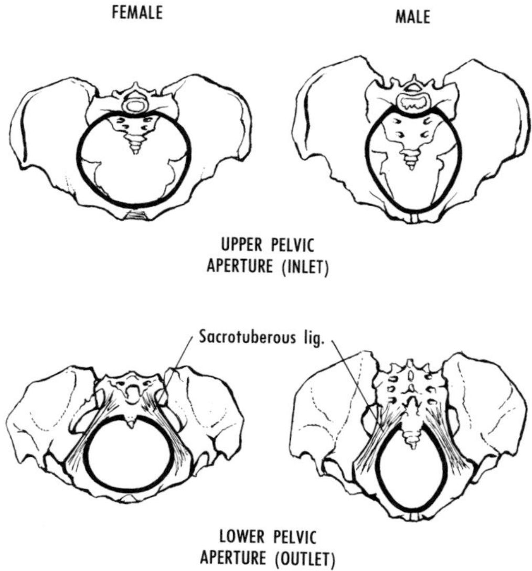

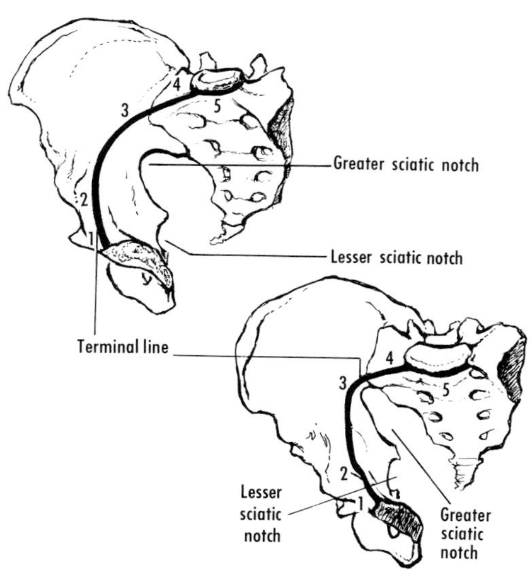

Figure 40–1 Skeletons of female and male pelves, showing inlet, outlet, and sciatic notches. The female pelvis is of the gynecoid type (fig. 40–6). Note the difference in size and shape of the greater sciatic notch. The terminal line (linea terminalis) consists of (1) pubic crest, (2) pectineal line, (3) medial border of ilium (lower half), (4) ala of sacrum, and (5) promontory. It frequently passes below the promontory. The details of the hip bone and of the sacrum are dealt with on pages 165 and 517.

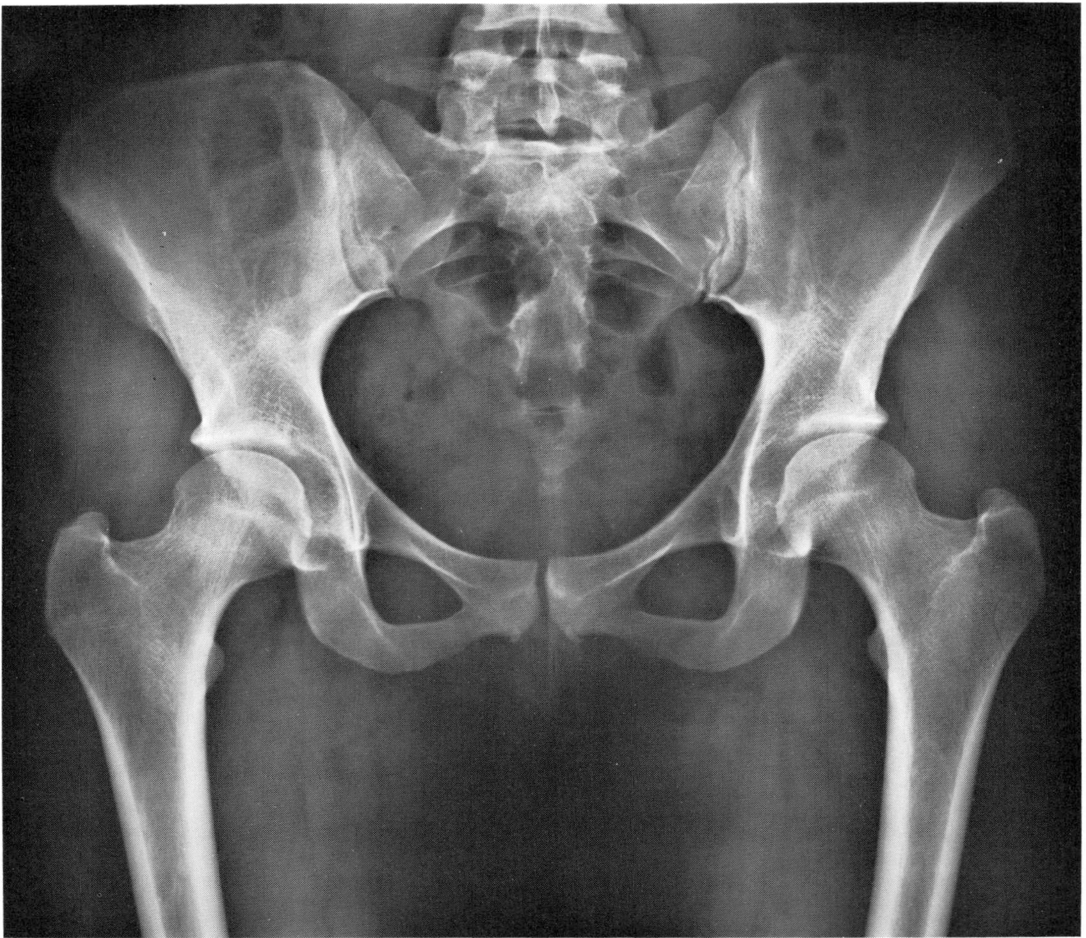

Figure 40–2 Female pelvis. Note especially the outlines of the sacroiliac joints, the subpubic angle, and the continuous curvature of the margin of the obturator foramen and the neck of the femur (Shenton's line).

conjugate diameter passes from the upper margin of the pubic symphysis to the middle of the sacral promontory. The *obstetrical conjugate diameter,* from the back of the pubic symphysis to the sacral promontory, is slightly shorter than the anteroposterior and is the minimal distance between the symphysis and the promontory. The *diagonal conjugate diameter* (fig. 40–4) is the only diameter that can be measured *per vaginam.* It is the distance between the lower margin of the pubic symphysis and the sacral promontory. When the sacral promontory cannot be reached *per vaginam,* the anteroposterior diameter of the inlet is considered to be adequate for successful parturition. When this promontory can be palpated, the pelvis is regarded as contracted. The *transverse diameter* passes across the widest part of the inlet. The *oblique diameter* extends from the sacroiliac joint of one side to the iliopubic eminence of the opposite side.

Cavity of Pelvis. The cavity of the pelvis passes backward and downward and extends from the superior to the inferior pelvic aperture. It is longer behind than in front. Its *anteroposterior* or *conjugate diameter*

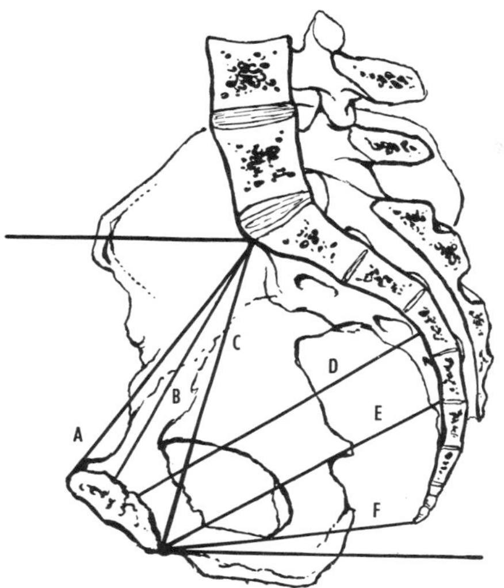

Figure 40–3 Median section of female bony pelvis, showing various diameters and planes. *A*, anteroposterior diameter of the inlet; this diameter is the true conjugate. It extends from the upper border of the public symphysis to the sacral promontory. *B*, obstetrical conjugate, the shortest diameter through which the fetal head must pass in its passage through the inlet. *C*, diagonal conjugate. This diameter can be measured during vaginal examination (fig. 40–4). *D*, the plane of greatest pelvic dimensions. *E*, the plane of the least pelvic dimensions. This plane is at the level of the ischial spines. *F*, the anteroposterior diameter of the outlet. The plane of the outlet usually makes an angle of 10 to 15 degrees with the horizontal. Based on Smout and Jacoby, cited on page 438.

passes from the middle of the back of the pubic symphysis to the center of the pelvic surface of the middle piece of the sacrum. The *transverse diameter* passes across the widest part of the cavity. The *oblique diameter* extends from the lower end of one sacroiliac joint to the center of the obturator membrane of the opposite side.

LOWER PELVIC APERTURE. The lower pelvic aperture (*pelvic outlet*) is diamond-shaped. It extends from the arcuate pubic ligament and the inferior rami of the pubis in front to the tip of the coccyx behind. It is bounded laterally by the ischial tuberosities and the sacrotuberous ligaments. The *anteroposterior* or *conjugate diameter* passes from the lower margin of the pubic symphysis to the tip of the coccyx. The *transverse diameter* extends between the ischial tuberosities. The *oblique diameter* extends from the junction of the ischial and pubic rami of one side to the point of crossing of the sacrotuberous and sacrospinous ligaments of the other side. The *plane of the outlet* forms an angle of about 10 to 15 degrees with the horizontal.

Pubic Arch. The pubic arch is formed by the conjoined rami of the pubis and ischium of the two sides. These rami meet at the symphysis to form the *subpubic angle* (fig. 40–5), which can be measured during a physical examination.

Classification of Pelves. **Two methods are used in classifying bony pelves. One of them depends upon the shape of the pelvic inlet, and the other depends upon measurements of its diameters.** These classifications are especially applicable in the female, because of the importance of the shape and size of the inlet in parturition.

The more recent classification,[3] based upon the shape of the inlet, has had wide acceptance among obstetricians and radiologists. **Four main pelvic inlet shapes are recognized (fig. 40–6). If the inlet resembles a long, narrow oval, the pelvis is said to be anthropoid. If it is rounded, the pelvis is gynecoid. An ovoid inlet having its long axis transverse is platypelloid, or flat. A pelvis having a heart-shaped inlet is android.** All four types may be found in the female; the gynecoid accounts for only about 50 per cent.[4] Furthermore, these types frequently overlap

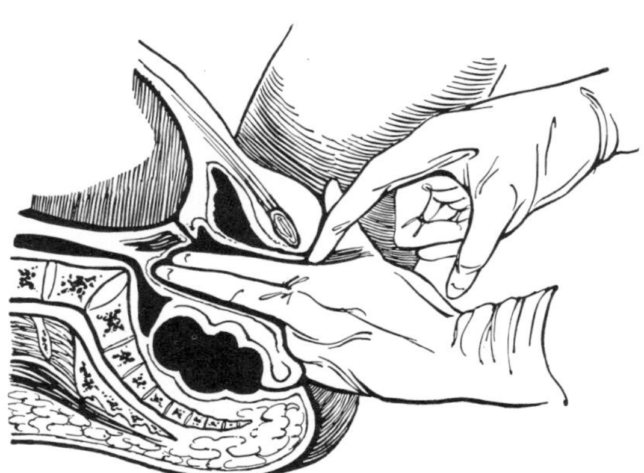

Figure 40–4 The middle finger measures the diagonal conjugate diameter. The "true" conjugate is about 1 to 2 cm less than the diagonal, and the obstetrical conjugate is about ½ cm less than the true. In the method illustrated, the indicated length on the index finger gives the true conjugate, because the index finger is about 1½ cm shorter than the middle. Based on Smout and Jacoby, cited on page 438. See also Moloy.[2]

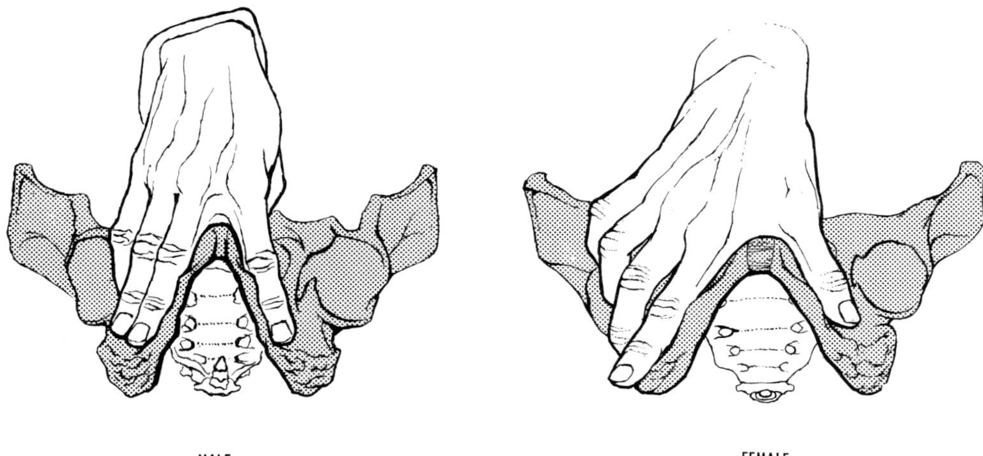

Figure 40–5 The subpubic angle in the female is nearly a right angle; that in the male is considerably less (about 60 degrees). When the vagina will admit three fingers placed side by side, the subpubic angle is adequate to permit proper extension of the fetal head after it has passed through the lower aperture.

one another, and a pelvis may be partly of one type and partly of another.[5]

The classification based upon the diameters of the inlet also distinguishes four main types.[6] A pelvis in which the anteroposterior diameter is longer than the transverse is *dolichopellic.* If these diameters are approximately equal, the pelvis is *mesatipellic.* If the transverse diameter is slightly greater than the anteroposterior, the pelvis is *brachypellic,* and if this ratio is increased, the pelvis is *platypellic.* The pelvic index is expressed as follows:

$$\frac{\text{anteroposterior diameter} \times 100}{\text{transverse diameter}}$$

Axis of Birth Canal. The axis of the birth canal (fig. 40–7) is the path followed by the fetal head in its course through the pelvic cavity, and is a guide to the direction of pull of obstetrical forceps. It extends downward and backward in the axis of the inlet (at a right angle to the plane of the inlet) as far as the ischial spines, which are at the level of the uterovaginal angle. Here, the axis of the birth canal turns forward and downward, at almost a right angle, and continues in the axis of the vagina, which is approximately parallel to the plane of the inlet. During parturition, the fetal head (usually the suboccipitobregmatic diameter, figure 40–8) occupies successively the transverse diameter of the inlet, the oblique diameter of the cavity, and the anteroposterior diameter of the outlet (fig. 40–9).

Radiographic Pelvimetry. A radiographic study provides information concerning the shape of the pelvis in all of its planes and makes possible the measurement of certain classical diameters. Pelvic shape can best be determined by a study of stereoscopic radiograms.

Some of the measurements made in pelvimetry are useful in predicting difficulties in parturition, although the size of the fetal head is equally important. **The following five measurements are especially important, and they depend upon radiographic measurements[7] for accuracy:**

1. The transverse diameter of the inlet.

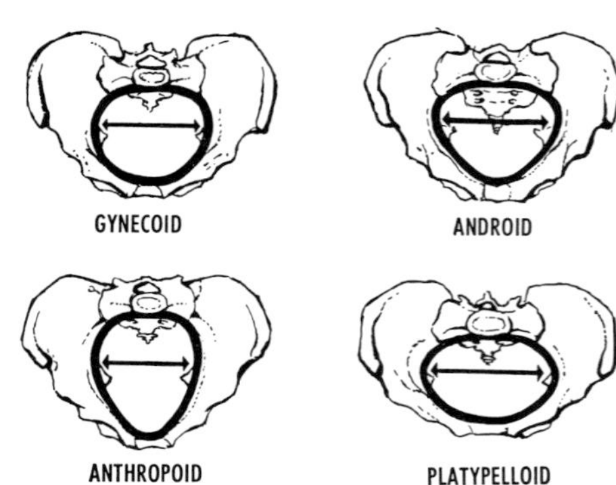

Figure 40–6 The four types of female pelves. Based on the classification of Caldwell and Moloy.[3]

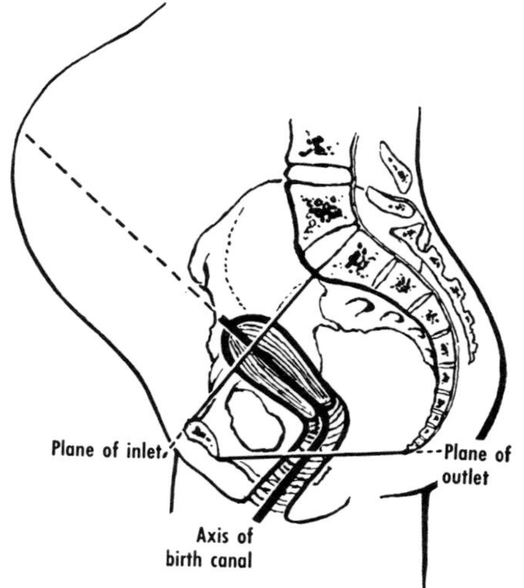

Figure 40-7 Bony pelvis and axis of birth canal. This axis, which turns at the uterovaginal angle, corresponds to the course taken by the fetal head in its passage through the pelvic cavity. The dotted line represents the upward and forward extension of the axis as the uterus enlarges during pregnancy. Based on Smout and Jacoby, cited on page 438.

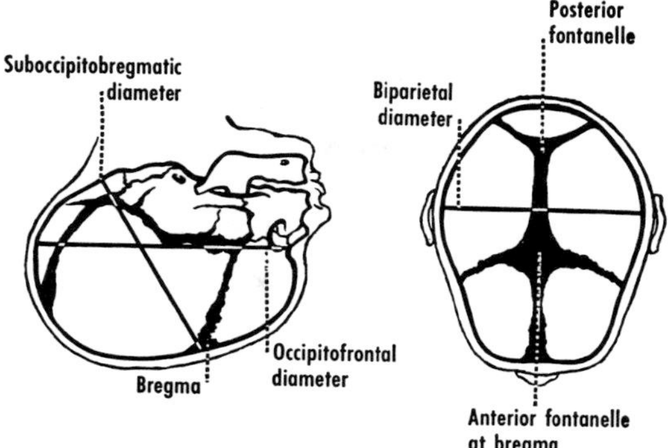

Figure 40-8 The fetal skull at the left is viewed upside down, in accordance with its position in the pelvis. The fetal skull at the right is viewed from above. Some common diameters at term are: suboccipitobregmatic, 9½ cm; occipitofrontal, 11½ cm; biparietal, 9½ cm. The greatest horizontal circumference averages about 33 to 36 cm. See page 578.

2. The obstetrical conjugate.*
3. The distance between the ischial spines.
4. The distance between the ischial tuberosities.
5. The posterior sagittal diameter (the distance between the midpoint of a line drawn between the ischial tuberosities and the apex of the sacrum).

When the fetal head is of average size, disproportion is likely to occur:

1. If the obstetrical conjugate is less than 10 cm.
2. If the distance between the ischial spines is less than 8 ½ cm.
3. If the distance between the ischial tuberosities is less than 8 cm.

Radiographic pelvimetry is indicated in only a small percentage of American women, usually in the following circumstances:

1. With clinical evidence that one or more pelvic diameters are smaller than average or that the pelvis has an unusual configuration.
2. When the fetal head remains above the level of the true pelvic cavity after the onset of labor.
3. In a nulliparous woman with breech presentation.
4. In women with histories of previous obstetric difficulties, pelvic fractures, or inflammatory disease of the pelvic bones.

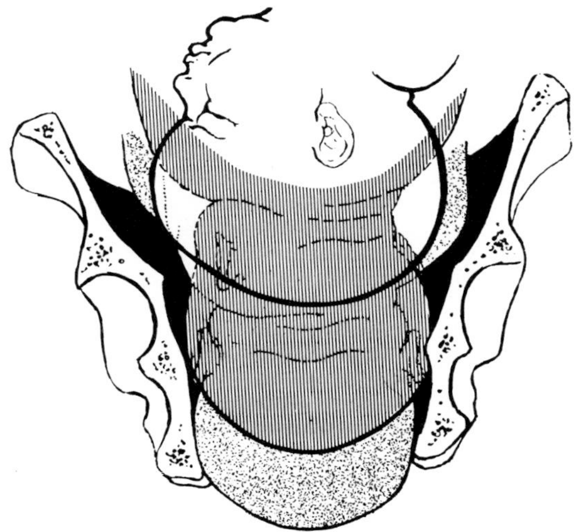

Figure 40-9 The bony pelvis and fetal head. Note how the head turns as it occupies first the inlet, then the cavity, and finally the outlet. Based on Smout and Jacoby (after Bumm), cited on page 438.

*This diameter of the female pelvis is very sensitive to nutrition, and the figures usually given are too low by about 13 mm.[8]

Although many features of the bony pelvis can be studied in the usual posterior radiogram, the shape of the inlet can be determined only from radiograms taken in such a way that the plane of the inlet is parallel to the film. The diameters of the inlet can be measured after corrections have been made for distortion. Certain measurements of the bony pelvis can be made only from lateral views.

A fetus is usually evident radiographically early in pregnancy, shortly after ossification of the fetal bones begins (see fig. 44–9). Many of the fetal bones may be obscured by the sacrum in posterior radiograms, and oblique views may be necessary to display them.

Sexual Differences.[9] In spite of terms such as android and gynecoid, relatively few pelves are typically male or typically female. In the female, the bones are usually thinner and lighter, and the muscular markings are not as prominent. The cavity is less funnel-shaped. The distances between the ischial spines and between the ischial tuberosities are greater, the greater sciatic notch is wider, and the surfaces of the sacrum for articulation with the ilium and with the fifth lumbar vertebra are smaller. The subpubic angle approximates a right angle in the female, but it is more acute in the male (fig. 40–5).

The inlet of the male pelvis is as variable as that of the female.[10] Also, in the male, the ischial spines are heavier than those of the female, and they project farther into the pelvic cavity.

Abnormal Types of Pelvis. A contracted bony pelvis is one in which one or more of the diameters are significantly decreased (by 1½ to 2 cm). Such a pelvis may be the result of a congenital anomaly, nutritional deficiencies, disease, or trauma, and it frequently interferes with normal parturition.

A high assimilation pelvis is a congenital anomaly in which the fifth lumbar vertebra is at least partially fused with the sacrum. In these types, the location of the sacral promontory is often altered, and the shape of the sacrum is frequently abnormal.

Nutritional deficiencies, such as rickets, occurring during the period of growth of the pelvis, often result in changes that are reflected by a widening of the pelvis transversely, a narrowing anteroposteriorly, and a forward displacement of the sacral promontory.

Diseases resulting in curvatures of the vertebral column often indirectly affect the shape of the pelvis. In kyphosis, the sacrum is usually narrow. This results in decreases in the subpubic angle and in the transverse diameter of the outlet. Furthermore, the sacrum is usually tilted in a way such that the anteroposterior diameter of the inlet is increased and that of the outlet is decreased. In scoliosis, the cavity of the pelvis is frequently oblique.

Growth of Bony Pelvis. Although some sexual differences are present in the fetal pelvis,[11] no marked differences are apparent at birth. In either sex, the inlet is ovoid, and its long axis is anteroposterior. Further differences are present during infancy and early childhood.The measurements relating to overall pelvic structure are larger in boys; those relating to internal structure, including the inlet, tend to be absolutely or relatively larger in girls.[12]

Profound sexual differences do not appear until puberty, when the pelvis grows rapidly, especially in girls.[13] The pubis grows mainly from the zone of ossification at its symphysial surface, and less from periosteum. Thus it increases in width in a frontal plane, and the pelvic cavity therefore becomes widened anteriorly.[14] The main growth of the sacrum is likewise largely in width and results in a widening of the cavity posteriorly. Such simultaneous increases in width in front and behind lead to a gynecoid pelvis. In males, at the same time, the bones become heavier, and the articular surfaces of the sacrum larger.

If, during puberty, all bones of the pelvis grow uniformly, the pelvis remains anthropoid in type. An accentuation of gynecoid growth results in the flat or platypelloid type, whereas a greater increase in growth posteriorly than anteriorly leads to the android type.

Growth at puberty is influenced by hormones (although in precocious puberty pelvic shape may not be altered), but is also very sensitive to nutritional and other environmental changes.[15]

JOINTS OF PELVIS

The joints of the pelvis include the lumbosacral, sacrococcygeal, and sacroiliac joints, and the pubic symphysis. The sacrotuberous and sacrospinous ligaments are associated with the joints of the pelvis, and the iliolumbar ligaments form an important connection between the vertebral column and the pelvis.

Lumbosacral and Sacrococcygeal Joints. The *lumbosacral joint* is formed by the body of the fifth lumbar vertebra and the sacrum. It is similar to other intervertebral joints, and includes an intervertebral disc, joints of the articular processes, and accessory ligaments.

The *sacrococcygeal joint* consists of an intervertebral disc between the sacrum and coccyx, reinforced by the *dorsal, ventral,*

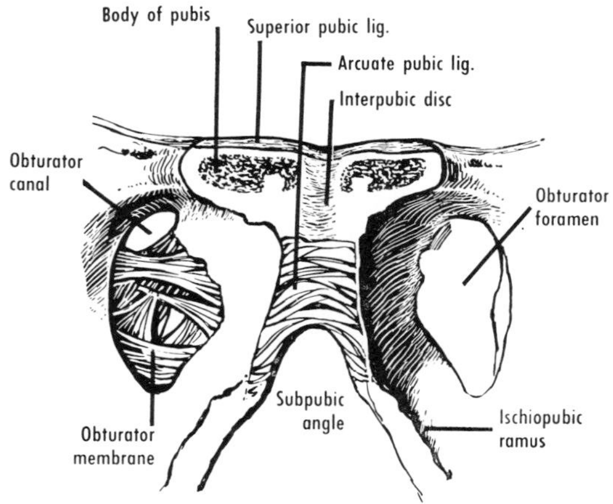

Figure 40–10 Diagram of pubic symphysis, cut in a coronal plane, and the obturator membrane. Drawn from a dissection.

and *lateral sacrococcygeal ligaments.* This joint often becomes partly or completely fused by bone.

Pubic Symphysis. The pubic symphysis (fig. 40–10) is a cartilaginous joint formed by the union of the bodies of the pubic bones in the median plane. The symphysial surface of the body of each pubic bone is covered by a thin layer of hyaline cartilage, which is united to that of the opposite side by a thick mass of fibrocartilage, the *interpubic disc.* A sagittal cleft is often present in this disc after childhood, but it has no synovial lining.

The *superior pubic ligament* consists of fibers passing transversely across the upper part of the joint. Extensions of the tendon of insertion of the rectus abdominis and of the aponeurosis of the external oblique reinforce the joint in front. The *arcuate pubic ligament* strengthens the joint below.

A relaxation of the ligaments and a loosening of the interpubic disc occur during pregnancy (p. 475), and facilitate the passage of the fetus.

Sacroiliac Joints (fig. 40–11). These are synovial joints, formed by union of the auricular surfaces of the sacrum and the ilium on either side. These surfaces are sometimes smooth and flat, but they are usually reciprocally curved and often have elevations and depressions that fit into corresponding irregularities of the opposed

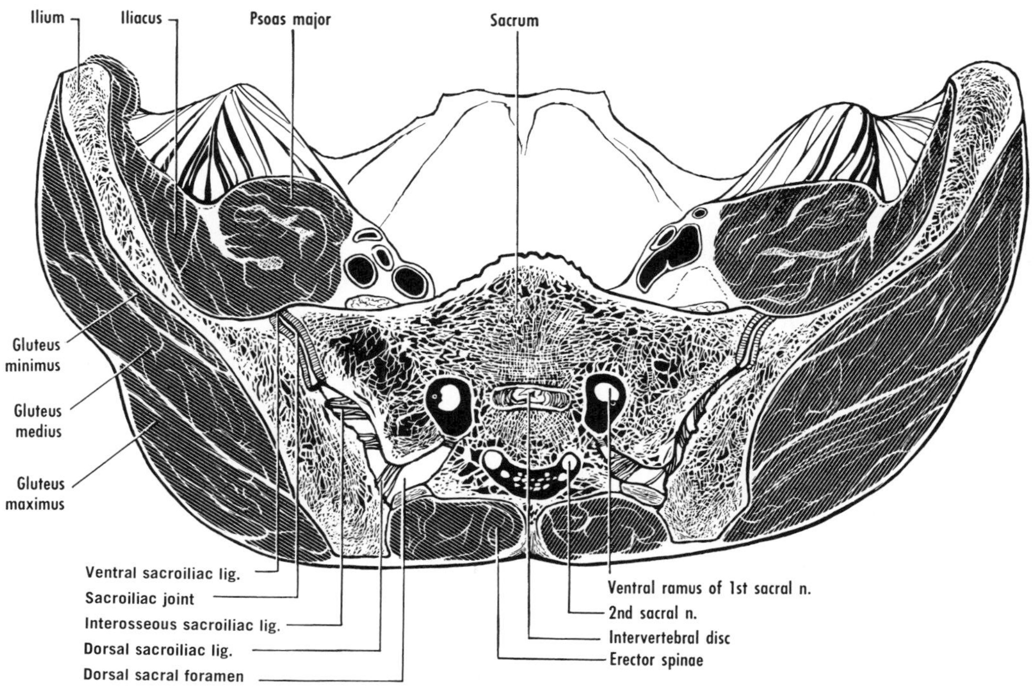

Figure 40–11 Diagram of sacroiliac joints. Drawn from horizontal section through upper part of sacrum, just above level of anterior superior iliac spines. The section was cut through the disc between the first and second sacral vertebrae, and through the first dorsal sacral foramina.

surface.[16] Greater stability of the joint results from such irregularities when they are present.

Supernumerary facets are commonly present behind the auricular surfaces (p. 521). These, too, aid in stabilizing the joint.

Hyaline cartilage covers the auricular surface of the sacrum, but columns of fibrocartilage separating islands of hyaline cartilage are found in the covering of the auricular surface of the ilium. After the third decade, both surfaces are found to be roughened, furred, and frayed.[17]

An articular capsule, lined by synovial membrane, connects the ilium with the sacrum at the periphery of the auricular surfaces.

The *interosseous sacroiliac ligaments* are the strongest of the ligaments of this joint, and are divided into upper and lower groups.[18] They are located behind the auricular surfaces and are attached to the tuberosities of the sacrum and ilium.

The *ventral sacroiliac ligaments* are thin bands that connect the ala and pelvic surface of the sacrum with the adjacent part of the ilium. Replacement of parts or all of these ligaments by bone often occurs after the fifth decade, especially in men.

The *dorsal sacroiliac ligaments* are attached to the tuberosity and posterior inferior spine of the ilium. They spread out to attach to the intermediate sacral crest and to areas of the sacrum adjacent to this.

MECHANICS OF PELVIS. The weight of the body is transmitted through the sacrum and ilia to the femora during standing, and to the ischial tuberosities in the sitting position. The wedge-shaped sacrum tends to be forced downward, but a displacement in this direction and a consequent separation of the hip bones are prevented by the reciprocal irregularities of the auricular surfaces and by the sacroiliac (mainly the interosseous) and the iliolumbar ligaments.

The two pubic bones and their connection in front act as a strut and prevent the sacroiliac joints from opening anteriorly and inferiorly, where their ligaments are weakest.

Apparently, very little movement occurs at the pubic symphysis and at the sacroiliac joints. However, these joints are somewhat more movable in women during the child-bearing age.

The sacrum has an angular movement around an axis that is most commonly from 5 to 10 cm vertically below the promontory. The promontory moves forward 5.6 ± 1.4 mm in changing from the recumbent to the standing position. During flexion and extension of the trunk, the movement is less constant and occurs over a shorter range. The changes in movement are somewhat more marked during pregnancy.[19]

Sacrotuberous and Sacrospinous Ligaments. These ligaments convert the greater and lesser sciatic notches into the greater and lesser sciatic foramina, respectively.

The *sacrotuberous ligament* (see fig. 40–12) is attached to the posterior iliac spines, to the lateral and lower part of the dorsal surface of the sacrum, and to the lateral margin of the upper part of the coccyx. From this broad attachment, the fibers converge to be attached to the medial margin of the ischial tuberosity. An extension of some of the fibers to the lower margin of the ramus of the ischium forms the *falciform process.*

The *sacrospinous ligament* (see fig. 40–12) is triangular in shape and lies in front of the sacrotuberous ligament. Its base is attached to the lateral margin of the lower part of the sacrum and the upper part of the coccyx. Its apex is attached to the ischial spine. The coccygeus muscle is more or less coextensive with its pelvic aspect.

The lateral margin of the sacrotuberous ligament forms the boundary that converts the sciatic notches into foramina, which are separated from each other by the sacrospinous ligament (figs. 20–1, p. 207, and 40–12).

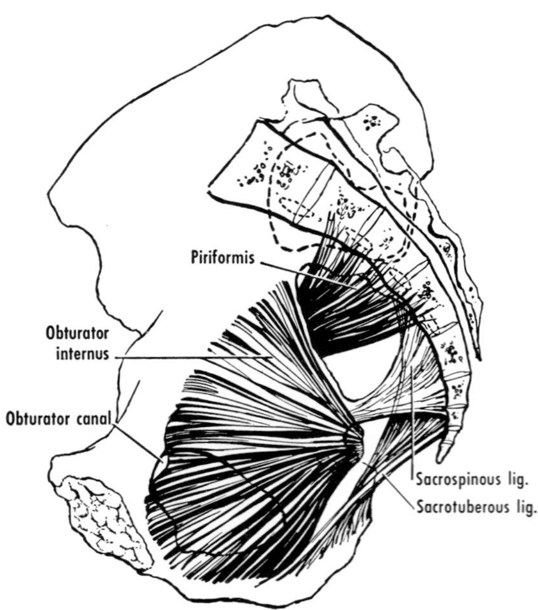

Figure 40–12 The muscles and ligaments of the lateral pelvic wall, pelvic aspect. Based on Shellshear and Macintosh.[20]

The *greater sciatic foramen* transmits the piriformis muscle, the superior and inferior gluteal vessels and nerves, the internal pudendal vessels, the pudendal nerve, the sciatic nerve, the posterior femoral cutaneous nerve, and the nerves to the obturator internus and quadratus femoris muscles.

The *lesser sciatic foramen* transmits the tendon of the obturator internus, the nerve to the obturator internus, the internal pudendal vessels, and the pudendal nerve.

WALLS OF PELVIS

The wall of the pelvic cavity, which is somewhat spherical in shape, has been described as having three planes: an intermediate, an external, and an internal.[20]

The intermediate plane consists of bones and ligaments. The bones are the sacrum, coccyx, and hip bones, and the ligaments are the obturator membrane and the sacrotuberous and sacrospinous ligaments.

The external plane comprises the muscles and fasciae that are superficial to the intermediate plane and includes the gluteal muscles.

The internal plane comprises the structures located on the pelvic aspect of the intermediate plane. It is formed by muscles and fasciae that are attached to the intermediate plane, by peritoneum, and by various blood vessels, nerves, and other structures located between the peritoneum and the fascial covering of the muscles.

Although the wall of the pelvic cavity is spherical and continuous, for purposes of description it is subdivided into two lateral walls, a posterior wall, and a floor.

Lateral Walls. Each lateral wall (figs. 40–12, 41–2, and 44–1) has as its bony framework the part of the hip bone below the terminal line. Most of the pelvic surface of this portion of the bone is covered by the obturator internus muscle and the obturator fascia. The obturator nerve and branches of the internal iliac vessels pass forward and downward medial to the obturator internus. The vessels are the umbilical artery, the obturator vessels, the superior vesical vessels, and, in the female, the uterine and vaginal vessels. The lateral wall is crossed in its posterior part by the ureter and in its anterior part by the round ligament in the female and the ductus deferens in the male. In the female, the ovary lies in a slight depression, the ovarian fossa, between the obliterated umbilical artery in front and the ureter and common iliac vessels behind.

A bony framework is lacking at the junction of the lateral and posterior walls. The space between the hip bone and the sacrum in this region is partially filled by the sacrotuberous and sacrospinous ligaments. The latter helps to divide the space into upper and lower parts, the greater and lesser sciatic foramina, respectively (p. 447).

Posterior Wall. The posterior wall is curved, and its upper part faces downward as well as forward. This wall is formed by the sacrum and coccyx, the lateral parts of which are covered by the piriformis and coccygeus and by the fasciae covering these muscles. The lumbosacral trunk, the sacral plexus of nerves, the sacral venous plexus, and some branches of the internal iliac vessels are situated in front of the piriformis. The median sacral artery and the sympathetic trunks pass downward on the sacrum. The glomus coccygeum is located near the tip of the coccyx.

MEDIAN SACRAL ARTERY. This arises from the back of the abdominal aorta, just above its bifurcation. It often arises in common with one or both fifth lumbar arteries. It passes downward in front of the lower lumbar vertebrae and the sacrum to the pelvic surface of the coccyx, where it ends by supplying the glomus coccygeum. In its course, it may supply part of the rectum and give small branches that anastomose with the lateral sacral arteries.

GLOMUS COCCYGEUM.[21] Commonly called the coccygeal body, this is a small cellular and vascular mass located in front of the tip of the coccyx. It receives the terminal branches of the medial sacral artery, and contains numerous arteriovenous anastomoses. Its functional significance is unknown.

Floor. The floor of the pelvis has been defined in various ways. Some authors define it as the pelvic diaphragm only, whereas others define it as both pelvic and urogenital diaphragms. **It seems**

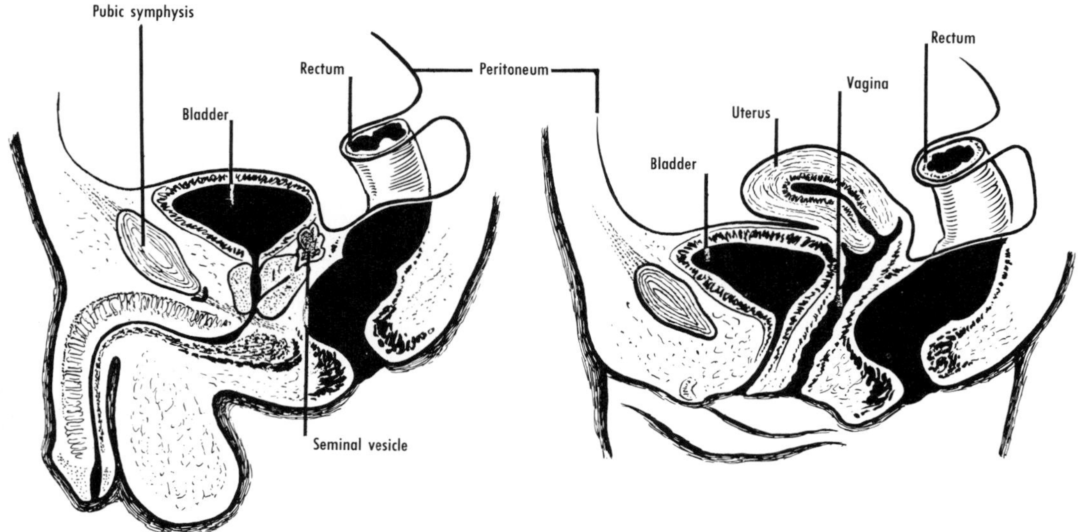

Figure 40–13 The peritoneal reflections from the pelvic viscera.

best, however, to include all structures that give support to the abdominal and pelvic viscera in the definition of the pelvic floor. These are the peritoneum above, the pelvic and urogenital diaphragms below, and various structures between the peritoneum and these diaphragms.

It has been stated that parts of the gluteus maximus and the sphincter ani externus should be included with the floor.[22] Some authors also include the pelvic viscera above the diaphragm as well as the skin and subcutaneous tissue of the perineum.[20]

The peritoneum (figs. 40–13 and 44–5) reaches its lowest level where it is reflected from the front of the rectum to the urinary bladder in the male to form the rectovesical pouch, or to the uterus and vagina in the female to form the rectouterine pouch. The lateral margins of these pouches are formed by elevations of the peritoneum called rectovesical and rectouterine folds, respectively. These are often called sacrogenital folds in both sexes. In the female, the uterus, vagina, and broad ligaments are located between the rectum and the urinary bladder. The anterior reflection of the peritoneum from the uterus to the bladder forms the uterovesical pouch.

The connective tissue between the peritoneum and the pelvic diaphragm varies in thickness in different locations. It contains the blood vessels that supply the viscera and the important nerve plexuses that innervate them. In addition, it contains the lower part of the ureter and the terminal part of the ductus deferens. Localized thickenings, containing numerous smooth muscle fibers, form ligaments that help to fix the various organs. Some of these ligaments are described with the pelvic diaphragm and fascia; others are described with the organs to which they are attached.

The lower part of the floor of the pelvis has two openings, both in the median plane. The posterior opening permits the rectum to pass through the pelvic diaphragm. The anterior opening, for the urethra in the male, and the urethra and vagina in the female, passes first through the pelvic and then through the urogenital diaphragm.

REFERENCES

1. A. Y. P. Garnett and J. B. Jacobs, Amer. J. Obstet. Gynec., *31*:388, 1936.
2. H. C. Moloy, *Clinical and Roentgenologic Evaluation of the Pelvis in Obstetrics*, Saunders, Philadelphia, 1951.
3. W. E. Caldwell and H. C. Moloy, Amer. J. Obstet. Gynec., *26*:479, 1933.

4. L. H. Garland, Amer. J. Roentgenol., *40*:359, 1938.
5. L. H. Garland, cited in reference 4. W. E. Caldwell, H. C. Moloy, and D. A. D'Esopo, Amer. J. Obstet. Gynec., *28*:482, 1934.
6. W. Turner, J. Anat., Lond., *20*:125, 1885. H. Thoms, Surg. Gynec. Obstet., *64*:700, 1937.
7. Personal communication from Charles E. McLennan, M.D. See also C. Nicholson, J. Obstet. Gynaec., Brit. Emp., *45*:950, 1938; J. G. H. Ince and M. Young, J. Obstet. Gynaec., Brit. Commonw., *47*:130, 1940.
8. C. Nicholson, J. Anat., Lond., *79*:131, 1945.
9. G. S. Letterman, Amer. J. phys. Anthrop., *28*:99, 1941. S. L. Washburn, Amer. J. phys. Anthrop., *6*:199, 1948; Amer. J. phys. Anthrop., *7*:425, 1949.
10. W. W. Greulich and H. Thoms, Anat. Rec., *75*:289, 1939.
11. B. J. Boucher, Amer. J. phys. Anthrop., *15*:581, 1957.
12. E. L. Reynolds, Amer. J. phys. Anthrop., *5*:165, 1947.
13. W. H. Coleman, Amer. J. phys. Anthrop., *31*:125, 1969.
14. W. W. Greulich and H. Thoms, Yale J. Biol. Med., *17*:91, 1944.
15. W. W. Greulich and H. Thoms, Anat. Rec., *72*:45, 1938. C. Nicholson, J. Anat., Lond., *79*:131, 1945.
16. H. Weisl, Acta anat., *22*:1, 1954.
17. G. B. Schunke, Anat. Rec., *72*:313, 1938.
18. H. Weisl, Acta anat., *20*:201, 1954.
19. H. Weisl, Acta. anat., *23*:80, 1955.
20. J. L. Shellshear and N. W. G. Macintosh, *Surveys of Anatomical Fields*, Grahame, Sydney, 1949.
21. W. H. Hollinshead, Anat. Rec., *84*:1, 1942.
22. A. W. Meyer, Calif. west. Med., *27*:1, 1927.

BLOOD VESSELS, NERVES, AND LYMPHATIC DRAINAGE

41

BLOOD VESSELS

Internal Iliac Artery

The internal iliac (hypogastric) artery (figs. 41–1 and 41–2) furnishes most of the blood supply to the pelvis. It arises from the common iliac in front of the sacroiliac joint, at the level of the intervertebral disc between the fifth lumbar vertebra and the sacrum. Its origin can be marked on the surface of the body by the upper point of trisection of a line extending between the anterior superior iliac spine and the pubic symphysis. It is usually about 4 cm long.

The internal iliac artery is crossed in front by the ureter. It is separated from the sacroiliac joint behind by the internal iliac vein and the lumbosacral trunk. In its upper part, the external iliac vein and psoas major are lateral to it; in its lower part the obturator nerve is lateral.

The internal iliac artery is commonly described as dividing into two main divisions, an anterior and a posterior. However, such a division may not be clear-cut. The various terminal branches may be close together at their origins, but they may arise in any one of the many different orders or combinations.[1] The major branches, direct and indirect, may be divided into parietal and visceral. The most common pattern is shown in figure 41–2. The superior and inferior gluteals sometimes arise by a common stem above the origin of the internal pudendal, the three vessels sometimes arise separately, and they occasionally arise by a common trunk.

The parietal branches include the iliolumbar, lateral sacral, obturator, superior gluteal, inferior gluteal, and internal pudendal.

The visceral branches are the umbilical, superior vesical, ductus deferens, inferior vesical, uterine, vaginal, and middle rectal.

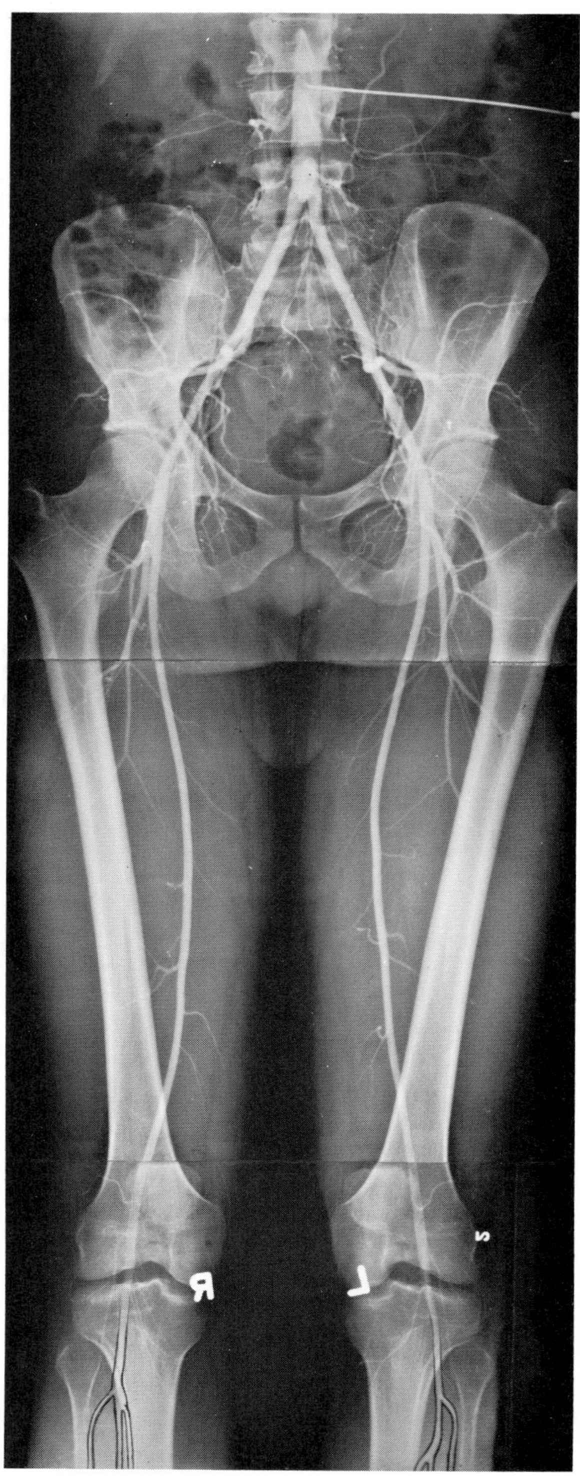

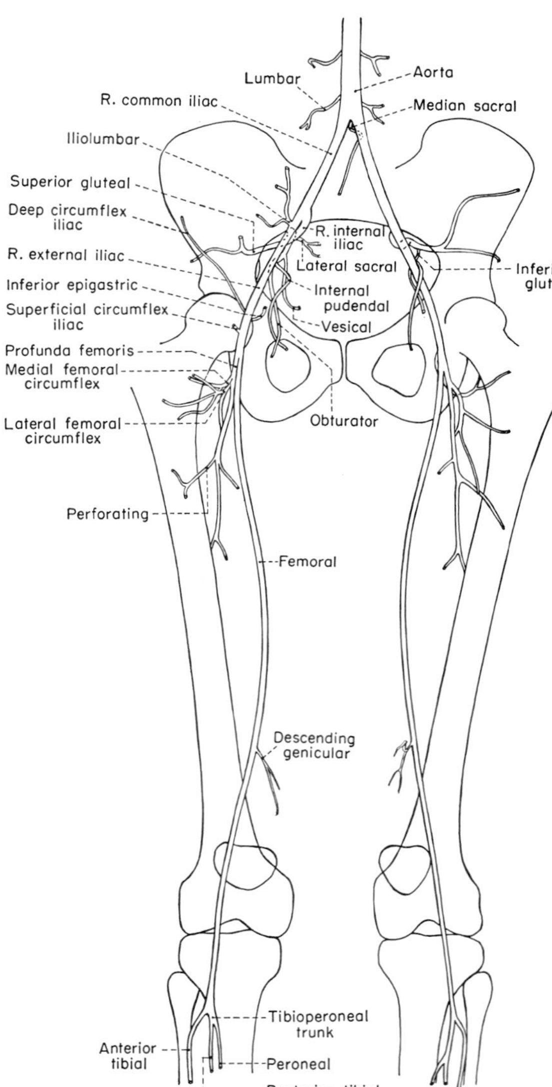

Figure 41–1 A normal, low lumbar aortogram in a 36 year old male. The arteries about the knee were visualized during an exposure made a few seconds after the initial exposure. From *Angiography*, H. L. Abrams, ed., Little, Brown, Boston, 1961; courtesy of S. M. Rogoff, M.D.

PARIETAL BRANCHES

Iliolumbar Artery. The iliolumbar artery runs upward and laterally to the iliac fossa, where it divides into an *iliac branch* that supplies the iliacus and the ilium, and a *lumbar branch* that supplies the psoas major and the quadratus lumborum. It sends a *spinal branch* through the intervertebral foramen between the fifth lumbar vertebra and the sacrum.

Lateral Sacral Arteries. Usually an upper and a lower, these may arise from a common trunk. The upper artery passes medially and enters the first or second pelvic sacral foramen. The lower artery descends in front of the piriformis and sacral nerves, lateral to the sympathetic trunk, and reaches the coccyx. Both lateral sacral arteries give off *spinal branches,* which, after passing through pelvic sacral foramina and supplying the contents of the sacral canal, may emerge through dorsal sacral foramina.

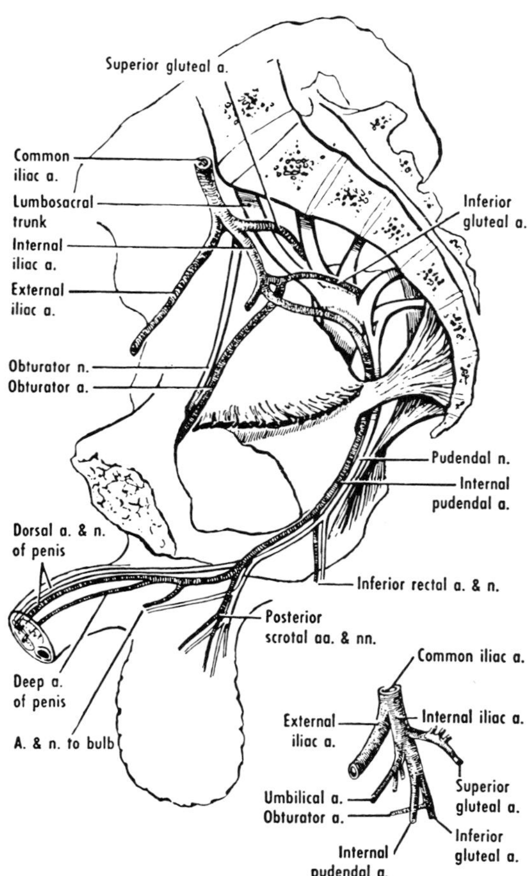

Figure 41–2 The sacral plexus and branches of the internal iliac artery as viewed from the medial side. *Lower right,* a diagram of the most common pattern of branches of the internal iliac artery. The superior gluteal artery arises proximal to a common stem for the umbilical, internal pudendal, and inferior gluteal arteries. The obturator arises from the inferior gluteal, and the middle rectal (not shown) from the internal pudendal. The iliolumbar and lateral sacral arteries (shown without labels) arise from the superior gluteal artery.

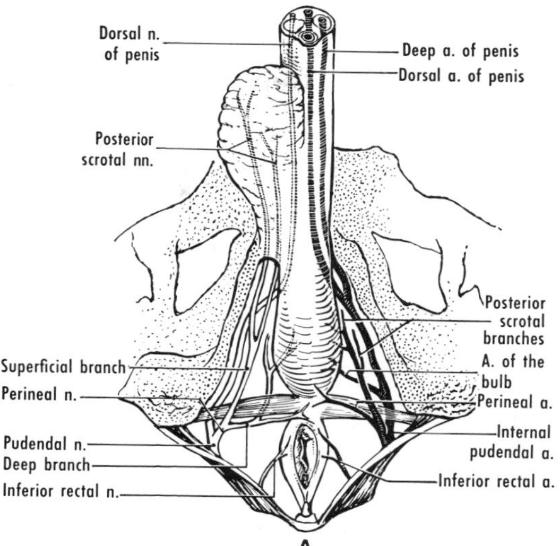

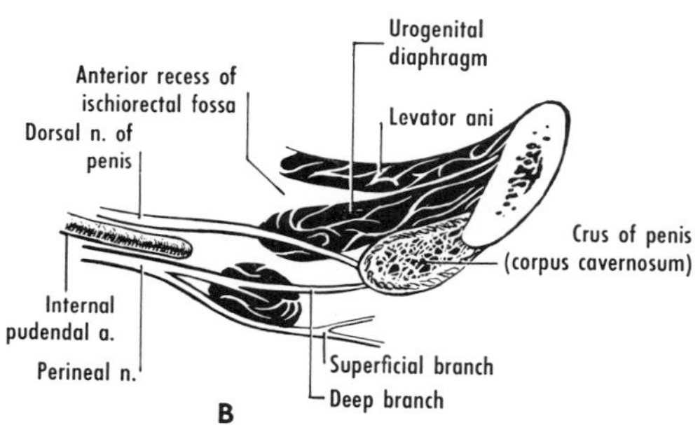

Figure 41–3 *A,* the internal pudendal artery and the pudendal nerve. The nerves are shown at the left. The dorsal nerve of the penis is more deeply placed. The arteries are shown at the right. *B,* a nearly sagittal section through the lateral part of the perineum.

Obturator Artery. This artery, the origin of which is variable,[2] passes forward and downward on the obturator fascia to the obturator foramen. The obturator nerve is above it, the obturator vein below it. It is crossed by the ureter near its origin. Within the pelvis, it gives off some muscular branches, a nutrient branch to the ilium, and a *pubic branch,* which ascends on the pelvic surface of the ilium. After passing through the upper part of the obturator foramen, it divides into *anterior* and *posterior branches,* which pass forward and backward, respectively, around the margin of the obturator foramen. They lie on the obturator membrane, deep to the obturator externus, and they supply adjacent muscles. The posterior gives off an *acetabular branch,* which supplies the fat in the acetabular fossa and the ligament of the head of the femur (medial epiphysial branches, p. 176).

The obturator artery arises from the inferior epigastric in about one-fifth of instances (fig. 33–7). It then passes on either the lateral or medial side of the femoral ring before reaching the obturator foramen. A medially placed obturator artery is susceptible to damage during operations for femoral hernia.

Superior Gluteal Artery. This passes backward, usually between the lumbosacral trunk and the first sacral nerve, and leaves the pelvis through the greater sciatic foramen, above the piriformis. Its distribution is described on page 208.

Inferior Gluteal Artery. This passes backward, between the first and second, or second and third sacral nerves, and leaves the pelvis through the greater sciatic foramen, below the piriformis. Its distribution is described on page 208.

Internal Pudendal Artery (fig. 41–3). This artery is larger in the male than in the female. It runs downward and lateralward to the lower margin of the greater sciatic foramen and leaves the pelvis by passing between the piriformis and the coccygeus. After crossing the back of the ischial spine, where it is medial to the nerve of the obturator internus, it enters the perineum through the lesser sciatic foramen. It then passes with the internal pudendal veins and the branches of the pudendal nerve through the pudendal canal in the lateral wall of the ischiorectal fossa. Continuing forward, it pierces the posterior margin of the urogenital diaphragm and runs in the deep perineal space, close to the inferior ramus of the pubis. Just before it reaches the pubic symphysis, it divides into its terminal branches, the deep and dorsal arteries of the penis (or clitoris).

BRANCHES. Small branches are distributed to the sacral plexus, to muscles within the pelvis, and to the muscles of the gluteal region. In addition to these, the internal pudendal artery gives off the inferior rectal artery, the posterior scrotal (or labial) branches, the perineal artery, the artery of the bulb of the penis (or vestibule), the urethral artery, and the deep and dorsal arteries of the penis (or clitoris).

The *inferior rectal artery* arises from the internal pudendal within the pudendal canal. It pierces the fascia of this canal and divides into several branches, which traverse the ischiorectal fossa and supply the muscles, fascia, and skin around the anal canal.

The posterior *scrotal* (or *labial*) *branches,* two in number, pierce the superficial and deep perineal fasciae, and pass forward in the superficial perineal space, between the ischiocavernosus and the bulbospongiosus. They help to supply these muscles, and are ultimately distributed to the scrotum in the male or to the labium majus and labium minus in the female.

The *perineal artery* passes below the superficial transversus perinei and supplies the tendinous center of the perineum and adjacent muscles.

The *artery of the bulb of the penis* arises from the internal pudendal artery within the deep perineal space. It passes medially through the urogenital diaphragm, pierces its inferior fascia, and supplies the erectile tissue of the bulb of the penis and the bulbourethral gland.

The *artery of the bulb of the vestibule,* after a similar course, supplies the bulb of the vestibule, the erectile tissue of the vagina, and the greater vestibular gland.

The *urethral artery* arises anterior to the origin of the artery of the bulb, and it, too, pierces the inferior fascia of the urogenital diaphragm. It enters the corpus spongiosum penis and ultimately reaches the glans penis.

The *deep artery of the penis* (or *clitoris*), one of two terminal branches, pierces the inferior fascia of the urogenital diaphragm and enters the crus of the penis (or clitoris). It runs near the center of the corpus cavernosum penis (or clitoris), which it supplies.

The *dorsal artery of the penis* (or *clitoris*), the other of the two terminal branches, also pierces the inferior fascia of the urogenital diaphragm. It passes first between the crus penis (or clitoris) and the pubic symphysis, and then between the two layers of the suspensory ligament of the penis (or clitoris). As it runs forward beneath the deep fascia on the dorsum of the penis (or clitoris), the dorsal nerve is lateral, and the deep dorsal vein is medial. Its terminal branches supply the glans and the prepuce.

VISCERAL BRANCHES

Umbilical Artery. This is usually the first visceral branch of the internal iliac. The umbilical arteries in the fetus are the main channels from the aorta to the placenta. After birth, when circulation to the placenta has ended, the part of each artery between its last branch and the umbilicus atrophies. The cordlike, fibrous remnant that is formed on each side is the medial (also called lateral) umbilical ligament (p. 462). The proximal portion of each artery remains patent. It runs forward along the lateral wall of the pelvis and along the inferolateral surface of the bladder, and gives rise usually to the superior vesical artery and the artery of the ductus deferens.

Superior Vesical Artery. This sometimes arises as a single vessel (in about one-fifth of instances) but more commonly as two or three vessels from the patent portion of the umbilical artery, just before the beginning of the medial umbilical ligament.[3] It supplies the upper part of the bladder and the medial umbilical ligament, and it may supply the lower part of the ureter.

Artery of the Ductus Deferens. This usually arises from the umbilical artery, supplies the seminal vesicles and the back of the bladder, and gives *ureteral* branches to the ureter. It accompanies the ductus deferens as far as the testis.

Inferior Vesical Artery. This usually arises from the common trunk for the internal pudendal and inferior gluteal arteries, or from a branch of this trunk, and passes medially to the lower part of the bladder. Most of its branches are distributed to the lower part of the bladder and to the prostate, but it also sends branches to the seminal vesicle, to the ductus deferens, and to the lower part of the ureter.

Uterine Artery. **The uterine artery (fig. 44–10, p. 484), which is homologous to the artery of the ductus deferens in the male, usually arises separately from the internal iliac, but may arise as a common trunk with the vaginal artery or with the middle rectal artery.** It runs downward, forward, and medially to the lower margin of the broad ligament, where it is lateral to the lateral fornix of the vagina. It passes in front of and above the ureter, to which it may give a small branch, and ascends between the two layers of the broad ligament along the body of the uterus. At the level of the uterine tube, it turns laterally and ends as the *ovarian branch,* which anastomoses with the ovarian artery. In addition to a variable number of branches to the uterus, it sends branches to the upper part of the vagina, to the medial part of the uterine tube (*tubal branch*), to the round ligament of the uterus, and to the ligament of the ovary.

Vaginal Artery (fig. 44–10, p. 484). This arises from the uterine artery, sometimes as several branches, and sometimes from the internal iliac in common with the uterine artery. It runs downward and medially to the side of the vagina, and divides into numerous branches that are distributed on the front and back of the vagina. The branches may anastomose in the median plane to form the anterior and posterior azygos arteries of the vagina. The vaginal artery also sends small branches to the bladder, rectum, and bulb of the vestibule.

Middle Rectal Artery. This runs medially to the rectum, to which most of its branches are distributed. However, some branches go to the prostate, the seminal vesicle, and the ductus deferens. The middle rectal artery is sometimes absent.

COLLATERAL CIRCULATION

The collateral circulation that develops after obstruction of an internal iliac artery results from anastomoses (1) with branches of the opposite internal iliac, (2) between parietal branches and branches of the femoral artery in the thigh (p. 217), and (3) between the superior and middle rectal arteries. The collateral circulation may be demonstrated by arteriography.[4] The collateral channels also supply the lower part of the abdomen if the abdominal aorta is obstructed, and the lower limb if the femoral artery is obstructed.

Internal Iliac Vein

The internal iliac (hypogastric) vein is a short trunk, which unites with the external iliac to form the common iliac vein. It lies behind the internal iliac artery and is crossed laterally by the obturator nerve. Its tributaries correspond in general to the branches of the internal iliac artery, with the exception of the umbilical and the iliolumbar arteries. Only the differences between the tributaries of the internal iliac vein and the branches of the internal iliac artery are discussed here.

The *superior* and *inferior gluteal veins* are each usually double, but commonly unite to form single trunks before emptying into the internal iliac vein. The *internal pudendal vein,* which in the male arises from the lower part of the prostatic plexus, is also double. However, it too usually empties into the internal iliac vein by a common trunk. The *deep dorsal vein of the penis* (or *clitoris*) runs in the median plane, between the left and right dorsal arteries. After passing through the inferior fascia of the urogenital diaphragm, it divides into two branches, which empty into the prostatic plexus in the male or into the vesical plexus in the female.

Each of the viscera within the pelvis is surrounded by a network of relatively large, thin-walled veins, which have few valves. These plexuses communicate freely with each other and give rise to the visceral tributaries of the internal iliac vein. They also communicate with the parietal tributaries, and thereby provide easy pathways for the spread of infections. The plexuses are named as follows: the *rectal venous plexus,* the *vesical venous plexus,* the *prostatic venous plexus,* the *uterine venous plexus,* and the *vaginal venous plexus.* The *sacral venous plexus,* located on the pelvic surface of the sacrum, is not associated with an organ, but it provides a pathway for blood to pass from the pelvic viscera to the azygos and vertebral venous systems (pp. 327, 328). Material injected into the deep dorsal vein of the penis has been found in the veins of the head, thorax, abdomen, pelvis, and thighs, and in the vertebral venous system.[5] During hysterosalpingography, radio-opaque material reached the veins of the uterus and was detected in the ascending lumbar veins.[6]

NERVES

The nerve supply to the pelvis is derived mainly from the sacral and coccygeal spinal nerves, and from the pelvic part of the autonomic nervous system.

Each of the five sacral nerves and the coccygeal nerve divide into a dorsal and a ventral ramus within the sacral canal.

The dorsal rami of the first four sacral nerves pass backward through the dorsal sacral foramina; those of the fifth sacral and the coccygeal emerge from the sacral canal through the sacral hiatus. Their distribution is described in the section on The Back (p. 532).

The ventral rami of the first four sacral nerves emerge from the sacral canal through the pelvic sacral foramina. The ventral ramus of the fifth sacral nerve enters the pelvis between the sacrum and the coccyx; that of the coccygeal nerve passes forward below the rudimentary transverse process of the first piece of the coccyx. The ventral rami of the first and second sacral nerves are the largest. Those of the lower sacral and the coccygeal decrease progressively in size from above downward.

Sacral Plexus (fig. 38–8, p. 426)

The ventral ramus of the fourth sacral nerve divides into an upper and a lower division; the upper division and the first three ventral rami combine with the lumbosacral trunk to form the sacral plexus[8] (figs. 41–2 and 41–4). Anterior and posterior divisions of the first three rami have been described, but, as in the case of the lumbar plexus, they are difficult to demonstrate. Each ramus contributing to the sacral plexus is connected to a single ganglion of the sacral sympathetic trunk by one or more rami communicantes.

The sacral plexus lies in front of the piriformis, and it is separated from the internal iliac vessels and the ureter anteriorly by the parietal pelvic fascia. The superior gluteal vessels usually pass between the lumbosacral trunk and the ventral ramus of the first sacral nerve. The inferior gluteal vessels pass between the ventral branches of the first and second, or second and third, sacral nerves. The internal pudendal vessels run between the sciatic and pudendal nerves.

The sacral plexus has 12 named branches. Seven of these are distributed to the buttock and the lower limb; the others supply structures belonging to the pelvis.

The branches that help to supply the buttock and the lower limb are as follows:

Superior Gluteal Nerve (L4 to S1). This passes backward through the greater sciatic foramen, above the piriformis. It accompanies the superior gluteal vessels into the buttock, where it is distributed (p. 208).

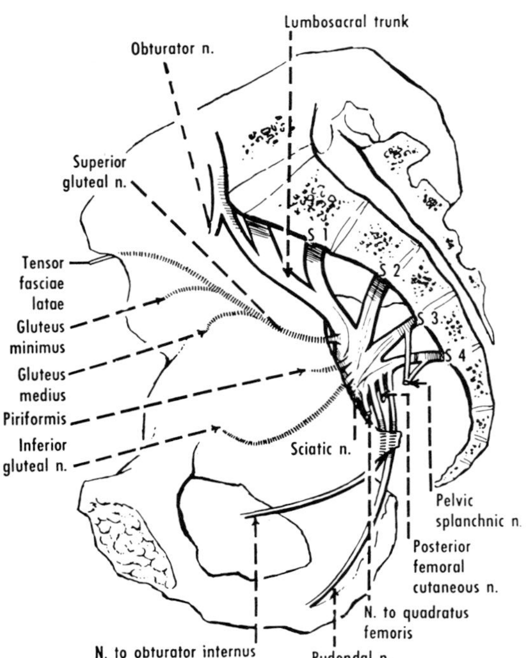

Figure 41–4 Schematic representation of the sacral plexus. Several branches are omitted, including the perforating cutaneous, branches to the pelvic and urogenital diaphragms, and twigs to the superior and inferior gemelli. The superior and inferior gluteal nerves and the nerve to the piriformis arise from the back of the plexus. A splanchnic nerve is shown arising from S3 and S4. Based on Pitres and Testut.[7]

Inferior Gluteal Nerve (L5 to S2). This passes backward through the greater sciatic foramen, below the piriformis. Its branches supply the gluteus maximus.

Nerve to the Quadratus Femoris (L4 to S1). This also leaves the pelvis below the piriformis, and then runs downward in front of the sciatic nerve. After giving a branch to the inferior gemellus, it enters the anterior aspect of the quadratus femoris. It also sends a branch to the hip joint.

Nerve to the Obturator Internus (L5 to S2). This passes through the sciatic foramen, below the piriformis. It gives off a branch that enters the posterior aspect of the superior gemellus, and, after crossing the ischial spine lateral to the internal pudendal vessels, it passes through the lesser sciatic foramen and then on the pelvic surface of the obturator internus, which it supplies.

Posterior Femoral Cutaneous Nerve (S1 to S3). This leaves the pelvis through the greater sciatic foramen, below the piriformis. Its branches and their distribution are described on page 209.

Perforating Cutaneous (Inferior Medial Clunial) Nerve (S2, 3). This pierces the sacrotuberous ligament and supplies the skin and subcutaneous tissue of the lower part of the buttock. It is sometimes absent.

Sciatic Nerve (L4 to S3). The largest nerve in the body, this consists of peroneal and tibial parts. It leaves the pelvis through the greater sciatic foramen, below the piriformis. In some cases, the two parts of the nerve are not bound together, and they leave the pelvis separately (p. 209). When this occurs, the peroneal portion pierces the piriformis, the tibial portion passes below it, and the two parts remain separate throughout their courses. The branches of the sciatic nerve are described on pages 213 and 220.

The following five branches of the sacral plexus are distributed to the pelvis.

Nerve to the Piriformis (S1, 2). This enters the anterior aspect of this muscle.

Nerves to the Levator Ani and Coccygeus (S3, 4). They descend on the pelvic aspect of these muscles and supply them.

Nerve to the Sphincter Ani Externus (perineal branch of S4). This passes either through the coccygeus or between this muscle and the levator ani. It passes forward in the ischiorectal fossa and supplies the sphincter ani externus, as well as the surrounding skin and subcutaneous tissue.

Pelvic Splanchnic Nerves[9] (S (2), 3, 4, (5)). These contain parasympathetic preganglionic and sensory fibers. They pass forward to take part in the formation of the inferior hypogastric plexus. The nerve supply of the sigmoid colon, which usually arises from the inferior hypogastric plexus (see p. 456), is derived from the pelvic splanchnic nerves and may arise directly from these nerves, either in part or entirely.

Pudendal Nerve. **The pudendal nerve (S2, 3, 4) (figs. 41–2 and 41–3) furnishes most of the innervation to the perineum.** It contains motor fibers, sensory fibers (including those carrying pain and taking part in reflexes), and postganglionic sympathetic fibers. It passes through the greater sciatic foramen, below the piriformis. It crosses the back of the ischial spine, where it is medial to the internal pudendal artery, and then enters the perineum with this artery through the lesser sciatic foramen. After entering the pudendal canal in the lateral wall of the ischiorectal fossa, it first gives off (1) the inferior rectal nerve, and then divides into (2) the perineal nerve and (3) the dorsal nerve of the penis (clitoris).

The *inferior rectal nerve,* which may arise independently from the sacral plexus (S3, 4), pierces the medial wall of the pudendal canal and divides into several branches, which traverse the ischiorectal fossa along with the corresponding vessels. They supply the sphincter ani externus, the skin around the anus, and the lining of the anal canal as far upward as the pectinate line.

The *perineal nerve* divides into superficial and deep branches while it is still in the pudendal canal. The deep branch gives off one or two twigs, which pierce the medial wall of the pudendal canal and help to supply the sphincter ani externus and the levator ani. It then pierces the superficial and deep perineal fasciae to enter the superficial perineal space, where it supplies the bulbospongiosus, the ischiocavernosus, the superficial transversus perinei, and the bulb of the penis. The superficial branch divides into two *posterior scrotal (labial) nerves,* which are medial and lateral. Both branches pierce the superficial and deep perineal fasciae and run forward with the corresponding arteries to be distributed to the scrotum in the male or to the labium majus in the female.

The *dorsal nerve of the penis* (or *clitoris*) pierces the posterior margin of the urogenital diaphragm. It supplies the deep transversus perinei and the sphincter urethrae as it runs forward on the lateral side of the internal pudendal artery. After piercing the inferior fascia of the urogenital diaphragm, it gives a branch to the corpus cavernosum penis (or clitoris) and then passes between the two layers of the suspensory ligament of the penis (or clitoris). It runs forward on the dorsum of the penis (or clitoris) and gives off branches, which are distributed to the skin, prepuce, and glans.

Coccygeal Plexus

The ventral rami of the fifth sacral and the coccygeal nerve pierce the coccygeus and join the lower division of the ventral ramus of the fourth sacral to form plexiform cords, which constitute the *coccygeal* (or *sacrococcygeal*) plexus. Fine filaments given off from this plexus supply the sacrococcygeal joint, the coccyx, and the skin over the coccyx.[10]

Pelvic Part of Autonomic Nervous System[11]

The sympathetic part of the autonomic nervous system reaches the pelvis by two different routes. One of these is the downward continuation of the sympathetic trunk (p. 423), and the other is the downward continuation of the aortic plexus (p. 424).

Sympathetic Trunk (fig. 64–15, p. 783). The sacral part of the sympathetic trunk[12] lies on the pelvic surface of the sacrum, medial to the upper three pelvic sacral foramina and usually in front of the fourth. It consists largely of preganglionic fibers; their levels of origin are referred to on page 335. It often ends by forming an enlargement, the *ganglion impar,* with the contralateral trunk in front of the coccyx. The number of ganglia interspersed along the sacral part of the trunk is variable, but there are usually three or four. Each ganglion tends to be connected by rami communicantes with only one spinal nerve.[13] The fibers in these rami are postganglionic, and most of them are distributed to the lower limb and perineum with branches of the sacral plexus. A variable number of fine fibers *(sacral splanchnic nerves)* passes forward from the trunk to join the inferior hypogastric plexus.

Autonomic Plexuses. After the aortic plexus continues downward in front of the fifth lumbar vertebra, it receives some fibers from the lower lumbar splanchnic nerves and is called the *superior hypogastric plexus* (or *presacral nerve)* (fig. 38–6, p. 424). The latter plexus divides in front of the sacrum into two elongated, narrow networks, which are sometimes collected into trunks, and which are termed the *right* and *left hypogastric nerves.* Each hypogastric nerve or plexus passes downward on the side of the rectum (or rectum and vagina in the female). At the level of the lower part of the front of the sacrum, each hypogastric nerve is joined by the pelvic splanchnic nerves of the corresponding side to form the right and left *inferior hypogastric* (or *pelvic) plexuses,* which consist of networks of interlacing nerves embedded in tough connective tissue. Small pelvic ganglia are scattered throughout these networks.

Subdivisions of the inferior hypogastric plexuses accompany the visceral branches of the internal iliac arteries and supply the pelvic organs. Although these subdivisions are named according to either the organs that they supply or the vessels that they accompany, they intercommunicate freely and their names are arbitrary.

A number of branches leave the inferior hypogastric plexus on each side and supply the rectum. One or two of these branches accompany the middle rectal artery and constitute the *middle rectal plexus,* which helps to supply the rectum (p. 489). A large part of the inferior hypogastric plexus forms the *prostatic plexus,* which supplies the prostate and parts of adjacent organs. It is continued forward as the *cavernous nerves of the penis* (p. 504). The *vesical plexus* supplies the bladder and parts of the ureter, ductus deferens, and seminal vesicle. The *uterovaginal plexus* passes with the uterine artery between the layers of the broad ligament. It supplies the uterus, ovary, vagina, urethra, and the erectile tissue of the vestibule. Fibers from the lowermost part of this plexus continue as the *cavernous nerves of the clitoris.*

FUNCTIONAL COMPONENTS. The inferior hypogastric plexus contains three types of fibers:

1. Postganglionic sympathetic fibers, some of which arise from the lumbar part of the sympathetic trunk and descend by way of the superior hypogastric plexus, and others of which arise from the sacral part of the trunk.

2. Preganglionic parasympathetic fibers, which arise from the sacral part of the spinal cord and reach the inferior hypogastric plexus by way of pelvic splanchnic nerves. They supply the descending colon, the sigmoid colon, and the pelvic viscera. The fibers that supply the descending and sigmoid colon may ascend directly to these organs, but usually they ascend in the plexus and then leave in a branch of the hypogastric nerve.[14] This branch gives a branch to the sigmoid colon and then ascends along the descending colon, sometimes as far as the left colic flexure.

3. Sensory fibers of various types. Some of these carry pain impulses, and

they enter the spinal cord in lumbar splanchnic nerves. They often ascend in the superior hypogastric plexus, but they also run by way of the pelvic splanchnic nerves. Other sensory fibers, which are concerned in various reflexes and with sensations from the urinary bladder, reach the sacral part of the cord in the pelvic splanchnic nerves.

Much more information in regard to the innervation of pelvic organs in man is needed. The types of fibers reaching a particular organ and the functions of these fibers are uncertain in many instances.

LYMPHATIC DRAINAGE

The lymphatic nodes of the pelvis (fig. 41–5) are variable in size, number, and location. Four main groups are located in or closely adjacent to the pelvis, and they receive most of the lymphatic vessels from the pelvis. They are named according to the arteries with which they are associated, but the division into definite groups is somewhat arbitrary. In addition to the nodes in the named groups, small nodes lie in the connective tissue along the pathways of various branches of the internal iliac artery.

Sacral Lymphatic Nodes. These lie in the hollow of the sacrum. They receive vessels from some of the pelvic organs and from the perineal and gluteal regions. They are frequently regarded as a part of the internal iliac group, and they drain either into this group or into the common iliac nodes.

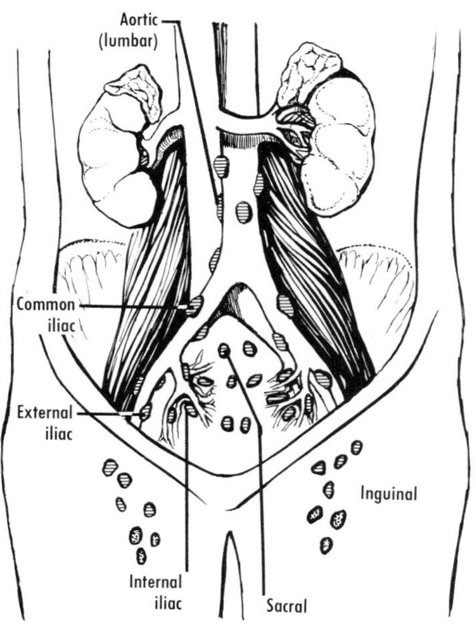

Figure 41–5 Schematic representation of the position of the major lymphatic nodes of the pelvis. The inguinal nodes are also shown in figure 19–5, p. 205.

Internal Iliac Lymphatic Nodes. These are arranged around the internal iliac artery and near the origins of the branches of this artery. They receive vessels from the pelvic viscera, the perineum, and the buttock. Their efferent vessels drain into the common iliac nodes.

External Iliac Lymphatic Nodes (fig. 41–6). These are arranged around the external iliac artery. They receive vessels from the superficial and deep inguinal nodes, from the deep part of the abdominal wall below the umbilicus, and from some of the pelvic viscera. Their efferent vessels drain into the common iliac nodes.

Common Iliac Lymphatic Nodes (fig. 41–6). These receive the drainage from the external and internal iliac and the sacral nodes. They drain into the lumbar group of nodes (fig. 38–5, p. 422).

There are many connections between the lymphatic vessels that drain the various pelvic organs. Because of these connections, no disturbance in drainage results from the removal of large numbers of nodes. Furthermore, cancer within the pelvis can spread to any pelvic or abdominal organ. Most of the lymphatic vessels follow the course of the arteries, but some do not.

The pelvic organs and the groups of lymphatic nodes into which their afferent lymphatic vessels drain are listed in table 41–1.

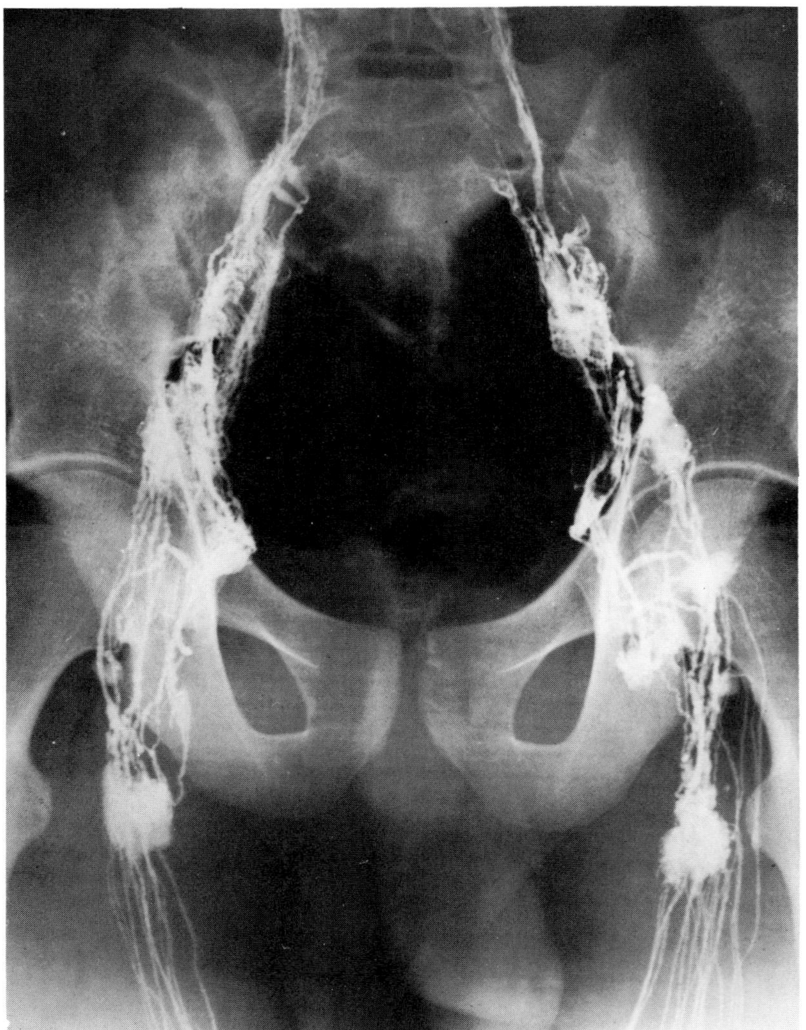

Figure 41–6 Normal iliac lymphogram, showing inguinal as well as iliac nodes. Large efferent vessels are present. From *The Lymphatics,* Edward Arnold, Ltd., London, 1972; courtesy of Prof. J. B. Kinmonth, F.R.C.S., and publisher.

TABLE 41–1 Lymphatic Drainage of Pelvic Organs

Organs	*Groups of Nodes Receiving Vessels Draining Pelvic Organs*
Ovary (along ovarian a.)	Lumbar
Uterine tube (except part near uterus) (along ovarian a.)	Lumbar
Uterus	
Upper part of body	Lumbar
Lower part of body	External iliac
Cervix	External iliac, internal iliac, and sacral
Region near uterine tube (along round ligament)	Superficial inguinal
Vagina	
Upper part (along uterine a.)	External and internal iliac
Middle part (along vaginal a.)	Internal iliac
Lower part	Sacral and common iliac
Part below hymen (with those from vulva and skin of perineum)	Superficial inguinal
Testis and epididymis (along testicular a.)	Lumbar
Seminal vesicle	External and internal iliac
Ductus deferens (pelvic portion)	External iliac
Prostate	Internal iliac mainly; sacral and external iliac
Scrotum	Superficial inguinal
Penis (clitoris)	
Skin and prepuce	Superficial inguinal
Glans	Deep inguinal and external iliac
Ureter (lower part)	External or internal iliac
Bladder	
Superior and inferolateral aspects	External iliac
Base	External iliac mainly; internal iliac
Neck	Sacral and common iliac
Urethra	
Female (along internal pudendal a.)	Internal iliac mainly; external iliac
Male	
Prostatic and membranous parts (along internal pudendal a.)	Internal iliac mainly; external iliac
Spongy part	Deep inguinal mainly; external iliac
Rectum	
Upper part	Inferior mesenteric
Lower part	Sacral, internal iliac, and common iliac
Anal canal	
Above pectinate line (along inferior rectal and internal pudendal aa.)	Internal iliac
Below pectinate line	Superficial inguinal

REFERENCES

1. F. L. Ashley and B. J. Anson, Amer. J. phys. Anthrop., *28*:381, 1941. J. L. Braithwaite, J. Anat., Lond., *86*:423, 1952. W. H. Roberts and G. L. Krishingner, Anat. Rec., *158*:191, 1967.
2. J. L. Braithwaite, J. Anat., Lond., *86*:423, 1952. J. W. Pick, B. J. Anson, and F. L. Ashley, Amer. J. Anat., *70*:317, 1942.
3. J. L. Braithwaite, Brit. J. Urol., *24*:64, 1952.
4. R. F. Muller and M. M. Figley, Amer. J. Roentgenol., *77*:296, 1957.
5. F. A. Beneventi and G. J. Noback, J. Urol., *62*:663, 1949.
6. T. N. A. Jeffcoate, J. Obstet. Gynaec., Brit. Emp., *62*:244, 1955.
7. A. Pitres and L. Testut, *Les nerfs en schémas,* Doin, Paris, 1925.
8. M. T. Horwitz, Anat. Rec., *74*:91, 1939. C. R. Bardeen and A. W. Elting, Anat. Anz., *19*:124, 209, 1901.
9. D. Sheehan, J. comp. Neurol., *75*:341, 1941.
10. A. Sicard and J. Bruézière, Arch. Anat., Strasbourg, *33*:43, 1950.
11. A. H. Curtis, B. J. Anson, F. L. Ashley, and T. Jones, Surg. Gynec. Obstet., *75*:743, 1942. F. L. Ashley and B. J. Anson, Surg. Gynec. Obstet., *82*:598, 1946. A. A. Pearson and R. W. Sauter, Amer. J. Anat., *128*:485, 1970.
12. A. Labbok, Anat. Anz., *85*:14, 1937.
13. J. Pick and D. Sheehan, J. Anat., Lond., *80*:12, 1946.
14. R. T. Woodburne, Anat. Rec., *124*:67, 1956.

URINARY BLADDER, URETER, AND URETHRA 42

URINARY BLADDER

The shape, size, position, and relations of the urinary bladder (fig. 42–1; figs. 47–2 and 47–3, pp. 497, 498) vary with age and with the amount of urine that it contains. The position and relations vary also with sex, but there is no significant difference between male and female bladders in size and shape.

Vesica is the Latin term for bladder. The noun vesicle and the adjective vesical are derived from it.

Position and Shape. **The empty urinary bladder in the living adult individual is somewhat rounded, although its roundness is altered by the pressure of and attachments to adjacent structures. It lies entirely, or almost entirely, within the pelvis, and rests on the pubis and the adjacent part of the floor of the pelvis. It is situated slightly lower in the female than in the male. As the bladder fills, it gradually rises into the abdomen and it may reach the level of the umbilicus.** During the early stages of filling, the transverse diameter of the bladder increases. With greater filling, the longitudinal diameter increases until, in the full bladder, the two diameters are about equal.[1]

At birth, the empty bladder is spindle-shaped (see fig. 44–8), and most of it lies in the abdomen proper. Its long axis extends from the anterior abdominal wall downward and backward. During childhood it gradually sinks to the position noted in the adult, and assumes its final shape.

Parts. The empty bladder of the living adult has four surfaces, or aspects: a superior, two inferolateral, and a posterior. The last is also called the *fundus,* or *base,* of the bladder. The superior and inferolateral surfaces meet in front at the *apex.* The inferolateral surfaces meet below at the *neck.* The part of the bladder between the

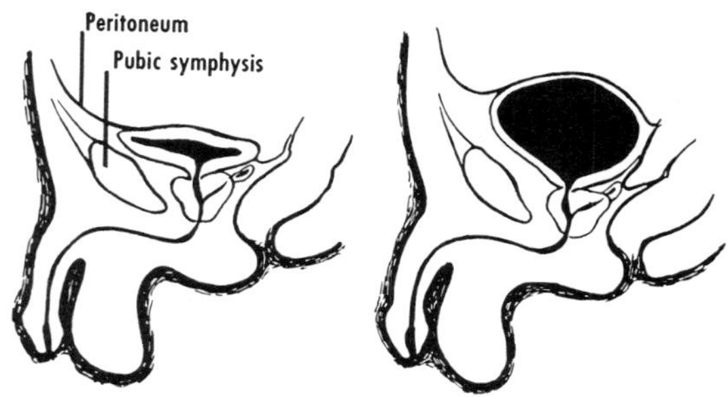

Figure 42–1 The urinary bladder, empty and full. Note that the peritoneum pulls away from the anterior abdominal wall as the bladder fills. Compare with the bladder at birth (fig. 44–8, p. 482).

apex in front and the base behind is the *body*.

Peritoneal Relations. The superior surface and the upper part of the base of the bladder are covered by peritoneum, which is reflected from the lateral wall of the pelvis and from the anterior abdominal wall just above the level of the pubic symphysis when the bladder is empty. As the bladder fills and rises into the abdominal cavity, the peritoneum is lifted away from the lower part of the anterior abdominal wall, and the reflection therefore becomes higher. Behind, the peritoneum is reflected onto the uterus in the female and onto the rectum in the male (fig. 40–13, p. 448).

Relations. The superior surface of the bladder is related through the peritoneum to the coils of the small intestine or to the sigmoid colon. In the female, the body of the uterus is above the bladder when the latter is empty.

The inferolateral surfaces and the rounded margin between them lie adjacent to the *retropubic space*. This space, which contains fat, loose fibrous tissue and a plexus of veins, lies between the umbilical prevesical fascia behind and the transversalis fascia in front (p. 369). Because of the predominance of fat in this region, the contained structures are collectively termed the *retropubic pad of fat*. The retropubic space[2] is **U**-shaped. The closed end of the **U** lies between the pubic symphysis and the bladder, and the open ends extend backward, on each side of the bladder, to the lateral ligaments of this organ. The space is limited above by the reflection of peritoneum to the anterior abdominal wall, and below by the puboprostatic ligaments. It is limited on each side by the parietal fascia covering the levator ani and the obturator internus. The retropubic space extends upward toward the umbilicus between the two medial umbilical ligaments and is therefore partly abdominal in position. Its upper limit varies with the level of reflection of the peritoneum from the anterior abdominal wall and consequently with the degree of fullness of the bladder.

The base of the bladder faces backward and slightly downward. In the male, it is closely related to the seminal vesicle in its lower, lateral part, to the ampulla of the ductus deferens just medial to the seminal vesicle, and to the rectum between the two ampullae. In the female, the base is connected by loose fibrous tissue with the anterior wall of the vagina below and with the supravaginal part of the cervix of the uterus above.

Attachments (fig. 44–5, p. 480). The neck is the least movable part of the bladder, and is firmly anchored to the pelvic diaphragm. In the male, it is continuous with the prostate, although a groove separates the two organs externally. The neck of the female bladder is lower than that of the male, and it rests on the pubococcygeal parts of the levatores ani.

Three ligaments help to fix the bladder. They are (1) the medial puboprostatic, (2) the lateral puboprostatic, and (3) the lateral ligament of the bladder. All three are localized thickenings of the superior fascia of the pelvic diaphragm.

The *medial puboprostatic* (or *pubovesical) ligament* is the forward continuation of the tendinous arch of the pelvic fascia. It fixes the prostate (or the neck of the bladder in the female) to the back of the body of the pubis. It contains bundles of smooth muscle fibers, collectively termed the *pubovesicalis*.

The *lateral puboprostatic* (or *pubovesical) ligament* is a smaller thickening, which extends from the prostate (or the neck of the bladder in the female) to the tendinous arch of the pelvic fascia.

The *lateral ligament* passes from the base of the bladder (and the seminal vesicle in the male) laterally and posteriorly, and is continued in the rectovesical fold in the male and in the rectouterine fold in the female. This fold contains visceral branches of the internal iliac vessels, the vesical plexus of nerves, the ureter, and, in the male, a part of the ductus deferens. It also contains some bundles of smooth muscle fibers, collectively termed the *rectovesicalis*.

In addition to the ligaments responsible for its fixation, three remnants of fetal structures are associated with the bladder.

They are (1) the median umbilical ligament and (2) the two medial umbilical ligaments. None of these ligaments is important in the fixation of the bladder.

The *median umbilical ligament* is a remnant of the *urachus,* and it extends from the apex of the bladder to the umbilicus. The part of the urachus nearest the bladder usually retains its lumen,[3] which sometimes communicates with that of the bladder.[4] The epithelial lining of the urachus occasionally gives rise to cysts.[5]

The *medial* (also called lateral) *umbilical ligaments* are the obliterated parts of the umbilical arteries. They extend from the level of the bladder to the umbilicus.

The median umbilical and the medial umbilical ligaments raise folds of peritoneum, called median and medial umbilical folds, respectively. These folds and the fossae associated with them are described on page 370.

Interior of Bladder. **The trigone of the bladder forms a triangle that is approximately equilateral and the angles of which are formed by the internal urethral orifice below and in front, and by the two orifices of the ureters at either side, above and behind.** The mucosa of the trigone is always smooth and flat. When examined with a cystoscope, it appears red when the bladder is empty and pale when it is full. Elsewhere, the lining of the bladder appears pale yellow through a cystoscope, and is wrinkled and folded when the bladder is empty.

An elevation, the interureteric ridge, extends between the two ureteral orifices, and is an indication of an underlying bundle of muscle fibers. The uvula is a median ridge above and behind the internal urethral orifice. It is formed by an underlying bundle of muscle fibers, by the median lobe of the prostate, or by both. It is usually more prominent in old men. The ureters, as they traverse the muscular coat of the bladder, produce folds, which are evident on the interior.

Structure. The bladder has the following four coats:

1. The *mucous membrane.*
2. The *submucosa,* which is absent in the region of the trigone.
3. The *muscular coat.*[6] The bundles of smooth muscle fibers forming this coat are collectively called the *detrusor urinae muscle.* In addition to these, a triangular sheet of muscle is present in the region of the trigone, between the detrusor and the mucosa. The fibers of this sheet that pass between the orifices of the two ureters are responsible for the interureteric ridge on the interior of the bladder, and a thickening of them in the median plane is responsible, at least in part, for the elevation called the uvula.

   Some of the fibers of the detrusor pass forward to form a *pubovesicalis* on each side, and others pass backward to form the *rectovesicalis.*
4. *The serous coat,* consisting of peritoneum, covers the superior surface and upper part of the base of the bladder. Elsewhere the bladder is covered by a *fibrous coat.*

Blood Supply.[7] Usually two or three superior vesical arteries arise from the patent part of the umbilical artery, and they supply the upper part of the bladder. In the male, the base is supplied by the artery of the ductus deferens. The lower part of the bladder, including the neck, is supplied by the inferior vesical artery (p. 453), and, in the female, by the vaginal artery also. In the female, the base is probably supplied by the inferior vesical and vaginal arteries.

The veins pass downward to join the prostatic (or vesical) plexus of veins, which drains into the internal iliac vein.

Lymphatic Drainage.[8] The lymphatic vessels from the superior and inferolateral surfaces of the bladder pass to the external iliac nodes. The vessels from the base drain into the external and internal iliac nodes. The vessels from the neck pass to the sacral and common iliac nodes.

Nerve Supply (fig. 42–2). The bladder is supplied by many nerve fibers from the vesical and prostatic plexuses, which are forward extensions from the inferior hypogastric plexuses to the sides and neck of the bladder. These fibers ramify throughout the wall of the bladder and include the following:

1. The motor supply of the bladder, namely, parasympathetic fibers to the detrusor.
2. The sensory supply of the bladder, namely, fibers that are stimulated by stretching of the detrusor and that activate various reflexes, fibers that are stimulated by stretching and give rise to the sensation of fullness (perhaps identical with the previous kind), and fibers that are concerned with a burning or spasmodic sensation which may be felt mainly in the hypogastric region, and with a sensation of urgency (as in urethral pain).
3. Sympathetic fibers, mostly for the supply of blood vessels. Some of these fibers may supply the detrusor muscle but have nothing to do with micturition. They may activate the detrusor so that it prevents the reflux of semen into the bladder during ejaculation.

Because the nerve supply to the bladder arises on each side of the rectum, it may be damaged during resection of the

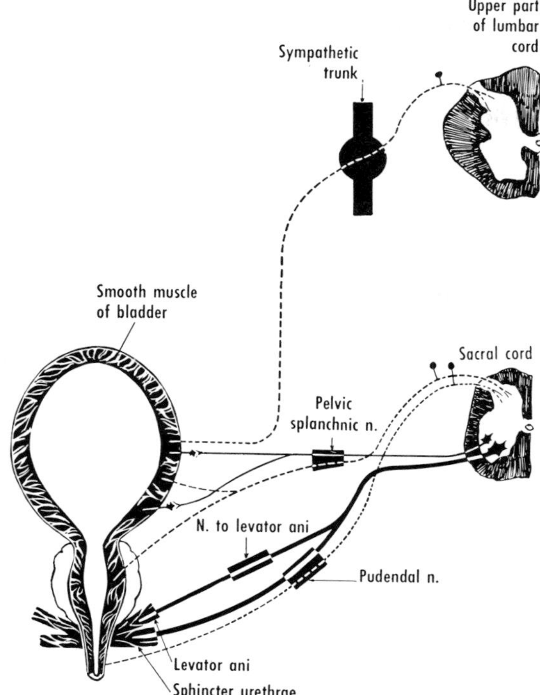

Figure 42–2 Schematic representation of nerve supply to urinary bladder and other structures associated with micturition. The preganglionic parasympathetic fibers to the smooth muscle of the bladder (detrusor urinae) synapse with ganglion cells in the wall of the bladder. Most of the afferent fibers from the bladder and urethra (see text) course in the pelvic splanchnics. A few pain fibers from the bladder ascend in the hypogastric plexuses and reach the upper lumbar part of the spinal cord. The sympathetic supply to the bladder is not shown.

Pain fibers from the urethra are shown coursing in the pelvic splanchnic and pudendal nerves. The heavy lines represent motor fibers to the levator ani and sphincter urethrae.

rectum. Urinary difficulties commonly follow such operations.[9]

Mechanism of Urination. Normal urination or micturition has been studied by fluoroscopy and cystometry in man, and by experimentation in lower animals. Unfortunately, the results of experimental studies of lower animals are often not applicable to man.

Fluoroscopic and cystometric studies have yielded different results, mainly concerning the question of voluntary control of the detrusor. The measurement of intravesical pressures by cystometry has made available valuable information on the process of filling, whereas fluoroscopy[10] has made it possible to study the onset and sequence of events during micturition.

Normal urination can occur only when the pelvic floor, the abdominal wall, and the diaphragm are normal. Before urination begins, the diaphragm and the muscles of the abdominal wall contract, the intra-abdominal pressure rises, and the pubococcygei relax. As the pubococcygei relax, the neck of the bladder moves downward.[11] This downward movement activates or initiates the contraction of the detrusor. At the same time, the contraction of the longitudinal fibers of the urethra, which are continuous with those of the detrusor, shortens the urethra and thereby widens and opens the internal urethral orifice. Urine is then expelled from the bladder. A contraction of the pubococcygei raises the neck of the bladder, the detrusor and the urethral musculature relax, the urethra lengthens, the internal urethral orifice narrows and closes, and urination stops. The smooth muscle of the detrusor is not under direct voluntary control, and its reflex contraction does not begin until the neck descends. If the pelvic floor is fixed so that the bladder cannot descend (e.g., by infiltration with cancer), the patient cannot voluntarily start or stop urination, which occurs automatically through the stretch reflex of the detrusor.

Urination may be interfered with by a variety of neurological disorders.[12] The most severe difficulties are those that result from the following:

1. Transection of the spinal cord above the sacral segments. Sensation and voluntary control are lost, but the reflex arcs are intact—the eventual result is an automatic "cord bladder."
2. Complete loss of the motor supply (e.g., by destruction of the sacral part of the spinal cord). In this condition, no reflex is possible; the detrusor acts independently and very inefficiently.
3. Loss of sensory supply (e.g., after destruction of the dorsal roots of the sacral nerves). Sensation and reflexes are lost, and the bladder becomes overdistended.

Cystometry may give a clue to the nature of these disorders. For example, if a patient does not experience any sensation or desire to urinate after a large amount of fluid is injected into his bladder, but the bladder nevertheless empties automatically and fairly completely, the diagnosis of a cord bladder is made.

Radiography (see also p. 435). The bladder can be viewed radiographically after the introduction of a radio-opaque substance into its cavity through a catheter, or after an intravenous injection of a suitable substance. When full, the bladder appears rounded.

Cystoscopy and Cystometry. The interior of the bladder can be seen with a cystoscope, which is a hollow tube having an electric lamp and a mirror at the end that is introduced through the urethra. Various landmarks may be identified, such as the internal urethral opening, the interureteric ridge, the orifices of the ureters, and the trigone.

Cystometry is the measurement of intravesical pressures after the introduction of a similar hollow tube. The reactions to instillation of measured amounts of fluid into the bladder may give valuable diagnostic information.

URETER

The ureter is the tube that conveys the urine from the kidney to the urinary bladder. It consists of an abdominal portion (p. 411) and a pelvic portion, each of which is about 12½ cm in length.

After entering the pelvis, the ureter passes downward on the lateral pelvic wall, where it is closely related to peritoneum. It is at first in front of and below the internal iliac artery, and then crosses the medial side of the umbilical artery and the obturator vessels and nerve. At the level of the ischial spine, it turns forward and medially. It reaches the back of the urinary bladder about 4 cm above the level of the pubic tubercle. The relations of the pelvic part of the ureter differ in the two sexes.

In the male, the ureter passes through the tissues of the sacrogenital fold and then those of the lateral ligament of the bladder. In this region, it is crossed on its medial side by the ductus deferens. As it approaches the bladder, it lies in front of the upper end of the seminal vesicle, on a plane anterior to that of the ductus deferens.

In the female, after entering the pelvis, the ureter is related to the free (posterior) border of the ovary. In the lower part of the pelvis, the ureter passes first in the tissue of the uterosacral ligament and then in that of the lateral cervical ligament, below the lower part of the broad ligament. It is accompanied in this part of its course by the uterine artery, which crosses above and in front of it. After reaching a point about 2 cm lateral to the cervix uteri, the ureter turns medially in the lateral ligament of the bladder and runs in front of the lateral margin of the vagina to the bladder. As a result of the common deviation of the upper end of the vagina to one side (usually the left), much more of one ureter than the other is situated in front of the vagina.[13] The close relationship of the ureter to the cervix uteri and the vagina is likely to be a special hazard in certain operations, such as the surgical removal of the uterus.[14]

In both sexes, the ureters are about 5 cm apart where they enter the back of the bladder. They are embedded in the wall of this organ for about 2 cm as they pass obliquely to its interior. They open by means of a pair of small, slitlike openings, the orifices of the ureters, which are about 2½ cm apart when the bladder is empty, but which may be 5 cm or more apart when it is distended. The muscular coat of the ureter is continued into the bladder, and its longitudinal layer forms the muscle underlying the trigone. The lumen is narrowest in the part of the ureter embedded within the wall of the bladder.

The ureters may be examined radiographically after the intravenous injection of an organic iodine compound, which is excreted by the kidneys (descending, or excretion, pyelography), or after the injection of sodium iodide through a ureteral catheter inserted through a cystoscope (ascending or retrograde pyelography). A greatly distended ureter may be palpated *per rectum* or *per vaginam*.

The structure, blood and nerve supply, and lymphatic drainage of the entire ureter are discussed on page 412.

URETHRA

The urethra is a fibromuscular tube that serves as a passage for urine from the urinary bladder to the exterior. In the male, it also serves as a passage for seminal fluid in most of its course. It is closed when fluid is not passing through it. The male and female urethra differ in many respects.

Male Urethra

The male urethra (figs. 43–1 and 47–3, pp. 467, 498) is about 20 cm in length. It begins at the neck of the bladder, and extends through the prostate, the pelvic and urogenital diaphragms, and the root and body of the penis to the tip of the glans. It is subdivided into three parts: prostatic, membranous, and spongy.

Prostatic Part. The prostatic part traverses the prostate, and extends from the base to the apex of this organ. It is approximately 3 cm in length, and is somewhat curved in an anteroposterior direction.[15] It is more dilatable than the membranous and spongy parts. When distended, it is the widest part of the entire urethra, and its cavity is fusiform in shape. When it is empty, the anterior and posterior walls are in contact, and the anterior and lateral walls are folded longitudinally.

The posterior wall (fig. 43–3, p. 471), sometimes called the "floor," is characterized by several markings. The *urethral crest* is a median ridge, which is sometimes continuous with the uvula of the bladder above, and which often extends downward into the membranous part of the urethra. The *colliculus seminalis (verumontanum)* is an ovoid enlargement of the crest, located at approximately the junction of the middle and lower thirds of the prostatic part. At the summit of the colliculus is the opening of a diverticulum, the *prostatic utricle,* which extends into the substance of the prostate for a short distance. This diverticulum is said to be the remains of the fused caudal ends of the paramesonephric ducts, which form the uterus and much of the vagina in the female. The minute openings of the ejaculatory ducts are located on each side of the opening of the utricle. The *prostatic sinus* is a groove on each side of the urethral crest. Most of the ducts of the prostate open into the floor of this groove, but a few, from the median lobe, open on the sides of the urethral crest.

Membranous Part. The membranous part extends downward and forward from the apex of the prostate to the bulb of the penis, and it passes through both pelvic and urogenital diaphragms. It is the shortest part of the urethra, and, except for the external opening, the narrowest and least dilatable. It is from 1 to 2 cm long, and is situated about 2½ cm behind the lower margin of the pubic symphysis. Within the urogenital diaphragm, it is surrounded by the sphincter urethrae. Immediately below the urogenital diaphragm, its posterior wall comes into contact with the bulb of the penis, but its anterior wall is not covered by the bulb until it has passed for a short distance below the diaphragm. As it enters the bulb, the urethra turns forward at almost a right angle. It is wider and its walls are thinner just below the urogenital diaphragm. Here it is most liable to rupture during injury, and also most liable to be penetrated during the passage of instruments. The mucosa of the membranous part is folded longitudinally when the urethra is empty.

Spongy Part. The spongy part lies in the corpus spongiosum. It traverses the bulb, body, and glans of the penis. In the first part of its course, it is fixed in position and is almost straight. When the lumen is closed, it is a sagittal slit in the glans and a transverse slit in the remainder of the spongy part. The lumen is somewhat wider where the urethra lies in the bulb *(intrabulbar fossa).* It is wider in the glans also, where the dilated portion is known as the *navicular fossa (fossa terminalis).*

The minute openings of the ducts of the bulbourethral glands are located on the lower wall of the urethra, shortly beyond the beginning of the spongy portion. The urethral lacunae are small pits in the walls, openings of which are usually directed toward the external opening.

Structure. The urethra consists of a mucous membrane and a muscular coat.

The *mucous membrane* varies in thickness in different parts of its course. It varies also in color. Through a urethroscope it appears red in the prostatic and membranous parts, but it is yellowish pink in the spongy part.

The ducts of numerous small *urethral glands* open in the inner surface of the urethra. Some ducts open into the urethral lacunae, but not all lacunae contain the openings of ducts.

The *muscular coat* of the prostatic and membranous parts of the urethra is continued downward from the bladder and can be regarded as a continuation of the detrusor muscle (p. 462). Fibers from the trigonal musculature pass downward on the posterior wall of the urethra.

Skeletal muscle fibers of the sphincter urethrae surround the membranous part.[16] Some of these pass upward along the anterior wall of the prostatic part for a short distance.

Blood Supply. The prostatic part is supplied mainly by the inferior vesical and middle rectal arteries. The membranous part is supplied by the artery of the bulb of the penis. The spongy part is supplied by the urethral artery, and also by some branches from the deep and dorsal arteries of the penis. The veins drain into the prostatic plexus and into the internal pudendal vein.

Nerve Supply (fig. 42–2, p. 463). The prostatic

part is supplied by the prostatic plexus, which continues as the cavernous nerves of the penis to supply the membranous part. Branches of the pudendal nerve supply the spongy part. The distribution and functions of these nerves are not agreed upon.

Lymphatic Drainage. The lymphatic vessels from the prostatic and membranous parts course along the internal pudendal vessels and drain mainly into the internal iliac nodes, but some pass to the external iliac. Most of the lymphatic vessels from the spongy part pass to the deep inguinal nodes. Some, however, drain into the external iliac nodes.

Rupture of Urethra. In some accidents resulting in crushing of the pelvis, the urethra may be ruptured at the junction of its prostatic and membranous parts. Attempts to urinate then result in the extravasation of urine into the extraperitoneal tissue around the bladder, and perhaps around the rectum also.

Rupture of the membranous part of the urethra occurs more frequently. Extravasated urine tends to be limited to the superficial perineal space at first, but it may pass through the deep perineal fascia and infiltrate the space between the superficial and deep perineal fascia, and spread from there to the scrotum, the penis, and the front of the abdomen.

Examination in the Living. The urethra may be examined radiographically after the introduction of a suitable radiopaque material into its lumen. The lining may be viewed with the aid of a urethroscope. When a catheter is passed through the entire length of the urethra, it can be palpated in the spongy part through the ventral aspect of the penis, in the membranous part through the perineum, and in the prostatic part *per rectum.*

Female Urethra

The female urethra is about 4 cm long. It is quite distensible and can be dilated to about 1 cm without damage to it. It extends downward and slightly forward from the neck of the bladder to the *external urethral orifice,* which is situated between the labia minora, in front of the opening of the vagina, and below and behind the glans of the clitoris. The margins of the external opening are slightly everted. In its course, the urethra passes through the urogenital and pelvic diaphragms. It is closed except during the passage of urine. When closed, its interior is marked by longitudinal folds, the most prominent of which is located on the posterior wall and is called the *urethral crest.*

The urethra is fused with the anterior wall of the vagina, and it can be palpated *per vaginam* between the anterior vaginal wall and the pubic symphysis. It is attached to the pubis by some of the fibers of the pubovesical ligament. A thickening of the superior fascia of the urogenital diaphragm also helps to fix it.[17]

Structure. The urethra consists of a mucous membrane and a muscular coat.

The *mucous membrane* contains the openings of many small *urethral glands* and numerous small pits, the *urethral lacunae.* Glands, tending to occur in four groups, drain through 6 to 31 ducts into the vestibule near the external urethral orifice.[18] These groups of glands are said to correspond to the prostate of the male.

The *muscular coat*[19] of the upper part of the urethra is a downward continuation of some of the bundles of the detrusor muscle. It is fusiform in the middle part of the urethra. The lower part of the urethra has no muscular coat.

Blood Supply. The upper part of the urethra is supplied by the inferior vesical artery, the middle part by the inferior vesical and uterine arteries, and the lower part by the internal pudendal artery.

The veins from the urethra drain into the vesical plexus and into the internal pudendal vein.

Lymphatic Drainage. The lymphatic vessels run along the internal pudendal artery and drain mainly into the internal iliac nodes. Some pass to the external iliac nodes.

Nerve Supply. The upper part of the urethra is supplied by the vesical and uterovaginal plexuses. The pudendal nerve supplies the lower part. The functions of the component autonomic and sensory fibers are uncertain.

REFERENCES

1. H.-J. F. H. von Lüdinghausen, Z. Anat. EntwGesch., *97*:757, 1932.
2. L. C. Jacobs and E. J. Caspar, Urol. cutan. Rev., *37*:729, 1933.
3. R. C. Begg, J. Anat., Lond., *64*:170, 1930.
4. G. Hammond, L. Yglesias, and J. E. Davis, Anat. Rec., *80*:271, 1941.
5. M. Douglass, Amer. J. Surg., *22*:557, 1933.
6. R. T. Woodburne, J. Urol., *100*:474, 1968. S. Gil Vernet, *Morphology and Function of Vesico-Prostato-Urethral Musculature,* Liberia Editrice Canova, Treviso, 1968. J. A. Hutch, *Anatomy and Physiology of the Bladder, Trigone, and Urethra,* Appleton-Century-Crofts, New York, 1972.
7. J. L. Braithwaite, Brit. J. Urol., *24*:64, 1952.
8. A. E. Parker, Anat. Rec., *65*:443, 1936. T. O. Powell, Surg. Gynec. Obstet., *78*:605, 1944.
9. V. F. Marshall, R. S. Pollack, and C. Miller, J. Urol., *55*:409, 1946.
10. C. E. Shopfner and J. A. Hutch, Radiol. Clin. N. Amer., *6*:165, 1968.
11. S. R. Muellner and F. G. Fleischner, J. Urol., *61*:233, 1949. S. R. Muellner, J. Urol., *65*:805, 1951. M. Emanuel, Surg. Clin. N. Amer., *45*:1467, 1965.
12. J. S. Ritter and A. Sporer, J. Urol., *61*:528, 1949. C. C. Prather, *Urological Aspects of Spinal Cord Injuries,* Thomas, Springfield, Illinois, 1949.
13. J. C. Brash, Brit. med. J., *2*:790, 1922.
14. J. Hawkins, Ann. R. Coll. Surg. Engl., *15*:326, 1954.
15. T. W. Glenister, J. Anat., Lond., *96*:443, 1962.
16. H. C. Rolnick and F. K. Arnheim, J. Urol., *61*:591, 1949.
17. A. H. Curtis, B. J. Anson, and C. B. McVay, Surg. Gynec. Obstet., *68*:161, 1939.
18. J. W. Huffman, Amer. J. Obstet. Gynec., *55*:86, 1948; Arch. Surg., *62*:615, 1951. R. L. Deter, G. T. Caldwell, and A. I. Folsom, J. Urol., *55*:651, 1946.
19. L. Beck, Z. Geburtsh, Gynäk., *169*:1, 1969.

MALE GENITAL ORGANS

43

The male genital organs (fig. 43–1) consist of the testes and epididymides, which are situated in the scrotum, the ductus deferentes (vasa deferentia), which are contained in the spermatic cords in a part of their course, the seminal vesicles, the ejaculatory ducts, the prostate, the bulbourethral glands, and the penis. All of these organs are paired, except the prostate and the penis, which are single. The scrotum and the penis, which are classified as external genital organs, are described on page 501.

The spermatozoa, which are formed in the testis, are the essential constituents of the seminal fluid. They pass from the testis to the epididymis, where they are stored. A mucoid secretion from the epididymis forms one of the constituents of the seminal fluid. After their emission from the epididymis, the spermatozoa pass through the ductus deferens and ejaculatory duct into the urethra, through which they reach the exterior.

The remaining constituents of the seminal fluid are produced in the seminal vesicles, the prostate, the bulbourethral glands, and the urethral glands. The secretions of these structures, which are sometimes called the glandular accessory genital organs, empty into the urethra.

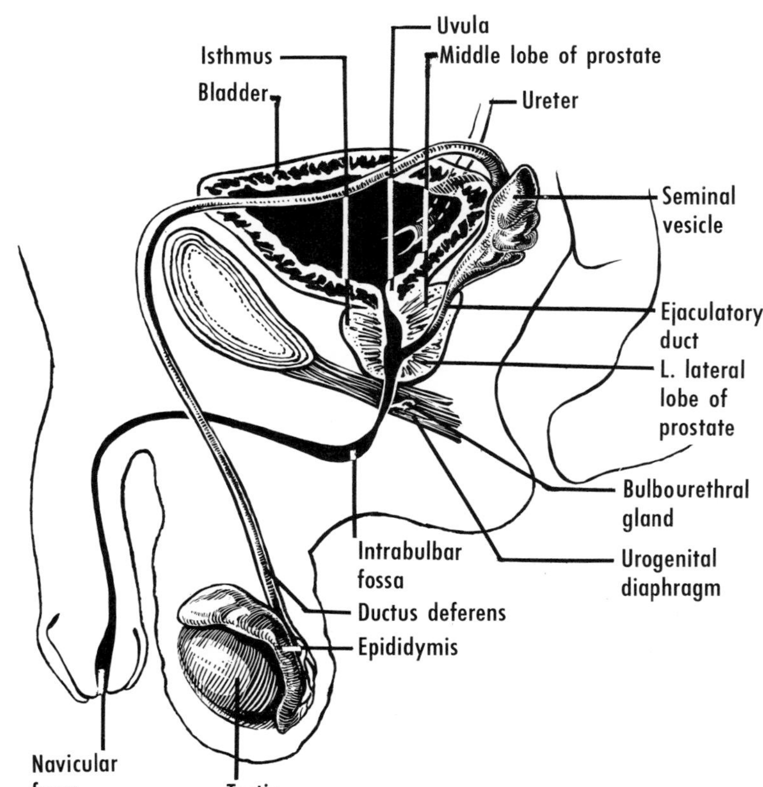

Figure 43–1 Scheme of the male genital system.

TESTIS AND EPIDIDYMIS

Testis

The testes (fig. 43–2) are paired, ovoid organs. After puberty they produce spermatozoa, and, inasmuch as they are in part endocrine glands, they secrete a hormone, which is responsible for the secondary sexual characteristics of the male. They are situated in the scrotum, where the left is usually at a lower level than the right. The right is lower than the left in cases of situs inversus totalis,[1] and is usually lower than the left in left-handed men.[2] In the adult, each testis weighs, on the average, 25 gm; in the majority of cases the right is heavier than the left.[2] The testis may weigh much less in old age. The Greek word for testis is *orchis;* such words as orchitis are derived from it.

Each testis has superior and inferior ends, medial and lateral surfaces, and anterior and posterior margins. Both surfaces are somewhat flattened. The posterior margin is covered by the epididymis and the lower part of the spermatic cord.

Structure. The *tunica albuginea* is the outer covering of the testis (fig. 43–2*C*). It lies beneath the visceral layer of the tunica vaginalis and consists mainly of dense, inelastic connective tissue. Delicate fibrous septa pass from its deep aspect into the interior and incompletely divide the testis into wedge-shaped lobules, between 250 and 400 in number, each with one to four tubules.[3] The bases of the wedges are at the deep aspect of the tunica albuginea; the apices

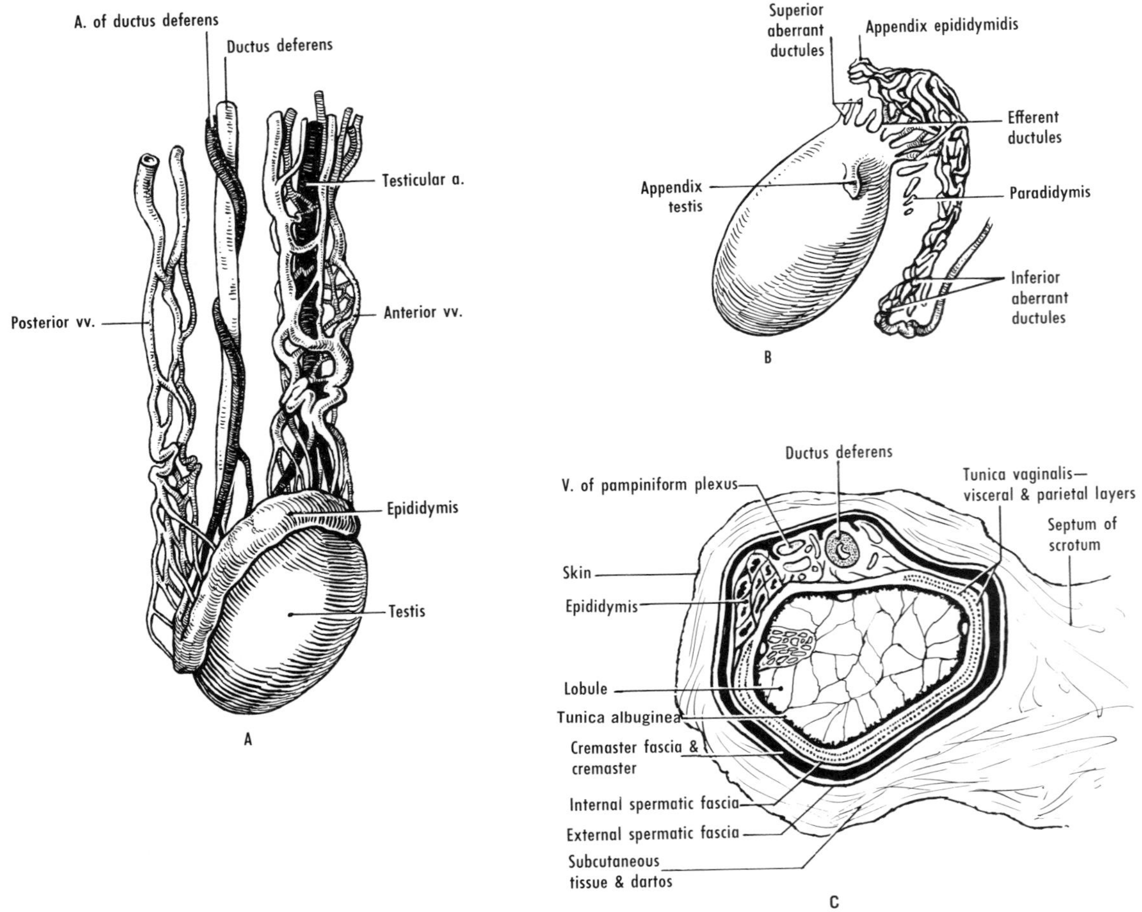

Figure 43–2 The right testis and epididymis from the lateral aspect (*A*) and from the anteromedial aspect (*B*). *C*, horizontal section of the right testis and scrotum. The posterior aspect is uppermost.

converge near the posterior margin of the testis, where the septa also converge and form the *mediastinum testis,* which is a mass of fibrous tissue continuous with the tunica albuginea.

The parenchyma of the testis is located within the lobules and consists of the *convoluted seminiferous tubules,* which resemble delicate, tortuous threads. It is estimated that more than 800 tubules are present in each testis. These tubules become less twisted and convoluted in their course backward. As they approach the mediastinum they unite to form between 20 and 30 *straight seminiferous tubules.* These in turn pass in the *rete testis,* an elaborate network of canals, which traverses the mediastinum. From the network are formed 15 to 20 channels, the *efferent ductules,* which enter the head of the epididymis.

The interstitial cells are located in the loose tissue under the tunica albuginea, in the septa, and in the stroma that surrounds the individual convoluted seminiferous tubules. They secrete testosterone, the male sex hormone.

Epididymis

The epididymis (fig. 43–2) is a C-shaped structure, which is applied to the posterior margin of the testis and overlaps the adjacent part of the lateral surface. The spermatozoa are stored in it until they are emitted. It is subdivided into three parts: a head, a body, and a tail.

The efferent ductules of the testis, which are at first straight, become very tortuous after they enter the head of the epididymis. Here they form wedge-shaped masses, the *lobules* (or *cones*) *of the epididymis,* the apices of which are directed toward the testis. After a twisted course, each ductule opens opposite the base of a lobule into a single tube, the *duct of the epididymis.* This duct is very greatly convoluted, and it makes up the main mass of the remainder of the epididymis. It is about 6 meters long.

The *head of the epididymis* is the upper, larger part, which lies on the superior end of the testis and overhangs it. The *body of the epididymis* is attached to the posterior margin of the testis. It is separated from the adjacent part of the lateral surface by the *sinus of the epididymis,* a space formed by an invagination of the visceral layer of the tunica vaginalis in this region. The *tail of the epididymis* is the lower, smaller part. In it the duct of the epididymis increases in thickness and diameter, and becomes the ductus deferens.

The *appendix testis* is a small body on the upper end of the testis. It is usually sessile, but may be pedunculated. It is a remnant of the upper end of the paramesonephric duct, and is homologous with the fimbriated end of the uterine tube of the female.

The *appendix of the epididymis* is a small appendage, usually pedunculated, on the head of the epididymis. It is regarded as a remnant of the mesonephros.

Blood Supply.[4] The testis is supplied by the testicular artery (fig. 43–2), which divides into a variable number of branches. These branches pass to the posterior border of the testis, medial to the epididymis. They penetrate the tunica albuginea and ramify in the underlying loose connective tissue, the tunica vasculosa. Small branches pass along the septa toward the mediastinum. The testicular artery, or one of its branches, anastomoses with the artery of the ductus deferens and with the external spermatic artery.[5] The veins of the testis pass backward to the posterior border, pierce the tunica albuginea, and join the pampiniform plexus.

The epididymis is supplied by the testicular artery or by one or more of its branches. Its veins drain into the pampiniform plexus.

Lymphatic Drainage. The lymphatic vessels from the testis and epididymis pass upward with the testicular vessels. They drain into the lumbar (aortic) nodes.[6]

Nerve Supply.[7] The testis is supplied by the testicular plexus, which receives additional fibers from the genitofemoral nerve and also, according to clinical evidence,[8] from the posterior scrotal nerves. The sympathetic fibers reaching the testis are probably mainly vasomotor. Testicular pain resulting from squeezing or swelling is severe, and often sickening or shocking, especially when it is acute. Under certain conditions it may be referred to the groin or to the lower part of the abdominal wall.[9]

The epididymis is supplied by fibers of the inferior hypogastric plexus that are continued along the ductus deferens. The importance of the autonomic supply to the smooth muscle of the ductus is uncertain.

DUCTUS DEFERENS, SEMINAL VESICLE, AND EJACULATORY DUCT

Ductus Deferens

The ductus deferens (vas deferens) (fig. 43–2) is the continuation of the duct of the epididymis, and carries the spermatozoa from the epididymis to the ejaculatory duct. It begins at the tail of the epididymis, where it is very tortuous. It becomes straighter as it ascends on the medial side of the epididymis near the posterior border of the testis. Here it is surrounded by the pampiniform plexus of veins and is incorporated into the spermatic cord. It continues upward from the superior end of the testis to the superficial inguinal ring, and in this part of its course it can be felt as a firm cord when held between the thumb and the index finger. After passing through the inguinal canal, it leaves the other structures in the spermatic cord by turning around the lateral side of the inferior epigastric artery and ascending in front of the external iliac artery for a short distance. It then turns backward and slightly downward, crosses the external iliac vessels, and enters the pelvis. As it continues backward, it is covered medially by peritoneum, and is related laterally to the umbilical artery, the obturator vessels and nerve, and the superior vesical vessels. After crossing the medial side of the ureter, it turns medially and downward to run in the sacrogenital fold. It reaches the posterior aspect of the bladder, and then runs downward and medially on the medial side of the seminal vesicle. In this location, the canal of the ductus is enlarged and tortuous, and this portion of the ductus is termed the *ampulla*. The canal is again small in caliber near the base of the prostate, where the ductus deferens joins the duct of the seminal vesicle to form the ejaculatory duct.

The *inferior ductulus aberrans* is a narrow, coiled tube, which is often connected with the first part of the ductus deferens, or with the lower part of the duct of the epididymis. It may be as long as 35 cm when uncoiled.

The *superior ductulus aberrans* is a narrow tube of variable length, which lies in the head of the epididymis and is connected with the rete testis.

The *paradidymis* lies above the head of the epididymis in the anterior part of the spermatic cord. It consists of a few tortuous tubules, and is considered to be a remnant of the mesonephros.

Structure. The ductus deferens consists of three layers: a mucous membrane, a muscular coat, and an adventitia.

Seminal Vesicle

The seminal vesicles (fig. 43–3) are two sacculated pouches that produce a large part of the seminal fluid. Each vesicle is usually about 5 cm long but it may be much shorter. The broad end is directed laterally, upward, and backward. Its narrow end closely approaches that of the contralateral vesicle. When the urinary bladder is distended, the seminal vesicles are more nearly vertical, but when the bladder is empty, they are more nearly horizontal.

The seminal vesicles are embedded in a dense sheath, consisting of smooth muscle and fibrous tissue and attached to the posterior aspect of the urinary bladder. Their upper parts, which are separated from the rectum by the rectovesical pouch, are covered by peritoneum. Their lower parts are separated from the rectum by the rectovesical septum. The terminal parts of the ureters and the ampullae of the ductus deferentes are medial to the vesicles, and the prostatic and vesical venous plexuses are lateral.

Each seminal vesicle consists of a coiled tube, which gives off several diverticula and which ends blindly above. Its lower end becomes narrow and straight to form a duct. This joins the corresponding ductus deferens, and the tube resulting from this union is the ejaculatory duct.

The seminal vesicles can be palpated *per rectum* when the urinary bladder is full. When the seminal vesicles are full, they are very sensitive to pressure. Each has a capacity of 1½ to 3 ml of fluid.

Ejaculatory Duct

The ejaculatory duct (fig. 43–3) is formed by the union of the ductus deferens and the duct of the seminal vesicle. After penetrating the base of the prostate, it passes downward and forward to enter the prostatic part of the urethra on the colliculus seminalis, just lateral to the prostatic utricle. (Openings of the ejaculatory ducts sometimes appear in the utricle, but they are alleged to be secondary.[10]) In its course through the prostate, each ejaculatory duct approaches its fellow of the opposite side. Also, its walls become thinner and it diminishes in size.

Blood Supply. The artery of the ductus deferens supplies the seminal vesicle and ejaculatory duct, and accompanies the ductus deferens as far as the testis, where it anastomoses with the testicular artery. It gives branches to the ductus throughout its course. Branches from the inferior vesical and from the middle rectal artery (when present) also help to supply the seminal vesicle and the adjacent part of the ductus deferens.

The veins from the ductus deferens, seminal vesicle, and ejaculatory duct join the prostatic and vesical venous plexuses.

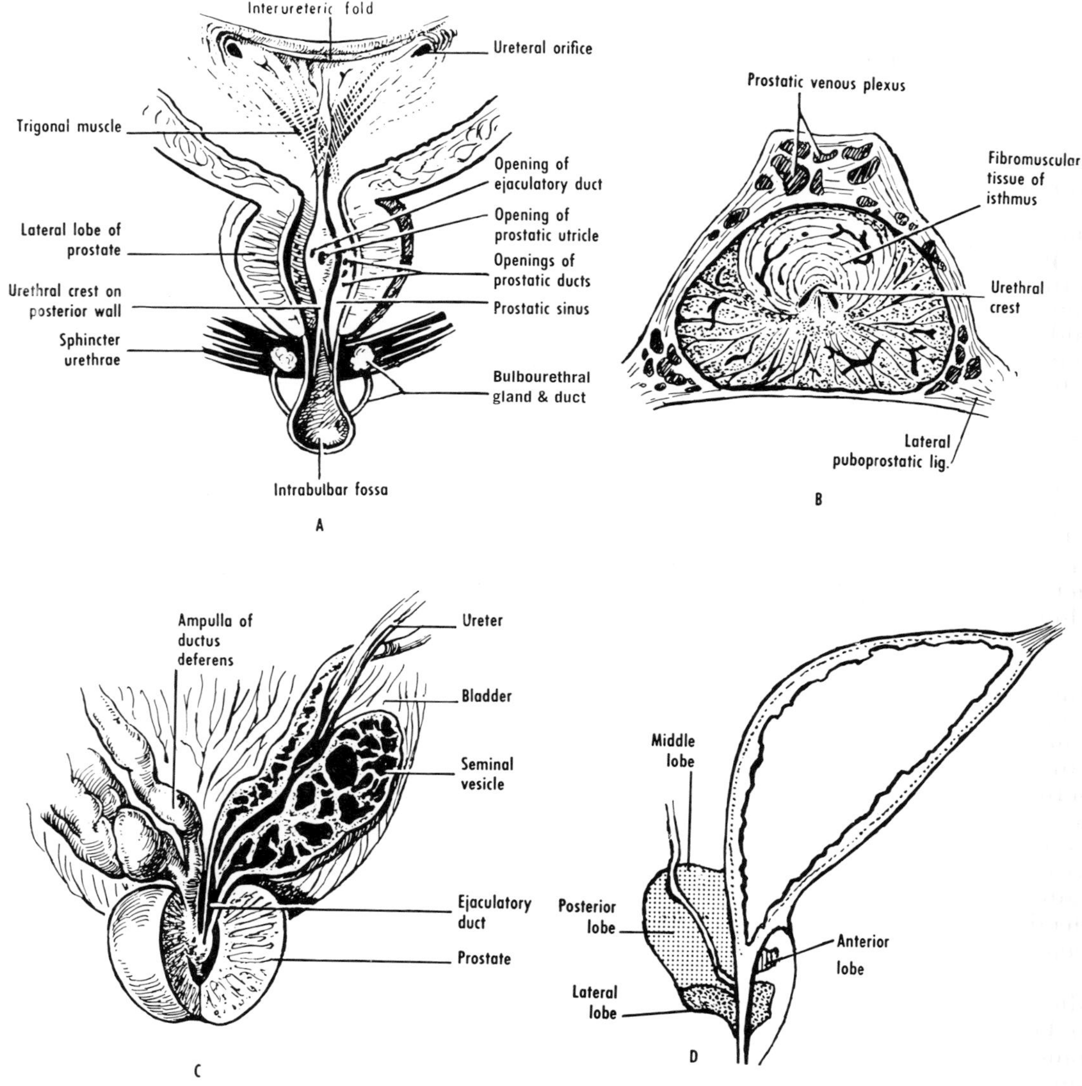

Figure 43–3 *A*, coronal section of bladder and prostate. *B*, horizontal section of prostate. *C*, seminal vesicles and ejaculatory ducts. *D*, diagram of the prostate of a newborn, showing the position of some of the "lobes." The only ones persisting in the adult are the lateral lobes. Based on O. S. Lowsley.[13]

Lymphatic Drainage. The lymphatic vessels from the ductus deferens drain into the external iliac nodes. Those from the seminal vesicles drain into the external and internal iliac nodes.

Nerve Supply. The ductus deferens is supplied by autonomic fibers from the superior and inferior hypogastric plexuses. The function of these fibers is not certain, and it is not definitely known whether sensory fibers are distributed with them. The seminal vesicles are supplied by nerves from the inferior hypogastric and prostatic plexuses. The functional significance of these nerves is also uncertain.

SPERMATIC CORD; COVERINGS OF SPERMATIC CORD, TESTIS, AND EPIDIDYMIS

Spermatic Cord

The spermatic cord is formed at the deep inguinal ring by the structures that accompany the testis and epididymis during their descent. It extends through the inguinal canal and into

the scrotum, where it ends along the posterior margin of the testis. The left spermatic cord is the longer in men in whom the left testis is lower than the right.

Below the superficial inguinal ring, the spermatic cord lies in front of the adductor longus. Here the superficial external pudendal artery crosses it anteriorly, and the deep external pudendal artery crosses it posteriorly.

The spermatic cord contains the following structures, all of which are embedded in a downward continuation of extraperitoneal tissue: (1) The ductus deferens, together with the closely associated artery and vein of the ductus deferens, and the nerves that pass to the epididymis. The ductus lies in the posterior part of the cord below the superficial inguinal ring, and in the lower part of the cord within the inguinal canal. (2) The testicular artery, which lies in front of the ductus deferens and is accompanied by the testicular plexus of nerves. (3) The *pampiniform plexus* of veins, which forms much of the bulk of the spermatic cord. This plexus is formed by the veins that drain the testis and epididymis and ascend as a number of anastomosing longitudinal vessels to the deep inguinal ring, where their number is reduced to two or three. These veins often become varicose, more frequently on the left side,[11] and the resulting condition is called varicocele. (4) Lymphatic vessels.[12] (5) The cremasteric artery. (6) The genital branch of the genitofemoral nerve. (7) Remnants of the processus vaginalis peritonei.

Coverings of Spermatic Cord, Testis, and Epididymis

The coverings of the spermatic cord, testis, and epididymis (fig. 43–2) are derived from several layers of the abdominal wall (fig. 33–6, p. 356). They are not easily separable from one another, either in the cadaver or in the living individual. Occasionally, one of the coverings can be separated into two or more layers.

Internal Spermatic Fascia. This is the thin, innermost covering and is derived from the transversalis fascia. It forms a loose investment for the spermatic cord and the associated extraperitoneal tissue.

Cremasteric Fascia. This is closely applied to the external aspect of the internal spermatic fascia. It can be recognized by the presence of many bundles of skeletal muscle fibers, collectively termed *cremaster muscle,* which are continuous above with the internal oblique. The cremaster muscle receives its blood supply from the cremasteric artery, and its nerve supply from the genital branch of the genitofemoral nerve. Contraction of its fibers can often be produced by a gentle stroking of the skin of the medial aspect of the thigh (cremasteric reflex). Its contraction results in a raising of the testis and epididymis to a higher position within the scrotum.

External Spermatic Fascia. This is the thin, outer covering. It is attached above to the crura of the superficial inguinal ring, and is continuous with the fascia covering the external oblique.

Tunica Vaginalis Testis. This is a double-layered serous membrane that covers the front and sides of the testis and epididymis. Covered by the internal spermatic fascia, it extends for a variable distance above the testis. During prenatal development, the tunica vaginalis is continuous with the peritoneum. This connection is usually lost, however, and most of the part above the testis disappears or becomes reduced to a strand of connective tissue lying in the anterior part of the spermatic cord. The layers of the tunica vaginalis are separated from one another by a small space, which contains serous fluid. The accumulation of an abnormally large amount of fluid in this space results in a condition called *hydrocele.* The inner or *visceral layer* of the tunica vaginalis is firmly attached to the front and sides of the testis and epididymis. Laterally it passes into the narrow interval between these organs to form the *sinus of the epididymis.* Posteriorly it is reflected from the testis and epididymis as the outer or *parietal layer.*

PROSTATE AND BULBOURETHRAL GLAND

Prostate

The prostate[13] (fig. 43–3) consists chiefly of smooth muscle and fibrous tissue.

It also contains glands, the secretion of which accounts for the characteristic odor of semen, and, together with the secretion of the seminal vesicles, forms the bulk of the seminal fluid. It is situated in the pelvis, behind the pubic symphysis, and on the medial margins of the pubococcygei. It is structurally continuous with the urinary bladder, which lies above it, but laterally and posteriorly a superficial groove marks a separation between the two organs. This groove is evident after the removal from it of a venous plexus embedded in fat and loose connective tissue. The size of the prostate is variable; the greatest diameters of a prostate considered to be free from disease are approximately as follows: transverse, 4 cm; vertical, 3 cm; anteroposterior, 2 cm.[14]

Parts. The *apex* is the lowermost part of the prostate, and it is located about 1½ cm behind the lower margin of the pubic symphysis. The *base* is in a horizontal plane that passes through the middle of the symphysis. It is structurally continuous with the wall of the urinary bladder except at its periphery, where a narrow rim forms the floor of a groove that separates it from the bladder. The internal urethral orifice is located approximately in the middle of the base. The *inferolateral surfaces* are convex, and are separated from the superior fascia of the pelvic diaphragm by a plexus of veins. The *anterior surface* is narrow. It is separated from the pubis by the retropubic pad of fat. The (medial) puboprostatic ligaments attach to its lower part. The urethra leaves the anterior surface of the prostate just above and in front of the apex. The *posterior surface* is flattened and triangular, and it presents a more or less prominent median groove. Its upper part is related to the seminal vesicles and the lower ends of the ductus deferentes, and, near the base, it presents small depressions for the entrance of the ejaculatory ducts. It can be palpated *per rectum* in the living individual.

The prostate has left and right lateral lobes and a middle or median lobe. Superficially the *lateral lobes* are not demarcated from each other. They are connected with each other in front of the urethra by the *isthmus of the prostate,* which consists mainly of smooth muscle tissue and is devoid of glands. The isthmus is not apparent from the exterior. The *median lobe,* variable in size, is the part of the prostate projecting inward from the upper part of the posterior surface between the ejaculatory ducts and the urethra. The enlargement of this lobe is at least partially responsible for the formation of the uvula, which, projecting into the wall of the bladder, may block the passage of urine. In structure, the median lobe is normally inseparable from the lateral lobes or from the wall of the bladder.

Fascia or Sheath of the Prostate. The superior fascia of the pelvic diaphragm is reflected upward as the visceral fascia of the pelvis to ensheathe the prostate and then to continue upward over the urinary bladder (fig. 47–2, p. 497). The part of this fascia that covers the prostate is dense and fibrous, and is termed the fascia (sheath) of the prostate. It is situated outside of the capsule of the prostate, and is separated from the capsule in front and at the side by loose connective tissue containing the prostatic plexus of veins. It is fused anteriorly with the tendinous arch of the pelvic fascia, which passes forward to the pubis as the medial puboprostatic ligament. Smooth muscle fibers are contained in this ligament, and they are collectively termed the puboprostatic muscle. The lateral puboprostatic ligament (fig. 47–2, p. 497) extends from the fascia of the prostate laterally to the tendinous arch of the pelvic fascia. Immediately below the puboprostatic ligaments, the prostate is closely associated with the medial margins of the pubococcygei. Here muscle fibers extend upward from the pubococcygei and fuse with the fascia of the prostate *(levator prostatae muscle).* Posteriorly the fascia of the prostate is separated from that covering the rectum by the rectovesical septum, which extends upward on the posterior aspects of the seminal vesicles and ductus deferentes and fuses with the peritoneum of the rectovesical pouch.

Structure. The *capsule* of the prostate lies inside the fascia of this organ. Numerous strands pass inward from the capsule, and incompletely divide the organ into about 50 poorly defined lobules. Skeletal muscle fibers from the sphincter urethrae pass upward into the prostate. They are located in front of the lower portion of the prostatic part of the urethra.

The *musculofibrous tissue* of the prostate, especially that lateral and posterior to the urethra, is broken up by as many as 50 branched tubuloalveolar glands. These drain into 20 to 30 minute *prostatic ductules*, which open into or near the prostatic sinuses in the posterior wall of the urethra.

Changes in the level of androgens affect the size and structure of the prostate. It is small at birth, but at puberty it rapidly increases in size, and, after one-half to one year of rapid growth, it is transformed into an organ resembling that of the adult. During the fifth decade, it commonly decreases in size, and the decrease is accompanied by an atrophy of the glandular tissue. In some men, however, the glandular tissue undergoes hyperplasia, and the size of the prostate increases with age.[15]

Blood Supply.[16] The main artery to the prostate usually arises in common with the inferior vesical from one of the branches of the internal iliac artery. Some of its branches ramify in the prostatic fascia, and the resulting subdivisions give off branches that penetrate the capsule and supply the outer portion of the prostate. The prostate often receives a branch from the superior rectal artery, and, when the middle rectal is present, it usually sends branches forward to reach this organ. Branches of the inferior vesical enter the prostate at its junction with the urinary bladder. These accompany the prostatic part of the urethra and supply adjacent portions of the prostate.

The veins from the prostate drain mainly into the prostatic plexus, an extensive network of thin-walled vessels lying in the fascia of the prostate. This network joins the vesical plexus in the groove that superficially separates the urinary bladder from the prostate, and the combined plexuses drain into the internal iliac vein.

Lymphatic Drainage. Most of the lymphatic vessels from the prostate pass to the internal iliac nodes, but some of them enter the external iliac group, and some others the sacral group.

Nerve Supply. The prostate is supplied by the prostatic plexus, which consists mainly of sympathetic nerves. These fibers presumably innervate the smooth muscle and blood vessels within this organ. A parasympathetic supply to the prostate has not been conclusively demonstrated. Some pain fibers may be present in the plexus, but in general the sensory innervation of the prostate is still unknown.

Bulbourethral Gland

The bulbourethral glands (fig. 43–3) are two rounded structures, 1/2 to 1 1/2 cm in diameter, situated on each side of the median plane. They are embedded in the substance of the sphincter urethrae, just behind the membranous part of the urethra. They secrete a mucus-like substance, the function of which is obscure.

The ducts of the bulbourethral glands pass through the inferior fascia of the urogenital diaphragm, enter the bulb of the penis, and traverse its substance. After a course of 2 1/2 to 4 cm, they end by opening into the lower aspect of the spongy part of the urethra.

Blood Supply. The bulbourethral glands are supplied by the arteries of the bulb of the penis.

Lymphatic Drainage. The lymphatic vessels drain into the internal iliac group of nodes.

REFERENCES

1. M. R. Cholst, Amer. J. Surg., *73*:104, 1947.
2. K. S. F. Chang *et al.*, J. Anat., Lond., *94*:543, 1960.
3. Ph. C. Sappey, *Traité d'anatomie descriptive*, Battaille, Paris, 1889.
4. R. G. Harrison and A. E. Barclay, Brit. J. Urol., *20*:57, 1948.
5. R. G. Harrison, J. Anat., Lond., *83*:267, 1949. H. Neuhof and W. H. Mencher, Surgery, *8*:672, 1940.
6. L. Wahlquist, L. Hultén, and M. Rosencranz, Acta chir. scand., *132*:454, 1966.
7. G. A. G. Mitchell, J. Anat., Lond., *70*:10, 1935. D. J. Gray, Anat. Rec., *98*:325, 1947.
8. H. H. Woollard and E. A. Carmichael, Brain, *56*:293, 1933.
9. H. H. Woollard and E. A. Carmichael, cited in reference 8.
10. S. McMahon, J. Anat., Lond., *72*:556, 1938.
11. H. L. Skinner, Ann. Surg., *113*:123, 1941.
12. M. Paul, Ann. R. Coll. Surg. Engl., *7*:128, 1950.
13. O. S. Lowsley, Surg. Gynec. Obstet., *20*:183, 1915. I. E. LeDuc, J. Urol., *42*:1217, 1939.
14. R. A. Moore, Amer. J. Path., *12*:599, 1936.
15. G. I. M. Swyer, J. Anat., Lond., *78*:130, 1944.
16. E. J. Clegg, J. Anat., Lond., *89*:209, 1955.

FEMALE GENITAL ORGANS

44

The female genital organs (figs. 44–1 and 44–2) consist of the ovaries, uterine tubes, uterus, vagina, and external genital organs. The paired ovaries and uterine tubes and the single uterus are situated in the pelvic cavity. The single vagina is located partly within the pelvic cavity and partly in the perineum. The external genital organs lie in front of and below the pubis. They are described in Chapter 47.

OVARY

The ovaries[1] (fig. 44–1) are paired organs, which produce oöcytes after puberty. In addition, parts of them function as endocrine glands and are responsible for the production of two main hormones. One of these is called the estrogenic, or follicular, hormone, and it is secreted by the ovarian follicle. It controls the development of the secondary sexual characteristics, such as the enlargement of the breasts, the deposition of fat over the hips and buttocks, and the growth of pubic and axillary hair. It also initiates the growth of the lining of the uterus during the menstrual cycle. The other endocrine secretion is called progesterone, or corpus luteum hormone, and it is secreted by the corpus luteum. It is indispensable for the implantation of the fertilized oöcyte and for the early development of the embryo. The secretion of both ovarian hormones is controlled by the gonadotropic hormone from the pars distalis of the hypophysis. The ovaries are homologous with the testes of the male.

A third hormone, or hormone-like substance, called relaxin, is secreted by the

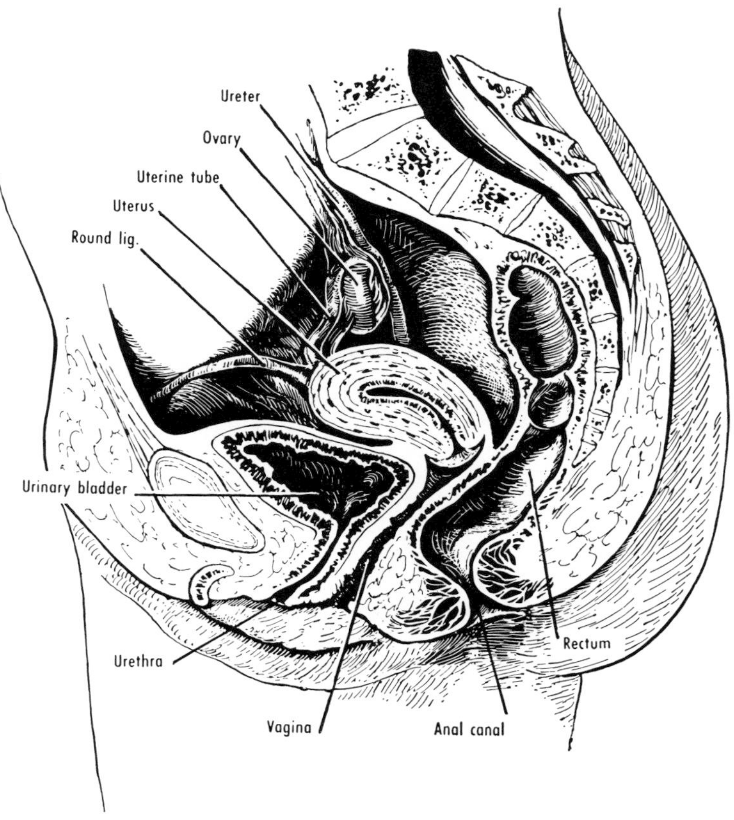

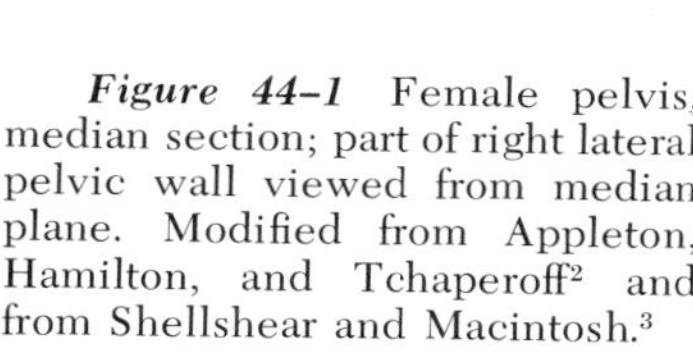

Figure 44–1 Female pelvis, median section; part of right lateral pelvic wall viewed from median plane. Modified from Appleton, Hamilton, and Tchaperoff[2] and from Shellshear and Macintosh.[3]

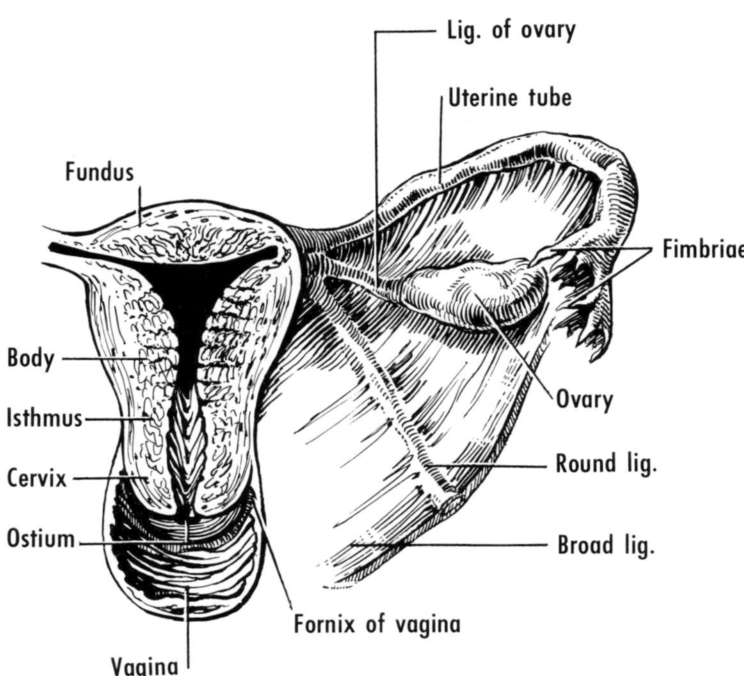

Figure 44–2 Schematic representation of female genital organs, posterior view.

ovary during pregnancy. It is said to inhibit premature contraction of the uterus during pregnancy, and, in certain mammals, is responsible for the relaxation of the sacroiliac joint and the pubic symphysis.

In a woman who has not borne children (nullipara), the ovary is situated on the lateral wall of the pelvis, at the level of the anterior superior spine, and just medial to the lateral plane, where it can be palpated bimanually. Its position may be altered by other pelvic organs, especially the uterus, to which the ovary is attached by ligaments. When the uterus ascends into the abdomen during pregnancy, the ovary is pulled away from its original position, which is usually regained after parturition.

Before the first ovulation, the ovary is smooth and pink, but later it becomes gray and puckered, because of the scars that follow the discharge of oöcytes from their follicles. It resembles a large almond in shape. Its size varies with age and also with the stage of the ovarian cycle. It is somewhat larger before than after pregnancy. After pregnancy, it is about 2½ to 4 cm long, and its average weight is 7 gm.[4] In old age it becomes further reduced in size.

When the ovary is in its usual position, its long axis is nearly vertical. It has medial and lateral surfaces, tubal and uterine ends, and mesovarian and free borders. It lies in a depression, the *ovarian fossa,* which is bounded in front by the obliterated umbilical artery and behind by the ureter and internal iliac artery.

The *lateral surface* is in contact with the parietal peritoneum lining the ovarian fossa and is separated by this peritoneum from the extraperitoneal tissue that covers the obturator vessels and nerve. Most of the *medial surface* is covered by the uterine tube; elsewhere this surface is related to the coils of the ileum.

The *mesovarian* or *anterior border* is attached to the mesovarium and faces the obliterated umbilical artery. The *hilus of the ovary,* through which blood vessels, lymphatic vessels, and nerves pass, is located on this border. The *free* or *posterior border* is related to the uterine tube and, behind this, to the ureter.

The *tubal* or *upper end* is closely connected to the uterine tube; the suspensory ligament of the ovary is attached to this end. The *uterine* or *lower end* gives attachment to the ovarian ligament.

Attachments. The *mesovarium* is a short, double-layered mesentery that extends backward from the posterior layer of the broad ligament to the mesovarian border of the ovary. Its two layers are attached one on each side of this border. The *suspensory ligament of the ovary,* or *infundibulopelvic ligament,* extends upward over the external iliac vessels and becomes lost in the connective tissue covering the psoas major. It contains the ovarian vessels and the ovarian plexus of nerves. The *ovarian ligament* passes from the uterine end of the ovary to the body of the uterus, just below and behind the entrance of the uterine tube. It is a rounded cord and contains some smooth muscle fibers.

Structure. The structure of the ovary varies with age and with the stage of the ovarian cycle. It is covered by a layer of cuboidal cells (germinal epithelium), which joins the mesothelium of the mesovarium at the hilus. The part of the ovary beneath the germinal epithelium is customarily divided into a cortex and a medulla.

Blood Supply. The ovary is supplied by the ovarian artery and by the ovarian branch of the uterine artery. After descending to the brim of the pelvis, the ovarian artery passes in the suspensory ligament and then between the two layers of the broad ligament until it reaches the mesovarium, in which it passes to reach the hilus of the ovary. The ovarian branch of the uterine artery passes laterally in the broad ligament to the mesovarium, where it ends by anastomosing with the ovarian artery.

The veins from the ovary begin as a plexus that communicates with the uterine plexus. Two veins arise from this plexus and become a single vein by the time they reach the abdomen (p. 419).

Lymphatic Drainage. The lymphatic vessels from the ovary pass upward with the ovarian vessels and drain into the lumbar (or aortic) nodes.

Nerve Supply.[5] The ovary is supplied by the ovarian plexus. Most of the fibers in this plexus are vasomotor.

UTERINE TUBES

The uterine tubes (fig. 44–2), two in number, convey the oöcytes from the ovaries to the cavity of the uterus. They transmit spermatozoa in the opposite direction, and fertilization of an oöcyte occurs usually within the tube.

The Greek word *salpinx,* which means a trumpet or a tube, is also used in referring to the uterine tubes. Such words as salpingitis, salpingography, and mesosalpinx are derived from it.

Each uterine tube is about 10 cm long and is located in the upper margin, and between the two layers, of the broad ligament. It runs laterally from the uterus to the uterine end of the ovary. It then passes upward on the mesovarian border, arches over the tubal end, and terminates on the free border and medial surface. It is subdivided into four parts, which, in passing from the uterus to the ovary, are: a uterine part, an isthmus, an ampulla, and an infundibulum.

The *infundibulum* is somewhat funnel-shaped. The *abdominal* or *pelvic opening of the uterine tube* is located at the bottom of the funnel, and the ovum enters the tube through it. This opening permits a communication of the peritoneal cavity with the exterior of the body. (In the male, no such communication exists, and the peritoneal cavity is closed.) The abdominal opening of the tube is about 2 mm in diameter when the muscles around it are relaxed. The *fimbriae* are a number of thin, irregular processes that project from the margins of the infundibulum. One of these, the *ovarian fimbria,* is longer than the rest, and is usually attached to the tubal end of the ovary.

The *ampulla* is the longest and widest part of the tube. It is slightly tortuous, and its walls are relatively thin. The *isthmus* is narrower and has thicker walls than the ampulla. The *uterine part* lies in the wall of the uterus; it ends in the cavity of the uterus as the *uterine opening.*

In passing medially, the lumen of the tube decreases in size. Its diameter is about 1 mm at the uterine opening.

When an oöcyte is discharged from the ovary, it is caught by the fimbriae and passes through the abdominal opening of the tube. Spermatozoa reach the infundibulum within hours after entering the cervix, and fertilization usually occurs here. Whether or not an oöcyte is fertilized, its movement through the tube to the uterus requires three to four days, and is probably influenced by both the ciliary action of the epithelial cells and the peristaltic action of the muscular coat. A fertilized oöcyte occasionally becomes embedded in the tube (usually the ampulla). The uterine tube is the commonest site of ectopic pregnancy.

The movement of spermatozoa and of oöcytes through each tube is obviously dependent upon its patency, which can be determined radiographically after the injection of a radio-opaque material into it by way of the vagina and uterus (fig. 44–3). It can also be tested by blowing air through the same route. If the tubes are patent, the air escapes through them into the peritoneal cavity. When the patient stands, the air ascends to the lower aspect of the diaphragm (usually the right dome), where it can be demonstrated radiographically. The air in this location may serve as a painful stimulus to the diaphragm, and the patient may experience pain in the region of the shoulder (p. 332).

Structure. Each uterine tube has three layers: a mucosa, a muscular coat, and a serosa. The serosa is the peritoneum of the broad ligament.

Blood Supply. The tubal branches of the uterine arteries and small branches of the ovarian arteries

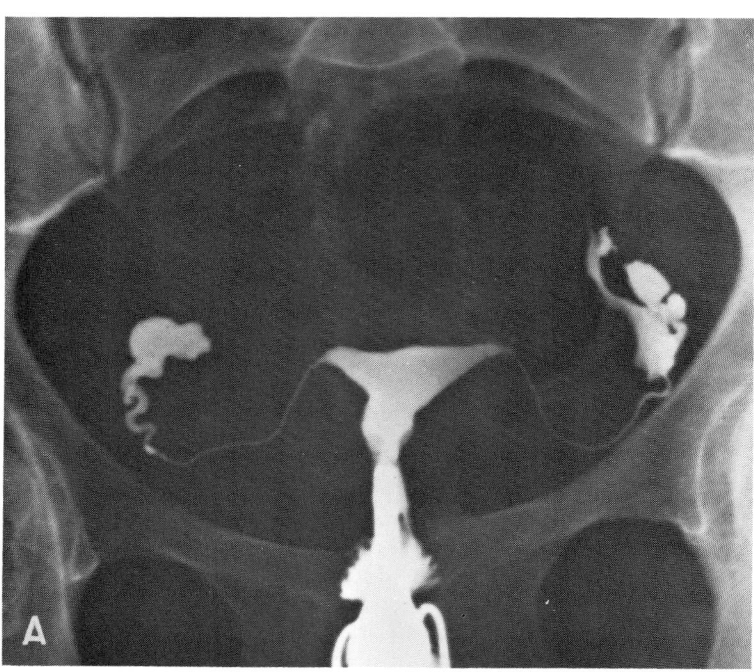

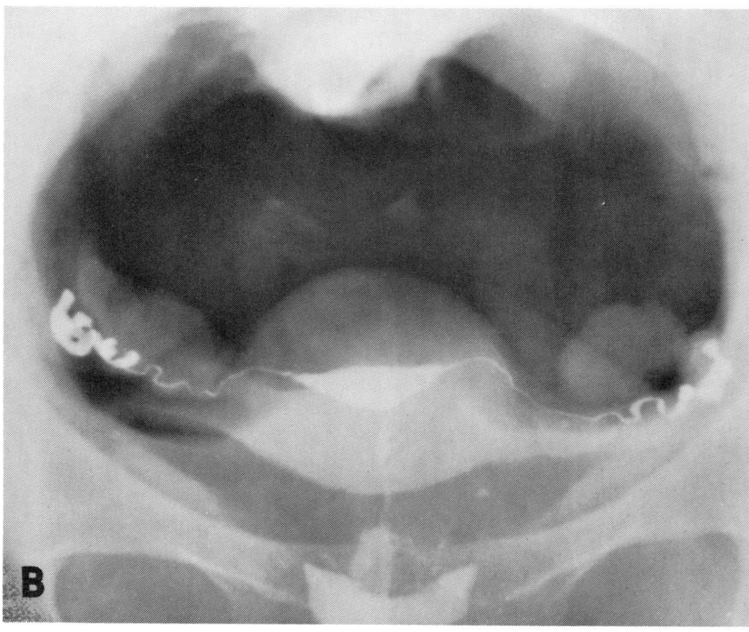

Figure 44–3 Uterus and uterine tubes. *A*, hysterosalpingogram. Note cavity of uterus, uterine tubes, and bilateral "spill" of radio-opaque medium into peritoneal cavity, thereby demonstrating patency of tubes. *B*, hysterosalpingogram, showing uterus and uterine tubes from in front and above. Note slitlike shape of uterine cavity in this view. Note also thickness of uterine wall. *A*, courtesy of Sir Thomas Lodge, The Royal Hospital, Sheffield, England. *B*, courtesy of Robert A. Arens, M.D., Chicago, Illinois.

supply the uterine tubes. The veins from the tubes have courses similar to those of the arteries.

Lymphatic Drainage. The lymphatic vessels from the uterine tubes follow the blood vessels and drain into the lumbar (or aortic) nodes.

Nerve Supply. The uterine tubes are supplied by the ovarian plexuses and by fibers from the inferior hypogastric plexuses. Some of the nerve fibers are sensory, others are autonomic for the supply of the muscular coat, and still others are vasomotor for the supply of blood vessels. The functional importance of these fibers is uncertain.

UTERUS

The uterus (figs. 44–1 and 44–2) is the organ in which the fertilized oöcyte normally becomes embedded and in which the developing organism grows and is nourished until its birth.[6] The cavity of the uterus and that of the vagina below it together form the "birth canal," through which the fetus passes at the end of its period of gestation. The uterine tubes open into the upper part of the uterine cavity.

The Greek word for uterus is *hystera.* Its combining form, hyster-, is used in words such as hysterogram and hysterectomy.

The uterus varies in shape, size, location, and structure. These variations are dependent upon age and upon other circumstances, such as pregnancy.

In the nulliparous woman, the walls of the uterus are thick and muscular. The entire organ is shaped like an inverted pear, and its narrow end, which is directed downward and backward, forms an angle of slightly more than 90 degrees with the vagina (angle of anteversion, fig. 44–4). The uterus lies within the pelvis, and its long axis is approximately in the axis of the upper pelvic aperture. It does not usually lie exactly in the median plane, but is inclined to one side or the other, usually to the right. Commonly it is also slightly twisted.[7] Its position is not fixed, however, and readily changes with the degree of fullness of the bladder, which is below and in front, and with the degree of fullness of the rectum, which is above and behind. The uterus is about 8 cm long, 4 cm wide in its upper part, and 2 cm thick. It is subdivided into a fundus, a body, an isthmus, and a cervix.

Parts and Relations. The *fundus* is the rounded part of the uterus that lies above and in front of the plane of the openings of the uterine tubes.

The *body* is the main part of the uterus, and it extends downward and backward to a constriction, the isthmus. It can be palpated bimanually (see p. 485). It has two surfaces, and two borders, or margins. The *vesical surface* is separated from the urinary bladder in front and below by the uterovesical pouch. The *intestinal surface* is separated from the sigmoid colon above and behind by the rectouterine pouch, which usually contains some coils of the ileum. The *left* and *right margins* are related to the respective broad ligaments, and to the structures contained between the two layers of each ligament.

The *isthmus* is the constricted part of the uterus and is about 1 cm or less in length. During pregnancy it becomes taken up by the body, and is therefore often referred to by obstetricians as the "lower uterine segment." The fetal membranes, however, do not usually become firmly attached to it. It resembles the body histologically but shows some differences in its musculature, epithelium, and number of glands. The changes that it undergoes dur-

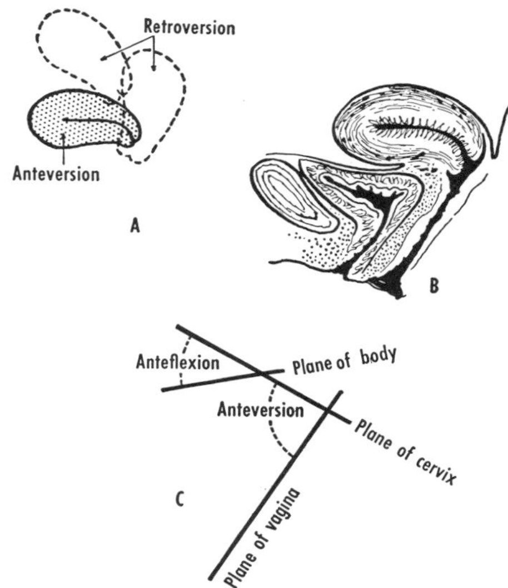

Figure 44–4 *A*, three outlines of the uterus, showing a normal position (anteversion) and a moderate and a more extreme degree of retroversion. The last is most often brought about by filling of the bladder. *B*, the normal adult uterus is anteverted, that is, at an angle with respect to the vagina, and anteflexed, that is, flexed somewhat on itself. *C*, the solid lines indicate the planes and the angles of anteversion and anteflexion. These are not fixed positions.

ing menstruation are not as marked as those in the body.

The *cervix* extends downward and backward from the isthmus to the opening within the vagina. It is the least freely movable part of the uterus and is divided into two parts by the anterior wall of the vagina, through which it passes. The *supravaginal part* is separated from the urinary bladder in front by loose connective tissue, and from the rectum behind by the rectouterine pouch. It is related laterally to the ureter and uterine artery. The *vaginal part* extends into the vagina. Its cavity communicates with that of the vagina by means of the *ostium of the uterus* (formerly called external os or orifice). This opening is a short depressed slit in the nullipara, but in women who have borne children it is larger and more irregular in outline. The ostium has *anterior* and *posterior lips*, which usually reach the posterior wall of the vagina.

The *cavity of the uterus* is wide above at the entrance of the uterine tubes, but it gradually decreases in width as it extends downward to the isthmus. It is very narrow in sagittal section, because the anterior and posterior walls are almost in contact.

The canal of the cervix is narrower at its ends than in its middle. A vertical fold is located on its anterior wall and another on its posterior wall. *Palmate folds* radiate obliquely from these in such a way that those on the anterior wall do not oppose those on the posterior wall. Instead, they fit each other so as to close the canal. They tend to disappear after pregnancy. The cavity of the uterus and the canal of the cervix can be viewed radiographically after the introduction through the vagina of a suitable radio-opaque material (hysterosalpingography).

Position of Uterus (fig. 44–4). In the adult, the entire uterus is usually anteverted. In this position, it extends forward and upward from the upper end of the vagina at an angle of about 90 degrees. The uterus is generally anteflexed also; that is, the body is bent downward at its junction with the isthmus. These positions are readily altered, especially during distention of the urinary bladder or the intestine. When the bladder is full, the uterus extends upward and backward (retroversion). In some women, the uterus is retroverted even when the bladder is empty, and the body may be bent backward on the isthmus (retroflexion).

Attachments and Peritoneal Relations[8] (fig. 44–5). The uterus gains much of its support by its direct attachment to the vagina. Indirect attachment to nearby structures, such as the rectum, urinary bladder, pelvic diaphragm, and bony pelvis, also helps to support it.

The peritoneum is reflected from the posterior aspect of the bladder to the isthmus of the uterus and then passes upward on the vesical surface of the body. This reflection forms the *uterovesical pouch.* After passing around the fundus of the uterus, the peritoneum passes downward on the intestinal surface of the body, and on the back of the cervix and the upper part of the vagina, from which it is reflected onto the front of the rectum. The

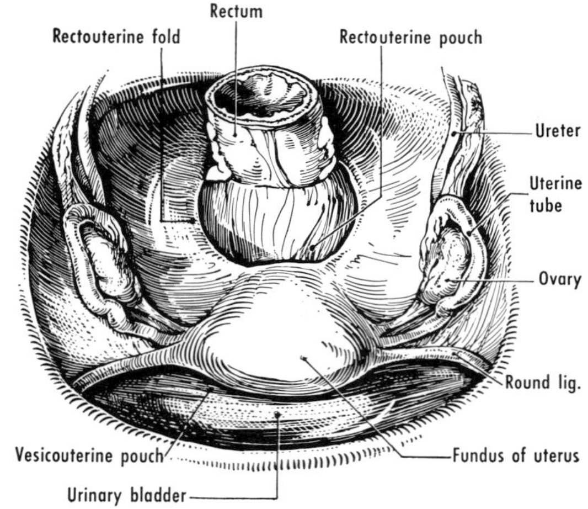

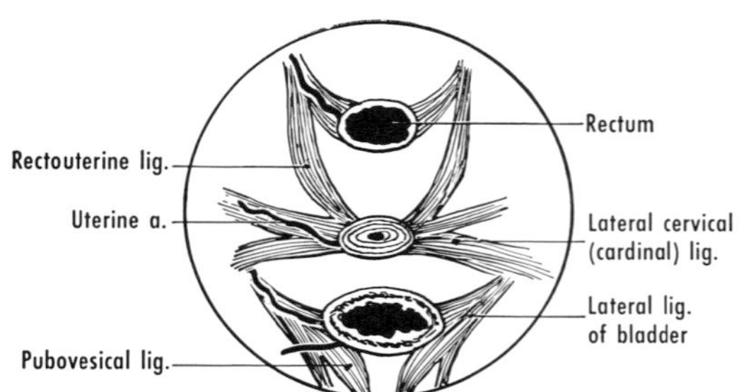

Figure 44–5 *Upper,* the pelvic viscera viewed from above, with peritoneum intact. *Lower,* a schematic representation of a horizontal section at the level of the cervix. The ligaments carry the blood vessels (indicated only on the left) to the organs.

recess formed by this reflection is the *rectouterine pouch.*

The *broad ligament* (fig. 44–6) is formed at the lateral margin of the uterus by the two layers of peritoneum that cover the vesical and intestinal surfaces. It extends to the lateral wall of the pelvis. The two layers are continuous with each other above, where they enclose the uterine tube. They are close to each other near the uterus, but they diverge laterally and below. The anterior layer passes forward to become continuous with the peritoneum covering the floor and lateral wall of the pelvis. The posterior layer extends backward from the cervix of the uterus as the *rectouterine fold.* This fold forms the lateral boundary of the rectouterine pouch, and, after passing along the side of the rectum, reaches the posterior wall of the pelvis. The plane of the broad ligament varies with the position of the uterus.

The *mesosalpinx* is the part of the broad ligament between the uterine tube and the line along which the broad ligament is drawn out to form the mesovarium. In addition to branches of the ovarian and uterine vessels, it contains two structures, called the epoöphoron and the paroöphoron.[9] The *mesometrium* is the part of the broad ligament below the mesosalpinx and mesovarium.

The *epoöphoron* consists of a duct, which runs parallel to and below the tube, and tubules, which run upward from the region of the ovary to join the duct at a right angle (fig. 44–7). It is the remains of a part of the mesonephric duct and some of its tubules. The *paroöphoron* lies medial to the epoöphoron and is a group of very small tubules. It usually cannot be recognized grossly in the adult. Both structures are important only in that cysts sometimes arise in them.[10]

The broad ligament encloses between its two layers some loose connective tissue and smooth muscle, collectively called the *parametrium.* Where the two layers are close together (near the uterus and near the uterine tube), the parametrium is not abundant, but laterally and below, where the layers diverge, it increases in amount. The broad ligament also encloses the uterine tube, the ovarian ligament, part of the round ligament, the uterine artery and venous plexus, the uterovaginal plexus of nerves, and a part of the ureter.

The *round ligament* is a narrow, flat band of fibrous tissue that is attached to the uterus just below and in front of the entrance of the uterine tube. It contains smooth muscle near this attachment. After passing laterally and forward across the umbilical artery and external iliac vessels, it hooks around the inferior epigastric artery. It then traverses the inguinal canal and becomes lost in the subcutaneous tissue of the labium majus. In the fetus, a tubular process of peritoneum, the *processus vaginalis peritonei,* accompanies the round ligament into the inguinal canal. This prolongation occasionally remains in the adult.

The visceral pelvic fascia at the side of the cervix and vagina is considerably thickened and contains numerous smooth muscle fibers. Part of this thickening passes laterally to merge with the upper fascia of the pelvic diaphragm and is called the *lateral* (or *transverse) cervical* or *cardinal ligament.* The uterine artery runs on its upper

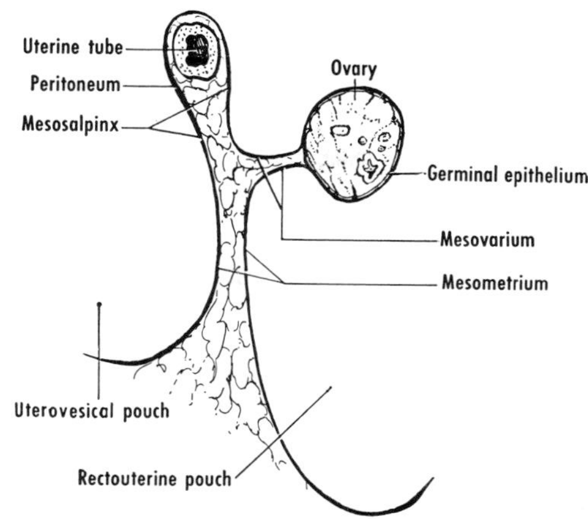

Figure 44–6 Diagram of a sagittal section, showing the broad ligament and its relations to the ovary and uterine tube. The anterior aspect is on the left side of the diagram.

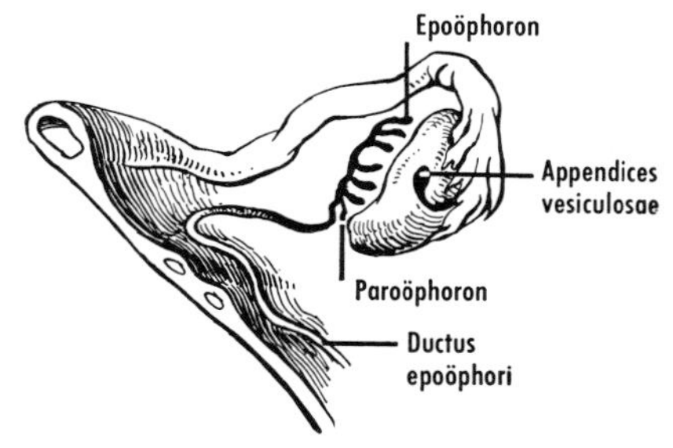

Figure 44–7 Diagram of a vertical section of the broad ligament, showing accessory elements in the ligament.

aspect. The rest of this thickening passes backward in the rectouterine fold and is attached to the front of the sacrum. This is the *uterosacral ligament,* and it can be palpated *per rectum.*

Changes with Age. At birth, the uterus reaches above the level of the pelvic inlet. The cervix is larger than the body, and the palmate folds extend into the upper part of the uterine cavity. The difference between the axis of the uterus and that of the vagina is relatively small (fig. 44–8). The growth of the uterus is slow until puberty, when it grows rapidly until its adult size and shape are reached. After the menopause, the uterus becomes smaller, more fibrous, and paler in color.

Changes during Pregnancy and after Parturition. The size of the uterus increases tremendously during pregnancy (fig. 44–9). The fundus rises above the level of the pubic symphysis in the third month. It reaches the supracristal plane in the sixth month and the level of the xiphisternal joint in the eighth month. It descends slightly in the ninth month, when the maximal circumference of the fetal head becomes engaged below the pelvic inlet. During this increase in size of the uterus, there is also a large increase in its weight, and the walls of the uterus become thinner.

After parturition, the uterus undergoes a process called involution. It gradually becomes reduced in size and weight until, after six to eight weeks, it reaches its resting state, in which it is about 1 cm larger in all dimensions than it was before pregnancy. It is also slightly heavier, its cavity is somewhat larger, and the lips of the opening into the vagina are irregular in outline.

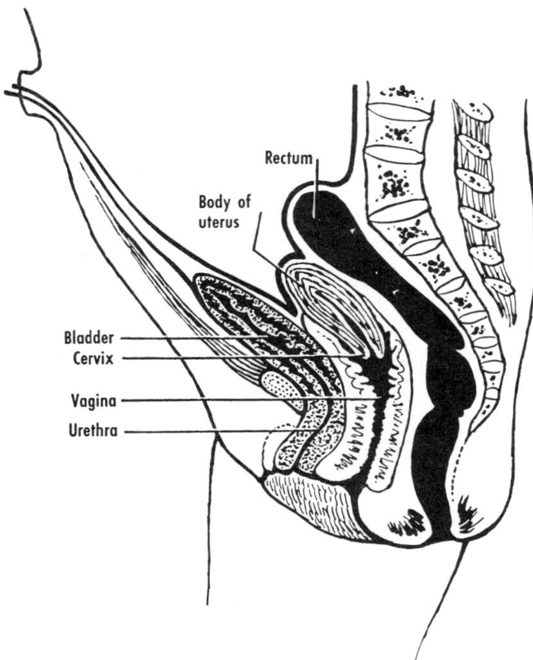

Figure 44–8 Median section of female pelvis, newborn. Drawn from a section. Note difference in shape and position of urinary bladder as compared with the adult (fig. 42–1). In an infant, a full bladder extends well into the abdomen. Note also shape and position of uterus. (Cf. fig. 44–1.)

Structure. The uterus has three layers: a mucosa, a muscular coat, and a serosa.

The *mucosa,* or *endometrium,* differs in structure with the stage of the menstrual or uterine cycle. It also undergoes changes during pregnancy. It contains numerous glands, which traverse the entire thickness of the lamina propria or endometrial stroma.

When an oöcyte is not fertilized, ovulation is usually followed by menstruation, which occurs every three to five weeks and which coincides fairly well with the degeneration of the corpus luteum and with a decrease in the level of estrogen and progesterone.

After a fertilized oöcyte is embedded in the wall of the uterus, the endometrium is usually called the decidua, which is divided into three parts. The part of the endometrium between the fertilized ovum and the cavity of the uterus is the *decidua capsularis* (or *reflexa*), the part between the ovum and the deeper portion of the wall of the uterus is the *decidua basalis* (or *serotina*), and the remainder of the endometrium is the *decidua parietalis* (or *vera*). The placenta forms at the site of the decidua basalis.

The *placenta* occupies approximately one-half of the uterine wall at about the fourth month of pregnancy, and from one-fourth to one-third at term, when it weighs about 500 gm.

The placenta is variable in shape, size, and degree of vascularity. It also varies in its site of formation. It is formed in the following sites in a decreasing order of frequency: posterior wall, anterior wall, sides, lower uterine segment, and fundus.

The *muscular coat,* or *myometrium,*[11] forms the greater part of the wall of the uterus. During parturition, the normal functioning of the uterine musculature is as important as the shape of the bony pelvis and the size of the fetal head. The muscle coat is continuous with that of the uterine tubes above and with that of the vagina below. Bundles of smooth muscle from its superficial part are prolonged into the various ligaments that are attached to the uterus. Its deeper part contains numerous blood vessels and nerves. There is relatively less muscle but more fibrous tissue in the isthmus and cervix than there is in the body and fundus.[12]

The *serosa,* or *perimetrium,* is formed by the peritoneum. It is firmly attached to the fundus and body except at the lateral margins; it is loosely attached to the back of the cervix.

Blood Supply. The uterine arteries (fig. 44–10) provide the main blood supply to the uterus. Each artery passes medially on the upper aspect of the lateral cervical ligament. It supplies the cervix and the upper part of the vagina and it then turns upward to run between the layers of the broad ligament, near the lateral margins of the body, and sends branches to both surfaces of the body. The uterine arteries become greatly enlarged during pregnancy and are tortuous after parturition.

The blood is returned from the uterus by way of a venous plexus that follows the uterine artery. An important anastomosis between the portal and systemic venous systems is formed by veins that run below the rectouterine pouch and connect the uterine venous plexus with the superior rectal vein.[13]

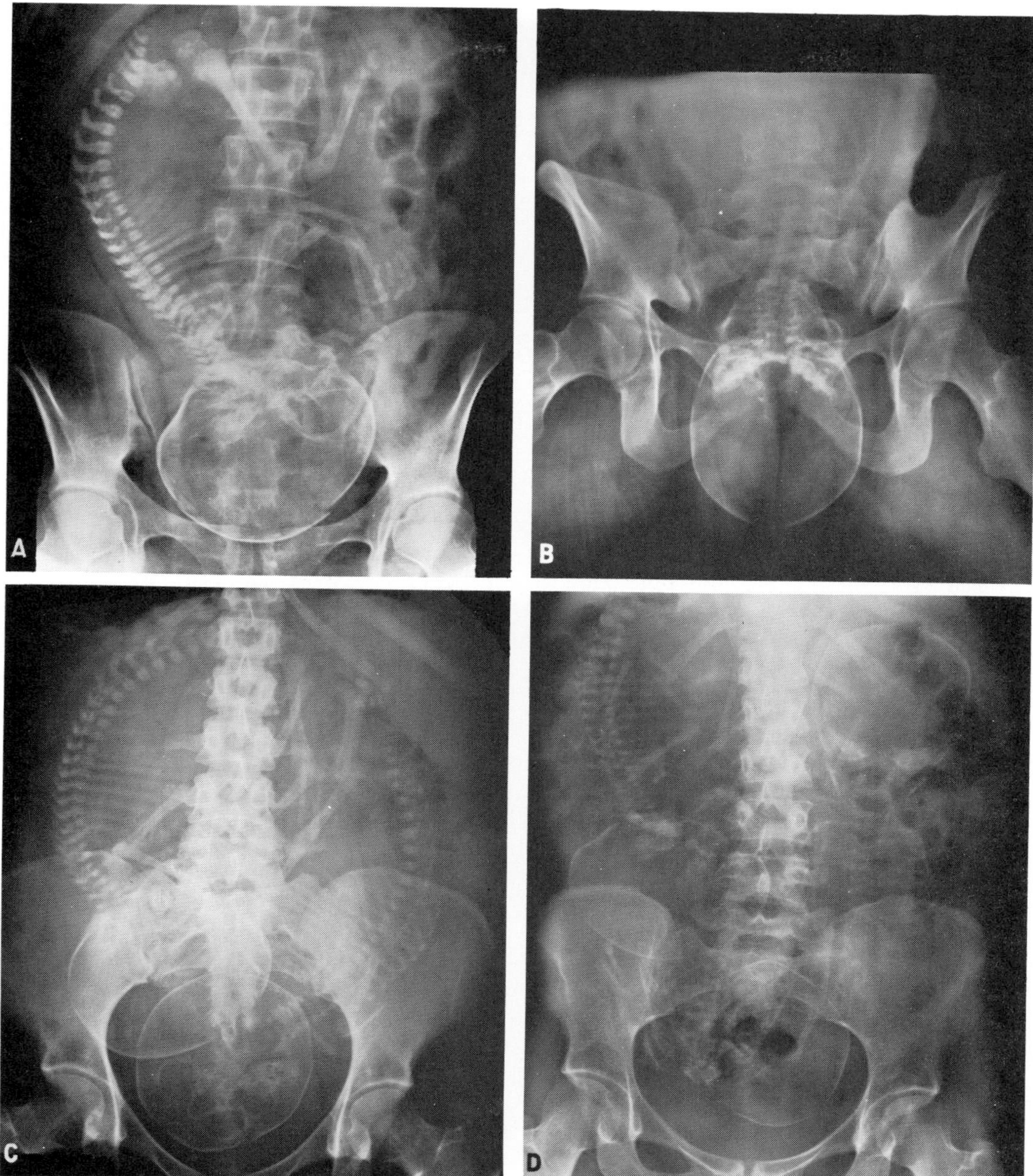

Figure 44–9 Pregnancy. *A*, fetus at term *in utero*. Cephalic presentation. Note fetal vertebrae and ribs, bones of limbs, and skull. Parietal bones have overlapped frontal bone at coronal suture. *B*, radiogram of child during birth. Note that the median plane of the infant's head coincides with the median plane of the mother's body at this stage of parturition. *C*, twins *in utero*. Both in cephalic presentation. Note orbits and nasal cavity of lower head. The right side of the mother's body is at the right in this print. *D*, triplets *in utero*. Two are in cephalic, the third in breech, presentation. *A*, courtesy of Robert A. Arens, M.D., Chicago, Illinois. *B*, courtesy of Robert P. Ball, M.D., Oak Ridge, Tennessee. *C*, courtesy of Herbert Pollack, M.D., Chicago, Illinois. *D*, from *Medical Radiography and Photography;* courtesy of Keith P. Bonner, M.D., Toronto, Canada.

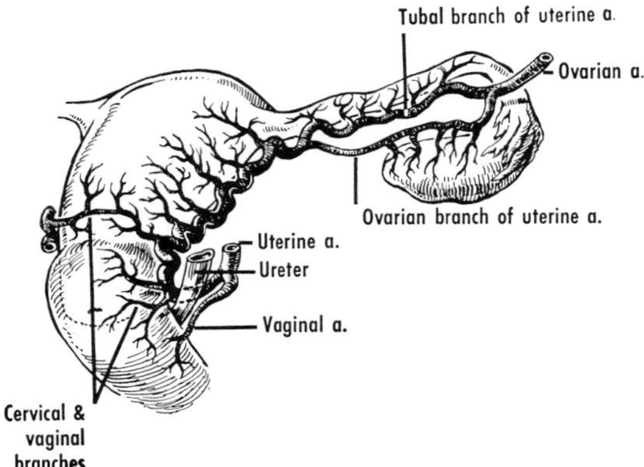

Figure 44–10 Schematic representation of the ovarian, uterine, and vaginal arteries, viewed from behind and slightly obliquely. Cervical branches of the uterine arteries anastomose across the median plane. The anastomoses between the uterine and ovarian arteries are very extensive. The vaginal artery may arise separately from the internal iliac, and the vagina also receives branches from the inferior vesical artery.

Lymphatic Drainage. The lymphatic vessels from the fundus and upper part of the body drain into the lumbar (or aortic) nodes, those from the lower part of the body into the external iliac nodes, and those from the cervix into the external iliac, internal iliac, and sacral nodes. Some vessels from the region of the uterus near the entrance of the uterine tube pass with the round ligament and drain into the superficial inguinal nodes.

Nerve Supply. The uterus receives autonomic and sensory fibers by way of the uterovaginal plexuses, which run along the uterine arteries. In spite of the extent of these plexuses, it is doubtful that the nerve supply is important for normal uterine function.

The uterus is insensitive to most stimuli, but pain may be felt when the cervix is grasped with a forceps or is dilated.[14] Some uterine disorders are painful, and pelvic pain may be felt in some phases of the menstrual cycle. There is some evidence that the fibers concerned ascend and enter the spinal cord by way of the lumbar splanchnic nerves. Resection of the superior hypogastric plexus has been performed to alleviate severe pain of this kind.

VAGINA

The vagina[15] is the female organ of copulation. It is also the lower end of the "birth canal," and it serves as the excretory duct for the products of menstruation. The cavity of the vagina communicates with that of the uterus above, and it opens into the vestibule of the vagina below. The vagina extends downward and forward in a plane parallel to that of the pelvic inlet. This plane is approximately 60 degrees from the horizontal. In the adult, when the urinary bladder is empty, the axis of the vagina forms an angle of a little more than 90 degrees with the axis of the uterus. However, this angle increases as the urinary bladder fills and pushes the fundus of the uterus upward and backward.

The vagina is highly dilatable, especially in the part above the pelvic diaphragm. When the cavity is empty, it is **H**-shaped in transverse section in most of its extent. Its anterior and posterior walls are in contact with each other below the entrance of the cervix. The *anterior wall,* which is pierced by the cervix, is about 7½ cm long; the *posterior wall* is about 9 cm long. These walls are especially distensible. The *lateral walls* are attached above to the lateral cervical ligament and, below this, to the pelvic diaphragm. They are, therefore, more rigid.

The recess between the vaginal part of the cervix and the walls of the vagina is termed the *fornix of the vagina.* Although it is continuous around the cervix, it is often subdivided into anterior, posterior, and lateral fornices. The posterior fornix is the deepest, and its wall is related to the peritoneum of the rectouterine pouch.

In most virgins, the opening of the vagina into the vestibule is partially closed by a fold called the *hymen.* This fold is variable in size and shape, but is often annular or crescentic. It usually has one opening, but it may be cribriform. When an opening is lacking, the fold is called an imperforate hymen. After the hymen has been torn or ruptured, small rounded fragments remain at the site of its attached margin. These are termed the *carunculae hymenales.*

Relations. Anteriorly, the upper part of the vagina is related to the cervix. Just below this, it is separated from the urinary bladder and ureters by loose connective tissue. Because the uterus is usually twisted and the upper part of the vagina is correspondingly deviated, much more of one ureter than the other is situated in front of the vagina. The urethra is fused with the lower two-thirds of the anterior wall of the vagina.[16]

Posteriorly, the upper part of the vagina is related to the rectouterine pouch, and, below this, it is separated from the rectum by relatively avascular connective tissue. The lower part of the vagina is fused with the tendinous center of the perineum.

Laterally, the upper part of the vagina is attached to the parametrium forming the lateral cervical ligament and the two layers of the broad ligament on either side of this. The ureter and uterine artery are also related to this part of the vagina. The pubococcygeal portions of the levatores ani embrace the vagina about 3 cm above its opening and act as a sphincter. Below the pelvic diaphragm, the vagina is related laterally to the greater vestibular gland, the bulb of the vestibule, and the bulbospongiosus muscle.

Structure. The vagina has three layers: a mucosa, a muscular coat, and a fibrous coat.

The *mucosa* is lined by stratified squamous epithelium, the appearance of which is subject to hormonal influences, and which differs in different stages of the ovarian cycle. Vaginal and cervical smears, containing desquamated cells from the mucosa of the uterus, are useful in the diagnosis of early carcinoma of the uterus.

The mucosa is thick and is marked by a number of transverse ridges, which are more prominent in the lower part of the vagina. These ridges, called *vaginal rugae*, tend to disappear in older women and in those who have borne children. A longitudinal ridge, termed the *anterior column of the rugae*, marks the anterior wall, and a similar ridge, termed the *posterior column of the rugae*, marks the posterior wall. Additional prominence is given to the lower part of the anterior column by the *urethral carina* or *ridge*, which is formed by the urethra.

The *muscular coat* consists of smooth muscle. Most of the fibers run longitudinally, and some of the bundles are continuous with the more superficial bundles of the uterus. The muscular coat is joined by skeletal muscle fibers (pubovaginalis) from the pubococcygeal part of the levator ani at the level of the pelvic diaphragm.

The *fibrous coat* is continuous with the part of the visceral pelvic fascia that surrounds the vagina. It contains a large venous plexus. A serous coat covers the upper part of the posterior wall of the vagina.

Blood Supply. The upper part of the vagina is supplied by branches of the uterine artery (fig. 44–10). The vaginal artery, sometimes arising as two or three branches from the internal iliac, gives branches that are distributed to the front and back of the vagina. These may anastomose in the median plane to form two longitudinal trunks, called the anterior and posterior azygos arteries of the vagina. A few branches from the artery of the bulb of the vestibule reach the lower part of the vagina.

The blood from the vagina drains into the vaginal venous plexus, which communicates with the uterine and vesical plexuses.

Lymphatic Drainage. The lymphatic vessels from the upper part of the vagina pass along the uterine artery and drain into the external and internal iliac nodes. Those from the middle part pass with the vaginal artery and drain into the internal iliac nodes, whereas those from the lower part pass to the sacral and common iliac nodes. Lymphatic vessels from the part of the vagina adjacent to the hymen pass to the superficial inguinal nodes.

Nerve Supply. Except for its lowermost part, which is supplied by the pudendal nerve, the vagina is supplied by the uterovaginal plexus. This plexus contains autonomic fibers for the supply of smooth muscle as well as vasomotor fibers, but both types are of doubtful significance. There is little sensation in the vagina, except in its lowermost part.

Examination of Pelvic Organs. Digital examinations *per vaginam* are made by placing one or two fingers in the vagina. In bimanual examinations, pelvic structures are palpated between these fingers in the vagina and the other hand placed on the anterior abdominal wall.

The following structures are palpable:

In front, the urethra and the vaginal part of the cervix, the urinary bladder when distended, and the body of the uterus (bimanually).

Behind, the rectum and any masses present in the rectouterine pouch, which is readily accessible. When the sacral promontory is felt, the diagonal conjugate diameter can be measured.

Laterally, the ureters, displaced or enlarged broad ligaments and lymphatic nodes, and displaced or enlarged ovaries and uterine tubes (bimanually).

A speculum introduced into the vagina permits visualization of the vagina and cervix, performance of certain minor operations on the cervix, and, under anesthesia, removal of the uterus.

REFERENCES

1. S. Zuckerman (ed.), *The Ovary*, Academic Press, New York, 1962.
2. A. B. Appleton, W. J. Hamilton, and I. C. C. Tchaperoff, *Surface and Radiological Anatomy*, Heffer, Cambridge, 4th ed. by W. J. Hamilton and G. Simon, 1958.
3. J. L. Shellshear and N. W. G. Macintosh, *Surveys of Anatomical Fields*, Grahame, Sydney, 1949.
4. F. W. Sunderman and F. Boerner, *Normal Values in Clinical Medicine*, Saunders, Philadelphia, 1949.
5. G. A. G. Mitchell, J. Anat., Lond., *72*:508, 1938.
6. S. R. M. Reynolds, *Physiology of the Uterus*, Hoeber, New York, 2nd ed., 1949. O. V. St. Whitelock (ed.), Ann. N. Y. Acad. Sci., *75*:385, 1959. H. J. Norris, A. T. Hertig, and M. R. Abell (eds.), *The Uterus*, Internat. Acad. Pathol. Monograph No. 14, Williams and Wilkins, Baltimore, 1973.
7. J. C. Brash, Brit. med. J., *2*:790, 1922.
8. V. Bonney, J. Obstet. Gynaec., Brit. Emp., *41*:669, 1934. A. H. Curtiss, B. J. Anson, and L. E. Beaton, Surg. Gynec. Obstet., *70*:643, 1940.
9. G. M. Duthie, J. Anat., Lond., *59*:410, 1925.
10. G. H. Gardner, R. R. Greene, and B. M. Peckham, Amer. J. Obstet. Gynec., *55*:917, 1948.
11. K. H. Renn, Z. Anat. EntwGesch., *132*:75, 1970.
12. H. Schwalm and V. Dubrausky, Amer. J. Obstet. Gynec., *94*:391, 1966.
13. E. G. Wermuth, J. Anat., Lond., *74*:116, 1939.
14. G. W. Theobald, chap. 8 in K. Bowes (ed.), *Modern Trends in Obstetrics and Gynaecology*, Hoeber, New York, 1956.
15. W. Shaw, Brit. med. J., *1*:477, 1947. An account of the surgical anatomy of the vagina.
16. B. H. Goff, Surg. Gynec. Obstet., *87*:725, 1948. J. V. Ricci, J. R. Lisa, C. H. Thom, and W. L. Kron, Amer. J. Surg., *74*:387, 1947. J. V. Ricci, C. H. Thom, and W. L. Kron, Amer. J.Surg., *76*:354, 1948.

RECTUM AND ANAL CANAL

45

RECTUM

The rectum (fig. 45–1) is the part of the large intestine between the sigmoid colon and the anal canal. The rectosigmoid junction is arbitrarily located at the level of the middle of the sacrum, and it may be marked by a constriction. The rectum is not otherwise well demarcated from the sigmoid colon, and changes in structure in passing from the one to the other are quite gradual. The lower limit of the rectum is at the upper aspect of the pelvic diaphragm. The anorectal junction is marked by the puborectalis muscle, which forms a sling behind it.

The rectum is about 15 cm in length. It is narrowest at its junction with the sigmoid colon. Its widest part, the *ampulla of the rectum,* is located just above the pelvic diaphragm and is capable of considerable distention. When the rectum is empty, its anterior and posterior walls are in contact.

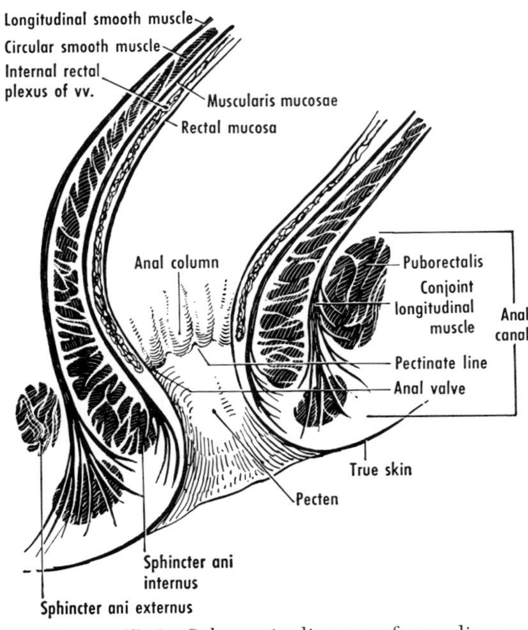

Figure 45–1 Schematic diagram of a median section of the rectum and anal canal. Based on Morgan and Thompson[13] and on Walls.[9] The anterior aspect is on the left side of the diagram.

The shape of the rectum depends upon whether it is full or empty. It also varies with the individual. The rectum is much straighter and relatively larger in the child than in the adult.

The rectum lies in the dorsal part of the pelvic cavity and follows the curvature of the sacrum and coccyx. The resulting anteroposterior curvature is called the *sacral flexure.* Another anteroposterior curvature is located at the junction of the rectum and the anal canal. This curvature, the *perineal flexure,* has an angle of 80 to 90 degrees. The puborectalis sling fits into its concavity (fig. 45–2). Here the puborectalis can be palpated *per anum.*

Usually three, sometimes more, lateral curvatures are present, and they are particularly evident during distention of the rectum. Their concavities coincide with semilunar *transverse rectal folds,* which project into the interior. In addition to mucosa and submucosa, these folds also contain a part of the inner circular layer of smooth muscle. They vary in location and in prominence and can be seen *per anum* with the aid of a speculum. They may impede the progress of instruments introduced into the rectum. Their function has been the subject of much speculation.

In external appearance, the rectum can be distinguished by the absence of a mesentery and of haustra.[1] Furthermore, the taeniae coli spread out on the rectum to form a more complete outer longitudinal coat of smooth muscle, which is thicker on the front and back than it is on the sides.

Peritoneal Relations. Peritoneum covers the front and sides of the upper part of the rectum, only the front of the middle part, and none of the lower part. In areas of the rectum covered by peritoneum, and especially on the sides, loose tissue and fat separate it from the muscular coat, and considerable expansion is therefore permitted.

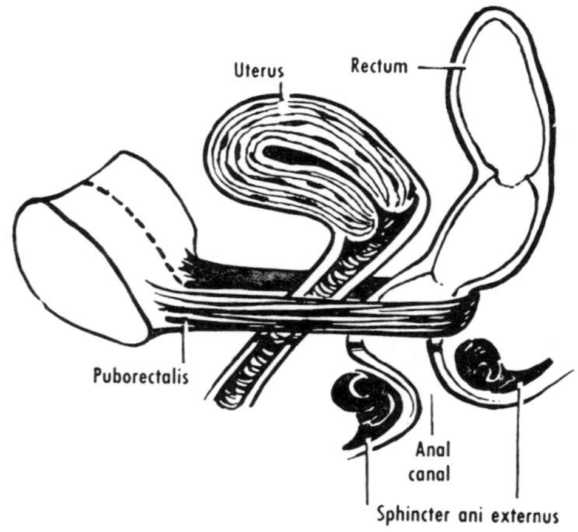

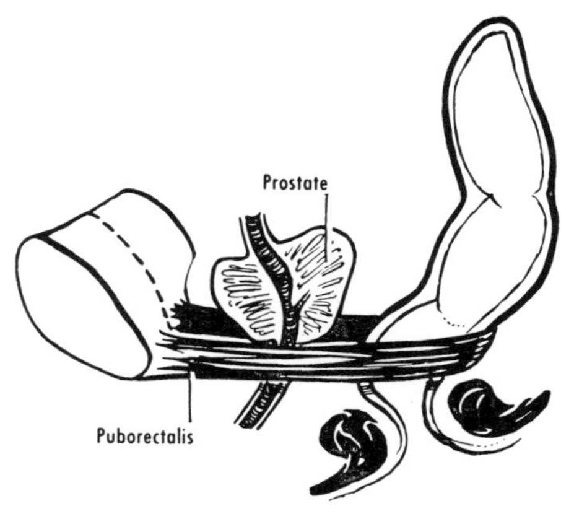

Figure 45–2 Schematic representations of puborectalis muscle.

The level at which the peritoneum leaves the front of the rectum varies with age, with sex, and with the individual. The level of reflection in the male is usually about 7 to 8 cm above the anus; in the female it is about 2 cm lower. The peritoneum passes from the front of the rectum to the urinary bladder in the male, and the floor of this reflection is called the *rectovesical pouch.* Below the floor of the pouch, the rectovesical septum, which is a membranous extension upward from the parietal pelvic fascia, separates the rectum from the prostate and urinary bladder (p. 494). In the female, the peritoneum passes from the front of the rectum to the back of the vagina, and the floor of this reflection is called the *rectouterine pouch.* A *rectovaginal septum* (p. 494), similar to the rectovesical septum, is said to be present below the floor of the rectouterine pouch.[2] The *pararectal fossa* is formed by the reflection of peritoneum from the side of the rectum to the posterior wall of the pelvis. It contains some of the ileum or sigmoid colon when the rectum is empty, but it is obliterated when the rectum is full.

Fascia of Rectum. The superior fascia of the pelvic diaphragm is reflected upward as visceral fascia to surround the rectum. The *rectosacral fascia* is an avascular band that passes backward and attaches to the pelvic surface of the sacrum. Two other condensations of connective tissue, in which are embedded the middle rectal arteries and plexuses, connect the rectum to the parietal pelvic fascia covering the sacrum at the levels of the second, third, and fourth pelvic sacral foramina. These condensations are termed the *lateral ligaments of the rectum,* or "rectal stalks." They divide the potential (pelvirectal) space surrounding the rectum and above the pelvic diaphragm into anterior and posterior divisions.

Relations. Posteriorly, from above downward, the rectum is related to the sacrum, coccyx, and pelvic diaphragm. When distended, it is also related to the sacral plexus and piriformis, which are covered by parietal pelvic fascia containing branches of the superior rectal vessels. In addition, the median sacral artery and vein, the sympathetic trunks, the lateral sacral vessels, and the sacral lymphatic nodes are behind the rectum.

Laterally, the ileum or sigmoid colon is related to the upper part of the rectum; the inferior hypogastric plexus of nerves and the pelvic diaphragm are related to the lower part.

Anteriorly, the relations differ in the two sexes. **In the male, the rectovesical pouch containing some coils of the small intestine separates the upper part of the rectum from the urinary bladder. The lower part is related to the posterior aspect of the bladder, to the posterior surface of the prostate, and, on either side of the prostate, to the seminal vesicle and the ductus deferens. In the female, the rectouterine pouch, which contains some of the small intestine, separates the upper part of the rectum from the uterus and the upper part of the**

vagina. The lower part of the rectum, below the reflection of peritoneum, is related to the back of the vagina.

Structure (fig. 45–1). The rectum has four layers, which are similar to those of the colon. The mucosa is somewhat redder in color, and, when the rectum is empty, it has many folds. The taeniae coli spread out on the rectum to form a more complete outer longitudinal muscular layer, which is thicker in front and behind than on the sides. Just above the perineal flexure, slips of smooth muscle pass backward to the coccyx as the *rectococcygeus,* and other slips pass forward to the urethra as the *rectourethralis.*[3]

ANAL CANAL

The anal canal (fig. 45–1) is defined anatomically as the part of the large intestine that extends from the level of the upper aspect of the pelvic diaphragm to the *anus.* **However, many surgeons prefer to regard only the part of the intestine below the pectinate line as the anal canal, because of differences in the nerve supply, the venous and lymphatic drainage, and, to some extent, the type of epithelium above and below this line.** The upper part of the canal is marked by the anorectal ring of muscle, which is formed mainly by the puborectalis sling. The anal canal is about 3 cm in length, and it passes downward and backward from the perineal flexure. Its cavity at this flexure is a small, anteroposterior slit.

Relations. As it passes through the pelvic diaphragm, the anal canal is surrounded by the levatores ani. Below this diaphragm, it is surrounded by the sphincter ani externus. The tendinous center of the peritoneum and the bulb of the penis are anterior to the anal canal in the male; the tendinous center of the perineum and the vagina are in front of it in the female. In both sexes, the anococcygeal ligament is posterior, and the ischiorectal fossa is lateral.

Sphincter Ani Externus.[4] The sphincter ani externus (figs. 45–1 and 47–5) surrounds the part of the anal canal that is located in the anal triangle, below the pelvic diaphragm. It is usually described in three parts, a subcutaneous, a superficial, and a deep, but the subdivision between the superficial and deep parts is often artificial.[5]

The *subcutaneous part* surrounds the lowermost portion of the anal canal, and its fibers decussate both in front of and behind the canal. The *superficial part* passes around the upper portion of the subcutaneous part. It is attached behind to the tip of the coccyx and to the *anococcygeal ligament,* a band that contains both muscle and connective tissue fibers and that passes from the coccyx to the anus. It is attached in front to the tendinous center of the perineum. The *deep part* surrounds the upper part of the anal canal. It is intimately associated with the puborectalis posteriorly; anteriorly some of its fibers pass into the tendinous center.

NERVE SUPPLY. The sphincter ani externus is supplied by the inferior rectal nerves and by a perineal branch of the fourth sacral nerve.

ACTION. This muscle is in a state of variable tonic contraction during waking hours, but its tone becomes minimal during sleep. Its tone increases when intra-abdominal pressure is increased, but decreases during straining for defecation. One can voluntarily cause the muscle to contract.[6]

Interior. The upper half of the anal canal is marked by a series of five to ten vertical folds of mucosa, the *anal* (formerly called *rectal*) *columns,*[7] which are well marked in children but may be poorly defined in adults. Each column contains a small artery and a small vein ("terminal" branches of the superior rectal vessels). An important plexus is formed by these veins, and its enlargement results in internal hemorrhoids. The lower ends of the anal columns are joined together by small crescentic folds of mucosa, the *anal valves.* A small recess, the *anal sinus,* lies external to each anal valve. The sinuous *pectinate line* marks the lower limit of the anal valves around the circumference of the canal. The lower half of the anal canal, from the pectinate line to the anus, is marked from above downward first by the *pecten,* a bluish white zone in the lining, and then by the *anal verge,* the lining of which merges with the skin of the anus. The term white line is employed for the intersphincteric interval, and is usually considered to be synonymous with pecten, but it has been used to designate the junction between the pecten and the anal verge. This junction is often very difficult to distinguish.

Ducts and glands may open into the anal sinuses and form fistulas. Infections in these ducts may rupture into an ischiorectal fossa and form an abscess there. Abscesses in an ischiorectal fossa may drain through these *fistulae in ano* into the anal canal.[8]

Structure. The anal canal consists of a mucosa, a submucosa in a part of its extent, and a muscular coat.

The *mucosa* has an epithelial lining, which differs according to level.[9] The mucosa of the pecten and of the anal verge is pink and moist, but lacks hairs and glands. The mucosa finally merges with the skin of the anus, which is pigmented and contains hair follicles and glands.

The *submucosa* of the upper half of the anal canal contains a plexus of veins (see below). The lamina muscularis mucosae becomes thickened at the level of the pecten. The close adherence of the epithelium to the lamina muscularis mucosae at the level of the lower margin of the internal sphincter appears to be responsible for a groove called the anal intermuscular septum.[10]

The *muscular coat*[11] of the anal canal consists of an inner circular and an outer longitudinal coat of smooth muscle. The fibers of the outer longitudinal coat become intermingled with skeletal muscle fibers of the puborectalis.

The inner circular layer, which is continued downward from the rectum, becomes thickened in the anal canal to form the *sphincter ani internus* (fig. 45–1). This sphincter extends from the level of the pelvic diaphragm to about the level of the lower part of the pecten. It determines the degree of permissible dilatation of the anal canal.[12]

The longitudinal coat of the rectum fuses with some of the fibers of the puborectalis. The conjoint muscle[13] thus formed passes downward in the anal canal between the internal and external sphincters, and it becomes increasingly fibroelastic as it descends. It divides into a number of septa, which separate the subcutaneous part of the external sphincter into circular bundles of muscle fibers. Some of these septa are inserted into the perianal skin and have been termed the *corrugator cutis ani;* it is doubtful that they wrinkle the skin.

Blood Supply.[14] The rectum and the anal canal are supplied by (1) the superior rectal artery, (2) the middle rectal arteries, (3) the inferior rectal arteries, and (4) the median sacral artery.

The *superior rectal artery* (figs. 35–10 and 35–14, pp. 387, 390) furnishes most of the blood supply to the rectum and anal canal. It is a continuation of the inferior mesenteric artery, and it divides into left and right branches, divisions of which pierce the muscular coat and then pass downward in the mucosa of the anal columns as far as the anal valves.

The *middle rectal arteries* help to supply the lower part of the rectum and the upper part of the anal canal.

Each of the *inferior rectal arteries* divides into several branches which, after traversing the ischiorectal fossa, supply the lower part of the anal canal as well as the surrounding muscles and skin.

The *median sacral artery* gives off small branches that supply the back of the rectum.

The anastomosis of the various arteries in the wall of the gut is so extensive that the middle and inferior rectal arteries can supply the entire rectum if the inferior mesenteric artery is ligated.[15]

The submucosal venous plexus drains in opposite directions from the level of the pectinate line. Veins above this line drain mainly into the superior rectal veins and thereby into the portal system (fig. 38–3, p. 418). Because the superior rectal veins lack valves and are subject to marked changes in pressure during straining and the like, they often become varicosed. Varicosities of these veins are termed internal hemorrhoids or piles.

Above the pelvic diaphragm, venous plexuses anastomose to form left and right middle rectal veins, which empty into the corresponding internal iliac veins. The communication between the superior and middle rectal veins forms an important anastomosis between the portal and systemic systems. In women, an additional anastomosis between these two systems is provided by a connection of the uterine plexus with the superior rectal veins.[16] The middle and inferior rectal veins have competent valves.

Below the pectinate line, the submucosal plexus drains into the small inferior rectal veins around the margin of the sphincter externus. The anastomosis of these veins with the middle and superior rectal veins (through plexuses in the wall of the gut) constitutes another important portal-systemic anastomosis. External hemorrhoids or piles, which may be very painful (see below), are varicosities of the inferior rectal veins.

Lymphatic Drainage.[17] The lymphatic vessels are arranged in three groups and, in general, follow the course of the blood vessels. Those from the upper part of the rectum pass to the inferior mesenteric nodes, whereas those from the lower part pass to the sacral, internal iliac, and common iliac nodes. The lowermost group drains in two directions. The lymphatic vessels from the part of the anal canal above the pectinate line drain into the internal iliac nodes; those below this line pass to the superficial inguinal nodes.

Nerve Supply (fig. 45–3). The nerves to the rectum and anal canal are derived from the superior and middle rectal plexuses and from the pudendal nerves by way of the inferior rectal nerves.

The plexuses that supply the rectum and the anal canal as far downward as the pectinate line contain the following: (1) Preganglionic parasympathetic fibers, which synapse with ganglion cells in the wall of the intestine. Postganglionic fibers from these cells supply the smooth muscle, including that of the sphincter ani internus (no parasympathetic fibers are present in the superior rectal plexuses, p. 425). (2) Postganglionic sympathetic fibers, some of which are vasomotor, and others of which supply smooth muscle but are of doubtful functional importance. (3) Sensory fibers, most of which are concerned with the reflex control of the sphincters, and others of which are concerned with pain. The sensory fibers are stimulated by distention of the wall of the rectum. They run centrally in the pelvic splanchnic nerves, although a few pain fibers may ascend with sympathetic fibers.

The inferior rectal nerves supply the lower half of the anal canal and contain the following: (1) motor fibers to the sphincter ani externus; (2) vasomotor

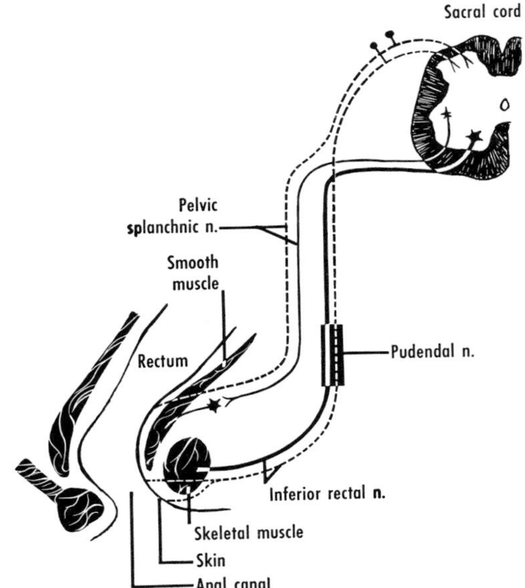

Figure 45-3 Schematic representation of nerve supply to rectum and anal canal. Sympathetic fibers are not shown. The fibers in the pelvic splanchnic nerves reach the intestine by way of plexuses that are described on page 456.

fibers; and (3) sensory fibers, most of which conduct pain, as well as others that may be involved in reflex sphincteric control.

The lower part of the anal canal is very sensitive,[18] but the part above the pectinate line is relatively insensitive. Consequently, internal hemorrhoids are usually painless. However, the overlap of the nerves is such that pain fibers extend for a short distance into the part of the canal above the pectinate line. Therefore, although the mucosa over the upper part of an internal hemorrhoid is insensitive, that over the lower part may be quite sensitive.

Fecal Continence and Mechanism of Defecation. Radiographic and proctoscopic studies of an adult indicate that the sigmoid colon normally contains fecal material and the rectum is empty.[19] Although an anatomical sphincter is not present at the rectosigmoid junction, a physiological sphincter is present here.

Fecal continence depends upon two factors: (1) Colonic control, or the mechanism by which the sigmoid colon retains fecal material until defecation. This mechanism involves a plastic adaption to an enlarging mass as a result of a relaxation or diminution of tone of the sigmoid colon. (2) Sphincteric control, or the reflex control of the sphincter ani externus and internus. The sphincter ani externus has a variable amount of tone (p. 488), but it fatigues rapidly during attempts to sustain its contraction. It is therefore desirable that the rectum be empty except during defecation. The sphincter ani externus is reflexly activated by sensory stimuli from the rectum (it relaxes during defecation, however). Sphincteric control is aided by the gluteus maximus (p. 206) and by the puborectalis (p. 492).

A variety of disorders may interfere with sphincteric control. Some of the more severe of these are: (1) Transection of the spinal cord above the sacral segments (the reflexes are maintained, and the gut empties automatically). (2) Destruction of the sacral part of the cord (a loss of all motor supply, and therefore a loss of sphincteric control ensues). (3) Loss of sensory fibers, from lesions of the dorsal root or resection of the rectum (no reflexes are possible and sphincteric control is therefore lost).

The particular stimuli responsible for the sensation of fullness or the desire to defecate are unknown. However, some evidence indicates that the sigmoid colon, which empties by a mass movement, is stimulated reflexly from the rectum. During increases in intra-abdominal pressure, resulting from the contraction of abdominal muscles, the puborectalis muscles and the sphincters relax, and the rectal musculature contracts. The relaxation of the puborectalis sling helps to decrease the angle of the perineal flexure. During evacuation, the colon and rectum move downward, and the rectum becomes narrow and elongated.[20] The puborectal sling and the sphincter ani externus act in closing the anal canal, and the sphincter ani internus completes the closure after the passage of each fecal mass.

Digital Examination. Valuable clinical information is often gained by introducing the index finger first into the anal canal and then into the rectum, and by palpating structures related to their walls. The finger encounters resistance offered first by the sphincters and then by the puborectalis. Although both sphincters can be palpated, under anesthesia the sphincter ani externus descends and is the one that is palpated.[21] During digital examination, the finger can usually reach the lowermost of the transverse rectal folds.

Anteriorly, in the male, the following structures can be felt: the membranous part of the urethra when catheterized, the prostate, the rectovesical fossa, the seminal vesicles when distended, the bladder when full, the bulbourethral glands when enlarged, and the ductus deferentes when displaced or enlarged. In the female, the following structures can be felt: the cervix and the ostium of the uterus, the vagina, the body of the uterus when retroverted, the rectouterine fossa, and, under certain pathological conditions, the ovary, the uterine tube, and the broad ligament.

Laterally, the ischial tuberosity, the ischial spine, and the sacrotuberous ligament may be palpated, as well as enlarged iliac lymphatic nodes and abnormal structures in the ischiorectal fossa.

Posteriorly, the pelvic surfaces of the sacrum and coccyx may be felt.

Proctoscopic and Sigmoidoscopic Examination. The interior of the rectum and anal canal may be examined with the aid of a proctoscope, an instrument which, when introduced *per anum,* can reach the lower part of the rectum. If the rectum is full of air, as it is in the knee-chest position, its red mucosa can be seen, and underlying veins in the submucosa are apparent. The anorectal ring, formed by the puborectalis sling of muscle, can be observed as the proctoscope is slowly withdrawn. Below this, internal hemorrhoids can be seen if present. Still farther below, the various markings on the inner wall of the anal canal can be observed (see Interior, p. 488). Complete closure of the canal by the sphincters occurs after withdrawal of the proctoscope.

With the aid of a sigmoidoscope, the lower part of the sigmoid colon and the rectosigmoid junction, as well as the rectum and the anal canal, may be studied.

REFERENCES

1. M. R. Ewing, Brit. J. Surg., *39*:495, 1952.
2. P. S. Milley and D. H. Nichols, Anat. Rec., *163*:443, 1969.
3. M. B. Wesson, J. Urol., *8*:339, 1922; Amer. J. Surg., *82*:714, 1951.
4. C. Oh and A. E. Kark, Brit. J. Surg., *59*:717, 1972.
5. H. Courtney, Surg. Gynec. Obstet., *89*:222, 1949.
6. W. F. Floyd and E. W. Walls, J. Physiol., *122*:599, 1953.
7. G. Ottaviani, Z. Anat. EntwGesch., *109*:303, 1939.
8. G. L. Kratzer and M. B. Dockerty, Surg. Gynec. Obstet., *84*:333, 1947. R. M. Burke, D. Zavela, and D. H. Kaump, Amer. J. Surg., *82*:659, 1951.
9. E. W. Walls, Brit. J. Surg., *45*:504, 1958.
10. E. S. R. Hughes, Aust. N.Z. J. Surg., *26*:48, 1956.
11. R. Fowler, Landmarks and Legends of the Anal Canal, chapter 4 in *Congenital Malformations of the Rectum, Anus, and Genito-Urinary Tracts,* F. D. Stephens (ed.), Livingston, Edinburgh, 1963. J. O. N. Lawson, Ann. R. Coll. Surg. Engl., *54*:288, 1974.
12. S. Eisenhammer, S. Afr. med. J., *27*:266, 1953.
13. F. R. Wilde, Brit. J. Surg., *36*:279, 1949. J. C. Goligher, A. R. Leacock, and J.-J. Brossy, Brit. J. Surg., *43*:51, 1955. A. G. Parks, Brit. J. Surg., *43*:337, 1956. C. N. Morgan and H. R. Thompson, Ann. R. Coll. Surg. Engl., *19*:88, 1956. G. L. Stonesifer, Jr., G. P. Murphy, and C. R. Lombardo, Amer. J. Surg., *100*:666, 1960.
14. O. Widmer, Z. Anat. EntwGesch., *118*:398, 1955. S. Sunderland, Aust. N.Z. J. Surg., *11*:253, 1942. A. Faller, Acta anat., *30*:275, 1957. N. A. Michels *et al.,* Dis. Colon Rectum, *8*:251, 1965.
15. J. C. Goligher, Brit. J. Surg., *37*:157, 1949. G. W. Ault, A. F. Castro, and R. S. Smith, Surg. Gynec. Obstet., *94*:223, 1952. J. D. Griffiths, Brit. med. J., *1*:323, 1961.
16. E. G. Wermuth, J. Anat., Lond., *74*:116, 1939.
17. J. B. Blair, E. A. Holyoke, and R. R. Best, Anat. Rec., *108*:635, 1950.
18. H. L. Guthie and F. W. Gairns, Brit. J. Surg., *47*:585, 1960.
19. E. A. Gaston, Surg. Gynec. Obstet., *87*:280, 1948; J. Amer. med. Ass., *146*:1486, 1951.
20. R. A. Rendich and L. A. Harrington, Amer. J. Roentgenol., *40*:173, 1938.
21. J. C. Goligher, A. G. Leacock, and J.-J. Brossy, Brit. J. Surg., *43*:51, 1955.

PELVIC DIAPHRAGM AND PELVIC FASCIA 46

PELVIC DIAPHRAGM

The pelvic diaphragm consists of the levatores ani and coccygei muscles and the fasciae that cover their upper and lower aspects. These fasciae are a part of the parietal pelvic fascia, which also includes the fascia covering the lateral and posterior walls of the pelvis. The visceral pelvic fascia is associated with the organs.

The levatores ani are by far the more important muscles of the pelvic diaphragm (figs.

46–1 and 46–2). The coccygei are relatively unimportant.

Levator Ani. The levator ani is variable in thickness and in strength. It lies almost horizontally in the floor of the pelvis.[1] The narrow gap between the medial edges of the left and right muscles transmits the vagina in the female and the urethra and rectum in both sexes. These organs, as well as those immediately above them, receive important support from these muscles.

The levator ani is usually divided into three parts (pubococcygeus, puborectalis, and iliococcygeus), designated according to the direction and attachment of their fibers, but this division is an oversimplification and tends to ignore important relationships with the urinary bladder, prostate, and vagina.

The main part of the levator ani arises from the back of the body of the pubis and runs backward in a sagittal plane toward the coccyx. This is the *pubococcygeus,* and it has several insertions. In the male, as it runs backward, some of its more medial fibers insert into the prostate *(levator prostatae).* In the female, some of the medial fibers insert into the urethra and vagina *(pubovaginalis),* and others, together with fibers from the contralateral muscle, encircle the urethra and vagina *(sphincter vaginae).* Behind the urethra in the male and the vagina in the female, some fibers are inserted into the tendinous center of the perineum, and a few (puboanal fibers) continue to the walls of the anal canal.

The most lateral fibers of the pubococcygeus arise from the tendinous arch of the levator ani when this arch is present. They pass behind the anal canal and reach the anococcygeal ligament.

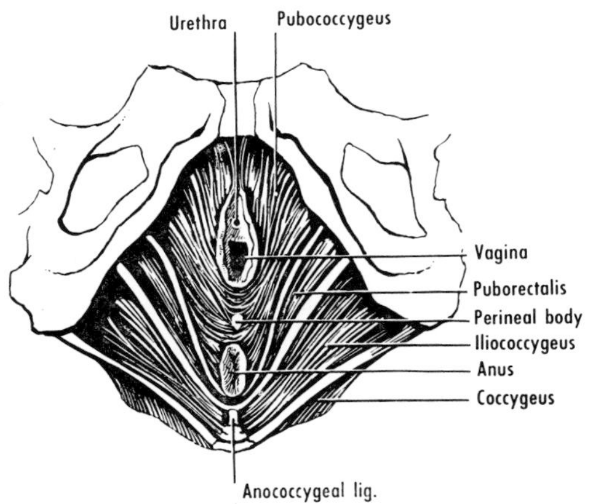

Figure 46–1 Muscles of pelvic diaphragm from below, female. Based on Milligan and Morgan.[2]

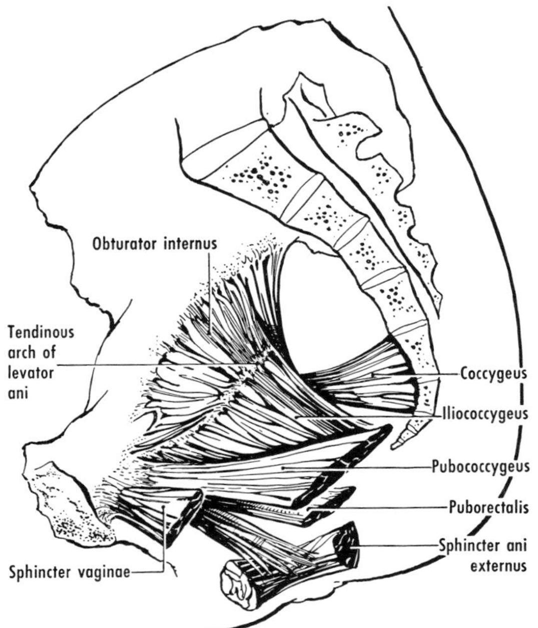

Figure 46–2 Muscles of pelvic diaphragm from their pelvic aspect, showing the different parts of the levator ani. Note that the pubococcygeus has several parts—sphincter vaginae, puborectalis, and pubococcygeus proper—depending on the direction and insertion of the fibers. Note that some fibers of the puborectalis pass toward the sphincter ani externus. These puboanal fibers are the only ones of the levator ani that can elevate the anus.

A conspicuous part of the levator ani, called the puborectalis, passes backward and unites with a corresponding part of the contralateral muscle to form a muscular sling behind the anorectal junction (fig. 45–2). Some of the fibers of the puborectalis blend with the sphincter ani externus and the longitudinal muscular coat of the rectum.

The *iliococcygeus,* the most posterior part of the levator ani, is often poorly developed and may be mostly aponeurotic. It is frequently deficient in some areas, and here the pelvic diaphragm is formed by the fused superior and inferior fasciae. It arises from the pelvic aspect of the ischial spine and from the tendinous arch of the levator ani (or obturator fascia) behind the level of the obturator canal. Its fibers pass obliquely and insert into the sides of the coccyx and into the anococcygeal ligament.

Coccygeus (Ischiococcygeus). This muscle is located behind the levator ani. Portions or all of it may be present as tendinous strands. It arises from the pelvic aspect of the ischial spine and spreads out to insert on the lateral margins of the lower part of the sacrum and upper part of the coccyx.

Nerve Supply. The levator ani and coccygeus are supplied by branches of the ventral rami of the third and fourth sacral nerves, which enter these muscles on their pelvic aspects. The anterior part of the levator ani is supplied by the perineal branch of the pudendal nerve.

Action. The pelvic diaphragm helps to support the pelvic viscera and resists increases in intra-abdominal pressure. Together with the anterior abdominal muscles, it enables the diaphragm to function effectively in all activities requiring increased intra-abdominal pressure.

Different parts of the levator ani have additional important functions. The levator prostatae in the male and the pubovaginalis in the female lie directly below the urinary bladder and are involved in the control of micturition.

The puborectalis is responsible for the flexure at the anorectal junction, and its relaxation during defecation permits a straightening at this junction. The puborectal sling, upon which the fetal head impinges during parturition, directs the head forward into the lower part of the birth canal. Although the pubococcygeus is capable of considerable relaxation during parturition, it is frequently torn or otherwise damaged. Defective support of the pelvic viscera may be a consequence of such injuries.

Fascia of Pelvic Diaphragm. These fasciae (see figs. 47–2, 47–4, and 47–5) are a part of the parietal pelvic fascia and are arranged in two layers.

The *superior fascia* covers the pelvic surface of the levatores ani and coccygei. The *tendinous arch of the pelvic fascia* is a thickening of this layer and extends from the ischial spine to the back of the body of the pubis near the symphysis. This arch is almost always present, at least in its anterior part, which forms the *medial puboprostatic (pubovesical) ligament*[3] (p. 462). The superior fascia is thin over the coccygeus, especially when this muscle is largely aponeurotic. It is fused with the sacrospinous ligament and often passes backward to cover the piriformis, where it is separated from this muscle and its delicate intrinsic fascia by the sacral plexus.

The *inferior fascia,* thinner than the superior, covers the lower surface of the levator ani and coccygeus. It forms the medial wall of the ischiorectal fossa (p. 500).

PELVIC FASCIA

The pelvic fascia[4] (see figs. 47–2, 47–4, and 47–5) comprises the parietal pelvic fascia and the visceral pelvic fascia.

Parietal Pelvic Fascia. This fascia is a part of the general layer that lines the inner aspect of the abdominal and pelvic walls (p. 353). Its continuity with the transversalis fascia and the fascia iliaca is usually interrupted by the fusion of these fasciae with the periosteum covering the terminal lines of the hip bone and the back of the body of the pubis. The parietal pelvic fascia forms a part of the floor of the pelvis (superior and inferior fasciae of the pelvic diaphragm described above), and it covers the lateral walls of the pelvis (obturator fascia). It covers the posterior wall incompletely, because it is absent on the median portion of the front of the sacrum.

The *obturator fascia* is the part of the parietal fascia covering the obturator internus muscle. It is attached around the margin of this muscle and covers its pelvic surface. A curved thickening of this fascia may be present below the level of the obturator canal. This thickening is the *tendinous arch of the levator ani.* It represents the line of fusion of the obturator fascia with the superior and inferior fasciae of the pelvic diaphragm, and it extends from the back of the body of the pubis to the ischial spine. It gives origin to a part of the levator ani. This arch is often absent, or it may be present in only part of its extent.

The part of the obturator internus below the level of the origin of the levator ani is extrapelvic, and the fascia covering it forms the lateral wall of the ischiorectal fossa. The *pudendal canal* is a tunnel in a special fascial sheath, the fascia lunata (p. 500), which is closely related to the obturator fascia. The internal pudendal vessels and the pudendal nerve pass through this canal.

Visceral Pelvic Fascia (see fig. 47–3). This fascia is formed by the extraperitoneal tissue that serves as a packing for organs and as sheaths for vessels. It lies between the peritoneum and parietal fascia, and is continuous above with the extraperitoneal tissue of the abdomen. It is variable in structure, and may be membranous, areolar, or fatty. It ensheathes the pelvic organs, and, where organs pass through the floor of the pelvis, it is continuous with the parietal fascia.

Special thickenings of the parietal and visceral fasciae form sheaths for blood vessels and nerves. Some of these sheaths are termed ligaments[5] and are described with the organs with which they are associated. They may contain large numbers of smooth muscle fibers.[6]

The *rectovesical septum* (see fig. 47–3) is a membranous partition between the rectum and the prostate and urinary bladder. It provides a cleavage plane during surgery. Its origin is disputed[7] and the existence of a similar membrane between the vagina and the rectum *(rectovaginal septum)* is frequently denied.[8]

REFERENCES

1. B. Berglas and I. C. Rubin, Surg. Gynec. Obstet., 97:677, 1953. B. Berglas, Wien med. Wschr., *4*:836, 1966.
2. E. T. C. Milligan and C. N. Morgan, Lancet, *2*:1150, 1934.
3. A. W. Meyer, Calif. west. Med., *27*:1, 1927.
4. D. L. Bassett, *A Stereoscopic Atlas of Human Anatomy,* section VI, The Pelvis, Sawyer's, Portland, Oregon, 1952–62. See also B. Berglas and I. C. Rubin, cited in reference 1.
5. A. H. Curtis, B. J. Anson, F. L. Ashley, and T. Jones, Surg. Gynec. Obstet., *75*:421, 1942. B. Berglas and I. C. Rubin, Surg. Gynec. Obstet., *97*:667, 1953.
6. R. M. H. Power, Amer. J. Obstet. Gynec., *38*:27, 1939.
7. M. B. Wesson, J. Urol., *8*:339, 1922. C. E. Tobin and J. A. Benjamin, Surg. Gynec. Obstet., *80*:373, 1945.
8. A. H. Curtis, B. J. Anson, and L. E. Beaton, Surg. Gynec. Obstet., *70*:643, 1940. J. F. Ricci and C. H. Thom, Quart. Rev. Surg. Obstet. Gynec., *2*:253, 1954.

PERINEAL REGION AND EXTERNAL GENITAL ORGANS 47

PERINEAL REGION

The perineal region (fig. 47–1) is the part of the trunk below the pelvic diaphragm. It is a diamond-shaped space, which has the same boundaries as the outlet of the pelvis (p. 441). The bony angles and sides can be palpated, but the sacrotuberous ligament usually cannot be felt, because it lies deep to the margin of the gluteus maximus. The perineal region is much deeper behind and at the sides than in front. The term *perineum* is usually restricted, especially in obstetrics and gynecology, to the region between the anal and vaginal orifices.

A median ridge, the *raphe,* runs forward from the anus. In the male, it is continuous with the raphe of the scrotum and penis. The bulb of the urethra can be palpated deep to the raphe, several centimeters in front of the anus. The central point of the perineum, located between the anus and the bulb of the urethra, is the superficial indication of the tendinous center of the perineum.

The *tendinous center of the perineum,* or *perineal body,* is a fibromuscular mass that is located in the median plane between the anal canal and the urogenital diaphragm, with which it is fused. It contains collagenous and elastic fibers and both skeletal and smooth muscle. Several muscles are attached to it, at least in part. These are the superficial and deep transversus perinei, the bulbospongiosus, the levator ani (levator prostatae), the sphincter ani externus, and smooth muscle from the longitudinal coat of the rectum (rectourethralis, p. 488). In addition, the superficial and deep perineal fasciae, and the superior and inferior fasciae of the urogenital diaphragm are attached to it. The tendinous center of the perineum is of special importance in the female, because it may be torn or otherwise damaged during parturition. In order that such damage may be avoided, the opening for the passage of the fetal head is often enlarged by making an incision through the posterior wall of the vagina and the adjacent part of the perineum. This operation is called *episiotomy.*

The perineal region is customarily divided into an anterior, urogenital region and a posterior, anal region by a line drawn transversely through the central point of the perineum, just in front of the ischial tuberosities (fig. 47–1*A*). The differences between the urogenital regions in the male and female are very pronounced.

UROGENITAL REGION

IN THE MALE

The male urogenital region is pierced by the urethra. From below upward, it comprises (1) skin, (2) the superficial perineal fascia, (3) the deep perineal fascia, (4) the superficial perineal space, which contains the root of the penis and the muscles associated with it, a part of the urethra, and branches of the internal pudendal vessels and the pudendal nerves, (5) the inferior fascia of the urogenital diaphragm, (6) the deep perineal space, which contains the urogenital diaphragm, the bulbourethral glands, and branches of the internal pudendal vessels and the pudendal nerves, and (7) the superior fascia of the urogenital diaphragm.

Superficial Perineal Fascia

The subcutaneous tissue of the urogenital region is called the superficial fascia[1] (figs. 47–1*B*, 47–2, and 47–3), and it consists of a superficial fatty and a deep membranous layer. The fatty layer, which contains some smooth muscle fibers, is continuous behind with a similar layer of the anal region. It loses its fat as it passes forward into the scrotum, where the fat is replaced by large numbers of smooth muscle fibers, which help to form the dartos. The fatty layer is continuous between the scrotum and the thighs with the subcutaneous tissue of the abdomen.

The membranous layer of the superficial perineal fascia is attached behind to the posterior margin of the urogenital diaphragm and to the tendinous center. At the sides, it is attached to the ischiopubic ramus along a line to which the fascia lata also is attached. In front, it is continuous with the dartos in the male, but at the sides of the scrotum it becomes continuous with the membranous layer of subcutaneous tissue that covers the front and sides of the lower half of the abdomen. It is fused with the perineal raphe below and with the raphe of the bulbospongiosus above. The membranous layer of the superficial perineal fascia is separated from the deep perineal fascia by a shallow cleft, which contains fat, loose connective tissue, and branches of the posterior scrotal (or labial) vessels and nerves.

Deep Perineal Fascia

The deep perineal fascia[2] (figs. 47–1*B*, 47–2, and 47–3) also is attached to the posterior margin of the urogenital diaphragm. At the sides, it is attached to the ischiopubic ramus, just above the attachment of the membranous layer of the superficial fascia. In front, it is fused with the suspensory ligament of the penis and is continuous with the fascia covering the external oblique and the sheath of the rectus.

Muscles of Superficial Perineal Space

The superficial perineal space, or pouch (figs. 47–1*D* and 47–2), is bounded below by the deep perineal fascia and above by the inferior fascia of the urogenital diaphragm. It contains the following muscles: the superficial transversus

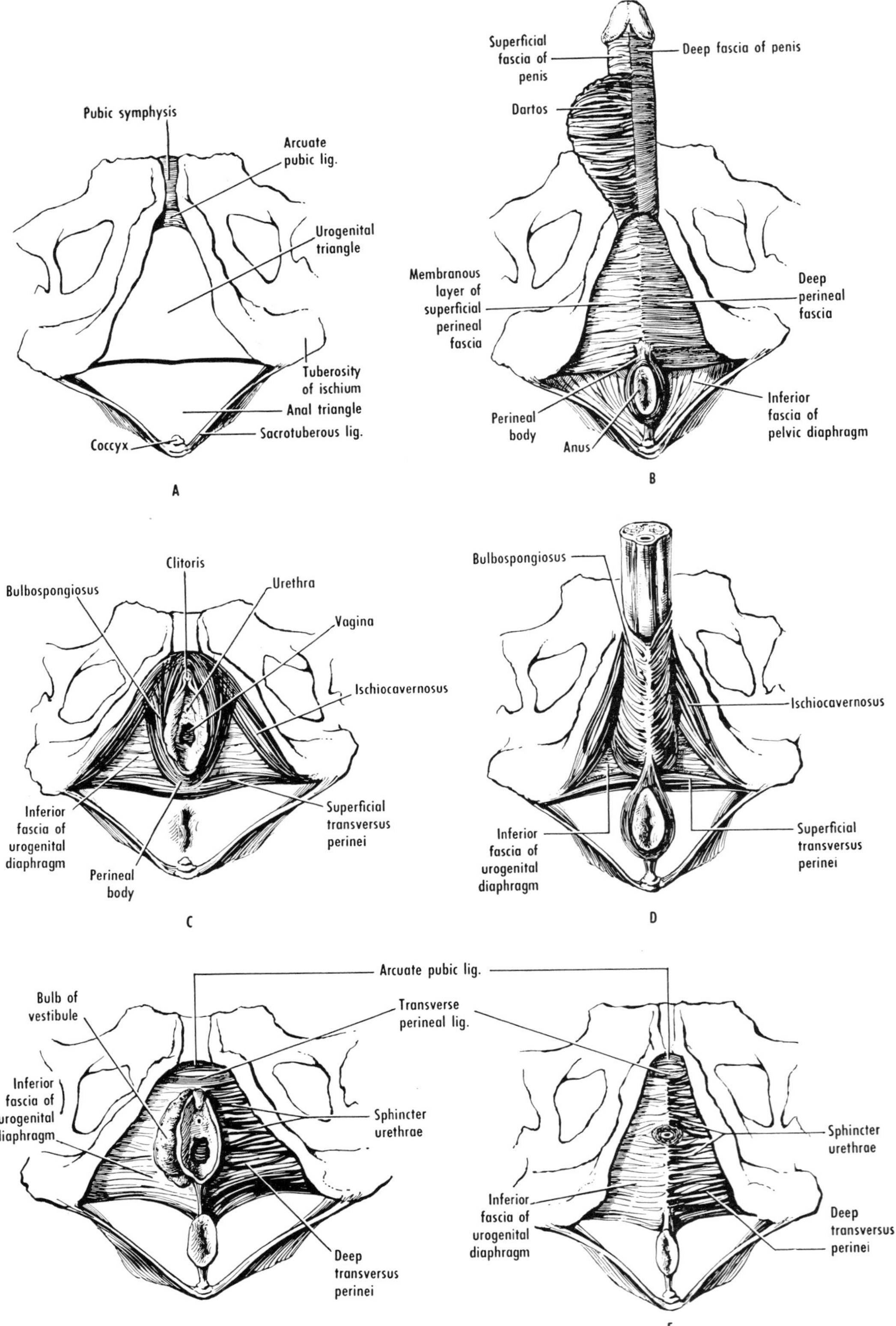

Figure 47–1 *See opposite page for legend.*

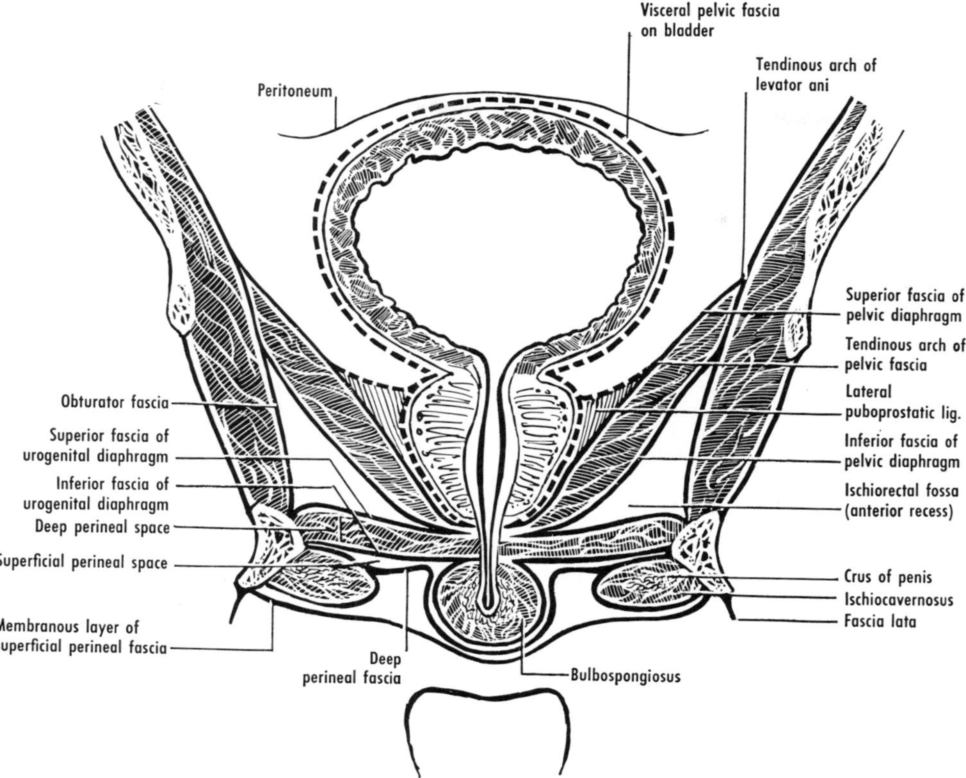

Figure 47–2 Fasciae of pelvis and urogenital region, male. Coronal section through prostatic part of urethra.

perinei, the ischiocavernosus, and the bulbospongiosus.

Superficial Transversus Perinei. This is usually poorly developed. It arises from the lower part of the inner surface of the ramus of the ischium, adjacent to the tuberosity, and inserts into the tendinous center of the perineum. It is supplied by a perineal branch of the pudendal nerve. Its action is insignificant.

Ischiocavernosus. This arises from the inner surface of the ramus of the ischium, just below the origin of the superficial transversus perinei. Its origin embraces the crus of the penis above, behind, and below. It spreads out to insert on the lower and medial aspects of the crus. It is supplied by perineal branches of the pudendal nerve. The ischiocavernosus may help to maintain erection of the penis by compressing the crus and thereby retarding the flow of blood from this organ.

Bulbospongiosus (Bulbocavernosus). This arises from the tendinous center of the perineum and from a median fibrous raphe on the lower aspect of the bulb of the penis. It passes upward and forward around the side of the bulb. Some of its fibers insert into the inferior fascia of the urogenital diaphragm, others into the upper aspect of the corpus spongiosum, and still others into the deep fascia on the

---

Figure 47–1 A, boundaries and subdivisions of the perineal region, inferior view. *B*, fasciae of the male perineal region, inferior view. The superficial fascia has been removed at the right. *C*, muscles of the superficial perineal space, female, inferior view, after removal of superficial and deep perineal fascia. *D*, muscles of the superficial perineal space, male, inferior view, after removal of superficial and deep fascia. *E*, muscles of the deep perineal space, female, inferior view. The inferior fascia of the urogenital diaphragm has been removed at the right. *F*, muscles of the deep perineal space, male, inferior view. The inferior fascia of the urogenital diaphragm has been removed at the right.

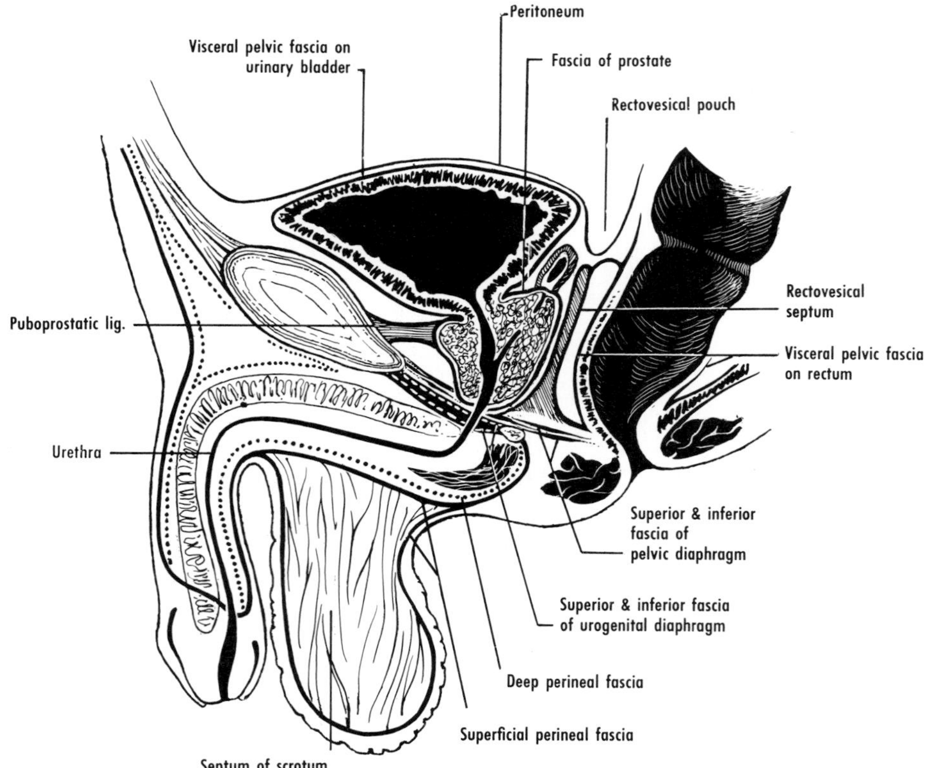

Figure 47–3 Median section, male pelvis. The bladder is shown as moderately full.

dorsum of the penis. The bulbospongiosus is supplied by perineal branches of the pudendal nerve. Acting together, the two muscles serve to expel the last drops of urine or semen from the urethra. Some of their fibers aid in retarding venous return from the penis and thereby help to maintain erection.

Muscles of Deep Perineal Space

The deep perineal space, or pouch, is filled by the *urogenital diaphragm* (figs. 47–1*F*, 47–2 and 47–3), in which the bulbourethral glands are located, and which consists of two muscles, the deep transversus perinei and the sphincter urethrae. The urogenital diaphragm is in contact through its fascia with the inferior fascia of the anterior part of the pelvic diaphragm above, and it is almost horizontal in the erect subject. It is pierced by the urethra, about 2½ cm behind the pubic symphysis.

Deep Transversus Perinei. This originates from the inner surface of the ramus of the ischium. Most of its fibers insert into the tendinous center of the perineum. If well developed, it may be fused with the posterior margin of the sphincter urethrae. If poorly developed, it may be separated from the sphincter urethrae by an interval in which the superior and inferior fasciae of the urogenital diaphragm are fused with each other, except where they are separated by the bulbourethral glands. It is supplied by the dorsal nerve of the penis. It helps to fix the tendinous center of the perineum.

Sphincter Urethrae. This arises from the inner surface of the inferior ramus of the pubis. Its fibers pass both in front of and behind the urethra, and some interdigitate with fibers of the opposite side.[3] It is supplied by the dorsal nerve of the penis. It is said that it constricts and expels the last drops of urine from the membranous part of the urethra.

The deep transversus perinei and the sphincter urethrae are covered above by the relatively thin and delicate *superior fascia of the urogenital diaphragm.* They are covered below by the strong and dense *inferior fascia of the urogenital diaphragm* (or *perineal membrane,* or *triangular ligament*). These fasciae are attached laterally to the ischiopubic ramus, at a level just above the crus of the penis. At the posterior limit of the urogenital region, they fuse with each other, with the membranous layer of the superficial perineal fascia, and, in the median plane, with the tendinous center of the perineum. In front, they fuse with each other to form the *transverse perineal ligament,* which is separated from the arcuate pubic ligament by a space through which the deep dorsal vein of the penis passes on its way to the prostatic plexus.

IN THE FEMALE

The female urogenital region[4] differs markedly from the corresponding region of the male. In addition to the urethra, it contains the lower end of the vagina and the female external genital organs. The fasciae, fascial spaces, muscles, blood vessels, and nerves resemble those of the male, but their anatomy is modified considerably by the presence of the genital organs.

Superficial Perineal Fascia

The fatty layer of the superficial perineal fascia continues forward into the labia majora, and from there into the mons pubis and the fatty layer of subcutaneous tissue covering the abdomen.

The deep, membranous layer of the superficial perineal fascia (fig. 47–4) passes through the deeper portions of the labia majora and becomes continuous with the membranous layer of the subcutaneous tissue covering the abdomen. It is fused with the perineal raphe below; its posterior and lateral attachments are the same as those in the male.

Deep Perineal Fascia

The deep perineal fascia (fig. 47–4) is fused in front with the suspensory ligament of the clitoris and

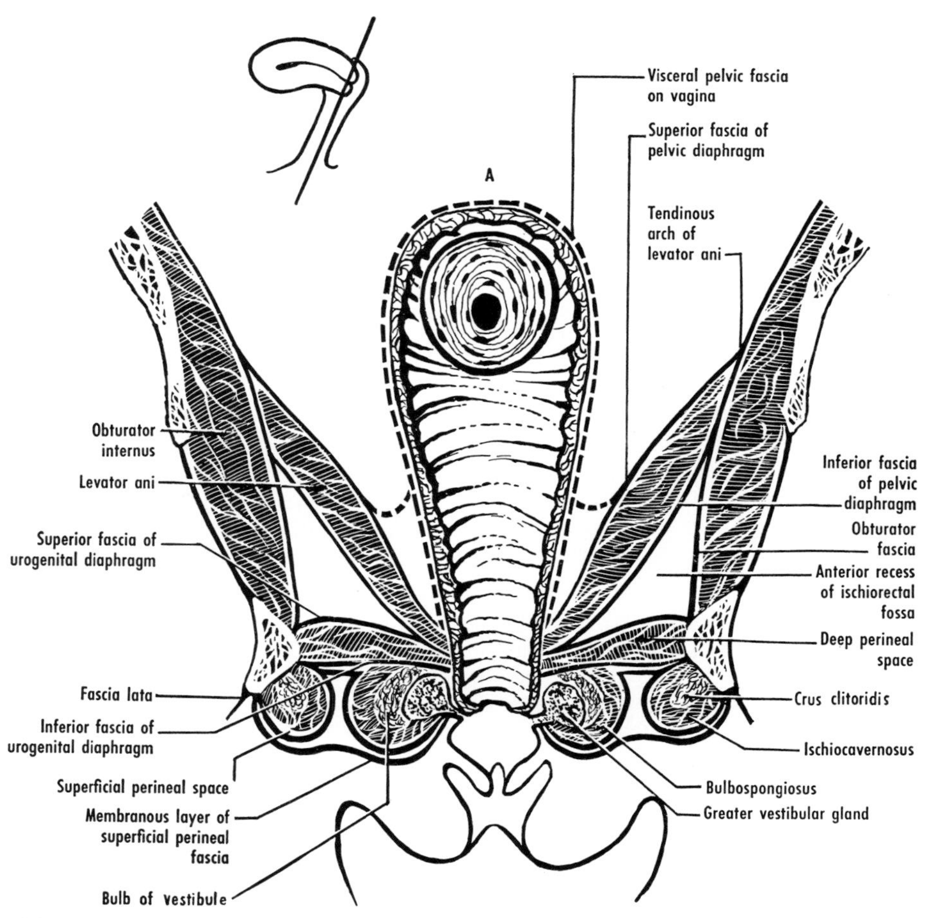

Figure 47–4 Fasciae of pelvis and urogenital region, female. The small figure inset above shows the plane of section. A indicates the uppermost part of the vagina.

is continuous with the fascia covering the external oblique and the sheath of the rectus. Its posterior and lateral attachments are also similar to those in the male.

Muscles of Superficial Perineal Space

The superficial perineal space, or pouch (fig. 47–1*C*), contains the following muscles: (1) the superficial transversus perinei, (2) the ischiocavernosus, and (3) the bulbospongiosus. The bulbospongiosus especially differs from its homologue in the male.

Superficial Transversus Perinei. This is similar to the corresponding muscle of the male.

Ischiocavernosus. This is smaller than its homologue in the male. At its origin from the inner surface of the ramus of the ischium, it embraces the crus of the clitoris. It spreads out to insert on the lower and medial aspects of the crus. It may help to maintain erection of the clitoris by compressing the crus and thereby retarding the flow of blood from this organ.

Bulbospongiosus (Bulbocavernosus). This differs from the homologous muscle in the male in being widely separated from the contralateral muscle by the lower part of the vagina. It arises from the tendinous center of the perineum and passes forward around the lower part of the vagina. In its course, it covers the bulb of the vestibule. It is inserted partly into the side of the pubic arch and partly into the root and dorsum of the clitoris. Acting together the two muscles weakly constrict the vagina.

Muscles of Deep Perineal Space

The *urogenital diaphragm*[5] (figs. 47–1*E* and 47–4) of the female is much less complete than that of the male, because it is almost divided into two halves by the vagina and the urethra.

The muscles of the female urogenital diaphragm, the deep transversus perinei and the sphincter urethrae, are usually much less well developed than those of the male. They are often described as a single muscle that is designated the deep transversus perinei.

Deep Transversus Perinei. This muscle arises from the inner surface of the ramus of the ischium. Its more posterior fibers insert into the tendinous center of the perineum, and its more anterior fibers into the lateral wall of the vagina. It helps to fix the tendinous center.

Sphincter Urethrae. This arises from the inner surface of the inferior ramus of the pubis. Most of its fibers insert into the lateral wall of the vagina, but a few pass in front of the urethra, and a few others may pass between the urethra and the vagina. In spite of its name, this muscle cannot act as a sphincter, because the urethra and the vagina are fused, and its fibers therefore do not surround the urethra. Furthermore, section of the perineal nerve, which supplies the sphincter urethrae, does not result in incontinence.[6]

The *inferior fascia of the urogenital diaphragm* is relatively strong and dense, but the *superior fascia* is indistinct.

ANAL REGION

The subcutaneous tissue of the anal region (fig. 47–1*A*) extends upward on both sides of the anus to fill the ischiorectal fossae. This tissue is called the *ischiorectal pad of fat,* and it contains many tough, fibrous septa. It gives support to the anal canal, but is readily displaced to allow the passage of feces.

Each ischiorectal fossa (figs. 47–2, 47–4, and 47–5) is a space located between the skin of the anal region below and the pelvic diaphragm above. It is triangular in coronal section. Its lateral wall, which is almost vertical, is formed by the obturator fascia, covering the obturator internus, and by the fascia lunata. The pudendal canal is located on the lateral wall between these two fasciae. The superomedial wall is formed by the inferior fascia of the pelvic diaphragm and by the sphincter ani externus. The lateral and superomedial walls meet above at the line of fusion of the obturator fascia with the inferior fascia of the pelvic diaphragm.

The ischiorectal fossa is bounded in front by the posterior margin of the urogenital diaphragm and the tendinous center of the perineum. However, an anterior diverticulum or recess can be followed for a variable distance between the urogenital and pelvic diaphragms, and sometimes reaches the retropubic space. Posteriorly, the fossa extends beyond the limit of the anal region, above the gluteus maximus, as far as the sacrotuberous ligament.

The *fascia lunata*[7] begins at the sacrotuberous ligament and forms the medial wall of the pudendal canal. Irregular strands leave it to pass medially across the ischiorectal fossa to join the inferior fascia of the pelvic diaphragm. They incompletely divide the fossa into upper and lower parts (supra- and infrategmental spaces).

In addition to the ischiorectal pad of fat, the ischiorectal fossa contains the internal pudendal vessels and pudendal nerve, which run on its lateral wall; the inferior

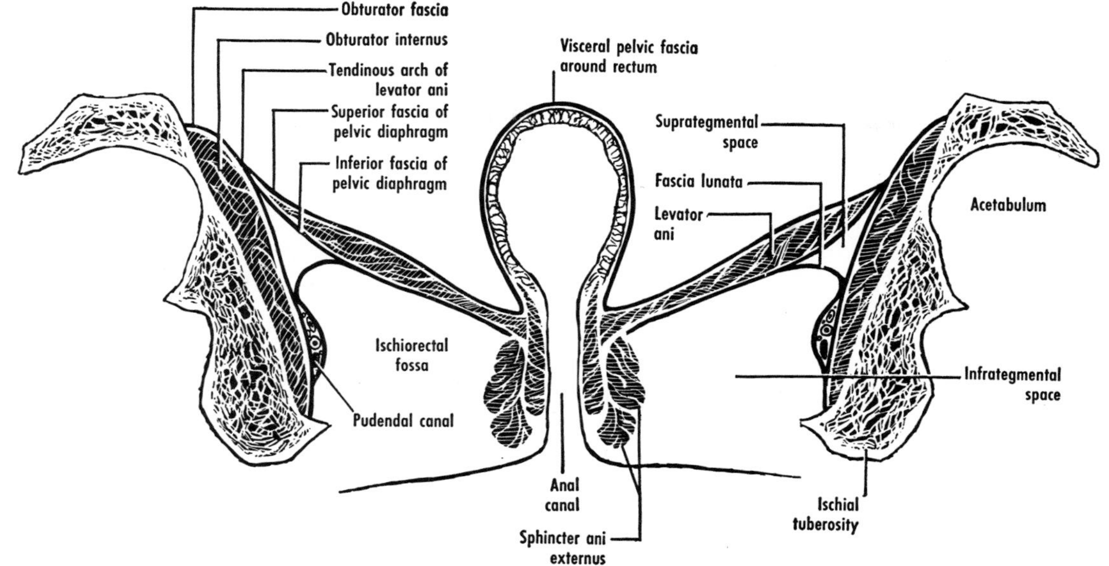

Figure 47–5 Ischiorectal fossae and pelvic diaphragm.

rectal vessels and nerves, which cross the fossa to reach the region around the anus; some muscle fibers continued downward from the outer longitudinal coat of the rectum; a branch of the posterior femoral cutaneous nerve; and the perforating cutaneous nerve.

The ischiorectal fossa is sometimes the site of an abscess that may be connected with the rectum or the anal canal. Because the two ischiorectal fossae communicate with each other behind the anal canal and around the sphincter ani externus, an abscess in one ischiorectal fossa may spread to the other.[8]

One muscle, the sphincter ani externus, is located in the anal triangle. It is described on page 488.

EXTERNAL GENITAL ORGANS

IN THE MALE

The *male external genital organs* comprise the scrotum and the penis. The scrotum is situated below and in front of the urogenital region. A part of the penis is located in the urogenital region, and the remainder is located in front of the scrotum.

Scrotum

The scrotum (figs. 43–2 and 47–3) is the pouch that is situated behind the penis and below the pubic symphysis. It is divided into two compartments, each of which contains a testis, an epididymis, and the lower part of the spermatic cord and its coverings. The left compartment usually hangs a little lower than the right (p. 468). The scrotum consists of skin and the closely associated underlying dartos.

The skin is relatively thin, and it contains more pigment than the skin adjacent to it. It contains few hairs, but many sebaceous and sweat glands. A median ridge, the raphe of the scrotum, is the superficial indication of the division of the scrotum into two compartments. This raphe is continuous anteriorly with the raphe of the penis and posteriorly with the raphe of the perineum.

The *dartos* consists largely of smooth muscle fibers, and it contains no fat. It is firmly united to the skin. It is continuous with the superficial perineal fascia and with the superficial fascia of the penis. Its superficial part is continuous around the scrotum, but its deeper part passes inward at the raphe to form the septum of the scrotum, which divides the scrotum into its two compartments. The dartos is separated by loose connective tissue from the external spermatic fascia, upon which the scrotum moves freely. This loose tissue is a common site for the collection of edematous fluid or blood.

The appearance of the scrotum varies with the state of contraction or relaxation of the smooth muscle of the dartos. It contracts under the influence of cold, exercise, or sexual stimulation, and the scrotum then appears short and wrinkled. It relaxes under the influence of warmth. It loses its tone in old men, in whom the scrotum is smooth and elongated.

Blood Supply. The front of the scrotum is supplied by the external pudendal arteries, whereas the back is supplied by the scrotal branches of the internal pudendal artery. Branches of the testicular and cremasteric arteries, which run in the spermatic cord, help to supply the scrotum. The veins accompany the arteries.

Lymphatic Drainage. Lymphatic vessels are especially numerous in the scrotum. They drain into the superficial inguinal nodes.

Nerve Supply. The anterior part of the scrotum is supplied by the ilioinguinal nerve and by the genital branch of the genitofemoral nerve. The posterior part is supplied by the medial and lateral scrotal branches of the perineal nerve and by the perineal branch of the posterior femoral cutaneous nerve.

Penis

The penis (figs. 47–3 and 47–6) is the male organ of copulation. Its erection and enlargement are due to its engorgement with blood. It consists of a root and a body.

Root of Penis. The root of the penis is the attached portion. It is situated in the superficial perineal space, between the inferior fascia of the urogenital diaphragm above and the deep perineal fascia below. It includes two crura and the bulb of the penis, all three of which are masses of erectile tissue.

Each *crus penis* is attached to the lower part of the internal surface of the ramus of the corresponding ischium, just in front of the ischial tuberosity (fig. 47–2). It grooves or lies close to the inferior ramus of the pubis as it passes forward, covered by the ischiocavernosus, and it joins the contralateral crus. Near the lower margin of the pubic symphysis, the conjoined crura turn downward. From here on they are termed the corpora cavernosa of the body of the penis.

The *bulb of the penis* is located between the two crura in the superficial perineal space. Superiorly it is flattened and attached to the inferior fascia of the urogenital diaphragm. Inferiorly and laterally it is rounded and covered by the bulbospongiosus. The enlarged posterior part of the bulb is penetrated above by the urethra, which extends forward in its substance. As the bulb passes forward, it becomes narrower, and it bends downward to continue as the corpus spongiosum of the body of the penis.

Body of Penis. The body of the penis is the free, pendulous part, and it is covered with skin. The dorsum of the penis is the aspect that faces forward when the organ is flaccid, and upward and backward when it is erect. The urethral aspect faces

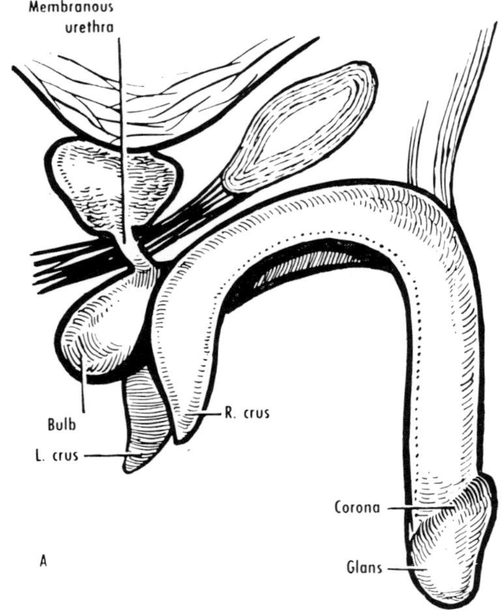

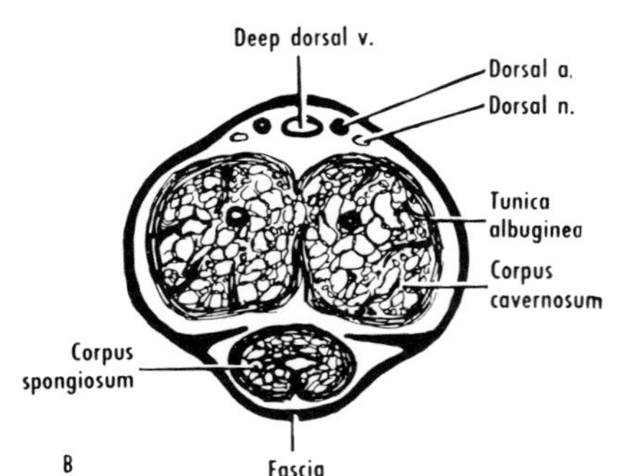

Figure 47–6 A, diagram of penis. *B*, cross section of penis.

in the opposite direction. A median ridge, the raphe of the penis, is located on this aspect and is continuous with the raphe of the scrotum. The body of the penis contains the paired corpora cavernosa, which are continuations of the crura, and the single corpus spongiosum, which is a continuation of the bulb.

The *corpora cavernosa* constitute the main bulk of the body of the penis, and they form its dorsum and sides. At their junction on the urethral aspect, they leave a wide median groove, in which the corpus spongiosum is situated. They end as blunt projections, which are covered by the glans penis.

The *corpus spongiosum* is smaller than either corpus cavernosum. In passing through the body of the penis it tapers slightly, but at the end it suddenly expands to form the *glans penis,* the concavity of which covers the free blunted ends of the corpora cavernosa.

The glans penis is separated superficially from the rest of the body of the organ by a constriction, the *neck of the glans.* The *corona glandis* is the prominent margin of the glans adjacent to the neck. A median slit near the tip of the glans is the *external urethral orifice.* A double layer of skin, the *prepuce,* passes from the neck to cover the glans for a variable extent. The *frenulum of the prepuce* is a median fold that passes from the deep layer of the prepuce to the part of the urethral aspect adjacent to the external urethral orifice.

Structure. The skin of the penis is thin, smooth, elastic, and dark in color. Near the root of the penis it contains a few hairs. It is loosely attached to the subcutaneous tissue except at the glans, which is firmly attached to the underlying erectile tissue. Many small *preputial glands* are located on the corona and neck of the glans. They are responsible for the sebaceous secretion called smegma, which has a characteristic odor.

The subcutaneous tissue, termed *superficial fascia of the penis,* consists of loosely arranged connective tissue characterized by the presence of some smooth muscle fibers and by an almost complete absence of fat. It is continuous with the dartos of the scrotum and with the superficial perineal fascia.

The *deep fascia of the penis* is a continuation of the deep perineal fascia. It is strong and membranous, and surrounds both corpora cavernosa and the corpus spongiosum as one sheath. It does not extend into the glans, but at the neck blends with the fibrous sheaths that surround the three erectile bodies.

The *tunica albuginea of the corpus cavernosum* is a dense fibrous envelope that lies under the deep fascia. Its superficial fibers are longitudinal in direction, and these form a sheath that surrounds both corpora cavernosa. Its deep fibers are circularly arranged around each corpus, and they meet in the median plane to form the *septum of the penis.* Near the root of the penis, this septum is thick and complete. Toward the free end, however, it becomes thinner and has deficiencies through which the corpora cavernosa communicate with each other. The *tunica albuginea of the corpus spongiosum* is thinner and more elastic than the corresponding sheath around the corpora cavernosa.

The corpora cavernosa and the corpus spongiosum are broken up into numerous blood-filled *cavernous spaces* by many *trabeculae,* which extend into them from the tunica albuginea and form the septum of the penis. These trabeculae run in all directions through the erectile tissue. They consist of collagenous, elastic, and smooth muscle fibers, and they are traversed by arteries and nerves.

Ligaments.[9] Two ligaments are attached to the penis near the junction of the body with the root. The elastic *fundiform ligament* arises from the lower part of the linea alba and the membranous layer of the subcutaneous tissue covering it. As it descends, it splits into left and right parts, which pass on the corresponding sides of the penis. They unite on the urethral side and pass into the septum of the scrotum. The *suspensory ligament,* deep to the fundiform, arises from the front of the pubic symphysis. It passes downward to attach to the deep fascia on each side of the penis.

Blood Supply (figs. 41–2 and 41–3, p. 451). The artery of the bulb of the penis passes through the erectile tissue of the bulb, and then continues through the corpus spongiosum. The deep artery of the penis, after entering the crus, gives off a branch that passes backward in the crus toward its bony attachment. It then passes through the crus and the corpus cavernosum, and supplies most of the blood to the erectile tissue of this body. The dorsal artery of the penis runs beneath the deep fascia on the dorsum of the penis, between the dorsal nerve on its lateral side and the deep dorsal vein on its medial side. Its branches help to supply the erectile tissue of the corpora cavernosa and the corpus spongiosum, and they anastomose with branches of the deep artery of the penis and of the artery of the bulb. The dorsal artery provides most of the blood supply to the glans.

The small branches of the arteries that supply the erectile tissue run in the trabeculae. Many of these have a helical appearance and are therefore termed *helicine arteries.* The capillaries arising from the small branches open into the cavernous spaces.

The unpaired dorsal vein of the penis drains most of the blood from the glans and prepuce, and from the corpus spongiosum and corpora cavernosa. In its course it divides into left and right veins, which drain into the prostatic plexus. The skin and subcutaneous tissues are drained by the superficial dorsal vein, and this empties into the great saphenous vein.

Lymphatic Drainage. The lymphatic vessels from the skin and prepuce drain into the superficial inguinal nodes, whereas those from the glans drain into the deep inguinal and external iliac nodes.

Nerve Supply (figs. 41–2 and 41–3, p. 451). The penis is innervated by (1) the dorsal nerves of the penis, which are branches of the pudendal nerves, and which are distributed mainly to the skin and especially to the glans; (2) the deep branches of the perineal nerves, which enter the bulb, continue through it and the corpus spongiosum, and supply mainly the urethra; (3) the ilioinguinal nerve, the branches of which are distributed to the skin near the root; (4) the cavernous nerves of the penis, which supply the erectile tissue of the bulb and crura, and then continue forward into the corpus spongiosum and corpora cavernosa. Many of the autonomic fibers in the cavernous nerves arise from lumbar sympathetic ganglia,[10] and some of their fibers join the dorsal nerves.

These fibers contain a large number of sensory fibers, which include pain fibers from the skin and urethra, as well as fibers from a variety of special receptors. They also contain many sympathetic and parasympathetic fibers, which are concerned with the control of circulation of blood in the penis.

Mechanism of Erection.[11] Stimulation of the parasympathetic fibers in the cavernous nerves produces a vasodilatation of the helicine and other small arteries in the trabeculae of the erectile tissue. The resulting flow of blood into the cavernous spaces produces a distention of both corpora cavernosa and of the corpus spongiosum. The egress of blood from these spaces is prevented by pressure on the veins that drain the erectile bodies. At the end of ejaculation, stimulation of the sympathetic nerves presumably produces a vasoconstriction of the arteries, the blood is permitted to enter the veins, and the penis returns to its flaccid state.

IN THE FEMALE

The *female external genital organs* (vulva; pudendum) (fig. 47–7) comprise the mons pubis, the labia majora, the labia minora, the vestibule of the vagina, the clitoris, the bulb of the vestibule, and the greater vestibular glands.

Mons Pubis

The mons pubis is a rounded, median elevation in front of the pubis symphysis. It consists mostly of an accumulation of fat. After puberty, the skin over it is covered by coarse hairs.

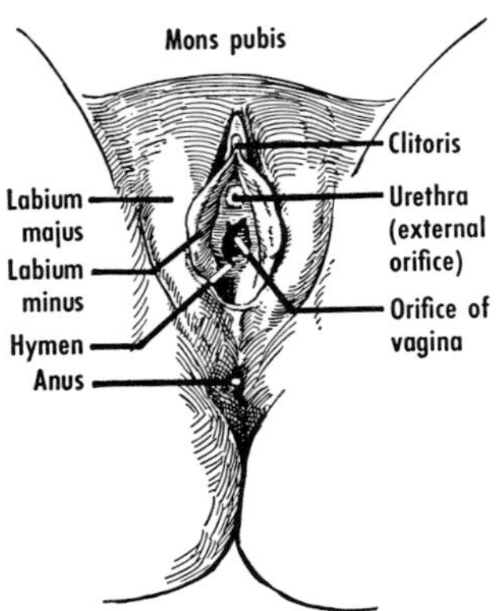

Figure 47–7 Female external genital organs. Schematic.

Labia Majora

The labia majora are two elongated folds that run downward and backward from the mons pubis and enclose between them the median *pudendal cleft.* Their outer aspects are overlaid by pigmented skin containing many sebaceous glands and covered with hair after puberty. Their inner aspects are smooth and hairless. The labia majora are usually united in front by an *anterior commissure.* They are not united behind, but the forward projection of the tendinous center of the perineum into the pudendal cleft sometimes gives the appearance of a *posterior commissure.*[12] The subcutaneous tissue of the labia majora consists mostly of fat. It is continuous behind with the subcutaneous tissue of the urogenital region and in front with that of the mons pubis and abdomen. The superficial perineal fascia passes through the deeper part of the labia majora and is continuous with the membranous layer of superficial fascia covering the abdomen. The labia majora also contain the terminations of the round ligaments, some bundles of smooth muscle fibers, nerves, and blood and lymphatic vessels. **The labia majora are homologous with the scrotum of the male.**

Labia Minora

The labia minora are two small folds of skin located between the labia majora, at either side of the opening of the vagina. They end behind by joining the medial aspects of the labia majora, and here, in the virgin, they are usually connected with each other by a transverse fold, called the *frenulum of the labia,* or *fourchette.* In front, each labium minus divides into a lateral and a medial part. The lateral part meets the corresponding one from the opposite labium minus to form a fold over the glans clitoridis, called the *prepuce of the clitoris.* The two medial parts unite below the clitoris to form the *frenulum of the clitoris.* The labia minora are devoid of fat, and the skin covering them is smooth, moist, and pink. They are hidden by the labia majora except in children and in women after the menopause, when the labia majora contain less fat and are smaller.

Vestibule of Vagina

The vestibule of the vagina is the cleft between the labia minora. It contains the openings of the vagina, the urethra, and the ducts of the greater vestibular glands. The *external urethral orifice* is situated behind the clitoris and immediately in front of the vaginal orifice. It is usually a median slit, the margins of which are slightly everted. The *vaginal orifice,* larger than the urethral orifice, is also a median cleft. Its size and appearance depend upon the condition of the hymen (see Vagina, in chapter 44). The ducts of the greater vestibular glands, two in number, open on each side of the vaginal orifice, between it and the labia minora. Smaller openings for the ducts of the lesser vestibular glands are located in the vestibule between the urethral and vaginal openings. The *vestibular* or *navicular fossa* is a shallow depression in the vestibule between the vaginal orifice and the frenulum of the labia.

Clitoris

The clitoris, like the penis with which it is homologous, consists mainly of erectile tissue and is capable of enlargement as a result of engorgement with blood. Unlike the penis, it is not traversed by the urethra. It is located behind the anterior commissure of the labia majora, and most of it is hidden by the labia minora.

The clitoris arises from the bony pelvis by two crura. Each *crus of the clitoris* is attached to the lower part of the internal surface of the ramus of the corresponding ischium, just in front of the ischial tuberosity. It grooves or lies close to the inferior ramus of the pubis as it passes forward in the superficial perineal space, where it is covered by the ischiocavernosus, and it joins the contralateral crus. Near the lower margin of the pubic symphysis the conjoined crura turn downward. From here on they are called the *corpora cavernosa,* which together form the *body of the clitoris.* The corpora cavernosa are enclosed by a densely fibrous envelope and are separated from each other by an incomplete septum. The *glans of the clitoris* is the small rounded elevation on the free end of the body. It also consists of erectile tissue and, like the glans of the penis, is highly sensitive. The *suspensory ligament of the clitoris* connects this organ to the front of the pubic symphysis.

Bulb of Vestibule

The bulb of the vestibule consists of paired elongated masses of erectile tissue, which lie at the sides of the vaginal opening under cover of the bulbospongiosus muscle. These masses are broad behind, but they become narrow in front, where they unite with each other to form a thin strand, which passes along the lower surface of the body of the clitoris to the glans. The bulb of the vestibule is the homologue of the bulb of the penis and the adjacent part of the corpus spongiosum.

Greater Vestibular Glands

The greater vestibular glands are two small rounded or ovoid bodies, located immediately behind the bulb of the vestibule or under cover of its posterior parts. The duct of each gland opens into the groove between the labium minus and the attached margin of the hymen. The greater vestibular glands are homologous with the bulbourethral glands of the male. They are compressed during coitus and secrete mucus, which serves to lubricate the lower end of the vagina.

Blood Supply, Lymphatic Drainage, and Nerve Supply

Blood Supply. The labia majora and minora are supplied by the anterior labial branches of the external pudendal arteries and by the posterior labial branches of the internal pudendal arteries. The crura and corpora cavernosa of the clitoris are supplied by the deep arteries of the clitoris; the glans is supplied by the dorsal arteries of the clitoris. The bulb of the vestibule and the greater vestibular gland receive their blood supply from the artery of the bulb of the vestibule and from the anterior vaginal artery.

Lymphatic Drainage. The lymphatic vessels

from the external genital organs drain into the superficial inguinal nodes.

Nerve Supply. The labia majora and minora are supplied by the anterior labial nerve (from the ilioinguinal nerve) and the posterior labial nerves (from the pudendal nerve). The bulb of the vestibule is supplied by the uterovaginal plexus, which is continued as the cavernous nerves of the clitoris. The clitoris is also supplied by the dorsal nerve of the clitoris.

These various nerves include (1) sensory fibers, some of which conduct pain and others of which arise from a variety of special receptors; (2) autonomic fibers, which supply the numerous blood vessels; and (3) autonomic fibers, which supply the various glands.

REFERENCES

1. C. E. Tobin and J. A. Benjamin, Surg. Gynec. Obstet., *79*:195, 1944; Surg. Gynec. Obstet., *88*:545, 1949.
2. D. L. Bassett, *A Stereoscopic Atlas of Human Anatomy,* Sawyer's, Portland, Oregon, 1952–1962. W. H. Roberts, J. Habenicht, and G. Krishinger, Anat. Rec., *149*:707, 1964.
3. H. C. Rolnick and F. K. Arnheim, J. Urol., *61*:591, 1949.
4. A. H. Curtis, B. J. Anson, and F. L. Ashley, Surg. Gynec. Obstet., *74*:709, 1942.
5. A. H. Curtis, B. J. Anson, and C. B. McVay, Surg. Gynec. Obstet., *68*:161, 1939.
6. J. R. Learmonth, H. Montgomery, and V. S. Counseller, Arch. Surg., Chicago, *26*:50, 1933.
7. D. E. Derry, J. Anat., Lond., *42*:107, 1907. G. E. Smith, J. Anat. Lond., *42*:198, 1908. A. R. Barnes, Anat. Rec., *22*:37, 1921.
8. H. Courtney, Surg. Gynec. Obstet., *89*:222, 1949.
9. E. D. Congdon and J. M. Essenberg, Amer. J. Anat., *97*:331, 1955.
10. P. Calabrisi, Anat. Rec., *125*:713, 1956.
11. G. Conti, Acta anat., *14*:217, 1952.
12. F. W. Jones, J. Anat., Lond., *48*:73, 1913.

Part Seven

THE BACK

ERNEST GARDNER
DONALD J. GRAY

Introduction

The back includes the muscles, fasciae, and bones of the posterior portions of the trunk. The back is of the utmost importance in posture, in the support of weight, in locomotion, and in the protection of the spinal cord and spinal nerves.

The bones of the back form the vertebral column, which consists of 24 movable presacral vertebrae, the sacrum, and the coccyx.* The vertebral column, with its muscles and joints, is the axis of the body, a pillar capable of rigidity or flexibility. The head pivots on it, and the upper limbs are attached to it. It completely surrounds or encases the spinal cord, and it partially shields the thoracic and abdominal viscera. It transmits the weight of the rest of the body to the lower limbs and the ground when standing.

---

*A short but excellent account of the vertebral column is given by D. B. Allbrook, E. Afr. med. J., *33*:9, 1956. See also L. A. Hadley, *Anatomico-Roentgenographic Studies of the Spine,* Thomas, Springfield, Illinois, 1964; R. E. M. Bowden, S. Abdullah, and M. R. Gooding, Anatomy of the Cervical Spine, Membranes, Spinal Cord, Nerve Roots and Brachial Plexus, in *Cervical Spondylosis,* ed. by Lord Brain and M. Wilkinson, Saunders, Philadelphia, 1967; D. von Torklus and W. Gehle, *The Upper Cervical Spine,* Grune & Stratton, New York, 1972; and B. S. Epstein, *The Spine,* Lea & Febiger, Philadelphia, 3rd ed., 1969.

VERTEBRAL COLUMN

48

The 24 movable presacral vertebrae comprise 7 cervical, 12 thoracic, and 5 lumbar. The five vertebrae immediately below the lumbar are fused in the adult to form the sacrum. The lowermost four, fused in later life, form the coccyx. The vertebrae of each group can usually be identified by special characteristics. Furthermore, individual vertebrae have distinguishing characteristics of their own.

The vertebral column is flexible because it is composed of many slightly movable parts, the vertebrae. Its stability depends largely upon ligaments and muscles. Some stability, however, is provided by the form of the column and its constituent parts. From the head to the pelvis the column carries progressively more weight. **The vertebrae become progressively larger down to the sacrum, and then become successively smaller. Each vertebra above the last lumbar is taller than the one immediately above it. The length of the vertebral column amounts to about two-fifths of the total height of the body.**

CURVATURES OF THE VERTEBRAL COLUMN

The adult vertebral column presents four sagittal curvatures: cervical, thoracic, lumbar, and sacral (fig. 48–1). These are evident in lateral radiograms. The thoracic and sacral curvatures are termed primary, because they are in the same direction as the curvature of the fetal vertebral column. The primary curves are due to differences in height between the fronts and backs of the bodies of the vertebrae. The secondary curves, cervical and lumbar, begin before birth, and are due mainly to differences in thickness between the anterior and posterior parts of the intervertebral discs. The secondary curves are concave posteriorly and thus compensate for and counteract the primary curvatures which persist in the thoracic and sacral regions. The cervical curve becomes prominent as the infant begins to support and pivot its head, and both the cervical and lumbar curves are accentuated with the assumption of the upright posture. The lumbar curve is said to be more prominent in women.

The *lumbosacral angle,* which is not one of the curvatures, is the angle between the long axis of the lumbar part of the vertebral column and that of the sacrum. It varies from 130 to 160 degrees.[1]

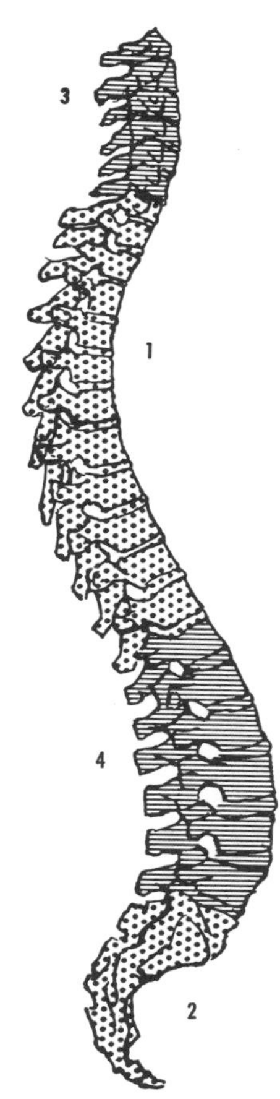

Figure 48–1 Diagrammatic representation of the primary and secondary curves of the vertebral column. The primary curves are (1) the thoracic and (2) the sacral. The secondary curves are (3) the cervical and (4) the lumbar.

An accentuated curvature of the vertebral column with the concavity anteriorly (primary curvature) is sometimes termed a *kyphosis* (hunchback), whereas one with the concavity posteriorly (secondary curvature) is termed a *lordosis*. These terms, however, are commonly used to refer to exaggerated curves resulting from pathological conditions.

A lateral curvature of the spine (to the right or left) is termed *scoliosis*. Scoliosis may be functional or structural. Functional or physiologic scoliosis occurs in the thoracic region, but is not present until middle or late childhood. Its concavity is usually to the left, with compensatory curves above and below. However, the curves are often reversed in left-handed people, and opposite scoliosis occurs in *situs inversus viscerum*.[2] It has been attributed to inequality of muscle actions in walking. However, differences in weight between the two halves of the body may also be a factor. Some degree of lateral curvature and torsion is usually present also in the lumbar region.

Structural scoliosis is abnormal. It appears during childhood and becomes progressively more severe, owing, apparently, to unequal growth of certain vertebrae. The deformity may be extremely pronounced. In most instances the cause of structural scoliosis is unknown.

COMPONENTS OF VERTEBRAL COLUMN

Parts of a Typical Vertebra

A typical vertebra consists of a body, a vertebral arch, and several processes for muscular and articular connections (fig. 48–2).

Each vertebra has three relatively short processes or levers (two transverse processes and one spinous process), and 12 vertebrae are connected with two long levers (ribs). Each lever is acted upon by several muscles or muscle slips. Thus, several hundred slips act upon the whole vertebral column.

The *body* of the vertebra is the part that gives strength and supports weight. It consists mostly of spongy bone that contains red marrow. The compacta on the margins of the upper and lower surfaces of the body is thicker than elsewhere and forms a ring (see ring epiphysis, page 522). The body is separated from the bodies of the vertebrae above and below by an intervertebral disc. Within the ring formed by the raised margin, the bone of the body is pierced by vascular foramina of varying sizes. Foramina are present elsewhere in the vertebral body also, especially on the posterior aspect where rather large ones serve as exits for basivertebral veins. Some of these veins are so large that they notch the vertebral body in front and behind, as seen in lateral radiograms, especially in infants.[3]

Posterior to the body is the *vertebral arch,* which, with the posterior surface of the body, forms the walls of the vertebral foramen. These walls enclose and protect the spinal cord. The vertebral arch is composed of right and left *pedicles* and right and left *laminae.* The lower parts of the inner surfaces of the laminae often show spicules or spurs of bone (p. 517). In intact vertebral columns, the series of vertebral foramina together form the *vertebral canal. A spinous process (spine)* projects backward from each vertebral arch, at the junction of the two laminae (fig. 48–3*A*). *Transverse processes* project on either side from the junction of the pedicle and the lamina (fig. 48–3*B*). *Superior* and *inferior articular processes* on each side bear *superior* and *inferior articular facets,* respectively.

A deep *vertebral notch* is present on the lower edge of each pedicle, and a shallow notch on the upper edge of each pedicle. Two adjacent notches, together with the intervening body and the intervertebral disc, form an *intervertebral foramen,* which transmits a spinal nerve and accompanying vessels.

Radiograms to show the vertebrae are usually taken in posterior and lateral views. In posterior views, the spinous processes appear as ovoid or somewhat elongated shadows, and the pedicles as ovoid shadows (fig. 48–4). The distances between the pedicles can be measured, and the width of the vertebral canal is thereby obtained. The bodies and their cancellous structure are clearly visible in lateral views, as are the translucent areas occupied by the intervertebral discs. Oblique views may be necessary to show the intervertebral foramina, the joints between articular facets, and the pars interarticularis (fig. 48–5). In posterior views, the upper cervical vertebrae are obscured by the shadow of the mandible when the mouth is closed. The atlas and axis, especially the dens and the atlantoaxial joints, may be shown more clearly in a radiogram taken through the open mouth (fig. 60–9, p. 687).

Cervical Vertebrae

The cervical vertebrae are those between the skull and the thorax. They are characterized by the presence of a foramen in each transverse process. This foramen,

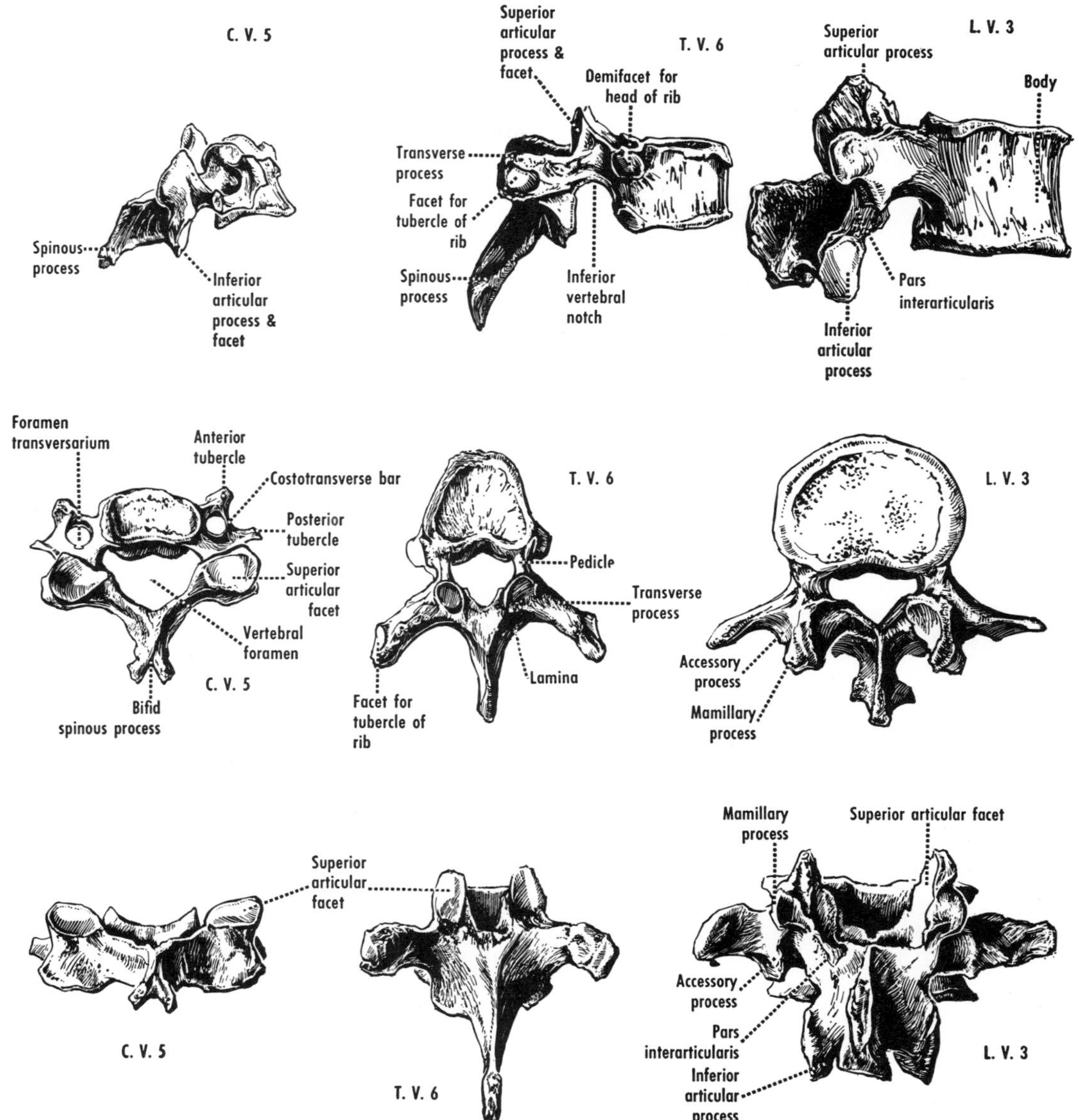

Figure 48–2 The fifth cervical, sixth thoracic, and third lumbar vertebrae, from lateral, superior, and posterior aspects.

termed the *foramen transversarium,* transmits the vertebral artery (except for C. V. 7), the vertebral veins, and a sympathetic plexus. **The first cervical vertebra is called the atlas, and the skull rests on it; it is named after Atlas, who, according to a Greek myth, was reputed to support the heavens. The second cervical vertebra is called the axis, because it forms a pivot around which the atlas turns and carries the skull. The atlas and axis are specialized cervical vertebrae, and the seventh is a transitional vertebra. The third to the sixth cervical vertebrae are regarded as typical.**

Atlas. The atlas[4] has neither spine nor body (fig. 48–6). It consists of two lateral masses, connected by a short anterior arch and by a long posterior arch. The atlas is the widest of the cervical vertebrae.

The *anterior arch,* about half as long as the posterior, presents an anterior *tubercle* in front for the attachment of the anterior

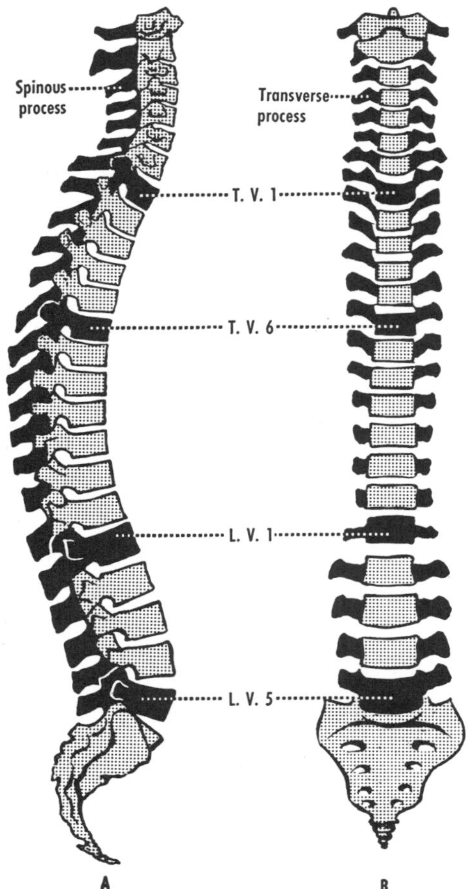

Figure 48–3 Diagrammatic representation of the positions, lengths, and directions of the *(A)* spinous processes and *(B)* transverse processes. The vertebrae indicated in black mark the levels at which a change in direction of curvature occurs.

longitudinal ligament. Behind, the arch has a facet *(fovea dentis)* for the dens of the axis. The transverse ligament of the atlas is attached on each side to a tubercle at the junction of the posterior surface with the lateral mass.

The *posterior arch,* which corresponds to the laminae of other vertebrae, presents a wide *groove for the vertebral artery* on its upper surface. The first cervical nerve also occupies this groove. The lower edge of the posterior atlanto-occipital membrane, which bridges the groove, may become ossified and thereby convert the groove into a foramen for the vertebral artery and first cervical nerve.[5] This trait is familial and genetic.[6] Behind, the posterior arch presents a small tubercle for the attachment of the ligamentum nuchae. The intervertebral foramen for the second cervical nerve is formed by a notch on the lower surface of the arch together with a corresponding notch on the axis.

Each *lateral mass* presents an upper, elongated facet for the corresponding occipital condyle of the skull, and a lower, circular facet for articulation with the axis. The upper facets, which enable nodding movements to take place at the atlanto-occipital joints, are usually constricted near their middle.

The transverse processes, which are related to the internal jugular veins and the accessory nerve, are long, and their ends correspond to the posterior tubercles of the transverse processes of the typical cervical vertebrae. The tip of each can be felt indistinctly through the skin by deep pressure midway between the tip of the mastoid process and the angle of the mandible (just below the auricle).

Axis. The axis, or *epistropheus* (fig. 48–7), is characterized by the *dens* or *odontoid process,* which projects upward from the body. The dens develops as the centrum of the atlas. It articulates in front with the anterior arch of the atlas. Behind, it is usually separated by a bursa from the transverse ligament of the atlas. The apical ligament anchors the tip of the dens to the front margin of the foramen magnum; the alar ligaments anchor it to the lateral margins. Lateral to the odontoid process, the body of the axis presents on each side a facet for the lower surface of the lateral mass of the atlas.

The lower aspect of the axis resembles that of a typical cervical vertebra. It presents two facets for articulation with the articular processes of the third cervical vertebra. They are directed forward and downward, similarly to those of the lower cervical vertebrae.

The thick bifid spinous process can be palpated just below the external occipital protuberance.

The transverse processes of the axis are the smallest of any belonging to cervical vertebrae; each has a tubercle at its end.

Third to Sixth Cervical Vertebrae. Each of these has a small, broad body and a large, triangular vertebral foramen (fig. 48–2). Their spines are short (fig. 48–3), and

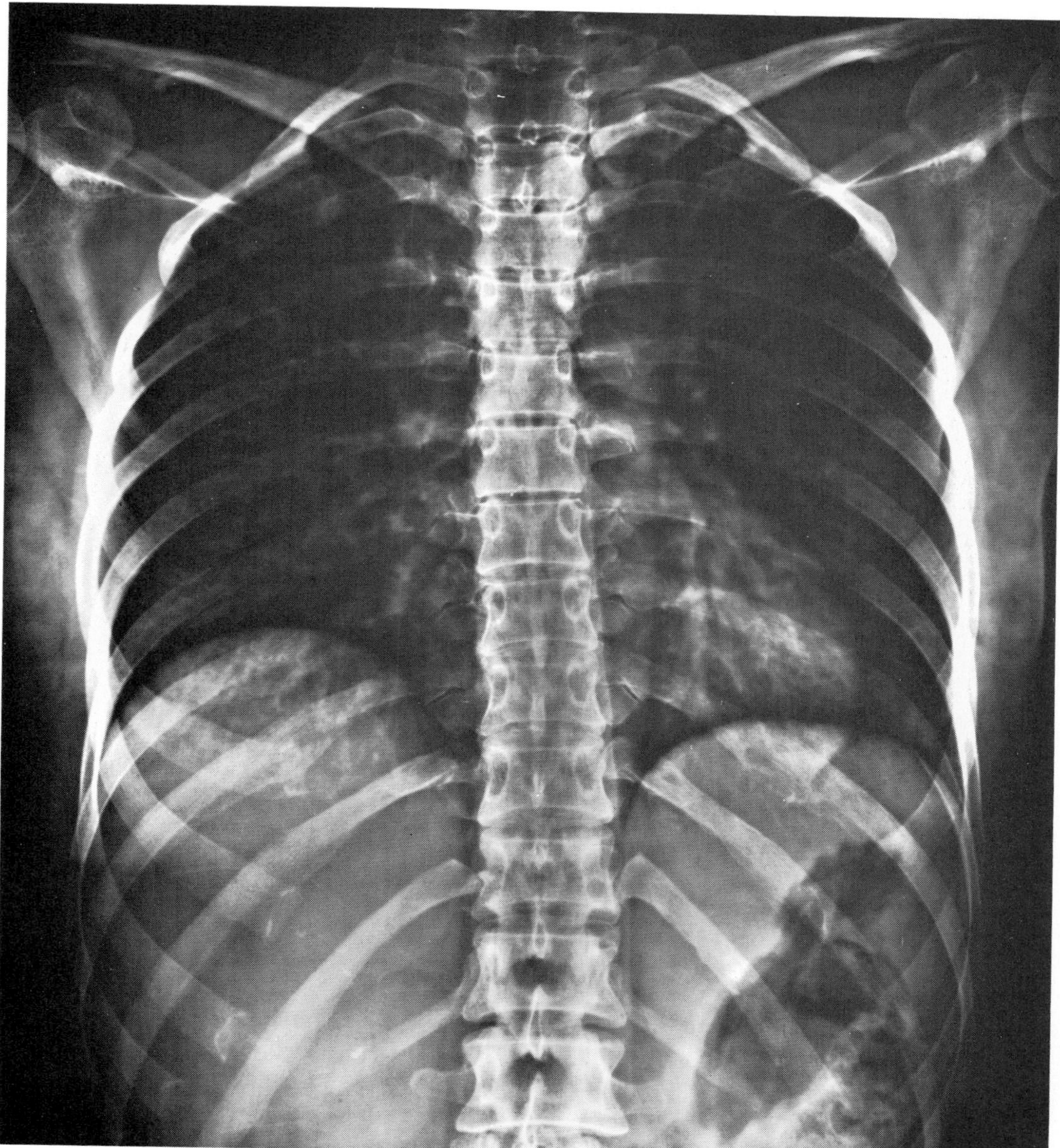

Figure 48–4 Thoracic vertebrae (C.V. 7 and L.V. 1 are included also). Note bodies, pedicles, transverse and spinous processes, and costotransverse joints. Courtesy of V. C. Johnson, M.D., Detroit, Michigan.

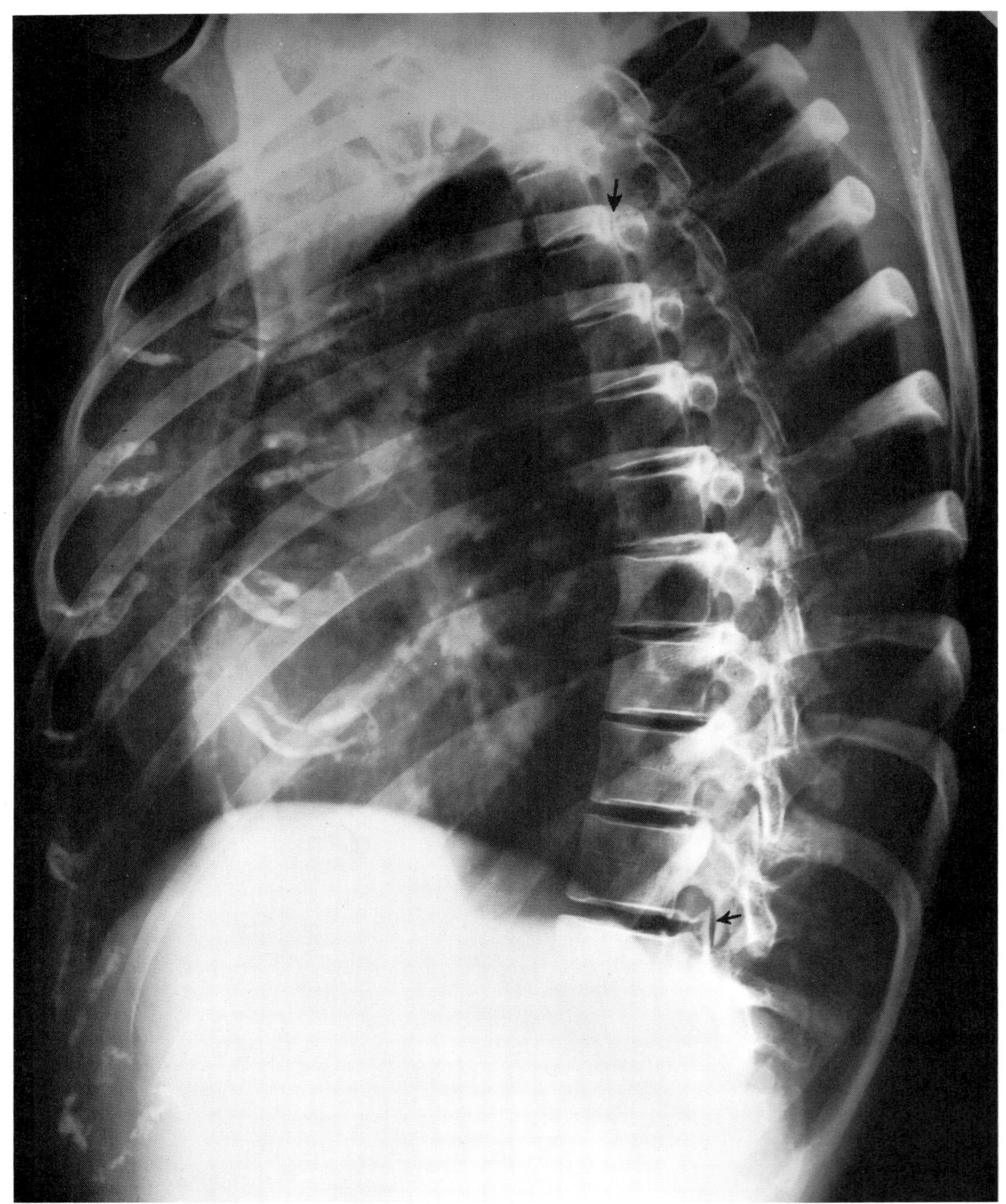

Figure 48–5 Oblique view of thoracic vertebrae. Note costotransverse joint *(upper arrow)* and joint between articular processes *(lower arrow)*. Courtesy of Claude Snead, M.D., Oak Park, Illinois.

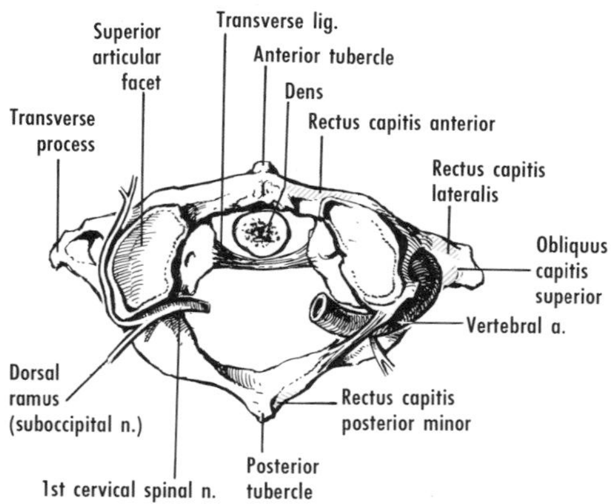

Figure 48–6 The atlas from above. Muscle origins and the vertebral artery are shown on the right side. Based on Frazer.[7]

the ends of the spines are bifid. The spines are usually palpable. At the junctions of pedicles and laminae, each vertebra presents pillars, which consist of the superior and inferior articular processes. These processes present facets, which are placed somewhat more horizontally than vertically. The superior facets are directed upward and backward, the inferior downward and forward.

Each transverse process is pierced by a foramen transversarium and ends laterally in two projections, the *anterior* and *posterior* (scalene) *tubercles.* These are connected by a grooved bridge of bone. Because the anterior tubercle corresponds to a thoracic rib, and the posterior tubercle to the entire transverse process of a thoracic vertebra, the connecting bridge is often called the costotransverse bar. The anterior tubercle of the sixth cervical vertebra is large and is termed the *carotid tubercle,* because the common carotid artery can be compressed against it. Grooves for the ventral rami of spinal nerves are located on the upper aspects of the costotransverse bars. The slope of the groove for the nerves is related to the direction of the ventral rami. The grooves on the third and fourth vertebrae slope more forward. Those on the fifth and sixth vertebrae slope downward.[8] Each of the costotransverse bars of the lower five cervical vertebrae often presents a middle scalene tubercle, for the attachment of a part of the scalenus medius.[9]

The upper borders of the bodies are raised behind, and especially at the sides, and are depressed in front. The raised margins are sometimes termed uncal processes.

The first cervical nerve emerges between the skull and the atlas, and each cervical nerve except the eighth leaves the vertebral canal above the correspondingly numbered vertebra. The eighth emerges above the first thoracic vertebra. The remaining spinal nerves emerge below the correspondingly numbered vertebrae. The ventral ramus of each cervical nerve passes behind the foramen transversarium (and therefore behind the vertebral artery also) of a typical cervical vertebra. The dorsal ramus winds around the front of the articular process.

Seventh Cervical Vertebra. This is characterized by a long spine that does not bifurcate, but ends in a tubercle that gives attachment to the ligamentum nuchae. This

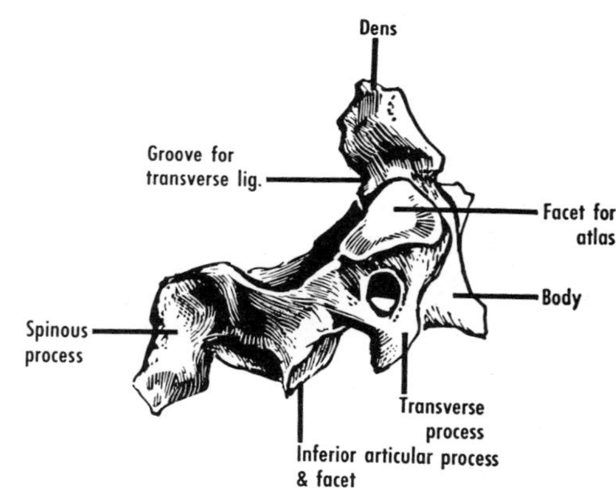

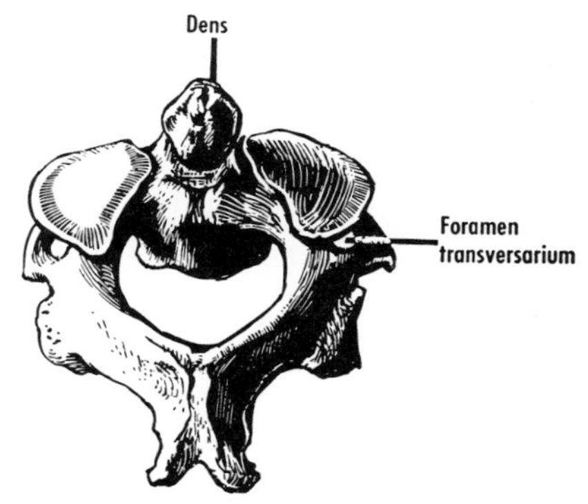

Figure 48–7 Lateral and posterosuperior views of the axis.

vertebra is known as the *vertebra prominens*, although it is only the spine that is prominent. However, three spines (C.V. 6, 7; T.V. 1) are usually visible, especially if the neck is flexed.[10]

The transverse process is large, the costal process small, the anterior tubercle often absent, and the foramen transversarium small and sometimes absent. The foramen transmits small veins, sometimes an accessory vertebral vein, and rarely the vertebral artery. The costal process may develop separately and form a cervical rib (p. 703).

Thoracic Vertebrae

The thoracic vertebrae (figs. 48–4 and 48–5) bear the ribs, and are usually 12 in number. The second to eighth thoracic vertebrae have similar characteristics, and may be regarded as typical. The first, and the ninth to twelfth, have special features that distinguish them from the typical vertebrae.

First Thoracic Vertebra. This resembles a cervical vertebra. The upper border of its body is raised posterolaterally, and forms the anterior boundary of the intervertebral foramen for the eighth cervical nerve.

The head of the first rib articulates with a complete, circular, superior articular facet on each side of the body near the upper border. The head of the second rib articulates with a small inferior facet on the lower border, and with a superior facet on the second thoracic vertebra.

The spinous process, which is more nearly horizontal than those of the typical thoracic vertebrae, may be more prominent than that of C.V. 7.

Second to Eighth Thoracic Vertebrae. These are the typical thoracic vertebrae (fig. 48–2). The outline of the body is kidney- or bean-shaped; that of the vertebral foramen is circular. An impression for the aorta is often evident on the side of the bodies of the fifth to the seventh or eighth vertebrae.

Superior and inferior costal facets of semilunar shape are located on the upper and lower borders of the junction of the body and the arch. The larger superior facet, together with the intervertebral disc above it and the inferior costal facet of the suprajacent vertebra, forms a socket for the head of the corresponding rib (p. 275).

The pedicles are short and compressed from side to side. The laminae slope downward and backward. Each overlaps the lamina of the subjacent vertebra. The spinous process, which is long and slender, slopes downward and backward, overlapping the spine of the vertebra below (fig. 48–3). It ends in a small, easily palpable tubercle, which lies at about the level of the disc below the subjacent vertebra. The transverse processes, which extend laterally, backward, and upward, are long, rounded, and strong. The costal facet on the front of the clubbed end of each articulates with the tubercle of the rib.

The articular facets lie mainly in a coronal plane; the superior facets lie more anteriorly than the inferior.

Ninth Thoracic Vertebra. This possesses only a single costal facet on each side. Each of these facets articulates with the lower half of the head of the ninth rib. A small, semilunar inferior facet is occasionally present.

Tenth Thoracic Vertebra. This presents a large, semicircular costal facet for the tenth rib, but none for the eleventh. The costal facet on the transverse process faces upward. The spine is sometimes short, so that a slight depression is located behind it.[11, *]

Eleventh Thoracic Vertebra. The body of this vertebra resembles that of a lumbar vertebra. Complete, circular costal facets are seen above, encroaching on the pedicle. No costal facets are found on the transverse processes.

Twelfth Thoracic Vertebra. This presents complete circular, costal facets, which are located mainly on the pedicle. The inferior articular processes resemble those of the lumbar vertebrae. The spinous process is short and horizontal. Three processes or tubercles are present in place of the transverse process: (1) a *mamillary process* above, (2) an *accessory process* (inferior tubercle) below, and (3) a *costal process* (lateral tubercle) on the pedicle.[13] All three processes correspond to similar processes on the lumbar vertebrae.

Owing to its transitional position, the twelfth thoracic vertebra, and sometimes the eleventh also, presents characteristics

---

*Sometimes it is the spine of T.V. 11 or 12 which is short.[12]

of both thoracic and lumbar vertebrae. The mamillary process of T.V. 12 is larger and in a different plane from that of L.V. 1, and projects backward behind the superior articular process.[14] Like the other mamillary processes, it serves for muscle attachments (p. 528). The articular facets of T.V. 11 and 12 are asymmetrical. Usually the right facet is flat, whereas the left is concave and resembles the lumbar articular facets.[15] One of the concave facets is usually shaped so that, in articulating with the facet of the subjacent vertebra, it forms a joint comparable in arrangement to a carpenter's mortise.[16]

Bony spurs are often present in the ligamenta flava connecting the thoracic vertebrae. These spurs, which project downward from the lamina, occur almost entirely on thoracic vertebrae, and chiefly on the tenth thoracic vertebra.[16, 17] They have been termed para-articular processes.

Lumbar Vertebrae

The lumbar vertebrae (figs. 48–8 and 48–9), which are the vertebrae between the thorax and the sacrum, are distinguished by their large size, by the absence of costal facets and foramina transversaria, by their thin transverse processes, and by their quadrilateral spinous processes (fig. 48–2). They account for much of the thickness of the trunk in the median plane (one-third to one-half in thin individuals).

The lumbar vertebrae present the following common characteristics. The bodies are kidney-shaped; the concavity faces the triangular vertebral foramen. The pedicles are short and thick. The laminae are also short, thick, and relatively uneven; they extend below the level of the pedicles. The part of the lamina between the superior and the inferior articular process is sometimes called the pars interarticularis. The quadrilateral, hatchet-shaped spinous processes extend horizontally backward, and their lower edges are at about the level of the lower surface of the body. The superior articular facets are concave medially, the inferior ones convex laterally. The joints formed by them are therefore in almost a sagittal plane. The *mamillary processes* project backward from the superior articular processes. The long thin *transverse* or *costal processes,* which are homologous to ribs, extend laterally and somewhat backward (fig. 48–3). Small *accessory processes* project downward from the lower aspect of the transverse processes at their junction with the pedicles.

It is relatively easy to distinguish lumbar vertebrae and to put them into their proper order when they are considered together, but it is much more difficult to identify a specific lumbar vertebra when it is isolated from the series.[18] The first has the most distinct accessory process, the smallest transverse process, and the narrowest pedicle. The fifth has a wedge-shaped body, thick, rough transverse processes, widely separated inferior articular processes, and a smaller, more rounded spine. From the second to the fourth, the pedicles become thicker, the mamillary processes become smaller, the width of the body increases, and the articular facets become more variable.[19, 20] Asymmetry of the facets, and disorders of the joints of these and other facets, may be factors in low back pain.

Fifth Lumbar Vertebra. This is usually the largest of the vertebrae. It is distinguished by strong, massive transverse processes, each of which is anchored to all of the adjacent pedicle and encroaches on the body. The body is thicker in front than behind, a shape that is associated with the prominence of the lumbosacral angle. The inferior articular processes are widely separated from each other, and their facets face forward and laterally. The joints formed between them and the superior articular facets of the sacrum are about in a coronal plane. Frequently, however, the facets are asymmetrical, facing approximately inward on one side and backward on the other.

Sacrum

The sacrum (figs. 48–10 and 48–11) is formed by five vertebrae* that fuse in the adult to form a wedge-shaped bone, which can be palpated below the "small of the back." The fusion is such that the transverse and costal elements of each vertebra meet the corresponding parts of adjoining vertebrae lateral to the intervertebral foramina and form sacral foramina, which are completely ringed by bone or cartilage. The sacrum articulates above with the fifth lumbar ver-

*In some races, the sacrum usually has six elements.[21]

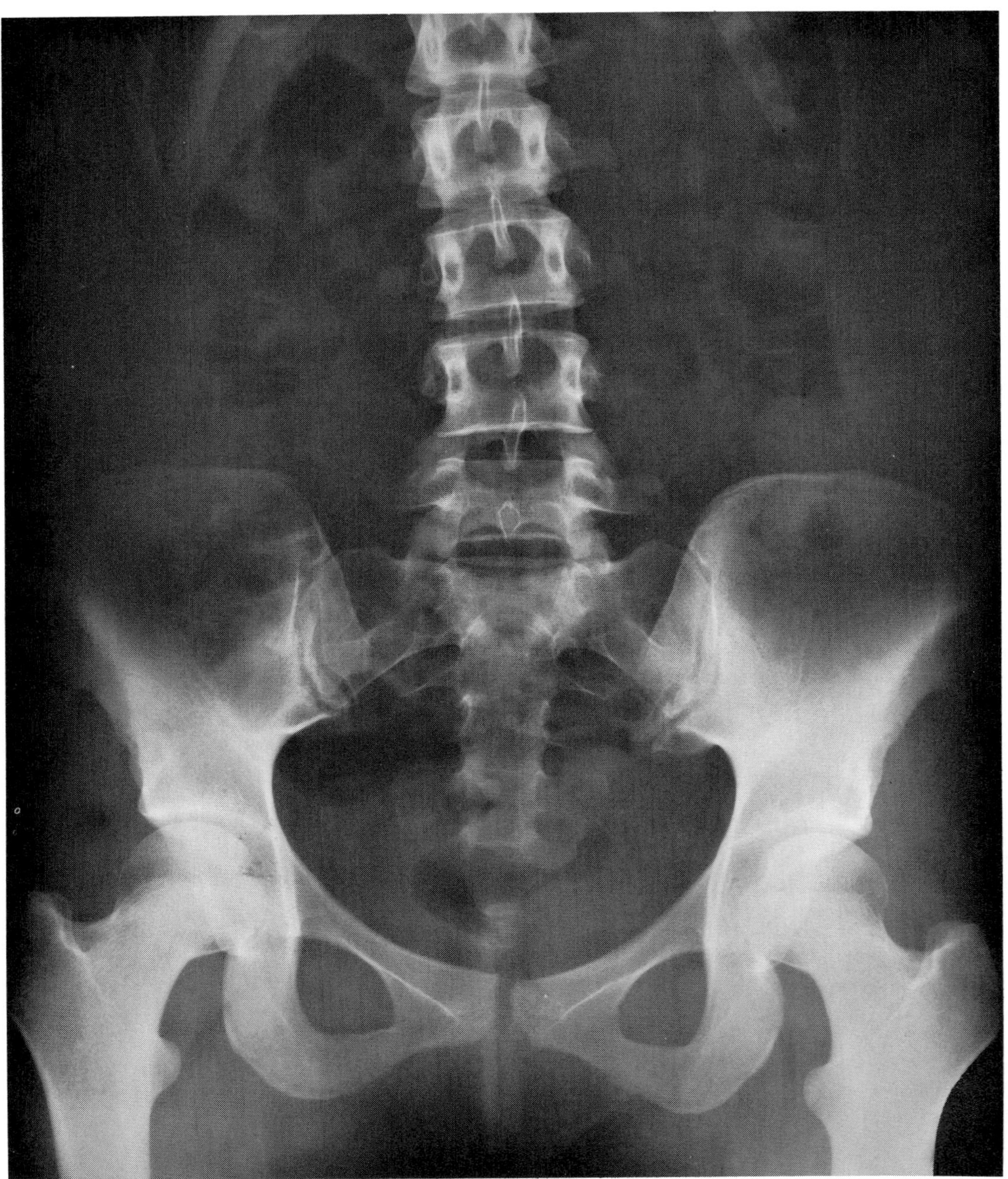

Figure 48–8 Lumbar vertebrae and pelvis of female.

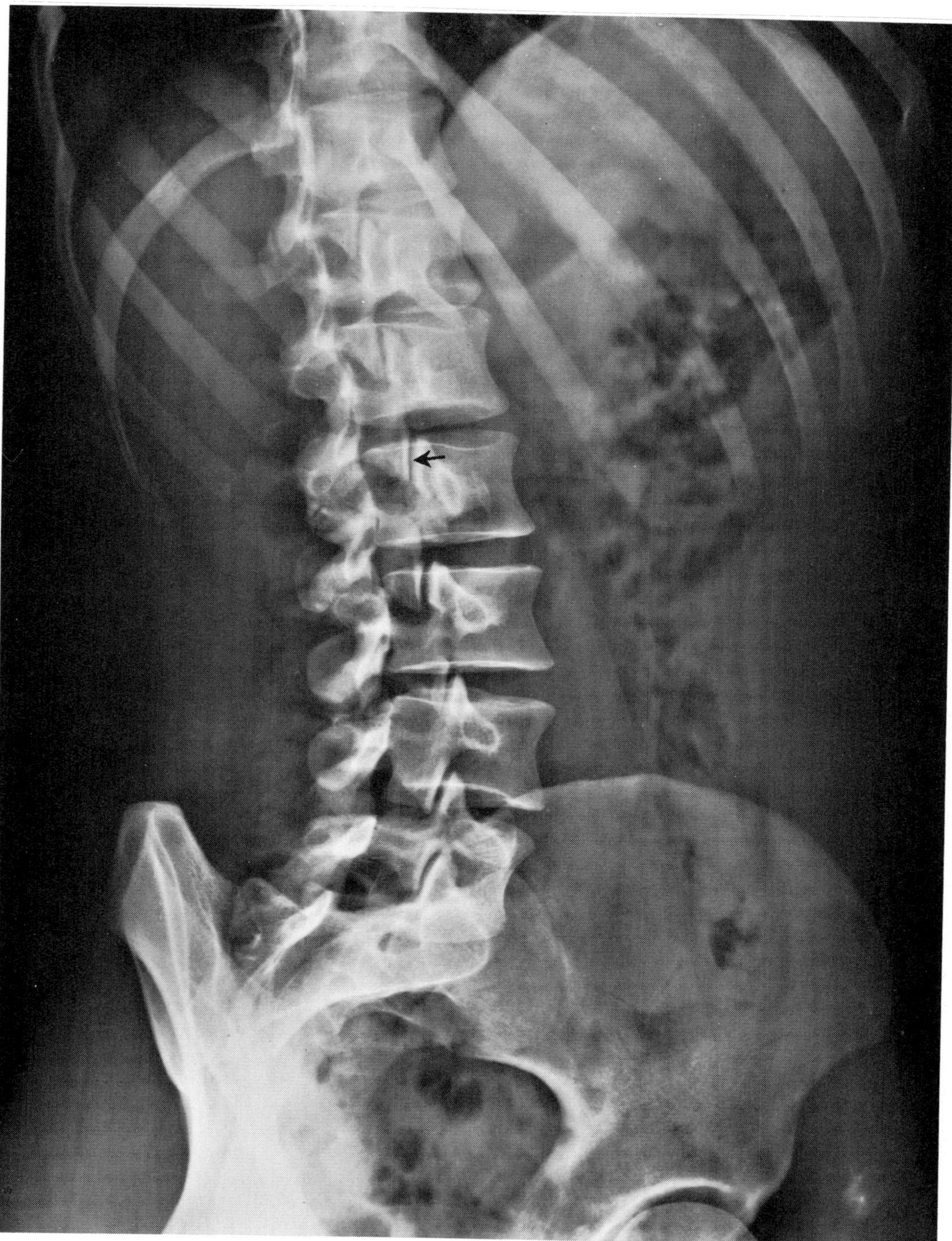

Figure 48–9 Oblique view of lumbar vertebrae. Note very small twelfth rib, joints between articular processes of lumbar vertebrae (arrow indicates the joint between the articular processes of L.V. 1 and L.V. 2), and sacrum.

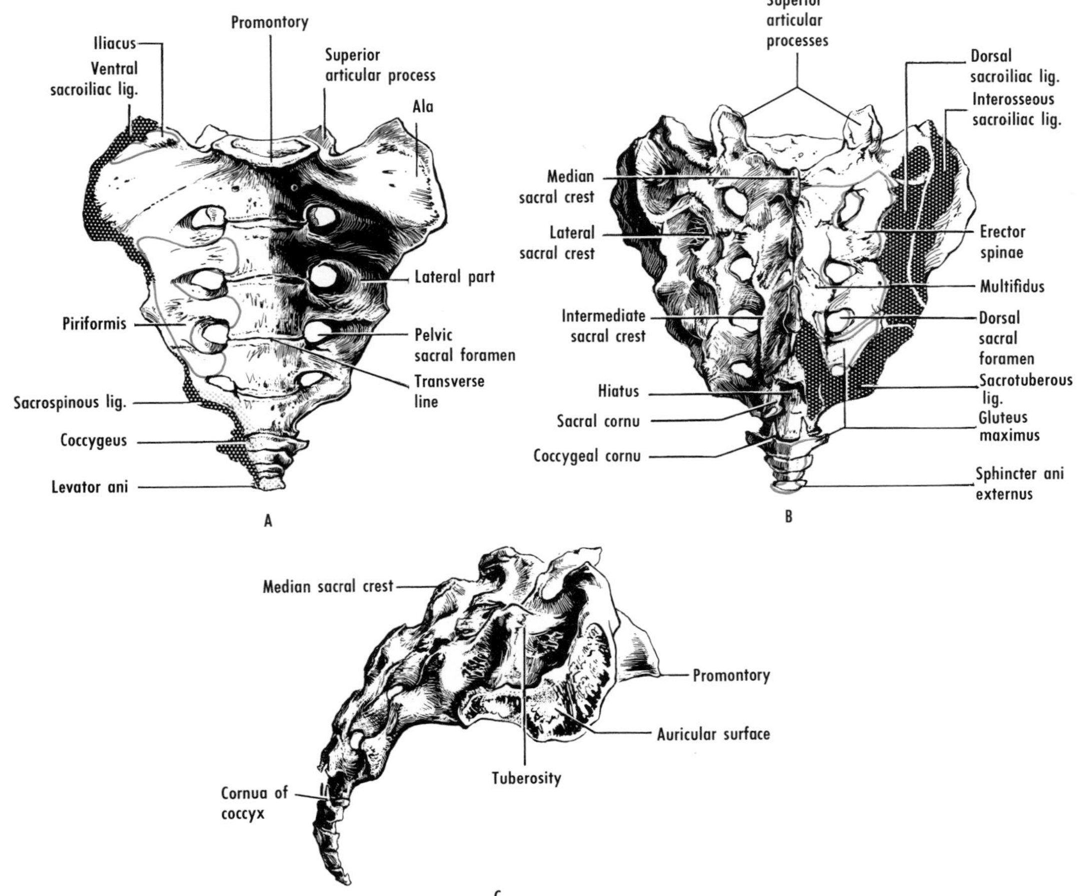

Figure 48–10 Female sacrum and coccyx. Pelvic *(A)* and dorsal *(B)* aspects, showing muscular and ligamentous attachments. The central part of the area indicated as the origin of the erector spinae is actually the origin of the multifidus, which is covered by the erector spinae. *(C)*, lateral aspect, in the anatomical position.

tebra and on the sides with the hip bones. It has pelvic and dorsal surfaces, and two lateral parts. The center of gravity of the body is about 1 cm behind the sacral promontory (range: from 1/2 cm in front to 4 cm behind).[22]

The *pelvic surface,* concave and smooth, faces downward and forward. Its median part, which represents the fronts of the fused bodies of the five vertebrae, is crossed by four *transverse lines.* These lines mark the fusions and indicate the levels of the intervertebral discs, remains of which can usually be seen in sagittal sections. Four pairs of *pelvic sacral foramina* at the ends of the lines transmit the ventral rami of the first four sacral nerves and the accompanying vessels.

The *dorsal surface,* rough and convex, faces backward and upward. The spines of the upper three or four sacral vertebrae are modified to form the *median sacral crest,* variable in prominence. The uppermost spine is usually the largest, the second is often fused with the third, whereas the fourth is rudimentary. The sacral groove on each side of the crest represents fused laminae. Fused articular processes form the *intermediate sacral crest,* just lateral to the groove. Four pairs of *dorsal sacral foramina* lateral to the intermediate crest transmit the dorsal rami of the sacral nerves and the accompanying vessels. The intermediate sacral crests project downward as the *sacral cornua* or *horns* (the articular processes of the fifth sacral vertebra), which bound the *sacral hiatus*. The hiatus, immediately below the median crest, is a gap shaped like an inverted **V** where no spine or lamina (of S.V. 5) forms and only a membrane is present. The cornua articulate with the coccygeal cornua. The sacral hiatus varies greatly in depth, owing in part to the fact that the laminae and spine of S.V. 4 may be incomplete or absent.

Anesthetics can be injected through the sacral hiatus, a procedure known as caudal analgesia.[23] Spreading upward and extradurally, the anesthetic acts directly on the spinal nerves. The height to which it ascends can be controlled by the amount injected and by the position of the patient. Anesthetics can also be injected through the dorsal sacral foramina.

The *lateral part* or mass of the sacrum is the part of the sacrum lateral to the foramina. It consists of the fused transverse processes together with their costal elements. The transverse processes are indicated by a series of elevations, which form the *lateral sacral crest*, immediately lateral to the dorsal sacral foramina. The upper part of the lateral mass presents the irregular, sometimes flat, often ear-shaped, *auricular surface*, which articulates with the ilium. The *sacral tuberosity*, to which the interosseous sacroiliac ligaments are attached, limits the auricular surface behind. Very often accessory facets are present above and behind the auricular surface.[24] These facets, the incidence of which appears to vary with age, form synovial joints with similar facets on the ilium.

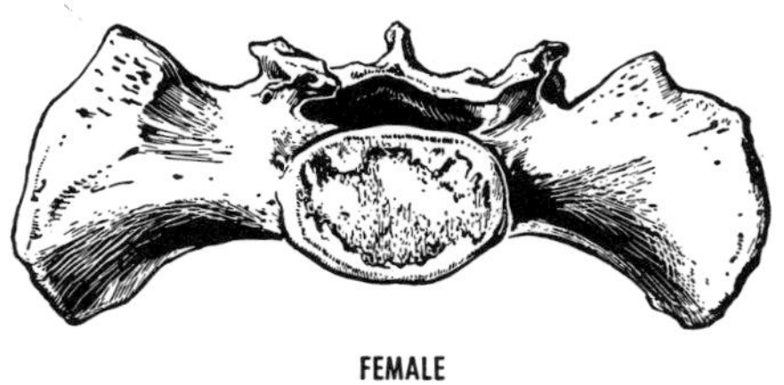

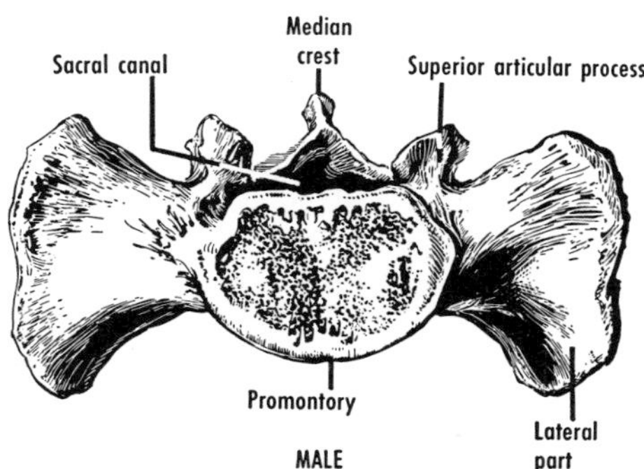

Figure 48–11 Female and male sacra from above. The superior aspect of the lateral part constitutes the ala.

The *base* or upper surface presents the *promontory* (anterior margin of the upper surface of the body of S.V. 1), the sacral canal, right and left *alae* (upper surfaces of the lateral parts), and two superior articular processes with mamillary processes. The promontory is the sacral portion of the terminal lines (fig. 40–1, p. 439). The articular processes and their facets vary, as do those of the fifth lumbar vertebra. The sacral canal, which begins at the base and ends at the hiatus, is enclosed by the modified bodies and arches. In transverse section, the canal is triangular in outline above, but it is flattened anteroposteriorly below. Four large openings on each side divide and end as the pelvic and dorsal sacral foramina. The sacral canal contains the dural sac and the filum of the dura mater, the lower part of the cauda equina, and the filum terminale.

The apex of the sacrum is separated from the coccyx by an intervertebral disc. The apex and the coccyx may be fused.

Congenital absence of the sacrum (sacral agenesis) is rare, and is usually associated with other skeletal abnormalities.[25] Less severe degrees of failure of development are more common. They may be associated with neurological deficits.

Sexual Differences (fig. 48–11). The sagittal curve of the male sacrum is quite uniform, whereas that of the female is sharper below. The female sacrum is wider and shorter, and its pelvic surface faces in a more downward direction. The auricular surface of the male sacrum includes S.V. 1 to 3, whereas that of the female may be limited to S.V. 1 and 2. The various differences that have been listed cannot, however, be completely relied upon for positive identification of the sex of the subject from whom the sacrum came.

Coccyx

The coccyx[26] lies slightly above the anus. Like the sacrum, it resembles a wedge in shape and has a base, an apex, pelvic and dorsal surfaces, and two lateral borders. It usually consists of four segments (vertebrae), sometimes five, and occasionally three. The first has short transverse processes that connect with the sacrum, and two *cornua* or horns that are connected to the sacral cornua. The first segment is often partly and sometimes entirely fused with the sacrum. The second segment, which has rudimentary horns, can move on the first. It retains this movement even if the first is fused with the sacrum, an important feature during parturition. The third and fourth segments are rudimentary. The joint between the second and third segments is sometimes movable; that between the third and fourth is occasionally so.

DEVELOPMENT[27] AND OSSIFICATION[28] OF VERTEBRAL COLUMN

The vertebrae begin to develop during the embryonic period as mesenchymal condensations around the notochord (fig. 48–12). Later, the mesenchymal condensations chondrify, and the cartilage thus formed is in turn replaced by bone. The pattern of ossification is subject to wide variation among individuals and also in different regions of the same vertebral column.

Most vertebrae begin to ossify during the fetal period, when three or more primary centers of ossification begin in each cartilaginous vertebra (usually one center in the centrum and one in each half of the neural arch). In spite of many statements in the literature regarding the presence of two centers in each half of the neural arch, it is clearly established that only one is present in each half.[29]

At birth, the last sacral and the coccygeal vertebrae may still be entirely cartilaginous (in which event they begin to ossify during infancy).

Bony union begins during early childhood and occurs at two sites: (1) at each neurocentral joint by union of the centers of the neural arch and centrum, and (2) at the junction of the two neural arch centers posteriorly, from where ossification spreads into the spinous process. The neural arches of the sacral vertebrae and the posterior arch of the atlas do not fuse with their respective centra until late childhood, and they sometimes fail to fuse. During early childhood, also, the costal elements that develop separately (cervical and sacral) fuse with the bodies of their respective vertebrae.

At about the time of puberty, a secondary center of ossification appears in the margin of each growth plate (the plate of hyaline cartilage on the upper and lower aspects of the body), and is termed a *ring epiphysis*. This ring epiphysis usually unites with the vertebral body early in adult life.* The union results in a characteristic smooth, raised margin around the edges of the upper and lower surfaces of the vertebral body.

Also, at about the time of puberty, secondary centers often appear in the cartilage of the tips of the major processes. These secondary centers are usually united by early adult life, but the times of their fusion are highly variable. A persistent epiphysis is sometimes mistaken in a radiogram for a fracture.[31]

The atypical vertebrae, atlas,[32] axis, and sacrum[33] show special patterns of ossification. Some 56 to 58 primary and secondary centers have been described for the sacrum.

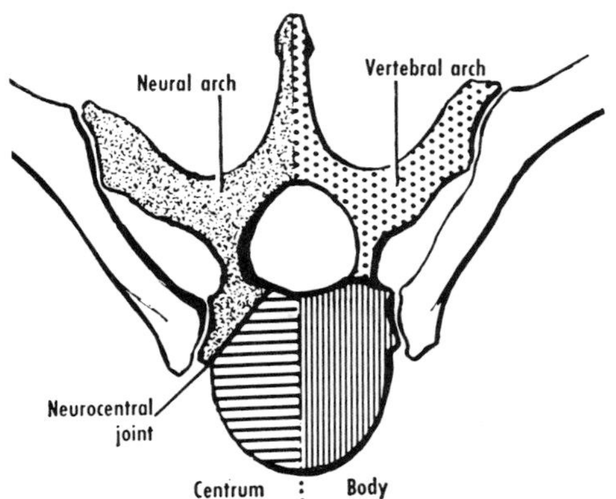

Figure 48–12 Schematic representation of the neural arch and centrum *(left half of figure)*, and the vertebral arch and body *(right half)*. The terms centrum and neural arch refer specifically to those parts of a vertebra ossified from primary centers. The terms vertebral arch and body are descriptive terms, generally applied to adult vertebrae. The body of a vertebra includes the centrum and part of the neural arch. The vertebral arch, therefore, is less extensive than the neural arch. Note that the rib articulates with the neural arch and not with the centrum.

SERIAL HOMOLOGIES

In a serial or repeating series of elements such as vertebrae, certain homologies may be obvious. For example, spinous processes are homologous. Transverse processes, however, are more difficult to homologize. It is generally assumed that during the early stages of development two processes form in the region of the transverse process. In the thorax, one process (costal element) becomes the rib; the other (transverse element) becomes the transverse process. The transverse process of a cervical vertebra is a composite process, its anterior part developing from the costal element and its posterior part from the transverse element (fig. 48–13). The "transverse process" of a lumbar vertebra is thought to develop from the costal element, and is therefore homologous with a rib. The mamillary and accessory processes are thought to develop from the transverse element. The lateral part of the sacrum is also composite, being developed from costal and transverse elements. The homologies of the various parts of the transverse processes have not yet been adequately verified by embryologic studies.

*A small but significant increase in the height of the vertebral bodies of men occurs between the ages of 20 and 45. Increase in stature may thus continue for a longer time than is ordinarily supposed.[30]

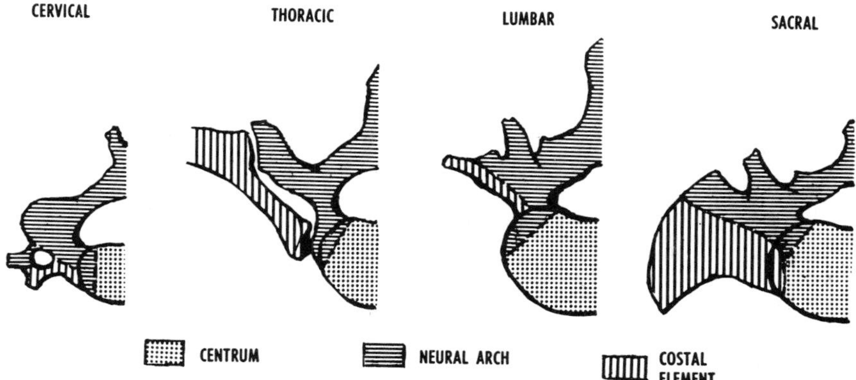

Figure 48–13 Schematic representation of horizontal sections of vertebrae, showing what are thought to be homologies. Note that the costal element forms a part of the transverse process of cervical vertebrae. It forms the rib in the thoracic region, most of the transverse process in the lumbar region, and the greater portion of the lateral part of the sacrum.

VARIATIONS[34] AND ANOMALIES

Variations in vertebrae are affected by race and sex, and by genetic and environmental factors. For example, columns with an increased number of vertebrae occur more often in males; those with a reduced number occur more often in females.

Many variations are congenital. They include, for example, variations in number, shape, and position (fig. 48–14). Among the striking congenital variations are those involving numbers of vertebrae. For in-

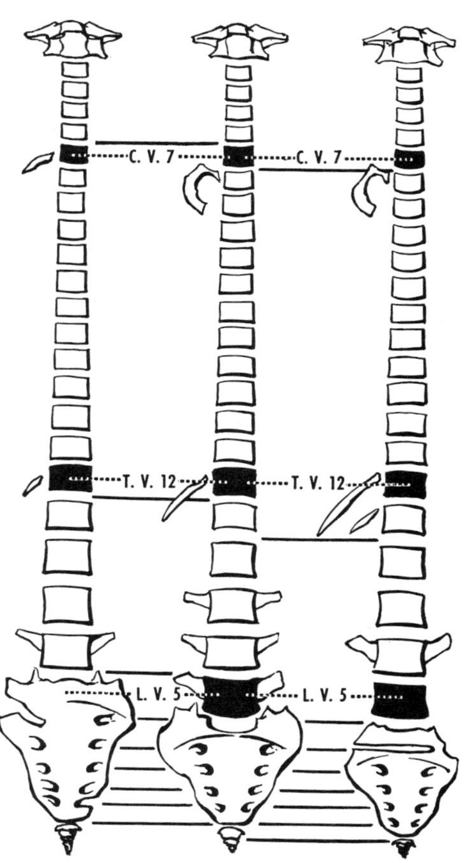

Figure 48–14 Variations in vertebrae. The middle diagram shows the common arrangement of vertebrae and the first and twelfth ribs. In the diagram on the left, which shows cranial variation, there are thirteen ribs. A cervical rib articulates with the seventh cervical vertebra, and the twelfth thoracic rib is reduced in size. The fifth lumbar vertebra is partially incorporated into the sacrum (sacralization of fifth lumbar vertebra). In some instances it is incorporated completely. There is partial segmentation of the lowest sacral segment. The diagram on the right shows caudal variation. The twelfth thoracic rib is increased in size and there is a small lumbar rib. The transverse processes of the fourth lumbar vertebra are increased in size; those of the fifth are greatly diminished. The first sacral segment is partially separated from the rest of the sacrum (lumbarization of first sacral segment). Sometimes it is completely separated. The upper segment of the coccyx is incorporated in the sacrum. Based on Schinz *et al.*[35]

stance, nearly all columns have the modal number of 24 presacral vertebrae. A few have 23, and some have 25. However, in columns with the modal number, there are some with more than five lumbar vertebrae (and a compensatory decrease in the number of thoracic vertebrae), and some with fewer than five lumbar vertebrae (and a compensatory increase in the number of thoracic vertebrae). The number of cervical vertebrae, however, is relatively stable. In characterizing a variation or in determining the total number of vertebrae, the thoracic and lumbar regions must be considered together. A radiogram of only the lumbar part of the vertebral column is inadequate in this regard.

Some races show much more variation in the vertebral column and have a greater tendency to an increased number of presacral vertebrae.[36] East Africans have the highest reported incidence of variation in the number of vertebrae; 15 per cent have other than the modal number of presacral vertebrae.[37] Population groups in North America are reported to have 11 per cent variation, with males tending to have more (25) presacral vertebrae, and females fewer (23).[38]

REFERENCES

1. W. H. Robinson and H. W. Grimm, Arch. Surg., Chicago, *11*:911, 1925.
2. A. Farkas, J. Bone Jt Surg., *23*:607, 1941.
3. G. P. Wagoner and E. P. Pendergrass, Amer. J. Roentgenol., *42*:663, 1939.
4. A. J. E. Cave, J. Anat., Lond., *68*:416, 1934.
5. W. F. Ossenfort, Amer. J. phys. Anthrop., *9*:439, 1926. B. G. H. Lamberty and S. Zivanović, Acta anat., *85*: 113, 1973.
6. S. Selby, S. M. Garn, and V. Kanareff, Amer. J. phys. Anthrop., *13*:129, 1955.
7. Frazer's *Anatomy of the Human Skeleton*, revised by A. S. Breathnach, Churchill, London, 6th ed., 1965.
8. For these and other details on the markings of the cervical vertebrae, see F. W. Jones, J. Anat., Lond., *46*:41, 1911. But for an interpretation of the cause of the differences in slope and for illustration of the relationships, see A. J. E. Cave, J. Anat., Lond., *71*:497, 1937.
9. A. J. E. Cave, J. Anat., Lond., *67*:480, 1933.
10. L. J. A. Di Dio, Anat. Anz., *120*:210, 1967.
11. E. F. Cyriax, J. Anat., Lond., *56*:147, 1922.
12. T. A. Willis, Amer. J. Anat., *32*:95, 1923.
13. A. J. E. Cave, J. Anat., Lond., *70*:275, 1936.
14. E. B. Kaplan, Surgery, *17*:78, 1945.
15. C. Whitney, Amer. J. phys. Anthrop., *9*:451, 1926.
16. P. R. Davis, J. Anat., Lond., *89*:370, 1955.
17. H. Nathan, Anat. Rec., *133*:605, 1959.
18. E. Fawcett, J. Anat., Lond., *66*:384, 1932.
19. J. F. Brailsford, Brit. J. Surg., *16*:562, 1929. P. N. B. Odgers, J. Anat., Lond., *67*:301, 1933.
20. V. Putti and D. Logròscino, Chir. Organi Mov., *23*:317, 1938. C. E. Badgley, J. Bone Jt Surg., *23*:481, 1941.
21. O. S. Heyns and J. E. Kerrich, Amer. J. phys. Anthrop., *5*:67, 1947.
22. B. Åkerblom, *Standing and Sitting Posture*, A.-B. Nordiska Bokhandeln, Stockholm, 1948.
23. Variations of the sacrum that are important in connection with caudal analgesia are described by M. Trotter and G. S. Letterman, Surg. Gynec. Obstet., *78*:419, 1944; by G. S. Letterman and M. Trotter, Surg. Gynec. Obstet., *78*:551, 1944; and by M. Trotter, Curr. Res. Anesth., *26*:192, 1947.
24. M. Trotter, Amer. J. Phys. Anthrop., *22*:247, 1937; J. Bone Jt Surg., *22*:293, 1940. L. A. Hadley, J. Bone Jt Surg., *34A*:149, 1952.
25. I. M. Zeligs, Arch. Surg., Chicago, *41*:1220, 1940. A. Lichtor, Arch. Surg., Chicago, *54*:430, 1947. J. F. Katz, J. Bone Jt Surg., *35A*:398, 1953.
26. A detailed study of the development, ossification, variations, and joints of the coccyx has been published by R. Dieulafé, Arch. Anat., Strasbourg, *16*:41, 1933.
27. E. C. Sensenig, Contr. Embryol. Carneg. Instn., *33*:21, 1949; *36*:141, 1957.
28. C. R. Noback and G. G. Robertson, Amer. J. Anat., *89*:1, 1951.
29. S. Friberg, Acta chir. scand., Suppl. 55, 1939. G. G. Rowe and M. B. Roche, J. Bone Jt Surg., *35A*:102, 1953. J. Mutch and R. Walmsley, Lancet, *1*:74, 1956.
30. D. B. Allbrook, Amer. J. phys. Anthrop., *14*:35, 1956.
31. W. Bailey, Amer. J. Roentgenol., *42*:85, 1937.
32. For details of the sequence of ossification in the cervical vertebrae, see D. K. Bailey, Radiology, *59*:712, 1952.
33. E. Fawcett, Anat. Anz., *30*:414, 1907. E. N. Cleaves, Amer. J. Roentgenol., *30*:450, 1937.
34. A.-F. LeDouble, *Traité des variations de la colonne vertébrale de l'homme*, Vigot, Paris, 1912. See also A. H. Schulz and W. L. Straus, Proc. Amer. phil. Soc., *89*:601, 1925, on the numbers of vertebrae in primates. A review of vertebral variations has been published by K. Theiler, Z. KonstLehre, *31*:271, 1952.
35. H. R. Schinz *et al.*, *Roentgen-Diagnostics*, English translation edited by J. T. Case, Grune & Stratton, New York, vols. 1 and 2, 1921.
36. L. R. Shore, J. Anat., Lond., *64*:206, 1930. T. D. Stewart, Amer. J. Phys. Anthrop., *17*:123, 1932.
37. D. B. Allbrook, Amer. J. phys. Anthrop., *13*:489, 1955; E. Afr. med. J., *33*:9, 1956.
38. P. E. Bornstein and R. R. Peterson, Amer. J. phys. Anthrop., *25*:139, 1966.

MUSCLES, VESSELS, NERVES, AND JOINTS OF BACK

49

MUSCLES

The muscles of the back are arranged into two main groups, anterior and posterior. Those on the anterior aspect of the vertebral column (prevertebral muscles) include muscles in the neck and in the abdomen; they are supplied by ventral rami of spinal nerves. **Those muscles on the posterior aspect of the vertebral column include (1) a superficial layer composed of the trapezius and the latissimus dorsi; the sternocleidomastoid is also seen at the back of the neck; (2) a deeper layer composed of the levator scapulae, the rhomboids, and the serrati posteriores; and (3) still deeper layers comprising the proper muscles of the back, which are supplied mostly by dorsal rami.** (The levatores costarum, which also belong to this group, are described with the thorax) (p. 266).

The subcutaneous tissue of the back is thick and, in spite of its contained fat, is very tough. The fascia of the back is attached in the median plane to the spines, supraspinous ligaments, and ligamentum nuchae. Extending laterally, it ensheathes the muscles of the back, including the superficial ones, and is continuous with the fasciae of the neck, axilla, thorax, and abdomen. Above, it is attached to the superior nuchal line, and below, to the iliac crest. The part of the fascia in the neck is termed the nuchal fascia; the remainder is called the thoracolumbar fascia. The noun *nucha* and the adjective *nuchal,* terms of Arabic origin, refer now to the nape or back of the neck.

The *thoracolumbar (lumbar)* fascia extends laterally from the vertebral spines and forms a retinaculum or retaining layer for the underlying muscles. In the thoracic region, it is attached to the angles of the ribs. In the lumbar region, it consists of several rather thick ensheathing layers. On each side, a strong glistening posterior layer extends laterally from the spines and, at the lateral edge of the underlying erector spinae, divides and encloses the latissimus dorsi (fig. 49–10). Between the latissimus dorsi and the external oblique it forms the roof of the lumbar triangle. Below, this layer reaches the iliac crest and the sacrum. Each of the intertransverse ligaments of the lumbar region divides and encloses the quadratus lumborum, and in so doing forms the middle and anterior layers of the thoracolumbar fascia (fig. 49–10). These two layers meet at the lateral edge of the quadratus lumborum, where they join the posterior layer and form a strong, more or less common aponeurotic sheet, to which the internal oblique and the transversus abdominis are attached, and with which their investing fasciae are continuous.

The glistening appearance of the posterior layer is due to the tendinous expansions into it of the latissimus dorsi and the serratus posterior inferior. The aponeurosis of the latter muscle is practically inseparable from the posterior layer of the fascia.

Serrati Posteriores.[2] These are two thin, partly membranous muscles that are supplied by ventral rami of spinal nerves. They perhaps serve as retinacula for the deep muscles. The *serratus posterior superior,* hidden by the rhomboids, extends from the ligamentum nuchae and the spines of the last cervical and several upper thoracic vertebrae to the second to fifth ribs. Nerve supply: C8 to T3. The *serratus posterior inferior,* lying deep to the latissimus dorsi, extends from the spines of the lower thoracic vertebrae to the lower four ribs. Nerve supply: T9 to 11.

Deep Muscles

These, the proper muscles of the back, form a complex mass of poorly defined bundles that, for the most part, lie in the "gutters" of the vertebral column. With few exceptions, the muscles are supplied by the dorsal rami of spinal nerves.

These muscles can be grouped according to the direction and attachments of their component bundles or slips. The more superficial bundles are fairly long and straight, whereas the deeper bundles are progressively shorter and more oblique. **In the lumbar region the deep muscles form a medial (transversospinalis) and a lateral (erector spinae) group.[3] At higher levels, the arrangement becomes increasingly complex.**

Erector Spinae and Splenius (fig. 49–1). These muscles form a longitudinal series that extends from the sacrum to the skull.

The erector spinae (*sacrospinalis*) begins at the sacrum (fig. 48–10, p. 520), the ilium, and associated ligaments. It thickens as it ascends alongside the lumbar spines, from which it receives additional attachments. At about the level of the last rib, it divides into three columns that ascend on the back of the chest, where they are inserted into ribs and vertebrae. Additional slips also arise from these bones and continue into the neck.

The narrow medial column of the three lies alongside the spines and consists of slips that extend between the upper lum-

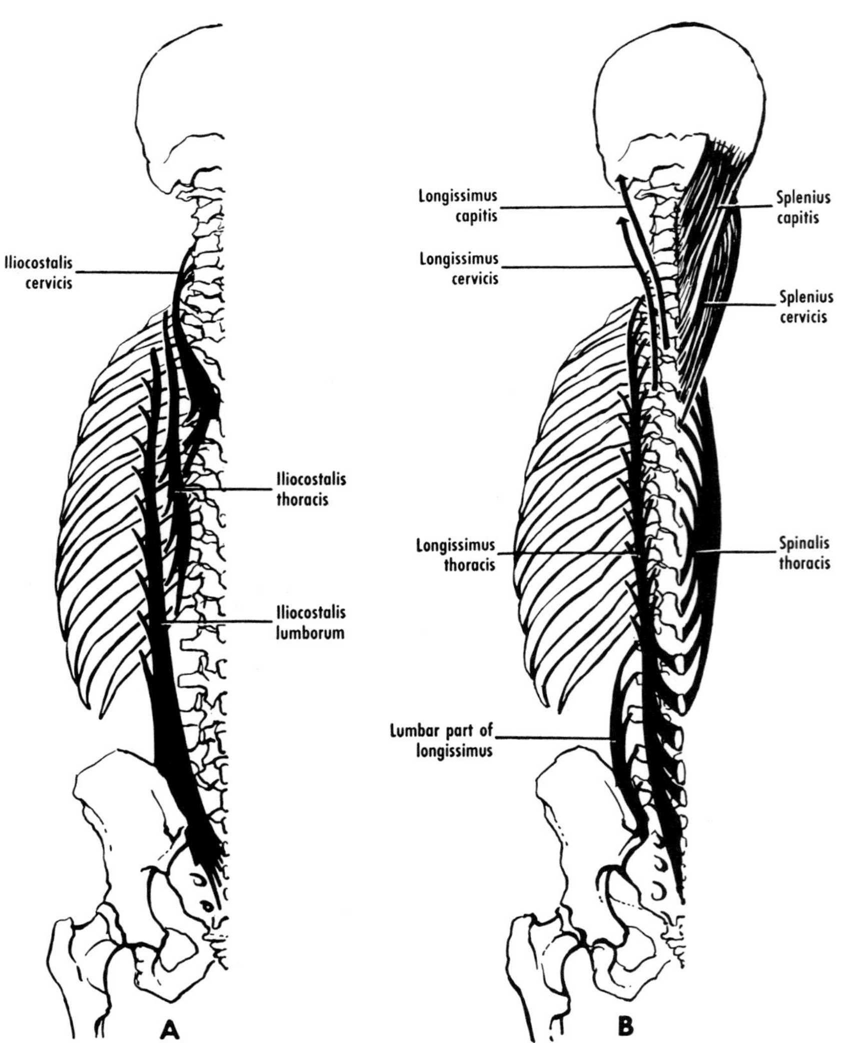

Figure 49–1 Diagrammatic representation of the parts of the erector spinae and the splenius. *A*, the parts of the iliocostalis; their separation is exaggerated. *B*, the parts of the longissimus and splenius, together with the spinalis thoracis. The insertion of the splenius cervicis is not shown. Based on Winckler.[1]

bar and upper thoracic spines *(spinalis thoracis)*. Extensions into the neck comprise the *spinalis cervicis*, an inconstant muscle that extends from the ligamentum nuchae and the spine of T.V. 7 to the spine of the axis, and the *spinalis capitis*, which is usually considered to be a part of the semispinalis capitis.

The large, long, intermediate group comprises the bulk of the erector spinae. It is attached to the accessory processes of the upper lumbar vertebrae and to the ribs and transverse processes of thoracic vertebrae *(longissimus thoracis)*, and, as the *longissimus cervicis*, it extends from the transverse processes of the upper thoracic vertebrae to the posterior tubercles of the transverse processes of the lower cervical vertebrae. Its upper medial part, the *longissimus capitis*, arises from the transverse and articular processes of the lower cervical and upper thoracic vertebrae and is inserted into the back of the mastoid process.

The lateral division of the erector spinae consists of an ascending series of slips *(iliocostalis lumborum, thoracis,* and *cervicis)* that are attached successively to the angles of the ribs and to the transverse processes of the lower cervical vertebrae. The iliocostalis cervicis is also known as the *costocervicalis*.

The splenius, a thick, flat muscle that is spinotransverse in arrangement, runs obliquely across the back of the neck and covers the underlying vertical muscles (L. *splenius*, a bandage). It consists of two parts. The *splenius capitis* arises from the lower part of the ligamentum nuchae and from the spines of the seventh cervical and the upper thoracic vertebrae. The fibers are directed upward and laterally to the mastoid portion of the temporal bone and the lateral third of the superior nuchal line on the occipital bone. The muscle arises under cover of the rhomboids and the trapezius and is inserted under cover of the sternocleidomastoid. The *splenius cervicis*, which is lateral to the splenius capitis, arises from the spines of the upper thoracic vertebrae and is inserted into the transverse processes (posterior tubercles) of the upper cervical vertebrae.

The direction of the erector spinae and its subdivisions is chiefly upward. Its longissimus division and the splenius also extend laterally, from spines to transverse processes, and may be termed a spinotransverse system, in contrast to the next deeper layer, the transversospinalis.

Transversospinalis (fig. 49–2). This is more deeply placed than the erector spinae. It consist chiefly of a large number of small muscles that run obliquely upward and medially, from transverse processes to spines.

The outermost muscle bundles span a number of segments (four to six) and comprise the *semispinalis*, which constitutes the chief component of the transversospinalis. The semispinalis is so called because it is located chiefly along the upper half of the vertebral column. However, the *semispinalis thoracis* consists of bundles that extend from the transverse processes of the lower thoracic vertebrae to the spinous processes of the upper thoracic and lower cervical vertebrae. The *semispinalis cervicis* is concealed by the semispinalis capitis. It arises from the transverse (and articular) processes of the upper thoracic vertebrae and is inserted into the spines of the cervical vertebrae.

The *semispinalis capitis* lies largely under cover of the splenius and conceals the suboccipital triangle and the semispinalis cervicis. Its medial border is free and is in contact with the ligamentum nuchae. This long thick muscle is responsible for the longitudinal bulge of the neck on each side of the median furrow. It arises from the transverse (and articular) processes of the lower cervical and upper thoracic vertebrae, and is inserted into the medial part of the area between the superior and inferior nuchal lines on the occipital bone. The upper part of the muscle is generally characterized by an imperfect tendinous intersection. Furthermore, the medial portion of the muscle is commonly more or less free and consists of upper and lower muscular parts with an intervening tendon. Hence it has been termed the *biventer cervicis*.

The *multifidus* consists of short, triangular bundles that lie deep to the semispinalis. The base of each triangle comprises many fascicles, which run upward and medially and converge to a single fasciculus that is inserted onto a spinous process. Massive bundles arise from the sacrum and from the mamillary processes of the lumbar vertebrae (and of T.V. 12), and smaller bundles proceed from the transverse pro-

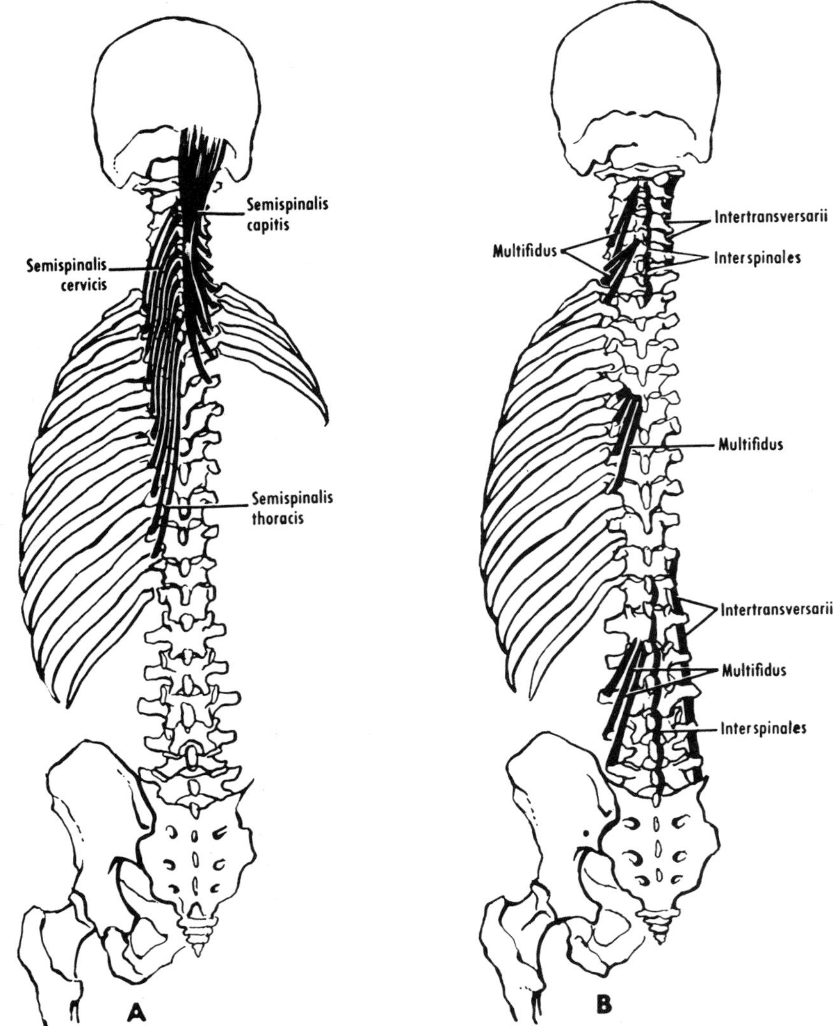

Figure 49–2 Diagrammatic representation of the transversospinalis, interspinales (interspinous), and intertransversarii (intertransverse), which are shown only in part in each region. *A*, the semispinalis, the thoracic and cervical parts of which are not separate muscles. *B*, the multifidus, interspinales, and intertransversarii. The sacral part of the multifidus is not shown. Based on Winckler.[1]

cesses of the thoracic vertebrae, and from the articular processes of the lower cervical vertebrae. The bundles cross several (two to five) vertebrae in ascending to their insertion on the spines. The deepest bundles of the group are the *rotatores,* which are inserted into the lamina immediately above or into the spine next above, and thus cross one or two segments. They are best developed in the thoracic region.

The size of the mamillary process of a vertebra varies according to the size of the muscular slip attached to it. The upper part of the lumbar multifidus consists of massive, clearly demarcated slips. Hence, the mamillary processes of the upper lumbar vertebrae, and especially those of T.V. 12, are large.[4]

Interspinales and Intertransversarii (fig. 49–2). These muscles connect spinous processes and transverse processes, respectively. They are poorly developed or absent in the thoracic region, and are of interest chiefly because of classifications into homologous subdivisions and because of homologies of nerve supply. This, depending upon location and group, may be from ventral rami, dorsal rami, or both.[5]

Suboccipital Triangle

The suboccipital triangle (fig. 49–3), which is in the suboccipital region, is bounded by the rectus capitis posterior major, the obliquus capitis inferior, and the obliquus capitis superior. It is roofed by the semispinalis capitis and the longissimus capitis. The floor is formed by the posterior atlanto-occipital membrane and by the posterior arch of the atlas. **The suboccipital triangle contains the vertebral artery and the suboccipital nerve, both of which lie in a groove on the upper surface of the posterior arch of the atlas.**

The posterior atlanto-occipital membrane connects the posterior arch of the atlas to the posterior margin of the foramen magnum. It arches over the vertebral artery (fig. 49–7) and the first cervical nerve. The free border of this arch is sometimes ossified.

The subarachnoid space can be tapped by inserting a needle at the back of the neck and penetrating the posterior atlanto-occipital membrane. This procedure is termed cisternal puncture.

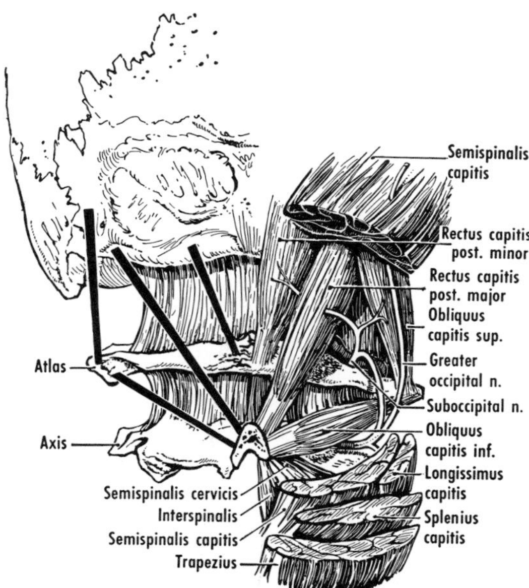

Figure 49–3 The suboccipital triangle. Most of the semispinalis capitis has been removed. Note the greater occipital nerve emerging at the lower border of the inferior oblique. The vertebral artery and the suboccipital nerve are seen in the triangle, but a mass of veins belonging to the suboccipital venous plexus has been omitted. On the left side, the thick black lines indicate schematically the directions and attachments of the muscles that bound the triangle.

Suboccipital Muscles. These comprise the recti capitis posteriores major and minor and the obliqui capitis inferior and superior. They are supplied chiefly by the suboccipital nerve. The suboccipital musles can act as extensors and rotators of the head but they function chiefly as postural muscles.

The *rectus capitis posterior major* runs from the spine of the axis to the lateral part of the inferior nuchal line on the occipital bone and to the area of bone below the line.

The *rectus capitis posterior minor* arises from the posterior tubercle of the atlas and ends on the medial part of the inferior nuchal line and on the area of bone below the line.

The *obliquus capitis inferior* extends from the spine of the axis to the transverse process of the atlas.

The *obliquus capitis superior* arises from the transverse process of the atlas and is inserted into the lateral part of the area between the superior and inferior nuchal lines.

Nerve Supply and Actions of Muscles of Back[6]

With the exception of certain intertransverse muscles, the muscles of the back are supplied by dorsal rami of spinal nerves, as are also the levatores costarum (p. 266).

As far as actions are concerned, very few of the muscles of the back have been studied directly. This is particularly true of the more deeply placed muscles. The rotatores, for example, are so named only because their direction and attachments suggest that they can rotate. Some of the actions of the muscles of the back are more or less self-evident, others have been determined from the study of patients with paralyses, and still others have been determined by electromyography (fig. 49–11). In many respects, however, the functions of the muscles of the back are not known or are poorly understood, particularly the way in which they are co-ordinated.

The functions of the muscles of the back can be outlined as follows: In the various movements of the trunk, a muscle may at times initiate movement, and at other times stabilize the trunk. **The more longitudinal the course of a muscle, the more it is concerned with flexion or extension of the vertebral column (head), and with lateral flexion. (Lateral flexion is especially important in balancing the trunk over the supporting limb during locomotion.)** The erector spinae is the chief extensor; it is aided by the suboccipital muscles, the splenii, and the semispinales capitis. **The more oblique the course of a muscle bundle, the more it is concerned with rotation.** The multifidus is the chief rotator of the trunk; it is aided by the external and internal abdominal obliques; the multifidus proper, however, seems to be more of a stabilizer than a prime mover. The splenius, semispinalis capitis, and sternocleidomastoid muscles are the chief rotators of the head, aided by the suboccipital muscles. The splenii of the two sides, acting together, extend the head. The splenii of one side draw the head to that side and turn the face toward the same side, and thus act with the sternocleidomastoid of the opposite side. Movements of the vertebral column are considered on page 537.

BLOOD VESSELS

Arteries

The structures of the back receive their arterial supply as follows: (1) in the neck, from the muscular branches of the occipital artery, and from the muscular and spinal branches of the ascending cervical, vertebral, and deep cervical arteries; (2) in the thorax and abdomen, from the muscular and spinal branches of the posterior intercostal, subcostal, and lumbar arteries; (3) in the pelvis, from the iliolumbar and lateral sacral branches of the internal iliac artery.

The branches that supply the spinal cord are described on page 541. The parts of the occipital and vertebral arteries that course through the suboccipital region are described below. Further details of the course and distribution of the various arteries are given in the appropriate sections.

Occipital Artery. This arises from the external carotid in the upper part of the neck. Its course may be considered in three portions: anterior, deep, and posterior to the sternocleidomastoid. The artery arises in the carotid triangle in front of the sternocleidomastoid. Deep to that muscle, the occipital artery occupies the occipital groove on the temporal bone, medial to the mastoid process. Here it is covered by the muscles attached to the mastoid process (sternocleidomastoid, splenius capitis, longissimus capitis, and digastric). It then lies on the obliquus capitis superior and the semispinalis capitis. Posterior to the sternocleidomastoid, the occipital artery pierces the trapezius, is accompanied by the greater occipital nerve, and divides into numerous branches on the scalp.

The most important branch of the occipital artery is its *descending branch*, which provides collateral circulation after ligation of the external carotid or subclavian artery. It arises on the obliquus capitis superior and divides into superficial and deep branches, which embrace the semispinalis capitis. The superficial branch passes deep to the splenius and anastomoses with the transverse cervical artery. The deep branch passes between the semispinalis capitis and cervicis, and anastomoses with the deep cervical artery from the costocervical trunk.

The other branches of the occipital artery are described on page 694.

Vertebral Artery. **The vertebral artery supplies chiefly the posterior part of the brain.** It arises from the first part of the subclavian artery and ascends through the foramina transversaria of the upper six cervical vertebrae. It then winds behind the lateral mass of the atlas and enters the cranial cavity through the foramen magnum. The course of the vertebral artery may be considered in four parts: cervical (p. 704), vertebral (p. 704), suboccipital, and intracranial (p. 605).

The *suboccipital part* winds backward around the lateral mass of the atlas and comes to lie in a groove on the upper surface of the posterior arch of the atlas (fig. 48–6). Here it forms a part of the contents of the suboccipital triangle (fig. 49–3) and is covered by the semispinalis capitis. The vertebral artery leaves the suboccipital triangle by going forward past the lateral edge of the posterior atlanto-occipital membrane and entering the vertebral canal. It then perforates the dura and the arachnoid, and passes through the foramen magnum. The vertebral artery forms a loop between the atlas and the skull.

The suboccipital part of the vertebral artery gives muscular branches to the suboccipital muscles and meningeal branches that ramify in the posterior cranial fossa.

Veins

The vertebral venous system[7] consists of a valveless plexiform network that is connected above with the cranial dural sinuses, below with pelvic veins, and in the neck and trunk with the azygos and caval systems. It provides for the flow of blood in either direction, according to variations in intrathoracic and intra-abdominal pressure (pp. 327, 420), and it provides a path for the spread of cancer cells, emboli, and infections. This venous network has three intercommunicating divisions:

1. A large plexus that surrounds the dura mater and drains the structures in the vertebral canal. This plexus, which is termed the *internal vertebral (epidural) venous plexus* (fig. 50–3, p. 542), is interrupted only at the levels of the intervertebral discs and between adjacent vertebrae posteriorly. The plexus is drained by segmental veins by way of *intervertebral veins,* which leave through the intervertebral and pelvic sacral foramina.

2. A network of veins in the marrow spaces of the bodies of the vertebrae that drains backward into the cranial network, and forward and laterally into the external plexus (fig. 37–6, p. 410). The *basivertebral veins* that issue from the posterior surfaces of the bodies of the vertebrae may be very large (p. 510).

3. An *external vertebral venous plexus,* the anterior part of which lies on the anterior aspects of the bodies of the vertebrae, and the posterior part of which lies on the external aspects of the vertebral arches, receives blood from the muscles of the back and from the bones. It is drained by segmental veins by way of the intervertebral veins.

The *suboccipital venous plexus* is a part of the external plexus that is notable for its extent and complexity. It lies on and in the suboccipital triangle, receives the occipital veins of the scalp, is connected with the transverse sinus by emissary veins, and communicates with the vertebral veins.

Each *vertebral vein* is formed by two roots that arise from the venous plexus around the foramen magnum. (This plexus is connected above with the basilar plexus.)[8] The vertebral receives tributaries from the suboccipital venous plexus and adjacent muscles, enters the foramen transversarium of the atlas, and descends through successive foramina. During the first part of its course, it forms a plexus around the vertebral artery, but at lower levels the plexus forms a single vein that leaves the foramen transversarium of the sixth cervical vertebra and empties into the brachiocephalic vein. The vertebral vein receives intervertebral and muscular veins throughout its course, and, just before it ends, it receives the deep cervical and the anterior vertebral veins. The *anterior vertebral vein* is a small vein that accompanies the ascending cervical artery. Sometimes an *accessory vertebral vein* arises from the plexus on the vertebral artery. It descends with the vertebral vein, but leaves through the foramen transversarium of the seventh cervical vertebra and ends in the brachiocephalic vein.

The *occipital vein,* which drains a plexus of veins in the scalp, accompanies the occipital artery through the trapezius and ends in the suboccipital venous plexus, where it communicates with the deep cervical vein. On the scalp, the occipital vein or one of its tributaries receives the parietal emissary vein and the mastoid emissary veins (p. 613). It sometimes continues with the occipital artery and ends in the internal jugular vein.

The *deep cervical vein* begins in the suboccipital region and descends between the splenius capitis and cervicis. It passes forward, between the transverse process of the seventh cervical vertebra and the neck of the first rib, and ends in the vertebral vein.

LYMPHATIC DRAINAGE

The lymphatic drainage of the deep structures of the back is by vessels that run mostly with the veins. Lymphatic vessels of the skin of the neck drain into cervical lymph nodes, those of the trunk above the umbilicus drain into the axillary nodes, and those arising below the umbilicus drain into superficial inguinal nodes.

NERVES

The nerve supply of the back is provided by the meningeal branches and dorsal rami of spinal nerves.

Meningeal Branches. Each spinal nerve gives off a meningeal branch, or sinuvertebral nerve,[9] which re-enters the vertebral canal and divides into fine filaments that anastomose with filaments from adjacent meningeal branches. These branches contain vasomotor and sensory fibers that supply the dura mater, posterior longitudinal ligament, and periosteum, and the epidural and intraosseous blood vessels. The meningeal branches of the first three cervical nerves give off branches that ascend through the foramen magnum and supply the dura mater on the anterior part of the floor of the posterior cranial fossa.[10]

Dorsal Rami. The dorsal rami, which contain motor, sensory, and sympathetic[11] fibers, pass backward and supply the muscles, bones, joints, and skin of the back. Most dorsal rami divide into medial and lateral branches, each of which descends as it runs dorsally.[12] Each anastomoses with nerves above and below and forms a plexus in the muscles of the back.

In the upper half of the trunk the medial branches supply skin, whereas in the lower half the lateral branches become cutaneous. However, the level of shift is variable (see Thoracic Dorsal Rami).

CERVICAL DORSAL RAMI. Connecting loops between the first three or four cervical nerves (fig. 49–3) form what is termed the *posterior cervical plexus.* The dorsal ramus of the first cervical nerve is known as the *suboccipital nerve.* It emerges above the posterior arch of the atlas, below the vertebral artery, and supplies the semispinalis capitis and the suboccipital muscles. The first cervical nerve often has no dorsal root; in such instances the nerve usually anastomoses with the accessory nerve.[13] Even if a dorsal root is present, the dorsal ramus usually has no cutaneous distribution.

The dorsal ramus of the second cervical nerve emerges below the obliquus capitis inferior, which it supplies, and then divides into medial and lateral branches. The lateral branch merely supplies adjoining muscles, but the larger medial branch, termed the *greater occipital nerve* (fig. 57–2, p. 653), has a more extensive distribution. It ascends under cover of the semispinalis capitis and the trapezius, both of which it pierces. It then accompanies the occipital artery and supplies the skin of the scalp as far forward as the vertex.

The medial branch of the dorsal ramus of the third cervical nerve gives off the *third occipital nerve,* which pierces the trapezius and ends in the skin on the back of the head.

The dorsal rami of C6, C7, and C8, as well as C1, usually have no cutaneous branches,[14] hence the C5 dermatome is adjacent to T1 and, with overlap, C4 meets T2.

THORACIC DORSAL RAMI. Each ramus passes backward, supplying the deeply placed muscles, and divides into a *medial* and a *lateral cutaneous branch,* which are separated by slips of the longissimus thoracis. The medial branches pass backward and downward, supplying the erector spinae and its divisions, as well as periosteum, ligaments, and joints. The medial branches of T1 to T3 also become cutaneous. The lateral branches supply the levatores costarum, the longissimus thoracis, and the iliocostalis thoracis. They have a long downward course, the lower ones (T9 to T12) piercing the latissimus dorsi and supplying the skin as far down as the gluteal region. In what might be termed a transitional zone, both medial and lateral branches of T4 to T8 give cutaneous twigs.

LUMBAR, SACRAL, AND COCCYGEAL DORSAL RAMI. Most of the medial branches of these dorsal rami supply the erector spinae. The lateral branches of the upper lumbar dorsal rami are the *superior clunial nerves,* which supply the skin of the buttock. The lateral branches of the lower lumbar dorsal rami, together with those of the first four sacral nerves (and usually a contribution from the fifth sacral), form a series of loops that comprise the *dorsal sacral plexus,* immediately posterior to the sacrum and coccyx.[15] These loops give off two or three *middle clunial nerves* that pierce the overlying gluteus maximus and supply the skin of the buttock.

The dorsal rami of the fifth sacral and coccygeal nerves lack medial and lateral branches. They communicate (often forming a single nerve) and supply adjacent ligaments and overlying skin.

JOINTS

The joints of the vertebral column include (1) those between the bodies of adjacent vertebrae, (2) the joints of the vertebral arches, (3) special joints, the atlanto-occipital and atlantoaxial, (4) costovertebral joints, and (5) the sacroiliac joints, with the iliolumbar ligaments.

Joints Between Bodies of Vertebrae

The bodies of adjacent vertebrae are joined together by longitudinal ligaments and by intervertebral discs (fig. 49–4).

Ligaments. The *anterior longitudinal ligament* is a fairly broad, thick band, which runs longitudinally on the front of the vertebral bodies and intervertebral discs, and fuses with the periosteum and anulus fibrosus, respectively. Above, it has a pointed attachment to the anterior tubercle of the atlas. Below, it spreads out on the pelvic surface of the sacrum. It is not uncommon to find bone in this ligament, particularly at the margins of vertebral bodies. This condition usually represents ossification in response to intermittent tension induced in the ligament by movement,[16] and is commonest at the summit of the lumbar curvature. Bony projections termed *osteophytes* are often present at the margins of the anterior and posterior parts of the vertebral bodies and extend into the adjacent ligaments.[17] These osteophytes, which are distinct from the para-articular processes, are said to increase in number with age.

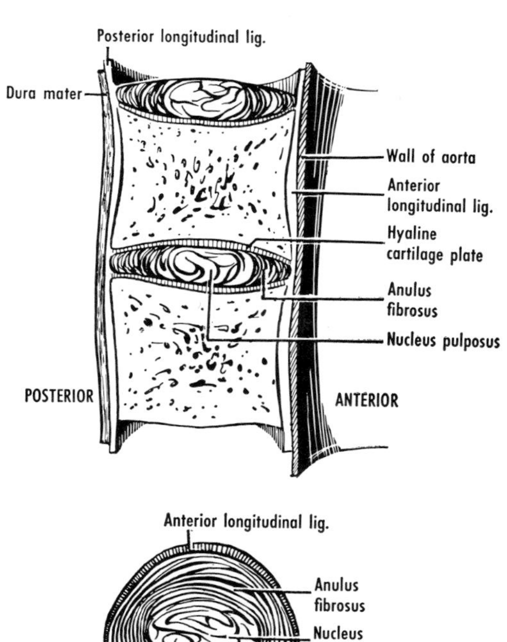

Figure 49–4 *Upper,* a median section of two vertebral bodies and an intervertebral disc of a young adult, and, *lower,* a horizontal section through a disc.

The *posterior longitudinal ligament* lies within the vertebral canal, on the posterior aspects of the vertebral bodies and intervertebral discs. Above, it is continuous with the membrana tectoria, and thus gains attachment to the occipital bone. As it descends, it narrows behind each vertebral body, but at the level of each disc it spreads laterally and fuses with the anulus fibrosus. Below, it continues into the sacral canal. This ligament is only loosely attached to the vertebral bodies, being separated from them by the basivertebral veins issuing from the spongiosa.

Intervertebral Discs.[18] The intervertebral discs are resilient pads that form fibrocartilaginous joints between the bodies of adjacent vertebrae (fig. 49–4). The uppermost disc is that between C.V. 2 and 3; the lowermost is within the coccyx. The structure and arrangement of discs vary with age.[19]

In young adults, each intervertebral disc consists of the nucleus pulposus and the anulus fibrosus, which contain few if any blood vessels or nerves, and which are separated from bone above and below by two thin hyaline cartilage plates. These plates cover the upper and lower aspects of the vertebral body. In growing bone, they form the zone from which the vertebral body grows in height. The nucleus pulposus, which occupies the center of the disc, is white, glistening, and semigelatinous. It contains fine bundles of collagenous fibers, connective tissue cells, cartilage cells, and much amorphous intercellular material. The nucleus is highly plastic, it behaves like a fluid, and it is held in shape by the cartilage plates and by the anulus fibrosus. The anulus fibrosus consists of a series of lamellae of collagenous bundles, which are arranged spirally. The bundles in adjacent lamellae lie at angles with respect to one another. Above and below, the fibers of the anulus are anchored to the ring epiphysis (p. 522) and to the margins of the hyaline plates. The outermost fibers of the anulus blend with the longitudinal ligaments; fibrocartilage is present in the innermost lamella. With advancing age, the entire disc tends to become fibrocartilaginous, and the structural differences between the nucleus and the anulus are often lost.

The nucleus pulposus subserves several functions: (1) It is a shock-absorbing mechanism; (2) it equalizes stresses; (3) it is important in the exchange of fluid between the disc and the capillaries in the vertebrae; and (4) the axis of movement between two adjacent vertebrae runs vertically through it. The anulus fibrosus also subserves several functions: (1) It binds the vertebral bodies together and provides stability; (2) it permits motion between vertebral bodies (because of the spiral arrangement of its fibers); (3) it acts as a check ligament; (4) it retains the nucleus pulposus; and (5) it is a shock-absorbing mechanism. The hyaline cartilage plates, in addition to serving as growth zones for the bodies of the vertebrae, probably protect the bodies to some extent, and they also permit the diffusion of fluid between the disc and the capillaries in the vertebrae.

The intervertebral discs have a high water content, which is maximal at birth and decreases with advancing age. Diurnal changes in water content probably account for the diurnal variation in height (1 to 2 cm); the height usually decreases during the day. The decrease in water content with age, together with other factors, results in a permanent thinning of the discs and a permanent decrease in stature. With advancing age, also, the fibers become coarse and hyalinized.

The discs are subject to pathological changes that may be followed by herniation of the nuclei and compression of adjacent nervous structures. Fissures that develop in the periphery of cervical discs, next to the margins of the bodies, have been termed uncovertebral joints on the erroneous assumption that they were synovial joints.[20]

SPECIAL FEATURES. **The discs account for about one-fourth of the length of the vertebral column.** They are thinnest in the thoracic region, and thickest in the lumbar region, where, also, disc disorders are most common. The lumbar and cervical discs are thicker in front than behind, and thus contribute to the secondary curves of these regions (p. 509). As discs become thinner with age, these curvatures are altered. In old age, for example, the cervical region of the vertebral column is usually straight.

The intervertebral disc forms one of the anterior boundaries of an intervertebral foramen (fig. 49–5). Cervical and thoracic nerves passing through a foramen lie directly behind the disc, in a position to be compressed by a posterolateral protrusion of a herniated nucleus pulposus. Most lumbar nerves, however, emerge above the disc (fig. 50–5, p. 543). Decrease in height of a disc results in narrowing of the foramen. Such narrowing is a potential cause of spinal nerve compression.

The *sacrococcygeal joint* consists of an intervertebral disc between the sacrum and coccyx, reinforced by *dorsal, ventral,* and *lateral sacrococcygeal ligaments.* The joint is often partly or wholly fused.

Joints of Vertebral Arches

The vertebral arches are connected by synovial joints between the articular processes, and by accessory ligaments that connect the laminae, the transverse processes, and the spinous processes.

Joints Between Articular Processes. These are mostly of the plane type that allows gliding between the facets. The loose capsule attached around the margins of the facets is supplied by medial branches of dorsal rami of spinal nerves.

Accessory Ligaments. These are the ligamenta flava, the ligamentum nuchae, and the supraspinous, interspinous, and intertransverse ligaments.

The *supraspinous ligament* connects the tips of the spinous processes. It is

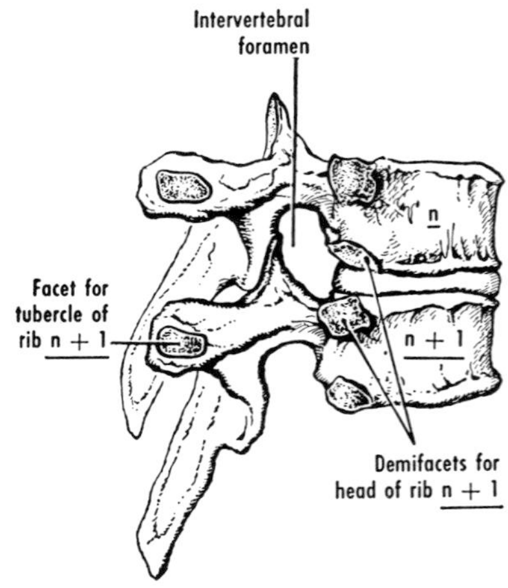

Figure 49–5 A lateral view of two thoracic vertebrae, showing the position of the intervertebral disc and its relation to the intervertebral foramen.

poorly developed in the lower lumbar region.[21] Tension in the ligament may result in the formation of small bony spurs at the tips of the spines, especially those of the thoracic vertebrae. Above, the supraspinous ligament merges with the *ligamentum nuchae,* a triangular membrane that forms a median fibrous septum between the muscles of the two sides of the neck (fig. 49–6).[22] The ligamentum nuchae presents (1) a superior border attached to the external occipital crest, (2) an anterior border attached to the spines of the cervical vertebrae, and (3) a free posterior border extending from the external occipital protuberance to the spine of the seventh cervical vertebra. In quadrupeds, the ligamentum nuchae is an elastic ligament that supports the head.

The *interspinous ligaments* are found between adjacent spines and are well developed only in the lumbar region.

The elastic *ligamenta flava* connect the borders of the laminae of adjacent vertebrae. Some fibers may lie on the anterior aspect of the lamina. Laterally, each ligamentum flavum extends to the capsule of the joint between the facets and thereby contributes to the posterior boundary of the intervertebral foramen. The two ligaments are separated by a narrow, median interval, through which pass the veins that connect the epidural and extravertebral venous plexuses. Bony spurs are often present in the ligamenta flava connecting thoracic vertebrae.

The *intertransverse ligaments* are insignificant except in the lumbar region, where they connect adjacent transverse processes.

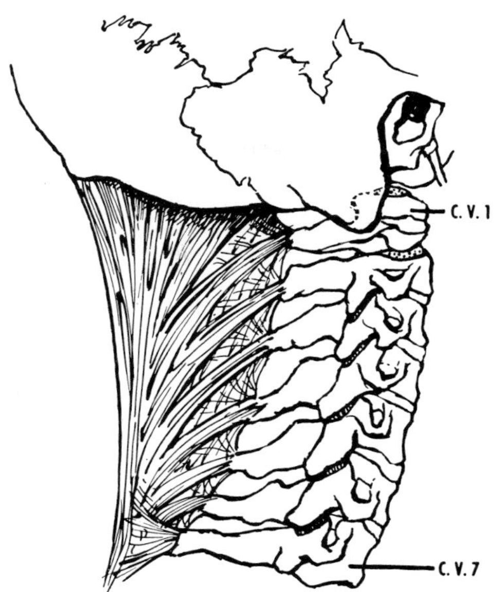

Figure 49–6 The ligamentum nuchae, right lateral aspect. The joints between the articular processes of the cervical vertebrae are indicated by cross-hatching. Note, however, that first, the joint between the superior articular facet of the axis and the lateral mass of the atlas, and second, the joint between the lateral mass of the atlas and the occipital condyle are located more anteriorly. These two joints do not correspond to the joints between the articular processes from C.V. 7 to C.V. 2. Associated with this is the fact that the first two cervical nerves emerge behind these two joints, respectively, whereas the remaining nerves emerge in front of the articular processes. Based on Poirier and Charpy.

Atlanto-occipital and Atlantoaxial Joints

Atlanto-occipital Joint. The atlanto-occipital joint of each side is that between the superior articular facet on the lateral mass of the atlas and the corresponding occipital condyle. Furthermore, these two bones are also connected by *anterior* and *posterior atlanto-occipital membranes,* which extend from the anterior and posterior arches of the atlas, respectively, to the anterior and posterior margins of the foramen magnum (figs. 49–7 and 49–8). The facet on the occipital condyle is generally reniform or hourglass-shaped, occasionally bipartite.[23]

Each atlanto-occipital joint is synovial in type and has a capsular ligament lined with synovial membrane.

Functionally, the right and left joints act together as an ellipsoidal articulation, the long axis of which is transverse in direction. The movements are nodding (flexion and extension) around a transverse axis, and tilting of the head sideways around an anteroposterior axis.

Atlantoaxial Joints and Occipitoaxial Ligaments. The atlantoaxial joints are synovial in type and three in number: two lateral and one medi n. Each *lateral atlantoaxial joint* unites the opposed articular processes and is a plane joint. It presents a synovial membrane and a capsule. The capsule is reinforced posteriorly by an accessory atlantoaxial ligament that extends from the body of the axis upward

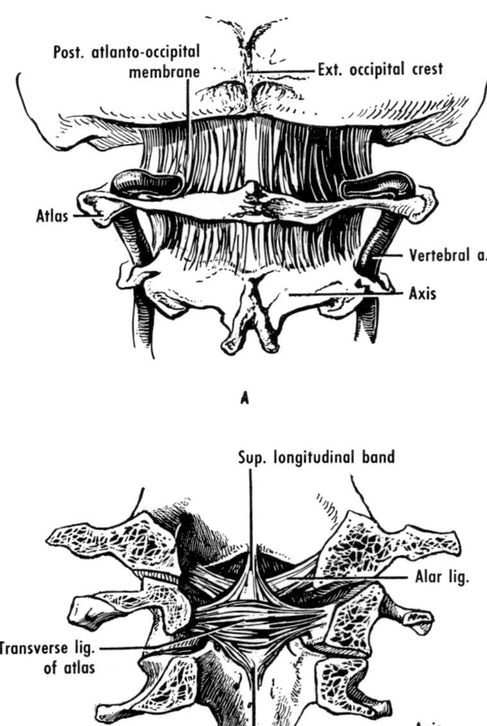

Figure 49–7 The ligaments of the atlas and the axis. *A*, posterior view, showing the vertebral arteries. *B*, posterior view, showing the interior of the vertebral canal after removal of portions of the skull and vertebrae.

and laterally to the lateral mass of the atlas. The *median atlantoaxial* is a pivot joint. It is formed by (1) the anterior arch and the transverse ligament of the atlas, and (2) the dens of the axis. Synovial cavities are present in front of and behind the dens, each with its own capsular ligament and synovial membrane (fig. 49–8).

The term *cruciform (cruciate) ligament of the atlas* is applied to (1) the *transverse ligament of the atlas* that unites the medial aspects of the lateral masses (fig. 48–6), and (2) the longitudinal bands that extend upward to the front edge of the foramen magnum and downward to the back of the body of the axis. The transverse ligament usually does not rupture in judicial hanging, but fracture of the axis occurs and the spinal cord is ruptured.[24]

In addition to the cruciform ligament, the axis is connected to the occipital bone by (1) the *apical ligament of the dens*, from the apex of the dens to the front edge of the foramen magnum (fig. 49–8); (2) the *alar ligaments*, from the apex of the dens to the medial side of each occipital condyle (fig. 49–7); and (3) the *membrana tectoria*, a broad sheet that extends from the posterior longitudinal ligament and the body of the axis to the upper surface of the basilar part of the occipital bone in front of the foramen magnum (fig. 49–8).

Costovertebral Joints

The costovertebral joints (p. 275) consist of the joints of the heads of the ribs with the bodies of the vertebrae, and the costotransverse joints, between the tubercles of the ribs and the transverse processes of the vertebrae.

Sacroiliac Joint and Iliolumbar Ligaments

The weight of the head, the upper limbs, and the trunk is transmitted through the sacrum and ilia to the femora when standing, and to the ischial tuberosities when sitting. The strength and stability of this region are due to the configuration of the bones, to the arrangement of the sacroiliac ligaments, and to strong ligaments that connect the lower lumbar vertebrae and ilia.

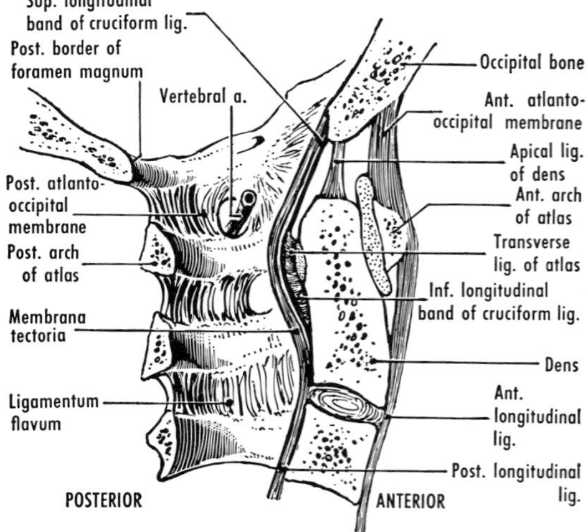

Figure 49–8 Median section of atlas and axis. Based on Poirier and Charpy.

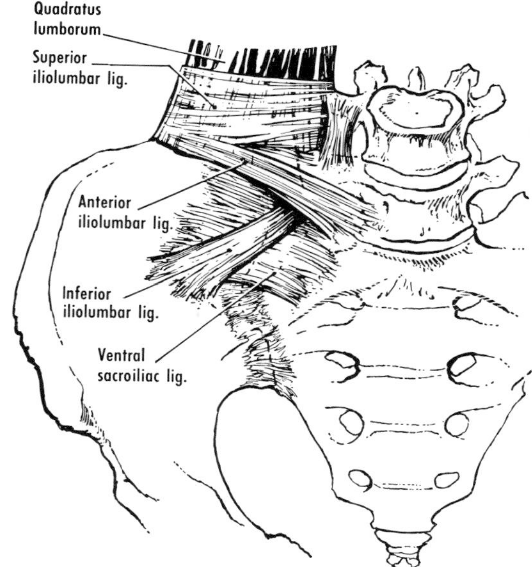

Figure 49–9 The iliolumbar and ventral sacroiliac ligaments. The posterior and vertical iliolumbar ligaments are not shown. Based on Shellshear and Macintosh.[25]

The *sacroiliac joint* is a plane synovial joint, formed by the union of the auricular surfaces of the sacrum and ilium. It is described with the joints of the pelvis (p. 445).

The *iliolumbar ligaments* are several strong ligaments (figs. 49–9 and 49–10), arranged so that they contribute to the stability of the lumbosacral region. These ligaments include: (1) The superior iliolumbar ligament, the medial part of which is the intertransverse ligament between the fourth and fifth lumbar vertebrae. Like the other lumbar intertransverse ligaments, it extends laterally to divide around the quadratus lumborum as the middle and anterior layers of the thoracolumbar fascia. In extending laterally, the ligament also reaches the ilium, and hence is an iliolumbar as well as an intertransverse ligament. (2) The anterior and posterior iliolumbar ligaments, which extend from the transverse process of L.V. 5 to the sacrum and ilium. (3) An inferior iliolumbar ligament, which extends from the transverse process of L.V. 5 to the iliac fossa, where it is fused with the ventral sacroiliac ligament (its fibers have a different direction). (4) A variable vertical iliolumbar ligament.

Movements of Vertebral Column

Movements of the vertebral column are flexion (forward bending) and extension (backward bending), both in the median plane; lateral flexion (side bending), to the right or left, in a coronal plane; and rotation around a longitudinal axis. The precise nature and axes of these movements in the living are very difficult to determine. Radiological studies have provided valuable data.

Each type of movement just listed can occur in each of the three movable regions of the vertebral column: cervical, thoracic, and lumbar.[26] The axis of each type of movement apparently runs through the nucleus pulposus, but shifts during the movement. During movement the articular facets glide or slip on one another. Their arrangement is such that lateral flexion is automatically accompanied by some degree of rotation. During each one-half of a walking cycle, rotation occurs in one direction above T.V. 6 to 8, and in the opposite direction below these vertebrae.[27]

The range of movement varies according to region and with the individual, being extraordinary in some individuals, such as acrobats and contortionists. The range is limited by the thickness and compressibility of the intervertebral discs, and

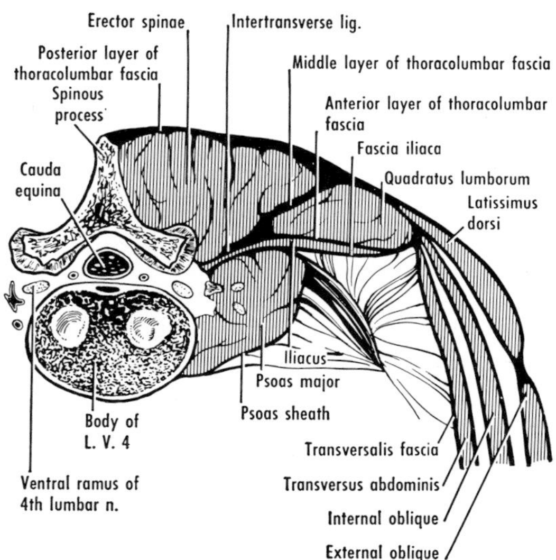

Figure 49–10 Horizontal section through the muscles of the back, at the level of the body of L.V. 4. Note the middle and anterior layers of the thoracolumbar fascia, formed by the division of the superior iliolumbar ligament (see fig. 49–9) and enclosing the quadratus lumborum.

by muscles and ligaments. During flexion of the vertebral column, the discs narrow in front and widen behind, and, in the lower part of the cervical region, there is a forward displacement of each vertebra over the one immediately below. The converse changes occur in extension. The cervical and lumbar regions are the most mobile. Extension can be carried out to a greater degree than flexion, especially in the lumbar region.[28] During full flexion, the lumbar column may become straight or slightly concave.[29]

Flexion and extension of the head occur mainly at the atlantoaxial and atlanto-occipital joints; less at other cervical joints.[30] Extension at the atlanto-occipital joint is freer than flexion, and both have a greater range than at the atlantoaxial joint.

The skull and the atlas together rotate on the axis at the three atlantoaxial joints, pivoting on the dens. They function as a ball-and-socket joint. The details of the movements are in dispute, however.[31] After about 45 degrees of rotation, additional rotation occurs between the remaining cervical vertebrae (for a total of about 90 degrees).

Muscles. The chief flexors of the vertebral column are the prevertebral muscles, the rectus abdominis, the iliopsoas, the scaleni, and the sternocleidomastoid. These muscles act bilaterally and flex the trunk against resistance. Flexion may also be brought about by gravity, and its rate and extent are controlled by the erectores spinae. The erectores spinae are the chief extensors of the vertebral column. In the beginning of extension from a position of full flexion, the gluteus maximus and the hamstrings are important extensors.

Lateral flexion or side-bending of the trunk is carried out by the oblique muscles of one side of the abdominal wall, aided by the quadratus lumborum and the psoas major. Activity in the erector spinae is related to extension during lateral bending.[32] Lateral flexion of the neck is carried out by the upward extensions of the erector spinae and by the semispinalis capitis and splenius capitis, aided perhaps by the sternocleidomastoid and trapezius.

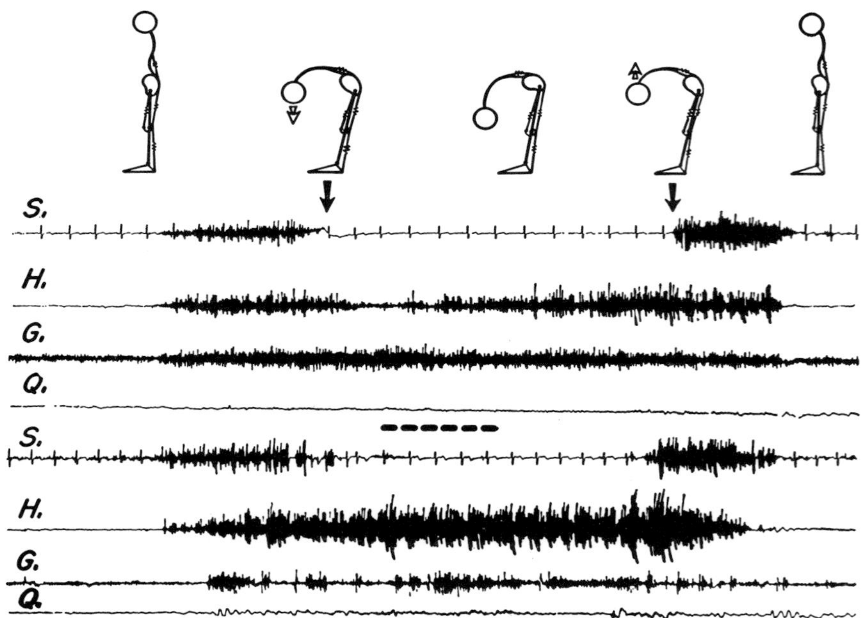

Figure 49–11 Electromyograms during standing and during flexion of the trunk and hips. In standing, no activity was recorded from the sacrospinalis *(S.)*, hamstrings *(H.)*, or quadriceps femoris *(Q.)*. The short vertical deflections in *S.* are the deflections recorded in the electrocardiogram. As the subject leaned forward, activity appeared in the sacrospinalis, in the hamstrings, and in the gastrocnemius *(G.)*. Activity in the sacrospinalis stopped at full flexion, and reappeared with extension of the trunk and a return to an easy standing position. The upper four records are from one subject, the lower four from another. From figure 2, H. Portnoy and F. Morin, Amer. J. Physiol., *186*:122, 1956. By permission of the authors and publishers.

Any obliquely running muscle of the back may take part in rotation, aided in the thoracolumbar region by the oblique muscles of the abdominal wall.

The muscles of the back are relatively inactive when one is standing at ease.[33] They serve mainly as lateral steadiers. The erector spinae controls the act of bending forward, and it acts powerfully in the return to the erect position (extension), being aided by the gluteus maximus and hamstrings. Surprisingly, during full flexion, as when an attempt is made to touch the floor with the fingertips without bending the knees, the erector spinae relaxes when a certain degree of flexion has been reached (fig. 49–11). The strain is then taken entirely by ligaments. On reversing the movement, the erector spinae is inactive at first, the initial phase of extension being carried out by the hamstrings and the gluteus maximus. The erector spinae then suddenly resumes activity. Lifting a load or moving rapidly from a bent over or fully flexed position is not only mechanically disadvantageous but potentially dangerous from the standpoint of injuries to muscles, ligaments, or intervertebral discs.

In addition to acting as prime movers, and acting paradoxically, the muscles of the back have important postural and synergistic functions. When one is standing on one foot, for example, the erector spinae contracts and helps to maintain balance. During walking, the alternating contractions of these muscles can easily be felt as weight shifts from one foot to the other.

The muscles of the back readily go into reflex spasm following injury to, or inflammation of, back structures. The spasm tends to get worse, and is itself painful.

REFERENCES

1. G. Winckler, Arch. Anat., Strasbourg, *31*:1, 1948.
2. J. Satoh, Okajimas Folia anat. jap., *46*:65, 1969; *47*:19, 1970.
3. B. Jonsson, Z. Anat. EntwGesch., *130*:177, 1970.
4. P. N. B. Odgers, J. Anat., Lond., *67*:301, 1933. For a study of the multifidus, see O. Kuran, Rev. Fac. med. Univ. Istanbul, *19*:193, 1956.
5. A. J. E. Cave, J. Anat., Lond., *71*:497, 1937. W. Langenberg and S. Jüschke, Z. Anat. EntwGesch., *130*:255, 1970. T. Sato, Z. Anat. EntwGesch., *143*:143, 1974.
6. E. Cyriax, J. Anat., Lond., *67*:178, 1932. B. Åkerblom, *Standing and Sitting Posture*, A.-B. Nordiska Bokhandeln, Stockholm, 1948. W. F. Floyd and P. H. S. Silver, Lancet, *1*:133, 1951; J. Anat., Lond., *85*:433, 1951. P. H. S. Silver, J. Anat. Lond., *88*:550, 1954. J. Joseph, *Man's Posture: Electromyographic Studies*, Thomas, Springfield, Illinois, 1960. J. M. Morris, G. Benner, and D. B. Lucas, J. Anat., Lond., *96*:509, 1962. R. L. Waters and J. M. Morris, J. Anat. Lond., *111*:191, 1972. E. W. Donisch and J. V. Basmajian, Amer. J. Anat., *133*:25, 1972.
7. O. V. Batson, Amer. J. Roentgenol., *78*:195, 1957. H. J. Clemens, *Die Venensysteme der menschlichen Wirbelsäule*, Walter de Gruyter, Berlin, 1961. M. H. von Lüdinghausen, Münch. med. Wschr., *110*:20, 1968.
8. E. Stolic and D. Mrvaljevic, C. R. Ass. Anat., *55*:1027, 1970.
9. G. Lazorthes and J. Gaubert, C. R. Ass. Anat., *43*:488, 1956. D. L. Stilwell, Jr., Anat. Rec., *125*:139, 1956. H. E. Pedersen, C. F. J. Blunck, and E. Gardner, J. Bone Jt Surg., *38A*:377, 1956. J. Molinsky, Acta anat., *38*:96, 1959. H. C. Jackson, R. K. Winkelman, and W. H. Bickel, J. Bone Jt Surg., *48A*:1272, 1966.
10. D. L. Kimmel, Neurol., *11*:800, 1961.
11. R. Dass, Anat. Rec., *113*:493, 1952.
12. H. M. Johnston, J. Anat., Lond., *43*:80, 1908. A. J. E. Cave, J. Anat., Lond., *71*:497, 1937.
13. A. A. Pearson, R. W. Sauter, and G. R. Herrin, Amer. J. Anat., *114*:371, 1964. G. Ouaknine and H. Nathan, J. Neurosurg., *38*:189, 1973.
14. A. A. Pearson, R. W. Sauter, and J. J. Bass, Amer. J. Anat., *112*:169, 1963. A. A. Pearson, R. W. Sauter, and T. F. Buckley, Amer. J. Anat., *118*:891, 1966. See also reference 12.
15. M. T. Horwitz, Anat. Rec., *74*:91, 1939.
16. D. Allbrook, J. Bone Jt Surg., *39B*:339, 1957.
17. H. Nathan, J. Bone Jt Surg., *44A*:243, 1962.
18. J. R. Armstrong, *Lumbar Disc Lesions*, Livingstone, Edinburgh, 2nd ed., 1958. E. J. Eyring, Clin. Orthop., *67*:16, 1969. A. F. DePalma and R. H. Rothman, *The Intervertebral Disc*, Saunders, Philadelphia, 1970.
19. A. Peacock, J. Anat., Lond., *86*:162, 1952. R. Walmsley, Edinb. med. J., *60*:341, 1953.
20. G. Töndury, Z. Anat. EntwGesch., *112*:448, 1943. E. E. Payne and J. D. Spillane, Brain, *80*:571, 1957. See also M. C. Hall, *Luschka's Joint*, Thomas, Springfield, Illinois, 1965.
21. P. M. Rissanen, Acta orthopaed. scand., Suppl. 46, 1960.
22. H. V. Vallois, C. R. Soc. Biol., Paris, *93*:1339, 1925.
23. A. J. E. Cave, J. Anat., Lond., *68*:416, 1934.
24. F. Wood-Jones, Lancet, *1*:53, 1913. V. Vermooten, Anat. Rec., *20*:305, 1921.
25. J. L. Shellshear and N. W. G. Macintosh, *Surveys of Anatomical Fields*, Grahame, Sydney, 1949.
26. S. N. Bakke, Acta radiol., Stockh., Suppl. 13, 1931.
27. G. G. Gregerson and O. B. Lucas, J. Bone Jt Surg., *49A*:247, 1967.
28. P. Wiles, Proc. R. Soc. Med., *28*:647, 1935. J. F. Elward, Amer. J. Roentgenol., *42*:91, 1939. J. F. Brailsford, *The Radiology of Bones and Joints*, Churchill, London, 5th ed., 1953.
29. J. J. Keegan, J. Bone Jt Surg., *35A*:589, 1953. D. Allbrook, J. Bone Jt Surg., *39B*:339, 1957.
30. J. W. Fielding, J. Bone Jt Surg., *39A*:1280, 1957. M. Hohl and H. R. Baker, J. Bone Jt Surg., *46A*:1739, 1964.
31. S. Werne, Acta orthopaed. scand., Suppl. 23, 1957; *28*:165, 1959.
32. J. E. Pauly, Anat. Rec., *155*:223, 1966.
33. J. Joseph, cited in Reference 6. J. V. Basmajian, *Muscles Alive*, Williams & Wilkins, Baltimore, 3rd ed., 1974.

SPINAL CORD AND MENINGES

50

SPINAL CORD

The *spinal cord*, which averages 45 cm in length, extends from the foramen magnum, where it is continuous with the medulla oblongata, to the upper part of the lumbar region. **The spinal cord ends most frequently at the level of the disc between L.V. 1 and L.V. 2, but the range is from the disc between T.V. 12 and L.V. 1 to that between L.V. 2 and L.V. 3.**[1] Below this level, the vertebral canal is occupied by meninges and spinal nerve roots. A thin, glistening fibrous strand, the *filum terminale*, continues downward from the spinal cord as a prolongation of pia mater. It fuses with dura mater at the apex of the dural sac and continues with it as the filum of the dura mater (p. 544). The filum terminale is about 15 to 20 cm long. The central canal continues into it for a variable distance. The filum terminale is derived from that portion of the spinal cord that is caudal to the second coccygeal segment (thirty-first segment) in the embryo. This portion dedifferentiates, leaving only a fibrous strand.[2] Subsequently, differential growth leads to an increasingly higher termination of the spinal cord. The adult level is reached about two months after birth.[3]

The spinal cord is almost cylindrical, but is slightly flattened from before backward. It presents a *cervical* and a *lumbar enlargement* at the levels of attachment of the nerves to the limbs. The lower end of the cord has a conical shape and is termed the *conus medullaris*. The filum terminale descends from the apex of the conus.

Behind, throughout its whole length, the spinal cord presents a slight longitudinal groove, the *posterior median sulcus*. *Dorsal root* filaments enter lateral to this sulcus at regular intervals (fig. 50–1). A variable number of small arteries and veins, derived from posterior spinal and medullary vessels, are present on the posterior surface of the cord.

In front, throughout its whole length, the spinal cord presents the *anterior median fissure*, occupied by the anterior spinal artery and small veins. *Ventral root* filaments leave the anterolateral aspect of the cord at regular intervals.

The fissures and sulci of the spinal cord are continuous above with corresponding ones of the medulla oblongata.

The part of the spinal cord to which a pair of dorsal roots and a pair of ventral roots are attached is termed a segment of the cord. Each dorsal root presents an ovoid swelling, the *spinal ganglion*, which lies near or within the intervertebral foramen. Distal to

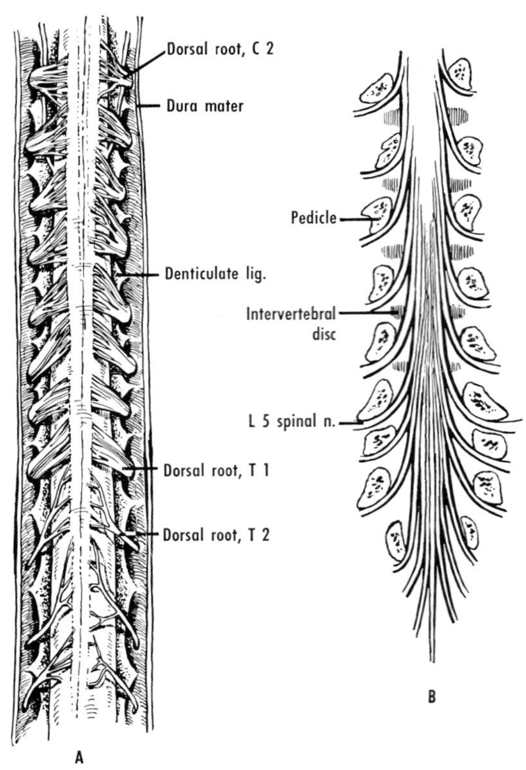

Figure 50–1 *A*, the dorsal roots of the cervical and upper thoracic parts of the spinal cord. Note that the filaments are more numerous in the lower cervical dorsal roots than in the upper cervical or upper thoracic. *B*, the cauda equina, after removal of the vertebral arches. Note how the lumbar nerves course close to the pedicles, above the intervertebral discs. Modified from Hovelacque.[4]

the ganglion, each dorsal root combines with the corresponding ventral root and forms a spinal nerve. **The first pair of spinal nerves emerges between the atlas and the skull, and the remaining cervical nerves, except the eighth, leave the vertebral canal above the correspondingly numbered vertebra. The eighth emerges below the seventh cervical vertebra. Thereafter, nerves leave below their correspondingly numbered vertebrae.** There are, then, eight cervical, twelve thoracic, five lumbar, and five sacral pairs of spinal nerves. Usually there is but one pair of coccygeal nerves. Additional ones, if any, are rudimentary.

The term *cauda equina* refers to the collection of spinal roots that descend from the lower part of the spinal cord and occupy the vertebral canal below the cord (fig. 50–1*B*). The term arises from the resemblance of this collection of roots to the tail of a horse.

Filaments that arise from the lateral aspect of the upper cervical part of the cord (midway between dorsal and ventral roots) form the spinal part of the accessory nerve. These filaments ascend through the foramen magnum and join the filaments of the accessory nerve that arise from the medulla oblongata.

Structure. The spinal cord has essentially the same arrangement throughout. The gray matter, which is shaped like the letter **H**, is modified locally because of differences in numbers and types of its contained nerve cells. Cells are more numerous in the cervical and lumbar enlargements because these regions supply the limbs. White matter similarly differs but little in its arrangement throughout the cord, and is contained within three funiculi on each side (fig. 50–2). The central canal is continuous above with that of the medulla oblongata, and thereby with the fourth ventricle. Below, it is often somewhat enlarged (terminal ventricle), just before tapering into the filum terminale, where it ends. In children, the central canal in the equinal portion of the cord is frequently forked.[5] In the adult, the canal is often obliterated at various levels by proliferating lining cells.

Blood Supply (fig. 50–3). **The pattern of arterial supply is formed by three longitudinal arterial channels that extend from the medulla oblongata to the conus medullaris, one channel in the anterior median position and the other two placed posterolaterally. All three channels are reinforced by "medullary feeders," which are branches of the segmental arteries.**[6]

The anterior channel is the *anterior spinal artery*, which is formed by the union of two branches of the vertebral artery. In its course in the anterior median fissure, it is reinforced by medullary arteries. The

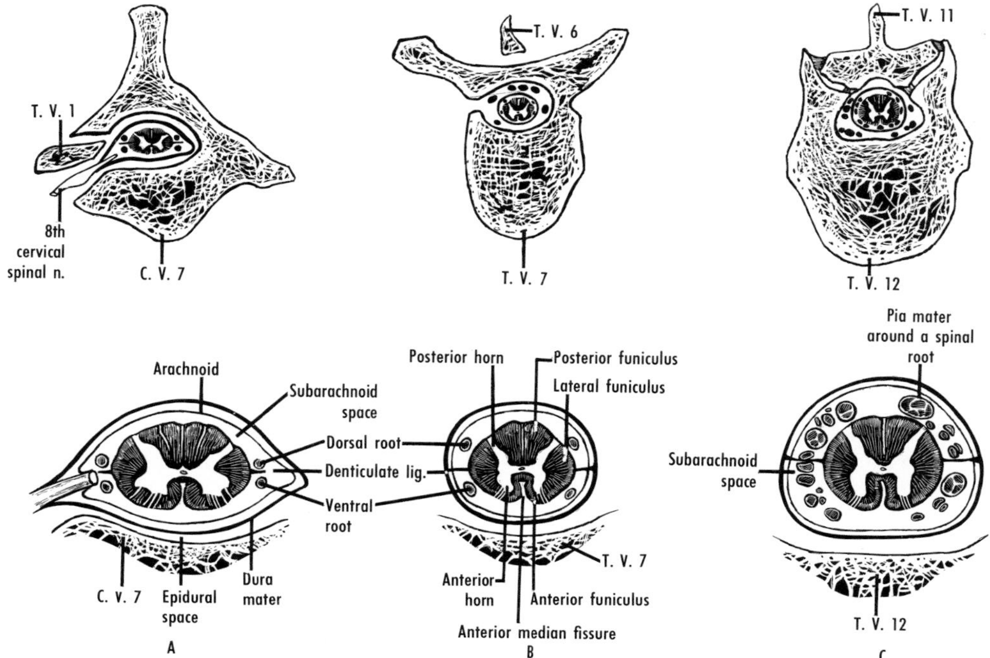

Figure 50–2 Diagrammatic representation of three levels of the vertebral column and spinal cord. *A*, section through body of C.V. 7; the spinal cord (about second thoracic segment) is shown below. Note the free edge of the denticulate ligament. *B*, section through body of T.V. 7; the spinal cord (about tenth thoracic segment) is shown below. Note that at this level (between exit of adjacent roots), the denticulate ligament is attached at both edges. *C*, section through body of T.V. 12; the spinal cord (upper sacral or lower lumbar) is shown below.

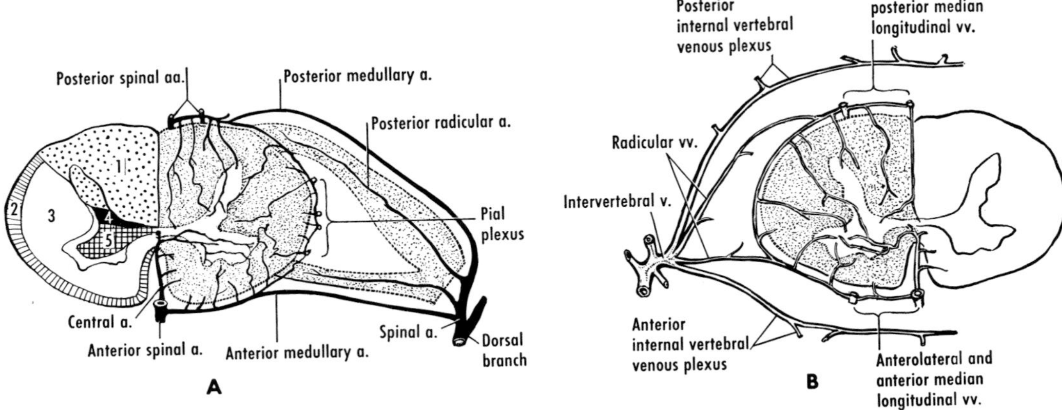

Figure 50–3 *A*, the arterial and *B*, the venous drainage of the spinal cord. (1) The area supplied by the posterior plexus; (2) the area supplied by the lateral and ventral parts of the pial plexus; (3) the area of common supply by the central artery and by the lateral and ventral pial parts of the pial plexus; (4) the area of common supply by the central artery and the posterior plexus; (5) the area supplied by the central artery. Based on Gillilan.[6, 9]

anterior spinal artery gives central (sulcal) branches that supply approximately the anterior four-fifths of the cord.

The two *posterior spinal arteries*, which are derived from the vertebral arteries or from the posterior inferior cerebellar arteries, commonly form plexiform channels in the pia mater on each side.[7] These channels are reinforced by medullary arteries.

The reinforcing segmental supply of the spinal cord is provided by the *medullary arteries*, which are derived primarily from the spinal branches of the ascending cervical, deep cervical, vertebral, posterior intercostal, and lateral sacral arteries. These medullary arteries (medullary feeders) are present chiefly where the need for blood supply is greatest, namely in the enlargements. There is an average of eight anterior and twelve posterior feeders. A particularly large anterior feeder is often present on the left side, commonly entering at thoracic levels 9 to 11; it turns downward to the lumbar enlargement. Another important region is the junction of the medulla oblongata and the spinal cord, where branches of the occipital artery also contribute.[8]

The ventral and dorsal roots are supplied by radicular branches that usually do not reach the spinal cord.

The veins that drain the spinal cord,[9] together with the internal vertebral venous plexuses, drain into intervertebral veins (fig. 50–3). These in turn drain into segmental veins.

REGIONAL FEATURES

Functional Features. The spinal cord contains the descending motor tracts and the ascending sensory tracts. Furthermore, different levels of the cord have different, more or less specific, functions. The cervical and lumbar enlargements, for example, contain the neurons that supply the limbs.

Special nerve fibers that arise from the cervical part of the spinal cord form the spinal part of the accessory nerve. In addition, the important cells that supply the diaphragm are located in the cervical part of the spinal cord. Injury to this part of the spinal cord is dangerous because of its interference with respiration.

The autonomic outflow is from the thoracic and upper lumbar parts of the cord, and also from the sacral part. This last is the parasympathetic outflow. The sacral part of the cord is therefore an important center for the control of urination and defecation. Injury to the conus medullaris or to the cauda equina commonly causes disturbances of bladder and intestinal function.

Morphological and Topographical Features. The spinal cord varies in shape according to level. It presents a cervical and a lumbar enlargement at the levels of attachment of the nerves that pass to the limbs. The spinal cord is almost cylindrical, being slightly flattened from before backward, especially in the cervical region.*

The relationships of the levels of attachment of spinal roots to vertebral levels, as compared with the levels of emergence of spinal nerves, are important. The level of attachment of roots is higher than the level of emergence

---

*The average maximal width of the cervical cord is 14 mm, the anteroposterior diameter, 6½ to 9 mm.[10]

from an intervertebral foramen (fig. 50–4). The discrepancy is greatest in the lumbosacral region. The correlation between levels of attachment and levels of emergence varies greatly.[11]

The manner in which roots attach to the cord differs according to the region. The length of the roots increases from above downward. Those of the lumbosacral region are not only the longest, but also the thickest. **The lumbar spinal nerves increase in size from above downward, whereas the lum-**

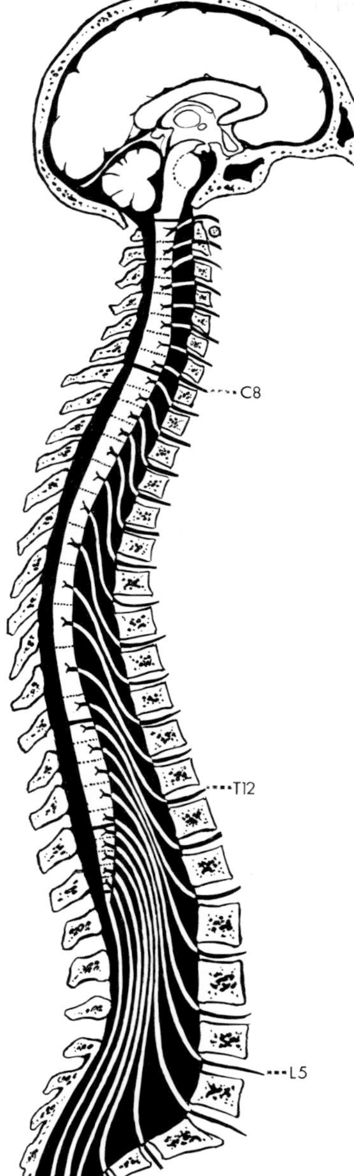

Figure 50–4 Diagrammatic representation of the brain and spinal cord to show the relation of cord, spinal nerves, and vertebrae. The brain is shown in the median plane. The dorsal and ventral roots are shown as single nerves emerging from each segment. Modified from Gardner.[12]

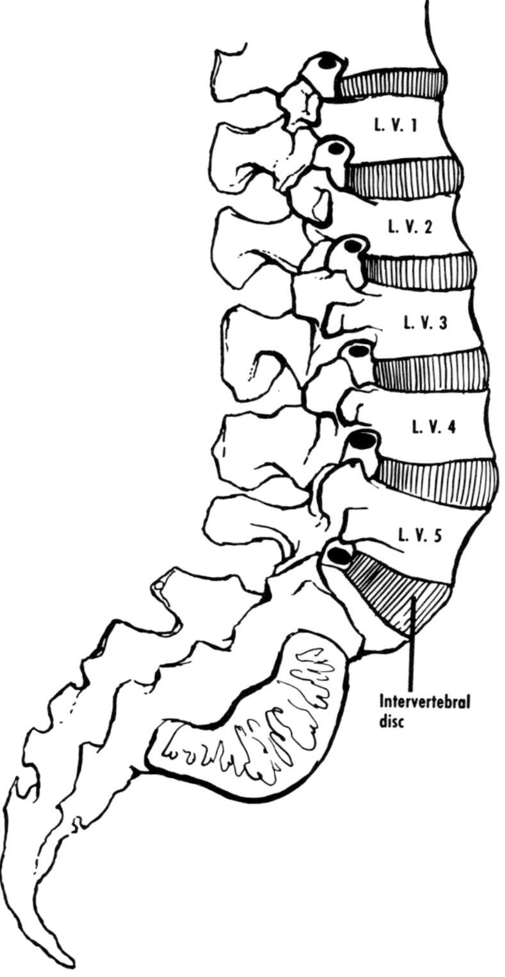

Figure 50–5 The emergence of the right lumbar spinal nerves through the intervertebral foramina. The drawing shows the relative sizes of lumbar intervertebral foramina and spinal nerves. Based on Danforth and Wilson.[13]

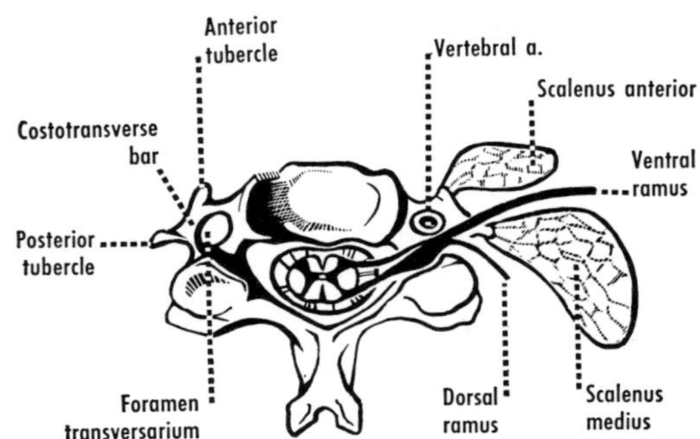

Figure 50–6 Superior aspect of the fourth cervical vertebra, showing the relations of a cervical nerve.

bar intervertebral foramina decrease in diameter. Thus, the fifth lumbar nerve is the thickest, and its foramen the narrowest (fig. 50–5).[13] This relationship increases the chance that the nerve will be compressed in the event of herniation of the nucleus pulposus.

In the cervical region, each spinal root passes through the intervertebral foramen, which is really a canal about 4 mm long, whereas the spinal ganglia and nerves lie outside the foramen, in the gutter or groove of the transverse process (fig. 50–6). Elsewhere, the spinal ganglia and nerves usually lie within the foramina.

MENINGES

Spinal Dura Mater

The spinal cord, like the brain, is surrounded and well protected by the three meninges. The outermost investment, the *spinal dura mater,* is a dense, tough, fibrous tube, which extends from the foramen magnum to the sacrum and coccyx. At the foramen magnum it is continuous with the dura mater of the brain. It is slightly adherent to the margins of the foramen,[14] and to the bodies of C.V. 2 and 3. Immediately below the foramen magnum, the dura mater is thick and vascular. In the lower part of the vertebral canal it narrows rapidly and ends blindly at about the level of S.V. 2 (caudal to the level of the middle of S.V. 2 in about half of instances[15]). It is prolonged as *the filum of the dura mater* to the back of the coccyx, where it blends with periosteum.

The dura mater is separated from the walls of the vertebral canal by the epidural or extradural space, which contains a quantity of semifluid fat and many thin-walled veins. These veins, which constitute the internal vertebral venous plexuses, are comparable in position to the venous sinuses of the cerebral dura mater, in that they lie between the periosteum lining the vertebral canal and the dura mater. (The outer part of the cranial dura mater serves as periosteum.) The epidural fat extends into and occupies the intervertebral foramina and extrudes alongside the vertebrae.[16] Owing to its semifluid nature at body temperature, the fat is readily squeezed into or out of the vertebral canal with variations in intrathoracic and intra-abdominal pressure. The vertebral canal is therefore not a rigid, closed canal. The veins of the canal can carry more or less blood, according to the intrathoracic and intra-abdominal pressure relationships (pp. 327, 420). Caudal analgesia involves injecting a solution of anesthetic into the sacral hiatus (p. 521), so that it diffuses upward into the epidural space.

As each spinal root approaches an intervertebral foramen, it enters a funnel-shaped or tubular prolongation of dura mater, the *dural sheath.*[17] At about the position of the spinal ganglion, the sheaths of the dorsal and ventral roots blend and form a single sheath that continues into the epineurium of the spinal nerve. Certain roots (especially the lower cervical and upper thoracic) descend intradurally and are then often angulated as they turn sharply upward to the intervertebral foramen.[18]

The *subdural space* is a potential space between the arachnoid and the deep surface of the dura mater. It contains only a film of fluid of capillary thinness. The space is prolonged within the dural sheaths and it communicates with lymphatic vessels in the nerves.

Spinal Arachnoid

This is a rather delicate, transparent membrane that forms a loose, wide investment for the spinal cord. Above, it is continuous with the cerebral arachnoid through the foramen magnum. Below, it ends together with the dural sac and is pierced by the filum terminale. At each side, it is prolonged within the dural sheaths as tubular investments for the spinal roots, but only for a short distance.

The subarachnoid space, which contains cerebrospinal fluid, is a wide interval between the arachnoid and the pia mater. A few connective strands cross the subarachnoid space, and a variable fibrous condensation, the subarachnoid septum, extends backward from the cord in the median plane. The subarachnoid space is partially subdivided by the denticulate ligaments.

Lumbar Puncture.[19] **Owing to the circumstance that the spinal cord ends at about L.V. 2, whereas the subarachnoid space continues to the level of S.V. 2, a needle may be introduced below the end of the cord and cerebrospinal fluid withdrawn.** Before fluid is withdrawn, its pressure can be measured. The fluid that is withdrawn can be examined for bacteria or other cells, can be subjected to serological and chemical tests, and can be replaced by a solution of an anesthetic agent. The fluid also can be replaced by contrast media, such as a gas or iodized oil, and the subarachnoid space visualized radiographically.

Spinal Pia Mater[20]

The spinal pia mater consists of reticular tissue and collagenous fibers. The reticular tissue closely invests the spinal cord and passes backward into the anterior median fissure. It forms the posterior median septum and other incomplete septa. It also forms an investment for the rootlets, is prolonged about them, and becomes continuous with the reticular tissue of the arachnoid. The collagenous fibers lie external to the reticular tissue and form a network of bundles. This network contains the vessels of the surface of the cord, is interrupted at the attachments of the rootlets, and gives a few collagenous strands that cross the subarachnoid space. The outermost bundles run longitudinally and, in the median plane in front, form a glistening band, the *linea splendens,* which ensheathes the anterior spinal artery. Below, the longitudinal fibers are contained as the filum terminale.

On each side, the collagenous layer sends a thin, longitudinal septum laterally. The lateral edge of this *denticulate ligament* (fig. 50–2) is free, except for a series of toothlike processes (usually 21 on each side) that fuse with the arachnoid and the dura mater and help to anchor the cord. The uppermost process is at the level of the foramen magnum. Each of the remaining processes is anchored in the interval between two adjacent dural sheaths, the lowermost being below the sheath for the last thoracic nerve. The two denticulate ligaments continue below into the filum terminale.

The attachment of the denticulate ligament to the spinal cord is about midway between the zones of attachment of dorsal and ventral roots. This attachment is a landmark for various surgical procedures.

REFERENCES

1. J. H. Needles, Anat. Rec., *63*:417, 1935. A. F. Reimann and B. J. Anson, Anat. Rec., *88*:127, 1944. The types and levels of endings, and the relation of the segments of the cord to the bodies and spinous processes of the vertebrae are graphically presented by R. Louis, Bull. Ass. Anat., *54*:272, 1970.
2. G. L. Streeter, Amer. J. Anat., *25*:1, 1919.
3. A. J. Barson, J. Anat., Lond., *106*:489, 1970.
4. A. Hovelacque, *Anatomie des nerfs craniens et rachidiens et du système grand sympathique chez l'homme,* Doin, Paris, 1927.
5. R. G. Lendon and J. L. Emery, J. Anat., Lond., *106*:499, 1970.
6. L. A. Gillilan, J. comp. Neurol., *110*:75, 1958. H. Garland, J. Greenberg, and D. G. F. Harriman, Brain, *89*:645, 1966. R. A. Henson and M. Parsons, Quart. J. Med., *36*:205, 1967. R. Djindjian *et al., Angiography of the Spinal Cord,* Masson, Paris, 1970. G. F. Dommisse, J. Bone Jt Surg., *56B*:225, 1974.
7. C. Maillot and J.-G. Koritké, C. R. Ass. Anat., *55*:837, 1970.
8. G. Lazorthes *et al.,* C. R. Ass. Anat., *52*:786, 1967. See also references 6 and 7.
9. L. A. Gillilan, Neurology, *20*:860, 1970.
10. R. M. Lowman and A. Finkelstein, Radiology, *39*:700, 1942. E. C. Porter, Amer. J. Roentgenol., *76*:270, 1956. Diameters of the vertebral canal at various levels are given by B. S. Wolf, M. Khilnani, and L. Malis, J. Mt Sinai Hosp., *23*:283, 1956; A. Delmas and H. Pineau, C. R. Ass. Anat., *51*:282, 1966; J. Minne *et al.,* C. R. Ass. Anat., *56*:1081, 1972.
11. R. W. Reid, J. Anat., Lond., *23*:341, 1889. E. Hintzsche and P. Gisler, Schweiz. Arch. Neurol. Psychiat., *35*:287, 1935. See especially R. Louis, reference 1.
12. E. Gardner, *Fundamentals of Neurology,* Saunders, Philadelphia, 6th ed., 1975.
13. M. S. Danforth and P. D. Wilson, J. Bone Jt Surg., *7*:109, 1925.
14. L. C. Rogers and E. E. Payne, J. Anat., Lond., *95*:586, 1961.
15. M. Trotter, Curr. Res. Anesth., *26*:192, 1947.
16. R. R. Macintosh and W. W. Mushin, Anaesthesia, *2*:100, 1947.
17. H. Swanberg, Med. Rec., N.Y., *87*:176, 1915. R. Frykholm, J. Neurosurg., *4*:403, 1947; Acta chir. scand., *101*:457, 1951. S. Kubik, Acta anat., *63*:324, 1966. The meningeal-neural relationships in the intervertebral foramen are described by S. Sunderland, J. Neurosurg., *40*:756, 1974.
18. H. Nathan and M. Feuerstein, J. Neurosurg., *32*:349, 1970.
19. R. R. Macintosh, *Lumbar Puncture and Spinal Analgesia,* Livingstone, Edinburgh, 2nd ed., 1957.
20. J. W. Millen and D. H. M. Woollam, Brain, *84*:514, 1961. *Idem, The Anatomy of the Cerebrospinal Fluid,* Oxford University Press, London, 1962.

SURFACE ANATOMY OF BACK 51

Some of the features of the surface anatomy of the back are shown in figure 51–1.

The spinous processes of the vertebrae are palpable in the median furrow of the back, a furrow rendered prominent in the neck by the semispinalis capitis on each side, and in the lumbar region by the erectores spinae. **The spines of C.V. 6 and 7 and T.V. 1 are usually prominent and palpable, and are made more conspicuous by flexion of the neck and trunk.**

The external occipital protuberance is palpable in adults (but not in children) in the median plane where the back of the head joins the back of the neck. The hollow below it lies at the level of the atlas. The spine of the axis is palpable about 5 cm below the external occipital protuberance, although this and the spines of C.V. 3 to 5 are obscured by the ligamentum nuchae.

The thoracic spines are usually palpable. The spine of the tenth (or eleventh) is often shorter than the others, so that a slight depression is found over it. In the thoracic region, the spine of each vertebra extends to the level of the body of the vertebra below; in the midthoracic region, a spine may reach the disc below the subjacent vertebra. A horizontal line between the inferior angles of the scapula crosses the spine of T.V. 7.

The lumbar spines are in the furrow between the wide erector spinae muscles on each side of the median plane. The spine of L.V. 5 is sometimes marked by a dimple. **A horizontal plane between the highest points of the iliac crests (supracristal plane)**

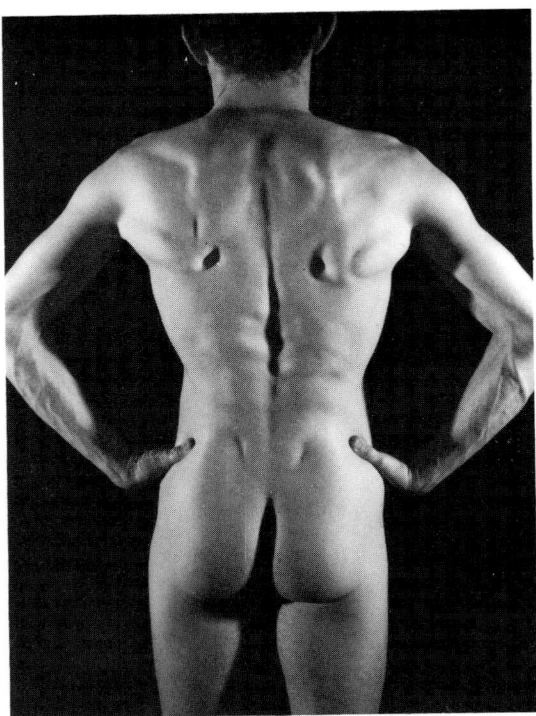

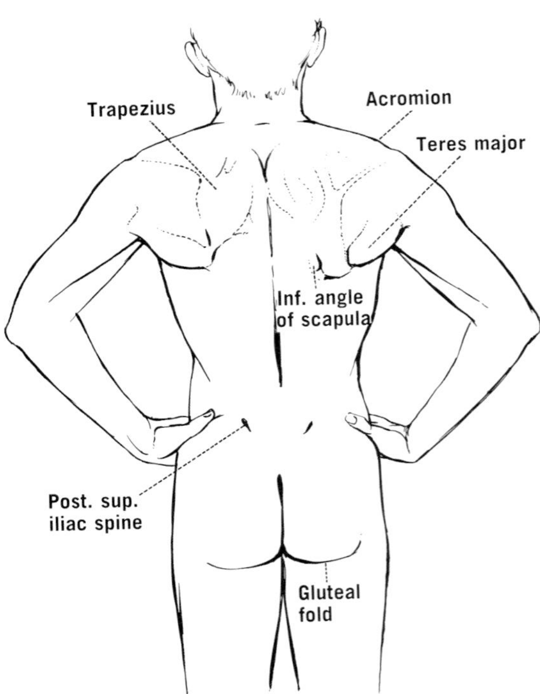

Figure 51–1 Some surface landmarks of the back. Courtesy of J. Royce, Ph.D. (cited on page 70), and Davis, Philadelphia.

passes through the lower part of the spine of L.V. 4, or just below it. This plane is therefore used as a landmark during lumbar puncture. A needle introduced in this plane enters the subarachnoid space at a depth of 4 to 6 cm (less than 2½ cm in infants), and, continued forward, will enter the disc between L.V. 4 and L.V. 5.

The spines of the sacrum are not readily distinguished. Each posterior superior iliac spine is commonly marked by a skin dimple. A line connecting the two iliac spines crosses the spine of S.V. 2.

The sacral hiatus can usually be palpated, as can the movable coccyx below it.

The body of each vertebra above the last lumbar is slightly taller than that of the suprajacent vertebra. As a rough estimate, however, one can take the height of a presacral vertebra and its disc as about 2½ cm.

Part Eight

HEAD AND NECK

Ronan O'Rahilly

Introduction

The cervical vertebrae and the back of the neck have already been described. The first chapter of this part deals with the skull and the hyoid bone. In the remaining chapters, the head and neck will be considered in an order that may be described as generally, but not exclusively, from above downward, beginning with the brain, ear, and eye, and ending with the mouth, nose, pharynx, and larynx.

Particular attention is directed to the following atlases.

R. C. Truex and C. E. Kellner, *Detailed Atlas of the Head and Neck,* Oxford University Press, New York, 1948. This work comprises colored regional drawings of dissections of the head and neck. The last third of the book consists of a series of coronal and a series of horizontal sections.

O. F. Kampmeier, A. R. Cooper, and T. S. Jones, *A Frontal Section Anatomy of the Head and Neck,* University of Illinois Press, Urbana, 1957. A series of 20 life-size photographs of coronal sections, made about 1 cm apart.

R. Aubaniac and J. Porot, *Radio-anatomie générale de la tête,* Masson, Paris, 1955. Key drawings and radiograms of coronal, sagittal, and horizontal sections through the head, made 1 cm apart. References to atlases of the radiological anatomy of the skull are given on page 580.

J. Symington, *An Atlas Illustrating the Topographical Anatomy of the Head, Neck and Trunk,* Oliver and Boyd, Edinburgh, reprinted in 1956. This series of life-size drawings of horizontal sections includes a dozen pertaining to the head and neck.

The head and neck together comprise a highly complicated area. Included in this region are a number of important structures, the diseases of which form the subject matter of certain medical and surgical specialties: neurology and neurosurgery (brain and nerves), ophthalmology (eye), otology (ear), rhinolaryngology (nose and throat), dentistry and oral surgery (teeth and associated structures). A vast amount of information concerning the anatomical basis of these specialties is available, and an introduction to the relevant literature can be gained by studying the special monographs cited among the following chapters.

SKULL AND HYOID BONE

52

The skeleton of the head and neck consists of the skull, the hyoid bone, and the cervical vertebrae. These bony parts should be reviewed throughout the study of the head and neck. The vertebrae have already been described (chapter 48); the skull and the hyoid bone will be discussed in this chapter.

SKULL

The skull provides (1) a case for the brain, (2) cavities for organs of special sensation (sight, hearing, equilibration, smell, and taste), (3) openings for the passage of air and food, and (4) the teeth and jaws for mastication.

The skull consists of a series of bones, which, for the most part, are united at immovable joints. One bone, the mandible, or lower jaw, is freely movable, being connected with the rest of the skull by a synovial articulation, the temporomandibular joint. Some of the bones of the skull are paired, whereas others are unpaired. The cranial bones consist of *external* and *internal tables* of compact substance, and a middle spongy layer, termed *diploë*. The inner table is thinner and more brittle than the outer. The skull is covered and lined by periosteum, the covering layer being termed *pericranium*, and the lining layer, *endocranium* (or endosteal layer of the dura mater).

The term *cranium* (Gk., skull) is sometimes restricted to mean the skull without the mandible. The word *calvaria* (calvarium is incorrect) usually refers to the skull-cap, that is, the top part or vault of the skull, without the facial bones.

Certain of the bones of the skull bound the cranial cavity, in which are situated the brain and its covering membranes (meninges). These bones are the frontal, ethmoid, sphenoid, occipital, temporal, and parietal; the last two are paired.

In addition to a portion of the frontal bone, the skeleton of the face is composed of several paired bones (nasal, lacrimal, zygomatic, and the maxillae), together with the mandible. Another single bone, the vomer, and two paired bones, the palatine and the inferior nasal conchae, are placed more deeply.

The various bones should be identified on a skull with the aid of figures 52–1 to 52–3, and 52–9.

The immovable joints between most of the bones of the skull are termed *sutures*. These have the appearance of irregular lines in the skulls of young adults. The connective tissue between the bones is usually termed a *sutural ligament*. A suture, however, comprises several layers.[1] Although growth of the calvaria occurs at sutures, expansion of the skull is due primarily to brain growth. With increasing age, many of the sutures become obliterated by osseous fusion between the adjacent bones. The closure of sutures, however, is not a reliable criterion of age.[2] Moreover, premature synostosis may occur abnormally; conversely, delayed closure is found in cretinism and in hydrocephalus.

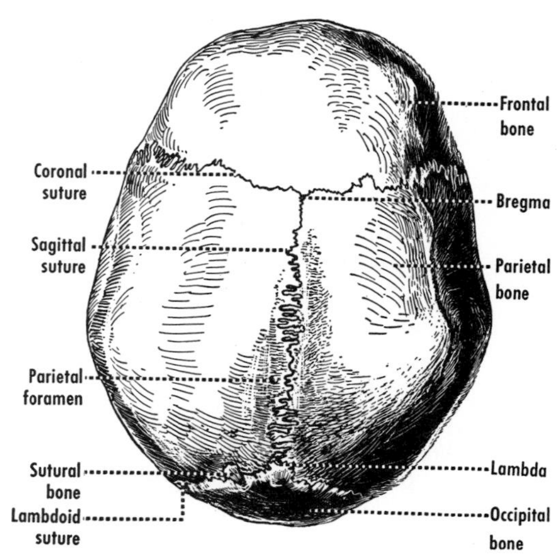

Figure 52–1 Superior aspect of the skull. Note that some portions of a suture are more serrated than others.

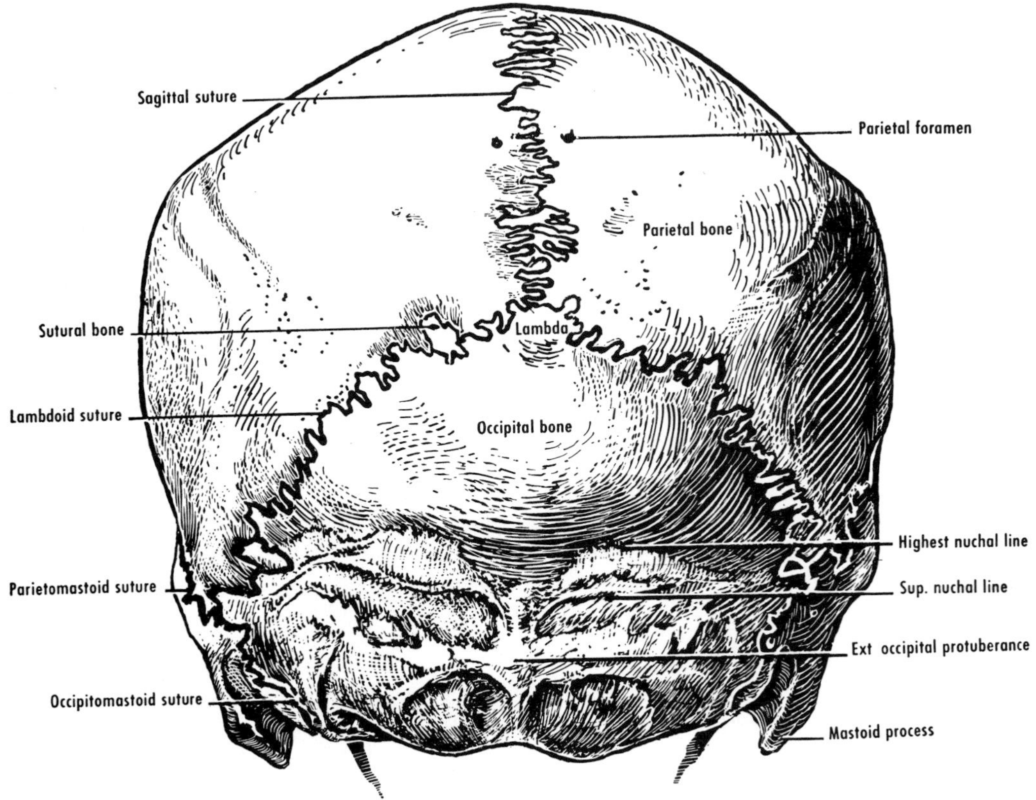

Figure 52–2 Posterior aspect of the skull. Sutural bones are shown; the most frequent site of occurrence of these variable ossicles is along the lambdoid suture.

Circumscribed bony areas, termed *sutural bones*,[3] are seen frequently in some sutures (fig. 52–2).

Although the contrary has been claimed, the position of the sutures seems to be "determined by the meeting of the individual bones, and not independently," as judged from experiments on rats *in utero*.[4] Many sutures are visible radiographically,[5] but closure as estimated radiographically is not contemporaneous with closure as determined anatomically.[6]

For convenience of description, the skull is oriented so that the lower margins of the orbits and the upper margins of the external acoustic meatuses are in the same horizontal plane, termed the *orbitomeatal plane*.

The orbitomeatal plane was accepted as a standard at an anthropological congress in Frankfurt in 1884. Strictly speaking, it passes through the upper margins of the external acoustic meatuses and the lower margin of the left orbit. The orbitomeatal plane corresponds fairly well with a natural horizontal plane through the cranium, that is, with the subject in the anatomical position and the gaze directed toward a vertical mirror on which he fixes his pupils.[7]

The following account is largely concerned with the skull as a whole. For a separate consideration of each individual bone of the skull, more detailed works should be consulted.[8] A good example of an undamaged, well-prepared, dried skull should be available for study and should be examined closely in conjunction with the ensuing description.

The skull presents approximately 85 normal, named foramina, canals, and fissures. Some of these are of little importance, and attention should be directed first to those openings through which a cranial nerve or one of its large branches passes. The openings for (1) the spinal cord and the vertebral arteries, (2) the internal jugular veins, and (3) the internal carotid arteries should be identified before proceeding further.

These openings can be seen clearly on the inferior aspect of the skull (see fig. 52–14). They are termed (1) the foramen magnum, a very large (about 35 mm in length), ovoid, median opening near the back of the skull; (2) the jugular foramen on each side, on a line with the front part of the foramen magnum and about 30 mm from the median plane; and (3) the carotid canal on each side, immediately in front of the jugular foramen.

Radiological Anatomy

The radiological anatomy of the skull is a highly specialized study, and details, when needed, should be sought in appropriate publications.[9]

The most frequently used views in radiography of the head are right and left lateral (fig. 52–4), postero-anterior (brow and nose against film; fig. 52–5), and anteroposterior (occiput against film, as in Towne's position; fig. 52–6). Additional views include a special postero-anterior (chin against film and nose raised from film, as in Water's projection; fig. 62–7A, p. 739) to show some of the paranasal sinuses.

Areas of calcification may sometimes be found normally in the habenular commissure, the pineal body, the choroid plexuses of the lateral ventricles, or in the falx cerebri.

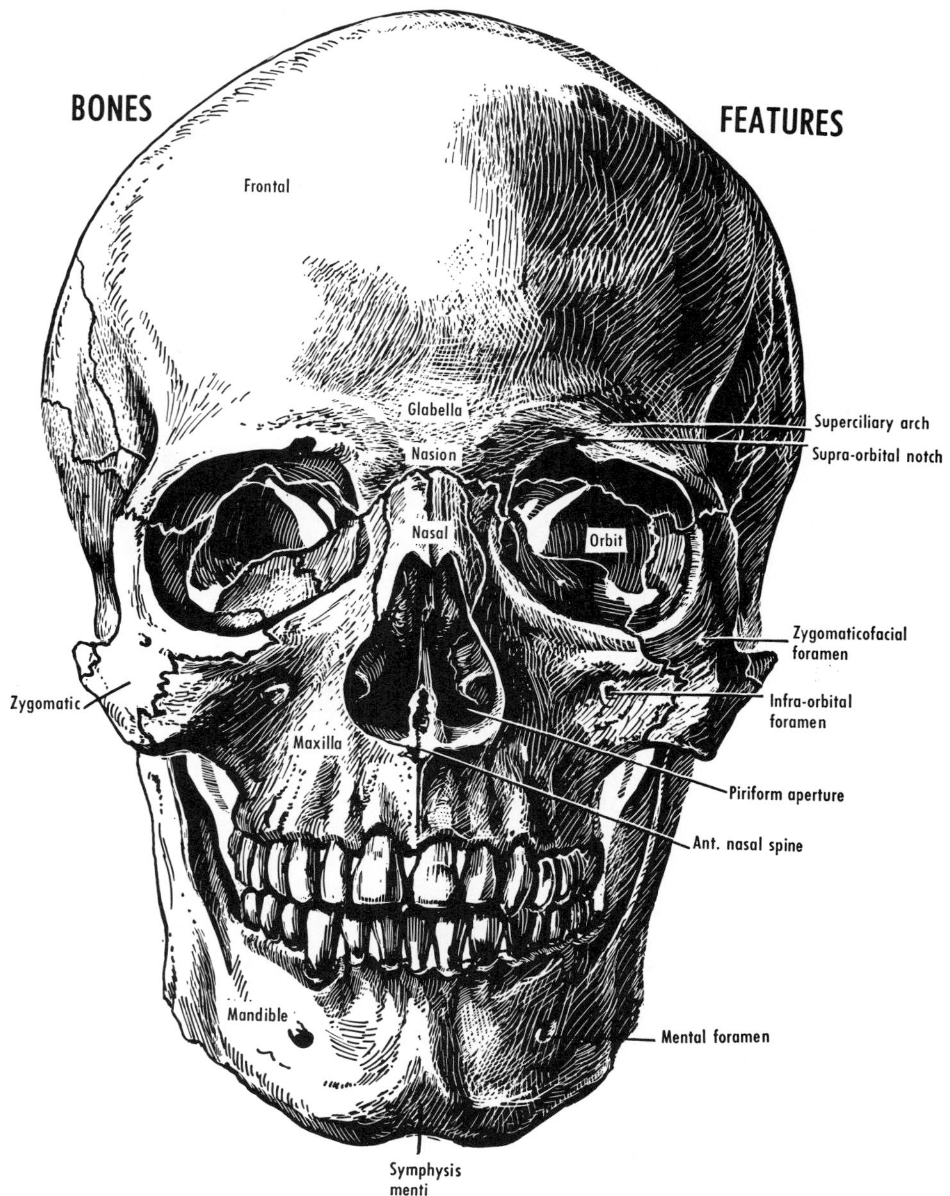

Figure 52–3 Anterior aspect of the skull. Observe that **the supra-orbital notch, the infra-orbital foramen, and the mental foramen are located approximately in a vertical line.**

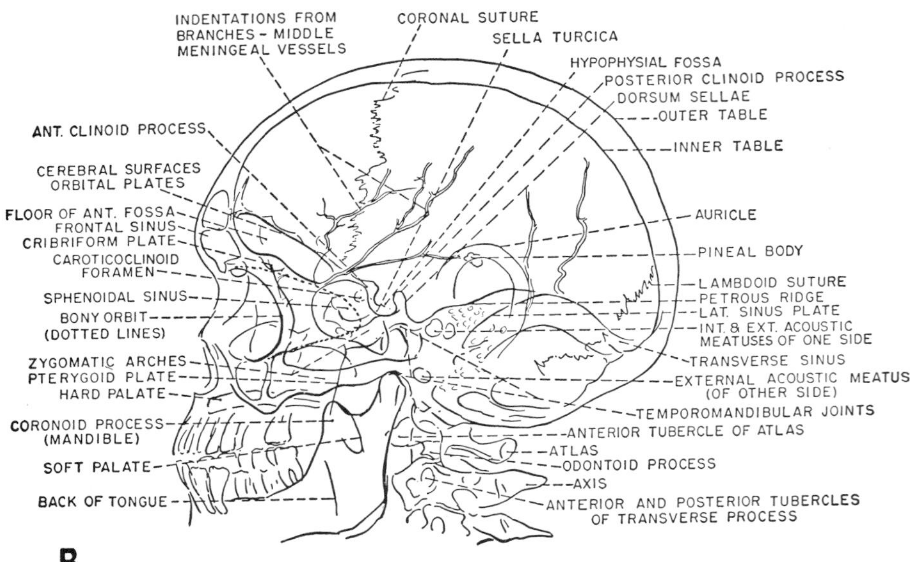

Figure 52–4 Head. Lateral radiogram, and key diagram. From I. Meschan, *Normal Radiographic Anatomy*, Saunders, Philadelphia, 2nd ed., 1959, courtesy of the author.

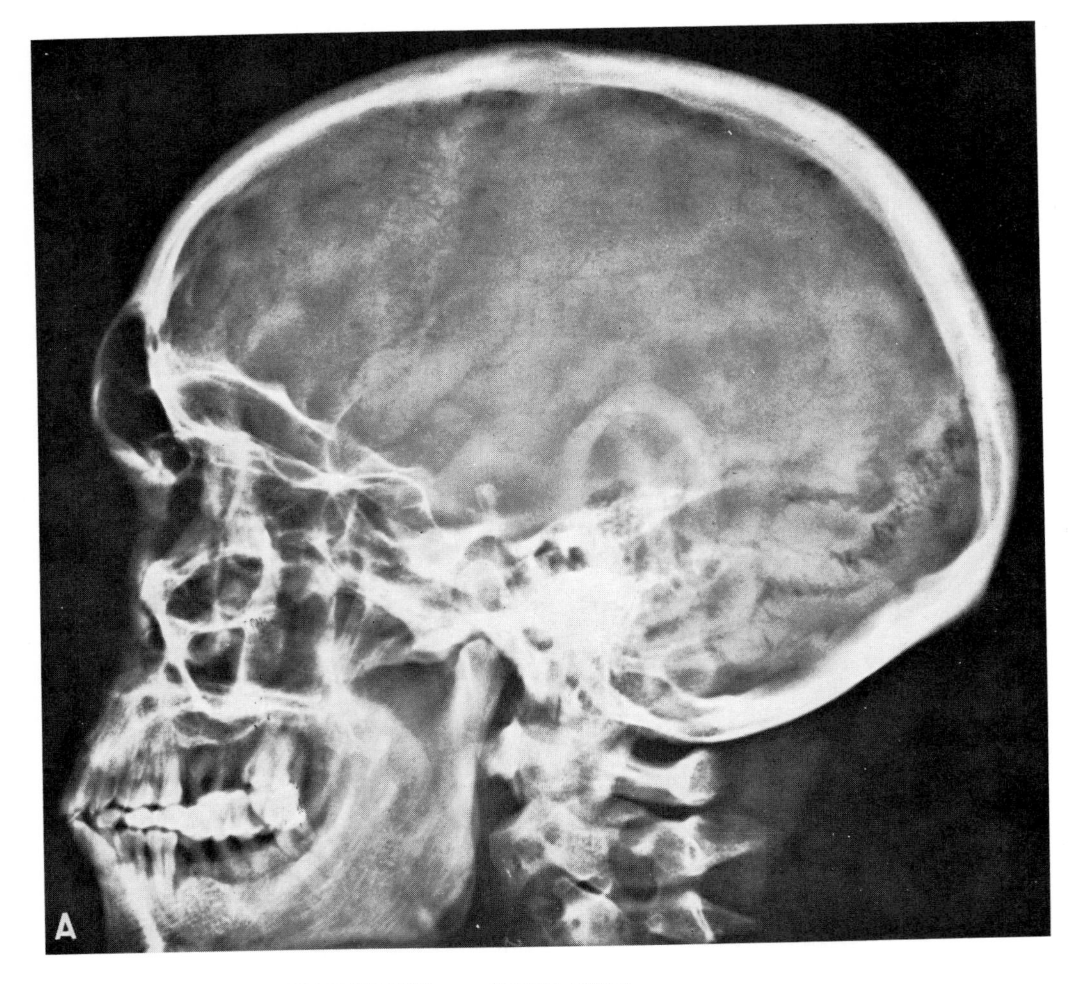

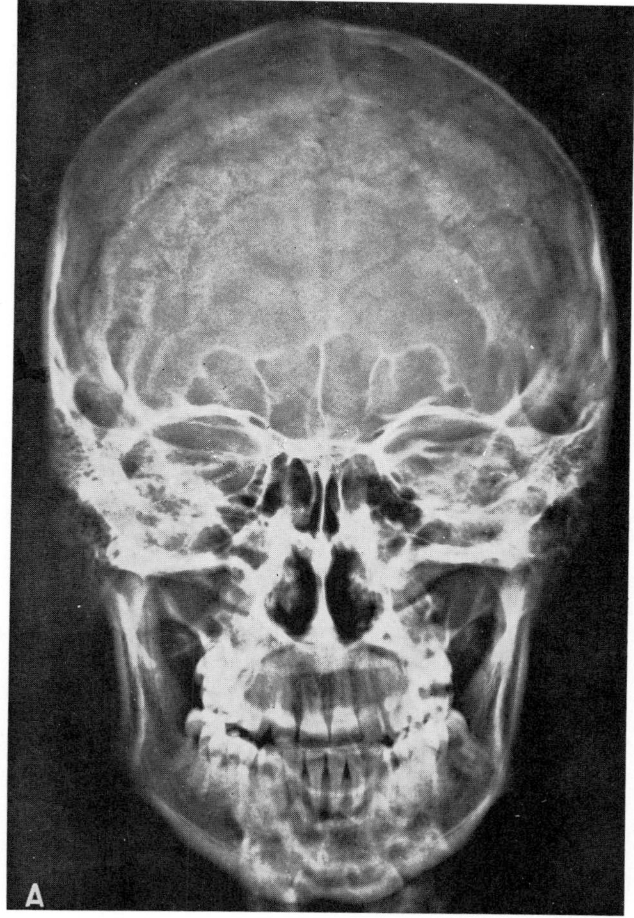

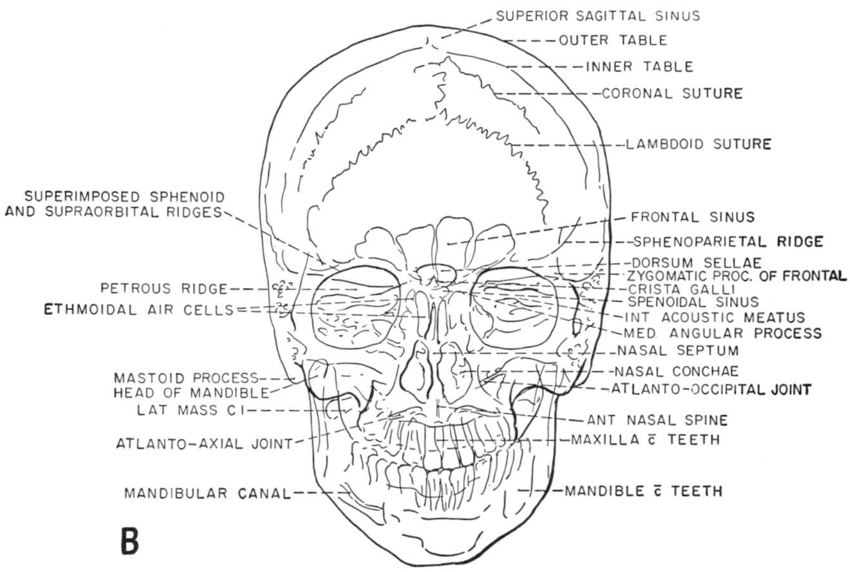

Figure 52–5 Head. Posteroanterior radiogram, and key diagram. From Meschan, cited with figure 52–4, courtesy of the author.

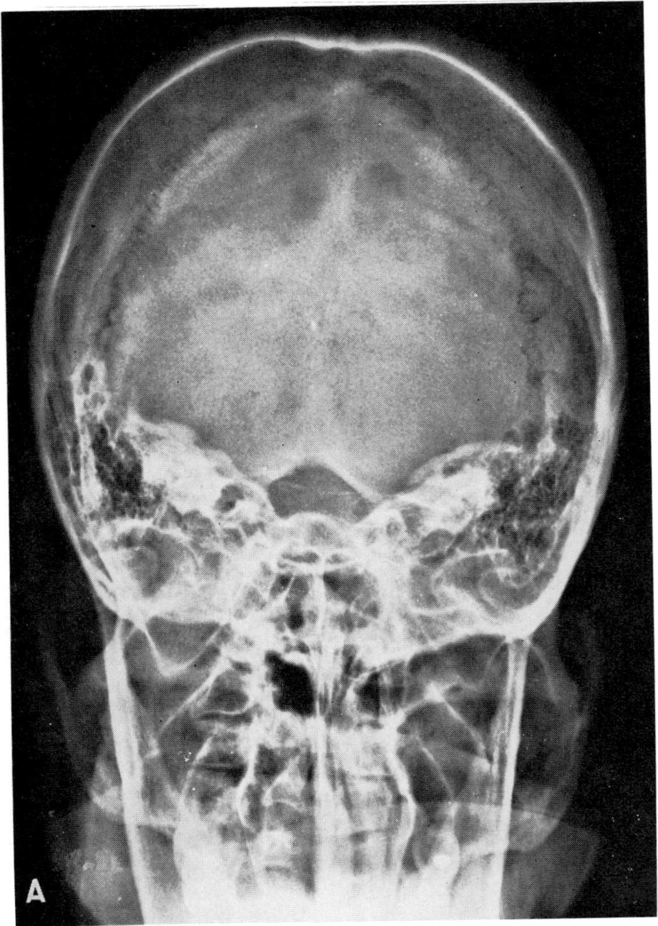

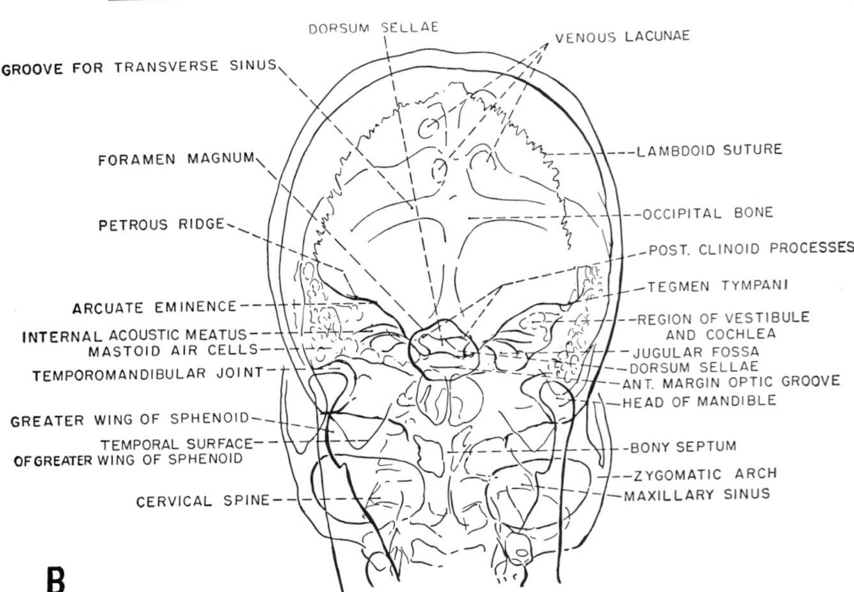

Figure 52–6 Head. Anteroposterior radiogram (Towne's projection), and key diagram. In this projection, in which the subject is supine and the back of the head rests on the film, a line from the lateral angle of the eye to the external occipital protuberance is perpendicular to the film. The central ray is directed slightly caudally, from the forehead to the external occipital protuberance. From Meschan, cited with figure 52–4, courtesy of the author.

Superior Aspect of Skull (fig. 52–1)

The skull is generally ovoid when viewed from above, and is wider behind than in front. Four bones can be identified: the frontal in front, the occipital behind, and the right and left parietal in between. These bones are united by sutures.

The suture between the two parietal bones is termed the *sagittal* (L., an arrow) *suture*. That between the parietals and the frontal is the *coronal* (L., a crown) *suture*, and that between the parietals and the occipital is the *lambdoid* (Gk., the letter L) *suture*. The intersection of the sagittal and coronal sutures is called the *bregma* and, in the fetus and infant, is the site of a membranous area, the anterior fontanelle. The intersection of the sagittal and lambdoid sutures is called the *lambda*. The *vertex* or highest point of the skull lies on the sagittal suture near its middle and is situated a few centimeters behind the bregma. The *parietal eminence* is the most convex portion of each parietal bone. A few centimeters in front of the lambda, a minute vascular opening, the *parietal foramen*, is sometimes found on one or both sides of the sagittal suture. It transmits an emissary vein. Rarely, the foramen may be very large.[10]

Posterior Aspect of Skull (fig. 52–2)

The back of the skull is composed of portions of the parietal bones, the occipital bone, and the mastoid parts of the temporal bones. The sagittal and lambdoid sutures meet at the *lambda*, which can sometimes be felt as a depression *in vivo*. The lower ends of the lambdoid suture meet the *parietomastoid* and *occipitomastoid sutures* on each side at a point known as the *asterion*. The occipitomastoid suture separates the occipital bone from the mastoid portion of the temporal. A vascular opening, the *mastoid foramen*, is frequently found near this suture; it transmits an emissary vein.

The *external occipital protuberance* is a median projection about midway between the lambda and the foramen magnum. It is palpable *in vivo* and is generally located a little below the most bulging part of the back of the head; therefore, it cannot be seen when the skull is viewed from above. Its center is termed the *inion*. On each side, a curved ridge, the *superior nuchal line*, arches laterally from the protuberance. It marks the upper limit of the neck. The *highest nuchal lines*, when present, lie about 1 cm above the superior nuchal lines, and are more arched.

ANTERIOR ASPECT OF SKULL
(figs. 52–3 and 52–5)

The anterior aspect of the skull presents the forehead, the orbits, the prominence of the cheek, the bony external nose, and the upper and lower jaws.

Forehead

The *frontal bone* forms the skeleton of the forehead. Below, on each side of the median plane, it articulates with the nasal bones. The intersection of the frontal and the two nasal bones is termed the *nasion*. The region above the nasion and between the eyebrows is called the *glabella*. The *superciliary arch* is an elevation that extends laterally on each side from the glabella. The two halves of the frontal bone are separated until about six years of age by the frontal suture. In some skulls the line of separation persists into adult life and is known as the *metopic suture*.[11]

Orbits

The orbits are the two bony cavities in which the eyes are situated. They are described in detail in chapter 55. At its junction with the face, each orbit presents superior, lateral, inferior, and medial margins.

The superior or *supra-orbital margin* is formed by the frontal bone. Its medial portion is characterized by the *supra-orbital notch* (or *foramen* in some skulls), which lodges the supra-orbital nerve and vessels. Medial to the notch, the margin is crossed by the supratrochlear nerve and vessels. Laterally, the supra-orbital margin ends in the *zygomatic process of the frontal bone*, which can be felt easily *in vivo*. At each supra-orbital margin, the frontal bone turns sharply backward as an *orbital plate* that forms the greater part of the roof of the corresponding orbit.

The lateral margin is formed by the zygomatic and frontal bones. The inferior margin is formed by the maxilla and the zygomatic. The medial margin of the orbit, which is not as clear-cut as the other boundaries, is formed by the maxilla, lacrimal, and frontal.

Below the inferior margin of the orbit, the maxilla presents an opening, the *infra-orbital foramen,* which transmits the infra-orbital nerve and artery.

Prominence of Cheek

The prominence of the cheek is formed by the *zygomatic (malar) bone* (fig. 52–7). The zygomatic bone is situated on the lower and lateral side of the orbit and rests on the maxilla. It presents (1) a lateral surface on the face, (2) an orbital surface, which contributes to the lateral wall of the orbit, and (3) a temporal surface located in the temporal fossa. A *frontal process* articulates with the zygomatic process of the frontal bone, and a *temporal process* articulates with the zygomatic process of the temporal bone. On the lateral aspect, the zygomatic bone is pierced by the small *zygomaticofacial foramen* for the nerve of the same name.

Bony External Nose

The bony part of the external nose is formed by the nasal bones and by the maxillae and it terminates in front as the *piriform aperture.* The pliable part of the external nose has a cartilaginous framework that is anchored to the piriform aperture by fibrous tissue. The two apertures are bounded above by the nasal bones, and laterally and below by the maxillae. Through the aperture the *nasal cavity* can be seen, divided by the nasal septum into right and left portions, each of which is frequently termed a nasal cavity. The front portion of the *nasal septum* is composed of cartilage, the back of bone (ethmoid and vomer). Each lateral wall of the nasal cavity presents three or four curled plates of bone termed *conchae* (or *turbinates*), the spaces beneath which are called *meatuses.* In the median plane the lower margin of the piriform aperture presents the *anterior nasal spine,* a sharp bony spur formed by the junction of the two maxillae.

The *nasal bones* lie between the frontal processes of the maxillae and meet each other medially. They articulate with the frontal bone above, whereas their inferior borders are anchored to nasal cartilages.

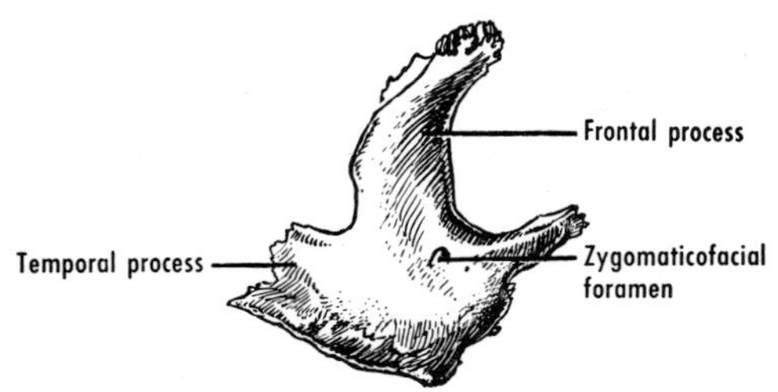

Figure 52–7 Lateral surface of the right zygomatic bone. The orbital and temporal surfaces cannot be seen in this view. The frontal and temporal processes articulate with the zygomatic processes of the frontal and temporal bones, respectively.

Upper and Lower Jaws

Maxillae. The upper jaw is composed of the two maxillae. **The growth of the maxillae is responsible for the vertical elongation of the face between six and twelve years of age.**

Each maxilla (fig. 52–8) consists of (1) a *body,* which contains the maxillary sinus; (2) a *zygomatic process,* which extends laterally and articulates with the zygomatic bone; (3) a *frontal process,* which projects upward and articulates with the frontal bone; (4) a *palatine process,* which extends horizontally to meet its fellow of the opposite side and form the greater part of the skeleton of the palate; and (5) an *alveolar process,* which carries the upper teeth.

The *body of the maxilla* is pyramidal and presents (1) a *nasal surface* or base, which contributes to the lateral wall of the nasal cavity; (2) an *orbital surface,* which forms most of the floor of the orbit; (3) an *infratemporal surface,* which forms the front wall of the infratemporal fossa; and (4) an *anterior surface,* which is covered by facial muscles. About 1 cm below the infra-orbital margin, the anterior surface of the maxilla presents the *infra-orbital foramen* (sometimes multiple[12]), which transmits the infra-orbital nerve and artery.

The upper teeth are carried by the alveolar processes of the maxillae. Vertical ridges corresponding to the roots of the teeth are frequently seen on the front of the bone. The two maxillae are united in the median plane at the *intermaxillary suture.* The portion of the maxillae that supports the incisor teeth is sometimes termed the *premaxilla.*

Mandible. The lower teeth are carried by the *alveolar part of the mandible.* Approximately below the second premolar

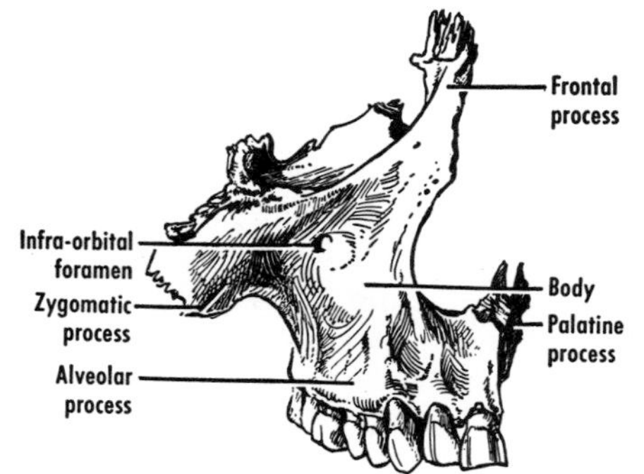

Figure 52–8 Anterior surface of the right maxilla. The maxilla consists of a body and four processes: zygomatic, frontal, palatine, and alveolar.

tooth the mandible displays the *mental foramen,* which transmits the mental nerve and vessels. The *symphysis menti* is the median region of the mandible where the two halves of the fetal lower jaw become fused. The mandible is described in detail later.

LATERAL ASPECT OF SKULL
(figs. 52–4 and 52–9)

The lateral aspect of the skull includes certain portions of the temporal bone, and the temporal and infratemporal fossae.

Certain Features of the Temporal Bone

The temporal bone comprises squamous, tympanic, styloid, mastoid, and petrous parts (fig. 52–10). Certain features of these are considered here, and the remaining features are presented in connection with the inferior aspect of the skull (p. 566) and with the cranial cavity (pp. 573 and 575).

1. Squamous Part. The parietal bone articulates below with the squamous part of

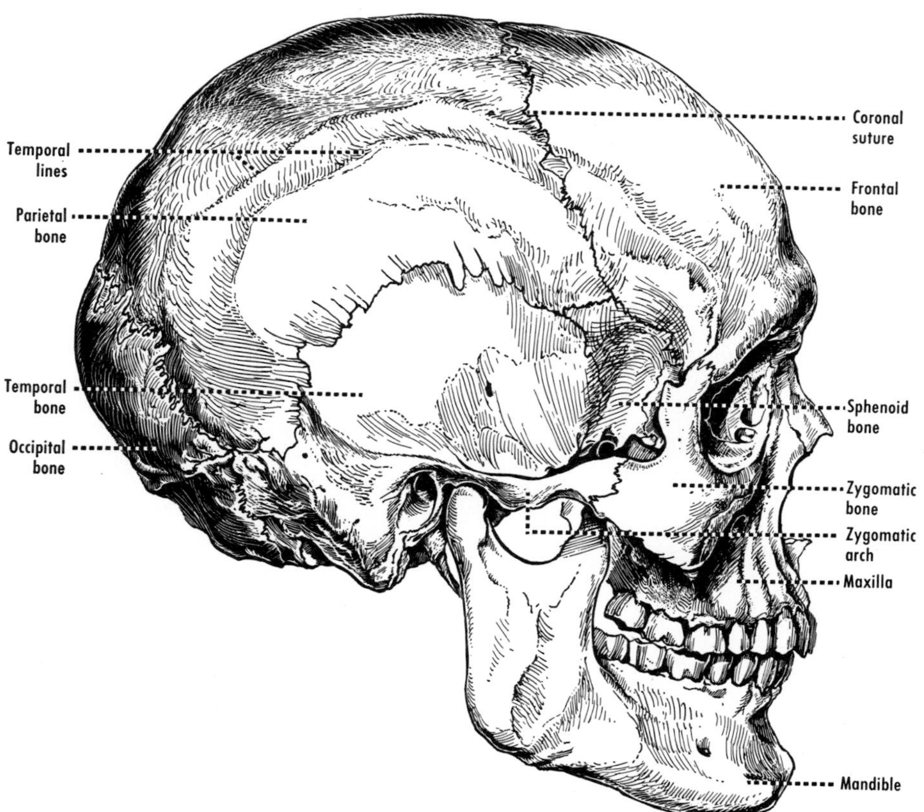

Figure 52–9 Lateral aspect of the skull. Observe the pterion, that is, the area where the parietal, frontal, greater wing of the sphenoid, and squamous part of the temporal approach one another. The temporal fossa is bounded by the temporal lines above and the zygomatic arch below.

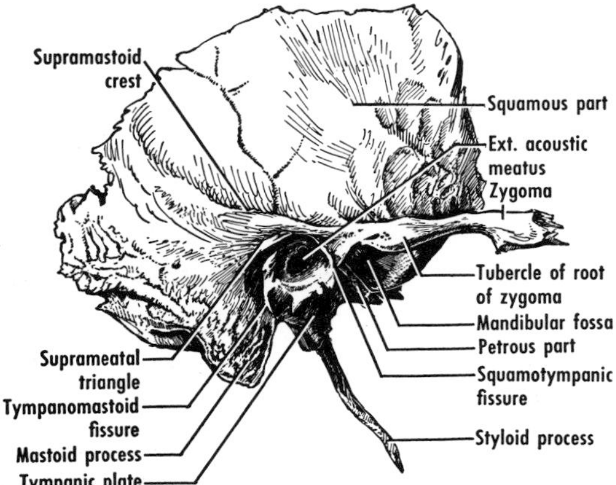

Figure 52–10 Lateral aspect of the right temporal bone. The styloid process is unusually long in this skull.

the temporal *(squamous suture,* fig. 52–21A). From the squamous portion, the *zygomatic process (zygoma)* projects forward to meet the zygomatic bone, thereby completing the zygomatic arch, which is readily palpable *in vivo.* **The upper border of the arch corresponds to the lower limit of the cerebral hemisphere,** and it gives attachment to the temporal fascia. The lower border and deep surface of the arch give origin to the masseter muscle.

The lower border of the zygomatic arch, traced backward, presents the *tubercle of the root of the zygoma,* for the attachment of the lateral ligament of the temporomandibular joint. Behind the tubercle, the head of the mandible is lodged in the mandibular fossa.

The *external acoustic (auditory) meatus,* situated behind the head of the mandible, leads from the exterior to the eardrum (tympanic membrane). It is about 3 cm in length, but the lateral third is cartilaginous, and therefore is not found in a dried skull. The roof and the adjacent part of the posterior wall of the bony meatus are formed by the squamous temporal,* whereas the other walls are formed by the tympanic part. The medial end of the meatus, *in vivo,* is separated by the tympanic membrane from the tympanic cavity (middle ear), which is a space in the temporal bone. In a dried skull, on looking into the meatus, one sees the medial wall of the middle ear, because the lateral wall (the tympanic membrane) has been removed.

A small depression, the *suprameatal triangle,* lies immediately above and behind the external acoustic meatus. **The mastoid antrum, one of the cavities in the temporal bone, lies about 1 cm medial to the suprameatal triangle.** The triangle corresponds to the uppermost part of the concha of the auricle.

2. Tympanic Part. The floor and the anterior wall of the external acoustic meatus are formed by a curved piece of the temporal bone known as the *tympanic plate.* In a child's skull this plate is merely an incomplete *tympanic ring.*

3. Styloid Part. The *styloid process,* a slender projection of variable length (sometimes as long as 8 cm;[13] fig. 52–10, and fig. 60–9*B*, p. 687), extends downward and forward from under cover of the tympanic plate. The hyoid bone in the neck is suspended from the skull by the stylohyoid ligament on each side. The styloid process gives origin to three muscles (styloglossus, stylopharyngeus, and stylohyoid), and provides attachment for the stylomandibular ligament. Laterally, the process is covered by the parotid gland. Its lower part may remain as a separate bone throughout life, and an elongated process may contain a cartilaginous joint.[14] The styloid process develops from the cartilage of the second pharyngeal arch.

4. Mastoid Part. The posterior portion of the temporal bone is termed its mastoid part, and it is fused with the squamous portion. **In the adult the mastoid part generally contains a number of air spaces, the *mastoid air cells,* which communicate with the middle ear by way of the mastoid antrum.** The mastoid part of the temporal is characterized by the downward-projecting *mastoid process,* easily felt *in vivo.* **The mastoid processes of the two sides of the head are in line with the foramen magnum.** The processes are absent at birth and develop gradually during early childhood. Each process gives attachment to several muscles (fig. 52–11). The front of the mastoid process is separated from the tympanic plate by the *tympanomastoid fissure,* which transmits the auricular branch of the vagus.

*Certain abbreviated expressions, such as squamous temporal and petrous temporal for the squamous and petrous parts, respectively, of the temporal bone, are in common usage, and it would be pedantic to refuse to employ them.

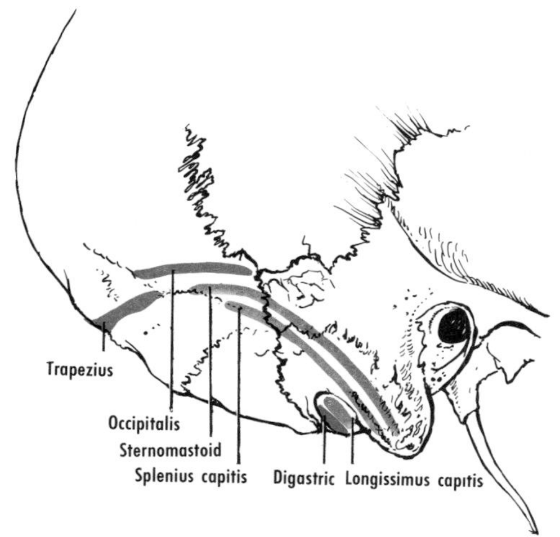

Figure 52–11 Lateral aspect of the occipitomastoid region of the skull, showing muscular attachments.

5. *Petrous Part.* The petrous part is placed deeply and is described later (p. 566).

Temporal Fossa

The *temporal line,* to which the temporal fascia is attached, begins at the zygomatic process of the frontal bone. It arches backward across the frontal and parietal bones, at a variable distance from the sagittal suture.[15] The indistinct posterior part of the temporal line joins a ridge on the temporal bone termed the *supramastoid crest.* The middle portion of the temporal line is usually double, and the lower line indicates the limits of the temporal muscle.

The temporal fossa, in which the temporal muscle is located, is bounded by the temporal line above and the zygomatic arch below. Its floor, which gives origin to the temporalis, is composed of portions of the parietal, frontal, greater wing of the sphenoid, and squamous part of the temporal. The area where these four bones approach each other is known as the *pterion* (fig. 52–9). **The pterion overlies the anterior branch of the middle meningeal artery on the internal aspect of the skull, and it corresponds also to the stem of the lateral sulcus of the brain. The center of the pterion is approximately 4 cm above the midpoint of the zygomatic arch and nearly the same distance behind the zygomatic process of the frontal.**

In the anterior wall of the temporal fossa, the zygomatic bone presents the small *zygomaticotemporal foramen,* for the nerve of the same name.

Infratemporal Fossa

The interval between the zygomatic arch and the rest of the skull is traversed by the temporal muscle and the deep temporal nerves and vessels. By means of this interval the temporal fossa above communicates with the infratemporal fossa below. The infratemporal fossa is an irregularly shaped space behind the maxilla. Medial to its communication with the temporal fossa, the roof of the infratemporal fossa is formed by the *infratemporal surface of the greater wing of the sphenoid* (fig. 52–18). Medially, the infratemporal fossa is limited by the *lateral pterygoid plate* of the sphenoid (fig. 52–12); laterally, it is limited by the ramus and the coronoid process of the mandible.

The infratemporal fossa contains the lower part of the temporalis, and the lateral and medial pterygoid muscles; the maxillary artery and its branches, and the pterygoid venous plexus; and the mandibular, maxillary, and chorda tympani nerves.

Above the posterior surface of the maxilla, between it and the greater wing of the sphenoid, the infratemporal fossa communicates with the orbit by means of the *inferior orbital fissure.* This fissure is continuous behind with the *pterygomaxillary*

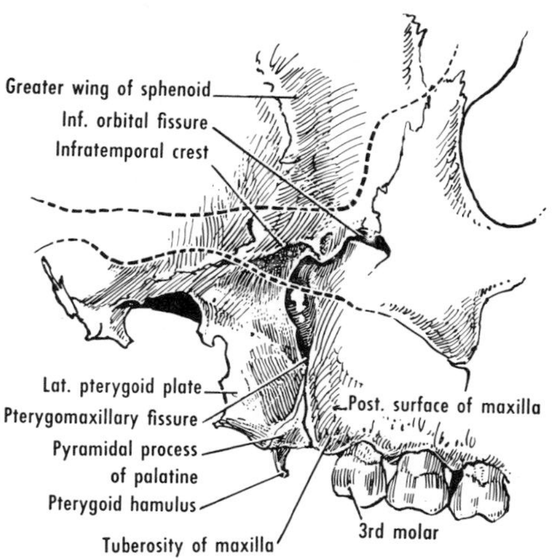

Figure 52–12 The medial wall of the right infratemporal fossa. The mandible has been removed. The zygomatic arch is shown as if transparent; its borders are indicated by interrupted lines.

fissure, a cleft located between the lateral pterygoid plate and the maxilla. The infratemporal fossa communicates with the pterygopalatine fossa by means of the pterygomaxillary fissure, which transmits the maxillary artery. The *pterygopalatine fossa* is so named because it lies between (1) the pterygoid plates of the sphenoid and (2) the palatine bone. It is located below the apex of the orbit, and it contains the maxillary nerve and artery, and the pterygopalatine ganglion. The fossa communicates with the nasal cavity through the sphenopalatine foramen. Below the pterygomaxillary fissure, the lateral pterygoid plate appears to meet the *tuber of the maxilla* but is actually separated from it by the *pyramidal process of the palatine bone.*

The communications just described are summarized in table 52–1.

The boundaries of, and openings from, both the infratemporal and the pterygopalatine fossa are indicated in table 52–2.

TABLE 52–1 Summary of Communications with Infratemporal Fossa

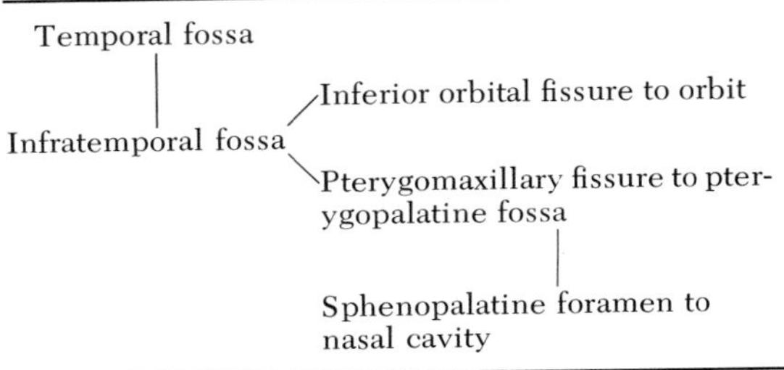

INFERIOR ASPECT OF SKULL
(figs. 52–13 and 52–14)

The features of this aspect of the skull will be considered from behind forward.

Occipital Bone (fig. 52–15)

The lower surface of the base of the skull is formed behind by the occipital bone, which is continuous in front with the sphenoid. The **occipital bone consists of four parts, arranged around the foramen magnum: (1) a squamous part behind, (2) a lateral part at each side, and (3) a basilar part in front. These four portions, separable at birth, become fused by about six years.**

The *foramen magnum* is located midway between, and on a level with, the mastoid processes. Through it the posterior cranial fossa communicates with the vertebral canal, and the brain becomes continuous with the spinal cord. The foramen contains: the junction between the medulla oblongata and the spinal cord, a small portion (the tonsils) of the cerebellum occasionally, the meninges, the spinal roots of the accessory nerves, meningeal branches of cervical nerves 1 to 3, the vertebral arteries and their sympathetic plexuses, the spinal arteries derived from the vertebral, and certain ligaments (apical ligament of the dens, cruciform ligament of the atlas, and membrana tectoria). The anterior and posterior margins of the foramen give attachment to the corresponding atlanto-

TABLE 52–2 Boundaries of and Openings from Infratemporal and Pterygopalatine Fossae

Aspect	*Boundaries of and Openings from Infratemporal Fossa*	*Boundaries of Pterygopalatine Fossa*	*Openings from Pterygopalatine Fossa*
Superior	Infratemporal surface of greater wing of sphenoid	Body of sphenoid and orbital process of palatine	Inferior orbital fissure to orbit
Inferior	Open	Anterior and posterior walls meet	Greater (and sometimes lesser) palatine canal to palate
Anterior	Posterior surface of maxilla and inferior orbital fissure	Posterior surface of maxilla	None
Posterior	Open	Lateral pterygoid plate and greater wing of sphenoid	Foramen rotundum to middle cranial fossa Pterygoid canal to foramen lacerum Palatovaginal canal to choana
Medial	Lateral pterygoid plate and pterygomaxillary fissure	Perpendicular plate of palatine	Sphenopalatine foramen to nasal cavity
Lateral	Ramus and coronoid process of mandible	Open	Pterygomaxillary fissure to infratemporal fossa

occipital membranes. The midpoint of the anterior margin of the foramen magnum is termed the *basion*.

1. Squamous Part. The squamous part of the occipital bone lies partly in the base and partly in the back of the skull, the line of demarcation being the *external occipital protuberance* and the superior nuchal lines. The *external occipital crest,* to which the ligamentum nuchae is attached, extends from the protuberance to the foramen magnum. About midway along the crest, an *inferior nuchal line* extends laterally on each side. The *superior nuchal lines* give attachment to the galea aponeurotica and several muscles (trapezius, occipitalis, splenius capitis, and sternomastoid). Certain muscles are inserted into the areas between the superior and inferior nuchal lines, and others in front of the inferior lines (fig. 52–15, p. 566).

2. Lateral (Condylar) Part. The lateral parts of the occipital bone present the *occipital condyles,* two large protuberances at the sides of the foramen magnum. The condyles articulate with the lateral masses of the atlas, and, through them, the weight of the head is transmitted to the vertebral column. The occipital condyles are on the level of the hard palate. Behind each condyle is a *condylar fossa,* which is frequently perforated by an opening *(condylar canal)* that transmits an emissary vein.

A short passage termed the *hypoglossal canal,*[16] frequently bipartite,[17] lies hidden above the front of each condyle (fig. 52–16, *arrow*). It transmits the hypoglossal nerve and some small vessels.

The *jugular process* extends laterally from each condyle to the temporal bone and presents a concave, anterior border termed the *jugular notch.* This notch forms the posterior boundary of the *jugular foramen,* a large opening between the occipital and the petrous temporal. The transverse process of the atlas lies immediately below the jugular process (fig. 52–16).

3. Basilar Part. The basilar part of the occipital *(basi-occiput)* is the wide bar of bone that joins the sphenoid; the union is cartilaginous until about puberty, when bony fusion takes place (12 to 16 years in the female, 13 to 18 in the male;[18] fig. 52–23). The pharynx (superior constrictor and pharyngeal raphe) is attached to the *pharyngeal tubercle,* an ill-defined elevation about 1 to 1½ cm in front of the foramen magnum. **The pharyngeal tubercle may be regarded as a point of division between the pharynx in front and the bones and muscles of the neck behind.** The longus capitis is inserted on each side in front of the pharyngeal tubercle. A jagged opening, the *foramen lacerum,* is seen on each side of the basilar part of the occipital. It is closed by cartilage *in vivo.*

Temporal Bone (figs. 52–10, 52–17, and 52–24)

The words temple and temporal are derived from the Latin *tempus,* time, and are used because gray hairs generally appear first in this area.

All the divisions of the temporal bone (petrous, mastoid, tympanic plate, styloid process, and squamous) can be identified on the lower aspect of the base (fig. 52–17). The boundaries of the temporal bone should be followed on a skull. **The temporal bone is important particularly because it contains the middle and internal portions of the ear.** The account of certain features of the temporal bone already given with the lateral aspect of the skull (p. 559) should be read again in conjunction with the following description.

1. Squamous Part (see also p. 559). The squamous part is a thin bony plate placed vertically in the side of the skull. It presents a cerebral surface medially and a temporal surface laterally.

The *zygomatic process (zygoma)* extends forward from the squamous part to articulate with the temporal process of the zygomatic bone and thereby form the *zygomatic arch.* The arch, traced backward, is considered to divide into two roots. The anterior root is continuous with the *articular tubercle,* a smooth elevation situated in front of a deep concavity known as the *mandibular fossa.* **The mandibular fossa and articular tubercle are fundamentally portions of the squamous part of the temporal.** Their margins give attachment to the capsule of the temporomandibular joint. The head of the mandible occupies the mandibular fossa when the mouth is closed, and rests on the articular tubercle when the mouth is open (fig. 58–5, *B* and *C,* p. 665). An articular disc intervenes, however, between

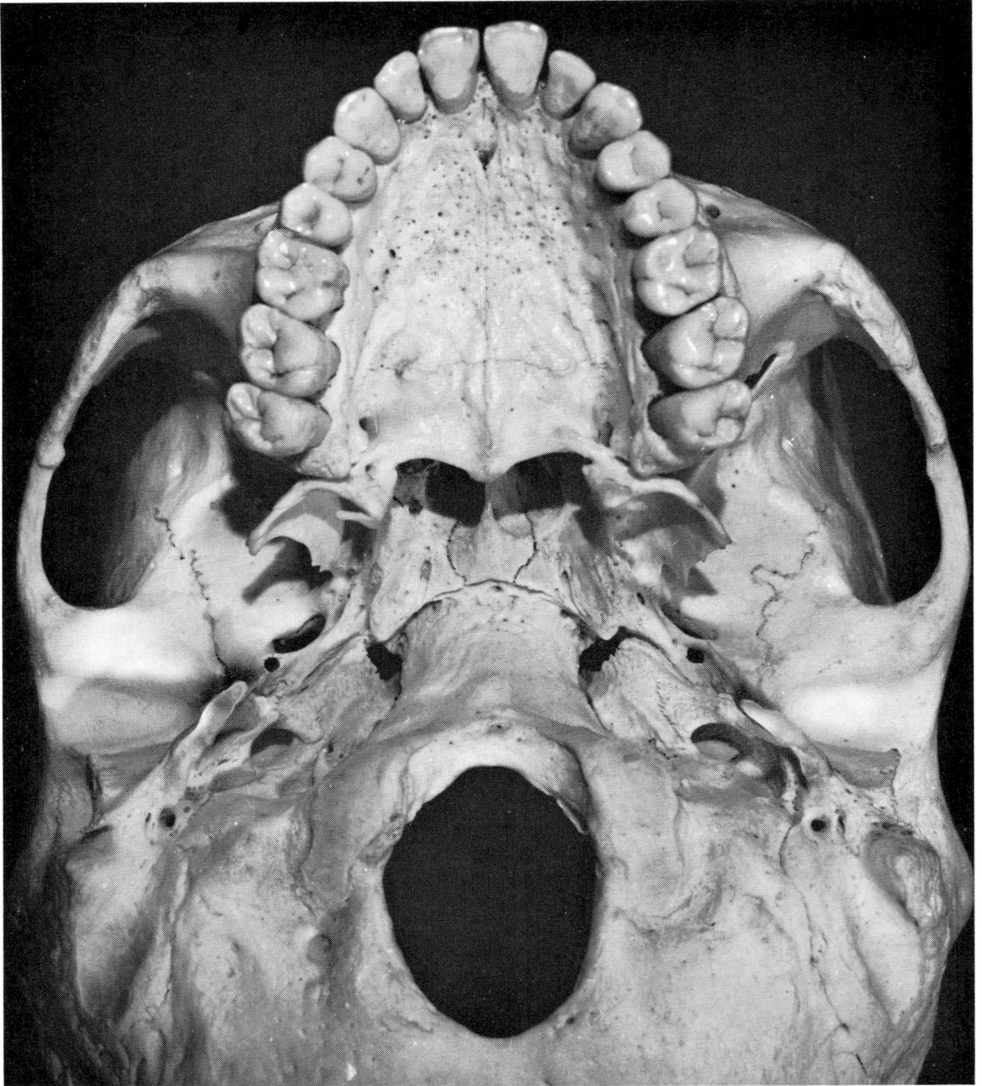

Figure 52–13 Photograph of the base of the skull. See figure 52–14 for the identification of the various features.

the base of the skull and the head of the mandible.

The posterior root of the zygomatic arch joins the *supramastoid crest.* The portion of the posterior root immediately in front of the external acoustic meatus is called the *postglenoid tubercle.*

2. *Tympanic Part.* The tympanic part of the temporal bone consists of a curved tympanic plate that is fused behind with the mastoid and petrous parts and forms a sheath for the styloid process. Its superior aspect forms the floor and the anterior wall of the external acoustic meatus. Its anterior surface is separated from the head and neck of the mandible (and the capsule) by a portion of the parotid gland. In the mandibular fossa the tympanic plate is separated from the squamous temporal by the *squamotympanic fissure.* The medial part of this fissure is generally occupied by a portion of the tegmen tympani (a part of the petrous temporal) that forms the anterolateral wall of the bony part of the auditory

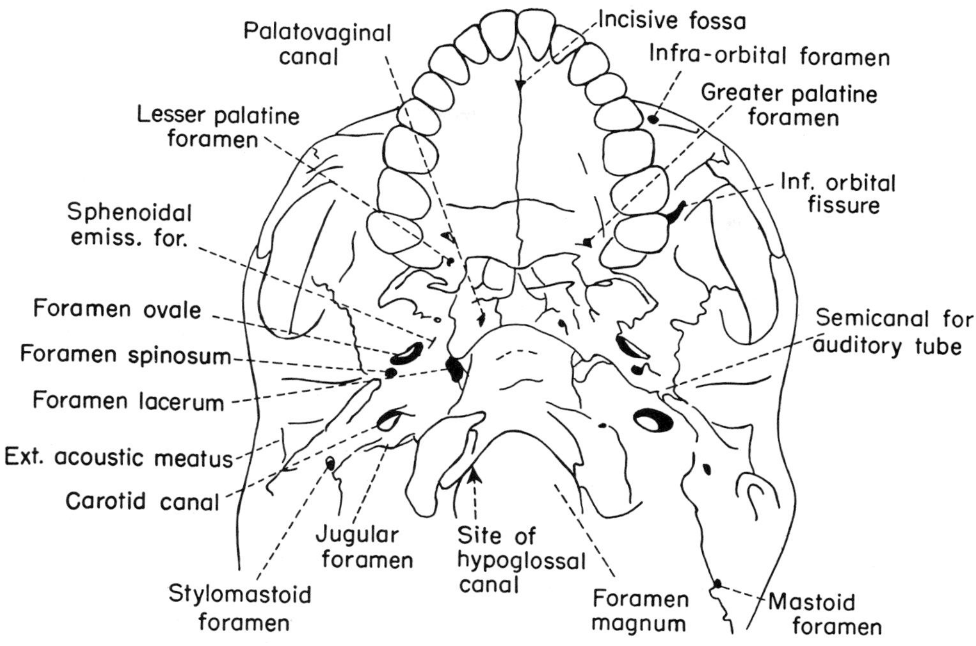

Palatine proc. of maxilla
Horizontal plate of palatine
Post. nasal spine
Choana
Sphenoidal proc. of palatine
Tuber of maxilla
Ala of vomer
Pterygoid hamulus
Infratemp. crest of sphenoid
Zygomatic arch
Vaginal proc.
Pterygoid fossa and lat. pterygoid plate
Quadrate area
Spine of sphenoid
Scaphoid fossa
Artic. tubercle
Styloid process
Mandib. fossa
Squamotympanic fissure
Spheno-occipital junction
Tympanomastoid fiss.
Mastoid proc.
Jugular proc. of occipital
Pharyngeal tubercle
Occipitomastoid suture
Mastoid notch
Occipital condyle
Sutural bone
Inf. nuchal line
Condylar fossa
Sup. nuchal line
Ext. occipital crest
Ext. occipital protuberance

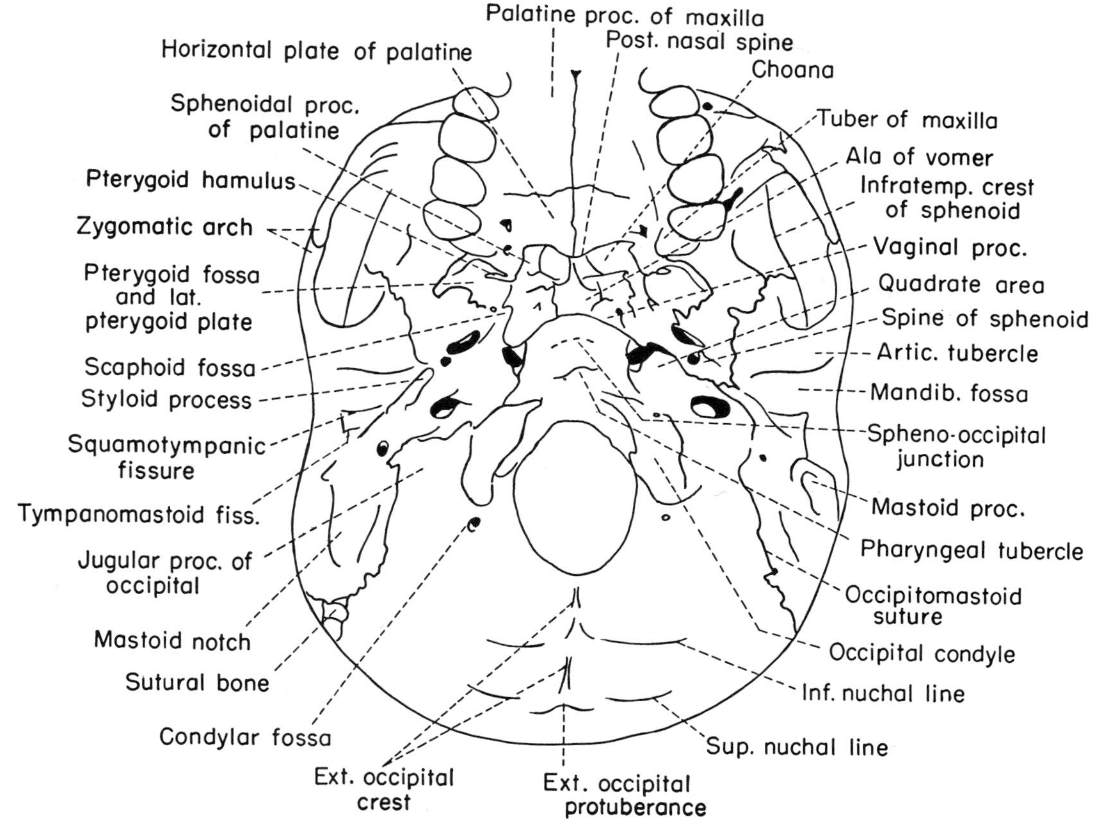

Figure 52–14 The base of the skull. Keys to figure 52–13. The upper drawing shows mainly foramina and canals.

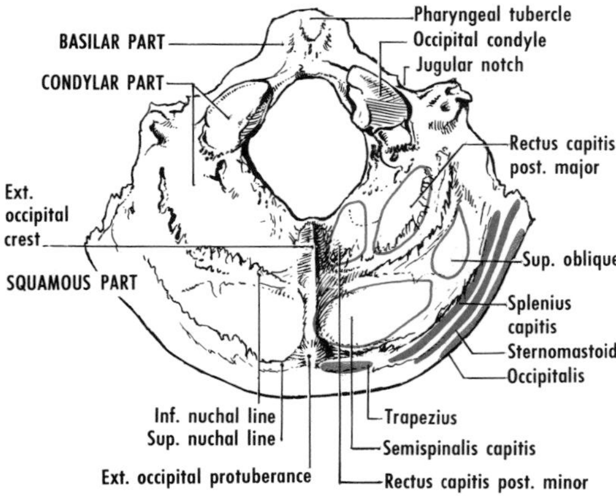

Figure 52–15 Inferior aspect of the occipital bone. The four chief parts of the bone can be seen around the foramen magnum: basilar, two lateral, and squamous. The muscular attachments are indicated on the left side of the bone. The insertions of the rectus capitis lateralis (lateral to the occipital condyle), rectus capitis anterior (lateral to the pharyngeal tubercle), and the fibrous raphe of the pharynx (to the pharyngeal tubercle) are not shown.

tube. The fissure is thereby divided into a *petrosquamous fissure* in front and a *petrotympanic fissure* behind (fig. 52–17). The latter allows the exit of the chorda tympani from the skull.

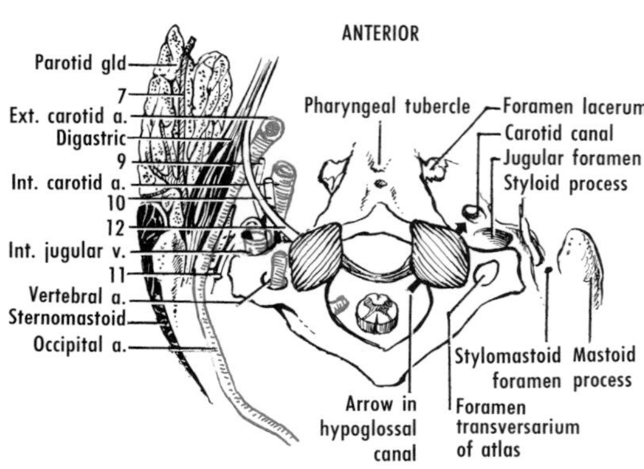

Figure 52–16 Inferior aspect of the base of the skull with the atlas in position. The interval in front of the transverse ligament of the atlas is for the dens of the axis. The transverse process of the atlas lies immediately behind the jugular foramen and immediately below the jugular process of the occipital bone. Cranial nerves 9, 10, and 11 emerge through the jugular foramen, which lies behind the carotid canal (occupied by the internal carotid artery). Cranial nerve 12, however, emerges more medially than the previous three nerves, namely, through the hypoglossal canal (*arrow*). Cranial nerve 7, by contrast, is "standing somewhat aloof," and it emerges through the more laterally placed stylomastoid foramen. Based on von Lanz and Wachsmuth and on Shellshear and Macintosh.

*3. **Styloid Part.*** The styloid part of the temporal bone consists of the styloid process, which has been described on page 560.

The *stylomastoid foramen,* through which the facial nerve emerges from the temporal bone, is located between the styloid and mastoid processes.

*4. **Mastoid Part*** (see also p. 560). The mastoid part is situated behind the squamous and tympanic parts, and is fused medially with the petrous part, from which it is developed. Below, it gives rise to the prominent *mastoid process,* which presents the *mastoid notch* on its medial side, for the origin of the posterior belly of the digastric muscle. The mastoid notch leads anteriorly to the stylomastoid foramen, through which the facial nerve emerges from the skull. Medial to the notch there is usually a *groove for the occipital artery.* The muscular attachments to the mastoid process are shown in figure 52–11.

*5. **Petrous Part.*** The petrous part of the temporal bone has the shape of a three-sided pyramid. It contains the internal ear and contributes to the boundaries of the middle ear. Its base, directed laterally, is fused with the other parts of the temporal bone. Its apex is directed medially and forward, between the sphenoid laterally and the occipital medially. The three sur-

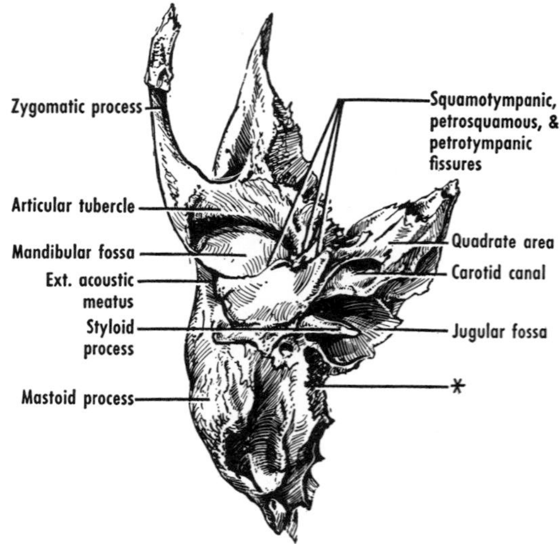

Figure 52–17 Inferior aspect of the right temporal bone. For the position of this bone in the base of the skull, see figure 52–13. The area marked with an asterisk articulates with a corresponding area (see fig. 52–22) on the jugular process of the occipital bone.

faces are: (1) anterior and (2) posterior, both of which face the cranial cavity (pp. 574 and 575), and (3) the inferior surface, which is considered here.

The *jugular fossa,* a depression situated medial to the styloid process, forms one boundary of the jugular foramen. The *jugular foramen* is located between the jugular notch of the occipital and the jugular fossa of the petrous temporal. **The jugular foramen is related to the carotid canal anteriorly, the transverse process of the atlas posteriorly, the styloid process laterally, and the hypoglossal canal medially** (fig. 52–16). The internal jugular vein (and thus the foramen) is usually larger on the right side of the head. The following structures are transmitted:

Posterior part of foramen: the internal jugular vein, a continuation of the sigmoid sinus. The vein is dilated at its commencement to form its superior bulb, which is lodged in a concavity, the jugular fossa, of the petrous temporal.

Middle part: the glossopharyngeal, vagus, and accessory nerves.

Anterior part: the inferior petrosal sinus, which is a tributary of the internal jugular vein.

The floor of the jugular fossa separates the superior bulb of the internal jugular vein from the middle ear (tympanic cavity). A minute opening, the *mastoid canaliculus,* is located on the lateral wall of the jugular foramen, and transmits the auricular branch of the vagus. Another minute opening, the *tympanic canaliculus,* is found on or near the ridge between the jugular foramen and the opening of the carotid canal; it transmits the tympanic nerve (from the glossopharyngeal) to the tympanic cavity.

The *carotid canal,*[19] a tunnel in the petrous temporal, transmits the internal carotid artery to the cranial cavity. The lower end of the carotid canal is found immediately in front of the jugular foramen, and thus the internal carotid artery enters the base of the skull anterior to the exit of the internal jugular vein. The canal is closely related to the internal ear, and the beating of the artery during excitement or after exertion can sometimes be heard as a thundering sound in the head. The carotid canal transmits the internal carotid artery and its associated sympathetic plexus.

The *quadrate area* of the petrous temporal lies between the carotid canal and the foramen lacerum. It gives origin to the levator veli palatini muscle. The *foramen lacerum* (compare the word lacerated) is a jagged opening between the petrous temporal, the body and the greater wing of the sphenoid, and the basilar part of the occipital. It is closed by cartilage *in vivo* and is related below to the cartilaginous part of the auditory tube. From its anterior margin the *pterygoid canal* passes forward to the pterygopalatine fossa. The canal transmits a nerve formed by the union of the deep petrosal nerve (from the carotid sympathetic plexus) with the greater petrosal nerve in the foramen lacerum. The pterygoid canal opens anteriorly in the posterior wall of the pterygopalatine fossa.

The groove between the quadrate area medially and the greater wing of the sphenoid laterally is occupied *in vivo* by the cartilaginous part of the auditory tube. The posterolateral end of the groove is continuous with two semicanals within the temporal bone. These lead into the middle ear (tympanic cavity). The lower is the bony part of the auditory tube; the upper and smaller is occupied by the tensor tympani muscle.

A number of transverse "datum lines" have been described for the base of the skull. It is instructive, for example, to consider some of the features that are situated medial to the external acoustic meatus, the mandibular fossa, and the articular tubercle, respectively:

Posterior transverse line:
- External acoustic meatus (posterior border)
- Stylomastoid foramen
- Jugular foramen
- Occipital condyle and hypoglossal canal
- Foramen magnum (anterior margin)

Middle transverse line:
- Mandibular fossa and head of mandible
- Petrotympanic and petrosquamous fissures
- Spine of sphenoid
- Junction of osseous and cartilaginous parts of auditory tube
- Pharyngeal tubercle

Anterior transverse line:
- Articular tubercle
- Foramen ovale
- Auditory tube
- Foramen lacerum
- Spheno-occipital junction

Sphenoid

The sphenoid bone (see also pp. 573 and 574) consists of a body and three pairs

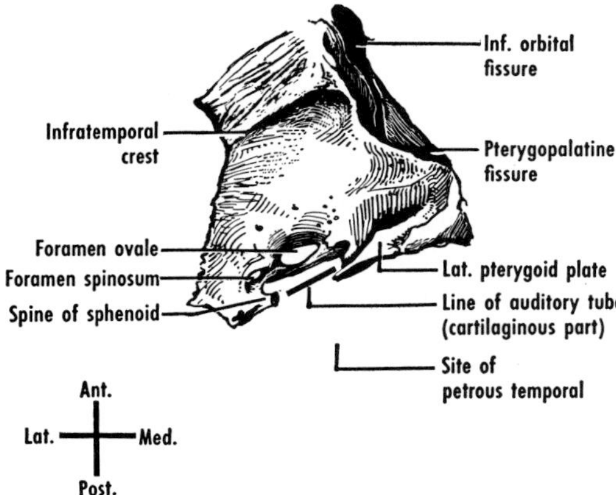

Figure 52–18 The infratemporal surface of the greater wing of the sphenoid. This area should be identified on an intact skull.

of processes, or "wings": greater wings, lesser wings, and pterygoid processes. The portions other than the lesser wings can all be identified on the inferior aspect of the skull. The greater wing and the pterygoid process are described here.

Greater Wing. The *infratemporal surface of the greater wing of the sphenoid* (fig. 52–18) is located lateral to the upper end of the lateral pterygoid plate and forms the roof of the infratemporal fossa. It gives origin to the upper head of the lateral pterygoid muscle. It faces downward and is roughly pentagonal. Anteriorly, it is limited by the inferior orbital fissure. Laterally, it is demarcated from the temporal surface of the greater wing by a variably developed *infratemporal crest.* Medially, it becomes continuous with the lateral surface of the lateral pterygoid plate. Posterolaterally, it articulates with the squamous temporal near the articular tubercle. Posteromedially, it presents two openings into the middle cranial fossa. The anterior and larger is the *foramen ovale,* which transmits the mandibular nerve and a few small vessels.[20] The posterior and smaller aperture is the *foramen spinosum,* which transmits the middle meningeal vessels and the meningeal branch of the mandibular nerve. This foramen is so named because a variably developed, sharp spur of bone, the *spine of the sphenoid,* is found behind it. The spine is related to the auriculotemporal nerve laterally and to the chorda tympani medially. It gives attachment to the tensor veli palatini and the sphenomandibular and pterygospinous ligaments. Occasionally there is a small opening (*canaliculus innominatus*) behind the foramen ovale for transmission of the lesser petrosal nerve. Frequently a small aperture, the *sphenoidal emissary foramen,* is found anteromedial to the foramen ovale. The sinus that passes through it connects the cavernous sinus and the pterygoid plexus. The pterygospinous ligament may be ossified.[21]

Medial to the foramen ovale and the foramen spinosum, the sphenoid is separated from the petrous temporal by a groove for the cartilaginous part of the auditory tube. Hence, at the base of the skull, the mandibular nerve and the middle meningeal artery lie on the lateral aspect of the tube (fig. 58–3*C*, p. 663).

Pterygoid Process. The pterygoid processes extend downward on each side from the greater wings of the sphenoid. They lie behind the maxillae and they separate the infratemporal fossae from the choanae. Each consists of a lateral and a medial pterygoid plate, separated from each other by the *pterygoid fossa.* Although it is readily evident only on young skulls, the lower ends of the lateral and medial plates are separated from each other by the pyramidal process of the palatine bone. The latter articulates with the maxillary tuber, that is, the portion of the maxilla behind the last molar tooth.

The posterior border of the *medial pterygoid plate* gives attachment to the cartilaginous part of the auditory tube, the pharyngobasilar fascia, and the superior constrictor of the pharynx. Its lower end is prolonged as a slender process, the *pterygoid hamulus,* to which the pterygomandibular ligament is anchored. The upper end of the posterior border of the medial pterygoid plate splits to enclose a depression, the *scaphoid fossa,* which gives origin to a part of the tensor veli palatini. The muscle descends in the pterygoid fossa and bends around the hamulus to pass medially to the soft palate. The *lateral pterygoid plate* gives origin, by its lateral and medial surfaces, to portions of the lateral and medial pterygoid muscles, respectively.

Choanae and Bony Palate

Choanae. **The nasal cavities are continuous with the cavity of the nasopharynx through the choanae.** These are two large openings above the posterior margin of the bony palate. Medially, they are separated from each other by the *vomer,* which here forms the back of the nasal septum. Laterally, each is bounded by the medial pterygoid plate. Superiorly, they are limited by the junction of the vomer with each medial pterygoid plate, immediately below the body of the sphenoid. This portion of the vomer is termed its *ala,* and it meets the *vaginal process* of each medial pterygoid plate. (A small vascular channel, the *vomerovaginal* canal, may be present at the junction.) The vaginal process is overlapped anteriorly by a portion *(sphenoidal process)* of the palatine bone. (A small *palatovaginal canal* may be identified at the junction.)

Bony Palate. **The bony palate, or skeleton of the hard palate, lies in the roof of the mouth and in the floor of the nasal cavity.** It is formed by the *palatine processes of the maxillae* in front, and by the *horizontal plates of the palatine bones* behind. These four processes are united by a *cruciform suture.*[22] The *torus palatinus* is a median bony elevation found occasionally on the palate.[23] Sometimes the two halves of the palate fail to meet in the median plane (cleft palate). The bony palate is covered below by the mucoperiosteum of the mouth.

In front, behind the incisor teeth, there is a depression, the *incisive fossa,* through which the nasopalatine nerves pass from the nose by way of a variable number of *incisive canals* and *foramina.*[24] The posterior border of the bony palate gives attachment to the soft palate (palatine aponeurosis). In the median plane, the posterior border presents the *posterior nasal spine.* Posterolaterally, the bony palate presents an opening on each side, the *greater palatine foramen.* This is the lower end of the *greater palatine canal,* which transmits the greater palatine nerve and vessels from the pterygopalatine fossa. One or more *lesser palatine foramina* and *canals* are found behind the greater, and contain the lesser palatine nerve and vessels.

Each palatine bone is **L**-shaped and consists of (1) a *perpendicular plate,* which is applied to the back part of the medial aspect of the maxilla, and (2) a *horizontal plate,* which projects medially to meet its fellow of the opposite side and form the posterior portion of the bony palate. At the junction of the two plates, a *pyramidal process* projects backward and laterally, and separates the maxilla from the pterygoid process of the sphenoid bone (fig. 52–12). Two small projections (*orbital* and *sphenoidal processes*) jut upward from the top of the perpendicular plate and help to bound the *sphenopalatine foramen.*[25]

CRANIAL CAVITY

The cranial cavity lodges the brain and the meninges, certain portions of the cranial nerves, and blood vessels. It is roofed over by the skull-cap (fig. 52–19), and its floor (fig. 52–20) is formed by the upper surface of the base of the skull. **The floor of the cranial cavity is divisible into three "steps" by two prominent bony ledges on each side, namely, the posterior border ("sphenoidal ridge") of the lesser wing of the sphenoid in front, and the superior border ("petrous ridge") of the petrous temporal behind. The three steps are known as the anterior, middle, and posterior cranial fossae.** The anterior is at the highest level, the posterior at the lowest. The floor of the fossae is irregular and reflects certain features of the brain; impressions of the cerebral gyri are evident in the anterior and middle fossae. The endocranium is firmly adherent to the base and is continuous with the pericranium through the various foramina and fissures.

Calvaria (fig. 52–19)

The skull-cap, or vault of the skull, forms the roof of the cranial cavity. Its external surface has been described on page

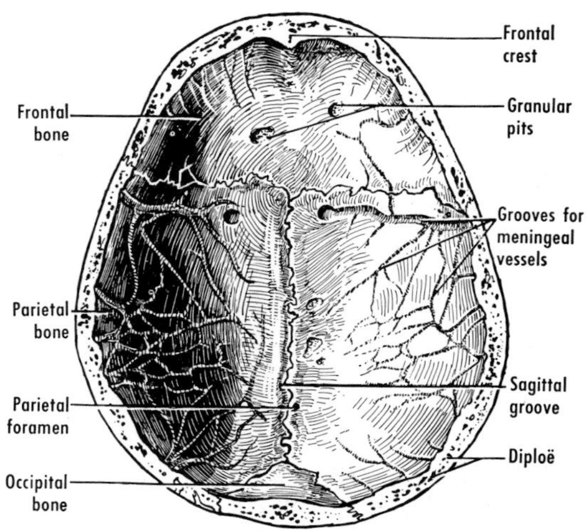

Figure 52–19 Internal aspect of the calvaria.

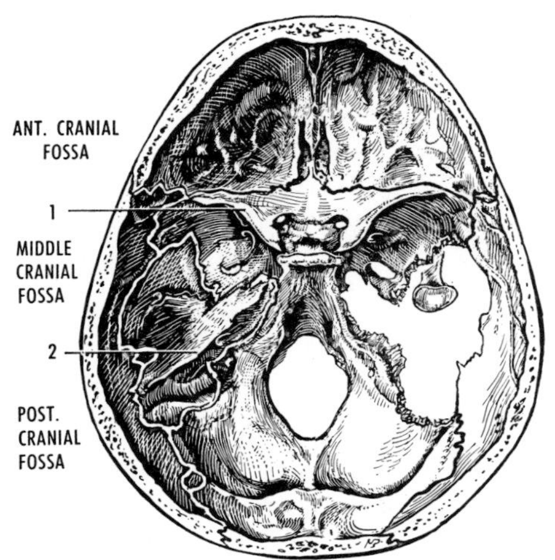

Figure 52–20 Superior aspect of the base of the skull. The right temporal bone has been removed, and is shown in figure 52–24. Notice the head of the mandible on the right-hand side. The details of the sphenoid and occipital bones are presented in figure 52–22. Note that the floor of the cranial cavity is divisible into three "steps" by two prominent bony ledges on each side. The three steps are known as the anterior, middle, and posterior cranial fossae. The ledges are formed by (1) the lesser wings of the sphenoid and (2) the upper border of the petrous part of the temporal bone on each side.

557. In young skulls its internal surface shows the coronal, sagittal, and lambdoid sutures, and a portion of the squamous suture (between the parietal and the squamous temporal) may be included. Moreover, in young skulls, markings *(digital impressions)* that correspond with some of the convolutions (gyri) of the brain may be visible. These markings may also be detected radiographically[26] (fig. 52–21).

In the median plane a shallow *sagittal groove* runs backward on the internal surface of the vault. It becomes increasingly wider as it is traced backward, and it lodges the superior sagittal sinus. A number of depressions, the *granular pits,* are found on each side of the sagittal groove.[27] They lodge lateral lacunae and arachnoid granulations (p. 603), and they are more numerous and more evident in old age.

One or more *parietal foramina* may be found, corresponding to those seen on the external aspect. They transmit emissary veins. The internal surface of the vault is characterized by numerous vascular grooves, which lodge meningeal vessels. The largest are on the parietal bones and are for the branches of the middle meningeal vessels. The terminal branches of the middle meningeal artery are separated from the bone by their accompanying veins.[28] Deeper grooves with sharper edges are for diploic veins.[29]

Anterior Cranial Fossa (fig. 52–20)

The anterior cranial fossa lodges the frontal lobes of the cerebral hemispheres. Its floor is composed of portions of three bones: ethmoid, frontal, and sphenoid.

The *crista galli* is a median process that extends upward from the ethmoid. Together with the frontal crest in front of it, the crista gives attachment to a fold of dura mater, the falx cerebri. A small pit in front of the crista, between the ethmoid and frontal bones, generally ends blindly and is known as the *foramen cecum.* Rarely, it transmits a vein from the nasal mucosa to the superior sagittal sinus. Behind and at each side of the crista galli, the *cribriform plate* of the ethmoid is characterized by a number of small apertures. They transmit the filaments of the olfactory nerves from the nasal mucosa to the olfactory bulbs. The cribriform plate supports the olfactory bulbs. The ethmoid articulates behind with the *jugum sphenoidale,* the part of the body of the sphenoid that roofs the sphenoidal air-sinus. The ethmoid is described with the nasal cavity (chapter 62).

Laterally, the greater part of the anterior cranial fossa is formed by the *orbital plates of the frontal bone.* Each plate is convex and presents impressions of cerebral sulci and gyri. It roofs the orbit and the ethmoidal (and the lower part of the frontal) air-sinuses. The orbital plate articulates behind with the *lesser wing of the sphenoid,* but the suture may be obliterated. The lesser wing presents a sharp posterior border, sometimes called the "sphenoidal ridge," which overhangs the middle cranial fossa and projects into the lateral sulcus of the cerebral hemisphere. It ends medially in the *anterior clinoid process,* which gives attachment to a fold of dura called the tentorium cerebelli.

Middle Cranial Fossa

The floor of the middle cranial fossa resembles a butterfly in that it comprises a smaller median part and an expanded lateral part on each side.

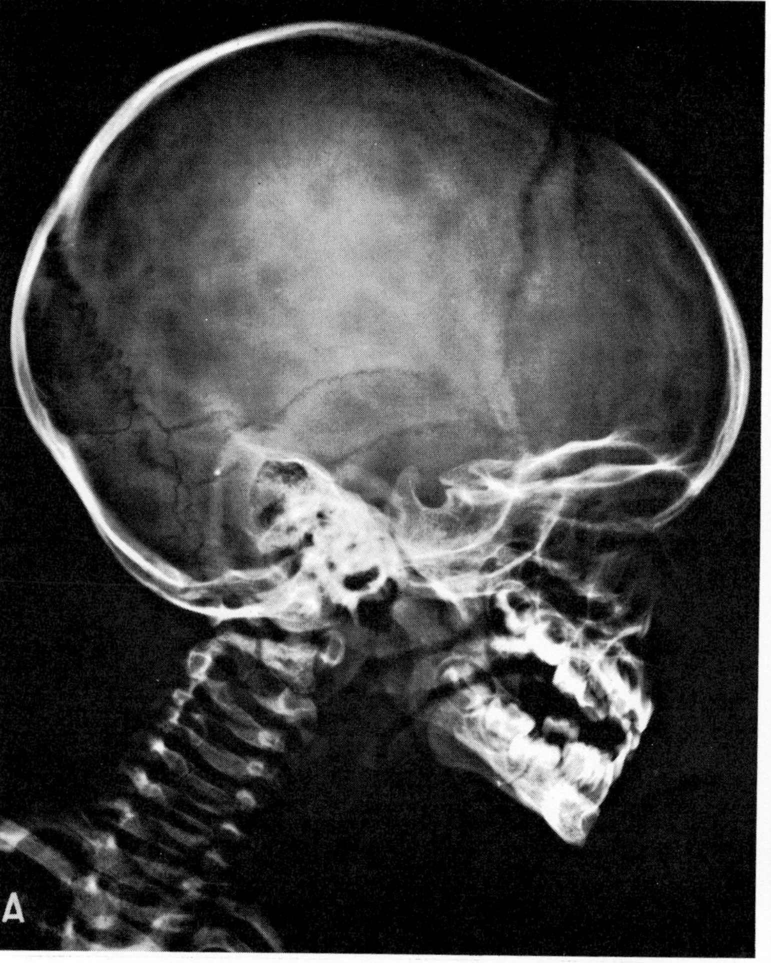

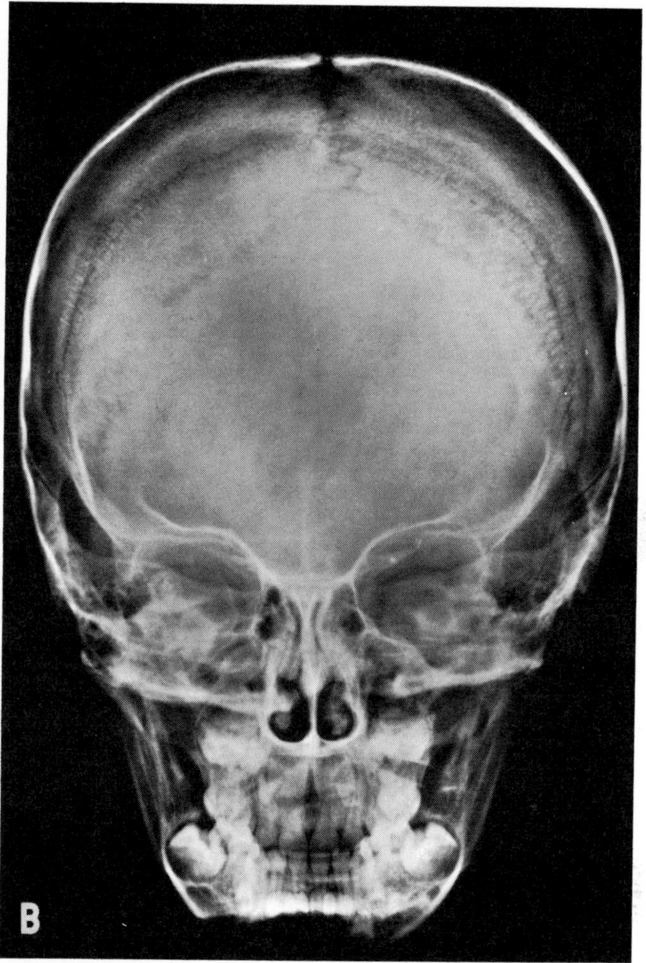

Figure 52–21 Head of child. *A*, lateral radiogram of the skull of a child between one and two years of age. Two adult phalanges overlie the body of the mandible. Note the clearly defined sutures, including both squamous. "Digital impressions" are clearly visible in the parietal region. Observe also the teeth and their large size in relation to the mandible. Neurocentral synchondroses are visible in the cervical vertebrae, which are rotated. *B*, anteroposterior view of the skull of a child aged three years. Note that the maxillary sinuses are small and the upper teeth are very close to the orbits. Observe the large, unerupted first permanent molar on each side in the mandible. Both films courtesy of E. S. Gurdjian, M.D., and J. E. Webster, M.D., Department of Neurosurgery, Wayne State University College of Medicine, Detroit, Michigan.

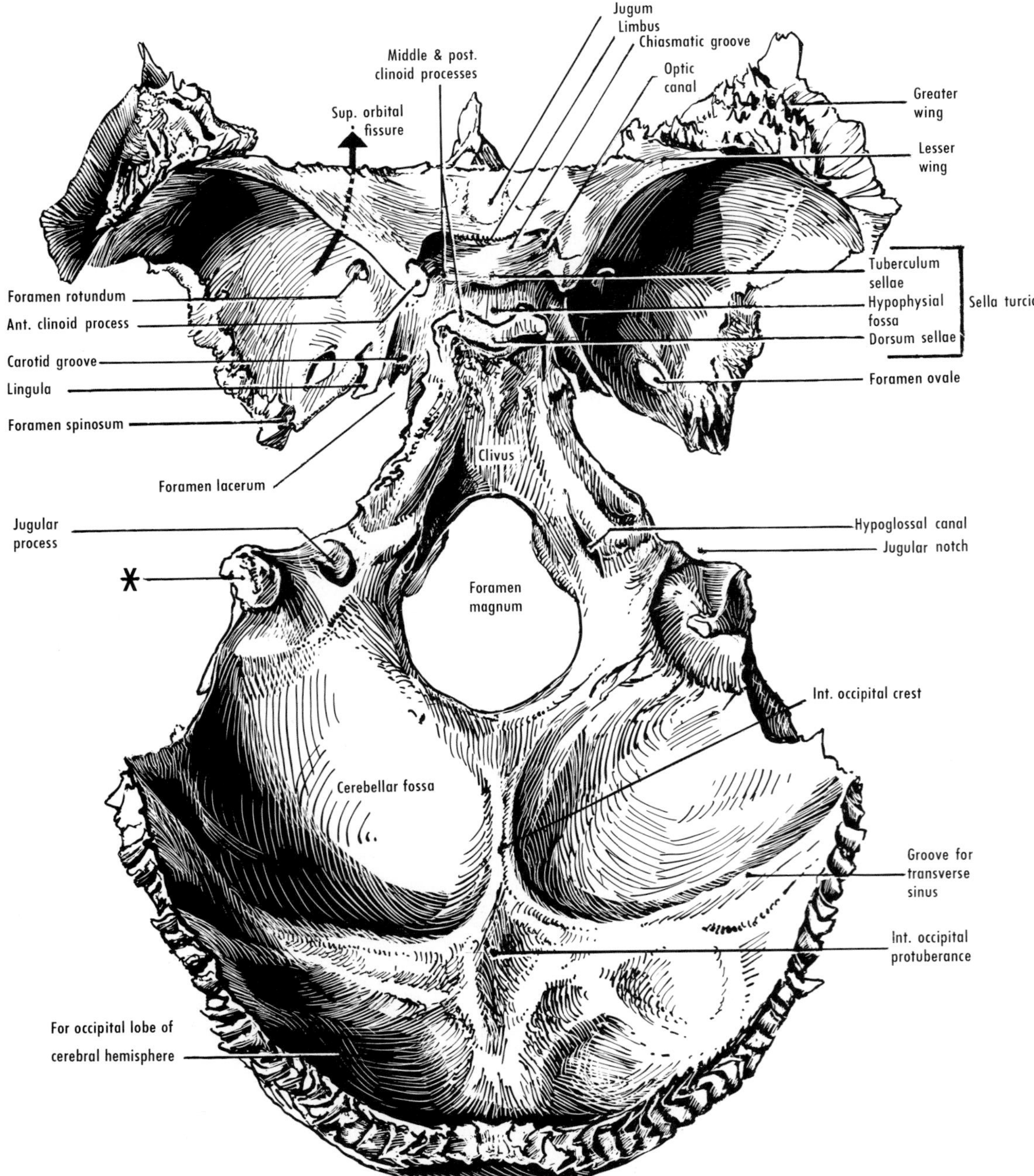

Figure 52–22 Superior aspect of the sphenoid and occipital bones. These two elements, separated by cartilage in the child, become united by bone at about puberty. The body, the paired greater wings, and the paired lesser wings of the sphenoid can be identified, but the pterygoid processes cannot be seen in this view. Note the crescentic row of openings in the greater wing: superior orbital fissure (*arrow*), foramen rotundum, foramen ovale, and foramen spinosum. Of these openings, only the last two can be seen on the inferior aspect of the base (figs. 52–14 and 52–18). The four chief parts of the occipital bone can be seen around the foramen magnum: basilar, two lateral, and squamous. The area marked with an asterisk articulates with a corresponding area (see fig. 52–17) on the temporal bone. Note the wide gap between the greater wing of the sphenoid and the basilar part of the occipital bone. In the intact condition this gap is occupied by the petrous part of the temporal bone. The apical area at the point of junction of the three bones, however, remains cartilaginous and, in the dried skull, is known as the foramen lacerum.

Median Part of Middle Cranial Fossa. The sphenoid bone (fig. 52–22) consists of a body and three pairs of processes: greater wings (alisphenoids), lesser wings (orbitosphenoids), and pterygoid processes. The pterygoid processes are found on the inferior aspect of the skull (p. 568), but the other portions of the sphenoid bone contribute to the middle cranial fossa. The median part of the fossa is formed by the body of the sphenoid.

The *body of the sphenoid* is more or less cuboidal. It presents, therefore: (1) lateral surfaces with which the greater wings and the pterygoid processes are fused; (2) an anterior surface, which contributes to the roof of the nose; (3) an inferior surface, which contributes to the roof of the pharynx; (4) a posterior surface, which is fused with the occipital in the adult; and (5) a superior surface on which the hypophysis cerebri, or pituitary gland, is lodged. Each lesser wing is attached to the upper and front part of the body of the sphenoid by two roots, between which lies the optic canal.

The median part of the middle cranial fossa (fig. 52–23) is limited in front by the *limbus sphenoidalis.* The limbus is the anterior edge of a shallow, transverse area, the *chiasmatic groove,* which leads into the optic canal on each side. The optic chiasma does not sit in the groove, however, but lies a short distance above it. The *optic canal,* directed forward and laterally, transmits the optic nerve and the ophthalmic artery to the orbit. It is bounded by the body of the sphenoid and the two roots of the lesser wing.

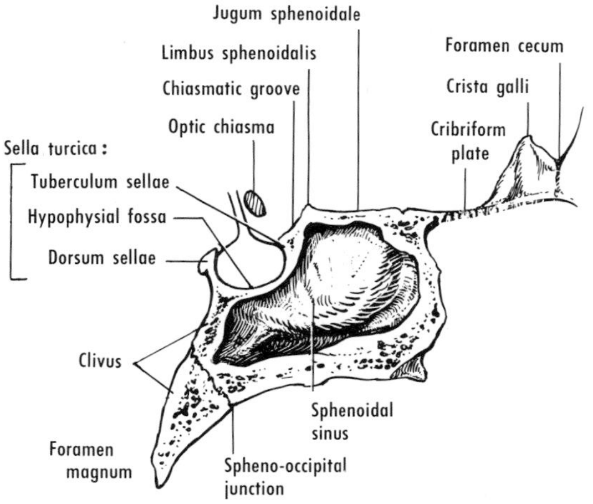

Behind the chiasmatic groove, the upper surface of the body of the sphenoid is termed the *sella turcica.* The sella is bounded in front by the *tuberculum sellae,* which is the posterior limit of the chiasmatic groove. The sella is bounded behind by the *dorsum sellae,* a square plate of bone that projects upward and presents a *posterior clinoid process* on each side. The two processes give attachment to the tentorium cerebelli. The seat of the saddle lodges the hypophysis cerebri, or pituitary gland, and is called the *hypophysial fossa.*[30] It roofs the sphenoidal air-sinuses. Occasionally a bead of notochordal tissue may be found beneath the dura over the dorsum sellae. The sella varies considerably in size, shape, and tilt.[31]

The *carotid groove* is a shallow furrow on the side of the body of the sphenoid, lateral to the hypophysial fossa. It begins in the foramen lacerum, runs upward, then forward, and finally upward again, medial to the anterior clinoid process. **The carotid groove contains the internal carotid artery, embedded in the cavernous sinus.** In many skulls a *middle clinoid process* arises near each end of the tuberculum sellae. In some instances the middle clinoid process is united by bone to the corresponding anterior clinoid process, thereby taking part in an opening *(caroticoclinoid foramen)* for the internal carotid artery.

A small median channel *(craniopharyngeal canal)* is found rarely between the center of the floor of the hypophysial fossa and the inferior aspect of the body of the sphenoid (behind the posterior border of the vomer and in front of the spheno-occipital junction). It is often described as the track of the developing adenohypophysis (craniopharyngeal pouch), but it is maintained that "no direct relationship exists between the two," and that the canal is a vascular conduit formed during osteogenesis.[32]

Figure 52–23 Median section through the base of the skull, to illustrate the terminology of some important features of the cranial fossae.

Lateral Part of Middle Cranial Fossa. The lateral part of the middle cranial fossa is formed by the greater wing of the sphenoid, together with the squamous and petrous portions of the temporal. It lodges the temporal lobe of the cerebral hemisphere. It is limited in front by the sharp posterior border of the lesser wing of the sphenoid, and behind by the prominent upper border of the petrous temporal. These two ridges are closely related to venous sinuses (the sphenoparietal and the superior petrosal, respectively).

The *superior orbital fissure* is a slit between the greater and lesser wings of the sphenoid. It transmits several important nerves (the oculomotor, trochlear, and abducent), including the branches of the ophthalmic nerve (which is a division of the trigeminal). It is described in detail with the orbit (p. 630, and fig. 55–5, p. 631).

The *foramen rotundum* is located immediately below the medial end of the superior orbital fissure. It transmits the maxillary nerve from the trigeminal ganglion to the pterygopalatine fossa. The foramen is quite commonly oval rather than round and a canal rather than a foramen.[33] The *foramen ovale* is found behind the foramen rotundum. It transmits the mandibular nerve from the trigeminal ganglion to the infratemporal fossa, where it has already been noted on the lower surface of the base. A *sphenoidal emissary foramen* may be present medial to the foramen ovale. The *foramen spinosum,* for the middle meningeal vessels, is posterior and lateral to the foramen ovale. A groove for the vessels extends laterally and forward from the foramen spinosum. After a short distance the groove divides into anterior and posterior grooves, which lodge the anterior and posterior branches of the vessels. The anterior groove continues to the pterion and then arches upward and backward across the parietal. At the pterion the groove may be converted into a tunnel, thereby increasing the possibility of vascular tearing when the skull is injured. The posterior groove passes backward across the squamous temporal and also reaches the parietal.

The superior orbital fissure, the foramen rotundum, the foramen ovale, and the foramen spinosum are arranged in a crescent on the greater wing of the sphenoid (fig. 52–22). Of these four openings, however, only the last two can be seen on the lower surface of the base.

It should be observed that the greater wing of the sphenoid presents (1) a cerebral surface in the middle cranial fossae, (2) an orbital surface, which forms a large portion of the lateral wall of the orbit, (3) a temporal surface in the temporal fossa, and (4) an infratemporal surface in the infratemporal fossa.

The boundaries of the temporal bone in the middle and posterior cranial fossae are shown in figure 52–20 and should be followed on a skull. The anterior surface of the petrous temporal (fig. 52–24) presents the *trigeminal impression* medially, near the apex of the bone. The trigeminal ganglion lies on this shallow depression. The abducent nerve bends sharply forward across the apex of the petrous temporal, medial to the trigeminal ganglion. A rounded elevation, the *arcuate eminence,* may be found on the anterior surface of the petrous temporal. It indicates the position of the underlying anterior semicircular canal. In front of the arcuate eminence a slit or *hiatus for the greater petrosal nerve* is continued as a groove to the foramen lacerum. The lesser petrosal nerve lies lateral to the greater and may occupy a small groove. The lateral part of the anterior surface of the petrous temporal roofs the tympanic cavity, the mastoid antrum, and the auditory tube. It is known as the *tegmen tympani.* Its anterior part turns downward into the squamotympanic fissure, where it can be identified on the inferior aspect of the skull.

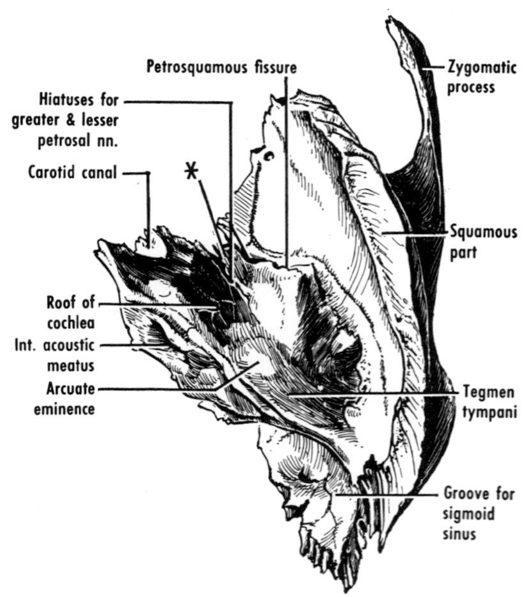

Figure 52–24 Superior aspect of the right temporal bone. For the position of this bone in the base of the skull, see figure 52–20. The asterisk indicates the portion of the petrous part (tegmen tympani) that turns downward into the squamotympanic fissure. The trigeminal impression, which lodges the trigeminal ganglion, is situated immediately behind the apex of the bone and the opening of the carotid canal.

The *foramen lacerum* has already been observed on the lower surface of the base. In the middle cranial fossa it is seen between the petrous temporal and the sphenoid. Its size depends on the forward extent of ossification in the apex of the petrous temporal. The carotid canal opens into the foramen, and the internal carotid artery crosses the foramen to reach the carotid groove on the sphenoid. The lateral wall of the foramen is formed by a variably developed spur *(lingula)* of the sphenoid bone. The lower part of the foramen lacerum is occupied by cartilage, and the only transmitted structures are minute vessels. The internal carotid artery lies above the cartilage, and the foramen contains several small nerves (deep petrosal and greater petrosal, which unite to form the nerve of the pterygoid canal).

The temporal bone consists of four parts at birth: squamous, tympanic, petromastoid, and styloid.

Posterior Cranial Fossa

The posterior cranial fossa lodges the hindbrain: cerebellum, pons, and medulla oblongata. It is formed by portions of the sphenoid, temporal (fig. 52–24), parietal, and occipital (fig. 52–22) bones. The posterior fossa is limited above by a wide fold of dura, the tentorium cerebelli, which intervenes between the occipital lobes of the cerebral hemispheres above and the cerebellum below. It is attached to the upper border of the petrous temporal and to the lips of a transverse groove on the internal surface of the occipital.

The lowest part of the posterior cranial fossa presents the *foramen magnum,* which has been considered already (p. 562). A little above the margin of the front part of the foramen magnum, the *hypoglossal canal* on each side transmits the hypoglossal nerve. The canal is sometimes divided. The *jugular tubercle* is an elevation above the hypoglossal canal, and between the jugular foramen and the foramen magnum. It is frequently grooved by cranial nerves 9, 10, and 11.

In front of the foramen magnum, the basilar part of the occipital bone ascends to meet the body of the sphenoid, with which it fuses at about puberty (p. 563 and fig. 52–23). This sloping bony surface, termed the *clivus* (fig. 52–23), is related to the pons and the medulla. It is continuous with the dorsum sellae above. The inferior petrosal sinus lies between the basilar part of the occipital and the petrous temporal.

Behind the foramen magnum a median ridge, the *internal occipital crest,* leads upward to the internal occipital protuberance. A median fold of dura, the falx cerebelli, is attached to it and intervenes between the two cerebellar hemispheres. The *internal occipital protuberance,* usually slightly higher than the external, gives attachment to the falx cerebri, the tentorium cerebelli, and the falx cerebelli. In this region the superior sagittal and straight sinuses end, and the right and left transverse sinuses begin. The arrangement at the confluence of the sinuses is variable (fig. 53–22, p. 608).

Each transverse sinus lies in the *groove for the transverse sinus,* which runs laterally from the internal occipital protuberance. Each transverse sinus then turns downward and becomes the sigmoid sinus. The *groove for the sigmoid sinus* can be traced medially and forward into the jugular foramen, which has been considered already (p. 567). The foramen transmits the sigmoid and inferior petrosal sinuses, and the glossopharyngeal, vagus, and accessory nerves. The sigmoid sinus becomes continuous with the internal jugular vein outside the skull. The upper part of the sigmoid sinus is closely related to the mastoid antrum. Vascular openings (*mastoid foramen* and *condylar canal*) may be present in or near the sigmoid groove. The *cerebellar fossa* on each side lies between the transverse and sigmoid grooves and the foramen magnum. The occipital lobes of the cerebral hemispheres are lodged above the transverse sinuses in two fossae on the occipital bone.

The posterior surface of the petrous temporal presents a conspicuous opening, the *internal acoustic (auditory) meatus.*[34] It lies almost directly medial to the external acoustic meatus. The internal meatus, about 1 cm in length, transmits the facial and vestibulocochlear nerves from the internal ear, and also the labyrinthine vessels.

An indistinct depression *(subarcuate fossa),* containing a dural fold and some vessels,[35] may be found lateral and superior to the internal acoustic meatus.

Behind the internal acoustic meatus, a

depression or slit, termed the *aqueduct of the vestibule,* may be seen. It transmits the endolymphatic duct of the internal ear (fig. 54–8, p. 622).

At the junction of the posterior and inferior surfaces of the petrous temporal, directly below the internal acoustic meatus, a notch termed the *cochlear canaliculus* may be identified. It lodges the aqueduct of the cochlea or perilymphatic duct.

MANDIBLE

The mandible, or lower jaw, is the largest and strongest bone of the face. It presents a body and a pair of rami (fig. 52–25). The junctional region, behind and below the lower third molar tooth, is described by some as a part of the ramus and by others as a part of the body. This region is marked by the *angle of the mandible,* which can readily be felt *in vivo.* Its most prominent, laterally directed point is called the *gonion.*

The angle of the mandible, which has a mean value of 125, varies from about 110 to 140 degrees.

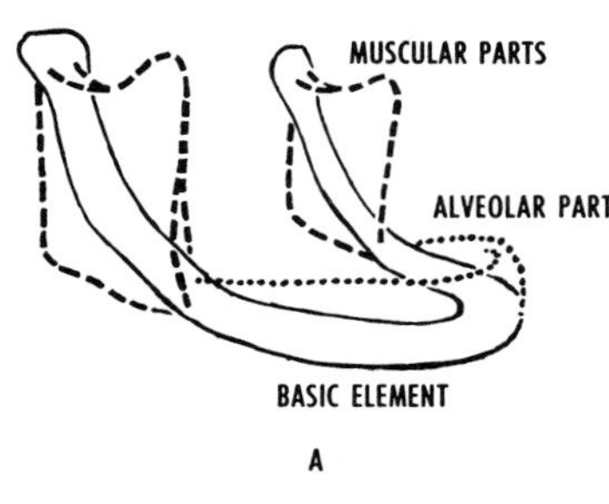

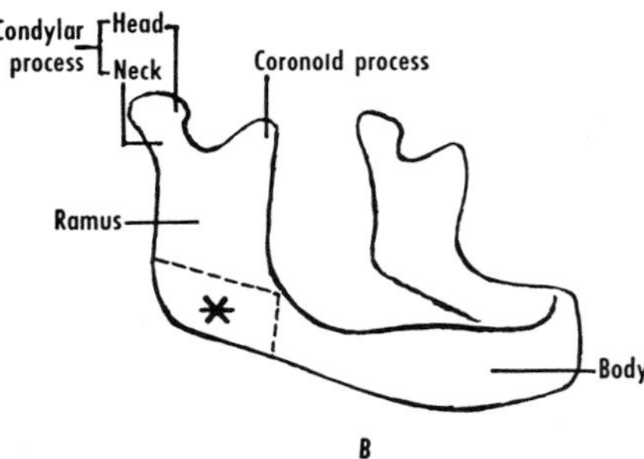

Figure 52–25 Schemes of the mandible. *A* shows the muscular and alveolar parts added to a basic element. *B* shows the main structural portions. The area marked by an asterisk may be classified as a part of either the ramus or the body of the bone. *A* is based on Symons.

The maxilla and the mandible[36] each include an alveolar part added to a basic element (fig. 52–25). These parts are at least partially separated by a groove.[37] In the maxilla, the alveolar part is dependent on the presence of the teeth, whereas, in the mandible, the inferior portion of the alveolar part is independent of the teeth, and persists in an edentulous jaw.[37]

Body of Mandible

The body of the mandible (fig. 52–26) is **U**-shaped, and it presents an external and an internal surface, a superior border or alveolar part, and an inferior border or base.

The *external surface* is generally characterized by a faint, median ridge, which marks the line of fusion of the two halves of the mandible at the *symphysis menti.* It expands below into a triangular elevation termed the *mental protuberance,* the base of which is limited on each side by the *mental tubercle.* Further laterally, frequently below the second premolar tooth,[38] the *mental foramen* can readily be seen. The mental nerve and vessels emerge from the foramen usually upward, backward, and laterally.[39] The *oblique line* is a faint ridge that runs backward and upward from the mental tubercle to the anterior border of the ramus.

The upper border of the body of the mandible is termed the *alveolar part* and contains the lower teeth, in sockets or *alveoli.* The edge of the alveolar part is the *alveolar arch.* The alveolar part is largely covered with the mucous membrane of the mouth.

The lower border of the mandible is termed its *base.* The *digastric fossa* is a rough depression on or behind the base, near the symphysis. Toward the back, about 4 cm anterior to the angle of the mandible, the base may show a faint groove for the facial artery. The pulsation of the artery can be felt against the base of the mandible.

The *internal surface* is characterized by an irregular elevation, the *mental spine,* at the back of the symphysis. It may consist of one to four parts, termed *genial tubercles,* which give origin to the geniohyoid and genioglossus muscles. Further back, the *mylohyoid line* can be distinguished as an oblique ridge that runs backward and upward from above the digastric fossa to a point behind the third molar tooth. It gives origin to the mylohyoid muscle. The *sub-*

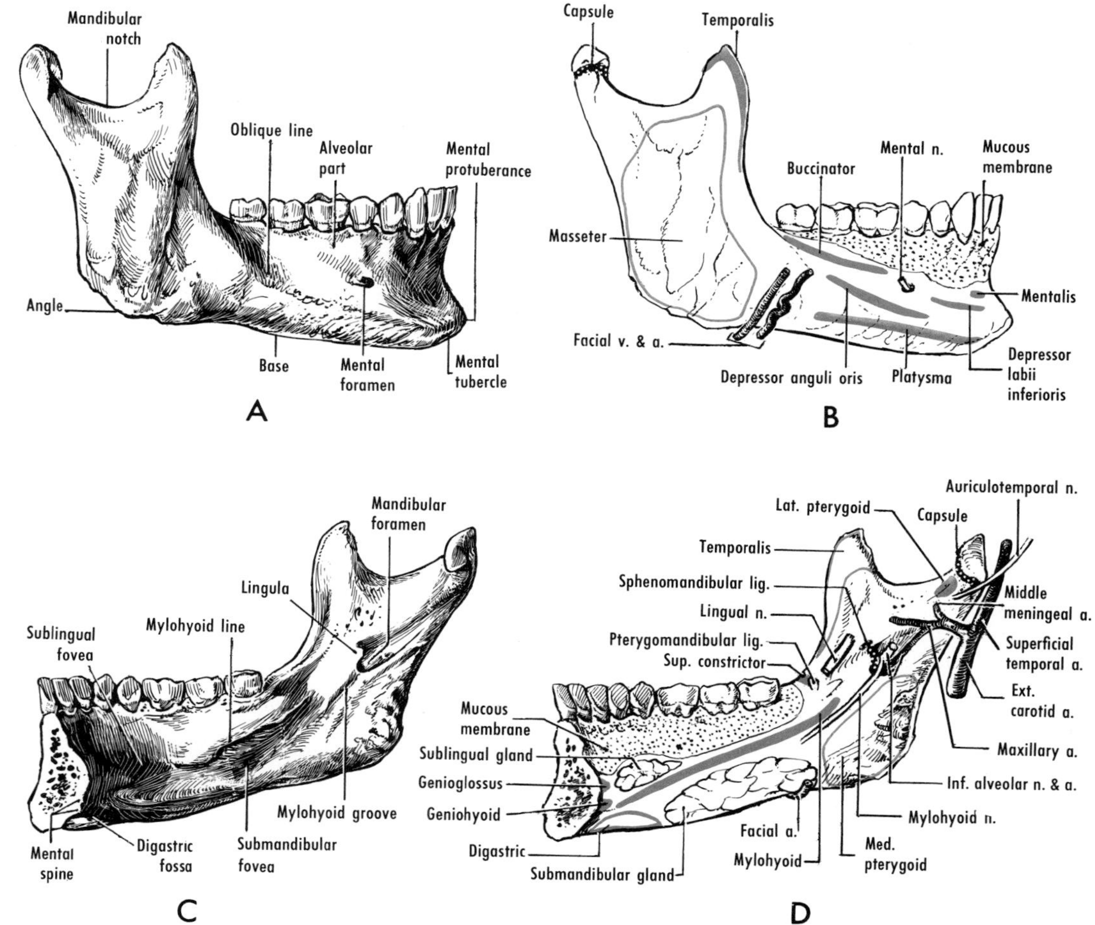

Figure 52–26 A and B, right lateral aspect of the mandible. C and D, medial aspect of the right half of the mandible. A and C show the main structural features. B and D show the attachments and relations. B and D are based on *Frazer's Anatomy of the Human Skeleton.*

mandibular fovea lies below the mylohyoid line and lodges a part of the submandibular gland. The *sublingual fovea* lies further forward, above the mylohyoid line, and lodges the sublingual gland. The anterior end of the mylohyoid groove reaches the body of the mandible below the posterior end of the mylohyoid line.

Ramus of Mandible

The ramus of the mandible (fig. 52–26) is a more or less quadrilateral plate of bone that presents lateral and medial surfaces, and anterior, superior, and posterior borders. The ramus and its attached muscles are in contact with the side of the pharynx.

The *lateral surface* is flat and gives insertion to the masseter. The *medial surface* is characterized by the *mandibular foramen,* which leads downward and forward into the mandibular canal, and transmits the inferior alveolar nerve and vessels. The foramen is limited medially by a projection[40] termed the *lingula,* to which the sphenomandibular ligament is attached. The *mandibular canal* runs as far as the median plane and, on its way, gives off a side canal that opens at the mental foramen. The *mylohyoid groove* begins behind the lingula and runs downward and forward to the submandibular fovea. It contains the mylohyoid nerve and vessels. Below and behind the mylohyoid groove, the medial surface is roughened and gives insertion to the medial pterygoid muscle.

The concave upper border of the ramus is the *mandibular notch*.[41] It is bounded in front by the *coronoid process*, into which the temporalis is inserted. The *condylar process* bounds the mandibular notch behind, and comprises the head and neck of the mandible. The *head (condyle)*, covered with fibrocartilage, articulates indirectly with the temporal bone to form the temporomandibular joint. The long axis of the head is directed medially and slightly backward. The lateral end of the head of the mandible can be felt *in vivo*. The *neck* gives attachment to the lateral ligament laterally and insertion to the lateral pterygoid muscle anteriorly.

The sharp anterior border of the ramus can be felt inside the mouth. It is continuous with the oblique line. The rounded posterior border is closely related to the parotid gland.

The mandible appears bilaterally in the embryo immediately external to the cartilage of the first pharyngeal arch. Bony union between the halves of the mandible takes place during the first postnatal year. A secondary cartilage in the condylar process is responsible for most of the growth in length of the mandible.

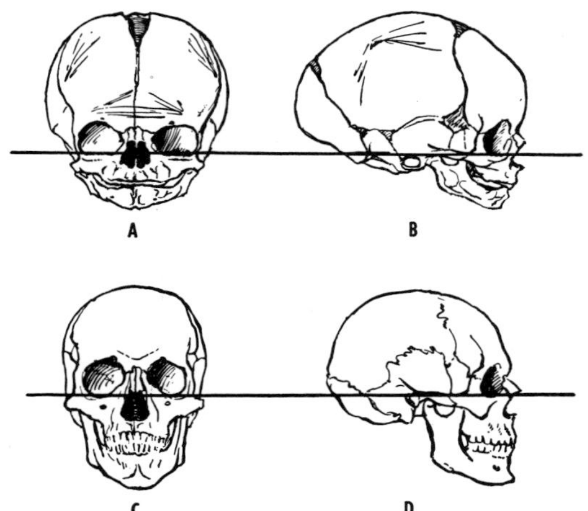

Figure 52–27 The growth of the skull. Skulls of (*A* and *B*) newborn infant and (*C* and *D*) adult, anterior and lateral aspects. The scale used here for the adult skull is half of the neonatal. The horizontal lines indicate the orbitomeatal plane. Note that, in the infant's skull, although the vault appears large, the facial region (chiefly below the horizontal line) is relatively small. The jaws, the nasal cavities, and the paranasal sinuses are all small; observe the proximity of the orbits to the teeth. In the adult, however, the horizontal line approximately bisects the vertical height of the skull. Note the fontanelles in *A* and *B*. Based chiefly on the work of J. C. Brash.

DEVELOPMENT OF SKULL

The bones of the skull are formed in the mesenchyme around the developing brain. In brief, the vault and portions of the base ossify intramembranously, whereas the greater part of the base undergoes chondrification (to form the *chondrocranium*) early in fetal life and, subsequently, undergoes endochondral ossification. Portions of the chondrocranium, however, retain their cartilaginous structure, namely, the front portion of the nasal septum and the foramina lacera.

The following bones of the skull ossify intramembranously: frontal, parietal, squamous part of the temporal, upper portion of the squamous part of the occipital, vomer, lacrimal, nasal, palatine, most of the pterygoid processes and greater wings of the sphenoid, zygomatic, maxilla, and mandible. The following ossify endochondrally: most of the occipital, petrous and mastoid parts of the temporal, body and lesser wings of the sphenoid, ethmoid, and inferior nasal concha.

Neonatal Skull (fig. 52–27, *A* and *B*)

At birth the portion of the skull above the orbitomeatal plane is very much larger than the portion below. The former is related to the growth of the brain, the eyes, and the organs of hearing and equilibration. The latter is concerned with the teeth, the tongue, the respiratory region of the nasal cavity, and the maxillary sinuses. In the adult, the upper portion of the skull, although still somewhat larger than the lower, has not increased nearly as much in size. The growth of the skull is intimately associated with that of the brain and eyes (up to 2 years), the teeth (up to 2 years, and again at 6 to 12 years), and the muscles of mastication (12 to 18 years). Radiograms of children's skulls are reproduced in figure 52–21.

Fontanelles (fig. 52–27, *A* and *B*)

The *fonticuli*, or *fontanelles*, are temporary membranous areas that bridge the gaps between the angles or margins of some of the ossifying bones of the skull. Usually six fontanelles are normally present at birth, and these are situated at the angles of the parietal bones. The anterior and posterior fontanelles are unpaired, and are located at the bregma and lambda, respectively. The sphenoidal and mastoid fontanelles are paired and are located at the pterion and the asterion, respectively. Accessory fontanelles may occasionally be found at various sites, particularly along the sagittal suture.[42]

The anterior fontanelle is the largest. It is commonly seen to pulsate (owing to the cerebral arteries), and it is readily palpable in an infant. It decreases in size postnatally and is usually obliterated by two years of age. The anterior fontanelle may be used (1) to determine during parturition, by palpation *per vaginam*, the position of the fetal head in a vertex presentation; (2) to estimate abnormal intracranial pressure in an infant; (3) to assess the degree of development of the skull; and (4) to obtain a sample of blood from the underlying superior sagittal sinus.

GROWTH OF SKULL

The growth of the skull[43] takes place by three methods. (1) Cartilage is replaced by bone. This occurs, in fetal life, in the base and, after birth, at the spheno-occipital junction, the condylar processes of the mandible, and the nasal septum. (2) Growth takes place at sutures. This occurs in the vault and the upper part of the face during fetal life and for several years (about seven) after birth. It accounts largely for the increase in the width of the head. (3) Surface deposition, associated with surface resorption internally, takes place in the face during late childhood and adolescence (about 7 to 21 years).

Analysis of the growth of the skull in any given individual indicates that growth proceeds in a discontinuous manner, that is, in spurts. Moreover, the pattern of growth varies from one individual to another.

HYOID BONE

The hyoid bone lies in the front portion of the neck between the mandible and the larynx, at the level of the third cervical vertebra. It does not articulate with any other bone, but is suspended from the styloid processes of the skull by the stylohyoid ligaments. The hyoid bone presents a body, and a pair of greater and a pair of lesser horns (fig. 52–28).

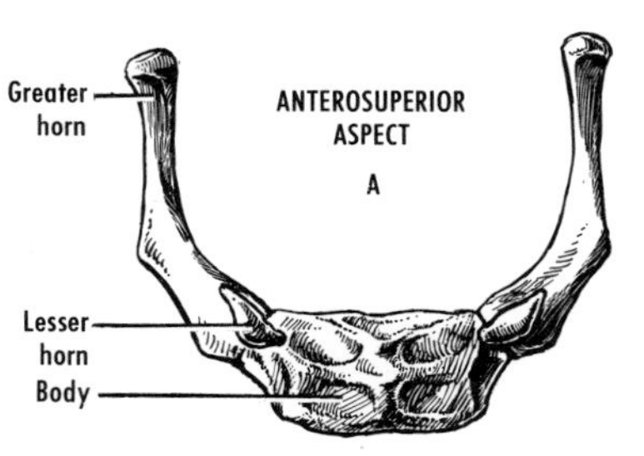

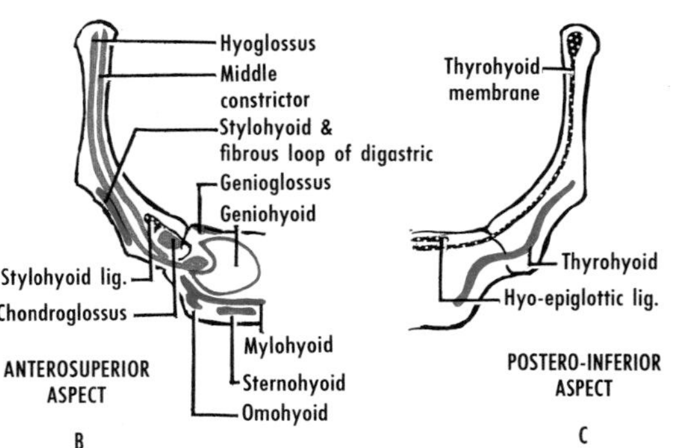

Figure 52–28 The hyoid bone. *A*, anterosuperior aspect, showing the main parts of the bone. *B*, anterosuperior aspect of the right half of the bone, showing attachments. *C*, postero-inferior aspect of the right half of the bone, showing attachments. *B* and *C* are based on Frazer.

The anterior surface of the *body* looks forward and upward. The posterior surface is separated from the thyrohyoid membrane by a bursa.

Each *greater horn* or *cornu* projects backward and upward from the side of the body of the bone, to which it is united by a cartilaginous joint or, later in life, by bony fusion. **When the neck is relaxed, the two greater horns can be gripped *in vivo* between the index finger and the thumb, and the hyoid bone can then be moved from side to side.** The tips of the greater cornua are overlapped by the sternomastoid muscles.

Each *lesser horn* or *cornu* is a small, upward projection that is attached at the junction of the body and the corresponding greater horn. The stylohyoid ligament is attached to the tip of the lesser horn and is sometimes ossified. The lesser cornua are anchored to the body of the bone by fibrous tissue and to the greater horns sometimes by synovial joints.

The hyoid bone develops from the cartilages of the second and third pharyngeal arches. It is ossified from three pairs of centers, some of which appear shortly before, and others after, birth. The hyoid bone is of little functional significance.[44]

REFERENCES

1. J. J. Pritchard, J. H. Scott, and F. G. Girgis, J. Anat., Lond., *90*:73, 1956.
2. R. Singer, J. forensic Med., *1*:52, 1953. See also S. T.

Brooks, Amer. J. phys. Anthrop., *13*:567, 1955, and L. Dérobert and G. Fully, Ann. Méd. lég., *40*:154, 1960.
3. L. Hess, Hum. Biol., *18*:61, 1946.
4. F. Girgis and J. J. Pritchard, Anat. Soc. G. B. and Ireland, April, 1955. See also M. L. Moss, Acta anat., *44*:263, 1961.
5. A. Crocellà, Arch. ital. Anat. Embriol., *60*:201, 1955. J. B. Christensen, E. Lachman, and A. M. Brues, Amer. J. Roentgenol., *83*:615, 1960.
6. E. Lachman, Amer. J. Roentgenol., *79*:721, 1958.
7. R. Bjerin, Acta odont. scand., *15*:1, 1957.
8. M. Augier, *Squelette céphalique*, in P. Poirier and A. Charpy, *Traité d'anatomie humaine*, Masson, Paris, vol. 1, fascicle 1, division 1, 1931. The standard works on variations are A.-F. LeDouble, *Traité des variations des os du crâne de l'homme*, Vigot, Paris, 1903, and *Traité des variations des os de la face de l'homme*, Vigot, Paris, 1906.
9. W. Bergerhoff, *Atlas of Normal Radiographs of the Skull*, Springer, Berlin, 1961. L. E. Etter, *Atlas of Roentgen Anatomy of the Skull*, Thomas, Springfield, Illinois, 1955. G. S. Schwarz and C. R. Golthamer, *Radiographic Atlas of the Human Skull*, Hafner, New York, 1965. R. Shapiro and A. H. Janzen, *The Normal Skull. A Roentgen Study*, Hoeber, New York, 1960.
10. R. O'Rahilly and M. J. Twohig, Amer. J. Roentgenol., *67*:551, 1952.
11. L. Hess, Hum. Biol., *17*:107, 1945. J. Torgersen, Acta radiol., Stockh., *33*:1, 1950.
12. A. Riesenfeld, Amer. J. phys. Anthrop., *14*:85, 1956.
13. T. Dwight, Ann. Surg., *46*:721, 1907. W. Arendt, Beitr. ges. Arb. Orthopäd., *6*:1, 1959. S. M. Kaufman, R. P. Elzay, and E. F. Irish, Arch. Otolaryng., Chicago, *91*:460, 1970.
14. J. Frommer, C. W. Monroe, and B. Spector, Anat. Rec., *142*:305, 1962, abstract.
15. A. Riesenfeld, Amer. J. phys. Anthrop., *13*:599, 1955.
16. B. E. Ingelmark, Acata anat., suppl. 6 = 1 ad vol. *4*:1, 1947.
17. M. A. Kirdani, Amer. J. Roentgenol., *99*:700, 1967.
18. B. Ingervall and B. Thilander, Acta odont. scand., *30*:349, 1972. B. Melsen, Acta anat., *83*:112, 1972.
19. D. K. von Brzezinski, Anat. Anz., *113*:164, 1963.
20. S. Sunderland, Aust. N.Z. J. Surg., *8*:170, 1938.
21. K. S. Chouke, Amer. J. phys. Anthrop., *4*:203, 1946, and *5*:79, 1947. J. Priman and L. E. Etter, Med. Radiogr. Photogr., *35*:2, 1959.
22. J.-K. Woo, Amer. J. phys. Anthrop., *7*:385, 1949.
23. J.-K. Woo, Amer. J. phys. Anthrop., *8*:81, 1950. B. Vidić, J. dent. Res., *45*:1511, 1966.
24. A. W. Meyer, Anat. Rec., *49*:19, 1931.
25. V. Nikolić, Acta anat., *68*:189, 1967.
26. D. Macaulay, Brit. J. Radiol., *24*:647, 1951.
27. A. Mayet and S. Heil, Anat. Anz., *128*:454, 1971.
28. F. W. Jones, J. Anat., Lond., *46*:228, 1912.
29. I. M. Thompson, Canad. med. Ass. J., *16*:1194, 1926.
30. For the radiographic mensuration and the growth of the hypophysial fossa, see R. M. Acheson and M. Archer, J. Anat., Lond., *93*:52, 1959.
31. J. Bull, Acta radiol., Stockh., *46*:72, 1956.
32. L. B. Arey, Anat. Rec., *106*:1, 1950.
33. S. Jovanović and V. Radojević, Bull. Acad. serbe Sci. méd., *23*:19, 1958.
34. A. H. Amjad, A. A. Scheer, and J. Rosenthal, Arch. Otolaryng., Chicago, *89*:709, 1969.
35. M. Bošković, Acta med. iugoslavica, 12, suppl. 1, 1958 (translated into English, 1962).
36. N. B. B. Symons, Brit. dent. J., *94*:231, 1953.
37. G. Inke in G.-H. Schumacher (ed.), *Morphology of the Maxillo-Mandibular Apparatus*, Thieme, Leipzig, 1972.
38. H. G. Tebo and I. R. Telford, Anat. Rec., *107*:61, 1950. A. P. S. Sweet, Dent. Radiogr., *32*:28, 1959.
39. R. Warwick, J. Anat., Lond., *84*:116, 1950. M. F. A. Montagu, Amer. J. phys. Anthrop., *12*:503, 1954.
40. R. Depreux, Arch. Anat. path., *6*:92, 1958, and *8*:42, 1960.
41. M. R. Simon and M. L. Moss, Acta anat., *85*:133, 1973.
42. F. L. Adair and R. E. Scammon, Amer. J. Obstet. Gynec., *14*:149, 1927.
43. J. H. Scott, *Dento-facial Development and Growth*, Pergamon, Oxford, 1967. D. H. Enlow, *The Human Face*, Hoeber, New York, 1968.
44. W. Lesoine, Z. Laryng. Rhinol., *49*:461, 1970.

BRAIN, CRANIAL NERVES, AND MENINGES 53

BRAIN

The nervous system in general has been described in chapter 5. The divisions of the brain are summarized here in table 53–1. The present chapter is limited to a brief description of the gross structure of the brain, an account of the ventricles, some general remarks on the cranial nerves, the meninges, and the blood supply. All these topics may be considered during a course in gross anatomy, whereas further details, including most of the internal structure of the brain, are generally

TABLE 53–1 Divisions of the Brain (Encephalon)

Divisions	*Cavities*
Prosencephalon (forebrain)	
Telencephalon (endbrain)	two lateral ventricles
Diencephalon (interbrain)*	third ventricle
Mesencephalon (midbrain)*	aqueduct
Rhombencephalon (hindbrain)	
Metencephalon (afterbrain)	fourth ventricle and part of central canal
Cerebellum	
Pons*	
Myelencephalon (marrow brain, or medulla oblongata)*	

*These divisions comprise the brain stem.

reserved for a special course in neuroanatomy, for which separate textbooks are employed.

GROSS STRUCTURE OF BRAIN

The brain, which is about 2 per cent of the body weight, receives about one-sixth of the cardiac output and consumes one-fifth of the oxygen utilized by the body at rest.

The brain comprises, from below upward, the hindbrain, the midbrain, and the forebrain.

In their attachment to the brain, the first two cranial nerves are associated with the forebrain, nerves 3 and 4 with the midbrain, and nerves 5 to 12 with the hindbrain (nerve 5 with the pons, nerves 6 to 8 with the junction between the pons and the medulla, and nerves 9 to 12 with the medulla).

The hindbrain, or rhombencephalon, consists of the medulla oblongata, the pons, and the cerebellum.

Medulla Oblongata (fig. 53–1)

The uppermost part of the spinal cord expands, passes through the foramen magnum, and becomes the medulla oblongata. The medulla rests anteriorly on the basilar part of the occipital bone, separated partially from the bone by the right and left vertebral arteries, which ascend and unite to form the basilar. Posteriorly, the medulla is largely covered by the cerebellum.

The lower half of the medulla contains the continuation of the central canal of the spinal cord, which, in the upper half of the medulla, widens to become the fourth ventricle.

The medulla presents an *anterior median fissure,* the lower part of which is interrupted by the *decussation of the pyramids,* where about two-thirds of the descending pyramidal fibers cross the median plane. The portion of the medulla adjacent to the upper part of the anterior median fissure on each side is termed the *pyramid.* It contains the fibers of the *pyramidal (corticospinal) tract.* Lateral to each pyramid, an elevation termed the *olive* is found. It is composed largely of gray matter. The olive is bounded medially by an *anterolateral sulcus* and laterally by a *posterolateral sulcus.*

The *hypoglossal (12th cranial) nerve* emerges from the medulla between the pyramid and the olive, whereas the *accessory (11th cranial), vagus (10th cranial),* and *glossopharyngeal (9th cranial) nerves* emerge posterolateral to the olive.

The dorsal aspect of the medulla presents a *posterior median sulcus.* On each side, two tracts from the spinal cord (the *fasciculus gracilis* medially, and the *fasciculus cuneatus* laterally) terminate in eminences known as the *gracile* and *cuneate tubercles* respectively. Higher up, the lower part of the fourth ventricle is bounded laterally by the *inferior cerebellar*

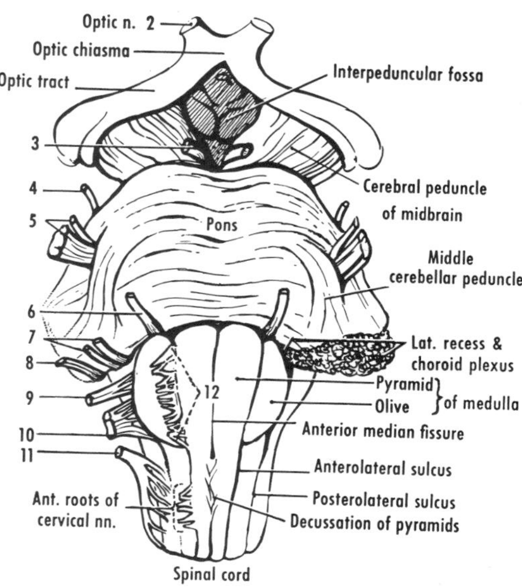

Figure 53–1 The brain stem, anterior aspect, showing the cranial nerves (indicated by numerals).

peduncle, which comprises fibers connecting the medulla and spinal cord with the cerebellum.

The medulla contains very important nerve centers associated with functions such as respiration and circulation.

Pons (fig. 53–1)

The pons lies between, and is quite sharply demarcated from, the medulla and the midbrain. It is situated in front of the cerebellum and appears superficially to bridge (hence its name) the two cerebellar hemispheres. As seen from the front, the transverse fibers of the pons form the *middle cerebellar peduncle* on each side and enter the cerebellum. The middle peduncle actually comprises fibers that connect one cerebellar hemisphere with the contralateral cerebellar hemisphere.

The front of the pons rests on the basilar part of the occipital bone and on the dorsum sellae. The pons is grooved longitudinally in front, and this groove is frequently occupied by the basilar artery.

The *vestibulocochlear (8th cranial), facial (7th cranial),* and *abducent (6th cranial) nerves* emerge in the groove between the pons and the medulla. Higher up, the *trigeminal (5th cranial) nerve* emerges from the side of the pons by a large sensory and a smaller motor root.

The back of the pons forms the floor of the upper part of the fourth ventricle (fig. 53–2), which is bounded laterally by the *superior cerebellar peduncles.* Each superior peduncle comprises fibers that connect the cerebellum with the midbrain. The floor of the fourth ventricle is discussed later (p. 592).

The *trigeminal (5th cranial) nerve* is large and complicated. It is sensory from the face, teeth, mouth, and nasal cavity, and motor to the muscles of mastication. It emerges from the side of the pons as a *sensory* and a *motor root,* generally with some accessory or intermediate fibers.[1] The two portions proceed from the posterior to the middle cranial fossa by passing beneath the attachment of the tentorium cerebelli to the petrous part of the temporal bone, and also by passing usually beneath the superior petrosal sinus.[2] The sensory root expands into a large, flat *trigeminal (semilunar) ganglion,* which contains the cells of origin of most of the sensory fibers. The ganglion

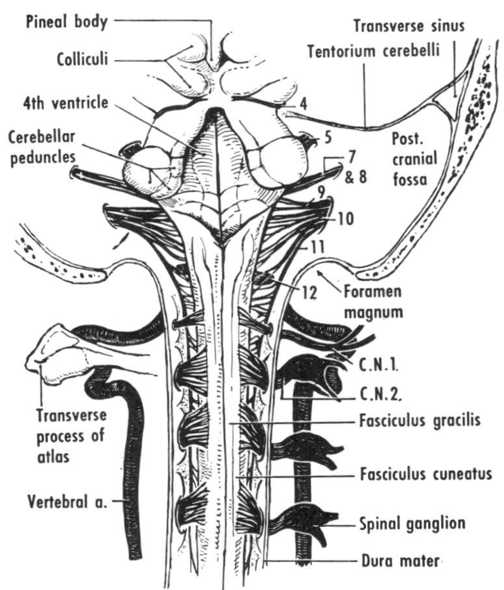

Figure 53–2 Posterior aspect of the brain stem and the upper part of the spinal cord, after removal of the cerebellum. The vertebral artery on each side is indicated, together with certain of the cranial and spinal nerves.

overlies the foramen lacerum, and the roots of the nerve occupy an impression on the anterior surface of the petrous part of the temporal bone near its apex. Most of the ganglion is enclosed in the cavity of the dura known as the cavum trigeminale (fig. 53–17). The ganglion gives rise to three large divisions: the ophthalmic, maxillary, and mandibular nerves. The motor root, which contains proprioceptive as well as motor fibers, continues beneath the ganglion and joins the mandibular nerve.

The ganglion can be "blocked" by passing a needle through the mandibular notch and the foramen ovale,[3] and injecting an anesthetic. The sensory root may be sectioned in the middle cranial fossa for the relief of trigeminal neuralgia (*tic douloureux*).

Cerebellum (figs. 53–4 and 53–7)

The cerebellum is situated on the back of the brain stem, to which it is attached by the three cerebellar peduncles on each side. The inferior peduncles connect the cerebellum with the medulla; the middle connect it with the pons; and the superior connect it with the midbrain. The cerebellum is located in the posterior cranial fossa. It comprises a median portion, termed the *vermis*, and two lateral parts known as *cerebellar hemispheres*. Like the cerebral hemispheres, the cerebellum has a cortex of gray matter. The *cerebellar cortex* is folded to form *folia*, which are separated from one another by *fissures*. The cerebellum is connected by tracts with the cerebral cortex and with the spinal cord. It is important in the coordination of muscular activities.

Midbrain (fig. 53–1)

The midbrain, or mesencephalon, connects the hindbrain with the forebrain. It is located in the tentorial notch of the dura mater (fig. 53–16). It consists of a ventral part, the cerebral peduncles, and a dorsal part, the tectum.

The *cerebral peduncles* are two large bundles that converge as they descend from the cerebral hemispheres, in which each is continuous with a band of white matter termed the internal capsule. The front portion of each peduncle is termed the *crus cerebri (basis pedunculi)*, whereas the back portion is the *tegmentum*. The upper part of each peduncle is crossed by the corresponding optic tract. The right and left optic tracts emerge from the optic chiasma, which is formed by the junction of the two optic nerves. The depression behind the chiasma, and bounded by the optic tracts and the cerebral peduncles, is termed the interpeduncular fossa (fig. 53–5).

The *interpeduncular fossa* contains, from before backward: (1) the tuber cinereum and the infundibular stem of the hypophysis, (2) the mamillary bodies, and (3) the posterior perforated substance.

The *oculomotor (3rd cranial) nerve* emerges at the upper border of the pons and at the medial border of the corresponding cerebral peduncle.

The *tectum*, or posterior part of the midbrain, consists of four hillocks known as the *superior* and *inferior colliculi (quadrigeminal bodies)* (fig. 53–2). The superior colliculi are concerned with visual, the inferior with auditory, functions. The pineal body is attached to the forebrain above the superior colliculi.

The *trochlear (4th cranial) nerve* decussates and emerges from the dorsal aspect of the midbrain, below the corresponding inferior colliculus (fig. 53–2).

The midbrain is traversed by the aqueduct, which is a channel that connects the fourth with the third ventricle (fig. 53–9).

The medulla, pons, midbrain, and diencephalon (that part of the forebrain that is adjacent to the midbrain) are collectively known as the brain stem (fig. 53–1).

Forebrain

The forebrain, or prosencephalon, comprises a smaller part, the diencephalon, and a massive portion, the telencephalon.

Diencephalon. The term diencephalon is applied to the part of the brain that largely bounds the third ventricle. A small portion of the third ventricle, however, is telencephalic.

The diencephalon includes (1) the thalami, (2) the medial and lateral geniculate bodies, (3) the pineal body and habenulae, and (4) the hypothalamus.

The *thalami* are two large masses of gray matter situated one on each side of the third ventricle. Each thalamus includes many nuclei, and functions as an important sensory correlation center.

The *medial* and *lateral geniculate bodies* are two elevations on each side of the colliculi, to which they are connected. They lie under cover of the posterior portion of the thalamus.

The *pineal body*, or *epiphysis*, is located below the splenium of the corpus callosum (fig. 53–18*A*) and is covered by a layer of the tela choroidea of the third ventricle. Calcified areas are often found in the pineal body, and hence the organ is frequently visible radiographically (fig. 52–4, p. 554).

The term *hypothalamus* is restricted functionally to the anterior part of the floor and the lower part of the lateral walls of the third ventricle. This region is concerned with autonomic and neuro-endocrinological functions. Anatomically, however, certain adjacent areas are generally included in the term hypothalamus: the optic chiasma, the *tuber cinereum* (a sheet of gray matter to which the infundibular stem of the hypophysis is attached), the hypophysis, and the *mamillary bodies* (two small masses covered by white matter) (fig. 53–5).

Telencephalon. The term telencephalon is virtually synonymous with *cerebral hemispheres* (figs. 53–4 and 53–7, 53–3 and 53–6). The term *cerebrum*, however, refers either to the brain as a whole or to merely

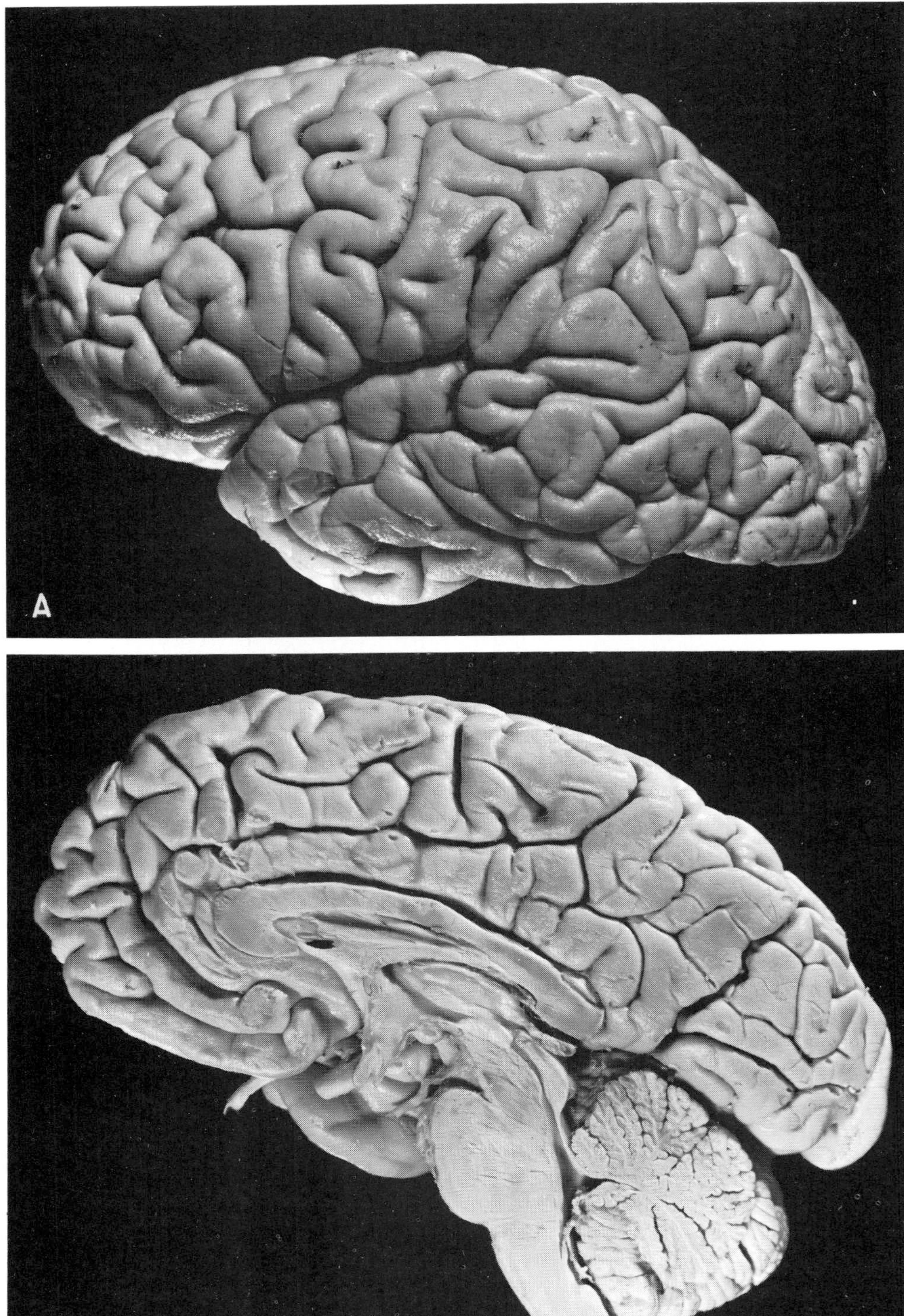

Figure 53–3 Photographs of the brain. *A*, left lateral aspect. In most right-handed people the left cerebral hemisphere is dominant, in the sense that it is concerned with the control of many functions that relate to speech and the integration of sensations. *B*, medial aspect of right half of brain. The arachnoid and pia mater were removed from both specimens. For identification of the various features, see figure 53–4. From Gardner.

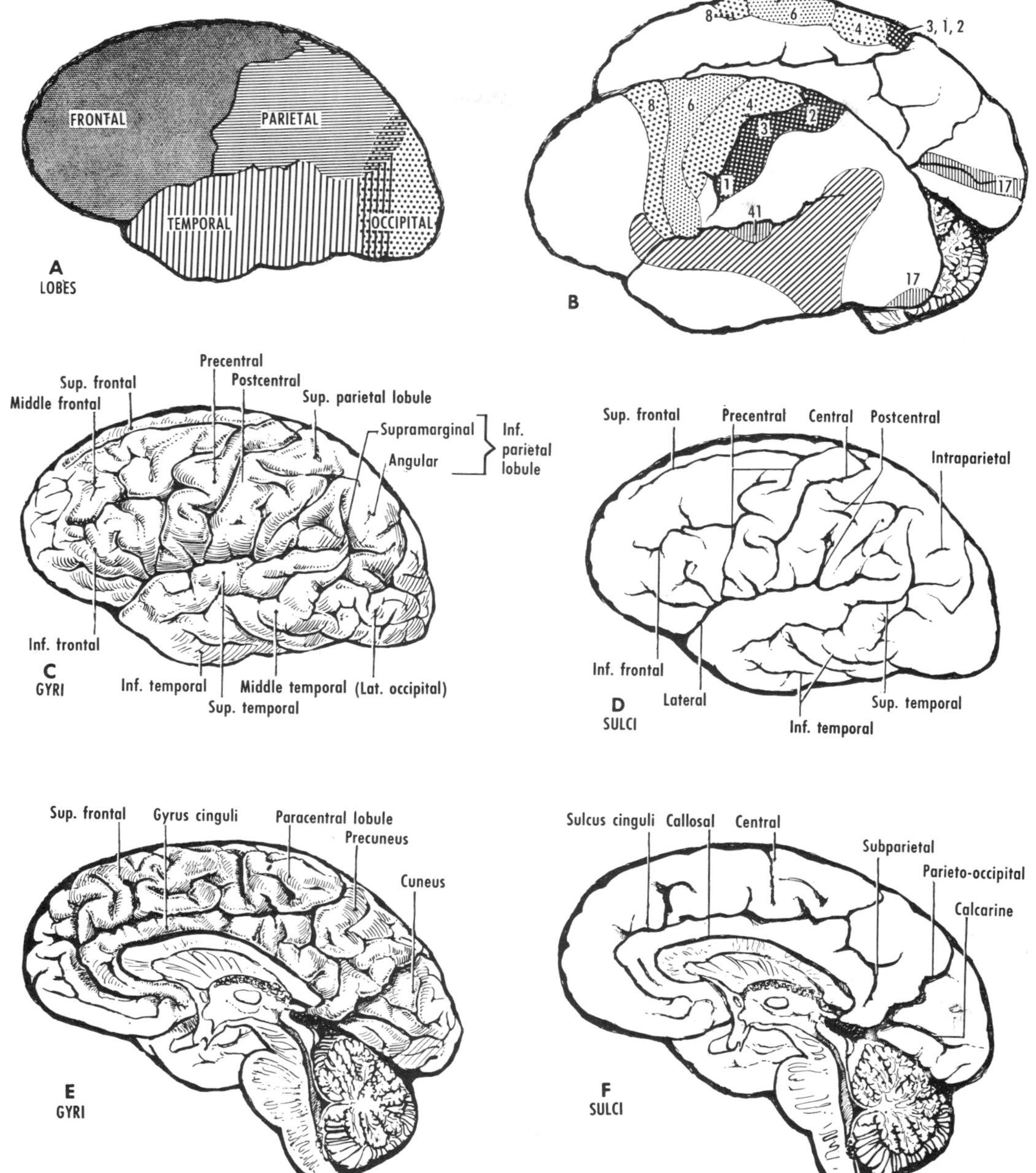

Figure 53–4 Features of the cerebral hemispheres. Compare figure 53–3. *A* shows the lobes of the brain, left lateral aspect. *B* shows certain functional areas on the lateral and medial aspects. The motor areas are marked 4, 6, and 8. The primary receptive areas are marked 1, 2, 3, 17, and 41. These numerals refer to Brodmann's system. The area indicated by diagonal lines is the region involved in aphasia. *C* shows the gyri, and *D* the sulci, on the left lateral aspect. *E* shows the gyri, and *F* the sulci, on the medial aspect of the right hemisphere.

the forebrain and the midbrain together. Each hemisphere contains a cavity known as the lateral ventricle.

As seen from above, the cerebral hemispheres conceal the other parts of the brain from view. Each hemisphere presents a superolateral, a medial, and an inferior surface. The right and left cerebral hemispheres are partly separated from each other by the *longitudinal fissure,* which is

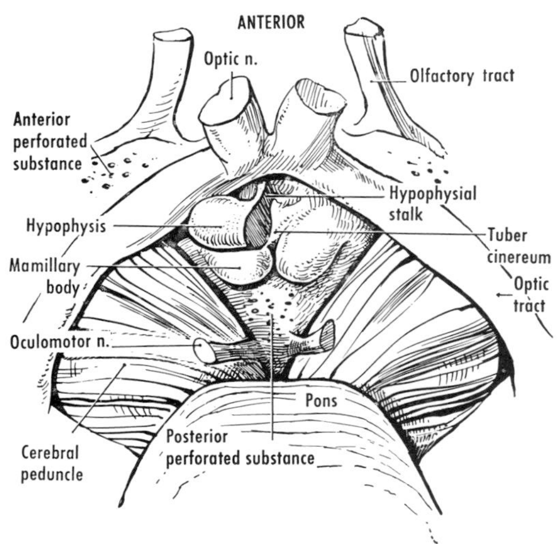

Figure 53–5 Close-up view of the interpeduncular fossa and adjacent region, antero-inferior aspect. The left half of the hypophysis cerebri has been removed. Based on a color transparency in Bassett, *A Stereoscopic Atlas of Human Anatomy,* section I, reel 4, views 5 and 6.

occupied by a fold of dura mater, the falx cerebri. The *corpus callosum* (fig. 53–18*A*, fig. 63–4, p. 743), found in the depths of the longitudinal fissure, is a bundle of fibers connecting the hemispheres. It forms the roof of the central part and of the anterior horn of the lateral ventricle of each side. It is curved sagittally and consists, from before backward, of the *rostrum, genu, trunk,* and *splenium.*

Each hemisphere presents *frontal, occipital,* and *temporal poles.* These poles are located, respectively, in the anterior, posterior, and middle cranial fossae, and are respectively related to the frontal and occipital bones and the greater wing of the sphenoid.

The gray matter of the surface of each hemisphere is termed the *cerebral cortex.* It is folded or convoluted into *gyri,* which are separated from each other by *sulci.* The pattern is variable, and it is necessary to remove the pia-arachnoid in order to identify individual gyri and sulci.

The *lateral sulcus* begins on the inferior surface of the brain. It extends laterally and, on gaining the superolateral surface of the hemisphere, proceeds backward, between (1) the frontal and parietal lobes and (2) the temporal lobe. (The *posterior ramus* is being referred to here; small *anterior* and *ascending rami* arise from the stem of the lateral sulcus where the sulcus reaches the superolateral surface of the hemisphere.) A portion of the cerebral cortex termed the insula lies buried in the depths of the lateral sulcus.

In disorders of speech (aphasia), a portion of the frontal lobe and a larger portion of the temporal lobe, adjacent to the lateral sulcus, are frequently involved. The involvement is usually on the left side of the brain (fig. 53–4*B*).

The *central sulcus* begins on the medial surface of the hemisphere and, on gaining the superolateral surface, descends between the frontal and parietal lobes. The area of cortex immediately in front of the central sulcus is known as the motor area, and is concerned with muscular activity, mostly in the opposite half of the body. The contralateral control may be demonstrated by artificial stimulation of this area, particularly of that part known as the *precentral gyrus,* or area 4, as a result of which movements of the opposite half of the body take place. Furthermore, the body is represented in an inverted position in the motor area. That is, stimulation of the upper part of the motor area gives rise predominantly to movements of the opposite lower limb; stimulation of the middle part to movements of the upper limb, and stimulation of the lower part to movements of the head and neck. The area of cortex immediately behind the central sulcus (the *post-central gyrus)* is an important primary receptive area, to which afferent pathways project by means of relays in the thalamus.

The cortex of each cerebral hemisphere is arbitrarily divided into frontal, parietal, occipital, and temporal lobes. The *frontal lobe* is bounded by the central and lateral sulci. It lies in the anterior cranial fossa. The *parietal lobe* extends from the central sulcus in front to an arbitrary line behind (between a groove above, the *parieto-occipital sulcus,* and an indentation below, the *pre-occipital notch*). The *occipital lobe* lies behind this line. The *temporal lobe* is situated in front of this line and below the lateral sulcus. It lies in the middle cranial fossa. The gyri and sulci characteristic of these lobes are shown in figures 53–4 and 53–7. The *calcarine sulcus* is located on the medial surface of the occipital lobe but may extend onto the superolateral surface of the hemisphere. When the cerebellum is displaced, portions of each of the four lobes can be seen also on the medial and inferior surfaces of the hemisphere. The occipital lobe is concerned especially with vision.

The *olfactory (1st cranial) nerves* are groups of nerve filaments which, on leaving the nose and passing through the base of the skull (cribriform plate of ethmoid), end in the olfactory bulbs. Each *olfactory bulb* lies on the inferior aspect of the corresponding frontal lobe, and gives rise to an *olfactory tract* (fig. 53–7) that runs backward and is attached to the brain.

The *optic (2nd cranial) nerves* leave the orbits through the optic canals and unite to form the *optic chiasma* (fig. 53–5). The chiasma gives rise to the right and left *optic tracts,* which proceed backward and around the cerebral peduncles. The optic chiasma and the interpeduncular fossa are contained within a very important arterial anastomosis, which is known as the circulus arteriosus. The infundibular stem of the neurohypophysis emerges from the tuber cinereum in the interpeduncular fossa, in front of the mamillary bodies. The area immediately anterolateral to each optic tract is pierced by branches of the anterior and middle cerebral arteries, and is known as the *anterior perforated substance.*

The term *basal ganglia,* or *basal nuclei,* is used for certain masses of gray matter found within the white substance of the cerebral hemispheres. These nuclei are the corpus striatum, the subthalamic nucleus, and the claustrum. The amygdaloid body is often included, and sometimes the thalamus also.

The *corpus striatum* comprises the caudate and lentiform nuclei. The *caudate nucleus* bulges into the lateral ventricle and presents a head, a body, and a tail. It has an arched form, for which reason it is often seen twice in a section. The *head* lies anteriorly, behind the genu of the corpus callosum; the *body* extends backward, above and lateral to the thalamus; and the *tail* of the nucleus curves downward and forward into the temporal lobe to end in the *amygdaloid body.* The *lentiform nucleus* lies lateral to the head of the caudate nucleus and to the thalamus. In front, it is connected with the head of the caudate nucleus by bars of gray matter, hence the name corpus striatum for the two nuclei. The lateral part of the lentiform nucleus, known as the *putamen,* is related laterally to the *claustrum* and the insula. The two medial parts of the lentiform nucleus are called the *globus pallidus.*

The *internal capsule* is a broad band of white matter situated between (1) the lentiform nucleus laterally and (2) the head of the caudate nucleus and the thalamus medially. The internal capsule consists of an *anterior limb* (between the lentiform and caudate nuclei), a *genu,* a *posterior limb* (between the lentiform nucleus and the thalamus), and *retrolentiform* and *sublentiform parts* (behind and below the lentiform nucleus, respectively).

The fibers of the internal capsule, on being traced upward, spread out in the hemisphere to form a fan-shaped arrangement termed the *corona radiata.* The fibers of the corona are intersected by those of the corpus callosum.

CRANIOCEREBRAL TOPOGRAPHY
(fig. 53–8)

Considerable variation occurs in the precise relations of the brain to the skull so that only an approximate localization of the parts of the brain is possible on the surface of the body.[4]

The inferior limit of the cerebral hemisphere lies above the eyebrow, the zygomatic arch, the external acoustic meatus, and the external occipital protuberance. Hence **the hemispheres lie above the orbitomeatal plane. A considerable portion of the cerebellum, however, lies below the level of that plane.**

The central sulcus begins at 1 cm behind the vertex, that is, behind the midpoint of a line on the head between the nasion and the inion. The sulcus runs downward, forward, and laterally for about 10 cm toward the midpoint of the zygomatic arch. The sulcus makes about three-quarters of a right angle with the median plane.

The lateral sulcus, on the superolateral surface of the hemisphere, extends from the pterion backward and slightly upward, and ends below the parietal eminence. The temporal pole of the hemisphere, from the point of view of surface anatomy, extends into the angle formed between the lateral sulcus and the zygomatic arch.

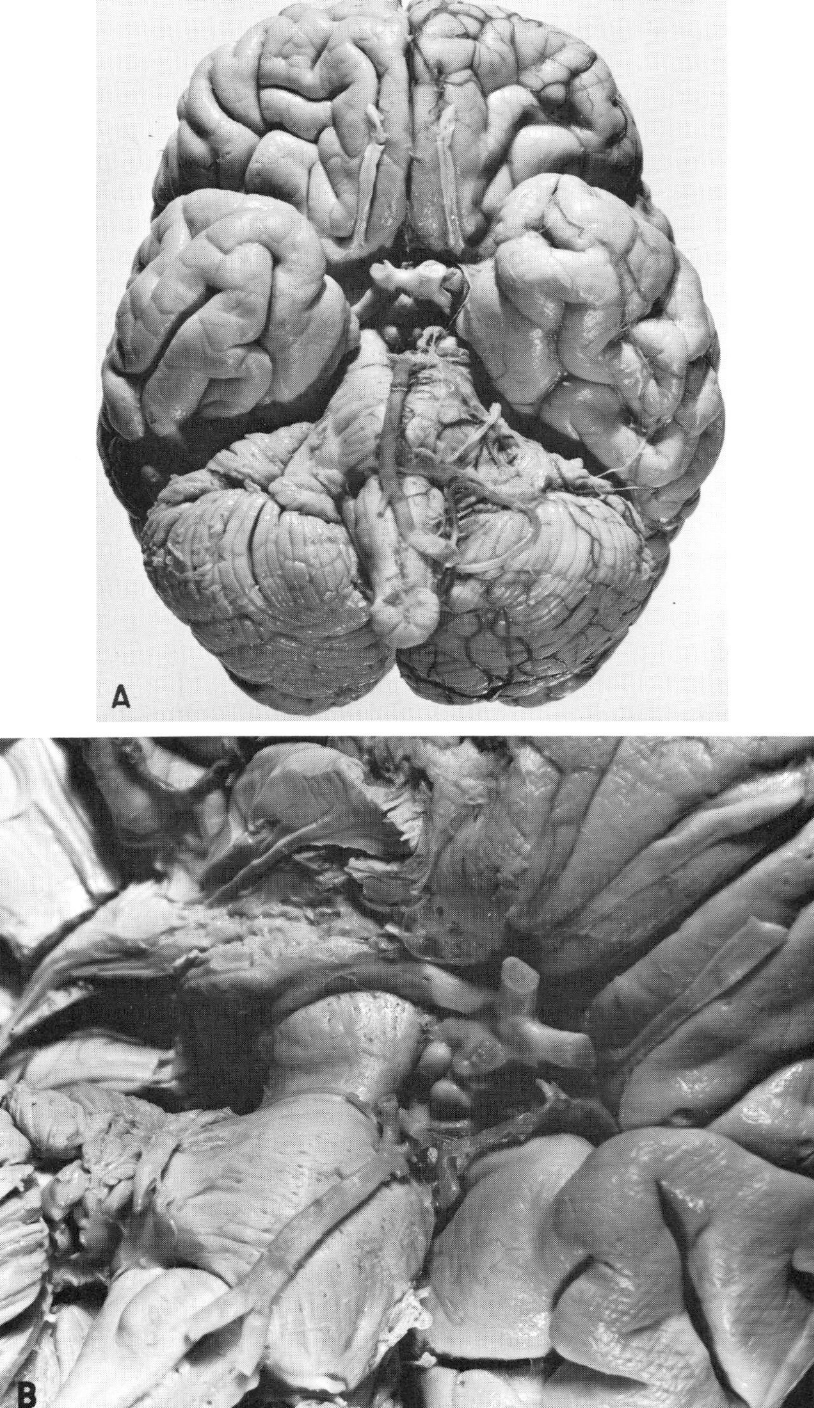

Figure 53–6 Photographs of the inferior aspect of the brain. In *A*, the arachnoid and pia mater have been removed from the right half of the brain. In *B*, most of the cranial nerves have been removed, as well as one olfactory bulb. The right temporal lobe has been cut away to expose the right optic tract. For identification of the various features, see figure 53–7. From Gardner.

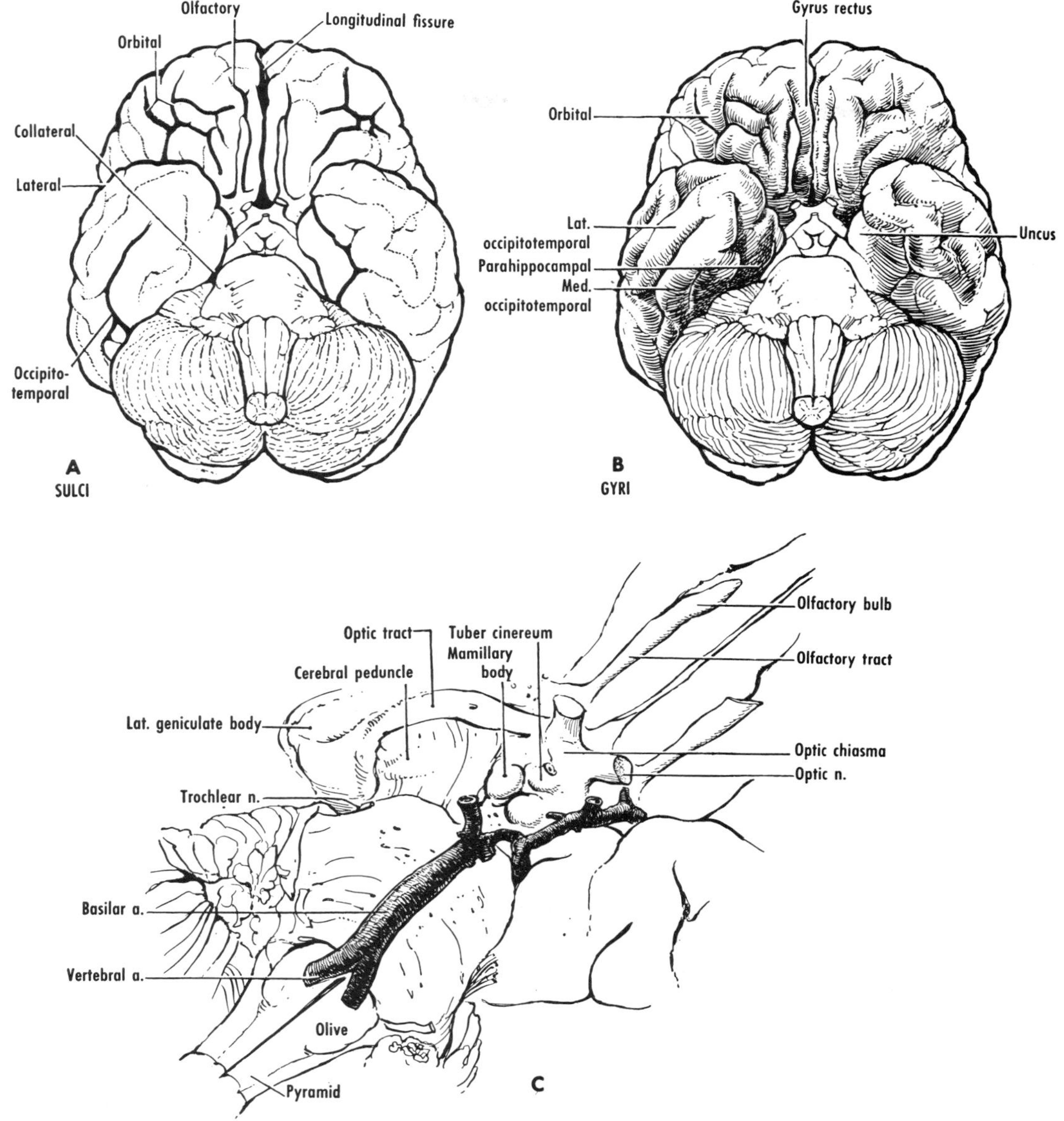

Figure 53–7 Inferior aspect of the brain. Compare figure 53–6. *A* shows the sulci. *B* shows the gyri. *C* is an enlarged view of the brain stem. The area above and to the right is anterior, that below and to the left is posterior.

VENTRICLES

The ventricles (fig. 53–9) will be described from above downward, that is, in the order in which they are numbered. The two lateral ventricles communicate with the third ventricle by an interventricular foramen on each side. The third ventricle communicates with the fourth ventricle through the aqueduct. The fourth ventricle becomes continuous with the central canal of the medulla and spinal cord, and opens by means of apertures into the subarachnoid space.

The neuroglia that lines the ventricles of the brain and the central canal of the spinal cord is termed *ependyma.* It is generally not ciliated in the adult. In the ventricles, vascular fringes of pia mater, known as the *tela choroidea,* invaginate their covering of modified ependyma and project into the ventricular cavities. This combina-

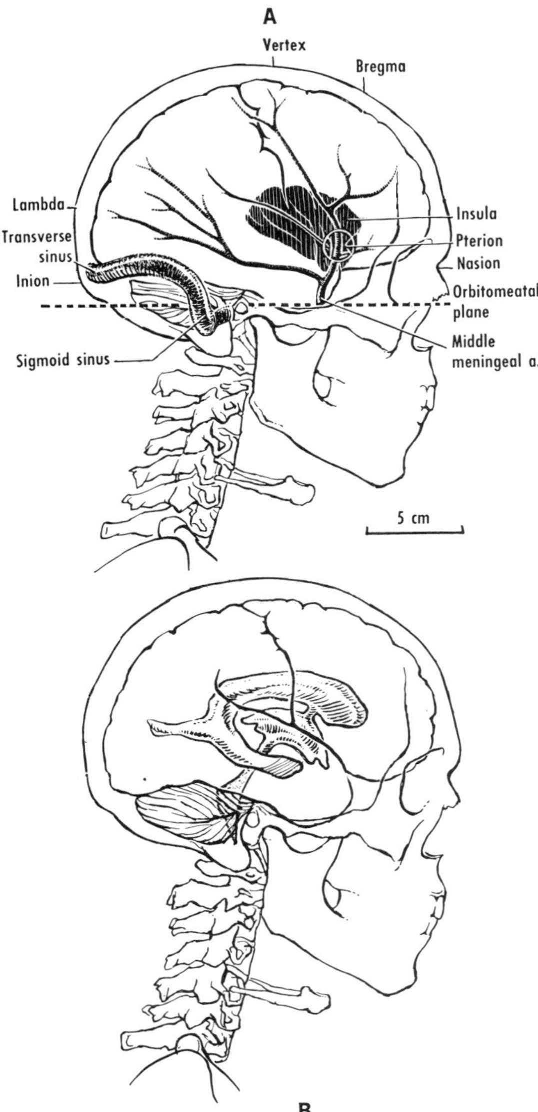

Figure 53–8 *A*, the surface anatomy of intracranial structures. *B*, the surface anatomy of the ventricles. *A* is based partly on Mettler.

tion of vascular tela and cuboidal ependyma is termed the *choroid plexus** (fig. 53–18). The plexuses are invaginated into the cavities of the lateral, third, and fourth ventricles and they are concerned with the formation of cerebrospinal fluid.

The term "blood-cerebrospinal fluid barrier" refers to the tissues that intervene between the blood and the cerebrospinal fluid. These include the capillary endothelium, several homogeneous and fibrillary layers (identified by electron microscopy), and the ependyma of the choroid plexus. The chief elements in the barrier appear to be tight junctions between the ependymal cells.

*An additional *i*, which used to be characteristic of the Latin forms (*plexus chorioideus, tela chorioidea*), although not of the English (choroid plexus), has been omitted in the *Nomina anatomica*.

Lateral Ventricles

Each lateral ventricle is a cavity in the interior of a cerebral hemisphere, and each communicates with the third ventricle by means of an interventricular foramen. The portion of the lateral ventricle in front of the foramen is termed its first part, or anterior horn. Behind this is the central part of the ventricle. The front, middle, and back portions of the central part are numbered second, third, and fourth parts, respectively. The fourth part of the ventricle divides into the fifth part, or posterior horn, and the sixth part, or inferior horn. The anterior, posterior, and inferior horns are found in the frontal, occipital, and temporal lobes of the cerebral hemisphere, respectively (fig. 53–8).

The anterior horn, central part, and inferior horn show the curve characteristic of the fetal as well as of the adult lateral ventricle. The posterior horn develops about halfway through fetal life as a backward extension and is highly variable in size in the adult.

The *anterior horn* of the lateral ventricle is bounded below by the rostrum, in front by the genu, and above by the trunk, of the corpus callosum. Laterally, it is limited by the bulging head of the caudate nucleus. Medially, it is separated from the lateral ventricle of the opposite side by a thin vertical partition, the *septum pellucidum*. Occasionally the septum contains a space *(cavity of the septum pellucidum)* between its two laminae.

The *central part* of the lateral ventricle lies beneath the trunk of the corpus callosum, and upon the thalamus and the body of the caudate nucleus. Medially, the two lateral ventricles are separated from each other by the posterior portion of the septum pellucidum.† In the angle between the diverging posterior and inferior horns, the floor of the cavity presents a triangular elevation, the *collateral trigone,* associated

†The details of certain other relations, e.g., the fornix, should be sought in texts on neuroanatomy.

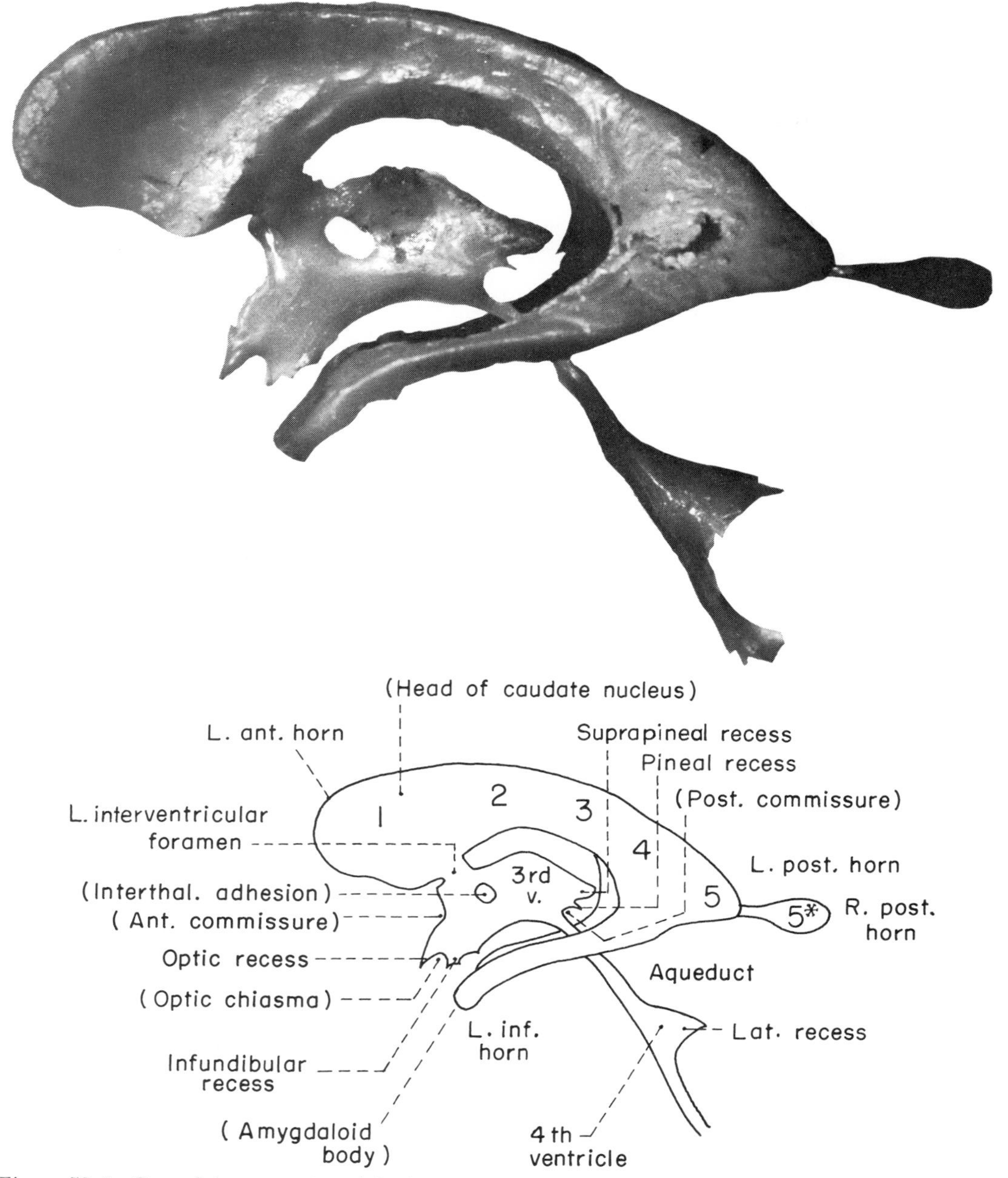

Figure 53–9 Cast of the ventricles of the brain, left lateral aspect. In this brain, the right posterior horn is considerably longer than the left. Key diagram indicates related, solid structures in parentheses. Courtesy of David Tompsett, M.D., Royal College of Surgeons of England, London.

with an underlying groove (generally the *collateral sulcus*).

The variable *posterior horn* tapers backward into the occipital lobe of the hemisphere. The two posterior horns are usually asymmetrical, and the back portion of one horn may appear as a separate vesicle in some instances.[5] Above, and on the lateral side, each posterior horn is bounded by a sheet of fibers (the *tapetum*) derived from the trunk and the splenium of the corpus callosum. Medially, two elevations may project laterally into the posterior horn. The upper elevation *(bulb of the posterior horn)* is produced by fibers *(forceps major)* derived from the splenium. The lower *(calcar avis)* is associated with a groove *(calcarine sulcus)* on the exterior of the hemisphere.

The *inferior horn* extends downward and forward behind the thalamus and into the temporal lobe of the hemisphere. It is bounded laterally by fibers (the tapetum) derived from the corpus callosum. Inferiorly, the most noticeable feature is an elevation known as the *hippocampus,* which is partly covered by the choroid plexus. Superiorly, the tail of the caudate nucleus runs forward to end in the amygdaloid body.

The choroid plexus of each lateral ventricle is invaginated along a curved line known as the *choroid fissure* (fig. 53–18). The fissure extends from the interventricular foramen in front, and in an arched manner around the posterior end of the thalamus, as far as the end of the inferior horn. The choroid plexus of the lateral ventricle is practically confined to the central part and the inferior horn. It is best developed at the junction of the central part with the inferior horn, and is there known as the *glomus choroideum.* Calcified areas (corpora amylacea) are frequent in the glomera. The vessels of the plexus are derived from the internal carotid (anterior choroid artery) and from the posterior cerebral (posterior choroid arteries). At the interventricular foramina the choroid plexuses of the two lateral ventricles become continuous with each other and with that of the third ventricle.

Third Ventricle

The third ventricle is a narrow cleft between the two thalami. Over a variable area the thalami are frequently adherent to each other, giving rise to the *interthalamic adhesion (massa intermedia).* The floor of the ventricle is formed by the hypothalamus. In front, the floor is crossed by the optic chiasma. The anterior wall is formed by the *lamina terminalis,* a delicate sheet that connects the optic chiasma to the corpus callosum. The thin roof consists of ependyma covered by two layers of pia (known as the *velum interpositum).*

The third ventricle communicates with the lateral ventricles by means of the interventricular foramina. Each *interventricular foramen* is situated at the upper and anterior portion of the third ventricle, at the front limit of the thalamus, and at the site of the outgrowth of the cerebral hemisphere in the embryo. From it a shallow groove, the hypothalamic sulcus, may be traced backward to the aqueduct. The sulcus marks the boundary between the thalamus above and the hypothalamus below.

The third ventricle presents several recesses (figs. 53–9 and 53–11): (1) the *optic recess* above the optic chiasma, (2) the *infundibular recess* in the infundibulum of the hypophysis, (3) a recess sometimes found in front of the mamillary bodies, (4) the *pineal recess* in the stalk of the pineal body, and (5) the *suprapineal recess.* Moreover, notches are produced in the outline of the third ventricle by the anterior and posterior commissures (bundles of white fibers that cross the median plane in front of and behind the third ventricle, respectively) and by the optic chiasma.

The choroid plexuses of the third ventricle invaginate the roof of the ventricle on each side of the median plane (fig. 53–18*B*). At the interventricular foramina they become continuous with those of the lateral ventricles. Their vessels (posterior choroid arteries) are derived from the posterior cerebral.

The *aqueduct* is the narrow channel in the midbrain that connects the third and fourth ventricles. It is about 1 cm in length and is widest in its central portion.[6]

Fourth Ventricle

The fourth ventricle is a rhomboidal cavity (fig. 53–2) located in the posterior portions of the pons and medulla. Above, it narrows to become continuous with the aqueduct of the midbrain. Below, it narrows and leads into the central canal of the medulla, which, in turn, is continuous with the central canal of the spinal cord. Laterally, the widest portion of the ventricle is prolonged on each side as the *lateral recess* (fig. 53–1). The superior and inferior cerebellar peduncles form the lateral boundaries of the ventricle.

The anterior boundary or floor (the *rhomboid fossa*) of the fourth ventricle is formed by the pons above and by the medulla below (fig. 53–2). It is related directly or indirectly to the nuclei of origin of the last eight cranial nerves. A *median groove* divides the floor into right and left halves. Each half is divided by a longitudinal groove (the *sulcus limitans*) into medial (basal) and lateral (alar) portions. The medial portion, known as the *medial emi-*

nence, overlies certain motor nuclei, for example, those of the abducent and hypoglossal nerves. The abducent nucleus, however, is covered by the motor fibers of the facial nerve, which loop backward around it in an elevation of the floor of the ventricle *(facial colliculus)*. The area lateral to the sulcus limitans overlies certain afferent nuclei, for instance, that of the vestibular part of the vestibulocochlear nerve.

The lowermost portion of the floor of the fourth ventricle is shaped like the point of a pen, and hence is sometimes referred to as the *calamus scriptorius*. This portion of the floor contains the important respiratory, cardiac, vasomotor, and deglutition centers.

The posterior boundary or roof of the fourth ventricle is extremely thin and is concealed by the cerebellum (fig. 53–18*A*). It consists of sheets of white matter (the *superior* and *inferior medullary vela*), which are lined by ependyma, and which stretch between the two superior and the two inferior cerebellar peduncles. The lower portion of the roof presents a deficiency, the *median aperture*[7] of the fourth ventricle, through which the ventricular cavity is in direct communication with the subarachnoid space. The ends of the lateral recesses have similar openings, the *lateral apertures*. There is little doubt that the various apertures are true openings and not artifacts. The median and lateral apertures are the only means by which cerebrospinal fluid formed in the ventricles enters the subarachnoid space. In the event of occlusion of the apertures, the ventricles become distended (hydrocephalus).

The choroid plexuses of the fourth ventricle invaginate the roof of the ventricle on each side of the median plane. A prolongation of each plexus protrudes through the corresponding lateral aperture (fig. 53–1). The vessels to the plexus are derived from cerebellar branches of the vertebral and basilar arteries.

VENTRICULOGRAPHY AND PNEUMO-ENCEPHALOGRAPHY[8] (figs. 53–10 and 53–11)

In ventriculography, gas (air, oxygen, or helium) is injected directly into a lateral ventricle through a hole trephined in the skull. The gas spreads through the ventricles, which then appear relatively radiolucent on radiography. Most of the gas is absorbed within a few days.

In pneumo-encephalography, gas is injected into the subarachnoid space by either lumbar or cisternal puncture. From the spinal subarachnoid space, the gas passes through the foramen magnum to the cerebellomedullary cisterna (fig. 53–12), and from there to (1) the fourth ventricle, aqueduct, third and lateral ventricles, and (2) the pontine and interpeduncular cisternae, and the subarachnoid space around the cerebral hemispheres, particularly in the various sulci. (It is seldom possible, however, to identify individual sulci.)

The diagnostic value of the use of contrast media in the ventricles depends on the deformity of the ventricular pattern produced by almost every expanding or contracting lesion of the brain. The localization of a tumor, for example, is facilitated by this type of radiography.

CEREBROSPINAL FLUID[9]

The cerebrospinal fluid (C.S.F.) may be examined by means of lumbar puncture (p. 545). The total volume of the fluid is thought to be about 100 to 150 ml, and its pressure about 150 mm of saline (normal range: 70 to 180) in the lateral recumbent position. The pressure is several times higher in the lumbar region when sitting, but is about atmospheric at the foramen magnum, and is negative in the ventricles. An abnormal increase in the quantity and pressure of the cerebrospinal fluid is termed hydrocephalus. It is caused usually by obstruction to the flow of the fluid.

It is generally held that the cerebrospinal fluid is formed chiefly by the choroid plexuses. The course of the cerebrospinal fluid is shown in figure 53–12. The arachnoid villi and arachnoid granulations (p. 603) seem to be responsible for the drainage of the cerebrospinal fluid into the venous sinuses of the cranial dura and the spinal veins.

The functions of the cerebrospinal fluid are not entirely clear. The liquid acts as a fluid buffer for the protection of the nervous tissue. It also compensates for changes in blood volume within the cranium, allowing the cranial contents to remain at a fairly constant volume. It has been stated (Monroe-Kellie doctrine) that no one element of the cranial contents (brain, blood, or cerebrospinal fluid) can increase except at the expense of the others.

HYPOPHYSIS CEREBRI

The hypophysis cerebri, or pituitary gland (fig. 53–13), is an important endo-

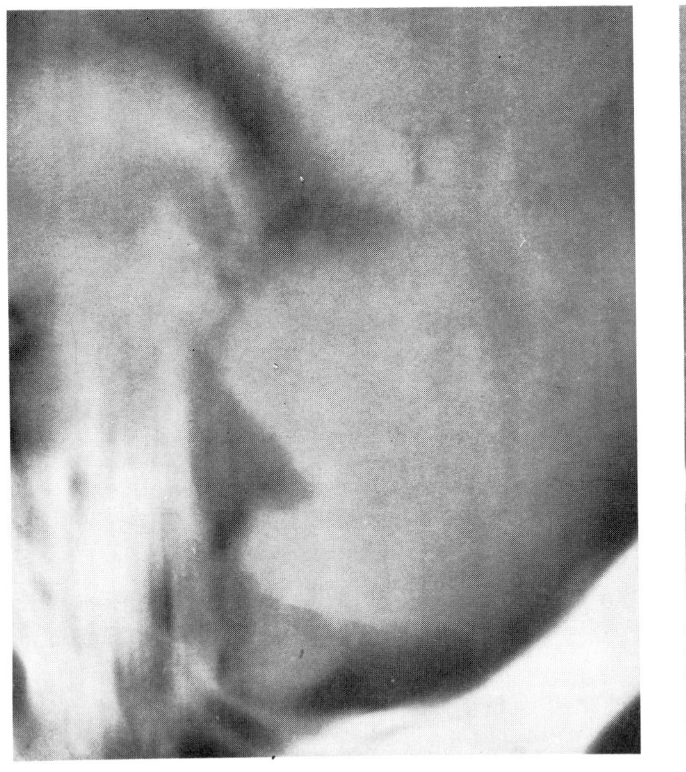

Figure 53–10 Pneumo-encephalographic tomograms. *Left-hand photograph:* pneumo-encephalographic tomogram showing the fourth ventricle. A key diagram is situated below the photograph. *Right-hand photograph:* pneumo-encephalographic tomogram showing the lateral ventricle of one side. The anterior horn is only partially filled. A key diagram is located below the photograph. Courtesy of G. Di Chiro, M.D., *An Atlas of Detailed Normal Pneumoencephalographic Anatomy,* 1961, and Charles C Thomas, Publisher, Springfield, Illinois.

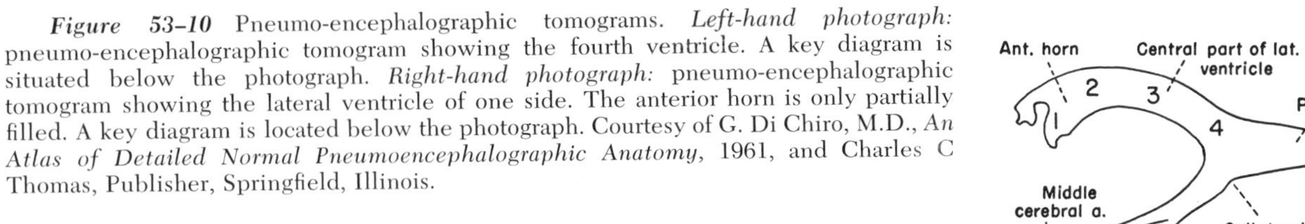

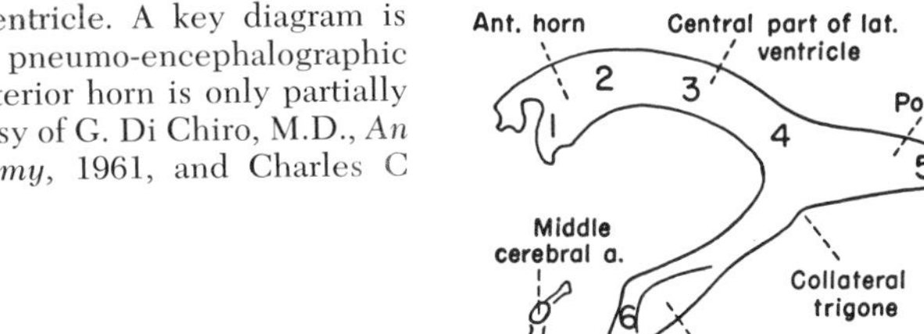

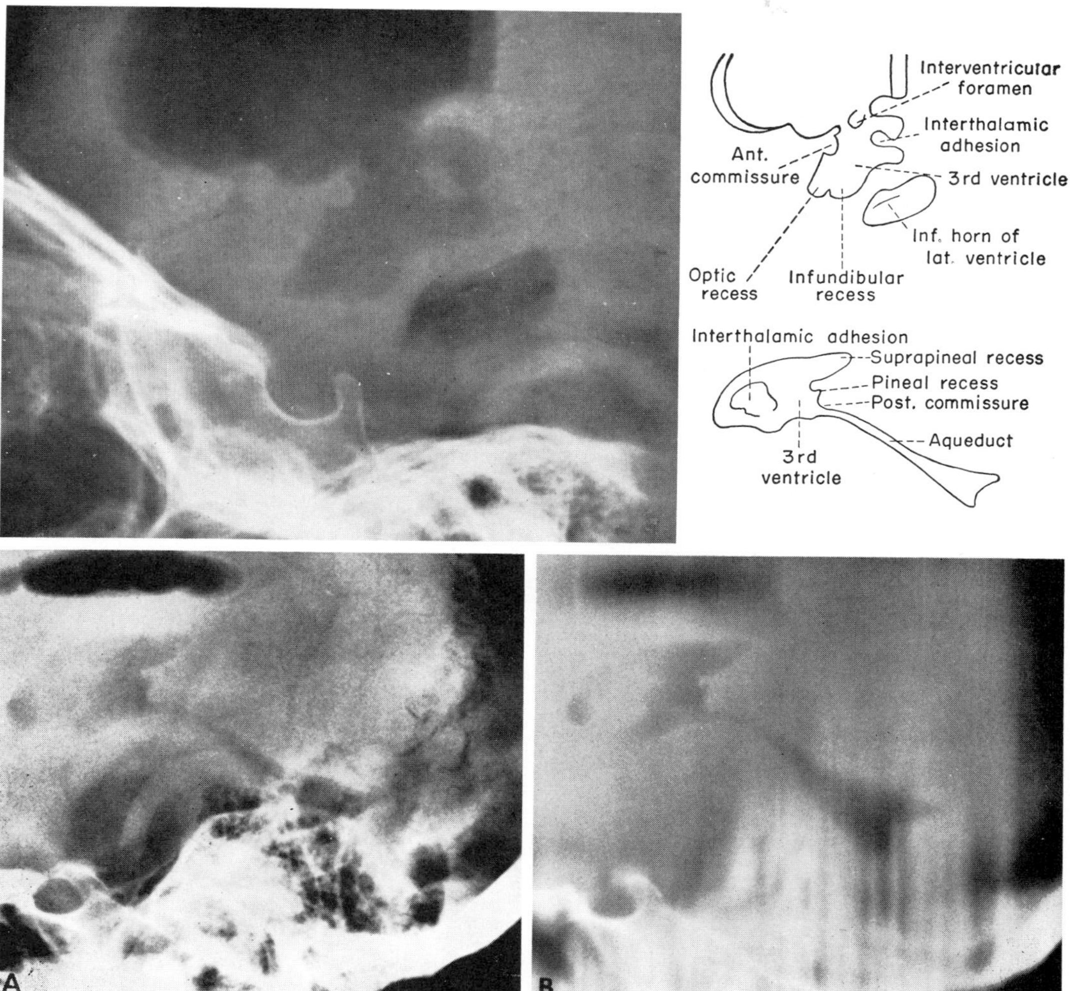

Figure 53–11 Pneumo-encephalograms and a tomogram. *Upper left:* pneumo-encephalogram showing filling of anterior portion of third ventricle. A key diagram is situated at the right. *Lower left:* pneumo-encephalogram showing filling of posterior portion of third ventricle and also of aqueduct. *Lower right:* a tomogram in the median plane. A key diagram is located above the photograph. Courtesy of G. Di Chiro, M.D., *An Atlas of Detailed Normal Pneumoencephalographic Anatomy,* 1961, and Charles C Thomas, Publisher, Springfield, Illinois.

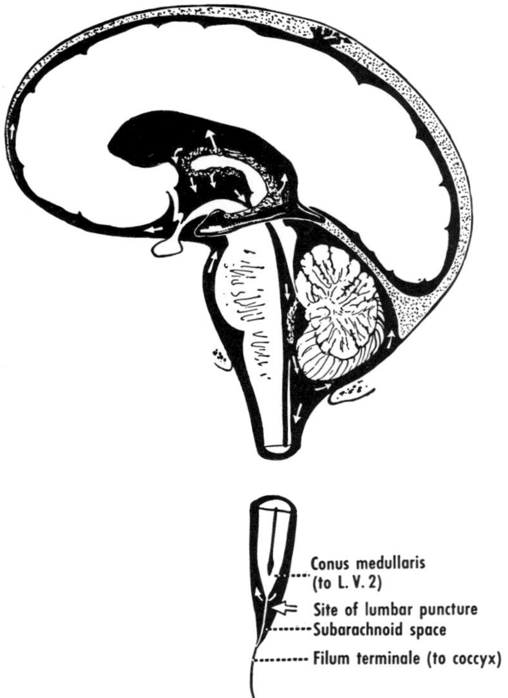

Figure 53–12 The course of the cerebrospinal fluid. The ventricles and the subarachnoid space are shown in black; the dura has been omitted. Arrows lead from the choroid plexuses of the lateral and third ventricles toward the aqueduct. The fluid formed in the lateral ventricles passes through the interventricular foramina and joins that produced in the third ventricle. The fluid then passes through the aqueduct and joins that formed in the fourth ventricle. It next passes through the median (occupied by an arrow in the figure) and lateral apertures in the roof of the fourth ventricle. From the cerebellomedullary cisterna (below the cerebellum and behind the medulla) the fluid (*arrows*) passes (1) upward around the brain and (2) downward around the spinal cord. An arachnoid granulation is represented at the top of the figure. Note that a lumbar puncture is performed in the part of the subarachnoid space that is below the termination of the spinal cord. Based chiefly on Rasmussen.

crine organ. It is an ovoid body the main portion of which is situated in the hypophysial fossa of the sphenoid bone, where it generally remains after removal of the brain. This main portion is connected to the brain by the infundibulum (fig. 53–5; fig. 63–4, p. 743). The diaphragma sellae (fig. 53–15*A*) forms a dural roof for the greater part of the hypophysis and is pierced by the infundibulum. In front of the infundibulum, the upper aspect of the gland is related directly to arachnoid and pia,[10] and the subarachnoid space here extends below the diaphragma.[11] The gland is surrounded in its fossa by a fibrous capsule that is fused with the endosteum.[12]

The hypophysis is related above to the optic chiasma, below to an intercavernous venous sinus and the sphenoidal air-sinus (through which it can be approached endonasally[13]), and laterally to the cavernous sinus and the structures contained therein (fig. 53–23). Hypophysial tumors, by causing pressure on the chiasma, commonly result in visual defects (e.g., superior temporal anopsia).

Terminology of Hypophysis.[14] The hypophysis* is best divided on embryological grounds into two main portions: the adenohypophysis and the neurohypophysis (fig. 53–13*A*). The *adenohypophysis* comprises the *pars infundibularis (pars tuberalis), the pars intermedia,* and the *pars distalis.*

The neurohypophysis comprises the *median eminence,* the *infundibular stem,* and the *infundibular process (neural lobe).* The median eminence is frequently classified also as a part of the tuber cinereum. The term *infundibulum (neural stalk)* is used for the median eminence and the infundibular stem. The term *hypophysial stalk* refers to the pars infundibularis and the infundibulum (fig. 53–13*B*).

Functional Considerations. The adenohypophysis, which develops as a diverticulum of the buccopharyngeal region, is an endocrine gland, the pars distalis of which secretes a number of hormones.

The neurohypophysis develops as a diverticulum of the floor of the third ventricle. Strictly speaking, it is not an endocrine gland, but is considered to be a storehouse for neurosecretions produced by the hypothalamus, and carried down the axons of the supra-opticohypophysial tract.

Blood Supply[15] ***and Innervation***[16] ***of Hypophysis.*** The hypophysis is supplied by a series of hypophysial arteries from the internal carotids (fig. 53–13*C*).

The maintenance and regulation of the activity of the adenohypophysis are dependent on the blood supply by way of the

*The terms "anterior and posterior lobes" are avoided because they are defined variously. Many physiologists include the pars intermedia in the posterior lobe, whereas many anatomists, together with the *Nomina anatomica,* include it in the anterior lobe.

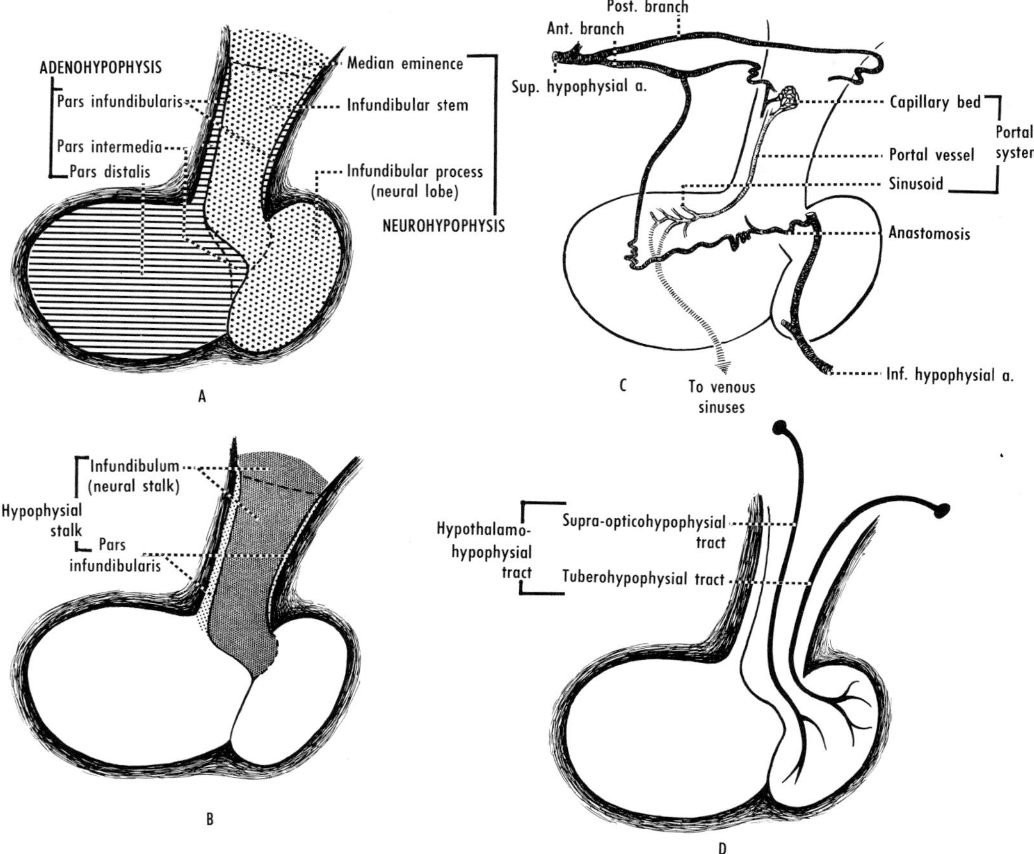

Figure 53–13 Human hypophysis cerebri. *A* and *B* illustrate the terminology. *C* shows the blood supply. Arteries in the hypophysial stalk break up into capillary loops, which drain into hypophysial portal vessels. These, on reaching the pars distalis, drain into sinusoids, which enter the venous sinuses around the gland. *D* is a schematic representation of the hypothalamohypophysial tract.

hypophysial portal system. It is likely that nerve fibers from the hypothalamus liberate releasing factors into the capillary bed in the infundibulum and that these substances are then carried by the portal vessels to the distal part of the gland, which they affect.

The neurohypophysis receives its main nerve supply from the hypothalamus by way of fibers known collectively as the *hypothalamohypophysial tract* (fig. 53–13*D*). This contains two sets of fibers, the *supra-opticohypophysial tract* and the *tuberohypophysial tract.*

CRANIAL NERVES

The cranial nerves, that is, the nerves attached to the brain (fig. 53–1; table 64–1, p. 776), are twelve on each side. They are numbered and named as follows. (The page numbers refer to the site of discussion of the specific nerves, and the figure numbers to illustrations in which they are shown.)

1. Olfactory nerve
   (page 735; fig. 62–4)
2. Optic nerve
   (page 637; figs. 53–5, 55–7*B*)
3. Oculomotor nerve
   (page 633; fig. 55–7*B*)
4. Trochlear nerve
   (page 634; fig. 55–7*A*)

5. Trigeminal nerve
   (page 582; fig. 53–23)
   (1) Ophthalmic nerve
   (page 630; fig. 55–6)
   (2) Maxillary nerve
   (page 668; fig. 58–7)
   (3) Mandibular nerve
   (page 670; fig. 58–9)
6. Abducent nerve
   (page 634; figs. 55–5, 55–7*B*)
7. Facial nerve
   (pages 619, 656; figs. 54–7, 57–5A)
8. Vestibulocochlear nerve
   (page 626; fig. 54–9)
9. Glossopharyngeal nerve
   (page 696; figs. 60–17 to 60–19)
10. Vagus nerve
   (page 698; figs. 60–18, 60–20, 60–21)
11. Accessory nerve
   (page 700; fig. 60–22)
12. Hypoglossal nerve
   (page 701; figs. 60–18, 60–23)

Functional Components

Some of the cranial nerves are exclusively or largely afferent (1, 2, and 8), others are largely efferent (3, 4, 6, 11, and 12), and still others are mixed, that is, contain both afferent and efferent fibers (5, 7, 9, and 10). The efferent fibers of the cranial nerves arise within the brain from groups of nerve cells termed motor nuclei. The afferent fibers arise outside the brain from groups of nerve cells, generally in a ganglion along the course of the nerve. The central processes of these nerve cells then enter the brain, where they end in groups of nerve cells termed sensory nuclei. The fibers of the cranial nerves begin to acquire their myelin sheaths during the fetal period.

The four functional types of fibers found in spinal nerves (see chapter 5) are present also in some of the cranial nerves: somatic afferent, visceral afferent, visceral efferent, and somatic efferent. These four types are termed "general." In certain cranial nerves, however, components that are "special" to the cranial nerves are present. The special afferent fibers comprise visual, auditory, equilibratory, olfactory, taste, and visceral reflex fibers (the first three are usually classified as somatic, and the last three as visceral). The special efferent fibers (which are classified as visceral) are those to skeletal muscles either known or thought to be derived from the pharyngeal arches (muscles of mastication, facial muscles, muscles of pharynx and larynx, sternomastoid and trapezius).

The cranial nerves, from the point of view of their chief functional components, may be grouped as follows:

Olfactory, Optic, and Vestibulocochlear Nerves (1, 2, and 8) pertain to organs of special sense (special afferent).

Oculomotor, Trochlear, Abducent, and Hypoglossal Nerves (3, 4, 6, and 12) supply skeletal muscle of specific regions of the head (eyeballs in the case of 3, 4, and 6; tongue in the case of 12). Nerve 3 also contains parasympathetic fibers to the smooth muscle of the sphincter pupillae and the ciliary muscle (general visceral efferent).

Trigeminal Nerve (5) contains motor fibers to the muscles of mastication (special visceral efferent) and sensory fibers from various parts of the head, for example, face, nasal cavity, tongue, and teeth (general somatic afferent).

Facial, Glossopharyngeal, Vagus, and Accessory Nerves (7, 9, and 10) contain several types of components. These are chiefly:

(a) Motor fibers to the muscles of facial expression (7) and the muscles of the pharynx and larynx (9 and 10) (special visceral efferent); many of the fibers to the pharynx and larynx are derived from nerve 11 (internal branch) and travel by way of nerve 10 (hence 11 is "accessory" to the vagus).

(b) Parasympathetic secretory fibers to the lacrimal and salivary glands (nervus intermedius of 7), the salivary glands (9), and certain glands associated with the respiratory and digestive systems (10) (general visceral efferent); nerve 10 also supplies most of the smooth muscle of the respiratory and digestive systems, as well as cardiac muscle.

(c) Taste fibers (nervus intermedius of 7; also 9 and 10) (special visceral afferent).

(d) Fibers from the mucous membrane of the tongue and pharynx (hence the name glossopharyngeal) and of much of the respiratory and digestive systems (general visceral afferent) are contained in nerves 9 and 10.

The spinal part of nerve 11 supplies the sternomastoid and the trapezius, two muscles of somewhat disputed development.

Parasympathetic Ganglia Associated with Cranial Nerves

The ciliary, pterygopalatine (sphenopalatine), otic, and submandibular ganglia are associated with certain of the cranial nerves. In these ganglia, parasympathetic fibers synapse, whereas sympathetic and other fibers merely pass through. The chief features of the ganglia are summarized in table 64–4, p. 782.

MENINGES

The brain, like the spinal cord, is surrounded by three membranes or meninges: the dura mater, the arachnoid, and the pia mater (fig. 53–14). The last two are known as the leptomeninges.

PACHYMENINX, OR DURA MATER

The part of the pachymeninx, or dura mater, that surrounds the brain is usually described as consisting of two layers: an external (endosteal) and an internal (meningeal) layer. The two layers are indistinguishable, however, except along certain strips where they are separated by venous sinuses. The *endosteal layer*, or endocranium (p. 551), is adherent to the inner aspect of the cranial bones, particularly at the sutures and at the base of the skull.[17] At the sutures and at the various foramina in the skull, the endosteal layer is continuous with the pericranium. At the foramina, the endosteal layer provides sheaths for the cranial nerves. The *meningeal layer* is lined internally by flat cells. The cranial dura is continuous at the foramen magnum with the spinal dura.[18]

Processes of Dura Mater

The meningeal layer sends four processes internally: the falx cerebri, the tentorium cerebelli, the falx cerebelli, and the diaphragma sellae (fig. 53–15).

1. Falx Cerebri. This process, shaped like a sickle, lies in the longitudinal fissure between the two cerebral hemispheres. It is attached to the crista galli in front and fuses with the tentorium cerebelli behind. The upper, convex border of the falx divides to enclose the superior sagittal sinus and is attached to the inner aspect of the skull (frontal, parietal, and occipital bones). The lower, concave border of the falx contains the inferior sagittal sinus and is free (fig. 53–14*C*); it lies above, and more or

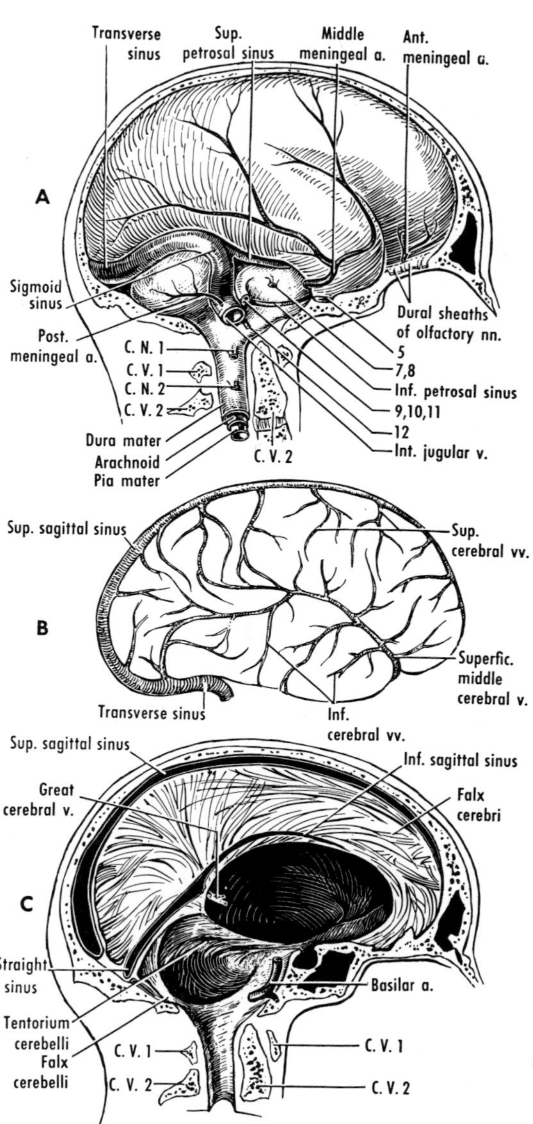

Figure 53–14 Meninges and associated vessels. *A*, lateral aspect of the intact dural sac. *B*, cerebral veins as seen through the arachnoid after removal of the dura. *C*, processes of dura mater after removal of the brain and the spinal cord. *A* is based on Strong and Elwyn.

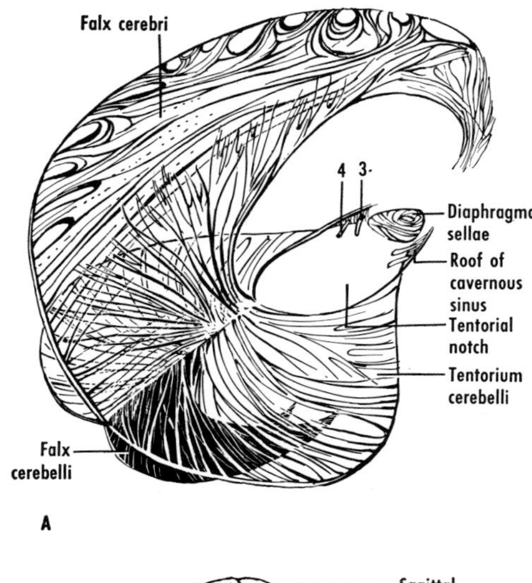

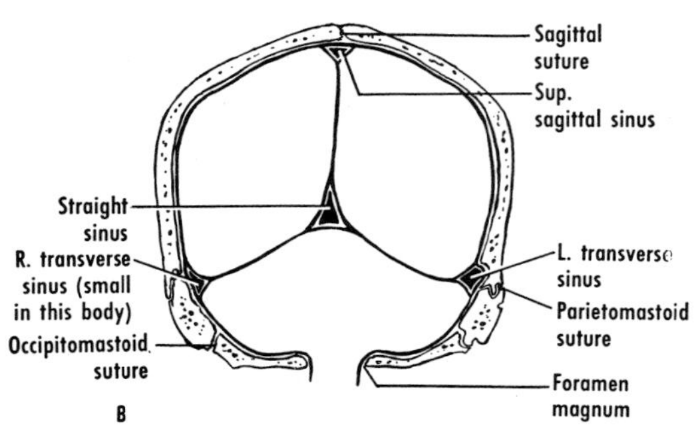

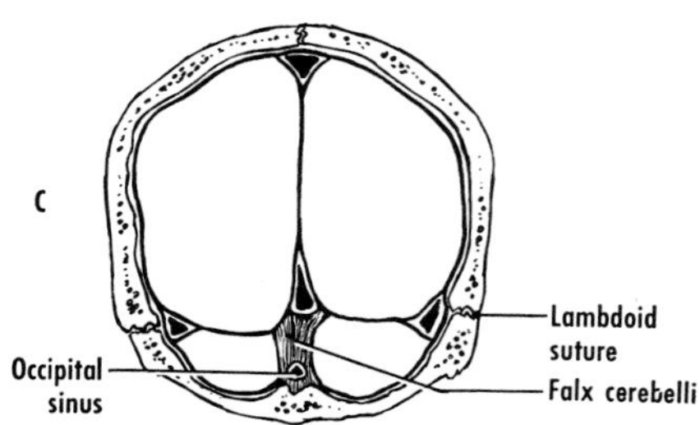

Figure 53–15 Processes of the dura mater. *A*, as seen from the right lateral aspect and also from above and behind. *B*, a coronal section through the foramen magnum. *C*, a coronal section about 1½ cm behind that shown in *B*.

less follows the curve of, the corpus callosum.

2. *Tentorium Cerebelli.* This process supports the occipital lobes of the cerebral hemispheres and covers the cerebellum. Its internal, concave border is free and, together with the dorsum sellae of the sphenoid bone, forms the boundary of the tentorial notch (fig. 53–16), which is occupied largely by the midbrain. The external, convex border encloses the transverse sinus behind and is attached to the inner aspect of the skull (occipital and parietal bones). Further forward, this margin encloses the superior petrosal sinus and is attached to the upper border of the petrous temporal. Near the apex of the petrous temporal, the two borders of the tentorium cross each other. The free border passes above and is anchored to the anterior clinoid process on each side; the attached border passes below and is attached to the posterior clinoid process. After the two borders have crossed each other, the triangular area of dura between them forms the roof of the cavernous sinus, which is pierced by the trochlear and oculomotor nerves. The roof is continuous medially with the diaphragma sellae.

Owing to the crossing over of the borders of the tentorium, its superior and inferior surfaces become lateral and superomedial, respectively, in front where, together with the body of the sphenoid bone, they enclose the cavernous sinus.

The *tentorial notch* (fig. 53–16) presents the following relations. The tip of the uncus (that part of the temporal lobe above which the optic tract disappears from view) lies above the cavernous sinus and is close to the oculomotor nerve, which pierces the roof of the sinus. The midbrain, surrounded by the subarachnoid space, lies in the tentorial notch. Miscellaneous veins, the posterior communicating and posterior cerebral arteries, the pineal body, and the splenium of the corpus callosum all lie within the subarachnoid space. A portion of the cerebellum usually rises through the notch. **Space-occupying intracranial lesions may cause herniation of the brain from one dural compartment to another through the tentorial notch, and distortion of the midbrain may ensue.**[19]

Near the apex of the petrous temporal, the meningeal layer of the dura of the posterior cranial fossa bulges forward and laterally like a three-fingered glove, beneath the superior petrosal sinus and the meningeal layer of the dura of the middle cranial fossa. The dural recess thereby formed is termed the *cavum trigeminale*[20] (fig. 53–17), because it contains the roots of the trigeminal nerve, together with most of the mandibular nerve and trigeminal ganglion (fig. 53–23). The cavum is fused anteriorly with the lateral wall of the cavernous sinus.

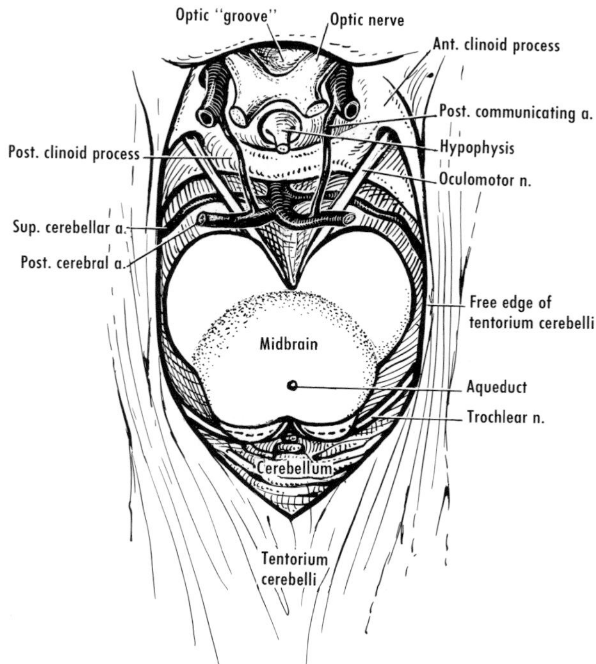

Figure 53–16 The tentorial notch, superior aspect. The dorsum sellae (between the two posterior clinoid processes) is the anterior boundary of the notch.

3. Falx Cerebelli. This sickle-shaped process lies below the tentorium. Its upper border is attached to the lower aspect of the tentorium. Its posterior border contains the occipital sinus and is attached to the occipital bone. Its anterior border is free and projects between the cerebellar hemispheres.

4. Diaphragma Sellae. This process, circular and horizontal, forms a dural roof for the sella turcica. It covers the hypophysis, and presents an opening for the infundibulum. The optic chiasma lies partly or completely above the diaphragma.[22]

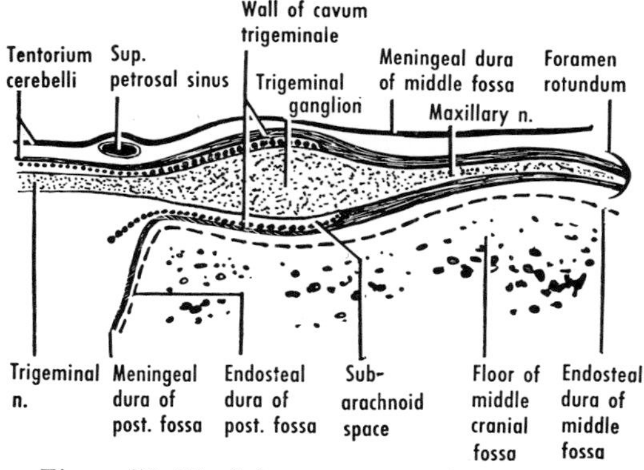

Figure 53–17 Schematic vertical section showing the cavum trigeminale. The anterior end of the section is at the right. The superior border ("petrous ridge") of the petrous temporal can be identified approximately below the superior petrosal sinus. See also figure 53–23.

Innervation of Dura Mater[23]

The dura, like the scalp, is supplied by the trigeminal and cervical nerves; autonomic fibers probably go mostly to vessels. The dura of the anterior cranial fossa is innervated by the ophthalmic nerve through branches of the anterior and posterior ethmoidal nerves. That of the middle fossa is supplied by the meningeal branches of the maxillary and mandibular nerves. The dura of the posterior fossa receives the meningeal branches of the vagus and hypoglossal nerves, both of which contain spinal fibers (C.N. 1, 2), and also meningeal branches of C.N. 1 to 3 that pass through the foramen magnum.[24] The tentorium is innervated by the tentorial branches of the ophthalmic nerve, which also supply the falx cerebri and related venous sinuses.[25] **The brain itself is normally quite insensitive, and headaches are commonly either of vascular (intracranial or extracranial) or of dural origin.**

Meningeal Vessels

The dura is supplied by anterior and posterior meningeal branches of various arteries (e.g., internal carotid, vertebral) and by the middle meningeal artery from the maxillary. The meningeal vessels are nutrient to the bones of the skull.

Middle Meningeal Artery. **This is clinically the most important branch of the**

maxillary artery because, in head injuries, tearing of this vessel may cause extradural hemorrhage. It arises in the infratemporal fossa, deep to the ramus of the mandible (fig. 58–3*C*, p. 663). It enters the cranial cavity through the foramen spinosum in the sphenoid bone. In the middle cranial fossa (fig. 53–14*A*) it runs forward and laterally for a variable distance, in a groove on the squamous part of the temporal bone. At a variable point[26] it divides into a frontal and a parietal branch. The frontal branch may lie in a bony canal and consequently be particularly susceptible to laceration when the skull is fractured. The middle meningeal artery may be torn in cranial injuries, however, even in the absence of a fracture of the skull. The resulting hemorrhage between the cranium and the dura may cause symptoms and signs of brain compression (e.g., contralateral paralysis), and necessitate trephining.

An accessory meningeal branch may arise from either the maxillary or the middle meningeal artery. It enters the middle cranial fossa through the foramen ovale and supplies the trigeminal ganglion and the dura.

SURFACE ANATOMY (fig. 53–8). **The middle meningeal artery divides at a variable point on a line connecting the midpoint of the zygomatic arch with the posterior end of the pterion.** The frontal branch passes upward and forward to the pterion,[27] then upward and backward toward the vertex. The parietal branch passes backward and upward toward the lambda.

BRANCHES. The branches of the middle meningeal artery arise in the middle cranial fossa. They contribute to the supply of the trigeminal ganglion, the tympanic cavity *(superior tympanic artery)*, and the orbit. (An anastomotic branch[28] passes through the superior orbital fissure and unites with the recurrent meningeal branch of the lacrimal artery.) The terminal branches, which are separated from the bone by their accompanying veins,[29] are as follows:

1. The *frontal branch* grooves the sphenoid and parietal bones and supplies the anterior portion of the dura. It may be enclosed in a bony canal for a part of its course.
2. The *parietal branch,* occasionally double, grooves the temporal and parietal bones and supplies the posterior portion of the dura.

Meningeal Veins. The meningeal veins, which are said to be actually sinuses, lie in and drain the dura mater (fig. 53–19). They communicate with *lateral lacunae* that are situated in the dura on each side of the superior sagittal sinus.[30] The lacunae are complicated venous meshworks that partly occupy the granular pits on the internal aspect of the calvaria. Arachnoid granulations project into the lacunae. The lacunae receive meningeal, diploic, and emissary veins, and they communicate with the superior sagittal sinus. They may also receive some cerebral veins.

LEPTOMENINGES

The leptomeninges (fig. 53–18) are frequently described as two membranes, the arachnoid and the pia mater, but the two are united by trabeculae of connective tissue, in the meshes of which (subarachnoid space) the cerebrospinal fluid circulates.[31] **The arachnoid and the pia are best regarded as one tissue, and the subarachnoid space may be considered as a space within the leptomeninges.**

Arachnoid

The arachnoid surrounds the brain loosely and is separated from the dura by a potential subdural space. It dips into the longitudinal fissure but not into the sulci. It is covered by flat cells.

At certain areas on the base of the brain, the arachnoid and the pia are separated widely by intervals known as *subarachnoid cisternae* (fig. 53–18*A*).[32] The *cerebellomedullary cisterna (cisterna magna)* is continuous below with the subarachnoid space around the spinal cord. It can be "tapped" by means of a needle inserted through the posterior atlanto-occipital membrane (cisternal puncture). The *pontine cisterna* is located on the front of the pons and contains the basilar artery. The *interpeduncular* and *chiasmatic cisternae* lie above the pons and between the temporal lobes. The interpenduncular cisterna contains the circulus arteriosus. The *cisterna of the lateral fossa* or *sulcus* is situated immediately in front of each temporal lobe. The *cisterna of the great cerebral vein* lies between the splenium of the corpus callosum and the cerebellum. It contains the great cerebral vein, and is connected around the brain stem (by the *cisterna ambiens*) with the interpeduncular cisterna.

The subarachnoid space communicates with the fourth ventricle by means of aper-

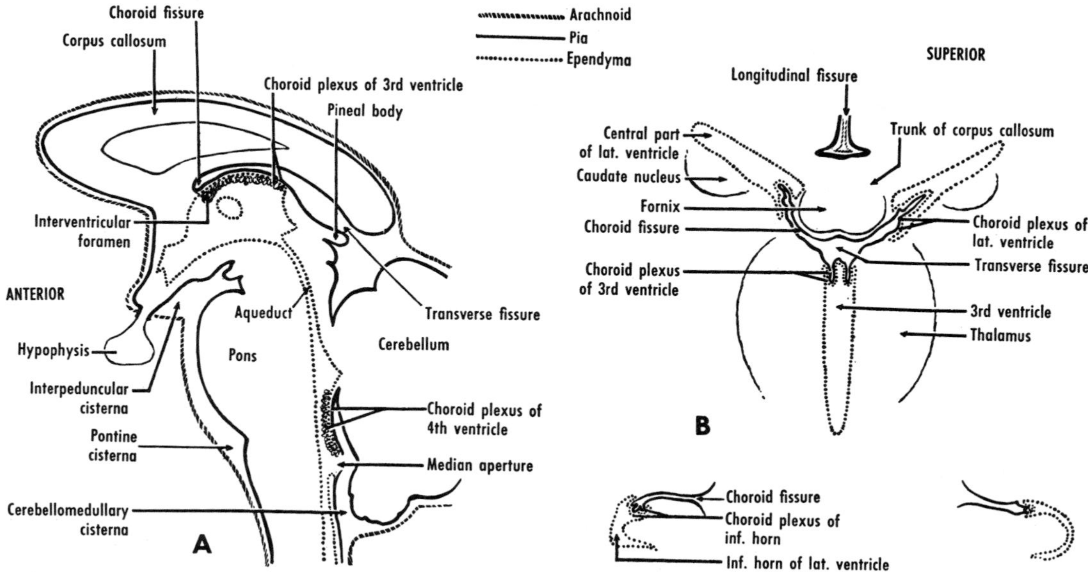

Figure 53–18 Schemes showing the arrangement of the leptomeninges and of the ependyma, the neuroglia lining the ventricles and the central canal. The choroid plexuses consist of vascular pia (tela choroidea) and ependyma. *A*, a median section through the brain stem. *B*, a coronal section through the forebrain. The inferior horns of the lateral ventricles, which are situated in the temporal lobes of the cerebral hemisphere, are shown in the lowermost part of this drawing.

tures, and it is continuous with the perineural space found around the olfactory and optic nerves.

In the region of many of the dural venous sinuses, the arachnoid presents externally a large number of microscopic projections, termed *arachnoid villi,* which are believed to be concerned with the absorption of the cerebrospinal fluid. The *arachnoid granulations*[33] are probably enlargements of the arachnoid villi and they are visible to the naked eye. They appear as small elevations that project into some of the venous sinuses, chiefly the superior sagittal (fig. 53–19) and the transverse, and also into the lateral lacunae found on each

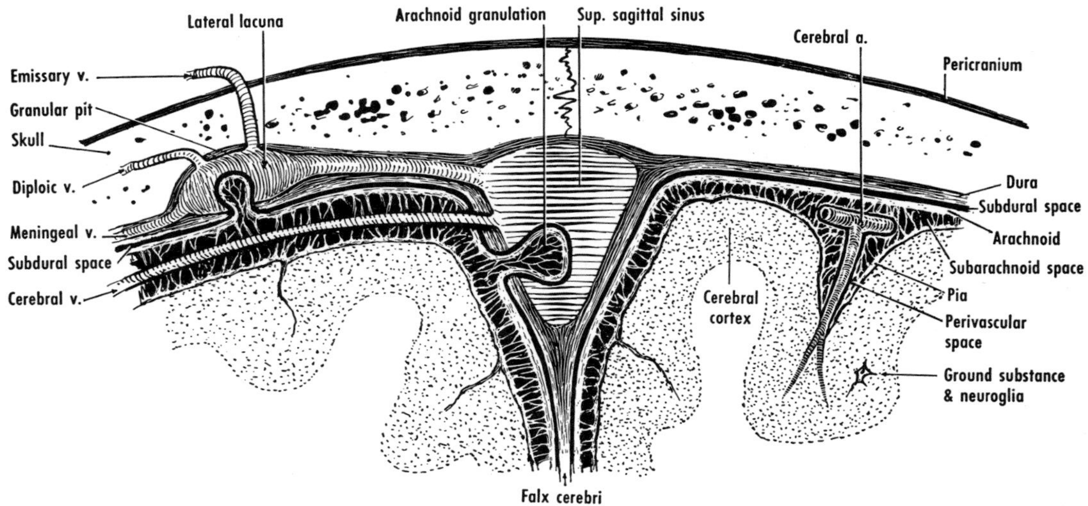

Figure 53–19 Schematic coronal section through the vault of the skull and the underlying brain. A neuron is represented (out of scale) at the lower right, so that its relationship to the adjacent blood vessels can be appreciated.

side of the superior sagittal sinus. The lacunae and the arachnoid granulations that project into them lie partly in depressions, termed granular pits, on the internal aspect of the calvaria. The granulations first appear during early childhood. The granulations are probably also responsible for absorption of the C.S.F., although some regard them as pathological.

Pia Mater

The pia covers the brain and dips in between the gyri of the cerebral hemispheres and between the folia of the cerebellum. It forms the tela choroidea of the ventricles. It consists of reticular and elastic fibers, covered superficially by the cerebral vessels within the subarachnoid space.

BLOOD SUPPLY OF BRAIN

ARTERIES OF BRAIN

The brain is supplied by the two internal carotid and the two vertebral arteries. The former supply chiefly the frontal, parietal, and temporal lobes, the latter the temporal and occipital lobes, together with the midbrain and the hindbrain. On the inferior surface of the brain the four arteries form an anastomosis, the circulus arteriosus.

It is important to realize that, in many of the domestic mammals used experimentally, the blood supply of the brain differs considerably from that found in the Primates.

The tissues that intervene between the blood and the neurons include (1) the vascular wall; (2) the internal and (3) the external layer of a leptomeningeal *perivascular sheath*, separated from each other by a *perivascular space* (fig. 53–19), which is probably continuous with the subarachnoid space; and (4) the neuroglia and the ground substance of the brain.[34] The inner and outer layers of the perivascular sheath become fused together at the level of the arterioles and venules. The sheath gradually fades away and, at the capillary level, is replaced by a neuroglial sheath. The capillary endothelial cells form a continuous layer and are united by tight junctions. These capillary cells and their externally applied basement membrane form the "blood-brain barrier" ("hemato-encephalic barrier"), which is a major factor in limiting the free movement of substances from the blood to the brain.

The cerebral arteries are believed to be involved in the production of many types of headaches.

INTERNAL CAROTID ARTERY (PETROUS, CAVERNOUS, AND CEREBRAL PARTS)

The cervical part of the internal carotid artery (chapter 60) enters the carotid canal in the petrous part of the temporal bone. The petrous part of the artery first ascends and then curves forward and medially. It is closely related to the cochlea, the middle ear, the auditory tube, and the trigeminal ganglion. The subsequent directions of the petrous, cavernous, and cerebral parts of the vessel are frequently numbered from 5 to 1, as follows (fig. 53–24, *inset*):

5. At the foramen lacerum, the petrous part of the internal carotid ascends to a point medial to the lingula of the sphenoid bone.

4. The artery then enters the cavernous sinus but is covered by the endothelial lining of the sinus (fig. 53–23). Hence this is termed the cavernous part of the artery. In the sinus the vessel passes forward along the side of the sella turcica.

3. It next ascends and pierces the dural roof of the sinus between the anterior and the middle clinoid process.

2. The cerebral part of the internal carotid turns backward in the subarachnoid space below the optic nerve. The **U**-shaped bend, convex forward, formed by parts 2, 3, and 4, is termed the carotid siphon.[35]

1. The artery finally ascends and, at the medial end of the lateral sulcus, divides into the anterior and middle cerebral arteries.

The internal carotid artery and its branches, including the cerebral arteries, are surrounded and supplied by a sympathetic plexus derived from the superior cervical ganglion.

The blood that enters the brain through one internal carotid artery is distributed almost wholly to the ipsilateral hemisphere and is drained predominantly by the internal jugular vein of the same side.[36]

When the collateral circulation is potentially good, one internal carotid artery can be occluded completely without any residual symptoms. The effects of occlusion of a major artery to the brain depend on the extent of the collateral circulation provided by the anterior and posterior communicating arteries, the extent of the collateral circulation between the cerebral arteries in the leptomeninges, the absence of vascular disease, and the level of the blood pressure.

Branches of Internal Carotid Artery (fig. 53–24). The internal carotid artery gives no named branches in the neck. Within the cranial cavity, however, it supplies the hypophysis, the orbit, and a major share of the brain.

The ophthalmic, posterior communicating, and anterior choroid arteries are branches of the carotid siphon.

The *ophthalmic artery* is described with the orbit. One of its branches (the dorsal nasal artery) anastomoses with branches of the facial artery, and another (the lacrimal artery) anastomoses with the middle meningeal artery. An anastomosis is thereby established ultimately between the internal and external carotid arteries.*

The *posterior communicating artery* is a slender vessel that connects the internal carotid with the posterior cerebral artery, and thereby forms a part of the circulus arteriosus. It may be very small or absent on one or both sides.

The *anterior choroid artery*[38] arises from the internal carotid near the termination of that vessel. It passes backward along the optic tract and enters the choroid fissure. Although its distribution is variable, it gives numerous small branches to the interior of the brain, including the choroid plexus of the lateral ventricle. The anterior choroid artery is frequently the site of thrombosis.

The terminal branches of the internal carotid are the anterior and middle cerebral arteries.

1. The *anterior cerebral artery*,[39] the smaller terminal branch, passes medially above the optic chiasma and enters the longitudinal fissure of the brain. Here it is connected with its fellow of the opposite side by the anterior communicating artery (which is frequently double and which sometimes gives off a median anterior cerebral artery[40]). It then runs successively forward, upward, and backward, usually lying on the corpus callosum.[41] Its precise course is subject to considerable variation.[42] The artery ends by turning upward on the medial surface of the hemisphere, just in front of the parieto-occipital sulcus.

2. The *middle cerebral artery*, the larger terminal branch of the internal carotid, is frequently regarded as the continuation of that vessel. It passes laterally in the lateral sulcus and gives rise to numerous branches on the surface of the insula.[43]

*The effectiveness of this collateral circulation seems clear in a woman of 48 years in whom the carotid arteries of both sides were ligated; a year later, "she was back at her work as a parochial school teacher . . . and had excellent vision in both eyes."[37]

Its branches (fig. 53–24) supply the motor and premotor areas, and the sensory and auditory areas. Occlusion of the middle cerebral artery causes a contralateral paralysis (hemiplegia) and a sensory defect. The paralysis is least marked in the lower limb (territory of anterior cerebral artery). When the dominant (usually left) side is involved, there are also disturbances of speech (aphasia).

The general distribution[44] of the cerebral arteries is shown in figure 53–20.

The chief branches of the internal carotid artery are summarized in table 53–2.

VERTEBRAL ARTERY (INTRACRANIAL PART) AND BASILAR ARTERY

The vertebral and basilar arteries and their branches[46] supply the upper part of the spinal cord, the brain stem, the cerebellum, and a large share of the postero-inferior portion of the cerebral cortex. The branches to the brain stem are functionally end-arteries.

Vertebral Arteries

The vertebral artery, a branch of the subclavian, has a complicated course that may be considered in four parts: cervical, vertebral, suboccipital, and intracranial. The suboccipital part of the vertebral artery (p. 530) perforates the dura and the arachnoid, and passes through the foramen magnum (fig. 53–2). The intracranial part of each vertebral artery ascends medially in front of the medulla oblongata and, at approximately the lower border of the pons, the two vertebrals unite to form the basilar

TABLE 53–2 Chief Branches of Internal Carotid Artery

Part	*Chief Branches*
Cervical	No named branches
Petrous	Caroticotympanic branches

Cavernous	Meningohypophysial trunk[45]
	  Tentorial branch
	  Meningeal branch
	  Inferior hypophysial artery
	Cavernous branch

Cerebral	Superior hypophysial arteries Ophthalmic artery Posterior communicating artery Anterior choroid artery Anterior cerebral artery Middle cerebral artery

artery (fig. 60–26). The two vertebral arteries are usually unequal in size, the left being larger than the right.

Branches of Intracranial Part of Vertebral Artery. The vertebral artery, which gives off muscular and spinal branches in the neck, supplies chiefly the posterior part of the brain, directly and, of greater importance, by way of the basilar.

1. The *anterior spinal artery* (p. 541) descends in front of the medulla and unites with the vessel of the opposite side to form a median trunk that contributes to the supply of the medulla and the spinal cord.

2. The *posterior inferior cerebellar artery*,[47] usually the largest branch of the vertebral, may be absent. It winds backward around the olive, between the roots of the hypoglossal nerve, and then behind the roots of the vagus and glossopharyngeal nerves. It gives branches to the medulla, the choroid plexus of the fourth ventricle, and the cerebellum. After a tortuous course it divides on the cerebellum into lateral and medial branches.

The *posterior spinal artery* (p. 542) is usually a branch of the posterior inferior cerebellar artery, but it may come directly from the vertebral. It descends on the side of the medulla and contributes to the supply of the spinal cord.

Basilar Artery

The basilar artery is formed by the union of the right and left vertebral arteries. It begins at approximately the lower border of the pons and ends at the upper border by dividing into the two posterior cerebral arteries (fig. 53–26). In its course it passes through the cisterna pontis and lies frequently in a longitudinal groove on the front of the pons. It is often curved toward one side, however. The basilar artery is so named because of its close relationship to the base of the skull.

Branches of Basilar Artery (fig. 53–24). The branches of the basilar artery are distributed to the pons, the cerebellum, the internal ear, and the cerebral hemispheres.

1. *Pontine branches* supply the pons.

2. The paired *anterior inferior cerebellar arteries*[48] pass backward on the lower surface of the cerebellum and anastomose with the posterior inferior cerebellar arteries of the vertebral. They are closely related to the facial and vestibulocochlear nerves. They supply the cerebellum and the pons.

The paired *labyrinthine (internal auditory)) arteries* may arise from either the basilar or the anterior inferior cerebellar, more commonly the latter. Their position is highly variable.[49] Each enters the corresponding internal acoustic meatus and is distributed to the internal ear.

3. The paired *superior cerebellar arteries* pass laterally below the oculomotor and trochlear nerves, and are distributed to the upper surface of the cerebellum. Anastomoses take place between the various cerebellar arteries.

4. The two *posterior cerebral arteries* are the terminal branches of the basilar. They supply much of the temporal and most of the occipital lobes (fig. 53–20). Each is connected with the corresponding internal carotid by a posterior communicating artery; occasionally the posterior cerebral arises as a branch of the internal carotid (an arrangement referred to as trifurcation of the internal carotid). The posterior cerebral runs backward, above and parallel to the superior cerebellar artery, from which it is separated by the oculomotor and trochlear nerves. The posterior cerebral arteries are distributed chiefly to the inferior and medial surfaces of the temporal and occipital lobes (fig. 53–20). They are surrounded by a sympathetic plexus derived from the vertebral and/or internal carotid plexuses. Among the branches of the posterior cerebral arteries are the *posterior choroid branches,* which supply the choroid plexuses of the third and lateral ventricles.[50]

ANASTOMOSES OF CEREBRAL ARTERIES

Branches of the three cerebral arteries to the cerebral cortex have important anastomoses with one another on the surface of the brain.

The arterial system in the brain, therefore, is not strictly terminal. However, in the event of occlusion, these microscopic anastomoses are not capable of providing a vicarious circulation for the ischemic brain tissue. The anastomoses are "practically always situated in the border area of the three main cerebral arteries, forming a communication between two arteries which originated in two different main cerebral arteries."[51]

CIRCULUS ARTERIOSUS
(fig. 53–21)

The circulus arteriosus[52] is an important polygonal anastomosis between the four arteries that supply the brain: the two verte-

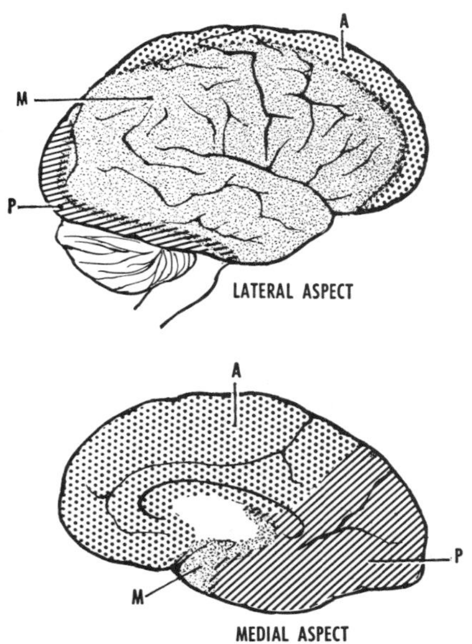

Figure 53–20 Approximate territories supplied by the cerebral arteries. Lateral aspect (*above*) and medial aspect (*below*) of right cerebral hemisphere. *A*, *M*, and *P* stand for the anterior, middle, and posterior cerebral arteries, respectively. Based on Beevor.

brals and the two internal carotids. It is formed by the posterior cerebral, posterior communicating, internal carotid, anterior cerebral, and anterior communicating arteries. **The circulus forms an important means of collateral circulation in the event of obstruction,**[53] but normally there is probably little mingling of the various blood streams[54] except perhaps during movements of the head. Variations in the size of the vessels that constitute the circulus are very common.[55] Absence of one or both posterior communicating arteries is sometimes noted. The anterior communicating artery may be double. The various components of the circulus give numerous minute branches to the brain.

VENOUS DRAINAGE

Veins of Brain

The veins of the brain[56] have thin walls and are devoid of valves. They pierce the arachnoid and the meningeal layer of the dura, and open into the venous sinuses of the dura.

The following veins[57] are not purely cortical but the surfaces of the cerebral hemispheres (fig. 53–14*B*) are drained by them:

1. Superior Cerebral Veins. These drain into the superior sagittal sinus. (The larger veins, situated posteriorly, are directed forward, that is, against the flow of blood in the sinus.)

2. Superficial Middle Cerebral Vein. This follows the lateral sulcus, sends *superior* and *inferior anastomotic veins* to the superior sagittal and transverse sinuses, re-

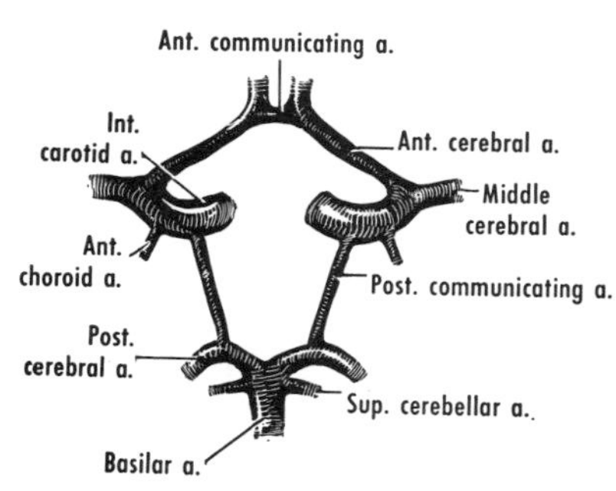

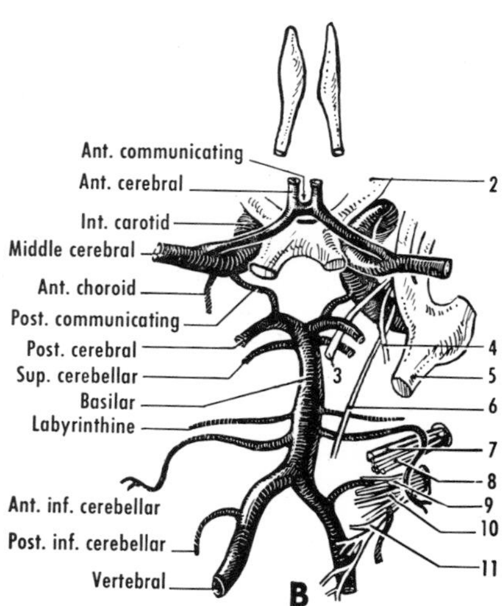

Figure 53–21 The circulus arteriosus and the origins of the cerebral arteries. *A*, the typical circulus arteriosus. Variations are frequent. *B*, the typical relationship of the cranial nerves to the intracranial arteries, superior aspect. The numerals indicate cranial nerves. Based on Padget.

spectively, and ends in the cavernous sinus.

3. Inferior Cerebral Veins. These drain the inferior aspect of the hemispheres and join nearby sinuses.

Basal Vein. The basal vein is formed by the union of several veins, including those that accompany the anterior and middle cerebral arteries. It winds around the cerebral peduncle and ends in the great cerebral vein.

Great Cerebral Vein. The single great cerebral vein (fig. 53–14*C*) is formed between the splenium and the pineal body by the union of two internal cerebral veins. It receives, directly or indirectly, a number of vessels from the interior of the cerebral hemispheres and also the basal veins.[58] It ends in the straight sinus.

The term *transverse fissure* (fig. 53–18) is used for a slit found below the splenium of the corpus callosum.[59] The layer of pia mater immediately on the lower aspect of the corpus callosum ascends on the back of the splenium and proceeds forward on the upper aspect of the corpus callosum. The layer of pia immediately on the roof of the third ventricle proceeds backward and covers the pineal body and the colliculi. The transverse fissure is the extracerebral space between these two layers of pia mater. It contains the internal cerebral veins.

Venous Sinuses of Dura Mater[60] (fig. 53–22)

The blood from the brain drains into sinuses that are situated between the meningeal and endosteal layers of the dura. The sinuses have thin walls, are devoid of valves, and are lined by endothelium continuous with that of the veins. They drain ultimately into the internal jugular veins.

Superior Sagittal Sinus (fig. 53–14*B, C*). The single superior sagittal sinus lies in the convex border of the falx cerebri. It begins in front of the crista galli, where, rarely, it may receive a small vein from the nasal cavity. The sinus runs backward in a groove on the frontal, parietal, and occipital bones. Near the internal occipital protuberance, it usually either enters a common pool with the straight sinus or divides into right and left limbs, which unite with corresponding limbs from the straight sinus and form the right and the left transverse sinus, respectively.[61] The superior sagittal sinus receives the superior cerebral veins and communicates with *lateral lacunae* in the adjacent dura (fig. 53–19). The sinus and the lacunae are partly invaginated by arachnoid granulations. The lacunae receive meningeal and diploic veins, and may be regarded as areas of fusion of these and other vessels.

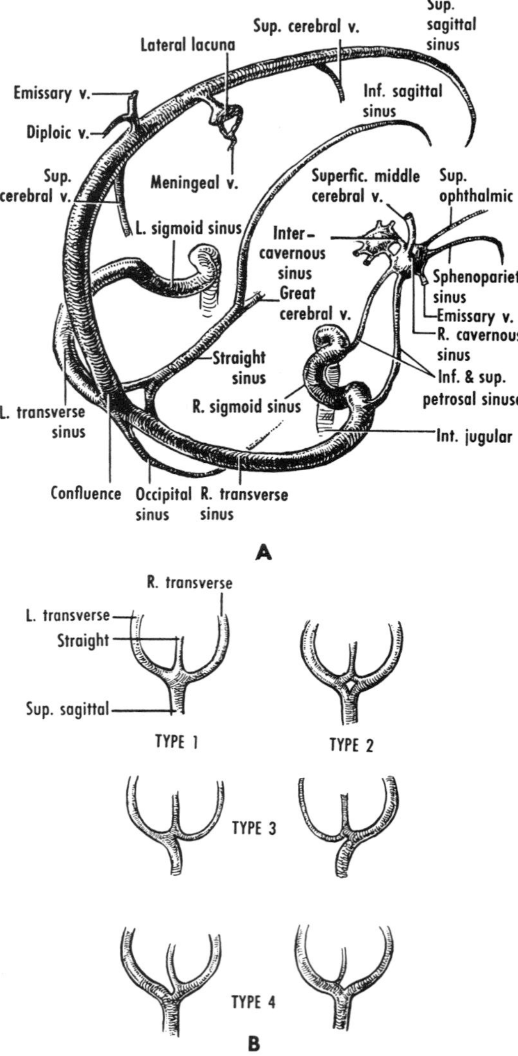

Figure 53–22 The venous sinuses of the dura mater. *A*, right lateral aspect and also from above and behind (as in fig. 53–15 *A*). *B*, schemes showing the chief types of the confluence of the sinuses. In type 1, the superior sagittal and straight sinuses drain into a common pool. In type 2, these two sinuses bifurcate. In type 3, the straight sinus bifurcates. In type 4, the superior sagittal sinus bifurcates. *B* is based on Browning.

SURFACE ANATOMY OF SUPERIOR SAGITTAL SINUS. The superior sagittal sinus extends from above the root of the nose, over the vault of the skull in the median plane, to the external occipital protuberance.

Confluence of Sinuses. The confluence of the sinuses (torcular) is the region where the superior sagittal and straight sinuses end and the right and left transverse sinuses begin. It is situated near the internal occipital protuberance. Several types of confluence occur (fig. 53–22*B*), and the grooves on the internal aspect of the skull are not necessarily an indication of the capacity of the individual sinuses.[62] Dominance, in the sense that one transverse sinus has a greater capacity than that of the opposite side, is usual, and right-sided dominance is more frequent than left. Asymmetries in the occipital region, however, are apparently not related to right- or left-handedness.[63] The occipital sinus (or sinuses) is a variable channel that arises near the margin of the foramen magnum and drains into the confluence.[64]

Inferior Sagittal Sinus. The single inferior sagittal sinus lies in the concave border of the falx cerebri. It ends in the straight sinus.

Straight Sinus. The single straight sinus lies at the junction of the falx cerebri with the tentorium cerebelli.[65] It receives the great cerebral vein and some of the cerebellar veins. It runs backward and downward, and joins the confluence.

Transverse Sinuses. The transverse sinus of each side begins in the confluence at the internal occipital protuberance. Each transverse sinus curves laterally and forward in the convex border of the tentorium cerebelli, and, on reaching the petrous temporal, becomes the sigmoid sinus (fig. 53–14*A*). Emissary veins connect the transverse sinuses with the suboccipital venous plexus (p. 531).

SURFACE ANATOMY OF TRANSVERSE SINUS (fig. 53–8). Each transverse sinus extends from the external occipital protuberance, laterally and with its convexity upward, to the base of the mastoid process. **The transverse sinus marks the boundary between the corresponding cerebral hemisphere and the cerebellum.**

Sigmoid Sinuses. The sigmoid sinus (formerly included as a part of the transverse or lateral sinus) of each side is a continuation of the transverse sinus where the latter leaves the tentorium cerebelli. The sigmoid sinus curves downward and medially in a deep groove on the mastoid part of the temporal bone. In the jugular foramen, it becomes continuous with the superior bulb of the internal jugular vein.

SURFACE ANATOMY OF SIGMOID SINUS (fig. 53–8). Each sigmoid sinus extends from the base of the mastoid process, downward, close to the posterior margin of the process, to within 1 cm of its tip.

Cavernous Sinus (fig. 53–23). The cavernous sinus extends from the superior orbital fissure in front to the apex of the petrous temporal behind. It comprises commonly one or more[66] main venous channels[67] (or, particularly in the newborn, a plexus), which lie in the periosteodural (or endosteomeningeal, fig. 53–23) compartment[68] bounded by the body of the sphenoid bone and the two surfaces of the front portion of the tentorium cerebelli. The compartment opens in front into the superior orbital fissure. **In addition to the main venous channel(s) of the cavernous sinus, the compartment contains the in-**

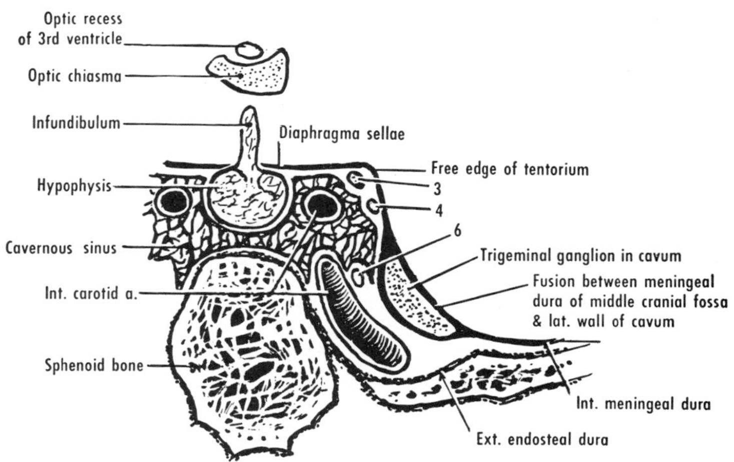

Figure 53–23 A coronal section through the cavernous sinus. Based on Truex and Kellner. See also fig. 53–17. In many instances, however, the trabeculation is negligible, or even non-existent (Bedford[68]).

ternal carotid artery and its sympathetic plexus, and the abducent nerve. Also in the compartment, but situated in the lateral wall of the sinus, are the oculomotor, trochlear, and ophthalmic nerves. These nerves are separated from the blood in the sinus by endothelium. It is believed that the pulsations of the internal carotid artery may aid in expelling blood from the sinus. The maxillary nerve is embedded in the dura lateral to the cavernous sinus.

The front part of the cavum trigeminale is fused with the lower and posterior part of the lateral wall of the sinus.

The cavernous sinus receives the superior ophthalmic vein, the superficial middle cerebral vein, and the sphenoparietal sinus. It communicates with the

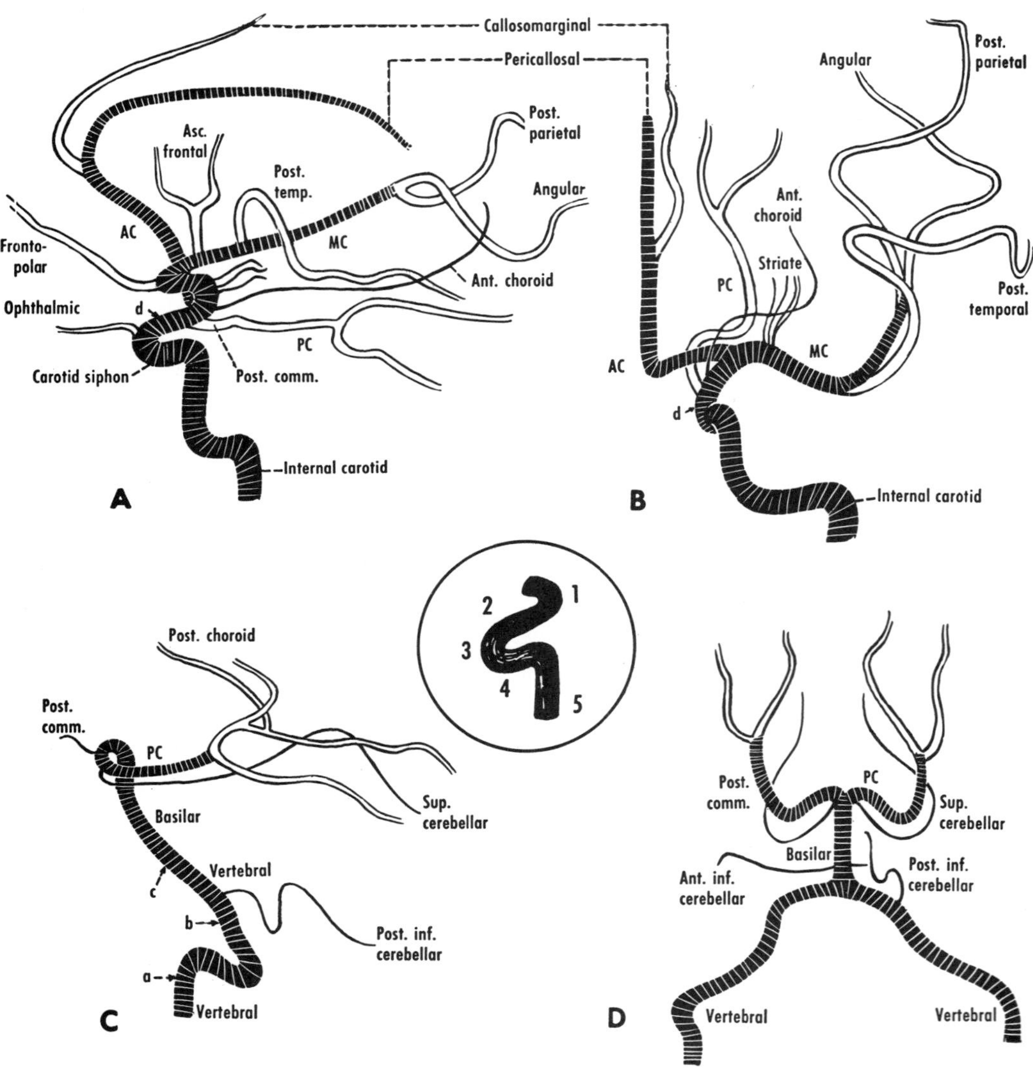

Figure 53–24 The chief arteries of the brain, as seen on angiography. Compare with figures 53–25 and 53–26. The main arteries are shaded. *A* and *B* are carotid arteriograms in lateral and anteroposterior projections, respectively. The internal carotid pierces the dura at *d*. The inset shows, in lateral view, the numbered portions of the internal carotid as described in the text. *C* and *D* are vertebral arteriograms in lateral and half-axial (fronto-occipital) projections, respectively. In *C*, *a* is the site of the foramen transversarium of the atlas, *b* is the foramen magnum, and *c* is at the junction of the right and left vertebral arteries, that is, the beginning of the basilar artery. Abbreviations: *AC*, anterior cerebral; *MC*, middle cerebral; *PC*, posterior cerebral. Based largely on Greitz and Lindgren.

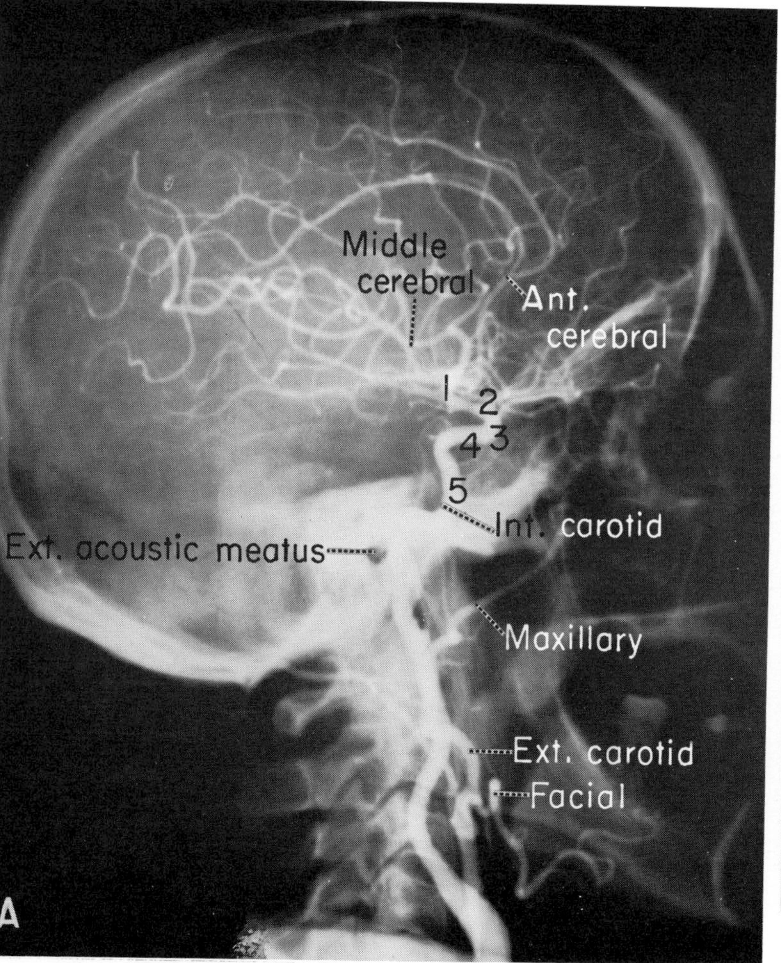

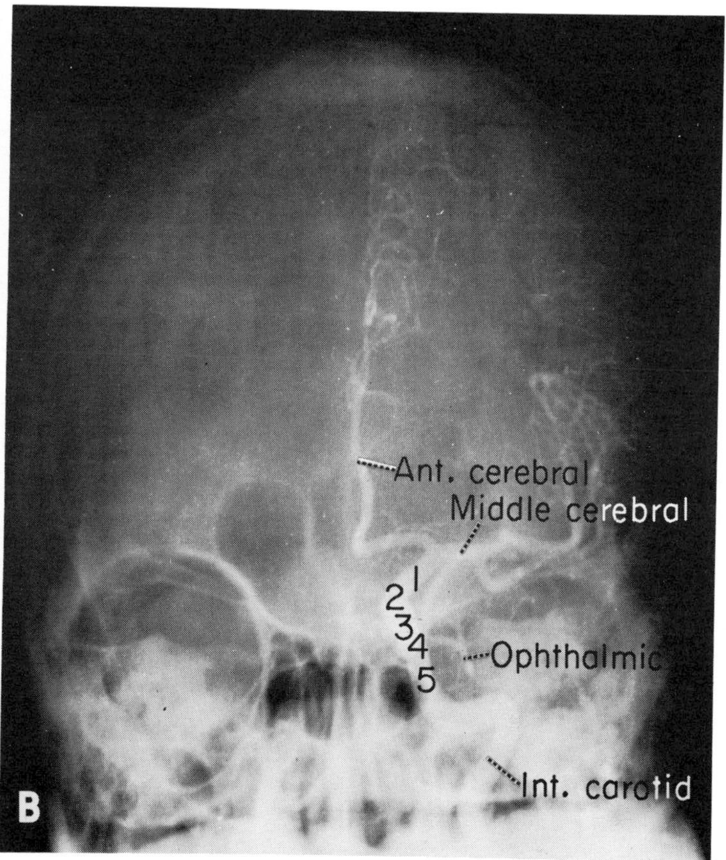

Figure 53-25 Internal carotid arteriograms *in vivo*. *A*, lateral view. *B*, postero-anterior view. The numerals from 1 to 5 refer to portions of the internal artery as described in the text. Parts 2, 3, and 4 constitute the carotid siphon. Courtesy of E. S. Gurdjian, M.D., and J. E. Webster, M.D., Department of Neurosurgery, Wayne State University School of Medicine, Detroit, Michigan.

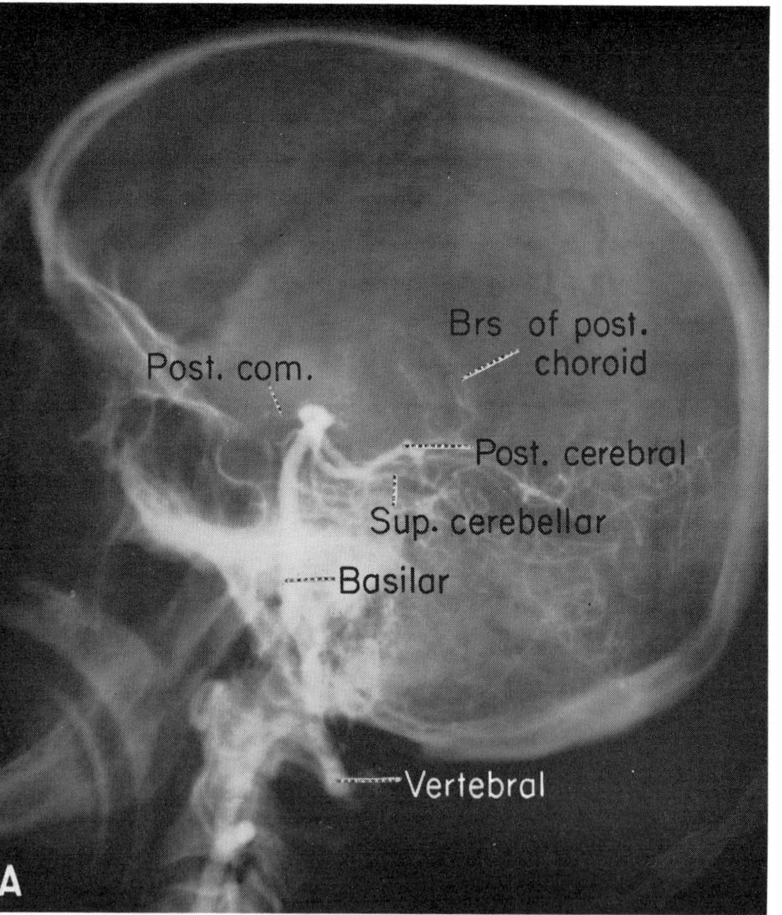

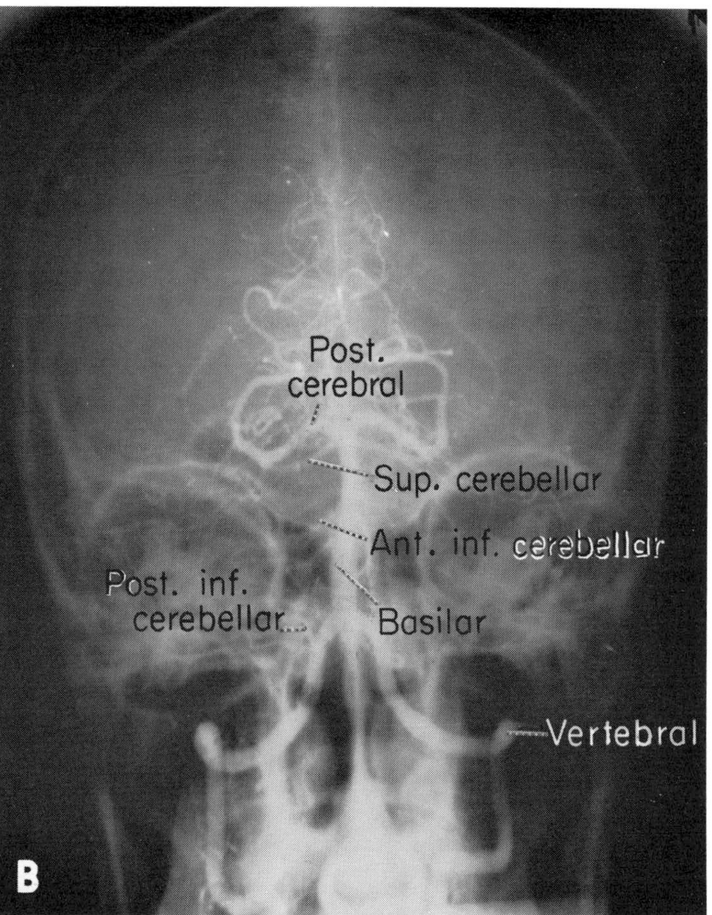

Figure 53–26 Vertebral arteriograms *in vivo*. *A*, lateral view. *B*, postero-anterior view. Courtesy of E. S. Gurdjian, M.D., and J. E. Webster, M.D., Department of Neurosurgery, Wayne State University School of Medicine, Detroit, Michigan.

transverse sinus and the internal jugular vein by way of the superior and inferior petrosal sinuses. It communicates also with the facial vein by way of the superior ophthalmic vein, with the pterygoid plexus by way of emissary veins, and with the opposite cavernous sinus by means of intercavernous sinuses.

Sphenoparietal, Superior Petrosal, and Inferior Petrosal Sinuses. The location of these sinuses can be appreciated from figure 53–22. The right and left inferior petrosal sinuses are connected by a basilar plexus on the dorsum sellae.

Diploic Veins

The diploic veins[70] lie in channels in the diploë of the skull (fig. 53–19). They are dilated at intervals along their course, they have thin walls, and they are devoid of valves. They develop after birth. The largest is situated in the occipital bone. Communications occur between the veins of the scalp, the diploic veins, the meningeal veins, and the venous sinuses of the dura mater.

Emissary Veins

The emissary veins (fig. 53–19) connect the venous sinuses of the dura with the veins of the scalp or with deep veins below the base of the skull. They pass through either special small apertures (e.g., parietal, mastoid, and sphenoidal emissary foramina) or some of the important large openings (e.g., foramen lacerum, hypoglossal canal, carotid canal) of the skull. The direction of blood flow in the emissary veins is normally away from the brain.

CEREBRAL ANGIOGRAPHY
(figs. 53–24 to 53–26)

Cerebral angiography[71] is the radiographic demonstration of the blood vessels of the brain. Practically all the vessels visualized lie on the surfaces of the hemispheres. Moreover, in the case of the brain, the arteries and veins do not run alongside each other. The anterior and middle (and frequently the posterior) cerebral vessels can be rendered radio-opaque by the percutaneous injection of an iodine compound into the common or the internal carotid artery. Carotid arteriography is of value in the diagnosis of cerebral tumors, particularly those in the anterior two-thirds of the hemisphere, because the growing mass may cause displacement of the anterior and/or the middle cerebral artery. It is also of use in the detection of vascular disorders, such as aneurysms. The arteriogram is taken about 2 seconds after the commencement of the injection. About 2 seconds later, the blood has passed to the veins of the brain, and a phlebogram (venogram) may be taken. After about another 2 seconds, the blood has passed to the venous sinuses of the dura mater, and a sinogram may be taken.

Vertebral angiography may be performed by percutaneous injection into the vertebral artery between two cervical transverse processes or between the skull and the atlas. Alternatively, the contrast medium may be instilled by means of a catheter passed through the brachial and into the subclavian artery, or through the femoral and into the aorta.

REFERENCES

1. J. Provost and J. Hardy, Neuro-chirurgie, *16*:459, 1970. K. Gudmundsson, A. L. Rhoton, and J. G. Rushton, J. Neurosurg., *35*:592, 1971.
2. A. E. Coates, J. Anat., Lond., *68*:428, 1934.
3. W. H. Sweet and J. G. Wepsic, J. Neurosurg., *40*:143, 1974.
4. L. P. Rowland and F. A. Mettler, J. comp. Neurol., *89*:21, 1948.
5. G. Lazorthes and J. Poulhès, C. R. Ass. Anat., 35th reunion, 1948.
6. D. H. M. Woollam and J. W. Millen, Brain, *76*:104, 1953. See also G. Flyger and U. Hjelmquist, Anat. Rec., *127*:151, 1957.
7. K. Peter, Z. Anat. EntwGesch., *106*:398, 1937. J. T. Wilson, J. Anat., Lond., *71*:423, 1937. M. L. Barr, Brain, *71*:281, 1948. W. Hewitt, J. Anat., Lond., *94*:549, 1960.
8. L. M. Davidoff and C. G. Dyke, *The Normal Encephalogram*, Lea & Febiger, Philadelphia, 3rd ed., 1951. G. Di Chiro, *An Atlas of Detailed Normal Pneumoencephalographic Anatomy*, Thomas, Springfield, Illinois, 1961.
9. H. Davson, *Physiology of the Cerebrospinal Fluid*, Churchill, London, 1967. T. H. Milhorat, *Hydrocephalus and the Cerebrospinal Fluid*, Williams & Wilkins, Baltimore, 1972.
10. S. Sunderland, J. Anat., Lond., *79*:33, 1945.
11. H. Ferner, Z. Anat. EntwGesch., *121*:407, 1960.
12. At least in the cow (W. H. Boyd, Anat. Rec., *137*:437, 1960).
13. S. Radojević, S. Jovanović, and N. Lotrić, Arch. Anat. path., *17*:274, 1969.
14. D. McK. Rioch, G. B. Wislocki, and J. L. O'Leary, Res. Publ. Ass. nerv. ment. Dis., *20*:3, 1940.
15. F. Morin, Arch. ital. Anat. Embriol., *45*:94, 1940. G. P. Xuereb, M. M. L. Prichard, and P. M. Daniel, Quart, J. exp. Physiol., *39*:199 and 219, 1954. J. P. Stanfield, J. Anat., Lond., *94*:257, 1960.
16. G. W. Harris, *Neural Control of the Pituitary Gland*, Arnold, London, 1955.
17. A. E. Walker, Anat. Rec., *55*:291, 1933.
18. L. C. Rogers and E. E. Payne, J. Anat., Lond., *95*:586, 1961.
19. S. Sunderland, Brit. J. Surg., *45*:422, 1958. H. F. Plaut, *Vertebral and Carotid Angiograms in Tentorial Herniations*, Thomas, Springfield, Illinois, 1961.
20. H. S. Burr and G. B. Robinson, Anat. Rec., *29*:269, 1925. R. D. Lockhart, J. Anat., Lond., *62*:105, 1927. L. Schwadron and B. C. Moffett, Anat. Rec., *106*:131, 1950.
21. M. Hasan and A. C. Das, Acta anat., *74*:624, 1969.
22. G. E. de Schweinitz, Trans. ophthal. Soc. U.K., *43*:12, 1923.
23. J. Grzybowski, Arch. Anat., Strasbourg, *14*:387, 1932. B. X. Ray and H. G. Wolff, Arch. Surg., Chicago, *41*:813, 1940. W. Penfield and F. McNaughton, Arch. Neurol. Psychiat., Chicago, *44*:43, 1940.
24. D. L. Kimmel, Neurol., *11*:800, 1961, and Chicago med. Sch. Quart., *22*:16, 1961.
25. W. Feindel, W. Penfield, and F. McNaughton, Neurol., *10*:555, 1960.
26. S. B. Chandler and C. F. Derezinski, Anat. Rec., 62:309, 1935.

27. G. Galli and E. Reggiani, Boll. Soc. med.-chir. Modena, *58*:75, 1958.
28. L. A. Gillilan, Arch. Ophthal., N.Y., *65*:684, 1961.
29. F. W. Jones, J. Anat., Lond., *46*:228, 1912.
30. J. Browder, A. Browder, and H. A. Kaplan, Anat. Rec., *176*:329, 1973.
31. A. Alvarez-Morujo, Acta anat., *74*:10, 1969.
32. C. E. Locke, and H. C. Naffziger, Arch. Neurol. Psychiat., Chicago, *12*:411, 1924. B. Liliequist, Acta radiol., *46*:61, 1956. J. Lang, Acta anat., *86*:267, 1973.
33. W. E. LeGros Clark, J. Anat., Lond., *55*:40, 1920. F. Kiss and J. Sattler, Anat. Anz., *103*:273, 1956.
34. L. Bakay, *The Blood-Brain Barrier,* Thomas, Springfield, Illinois, 1956.
35. W. Platzer, Fortschr. Röntgenstr., *84*:200, 1956.
36. H. A. Shenkin, M. H. Harmel, and S. S. Kety, Arch. Neurol. Psychiat., Chicago, *60*:240, 1948.
37. W. B. Hamby, *Intracranial Aneurysms,* Thomas, Springfield, Illinois, 1952.
38. L. H. Herman, O. U. Fernando, and E. S. Gurdjian, Anat. Rec., *154*:95, 1966. J. Furlani, Acta anat., *85*:108, 1973.
39. D. S. Ahmed and R. H. Ahmed, Anat. Rec., *157*:699, 1967. W. Firbas and H. Sinzinger, Acta anat., *83*:81, 1972.
40. A. G. Baptista, Neurol., *13*:825, 1963.
41. R. Dufour *et al.,* C. R. Ass. Anat., *44*:258, 1958.
42. A. A. Morris and C. M. Peck, Amer. J. Roentgenol., *74*:818, 1955.
43. B. A. Ring, Acta radiol., Stockh., *57*:289, 1962. K. K. Jain, Canad. J. Surg., *7*:134, 1964.
44. C. E. Beevor, Phil. Trans., *200B*:1, 1909.
45. D. Parkinson, Canad. J. Surg., *7*:251, 1964. C. Manelfe, M. Tremoulet, and J. Roulleau, Neuro-chir., *18*:581, 1972.
46. J. S. B. Stopford, J. Anat., Lond., *50*:131 and 255, 1916.
47. B. S. Wolf, C. M. Newman, and M. T. Khilnani, Amer. J. Roentgenol., *87*:322, 1962.
48. W. J. Atkinson, J. Neurol. Psychiat., *12*:137, 1949.
49. J. C. Watt and A. N. McKellop, Arch. Surg., Chicago, *30*:336, 1935.
50. J. R. Galloway and T. Greitz, Acta radiol., Stockh., *53*:353, 1960.
51. H. M. Vander Eecken, *Anastomoses between the Leptomeningeal Arteries of the Brain,* Thomas, Springfield, Illinois, 1959. See also L. A. Gillilan, J. comp. Neurol., *112*:55, 1959.
52. D. H. Padget, in W. E. Dandy, *Intracranial Arterial Aneurysms,* Comstock, Ithaca, New York, 1944. A. G. Baptista, Acta neurol. scand., *40*:398, 1964.
53. C. J. Dickinson, Brit. med. J., *1*:858, 1961.
54. L. Rogers, Brain, *70*:171, 1947. R. A. Kuhn, J. Amer. med. Ass., *175*:769, 1961.
55. G. Lazorthes and A. Gouazé, C. R. Ass. Anat., *53*:1, 1968, and *55*:826, 1970.
56. C. Johansson, Acta radiol., Stockh., Suppl. 107, 1954. O. Hassler, Neurol., *16*:505, 1966.
57. G. Di Chiro, Amer. J. Roentgenol., *87*:308, 1962.
58. M. Banna and J. R. Young, Brit. J. Radiol., *43*:126, 1970.
59. A. Bouchet, A. Goutelle, and G. Fischer, Arch. Anat., Strasbourg, *49*:453, 1966.
60. J. G. Waltner, Arch. Otolaryng., Chicago, *39*:307, 1944.
61. H. Browning, Amer. J. Anat., *93*:307, 1953. A. Elmohamed and K.-J. Hempel, Frankfurt. Z. Path., *75*:321, 1966.
62. B. Woodhall, Laryngoscope, St Louis, *49*:966, 1939.
63. W. E. LeGros Clark, Man, *34*:35, 1934.
64. A. C. Das and M. Hasan, J. Neurosurg., *33*:307, 1970.
65. G. Leutert and E. Lais, Anat. Anz., *120*:18, 1967.
66. H. Viallefont, R. Paleirac, and C. Boudet, Bull. Soc. franç. Ophtal., *72*:208, 1959.
67. H. T. Green, Amer. J. Anat., *100*:435, 1957.
68. M. A. Bedford, Brit. J. Ophthal., *50*:41, 1966. R. Mercier *et al.,* C. R. Ass. Anat., *55*:877, 1970. P. Patouillard and G. Vanneuville, Neuro-chir., *18*:551, 1972.
69. W. R. Henderson, J. Anat., Lond., *100*:95, 1966.
70. G. Jefferson and D. Stewart, Brit. J. Surg., *16*:70, 1928.
71. A. Ecker, *The Normal Cerebral Angiogram,* Thomas, Springfield, Illinois, 1951. T. Greitz and E. Lindgren, in H. L. Abrams (ed.), *Angiography,* Little, Brown, Boston, vol. 1, 1961.

GENERAL READING

Blinkov, S. M., and Glezer, I. I. (trans. by B. Haigh), *The Human Brain in Figures and Tables,* Basic Books, New York, 1968.

Bossy, J., *Atlas du système nerveux. Aspects macroscopiques de l'encéphale,* Éditions Offidoc, Paris, 1972.

Brodal, A., *The Cranial Nerves,* Blackwell, Oxford, 2nd ed., 1965.

Davson, H., *Physiology of the Cerebrospinal Fluid,* Churchill, London, 1967.

Ford, D. H., and Schadé, J. P., *Atlas of the Human Brain,* Elsevier, Amsterdam, 2nd ed., 1971. This is one of many good atlases of the human brain.

Gardner, E., *Fundamentals of Neurology,* Saunders, Philadelphia, 6th ed., 1975. An up-to-date introduction to the nervous system.

Kaplan, H. A., and Ford, D. H., *The Brain Vascular System,* Elsevier, Amsterdam, 1966.

Lazorthes, G., *Vascularisation et circulation cérébrales,* Masson, Paris, 1961.

Purves, M. J., *The Physiology of the Cerebral Circulation,* Cambridge University Press, London, 1972.

de Ribet, R.-M., *Les nerfs crâniens,* Doin, Paris, 1952.

Stephens, R. B., and Stilwell, D. L., *Arteries and Veins of the Human Brain,* Thomas, Springfield, Illinois, 1969.

THE EAR

54

The ears are vestibulocochlear organs, that is, they are concerned with equilibration and hearing.

The word *ear* is akin to the Latin *auris;* hence aural means pertaining to the ear. The Greek word for ear is *ous, otos,* and hence the study of the ear and its diseases is termed otology. The Latin word *audire* means to hear; hence auditory means pertaining to the sense of hearing. Acoustic is derived from a corresponding Greek word.

Each ear comprises three portions (fig. 54–1): external, middle, and internal. The external ear consists of the auricle and the external acoustic meatus. The middle ear, or tympanic cavity, is an air space in which the auditory ossicles are located. The internal ear comprises a series of complicated, fluid-filled spaces known as the labyrinth. Both the middle and the internal components of the ear are enclosed in the temporal bone. The temporal bone, which has already been described (pp. 559, 563, 573, and 575), should be reviewed in conjunction with the study of the ear.

EXTERNAL EAR

The external ear conducts sound toward the middle and internal components of the ear and protects those portions from outside damage.

The auricle is described in chapter 57.

External Acoustic Meatus

The external acoustic (auditory) meatus, about 25 mm or more in length, extends from the concha to the tympanic membrane. The lateral part of the canal is largely cartilaginous; the longer medial portion is bony. The cartilage of the meatus is continuous with that of the auricle. **The cartilaginous part of the meatus is slightly concave anteriorly; a speculum may be inserted more readily, therefore, when the auricle is pulled backward.** The meatus is lined by the skin of the auricle, and, in its cartilaginous part, it presents hair follicles, and sebaceous and ceruminous glands. Ceruminous glands[1] are simple, coiled tubules that resemble the apocrine sweat glands of the axilla. Cerumen, or ear wax, is a mixture of the secretions of sebaceous and ceruminous glands. The skin of the meatus is closely adherent to the perichondrium or the periosteum, a fact that accounts for the severe pain of a furuncle within the meatus. The meatus is related in front and below to the parotid gland, above and behind to the epitympanic recess and the mastoid air cells. The meatus is slightly constricted (1) at the junction of its cartilaginous and bony parts, and (2) near its medial end or *isthmus.* The meatus serves for protection and as a pressure amplifier.

Sensory Innervation and Blood Supply of External Ear

The sensory innervation of the external ear (fig. 57–6) is chiefly from the auriculotemporal nerve (5th cranial) and the cervical plexus (great auricular nerve, C2, 3). The region of the concha probably receives contributions also from cranial nerves 7, 9, and 10. The blood supply of the external ear is mainly from the posterior auricular and superficial temporal arteries (of the external carotid).

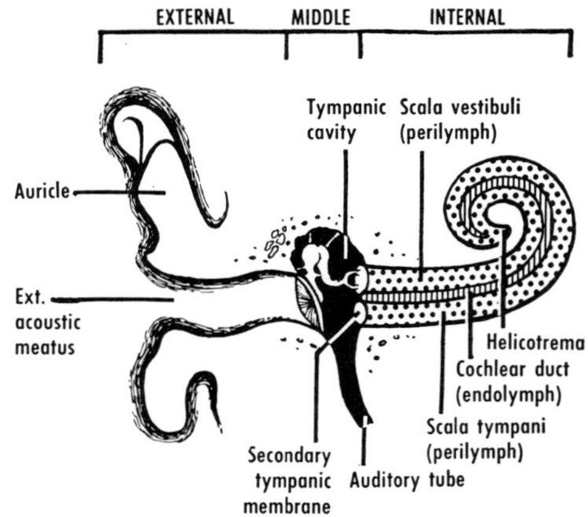

Figure 54–1 Scheme showing basic structure of the ear. The tympanic membrane and the auditory ossicles do not have labels. Note the base of the stapes closing the fenestra vestibuli, and the secondary tympanic membrane closing the fenestra cochleae.

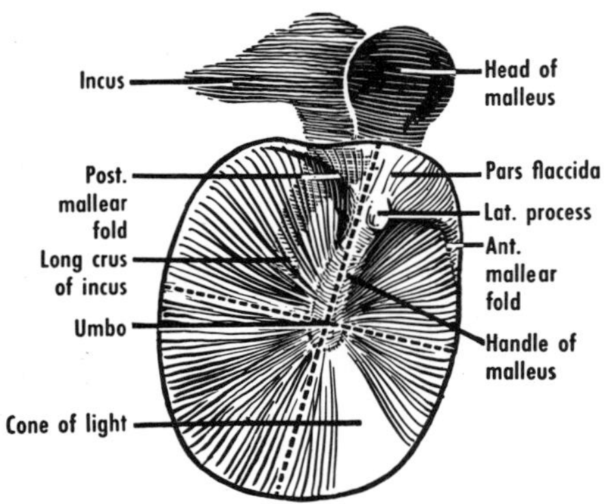

Figure 54–2 Right tympanic membrane. Lateral aspect. An imaginary line down the handle of the malleus and one at a right angle to it are used to delimit quadrants on the tympanic membrane. The umbo is approximately opposite the promontory of the middle ear. The head of the malleus and the body and short crus of the incus are in the epitympanic recess, above the level of the tympanic membrane. Compare figure 63–12*A*, p. 754.

Tympanic Membrane

The *tympanic membrane,* or ear drum (fig. 54–2), about 1 cm in diameter, forms a partition between the external acoustic meatus and the tympanic cavity. Its lateral surface is covered by epidermis, its medial surface by the mucous membrane of the middle ear. The middle fibrous basis[2] of the membrane, except anterosuperiorly, is attached to the tympanic plate of the temporal bone. This larger portion of the membrane is termed its *tense part.* The fibrous basis, however, is thinner in the anterosuperior portion of the membrane, and this part, limited by *anterior* and *posterior mallear folds,* is termed the *flaccid part.*

The tympanic membrane is placed very obliquely in the meatus (fig. 54–1). Its lateral surface is concave, and the center of the concavity is called the *umbo.* **The handle and the lateral process of the malleus are attached to the medial surface of the tympanic membrane, and this surface is related closely to the chorda tympani. Incisions through the membrane are usually made in its postero-inferior quadrant, thereby avoiding the ossicles and the chorda tympani.** The tympanic membrane is highly sensitive (cranial nerves 5 and 10 supply its lateral surface; 9 supplies its medial surface).

The tympanic membrane faces laterally, forward, and downward, "as though to catch sounds reflected from the ground as one advances" (J. C. B. Grant); this mnemonic enables one to remember that the anterior and inferior walls of the meatus are longer than the posterior and superior.

Examination of External Ear (Otoscopy)

The external acoustic meatus and the tympanic membrane can be examined *in vivo* by means of a speculum, termed an otoscope or auriscope.

The auricle is pulled backward and upward in order to straighten the cartilaginous part of the meatus. The normal tympanic membrane has a pearl-gray color and reflects a "cone of light" in its antero-inferior quadrant (fig. 63–12*A*, p. 754). Certain structures of the middle ear can generally be identified through the tympanic membrane: handle and lateral process of malleus, anterior and posterior mallear folds, occasionally long process of incus. **The deep relations of the tympanic membrane are as follows: Antero-inferior quadrant: carotid canal. Anterosuperior quadrant: tympanic opening of auditory tube. Posterosuperior quadrant: long process of incus, stapes, and fenestra vestibuli. Postero-inferior quadrant: promontory and fenestra cochleae.**

MIDDLE EAR

The middle ear consists largely of an air space, the *tympanic cavity* (fig. 54–3), in the temporal bone. The cavity contains the auditory ossicles (fig. 54–4) and is in communication with (1) the mastoid air cells and the mastoid antrum by means of the aditus, and (2) the nasopharynx by means of the auditory tube. The mucous membrane of the middle ear covers the structures in the tympanic cavity. It presents a cuboidal epithelium which, in the cartilaginous part of the auditory tube,

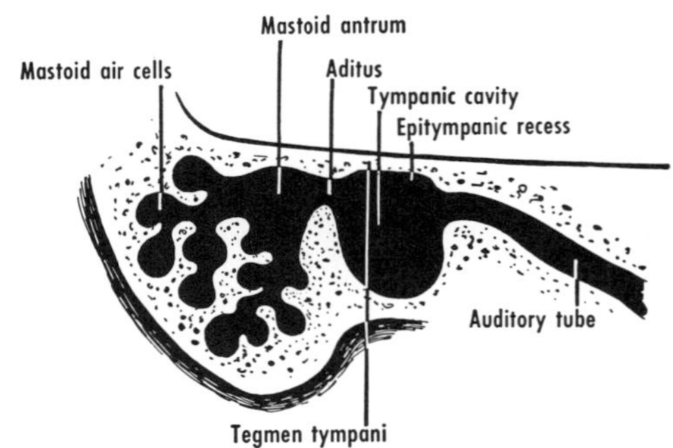

Figure 54–3 Scheme of the tympanic cavity and the air spaces with which it communicates. Right lateral aspect.

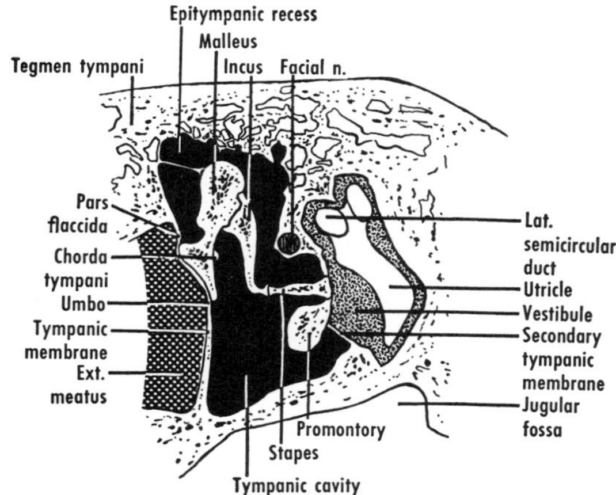

Figure 54–4 Schematic vertical section through the ear, based on several sections cut at a right angle to the long axis of the petrous part of the temporal bone. Based on Wolff, Bellucci, and Eggston.

becomes respiratory in type, that is, pseudostratified ciliated columnar. The tympanic cavity and the auditory tube develop as a recess of the embryonic pharynx.

Boundaries (fig. 54–5)

Lateral or Membranous Wall. The tympanic cavity is bounded laterally by the tympanic membrane. However, **a portion of the tympanic cavity is situated above the level of the membrane and is known as the attic, or *epitympanic recess.*** The main portion is sometimes called the *mesotympanum* and the lowest portion the *hypotympanic recess.* **The epitympanic recess contains the head of the malleus and the body and short crus of the incus. It communicates with the aditus.** A small pouch of the tympanic cavity lies between the flaccid part of the membrane and the neck of the malleus. The lateral and medial walls of the tympanic cavity are only 2 to 6 mm apart.

Roof or Tegmental Wall. The tympanic cavity is limited above by a portion of the petrous temporal, termed the *tegmen tympani.* The tegmen intervenes between the middle ear and the middle cranial fossa.

Floor or Jugular Wall. The middle ear is limited below by the *jugular fossa,* in which is lodged the superior bulb of the internal jugular vein.

Anterior or Carotid Wall. Anteriorly, the tympanic cavity communicates with the *semicanal for the tensor tympani muscle* (which runs backward on the medial wall) and, below this, with the nasopharynx by means of the auditory tube. The tube is described with the pharynx (p. 744). Below the tubal opening, the middle ear is related in front to the *carotid canal,* in which is lodged the internal carotid artery.

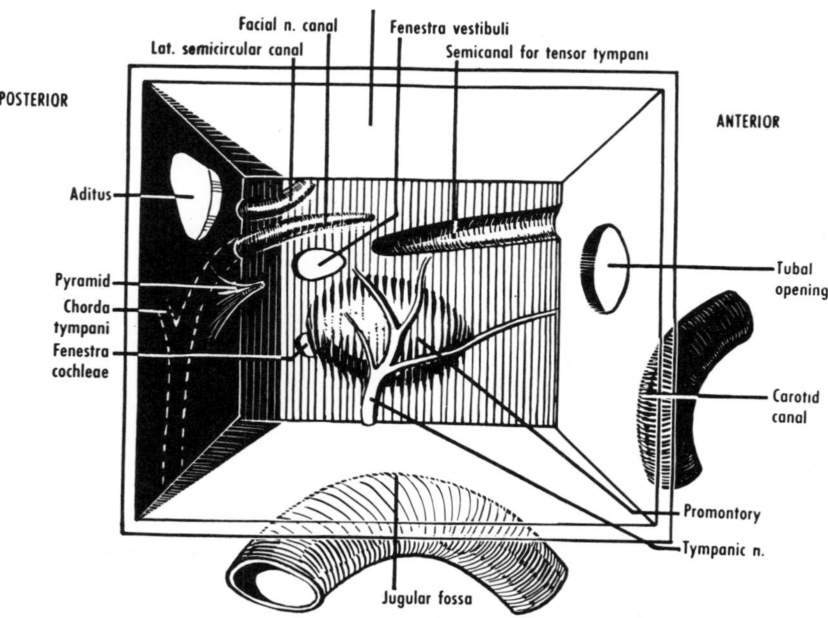

Figure 54–5 Highly schematic view of the medial wall of the right tympanic cavity, lateral aspect. Note the tympanic plexus on the promontory. Based on Maisonnet and Coudane.

Posterior or Mastoid Wall. Posteriorly, the tympanic cavity communicates with the *mastoid antrum* by means of the *aditus ad antrum.* Below the aditus is a small projection, the *pyramidal eminence (pyramid),* which contains the stapedius muscle.[3]

The mastoid portion of the temporal bone presents a variable arrangement of air cells in the adult. In most instances the greater part of the mastoid process is hollowed out (pneumatic mastoid), but in some cases a considerable amount of compact bone (diploic mastoid) may be present. The air cells communicate with one another and with the mastoid antrum, and are lined by a mucoperiosteum that includes a layer of squamous epithelium but no glands. Various groups of air cells have been described.[4] Pneumatization may involve other sites also, for instance, the zygomatic arch and the apex of the petrous temporal.

Medial or Labyrinthine Wall (fig. 54–5). The medial wall presents the following features from above downward: (1) the *prominence of the lateral semicircular canal* and the *prominence of the facial nerve canal;* (2) the *fenestra vestibuli (oval window),* closed by the base of the stapes, and the *processus cochleariformis,* a bony prominence above the window; (3) the *promontory,* formed by the basal turn of the cochlea and covered by the tympanic plexus; (4) the *fenestra cochleae (round window),* closed by the mucous membrane of the middle ear.

The *tympanic plexus*[5] of nerves lies on the promontory. It is formed chiefly by the tympanic nerve (from the 9th cranial), which gives sensory fibers to the middle ear and secretomotor fibers to the parotid gland. The latter emerge as the lesser petrosal nerve and synapse in the otic ganglion (fig. 58–10).

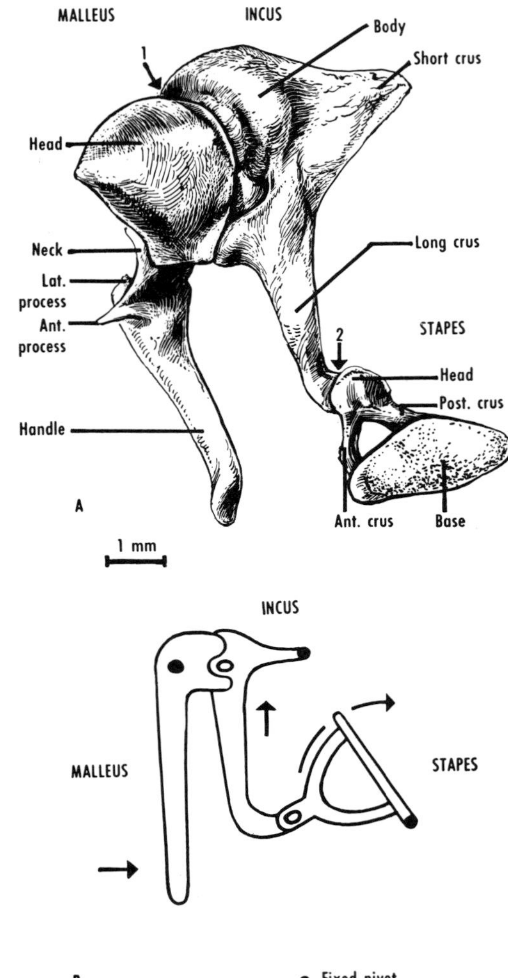

Figure 54–6 The ossicles of the right ear. Medial aspect. *A,* drawing showing the processes and the articulations (*1,* incudomallear; *2,* incudostapedial). *B,* diagram showing the lever-like action of the ossicles. For purposes of the illustration, the movements (indicated by arrows) have been shown as though they took place in one plane. The fixed pivots are anchored to the bony walls of the tympanic cavity. *A* is based on a photograph in Bassett's Atlas, section 2, 1954, reel 62, view 7. *B* is based on Guelke and Keen.

Auditory Ossicles, Joints, and Muscles

Ossicles. The auditory ossicles (fig. 54–6) are three small[6] bones termed the malleus (hammer), incus (anvil), and stapes (stirrup), respectively. The mucous membrane of the middle ear is wrapped around them.

MALLEUS. The malleus presents a head and neck, a handle, and two processes. The *handle,* or *manubrium,* and the *lateral process* are embedded in the fibrous layer of the tympanic membrane, whereas the *head* lies in the epitympanic recess. The *anterior process* is connected by an *anterior ligament* to the petrotympanic fissure and thence to the sphenomandibular ligament.

INCUS. The incus presents a body and two crura or processes. The *body* and the *short crus* lie in the epitympanic recess. The *long crus* lies parallel to and behind the handle of the malleus. Its lower end *(lenticular process)* projects medially to articulate with the stapes.

STAPES. The stapes presents a head and a base, united together by two limbs or crura. The *base* or footplate is attached by an annular ligament to the margin of the fenestra vestibuli. Abnormal ossification between the base of the stapes and the margin of the fenestra (otosclerosis, p. 627) causes deafness.

Joints. The *incudomallear* and *incudostapedial joints* are synovial in type: the former, a saddle joint; the latter, a ball-and-socket. The joints possess elastic capsules[7] and a complicated ligamentous apparatus, and they may have intra-articular cartilages.[8] The stapediovestibular junction is either a fibrous *(tympanostapedial syndesmosis)* or a synovial joint.[9] The chain of auditory ossicles acts as a system of levers (fig. 54–6*B*).

Muscles. The *tensor tympani* arises chiefly from the cartilaginous portion of the auditory tube. It enters a semicanal above the tube. Its tendon turns laterally around the processus cochleariformis and is inserted on the handle of the malleus. The muscle is supplied by the mandibular nerve[10] and by the tympanic plexus[11] (from the 5th cranial) and perhaps by autonomic fibers. It draws the handle of the malleus medially and thereby tightens the tympanic membrane; hence its name.

The *stapedius* is a small muscle located in the pyramidal eminence on the posterior wall of the tympanic cavity. Its tendon passes through an opening at the apex of the eminence and is inserted on the neck of the stapes (hence its name) and frequently into the incudostapedial capsule.[12] The muscle is supplied by the 7th cranial nerve. It draws the stapes laterally and perhaps rotates the incus.

Both the tensor tympani and the stapedius attenuate sound transmission through the middle ear. They also take part in non-acoustic activities, for example, facial movements, yawning, swallowing.[13]

Sensory Innervation and Blood Supply of Middle Ear

The sensory innervation is by means of the auriculotemporal (5th cranial) and tympanic (9th cranial) nerves, and the auricular branch of the vagus. The chief blood supply to the middle ear is from the external carotid (stylomastoid artery from posterior auricular artery) and the maxillary (anterior tympanic artery). Many small vessels contribute.[14]

FUNCTIONAL CONSIDERATIONS

The functioning of the middle ear is not well understood. The tympanic membrane, which seals the air-filled tympanic cavity, is set in motion by the sound waves that impinge on it. The vibrations of the membrane are converted into intensified movements of the stapes by means of the lever-like action of the auditory ossicles. The chain of ossicles acts as a mechanical transformer, and delivers the signal to the fenestra vestibuli, leaving the fenestra cochleae free to execute compensatory movements in the opposite direction.[15] The main axis of rotation of the ossicles probably passes between (1) a point below the anterior process of the malleus and (2) the short process of the incus. Actually the ossicles appear to have two movable pivots (the incudomallear and incudostapedial joints) and three fixed pivots (anterior process of malleus, short process of incus, and inferior and posterior edges of stapes) (fig. 54–6*B*).[16] In normal use, however, the malleus and the incus "vibrate as a rigid body."[17] The movement of the stapes in the fenestra vestibuli has been compared to tapping the foot while the heel rests on the ground.

Sound vibrations are transmitted to the inner ear (1) by the auditory ossicles and the fenestra vestibuli, (2) by the air in the tympanic cavity and by the fenestra cochleae, and (3) as a result of bone conduction, that is, through the bones of the skull. The first method normally predominates in importance. Defects in conduction through the ossicles result in partial deafness. Difficulty in hearing air-borne sound can be corrected by a hearing aid.

FACIAL NERVE

The facial or 7th cranial nerve (fig. 54–7) has a complicated course in the temporal bone, in which it gives off many of its branches. It is described here because of its close topographical relationship to the middle ear.

The facial nerve consists of a larger part, which supplies the muscles of facial expression,

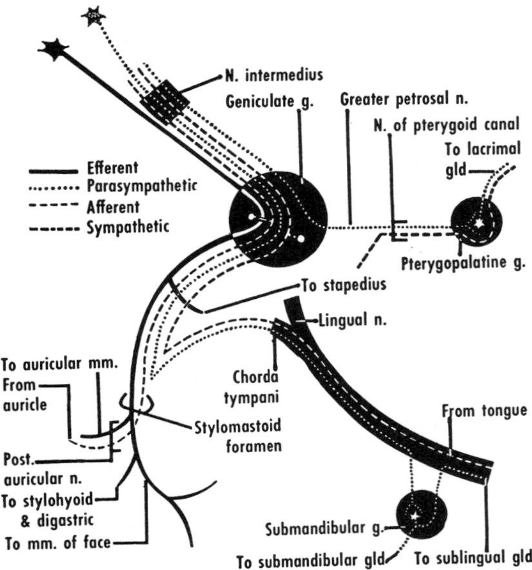

Figure 54–7 Scheme of the facial nerve and its components.

and a smaller part, termed the *nervus intermedius*,[18] which contains taste fibers for the anterior two-thirds of the tongue, secretomotor fibers for the lacrimal and salivary glands, and some pain fibers. The two portions of the facial nerve leave the brain at the lower border of the pons (in the cerebellopontine angle) and, together with the vestibulocochlear nerve, enter the internal acoustic meatus. The facial nerve proceeds laterally in the meatus and then enters the *facial canal* in the temporal bone.[19] Above the promontory on the medial wall of the middle ear, the nerve is expanded to form the *geniculate (facial) ganglion*[20] (fig. 54–9), which contains the cells of origin of its taste fibers. The facial nerve next turns sharply backward, the bend being termed the *geniculum (genu)*. The nerve then sweeps downward behind the middle ear (fig. 54–5) and emerges from the skull at the stylomastoid foramen. In infants the mastoid process is poorly developed, and hence the facial nerve is very superficial at its exit from the skull. Finally, the facial nerve enters the parotid gland, forms the parotid plexus, and gives rise to its terminal branches for the facial muscles (fig. 57–5A).

In its course, the facial nerve traverses in succession (1) the posterior cranial fossa, (2) the internal acoustic meatus, (3) the facial canal in the temporal bone, and (4) the parotid gland and the face.

Surface Anatomy

The main point of division of the facial nerve in the parotid gland lies 1/2 cm behind the ramus of the mandible and about 3 cm above the angle of that bone.[21]

Branches of Facial Nerve

In the internal acoustic meatus the facial communicates with the vestibulocochlear nerve.

In the facial canal the geniculate ganglion gives rise to several branches:

1. Greater Petrosal Nerve. The greater petrosal nerve passes forward in a groove toward the foramen lacerum. There it is joined by the deep petrosal nerve (from the sympathetic plexus on the internal carotid artery) to form the *nerve of the pterygoid canal*, which runs forward in the pterygoid canal and reaches the pterygopalatine ganglion. **The greater petrosal nerve contains secretomotor fibers for the lacrimal and nasal glands,** and perhaps vasodilator fibers for the middle meningeal arteries. It also contains a number of afferent fibers (the cells of origin being in the geniculate ganglion); their distribution and function are uncertain, but some of the fibers may subserve general sensation from the nasal mucosa. Others are believed by some workers to be taste fibers from the anterior two-thirds of the tongue and from the soft palate.

2. Communicating Branch. This joins the lesser petrosal nerve (p. 697).[22]

3. External Petrosal Nerve. This inconstant twig may join the sympathetic plexus on the middle meningeal artery.

Also in the facial canal, but arising from the facial nerve as it descends, several branches originate:

4. Nerve to the Stapedius. This supplies the stapedius muscle.

5. Chorda Tympani. The chorda tympani enters the tympanic cavity (although covered by a reflection of the mucous membrane), passes medial to the tympanic membrane and the handle of the malleus (i.e., between the malleus and the incus), and again enters the temporal bone. It leaves the skull through the petrotympanic fissure and descends in the infratemporal fossa. Medial to the lateral pterygoid muscle, **it joins the lingual nerve, with which it is distributed to the anterior two-thirds of the side and dorsum of the tongue.** The chorda tympani contains *(a)* fibers associated with taste[24] from the anterior two-thirds of the tongue and from the soft palate (the cell bodies being in the geniculate ganglion); and *(b)* preganglionic secretory and vasodilator fibers, which synapse in the submandibular ganglion, the postganglionic fibers then supplying the submandibular, sublingual and lingual glands. Below the base of the skull the chorda tympani communicates with the otic ganglion. It is possible that by this means the facial nerve sends secretory fibers to the parotid gland.[25]

Just below the base of the skull the facial nerve gives off several branches:

6. *Muscular Branches.* These pass to the stylohyoid and the posterior belly of the digastric.

7. ***Communicating Branches.*** These pass to the 9th and 10th cranial nerves, the auricular branch of 10, and the auriculotemporal, great auricular, and lesser occipital nerves.

8. *Posterior Auricular Nerve.* This nerve accompanies the posterior auricular artery and supplies the muscles of the auricle (the anterior and superior auricular muscles, however, are innervated by temporal branches of the facial nerve), together with the occipitalis. It also sends sensory fibers to the auricle.

9. *Terminal Branches.* In the parotid gland (p. 660), the facial nerve usually divides into two chief trunks (temporofacial and cervicofacial), the branches of which anastomose with each other in a variable manner[26] to form the *parotid plexus (pes anserinus).* The terminal branches emerge from the gland under cover of its lateral surface and radiate forward in the face, communicating with the terminal branches of the trigeminal nerve. They supply the anterior and superior auricular muscles, the frontalis, orbicularis oculi, buccinator, orbicularis oris, and other muscles of facial expression, including the platysma.

The terminal branches are variable in their arrangement but are commonly classified as *temporal, zygomatic* (which unite with the infra-orbital nerve to form a plexus), *buccal* (supplying the buccinator and other muscles of the mouth), *marginal branch of the mandible,* and *cervical* (lying deep to, and supplying, the platysma). Numerous anastomoses occur between these branches.

The branches of the facial nerve are summarized in table 54–1.

The terminal branches of the facial nerve contain afferent as well as motor fibers.[27] The afferent fibers are thought to be proprioceptive from the muscles of facial expression and/or concerned with deep pain in the skin, muscles, and bones of the face.

Examination

The facial nerve is tested chiefly in regard to the facial muscles. The subject is asked in turn to show his teeth, puff out his cheeks, whistle, wrinkle his forehead by looking upward, frown, and close his eyes tightly.

Lacrimation is tested by irritating the nasal mucosa with ammonia fumes. Sensitivity to sounds is ascertained by an instrument termed an audiometer. Taste on the anterior two-thirds of the tongue is tested by placing sweet, salt, sour, and bitter substances (e.g., sugar, salt, vinegar, and quinine) in turn on one-half of the protruded tongue; the subject is asked to write down the name of each substance that he tastes.

Facial Paralysis

Paralysis of the entire facial musculature on one side results only from lesions of the facial nucleus (in the pons) or of the facial nerve. Unilateral lesions at a higher (supranuclear) level spare the orbicularis oculi and frontalis because the portion of the facial nucleus concerned with these muscles is controlled by fibers from the cerebral cortex of both sides.

TABLE 54–1 Branches of Facial Nerve

Region	*Branches*
In internal acoustic meatus	Communicates with 8th cranial nerve
In facial canal, from geniculate ganglion	Greater petrosal nerve
	Communicating branch to tympanic plexus External petrosal nerve
In facial canal, beyond geniculate ganglion	Nerve to stapedius
	Chorda tympani
Below base of skull	Stylohyoid and digastic branches Communicating branches, e.g., with 9th and 10th cranial nerves Posterior auricular nerve
On face	Temporal branches Zygomatic branches Buccal branches Marginal branch of mandible Cervical branch

Although central lesions of the descending motor pathway abolish voluntary control of the lower facial muscles, the muscles may still take part in reflex emotional responses, such as smiling. Conversely, in certain conditions, emotional responses may be lost while voluntary control is retained.

The facial nerve in the facial canal may be involved by infection of the middle ear, in surgical exposure of the mastoid air cells, or in so-called idiopathic facial paralysis (Bell's palsy). In these paralyses, the lesions of which are below the level of the facial nucleus in the brain (i.e., infranuclear in type), the musculature of both the upper and the lower part of the face is affected. The most conspicuous feature of unilateral facial paralysis is the displacement of the mouth produced by the unopposed contraction of the muscles of the sound side (fig. 57–4 *D*). The corner of the mouth droops and the affected side cannot take part in smiling, whistling, or blowing. The nasolabial furrow is less pronounced. The upper eyelid droops, the lower lid is everted, the eye cannot be closed, blinking does not occur, and the eye is rendered readily susceptible to inflammation (conjunctivitis). The wrinkles of the forehead become obliterated.

The level of a lesion of the facial nerve is inferred from effects that depend on whether or not specific branches are intact.[28] Thus, (1) if the greater petrosal nerve is involved, decreased lacrimation may be found on testing; (2) if the nerve to the stapedius is affected, painful sensitivity to sounds (hyperacusis) may result; (3) if the chorda tympani is implicated, there may be a loss of taste for the anterior two-thirds of the tongue; (4) if the branch to the digastric is interfered with, the lower jaw and the tongue deviate to the sound side on maximal opening of the mouth.

INTERNAL EAR

The internal ear is located within the petrous part of the temporal bone.[29] It consists of a complex series of fluid-filled spaces, the membranous labyrinth, lodged within a similarly arranged cavity, the bony labyrinth. The terminology is very confusing, and a simplified scheme is given in figure 54–8.

The cochlea functions as the essential organ of hearing. Those portions (utricle and semicircular ducts) of the internal ear other than the cochlea constitute the vestibular apparatus. Equilibrium, however, is maintained by vision and by proprioceptive impulses as well as by the vestibular apparatus.

Osseous Labyrinth

The osseous labyrinth comprises a layer of dense bone (*otic capsule*) in the petrous temporal and the perilymphatic space surrounded by that bone. The term osseous labyrinth, however, is used by some authors for only the perilymphatic space, by others for only the surrounding bone. The space is lined by endosteum, is crossed in most places by delicate trabeculae, and contains perilymph, a liquid that resembles cerebrospinal fluid in composition but has a higher content of protein. The perilymphatic space of the osseous labyrinth comprises a series of continuous cavities: the cochlea, the vestibule, and the semicircular canals (fig. 54–9).

1. Semicircular Canals. There are three semicircular canals: *anterior, posterior,* and *lateral.* The axes of the anterior and posterior canals, although both are arranged in vertical planes, are at a right angle to each other (fig. 54–10).[30] The lateral canal is horizontal when the head is flexed about 30 degrees.

2. Vestibule. The term vestibule, in the official terminologies, is used to include the semicircular canals. In a more limited sense, however, the vestibule is the middle part of the bony labyrinth, and is located immediately medial to the tympanic cavity (fig. 54–4). It contains the utricle and saccule of the membranous labyrinth, lodged in the *elliptical* and *spherical recesses*, respectively. The fenestra vestibuli, situated laterally, is closed by the

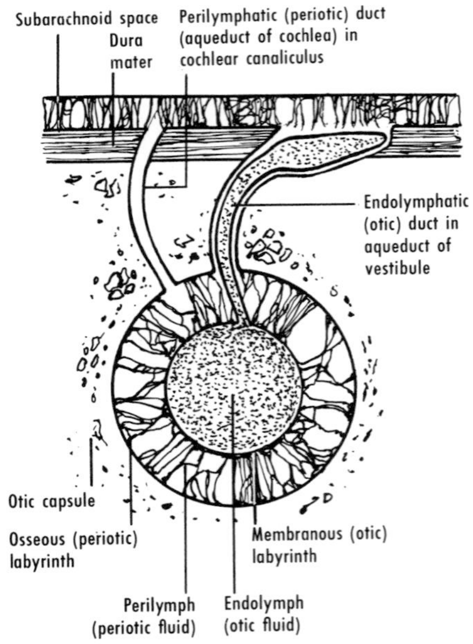

Figure 54–8 Scheme showing basic arrangement and terminology of the internal ear. That the aqueduct of the cochlea communicates with the subarachnoid space is questionable.

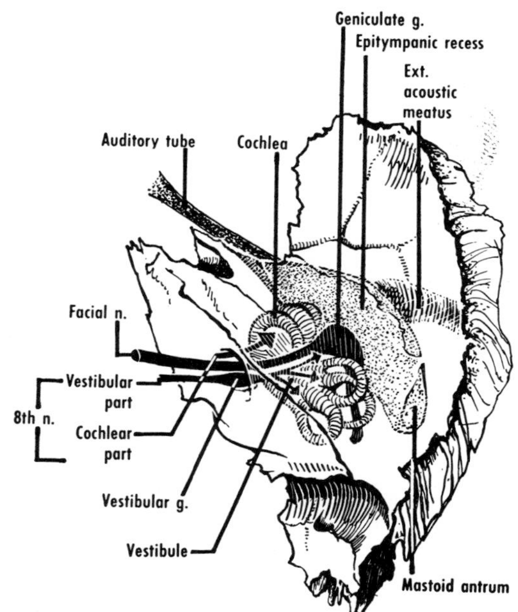

Figure 54–9 Right temporal bone viewed from above. The osseous labyrinth, and the tympanic cavity and the air spaces with which it communicates, are shown as if the surrounding bone were transparent.

At the apex of the cochlea, the modiolus and the cochlear duct end blindly, and the two scalae communicate with each other, the communication being termed the *helicotrema* (fig. 54–1). The *scala vestibuli* begins in the vestibule (hence its name), and the *scala tympani* ends blindly near the fenestra cochleae, which is closed by the secondary tympanic membrane.[32]

The *perilymphatic duct,* or *aqueduct of the cochlea,* is situated in a bony channel, the *cochlear canaliculus.** The location of the cochlear canaliculus can be identified on a skull by a notch on the inferior surface of the petrous temporal, in front of the medial side of the jugular fossa and directly below the internal acoustic meatus. The aqueduct of the cochlea is frequently said to connect the scala tympani with the subarachnoid space (fig. 54–8), although this is disputed. The endo-

*Terminological confusion abounds. Some use aqueduct of cochlea as a synonym of cochlear canaliculus. Moreover, some use aqueduct of vestibule as a synonym of endolymphatic duct.

base of the stapes. Irregular clefts *(fissula ante fenestram)* extend laterally from the vestibule toward the tympanic cavity. The *aqueduct of the vestibule* transmits the endolymphatic duct and perilymphatic tissue[31] (fig. 54–8). The location of the aqueduct of the vestibule can be identified on a skull by a slit on the posterior surface of the petrous temporal, behind the internal acoustic meatus.

3. Cochlea (fig. 54–11). The cochlea, named from its resemblance to the shell of a snail, is a helical tube of about 2½ turns. The base of the cochlea lies against the lateral end of the internal acoustic meatus, and the basal coil forms the promontory of the middle ear. The apex of the cochlea is directed anterolaterally (fig. 54–9).

The cochlea possesses a bony core, the *modiolus,* which transmits the cochlear nerve and contains the spiral ganglion (fig. 54–11). A winding shelf, the *osseous spiral lamina,* projects from the modiolus like the flange of a screw. The cochlear duct extends from this lamina to the wall of the cochlea. Thus the space in the cochlea is divided by the spiral lamina and the cochlear duct into two portions, a scala vestibuli anteriorly, and a scala tympani posteriorly.

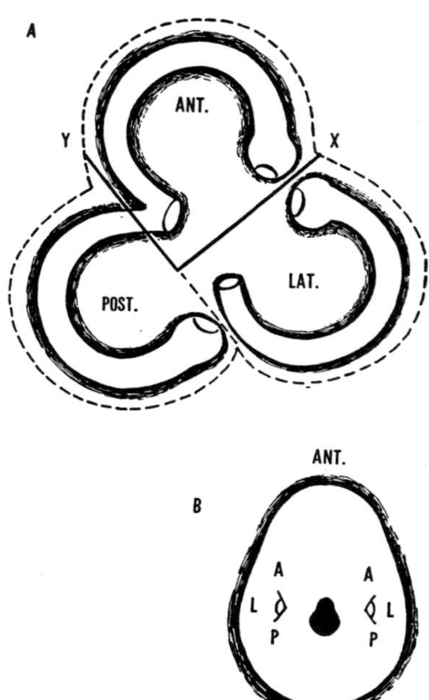

Figure 54–10 *A,* scheme demonstrating relative positions of the semicircular canals of the right ear. Trace this diagram on a card. Cut along the interrupted line. Fold the anterior canal up away from the plane of the page, along line *X.* Then fold the posterior canal toward the lateral, along line *Y. B,* scheme of the base of the skull, viewed from above, showing planes of the semicircular canals. *A* is based on Lithgow.

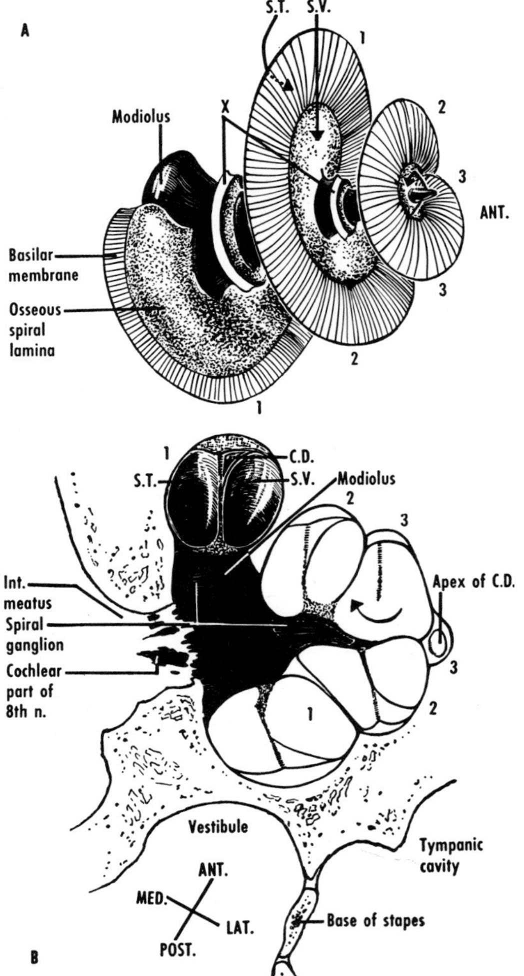

Figure 54–11 Cochlea. *A*, modiolus, osseous spiral lamina, and basilar membrane of the right ear, lateral aspect. The basal, middle, and apical coils of the cochlea are marked 1, 2, and 3, respectively. The coils are separated from one another by a bony, spiral partition, which is not shown; it is attached to the modiolus at *X*. Note that the fibers of the basilar membrane in general increase in length from the basal toward the apical coil of the cochlea; the reverse holds for the width of the osseous spiral lamina. An arrow is placed at the helicotrema. *S.T.*, site of scala tympani; *S.V.*, site of scala vestibuli. *B*, horizontal section through the cochlea. An arrow is placed in the helicotrema. *C.D.*, cochlear duct; *S.T.*, scala tympani; *S.V.*, scala vestibuli. *A* is modified from Bast and Anson. *B* is based on Wolff, Bellucci, and Eggston.

lymph, the blood, and the cerebrospinal fluid have each been suggested as a source of the perilymph. Perilymph is probably resorbed into the capillaries of the spiral ligament (fig. 54–13).

Membranous Labyrinth

The membranous labyrinth (fig. 54–12) lies within the bony labyrinth and contains endolymph, a liquid that differs from perilymph in its composition. The endolymph is probably secreted by specialized cells of the membranous labyrinth, for example, in the stria vascularis (fig. 54–13) or in the endolymphatic sac (fig. 54–12).[33] The wall of the membranous labyrinth is composed of fibrous tissue lined by epithelium, mostly simple squamous in type. The membranous labyrinth consists of a series of continuous cavities: the cochlear duct, the saccule and utricle, and the semicircular ducts.

1. Semicircular Ducts. The arrangement of the three semicircular ducts is similar to that of the semicircular canals, and they are similarly named. Each is situated eccentrically in its corresponding canal, to the convex wall of which it is attached.[34] One end of each duct is enlarged to form its ampulla. Each *ampulla* possesses an *ampullary crest* (fig. 54–12), which consists of neuro-epithelial hair cells surmounted by a gelatinous *cupola.* The hair cells are

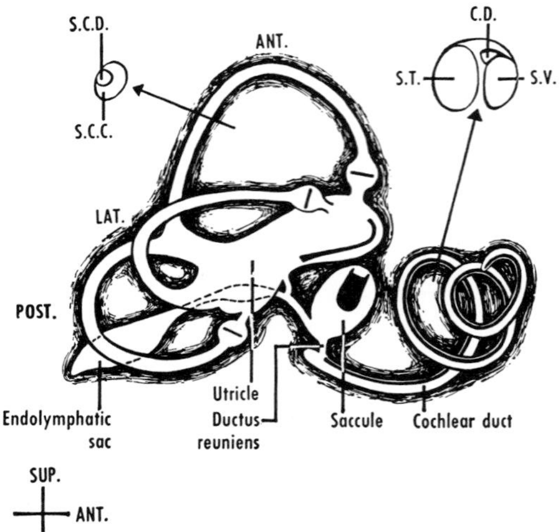

Figure 54–12 Diagram of the membranous labyrinth. The anterior, lateral, and posterior semicircular ducts are indicated; note their ampullae. The black lines or areas in the ampullae, in the utricle and saccule, and in the cochlear duct indicate the sites of the neuro-epithelium (ampullary crests, maculae, and spiral organ respectively). Two cross-sections, which show the relation of the membranous to the bony labyrinth, are shown above. *S.C.D.*, semicircular duct; *S.C.C.*, semicircular canal; *C.D.*, cochlear duct; *S.T.*, scala tympani; *S.V.*, scala vestibuli.

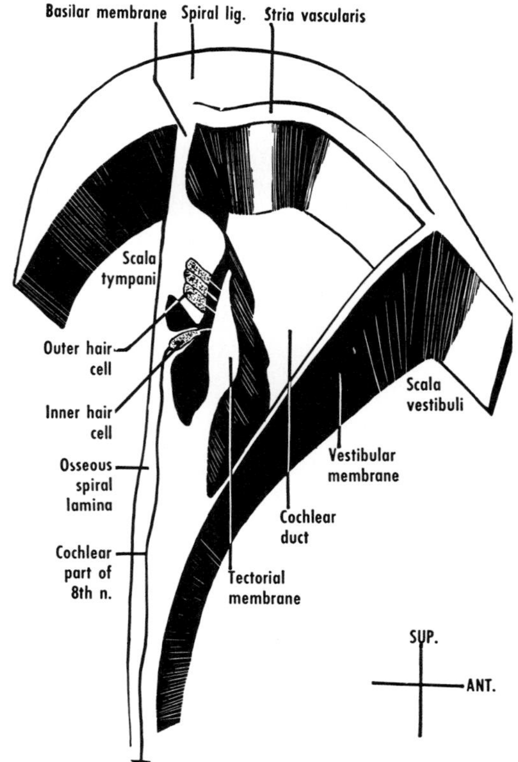

Figure 54–13 Diagram of a cross-section through the cochlear duct. Based on Stöhr.

closely related to the fibers of the vestibular nerve. The cupola may be bent by the pressure (due to movement) of the endolymph, and this in some way stimulates the nerve endings around the hair cells. In man, however, section of the vestibular nerve causes only a transient disturbance of posture because muscle-joint sense and vision are more important for posture than is the labyrinth. Motion sickness appears to be associated with stimulation of the semicircular ducts, but other factors are also involved.

2. Utricle and Saccule. The utricle and the saccule lie in the vestibule and are in communication with each other by the utricular and saccular ducts.* The utricle receives the five openings of the semicircular ducts; the anterior and posterior ducts have one opening in common. The saccule is continuous with the cochlear duct by means of the *ductus reuniens.* The utricle and the saccule each present a thickening termed a *macula* (fig. 54–12), which consists of neuro-epithelial hair cells surmounted by a gelatinous *otolithic membrane.* The membrane contains crystals of calcium carbonate *(otoliths,* or *otoconia).* The maculae are stimulated by gravity, but many authors maintain that the saccule may not be vestibular in function.

The *endolymphatic duct* arises from the utricular and saccular ducts and is transmitted by the aqueduct of the vestibule. The duct ends in the *endolymphatic sac* (fig. 54–12), which is located under cover of the dura on the posterior surface of the petrous temporal.[35]

3. Cochlear Duct. The cochlear duct (*scala media*), about 32 mm in length,[35] winds from the saccule to the apex of the cochlea, where it ends blindly (fig. 54–1). It extends from the osseous spiral lamina to the wall of the cochlea (fig. 54–11). Its posterior wall is formed by the *basilar membrane,* which is anchored to the wall of the cochlea by a thickening of the endosteum, the *spiral ligament* (fig. 54–13). The basilar membrane separates the cochlear duct from the scala tympani. The anterior wall of the cochlear duct is the thin *vestibular membrane,* which separates it from the scala vestibuli. The "cochlear partition" comprises the vestibular and basilar membranes, together with the structures in the cochlear duct. Both the basilar and the vestibular membrane extend from the osseous spiral lamina to the endosteum (spiral ligament) of the wall of the cochlea. A portion of the spiral ligament, the *stria vascularis,* may be the source of secretion of the endolymph. The endolymphatic duct and sac, and perhaps also the stria vascularis, may be the sites of absorption of the endolymph.

The *spiral organ,*[37] the organ of hearing, lies against the basilar membrane (fig. 54–13). It contains neuro-epithelial hair cells that are attached anteriorly to a gelatinous mass, termed the *tectorial membrane.* The basilar membrane is said to contain 24,000 fibers; the fibers in general increase in length,[38] whereas the membrane decreases in stiffness, from the base toward the apex of the cochlea. It has often been maintained that the basilar membrane vibrates, perhaps acting as a resonator, but

*The term *utriculosaccular duct,* not used here, is employed for the utricular duct by those who regard the endolymphatic duct as arising directly from the saccule.

it is now known that the fibers of the membrane are not under tension. Moreover, it has been suggested that the tectorial membrane is the vibrator.

The chief components of the internal ear, together with their openings, are summarized in table 54–2.

TABLE 54–2 Components of Internal Ear, and Their Openings

Osseous Labyrinth	*Membranous Labyrinth*
3 SEMICIRCULAR CANALS	3 SEMICIRCULAR DUCTS
VESTIBULE	UTRICLE
5 openings of semicircular canals	5 openings of semicircular ducts
aqueduct of vestibule containing endolymphatic duct	SACCULE
	endolymphatic duct and sac arise via utricular and saccular ducts
fenestra vestibuli closed by base of stapes	
COCHLEA	COCHLEAR DUCT
cochlear canaliculus containing perilymphatic duct (aqueduct of cochlea)	continuous with saccule via ductus reuniens
fenestra cochleae closed by secondary tympanic membrane	

Innervation and Blood Supply of Internal Ear

The nerve of special sensation from the internal ear is the vestibulocochlear. In addition, the internal ear receives sympathetic and parasympathetic fibers. The internal ear is supplied by the labyrinthine artery, a branch of either the anterior inferior cerebellar or the basilar. The blood drains mostly into the petrosal sinuses.

VESTIBULOCOCHLEAR NERVE

The *vestibulocochlear (acoustic,* or *auditory)* or 8th cranial nerve[39] emerges between the pons and the medulla oblongata, at the cerebellopontine angle and behind the facial nerve. It comprises afferent fibers from the internal ear. The nerve travels laterally through the internal acoustic meatus, in which it receives a communication from the facial nerve.

The vestibulocochlear nerve contains two sets of fibers:

1. The *vestibular part* (or nerve), concerned with equilibration, is distributed to the maculae of the utricle and saccule, and to the ampullary crests of the semicircular ducts. The vestibular fibers arise in bipolar cells in the *vestibular ganglion,* which is situated in the internal acoustic meatus. The branches of the nerve pierce the lateral end or *fundus* of the internal acoustic meatus, thereby gaining the labyrinth. The saccule appears to receive cochlear as well as vestibular fibers.[40]

2. The *cochlear part* (or nerve), concerned with hearing, is distributed to the hair cells of the spiral organ. A vestibulocochlear anastomosis takes place. The cochlear fibers pierce the fundus of the internal acoustic meatus, thereby gaining the modiolus of the cochlea. The fibers arise in bipolar cells in the *spiral ganglion* (fig. 54–11), which is lodged in a spiral canal in the modiolus.[41] The peripheral processes of the ganglion cells are distributed to the spiral organ by way of the osseous spiral lamina (fig. 54–13).

Examination

Examination of the vestibulocochlear nerve necessitates separate testing of the vestibular and cochlear functions. Irritative lesions produce exaggerated functional effects, whereas destructive lesions result in a loss of function.

Symptoms and signs that may suggest disease of the vestibular part include a sensation of giddiness or a loss of sense of position (vertigo). Tests of labyrinthine function depend on observing the response to labryinthine stimulation induced by rapid rotation of the body or by irrigation of the external acoustic meatus with hot and cold water in turn.

The chief symptom and sign of disease of the cochlear part of the 8th nerve is deafness. Deafness can be tested in one ear by finding the maximum distance at which the ticking of a watch can be heard while the opposite external meatus is plugged by wool. **When deafness is found, it is important to determine whether it is due to (1) a lesion of the cochlea or cochlear nerve (neural deafness), or (2) disease of the middle ear (conductive deafness).**

Partial neural deafness, in which appreciation of chiefly higher frequencies is lost, occurs in the majority of elderly persons.

An irritative lesion of the vestibulocochlear nerve may produce a ringing or buzzing in the ear (tinnitus).

FUNCTIONAL CONSIDERATIONS

The details of the mechanism of functioning of the internal ear are poorly understood. As mentioned in regard to the middle ear (p. 619), vibrations of the perilymph may arise from (1) movements of the tympanic membrane and the auditory ossicles, (2) air waves reaching the fenestra cochleae, and (3) vibrations of the walls of the bony labyrinth. Movement of the stapes in the fenestra vestibuli results in movement of the secondary tympanic membrane in the fenestra cochleae (fig. 54–1).[42] Vibrations in the perilymph affect the cochlear duct, and, owing to relative motion between the basilar and tectorial membranes, the hair cells of the spiral organ are stimulated.

Disease (otosclerosis) of the walls of the inner ear may involve the fenestra vestibuli, limit the movement of the stapes, and so cause conductive deafness. Provided the cochlea is normal, a new opening can be made (fenestration) from the middle into the internal ear. The surgical fenestra, which takes the place of the fenestra vestibuli, is cut into the lateral semicircular canal, thereby re-establishing a sound-pathway to the labyrinth.

Equilibrium is maintained by vision, by proprioceptive impulses, and by the vestibular apparatus. Labyrinthine equilibrium depends on the perception of (1) linear acceleration and movement at a constant speed (otolith organs), and (2) angular acceleration (semicircular canals and otolith organs).[43]

REFERENCES

1. E. T. Perry, *The Human Ear Canal,* Thomas, Springfield, Illinois, 1957.
2. C. D. Schneck and S. Kendall, Anat. Rec., *157*:317, 1967, abstract.
3. W. Platzer, Mschr. Ohrenheilk, Laryng.-Rhinol., *95*:553, 1961.
4. G. P. Stelter, T. H. Bast, and B. J. Anson, Quart. Bull. Northw. Univ. med. Sch., *34*:23, 1960.
5. G. Portmann and E. Puig, Rev. Laryng., Paris, *70*:1, 1949. K. M. Rushton, J. Laryng., *71*:100, 1957.
6. I. C. Heron, Amer. J. phys. Anthrop., *6*:11, 1923. I. de Vincentiis and A. Cimino, Riv. Biol., Perugia, *49*:181, 1957.
7. M. Harty, Z. mikr.-anat. Forsch., *71*:24, 1964.
8. D. Wolff and R. J. Bellucci, Ann. Otol., etc., St Louis, *65*:895, 1956.
9. H. Engström, Acta anat., *6*:283, 1948. E. A. Bolz and D. J. Lim, Acta otolaryng., Stockh., *73*:10, 1972.
10. R. Girardet, Arch. Anat., Strasbourg, *43*:121, 1960.
11. M. Lawrence, Ann. Otol., etc., St. Louis, *71*:705, 1962.
12. G. Djupesland and H. E. Gronas, Canad. J. Otolaryng., *2*:119, 1973.
13. G. Salomon and A. Starr, Acta neurol. scand., *39*:161, 1963.
14. C. T. Nager and M. Nager, Ann. Otol., etc., St Louis, *62*:923, 1953.
15. H. G. Kobrak, *The Middle Ear,* University of Chicago Press, 1959.
16. R. Guelke and J. A. Keen, J. Physiol., *116*:175, 1952.
17. I. Kirikae, *The Structure and Function of the Middle Ear,* University of Tokyo Press, 1960.
18. A. L. Rhoton, S. Kobayashi, and W. H. Hollinshead, J. Neurosurg., *29*:609, 1968.
19. D. R. Haynes, Ann. R. Coll. Surg. Engl., *16*:175, 1955, G. Botros, Ann. Otol., etc., St Louis, *66*:173, 1957. B. J. Anson, D. G. Harper, and R. L. Warpeha, *ibid.*, *72*:713, 1963.
20. G. M. Hall, J. L. Pulec, and A. L. Rhoton, Arch. Otolaryng., *90*:568, 1969.
21. L. J. McCormack, E. W. Cauldwell, and B. J. Anson, Surg. Gynec. Obstet., *80*:620, 1945.
22. B. Vidic and P. A. Young, Anat. Rec., *158*:257, 1967. B. Vidic, *ibid.*, *162*:511, 1968.
23. B. J. Anson, J. A. Donaldson, and B. B. Shilling, Ann. Otol., etc., St Louis, *81*:616, 1972. M. Maurizi, E. de Campora, and A. Frenguelli, Il Valsalva, *48*:186, 1972.
24. Y. Zotterman and H. Diamant, Nature, Lond., *183*:191, 1959.
25. H. Diamant and A. Wiberg, Acta otolaryng., Stockh., *60*:255, 1965.
26. R. A. Davis *et al.*, Surg. Gynec. Obstet., *102*:385, 1956. E. Nesci and P. Motta, Anat. Anz., *131*:82, 1972.
27. C. P. G. Wakeley and F. H. Edgeworth, J. Anat., Lond., *67*:420, 1933.
28. K. Tschiassny, Ann. Otol., etc., St Louis, *62*:677, 1953.
29. For special methods of removing and studying the temporal bone, see Trans. Amer. Acad. Ophthal. Otolaryng., *62*:601, 1958.
30. J. D. Lithgow, J. Laryng., *35*:81, 1920.
31. B. J. Anson *et al.*, Laryngoscope, St Louis, *74*:945, 1964. Y. Ogura and J. D. Clemis, Ann. Otol., etc., St Louis, *80*:813, 1971.
32. M. Harty, Z. mikr.-anat. Forsch., *70*:484, 1963.
33. L. Citron, D. Exley, and C. S. Hallpike, Brit. med. Bull., *12*:101, 1956.
34. M. Harty, J. Laryng., *62*:36, 1948.
35. B. J. Anson, R. L. Warpeha, and M. J. Rensink, Ann. Otol., etc., St Louis, *77*:583, 1968.
36. M. Hardy, Amer. J. Anat., *62*:291, 1938. J. A. Keen, J. Anat., Lond., *74*:524, 1940.
37. D. Wolff, Arch. Otolaryng., *56*:588, 1952.
38. E. G. Wever, Ann. Otol., etc., St Louis, *47*:37, 1938.
39. C. C. D. Shute, Proc. R. Soc. Med., *44*:1013, 1951.
40. M. Hardy, Anat. Rec., *59*:403, 1934.
41. D. Wolff, Amer. J. Anat., *60*:55, 1936.
42. L. M. Sellers, Laryngoscope, St Louis, *71*:237, 1961.
43. O. Lowenstein, Brit. med. Bull., *12*:114, 1956.

GENERAL READING

Anson, B. J., and Donaldson, J. A., *Surgical Anatomy of the Temporal Bone and Ear,* Saunders, Philadelphia, 2nd ed., 1973. Largely an atlas of detailed illustrations and photomicrographs of the ear.

Bast, T. H., and Anson, B. J., *The Temporal Bone and the Ear,* Thomas, Springfield, Illinois, 1949. The internal ear and the middle ear are described on a developmental basis.

Dallos, P., *The Auditory Periphery. Biophysics and Physiology,* Academic Press, New York, 1973.

Silverstein, H., *Atlas of the Human and Cat Temporal Bone,* Thomas, Springfield, Illinois, 1972. Photographs and photomicrographs of horizontal sections.

Vidić, B., and O'Rahilly, R., *An Atlas of the Anatomy of the Ear,* Saunders, Philadelphia, 1971. Color slides and key drawings.

Wolff, D., Bellucci, R. J., and Eggston, A. A., *Surgical and Microscopic Anatomy of the Temporal Bone,* Hafner, New York, 1971. Photomicrographs of serial sections of temporal bones cut in horizontal and vertical planes.

THE ORBIT

55

BONY ORBIT

The orbits (figs. 55–1 and 55–2) are two bony cavities that contain the eyes, together with their associated muscles, nerves, blood vessels, fat,[1] and much of the lacrimal apparatus. Each orbit is shaped roughly like a pear, or a four-sided pyramid, with its *apex* posteriorly and its *aditus (base)* anteriorly. The base is about 35 mm high by 40 mm wide. The sides of the orbit are termed the roof, lateral wall, floor, and medial wall. The periosteum of the walls is continuous with the dura. **Each orbit is related (1) above to the anterior cranial fossa and usually to the frontal sinus, (2) laterally to the temporal fossa in front and to the middle cranial fossa behind, (3) below to the maxillary sinus, and (4) medially to the ethmoidal air cells and generally to the sphenoidal sinus.** The orbit communicates with the cranial cavity by several apertures.

Margins

The margin of the orbit is readily palpable *in vivo.* It is formed chiefly by three bones (frontal, zygomatic, and maxilla) separated by three sutures (fig. 55–2A). The lacrimal also contributes to its composition. The margin may be subdivided into four continuous parts (supra-orbital, lateral, infra-orbital, and medial), each of which is frequently referred to individually as a margin.

Supra-Orbital Margin. The supra-orbital margin is formed by the frontal bone and, at the junction of its lateral two-thirds and medial one-third, presents the palpable *supra-orbital notch.* The notch is bridged *in vivo* by fibrous tissue that is sometimes ossified. The foramen, whether completed by fibrous tissue or by bone, transmits the supra-orbital nerve and vessels to the forehead. More medially, a *frontal notch* may be found; it transmits branches of the supra-orbital nerve and vessels. Still farther medially, the margin is crossed by the supratrochlear nerve and vessels. Only the medial part of the supra-orbital margin is covered by the eyebrow; laterally, the eyebrow runs above the margin.

Lateral Margin. The lateral margin is formed by the zygomatic process of the frontal and the frontal process of the zygomatic. This margin is concave forward, thereby increasing the extent of the visual field on the temporal side. A small *orbital tubercle* on the zygomatic bone gives attachment to the lateral palpebral ligament and may be palpable *in vivo.*[2]

Infra-orbital Margin. The infra-orbi-

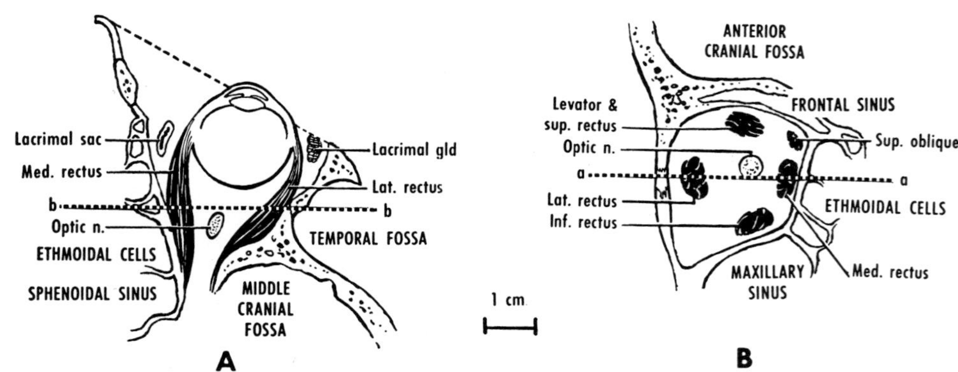

Figure 55–1 A, horizontal section through the orbit at the level of the palpebral fissure. Note that the eye projects in front of the interrupted line from the bridge of the nose to the lateral margin of the orbit. The line *bb* indicates the plane of section of *B. B,* coronal section through the orbit. The line *aa* indicates the plane of section of *A. A* is based on Truex and Kellner, *B* on Kampmeier, Cooper, and Jones.

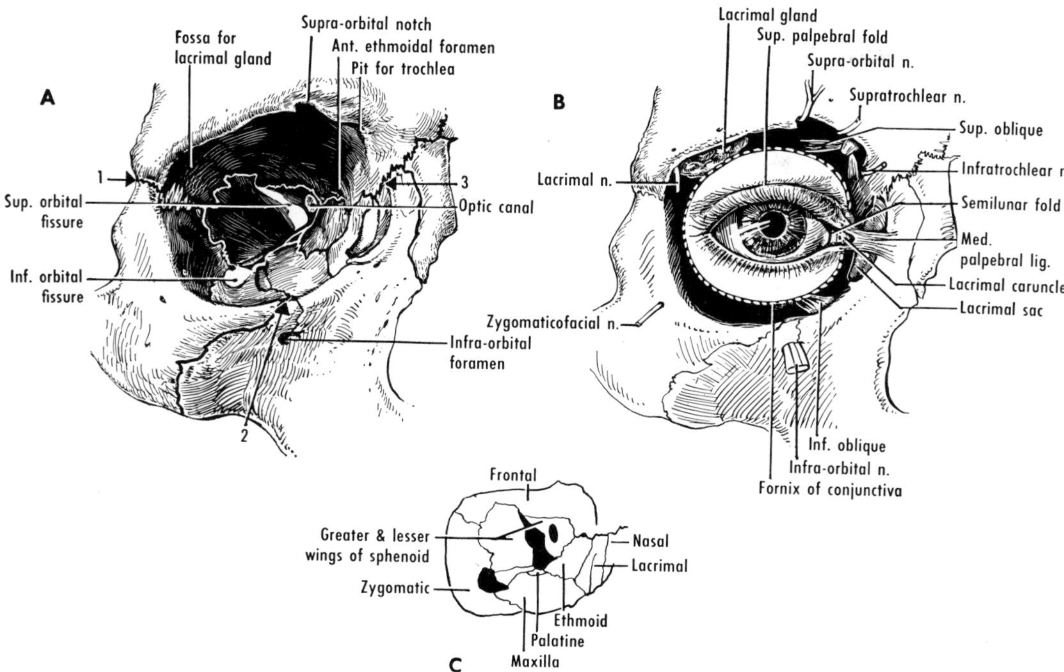

Figure 55–2 *A*, anterior aspect of the right bony orbit. The numerals *1*, *2*, and *3* indicate the three main sutures separating the three chief bones of the margin of the orbit. Note also the infra-orbital suture in the maxilla (above the infra-orbital foramen); it indicates the line of closure of the infra-orbital canal. *B*, surface anatomy of the right eye and orbit. Compare figure 56–5, p. 647. *C*, diagram showing the bones that form the walls of the orbit.

tal margin is formed by the zygomatic bone and the maxilla. An infra-orbital suture may be visible in the maxilla on this margin. It indicates the line of closure of the infra-orbital canal. The *infra-orbital foramen* opens ½ to 1 cm below the margin[3] and transmits the nerve and artery of the same name to the face.

Medial Margin. The medial margin is formed by the maxilla, lacrimal, and frontal. The infra-orbital margin can be traced upward to become continuous with the *anterior lacrimal crest* on the frontal process of the maxilla. The supra-orbital margin can be traced downward toward the *posterior lacrimal crest* on the lacrimal bone. The *fossa for the lacrimal sac* is formed by the expanded medial margin of the orbit (maxilla and lacrimal bone), between the anterior and posterior lacrimal crests. The fossa is continued downward through the floor of the orbit as the *nasolacrimal canal.* The canal is formed by the maxilla laterally and by the lacrimal bone and the inferior nasal concha medially. It transmits the nasolacrimal duct from the lacrimal sac to the inferior meatus of the nasal cavity. The anterior lacrimal crest gives attachment to the medial palpebral ligament and the orbicularis oculi; the posterior crest to the lacrimal part of the orbicularis, the orbital septum, and the medial check ligament (fig. 55–4).

Walls

The orbit possesses four walls (fig. 55–2, *A* and *C*), representable on a cardboard model.[4]

Roof or Superior Wall. Triangular in shape, the superior wall is formed by the orbital plate of the frontal bone and by the lesser wing of the sphenoid. The *fossa for the lacrimal gland* is at the anterolateral angle of the roof. The *trochlear pit* is a small fossa at the anteromedial angle of the roof. It indicates the attachment of the pulley of the superior oblique muscle, and is sometimes marked by a *spine.* **The optic canal is situated in the extreme back part of the roof. It lies between the two roots of the lesser wing of the sphenoid, and transmits the optic nerve and its meningeal coverings,**

together with the ophthalmic artery, from the middle cranial fossa.

Lateral Wall. The lateral wall is triangular in shape. **The lateral walls of the two orbits are set at approximately a right angle** (fig. 55–3). The posterior part of the lateral wall is demarcated above and below by the superior and inferior orbital fissures. This wall is formed by the zygomatic bone and the greater wing of the sphenoid, together with a small portion of the frontal bone. The lateral wall presents several small foramina (one or two for the zygomatic nerve and one for the orbital branch of the middle meningeal artery). **The superior orbital fissure communicates with the middle cranial fossa. It lies between the greater and lesser wings of the sphenoid and is closed laterally by the frontal bone.[5] Its wider, medial part transmits chiefly cranial nerves 3, 4, and 6, the three branches of the ophthalmic nerve, and the ophthalmic veins** (fig. 55–5). The *inferior orbital fissure* communicates with the infratemporal and pterygopalatine fossae. It lies between the greater wing of the sphenoid above and the maxilla and palatine bone below. It is frequently limited in front by the zygomatic bone. It transmits chiefly the maxillary or infra-orbital nerve, the zygomatic nerve, and the infra-orbital artery.

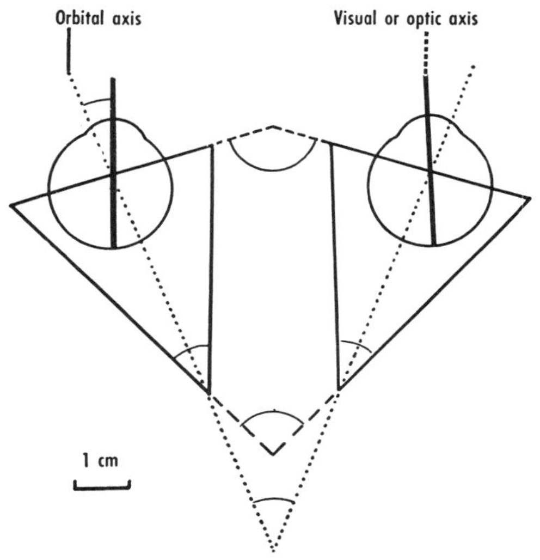

Figure 55–3 Diagrammatic horizontal section through orbits, showing angles formed by the walls. The dotted lines indicate the longitudinal axes of the orbits. The visual axes; here shown directed toward a distant object, are nearly parallel. Based on Whitnall.

The *orbital muscle,* or *orbitalis,* consists of smooth muscle fibers that incompletely close the inferior orbital fissure.[6]

Floor or Inferior Wall. The inferior wall, triangular in shape, extends backward only two-thirds of the depth of the orbit. It is formed by the maxilla, and by the zygomatic and palatine bones. It presents the *infra-orbital groove* and *canal,* which transmit the nerve and artery of the same name from the inferior orbital fissure to the infra-orbital foramen. The nasolacrimal canal has been mentioned with the medial margin. The inferior oblique muscle arises from the anteromedial angle of the floor of the orbit, just lateral to the opening of the nasolacrimal canal.

Medial Wall. The medial wall, quadrilateral in shape, is the thinnest wall of the orbit, and is almost parallel to the median plane. Thus **the medial walls of the two orbits are nearly parallel to each other** (fig. 55–3). The medial wall is formed by the ethmoid (orbital plate), lacrimal, and frontal bones, together with a small portion of the body of the sphenoid. If the fossa for the lacrimal sac is regarded as an expanded portion of the medial margin of the orbit, then the medial wall is limited in front by the posterior lacrimal crest. The *anterior* and *posterior ethmoidal foramina* are small openings at the junction of the medial wall with the roof of the orbit. They transmit the nerves and arteries of the same name to the anterior cranial fossa.

The dura mater of the middle cranial fossa passes through the optic canal and divides into two layers, which enclose the common tendinous ring between them (fig. 55–4). The external layer is the periosteum that lines the orbit, known as the *periorbita.* It can be detached readily. The internal layer forms the outer sheath of the optic nerve and is continuous with the fascia bulbi (p. 644).

OPHTHALMIC NERVE

The ophthalmic nerve (first division of the trigeminal) is an afferent nerve that supplies the eyeball and conjunctiva, the lacrimal gland and sac, the nasal mucosa and frontal sinus, the external nose, the upper eyelid, forehead, and scalp. It arises from the trigeminal ganglion and lies in the dura of the lateral wall of the cavernous sinus.

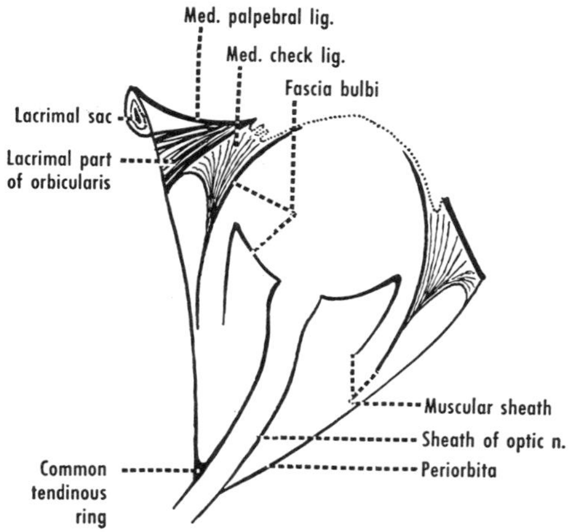

Figure 55–4 Diagramatic horizontal section of the right orbit, showing the fascia. Note the sheaths of the recti and the connective tissue (check ligaments) between these sheaths and the walls of the orbit. Based on Whitnall.

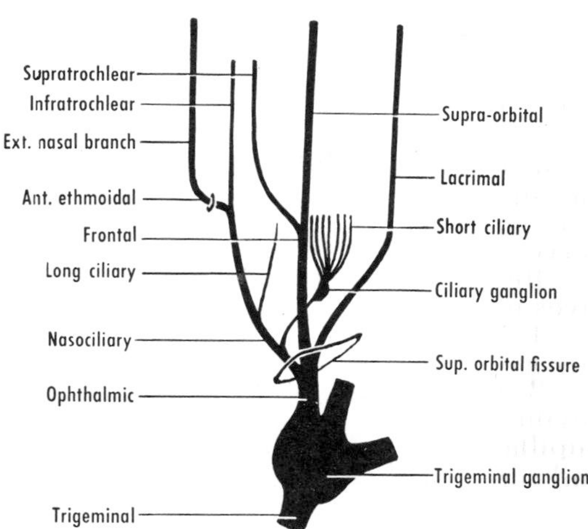

Figure 55–6 Scheme of the chief branches of the ophthalmic nerve. Superior aspect.

Branches

The ophthalmic nerve divides, near the superior orbital fissure, into three branches: lacrimal, frontal, and nasociliary. These three nerves pass through the superior orbital fissure (fig. 55–5) and traverse the orbit, where they give rise to branches (fig. 55–6).

Lacrimal Nerve. The lacrimal nerve enters the orbit through the superior orbital fissure, above the muscles of the eyeball. It proceeds along the upper border of the lateral rectus and ends at the front of the orbit by giving branches to the lacrimal gland, the conjunctiva, and the skin of the upper eyelid. The lacrimal nerve communicates in the orbit with the zygomatic nerve and, by this means, some secretory fibers are brought to the lacrimal gland.

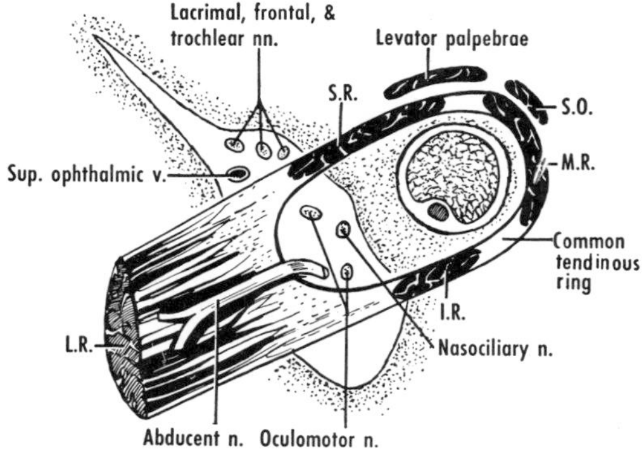

Figure 55–5 Superior orbital fissure and optic canal. Anterior aspect. Note that the optic canal and the adjacent part of the fissure are surrounded by the common tendinous ring, from which the four recti arise. The structures that pass through the common tendinous ring all lie at first within the cone formed by the muscles of the globe. The lacrimal, frontal, and trochlear nerves, however, enter the orbit above the ring and lie, therefore, above the muscular cone. The optic nerve and the ophthalmic artery do not have labels. Compare figure 55–7. Based on Wolff and on Whitnall.

Frontal Nerve. The frontal nerve enters the orbit through the superior orbital fissure, above the muscles of the eyeball, and proceeds directly forward on the levator palpebrae superioris.

BRANCHES. At an extremely variable point the frontal nerve divides into the supra-orbital and supratrochlear nerves.

The *supra-orbital nerve* is the direct continuation of the frontal. It leaves the orbit through the supra-orbital notch or foramen and is distributed to the forehead and scalp, the upper eyelid, and the frontal sinus.

The *supratrochlear nerve,* much smaller, leaves the orbit at the medial end of the supra-orbital margin. It supplies the forehead and upper eyelid.

Nasociliary Nerve. **The nasociliary nerve is the sensory nerve to the eye.** It enters the orbit through the superior orbital fissure, inside the cone formed by the muscles of the globe. It is on a lower plane, therefore, than the lacrimal and frontal nerves. It lies between the two divisions of the oculomotor nerve. It courses forward below the superior rectus and crosses the optic nerve with the ophthalmic artery. At the medial side of the orbit it lies between the superior oblique and the medial rectus. It is continued as the anterior ethmoidal nerve.

BRANCHES. The nasociliary nerve gives off:

1. A *communicating branch* to the ciliary ganglion.
2. One or two *long ciliary nerves* (conveying sympathetic fibers to the dilator pupillae and afferent fibers from the uvea and cornea).
3. The *infratrochlear nerve* to the eyelids, skin of the nose, and lacrimal sac.
4. The *posterior ethmoidal nerve,* frequently absent, to the sphenoidal and ethmoidal sinuses.
5. The *anterior ethmoidal nerve,* which is regarded as the continuation of the nasociliary. The anterior ethmoidal nerve passes through the foramen of the same name and enters the anterior cranial fossa. It then gains the nasal cavity and divides into *internal nasal branches,* which supply the walls of the nasal cavity. One of the branches reaches the skin of the nose as an *external nasal branch.*

In its course, the nasociliary, together with its continuation, the anterior ethmoidal nerve, traverses in succession the middle cranial fossa, the orbit, the anterior cranial fossa, the nasal cavity, and the external aspect of the nose.

Examination

The area of skin supplied by the ophthalmic nerve (fig. 57–6*B*) is tested for sensation by the use of cotton wool and a pin. Blowing on the cornea, or touching it with a wisp of cotton wool, results in "closure of the eyes" due to bilateral contraction of the orbicularis oculi muscles. This is termed the corneal reflex. The afferent limb of the reflex arc includes the nasociliary nerve; the efferent limb is the facial nerve.

OPHTHALMIC VESSELS

Ophthalmic Artery

The ophthalmic artery[7] (fig. 55–7*B*) is the chief vessel to the orbit. (The infra-orbital, the continuation of the maxillary, also contributes to the supply of this region.) It is a branch of the cerebral part of the inter-

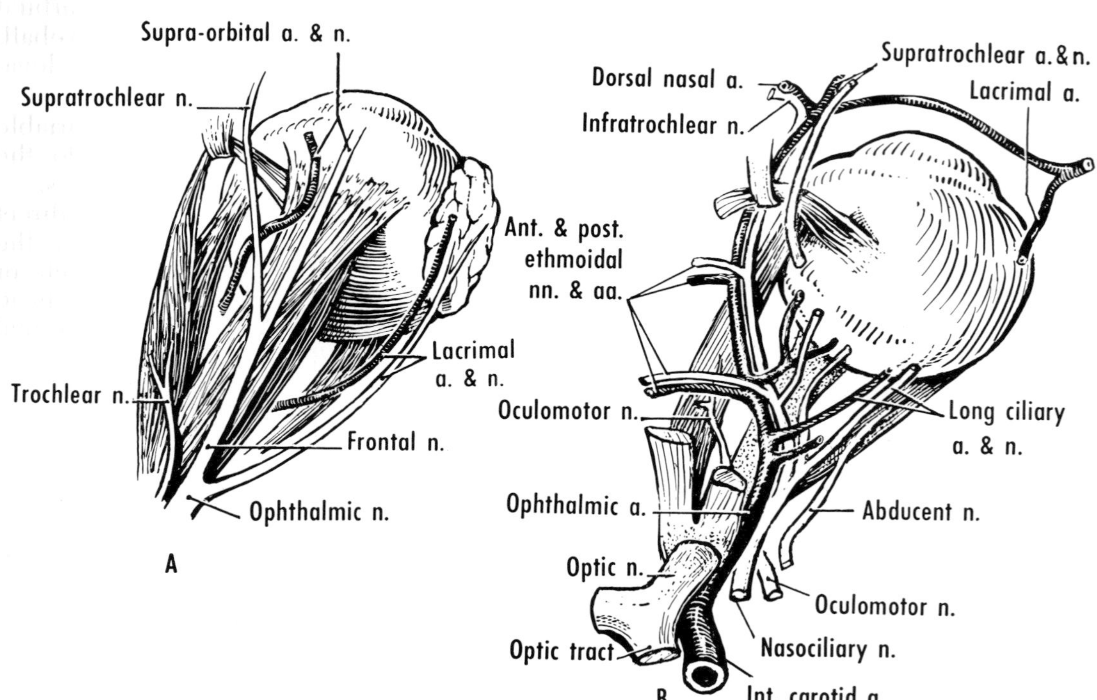

Figure 55–7 Right orbit from above. *A*, showing the three nerves that enter the orbit above the cone of muscles (see fig. 55–5). In *B*, the levator and the superior rectus have been removed, showing the artery and the nerves that enter the orbit within the cone of muscles (see fig. 55–5).

nal carotid artery and arises medial to the anterior clinoid process. It passes forward and laterally, through the optic canal, below the optic nerve. In the orbit it lies between the lateral rectus and the optic nerve. It then turns medially, usually between the superior rectus and the optic nerve, but sometimes below the nerve. At the medial wall of the orbit it turns forward between the superior oblique and the medial rectus. It is accompanied first by the nasociliary and then by the infratrochlear nerve. Near the front of the orbit it divides into the supratrochlear and the dorsal nasal artery.

Branches. **The more important of the numerous branches of the ophthalmic artery are the central artery, the posterior ciliary, and the anterior ciliary (derived from the muscular branches).**

1. The *central artery of the retina*[8] pierces the inferomedial aspect of the optic nerve and thereby reaches the retina, where it divides in a variable pattern into upper and lower temporal, and upper and lower nasal, branches. It anastomoses with the ciliary arteries,[9] but **its terminal branches are virtually end-arteries.**

2. Two *long posterior ciliary arteries* pierce the sclera and supply the ciliary body and the iris.

3. Several *short posterior ciliary arteries* pierce the sclera and supply the choroid.

4. The *lacrimal artery* runs forward along the upper border of the lateral rectus, and supplies the lacrimal gland, and the conjunctiva and eyelids. It gives off: *(a) A recurrent meningeal branch,* which passes backward through or near the superior orbital fissure and anastomoses with the middle meningeal artery. **This anastomosis is ultimately between the internal and external carotid arteries:** it may be large, and, in some cases, the lacrimal (or even the ophthalmic[10]) may be a branch of the middle meningeal. *(b)* Two *lateral palpebral arteries,* which contribute to arcades in the upper and lower lids.

5. *Muscular branches* give origin to the *anterior ciliary arteries,* which pierce the sclera, give off the *anterior conjunctival arteries,* and supply the iris (fig. 56–7).

6. The *supra-orbital artery* may or may not[11] pass through the supra-orbital notch or foramen, and it supplies the upper eyelid and the scalp.

7. A *posterior ethmoidal artery* may be present.

8. The *anterior ethmoidal artery* accompanies the nerve of the same name through the anterior ethmoidal canal to the anterior cranial fossa, nasal cavity, and external nose.

9. The two *medial palpebral arteries* contribute to the arcades in the upper and lower eyelids, and give off *posterior conjunctival arteries* (fig. 56–7).

10. The *supratrochlear artery* crosses the supra-orbital margin, and supplies the forehead and the scalp.

11. The *dorsal nasal artery* leaves the orbit above the medial palpebral ligament. It supplies the root of the nose and the lacrimal sac, and anastomoses with branches of the facial artery. **This is another example of an anastomosis ultimately between the internal and external carotid arteries.**

Ophthalmic Veins

The orbit is drained by the superior and inferior ophthalmic veins, which have no valves. They establish important communications with the facial vein, the pterygoid plexus, and the cavernous sinus. The *superior ophthalmic vein* is formed near the root of the nose by the union of the supra-orbital and angular veins **(hence the possibility of the spread of superficial infections of the face to the cavernous sinus).** It accompanies the ophthalmic artery, receives corresponding tributaries, passes through the superior orbital fissure, and ends in the cavernous sinus. The *central vein of the retina* usually enters the cavernous sinus directly but it may join one of the ophthalmic veins. The *inferior ophthalmic vein* begins as a plexus on the floor of the orbit, and ends directly or indirectly in the cavernous sinus. The uvea is drained by four veins, termed *venae vorticosae* (fig. 56–7), which pierce the sclera obliquely and end in the ophthalmic veins.

OCULOMOTOR, TROCHLEAR, AND ABDUCENT NERVES

Oculomotor Nerve

The oculomotor (3rd cranial) nerve, so named because it is the chief motor nerve to the ocular muscles, supplies all the muscles of the eyeball except the superior oblique and the

lateral rectus. It emerges from the brain stem medial to the cerebral peduncle. It passes between the posterior cerebral and superior cerebellar arteries, and runs forward in the interpeduncular cisterna of the subarachnoid space, on the lateral side of the posterior communicating artery. It pierces the dura lateral to the posterior clinoid process (which it may groove) and traverses the cavernous sinus. It divides into a superior and an inferior division, which traverse the superior orbital fissure within the common tendinous ring.

The *superior division* supplies the superior rectus and the levator palpebrae superioris. The *inferior division* supplies the medial rectus, inferior rectus, and the inferior oblique. **A parasympathetic communication from the branch to the inferior oblique joins the ciliary ganglion. This communication contains the motor fibers to the sphincter pupillae and the ciliary muscle.**

The oculomotor nerve contains motor, proprioceptive, parasympathetic (preganglionic), and sympathetic (postganglionic) fibers. **In focusing the eyes on a near object, the oculomotor nerves are involved in adduction (medial recti), accommodation (ciliary muscle), and miosis (sphincter pupillae).**

Examination. Paralysis of the oculomotor nerve results in ptosis (paralysis of levator), and abduction (unopposed action of lateral rectus and superior oblique) of the eyeball. There may also be limitation of movement, double vision (diplopia), dilatation of the pupil (mydriasis; due to paralysis of sphincter), and inability to accommodate (cycloplegia; due to paralysis of ciliary muscle).

Trochlear Nerve

The trochlear (4th cranial) nerve supplies only the superior oblique muscle of the eyeball and is named from the trochlea or pulley of that muscle. The fibers of the trochlear nerve of each side decussate across the median plane and then emerge from the back of the brain stem, below the corresponding inferior colliculus. It is the only motor nerve, cranial or spinal, that arises from the dorsal aspect of the central nervous system. The nerve is very long and slender. It passes laterally and winds forward around the cerebral peduncle, between the posterior cerebral and the superior cerebellar artery. It runs below the free edge of the tentorium cerebelli and pierces the dura. It passes forward in the lateral wall of the cavernous sinus, crosses the oculomotor nerve, and traverses the superior orbital fissure. It then lies above the levator palpebrae superioris and enters the upper aspect of the superior oblique.

Examination. When the subject's eye is in adduction, he is asked to look downward. Diplopia and limitation of movement are then found if the superior oblique muscle is paralyzed.

Abducent Nerve

The abducent (6th cranial) nerve supplies only the lateral rectus muscle of the eyeball, and is named from its role in abduction of the eye. The nerve emerges from the brain stem between the pons and the medulla oblongata. It runs upward, forward, and slightly laterally through the cisterna pontis, crossing usually dorsal to the anterior inferior cerebellar artery. It pierces the dura below the posterior clinoid process and crosses (sometimes perforates) the inferior petrosal sinus. **It then bends sharply forward, almost at a right angle, across the superior border of the apical portion of the petrous temporal.** It lies below the petroclinoid ligament (between the dorsum sellae and the petrous temporal).[12] It runs forward in the cavernous sinus on the lateral (and then on the inferolateral) aspect of the internal carotid artery. It traverses the superior orbital fissure within the common tendinous ring, and enters the medial aspect of the lateral rectus.

Like the trochlear nerve, the abducent contains motor, proprioceptive, and sympathetic (postganglionic) fibers.

Examination. If the lateral rectus muscle is paralyzed, the subject is unable to abduct the eye beyond the middle of the palpebral fissure. Diplopia is present on attempting to look laterally.

The abducent nerve, perhaps owing to stretching against the sharp upper border of the petrous temporal,[13] is affected in almost any cerebral lesion that is accompanied by increased intracranial pressure, and hence has been termed "the weakling of the cranial contents." Involvement of the abducent nerve alone, therefore, has no localizing value.

CILIARY GANGLION
(fig. 55–8)

The ciliary ganglion is "the peripheral ganglion of the parasympathetic system of the eye."[14] It is situated toward the back of the orbit, lateral to the optic nerve, medial to the lateral rectus muscle, and anterior or lateral to the ophthalmic artery. It

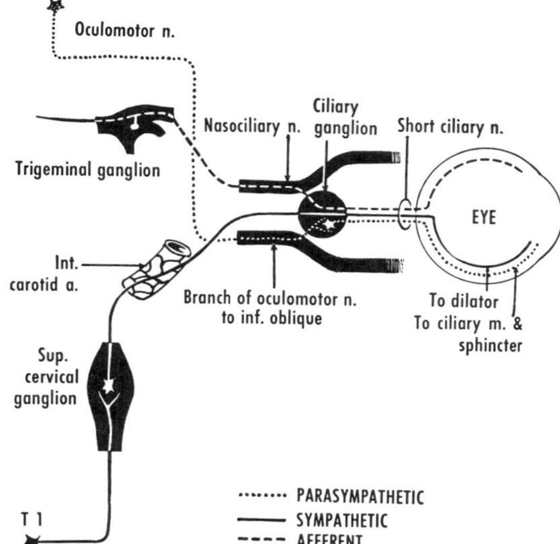

Figure 55–8 Scheme of ciliary ganglion and its connections. Lateral aspect.

is very small. The ganglion is frequently connected with the nasociliary nerve by communicating branches (sensory root). The *short ciliary nerves* are numerous branches of the ganglion that are distributed to the eyeball. Some of the afferent fibers from the eye (choroid, iris, and cornea) travel by the short ciliary nerves, pass through the ciliary ganglion, and reach the nasociliary nerve. The fibers connected with the ganglion are generally described as its roots. A parasympathetic (motor) root or roots come from the branch of the oculomotor nerve to the inferior oblique muscle. (The parasympathetic fibers in the oculomotor nerve are the axons of cells in one of the parts of the oculomotor nucleus in the midbrain.) These fibers synapse in the ganglion and are the only fibers that do so. **The postganglionic fibers pass to the short ciliary nerves and supply the ciliary muscle and the sphincter pupillae.** The sympathetic fibers are derived from the internal carotid plexus and reach the ganglion either directly or by means of the nasociliary nerve, a separate sympathetic root seldom being found. These fibers are postganglionic (arising in the superior cervical ganglion). They merely pass through the ciliary ganglion and, **by way of the short ciliary nerves, they supply the dilator pupillae, the orbitalis, the palpebral or tarsal muscles, and the blood vessels of the eyeball.**

MUSCLES OF EYEBALL

The chief muscles that move the eyeball (the ocular or extrinsic muscles of the eye) are the four recti and the two obliques (fig. 55–9). These six skeletal muscles are inserted into the sclera mostly by glistening tendons. With the exception of the inferior oblique, they all arise from the back of the orbit.

Rectus Muscles. **The four recti arise from a common tendinous ring, which surrounds the optic canal and a part of the superior orbital fissure (fig. 55–5). All the structures that enter the orbit through the optic canal and the adjacent part of the superior orbital**

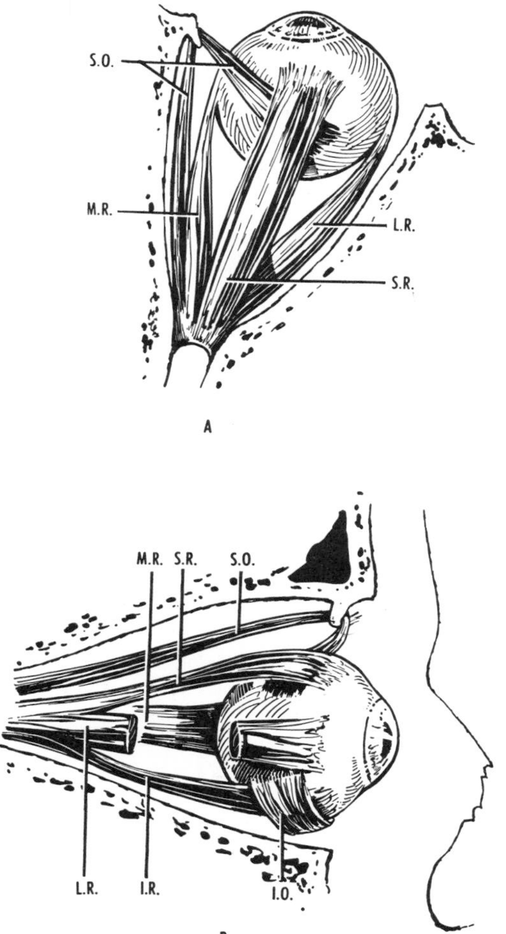

Figure 55–9 Muscles of eyeball. *A*, superior aspect. *B*, lateral aspect. The optic nerve has been omitted. Note that the two oblique muscles pass below the corresponding recti. *A* is based on Krimsky, *B* on Cogan.

fissure lie at first within the cone of the recti. The four muscles then run forward close to the walls of the orbit and are inserted into the front portion of the sclera. (Compare the oblique muscles, which are inserted into the back portion of the sclera.) The lateral rectus arises from the part of the common tendinous ring that spans the superior orbital fissure. (The continuous origin is sometimes described as comprising two heads.)

The lateral and medial recti lie in the same horizontal plane, whereas the superior and inferior recti lie in the same vertical plane (fig. 55–10*B*).

Oblique Muscles. The *superior oblique* arises from the sphenoid bone above and medial to the optic canal. It passes forward between the roof and the medial wall of the orbit, above the medial rectus. Its tendon then reaches a hyaline-cartilaginous sling known as the *trochlea,* or pulley, which is attached to the frontal bone. The tendon passes through the trochlea and then turns sharply laterally, backward, and downward, and is inserted into the posterolateral aspect of the sclera, largely under cover of the superior rectus. The trochlea indicates the "functional origin" of the muscle and, from there to its insertion, the superior oblique lies in the same vertical plane as the inferior oblique muscle (fig. 55–10*B*).

The *inferior oblique* arises in the front of the orbit, from a depression on the upper surface of the maxilla, lateral to the nasolacrimal canal. The muscle passes laterally and backward below the inferior rectus. It then winds upward, under cover of the lateral rectus, and is inserted into the posterolateral aspect of the sclera.

Innervation of Muscles of Eyeball

All three cranial nerves supplying the muscles of the globe enter the orbit through the superior orbital fissure. The superior oblique is supplied by the trochlear nerve,the lateral rectus by the abducent, and the superior, medial, and inferior recti, together with the inferior oblique, by the oculomotor nerve. **This may be summarized by the formula: SO_4, LR_6, remainder$_3$.**

Actions of Muscles of Eyeball

The eye is delicately poised in the fascia and fat of the orbit (fig. 55–4). Usually the globe and the fascia bulbi move

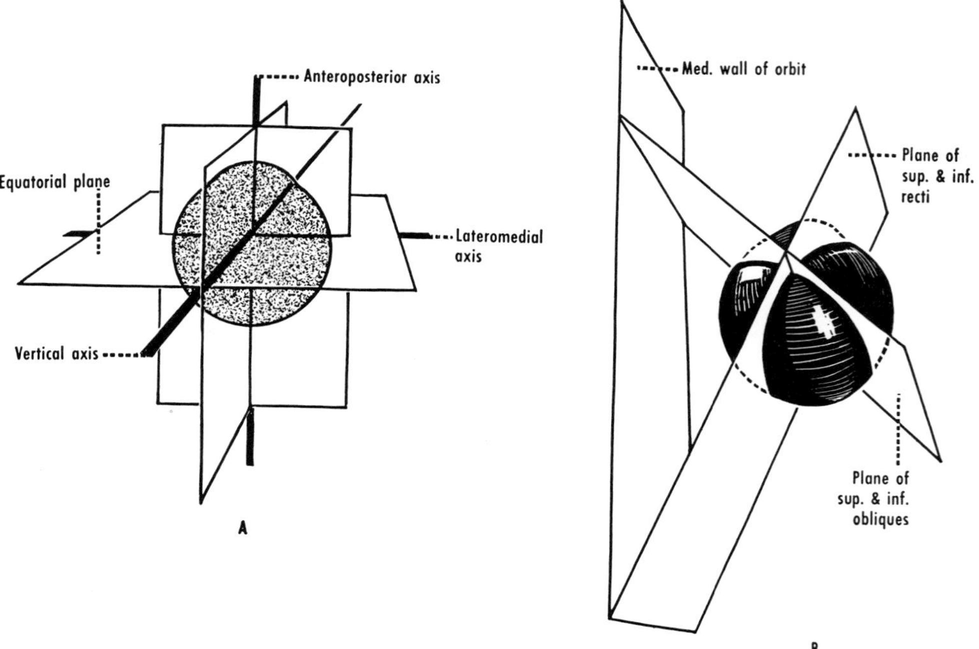

Figure 55–10 Axes of globe and orbit. Superior aspect. In *B* the planes of the muscles of the right eyeball are indicated. The lateral and medial recti lie in a horizontal plane (parallel with this page), which is not shown. *A* is based on Krimsky, *B* on Cogan.

together upon the orbital fat. The eyeball undergoes little shifting in position but rather oscillates around a center of movement about 2 mm behind the mathematical center of the globe. The position of rest of the eyes is that in which the gaze is straight ahead (the primary position). The muscles prevent protrusion of the eyeball.[15] Equilibrium is maintained by all the eyeball muscles, and one of the muscles never acts alone. Indeed, probably all the muscles participate in each movement of the eyes. Movements of the eyes are brought about by an increase in the tone of one set of muscles and a decrease in the tone of the antagonistic muscles.[16] The movements of the head itself are also important in vision.

The movements of the eyeball are commonly resolved into those taking place around three primary axes at right angles to one another (fig. 55–10*A*). The center of the cornea moves laterally (abduction) or medially (adduction) around a vertical (i.e., supero-inferior) axis, and upward (elevation) or downward (depression) around a lateromedial axis. These two axes are located in the equatorial plane of the eye. Furthermore, in combination with other movements, the eye may rotate around an anteroposterior axis, so that the upper part ("12 o'clock") of the cornea moves laterally ("extorsion") or medially ("intorsion"). The inferior rectus and inferior oblique are responsible for extorsion, the superior rectus and superior oblique[17] for intorsion. The lateral and medial recti are purely an abductor and an adductor, respectively; the actions of the other four muscles are more complicated, and are shown in table 55–1 and in figures 55–11 and 55–12.

Paralysis of a muscle of the globe is noted by (1) limitation of movement of the eye in the field of action of the paralyzed muscle, and (2) the production of two images that are separated maximally when an attempt is made to use the paralyzed muscle.

The normal and abnormal movements of the eyes are complicated. The two eyes may be moved in the same direction, e.g., to the right by the right lateral rectus and the left medial rectus. But the eyes may also be moved in opposite directions, e.g., in convergence.

Summary of Muscles of Eyeball
(fig. 55–12)

Innervation: SO_4, LR_6, $remainder_3$. The four recti extend from the back of the orbit to the front of the sclera. Therefore, L.R. and M.R. abduct and adduct, respectively, whereas S.R. and I.R. elevate and depress, respectively. The trochlea of S.O. serves as its functional origin. Hence, the two obliques may be said to extend from the front of the orbit to the back of the sclera. Therefore, S.O. and I.O. have a tendency to depress and elevate, respectively.

Furthermore, owing to their lateral course, the axes of the two recti (S.R. and I.R.) coincide with the anteroposterior axis of the *abducted* (L.R.) eye. In this position, S.R. and I.R. elevate and depress, respectively, whereas I.O. and S.O. cannot do so. By contrast, the axes of the two obliques (I.O. and S.O.) coincide with the anteroposterior axis of the *adducted* (M.R.) eye. In this position, I.O. and S.O. now elevate and depress, respectively, whereas S.R. and I.R. cannot do so.

TABLE 55–1 Main Actions of Superior and Inferior Muscles of Eyeball

Muscle	*Eye Abducted*	*Eye in Primary Position*	*Eye Adducted*
Sup. oblique	Intorsion	Depression Abduction Intorsion	Depression
Inf. oblique	Extorsion	Elevation Abduction Extorsion	Elevation
Sup. rectus	Elevation	Elevation Adduction Intorsion	Intorsion
Inf. rectus	Depression	Depression Adduction Extorsion	Extorsion

OPTIC NERVE

The optic (2nd cranial) nerve is the nerve of sight. It is about 5 cm in length and extends between the optic chiasma and the eyeball. **Developmentally, the optic nerve may be considered as a tract of fibers that connect the retina (a derivative of the brain) with the brain. About 90 per cent[18] of the fibers of the optic nerve are afferent and arise in the layer of ganglion cells in the retina.** The fibers, which are the axons of the ganglion cells, lie internally in the retina and converge on the optic disc. There they pierce

ELEVATION OF ADDUCTED EYE BY I.O.

ELEVATION OF ABDUCTED EYE BY S.R.

ADDUCTION BY M.R.

I.O.

S.O.

S.R.

I.R.

PRIMARY POSITION

ABDUCTION BY L.R.

DEPRESSION OF ADDUCTED EYE BY S.O.

DEPRESSION OF ABDUCTED EYE BY I.R.

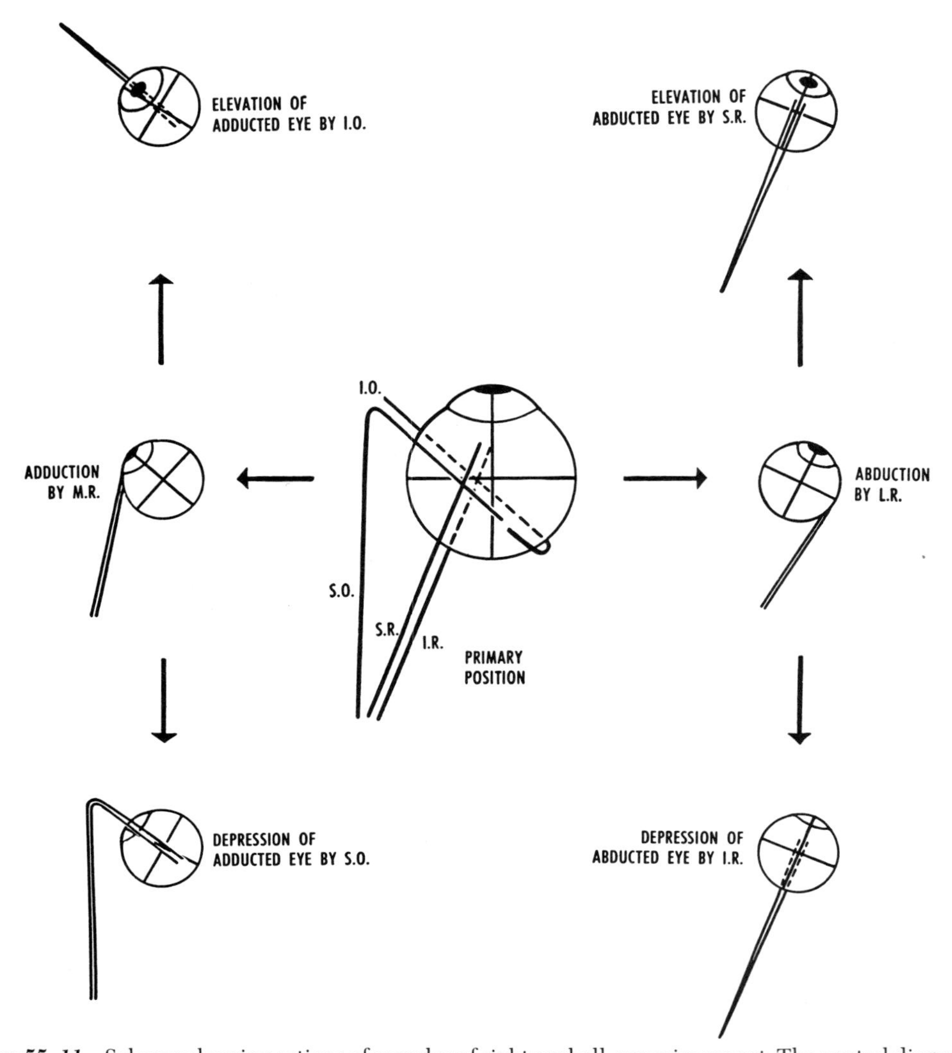

Figure 55–11 Scheme showing actions of muscles of right eyeball, superior aspect. The central diagram shows the primary position. The three diagrams at the left show the adducted eye, the three at the right, the abducted eye. To interpret the scheme, start at the primary position and follow the arrows.

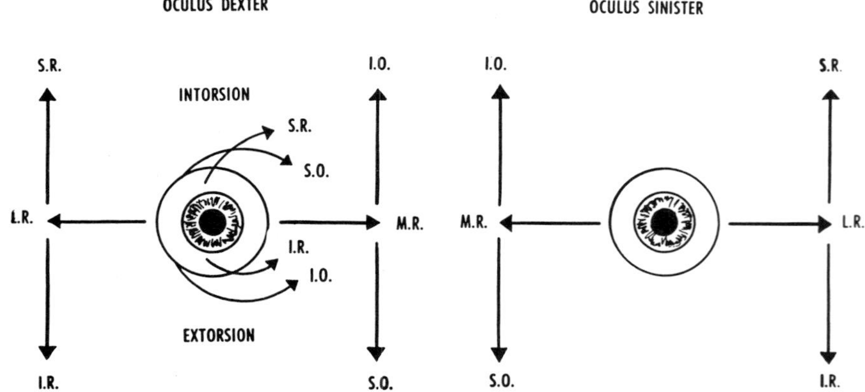

Figure 55–12 Scheme summarizing muscles responsible for vertical movements of abducted and adducted eyes, anterior aspect. The arrows must be followed in strict sequence: first, an abducting or an adducting arrow, and only then an elevating or depressing arrow. For the right eye, an indication of muscles responsible for intorsion and extorsion has been included.

the remaining layers of the retina, the choroid, and the lamina cribrosa of the sclera, and receive myelin sheaths. The fibers of the optic nerve do not possess a neurilemma.

The optic nerve proceeds backward and medially from the eyeball in a sinuous manner. In the orbit, it lies within the cone formed by the recti. It is crossed superiorly by the ophthalmic artery and the nasociliary nerve. It is pierced inferomedially by the central artery and vein of the retina, and these vessels reach the optic disc by running within the optic nerve.

The optic nerve leaves the orbit and gains the middle cranial fossa by passing through the optic canal. The canal is related medially to the sphenoidal sinus or to an ethmoidal air cell. **The optic nerve is related below to the internal carotid and ophthalmic arteries, and to the hypophysis. The nerve ends in the optic chiasma, where the medial fibers decussate. (The decussating fibers are those that come from the medial or nasal side of the retina and represent, therefore, the lateral or temporal side of the visual field.)** From the chiasma, the fibers are continued in the optic tracts to the lateral geniculate bodies and the midbrain.

The optic nerve is surrounded in the orbit by three sheaths, which are continuous with the three meninges, and a prolongation of the subarachnoid space is present.

Examination. The most commonly used testing procedures concerned with vision are: ophthalmoscopy (p. 649), testing of visual acuity, and plotting of visual fields (roughly by means of a finger, and more precisely by an instrument termed a perimeter).

EYELIDS

The eyelids *(palpebrae)* (fig. 55–13) are two movable, musculofibrous folds placed in front of each orbit. They protect the globe and rest the eye from light. The upper eyelid, more extensive and more mobile than the lower, meets the latter at the medial and lateral *angles (canthi)* of the eye (fig. 56–5). The *epicanthus* is a fold of skin that covers the medial canthus in some people, chiefly Orientals, who may also have a "Mongolian fold" of skin over the lower part of the upper eyelid. These folds are due to a less extensive insertion of the levator muscle.[19] The *palpebral fissure* is

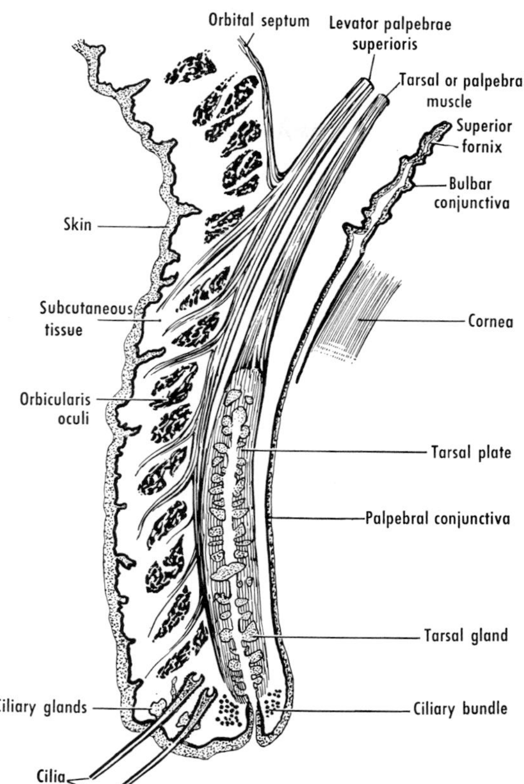

Figure 55–13 Diagram of sagittal section through the upper eyelid.

the opening bounded by the upper and lower eyelids. It is the mouth of the conjunctival sac and it varies in size according to the degree to which the "eye is open." When the "eye is closed," the fissure is merely a slit, and the cornea is completely covered by the upper eyelid (fig. 56–2).

The free margin of each eyelid possesses two or three rows of hairs termed *cilia,* or *eyelashes.* The nearby *ciliary glands* are both sweat and sebaceous in type. Infection of a ciliary gland may result in a sty(e). Near its medial end, the free margin of each lid presents an opening, the lacrimal punctum. The medial ends of the upper and lower lids delimit an area termed the *lacrimal lake,* the floor of which presents a small, red "fleshy" mass, the *lacrimal caruncle* (fig. 56–5). The caruncle lies on a fold of conjunctiva called the plica semilunaris, or semilunar fold.

Structure. Each eyelid consists of a series of layers (fig. 55–13). From before backward, in the case of the upper lid, these are:

1. Skin and subcutaneous tissue. The skin of the eyelids is very thin. **The subcutaneous tissue usually contains no fat, and fluid can readily accumulate there.**

2. The muscular plane includes portions of the orbicularis and levator muscles.

The *palpebral part of the orbicularis oculi* (p. 654) arises from the medial palpebral ligament, sweeps across the eyelids, and forms the lateral palpebral raphe. A small bundle close to the margin of each eyelid is termed the *ciliary bundle*. The eyelid can be split surgically into an anterior and a posterior portion along the plane of the submuscular tissue.

The *levator palpebrae superioris* (fig. 55–14) arises in the orbit from the lesser wing of the sphenoid, above the optic canal. It passes forward, above the superior rectus, and ends in an aponeurosis. This expands downward into the upper eyelid and is inserted chiefly into *(a)* the skin of the upper eyelid (which it reaches by passing through the fibers of the orbicularis) and the front of the tarsal plate, and *(b)* the upper border of the tarsal plate by means of the superior tarsal muscle. The lower eyelid has no special elevator or depressor. The levator is supplied by the oculomotor nerve; the tarsal muscle has a sympathetic innervation. The levator raises the upper eyelid, thereby uncovering the cornea and a portion of the sclera. Its antagonist is the orbicularis. **Paralysis of the levator results in drooping (ptosis) of the upper lid.**

3. The fibrous layer comprises the orbital septum and the tarsal plate.

The *orbital septum (palpebral fascia)* is a thin fibrous membrane attached to the entire margin of the orbit, where it is anchored to the periorbita. It extends to the levator in the upper lid, and to the tarsal plate in the lower lid.

A *tarsal plate* strengthens each lid. It consists of dense fibrous tissue, together with some elastic fibers. Each plate is extensively hollowed out (or perhaps grooved posteriorly) by *tarsal glands*, sebaceous in type. The glands can be seen as yellow streaks through the conjunctiva. The lateral and medial ends of the upper and lower tarsal plates are anchored to the margin of the orbit by the *lateral* and *medial palpebral ligaments*. The medial ligament presents a prominent inferior border that can be seen and felt through the skin on drawing the lids laterally *in vivo*. **The medial palpebral ligament is placed in front of the upper part of the lacrimal sac, to which it serves as a guide** (fig. 55–2*B*). It is attached to at least the anterior lacrimal crest and is frequently described as splitting to embrace the lacrimal fossa.

The superior and inferior *tarsal muscles* are thin sheets of smooth muscle found in the upper and lower eyelids, respectively. The superior tarsal muscle connects the levator with the upper border of the tarsal plate. The inferior tarsal muscle is poorly developed. Both muscles are supplied by sympathetic fibers. They widen the palpebral fissure. **A lesion of the cervical sympathetic trunk may result in ptosis of the upper eyelid (Horner's syndrome, p. 706).**

4. Mucous membrane. The mucosa of the eyelids is termed the palpebral part of the conjunctiva (see below). At the free margin of each lid it is continuous with the skin. In order to examine the palpebral conjunctiva of the upper eyelid, the eyelid should be everted around either a finger or a cotton applicator.

Sensory Innervation of Eyelids. The upper and lower lids are supplied chiefly by the supra-orbital and infra-orbital nerves, respectively; that is, by the first and second divisions of the trigeminal nerve, respectively.

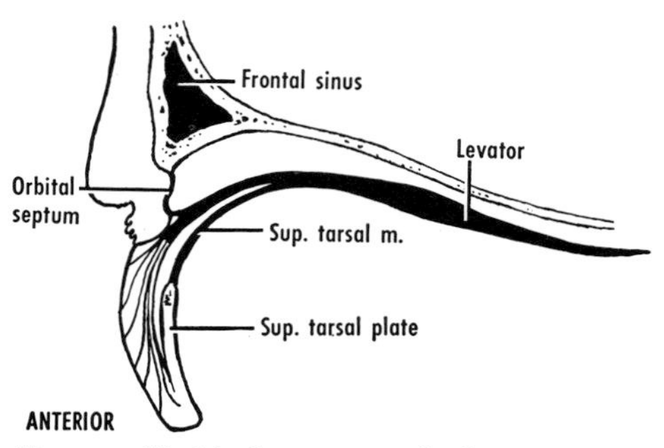

Figure 55–14 Levator palpebrae superioris. Sagittal section through eyelid and roof of orbit. Based on Whitnall.

CONJUNCTIVA

The conjunctiva (figs. 55–13 and 56–2) is the thin mucous membrane that lines the back of the eyelids and the front of the globe. The *conjunctival sac* is the capillary interval, lined by conjunctiva, between the eyelids and the globe (fig. 56–2). The palpebral fissure is the mouth of the conjunctival sac, and it varies in size according to the degree to which the "eye is open." For purposes of description, the conjunctiva is divided into palpebral and bulbar parts.

Palpebral Conjunctiva. **The palpebral conjunctiva lines the back of the eyelids. It contains the openings of the lacrimal canaliculi, as a result of which the conjunctival sac ultimately communicates with the inferior meatus of the nose. The palpebral conjunctiva is red and very vascular; it is examined in cases of suspected anemia.** The superior and inferior *fornices* of the conjunctiva are the culs-de-sac formed by the reflection of the conjunctiva from the upper and lower lids, respectively, to the eyeball. The superior fornix receives the openings of the ducts of the lacrimal glands.

Bulbar Conjunctiva. **The bulbar conjunctiva is transparent, thereby allowing the sclera to show through it as the "white of the eye." It is colorless, except when its vessels are dilated as a result of inflammation. Its peripheral part is loose, thereby allowing free movement of the eyeballs. Its central part is continuous at the limbus with the anterior epithelium of the cornea.** The *plica semilunaris,* or *semilunar fold,* is a fold of conjunctiva at the medial angle of the eye, deep to the lacrimal caruncle. The fold intercepts foreign bodies on the cornea and passes them to the region of the lacrimal caruncle. It does not correspond to the nictitating membrane of birds.[20]

Innervation of Conjunctiva. The conjunctiva is supplied by the infratrochlear, lacrimal, and ciliary nerves. The majority of the fibers terminate in free nerve endings.[21]

Blood Supply of Conjunctiva (fig. 56–7). **The vessels of the bulbar conjunctiva are clearly visible *in vivo* and the circulation in them can be studied microscopically.**[22] The bulbar conjunctiva is supplied by (1) a peripheral palpebral arcade (peripheral to the tarsal plate), from which posterior conjunctival arteries emerge and bend around the fornix; and (2) the anterior ciliary arteries, which arise from the branches to the recti, pass forward and give off anterior conjunctival arteries, and join the greater arterial circle of the iris. The anterior and posterior conjunctival arteries form a plexus around the cornea. The palpebral conjunctiva is supplied by the marginal palpebral arcade in the eyelid.

The superficial vessels of the pericorneal plexus, derived from the posterior conjunctival, are dilated on rubbing the eyelids, or by the wind, or in superficial affections of the cornea. **In conjunctivitis the bulbar conjunctiva becomes brick-red. The redness increases toward the fornices, and does not fade on pressure; the vessels move with the conjunctiva.**

The deep vessels of the pericorneal plexus, derived from the anterior conjunctival, are dilated in diseases of the deep portion of the cornea, the iris, or the ciliary body. **The result of such dilation is a rose-pink band of "ciliary injection." The redness disappears on pressure, but the vessels do not move with the conjunctiva.**

Lacrimal Apparatus

The lacrimal apparatus comprises (1) the lacrimal gland and its ducts, and (2) the lacrimal passages: the lacrimal canaliculi and sac, and the nasolacrimal duct (fig. 55–15).

Lacrimal Gland. The lacrimal gland is lodged in a fossa at the anterolateral angle of the roof of the orbit (frontal bone). It rests on the lateral rectus and the levator. The main part of the gland, limited in front by the orbicularis and the orbital septum, is termed the *orbital part.* A process, called the *palpebral part* of the gland, projects into the lateral part of the upper lid, where it rests on the conjunctiva of the superior fornix. The two parts are continuous with each other around the lateral border of the aponeurosis of the levator. Accessory lacrimal glands are found near the superior fornix.

The lacrimal gland is drained by means of a dozen *lacrimal ducts,* all of which traverse the palpebral part of the

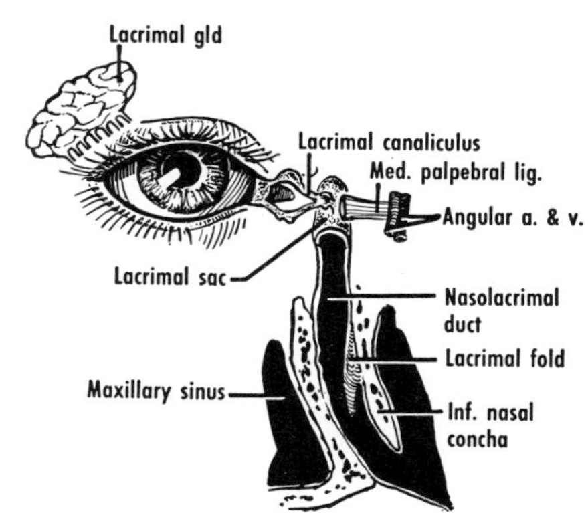

Figure 55–15 Diagram of lacrimal apparatus.

gland and open into the superior conjunctival fornix.

Tears[23] are secreted by the lacrimal and the accessory lacrimal glands. About half the quantity of tears secreted evaporates and the remainder drains into the lacrimal sac. Tears keep the front of the eye moist, thereby preventing drying of the anterior epithelium of the cornea.

The secretory fibers to the lacrimal gland are derived from the greater petrosal nerve (a branch of the facial) and the nerve of the pterygoid canal. The fibers synapse in the pterygopalatine ganglion[24] and are conveyed to the gland (1) as orbital branches of the ganglion and (2) by a filament given by the zygomatic to the lacrimal nerve. Sympathetic fibers from the superior cervical ganglion reach the gland by way of the lacrimal artery and nerve.

Lacrimal Canaliculi. The lacrimal canaliculi, one in each eyelid, are about 1 cm in length. Each commences at a *lacrimal punctum,* which is situated on a slight *papilla* and can be seen on everting the eyelid (fig. 56–5, p. 647). Each canaliculus presents a bend in its course (where a dilatation or *ampulla* may be found) and opens into the lacrimal sac. Tears enter the canaliculi at least partly by capillarity. The lacrimal part of the orbicularis oculi perhaps plays a role in the drainage of tears.[25]

Lacrimal Sac. The lacrimal sac, about 1 to 1½ cm in length, is continuous with the upper end of the nasolacrimal duct. It is lodged in a fossa at the medial margin of the orbit (lacrimal bone and maxilla). The upper part of the sac is covered in front by the medial palpebral ligament (fig. 55–2*B*). The lacrimal canaliculi open into the lateral wall of the sac, usually by means of a common sinus. The fossa for the sac is bounded by the anterior and posterior lacrimal crests, to which is attached the lacrimal fascia, forming a roof and a lateral covering for the sac. The sac is related posteriorly to the lacrimal part of the orbicularis oculi and medially to the ethmoidal air cells and the middle meatus. The angular vessels are situated anteromedial to the sac.

Nasolacrimal Duct. The nasolacrimal duct, about 2 cm in length, extends from the lower end of the lacrimal sac to the inferior meatus of the nose. The lumen of the duct is frequently irregular, and various folds of valve-like appearance may be encountered. The opening of the duct into the meatus is sometimes marked by a fold of mucous membrane termed the *lacrimal fold.* The nasolacrimal duct is situated in a bony canal formed by the lacrimal bone, the maxilla, and the inferior nasal concha.

REFERENCES

1. M. Neiger, Acta anat., Suppl. 39, 1960.
2. L. J. A. DiDio, Anat. Rec., *142*:31, 1962.
3. P. Keros and G. Nemanić, Folia morph., *15*:79, 1967.
4. R. O'Rahilly, Anat. Rec., *141*:315, 1961.
5. C. D. Ray, Amer. J. phys. Anthrop., *13*:309, 1955.
6. C. Vermeij-Keers, Z. Anat. EntwGesch., *141*:77, 1973.
7. S. S. Hayreh, Brit. J. Ophthal., *46*:212, 1962, and Brit. J. Surg., *50*:938, 1963. S. S. Hayreh and R. Dass, Brit. J. Ophthal., *46*:65 and 165, 1962.
8. S. Singh and R. Dass, Brit. J. Ophthal., *44*:193 and 280, 1960.
9. K. C. Wybar, Brit. J. Ophthal., *40*:65, 1956.
10. J. Brucher, Radiol., *93*:51, 1969. See also O. F. Gabriele and D. Bell, Radiol., *89*:841, 1967.
11. N. Kato and H. Outi, Okajimas Folia anat. jap., *38*:411, 1962.
12. K. M. Houser, Arch. Otolaryng., Chicago, *16*:488, 1932.
13. E. Wolff, Brit. J. Ophthal., *12*:22, 1928.
14. L. Devos and R. Marcelle, Arch. Anat., Strasbourg, *27*:277, 1939.
15. B. Tengroth, Acta ophthal., Kbh., *38*:698, 1960.
16. G. M. Breinin and J. Moldaver, Arch. Ophthal., *54*:200, 1955. A Björk and E. Kugelberg, Electroenceph. clin. Neurophysiol., *5*:595, 1953. G. M. Breinin, Amer. J. Ophthal., *72*:1, 1971.
17. R. S. Jampel, Arch. Ophthal., Chicago, *75*:535, 1966.
18. J. R. Wolter and R. R. Knoblich, Brit. J. Ophthal., *49*:246, 1965.
19. B. T. Sayoc, Amer. J. Ophthal., *42*:298, 1956.
20. E. P. Stibbe, J. Anat., Lond., *62*:159, 1928. But see also T. Arao and E. Perkins, Anat. Rec., *162*:53, 1968.
21. D. R. Oppenheimer, E. Palmer, and G. Weddell, J. Anat., Lond., *92*:321, 1958.
22. E. H. Bloch, Anat. Rec., *120*:349, 1954. R. Landesman *et al.*, Amer. J. Obstet. Gynec., *66*:988, 1953.
23. E. R. Veirs, *The Lacrimal System. Clinical Application,* Grune & Stratton, New York, 1955.
24. S. L. Ruskin, Arch. Ophthal., Chicago, *4*:208, 1930.
25. But see J. A. Brienen and C.A.R.D. Snell, Ophthalmologica, *159*:223, 1969.

GENERAL READING

Duke-Elder, S., and Wybar, K. C., *The Anatomy of the Visual System,* vol. 2 of S. Duke-Elder (ed.), *System of Ophthalmology,* Kimpton, London, 1961. A superb work of reference for the orbit and the eye.

Whitnall, S. E., *The Anatomy of the Human Orbit and Accessory Organs of Vision,* Oxford University Press, London, 2nd ed., 1932. The classical study of the orbit.

Wolff, E., *The Anatomy of the Eye and Orbit,* Lewis, London, 6th ed. revised by R. J. Last, 1968. A well-known and well-illustrated text.

THE EYE

56

The eye (L., *oculus;* Gk., *ophthalmos*) (fig. 56–1) occupies one-third or less of the cavity of the orbit. It measures 24 mm in diameter.[1] It comprises portions of two spheres: a posterior five-sixths and an anterior one-sixth. The optic nerve emerges from the eyeball a little medial to the posterior pole of the globe. The midpoints of the two pupils lie about 60 mm apart. The anteroposterior diameter of the globe may be greater (as in myopia) or less (as in hypermetropia) than the normal (emmetropia).

Certain terms of reference may most easily be understood by comparing the eye with the earth turned in a way such that one pole faces anteriorly and the other posteriorly. The *anterior* and *posterior poles* of the globe are the central points of the corneal and scleral curvatures, respectively. The geometrical or anteroposterior axis connects the two poles. The *equator* is an imaginary line about the globe, everywhere equidistant from the two poles. Hence the equator is in a coronal plane. A *meridian* is any imaginary line on the globe from pole to pole, cutting the equator at a right angle. Two opposite meridians form a circle, the plane of which is described as meridional. Hence meridional sections through the eye may be horizontal, sagittal, or oblique.

TUNICS OF EYE

The *eyeball (globe, bulb)* has three concentric coverings (fig. 56–2): (1) an external, protective, fibrous tunic, comprising the cornea and sclera; (2) a middle, vascular, pigmented tunic, comprising the iris, ciliary body, and choroid; and (3) an internal tunic termed the retina.

EXTERNAL FIBROUS TUNIC

Cornea

The cornea[2] is the anterior, transparent part of the external tunic of the eye. It is ½ mm in thickness at the center[3] and 1 mm at the periphery. Its main bulk, the substantia propria, is continuous with the sclera. The transparency of the substantia propria of the cornea is dependent on its dehydration, and this is believed to be maintained by its limiting epithelia. When the cornea is not a part of a sphere but is more curved in one meridian than another, the condition is termed astigmatism. How-

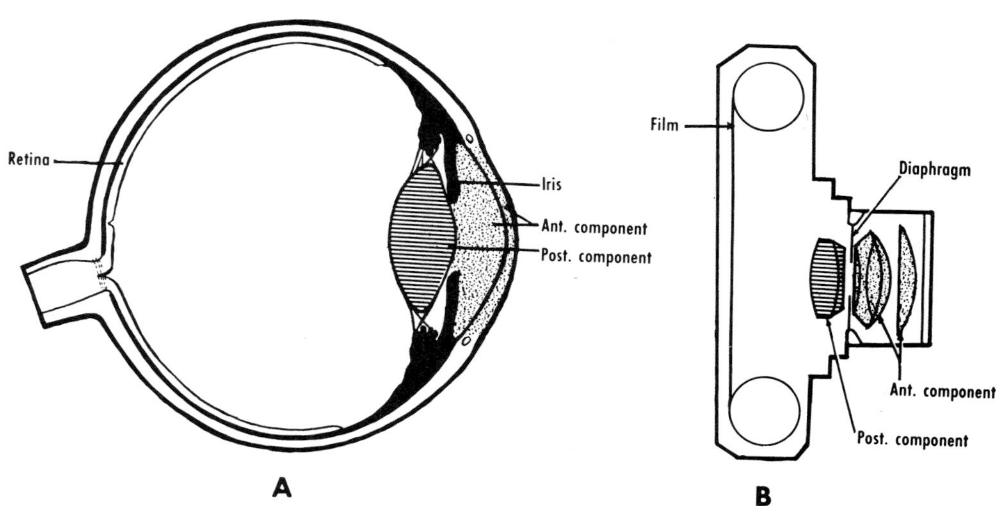

Figure 56–1 Diagram showing comparison between the optics of *(A)* the eye, and *(B)* a modern miniature camera. Note that, in each case, the refractive part consists of two components: (1) an anterior (cornea and aqueous humor, in the case of the eye), and (2) a posterior (lens, in the case of the eye), in each instance separated by an iris diaphragm. The lens of this camera has a focal length of 5 cm, and an aperture range of f 2–f 22. The lens system of the eye has a focal length of 2 cm, and an aperture range f 2.5–f 11. It should be noted, however, that the visual image is not imprinted as on a film but is coded and transmitted somewhat as in television.

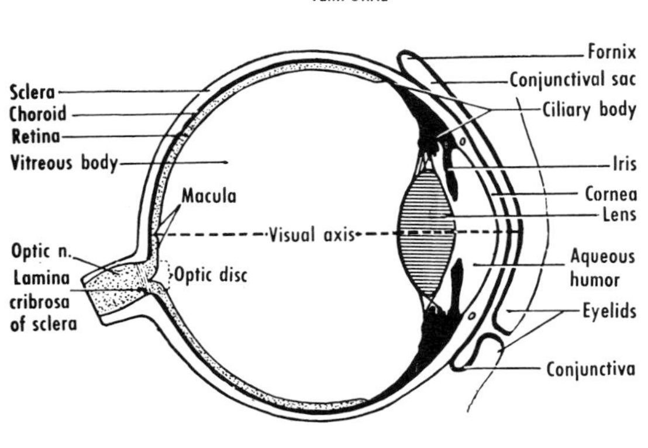

Figure 56–2 Horizontal section of eyeball. Note the three tunics and the visual axis.

ever, **most of the refraction by the eye takes place not in the lens but at the surface of the cornea.**

The *limbus* of the cornea comprises the region of the conjunctivocorneal and sclerocorneal junctions (fig. 56–3). The periphery of the cornea frequently displays a whitish ring *(arcus senilis)* in older persons, owing to fatty degeneration.[4]

The cornea is avascular and is nourished by permeation from the periphery along the substantia propria. **The cornea is supplied by the ophthalmic nerve (from the 5th cranial) by means of its ciliary branches.** The free nerve endings in the anterior epithelium subserve the sensations of pain, touch, and perhaps warmth and cold.[5] The nerves of the cornea are the afferent limb of the corneal reflex (closure of the eyelids on stimulation of the cornea).

The cornea consists, from before backward, of (1) anterior epithelium (continuous with that of the conjunctiva), (2) anterior limiting lamina, (3) substantia propria (connective tissue), (4) posterior limiting lamina, and (5) mesothelium (frequently and inappropriately termed endothelium).

The anterior epithelium, substantia propria, mesothelium, and the nerves in the cornea can be seen *in vivo* by means of the slit-lamp, an instrument that comprises a stereoscopic microscope mounted on a stand, together with an illuminating system with a slit-like aperture in its diaphragm.

Sclera

The sclera is the posterior, opaque part of the external tunic of the eye. The front part of the sclera can be seen through the conjunctiva as "the white of the eye." The sclera consists of a feltwork of collagenous fibers. Externally it is loosely connected with the fascia bulbi by episcleral tissue and receives the tendons of the muscles of the eyeball. Furthermore, it is pierced by the ciliary arteries and nerves, and by the venae vorticosae. Posteriorly, the sclera is perforated by the optic nerve. The sieve-like part of the sclera through which the nerve fibers pass is termed the *lamina cribrosa.* Fibers from the sclera form a sheath for the optic nerve, and the sheath is continuous with the dura mater.

In inflammation of the cornea (keratitis), and of the iris and ciliary body (iridocyclitis), the anterior ciliary vessels become dilated (ciliary injection) in the episcleral tissue. These vessels do not move when the conjunctiva is moved. By contrast, in inflammation of the conjunctiva (conjunctivitis), the posterior conjunctival vessels (from the palpebral) become dilated (p. 641).

The *fascia bulbi*[6] is a thin, fibrous membrane that envelops the globe from near the margin of the cornea

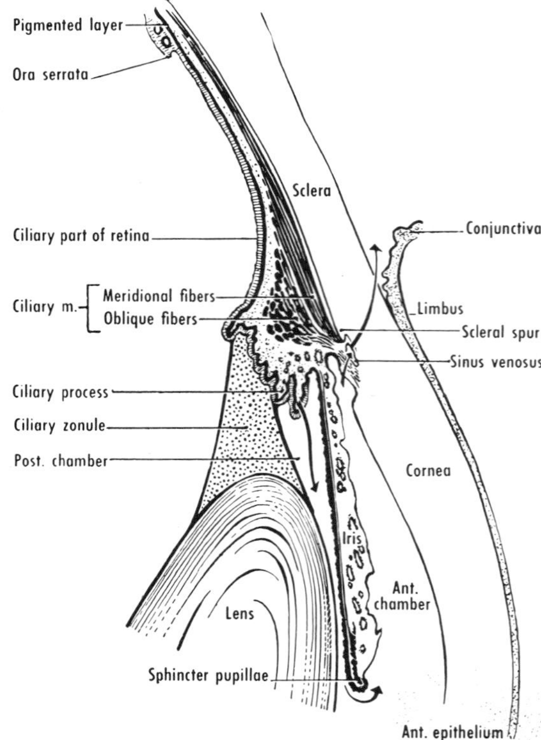

Figure 56–3 Meridional section to show the ciliary region and the iridocorneal angle. The trabecular meshwork can be seen between the sinus venosus and the iridocorneal angle. The four arrows indicate the formation and drainage of the aqueous humor. A narrow line in front of the pigmented epithelium of the iris indicates the approximate position of the dilator pupillae.

in front to the optic nerve behind. The fascia bulbi is pierced by the tendons of the muscles of the eyeball and sends a tubular reflection around each of the muscles (fig. 55–4). The globe is suspended by a blending of the sheaths of the inferior rectus and inferior oblique. The orbital fat is found outside the fascia bulbi. After surgical removal of an eye, the fascia bulbi forms a socket for the prosthesis.

The eyeball is marked at the sclerocorneal junction by a slight furrow internally. The sclera projects partly behind the furrow to form a roll termed the *scleral spur* (fig. 56–3). Adjacent to the spur is a canal termed the *sinus venosus sclerae.* The sinus runs around the eye and is frequently divided in a part of its course. It is separated from the iridocorneal angle by a meshwork of trabeculae and spaces,[7] which can be seen *in vivo* by means of a slit-lamp. **The aqueous humor, formed by the ciliary processes (p. 650), filters through channels leading from the anterior chamber to the sinus venosus.[9] The sinus is lined by endothelium and drains, by means of *aqueous veins*, into the scleral plexuses.[10]** The sinus contains either blood (under conditions of venous stasis) or aqueous humor.[11]

The *iridocorneal angle* (between the iris and the cornea), otherwise known as the *angle of the anterior chamber*, or the *filtration angle*, is very important physiologically and pathologically. It can be examined *in vivo* by a special instrument, termed a gonioscope, which comprises a corneal microscope and a special contact lens.

MIDDLE VASCULAR TUNIC

The middle, vascular tunic of the eye, frequently termed the *uvea*, comprises the choroid, the ciliary body, and the iris, from behind forward.

Choroid

The choroid is a brown coat that lines the greater part of the sclera. It contains pigment cells *(suprachoroid lamina)*, arteries (from the short posterior ciliary) and veins (to the venae vorticosae) *(vascular lamina)*, and capillaries *(choriocapillary lamina)*, adjacent to the basement membrane of the pigmented layer of the retina *(basal lamina)*.

Ciliary Body

The ciliary body is a thickening in the vascular tunic. It is situated in front of the ora serrata of the retina (fig. 56–3), it connects the choroid with the iris, and it contains the ciliary muscle and the ciliary processes.

The *ciliary muscle* comprises two main sets of smooth muscle fibers (fig. 56–3): *(a)* meridional or longitudinal fibers extend from the scleral spur in front to the suprachoroid lamina behind; and *(b)* oblique fibers enter the base of the ciliary processes. The ciliary muscle is supplied by parasympathetic fibers by way of the ciliary nerves (and also by sympathetic fibers). The details of the architecture and action of the muscle are disputed. **When it contracts, the ciliary body moves forward. This presumably decreases the tension on the fibers of the ciliary zonule (p. 650). The central part of the lens then becomes more curved (mostly anteriorly), and the eye can be focused on near objects, a process known as accommodation.** The tone of the ciliary muscle can be abolished by atropine.

The *ciliary processes*, about 70 in number, are arranged meridionally in a circle behind the iris (fig. 56–4). They may be regarded as localized, whitish thickenings of the vascular lamina.

The ciliary body is lined by the *ciliary part of the retina*, consisting of two layers of epithelium, the outer of which is heavily pigmented.

Iris

The iris is a circular, pigmented diaphragm that lies in front of the lens in a more or less coronal plane. Its peripheral border, or root, is attached to the ciliary body, whereas its central border is free and bounds an aperture known as the pupil. The pupil appears black because the rays of light reflected from the retina are refracted by the lens and the cornea, and go back to the source of the light. The central border corresponds to the rim of the embryonic optic cup. The iris divides the space between the cornea and the lens into an anterior and a posterior chamber (fig. 56–3). **The *anterior chamber* is bounded by the cornea and iris, and by portions of the sclera, ciliary body, and lens. The *posterior chamber* is bounded by the iris, ciliary processes, ciliary zonule, and lens. Both chambers are filled with aqueous humor.**

The anterior surface[12] of the iris presents excavations termed *crypts*, and also an irregular fringe known as the *collarette* (fig. 56–5). The collarette

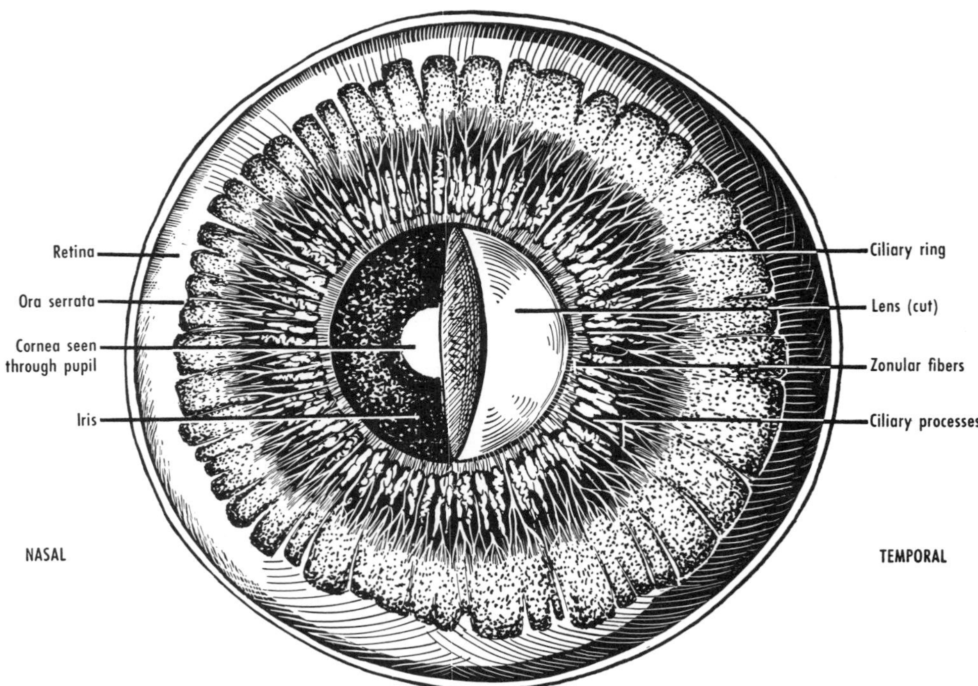

Figure 56–4 Anterior half of right eye seen from behind. The ciliary ring is that portion of the ciliary body situated between the ora serrata and the ciliary processes. Based on Wolff.

marks the line of attachment of the pupillary membrane in the fetus.

The *pectinate ligament* of the iris comprises some strands at the apex of the iridocorneal angle that are generally ill-developed and are seldom found.[13] The strands enclose the *spaces of the iridocorneal angle*.

A defect of a part of the iris is termed a coloboma.

The *stroma* of the iris contains collagenous fibers, tissue spaces,[14] vessels, nerves, pigment cells (chromatophores), and the sphincter pupillae.

The *sphincter pupillae* is in the posterior part of the stroma near the pupil. It consists of smooth muscle but develops from ectoderm. It is supplied by parasympathetic fibers by way of the ciliary nerves. Its contraction results in constriction of the pupil (miosis). **The iris contracts reflexly when light reaches the retina (the light reflex) and during focusing on a near object (the accommodation reaction).** The pathway for the light reflex involves the retina, optic nerve, midbrain, oculomotor nerve, ciliary ganglion, short ciliary nerves, and sphincter pupillae. **A drop of atropine placed on the eye annuls the actions of the ciliary muscle and the sphincter pupillae, both of which are under parasympathetic control.** The result is inability to accommodate, and also a dilatation of the pupil due to overaction of the dilator pupillae. Thus atropine is of use in the examination of the eye.

The color of the iris depends on the arrangement and type of pigment and on the texture of the stroma. The stroma contains little or no pigment in gray and blue irides, whereas melanophores are numerous in brown irides. The blue color is due to diffraction and arises in a manner similar to that of the blue sky. "It is the colour of a cloudy emulsion on a dark background" (A. Vogt). In terms of heredity, brown is dominant and blue is recessive. The pigment is relatively less at birth, for which reason the iris in infants is generally blue. In albinos, pigment is absent from both the stroma and the epithelium, and the pink color of the iris is due to blood.

The *dilator pupillae* consists of smooth muscle fibers derived probably from myoepithelial cells that form a part of the underlying pigmented epithelium and hence are ectodermal in origin. **The dilator pupillae is supplied by sympathetic fibers (roots from C8 to T4[15] by way of the ciliary nerves).** Contraction results in dilatation of the pupil (mydriasis).

The *iridial part of the retina* consists

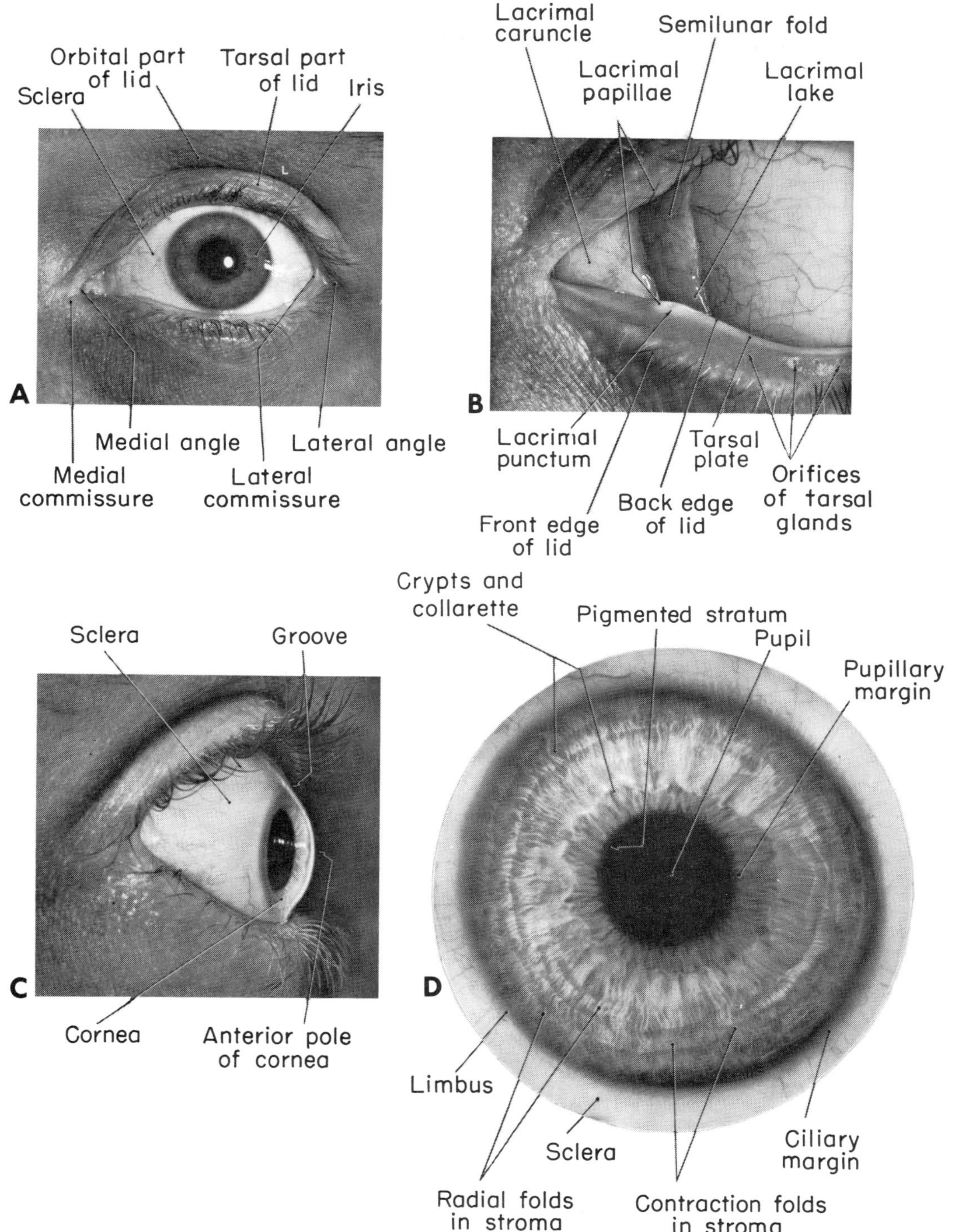

Figure 56–5 Photographs of the eye *in vivo*. Courtesy of Mr. H. L. Gibson, Eastman Kodak Company; from *Medical Radiography and Photography*, 28:123, 1952.

of two layers of pigmented epithelium and is derived from the most anterior part of the embryonic optic cup.

Autonomic Innervation of Eye

The autonomic innervation of the eye may be summarized as follows:

1. **Parasympathetic (synapses in ciliary ganglion):**
   - **sphincter pupillae** (p. 646);
   - **ciliary muscle** (p. 645).
2. **Sympathetic (synapses in superior cervical ganglion):**
   - **dilator pupillae** (p. 646);
   - **orbitalis (smooth muscle fibers in the region of the inferior orbital fissure,** p. 630**);**
   - **superior tarsal muscle (smooth muscle in eyelid,** p. 640**);**
   - **blood vessels of choroid and retina.**

INTERNAL NERVOUS TUNIC OR RETINA

The internal nervous tunic of the eye, or *retina*,[16] contains special receptors on which is projected an inverted image of objects seen. There is still disagreement as to how one adjusts to the inversion of the retinal image. **As a result of the partial crossing of nerve fibers at the optic chiasma, the retina of each eye is connected with both right and left visual areas of the forebrain.** The retina is shaped like a sphere that has had its anterior segment removed. The margin is irregular in outline and is termed the ora serrata.

The *ora serrata* (figs. 56–3, 4) is the dentate fringe that marks the anterior termination of the retina proper. Most of the cellular layers either end or become thinner and fuse at the ora. The very thin continuation of the retina in front of the ora constitutes the ciliary and iridial parts of the retina (pp. 645 and 646, respectively). Cystic degeneration is frequently seen at the ora.

Stratification of Retina

The retina comprises two strata: (1) an external, pigmented stratum, and (2) an internal, transparent, cerebral stratum,* which consists of several layers.

*It should be pointed out that until recently the term cerebral stratum was applied to only the more internal layers (numbers 5 to 10) derived from the inverted lamina of the optic cup.

Pigmented Stratum. The pigmented stratum develops from the external lamina of the optic cup and is adherent to the choroid. It contains granules of a pigment termed fuscin. In albinos, however, the granules are colorless.

Cerebral Stratum. The cerebral stratum develops from the inverted lamina of the optic cup. It has a purplish red color in the living, dark-adapted eye, owing to the presence of visual purple in the rods. **The cerebral stratum of the retina may become detached pathologically (or in the course of preparing histological sections) from the pigmented layer along a plane that indicates the cavity of the optic vesicle in the embryo.** The cerebral stratum of the retina consists essentially of three sets of neurons.

Macula and Optic Disc. Two special areas of the retina require particular mention: the macula and the optic disc. The *macula* or yellow spot *(macula lutea)* is a pigmented area of the retina on the temporal side of the optic disc. It presents a pit, the *fovea centralis,* the central depression or *foveola* of which is situated lateral to, and usually near the level of the lower margin of, the optic disc.[17] The fovea is avascular and is nourished by the choroid. **Cones, but not rods, are present at the foveola, and each cone here is connected with only one ganglion cell. The foveola functions in detailed vision, that is, when an object is looked at specifically.** A line joining the viewed object and the foveola indicates the *visual axis* of the eye.

The optic disc, or blind spot, has no receptors and consists merely of the optic nerve fibers. Therefore, it is insensitive to light. It is situated nasal to the posterior pole of the eye and to the fovea centralis. Normally it is flat and does not form a papilla. Near its center, however, a variable depression, termed the "physiological cup," is present.[18]

Blood Supply of Retina

The outer part of the cerebral stratum, including the rods and cones, is nourished by the choriocapillary lamina of the choroid, whereas the inner part is supplied by the central artery of the retina, a branch of the ophthalmic. The central artery travels in the optic nerve and, at the optic disc, it divides into superior and inferior branches, each of which then divides into temporal and nasal branches. The branches of the central artery do not anastomose with each other or with any other vessels. **Functionally, the central artery is an end-artery in the sense that no direct communications occur between the arterioles and venules, the junction being by way of the capillary network only. Occlusion of the central artery results in blindness.** The retinal veins more or less follow the arteries, and the central vein ends in the cavernous sinus.

Examination of Retina (Ophthalmoscopy)

The *fundus oculi* (fig. 56–6) is the back portion of the interior of the eye-

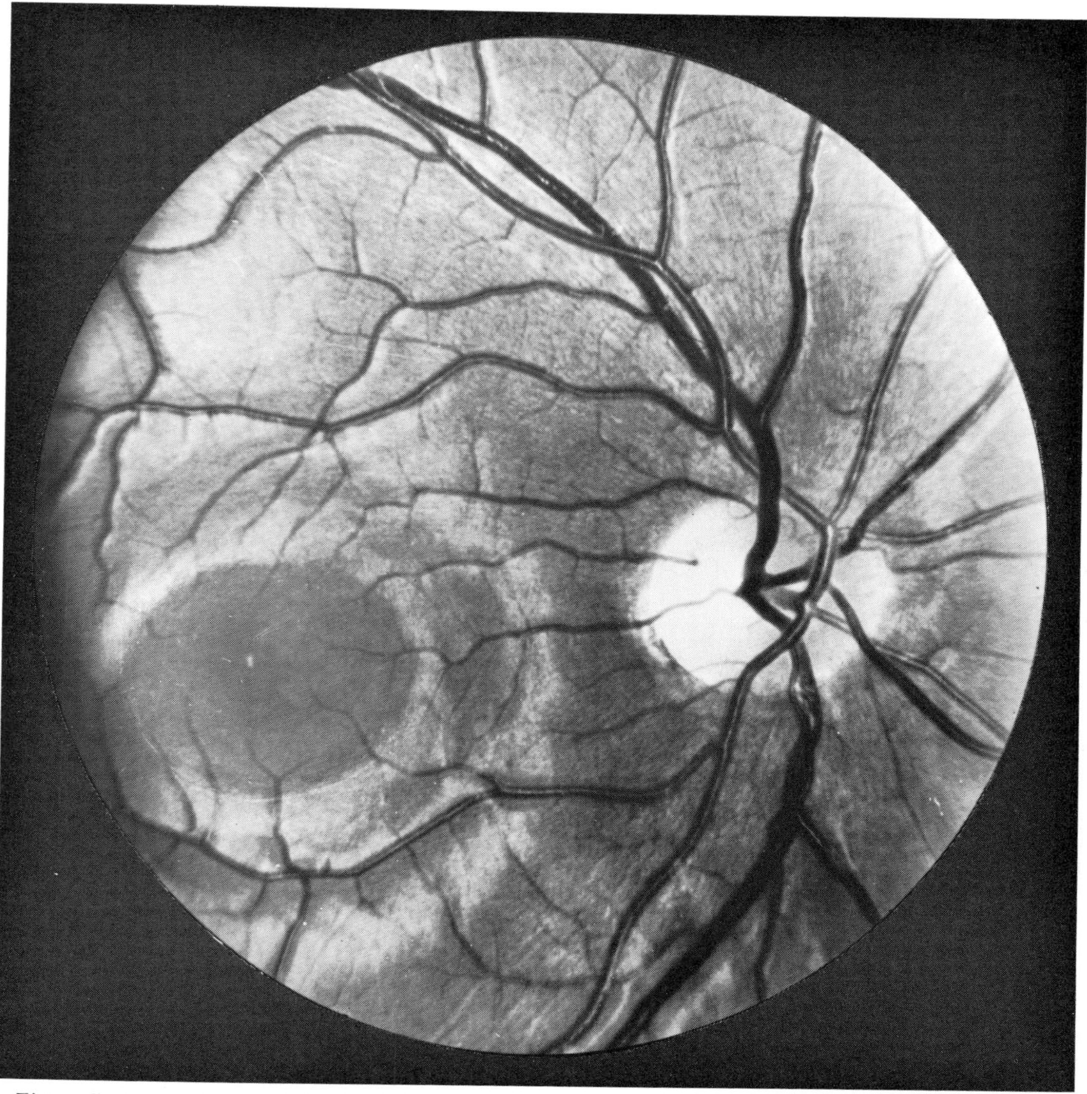

Figure 56–6 Fundus oculi *in vivo*. Photograph taken with the Fundus Camera of Carl Zeiss, Oberkochen. Note the optic disc on the right side of the photograph. The whitish component of its (pink) color is due to the lamina cribrosa. The lateral border of the disc is sharper than the medial. The retinal vessels radiate over the fundus from the disc. The arteries present a light streak along their middle. The veins are darker and wider than the arteries. The central vein is lateral to the central artery at the disc. The macula, situated lateral (on the left side of the photograph) to the optic disc, appears as a dark (red) oval area. The foveola appears here as a whitish spot in the macula. Striations due to the nerve fibers in the retina can be seen proceeding downward and medially, and upward and medially, toward the disc. Courtesy of Dr. Hans Littmann and Carl Zeiss, Inc.

ball as seen on ophthalmoscopy, that is, by use of an ophthalmoscope.

The three tunics of the eye are summarized in table 56–1.

DIOPTRIC MEDIA OF EYE

The dioptric or refractive apparatus of the eye comprises the cornea, the aqueous humor, the lens, and the vitreous body. Most of the optical power of the eye is contributed by the front surface of the cornea. The cornea has been described (p. 643).

Aqueous Humor

The aqueous humor fills the anterior and posterior chambers of the eye. Its composition is approximately that of protein-free plasma. It is formed probably by the ciliary processes[19] and is passed into the posterior chamber, and then through the pupil into the anterior chamber (fig. 56–3). It traverses the iridocorneal angle and the sinus venosus, and thereby reaches the ciliary veins. **Interference with resorption results in an increased intra-ocular pressure (glaucoma).** The intra-ocular pressure, normally[20] about 15 (8 to 21) mm of mercury, can be calculated from measurements of the impressibility of the anesthetized cornea (tonometry).

Lens

The lens[21] of the eye is biconvex and has a diameter of 1 cm. It presents anterior and posterior surfaces, separated by a rounded border termed the *equator*. The posterior surface is more convex than the anterior.

The lens consists of: (1) a *capsule*, which forms an elastic envelope with which the fibers of the ciliary zonule are fused; (2) an *epithelium*, cuboidal in type and confined to the front of the lens; and (3) *lens fibers*, which are long bands derived from the epithelium. The laminated structure of the lens is due to the continuous laying down of fibers in the region of the equator. Thus the lens grows slightly throughout life. The central part *(nucleus)* of the lens is harder than the external part *(cortex)*. The lens absorbs much of the violet light and it becomes increasingly yellow with age. **The lens also becomes harder with age, as a result of which the power of accommodation is lessened (presbyopia). This defect may be overcome by the use of convex spectacles. An opacity of the lens is termed a cataract.**

The *ciliary zonule* (fig. 56–3), or *suspensory ligament of the lens*, anchors the lens capsule to the ciliary body and the retina.[22] The fibers of the zonule are fine, viscous strands that enclose a series of spaces.

When distant objects are being looked at, it is believed that the elastic fibers of the suprachoroid lamina pull on the ciliary body, which, in turn, keeps the zonular fibers under tension. This results in tension on the lens capsule, and the curvatures of the lens are at a minimum. A more or less opposite train of events takes place when near objects are viewed (accommodation, p. 645).

Vitreous Body

The vitreous body is a transparent, gelatinous mass that fills the posterior four-fifths of the globe and is adherent to the ora serrata. Its composition resembles that of the aqueous humor, but it contains a meshwork of collagenous fibrils and a mucopolysaccharide, termed hyaluronic acid. The *hyaloid canal* extends from the optic disc to the lens. It sags downward in the vitreous body. The canal marks the site of the

TABLE 56–1 Tunics of Eye

Tunic	*Component Parts*		
External	Cornea	Sclera	
Middle	Iris { Uveal layers of iris	Ciliary body { Uveal layers of ciliary body	Choroid
Internal	Ext. pigmented layer, Int. pigmented layer } Iridial part of retina	Ext. pigmented layer, Int. layer } Ciliary part of retina	Pigmented stratum, Cerebral stratum } Optic part of retina

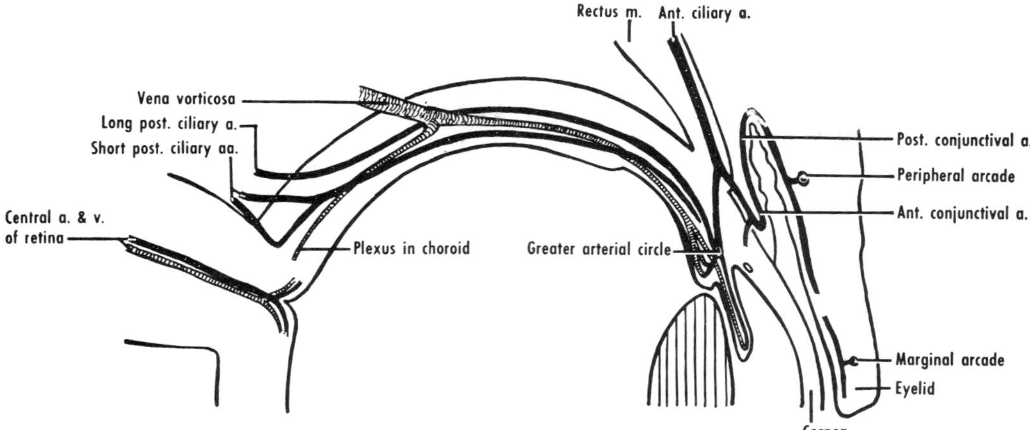

Figure 56–7 Blood supply of the eye. Based on Wolff.

hyaloid artery in the fetus. The movement of specks (possibly red blood corpuscles) in the vitreous body is sometimes seen as *muscae volitantes* (L., flitting flies).

GENERAL SENSORY INNERVATION AND BLOOD SUPPLY OF EYE

Sensory fibers from the cornea, iris, and ciliary body reach the nasociliary nerve (of the ophthalmic) by way of the short and long ciliary nerves. The eye receives its blood supply (fig. 56–7) from the ophthalmic artery by way of the central artery of the retina, the short and long posterior ciliary arteries, and the anterior ciliary arteries (from the muscular branches of the ophthalmic). The veins drain into the cavernous sinus by way of the ophthalmic veins. The veins from the eye accompany the arteries, except that the vessels corresponding to the posterior ciliary arteries are the venae vorticosae.

REFERENCES

1. H.-L. Thiel, v. Graefes Arch. Ophthal., *156*:590, 1955.
2. S. Duke-Elder and E. S. Perkins (eds.) *The Transparency of the Cornea,* Blackwell, Oxford, 1960.
3. G. von Bahr, Amer. J. Ophthal., *42*:251, 1956.
4. J. A. Pratt-Johnson, Amer. J. Ophthal., *47*:478, 1959.
5. P. P. Lele and G. Weddell, Brain, *79*:119, 1956. But see D. R. Kenshalo, J. app. Physiol., *15*:987, 1960.
6. L. T. Jones, Trans. Amer. Acad. Ophthal. Otolaryng., *72*:755, 1968.
7. M. Flocks, Arch. Ophthal., *56*:708, 1956. N. Ashton, A. Brini, and R. Sith, Brit. J. Ophthal., *40*:257, 1956.
8. H. M. Burian, A. E. Braley, and L. Allen, Arch. Ophthal., *53*:767, 1955.
9. J. Speakman, Brit. J. Ophthal., *43*:129, 1959, and *44*:513, 1960. A. Bill and C. I. Phillips, Exp. Eye Res., *12*:275, 1971.
10. K. W. Ascher, *The Aqueous Veins,* Thomas, Springfield, Illinois, 1961.
11. R. Smith, Brit. J. Ophthal., *40*:358, 1956.
12. V. Eskelund, *Structural Variations of the Human Iris and Their Heredity,* Nyt Nordisk, Copenhagen, 1938.
13. L. Allen, H. M. Burian, and A. E. Braley, Arch. Ophthal., *53*:799, 1955.
14. E. Gregersen, *Studies on the Spongy Structure of the Iris and its Imbibition with the Aqueous Humour,* Munksgaard, Copenhagen, 1960.
15. B. S. Ray, J. C. Hinsey, and W. A. Geohegan, Ann. Surg., *118*:647, 1943.
16. S. L. Polyak, *The Retina,* University of Chicago Press, 1941.
17. J. Fison, Brit. J. Ophthal., *40*:234, 1956.
18. D. Snydacker, Amer. J. Ophthal., *58*:958, 1964.
19. H. Davson, *Physiology of the Ocular and Cerebrospinal Fluids,* Churchill, London, 1956.
20. W. Leydhecker, K. Akiyama, and H. G. Neumann, Klin. Mbl. Augenheilk., *133*:662, 1958.
21. J. Nordmann, *Biologie du cristallin,* Masson, Paris, 1954.
22. C. McCulloch, Trans. Amer. ophthal. Soc., *52*:525, 1955.

GENERAL READING

Adler's Physiology of the Eye. Clinical Application, Mosby, St. Louis, 5th ed. by R. A. Moses, 1970. An attractive text on the functional aspects.

Davson, H., *The Eye,* Academic Press, New York, 4 vols., 1969–.

Davson, H., *The Physiology of the Eye,* Churchill, London, 3rd ed., 1972. Includes visual optics.

Duke-Elder, S., and Wybar, K. C., *The Anatomy of the Visual System,* vol. 2 of S. Duke-Elder (ed.), *System of Ophthalmology,* Kimpton, London, 1961. A superb work of reference for the orbit and the eye.

Kestenbaum, A., *Applied Anatomy of the Eye,* Grune & Stratton, New York, 1963. A brief although detailed account of applied anatomy, with many schematic drawings and regrettable terminology.

Wolff, E., *The Anatomy of the Eye and Orbit,* Lewis, London, 6th ed. revised by R. J. Last, 1968. A well-known and well-illustrated text.

SCALP, AURICLE, AND FACE

57

SCALP

LAYERS OF SCALP

The scalp (fig. 57–1) consists of five layers, the first three of which are connected intimately and move as a unit. A useful mnemonic is that the initial letters of the names of the layers form the word SCALP:

1. **S**kin, usually presenting numerous, long hairs.

2. **C**lose subcutaneous tissue consisting of a fatty, avascular, and a deeper, membranous, vascular stratum. The latter contains the larger blood vessels and nerves. **The scalp gapes when cut, and the blood vessels do not contract, resulting in considerable bleeding, which should be arrested by pressure.**

3. **A**poneurosis (galea aponeurotica) and occipitofrontalis muscle. The *galea aponeurotica (epicranial aponeurosis)* is a sheet of fibrous tissue that covers the vault of the skull between the occipitalis, auricularis superior, and frontalis muscles of the two sides. It is anchored to the external occipital protuberance, the highest nuchal line, and it extends over the temporal fascia to reach the zygomatic arch. It is sensitive to pain.

4. **L**oose subaponeurotic tissue containing the emissary veins. This layer allows free movement of the first three layers and is easily torn in deep wounds of the scalp. **The layer of subaponeurotic tissue has been termed a "dangerous area" because infection can spread easily in it, and also because of the possibility of the spread of infection from the scalp, by way of the emissary veins, to intracranial structures.**

5. **P**ericranium, that is, the periosteum on the outside of the skull. The pericranium has poor osteogenetic qualities, so that scant regeneration takes place if a bone flap is not replaced. The skull is insensitive and the pericranium is relatively so.

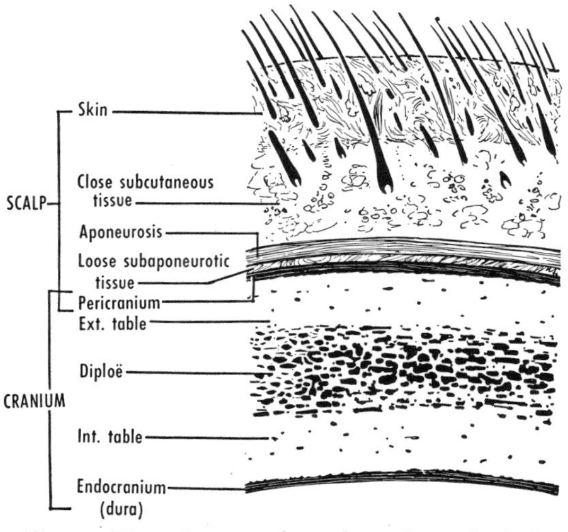

Figure 57–1 Section through scalp and skull.

INNERVATION AND BLOOD SUPPLY OF SCALP

The nerves and blood vessels of the scalp (fig. 57–2) enter it from below and ascend in its second layer. Hence surgical flaps of the scalp are cut in a way such that they remain attached inferiorly.

Sensory Innervation

The sensory innervation of the scalp, from before backward, is by the ophthalmic (supratrochlear and supra-orbital nerves), maxillary (zygomaticotemporal branch), and mandibular (auriculotemporal nerve) divisions of the 5th cranial nerve, and by the cervical plexus (lesser occipital nerve) and dorsal rami (greater occipital and 3rd occipital nerves) of the cervical spinal nerves. **The trigeminal and cervical territories are commonly equal in area.**[1]

Arterial Supply

The arterial supply of the scalp is partly by the internal carotid (supratrochlear and supra-orbital arteries) but chiefly by the external carotid (superficial temporal, posterior auricular, and occipital arteries). Anastomoses are abundant, so that

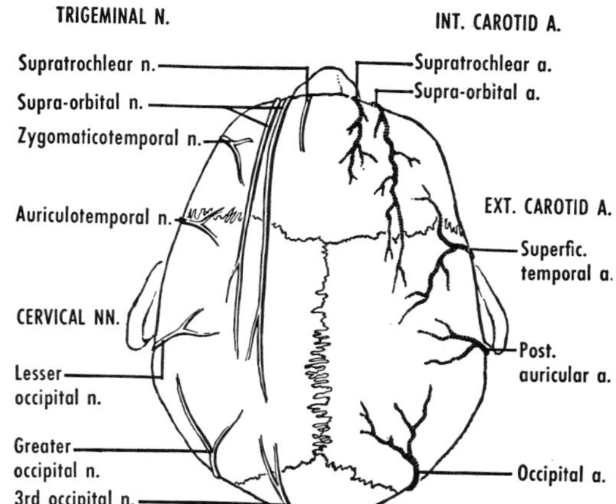

Figure 57–2 Innervation and blood supply of scalp. Superior aspect. In this instance the trigeminal territory extends behind the vertex of the head. For the purpose of simplification, the abundant arterial anastamoses have not been shown. Based on Grant's *Atlas*.

partially detached portions of the scalp may be replaced successfully. The veins parallel the arteries.

Superficial Temporal Artery. The superficial temporal artery is one of the terminal branches of the external carotid. It arises in the parotid gland, behind the neck of the mandible. It crosses the zygomatic process of the temporal bone and divides into frontal and parietal branches. The auriculotemporal nerve accompanies, and is posterior to, the superficial temporal artery. **The pulsations of the artery can be felt readily over the zygomatic process.**

Posterior Auricular Artery. The posterior auricular artery arises from the external carotid in the upper part of the neck and runs upward and backward, under cover of the parotid gland. It ends between the mastoid process and the back of the auricle, by dividing into auricular and occipital branches. The occipital branch supplies the scalp above and behind the ear.

Occipital Artery. The occipital artery arises from the external carotid in the upper part of the neck. It runs backward and upward. It lies at first anterior, then deep, and finally posterior, to the sternomastoid muscle. After it emerges from under cover of the sternomastoid, it pierces the trapezius, is accompanied by the greater occipital nerve, and divides into numerous occipital branches on the scalp. These branches, very tortuous in appearance, are accompanied by branches of the greater occipital nerve. The occipital vein ends variably in the vertebral, posterior auricular, external jugular, or internal jugular vein.

MUSCLES OF SCALP

Epicranius

The epicranius (fig. 57–4*C*) consists chiefly of the occipitofrontalis.

Occipitofrontalis. The occipitofrontalis consists of two occipital bellies (occipitalis) and two frontal bellies (frontalis) united by an intervening aponeurosis, the galea aponeurotica.

The *occipitalis* arises from about the lateral two-thirds of either the superior or the highest nuchal line on the occipital bone and from the mastoid temporal. It ends in the galea. The *frontalis* has no bony attachments. It arises from the galea aponeurotica and ends in adjacent muscles, and in the skin at the root of the nose and along the eyebrow.

The occipitofrontalis is supplied by the facial nerve.

The occipitalis, by pulling on the galea, provides a support for the frontalis. Alternate action of the occipitalis and frontalis moves the scalp backward and forward. The two frontalis muscles elevate the eyebrows, as in surprise. They are the antagonists of the orbicularis oculi muscles.

AURICLE

The auricle is a portion of the external ear. It consists of a plate of elastic cartilage covered by skin. It presents a number of depressions, the deepest of which is termed the *concha*. The margin of the auricle is called the *helix*. The names of the chief depressions and elevations of the auricle are given in figure 57–3. The lobule, which is devoid of cartilage, consists of fibrous tissue and fat. It is sometimes used as a source of blood for a blood cell count.

The sensory innervation of the auricle (fig. 57–6*B*) is from the auriculotemporal, lesser occipital, and great auricular nerves. Vagal fibers reach the auricle by way of the auriculotemporal nerve, and the facial nerve is believed to contribute to the external acoustic meatus. Although the skin of the auricle does not contain organized nerve endings, nevertheless touch, pain, cold, and warmth can be felt there.[2] The blood supply is from the superficial temporal and posterior auricular arteries.

MUSCLES OF AURICLE

Several unimportant intrinsic muscles have been described for the auricle.[3] Three extrinsic muscles, anterior, superior, and posterior, connect the auricle with the fascia on the side of the skull (fig. 57–4C). All the auricular muscles are supplied by the facial nerve.

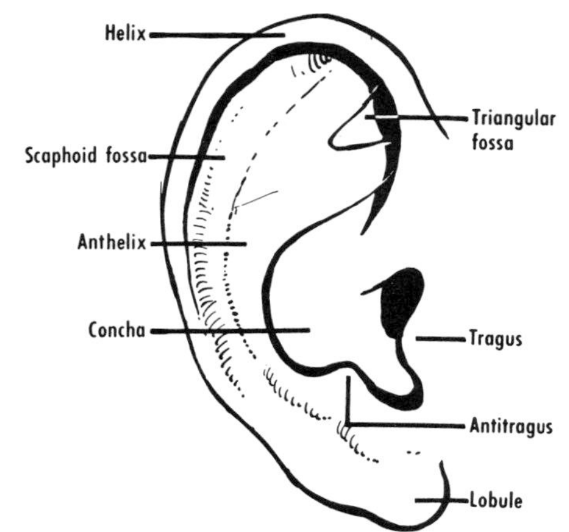

Figure 57–3 Lateral aspect of right auricle, showing chief landmarks.

FACE

The height of the adult body is usually seven and one-half head lengths, measured from the vertex of the head to the chin. The height of a one year old infant, however, is only four head lengths.

The eyelids (p. 639), the nose (p. 732), and the lips (p. 717) are dealt with elsewhere.

MUSCLES OF FACIAL EXPRESSION

The muscles of facial expression[4] have in common: (1) a very superficial location and an attachment to, or influence on, the skin; (2) great variability in degree of development and in shape and strength; and (3) an innervation by the facial nerve. Dissection of the facial muscles is rendered difficult because adjacent muscles are often fused, because the fibers of insertion are frequently intermingled, and because they are inserted into the skin by thin and isolated strands.

The muscles of facial expression may be grouped as (1) the muscles of the scalp and auricle (pp. 653 and 654), (2) the muscles around the opening of the orbit, (3) the muscles of the nose, (4) the muscles of the mouth, and (5) the platysma (p. 683).

Muscles around Opening of Orbit (fig. 57–4A)

This group includes the orbicularis oculi and an associated slip termed corrugator supercilii.

Orbicularis Oculi. The orbicularis oculi[5] is a thin, flat, elliptical sphincter that surrounds the rim of the orbit. It consists of three parts: orbital, palpebral, and lacrimal.

The *orbital part* is attached to the medial margin of the orbit (frontal bone and maxilla) and to the medial palpebral ligament. Its fibers may form complete ellipses, perhaps without interruption laterally, but this is disputed.

The *palpebral part* is contained in the eyelids. Its fibers arise from the medial palpebral ligament and sweep laterally, in front of the tarsal plate and orbital septum of each eyelid. The pretarsal fibers from the two eyelids form a common tendon (lateral palpebral ligament) that is inserted into the orbital tubercle of the zygomatic bone.[6] The preseptal fibers intermingle to form the lateral palpebral raphe. A small bundle close to the margin of each eyelid is termed the *ciliary bundle.*

The *lacrimal part* (fig. 55–4) lies behind the lacrimal sac. It arises from the crest of the lacrimal bone, passes across the tarsal plate of each eyelid, and is inserted mostly into the lateral palpebral raphe.

Actions. The orbicularis protects the eye from intense light and from injury. The palpebral part brings the eyelids together gently, as in blinking and in sleep. Strong closure of the lids is effected by the cooperation of the orbital part, whereby the skin of the forehead, temple, and cheek is drawn toward the medial angle of the eyelids. This results in radiating folds of skin at the lateral angle of the eyelids; these become permanent in older people ("crow's feet"). The effect of the lacrimal part of the orbicularis on the lacrimal sac is disputed. The levator palpebrae superioris (described with the orbit) and the frontalis are the antagonists of the orbicularis. **Paralysis of the orbicularis results in drooping of the lower eyelid (ectropion) and spilling of tears (epiphora), as occurs in facial palsy.**

Muscles of Nose

This group comprises the procerus, nasalis, and depressor septi, as shown in figure 57–4*A*. The alar part of the nasalis aids in widening the nostril, an action that becomes prominent when respiration becomes difficult.

Muscles of Mouth (fig. 57–4*B*)

This group comprises the risorius, depressor anguli oris, zygomaticus major, levator anguli oris, zygomaticus minor, leva-

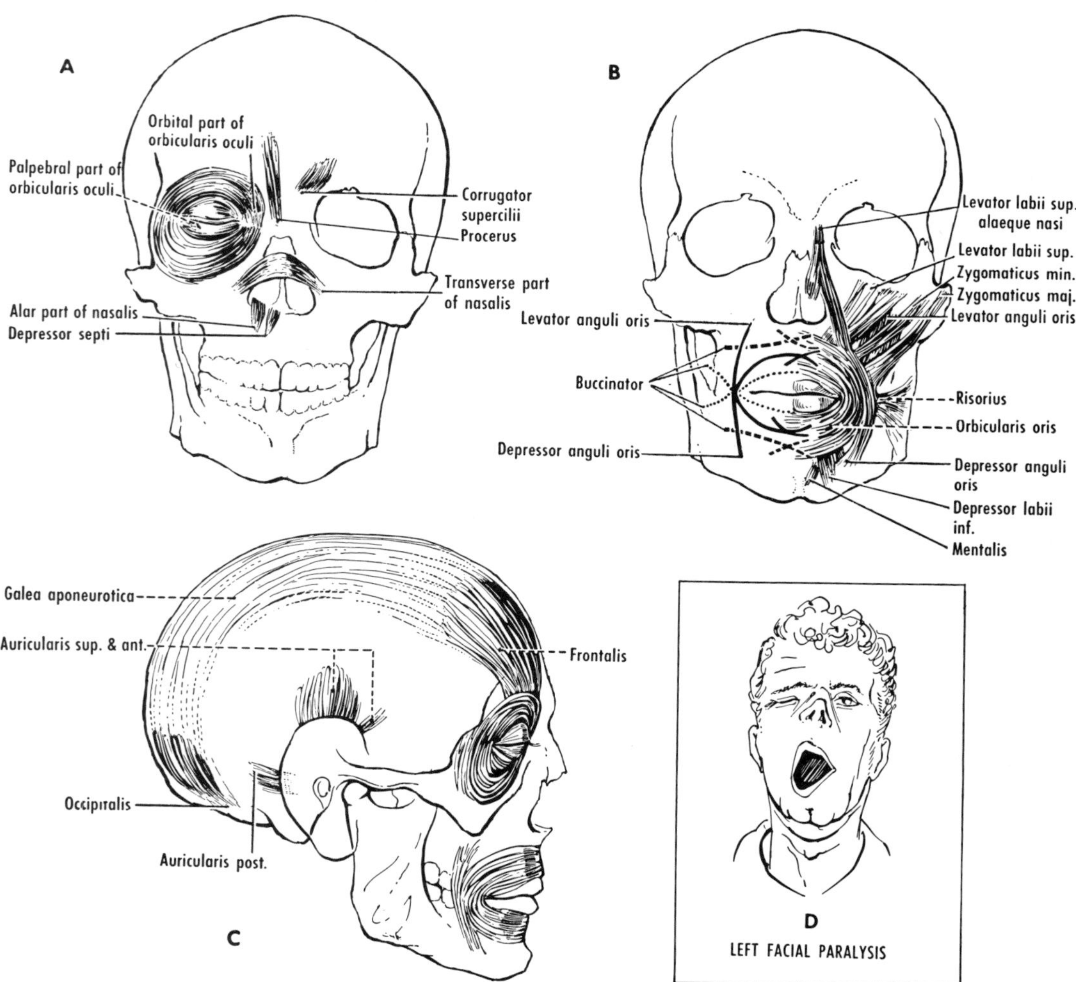

Figure 57–4 Muscles of facial expression. *A*, anterior aspect, showing muscles around openings of orbit and nose. *B*, muscles of mouth; on the right half of this face the course of the fibers that constitute the orbicularis oris is shown schematically. *C*, lateral aspect, showing muscles of scalp and auricle. (The orbicularis oculi and the orbicularis oris are shown, but do not have labels.) In *A*, *B*, and *C* the unbroken guide lines indicate the bony attachments of the muscles. *D*, results of left-sided facial paralysis, due to a lesion of the facial nerve at its exit from the skull. The patient has been asked to "shut his eyes" tightly and to open his mouth. Note deviation of the lips, characteristic triangular shape of the mouth, and failure to close the affected eye. From a photograph by Pitres and Testut.

tor labii superioris, levator labii superioris alaeque nasi, and depressor labii inferioris. These eight muscles are inserted into the skin at the angle of the mouth. Also included in the group are the buccinator, oribcularis oris, superior and inferior incisive muscles, and the mentalis.

Buccinator. The buccinator is a thin, quadrilateral muscle that occupies the back of the interval between the maxilla and the mandible. It arises from the alveolar processes of the maxilla and the mandible and from the pterygomandibular raphe (which separates it from the superior constrictor of the pharynx). It is inserted in a complicated manner into the orbicularis oris and into the lower and upper lips. The muscle is covered by the buccopharyngeal fascia and is pierced by the parotid duct. Its deep surface is lined by the mucous membrane of the mouth. A gap between the maxillary fibers and those from the pterygomandibular raphe transmits the tendon of the tensor veli palatini.

ACTIONS. The buccinator keeps the cheek taut, thereby preventing folding and injury from the teeth. It is also said to be concerned in whistling and blowing. It is active in smiling.[7]

Orbicularis Oris. The orbicularis oris is a complicated sphincter that contains fibers from other facial muscles as well as fibers proper to the lips. The deepest fibers are derived chiefly from the buccinator. More superficially, the levator et depressor anguli oris enter the lips. The fibers proper to the lips pass obliquely from the skin to the mucous membrane.

ACTIONS. The orbicularis oris closes the lips and can also protrude them.

Innervation of Muscles of Facial Expression

All the muscles of facial expression develop in the embryo from the second pharyngeal arch and are supplied by the facial nerve.

FACIAL NERVE (FACIAL PART)

The facial (7th cranial) nerve has already been described in detail (p. 619). It supplies all the muscles of facial expression. **In its course, the facial nerve traverses in succession (1) the posterior cranial fossa, (2) the internal acoustic meatus, (3) the facial canal in the temporal bone, and (4) the parotid gland and the face.**

Branches. Those concerned with the face and scalp (fig. 57–5A) are:

1. The *posterior auricular nerve* is given off just below the base of the skull. It accompanies the posterior auricular artery and supplies most of the muscles of the auricle together with the occipitalis. It also sends sensory fibers to the auricle.

2. The terminal branches of the facial nerve arise in the parotid gland and form the *parotid plexus.* The branches emerge from the gland under cover of its lateral surface and radiate forward in the face, communicating with the terminal branches of the trigeminal nerve. They supply the anterior and superior auricular muscles, the frontalis, orbicularis oculi, buccinator, orbicularis oris, and the other muscles of facial expression, including the platysma.

The terminal branches are variable in their arrangement but are commonly classified as *temporal,* several *zygomatic* (which unite with the infra-orbital nerve to form a plexus), *buccal* (supplying the buccinator and other muscles of the mouth), *marginal branch of the mandible,*[6] and *cervical* (lying deep to, and supplying, the platysma). Numerous anastomoses between these branches are common.

The terminal branches of the facial nerve contain afferent as well as motor fibers. The afferent fibers are thought to be proprioceptive from the muscles of facial expression and/or concerned with deep pain in the skin, muscles, and bones of the face.

FACIAL VESSELS (FACIAL PART)

Facial Artery

The facial artery is a branch of the external carotid. After a brief course in the neck (cervical part, p. 694), the facial artery winds around the lower border of the mandible at the anterior margin of the masseter, and proceeds upward and forward on the face (facial part, fig. 57–5*B*). It ends at the medial angle of the eye by anastomosing with branches of the ophthalmic artery. The facial artery is very tortuous, and it takes part in numerous anastomoses, including some across the median plane. These latter aid in the collateral circulation

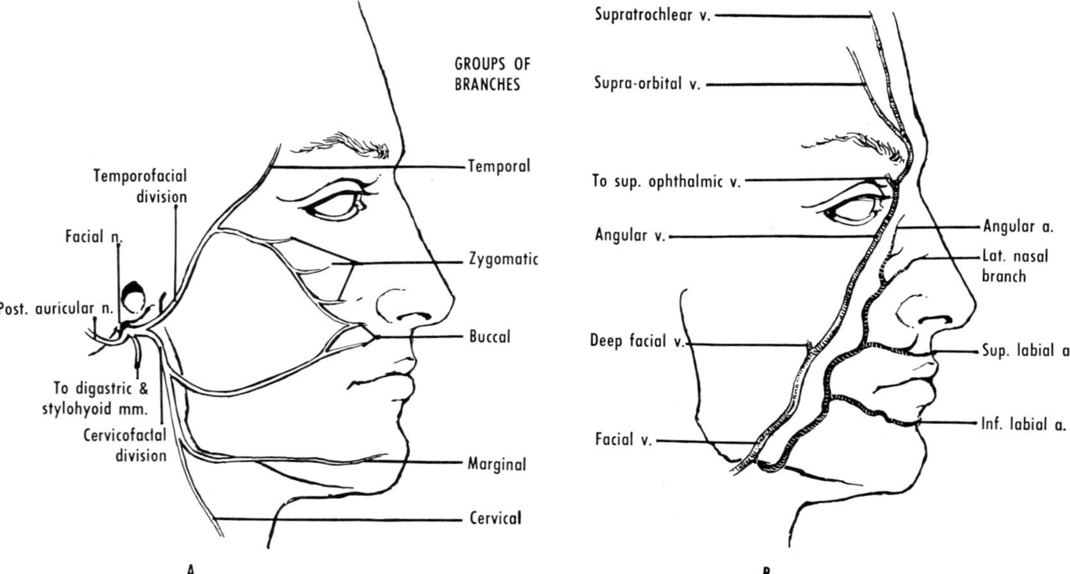

Figure 57–5 A, schematic representation of the facial nerve in the face. Although variations are numerous and frequent, two chief divisions (temporofacial and cervicofacial), together with an anastomosis between them, are generally found. For the purpose of simplification, the intricacies of the parotid plexus have been omitted. *B*, facial vessels in the face. Note that the vein is posterior to, more superficial than, and not as tortuous as, the artery.

after ligation of the common or the external carotid artery on one side.

Branches of Facial Part of Facial Artery. The facial artery supplies the lips and the external nose by means of the following:

1. The *inferior labial artery* (frequently double on each side) penetrates the orbicularis oris, supplies the skin, muscles, and mucous membrane of the lower lip, and anastomoses with its fellow of the opposite side.

2. The *superior labial artery,* larger and more tortuous than the inferior, has a similar course and distribution in the upper lip. It gives septal and alar branches to the nose. **Hemorrhage is controlled by compressing both parts of a cut lip between the index fingers and the thumbs.**

3. The *lateral nasal branch* supplies the ala and the dorsum of the nose.

4. The *angular artery* is the termination of the facial. At the medial angle of the eye it anastomoses with the dorsal nasal and palpebral branches of the ophthalmic artery, and thereby **establishes a communication ultimately between the external and internal carotid arteries.**

Facial Vein

The *facial vein* lies behind the facial artery and pursues a straighter course across the face. It begins at the medial angle of the eye as the *angular vein,* by the union of the supra-orbital and supratrochlear veins. It communicates freely with the superior ophthalmic vein and thereby with the cavernous sinus. The facial vein descends behind the facial artery and usually ends directly or indirectly in the internal jugular vein. In the cheek, the facial vein receives the *deep facial vein* from the pterygoid plexus. Its other tributaries correspond to the branches of the facial artery. The facial vein has no valves. **Because of its connections with the cavernous sinus and the pterygoid plexus, and the consequent possibility of spread of infection, the territory of the facial vein around the nose and the upper lip is frequently termed the "danger area" of the face.**

The vessels known formerly as the anterior and posterior facial veins are now termed the facial and the retromandibular veins (see fig. 60–6, p. 682). Sometimes the retromandibular vein, or a division of it, connects with the facial vein, in which event the common vessel so formed (which empties into the internal jugular) has long been known as the *common facial vein.*

CUTANEOUS INNERVATION OF HEAD AND NECK

The sensory innervation of the face (fig. 57–6) is largely through the branches of the trigeminal nerve, that of the front of the neck through the cervical plexus, and that of the back of the head and neck through the dorsal rami of the cervical nerves. **The "vertex-ear-chin line" (fig. 57–6A) indicates the approximate boundary between the cranial and spinal modes of innervation.**

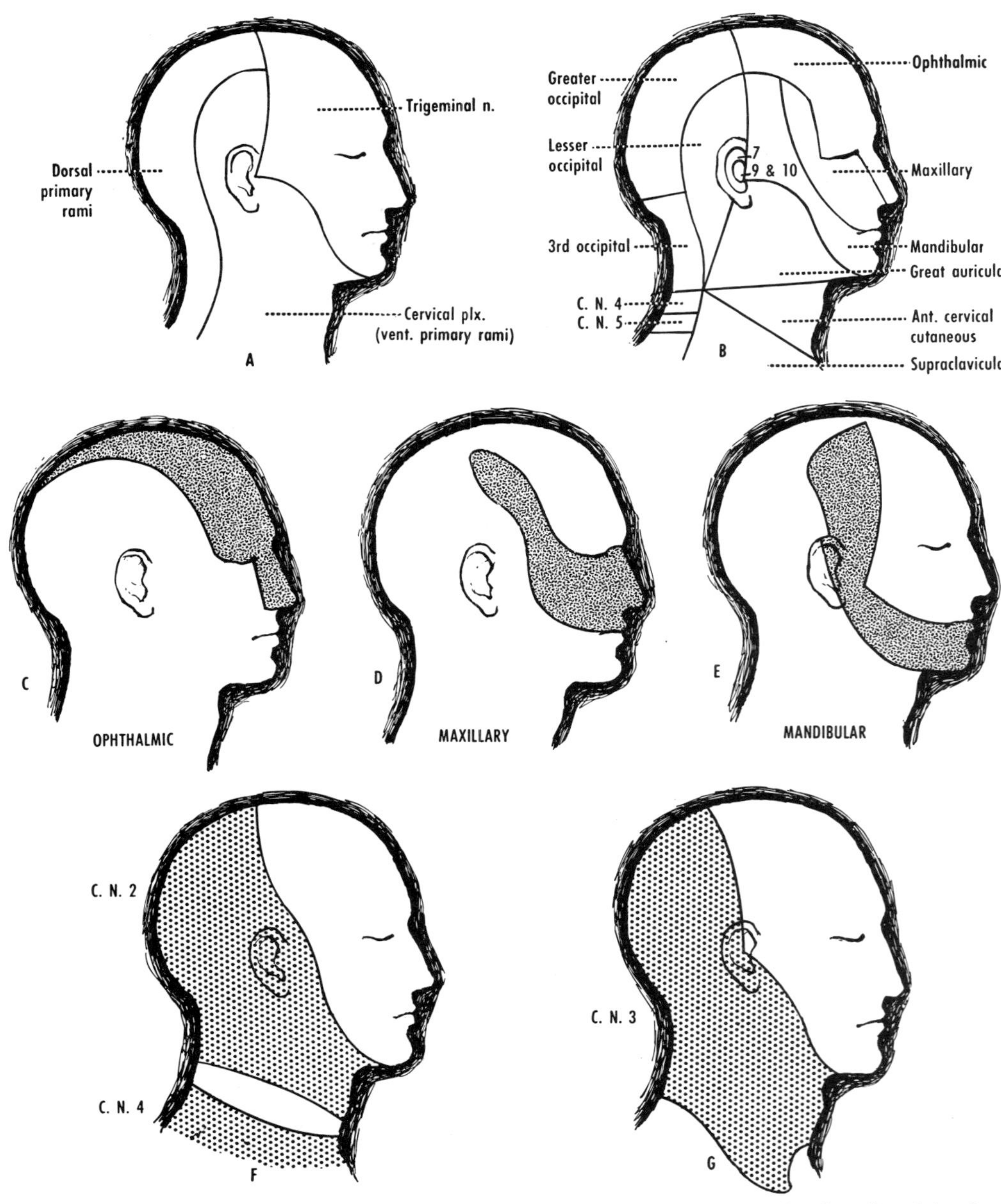

Figure 57–6 Cutaneous innervation of head and neck. *A*, showing "vertex-ear-chin line," and the three main areas of cutaneous innervation. *B*, showing approximate cutaneous territories of some of the individual nerves. C.N. 4 and 5 indicate the areas supplied by the dorsal rami of the respective cervical nerves. *C*, *D*, *E*, showing territories of the ophthalmic, maxillary, and mandibular nerves, respectively, as given by Corning. *F*, *G*, showing territories of cervical nerves 2 to 4. Note the overlapping of the territories of adjacent nerves.

Cranial Innervation

The three main divisions of the trigeminal nerve, namely, the ophthalmic, maxillary, and mandibular nerves, separate before emerging from the base of the skull. Thereafter, their courses differ widely, and, as a result, their cutaneous distributions must be tested separately. This is usually done by testing for sensation over the forehead, the prominence of the cheek, and the chin. Other cranial nerves (10 and 7) contribute to the auricle and external acoustic meatus.

Spinal Innervation

The spinal innervation of the skin may be considered in two ways (fig. 5–2, p. 34): (1) the area of distribution may be mapped out for each spinal nerve, including both its ventral and its dorsal ramus (fig. 57–6, *F* and *G*); (2) because the ventral rami combine to form plexuses (e.g., the cervical plexus) in which the individual rami become regrouped to constitute the named nerves of the body (e.g., the lesser occipital nerve), the area of distribution may be mapped out for each named nerve (fig. 57–6*B*). The named nerves usually contain fibers from more than one spinal nerve; thus the lesser occipital contains fibers from C.N. 2 and 3. In the case of either method of mapping out areas of cutaneous innervation, there is found to be considerable overlapping in the distributions of adjacent nerves. Moreover, a very considerable overlap occurs between the cutaneous areas supplied by cranial (5, 10, and 7) and spinal (C.N. 2, 3) nerves.[9]

REFERENCES

1. G. Lazorthes and G. Bastide, C. R. Ass. Anat., *43*:479, 1957.
2. D. C. Sinclair, G. Weddell, and E. Zander, J. Anat., Lond., *86*:402, 1952.
3. G. Winckler, Arch. Anat., Strasbourg, *43*:237, 1960.
4. G. S. Lightoller, J. Anat., Lond., *60*:1, 1925; *62*:319, 1928. E. Huber, *Evolution of Facial Musculature and Facial Expressions*, Johns Hopkins Press, Baltimore, 1931.
5. G. Winckler, Arch. Anat., Strasbourg, *24*:183, 1937.
6. L. T. Jones, Amer. J. Ophthal., *49*:29, 1960.
7. C. L. Isley and J. V. Basmajian, Anat. Rec., *176*:143, 1973.
8. E. Alajmo and T. Ricci, Il Valsalva, *41*:223, 1965.
9. D. Denny-Brown and N. Yanagisawa, Brain, *96*:783, 1973.

PAROTID, TEMPORAL, AND INFRATEMPORAL REGIONS

58

The parotid region (fig. 60–3*C*, p. 680) comprises the parotid gland and its bed. The bony areas that form a part of the bed for the gland include the ramus of the mandible anteriorly, the styloid process medially, and the mastoid process posteriorly. These structures, together with certain attached muscles, groove the parotid gland. The parotid region is limited behind by the sternomastoid, and below by the digastric.

The temporal region is that on the side of the head (the temple).

The infratemporal region is below the temporal region and medial to the ramus of the mandible. The masseter muscle, however, is conveniently described with this region, although it is on the lateral aspect of the ramus of the mandible.

The bony boundaries of the temporal, infratemporal, and pterygopalatine fossae (p. 561) should be reviewed at this time.

PAROTID GLAND

The parotid gland[1] is the largest of the three large, paired glands (the parotid, submandibular, and sublingual) which, with numerous small glands in the tongue, lips, cheeks, and palate, constitute the salivary glands (fig. 58–1). The combined secretion of all these glands is termed saliva.

The parotid gland is a compound tubulo-alveolar gland and is purely serous in type. It is a yellowish, lobulated structure of irregular shape. It occupies the interval between the sternomastoid and the mandible.

Surface Anatomy

The parotid gland lies below the zygomatic arch, below and in front of the external acoustic meatus, in front of the mastoid process, on the masseter, and behind the ramus of the mandible. Its inferior end or apex is below and behind the angle of the mandible. The very close relationship between the parotid gland and the mandible is emphasized by the pain on mastication that occurs in viral inflammation of the parotid gland (mumps).

Relations

The parotid gland is enclosed within a sheath (the parotid fascia) derived superficially from the investing layer of the deep cervical fascia and deeply from the fascia over the masseter. A fascial extension separates the parotid from the submandibular gland.[2]

The parotid gland is shaped roughly like an inverted pyramid, with either three or four sides (fig. 58–2*A*). It presents an apex, a base, and lateral, anterior, and posterior surfaces. The posterior surface is sometimes considered as two surfaces: posterior and medial.

The *apex* lies between the sternomastoid and the angle of the mandible. The *base* (superior aspect) is related to the root of the zygoma and the neck of the mandible. The superficial temporal vessels emerge from the base, with the auriculotemporal nerve.

The *lateral (superficial) surface* is characterized by lymph nodes embedded in the gland. It is covered by skin.

The *anterior surface* is grooved by the ramus of the mandible and the masseter muscle (fig. 58–2*B*). Thus it presents lateral and medial lips. The lateral lip frequently presents a detached portion, known as the *accessory parotid gland* (fig. 58–1). The parotid duct, the branches of the facial nerve, and the transverse facial artery emerge from under cover of the lateral lip. The medial lip of the anterior surface may pass between the two pterygoid muscles, and the maxillary artery emerges from this part of the gland.

The *posterior surface* is related above to the external acoustic meatus. It is grooved by (1) the mastoid process, and the sternomastoid and digastric muscles, and (2) the styloid process and its attached muscles. This second part of the posterior surface is frequently described as a *medial surface*.[3] In front of the styloid process, the medial border of the gland is related to the internal carotid artery. The medial border may make contact with the lateral wall of the pharynx. The part of the gland more or less behind the styloid process is related to the internal jugular vein and the last four cranial nerves. This portion of the gland is pierced by the facial nerve above and by the external carotid artery below.

The following structures lie partly within the parotid gland, from superficial to deep:

1. The facial nerve enters the posterior surface of the gland and forms the parotid plexus *(pes anserinus)* within the gland.

The parotid gland consists of a superficial and a deep portion ("lobe"), or layer, wrapped around the

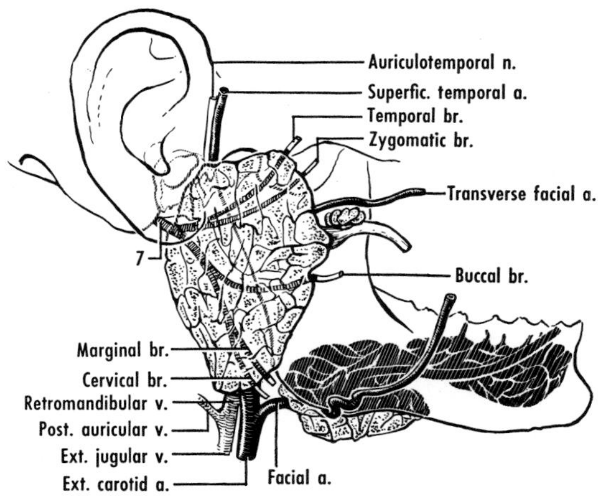

Figure 58–1 Diagram showing the location of the three main salivary glands and their ducts: parotid, submandibular, and sublingual. Note also the accessory parotid gland above the parotid duct. Although the details of the parotid plexus of nerves have not been shown, the chief branches of the facial nerve have been indicated.

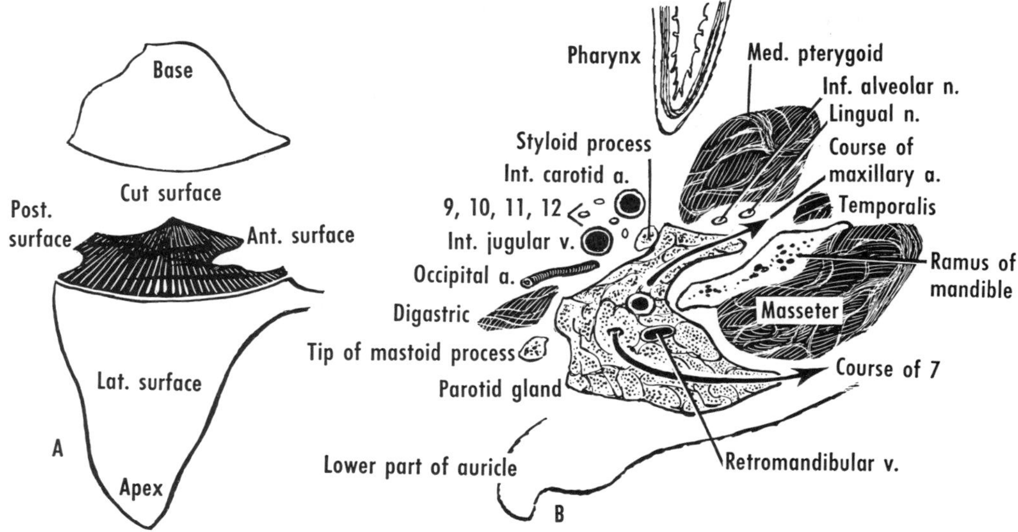

Figure 58–2 A, diagram of the lateral aspect of the parotid gland, sectioned horizontally, showing the surfaces and the plane of section of B. B, horizontal section through the parotid gland and adjacent structures at the level of the atlas. B is based on Truex and Kellner, and on Parsons.

branches of the facial nerve,[4] and connected by one[5] or more isthmuses.[6] The task of surgical excision of the parotid gland (e.g., for a tumor) remains a delicate procedure if damage to the facial nerve ("the hostage of parotid surgery") is to be kept minimal.

2. The superficial temporal and maxillary veins enter the parotid gland with their corresponding arteries and unite within the gland to form the retromandibular vein. This last emerges near the apex of the gland and contributes in a variable manner to the formation of the external jugular vein (fig. 60–6, p. 682). The latter begins just below, or occasionally within, the parotid gland.

3. The external carotid artery enters the posterior surface of the gland below and frequently gives off the posterior auricular artery within the gland. The latter vessel then emerges from the posterior surface. **The external carotid divides within the gland into its terminal branches:** *(a)* the superficial temporal artery, which gives off the transverse facial artery and then emerges from the base of the gland; and *(b)* the maxillary artery, which emerges from the medial lip of the anterior surface and runs forward, deep to the neck of the mandible.

Parotid Duct. The parotid duct, about 5 cm in length, emerges from under cover of the lateral surface of the gland. It proceeds forward on the masseter and, after turning medially at nearly a right angle, it pierces the buccal pad of fat and the buccinator muscle. After a brief course between the buccinator and the mucous membrane of the mouth, **the parotid duct opens into the oral cavity opposite the crown of the upper second molar tooth.** The opening may be marked by a projection termed the *parotid papilla* (fig. 61–1A, p. 718). The parotid duct can be felt *in vivo* by a finger inside the mouth.

With regard to surface anatomy, the parotid duct corresponds to the posterior half of a line from (1) the junction of the ala of the nose with the face, to (2) a fingerbreadth above the angle of the mandible.[7]

The pattern of branching of the parotid duct can be examined radiographically after the injection of a radio-opaque medium (iodized oil) into the opening of the duct. This procedure is termed sialography. See figure 58–5A.

Innervation of Parotid Gland
(fig. 58–10)

Each of the salivary glands is innervated by both parasympathetic and sympathetic fibers. In the case of the parotid gland, parasympathetic, preganglionic, secretomotor fibers pass through the glossopharyngeal, tympanic, and lesser petrosal nerves to reach the otic ganglion, where they synapse. Postganglionic fibers then pass to the parotid gland by way of the auriculotemporal nerve. Owing to communications between the facial and glossopharyngeal nerves (e.g., between the chorda tympani and the otic ganglion), it is possible that the facial nerve also sends secretory fibers to the parotid gland. Indeed, **secretory fibers to each of the three major salivary glands may travel in both the facial and the glossopharyngeal nerve.**[8] The sympathetic supply to the salivary glands includes vasomotor fibers.[9]

SUPERFICIAL TEMPORAL ARTERY

The superficial temporal artery, the smaller terminal branch of the external carotid, arises in the parotid gland, behind the neck of the mandible. It crosses the zygomatic arch and divides into frontal and parietal branches (fig. 60–15*C*, p. 691). The auriculotemporal nerve accompanies, and is posterior to, the superficial temporal artery. Pulsations of the artery can be felt against the zygomatic arch. The anastomoses between the various arteries to the scalp are very free, so that partially detached portions of the scalp may be replaced successfully. When a lateral flap of the scalp is made surgically, the incision is shaped like a horseshoe with its convexity upward, so that the flap contains the intact superficial temporal artery.

Branches. The superficial temporal artery gives branches to the parotid gland, the auricle, and the temporal fossa, in addition to the transverse facial artery.

The *transverse facial artery* arises in the parotid gland and runs forward across the masseter between the zygomatic arch above and the parotid duct below, accompanied by zygomatic branches of the facial nerve. It supplies the parotid gland and duct, the masseter, and skin, and anastomoses with branches of the facial artery.

The terminal branches of the superficial temporal artery are as follows:

1. The *frontal branch* supplies the muscles and skin of the frontal region. It is very tortuous. It anastomoses with branches of the ophthalmic artery.
2. The *parietal branch* supplies skin and the auricular muscles.

MUSCLES OF MASTICATION
(figs. 58–3 and 58–4)

The muscles of mastication are the (1) masseter, (2) temporalis, (3) medial pterygoid, and (4) lateral pterygoid. They are developed from the mesoderm of the mandibular arch and are all supplied by the mandibular nerve (motor root), a division of the trigeminal.

The *buccal pad of fat*[10] overlies the buccinator and masseter muscles and possesses several extensions. Like the buccinator, it is pierced by the parotid duct. The buccal pad contributes to the roundness of a child's cheek. It has been suggested that it may prevent the cheeks from being sucked inward while an infant suckles.

Masseter

The masseter (fig. 58–3*F*) is covered by the masseteric fascia. It is a thick, quadrate muscle that arises from the inferior border and medial surface of the zygomatic arch and is inserted on the lateral aspect of the ramus of the mandible. It can be divided partially into superficial, middle, and deep portions.[11]

Innervation. A branch (masseteric nerve) of the anterior trunk of the mandibular nerve reaches the deep surface of the muscle by passing through the mandibular notch.

Action. The masseter is a powerful elevator of the mandible. It can be palpated during clenching of the teeth.

Temporal and Infratemporal Fossae

The *temporal fossa* is bounded by the temporal line, the frontal process of the zygomatic bone, and the zygomatic arch. It communicates with the infratemporal fossa deep to the zygomatic arch. The temporal muscle arises from the floor of the temporal fossa and, by passing deep to the zygomatic arch, enters the infratemporal fossa.

The *infratemporal fossa* is bounded anteriorly by the posterior surface of the maxilla, superiorly by the infratemporal surface of the greater wing of the sphenoid (fig. 52–18, p. 568), medially by the lateral pterygoid plate (fig. 52–12, p. 561), and laterally by the ramus and the coronoid process of the mandible. **The infratemporal fossa contains a part of the temporalis, most of the two pterygoid muscles, the maxillary ar-**

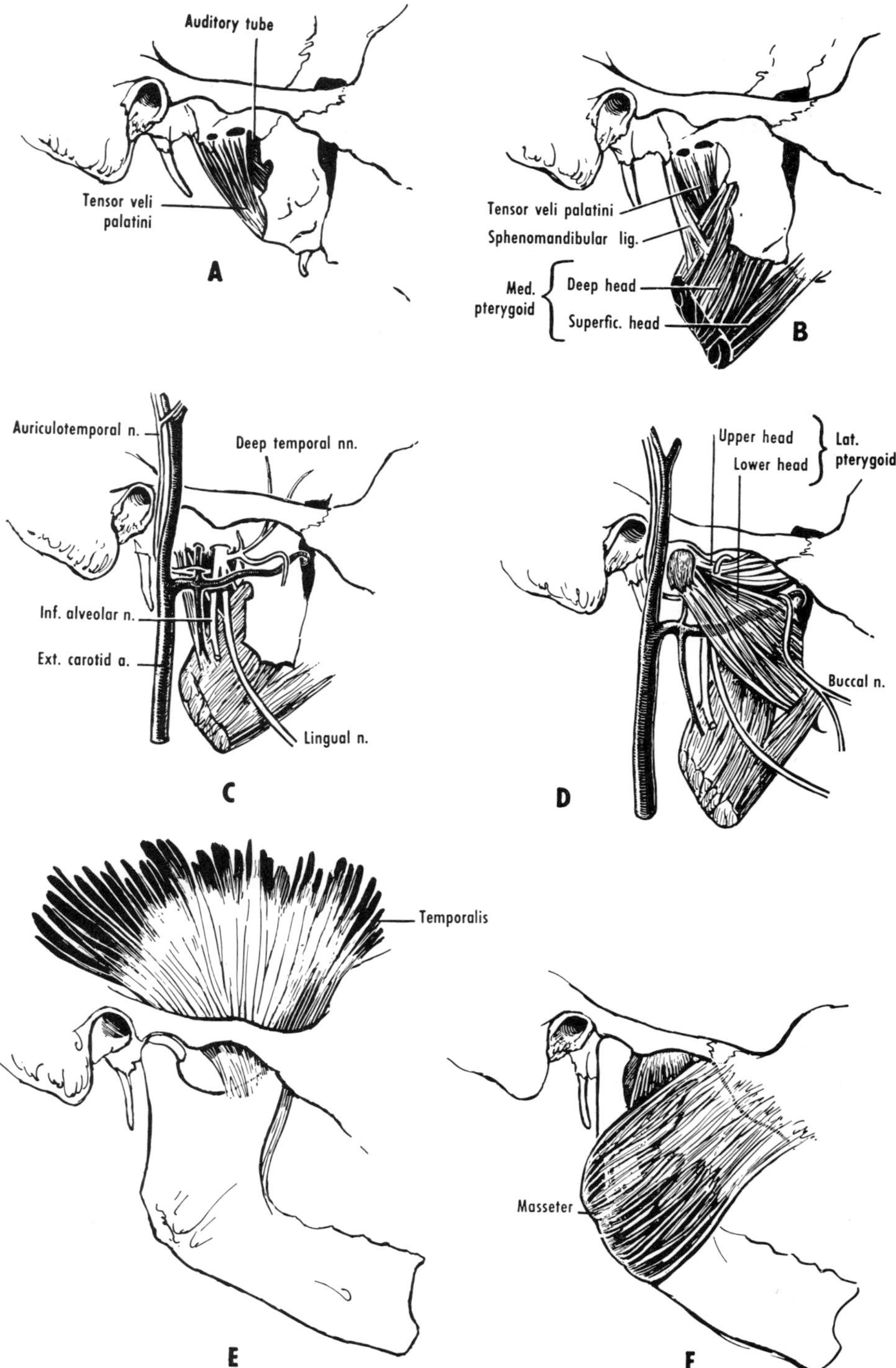

Figure 58–3 Six successively more superficial planes in the infratemporal region. *A*, medial wall of infratemporal fossa (compare fig. 52–12) and tensor veli palatini. *B*, medial pterygoid muscle. *C*, mandibular nerve and maxillary artery. The chorda tympani and the middle meningeal artery do not have labels. *D*, lateral pterygoid muscle. The second part of the maxillary artery is shown deep to the lower head of the lateral pterygoid; it is frequently superficial. Note the buccal nerve emerging between the heads of the lateral pterygoid, and the masseteric nerve above the upper head of the muscle. In this view the head of the mandible is shown in the mandibular fossa but the mandible has been cut away at its neck. *E*, temporalis. Note the anterior fibers attached to the anterior border of the ramus of the mandible. *F*, masseter. Note the deeper fibers proceeding directly downward.

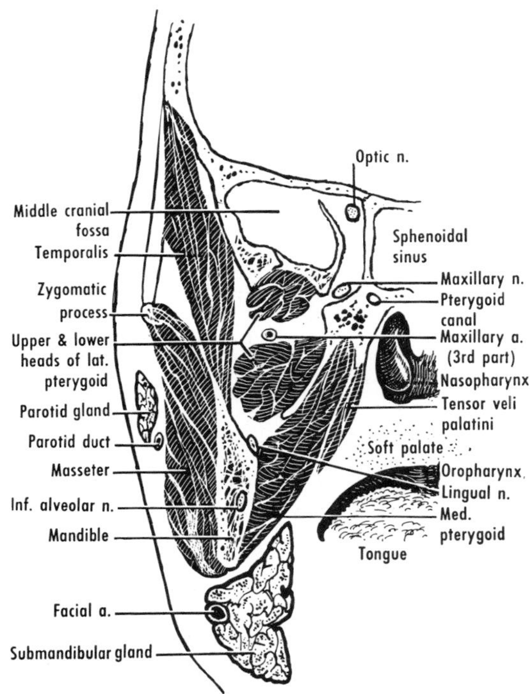

Figure 58–4 Coronal section through the ramus of the mandible. Note the four muscles of mastication and their relations. Based on Truex and Kellner.

tery and the pterygoid venous plexus, and the mandibular and chorda tympani nerves.

The *pterygoid venous plexus* is located partly between the temporalis and the lateral pterygoid, and partly between the two pterygoid muscles. Its numerous tributaries include (1) the deep facial vein, which connects it with the facial vein, and (2) veins that pass through the sphenoidal emissary foramen and the foramen ovale and connect it with the cavernous sinus.

The *temporal fascia* covers the temporal muscle above the zygomatic arch. Its upper part is thin and aponeurotic. It is attached above to the superior temporal line. Below, it consists of two layers, which are attached to the upper border of the zygomatic arch. They are separated below by fatty tissue. Its deep surface gives attachment to the temporalis.

Temporalis

The temporalis (fig. 58–3*E*) is a fan-shaped muscle that lies in the temporal fossa. It arises from the floor of the fossa below the inferior temporal line (frontal, parietal, sphenoid, temporal, and sometimes zygomatic bone) and from the deep surface of the temporal fascia. The cranial and fascial (aponeurotic) origins give the muscle a bipennate arrangement. The tendon of insertion passes deep to the zygomatic arch and is inserted into the coronoid process (medial surface, apex, and anterior border) and the anterior border of the ramus of the mandible.

Innervation. Deep temporal branches of the anterior trunk of the mandibular nerve.

Action. The temporalis maintains mandibular posture at rest and elevates the mandible in molar occlusion. The posterior fibers pull the head of the mandible backward from the articular tubercle into the mandibular fossa during closing of the mouth.

Medial Pterygoid

The medial pterygoid (fig. 58–3*B*) lies on the medial aspect of the ramus of the mandible. It possesses two heads of origin. The larger, deep head arises from the medial surface of the lateral pterygoid plate and from the pyramidal process of the palatine bone. The superficial head arises from the pyramidal process of the palatine and from the tuber of the maxilla. The two heads embrace the lower head of the lateral pterygoid and unite. The muscle passes downward and backward to be inserted into the medial surface of the mandible near its angle.

Innervation. A branch from the mandibular nerve.

Action. It acts as a synergist of the masseter in elevating the mandible. The lateral and medial pterygoids, acting together, protrude the mandible.

Lateral Pterygoid

The lateral pterygoid (fig. 58–3*D*) occupies the infratemporal fossa. It possesses two heads of origin. The upper head arises from the infratemporal surface and crest of the greater wing of the sphenoid bone. The larger, inferior head arises from the lateral surface of the lateral pterygoid plate. The muscle passes backward and the fibers converge to be inserted into the capsule of the temporomandibular joint, the articular disc,[12] and a pit on the front of the neck of the mandible.

Innervation. A branch from the anterior trunk of the mandibular nerve, which may come from the masseteric or buccal nerves.

Action. The lateral pterygoid, because of its attachment to the articular disc, is the chief protractor of the jaw. Furthermore, when the mouth is open, it prevents backward displacement of the articular disc and the head of the mandible. The mouth is opened by the rotational pull of the lateral pterygoid and digastric muscles.[13] Alternative, or complementary, factors in opening the mouth are relaxation of the muscles of mastication and gravity. In other words, according to the latter theory, the muscles of mastication are antigravity or postural muscles. The two heads can act independently.[14]

TEMPOROMANDIBULAR JOINT
(figs. 58–5 and 58–6)

The temporomandibular joint[15] is a synovial joint between (1) the articular tubercle, the mandibular fossa, and the postglenoid tubercle[16] of the temporal bone above, and (2) the head of the mandible below. The articular surfaces are covered with avascular, fibrous tissue, which may contain a variable number of cartilage cells. An articular disc divides the joint into two compartments. **The joint is subcutaneous laterally; medially it is related to the spine of the sphenoid and the foramen spinosum, anteriorly to the lateral pterygoid, and posteriorly to the parotid gland, the auriculotemporal nerve, and the superficial temporal vessels.**

The loose *articular capsule* is attached

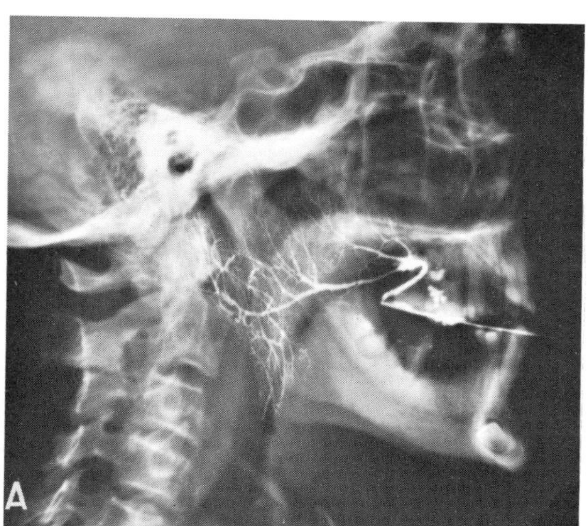

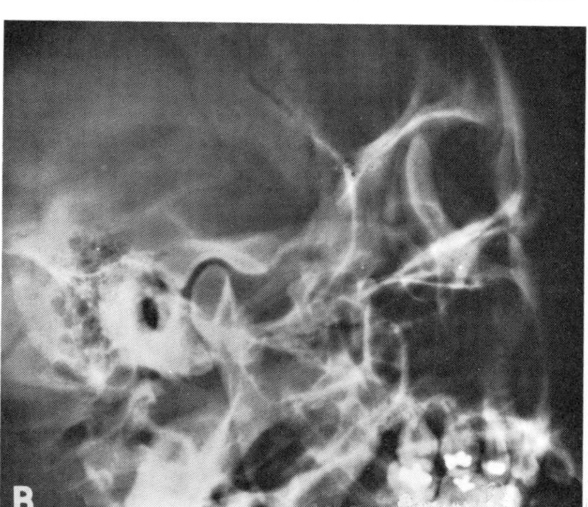

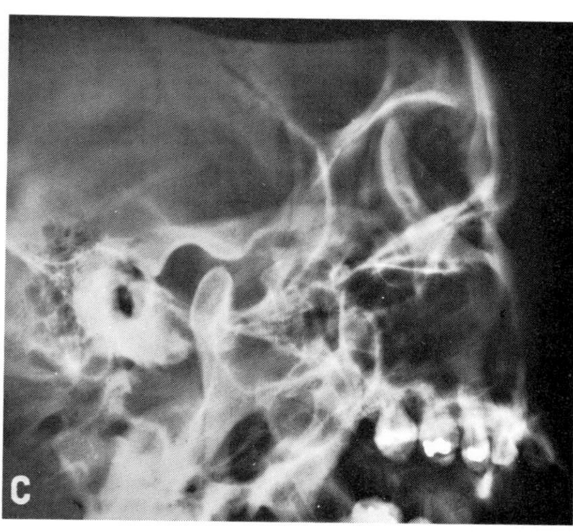

Figure 58–5 Parotid, temporal, and infratemporal regions. *A*, parotid sialogram *in vivo.* Iodized oil has been injected through the parotid duct. It can be seen that the parotid gland occupies an extensive area. *B* and *C*, radiograms showing the temporomandibular joint, (*B*) with the mouth closed and (*C*) with the mouth open. As the mouth is opened, the head of the mandible slides forward out of the mandibular fossa. The black oval area behind the head of the mandible is the external acoustic meatus. *A*, courtesy of Robert S. Sherman, M.D., New York, New York. *B* and *C*, courtesy of Mr. John A. Hill, Superintendent Radiographer, Birkenhead General Hospital, Birkenhead, England.

to the articular tubercle, the squamotympanic fissure, and the margins of the mandibular fossa between these two attachments (fig. 58–6*D*). Below, it is attached to the neck of the mandible. In front, it receives a part of the insertion of the lateral pterygoid. A portion of the neck of the mandible posteriorly is intracapsular.

The *articular disc*[17] is an oval plate of fibrous tissue (sometimes containing areas of fibrocartilage) the circumference of which is connected to the articular capsule. It is lost behind in elastic fibers and a retroarticular venous plexus.[18] In front, the disc is anchored to the tendon of the lateral pterygoid. The disc is tightly attached to the condyle, so that it follows the jaw in sliding movements. The disc divides the joint into two separate compartments, an upper between the temporal bone and the disc, and a lower between the disc and the mandible. The upper surface of the disc is concavoconvex, the lower concave. The disc is irregular in thickness,[19] but is only rarely perforated.

A separate synovial membrane lines the capsule in each of the two compartments of the joint but does not cover either

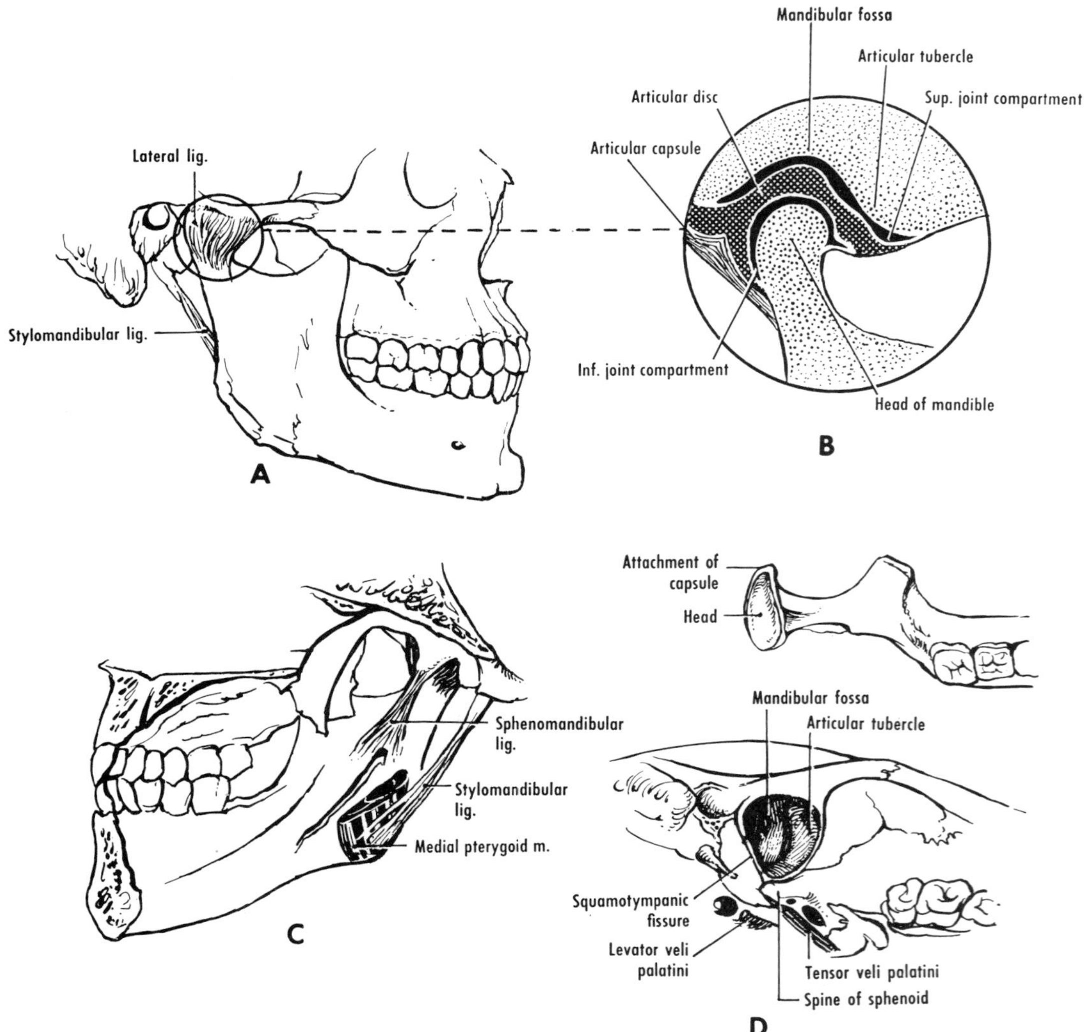

Figure 58–6 The temporomandibular joint. *A*, lateral aspect. *B*, lateral aspect after partial removal of the capsule. *C*, medial aspect. *D*, superior aspect of condylar process of mandible and inferior aspect of mandibular fossa and articular tubercle. Note the attachment of the capsule in both drawings, and the origins of the levator et tensor veli palatini in the lower of the two. Based on Sicher and Tandler.

the articular surfaces or the articular disc. The membrane presents folds and villi.

The *lateral (temporomandibular) ligament* extends from the tubercle on the root of the zygoma to the lateral surface of the neck of the mandible.

The *sphenomandibular ligament* is a thin band that lies medial to the joint. It extends from the anterior process and ligament of the malleus, the lips of the petrotympanic fissure, and the spine of the sphenoid bone to the lingula of the mandible.[20] It is related laterally to the lateral pterygoid muscle and the auriculotemporal nerve above, then to the maxillary vessels and the neck of the mandible, and below to the inferior alveolar nerve and vessels, and a portion of the parotid gland. It is related medially to the pharynx above and to the medial pterygoid muscle below. The sphenomandibular ligament apparently develops from the sheath of the cartilage of the first pharyngeal arch.

The *stylomandibular ligament* extends between the styloid process above and the angle and posterior border of the ramus of the mandible below.

Innervation and Blood Supply. Branches from the auriculotemporal and masseteric and/or deep temporal nerves of the mandibular supply the joint.[21] The blood supply is derived from the superficial temporal and maxillary arteries of the external carotid.

MOVEMENTS OF THE MANDIBLE (fig. 58–5, *B* and *C*)

The movements of the mandible are controlled more by the play of muscles than by either the shape of the articular surfaces or the ligaments. The movements are characteristic for each individual and depend on a pattern, in a manner similar to that which gives an individual a characteristic gait.

The chief movements of the mandible are depression, elevation (occlusion), protrusion (protraction), retraction, lateral movement, and circumduction.

The chief factors responsible for these movements may be summarized as follows:

Depression: lateral pterygoid, digastric, gravity.

Elevation: temporalis, masseter, medial pterygoid.

Protrusion: lateral and medial pterygoids, masseter.

Retraction: temporalis (posterior fibers).

Lateral movement: temporalis and masseter (ipsilateral), medial and lateral pterygoids (contralateral).

The temporomandibular articulation is the only normal joint that can be dislocated without the action of an external force. Dislocation of the jaw is almost always bilateral, and the displacement is anterior. **The head of the mandible may slip forward into the infratemporal fossa when the mouth is open, that is, when the head is situated on the articular tubercle. Reduction is accomplished by depressing the back of the jaw and elevating the chin.**

MAXILLARY ARTERY

The maxillary (formerly internal maxillary) artery, the larger terminal branch of the external carotid, arises in the parotid gland, behind the neck of the mandible. It has an extensive distribution to the upper and lower jaws, the muscles of mastication, the palate, and the nose. Its course may be considered in three parts: mandibular, pterygoid, and pterygopalatine.

1. The *mandibular part* runs forward between the neck of the mandible and the sphenomandibular ligament. It courses along the lower border of the lateral pterygoid muscle. Most of the branches of the first and second parts of the maxillary artery accompany branches of the mandibular nerve.

2. The *pterygoid part* runs forward and upward under cover of the temporalis. It lies either superficial or deep (fig. 58–3*D*) to the lower head of the lateral pterygoid muscle.[22] In its superficial location, it is placed between the temporalis and the lateral pterygoid; in its deep location, it lies between the lateral pterygoid and the branches of the mandibular nerve. The branches of the second part of the maxillary artery supply the muscles of mastication and the buccinator.

3. The *pterygopalatine part* of the maxillary artery passes between the upper and lower heads of the lateral pterygoid (fig. 58–4), and then through the pterygopalatine fossa. It supplies in part the orbit, face, upper teeth, palate, nasal cavity, paranasal sinuses, and nasopharynx. Its most important branch is the sphenopalatine artery. The nerves that accompany the branches of the third part of the maxillary artery are derived from the maxillary nerve either directly or through the pterygopalatine ganglion.

The *maxillary vein* is formed by the union of vessels in the pterygoid plexus (p. 664). It accompanies the first part of the maxillary artery and unites with the superficial temporal vein to form the retromandibular vein.

Branches of First Part. These branches supply chiefly the tympanic membrane, the dura and the skull, and the lower teeth.

1 and 2. The *deep auricular* and *anterior tympanic arteries* supply the tympanic membrane.

3. The *middle meningeal artery* is clinically the most important branch of the maxillary (fig. 58–3*C*). It ascends between the sphenomandibular ligament and the lateral pterygoid, and lies in the tensor veli palatini. It passes between the two roots of the auriculotemporal nerve and lies behind the mandibular nerve. It enters the cranial cavity by passing through the foramen spinosum in the sphenoid bone (p. 601).

4. An *accessory meningeal branch* may arise from[23] either the maxillary or the middle meningeal artery. It traverses the foramen ovale.

5. The *inferior alveolar artery* descends between the sphenomandibular ligament and the ramus of the mandible. The corresponding nerve lies in front of it and both enter the mandibular canal through the mandibular foramen. The inferior alveolar artery supplies the mucous membrane of the cheek, the chin, and the lower teeth.

Branches of Second Part. These branches supply chiefly the muscles of mastication and are named accordingly: *anterior* and *posterior deep temporal, pterygoid, masseteric,* and *buccal.*

Branches of Third Part. This part of the artery has an extensive distribution that includes the upper teeth, portions of the face and of the orbit, the palate, and the nasal cavity.

1. The *posterior superior alveolar artery* descends in the infratemporal fossa on the posterior surface of the maxilla, and supplies the molar and premolar teeth.

2. The *infra-orbital artery* arises in the pterygopalatine fossa, enters the orbit through the inferior orbital fissure, and runs along the infra-orbital groove and canal, and through the infra-orbital foramen. In addition to the orbit, it supplies the lower eyelid, the lacrimal sac, the upper lip, and the cheek.

Anterior and *middle superior alveolar arteries* give dental branches to the canine and incisor teeth.

3. The *descending palatine artery* descends through the pterygopalatine fossa and the greater palatine canal, and gives off the *greater* and *lesser palatine arteries* to the palate.

4. The *artery of the pterygoid canal,* which frequently arises from one of the palatine arteries, runs backward through the pterygoid canal.

5. A *pharyngeal branch* runs backward through the palatovaginal canal and is distributed to the roof of the nose and pharynx.

6. The *sphenopalatine artery* may be regarded as the termination of the maxillary. It enters the nasal cavity through the sphenopalatine foramen. It supplies the conchae, meatuses, and paranasal sinuses, and ends on the nasal septum. The sphenopalatine artery is important in bleeding from the nose (epistaxis).

MAXILLARY NERVE
(fig. 58–7)

The maxillary nerve (second division of the trigeminal) arises from the trigeminal ganglion and lies in the dura lateral to the cavernous sinus. It passes through the foramen rotundum and enters the pterygopalatine fossa (where it can be "blocked" by passing a needle through the mandibular notch and injecting a local anesthetic). Then, as the infra-orbital nerve, it gains the orbit through the inferior orbital fissure. It ends on the face by emerging through the infra-orbital foramen. Therefore, **in its course, the maxillary nerve traverses in succession the middle cranial fossa, the pterygopalatine fossa, the orbit, and the face.**

Examination. The area of skin supplied by the maxillary nerve is tested for sensation by the use of cotton wool and a pin (fig. 57–6*D*, p. 658).

Branches. The maxillary nerve gives off the following branches:

1. A *meningeal branch* arises in the middle cranial fossa.

2. Communicating branches are given to the pterygopalatine ganglion.

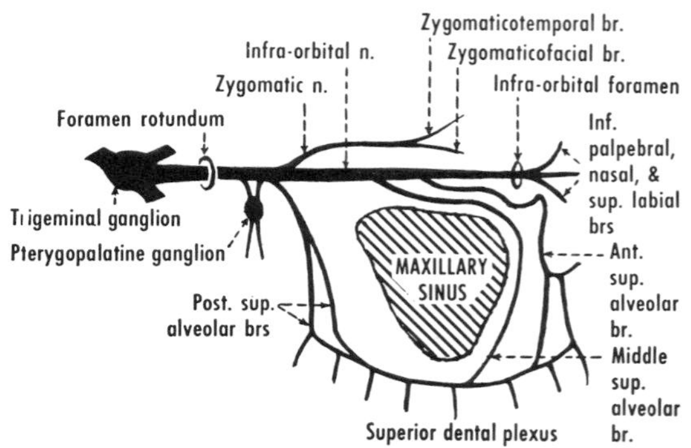

Figure 58–7 Scheme of the maxillary nerve. Lateral aspect.

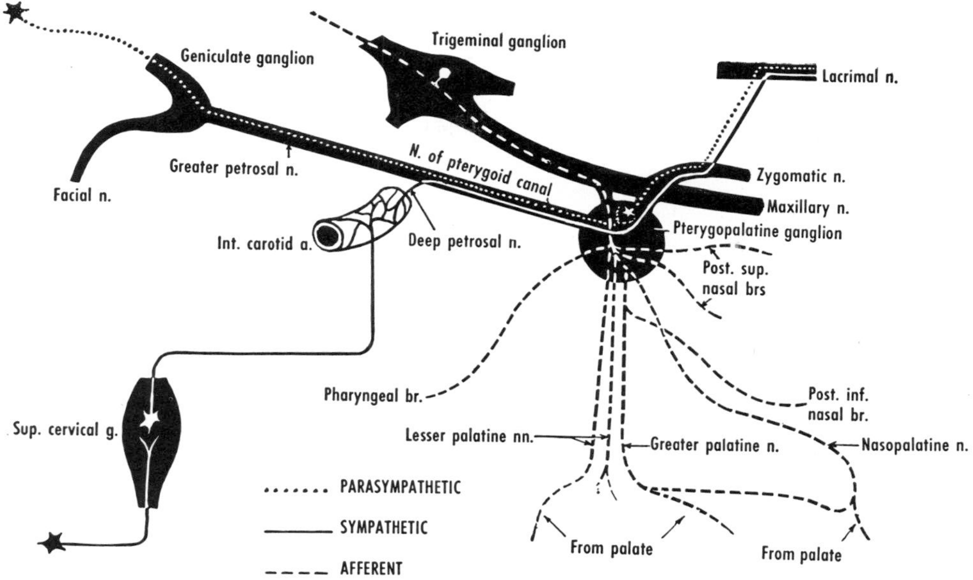

Figure 58–8 Scheme of the pterygopalatine ganglion and its connections. Lateral aspect.

3. *Posterior superior alveolar branches* emerge through the pterygopalatine fissure, enter canals on the back of the maxilla, and supply the maxillary sinus, cheek, gums, and the molar and premolar teeth. They take part in the superior dental plexus.

4. The *zygomatic nerve* enters the orbit through the inferior orbital fissure and divides on the lateral wall of the orbit into: *(a)* a *zygomaticotemporal branch,* which pierces the zygomatic bone and supplies the skin of the temple; and *(b)* a *zygomaticofacial branch,* which pierces the zygomatic bone and supplies the overlying skin of the face. The zygomatic nerve communicates in the orbit with the lacrimal nerve, by which means secretory fibers are probably conveyed to the lacrimal gland (fig. 58–8). Direct branches from the maxillary to the lacrimal nerve have also been described.[24]

5. The *infra-orbital nerve,* regarded as the continuation of the maxillary, enters the orbit through the inferior orbital fissure and occupies in succession the infra-orbital groove, canal, and foramen. It ends on the face by dividing into several branches: *inferior palpebral* (to conjunctiva and skin of lower eyelid), *nasal* (to skin of nose), and *superior labial* (to mucous membrane of mouth and skin of lip). A *middle superior alveolar branch* commonly arises from the infra-orbital nerve, runs in the anterior, lateral, or posterior wall of the maxillary sinus, and passes to the premolar part of the superior dental plexus.[25] An *anterior superior alveolar branch* arises from the infra-orbital nerve in the infra-orbital canal, and, by means of a sinuous canal,[26] descends along the anterior wall of the maxillary sinus. It takes part in the superior dental plexus and gives branches to the canine and incisors. Its terminal twigs emerge close to the nasal septum and supply the floor of the nose. The superior dental plexus is located in part on the posterior surface of the maxilla and in part in bony canals in the lateral and anterior aspects of the maxilla.[27] It is formed by the anterior and posterior and, when present, the middle, alveolar nerves.

The branches of the maxillary and infra-orbital nerves are summarized in table 58–1.

PTERYGOPALATINE GANGLION
(fig. 58–8)

The pterygopalatine (sphenopalatine) ganglion is situated in the pterygopalatine fossa, lateral to the sphenopalatine foramen, below the maxillary nerve, in front of the pterygoid

TABLE 58–1 Summary of Branches of Maxillary and Infra-orbital Nerves

Location	*Branches*
In middle cranial fossa	Meningeal branch
In pterygopalatine fossa	Communicating branches (pterygopalatine nerves) to pterygopalatine ganglion Posterior superior alveolar branches Zygomatic nerve Zygomaticotemporal branch Zygomaticofacial branch
In infra-orbital canal	Middle superior alveolar branch Anterior superior alveolar branch
On face	Inferior palpebral branches Nasal branches Superior labial branches

canal, and behind the middle nasal concha. The ganglion can be injected through the mandibular notch and the pterygopalatine fossa. The fibers connected with the ganglion are generally described as its roots. A parasympathetic (motor) root comes by way of the greater petrosal nerve and the nerve of the pterygoid canal. These fibers, derived from the facial nerve, synapse in the ganglion, and they are probably the only fibers that do so. The postganglionic fibers pass to the lacrimal gland (by way of the maxillary, zygomatic, and lacrimal nerves; and also by orbital branches of the ganglion, which pass through the inferior orbital fissure). Parasympathetic fibers also pass to the nasal and palatine glands. A sympathetic root, derived from the internal carotid plexus, travels by way of the deep petrosal nerve and the nerve of the pterygoid canal. These fibers are postganglionic (arising in the superior cervical ganglion). They merely pass through the pterygopalatine ganglion and are distributed with the parasympathetic fibers.

The afferent or sensory root consists of fibers that connect the pterygopalatine ganglion with the maxillary nerve. These fibers have reached the ganglion from the periphery (orbit, nasal cavity, palate, and nasopharynx) by way of the so-called branches of the pterygopalatine ganglion, which are predominantly fibers of the maxillary nerve. The branches referred to are as follows:

1. *Orbital branches* pass to the periosteum of the orbit and to the posterior ethmoidal and sphenoidal sinuses.
2. *Posterior superior nasal branches* supply the nasal cavity.
3. The *nasopalatine nerve* passes through the sphenopalatine foramen, descends along the nasal septum, and gains the hard palate by passing through a median incisive foramen.
4. The palatine nerves descend through the palatine canals. The *greater palatine nerve* gives off posterior inferior nasal branches, emerges through the greater palatine foramen, and supplies the palate. It may contain fibers from the facial as well as from the maxillary nerve. The *lesser palatine nerves* emerge through the lesser palatine foramina and supply the soft palate and the tonsil.
5. The *pharyngeal* branch passes backward through the palatovaginal canal and supplies the mucosa of the roof of the pharynx and the sphenoidal sinus.

The nasal, nasopalatine, and palatine nerves contain, in addition to sensory fibers, secretory fibers to the nasal and palatine glands, and also vasomotor fibers. Furthermore, the palatine nerves contain some fibers associated with taste; these reach the facial nerve by way of the greater petrosal nerve.

MANDIBULAR NERVE
(figs. 58–3*C* and 58–9)

The mandibular nerve (third division of the trigeminal) arises from the trigeminal ganglion and, together with the motor root of the trigeminal nerve, passes through the foramen ovale to the infratemporal fossa (where it can be "blocked" by passing a needle

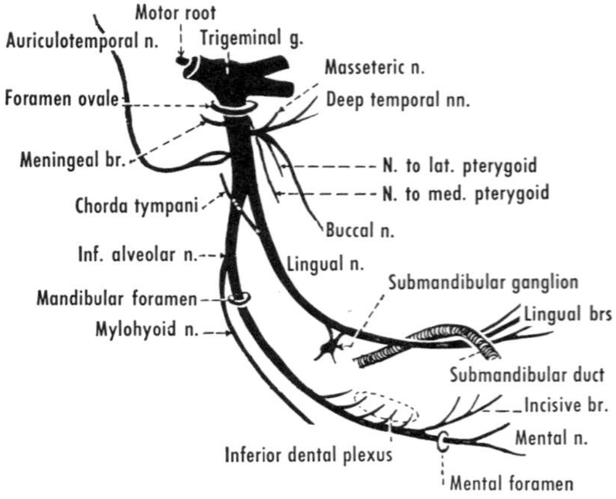

Figure 58–9 Scheme of the mandibular nerve. Lateral aspect.

through the mandibular notch and injecting a local anesthetic). As it passes through the base of the skull, the mandibular nerve is joined by the motor root. The trunk thereby formed divides almost immediately into a number of branches, which are classified into an anterior and a posterior division. The mandibular nerve is related at the base of the skull to the middle meningeal artery posteriorly, the lateral pterygoid muscle laterally, and the tensor veli palatini medially.

Examination. The area of skin supplied by the mandibular nerve is tested for sensation by the use of cotton wool and a pin (fig. 57–6*E*, p. 658). The muscles of mastication are tested by palpating the temporalis and the masseter while the subject clenches his teeth.

Branches. Two branches arise from the undivided trunk of the mandibular nerve: (1) a *meningeal branch (nervus spinosus),* which accompanies the middle meningeal artery upward through the foramen spinosum; and (2) the *nerve to the medial pterygoid,* which is commonly said to supply also the tensor tympani and tensor veli palatini by way of the otic ganglion.

The anterior division of the mandibular nerve comprises several small branches:

1. The *buccal nerve* passes between the two heads of the lateral pterygoid and becomes embedded temporarily in the anterior border of the temporalis.[28] Its branches spread out on the lateral surface of the buccinator. It supplies sensory fibers to the skin and the mucous membrane of the cheek, to the gums,[29] and perhaps also to the first two molars and the premolar teeth.[30] Some of its branches unite with those of the buccal branch of the facial nerve.

2. The *masseteric nerve* passes above the lateral pterygoid, behind the temporalis, and through the mandibular notch to supply the masseter.

3. Several *deep temporal nerves* supply the temporalis.

4. The *nerve to the lateral pterygoid* supplies that muscle.

The posterior division of the mandibular nerve is chiefly sensory. It gives off the auriculotemporal nerve and divides into the lingual and inferior alveolar nerves.

1. The *auriculotemporal nerve* arises generally by two roots that encircle the middle meningeal artery. The nerve breaks up immediately into a spray of branches,[31] the largest of which proceeds backward, deep to the lateral pterygoid, and between the sphenomandibular ligament and the neck of the mandible. It is closely related to the parotid gland and passes upward, behind the temporomandibular joint. It crosses the zygoma and lies behind the superficial temporal artery. Its terminal twigs are distributed to the scalp.

The auriculotemporal nerve receives communications from the otic ganglion (conveying secretory fibers from the glossopharyngeal nerve for the parotid gland), and it supplies the parotid gland, the temporomandibular joint, the tympanic membrane, the external ear, and the scalp.

Pain from disease of a tooth or of the tongue is sometimes referred to the distribution of the auriculotemporal nerve to the ear.

2. The *lingual nerve* descends medial to the lateral pterygoid and is there joined by the chorda tympani, a branch of the facial nerve that contains fibers associated with taste. The lingual nerve is located in front of the inferior alveolar nerve (with which it sometimes communicates) and it passes between the medial pterygoid and the ramus of the mandible. It then lies under cover of the mucous membrane of the mouth, being palpable against the mandible about 1 cm below and behind the third molar tooth. Next it crosses the lateral surface of the hyoglossus, passes deep to the mylohyoid and lies above the submandibular duct. It crosses downward on the lateral side of the duct and winds upward on its medial side, lying on the genioglossus. It proceeds forward along the side of the tongue and supplies that structure with sensory fibers. Its terminal branches communicate with those of the hypoglossal nerve.

In addition to the communication from the chorda tympani, the lingual nerve gives branches to the isthmus of the fauces, the submandibular ganglion, and the mucous membrane over the side and dorsum of the anterior two-thirds of the tongue, also supplying the mucous membrane of the mouth, the gums, and the first molar and the premolar teeth.

3. The *inferior alveolar nerve* descends in front of its companion artery, deep to the lateral pterygoid. It passes between the sphenomandibular ligament and the ramus of the mandible, and then through the mandibular foramen and canal. **Above its entry into the mandibular foramen, the inferior alveolar nerve can be "blocked" intra-orally with a local anesthetic.**

The inferior alveolar nerve gives off the following: (1) The *mylohyoid nerve* arises immediately before the inferior alveolar nerve enters the mandibular foramen. It pierces the sphenomandibular ligament, runs along a groove on the ramus of the mandible, lies on the lower surface of the mylohyoid, and supplies the mylohyoid and the anterior belly of the digastric. (2) *Inferior dental branches* arise in the mandibular canal, form the inferior dental plexus, and supply the lower teeth. (3) *Gingival branches* to the gums. (4) The *mental nerve* emerges through the mental foramen and supplies the skin of the chin and lower lip. (5) The *incisive branch* (as the terminal part of the inferior alveolar nerve, after it has given off the mental nerve, is sometimes called) forms a plexus that supplies the canine (sometimes) and the incisor teeth, and frequently also the incisors of the opposite side.[32]

OTIC GANGLION
(fig. 58–10)

The otic ganglion is situated in the infratemporal fossa, immediately below the foramen ovale, medial to the mandibular nerve, lateral to the tensor veli palatini, in front of the middle meningeal artery, and behind the medial pterygoid muscle. The fibers connected with the ganglion are generally described as its roots. The parasympathetic (motor) root is the lesser petrosal nerve. These preganglionic fibers, derived from the glossopharyngeal nerve, synapse in the ganglion, and they are the only fibers that do so. The postganglionic fibers pass to the auriculotemporal nerve. They are secretory to the parotid gland. A sympathetic root is derived from the plexus on the middle meningeal artery. These fibers are postganglionic (arising in the superior cervical ganglion). They merely pass through the otic ganglion and, by way of the auriculotemporal nerve, supply the blood vessels of the parotid gland. A so-called efferent root comes from the nerve to the medial pterygoid muscle. These fibers merely pass through the ganglion and are commonly said to supply the tensor tympani and the tensor veli palatini. Some taste fibers from the anterior two-thirds of the tongue may pass through the otic ganglion, which they reach by a communication from the chorda tympani and leave by a communication to the nerve of the pterygoid canal.

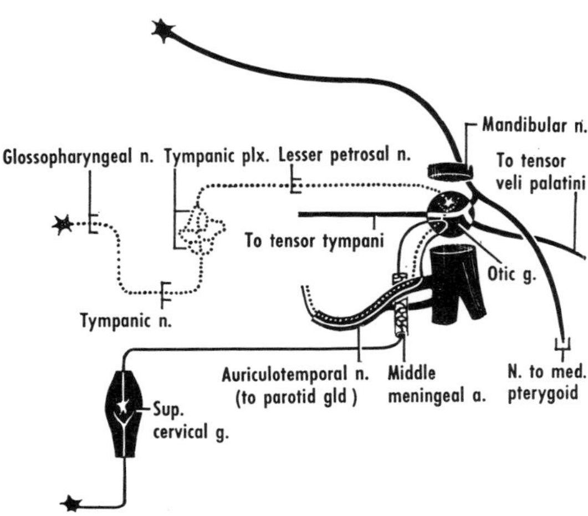

Figure 58–10 Scheme of the otic ganglion and its connections. Lateral aspect. Note parasympathetic (dotted lines) and sympathetic (continuous, narrow lines) fibers.

REFERENCES

1. S. Rauch, *Die Speicheldrüsen des Menschen,* Thieme, Stuttgart, 1959.
2. G. R. L. Gaughran, Ann. Otol., etc., St Louis, *70*:31, 1961.
3. F. G. Parsons, J. Anat., Lond., *45*:239, 1911.
4. H. Brünner, Anat. Anz., *110*:327, 1962.
5. G. L. McWhorter, Anat. Rec., *12*:149, 1917. R. A. Davis *et al.,* Surg. Gynec. Obstet., *102*:385, 1956.
6. J. McKenzie, J. Anat., Lond., *82*:183, 1948. D. H. Patey and I. Ranger, Brit. J. Surg., *45*:250, 1957.
7. H. Oppenheim and M. Wing, Arch. Otolaryng., *71*:80, 1960.
8. F. L. Reichert, Arch. Neurol. Psychiat., Chicago, *32*:1030, 1934.
9. A. Kuntz and C. A. Richins, J. comp. Neurol., *85*:21, 1946.
10. R. E. Scammon, Anat. Rec., *15*:257, 1919. G. R. L. Gaughran, Anat. Rec., *129*:383, 1957. C. Argenson *et al.,* C. R. Ass. Anat., *56*:1185, 1972.
11. J. D. B. MacDougall, Brit. dent. J., *98*:193, 1955.
12. M. R. Porter, J. prosthet. Dent., *24*:555, 1970.
13. R. J. Last, Proc. R. Soc. Med., *47*:571, 1954.
14. P. G. Grant, Amer. J. Anat., *138*:1, 1973. J. A. McNamara, Amer. J. Anat., *138*:197, 1973.
15. B. G. Sarnat (ed.), *The Temporomandibular Joint,* Thomas, Springfield, Illinois, 2nd ed., 1964.
16. T. R. Murphy, Brit. dent. J., *118*:163, 1965.
17. B. Thilander, Acta odont. Scand., *22*:135, 1964.
18. W. Zenker, Z. Anat. EntwGesch., *119*:375, 1956.
19. L. A. Rees, Brit. dent. J., *96*:125, 1954.
20. J. Cameron, J. Anat., Lond., *49*:210, 1915. J. Bossy and L. Gaillard, Acta anat., *52*:282, 1963. J. G. Burch, J. prosthet. Dent., *24*:621, 1970.
21. B. Thilander, Trans. roy. Sch. Dent., Stockh. Umeå, 2, 1961.
22. Z. Križan, Acta anat., *41*:319 and *42*:71, 1960. C. Skopakoff, Anat. Anz., *123*:534, 1968.
23. J. J. Baumel and D. Y. Beard, J. Anat., Lond., *95*:386, 1961.
24. T. H. Evans, Amer. J. Ophthal., *47*:225, 1959.
25. M. J. T. Fitzgerald, J. Anat., Lond., *90*:520, 1956. W. L.

McDaniel, J. dent. Res., *35*:916, 1956. W. Graf and G. Martensson, Acta otolaryng., Stockh., *47*:114, 1957. M. F. Gaballah, M. T. Rakhawy, and Z. H. Badawy, Acta anat., *86*:151, 1973.
26. F. W. Jones, J. Anat., Lond., *73*:583, 1939.
27. M. J. T. Fitzgerald and J. H. Scott, Brit. dent. J., *104*:205, 1958.
28. E. G. Sloman, J. Amer. dent. Ass., *26*:428, 1939.
29. P. Dziallas, Z. Anat. EntwGesch., *120*:466, 1958.
30. D. Stewart and S. L. Wilson, Lancet, *2*:809, 1928.
31. J. J. Baumel, J. P. Vanderheiden, and J. E. McElenney, Amer. J. Anat., *130*:431, 1971.
32. C. Starkie and D. Stewart, J. Anat., Lond., *65*:319, 1931. A. Brunetti, Stomatologia, Milano, *29*:85, 1931.

SUBMANDIBULAR REGION 59

The submandibular region is that under cover of the body of the mandible and between the mandible and the hyoid bone. It contains the submandibular and sublingual glands, the suprahyoid muscles, the submandibular ganglion, and the lingual artery. The lingual (p. 671) and hypoglossal (p. 701) nerves, and the facial artery (pp. 694 and 656) are discussed elsewhere.

Submandibular Gland

The submandibular gland is one of the three large paired salivary glands. It is predominantly serous in type. The gland comprises a larger superficial part, or body, and a smaller deep process. **The two parts are continuous with each other around the posterior border of the mylohyoid** (fig. 59–1*A*).

Surface Anatomy. The submandibular gland lies partly above and partly below the posterior half of the base of the mandible. Usually the gland is scarcely palpable.

Relations. **The body of the gland is located in and below the digastric triangle, and also partly under cover of the mandible.** It presents three surfaces: inferior, lateral, and medial (fig. 59–1*B*). The *inferior surface* is covered by the skin, platysma, and fascia, and is related to the facial vein and the submandibular lymph nodes. The *lateral surface* is related to the submandibular fovea on the medial surface of the mandible and also to the medial pterygoid muscle. The *medial surface* is related to the mylohyoid, hyoglossus, and digastric. The inferior and medial surfaces are covered by the cervical fascia.

The *deep process* of the submandibular gland lies between the mylohyoid laterally and the hyoglossus medially, and between the lingual nerve above and the hypoglossal nerve below (fig. 59–3).

Submandibular Duct. The submandibular duct, about 5 cm in length, emerges from the deep process of the gland (fig. 59–1A). It continues between the mylohyoid and the hyoglossus, where it is crossed laterally by the lingual nerve, and then between the sublingual gland and the genioglossus. The terminal branches of the lingual nerve ascend on the medial side of the duct. **The submandibular duct opens by one to three orifices into the oral cavity on the sublingual papilla, at the side of the frenulum of the tongue.**

The branches of the submandibular duct can be examined radiographically after the injection of a radio-opaque medium (iodized oil) into the opening of the duct. This procedure is termed sialography.

Innervation and Blood Supply. The submandibular gland is supplied by parasympathetic, secretomotor fibers, derived mostly from the submandibular ganglion (fig. 59–4). The preganglionic fibers are derived from the chorda tympani, a branch of the facial, and reach the ganglion by way of the lingual nerve. The postganglionic fibers are given to the gland directly from the ganglion. Owing to the presence of a

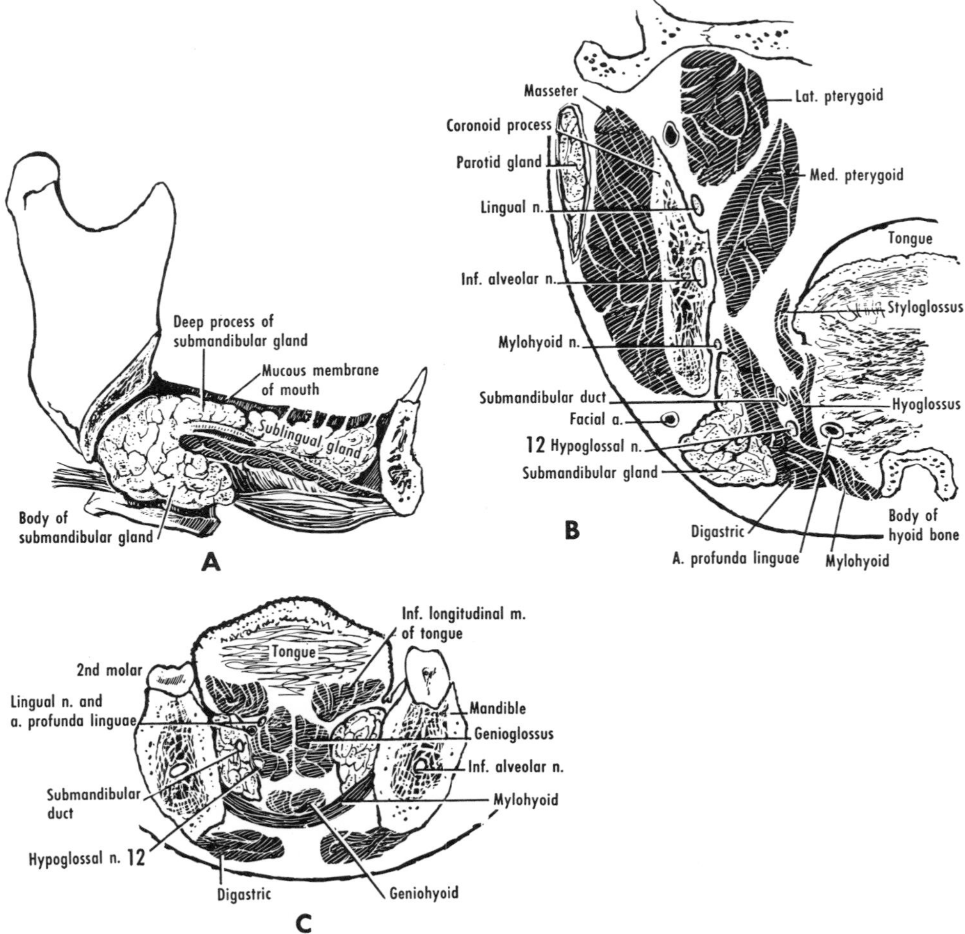

Figure 59–1 The submandibular and sublingual glands. *A*, right lateral aspect, after removal of a portion of the mandible. *B*, coronal section through ramus of mandible and hyoid bone, to show relations of submandibular gland. *C*, coronal section through body of mandible, to show relations of sublingual gland. Based on Kampmeier, Cooper, and Jones.

communication between the glossopharyngeal and facial nerves, it is possible that the submandibular gland is supplied by fibers associated with both of these cranial nerves. Sympathetic fibers also supply the gland and appear to be secretomotor.[1]

Sublingual Gland

The sublingual gland is the smallest of the three chief, paired salivary glands, and is predominantly mucous in type. It is related (fig. 59–1*C*) superiorly to the mucous membrane (sublingual fold) of the floor of the mouth, inferiorly to the mylohyoid, anteriorly to the gland of the opposite side, posteriorly to the deep process of the submandibular gland, laterally to the sublingual fovea on the medial surface of the mandible, and medially to the genioglossus, from which it is separated by the lingual nerve and the submandibular duct.

Sublingual Ducts. The sublingual ducts, about 10 to 30 in number, open mostly separately into the oral cavity on the sublingual fold (fig. 59–1*A*), but some enter the submandibular duct.

Innervation and Blood Supply. The sublingual gland is supplied by parasympathetic, secretomotor fibers, derived mostly from the submandibular ganglion (fig. 59–4). The preganglionic fibers come from the chorda tympani, a branch of the facial nerve, and reach the ganglion by way of the lingual nerve. The postganglionic fibers join the lingual nerve and thereby reach the sublingual gland. Owing to the presence of a communication between the glossopharyngeal and facial nerves, it is possible that the sublingual gland is supplied by fibers associated with both of these cranial nerves.

Suprahyoid Muscles

The suprahyoid muscles (figs. 59–2 and 60–7) connect the hyoid bone to the skull. They comprise the digastric, the stylohyoid, the mylohyoid, and the geniohyoid. The genioglossus and the hyoglossus are described with the tongue (p. 722).

Digastric (fig. 60–3, *C* and *D*). The digastric consists of two bellies united by an intervening tendon. The *posterior belly* arises from the mastoid notch on the temporal bone and is directed forward and downward toward the hyoid bone. The shorter *anterior belly* is attached to the digastric fossa on the lower border of the mandible close to the symphysis. It is directed backward and downward. The middle tendon is attached to the body and the greater horn of the hyoid bone by aponeurotic fibers from the cervical fascia. It commonly passes through the stylohyoid muscle.

The posterior belly of the digastric and the stylohyoid are crossed superficially by the facial vein, the great auricular nerve, and the cervical branch of the facial nerve. The external and internal carotid arteries, the internal jugular vein, the last three cranial nerves, and the sympathetic trunk lie deep to the posterior belly and the stylohyoid.

INNERVATION. The anterior belly is supplied by the mylohyoid branch of the inferior alveolar nerve, the posterior by the facial nerve.

ACTION. The digastric pulls the chin backward and downward in opening the mouth, thereby assisting the lateral pterygoid to rotate the mandible into the open-mouth position. The anterior bellies support the hyoid bone.[2]

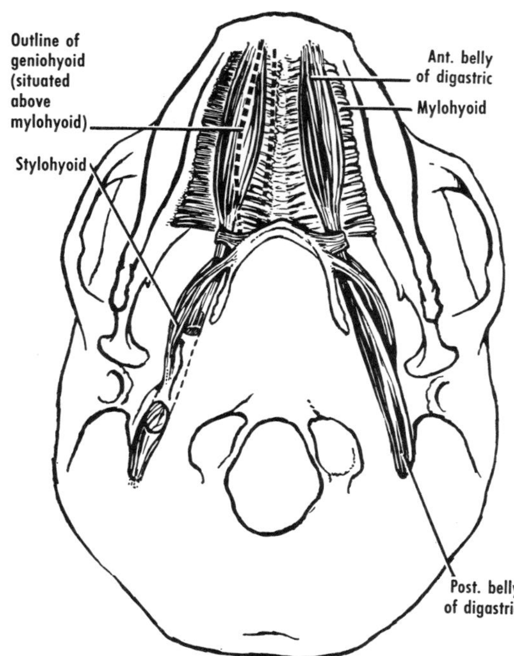

Figure 59–2 Inferior aspect of skull and hyoid bone, showing the suprahyoid muscles. A portion of the posterior belly of the digastric muscle has been removed on the right-hand side of the body, thereby revealing the origin of the stylohyoid muscle from the styloid process. The outline of the geniohyoid muscle is also indicated on this side. Most of the suprahyoid muscles are shown again in figure 60–7. Based on von Lanz and Wachsmuth.

Stylohyoid (fig. 60–7). This is a slender muscle located along the upper border of the posterior belly of the digastric. It arises from the back of the styloid process and is inserted into the hyoid bone at the junction between the body and the greater horn. The stylohyoid is commonly split near its insertion by the tendon of the digastric.

INNERVATION. Facial nerve.

ACTION.[3] The stylohyoid draws the hyoid bone backward and elongates the floor of the mouth. The anteroposterior position of the hyoid bone is determined by the stylohyoid, the geniohyoid, and the infrahyoid muscles.

Mylohyoid (fig. 59–3). This muscle is located above the anterior belly of the digastric. It arises from the mylohyoid line on the internal surface of the mandible, extending from the last molar tooth almost to the symphysis menti. The fibers are directed toward the median plane, where they end mostly upon a median, tendinous raphe. The posterior fibers, however, are inserted into the body of the hyoid bone. The two mylohyoid muscles together form a muscular floor (the *diaphragma oris*) beneath the front of the mouth. The mylohyoid partly covers the hyoglossus.

INNERVATION. Mylohyoid branch of the inferior alveolar nerve.

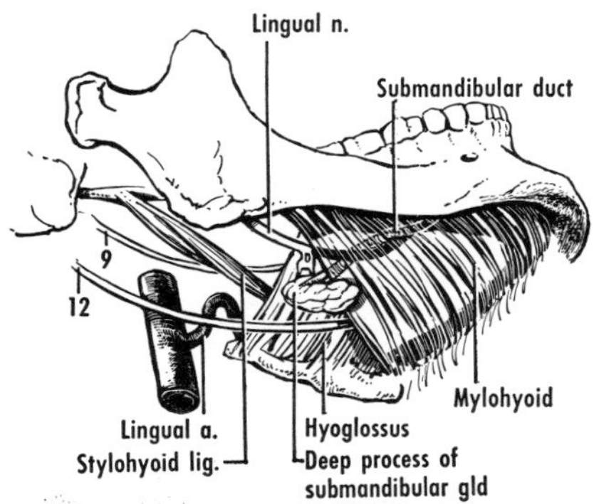

Figure 59–3 Relations of mylohyoid and hyoglossus muscles. **The lingual nerve, the deep process of the submandibular gland and the submandibular duct, and the hypoglossal nerve pass deep to the posterior border of the mylohyoid.** The submandibular ganglion is shown suspended from the lingual nerve. **The glossopharyngeal nerve, the stylohyoid ligament, and the lingual artery pass deep to the posterior border of the hyoglossus.**

ACTION. The two mylohyoids form a muscular diaphragm or sling, which supports the tongue. Contraction of the mylohyoids elevates, and thereby makes more shallow, the floor of the mouth.[4] This elevates the tongue and, if the teeth are held in occlusion, produces an increased pressure on the tongue, forcing it backward, as occurs during swallowing. The mylohyoids force either solids or liquids from the oropharynx to the laryngopharynx.[5]

Geniohyoid (figs. 59–5; 63–4, p. 743). The geniohyoid is situated above the mylohyoid. It arises from the inferior genial tubercle behind the symphysis of the mandible and is inserted into the front of the body of the hyoid bone. It is in contact or fused with the muscle of the opposite side (fig. 59–1*C*).

INNERVATION. A branch from the hypoglossal nerve. This branch consists of fibers of the first (and perhaps the second) cervical nerve.

ACTION. The geniohyoid protrudes the hyoid bone, thereby shortening the floor of the mouth.

Submandibular Ganglion
(fig. 59–4)

The submandibular ganglion is situated on the lateral surface of the hyoglossus muscle, medial to the mylohyoid, below the lingual nerve, and above the submandibular duct and the hypoglossal nerve. The ganglion is suspended from the lingual nerve by several communicating branches. The fibers connected with the ganglion are generally described as its roots. A parasympathetic (motor) root comes from the lingual nerve. These preganglionic fibers, derived from the chorda tympani, synapse in the ganglion, and they are the only fibers that do so. The postganglionic fibers pass to the submandibular gland, to which they are secretory, by way of glandular branches from the submandibular ganglion. Some of the branches join the lingual nerve and reach the sublingual and lingual glands. Portions of the submandibular ganglion may be found in the submandibular gland. A sympathetic root is derived from the plexus on the facial artery. These fibers are postganglionic (arising in the superior cervical ganglion). They merely pass through the submandibular ganglion and are distributed with the parasympathetic fibers. Thus, the submandibular ganglion is essentially a relay station for preganglionic secretomotor fibers of the facial nerve.

Lingual Artery
(fig. 59–5)

The lingual artery arises from the front of the external carotid at or above the level of the hyoid bone. Its course may be consid-

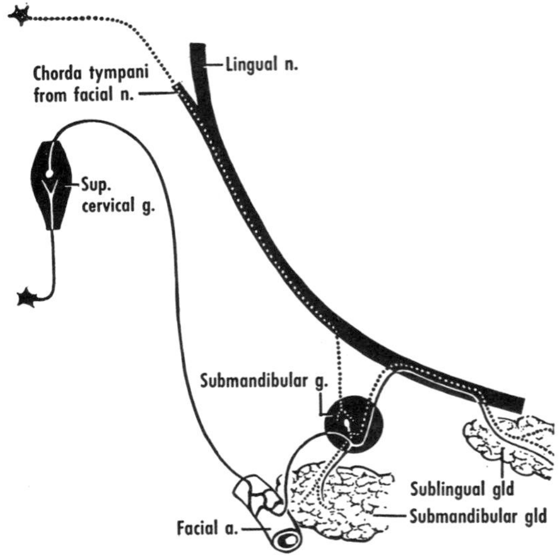

Figure 59–4 Scheme of submandibular ganglion and its connections. Lateral aspect. Note parasympathetic (dotted lines) and sympathetic (continuous, narrow lines) fibers.

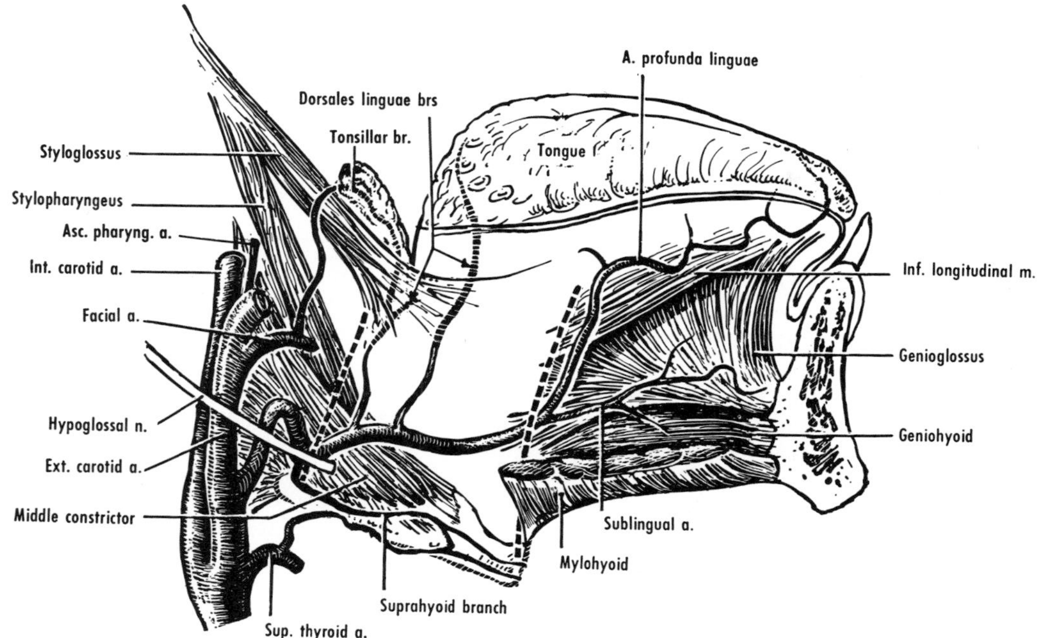

Figure 59–5 The lingual artery. The interrupted lines indicate the borders of the removed hyoglossus muscle.

ered in three parts: posterior, deep, and anterior to the hyoglossus muscle, respectively. The *first part* of the lingual artery lies mainly in the carotid triangle and forms a loop on the middle constrictor of the pharynx. The loop is crossed by the hypoglossal nerve. The *second part* of the artery passes deep to the hyoglossus and runs along the upper border of the hyoid bone. It lies on the middle constrictor. The *third part* of the artery, the arteria profunda linguae, is described below.

Branches. The lingual artery gives off: (1) a *suprahyoid branch,* which anastomoses with its fellow of the opposite side; (2) the *dorsales linguae branches* to the dorsum of the tongue; (3) the *sublingual artery* to the sublingual gland; (4) the *arteria profunda linguae (ranine artery),* the termination of the lingual, which ascends between the genioglossus and the inferior longitudinal muscle of the tongue. It runs along the lower surface of the tongue and anastomoses with its fellow of the opposite side.

REFERENCES

1. J.-E. Laage-Hellman and B. C. R. Strömblad, J. appl. Physiol., *15*:295, 1960.
2. S. Carlsöö, Acta anat., *26*:81, 1956.
3. R. J. Last, Int. dent. J., *5*:338, 1955.
4. R. J. Last, Proc. R. Soc. Med., *47*:571, 1954.
5. J. Whillis, J. Anat., Lond., *80*:115, 1946.

THE NECK

60

SUPERFICIAL STRUCTURES OF NECK

STERNOMASTOID AND TRAPEZIUS

Sternomastoid. The sternocleidomastoid (*cleid-* refers to the clavicle) or, more simply, the sternomastoid, extends obliquely up the neck from the sternoclavicular joint to the mastoid process (figs. 60–2 and 60–3*B*). It has two heads of origin. The rounded, tendinous sternal head arises from the front of the manubrium (fig. 26–2), the flat clavicular head from the upper surface of the medial third of the clavicle (figs. 11–2 and 13–3). The clavicular head varies greatly in width, and a variable interval is found between the two heads. The muscle is inserted on the lateral surface of the mastoid process (fig. 52–11) and into the lateral half or two-thirds of the superior nuchal line on the occipital bone (fig. 52–15).

The sternomastoid is crossed by the platysma, the external jugular vein, and the great auricular and transverse cervical nerves. The sternomastoid covers the great vessels of the neck, the cervical plexus, portions of a number of other muscles (splenius, digastric, levator scapulae, scaleni, sternohyoid, sternothyroid, and omohyoid), and the cupola of the pleura.

The sternomastoid is the "key muscle" of the neck because it divides the quadrilateral area on the side of the neck into anterior and posterior triangles (fig. 60–3*B*).

Trapezius. The trapezius (fig. 13–2) arises from the medial third of the superior nuchal line, the external occipital protuberance, the ligamentum nuchae, and the spines of the last cervical and all of the thoracic vertebrae, as well as the supraspinous ligament. The fibers from the occipital bone and the ligamentum nuchae are inserted into the posterior border and the upper surface of the lateral third of the clavicle. The remaining fibers are inserted on the acromion and the spine of the scapula, as described with the upper limb (p. 105).

Innervation of Sternomastoid and Trapezius. Both muscles are supplied mainly by the accessory (11th cranial) nerve.

There is also an innervation (perhaps

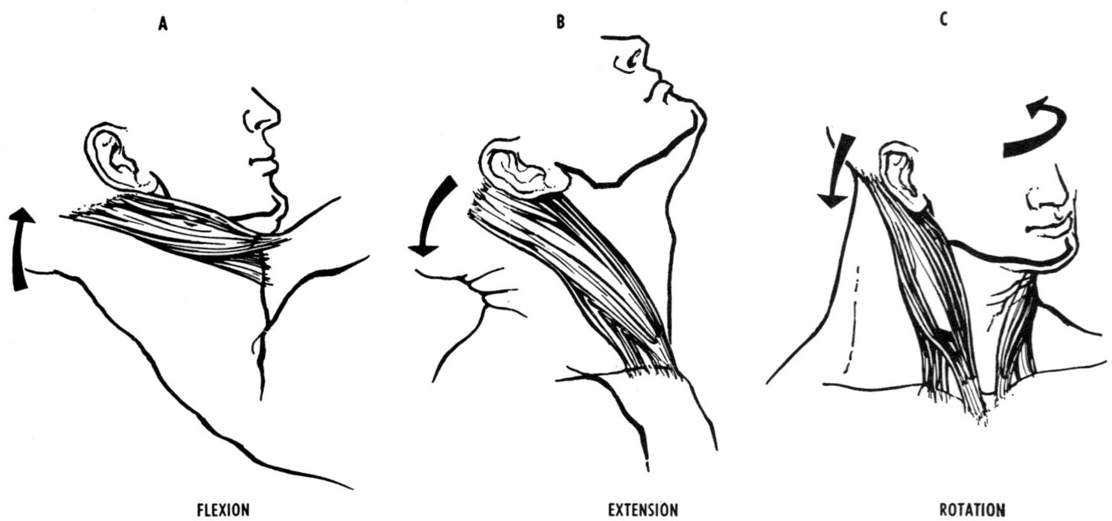

Figure 60–1 Actions of the sternomastoid. *A*, flexion from the recumbent position. *B*, extension. *C*, lateral flexion and contralateral rotation of the face.

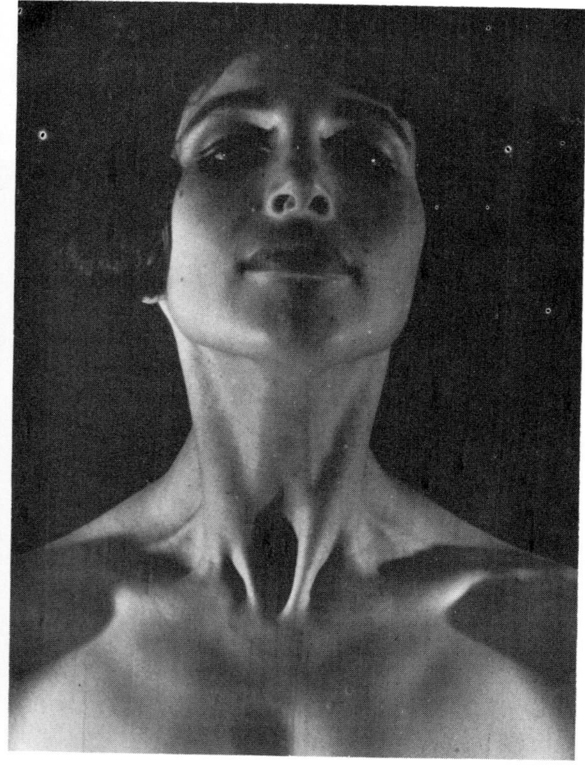

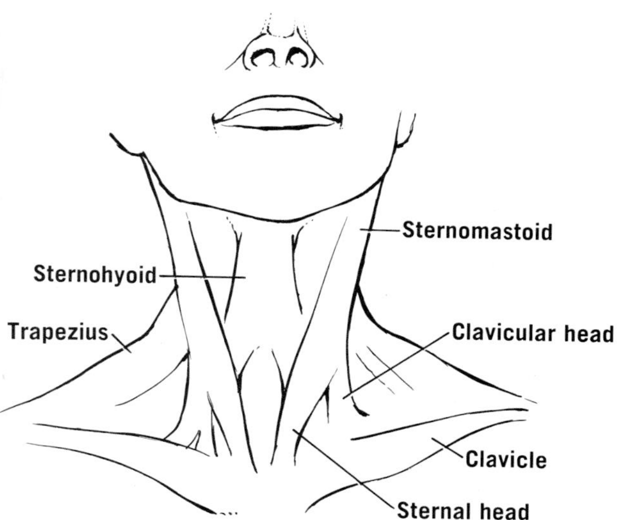

Figure 60–2 Surface anatomy of the neck. The sterno(cleido)mastoid muscles, including their sternal and clavicular heads of origin, are clearly visible. On each side, the anterior triangle of the neck is bounded by the anterior border of the sternomastoid, the anterior median line of the neck, and the lower border of the mandible. From J. Royce, Ph.D., *Surface Anatomy*, Davis, Philadelphia, 1965; courtesy of author and publisher.

proprioceptive[1] as well as motor) from the ventral rami of the cervical nerves (sternomastoid, C2, 3; trapezius, C3, 4).

Actions of Sternomastoid and Trapezius. The trapezius elevates and rotates the scapula, as explained in the section on the upper limb. The trapezius is probably the only muscle in the neck that can be relaxed.[2]

The sternomastoid muscles, acting together, bend the head forward against resistance (fig. 60–1*A*). Although the sternomastoids pull the cervical part of the vertebral column forward into flexion, their posterior fibers probably extend at the atlanto-occipital joints (fig. 60–1*B*). In any event, the sternomastoids are active during extension at those joints.[3] The sternomastoids are of importance in respiration only when the rate of ventilation is elevated and the ordinary muscles of inspiration are operating at a disadvantage.[4] When one of the muscles contracts, the head is inclined laterally toward that side and the face is rotated to the opposite side (fig. 60–1*C*). In rotation without resistance, the sternomastoids are usually active only toward the end of the movement.

Spasm of a sternomastoid, which may arise from various causes, produces wry neck (torticollis).

Flexion of the head is performed ordinarily by gravity and is carried out by a regulated relaxation of the extensor muscles. Active flexion is performed mainly by the sternomastoid and longus capitis muscles. The sternomastoids are the chief flexors, and their action is best seen when a person lying flat on his back raises his head (fig. 60–1*A*).

TRIANGLES OF NECK

The neck, when viewed from the side (fig. 60–3*B*), presents a roughly quadrilateral outline that has the following boundaries:

Superior: inferior border of the mandible and a line drawn from the angle of the mandible to the mastoid process.

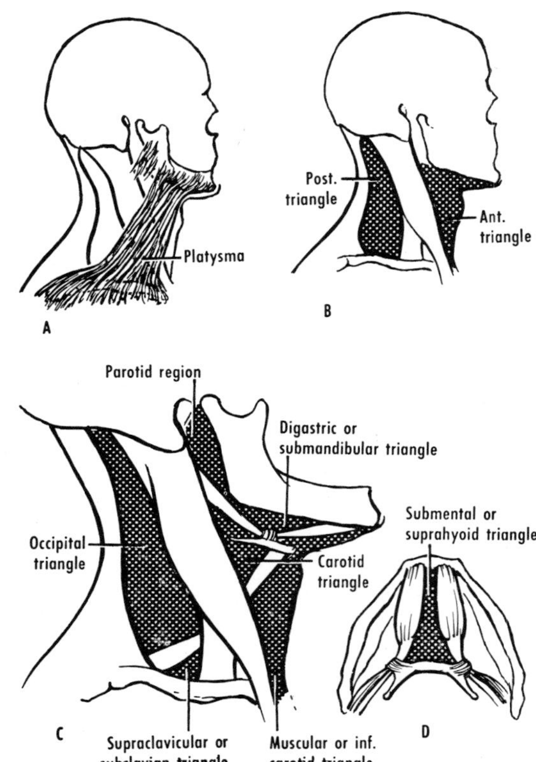

Figure 60–3 Triangles of the neck. *A*, showing the platysma, which forms a part of the roof of both the anterior and the posterior triangle. *B*, showing the general subdivision by the sternomastoid into anterior and posterior triangles. *C* and *D*, showing the smaller divisions of the triangles.

Inferior: superior aspect of the clavicle.
Anterior: anterior median line of the neck.
Posterior: anterior border of the trapezius.

This area is divided by the sternomastoid muscle into two triangles, an anterior in front of the muscle, and a posterior behind it.

POSTERIOR TRIANGLE OF NECK

The boundaries of the posterior triangle are as follows:

Inferior: superior aspect of the intermediate third of the clavicle.
Posterior: anterior border of the trapezius.
Anterior: posterior border of the sternomastoid.

The posterior triangle is crossed by the inferior belly of the omohyoid, and this muscle has been used to divide the posterior triangle into two smaller areas, namely, the occipital and supraclavicular triangles, as shown in figure 60–3*C*.

Roof of Posterior Triangle. The roof of the posterior triangle is formed by the investing layer of the fascia and the platysma. The fascia is pierced by the external jugular vein and the supraclavicular nerves.

Contents of Posterior Triangle (fig. 60–4*B*). **The most important contents of the posterior triangle are the accessory nerve, lymph nodes, the brachial plexus, and the third part of the subclavian artery.** The accessory nerve crosses the middle of the posterior triangle and lies on the levator scapulae. It is located in the fascial roof of the triangle, whereas the brachial plexus lies deep to both the investing and the prevertebral layer of the fascia. Other contents of the posterior triangle include the dorsal scapular nerve to the rhomboids, the long thoracic nerve to the serratus anterior, the nerve to the subclavius, the suprascapular nerve, and the transverse cervical artery.

The brachial plexus (figs. 13–7 to 9), formed by the ventral rami of the lower four cervical nerves (C5, 6, 7, 8) and the first thoracic nerve, should be reviewed at this time (p. 109). The nerves that form the plexus are sandwiched between the scalenus anterior and the scalenus medius. The brachial plexus is found in the posterior triangle of the neck, below a line from the posterior margin of the sternomastoid at the level of the cricoid cartilage to the midpoint of the clavicle. **In this area the plexus can be "blocked" by the injection of a local anesthetic between the first rib and the skin above the clavicle. Brachial plexus block is useful surgically because it renders insensitive all the deep structures of the upper limb and the skin distal to the middle of the arm.**

Floor of Posterior Triangle. The floor of the posterior triangle (fig. 60–4*A*) is formed by the splenius capitis, levator scapulae, scaleni medius et posterior, and the first digitation of the serratus anterior. These muscles are covered by the prevertebral layer of fascia. Sometimes a small part of the semispinalis capitis is visible superiorly at the apex of the triangle.

Accessory Nerve (External Branch)

The accessory (11th cranial) nerve comprises the cranial and the spinal portion, both of which traverse the jugular foramen, where they interchange fibers or unite for a short distance. The two portions separate below the foramen, and the cranial portion (internal branch) joins and is distributed with the vagus (p. 700).

The *external branch* of the accessory nerve runs backward and downward to be distributed to the sternomastoid and the trapezius. It crosses the transverse process of the atlas, and passes deep to the styloid process and the posterior belly of the digastric. It usually pierces the deep surface of the sternomastoid but sometimes remains deep to that muscle. It supplies the sternomastoid and communicates with branches from the second cervical nerve. **Above the middle of the posterior border of the sternomastoid, the accessory nerve crosses the posterior triangle of the neck obliquely, lying on the levator scapulae and being related to lymph nodes** (fig. 60–4*B*). It communicates with branches of C.N. 2 and 3. About 5 cm above the clavicle it passes deep to the anterior border of the trapezius and forms a plexus with branches from C.N. 3 and 4. It supplies the trapezius.

The external branch of the accessory nerve is tested by asking the subject to shrug his shoulders (trapezius) and then to rotate his head (sternomastoid).

Superficial Branches of Cervical Plexus

The *cervical plexus* is located deeply in the upper part of the neck, under cover of the internal jugular vein and the sternomastoid. It is formed by the ventral rami of the upper four cervical nerves (p. 711). The superficial or cutaneous branches of the plexus (fig. 60–5) are the lesser occipital, great auricular, and transverse cervical nerves (C2, 3), and the supraclavicular nerves (C3, 4). **All these superficial branches emerge near the middle of the posterior border of the sternomastoid.**

Lesser Occipital Nerve. The lesser occipital nerve hooks around the accessory nerve, ascends along the posterior border of the sternomastoid, runs behind the auricle, and supplies some of the skin on the side of the head and on the cranial surface of the auricle.

Great Auricular Nerve. The great auricular nerve winds around the posterior border of the sternomastoid (where it is occasionally palpable as a small nodule[5] and ascends obliquely across that muscle onto

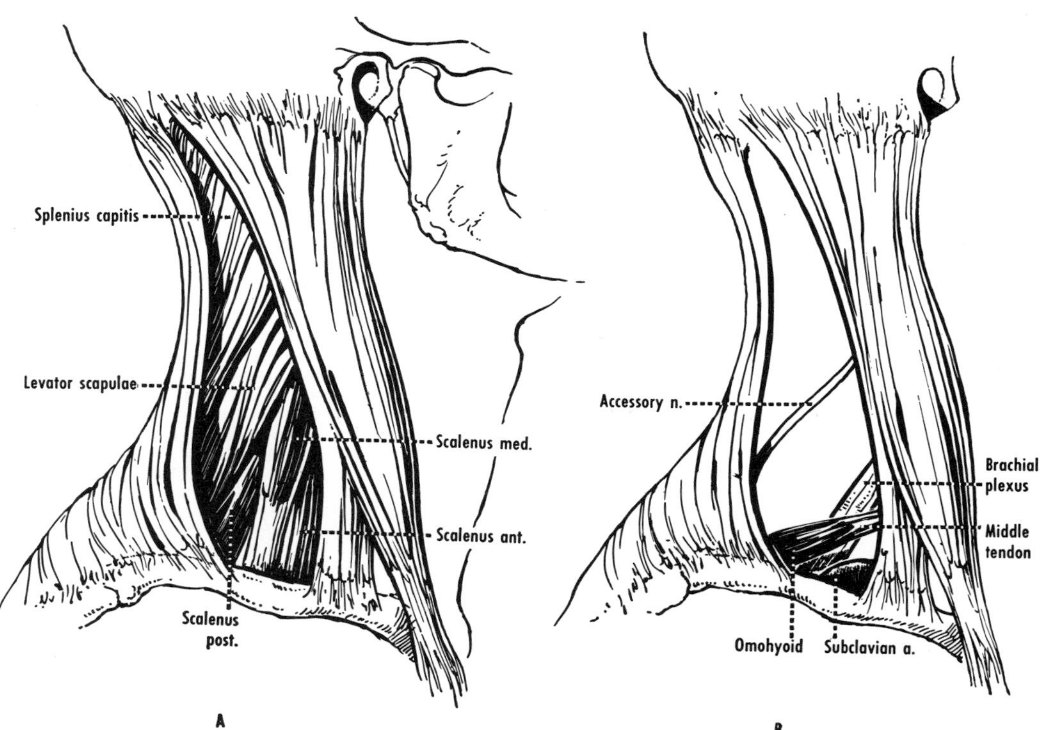

Figure 60–4 Posterior triangle of the neck. *A*, floor. *B*, chief contents. The brachial plexus meets and follows the route of the subclavian artery, but it lies a little more posteriorly and laterally. **The portion of the third part of the subclavian artery shown here is the site for compression.**

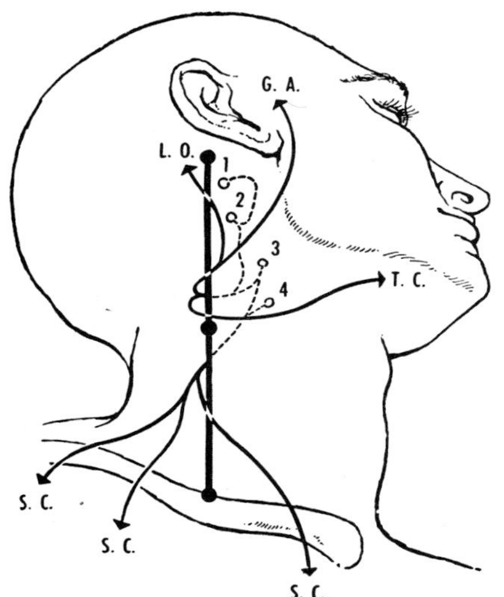

Figure 60–5 Scheme of the cutaneous branches of the cervical plexus. The vertical line indicates the posterior border of the sternomastoid, and the branches of the plexus emerge at the middle of this border. *G.A.*, great auricular nerve; *L.O.*, lesser occipital nerve; *S.C.*, supraclavicular nerves; *T.C.*, transverse cervical nerve. The numerals 1, 2, 3, and 4 indicate the first four cervical nerves. Based on von Lanz and Wachsmuth.

the parotid gland, where it divides to supply the skin over the gland and over the mastoid process, together with both surfaces of the auricle.

Transverse Cervical Nerve. The transverse cervical nerve turns around the middle of the posterior border of the sternomastoid and crosses that muscle deep to the platysma. It divides into branches that supply the skin on the side and front of the neck.

Supraclavicular Nerves. These emerge as a common trunk from under cover of the posterior border of the sternomastoid. The trunk divides into *anterior, middle,* and *posterior supraclavicular nerves,* which descend under cover of the platysma in the posterior triangle, cross the clavicle superficially, and supply the skin over the shoulder as far forward as the median plane.

External Jugular Vein
(fig. 60–6)

The external jugular vein drains the greater part of the face and the scalp, and also contains a significant fraction of cerebral blood. It begins just below, or occasionally within, the parotid gland. It is formed most frequently by the union of the posterior auricular vein and the retromandibular vein, but its mode of formation is highly variable.[6] It runs downward and backward, crossing the sternomastoid obliquely, and under cover of the platysma. It pierces the fascia in the posterior triangle of the neck and ends in the subclavian, or sometimes in the internal jugular vein. Its size varies inversely with that of other veins of the neck. It is provided with two valves which, however, do not prevent regurgitation of blood.

Surface Anatomy. The external jugular vein extends downward and backward from the angle of the mandible to the midpoint of the clavicle. **It is frequently visible on the sternomastoid** and can be made more prominent by having the subject blow while the mouth is closed.

Tributaries. The tributaries, which vary greatly in arrangement, include the *posterior auricular* and *ret-*

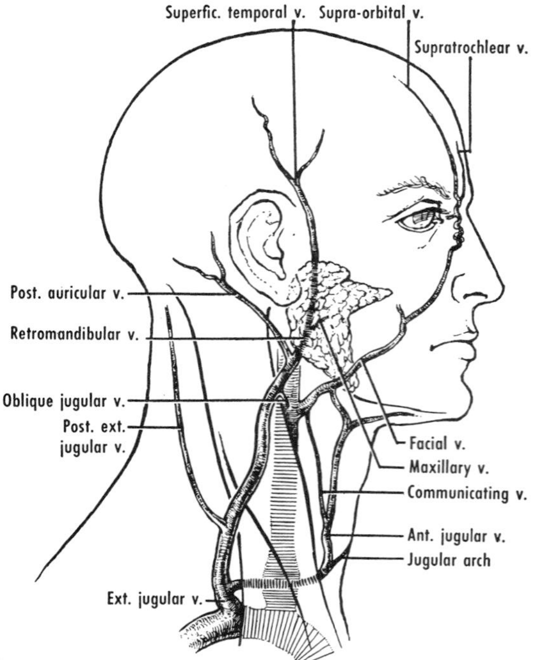

Figure 60–6 Scheme of the superficial veins of the head and neck. Only one pattern is shown, but it should be stressed that the arrangement of these veins is highly variable. The internal jugular, subclavian, and brachiocephalic veins do not have labels. See also figure 57–5*B* for the facial vessels.

romandibular veins, and communications with the internal jugular vein. An *anterior jugular vein* may descend on the front of the neck, turn laterally, pass deep to the sternomastoid, and enter either the external jugular or the subclavian vein. An inconstant jugular arch may connect the right and left anterior jugular veins above the sternum.

ANTERIOR TRIANGLE OF NECK (fig. 60–3, *C* and *D*)

The boundaries of the anterior triangle (fig. 60–3*B*) are as follows:

Superior: inferior border of the mandible and a line drawn from the angle of the mandible to the mastoid process.
Anterior: anterior median line of the neck.
Posterior: anterior border of the sternomastoid.

The anterior triangle is crossed by the digastric and the stylohyoid, and by the superior belly of the omohyoid. Dependent on these muscles, various schemes have been proposed to subdivide the anterior triangle into a series of smaller triangles. The names and boundaries of the four chief areas are as follows:

1. Digastric (Submandibular) Triangle. This is bounded by the inferior border of the mandible and the two bellies of the digastric. The term submandibular triangle has also been used for the area bounded by the inferior border of the mandible, the posterior belly of the digastric, the hyoid bone, and the anterior median line of the neck.

2. Submental (Suprahyoid) Triangle. This area is bounded by the body of the hyoid bone and the anterior belly of the digastric of each side (fig. 60–3*D*). It therefore extends across the anterior median line of the neck. The mylohyoid muscle forms its floor.

3. Carotid Triangle. **The carotid triangle is bounded by the posterior belly of the digastric, the superior belly of the omohyoid, and the anterior border of the sternomastoid.**

4. Muscular (Inferior Carotid) Triangle. This area is bounded by the superior belly of the omohyoid, the anterior border of the sternomastoid, and the anterior median line of the neck. The sternohyoid and the sternothyroid form its floor.

Roof of Anterior Triangle

The roof of the anterior triangle is formed by the fascia and the platysma. The cervical branch of the facial nerve and the transverse cervical nerve lie deep to the platysma.

Platysma. The platysma (fig. 60–3*A*) is a quadrilateral muscular sheet extending over the front and side of the neck and located in the subcutaneous tissue, that is, superficial to the cervical fascia. It arises from the subcutaneous tissue and the skin over the upper part of the deltoid and the pectoralis major. It is inserted mainly on the lower border of the mandible, but also into the skin and muscles around the mouth. Moreover, the most anterior fibers may cross the median plane and decussate with those of the opposite side. The platysma is extremely variable in its degree of development, however, and some people can contract one or both platysmata voluntarily.[7] It forms a part of the roof of the anterior and posterior triangles of the neck.

The platysma is supplied by the cervical branch of the facial nerve.

The platysma raises and carries forward the skin of the neck and shoulder, and diminishes the concavity between the jaw and the side of the neck. Thereby it probably relieves pressure on the underlying veins.[8]

Floor and Contents of Anterior Triangle

The floor of the digastric triangle is formed by the mylohyoid and the hyoglossus. The chief contents are the submandibular gland, the facial artery (deep to the gland), and the facial vein (superficial to the gland). Posteriorly, in the parotid region, portions of the parotid gland and the external carotid artery are found. The internal carotid artery, the internal jugular vein, and the glossopharyngeal and vagus nerves are situated more deeply.

The floor of the carotid triangle is formed by parts of the thyrohyoid, hyoglossus, and inferior and middle constrictors of the pharynx. **The chief contents are portions of the common, external, and internal carotid arteries, and the internal jugular vein, all overlapped by the anterior border of the sternomastoid.** Some of the branches of the external carotid artery (e.g., superior thyroid, lingual, and facial arteries), the corresponding tributaries of the internal jugular vein, and portions of the last three cranial nerves are also found in the carotid triangle. The larynx and the pharynx, and the internal and external laryngeal nerves are placed deeply in this area.

Infrahyoid Muscles

The infrahyoid muscles (fig. 60–7) are four ribbon-like muscles that anchor the hyoid bone to the sternum, the clavicle, and the scapula. They are arranged in (1) a superficial plane, comprising the sternohyoid and the omohyoid, and (2) a deep plane, comprising the sternothyroid and the thyrohyoid.

Sternohyoid. The sternohyoid arises from the back of the manubrium and/or from the medial end of the clavicle. It is inserted into the lower border of the body of the hyoid bone.

Omohyoid. The omohyoid consists of two bellies united by an intervening tendon. The inferior belly arises from the upper border of the scapula near the suprascapular notch and occasionally from the suprascapular ligament.[9] The inferior belly then passes forward and upward, under cover of the sternomastoid, to end in the middle tendon. From there the superior belly proceeds upward to be inserted into the lower border of the body of the hyoid bone. The middle tendon, situated deep to the sternomastoid, is attached to the manubrium and the first costal cartilage by a fascial sling, and is also anchored to the clavicle.

Sternothyroid. The sternothyroid lies under cover of the sternohyoid. It arises from the back of the manubrium and variably from the upper costal cartilages. It is inserted into the oblique line on the lamina of the thyroid cartilage.

Thyrohyoid. The thyrohyoid may be regarded as an upward continuation of the sternohyoid. It arises from the oblique line of the thyroid cartilage and is inserted into the lower border of the greater horn of the hyoid bone.

Innervation. The sternohyoid, the omohyoid, and the sternothyroid are supplied by the ansa cervicalis (p. 711) and its superior root. The thyrohyoid is innervated directly by a branch from the hypoglossal nerve. All the fibers to the infrahyoid muscles, however, are believed to be derived

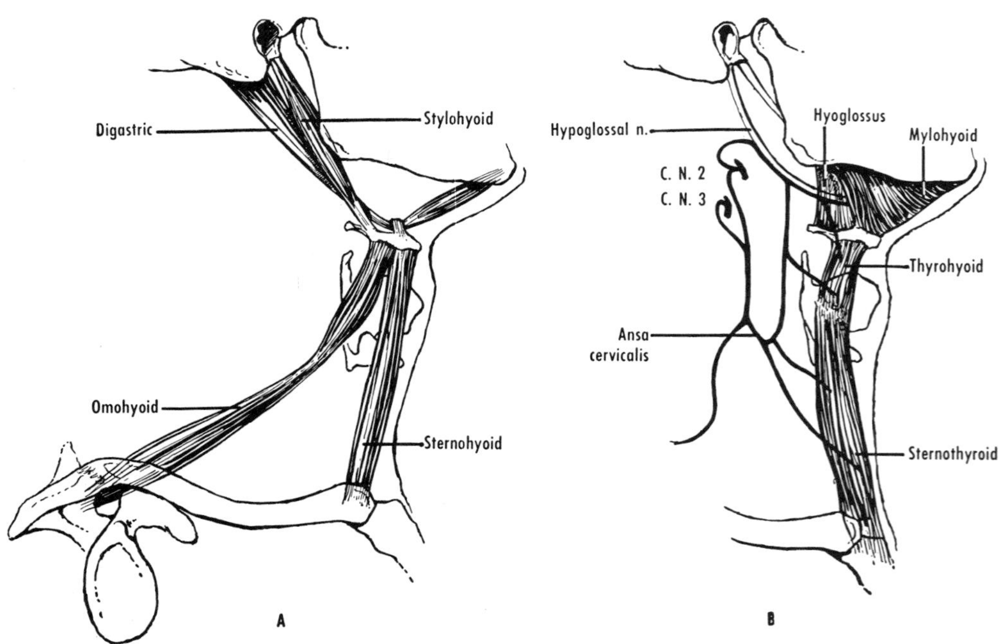

Figure 60–7 Suprahyoid and infrahyoid muscles. *A* shows the more superficial muscles, *B* the deeper muscles. The geniohyoid is not shown (See fig. 63–4). The suprahyoid muscles are also shown in figure 59–2. The infrahyoid muscles are innervated chiefly by the ansa cervicalis, as shown in *B*.

ultimately from the first three cervical nerves.

Actions. The infrahyoid muscles act as a group to depress the larynx, the hyoid bone, and the floor of the mouth, or to resist their elevation, according to circumstances. The vertical level of the hyoid bone is determined by the mylohyoid and infrahyoid muscles. The anteroposterior position of the hyoid bone is determined by the stylohyoid, the geniohyoid, and the infrahyoid muscles.[10]

The infrahyoid muscles show activity when the mouth is opened against resistance.[11]

The omohyoid probably relieves pressure on the apices of the lungs and on the internal jugular vein (which is attached to the fascial layer connecting the middle tendon to the clavicle).

DEEP STRUCTURES OF NECK

The cervical vertebrae should be studied in conjunction with the deep structures of the neck (p. 510, and figs. 60–8 and 60–9).

The lower part of the neck is a junctional region between the thorax and the upper limbs.

The *superior aperture (inlet) of the thorax* is a reniform interval bounded by the first thoracic vertebra, the first ribs and their costal cartilages, and the manubrium sterni. The neck communicates with the cavity of the thorax through the inlet. The chief structures that pass through the inlet (fig. 60–10) are vessels (brachiocephalic trunk and left common carotid, left subclavian, and both internal thoracic arteries), nerves (phrenic, vagus, recurrent laryngeal, and sympathetic trunk), trachea and esophagus, cupola of pleura and apex of lung, and the thymus.

The *apex of the axilla* corresponds to the interval bounded by the superior border of the scapula, the external border of the first rib, and the posterior surface of the clavicle. The neck communicates with the axillary region through this space. The chief structures that pass through the interval are the brachial plexus and the axillary artery and vein. The axillary vessels change their name to subclavian where they lie medial to the external border of the first rib.

THYMUS

The thymus, which has been described with the thorax (p. 331), possesses a cervical part on the front and sides of the trachea, behind the sternohyoid and sternothyroid. It is connected by fibrous strands with the tissues around the thyroid gland.

THYROID GLAND

The thyroid gland is an endocrine gland situated in the neck, opposite C.V. 5 to 7. It develops largely as a median diverticulum from the floor of the pharynx.

The immediate coverings of the thyroid gland are (1) a fibrous *capsule,* intimately adherent to the underlying gland, and (2) a *sheath* (the so-called "false capsule"), derived from the pretracheal layer of the deep cervical fascia. The anterior layer of the sheath encloses the infrahyoid muscles, whereas the posterior layer encloses the trachea and esophagus and the recurrent laryngeal nerves.

The thyroid gland, as seen from the front, is roughly H-shaped or U-shaped. It consists of two lobes, right and left, connected by an isthmus (figs. 60–11 to 13). The lobes are freely movable. Each *lobe* presents an apex, a base, and three surfaces. The *apex,* directed upward and backward, lies sandwiched in the interval between the sternothyroid and the inferior constrictor of the pharynx. The *base* is directed inferiorly and medially. The *lateral surface* is covered by infrahyoid muscles (sternothyroid, sternohyoid, and omohyoid). The *medial surface* is related to the larynx (cricothyroid muscles) and trachea, the pharynx (inferior constrictor) and esophagus, and the external and recurrent laryngeal nerves. The *posterior surface* (fig. 60–11*B*) is related to the carotid sheath and its contents (and also to the prevertebral

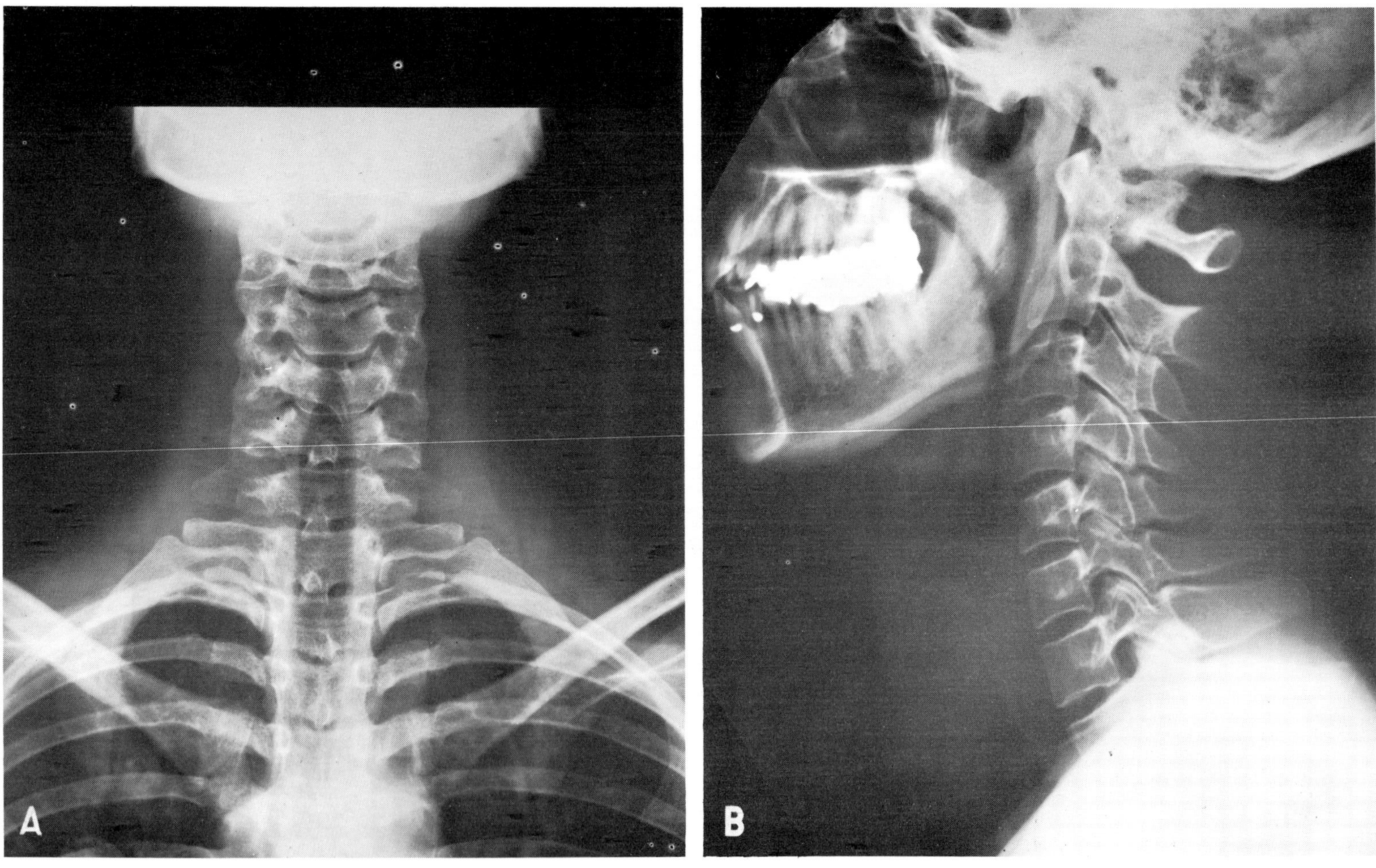

Figure 60–8 Cervical vertebrae. *A*, anteroposterior radiogram. Note the translucent area formed by the larynx and the trachea. *B*, lateral radiogram. Note the anterior and posterior arches of the atlas, the curve of the cervical vertebrae, and the slopes of the articular facets. Observe also the metallic fillings in the teeth.

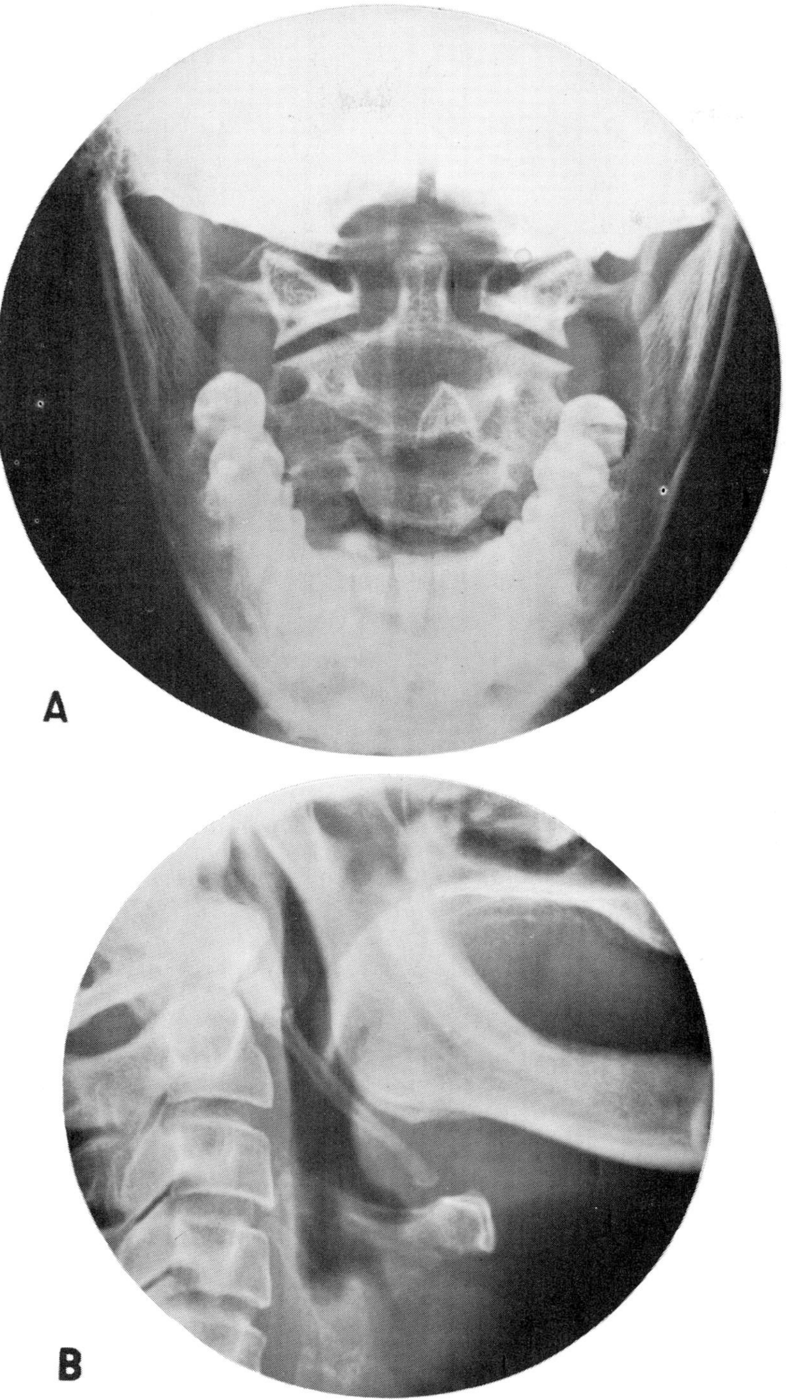

Figure 60–9 Cervical vertebrae. *A*, the first three cervical vertebrae visualized with the mouth open. Note the atlanto-axial joint on each side of the dens. Note also the foramen magnum above, the styloid processes and the transverse processes of the atlas laterally, and the spine of the axis, shaped like an inverted **V**. *B*, lateral radiogram, showing an exceptionally long styloid process. Note the hyoid bone and the edentulous mandible. Both films courtesy of the Department of Radiology, University of Rochester School of Medicine and Dentistry, Rochester, New York.

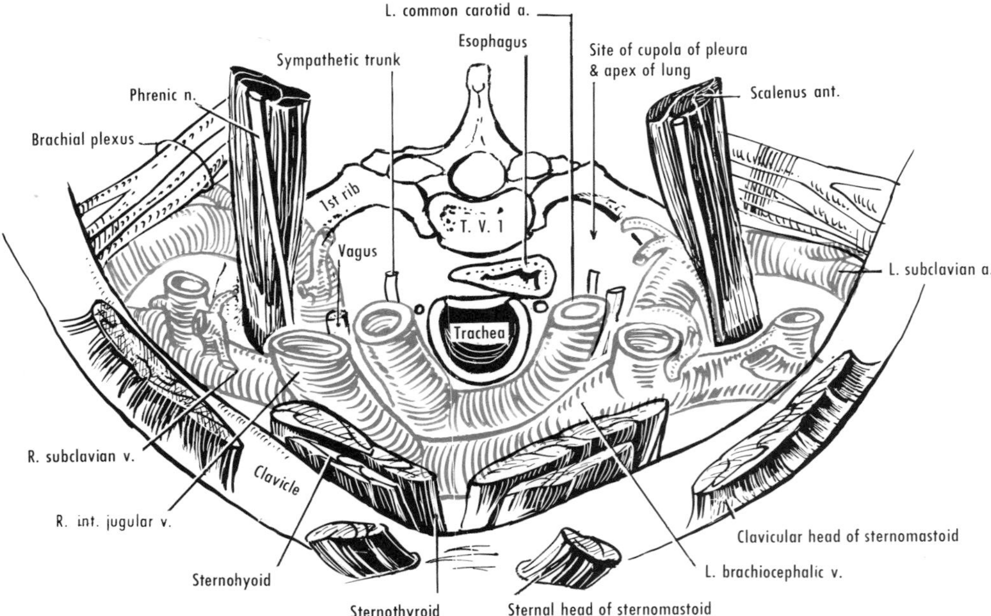

Figure 60–10 The main structures that cross the thoracic inlet (See text). In addition to the various vessels, note the recurrent laryngeal nerves and the thoracic duct. Based on von Lanz and Wachsmuth.

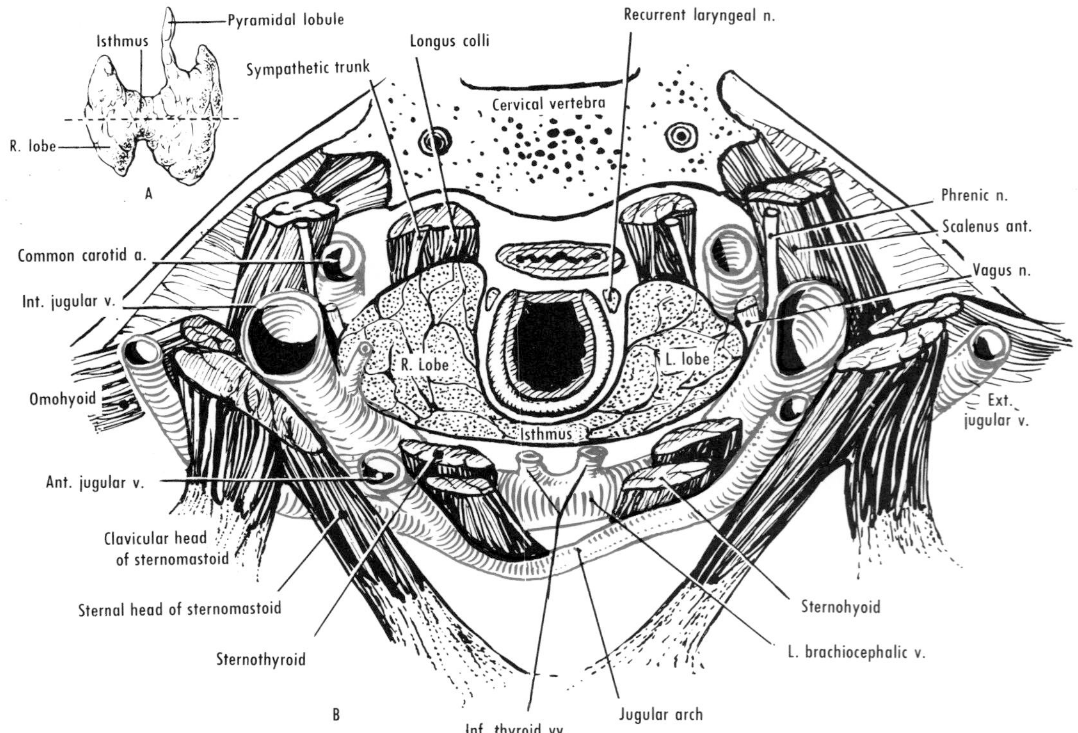

Figure 60–11 *A*, the thyroid gland, anterior aspect. The horizontal line indicates the plane of section shown in *B*. *B*, a horizontal section to show the relations of the thyroid gland. Based on von Lanz and Wachsmuth.

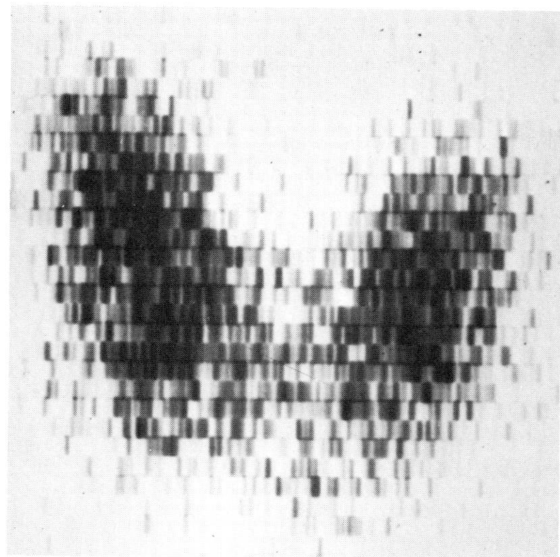

Figure 60–12 Scintigram (radio-isotope scan), showing thyroid gland *in vivo*. The right and left lobes, connected by the isthmus, are clearly visible. From F. H. DeLand, M.D., and H. N. Wagner, M.D., *Atlas of Nuclear Medicine,* volume 3, *Reticuloendothelial System, Liver, Spleen and Thyroid,* Saunders, Philadelphia, 1972; courtesy of the authors.

muscles and the sympathetic trunk) and, medially, to the parathyroid glands.

To palpate a lobe *in vivo,* the right lobe, for example, the chin of the subject is elevated (to bring the larynx forward) and rotated to the right (to relax the right sternomastoid). The thyroid cartilage is pressed toward the right by the right thumb, and the right lobe of the thyroid gland, together with the right sternomastoid and carotid sheath, is caught between the left thumb in front and the remaining fingers of the left hand behind.[12]

The *isthmus* is a variable band of glandular tissue that unites the lower parts of the right and left lobes. **The isthmus generally covers the second, third, and fourth rings of the trachea,** but is sometimes absent. An anastomosis between the right and left superior thyroid arteries takes place along the upper border of the isthmus. The *pyramidal lobe (lobule)* is an inconstant portion of the thyroid gland that extends upward from the isthmus, and may be anchored to the hyoid bone by fibrous or muscular tissue. When a muscular band is present, it is generally known as the *levator glandulae thyroideae,* although not all these muscles are inserted into the pyramidal lobule as such.[13]

Each anatomical lobe of the thyroid gland is composed of a large number of smaller structural lobes, which are irregular discs of tissue. Each structural lobe consists of many lobules. A *lobule* is made up of 20 to 40 vesicles or *follicles,* bound together by fine connective tissue. Each lobule receives its own artery.[14]

Blood Supply (figs. 60–13, 14). The thyroid gland is a highly vascular organ that easily becomes enlarged, for example, during menstruation and pregnancy. It is supplied chiefly by the superior and inferior thyroid arteries.

1. The superior thyroid artery is a branch of either the external or the common carotid (p. 694).
2. The inferior thyroid artery is a branch of the thyrocervical trunk of the subclavian (p. 704).
3. The arteria thyroidea ima is an inconstant branch from the brachiocephalic trunk, the right common carotid artery, the aortic arch, or another source. It ascends to the lower border of the isthmus, where it breaks up into several branches.

The main venous trunks are highly variable. The thyroid veins form a plexus on the surface of the gland and on the front of the trachea. The superior and middle thyroid veins drain this plexus on each side into the internal jugular vein. The inferior thyroid veins form a plexus in front of the trachea and drain into the brachiocephalic veins.

Lymphatic Drainage. The lymph vessels[15] drain (1) upward to the lower deep cervical nodes, and (2) downward to the paratracheal nodes. Lymph vessels from the isthmus drain upward to the prelaryngeal nodes and downward to the pretracheal nodes.

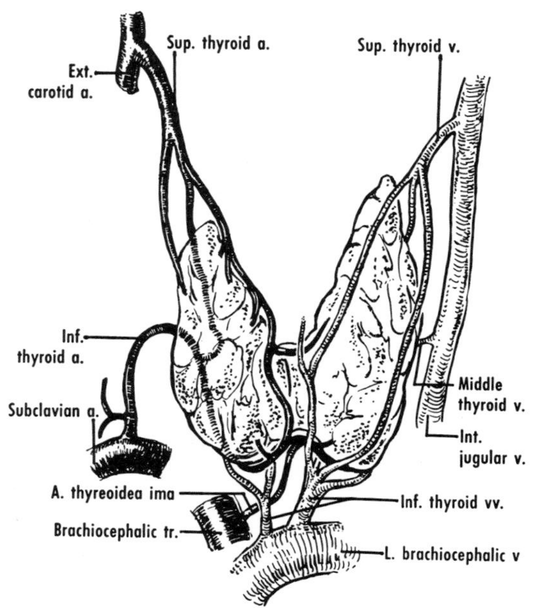

Figure 60–13 The blood supply of the thyroid gland. To simplify the illustration, only the arteries have been shown on the right side of the specimen, and only the veins on the left. Moreover, most of the smaller anastomotic vessels have been omitted.

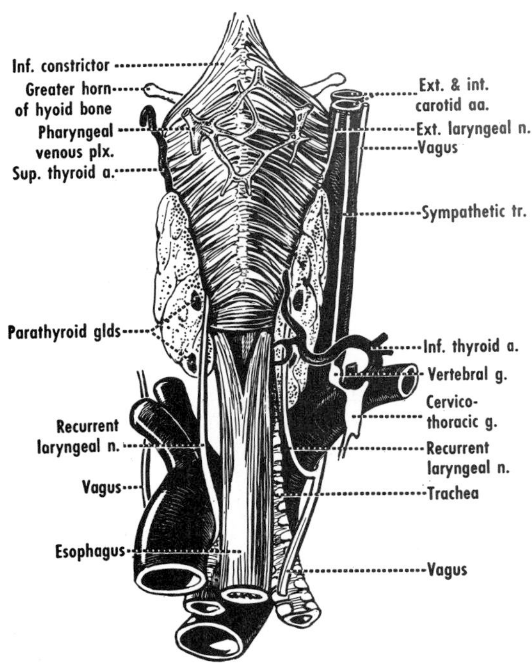

Figure 60–14 The lower portion of the pharynx and the upper portion of the esophagus. Posterior aspect. To simplify the illustration, certain structures have been drawn on only one side, for example, the superior thyroid artery on the left of the specimen, and the inferior thyroid artery on the right. Based chiefly on von Lanz and Wachsmuth.

Innervation. Branches from the cervical sympathetic (vasoconstrictor)[16] and the vagus (of obscure function) reach the thyroid gland.

Additional Considerations. **A non-neoplastic and non-inflammatory enlargement of the thyroid gland is termed a goiter.** Goiter is endemic in certain regions of the world, when the soil and water are deficient in iodine.

Portions of the embryonic thyroglossal duct may remain to form cysts,[17] fistulae, or the pyramidal lobe. The site of origin of the duct is marked by the foramen cecum on the tongue. Portions of the developing thyroid may become detached and form accessory thyroid glands. **Accessory thyroid tissue may be found anywhere along the course of the thyroglossal duct and also in the thorax.**

PARATHYROID GLANDS

The parathyroid glands are endocrine organs and are essential to life. They are small, pinkish yellow or brownish yellow bodies that lie most frequently on the medial half of the posterior surface of each lobe of the thyroid gland (hence their name). The greatest diameter of each measures only 6 mm approximately. They vary from two to six in number, the most frequent number being four.[18] They generally lie outside the capsule of the thyroid gland, but their position varies.[19] They possess delicate connective tissue capsules and septa, but no distinct lobules. According to their position, the parathyroid glands are usually named *superior* and *inferior* on each side. The superior are usually dorsal, the inferior ventral, to the recurrent laryngeal nerve.[20] Lymph nodes, fat lobules, or accessory thyroid tissue may be mistaken for parathyroid glands. Microscopical demonstration of parathyroid tissue is the only certain means of identification.

Blood Supply. The blood supply is from branches of the inferior thyroid.

CERVICAL PARTS OF TRACHEA AND ESOPHAGUS

Trachea

The trachea is characterized by a series of incomplete rings of hyaline cartilage (figs. 60–10 and 63–11). It lies partly in the neck and partly in the thorax. It is continuous with the larynx above, and it divides below into the right and left main bronchi. **The trachea extends from the level of C.V. 6 to that of about T.V. 6 or 7 *in vivo*.**

The cervical part of the trachea (fig. 60–11) is related anteriorly to the jugular venous arch, the sternohyoid and sternothyroid, the isthmus of the thyroid gland (which covers generally the second, third, and fourth rings of the trachea), the inferior thyroid veins (which form a plexus), the thymus, the arteria thyroidea ima, and, in the child, the brachiocephalic trunk just above the jugular notch of the sternum. The trachea is related posteriorly to the esophagus and the recurrent laryngeal nerves, and laterally to the lobes of the thyroid gland and the common carotid arteries. It is supplied chiefly by the inferior thyroid vessels and the recurrent laryngeal nerves.

Tracheostomy, the operation whereby an artificial opening is made in the trachea, is sometimes necessary for respiratory obstruction, for instance, that caused by spasm at the glottis due to a foreign body in the larynx. In an emergency, a rolled-up towel is placed beneath the shoulders, and the head is extended and held by an assistant. A vertical incision is made in the skin, from the thyroid notch of the thyroid cartilage to a point just above the jugular notch of the manubrium sterni. The incision, which is kept strictly median, is deepened, and the third and fourth rings of the trachea are incised. **For non-surgeons, however, cricothyrotomy is preferable** (p. 756).

Esophagus

The upper part of the esophagus lies in the neck, whereas the lower portion lies in the thorax (p. 279) and abdomen (p. 376). A muscular tube connecting the pharynx above with the stomach below, **the esophagus extends from the level of the cricoid cartilage (C.V. 6) to that of about T.V. 11. It presents several constrictions, one of which is in the neck, namely, that adjacent to, and partly due to, the inferior constrictor of the pharynx,[21] about 15 cm from the upper incisor teeth. This is the narrowest point in the esophagus.** The external and internal muscular layers of the upper portion of the esophagus consist of skeletal muscle that is attached by the crico-esophageal tendon to the back of the lamina of the cricoid cartilage. The cricopharyngeal fibers of the inferior constrictor of the pharynx act as a sphincter for the esophagus (fig. 60–14).

The cervical part of the esophagus, about one-fifth of its total length, is related anteriorly to the trachea and the recurrent laryngeal nerves, posteriorly to the longus colli and the vertebral column, and laterally to the lobes of the thyroid gland and the common carotid arteries. The trachea covers the right margin of the esophagus, whereas the left margin projects laterally from behind the trachea (fig. 60–14). Thus the cervical part of the esophagus can be exposed surgically more easily from the left side. The esophagus is supplied chiefly by the inferior thyroid vessels[22] and the recurrent laryngeal nerves.

Apart from radiography (fig. 28–4, p. 282), the esophagus can be examined *in vivo* by using an electrically lit tube termed an esophagoscope. Tissue may be obtained for biopsy, or a foreign body removed, by esophagoscopy.

COMMON, EXTERNAL, AND INTERNAL CAROTID ARTERIES

The main arteries of the head and neck (figs. 60–15*C* and 60–16) are the right and left common carotid arteries, each of which divides into (1) an external carotid, which supplies the structures external to the skull, as well as the face and the greater part of the neck, and (2) an internal carotid, which supplies the structures within the cranial cavity and the orbit.

The term carotid comes from a Greek word meaning heavy sleep. The finding that compression of the carotid arteries may result in a deep sleep or stupor dates from antiquity. In persons who are free of cerebrovascular disease, however, unilateral carotid compression rarely results in either syncope or changes in the pulsation of the radial artery at the wrist.[23]

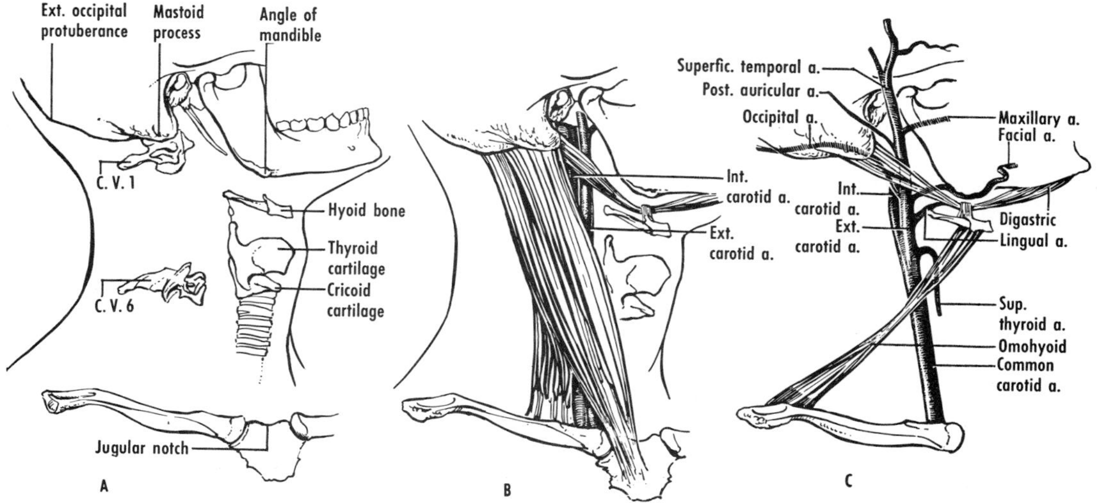

Figure 60–15 Diagrams illustrating the location of the carotid arteries in the neck. *A*, important bony landmarks in the neck. *B*, the sternomastoid and the underlying great vessels. Note the internal jugular vein in the interval between the heads of the sternomastoid below. *C*, the carotid arteries. Note the branches of the external carotid artery.

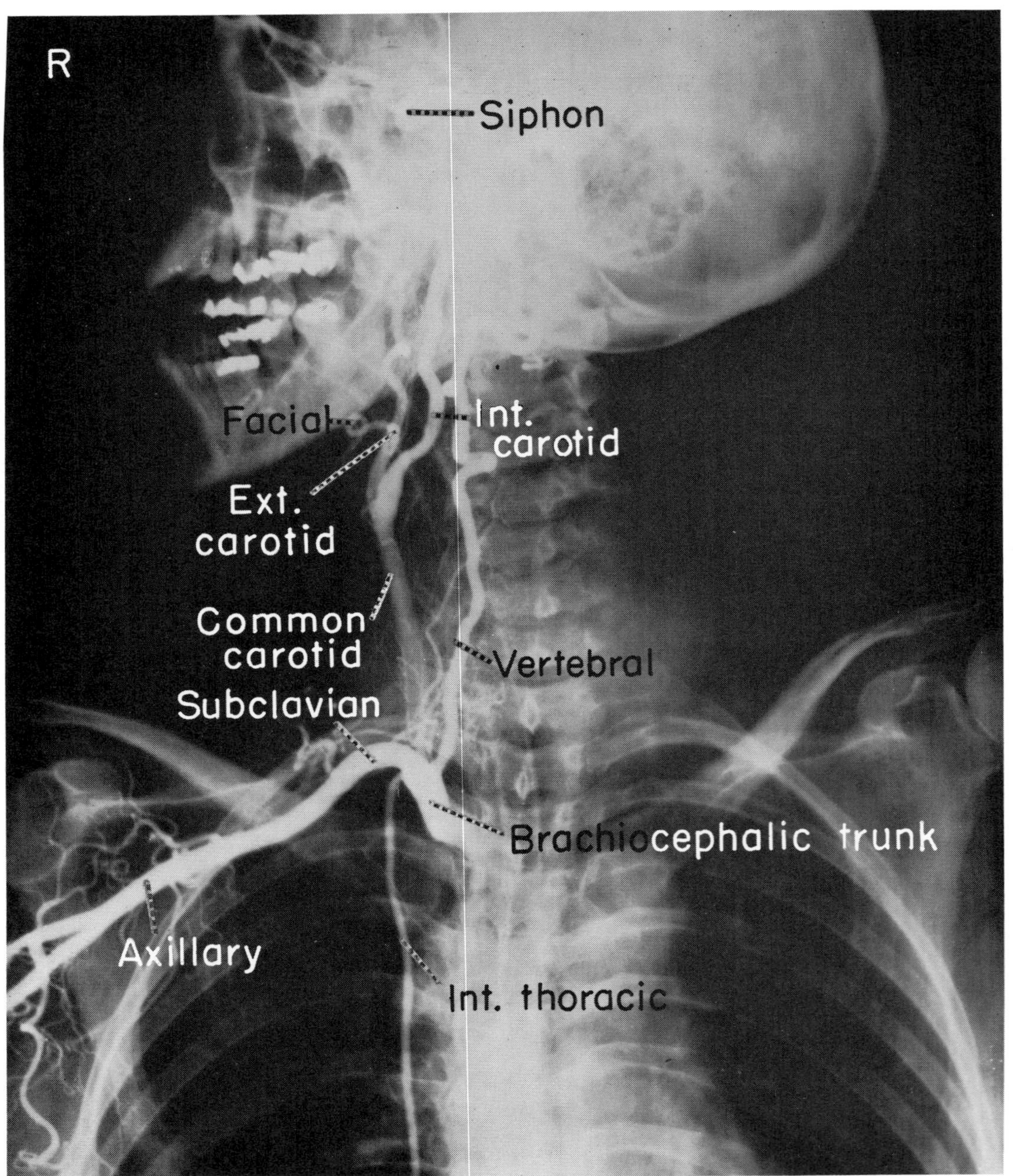

Figure 60–16 Carotid and subclavian arteriogram *in vivo*. Courtesy of E. S. Gurdjian, M.D., and J. E. Webster, M.D., Department of Neurosurgery, Wayne State University School of Medicine, Detroit, Michigan.

The common and internal carotid arteries of each side lie in a cleft bounded by (1) the cervical vertebrae and their attached muscles, (2) the pharynx, esophagus, larynx, trachea, and the thyroid gland, and (3) the sternomastoid, together with certain of the suprahyoid and infrahyoid muscles.

Common Carotid Artery

The *right common carotid artery* arises at the point of division of the brachiocephalic trunk (behind the right sternoclavicular joint), whereas the *left common carotid* is a branch of the arch of the aorta. The left artery, therefore, has a thoracic portion before it reaches the neck behind the left sternoclavicular joint.

Occasionally the left common carotid may arise from the brachiocephalic trunk; rarely right and left brachiocephalic trunks may be present, or the right common carotid may arise directly from the arch of the aorta.[24]

The common carotid artery commonly gives off no named branches in the neck, but the superior thyroid artery may arise from it. Moreover, the right common carotid may give off the arteria thyroidea ima.

Each common carotid artery divides usually at the level of the upper border of the lamina of the thyroid cartilage, that is, at the level of C.V. 4, occasionally one vertebra higher or lower. The point of division is generally 3 cm or less below the lower border of the mandible.[25] A dilatation, the carotid sinus,[26] is generally found on the internal carotid, or on the common and internal carotid arteries, near the point of division.

Collateral Circulation. An adequate collateral circulation can frequently be maintained after ligature of one common carotid artery, but signs of cerebral disturbance may occur. The chief pathways (fig. 60–24) are: (1) outside the skull, between the superior and inferior thyroid arteries (subclavian), and between the descending branch of the occipital and the deep cervical (subclavian); (2) inside the cranial cavity, the vertebral arteries may take the place of the common carotid.

Superficial Relations. The common carotid artery is crossed by the omohyoid at the level of the cricoid cartilage (C.V. 6) (fig. 60–15*C*). Below this muscle, the artery is placed deeply under cover of the sternothyroid, sternohyoid, sternomastoid, and platysma. In its upper part, the common carotid lies under cover of the anterior border of the sternomastoid. The superior root of the ansa cervicalis lies superficial to the artery, usually outside the carotid sheath. The sternomastoid branch of the superior thyroid artery, and the superior and middle thyroid veins, cross the common carotid. The internal jugular vein lies on the lateral side. The thyroid gland overlaps the artery anteromedially.

Deep Relations. The artery is related behind to the sympathetic trunk, the prevertebral muscles, and the transverse processes of the cervical vertebrae (C.V. 4 to 6). **The common carotid can be compressed against the transverse processes of the cervical vertebrae by pressing medially and posteriorly with the thumb.** (The carotid tubercle of C.V. 6 is about 4 cm above the sternoclavicular joint.) Below the level of the carotid tubercle the common carotid artery lies in the interval between the scalenus anterior and the longus colli, in front of the vertebral artery and the sympathetic trunk; the common carotid is crossed by the inferior thyroid artery and, on the left side, by the thoracic duct. The vagus is posterolateral. The pharynx and esophagus, and the larynx and trachea are medial to the artery, as is also the recurrent laryngeal nerve.

External Carotid Artery

The external carotid artery extends from the level of the upper border of the lamina of the thyroid cartilage to a point behind the neck of the mandible, midway between the tip of the mastoid process and the angle of the mandible. It divides in the substance of the parotid gland into the superficial temporal and maxillary arteries.

The external carotid is at first generally anteromedial to the internal carotid. As it ascends, however, it inclines backward and comes to lie lateral to the internal carotid.

Collateral Circulation. An adequate collateral circulation can usually be maintained after ligature of an external carotid artery. The chief pathways are between the large branches of the external carotid (superior thyroid, lingual, facial, and occipital) and the corresponding branches of the opposite side.

Superficial Relations. The external carotid artery begins in the carotid triangle, where it is partly overlapped by the sternomastoid, and crossed by the hypoglossal nerve and the lingual and facial veins. The external carotid passes deep to the posterior belly of the digastric and the stylohyoid (fig. 60–15*C*). It enters the substance of the parotid gland, where it is crossed by the facial nerve or its branches (fig. 58–1).

Deep Relations. The constrictors of

the pharynx and the superior (or internal and external) laryngeal nerve lie medially.

The following structures are found partly between the external and internal carotid arteries: the styloid process (fig. 60–18) or the stylohyoid ligament, the styloid muscles, the glossopharyngeal nerve and the pharyngeal branch of the vagus, and a portion of the parotid gland.

Branches. The branches of the external carotid artery are generally as follows.

1. The *superior thyroid artery* (fig. 60–13) arises either from the front of the external carotid or from the common carotid. Its origin is below the level of the tip of the greater horn of the hyoid bone and under cover of the sternomastoid. It runs downward and forward in the carotid triangle and passes deep to the omohyoid, sternohyoid, and sternothyroid. It lies on the inferior constrictor and is related to the external laryngeal nerve. At the apex of the corresponding lobe of the thyroid gland it divides into glandular branches.

The branches of the superior thyroid artery include *(a)* an *infrahyoid branch,* which anastomoses with its fellow on the opposite side; *(b)* a *sternomastoid branch,* frequently a direct branch of the external carotid, which runs across the carotid sheath and enters the sternomastoid; *(c)* the *superior laryngeal artery,* frequently a direct branch of the external carotid, which accompanies the internal laryngeal nerve, passes deep to the thyrohyoid, and pierces the thyrohyoid membrane to supply the larynx; *(d)* a *cricothyroid branch,* which anastomoses with its fellow of the opposite side; *(e)* several *glandular branches,* one of which anastomoses with its fellow of the opposite side along the upper border of the isthmus.

2. The *lingual artery* (fig. 59–5) arises from the front of the external carotid at or above the level of the hyoid bone. It may arise in common with the facial artery. Its course may be considered in three parts: posterior, deep, and anterior to the hyoglossus muscle, respectively.

The *first part* of the lingual artery lies mainly in the carotid triangle and forms a loop on the middle constrictor of the pharynx. The loop is crossed by the hypoglossal nerve.

The *second part* of the artery passes deep to the hyoglossus and runs along the upper border of the hyoid bone. It lies on the middle constrictor.

The *third part* of the artery, named the *arteria profunda linguae,* ascends between the genioglossus and the inferior longitudinal muscle of the tongue. It runs along the lower surface of the tongue and anastomoses with its fellow of the opposite side.[27]

The first part of the lingual artery gives off a *suprahyoid branch,* which anastomoses with its fellow of the opposite side. The branches of the second and third parts of the lingual artery have already been described (p. 677).

3. The *facial* (formerly *external maxillary) artery* arises from the front of the external carotid, frequently in common with the lingual (as a *linguofacial trunk*). It ascends in the carotid triangle and enters a groove on the posterior border of the submandibular gland. It then turns downward and forward between the submandibular gland and the medial pterygoid muscle. Next it winds around the lower border of the mandible at the anterior margin of the masseter, and proceeds upward and forward on the face. It ends at the medial angle of the eye by anastomosing with branches of the ophthalmic artery. The facial artery is very tortuous and it takes part in numerous anastomoses. Its course may be considered in two portions: cervical and facial.

The *cervical part of the facial artery* (figs. 59–5 and 58–1) is covered at its origin by the platysma. The artery ascends deep to the digastric and stylohyoid, to reach the back of the submandibular gland. It lies at first on the middle and superior constrictors of the pharynx and then on the lateral surface of the submandibular gland. The superior constrictor separates it from the tonsil.

The *facial part of the facial artery* (fig. 57–5*B*) begins where the facial artery winds around the lower border of the mandible at the anterior border of the masseter (fig. 52–26*B*). It supplies the muscles of facial expression and presents a variable relationship to them. The facial vein lies behind the artery and pursues a straighter course across the face.

The cervical part of the facial artery gives off the following branches:

(a) The *ascending palatine artery,* which ascends and accompanies the levator veli palatini, passes over the upper border of the superior constrictor, and supplies the soft palate, as well as portions of the pharynx.

(b) The *tonsillar branch* (fig. 59–5), the main vessel to the tonsil, pierces the superior constrictor, and ends in the tonsil.

(c) Glandular branches supply the submandibular gland.

(d) The submental artery runs forward on the mylohyoid and supplies nearby muscles.

The facial part of the facial artery gives off the inferior and superior labial arteries and the lateral nasal branch, and ends as the angular artery. These vessels have already been described (p. 657).

4. The *occipital artery* (fig. 60–15*C*)

arises from the back of the external carotid. Its course may be considered in three portions: anterior, deep, and posterior to the sternomastoid, respectively.

(*a*) In the carotid triangle, the hypoglossal nerve winds around the occipital artery at its origin (fig. 60–18). The artery passes backward and upward, deep to the lower border of the posterior belly of the digastric. It crosses the internal carotid artery, the internal jugular vein, and the last three cranial nerves.

(*b*) Deep to the sternomastoid, the occipital artery occupies the occipital groove on the temporal bone, medial to the mastoid process. There the artery is covered by the muscles attached to the mastoid process (sternomastoid, splenius capitis, longissimus capitis, and digastric). It then lies on the obliquus capitis superior and the semispinalis capitis.

(*c*) Posterior to the sternomastoid, the occipital artery pierces the trapezius, is accompanied by the greater occipital nerve, and divides into branches on the scalp.

The ascending pharyngeal artery sometimes arises from the occipital. Apart from this, the occipital gives rise to *sternomastoid, meningeal,* and *occipital branches,* and an important descending branch. **The *descending branch* provides the chief collateral circulation after ligation of the external carotid or the subclavian artery.** It arises on the obliquus capitis superior and divides into superficial and deep branches that embrace the semispinalis capitis. The superficial branch passes deep to the splenius and anastomoses with the superficial branch of the transverse cervical artery. The deep branch passes between the semispinalis capitis and the semispinalis cervicis, and anastomoses with the deep cervical artery from the costocervical trunk (fig. 60–24).

5. The *posterior auricular artery* arises from the back of the external carotid, immediately above the posterior belly of the digastric. It runs upward and backward, superficial to the styloid process and under cover of the parotid gland. It ends between the mastoid process and the auricle by dividing into auricular and occipital branches. In addition, it gives rise to the *stylomastoid artery* (to the middle and internal ear), which in turn may give off the *posterior tympanic artery.*

6. The *ascending pharyngeal artery* is a small vessel that arises from the medial side of the lower part of the external carotid. It ascends between the internal carotid artery and the wall of the pharynx. It may arise as a branch of the occipital artery. The branches of the ascending pharyngeal are irregular and inconstant: *pharyngeal* and *meningeal branches,* and the *inferior tympanic artery* (to the wall of the tympanic cavity).

7 and 8. The *superficial temporal artery* and the *maxillary artery* have already been described (pp. 662 and 667).

The branches of the external carotid artery are summarized in table 60–1.

Internal Carotid Artery (Cervical Part)

The internal carotid artery begins at the level of the upper border of the lamina of the thyroid cartilage. **It enters the skull through the carotid canal of the temporal bone, and it ends in the middle cranial fossa by dividing into the anterior and middle cerebral arteries.**

The *carotid sinus*[29] (fig. 60–17) is a more or less spindle-shaped dilatation of the internal carotid artery, or of the common and internal carotid arteries near the point of division. It is more evident *in vivo.* The wall of the sinus contains pressoreceptors (baroreceptors) that are stimulated by changes in blood pressure.

The *carotid body*[30] is a small structure that lies in the angle of bifurcation of the common carotid artery, usually medial (that is, deep) to the external and internal carotid arteries. It probably functions as a chemoreceptor, being stimulated in anoxemia. The result is an increase in blood pressure, in cardiac rate, and in respiratory movements. The carotid body may be an endocrine gland.[31]

TABLE 60–1 Summary of Branches of External Carotid Artery

Aspect	*Branches*
Anterior	1. Superior thyroid 2. Lingual 3. Facial
Posterior	4. Occipital 5. Posterior auricular
Medial	6. Ascending pharyngeal
Terminal	7. Superficial temporal 8. Maxillary

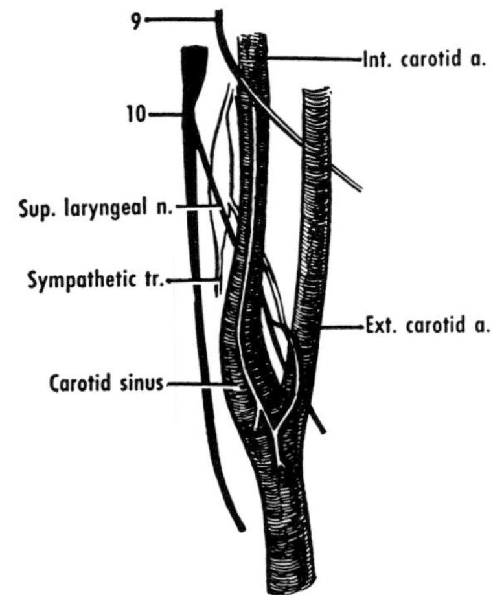

Figure 60–17 The carotid sinus and its innervation from the glossopharyngeal and vagus nerves, and from the sympathetic trunk.

The internal carotid gives no named branches in the neck. Its course may be considered in four parts: cervical, petrous, cavernous, and cerebral. The last three parts have already been described (p. 604). Kinking in the cervical part is not uncommon.[32]

Superficial Relations. The internal carotid artery begins in the carotid triangle, where it is partly overlapped by the sternomastoid and crossed by the hypoglossal nerve. The internal carotid passes deep to the posterior belly of the digastric and the stylohyoid (fig. 60–15*C*), and is crossed by the posterior auricular and occipital arteries. Higher up, the internal carotid is related superficially to the styloid process and its attached muscles, and to the glossopharyngeal nerve and the pharyngeal branch of the vagus. The external carotid artery is at first generally anteromedial[33] or anterior (less commonly anterolateral or lateral) to the internal carotid; higher up it becomes lateral. The internal jugular vein and the vagus lie mostly lateral to the internal carotid artery. At the base of the skull, however, the vein lies posterior to the artery, being separated by the last four cranial nerves (fig. 52–16).

Deep Relations. The internal carotid artery lies posteriorly on the superior cervical ganglion and the sympathetic trunk, the prevertebral layer of fascia and the prevertebral muscles, and the transverse processes of the upper cervical vertebrae (C.V. 1 to 3). Medially the artery is related to the wall of the pharynx, especially when the artery is tortuous,[34] and to the external and internal laryngeal nerves.

SURFACE ANATOMY OF CAROTID ARTERIES

The line of the carotid arteries extends upward from (1) the sternoclavicular joint, along the anterior border of the sternomastoid, to (2) the midpoint between the tip of the mastoid process and the angle of the mandible (fig. 60–15*B*). Point 2 is at the transverse process of the atlas and is approximately medial to the lobule of the auricle. The uppermost parts of the internal and external carotids extend upward to a point behind the neck of the mandible. The left common carotid artery has also a thoracic part. **Each common carotid artery is crossed by the corresponding omohyoid opposite the cricoid cartilage (C.V. 6), and this is the site for compression. The common carotid divides usually at the level of the upper border of the lamina of the thyroid cartilage,** and the point of division is generally 3 cm or less below the lower border of the mandible. **The pulsation of the common and external carotid arteries is palpable along the anterior border of the sternomastoid.** The superior thyroid artery arises below the level of the tip of the greater horn of the hyoid bone; the lingual and facial arteries arise at or just above the level of the hyoid.

GLOSSOPHARYNGEAL NERVE
(figs. 60–18 and 60–19)

The glossopharyngeal (9th cranial) nerve is afferent from the tongue and the pharynx (hence its name), and efferent to the stylopharyngeus and the parotid gland. It emerges from the medulla oblongata and rests on the jugular tubercle of the occipital bone. It passes through the middle part of the jugular foramen, where it presents two ganglia, a *superior (jugular)* and an *inferior (petrous)*. These two ganglia contain the cell bodies of the afferent fibers. The glossopharyngeal nerve next passes between the internal jugular vein and the internal carotid artery. It descends in front of the latter vessel, deep to the styloid process and the styloid muscles. **It curves forward around the stylopharyngeus,** runs deep to the posterior border of the hyoglossus (fig. 59–3), and passes between the superior and middle constrictors of the pharynx.

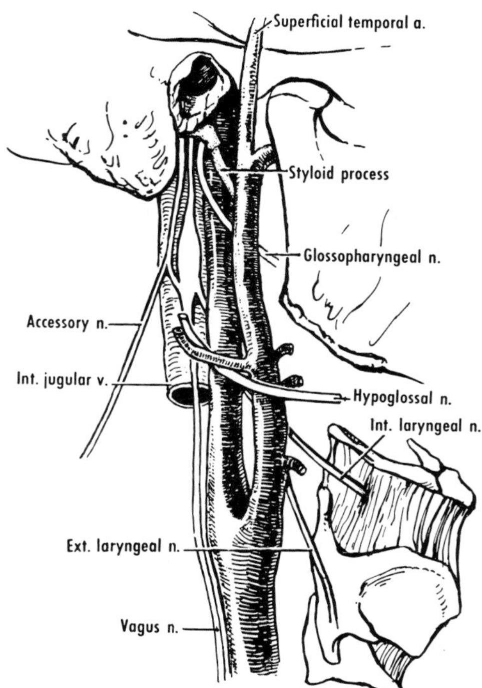

Figure 60–18 The last four cranial nerves below the base of the skull. Note the glossopharyngeal nerve passing between the carotid arteries; the vagus descending between the internal jugular vein and the internal and common carotid arteries; the accessory nerve crossing the internal jugular vein; and the hypoglossal nerve, superficial to the great vessels and winding around the origin of the occipital artery. Note the maxillary artery running forward, deep to the neck of the mandible.

Branches. The glossopharyngeal nerve gives rise to the following branches.

1. The *tympanic nerve* contains secretomotor and vasodilator fibers for the parotid gland. It appears[35] to arise from the inferior ganglion of the glossopharyngeal nerve. It passes through the tympanic canaliculus (between the jugular foramen and the carotid canal) to reach the tympanic cavity. There it divides into branches that form the tympanic plexus on the promontory on the medial wall of the tympanic cavity (fig. 54–5). The plexus gives branches to the mucous membrane of the tympanic cavity, the mastoid air cells, and the auditory tube.[36] Some fibers of the plexus reunite to form the *lesser petrosal nerve,* which contains secretomotor fibers to the parotid gland. The lesser petrosal nerve enters a small canal in the temporal bone. On its emergence from the bone, it passes through the foramen ovale or the adjacent canaliculus innominatus and joins the otic ganglion. Postganglionic fibers arise there and are conveyed to the parotid gland by the auriculotemporal nerve (fig. 60–19).

2. The *communicating branch* unites with the auricular branch of the vagus. A communication between the glossopharyngeal and facial nerves is sometimes present.

3. The *branch to the carotid sinus* (fig. 60–17) descends on the anterolateral aspect of the internal carotid artery and supplies the carotid sinus and the carotid body with afferent fibers from the pressoreceptors and chemoreceptors present in these structures. The carotid branch is frequently joined by twigs from the vagus and the sympathetic trunk.

4. The *pharyngeal branch* or branches unite on the middle constrictor with the pharyngeal branch of the vagus and with twigs from the sympathetic trunk. They

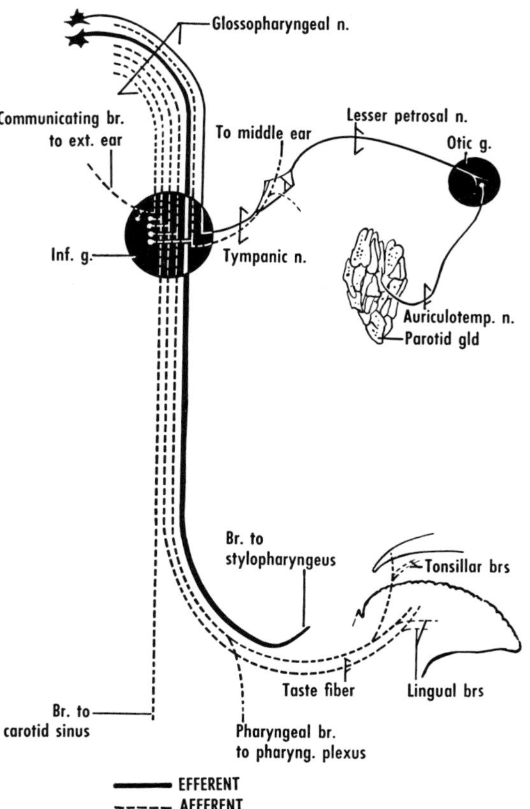

Figure 60–19 Scheme of the glossopharyngeal nerve.

supply general sensory fibers to the mucous membrane of the pharynx.

5. The motor *branch to the stylopharyngeus* is given off as the glossopharyngeal nerve crosses the muscle.

6. *Tonsillar branches* supply general sensory fibers to the mucous membrane over the tonsil and the soft palate.

7. *Lingual branches* supply taste and general sensory fibers to the posterior third of the tongue and the vallate papillae.

Examination of Glossopharyngeal Nerve. The sensation of taste is tested on the posterior third of the tongue (p. 724).

VAGUS NERVE
(figs. 60–18, 20, and 21)

The vagus (10th cranial) nerve is predominantly afferent. It has an extensive distribution (L. *vagus*, wandering) in the head and neck, and in the thorax and abdomen. It supplies afferent and efferent fibers to the pharynx and the larynx. From the neck downward, the vagus intercommunicates freely with sympathetic ganglia and branches.

The vagus emerges from the medulla oblongata and passes through the middle part of the jugular foramen. In the foramen it presents a *superior (jugular) ganglion* and, below the foramen, an *inferior (nodose) ganglion.* These ganglia contain the cell bodies of the afferent fibers and they have numerous connections (with cranial nerves 7, 9, 11, 12, cervical nerves 1, 2, and the sympathetic). Below the inferior ganglion, the vagus is joined by the internal branch of the accessory nerve, the fibers of which are distributed with branches of the vagus.

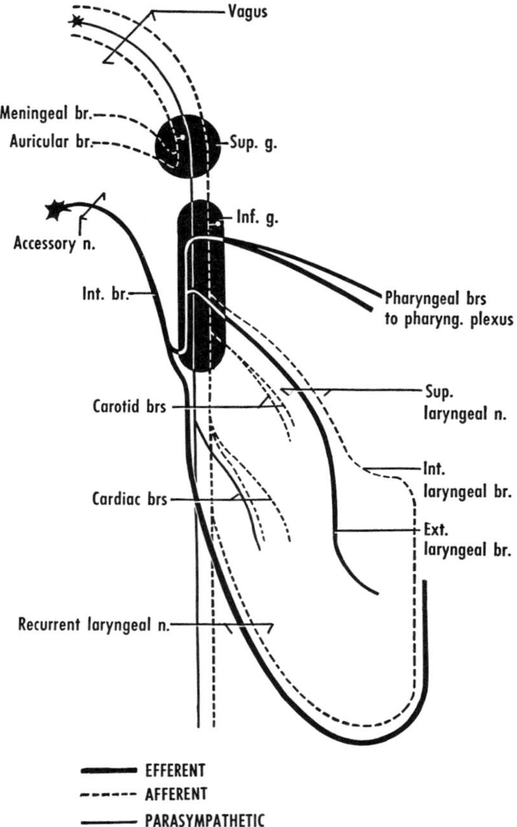

Figure 60–21 Scheme of the vagus nerve.

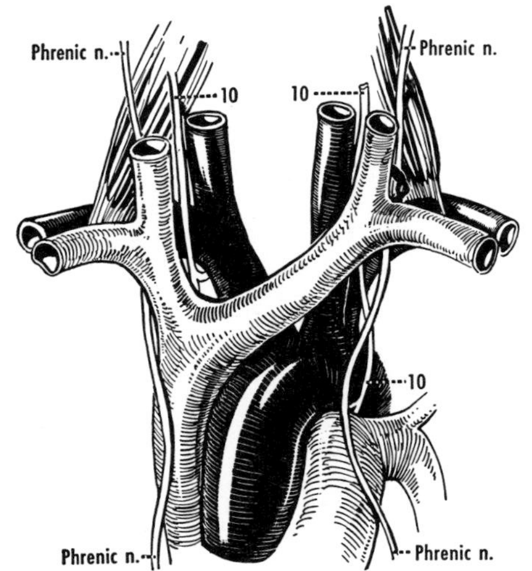

Figure 60–20 The course of the vagus and phrenic nerves, and their relationship to the great vessels. Anterior aspect. Note the different levels of origin of the right and left recurrent laryngeal nerves. The scalenus anterior muscle is depicted on each side. The termination of the thoracic duct on the left side of the body is shown.

The vagus descends within the carotid sheath, between the internal jugular vein and, successively, the internal and common carotid arteries. The right vagus passes between the internal jugular vein and the first part of the subclavian artery. The left vagus passes between the left common carotid artery and the first part of the subclavian artery. The subsequent course and distribution of both vagus nerves have been discussed (pp. 333 and 422).

Branches in Head and Neck. The vagus gives rise to the following branches:

The first two branches arise from the superior ganglion, in which may be located the cells of origin of their fibers.

1. The *meningeal branch* arises from the superior ganglion and supplies the dura of the posterior cranial fossa. It contains spinal fibers (C.N. 1, 2). Stimulation of meningeal arteries, venous sinuses, or dura in the posterior fossa produces pain referred to the cutaneous distribution of cranial nerves 9 and 10, and cervical nerve 1.

2. The *auricular branch* arises from the superior ganglion and is joined by a communication from the glossopharyngeal. It traverses the mastoid canaliculus on the lateral wall of the jugular fossa, communicates with the facial nerve, and passes through the tympanomastoid fissure. It supplies the cranial surface of the auricle, the floor of the external acoustic meatus, and the adjacent portion of the tympanic membrane.

The next three branches arise from either the inferior ganglion or the trunk of the vagus. Just below the ganglion, the trunk receives the internal branch of the accessory nerve.

3. The pharyngeal branches, several in number, are the chief motor nerves to the pharynx and the soft palate. Most of the fibers are derived from the internal branch of the accessory nerve. The pharyngeal branches arise from the inferior ganglion of the vagus and pass between the external and internal carotid arteries. They divide into branches which, together with branches from the glossopharyngeal and the sympathetic, form the pharyngeal plexus on the constrictors of the pharynx. The plexus supplies the muscles of the pharynx (except the stylopharyngeus) and of the soft palate (except the tensor veli palatini).

4. The *superior laryngeal nerve* arises from the inferior ganglion of the vagus (fig. 60–18). It descends along the side of the pharynx behind and medial to the internal carotid artery. It divides at a variable level into a larger internal and a smaller external branch.

The *internal branch,* commonly called the *internal laryngeal nerve,* is afferent from the mucous membrane of the larynx. The area of mucosa supplied extends from the epiglottis and the back of the tongue down to the vocal folds. Stimulation of the internal laryngeal nerve results in sensations of touch and pain.[37] The nerve pierces the thyrohyoid membrane above the superior laryngeal artery and divides into terminal branches. A twig is given to the transverse arytenoid muscle, but whether these fibers are motor or proprioceptive is disputed (p. 759). The internal laryngeal nerve ends by joining branches of the recurrent laryngeal nerve. The anastomosis may take place behind or in the substance of the posterior crico-arytenoid,[38] and the connection may pierce the inferior constrictor of the pharynx.[39]

The *external branch,* commonly called the *external laryngeal nerve,* descends under cover of the sternothyroid and lies deep to the superior thyroid artery. It pierces the inferior constrictor of the pharynx and enters the cricothyroid, supplying both of these muscles.

5. The *depressor nerves* or *carotid branches* are inconstant twigs that assist the glossopharyngeal nerve in supplying the carotid sinus and the carotid body.

6. A variable number of *cardiac branches* arise from the vagus in the neck and in the thorax. They are often classified into superior, middle, and inferior groups, but their arrangement is quite variable. They are closely related to, and often join, the cardiac branches of the sympathetic ganglia. Cardiac branches also arise from the recurrent laryngeal nerves. They all proceed to the cardiac plexus (p. 336).

7. The *recurrent laryngeal nerve* supplies the mucous membrane of the larynx below the vocal folds, and all the muscles of the larynx except the cricothyroid. Most of its fibers are derived from the cranial part of the accessory nerve. The recurrent laryngeal nerve arises at different levels on the two sides of the body (fig. 60–20), an arrangement that is correlated with the development of the aortic arches in the embryo. (Rarely the nerve may pass directly to the larynx without recurring.[40]) **The right nerve arises in front of the first part of the subclavian artery and winds around that vessel. The left nerve arises in the thorax on the left side of the arch of the aorta, and winds around that vessel behind the attachment of the ligamentum arteriosum. Both recurrent nerves ascend in or near the groove between the trachea and the esophagus** (the nerve may be lateral to the trachea, however), medial to the corresponding lobe of the thyroid gland. Opposite the first or second ring of the trachea, the recurrent nerve gives off a sensory branch to the laryngopharynx.[41] The recurrent nerve may be very closely related to the thyroid gland but is probably never embedded in normal glandular substance. However, **there is considerable danger of damage to the recurrent nerve in operations on the thyroid gland.[42] The recurrent laryngeal nerve is closely related to the inferior thyroid artery, but the relationship is variable:** the nerve or its branches may be posterior or anterior to, or in between, the branches of the artery. Furthermore, the arrangements on the two sides of the body are seldom similar. Before it enters the larynx, the nerve divides into two or more branches, but the arrangement of fibers[43] in these branches is variable; the branch to the posterior cricoarytenoid is given off within the larynx.[44] The recurrent nerve passes deep to the lower border of the inferior constrictor of the pharynx, in company with the inferior laryngeal artery, and enters the larynx behind the cricothyroid joint. It communicates with the internal laryngeal nerve.

The term *inferior laryngeal nerve* is sometimes used for the terminal portion of the recurrent laryngeal.

The branches of the recurrent laryngeal nerve are: cardiac branches to the cardiac plexus, communications to the sympathetic, tracheal and esophageal branches (to the mucous membrane and the musculature), twigs to the inferior constrictor, and several (4 to 12) terminal branches to the larynx.

That damage to the recurrent laryngeal nerves results in aphonia and respiratory distress was known to the Greeks of the first two centuries A.D. The paralysis, being of the lower motor neuron type, is flaccid, and is followed by atrophy, fibrosis, and contracture. The effects on the vocal folds of injury to the recurrent laryngeal nerve are discussed later (p. 759).

The branches of the vagus in the head and neck are summarized in table 60–2.

Examination of Vagus. The accessory fibers (internal branch) in the pharyngeal branches of the vagus can be tested by asking the subject to say "ah." The uvula should proceed backward in the median plane. In unilateral vagal paralysis the uvula becomes deviated toward the normal side. The condition of the laryngeal muscles can be ascertained by laryngoscopy.

TABLE 60–2 Summary of Branches of Vagus Nerve in Head and Neck

Part	*Branches*
From superior ganglion	1. Meningeal branch 2. Auricular branch
From inferior ganglion or from trunk of vagus	3. Pharyngeal branches 4. Superior laryngeal nerve (*a*) Internal laryngeal branch or nerve (*b*) External laryngeal branch or nerve 5. Depressor nerves or carotid branches (inconstant)
From trunk of vagus	6. Cardiac branches 7. Right recurrent laryngeal nerve

ACCESSORY NERVE
(fig. 60–22)

The accessory (11th cranial) nerve is formed by the union of a cranial and a spinal portion. The cranial roots emerge from the side of the medulla oblongata, below the roots of the vagus. The spinal roots emerge from the side of the spinal cord (as far down as a level between C.N. 3 and C.N. 7). Their cells of origin are in the gray matter of the spinal cord. The spinal roots unite to form a trunk that ascends in the vertebral canal, usually communicates with C.N. 1,[45] and passes through the foramen magnum. Both portions, cranial and spinal, traverse the jugular foramen, where they interchange fibers or unite for a short distance.

Below the jugular foramen, the *cranial* or *bulbar portion*, or *internal branch*, joins the

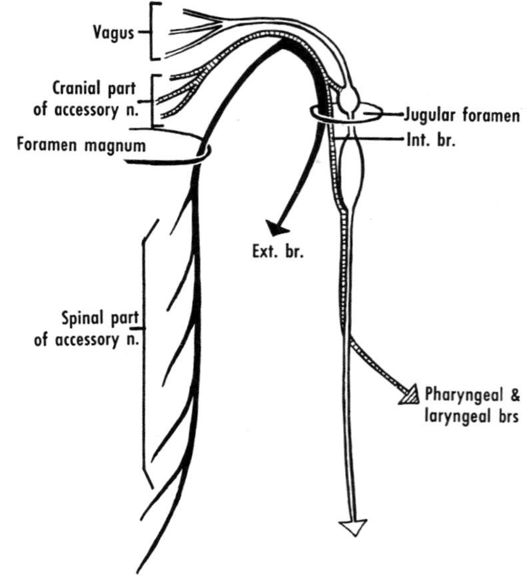

Figure 60–22 Scheme of the vagus and accessory nerves. The superior and inferior ganglia of the vagus do not have labels, nor has a small branch from the cranial part of the accessory nerve to the superior ganglion.

vagus at or just beyond the inferior ganglion of the vagus. The cranial portion contains motor fibers to skeletal muscles, and is best regarded as a part of the vagus.[46] By means of the pharyngeal and recurrent laryngeal branches of the vagus, it is distributed to the soft palate, the constrictors of the pharynx, and the larynx. Some fibers may enter the cardiac branches of the vagus.

The *spinal portion, or external branch,* of the accessory nerve is distributed to the sternomastoid and the trapezius. It runs backward, usually superficial to the internal jugular vein. It crosses the transverse process of the atlas and is crossed by the occipital artery (the upper sternomastoid branch of which accompanies it). The accessory nerve descends deep to the styloid process and the posterior belly of the digastric. It usually pierces the deep surface of the sternomastoid but sometimes remains deep to that muscle. It supplies the sternomastoid and communicates with branches from C.N. 2. **Above the middle of the posterior border of the sternomastoid, the accessory nerve crosses the posterior triangle of the neck obliquely (fig. 60–4*B*), lying on the levator scapulae and in relation to lymph nodes.** It communicates with branches of C.N. 2 and 3. About 5 cm above the clavicle it passes deep to the anterior border of the trapezius and forms a plexus with branches from C.N. 3 and 4. It supplies the trapezius.

Examination of Accessory Nerve. **The spinal part is tested by asking the subject to shrug his shoulders (trapezius) and then to rotate his head (sternomastoid).** For examination of the cranial part of the accessory nerve, see under Vagus.

HYPOGLOSSAL NERVE
(fig. 60–23)

The hypoglossal (12th cranial) nerve is principally the motor nerve to the tongue. Its roots emerge from the medulla oblongata between the pyramid and the olive. The roots unite to form two bundles, which pierce the dura and pass through the hypoglossal canal of the occipital bone. The bundles unite in or below the canal. The canal and the nerve are situated medial to the jugular foramen and its contents (fig. 52–16). The hypoglossal nerve descends behind the internal carotid artery and the glossopharyngeal and vagus nerves. It then descends between the internal carotid artery and the internal jugular vein, in front of the vagus and deep to the posterior belly of the digastric. **The hypoglossal nerve loops forward around the occipital artery** (fig. 60–18) and its lower sternomastoid branch, and receives a twig from the pharyngeal plexus. **It crosses the internal carotid, external carotid, and lingual arteries.** It lies on the hyoglossus and passes deep to the digastric and mylohyoid. On the hyoglossus it lies below the submandibular duct and the lingual nerve (fig. 59–3). It divides into terminal branches which continue forward between the mylohyoid and the genioglossus, and which communicate with the lingual nerve. The right and left hypoglossal nerves may unite.[47]

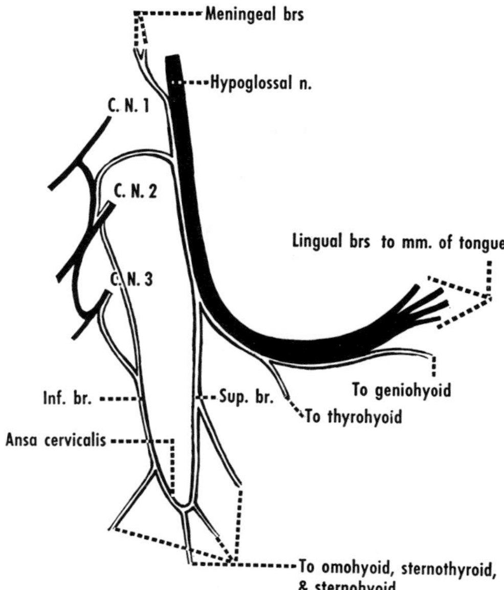

Figure 60–23 Scheme of the hypoglossal nerve.

Branches. The fibers of some of the branches of the hypoglossal nerve are proper to the nerve itself (intracranial in origin), whereas other fibers are merely being carried from cervical nerves (spinal in origin).

The meningeal branches, the superior root of the ansa cervicalis, the nerve to the thyrohyoid, and that to the geniohyoid are composed of fibers of the 2nd or 1st cervical nerve.

1. *Meningeal branches* course upward through the hypoglossal canal and supply the dura of the posterior cranial fossa.

2. The *superior root of the ansa cervicalis (descending branch of the hypoglossal nerve)* connects the hypoglossal nerve with the ansa cervicalis (p. 711). It lies superficial to, or in the substance of, the carotid sheath. The ansa and its superior root supply the infrahyoid muscles (fig. 60–7*B*).

3. The *thyrohyoid branch* arises in the carotid triangle. It supplies the thyrohyoid muscle.

4. The terminal *lingual branches* supply the extrinsic (styloglossus, hyoglossus, geniohyoid, and genioglossus) and the intrinsic muscles of the tongue. Plexiform communications occur between the terminal branches of the lingual and hypoglossal nerves.

Examination of Hypoglossal Nerve. The subject is asked to protrude his tongue. A lesion of one hypoglossal nerve results in deviation of the protruded tongue toward the affected side. The protrusion is due to the genioglossus and the intrinsic muscles of the normal side; the deviation is due probably to the weight or lag produced by the paralyzed half.[48]

SUBCLAVIAN ARTERY
(fig. 60–24)

The main artery carrying blood to the upper limb is called by various names (subclavian, axillary, and brachial) in different parts of its course. The territory supplied by the subclavian artery extends as far as the forebrain, the abdominal wall, and the fingers. The *left subclavian artery* arises directly from the arch of the aorta, whereas the *right subclavian* is a branch of the brachiocephalic trunk. The left subclavian artery, therefore, has a thoracic portion. Rarely the right subclavian artery arises from the arch of the aorta and passes behind the esophagus.

The course of each subclavian artery may be considered in three portions: the first part extends from the origin of the vessel to the medial border of the scalenus anterior, the second lies behind that muscle, and the third passes from the lateral border of the muscle to the external border of the first rib, where the subclavian is renamed the axillary artery.

The subclavian vein, which is the continuation of the axillary, passes in front of the scalenus anterior, and unites at the medial border of that muscle with the internal jugular to form the brachiocephalic vein. The subclavian vein, therefore, is co-extensive with only the second and third parts of the subclavian artery. Moreover, it seldom rises above the level of the clavicle.

The *first part* of the right subclavian artery and the cervical portion of the first part of the left subclavian artery arch upward and laterally from behind the sternoclavicular joint. The chief relations are as follows. Anteriorly: with the sternomastoid, sternohyoid, sternothyroid; internal jugular vein, thoracic duct (left side only); vagus, cardiac branches of vagus and of sympathetic, ansa subclavia (which encircles the subclavian artery), left phrenic nerve (left side only). Posteriorly: with the apex of the lung, cupola of the pleura, suprapleural

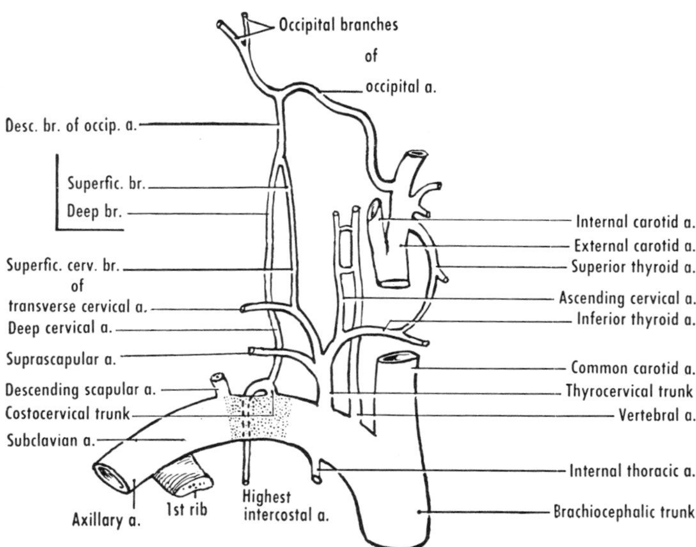

Figure 60–24 The subclavian artery and its branches: vertebral, internal thoracic, thyrocervical trunk, costocervical trunk, and descending scapular. The second part of the subclavian artery is shaded. Various anastomoses with other vessels are also shown.

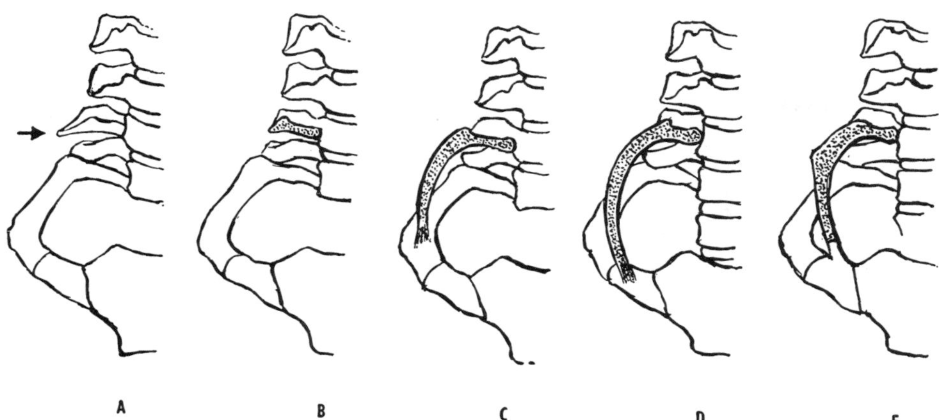

Figure 60–25 Cervical ribs. *A*, an unusually long transverse process of the seventh cervical vertebra (*arrow*). *B*, a minute cervical rib with head, neck, and tubercle. *C*, a cervical rib that is bound to the first rib; in other instances it may end freely. *D*, a cervical rib that reaches the first costal cartilage. *E*, a cervical rib that reaches the manubrium sterni, as do the upper thoracic ribs. From von Lanz and Wachsmuth.

membrane; sympathetic trunk and inferior cervical ganglion, right recurrent laryngeal nerve (which winds around the right subclavian artery).

The *second part* of the subclavian artery, which is normally narrowed between the scaleni,[49] extends a variable distance (up to 4 cm but commonly about 2 cm) above the clavicle. The chief relations are as follows. Anteriorly: with the scalenus anterior, sternomastoid; right phrenic nerve (right side only, and separated by the scalenus anterior); subclavian vein (separated by the scalenus anterior). Posteriorly: with the apex of the lung, cupola of the pleura, suprapleural membrane; scalenus medius.

The *third part* is the most superficial, and its pulsations can be felt on deep pressure. It is located mainly in the supraclavicular triangle, where it lies on the first rib (fig. 13–7). It can be compressed against the first rib by pressing downward, backward, and medially in the angle (fig. 60–4*B*) between the clavicle and the posterior border of the sternomastoid. (The posterior border of the sternomastoid suffices for the surface marking of the lateral border of the scalenus anterior.) It can also be ligated conveniently at this site, and the collateral circulation to the upper limb is generally adequate after ligation of any of the three parts of the subclavian artery. The chief relations are as follows. Anteriorly: with the external jugular vein and some of its tributaries; clavicle and subclavius; subclavian vein. Posteriorly: with the lower trunk of the brachial plexus; scalenus medius.

Neurovascular Compression.[50] **Abnormal compression of the subclavian or the axillary vessels and/or the brachial plexus produces a series of signs and symptoms that may be termed the "neurovascular compression syndromes" of the upper limb.** The features, which may include pain, paresthesia (e.g., prickling), numbness, weakness, discoloration, swelling, ulceration, and gangrene, may be produced also by other causes.

The neurovascular bundle to the upper limb is most frequently compressed: (1) in the interval between the scalenus anterior and the scalenus medius (fig. 60–32*B*), where compression may be produced or accentuated by a cervical rib (see below) or its fibrous extension, or by variations in the insertion of the scalene muscles; (2) in the interval between the first rib and the clavicle (fig. 13–3); (3) where the neurovascular bundle is crossed by the pectoralis minor and is related to the coracoid process (fig. 13–7). Compression at these three sites is accentuated, respectively, by (1) maintaining a deep inspiration while the neck is fully extended; (2) bracing the shoulders downward and backward; (3) placing the hands on the back of the head (commonly termed "hyperabduction" of the arms). It should, however, be emphasized that, even in normal people, one of these three procedures may result in weakening of the radial pulse. Moreover, factors such as a reduced muscle mass and reduced tone in middle age may be the precipitating cause of the compression in one of the neurovascular syndromes.

Cervical Rib (fig. 60–25). The head, neck, and tubercle of a mesenchymal or a cartilaginous rib are present on the seventh cervical vertebra *in fetu*. In the adult, although an unusually long transverse process is not infrequent, a separate cervical rib is rare. When present, a cervical rib may end freely, be bound to the first thoracic rib, be anchored to the first costal cartilage (by connective tissue, cartilage, or bone), or possess a costal cartilage that reaches the manubrium. An osseous cervical rib can be detected radiographically.

Branches. The branches of the subclavian artery[51] (fig. 60–24) arise generally from the first part of the artery, that is, medial to the scalenus anterior, with the exception of (1) the costocervical trunk of the right side, which usually springs from the second part, and (2) the descending scapular artery, which comes from either the second or the third part. The branches of the subclavian artery are the vertebral, internal thoracic, and descending scapular arteries, and the thyrocervical and costocervical trunks. The first three branches arise close together at the medial border of the scalenus anterior and in front of the cupola of the pleura.

1. The *vertebral artery* (fig. 60–26), in spite of its name, supplies chiefly the posterior part of the brain. It arises from the first part of the subclavian, that is, medial to the scalenus anterior. Rarely, the left vertebral artery may arise from the arch of the aorta[52] or from the brachiocephalic trunk. **The vertebral artery ascends through the foramina transversaria of the upper six cervical vertebrae, winds behind the lateral mass of the atlas (fig. 48–6), and enters the cranial cavity through the foramen magnum. At the lower border of the pons it unites with the vessel of the opposite side to form the basilar artery (which divides into the two posterior cerebral arteries). The course of the vertebral artery may be considered in four parts: cervical, vertebral, suboccipital, and intracranial.** The cervical and vertebral are considered here; the suboccipital and intracranial parts have already been described (pp. 530 and 605).

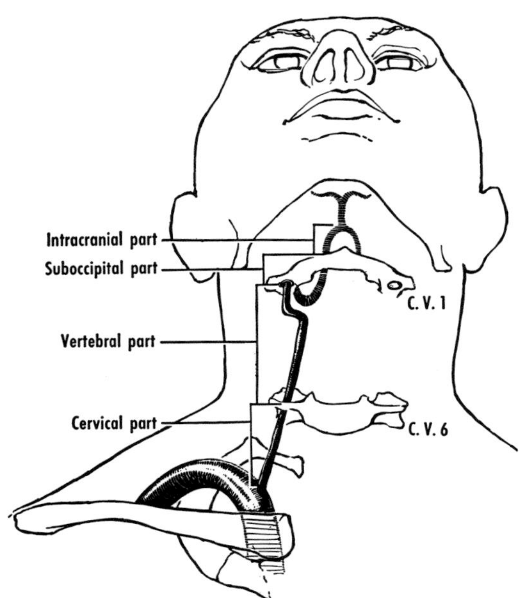

Figure 60–26 The course of the vertebral artery. The vertebral artery, a branch of the subclavian, presents cervical, vertebral, suboccipital, and intracranial parts. It unites with its fellow of the opposite side to form the basilar artery. The basilar divides into the right and left posterior cerebral arteries.

(a) The *cervical part of the vertebral artery* ascends behind the common carotid in the pyramidal space between the longus colli and the scalenus anterior (fig. 60–32*B*). The vertebral vein, a tributary of the brachiocephalic, lies anteriorly. The vertebral artery is crossed by the inferior thyroid artery and, on the left side, by the thoracic duct. It is closely related to the inferior cervical ganglion and lies in front of the ventral rami of C.N. 7 and 8, and the transverse process of C.V. 7. Muscular branches are given to the deep muscles of the neck.

(b) The *vertebral part of the vertebral artery* ascends through the foramina transversaria (from C.V. 6 to 1, occasionally from one vertebra higher or lower[53]), anterior to the ventral rami of the cervical nerves (C6 to 2). It is accompanied by a venous plexus and by sympathetic filaments. Spinal branches are given to the spinal cord and the vertebrae. These branches enter the vertebral canal in company with the roots of the spinal nerves; they form a plexus under cover of the posterior longitudinal ligament.[54]

2. The *internal thoracic artery* lies at first on the cupola of the pleura and is covered by the subclavian and internal jugular veins. It is crossed obliquely by the phrenic nerve. Its further course is described with the thorax (p. 270).

3. The *thyrocervical trunk* arises from the first part of the subclavian artery and divides almost at once into three branches: the inferior thyroid, transverse cervical, and suprascapular arteries. A simple trifurcation, however, occurs in less than one-half of instances.

(a) The *inferior thyroid artery* (fig. 60–14) ascends in front of the scalenus anterior and then arches medially in front of the vertebral vessels and behind the carotid sheath. It crosses either behind or in front of the sympathetic trunk at the level of C.V. 7 or T.V. 1 and may be related to the middle cervical ganglion. Its anterior relations are the carotid sheath and its contents, and the sympathetic trunk; the middle cervical ganglion is close to the artery. Posteriorly, the inferior thyroid artery is related to the

scalenus anterior, the vertebral vessels, and the longus colli. The relationship of the recurrent laryngeal nerve is variable: the nerve or its branches may be posterior or anterior to, or in between, the branches of the inferior thyroid artery. On reaching the lower part of the posterior surface of the lobe of the gland, the inferior thyroid artery pierces the sheath of the gland and breaks up into several glandular branches.

The branches of the inferior thyroid artery are as follows:

The *ascending cervical artery* ascends on the transverse processes of the cervical vertebrae in company with and medial to the phrenic nerve. It gives branches to the vertebrae and the spinal cord.

The *inferior laryngeal artery* accompanies the recurrent laryngeal nerve. It passes deep to the lower border of the inferior constrictor and supplies the larynx.

Tracheal, pharyngeal, and *esophageal branches* supply the structures indicated by their names.

The terminal *glandular branches* are distributed to the thyroid and parathyroid glands. A branch anastomoses with its fellow of the opposite side along the lower border of the isthmus (fig. 60–13).

(*b*) The *suprascapular (transverse scapular) artery* arises usually from the thyrocervical trunk and ends on the dorsal surface of the scapula, where it takes part in the anastomoses around that bone. It first passes laterally across the scalenus anterior and the phrenic nerve. It then crosses the subclavian artery and the cords of the brachial plexus, and runs behind and parallel with the clavicle and the subclavius. It reaches the upper border of the scapula, where it anastomoses with other vessels in relation to that bone. The suprascapular artery gives off: a *suprasternal branch,* which crosses the clavicle and supplies the skin over the front of the chest; an *acromial branch,* which pierces the trapezius and supplies the skin over the acromion; and *articular twigs* to the sternoclavicular, acromioclavicular, and shoulder joints.

(*c*) The *arteria transversa colli,* or *transverse cervical artery,* arises usually from the thyrocervical trunk, frequently in common with the suprascapular artery. It passes laterally across the scalenus anterior and the phrenic nerve but at a higher level than the suprascapular artery. It then crosses the trunks of the brachial plexus in the posterior triangle of the neck and passes deep to the trapezius, which it supplies (as the *superficial cervical artery*). It anastomoses with the superficial branch of the descending branch of the occipital artery. Alternatively, at the anterior border of the levator scapulae, the transverse cervical artery may divide into superficial and deep branches. In this event the latter vessel takes the place of the descending scapular artery, which is generally a direct branch of the subclavian.

4. The *costocervical trunk* arises from the back of the first (on the left side of the body) or second (on the right side) part of the subclavian artery. It arches backward over the cupola of the pleura to reach the neck of the first rib, where it divides into the deep cervical and the highest intercostal arteries.

The *deep cervical artery* passes backward, usually between the transverse process of C.V. 7 and the neck of the first rib. It then ascends between the semispinalis capitis and the semispinalis cervicis, and anastomoses with the deep branch of the descending branch of the occipital artery. It gives off a spinal branch to the vertebral canal. The deep cervical artery may arise directly from the subclavian.

The *highest intercostal (superior intercostal) artery* descends behind the pleura and in front of the necks of the first two ribs. It usually gives off the first two posterior intercostal arteries, that is, those to the first two intercostal spaces.

5. The *descending scapular (dorsal scapular) artery* usually arises from the subclavian (second or third part)[55] but may be the deep branch of the transverse cervical artery. As a direct branch of the subclavian, the dorsal scapular artery generally passes between either the upper and middle or the middle and lower trunks of the brachial plexus. On reaching the medial border of the scapula, the dorsal scapular artery descends in company with the dorsal scapular nerve and gives branches to the rhomboid muscles.

The branches of the subclavian artery are summarized in table 60–3.

CUPOLA OF PLEURA

The *cupula pleurae*[56] *(cervical pleura)* (fig. 60–32A) is the continuation of the costal and mediastinal pleura over the apex of the lung. It begins at the inlet of the thorax, that is, along the sloping internal border of the first rib (fig. 13–3), and projects upward

TABLE 60–3 Summary of Branches of Subclavian Artery

Branch of Subclavian A.	*Branches*
1. Vertebral a.	
Cervical part	Muscular brs.
Vertebral part	Spinal brs.
Suboccipital part	See p. 530
Intracranial part	See p. 606
2. Internal thoracic a.	See p. 270
3. Thyrocervical trunk	
(*a*) Inferior thyroid a.	Ascending cervical a. Inferior laryngeal a. Tracheal, pharyngeal, & esophageal brs. Glandular brs.
(*b*) Suprascapular a.	Suprasternal br. Acromial br. Articular brs.
(*c*) Transverse cervical a.	Superficial cervical a.
4. Costocervical trunk	Deep cervical a. Highest intercostal a. Posterior intercostal aa.
5. Descending or dorsal scapular a.	

into the root of the neck. Its superior limit is the neck of the first rib (level of spinous process of C.V. 7), or about 3 cm above the medial third of the clavicle. The cupola is covered by a fascial thickening termed the *suprapleural membrane,* which is strengthened by bands attached to the internal border of the first rib and the vertebrae (bodies of C.V. 7 and T.V. 1, transverse process of C.V. 7).[57] The *scalenus minimus (scalenus pleuralis)* is not constantly present (p. 712).

The cupola and the apex of the lung are related anteriorly to the subclavian artery and its branches, the scalenus anterior, the subclavian vein, and the phrenic and vagus nerves; posteriorly, to the sympathetic trunk, the first thoracic nerve, and the superior intercostal artery; laterally, to the scalenus medius; and medially, to the brachiocephalic trunk, right brachiocephalic vein, and trachea on the right side of the body, and to the left subclavian artery and left brachiocephalic vein on the left. The cupola and the apex occupy the lower part of the pyramidal interval between the scaleni and the longus colli.

Surface Anatomy of Cupola and Apex of Lung. **Their limit is indicated by a curved line from the sternoclavicuar joint to the junction of the medial and middle thirds of the clavicle. The summit of the curve should be about 3 cm above the medial third of the clavicle. The cupola and the apex lie entirely under cover of the sternomastoid.**

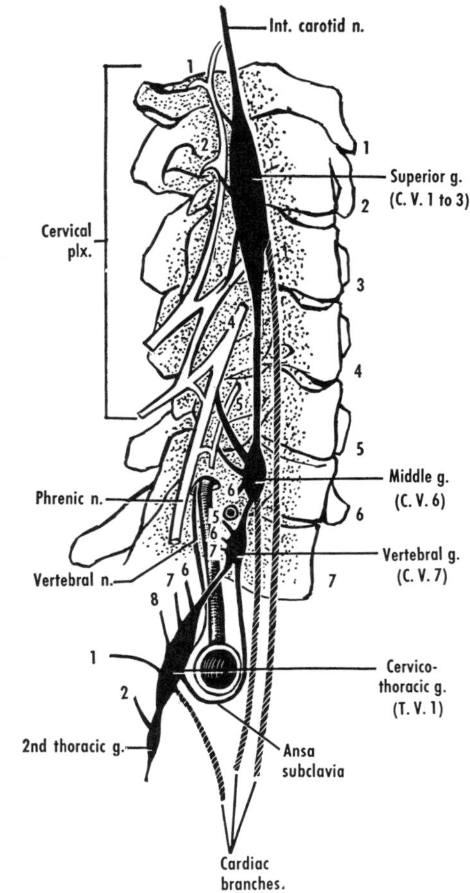

Figure 60–27 The right sympathetic trunk in the neck, lateral aspect. Only the first five cervical nerves are shown. The numerals on the left side of the drawing refer to those cervical and thoracic nerves to which rami communicantes (postganglionic fibers) are given. The numerals on the right side refer to the cervical vertebrae. The subclavian (in transverse section) and vertebral arteries are shown but do not have labels.

SYMPATHETIC TRUNK (CERVICAL PART) (fig. 60–27)

The sympathetic supply of the various structures in the head and neck arises in the first two segments of the thoracic part of the spinal cord, and sometimes in the eighth cervical segment also. The preganglionic fibers leave by way of ventral roots and pass through rami communicantes to the thoracic part of the sympathetic trunk. They then ascend to the cervical part of the sympathetic trunk, where they synapse, and from which postganglionic fibers are distributed to the blood vessels, smooth muscle, and glands of the head and neck.

The cervical part of the sympathetic trunk consists of three or four ganglia connected by an intervening cord or cords. The postganglionic fibers that leave the trunk do so by rami communicantes; these vary in number and arrangement, and they may contain ganglia and ganglion cells along their course.[58] Postganglionic fibers also leave in branches that are given directly to adjacent blood vessels, and in branches that go directly to certain viscera.

The surface anatomy of the sympathetic trunk is similar to that of the carotid arteries (p. 696).

Interruption of the cervical part of the sympathetic trunk cuts off impulses to the superior cervical ganglion and is followed by a characteristic group of signs (Horner's syndrome): constriction of the pupil (miosis, owing to unopposed action of the parasympathetic), drooping of the upper eyelid (ptosis, owing to paralysis of smooth muscle), an illusion that the eye has receded (enophthalmos), redness and increased temperature of the skin (vasodilatation), and absence of sweating (anhidrosis). It is probable that the preganglionic fibers for the eye and the orbit are derived mostly from T.N. 1. (The range[59] is from C8 to T4.) The fibers probably enter the upper portion of the cervicothoracic ganglion by a paravertebral route.[60] Section of the sympathetic trunk below the first thoracic ganglion leaves the eye unaffected.

A local anesthetic injected in the vicinity of the cervicothoracic ganglion will "block" the cervical and the upper three or four thoracic ganglia (stellate ganglion block).[61] This procedure is undertaken chiefly for vascular spasm or occlusion involving the brain or an upper limb.

Cervical Ganglia

1. Superior Cervical Ganglion. This extends from C.V. 1 to C.V. 2 or 3. It lies behind the internal carotid artery and in front of the longus capitis muscle.

The superior cervical ganglion gives rise to rami communicantes (postganglionic fibers) to the upper cervical (C1 to 3 or 4) and the last four cranial nerves. It also gives twigs to the carotid body and sinus, to the pharyngeal plexus and thence to the larynx and the pharynx, and cervical cardiac nerves to the heart (p. 335). Several branches arise and form a plexus in the adventitia of the external carotid artery.[62] Branches of the plexus continue along the branches of the external carotid artery, and some ultimately reach the salivary glands. Finally, a rather large branch (sometimes several) continues upward from the ganglion. This branch is termed the *internal carotid nerve*,[63] and it accompanies the internal carotid artery. As the artery ascends through the base of the skull, the nerve breaks up into branches that form the internal carotid plexus. The plexus gives twigs to the tympanic nerve (caroticotympanic branches), to the greater petrosal nerve (to form the nerve of the pterygoid canal, which is connected to the pterygopalatine ganglion), to various cranial nerves (3, 4, 5, and 6), and to the ciliary ganglion. Moreover, it forms subsidiary plexuses along the branches of the internal carotid artery, for example, the anterior and middle cerebral. All of the branches of the plexus consist mostly of postganglionic fibers. Those that are given to the ciliary ganglion are the dilator fibers for the pupil.

2. Middle Cervical Ganglion. The middle cervical ganglion is extremely variable and is sometimes fused with either the superior or the vertebral ganglion.[64] It is usually at the level of C.V. 6, above the arch formed by the inferior thyroid artery.

It gives postganglionic rami to some of the cervical nerves (e.g., C4 to 6), a branch to the heart, and small twigs that form a plexus along the inferior thyroid artery.

3. Vertebral Ganglion. The vertebral ganglion is commonly found at the level of C.V. 7. It lies generally in front of the vertebral artery and below the arch formed by the inferior thyroid artery, but it may be intimately related to the latter vessel. The middle and vertebral ganglia are frequently present together on the one sympathetic trunk.[65]

The ganglion gives postganglionic rami to some of the cervical nerves (e.g., C6) and thence to the brachial plexus for the supply of the upper limb. Branches are also given to the plexus along the vertebral artery. Cords connect the vertebral ganglion with the cervicothoracic or with the inferior cervical ganglion. These cords pass on each side of the vertebral artery and constitute a part of the sympathetic trunk. An additional cord loops around the first part of the subclavian artery to form the *ansa subclavia.* The ansa contributes fibers to the plexus on the subclavian artery.

4. Cervicothoracic (Stellate) Ganglion. This has two components: the inferior cervical and the first (and occasionally the second and third) thoracic ganglion. The two components may appear completely fused or partially or entirely separate.[66] The cervicothoracic ganglion is generally at the level of C.V. 7 to T. V. 1, and is placed anterior to C.N. 8 and T.N. 1. It lies behind the vertebral artery and in front of the seventh cervical transverse process and the neck of the first rib.

The cervicothoracic ganglion receives preganglionic rami from T.N. 1, or from T.N. 1 and 2. It gives postganglionic rami to the lower cervical and upper thoracic nerves (C6 to 8, and T1, and frequently T2). These fibers enter the brachial plexus and are distributed to the upper limb. It gives a cervicothoracic branch or branches to the heart, and a variable number of branches to the subclavian and vertebral arteries.

The fibers from the vertebral and cervicothoracic ganglia to the vertebral artery form a *vertebral plexus* that accompanies the artery, enters the cranial cavity, and is distributed along the basilar artery. Whether the plexus on the posterior cerebral artery is derived mainly from the plexus on the basilar artery or from the internal carotid plexus is disputed. The fibers in the vertebral plexus are mostly postganglionic, and many reach the lower cervical nerves (e.g., C5 to 7)[67] by way of small branches of the plexus. Small collections of ganglion cells are present in the vertebral plexus.

Some of the fibers from the vertebral and cervicothoracic ganglia ascend separately from the plexus and form a *vertebral nerve,* which is situated dorsal to the vertebral artery and ascends to the level of the axis or the atlas. The postganglionic fibers in it are given to the cervical nerves,[68] and it also sends branches to the spinal meninges.[69]

INTERNAL JUGULAR VEIN
(fig. 60–28)

The internal jugular vein drains the brain, the neck, and the face. At the level of its superior bulb, however, the internal jugular vein is almost free of extracerebral blood.

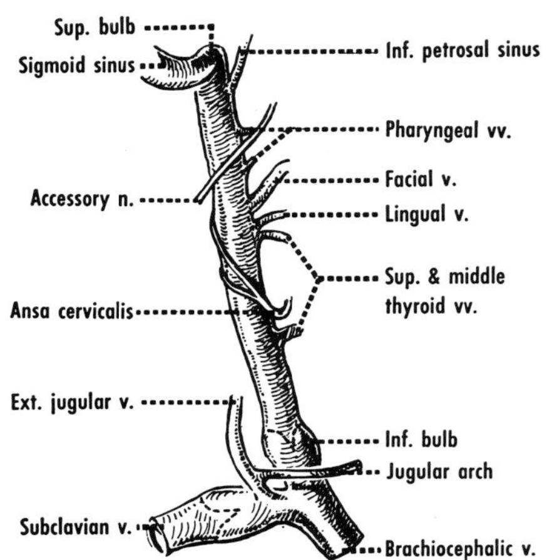

Figure 60–28 The internal jugular vein and its tributaries. Note the valves at the terminations of the subclavian and internal jugular veins; these are the last valves before the blood reaches the heart. From Grant's *Atlas.*

The internal jugular commences in the jugular foramen at the base of the skull, and is a continuation of the sigmoid sinus. A dilatation at its origin is known as the *superior bulb.* A structure similar to the carotid body is generally present near or in the bony wall of the bulb.[70] The vein descends in the carotid sheath and ends behind the medial end of the clavicle by uniting with the subclavian vein to form the brachiocephalic vein. The internal jugular usually possesses one bicuspid valve.[71] A dilatation near its termination is known as the *inferior bulb.*

The surface anatomy is similar to that of the carotid arteries (p. 696). **The internal jugular vein descends from behind the neck of the mandible to the interval between the sternal and clavicular heads of the sternomastoid** (fig. 60–15*B*). A pulsation of the internal jugular vein, due to contraction of the right atrium of the heart, is sometimes demonstrable in the lower part of the neck.

Superficial Relations. The sternomastoid overlaps the upper part and covers the lower part of the internal jugular vein. Other superficial relations include the posterior belly of the digastric, the superior belly of the omohyoid, and the other infrahyoid muscles. Also superficial are the facial nerve, the accessory nerve, usually the inferior root of the ansa cervicalis, and the anterior jugular vein. **The deep cervical lymph nodes lie along the course of the internal jugular vein.**

Deep Relations. These include the transverse processes of the cervical vertebrae, the levator scapulae, scalenus medius, cervical plexus, scalenus anterior, phrenic nerve, thoracic duct, the thyrocervical trunk and the first part of the subclavian artery, and the cupola of the pleura. The internal and common carotid arteries accompany the internal jugular vein medially, and the vagus lies behind and between the vein and the arteries. **At the base of the skull the internal carotid artery lies in front of the internal jugular vein (the carotid canal lies in front of the jugular foramen), and the two vessels are separated there by the last four cranial nerves** (figs. 52–16 and 58–2*B*).

Tributaries. The tributaries, which are variable, include the inferior petrosal sinus, and the pharyngeal, lingual, superior and middle thyroid veins. In addition, the *right lymphatic duct* or, on the left side, the *thoracic duct,* opens usually into the internal jugular vein, or into the junction between the internal jugular and the subclavian vein.

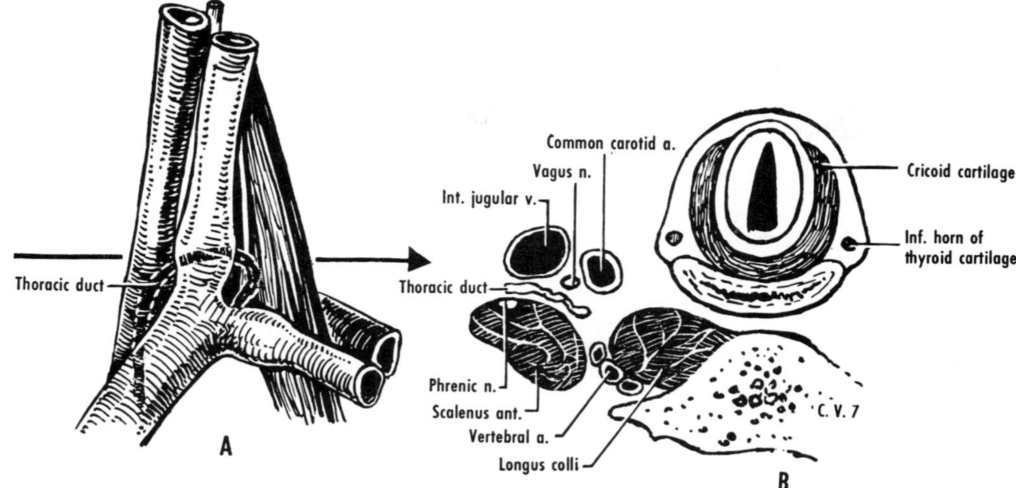

Figure 60–29 *A*, the termination of the thoracic duct. *B*, a horizontal section in which the arch formed by the thoracic duct is seen between the scalenus anterior muscle behind and the internal jugular vein and the common carotid artery in front.

THORACIC DUCT AND LYMPHATIC DRAINAGE OF HEAD AND NECK

Thoracic Duct

The thoracic duct (fig. 60–29), when it leaves the thorax (p. 330), arches laterally at the level of C.V. 7. The duct passes in front of the left sympathetic trunk, the left vertebral artery, and the left phrenic nerve and scalenus anterior (separated by the prevertebral layer of fascia). The duct is placed behind the left common carotid artery, the vagus, and the internal jugular vein (fig. 60–29*B*). It receives the left jugular trunk and ends in front of the first part of the left subclavian artery, by opening into one of the following: the left internal jugular, the angle between the left internal jugular and the left subclavian, the left subclavian, or the left brachiocephalic vein.[72] The duct is provided with a pair of valves at its termination. **The thoracic duct receives the lymph from most of the body, including the left side of the head and neck.** Some of its tributaries, however, e.g., the left transverse cervical trunk, frequently open independently into the venous junction.

Right Lymphatic Duct (fig. 60–30*B*). This duct, about 1 cm in length, is seldom present as a single structure; the chief vessels that unite to form it, namely, the right jugular, right subclavian, and right bronchomediastinal trunks, usually open separately into the right internal jugular and/or the right subclavian vein. **The right lymphatic duct receives the lymph from the right side of the head and neck, the right upper limb, and the right side of the thorax.**

Lymphatic Drainage of Head and Neck

Cervical Lymph Nodes. **All the lymphatic vessels from the head and neck drain into the deep cervical nodes, either (1) directly from the tissues or (2) indirectly after traversing an outlying group of nodes.**

Several outlying groups of lymph nodes form a "pericervical collar" (fig. 60–30*A*) at the junction of the head with the neck: *occipital, retro-auricular (mastoid), parotid, submandibular, buccal (facial),* and *submental.* The superficial tissues of the head and the neck drain into these groups and also into the superficial cervical nodes. Thus, the scalp drains into the various nodes of the pericervical collar; the auricle into the retro-auricular and parotid nodes; the eyelids and the cheek into the parotid and submandibular; the external nose into the submandibular; the lips into the submandibular and (in the case of the lower lip) submental nodes.

SUPERFICIAL CERVICAL NODES. **The superficial cervical nodes are found (1) in the**

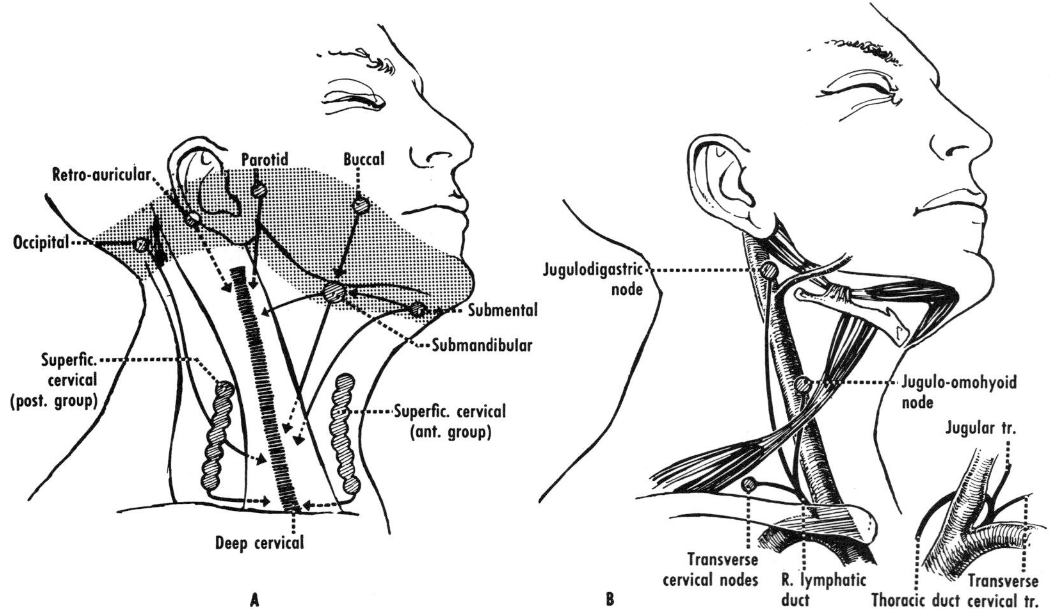

Figure 60–30 The lymphatic drainage of the head and neck. The arrows indicate the direction of drainage. *A*, the superficial groups of cervical lymph nodes. The wide, shaded band indicates the "pericervical collar" of nodes. Each circle represents a group of nodes. *B*, the deep cervical lymph nodes. The inset at the lower right shows one of the many patterns that may be found on the left side of the body.

posterior triangle along the course of the external jugular vein, and (2) in the anterior triangle along the course of the anterior jugular vein. The superficial tissues of the neck drain into the occipital, submandibular, and submental nodes, and into the superficial and deep cervical nodes.

DEEP CERVICAL NODES (fig. 60–30*B*). **The deep cervical nodes include several groups. The main group forms a chain along the internal jugular vein mostly under cover of the sternomastoid.** For purposes of description, the group is frequently subdivided into superior and inferior, or into superior, middle, and inferior groups. The *jugulodigastric node* lies on the internal jugular vein at the level of the greater horn of the hyoid bone, that is, just below the posterior belly of the digastric. It receives numerous afferents from the posterior third of the tongue and from the palatine tonsil; it derives its importance from the fact that it is frequently enlarged in carcinoma of either organ. The *jugulo-omohyoid node* lies on the internal jugular vein just above the middle tendon of the omohyoid muscle. It receives afferents from the tongue directly, and also indirectly by way of the submental, submandibular, and upper deep cervical nodes.

A group of deep nodes leaves the region of the internal jugular vein above and descends across the posterior triangle in company with the accessory nerve. These nodes drain chiefly into a group found along the transverse cervical artery which, in turn, drain forward into the lower nodes on the internal jugular vein. The nodes along the transverse cervical artery, which are sometimes termed the supraclavicular nodes, may be involved, presumably by retrograde extension, in carcinoma in the thorax or the abdomen.

Other groups of nodes that are placed deeply in the neck include the *prelaryngeal, pretracheal, paratracheal,* and *retropharyngeal nodes.* These take part in the drainage of some of the deeper structures of the head and neck. Thus, the middle ear drains into the retropharyngeal and upper deep cervical nodes; the nasal cavity and the paranasal sinuses into the submandibular, retropharyngeal, and upper deep cervical; the palate and the tonsil into the upper deep cervical; the tongue into the submental, submandibular, and upper and lower deep cervical (p. 723); the larynx into the upper and lower deep cervical; the pharynx into the upper and lower deep cervical, and into the retropharyngeal; and the thyroid gland into the lower deep cervical, prelaryngeal, pretracheal, and paratracheal.

The efferent vessels from the deep cervical nodes on either side form the *jugular trunk.* The trunk on the left side usually joins the thoracic duct; that on the right side ends in the junction of the internal jugular with the subclavian vein, or in the right lymphatic duct when that vessel is present.

CERVICAL PLEXUS

The ventral rami of the upper four cervical nerves unite to form the cervical plexus, whereas those of the lower four, together with the greater part of that of the first thoracic nerve, join to form the brachial plexus. Each of the ventral rami receives one or more rami communicantes from a cervical ganglion on the sympathetic trunk. These rami communicantes contain postganglionic fibers. The ventral rami of C.N. 3 to 7 lie in grooves on the costotransverse bars of C.V. 3 to 7.[73]

The cervical plexus, formed by the ventral rami of the upper four cervical nerves, is located in front of the levator scapulae and the scalenus medius, under cover of the internal jugular vein and the sternomastoid. **The cervical plexus is arranged as an irregular series of loops (fig. 60–27), from which the branches arise. This arrangement results in the supply of cutaneous areas by branches of more than one spinal nerve.** The branches supply the skin of the back of the head, the neck (fig. 57–6*B*), and the shoulder, and also certain muscles of the neck, together with the diaphragm. **The cutaneous branches (p. 659) all emerge near the middle of the posterior border of the sternomastoid** (fig. 60–5). They may communicate among themselves and also with branches of the dorsal rami and some of the cranial nerves. The branches of the cervical plexus are summarized in table 60–4.

TABLE 60–4 Summary of Branches of Cervical Plexus

Superficial branches:		
	Lesser occipital Great auricular Transverse cervical (anterior cervical cutaneous)	C. N. 2, 3
	Supraclavicular	C. N. 3, 4
Deep branches:		
To	Sternomastoid	C. N. 2, 3
	Trapezius Levator scapulae Scaleni	C. N. 3, 4
	Prevertebral muscles	C. N. 1 to 4
	Infrahyoid muscles by ansa cervicalis	C. N. 1 to 3
	Diaphragm by phrenic nerve	C. N. 3 to 5
Communicating to cranial nerves (10, 11, and 12) and to sympathetic		

Ansa Cervicalis

The ansa cervicalis *(ansa hypoglossi)* is a loop that lies superficial to (or in the substance of) the carotid sheath and that is formed by fibers of C.N. 1 to 3, or C.N. 2 and 3 (figs. 60–7*B* and 60–23). It presents a *superior root* (the so-called *descending branch of the hypoglossal nerve),* which connects it with the hypoglossal nerve (p. 702) but consists of fibers of C.N. 2 or 1, and an *inferior root (nervus descendens cervicalis),* which connects it with branches from C.N. 2 and 3. The inferior root usually passes lateral (occasionally medial) to the internal jugular vein. The ansa and its superior root supply the infrahyoid muscles; the thyrohyoid, however, receives its cervical fibers directly from the hypoglossal nerve.

Phrenic Nerve

The *phrenic nerve* supplies the diaphragm and the serous membranes of the thorax and abdomen. It arises chiefly from C.N. 4 (fig. 60–27). Most commonly it has two roots of origin, a principal root from C.N. 4 and an accessory root from C.N. 5.[74] Occasionally it may have a root from C.N. 3. The fibers from C.N. 5 sometimes reach the phrenic by way of the nerve to the subclavius, through what is termed an accessory phrenic nerve (see below).

The phrenic nerve is formed at the lateral border of the scalenus anterior. It then runs vertically downward across the front of that muscle (fig. 60–10), under cover of the internal jugular vein and the sternomastoid. It is situated behind the prevertebral fascia and is crossed by the transverse cervical and suprascapular arteries (fig. 60–32*B*). It is accompanied medially by the ascending cervical artery (a branch of the inferior thyroid). It passes between the subclavian artery and vein (fig. 60–20), and crosses the internal thoracic artery lateromedially. Its further course to the diaphragm is described with the thorax (p. 332). The phrenic contains afferent as well as efferent fibers.

In its course in the neck, the left phrenic nerve crosses the first part of the left subclavian artery, whereas the right nerve crosses the second part of the right subclavian artery.

Surgical interruption of the phrenic nerve on the scalenus anterior is sometimes performed, to aid in the collapse of a lung, for example. The diaphragm becomes paralyzed, and is therefore elevated, on the side of the interruption.

An *accessory phrenic nerve* is present in nearly one-third of instances. It arises variably from the brachial plexus (e.g., through the nerve to the subclavius), the cervical plexus (e.g., through the inferior root of the ansa), or by passing through one of the cardiac branches of the cervical sympathetic ganglia. Its fibers are derived from one or more of the spinal nerves from C3 to T1.

SCALENE MUSCLES

The scaleni (figs. 60–32, 33) are chiefly the anterior, medius, and posterior.

Scalenus Anterior. **The scalenus anterior (except at its insertion) lies entirely under cover of the sternomastoid.** It arises from the anterior tubercles of the transverse processes of the lower cervical vertebrae (C.V. 3 to 6), and is inserted into the scalene tubercle on the internal border of the first rib (and on the adjacent part of the upper surface) (fig. 26–9). **The subclavian artery passes behind the scalenus anterior, whereas the phrenic nerve lies on the muscle.**

Scalenus Medius. The scalenus medius arises from the posterior tubercles[75] of the transverse processes of the cervical vertebrae (most frequently C.V. 1 to 7), and is inserted on an impression on the superior surface of the first rib (See fig. 26–9). **The brachial plexus emerges between the scalenus anterior and the scalenus medius.**

Scalenus Posterior. The scalenus posterior, not infrequently absent or blended with the medius, arises from the posterior tubercles of the transverse processes of the lower cervical vertebrae (C.V. 4 to 6), and is inserted into the external surface of the second rib.

Scalenus Minimus (Scalenus Pleuralis). The scalenus minimus,[76] when present, runs from the transverse process of C.V. 7 (or C.V. 6 and 7) to the internal border of the first rib and usually to the cupola of the pleura also. Its place may be taken by, or it may give rise to, fibrous expansions associated with the suprapleural membrane.

A pyramidal space occurs between the scaleni laterally and the longus colli medially. The pleura and the apex of the lung project upward into this space (fig. 60–32A).

Innervation of Scalene Muscles. By branches from the ventral rami of the cervical nerves.

Actions. The scaleni flex the cervical part of the vertebral column laterally. The scaleni may act as muscles of inspiration in normal persons even during quiet breathing. They become active during moderately severe voluntary expiratory efforts, and may be important in coughing and straining.[77]

CERVICAL FASCIA[78]

Descriptions of the fascia vary in details because the arrangement of fasciae is complicated and variable, and because different appearances are produced according to the technique used, for example, gross dissection, microscopic examination of sections, injection of fluids, or the study of the spread of infections. Moreover, the degree of condensation necessary before a connective tissue is designated as a fascial layer is arbitrary.

The cervical fascia "affords that slipperiness which enables structures to move and pass over one another without difficulty, as in swallowing, and allowing twisting of the neck without it creaking like a manilla rope—a looseness, moreover, that provides the easiest pathways for vessels and nerves to reach their destinations."[79]

The fascia of the neck (fig. 60–31) is generally described as comprising three layers: investing, pretracheal, and prevertebral.

1. The *investing layer* is attached behind to the external occipital protuberance and the superior nuchal line, and to the ligamentum nuchae and the spines of the cervical vertebrae. Its other bony attachments are the mastoid process, lower border of the mandible, zygomatic arch, styloid process (by the stylomandibular ligament), hyoid bone, acromion, clavicle, and manubrium sterni. The investing layer surrounds the trapezius, roofs the posterior triangle of the neck (enclosing the omohyoid),

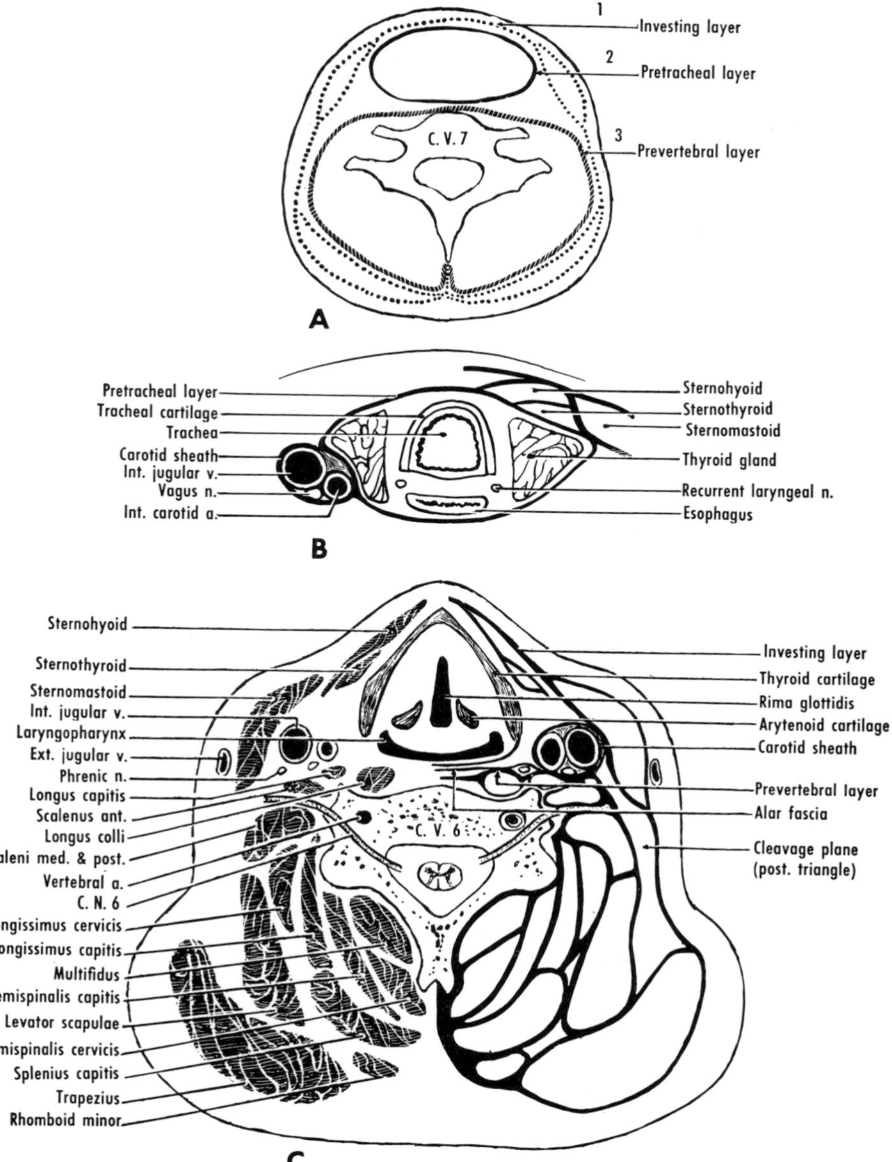

Figure 60–31 Horizontal sections illustrating the cervical fascia. *A* presents the general arrangement of the three layers. *B* shows the pretracheal layer at the level of the seventh vertebra. *C* shows the three layers at the level of the sixth cervical vertebra. In *C* the muscles are indicated on the left side, the fasciae on the right side. *C* is based on Truex and Kellner.

surrounds the sternomastoid, and roofs the anterior triangle, where it covers the infrahyoid muscles. Immediately above the sternum, the investing layer divides into two layers, which are anchored to the front and back of the manubrium, respectively. The interval between these two layers is termed the *suprasternal space.* It contains the sternal heads of the sternomastoids, the jugular venous arch, and an occasional lymph node. Above, the investing layer forms the sheaths of the parotid and submandibular glands.

2. The *pretracheal layer* is limited to the front part of the neck, but is more extensive than its name suggests. It lies below the hyoid bone and is attached to the oblique lines of the thyroid cartilage

and to the cricoid cartilage. It surrounds the thyroid gland, forming its sheath, and it invests the infrahyoid muscles and the air and food passages. Infections from the head and neck can spread in front of the trachea or behind the esophagus and reach the superior mediastinum in the thorax. Above, the pretracheal layer behind the esophagus becomes continuous with the buccopharyngeal fascia, which cover the constrictor muscles of the pharynx and the buccinator.

3. The *prevertebral layer* is attached to the base of the skull and to the transverse processes of the cervical vertebrae. It covers the prevertebral muscles, the scaleni (and the phrenic nerve), and the deep muscles of the back. It covers, therefore, the floor of the posterior triangle of the neck. In front of the subclavian artery, it is prolonged laterally as the axillary sheath, which invests the brachial plexus in addition to the vessel. In front of the bodies of the cervical vertebrae, an additional layer, the *alar fascia,* is found between the pretracheal and prevertebral layers. It is attached to the transverse processes.

The *carotid sheath* is a condensation of the fascia that surrounds the common and internal carotid arteries, the internal jugular vein, and the vagus nerve. It is fused with all three layers of the cervical fascia.

PREVERTEBRAL MUSCLES

The prevertebral muscles (figs. 60–32 and 60–33) are the longi capitis et colli, and the recti capitis anterior et lateralis.

Longus Capitis (fig. 60–33). The longus capitis covers the upper oblique fasciculi of the longus colli. It arises from the

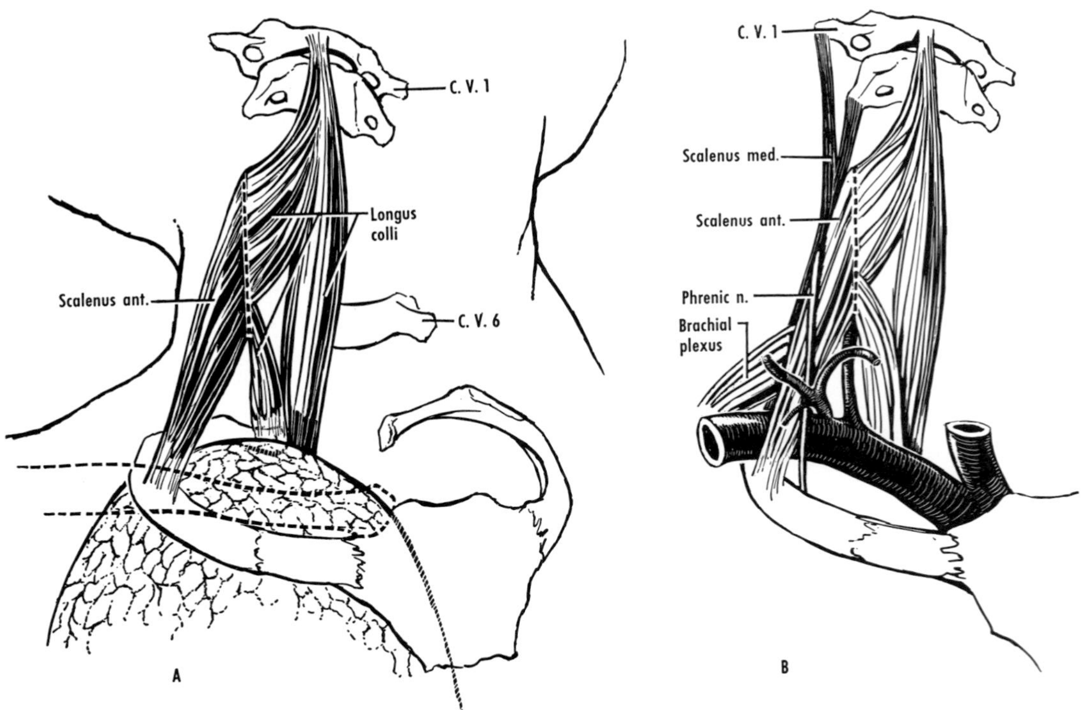

Figure 60–32 *A,* the cupola of the pleura, the scalenus anterior, and the longus colli. Anterior aspect, and slightly from the right side. Note that the cupola, which projects upward between the scaleni and the longus colli, rises about 3 cm above the medial third of the clavicle (interrupted line). *B,* the third part of the subclavian artery and the brachial plexus, between the scalenus anterior and the scalenus medius; the lower trunk of the plexus lies behind the artery. The phrenic nerve, which descends almost vertically on the obliquely set scalenus anterior, is bound down to the front of that muscle by the transverse cervical and suprascapular arteries (no labels). Note the triangle bounded by the scalenus anterior laterally, the longus colli medially, and the first part of the subclavian artery below. The carotid tubercle of C.V. 6 lies at the apex. The vertebral artery (no label) ascends through the triangle to reach the foramen transversarium of C.V. 6. *A* is based on von Lanz and Wachsmuth.

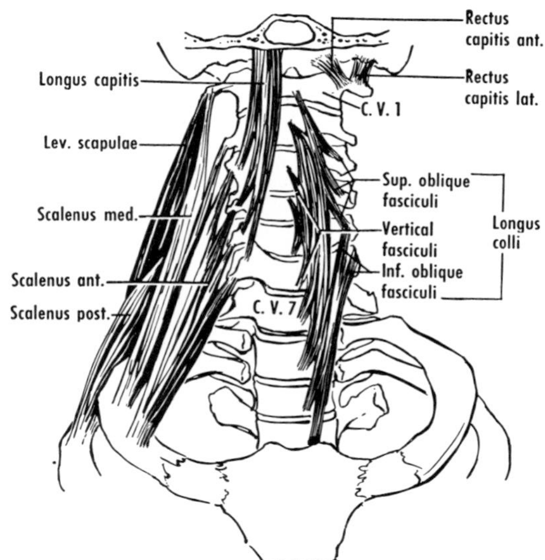

Figure 60–33 The scalene and prevertebral muscles. The three chief scalene muscles are shown on the right side of the body, together with the levator scapulae and the longus capitis. Observe that the scalenus anterior and the longus capitis arise from the same vertebrae but proceed in opposite directions. The longus colli is shown on the left side of the body, together with the recti capitis anterior et lateralis. Note that about 2½ thoracic vertebrae are visible above the level of the jugular notch of the manubrium sterni.

transverse processes (anterior tubercles) of the lower cervical vertebrae (C.V. 3 to 6) and is inserted into the lower surface of the basilar part of the occipital bone.

Longus Colli (Longus Cervicis) (figs. 60–32 and 60–33). The muscle of each side consists of a vertical and an oblique fasciculus. The former extend from the bodies of the upper thoracic and lower cervical vertebrae to the bodies of the upper cervical vertebrae. The lower oblique fasciculi run from the bodies of the upper thoracic vertebrae to the transverse processes (anterior tubercles) of the lower cervical vertebrae. The upper oblique fasciculi run from the transverse processes (anterior tubercles) of the upper cervical vertebrae to the anterior arch of the atlas. The longus colli muscles of the two sides are separated by a gap in which the anterior longitudinal ligament of the vertebral column is visible.

Rectus Capitis Anterior. This extends from the transverse process and the lateral mass of the atlas to the basilar part of the occipital bone.

Rectus Capitis Lateralis. This extends from the transverse process of the atlas to the jugular process of the occipital bone.

Innervation. All the prevertebral muscles are supplied by branches from the ventral rami of the cervical nerves.

Actions. The longus colli flexes the cervical part of the vertebral column, and is active during talking, coughing, and swallowing;[80] the other prevertebral muscles flex the head. The prevertebral muscles, together with the sternomastoids, act in conjunction with, and as antagonists to, the upper deep muscles of the back.

REFERENCES

1. K. B. Corbin and F. Harrison, Brain, *62*:191, 1939.
2. G. Weddell, B. Feinstein, and R. H. Pattle, Brain, *67*:178, 1944.
3. G. Causey and D. Slome, Anat. Soc. G. B. and Ireland, Feb., 1954. O. Machado de Sousa, J. Furlani, and M. Vitti, Electromyogr. clin. Neurophysiol., *13*:93, 1973.
4. E. J. M. Campbell, J. Anat., Lond., *89*:378, 1955.
5. A. Thévenard and H. Berdet, Pr. méd., *66*:529, 1958.
6. S. Brown, Amer. J. phys. Anthrop., *28*:213, 1941. M. Janský, B. Plucnar, and Z. Svoboda, Acta anat., *37*:298, 1959.
7. J. A. Pires de Lima, J. Anat., Lond., *59*:108, 1924.
8. O. Machado de Sousa, Folia clin. biol., *33*:42, 1964.
9. C. L. Langsam, Amer. J. phys. Anthrop., *28*:249, 1941.
10. R. J. Last, Proc. R. Soc. Med., *47*:571, 1954.
11. J. D. B. MacDougall and B. L. Andrews, J. Anat., Lond., *87*:37, 1953.
12. F. H. Lahey, J. Amer. med. Ass., *86*:813, 1926.
13. J. Cayotte and A. Brulé, *Recueil des travaux du laboratoire d'anatomie de Nancy*, 1952.
14. N. Johnson, Aust. N.Z. J. Surg., *23*:95, 1953, and Brit. J. Surg., *42*:587, 1955.
15. J.-P. Lassau, G. Hidden, and J. Hureau, Arch. Anat. path., *15*:107, 1967.
16. W. J. Cunliffe, Acta anat., *46*:135, 1961.
17. A. H. Bill, Surg. Clin. N. Amer., *36*:1599, 1956. J. B. Dalgaard and P. Wetteland, Acta chir. scand., *111*:444, 1956.
18. E. Hintzsche, Anat. Anz., *84*:18, 1937. J. R. Gilmour, J. Path. Bact., *46*:133, 1938. A. D. Vail and F. C. Coller, Missouri Med., *63*:347, 1966.
19. A. J. Walton, Brit. J. Surg., *19*:285, 1931.
20. L. J. Pyrtek and R. L. Painter, Surg. Gynec. Obstet., *119*:509, 1964.
21. C. Zaino *et al.*, *The Pharyngoesophageal Sphincter*, Thomas, Springfield, Illinois, 1970.
22. A. L. Shapiro and G. L. Robillard, Ann. Surg., *131*:171, 1950. LaV. L. Swigart *et al.*, Surg. Gynec. Obstet., *90*:234, 1950.
23. E. S. Gurdjian *et al.*, Neurology, *8*:818, 1958.
24. J. D. Liechty, T. W. Shields, and B. J. Anson, Quart. Bull. Northw. Univ. med. Sch., *31*:136, 1957.
25. D. K. McAfee, B. J. Anson, and J. J. McDonald, Quart. Bull. Northw. Univ. med. Sch., *27*:226, 1953.
26. G. Muratori, Anat. Anz., *119*:466, 1966.
27. H.-J. Hübner, Anat. Anz., *122*:133, 1968.
28. H. Pakula and J. Szapiro, J. Neurosurg., *32*:171, 1970.
29. D. Sheehan, J. H. Mulholland, and B. Shafiroff, Anat. Rec., *80*:431, 1941. J. D. Boyd, Anat. Anz., *84*:386, 1937.
30. W. E. Adams, *The Comparative Morphology of the Carotid Body and Carotid Sinus*, Thomas, Springfield, Illinois, 1958.
31. P. N. Karnauchow, Canad. med. Ass. J., *92*:1298, 1965.
32. H. Metz *et al.*, Lancet, *1*:424, 1961.
33. A. Faller, Schweiz. med. Wschr., *76*:1156, 1946. H.-J. Hübner, Anat. Anz., *121*:489, 1967.
34. J. Cairney, J. Anat., Lond., *59*:87, 1924.
35. G. Godlewski and J. Bossy, Bull. Ass. Anat., *57*:325, 1973.

36. B. Cochet, Arch. Anat., Strasbourg, *40*:1, 1967.
37. J. H. Ogura and R. L. Lam, Laryngoscope, St Louis, *63*:947, 1953.
38. F. Lemere, Anat. Rec., *54*:389, 1932.
39. T. F. M. Dilworth, J. Anat., Lond., *56*:48, 1921.
40. G. R. Stewart, J. C. Mountain, and B. P. Colcock, Brit. J. Surg., *59*:379, 1972.
41. H. Pichler and A. Gisel, Laryngoscope, St Louis, *67*:105, 1957.
42. W. H. Rustad, *The Recurrent Laryngeal Nerves in Thyroid Surgery*, Thomas, Springfield, Illinois, 1956. P. Blondeau, J. Chir., Paris, *102*:397, 1971. P. Vuillard, Arch. Anat. path., *19*:449, 1971.
43. S. Sunderland and W. E. Swaney, Anat. Rec., *114*:411, 1952.
44. A. F. Williams, J. Laryng., *68*:719, 1954. Pichler and Gisel, Laryngoscope, St Louis, *67*:105, 1957.
45. G. Ouaknine and H. Nathan, J. Neurosurg., *38*:189, 1973.
46. A. A. Pearson, R. W. Sauter, and G. R. Herrin, Amer. J. Anat., *114*:371, 1964.
47. W. Platzer, Arch. Psychiat. Nervenkr., *199*:372, 1959.
48. G. A. Bennett and R. C. Hutchinson, Anat. Rec., *94*:57, 1946.
49. S. Sunderland and G. M. Bedbrook, Anat. Rec., *104*:299, 1949.
50. L. M. Rosati and J. W. Lord, *Neurovascular Compression Syndromes of the Shoulder Girdle*, Grune & Stratton, New York, 1961.
51. E. H. Daseler and B. J. Anson, Surg. Gynec. Obstet., *108*:149, 1959. I. G. Schraibman, Leech, *27*:35, 1957.
52. H. Maisel, S. Afr. med. J., *32*:1141, 1958.
53. R. H. Bell, LaV. L. Swigart, and B. J. Anson, Quart. Bull. Northw. Univ. med. Sch., *24*:184, 1950.
54. R. S. Harris and D. M. Jones, J. Bone Jt Surg., *38B*:922, 1956.
55. D. F. Huelke, Anat. Rec., *132*:233, 1958, and *142*:57, 1962.
56. A. Hafferl, *Die Anatomie der Pleurakuppel*, Springer, Berlin, 1939.
57. G. R. L. Gaughran, Anat. Rec., *148*:553, 1964.
58. T. Skoog, Lancet, 2:457, 1947.
59. B. S. Ray, J. C. Hinsey, and W. A. Geohegan, Ann. Surg., *118*:647, 1943.
60. L. T. Palumbo, Surgery, *42*:740, 1957.
61. D. C. Moore, *Stellate Ganglion Block*, Thomas, Springfield, Illinois, 1954.
62. E. Gardner, Arch. Surg., Chicago, *46*:238, 1943.
63. A. Kuntz, H. H. Hoffman, and L. M. Napolitano, Arch. Surg., *75*:108, 1957.
64. M. Wrete, J. Anat., Lond., *93*:448, 1959.
65. R. F. Becker and J. A. Grunt, Anat. Rec., *127*:1, 1957.
66. R. W. Jamieson, D. B. Smith, and B. J. Anson, Quart. Bull. Northw. Univ. med. Sch., *26*:219, 1952.
67. S. Sunderland and G. M. Bedbrook, Brain, *72*:297, 1949.
68. S. A. Siwe, Amer. J. Anat., *48*:479, 1931. H. H. Hoffman and A. Kuntz, Arch. Surg., *74*:430, 1957.
69. D. L. Kimmel, J. comp. Neurol., *112*:141, 1959.
70. S. R. Guild, Ann. Otol., etc., St Louis, *62*:1045, 1953.
71. H. T. Weathersby, Anat. Rec., *124*:379, 1956, abstract.
72. D. A. Jdanov, Acta anat., *37*:20, 1959. P. Kinnaert, J. Anat., Lond., *115*:45, 1973.
73. F. W. Jones, J. Anat., Lond., *46*:41, 1912.
74. V. Fontes, C. R. Ass. Anat., *42*:518, 1956.
75. B. S. Nat, J. Anat., Lond., *58*:268, 1924. A. J. E. Cave, J. Anat., Lond., *67*:480, 1933.
76. L. Lazorthes and A. Haumont, C. R. Ass. Anat., *39*:312, 1953.
77. E. J. M. Campbell, J. Anat., Lond., *89*:378, 1955.
78. G. R. L. Gaughran, Ann. Otol., etc., St Louis, *68*:1082, 1959, and *70*:31, 1961.
79. S. E. Whitnall, *The Study of Anatomy*, Arnold, London, 4th ed., 1939.
80. F. P. Fountain, W. L. Minear, and R. D. Allison, Arch. phys. Med., *47*:665, 1966.

MOUTH, TONGUE, AND TEETH 61

ORAL CAVITY

The oral cavity (L. *os, oris,* mouth) is lined by a mucous membrane, the epithelium of which is stratified squamous in type. Although the epithelium is keratinizing, cornified cells are found only on the dorsum of the tongue, the hard palate, and the gums. Oral smears may be used for chromosomal sex determination. The temperature is commonly taken by inserting a clinical thermometer into the oral cavity. The normal mean temperature is 37 degrees C. or 98.6 degrees F. The normal range is approximately from 36 to 37.5 degrees C. Everyone should be acquainted with artificial respiration by mouth-to-mouth or mouth-to-nose, with the victim's neck fully extended and the chin pressed upward.[1] The cavity of the mouth comprises a smaller external portion, the vestibule, and a larger internal portion, the oral cavity proper.

VESTIBULE

The vestibule is the cleft between the lips and cheeks externally and the teeth and gums internally. The roof and floor of the vestibule are formed by the reflection of the mucous membrane from the lips and cheeks to the gums. The vestibule receives the minute openings of the labial glands. **The parotid duct opens into the vestibule opposite the upper second molar tooth. When the teeth are in contact, the vestibule communicates with the mouth cavity proper only by a variable gap between the last molars and the ramus of the mandible.**

ORAL CAVITY PROPER

The oral cavity proper (figs. 61–1 and 61–2) is bounded in front and on each side by the alveolar arches, the teeth, and the gums. It communicates behind with the oropharynx by means of an opening termed the faucial (oropharyngeal) isthmus, which is demarcated on each side by the palatoglossal arches. The roof of the oral cavity is the palate. The floor is largely occupied by the tongue, which is supported by muscles and other soft tissues in the interval between the halves of the body of the mandible. These soft structures are collectively termed the *floor of the mouth.* They include particularly the two mylohyoid muscles, which form the *diaphragma oris.* The lower surface of the tongue is connected to the floor of the mouth by a median fold of mucous membrane, termed the *frenulum of the tongue* (fig. 61–1*B*). **The lower end of the frenulum presents an elevation on each side, the *sublingual papilla,* on which the submandibular duct opens.** The sublingual gland produces an elevation, the *sublingual fold,* in the mucous membrane on each side of the frenulum. Most of the sublingual ducts open on the sublingual fold.

LIPS AND CHEEKS

The lips are two mobile, musculofibrous folds that bound the opening of the mouth. They meet laterally at the angle of the mouth. The median part of the upper lip is marked externally by a shallow groove, the *philtrum.* The internal aspect of each lip is connected to the corresponding gum by a median fold of mucous membrane, the *frenulum of the lip.* The lips are covered with skin, contain the orbicularis oris muscles and labial glands, and are lined by mucous membrane. **Harelip is most commonly found in the upper lip and in a paramedian position. It is frequently associated with cleft palate.**

The cheeks are similar in structure, and contain the buccinator and the buccal glands. The *buccal pad of fat* overlies the buccinator and masseter muscles. The parotid duct pierces the pad of fat and the buccinator, and opens opposite the upper second molar tooth. The junction between the cheeks and the lips may be marked externally on each side by a *nasolabial groove,* which extends downward and laterally from the nose to the angle of the mouth.

PALATE

The palate constitutes the roof of the mouth and the floor of the nasal cavity. It extends backward so as to form a partial division between the oral and nasal portions of the pharynx (fig. 63–4). The palate is arched both transversely and anteroposteriorly. **It consists of two parts: an anterior two-thirds, the hard palate, and a posterior third, the soft palate.**

Hard Palate

The hard palate is at the level of the axis in the adult, higher in the infant. It is characterized by an osseous framework, the *bony palate,* formed by the palatine processes of the maxillae in front and the horizontal plates of the palatine behind (fig. 62–2). The bony palate is covered superiorly by the mucous membrane of the nasal cavity and inferiorly by the mucoperiosteum of the hard palate. The mucoperiosteum contains blood vessels and nerves and, posteriorly, a large number of *palatine glands,* mucous in type. Its epithelium is keratinizing, stratified squamous in type, and highly sensitive to touch. The mucoperiosteum presents a median *raphe* that ends in front in the *incisive papilla.* Several *transverse palatine folds,* or *rugae,*[2] extend laterally and aid in gripping food against the tongue during mastication. A median bony protuberance, the *torus palatinus,* is sometimes present on the lower aspect of the hard palate.

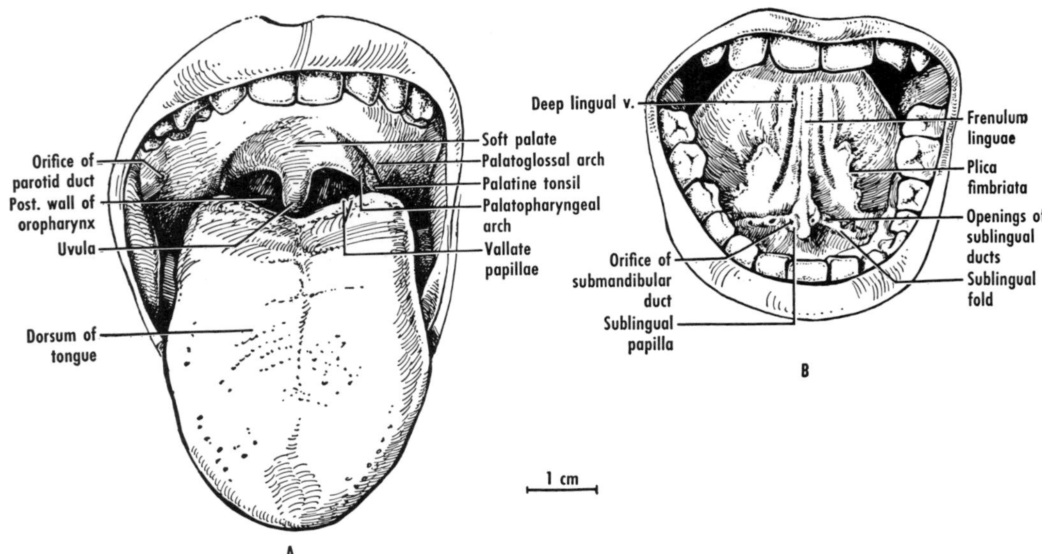

Figure 61–1 View of open mouth. *A*, with tongue protruded. *B*, with tip of tongue raised. These drawings serve as key diagrams to figure 61–2, on facing page.

Soft Palate

The soft palate, or *velum palatinum*, is a movable, fibromuscular fold, suspended from the posterior edge of the hard palate. It forms a partial partition between the nasopharynx above and the oropharynx below. It functions in closing the pharyngeal isthmus in swallowing and during speech. It is covered chiefly by stratified squamous epithelium, and numerous palatine glands are present on its anterior aspect.[3] Lymphatic follicles may also be present.[4] Taste buds are found posteriorly.[5] The free, inferior border of the soft palate presents, in the median plane, a projection of variable length, the *uvula* (fig. 61–1A). **The soft palate is continuous laterally with two folds, termed the palatoglossal and palatopharyngeal arches.**

The soft palate and palatopharyngeal folds may be considered as intervening between the nasopharynx (regarded as the back portion of the nasal cavity) and the oropharynx. These two cavities are separated by a gap, the pharyngeal isthmus, bounded in front by the posterior margin of the soft palate, laterally by a palatopharyngeal fold, and posteriorly by the pharyngeal ridge.[6] More recent studies in the living, however, have shown that the pharyngeal isthmus is situated above the pharyngeal ridge and is well above the level of the arch of the atlas during speech.[7]

Vessels and Sensory Nerves of Palate. The palate has "an extravagant arterial supply." The chief source on each side is the greater palatine artery, a branch of the descending palatine from the maxillary. The sensory nerves, branches of the pterygopalatine ganglion, include the palatine and nasopalatine nerves. The nerve fibers probably belong to the maxillary nerve.

Muscles of Soft Palate. The muscles of the soft palate are the palatoglossus, palatopharyngeus, musculus uvulae, levator veli palatini, and tensor veli palatini.

The *palatoglossus* occupies the palatoglossal fold. It arises from the palatine aponeurosis and is inserted into the side of the tongue.

The *palatopharyngeus* occupies the palatopharyngeal fold. It arises from the posterior border of the bony palate and from the palatine aponeurosis. In the soft palate it is arranged in two strands, medial and lateral, separated by the levator veli palatini.[8] These strands unite together, and the palatopharyngeus is then inserted into the posterior border of the thyroid cartilage *(palatothyroideus)* and into the side of the pharynx and the esophagus *(palatopharyngeus proper).*

The *musculus uvulae* arises from the posterior nasal spine of the palatine bones and from the palatine aponeurosis. It is inserted into the mucous membrane of the uvula.

The *levator veli palatini (levator palati)* arises[9] from the under surface of the petrous portion of the temporal bone, in

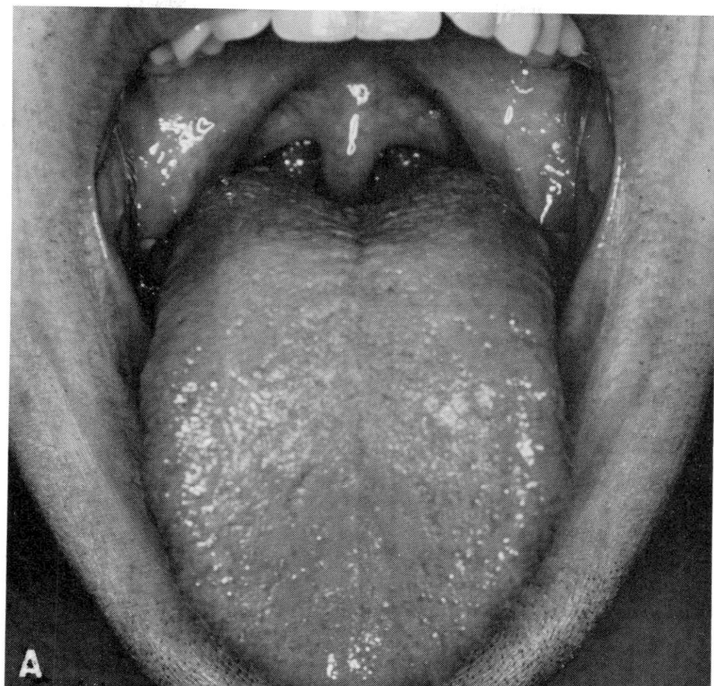

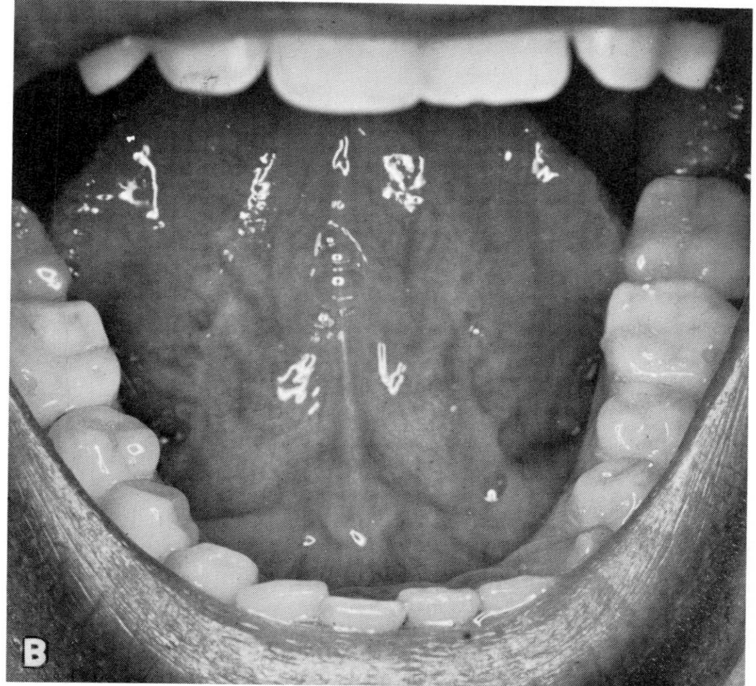

Figure 61–2 Oral cavity of a young man. In *A*, the tongue is protruded. In *B*, the tongue is elevated and its tip lies behind the upper incisors. For identification of the various features, see figure 61–1. Stereoscopic color transparencies of these views can be found in D. L. Bassett, *A Stereoscopic Atlas of Human Anatomy*, Sawyer's Inc., Portland, Oregon, 1954, section 2, reel 70, views 5 and 4. Courtesy of David L. Bassett, M.D., University of Washington, Seattle, Washington. Copyright 1954, Sawyer's Inc., U.S.A.

front of the carotid canal, from the carotid sheath, and from the cartilage of the auditory tube. It is inserted into the palatine aponeurosis and into the muscle of the opposite side. The levatores and the palatopharyngei form, respectively, an upper sling attached to the skull and a lower sling attached to the larynx.

The *tensor veli palatini (tensor palati)* arises from the scaphoid fossa at the root of the medial pterygoid plate, from the spine of the sphenoid bone, and from a crest between these attachments. It ends in a tendon that turns around the pterygoid hamulus of the medial pterygoid plate, passes through a gap in the origin of the buccinator, and is inserted into the palatine aponeurosis. Deeper fibers of the tensor merely connect the pterygoid hamulus with the cartilage and membranous wall of the auditory tube. The *palatine aponeurosis* is an expansion in the anterior two-thirds of the soft palate to which all the muscles of the palate are attached. It is formed by the expanded tendon of the tensor and is anchored to the posterior border of the hard palate.

Innervation of Muscles of Soft Palate. **With the exception of the tensor, all the muscles of the soft palate are generally said to be supplied through the pharyngeal plexus by fibers derived from the internal branch of the accessory nerve.** Other contributions mentioned by some workers include cranial nerves7,[10] 9, and 12. The tensor appears to be supplied largely by the mandibular nerve (perhaps through the branch to the medial pterygoid and by way of the otic ganglion).

Actions of Muscles of Soft Palate. The palatoglossi approximate the palatoglossal folds, thereby shutting off the oral from the pharyngeal cavity. The palatopharyngei approximate the palatopharyngeal folds, thereby separating the oropharynx from the nasopharynx. The musculus uvulae raises the uvula. The levator veli palatini elevates the soft palate and draws it backward, as in phonation and in sucking liquids. The levator is not only the principal mover of the soft palate but (because of its close relation to the auditory tube) is also the key elevator of the pharynx.[11] The tensor veli palatini tightens the soft palate as in blowing, and perhaps is responsible for opening the auditory tube.[12] It has little to do with speech but is active during swallowing.[13]

TONGUE

GENERAL FEATURES OF TONGUE
(figs. 61–1 to 61–3)

The tongue (L. *lingua*, Gk. *glossa*) is a muscular organ in the floor of the mouth. It is attached by muscles to the hyoid bone, mandible, styloid processes, and pharynx. The tongue is important in taste, mastication, swallowing, and speech. It is composed chiefly of skeletal muscle and is partly covered by mucous membrane. The tongue presents (1) a tip and a margin, (2) a dorsum, (3) an inferior surface, and (4) a root (fig. 61–3, *B* and *C*).

1. Tip. The tip, or apex, of the tongue usually rests against the incisor teeth.

Margin. The margin of the tongue is related on each side to the gums and the teeth.

2. Dorsum. The dorsum of the tongue (fig. 61–3*A*) is situated partly in the oral cavity and partly in the oropharynx. It is convex in shape and is related to the palate. It is characterized by a **V**-shaped groove, the *sulcus terminalis*, which runs laterally and forward on each side from a small pit, the foramen cecum. **The sulcus terminalis may conveniently be taken as the boundary between *(a)* the oral part, or anterior two-thirds, and *(b)* the pharyngeal part, or posterior third, of the tongue. The *foramen cecum*, not infrequently absent,[14] indicates the site of origin of the thyroglossal duct in the embryo.**

ORAL PART. The dorsum of the oral part of the tongue may present a shallow *median groove*. The mucous membrane is generally moist and pink, and appears velvety owing to the presence of numerous minute papillae. "Furring" or "coating" of the tongue bears no relation to digestive disturbances, and is usually due to smok-

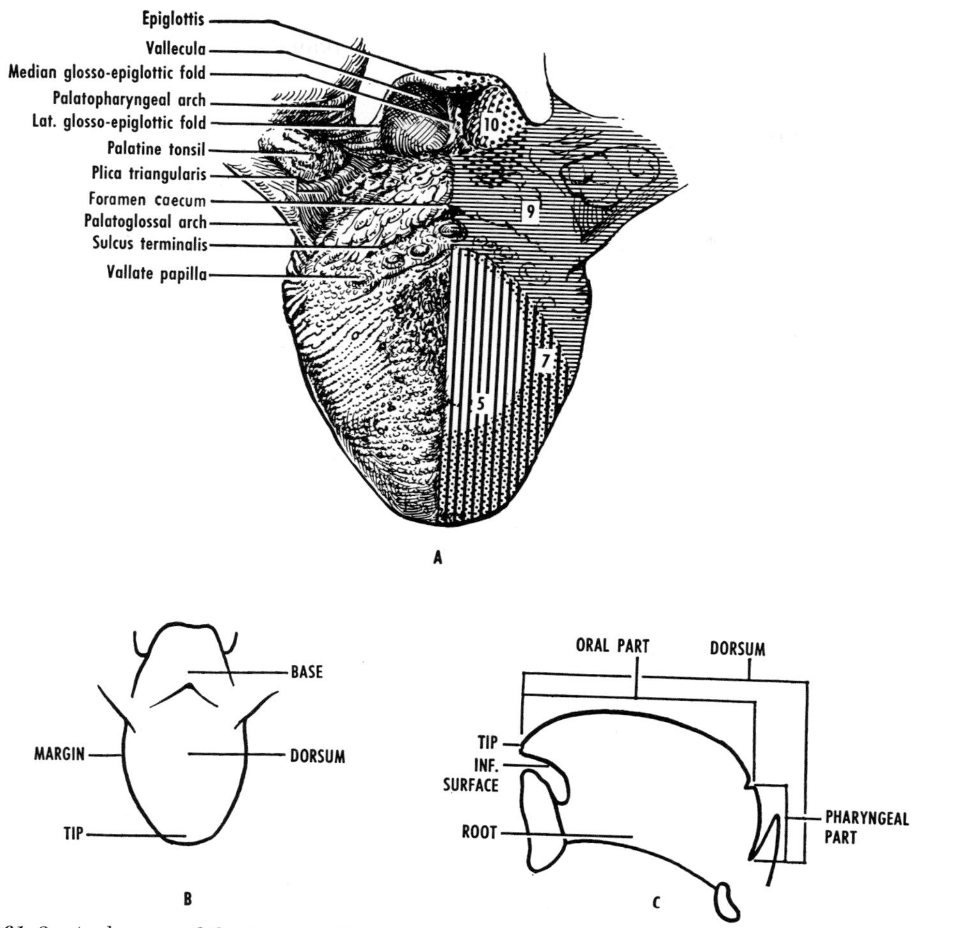

Figure 61–3 A, dorsum of the tongue, showing the sensory innervation on one side. The numbers refer to cranial nerves. *B* and *C*, diagrams showing the parts of the tongue.

ing, respiratory infection, fever, or oral infection.[15]

The *lingual papillae* are projections of the lamina propria or corium of the mucous membrane, covered with epithelium. Four chief types are found: *(a)* The *filiform papillae*, the narrowest and most numerous, are conical projections with sharply pointed tips. They are located abundantly on the dorsum of the oral part of the tongue. *(b)* The *fungiform papillae* have each a red, rounded head and a narrower base, and they usually contain taste buds. They are found mostly at the apex and the margin of the tongue. *(c)* The *vallate* (known formerly as *circumvallate) papillae* are the largest. They vary from 3 to 14 in number (perhaps depending on hereditary factors[16]), and are arranged in a **V**-shaped row in front of the sulcus terminalis. Each vallate papilla is shaped like a round castle, surrounded by a deep moat, which is bounded on its periphery by a wall, or "vallum." The ducts of serous glands open into the moat, and taste buds are found in the papilla and its vallum. The taste buds of the vallate papillae atrophy in old age,[17] but apparently there is little or no decrease in taste sensitivity with age.[18] *(d)* The *folia*, or *foliate papillae*, of the tongue consist of inconstant grooves and ridges near the posterior part of the margin.

Pharyngeal Part. The dorsum of the pharyngeal part of the tongue faces posteriorly, whereas that of the oral part faces superiorly. **The *base* of the tongue constitutes the anterior wall of the oropharynx and can be inspected only by the use of a**

mirror or by downward pressure on the tongue with a spatula. The mucous membrane over the base is devoid of visible papillae, contains numerous serous glands, and is made uneven by the presence of lymphatic follicles in the underlying submucosa. These follicles are collectively termed the *lingual tonsil*. The submucosa also contains mucous glands. The mucous membrane is continuous with that covering the palatine tonsils and the pharynx. Posteriorly, it is reflected onto the front of the epiglottis (as the *median glosso-epiglottic fold*) and onto the lateral wall of the pharynx (as the *lateral glosso-epiglottic*, or the *pharyngo-epiglottic, fold*). **The space on each side of the median glosso-epiglottic fold is termed the *epiglottic vallecula*.**

3. Inferior Surface. The inferior surface of the tongue (figs. 61–1*B* and 61–2*B*) is situated in the oral cavity only. It is thin, smooth, devoid of papillae, and purplish in color. It is connected to the floor of the mouth by a median fold of mucous membrane, the *frenulum of the tongue*. A short frenulum gives rise to the condition of tongue-tie, but is rarely a major factor in defective articulation. The profunda linguae vein can be seen through the mucous membrane on each side of the frenulum. A fringed fold of mucous membrane, the *plica fimbriata*, or *fimbriated fold*, lies on the lateral side of the vein. The *anterior lingual glands* are embedded in the musculature of the tongue on each side, close to the inferior surface and a little behind the apex. They are mixed in type, that is, both serous and mucous, and their minute ducts open on the inferior surface of the tongue.

4. Root. The root of the tongue is the part that rests on the floor of the mouth (the geniohyoid and mylohyoid muscles). It is attached by muscles to the mandible and the hyoid bone. The term "root" of the tongue, however, is sometimes used for the pharyngeal part of the organ, the oral part being called the "body" of the tongue. **The nerves, vessels, and extrinsic muscles enter or leave the tongue through its root, which is not covered by mucous membrane.**

MUSCLES OF TONGUE[19]

The muscles of which the tongue is largely composed comprise fibers peculiar to itself, the intrinsic muscles, and also fibers that arise from nearby parts, the extrinsic muscles. All the muscles of the tongue are bilateral, those of one side being partially separated from those of the opposite side by a median septum, which is not a fibrous dividing wall but a complicated linkage of the transverse muscles.[20]

Intrinsic Muscles

The intrinsic muscles of the tongue are arranged in several planes. They are generally classified as the *superior* and *inferior longitudinal*, the *transverse*, and the *vertical*.

Extrinsic Muscles

The extrinsic muscles of the tongue (fig. 61–4) are the genioglossus, hyoglossus, chondroglossus, styloglossus, and palatoglossus.

Genioglossus. The genioglossus is a fan-shaped muscle placed vertically and in contact with its fellow medially (fig. 59–1*C*). It constitutes the bulk of the posterior part of the tongue. It arises from the superior genial tubercle (and adjacent area[21]) behind the symphysis of the mandible. It is inserted into the inferior aspect of the tongue and the front of the body of the hyoid bone.

Hyoglossus. The hyoglossus is a flat, quadrilateral muscle that is largely concealed by the mylohyoid. It arises from the greater horn and the body of the hyoid bone. It passes upward and forward to be inserted into the side and the inferior aspect of the tongue. **The glossopharyngeal nerve, the stylohyoid ligament, and the lingual artery (second part) pass deep to the posterior border of the hyoglossus** (fig. 59–3).

Chondroglossus. The chondroglossus is a variable slip between the hyoid bone and the dorsum of the tongue.

Styloglossus. The styloglossus arises from the front of the styloid process and from the stylomandibular ligament. It is inserted into the side and the inferior aspect of the tongue.

Palatoglossus. The palatoglossus has been described with the muscles of the soft palate (p. 718).

Innervation of Muscles of Tongue

All the muscles of the tongue (except the palatoglossus, p. 720) are supplied by the

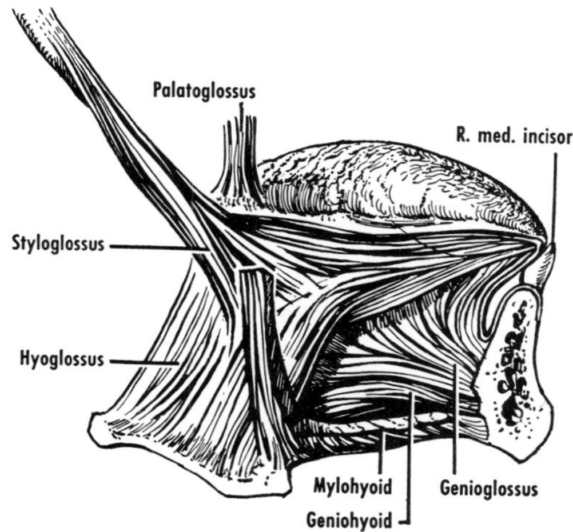

Figure 61–4 Extrinsic muscles of the tongue. Right lateral aspect. Most of the right half of the mandible and of the mylohyoid muscle has been removed.

hypoglossal nerve (p. 702). Proprioceptive afferents from the tongue probably travel in the hypoglossal rather than in the lingual nerve.[22]

Actions of Muscles of Tongue

The form of the tongue depends on its intrinsic and extrinsic muscles. The position of the tongue depends on its extrinsic muscles and also on the muscles attached to the hyoid bone.

The genioglossus is mainly a depressor of the tongue. Its posterior part pulls the tongue forward, that is, protrudes it. The hyoglossus and the styloglossus retract the tongue.

The attachment of the genioglossi to the mandible prevents the tongue from falling backward and obstructing respiration. Anesthetists keep the tongue forward by pulling the mandible forward.

BLOOD SUPPLY OF TONGUE

The main artery to the tongue is the lingual (fig. 59–5), a branch of the external carotid. The branches that supply the tongue are chiefly the dorsales linguae branches (to the pharyngeal part) and the arteria profunda linguae (p. 677).

The tongue is drained by (1) *lingual veins,* which act as venae comitantes for the lingual artery and which receive several *dorsales linguae veins;* and (2) the *vena profunda linguae,* or *ranine vein,* which runs backward under cover of the mucous membrane at the side of the frenulum (where it can be seen *in vivo*) and, after crossing the lateral surface of the hyoglossus, unites with the *sublingual vein* (from the sublingual salivary gland) to form the *vena comitans nervi hypoglossi.* This last ends in the facial, lingual, or internal jugular. All these veins terminate, directly or indirectly, in the internal jugular vein.

LYMPHATIC DRAINAGE OF TONGUE

The lymphatic drainage[23] is important because of the early spread of carcinoma of the tongue. The drainage is to the submental, submandibular, and deep cervical (including the jugulodigastric and jugulo-omohyoid) nodes. The arrangement is shown in fig. 61–5. Extensive communications occur across the median plane.

SENSORY INNERVATION OF TONGUE
(fig. 61–3A)

The anterior two-thirds of the tongue is supplied by (1) the lingual nerve (of the mandibular) for general sensation, and (2) the

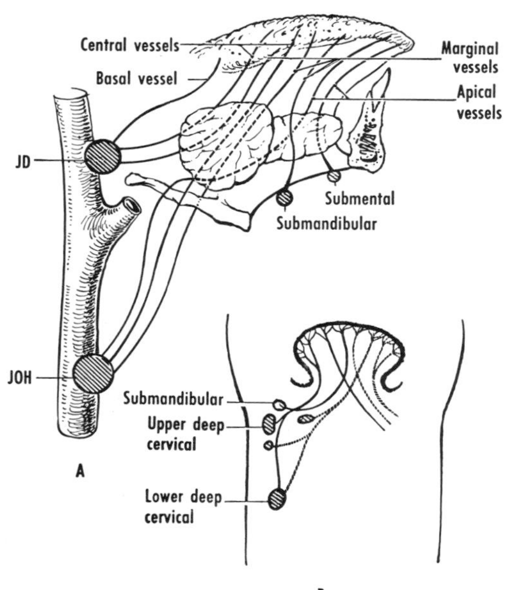

Figure 61–5 Lymphatic drainage of the tongue. *A,* right lateral aspect. The submandibular and sublingual salivary glands do not have labels. The cross-hatched circles indicate the sites of groups of nodes. *JD,* jugulodigastric; *JOH,* jugulo-omohyoid. *B,* schematic coronal section. *A* is from Rouvière. *B* is from Jamieson and Dobson.

chorda tympani (a branch of the facial that runs in the lingual nerve) for taste.

The posterior third of the tongue and the vallate papillae are supplied by the lingual branch of the glossopharyngeal nerve for both general sensation and taste. Other contributions come from the lingual branch of the facial nerve (taste), and, near the epiglottis, from the internal laryngeal branch of the vagus (general sensation and taste). Hence **the cranial nerves concerned with taste are nerves 7, 9, and 10.**

It has been claimed that all impulses concerned with taste from the anterior part of the tongue pass from the lingual nerve to the chorda tympani and thence to the nervus intermedius of the facial nerve.[24] Certain workers believe that, in some people, taste fibers leave the chorda tympani, pass through the otic ganglion and the greater petrosal nerve, and thereby reach the nervus intermedius.[25]

TEETH

The study of the teeth, strictly speaking, forms the subject of odontology (Gk. *odous, odontos*), whereas the specialty concerned with the diagnosis and treatment of diseases of the teeth and associated structures is dentistry (L. *dens, dentis,* tooth). For the many details necessary for the study of dentistry, including a description of individual teeth, special books are listed at the end of this section.

Functions of Teeth

The chief functions of the teeth are (1) to "incise and reduce food material during mastication," and (2) to help to "sustain themselves in the dental arches by assisting in the development and protection of the tissues that support them" (Wheeler).

STRUCTURE OF TEETH
(fig. 61–6)

Each tooth is composed of a specialized connective tissue, the pulp, covered by three calcified tissues: dentin(e), enamel, and cement(um). Localized disintegration of one or more of the dental tissues is termed caries. The dental calculus, or "tartar," frequently found on teeth is a layer of calcium salts derived from the saliva.

The *periodontium (periodontal membrane; periodontal ligament)* connects the cementum of a tooth to the alveolar bone, thereby establishing a fibrous joint between the tooth and its socket. The term periodontium is sometimes used in a more general sense to include the cementum, periodontal ligament, gums, and alveolar bone. The periodontium may be regarded as a modified alveolar periosteum, but it has more of the qualities of a ligament than of a membrane.

The *gums (gingivae)* are composed of dense fibrous tissue covered by oral mucous membrane (including keratinizing stratified squamous epithelium).

PARTS OF A TOOTH
(fig. 61–6)

The **anatomical crown** is the part of a tooth that is covered by enamel, whereas the **clinical crown** is the part that projects into the oral cavity. More of the crown becomes exposed with increasing age.

The **root** of a tooth is the part that is covered by cement. The **neck** is the part of the root that is adjacent to the crown. Some teeth (e.g., molars) have more than one root. The teeth are supported in that part of the jaw known as the *alveolar process.* Each tooth lies in a bony socket, or *alveolus.*

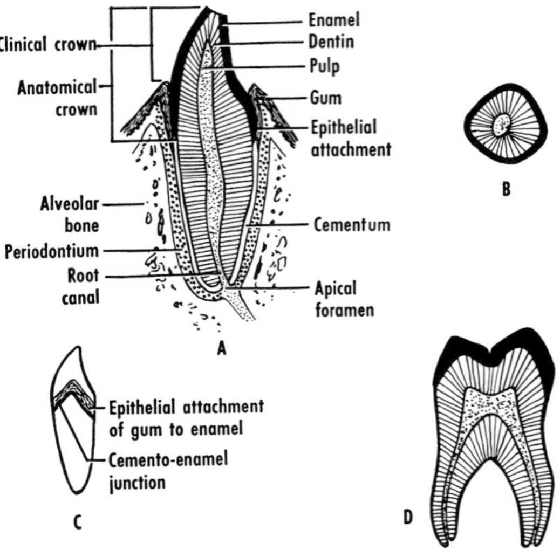

Figure 61–6 *A,* schematic drawing of a longitudinal section of a permanent lower incisor and its supporting tissues. *B,* cross-section of the crown of an incisor, showing enamel, dentin, and pulp. *C,* side view (mesial aspect) of an incisor, showing the area of the epithelial attachment and the line of the cemento-enamel junction. *D,* longitudinal section of a molar, showing the bifurcation of the pulp cavity.

Each tooth possesses a cavity occupied by pulp. The **pulp cavity** comprises a *pulp chamber* in the crown and one or more *root canals* in the root(s). Each root canal opens by one or more *apical foramina* at the tip of the root. The nerves, and blood and lymphatic vessels that supply the pulp enter or leave a tooth through its apical foramen.

TYPES OF TEETH

The teeth are classified as incisors, canines, premolars, and molars.

Incisors. The incisors incise, that is, cut the food by means of their cutting edges. The lingual surface of the crown is triangular and the apex of the triangle, directed toward the root, usually presents an elevation termed a *cingulum.* The two incisors on one side of each jaw are distinguished as (1) *medial (central),* and (2) *lateral.* The four upper incisors are carried by the premaxillary portion of the maxilla. Supernumerary teeth are sometimes found between or behind the upper incisors. The upper lateral incisors are extremely variable and often appear reduced in size. The lower incisors may be crowded.

Canines. The canines are so named because they are prominent in dogs. They are sometimes referred to as "cuspids," or "eye-teeth." The canines are long teeth, each of which presents a prominent cusp on its crown. They are usually the last deciduous teeth lost. Like the incisors, they assist in cutting the food. The canines are important also in maintaining natural facial expression.

Premolars. The premolars, sometimes termed "bicuspids," are those teeth that replace the deciduous molars. Each usually presents two *tubercles,* or *cusps,* on its crown. The premolars assist in crushing the food, but their crowns are not as complex as those of the molars.

Molars. The molars (L. *molar(es),* grinders) crush and grind the food. They possess three to five *tubercles* or *cusps* on their crowns, but the cusps become worn away with use, so that their enamel is lost and the underlying substance (dentin) may become exposed. Each upper molar generally has three roots, each lower molar two. **The roots of the upper molars are closely related to the floor of the maxillary sinus. Hence pulpal infection may cause sinusitis, or sinusitis may cause pain that is referred to the teeth.** The permanent molars have no deciduous predecessors. The first molars are usually the largest teeth. The third molar is known as the "wisdom tooth" *(dens serotinus).* The third molars are highly variable in form and position, and they may be lacking or impacted.

DENTAL TERMINOLOGY
(fig. 61–7)

Because of the curvature of the dental arches, a specialized system of nomenclature is employed to describe the surfaces of the teeth. The term "anterior teeth" is used for the incisors and canines; the term "posterior teeth" is used for the premolars and molars. **Most of the teeth in an adult are "successional," that is, they have succeeded a corresponding number of milk teeth. The permanent molars, however, are "accessional," that is, they have been added behind the milk teeth during development.**

(1) Mesial and Distal

The **mesial surface** is medial in the anterior teeth, but anterior in the posterior teeth.

The **distal surface** is lateral in the an-

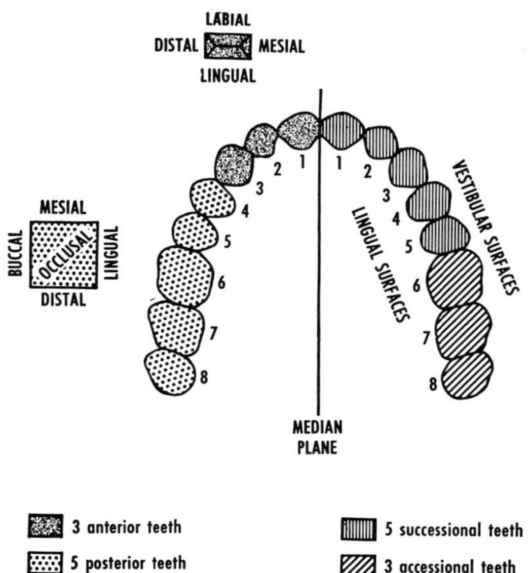

Figure 61–7 Dental terminology. Scheme of upper permanent teeth, viewed from below.

terior teeth, but posterior in the posterior teeth.

In the case of two adjacent teeth in a dental arch, the adjacent mesial and distal surfaces may be called *contiguous, proximate,* or *contact surfaces.* In each dental arch, the portions of adjacent clinical crowns that actually touch each other are called *contact areas.* The space between two contiguous teeth and immediately surrounding their contact area is termed an *embrasure.*

(2) Vestibular and Lingual

The **vestibular surface,** that is, the surface facing the vestibule of the oral cavity, is *labial* (facing the lips) in the anterior teeth, but *buccal* (facing the cheek) in the posterior teeth.

The **lingual surface** is that facing the tongue.

(3) Masticatory or Occlusal

The **masticatory** or **occlusal surface** of a tooth is the surface that comes into contact with its opposite number in the other jaw when the jaws are closed. In the anterior teeth the occlusal surfaces are merely narrow borders.

PRIMARY OR DECIDUOUS DENTITION
(figs. 61–8 and 61–9)

No functioning teeth are present in the oral cavity at birth. The *primary* or *deciduous teeth* ("milk teeth") appear in the oral cavity between one-half and two and one-half years. The first teeth to erupt are the lower medial incisors at about six months. The lower teeth frequently precede the upper in eruption.

The deciduous teeth are 20 in number, that is, five in each quadrant: two incisors, one canine, and two molars (I.2; C.1; M.2). The teeth in each quadrant can conveniently be given letters from A mesially to E distally:

E	D	C	B	A	A	B	C	D	E
E	D	C	B	A	A	B	C	D	E

By means of this scheme, the right lower lateral incisor, for example, can be indicated as B|. Similarly, the first left upper molar can be designated as |D.

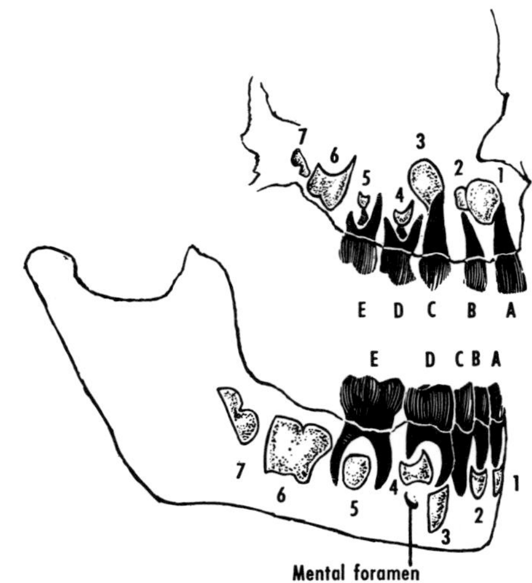

Figure 61–8 Right lateral aspect of maxilla and mandible of a five year old child, showing position of deciduous and of permanent teeth.

In an alternative method, the teeth are merely given the letters A (E|) to J (|E) and then K (|E) to T (E|).

The deciduous teeth are smaller and whiter than the permanent teeth.

The term "eruption" is used clinically for the appearance of a tooth in the oral cavity. The usual order of eruption is A, B, D, C, E. Movement of a tooth, however, begins at the time of root formation and continues through the life span of the

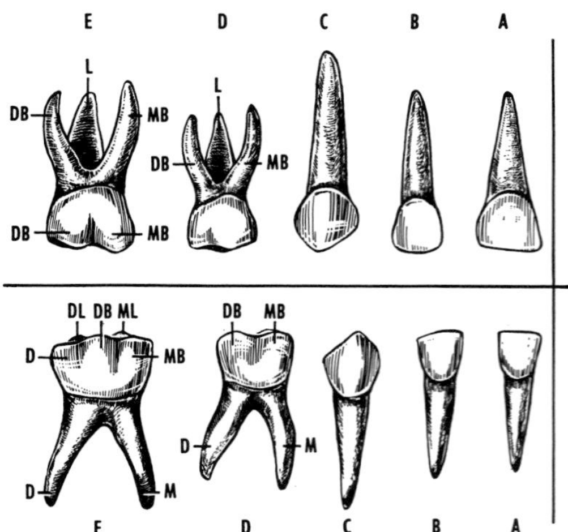

Figure 61–9 Vestibular surfaces of right deciduous teeth. **D,** distal; **DB,** distobuccal; **DL,** distolingual; **L,** lingual; **M,** mesial; **MB,** mesiobuccal; **ML,** mesiolingual. Based on Wheeler.

tooth. The emergence through the gum is merely an incident in a continuous and complicated process.

By about twelve years of age all of the deciduous teeth have been shed, mainly due to the resorption of their roots, associated with the eruption of the permanent teeth.

Various data concerning the deciduous teeth are summarized in table 61–1.

TABLE 61–1 Dental Data

A. *DECIDUOUS TEETH*

Calcification begins during fourth month of intra-uterine life (in the sequence A, D, B, C, E).

Extent of calcification at birth:

cusps	occlusal surf.	⅓ crown	⅔ crown	⅔ crown
similar to upper teeth				

Enamel of crowns is completed during first year.

Teeth erupt into mouth cavity (median age in years):

2½	1½	1¾	1	½
similar to upper teeth				

Roots are completed about 1 to 1½ years after eruption.

Resorption of roots begins about 5 years after eruption.

Resorption of roots ends, and crowns are shed, between 5 and 15 years.

Usual number of cusps of deciduous teeth:

4–5	2–4	1	—	—
5	4	1	—	—

Usual number of roots of deciduous teeth:

3	3	1	1	1
2	2	1	1	1

B. *PERMANENT TEETH*

Calcification begins (in years; B = birth)

7–9	3	B	2	2	½	1	⅓
8–10	3	B	2	2	½	⅓	⅓

Enamel of crowns completed (in years):

12–16	7–8	3	6–7	5–6	6–7	4–5	4–5
similar to upper teeth							

Teeth erupt into mouth cavity (median age in years):

*	12	7	12	11	11	8	7
*	12	7	11	11	10	7	6

*The highly variable third molar may erupt from 17 years onward, or not at all.

Roots are completed about 2 to 3 years after eruption.

Usual number of cusps:

3	4	4–5	2	2	1	—	—
4–5	4–5	5	2–3	2	1	—	—

Usual number of roots:

1–3	3	3	1–2	1–2	1	1	1
1–2	2	2	1	1	1	1	1

Frequent pattern of innervation of teeth:

Post. sup. alveolar					Ant. sup. alveolar			
8	7	6	5	4	3	2	1	
8	7	6	5	4	3	2	1	1
Inf. alveolar					Incisive br. of inf. alveolar			

Additional innervation of gums, alveolar bone, and periodontium:

(*a*) Vestibular aspect—

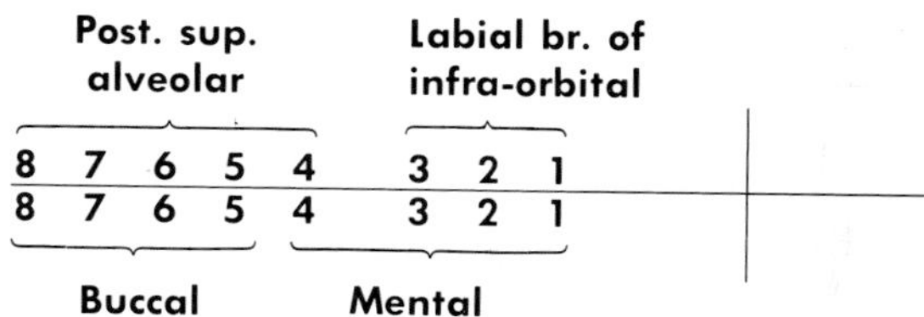

(*b*) Lingual aspect—

Greater palatine					Nasopalatine		
8	7	6	5	4	3	2	1
8	7	6	5	4	3	2	1
Lingual							

PERMANENT DENTITION
(figs. 61–10 and 61–11)

The *"permanent" teeth* begin to appear in the oral cavity at about six years and have replaced the deciduous teeth by about twelve years. The permanent teeth are 32 in number, that is, eight in each quadrant: two incisors, one canine, two premolars, and three molars (I.2; C.1; P.2; M.3). The teeth in each quadrant can conveniently be numbered from 1 mesially to 8 distally:

8 7 6 5 4 3 2 1	**1 2 3 4 5 6 7 8**
8 7 6 5 4 3 2 1	**1 2 3 4 5 6 7 8**

By means of this scheme, the right lower first molar, for example, can be indicated as $\overline{6}|$. Similarly, the left upper canine can be designated as $|\underline{3}$.

An alternative to this "symbolic system" is the so-called "universal system," in which the teeth are merely numbered from 1 ($\underline{8}|$) to 16 ($|\underline{8}$) and then from 17 ($|\overline{8}$) to 32 ($\overline{8}|$).[26]

The first five teeth in each quadrant are "successional," that is, they are preceded by the five deciduous teeth. The latter are shed as the permanent teeth erupt. The permanent molars, however, have no deciduous predecessors and may, therefore, be called "accessional." **The first permanent tooth to erupt is the sixth tooth in the arch (first molar) at about six years, before any of the deciduous teeth have been lost. Hence it is known as the six year molar. The adjacent second molar is known, for a similar reason, as the twelve year molar.** The third molar is highly variable in its time of eruption, and is sometimes impacted, that is, its crown is directed toward the side of the second molar. The mean eruption time of a given tooth is generally several months earlier in girls. The order of eruption is variable, but the teeth appear[27] usually in the following sequence: 6 and 1; 2; 4, 3, 5, 7; 8.

In forensic medicine and in archeology the teeth and the bones are used to estimate the age of an individual.[28]

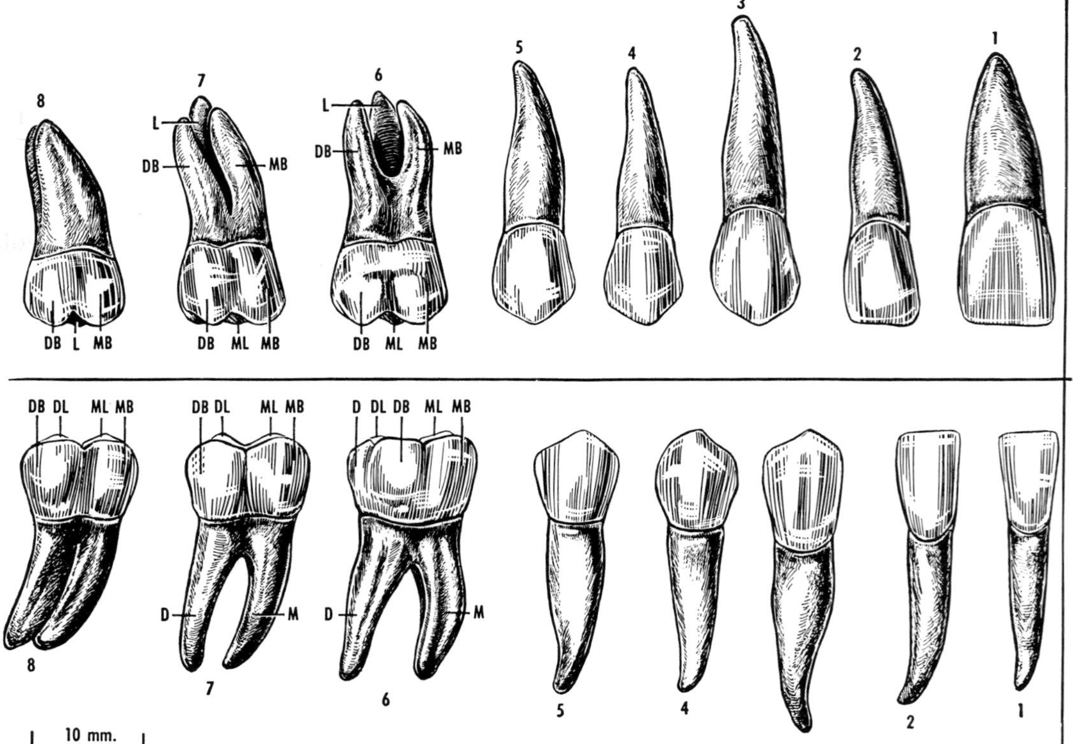

Figure 61–10 Vestibular surfaces of right permanent teeth. **B,** buccal; **D,** distal; **DB,** distobuccal; **DL,** distolingual; **L,** lingual; **M,** mesial; **MB,** mesiobuccal; **ML,** mesiolingual. Based on Wheeler.

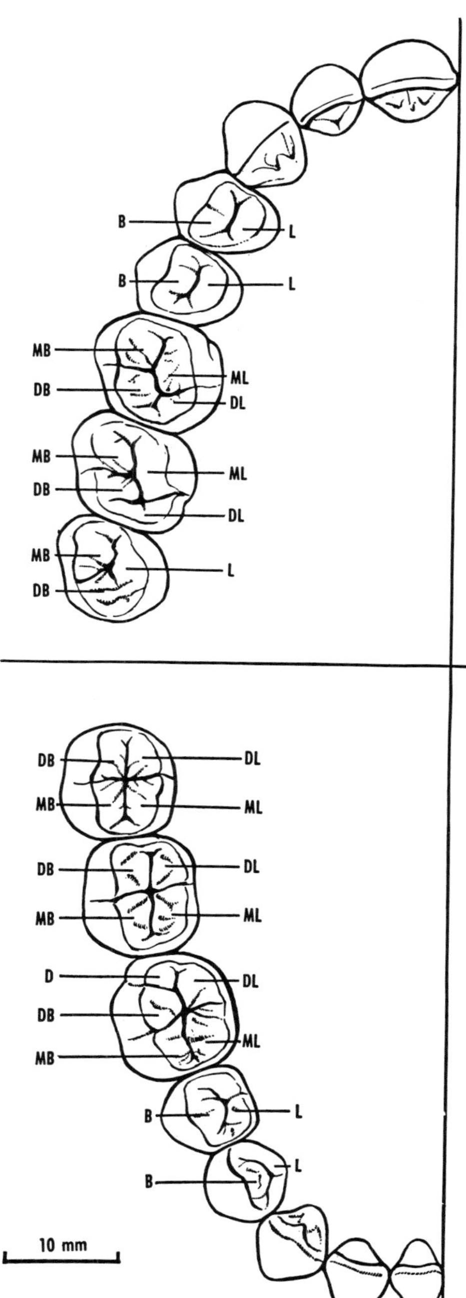

Figure 61–11 Occlusal aspect of upper (*above*) and lower (*below*) permanent teeth of right side. Abbrevations as in figure 61–10. Based on Wheeler.

Various data concerning the permanent teeth are summarized in table 61–1.[29] For the innervation,[30] see also the mandibular nerve (pp. 671 and 672).

ALIGNMENT AND OCCLUSION
(fig. 61–11)

The teeth are arranged in two *arches* or *arcades*, one in each jaw. The lower arch is movable. The way in which the teeth are arranged in an arch is termed their "alignment."

The term "occlusion" is used for any functional relation established when the upper and lower teeth come into contact with each other. "Centric occlusion" is the sum of the articular relations formed between the upper and lower teeth when the jaws are closed and the heads of the mandible are resting in the mandibular fossae of the skull. Normal occlusion depends on normal development and normal form of the dentition.[31] Abnormal occlusion is termed malocclusion. Orthodontics is that branch of dentistry that deals with the prevention and correction of malocclusion and other positional anomalies of the teeth and jaws.

CLINICAL EXAMINATION AND RADIOLOGICAL ANATOMY OF TEETH

In examinations of the teeth, the number and, as far as possible, the types of teeth present should be noted. The individual teeth should be inspected for discoloration, cavities, fillings, etching of enamel, erosion, abrasion, mobility, and occlusion. The color and form of the gums should be observed, and the lips, tongue, and oral mucosa should be examined. Anomalies of the teeth include congenital absence, the presence of supernumerary teeth, and fusion of teeth.

Apart from metallic fillings (fig. 60–8), enamel is the most radio-opaque portion of a tooth (fig. 61–12). The dentin and the cementum are of equal radio-opacity. The pulp cavity and the periodontium are both radiolucent. The alveolar bone is not as radio-opaque as the tooth. It presents a network of bony trabeculae (cancellous bone), bordered by a narrow layer of cortical bone, the *lamina dura,* which forms the tooth socket. The part of the alveolar bone between two adjacent teeth is termed an *interdental septum.*

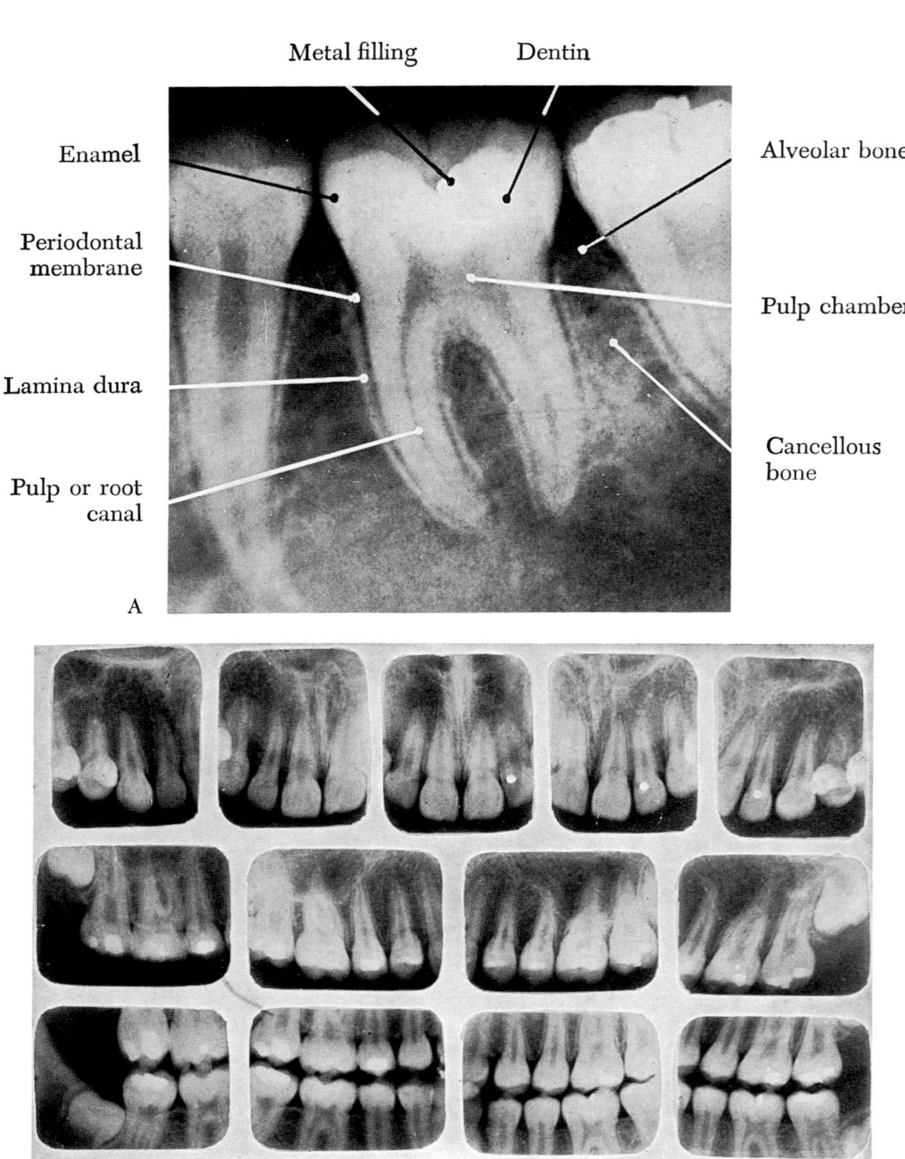

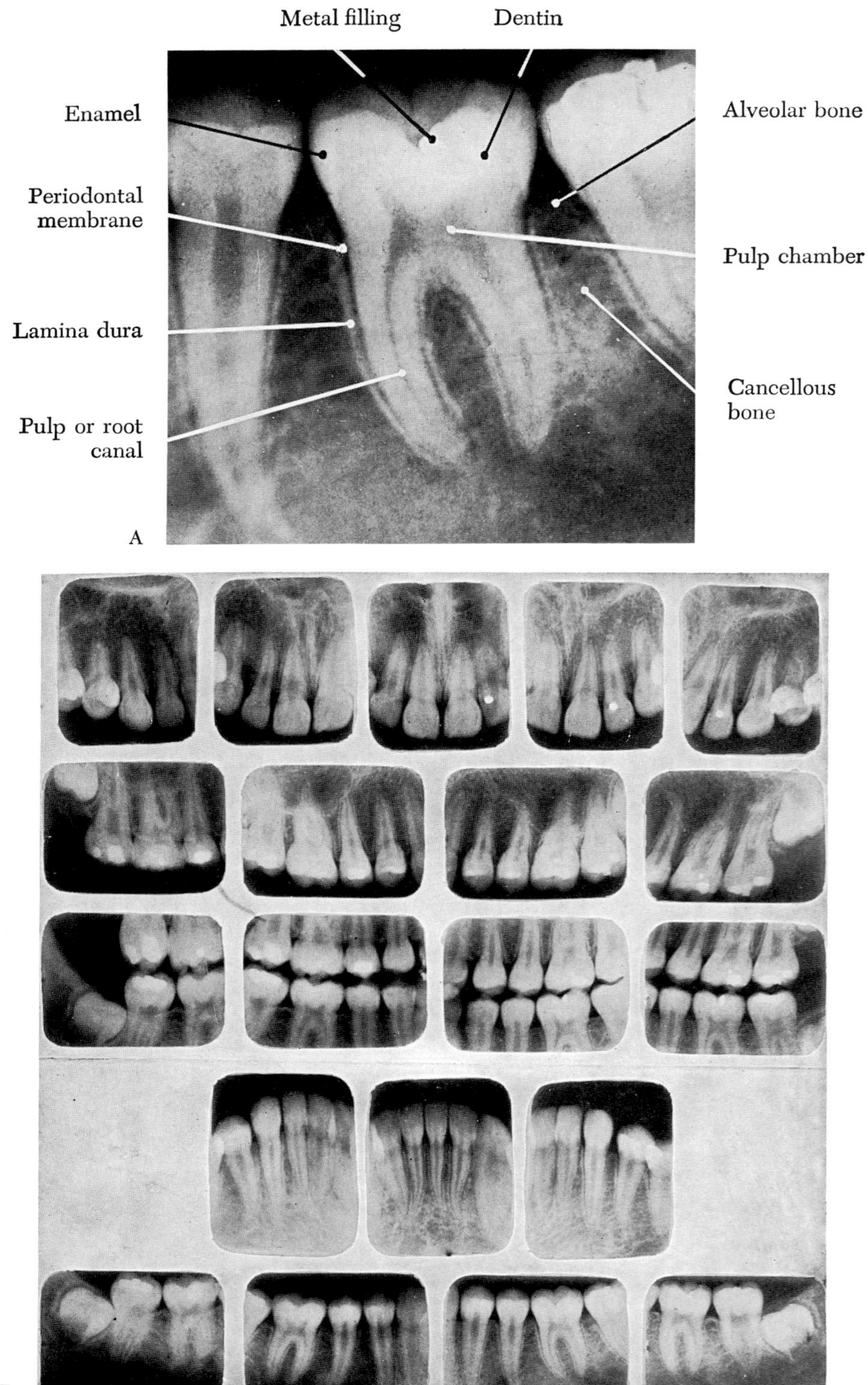

Figure 61–12 *See opposite page for legend.*

REFERENCES

1. J. O. Elam *et al.*, J. Amer. med. Ass., *172*:812, 1960.
2. L. Lysell, Acta odont. scand., *13*, suppl. 18, 1955.
3. K. Paulsen and L. Kleine, Z. Anat. EntwGesch., *139*:195, 1973.
4. M. J. Knapp, Oral Surg. oral Med. oral Path., *29*:155, 1970.
5. E. R. Lalonde and J. A. Eglitis, Anat. Rec., *140*:91, 1961.
6. F. W. Jones, J. Anat., Lond., *74*:147, 1940.
7. J. S. Calnan, Brit. J. plast. Surg., *5*:286, 1955.
8. J. Whillis, J. Anat., Lond., *65*:92, 1930.
9. R. F. Rohan and L. Turner, J. Anat., Lond., *90*:153, 1956.
10. W. Moritz, Z. Anat. EntwGesch., *109*:197, 1939. S. Podvinec, J. Laryng., *66*:452, 1952.
11. J. F. Bosma, Ann. Otol., etc., St Louis, *62*:51, 1953.
12. F. Korner, Z. Anat. EntwGesch., *111*:508, 1942.
13. W. E. M. Wardill and J. Whillis, Surg. Gynec. Obstet., *62*:836, 1936.
14. C. F. Marshall, J. Anat., Lond., *29*:234, 1895.
15. I. S. L. Loudon, Brit. med. J., *1*:18, 1956.
16. J. N. Spuhler, Cold Spr. Harb. Symp. quant. Biol., *15*:175, 1950.
17. L. B. Arey, M. J. Tremaine, and F. L. Monzingo, Anat. Rec., *64*:9, 1935.
18. E. Byrd and S. Gertman, Geriatrics, *14*:381, 1959.
19. S. Abd-el-Malek, J. Anat., Lond., *73*:201, 1939.
20. R. Dabelow, Morph. Jb., *91*:33, 1951.
21. G. A. Doran and H. Baggett, Acta anat., *83*:403, 1972.
22. A. K. Adatia and E. N. Gehring, J. Anat., Lond., *110*:215, 1971.
23. J. K. Jamieson and J. F. Dobson, Brit. J. Surg., *8*:80, 1920.
24. B. Krarup, Neurology, 9:53, 1959.
25. H. G. Schwartz and G. Weddell, Brain, *61*:99, 1938.
26. N. J. Paquette, Dent. Survey, 1960.
27. K. Koski and S. M. Garn, Amer. J. phys. Anthrop., *15*:469, 1957.
28. For reprints of several papers on the ages of calcification and eruption of teeth, see T. D. Stewart and M. Trotter (eds.), *Basic Readings on the Identification of Human Skeletons: Estimation of Age,* Wenner-Gren Foundation, New York, 1954. See also A. E. W. Miles, Proc. R. Soc. Med., *51*:1057, 1958.
29. For dates of eruption see E. M. B. Clements, E. Davies-Thomas, and K. G. Pickett, Brit. med. J., *1*:1421, 1953; R. S. Nanda, Amer. J. Orthodont., *46*:363, 1960. For calcification in the deciduous teeth see B. S. Kraus, J. Amer. dent. Ass., *59*:1128, 1959.
30. The innervation of the gums is based on D. Mongkollugsana and L. F. Edwards, J. dent. Res., *36*:516, 1957.
31. S. Friel, Int. J. Orthod., *13*:322, 1927.

GENERAL READING

Jenkins, G. N., *The Physiology of the Mouth,* Blackwell, Oxford, 3rd ed., 1966.

Manley, E. B., Brain, E. B., and Marsland, E. A., *An Atlas of Dental Histology,* Blackwell, Oxford, 2nd ed., 1955.

Orban's *Oral Histology and Embryology,* ed. by H. Sicher and S. N. Bhaskar, Mosby, St. Louis, 7th ed., 1972.

Scott, J. H., and Symons, N. B. B., *Introduction to Dental Anatomy,* Livingstone, Edinburgh, 6th ed., 1971.

Wheeler, R. C., *An Atlas of Tooth Form,* Saunders, Philadelphia, 4th ed., 1969.

Wheeler, R. C., *Dental Anatomy, Physiology and Occlusion,* Saunders, Philadelphia, 5th ed., 1974.

Figure 61–12 Teeth. *A,* enlargement from a radiogram of the right side of the mandible of an adolescent, showing the normal dental and periodontal tissues. *B,* Twenty intra-oral radiograms of the permanent teeth. The upper teeth are shown in the upper three rows, the lower teeth in the lower three rows. The teeth are being viewed as they would appear from within the oral cavity, that is, the teeth of the left side of the body are reproduced on the left side of the page. This convention is usual in dental radiography. From J. O. McCall and S. S. Wald, *Clinical Dental Roentgenology,* Saunders, Philadelphia, 4th ed., 1957; courtesy of the authors.

NOSE AND PARANASAL SINUSES

62

NOSE

The term *nose* includes the external nose, visible on the face, and the nasal cavity, which extends considerably further back. The word nose is akin to the Latin *nasus;* hence nasal means pertaining to the nose. The Greek word for nose is *rhis, rhinos,* from which is derived a series of words (e.g., rhinoceros). Thus the study of the nose and its diseases is termed rhinology.

The functions of the nose are (1) to subserve the sense of smell, (2) to provide an airway for respiration, (3) to filter, warm, and moisten the inspired air, that is, to serve for air conditioning, and (4) to cleanse itself of foreign matter which it extracts from the air.

EXTERNAL NOSE

The external nose presents a free *tip* or *apex,* and is attached to the forehead by the *root* or *bridge* of the nose. The rounded border between the apex and the root is the *dorsum* of the nose. The external nose is perforated inferiorly by the two *nostrils,* or *nares.* Each nostril is bounded medially by the nasal septum and laterally by the *ala* of the nose. The upper part of the external nose is supported by the nasal and frontal bones, and by the maxillae. The lower part has a hyaline cartilaginous framework. This consists of a *septal cartilage* that includes lateral expansions termed *lateral cartilages.*[1] A greater alar cartilage[2] (fig. 62–2*B*) is placed below each lateral cartilage, and several smaller cartilages may be found. A variable degree of fusion may be observed between the different cartilages of the nose. The muscles of the external nose have been mentioned with the face (p. 655). The chief arterial supply of the external nose is from branches of the facial and ophthalmic arteries. The cutaneous innervation is from branches of the ophthalmic and maxillary nerves.

NASAL CAVITY

RELATIONS AND OPENINGS

The nasal cavity extends from the nostrils, or nares, in front, to the choanae behind. It is related superiorly to the frontal sinus, the anterior cranial fossa, and the sphenoidal sinus and middle cranial fossa. Below, it is separated from the oral cavity by the hard palate. **Posteriorly, the nasal cavity communicates with the nasopharynx, which, in many respects, may be regarded as the back portion of the cavity** (p. 742). Laterally, it is related to the exterior in front, and, further back, to the orbit, the maxillary and ethmoidal sinuses, and the pterygopalatine and pterygoid fossae.

The *piriform aperture* of the nose is bounded above by the nasal bones, and laterally and below by the maxillae (fig. 52–3, p. 553).

The *choanae* (posterior apertures) are bounded medially by the vomer, inferiorly by the horizontal plate of the palatine bone, laterally by the medial pterygoid plate, and superiorly by the body of the sphenoid (covered by the ala of the vomer) and by the medial pterygoid plate (figs. 52–13 and 52–14, pp. 563, 564). The choanae are larger than the nostrils.

The openings that lead into or out of the nasal cavity are: the nostrils, the choanae, the openings of the maxillary, frontal, sphenoidal, and ethmoidal air sinuses, and the nasolacrimal duct. In the dried skull the sphenopalatine foramen, the incisive canal, and the foramina in the cribriform plate of the ethmoid also open into the nasal cavity, but these are covered with mucous membrane in the intact state.

The nasal cavity is divided into right and left halves (formerly called *nasal fossae*) by a median partition, the nasal septum. The term nasal cavity refers to either the entire cavity or one of its halves, depending on the context. Each half has a roof, a floor, and medial and lateral walls.

BOUNDARIES

Roof

The roof is formed, from before backward, by nasal cartilages and the following bones: nasal, frontal, cribriform plate of ethmoid, and the body of the sphenoid covered by parts of the vomer and palatine. The ethmoidal part (fig. 62–1) is more or less horizontal; the parts in front and behind slope downward. The roof is very narrow from side to side.

Floor

The floor of the nasal cavity is smooth, almost horizontal from front to back, and concave from side to side. It is wider than the roof. It is formed by the palatine process of the maxilla in front, and by the horizontal plate of the palatine bone behind. It intervenes between the nasal and oral cavities.

Medial Wall, or Nasal Septum

The medial wall (fig. 62–2*B*) is the partition between the two halves of the nasal cavity. **The *nasal septum* is formed, from before backward, by (1) the septal cartilage**

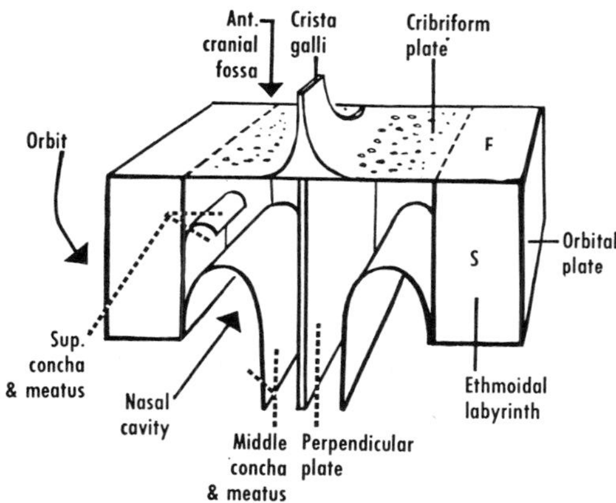

Figure 62–1 Scheme of the ethmoid bone, viewed from behind. Note the two ethmoidal labyrinths united by the cribriform plate. The perpendicular plate, which forms the upper part of the nasal septum, is set at a right angle to the horizontally placed cribriform plate. The lateral surface of each labyrinth forms a part of the medial wall of the orbit and is termed the orbital plate of the ethmoid. *F* indicates the portion of the labyrinth completed by the frontal bone; *S*, the portion completed by the sphenoid. The ethmoidal labyrinth contains ethmoidal air cells known collectively as the ethmoidal sinus. Modified from Grant's *Atlas*.

(destroyed in a dried skull), (2) the perpendicular plate of the ethmoid, and (3) the vomer. In the region of the apex of the nose, the septum is completed by skin, subcutaneous tissue, and the greater alar cartilages. (This is the *mobile* or *membranous part of the septum*, or *columna*, or *columella*.) **The septum is usually deviated to one or the other side.**

The *vomeronasal organ* is a small pouch that opens on the front and lower part of the nasal septum. It is sometimes found in the adult, and has been detected *in vivo*.[3]

Lateral Wall

The lateral wall (fig. 62–2*A*), uneven and complicated, is formed by parts of the nasal, maxilla, lacrimal, ethmoid (labyrinth and conchae), inferior nasal concha, palatine (perpendicular plate), and sphenoid (medial pterygoid plate). **The lateral wall is characterized by the medial projection of the nasal conchae and their underlying meatuses. (The plural of meatus is meatus in Latin and meatuses in English.) The highest (inconstant), superior, and middle conchae are portions of the ethmoid, whereas the inferior concha is a separate bone.**

The small space above and behind the superior concha is termed the **spheno-ethmoidal recess,** and it receives the opening of the sphenoidal sinus. Frequently a **highest concha** and **highest meatus** (fig. 62–3*A*) are present in this region.

The space under cover of the **superior concha** is the **superior meatus,** and it receives the openings of the posterior group of ethmoidal cells and, in the dried skull, the sphenopalatine foramen.

The **middle meatus** lies under cover of the **middle concha** and is continuous in front with a depression termed the *atrium*. The atrium lies above the *vestibule* (that part of the nasal cavity adjacent to the nostril) and is limited above by a ridge (the *agger nasi*). **The middle meatus receives the openings of the maxillary and frontal sinuses,** and the anterior group of ethmoidal cells. The *ethmoidal bulla* (fig. 62–3*B*) is an elevation of the ethmoidal labyrinth that projects medially from the lateral wall into the middle meatus. It overlies some of the anterior ethmoidal cells, which open on it. The *hiatus semilunaris* is a curved slit below and in front of the bulla. It receives the opening of the maxillary sinus. The *ethmoidal infundibulum* is a narrow passage that runs up-

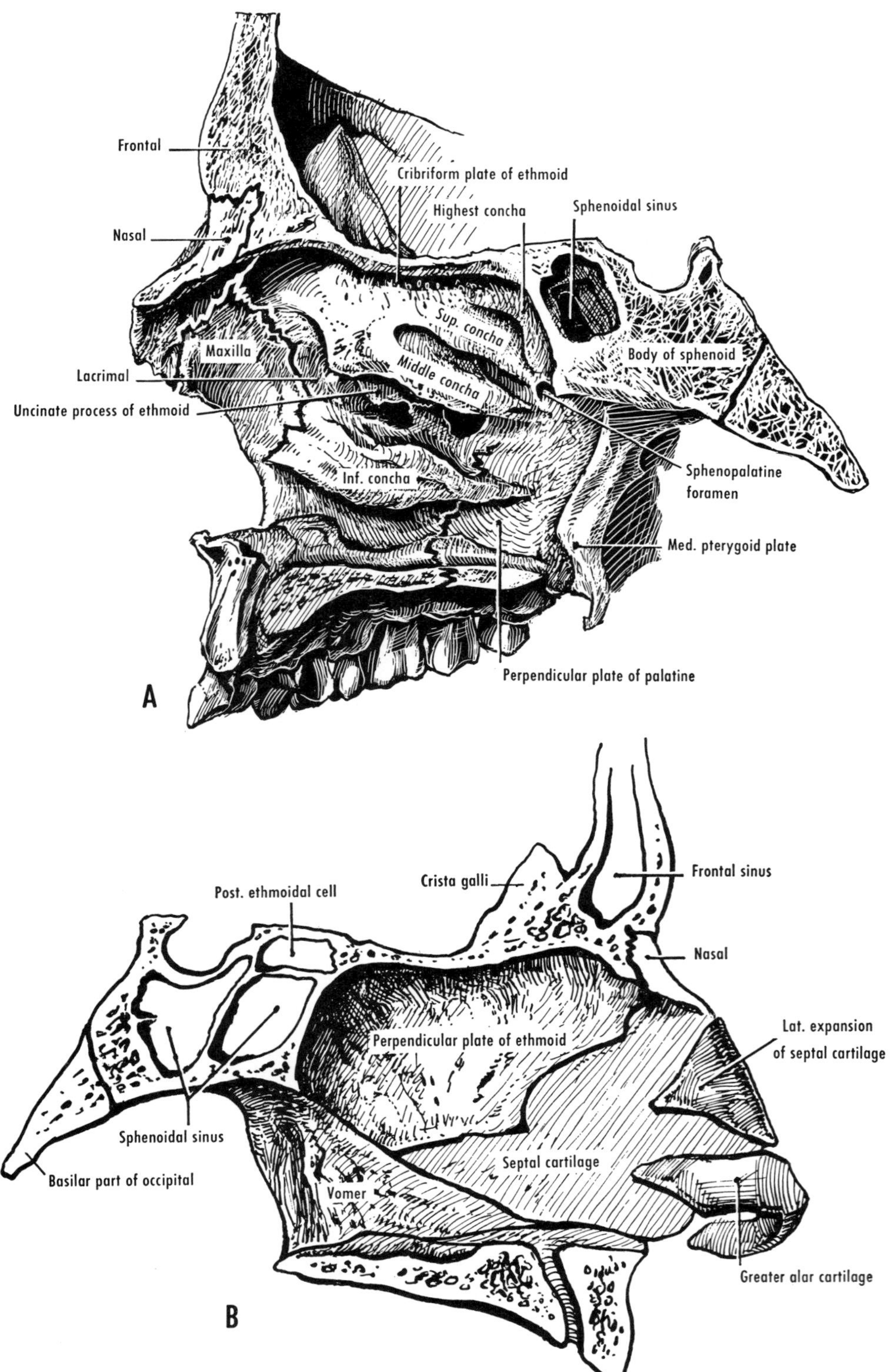

Figure 62–2 *A*, medial aspect of the bony framework of the lateral wall of the right nasal cavity. Note that the lateral boundary of the piriform aperture is formed by the nasal bone and the maxilla, and that of the choana is formed by the medial pterygoid plate of the sphenoid bone. Note also the line of the spheno-occipital junction. *B*, lateral aspect of the medial wall (nasal septum) of the right nasal cavity. The inferior limit of the ethmoid contribution to this wall varies considerably. The septal cartilage is attached to the vomer and the maxilla in a way such that considerable movement is possible without dislocation. The cartilages are based on Schaeffer and on Aymard.

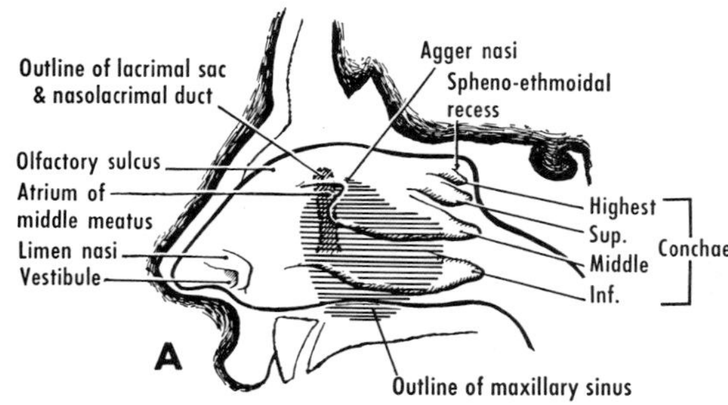

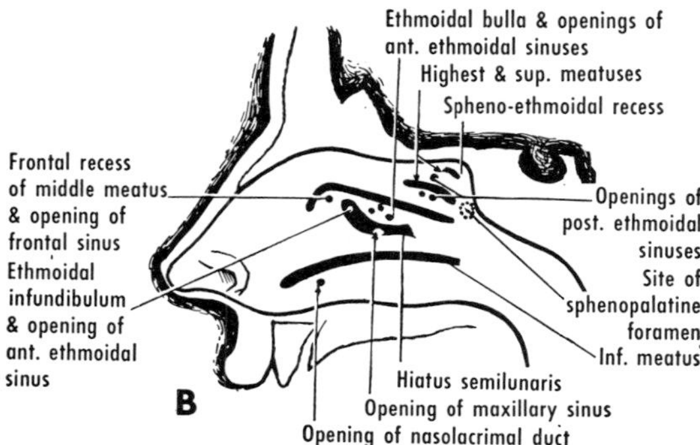

Figure 62–3 Medial aspect of the lateral wall of the right nasal cavity. *A*, showing the four conchae. **Each meatus is named after the concha that forms its roof.** In *B*, the conchae have been largely cut away. The frontal sinus may open into (1) the frontal recess (as shown here) or (2) the ethmoidal infundibulum.

ward and forward from the hiatus semilunaris and receives the openings of the frontal sinus and some of the anterior ethmoidal cells. These sinuses, however, may open into the *frontal recess* of the middle meatus in front of the infundibulum.

The **inferior nasal concha** is a separate bone that lies along the lower part of the lateral wall of the nasal cavity on each side of the body. Its lower border is free, whereas its upper border articulates with the maxilla, lacrimal, ethmoid, and palatine bones. **The *inferior meatus*, that between the inferior concha and the bony palate, receives the termination of the *nasolacrimal duct*.** The bony canal in which the duct lies is bounded by the maxilla, the lacrimal, and the inferior concha. The duct is in some cases protected at its termination by a *lacrimal fold* of mucous membrane.

SUBDIVISIONS AND MUCOUS MEMBRANE

The nasal cavity may be divided into a vestibule, a respiratory region, and an olfactory region.

*1. **Vestibule.*** The vestibule (fig. 62–3*A*) is a slight dilatation inside the opening of each nostril. It is lined largely with skin that presents hairs, and sebaceous and sweat glands. The vestibule is limited above and behind by a ridge (the *limen nasi*), over which the skin becomes continuous with the nasal mucosa. The junction between the vestibule and the respiratory region proper is constricted.

*2. **Respiratory Region.*** The respiratory region is covered by mucous membrane, which is closely adherent to the underlying periosteum or perichondrium, thereby constituting a mucoperiosteum or mucoperichondrium, respectively. It is continuous with the skin of the vestibule, the mucous membrane of the nasopharynx and of the paranasal sinuses, the lining of the nasolacrimal duct, and hence with the conjunctiva.

The anterior third of the nasal cavity is relatively inactive, as far as drainage is concerned, whereas the posterior two-thirds constitutes an area of active ciliary motion where rapid drainage takes place backward and downward into the nasopharynx.[5]

The nasal mucosa is highly vascular and, particularly over the conchae (which greatly increase its surface area), warms and moistens the incoming air. Because this is important for olfaction, the so-called respiratory region of the nasal cavity also has olfactory significance. The inferior concha, owing to the manner in which it conserves heat and moisture, has been likened to "an air conditioning plant."[6] The lamina propria or submucosa over the middle and inferior nasal conchae is characterized by large venous-like spaces ("swell bodies"), usually seen collapsed. These spaces are connected to arterioles by arteriovenous anastomoses.[7] The spaces may become congested with blood during a "cold in the nose" (coryza) or sometimes during menstruation.[8] The coverings of the conchae are distended considerably *in vivo*, so that the channels for breathing are slitlike (fig. 62–6).

*3. **Olfactory Region.*** The olfactory region of the nasal cavity is bounded by the superior nasal concha and the upper third of the nasal septum.[9] It is supplied by bundles of nerve fibers termed collectively the olfactory nerve. These bundles pierce the cribriform plate of the ethmoid bone and end in the olfactory bulb.

Olfactory Nerve

The mucous membrane of the olfactory region of the nasal cavity is yellowish (owing to pigment) rather than pink in color. The mucosa presents a thick, nonciliated pseudostratified columnar epithelium, including olfactory cells, which are bipolar neurons. Their dendrites reach the surface, where they become expanded and give rise to minute hairlike processes. Their non-myelinated axons (directed centrally) are the olfactory nerve fibers in the lamina propria. **The nerve fibers are collected into bundles, about 20 in number, which pass through the foramina in the cribriform plate of the ethmoid bone and are collectively known as the *olfactory* (1st cranial) *nerve*.** The nerve fibers enter the olfactory bulb, where they synapse. Degeneration of the olfactory nerve fibers increases with age.

Nerve filaments, collectively termed the *nervus terminalis*,[11] may be found between the olfactory bulb and the crista galli. They pass through the cribriform plate and are distributed to the mucosa of the nasal cavity. Their significance is obscure.

Examination. The olfactory nerve is tested by closing one nostril of the subject and presenting the test substance (e.g., peppermint or oil of cloves) to the other nostril.

GENERAL SENSORY INNERVATION AND BLOOD SUPPLY

Innervation

The nerves of ordinary sensation are derived from the first two divisions of the trigeminal nerve (fig. 62–4, *A*, *B*, and *C*). Although organized nerve endings are not found on the inferior concha, touch, pain, cold, and warmth can be felt there.[12] The nerves for the front portion of the nasal cavity come from the anterior ethmoidal nerve, derived (through the nasociliary) from the

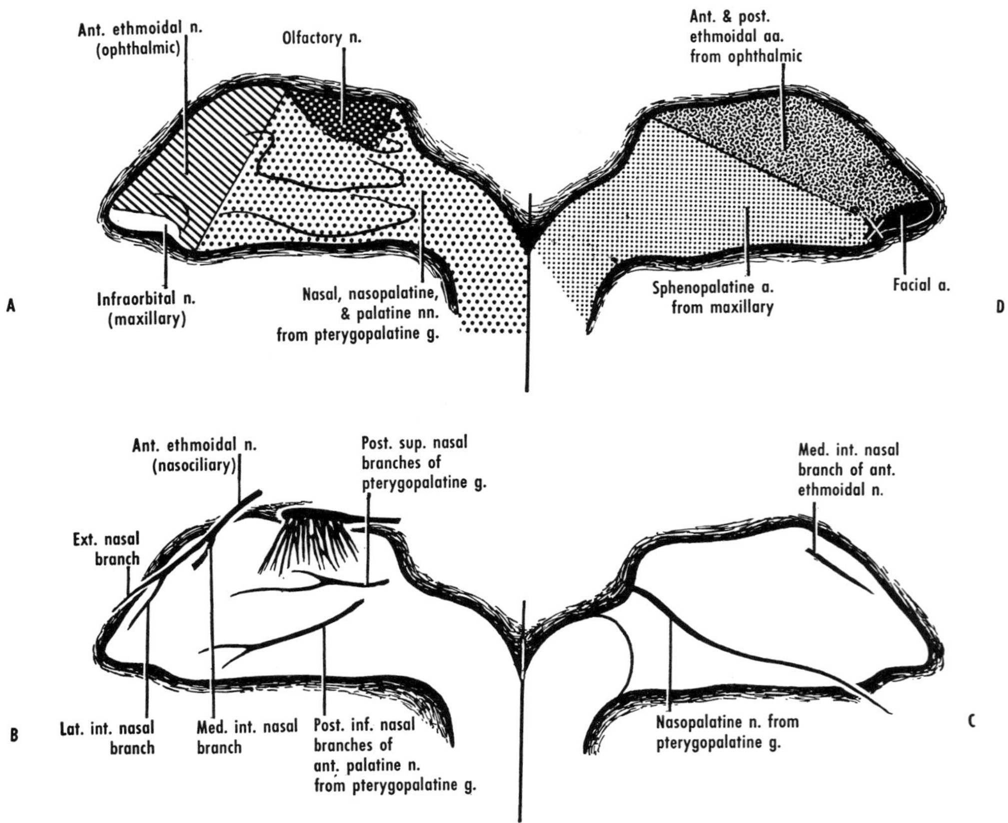

Figure 62–4 Innervation and blood supply of the right nasal cavity. *A*, showing the major nerve territories on the lateral wall; those on the medial wall are similar. **Apart from the olfactory region, the chief innervation of the nasal cavities is from the ophthalmic and maxillary nerves, the latter by way of the pterygopalatine ganglion.** *B* and *C*, showing the nerves in the lateral and in the medial wall, respectively. *D*, showing the major arterial territories on the medial wall; those on the lateral wall are similar. On the nasal septum the point **X** is the site of anastomosis between a septal branch of the superior labial artery (from the facial) and septal branches of the sphenopalatine artery; this is the chief area from which bleeding from the nose occurs.

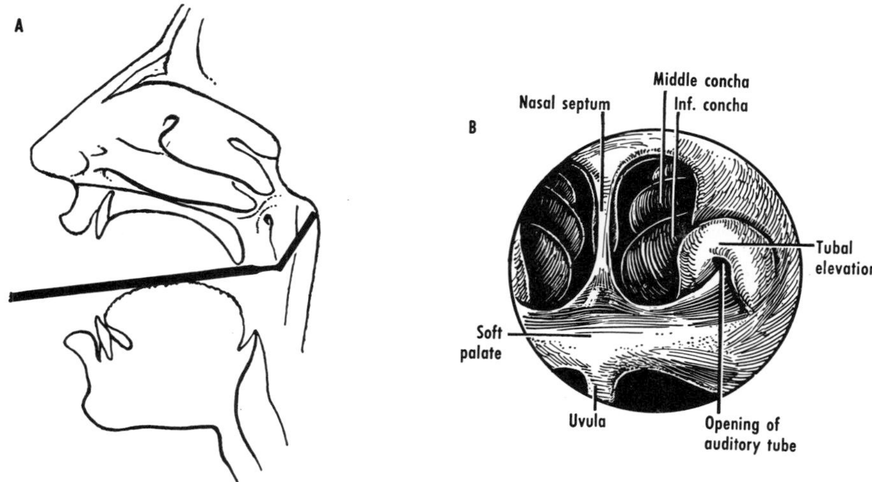

Figure 62–5 Posterior rhinoscopy. *A*, showing placement of mirror in the nasopharynx. *B*, details of structures seen in the mirror (Compare fig. 63–12*B*, p. 754).

ophthalmic. The nerves for the posterior and larger portion of the cavity come from the nasal, nasopalatine, and palatine branches of the pterygopalatine ganglion. These fibers are derived from the maxillary nerve and perhaps also from the facial.

Anesthetization of the walls of the nasal cavity may be achieved by injections through the mandibular notch and pterygopalatine fossa (maxillary nerve and pterygopalatine ganglion) and along the medial wall of the orbit (nasociliary nerve).

The chief sympathetic and parasympathetic innervation of the nasal cavity is from the branches of the pterygopalatine ganglion, but sympathetic fibers are carried along the walls of arteries also. Parasympathetic fibers from the facial nerve (nervus intermedius) travel in the greater petrosal nerve to the pterygopalatine ganglion, where they synapse. The postganglionic fibers from the ganglion are vasodilator and secretory in function. Sympathetic fibers, presumably from the upper thoracic segments of the spinal cord, synapse in the superior cervical ganglion. Postganglionic fibers travel in the internal carotid plexus, the deep petrosal nerve, and the nerve of the pterygoid canal, to reach the pterygopalatine ganglion. They are probably chiefly vasoconstrictor.

Blood Supply and Lymphatic Drainage

The most important arteries to the nasal cavity are the sphenopalatine (a branch of the maxillary) and the anterior ethmoidal[13] (a branch of the ophthalmic) (fig. 62–4*D*). The vast majority of instances of bleeding from the nose (epistaxis) occur from the junction of a septal branch of the superior labial artery with a septal branch of the sphenopalatine artery. The veins form a plexus beneath the mucosa and in general accompany the arteries. The lymph vessels drain into the deep cervical nodes.

It has been shown in the cat that India ink can pass readily from the subarachnoid space to the lymphatics of the nasal mucosa, probably by way of the perineural sheath of the olfactory nerve.[14] Similar results have been obtained in the rabbit.[15]

EXAMINATION OF NASAL CAVITY (RHINOSCOPY)

The nasal cavity can be examined in the living either through a nostril (anterior rhinoscopy) or through the pharynx (posterior rhinoscopy).

Anterior Rhinoscopy. In anterior rhinoscopy the front part of the nasal cavity is inspected by inserting a nasal speculum through a nostril. By this means the middle and inferior conchae and meatuses, the nasal septum, and the floor of the nose can be observed.

Posterior Rhinoscopy. In posterior rhinoscopy the back part of the nasal cavity is inspected through the choanae by inserting a postnasal mirror (fig. 62–5) through the mouth and pharynx. The posterior edge of the nasal septum, formed by the vomer, is a prominent landmark, and the opening of the auditory tube can be identified (fig. 63–12*B*, p. 754).

PARANASAL SINUSES

The paranasal sinuses are cavities found in the interior of the maxilla, frontal, sphenoid, and ethmoid bones. Their walls, composed of compact bone, are lined by a muco-endosteum that is continuous with the respiratory mucosa of the nasal cavity and is similar in type (ciliated pseudostratified columnar epithelium). Mixed glands are present. **The paranasal sinuses are supplied by branches of the ophthalmic and maxillary nerves. The sinuses develop as outgrowths from the nasal cavity, and hence they all drain, directly or indirectly, into the nasal cavity** (fig. 62–6). The drainage is by means of ciliary action and perhaps also by suction set up during blowing of the nose.[16] **Nasal infection (rhinitis), such as occurs during a "cold in the head," may spread to the lining of the sinuses, producing sinusitis.** The paranasal sinuses are very small at birth, but they enlarge greatly between puberty and adulthood. The degree of development of the individual sinuses is very variable. The functional significance[17] of the paranasal sinuses remains obscure.

Transillumination and Radiography. Information regarding the clarity of the maxillary sinuses can be obtained in a dark room by means of a small, strong light placed inside the mouth, the sinuses being observed through the face. The paranasal sinuses can be examined radiographically with or without (fig. 62–7) the injection of a radio-opaque medium (iodized oil).

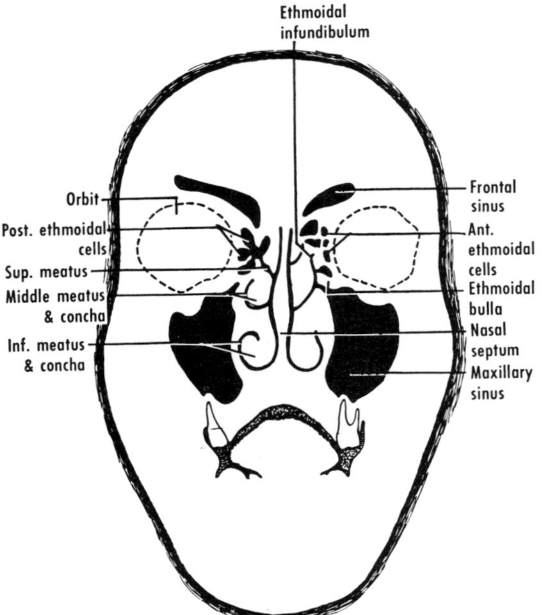

Figure 62–6 Coronal scheme of the nasal cavity, showing the conchae and the meatuses, and some of the paranasal sinuses. The size of the nasal passages is based on a most instructive laminogram reproduced by Proetz.

Maxillary Sinus (fig. 62–7A)

The maxillary sinus, the largest of the paranasal sinuses, is situated in the body of the maxilla. It is usually described as being shaped like a pyramid (fig. 62–8) laid on one side, the base being medial and the apex being in the zygomatic process of the maxilla, but the base of the pyramid is frequently superior. The medial wall is the lateral wall of the nasal cavity. The roof is the floor of the orbit. The floor is the alveolar process of the maxilla. **The floor of the maxillary sinus is usually 1/2 to 1 cm below the level of the floor of the nasal cavity, and it presents elevations produced by the first and second molar teeth. The teeth related to the floor of the sinus vary from the three molars to the molars, premolars, and canine. Maxillary sinusitis is frequently accompanied by toothache.** The posterior wall separates the maxillary sinus from the infratemporal and pterygopalatine fossae. The anterior wall is related to the face. Ridges and septa in the walls of the sinus are not infrequent.[19] **Infection may spread fairly readily among the frontal sinus or the anterior ethmoidal cells, the nasal cavity, the teeth, and the maxillary sinus.** The maxillary sinus is supplied by the anterior and posterior superior alveolar and infra-orbital nerves.

The maxillary sinus drains by one or more openings into the middle meatus of the nasal cavity by means of the hiatus semilunaris.

The opening is generally into the posterior third of the hiatus.[20] It is usually a short canal rather than an orifice,[21] and it may be placed horizontally, vertically, or obliquely.[22] The opening is in the anterosuperior part of the medial wall of the sinus and can frequently be cannulated *in vivo* through the nostrils. An accessory opening is often found, usually behind and below the main opening. Although the main opening appears large in a disarticulated maxilla, it is narrowed in the intact state by the surrounding bones (ethmoid, palatine, lacrimal, and inferior nasal concha) and by the mucous membrane. Drainage of the maxillary sinus is effected by ciliary activity, involving a spiral motion centered on the opening,[23] and by negative pressure produced in the sinus during inspiration.[24]

Ethmoidal Sinus

The ethmoidal sinus[25] comprises numerous (4 to 17 on each side) small cavities in the eth-

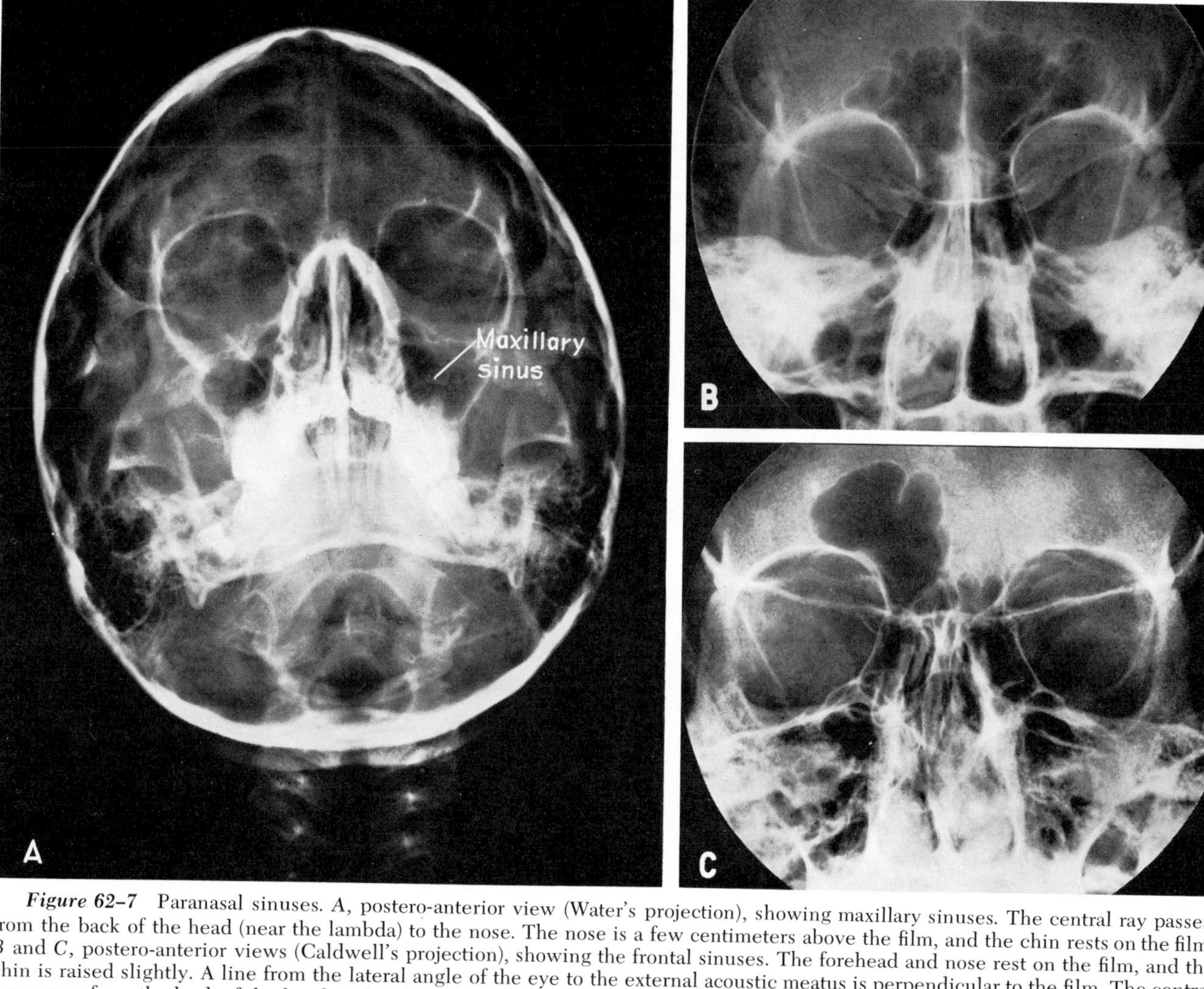

Figure 62–7 Paranasal sinuses. *A*, postero-anterior view (Water's projection), showing maxillary sinuses. The central ray passes from the back of the head (near the lambda) to the nose. The nose is a few centimeters above the film, and the chin rests on the film. *B* and *C*, postero-anterior views (Caldwell's projection), showing the frontal sinuses. The forehead and nose rest on the film, and the chin is raised slightly. A line from the lateral angle of the eye to the external acoustic meatus is perpendicular to the film. The central ray passes from the back of the head to the glabella, and is angled slightly caudally. In *B*, the horizontal white line between the orbits is the anterior edge (limbus sphenoidalis) of the optic groove. In *C*, the frontal sinus on one side is extremely small. The horizontal white lines crossing the upper portions of the orbits are the posterior edges (sphenoidal ridges) of the lesser wings of the sphenoid. *A*, from I. Meschan, cited with figure 52–4. *B*, courtesy of Eastman Kodak Company, Rochester, New York. *C*, courtesy of Earl L. Warren, M. D., Paterson, New Jersey.

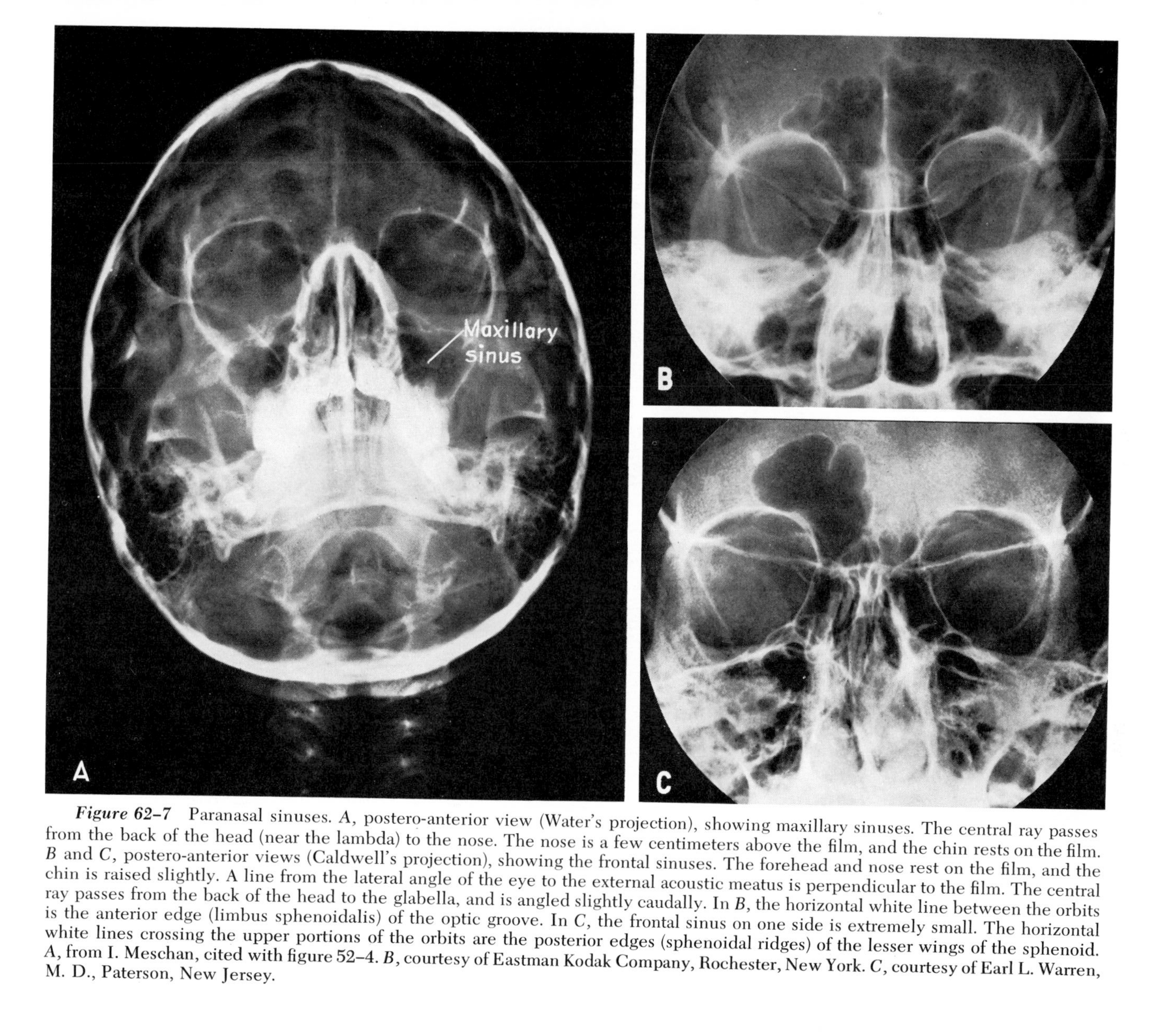

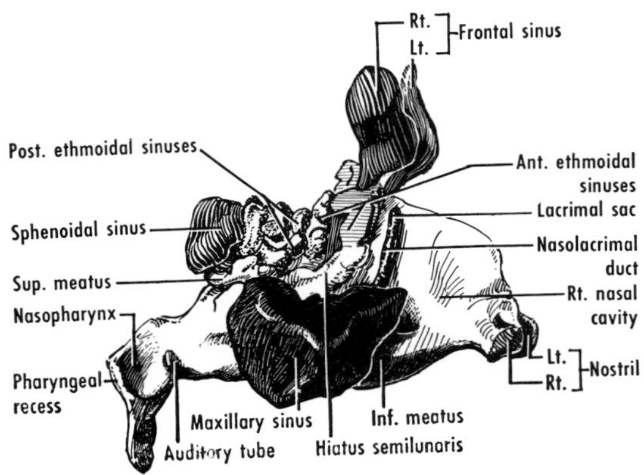

Figure 62–8 Drawing of a cast of the nasal cavity and the paranasal sinuses. Right lateral aspect. Note the orbital, facial, and infratemporal aspects of the maxillary sinuses. A description of this cast has been published.[18]

moidal labyrinth, between the orbit and the nasal cavity. The parts of the sinus are termed *ethmoidal cells.* Their walls are completed by the frontal, maxillary, lacrimal, sphenoid, and palatine bones, and their tendency to be found partly in adjacent bones has been graphically described as "the struggle of the ethmoids" (Seydel). On the basis of their openings, the ethmoidal cells can be classified into (1) anterior and (2) posterior groups, which drain into (1) the middle and (2) the superior and highest meatuses, respectively.

Frontal Sinus (fig. 62–7, *B* and *C*)

The frontal sinus may be regarded as an anterior ethmoidal cell that has invaded the frontal bone (after birth). It is separated from its fellow of the opposite side by a bony septum that is usually deviated to one side. The sinuses vary greatly and frequently are of different size on the two sides of the body. The shape of the brow *in vivo* gives virtually no indication of the size of the sinus. The sinus commonly extends backward in the roof of the orbit.[26] In many instances it may be related closely to the orbit, therefore, as well as to the anterior cranial fossa (fig. 55–14). **The frontal sinus opens into the middle meatus, directly or by means of a passage** (the *frontonasal duct,* which may or may not be continuous with the ethmoidal infundibulum[27]), at one of two sites as follows: (1) into the front of the middle meatus (frontal recess), in front of or above the ethmoidal infundibulum, or (2) into the ethmoidal infundibulum. The frontal sinus is supplied by the supra-orbital nerve from the ophthalmic.

Sphenoidal Sinus (fig. 62–2)

The sphenoidal sinus is situated in the body of the sphenoid bone. It is extremely variable in size,[28] and it may extend into the occipital bone.[29] It opens by an orifice in the upper part of its front wall into the spheno-ethmoidal recess of the nasal cavity. It can frequently be cannulated *in vivo* through the nostrils. The sphenoidal sinus is divided into right and left parts (each of which may be termed a sphenoidal sinus) by a bony septum, which is usually deviated to one side.[30] The anterior wall of the sinus is formed by two thin, curved plates (the *sphenoidal conchae*), which are usually injured or destroyed when a skull is disarticulated. The sphenoidal sinus is related posteriorly to the pons and the basilar artery; superiorly to the optic chiasma (above the chiasmatic groove) and optic nerves, and the hypophysis cerebri; anteriorly to the nasal cavity; inferiorly to the nasal cavity and the nasopharynx; and laterally to the optic nerve, the cavernous sinus and internal carotid artery, and the ophthalmic and maxillary nerves. The sphenoidal sinus is supplied chiefly by branches from the maxillary nerve.

REFERENCES

1. J. L. Aymard, J. Anat., Lond., *51*:293, 1917. J. M. Converse, Ann. Otol., etc., St Louis, *64*:220, 1955. M.-J. Peyrus, Lyon méd., *200*:871, 1958.
2. G. W. Drumheller, Anat. Rec., *176*:321, 1973.
3. S. J. Pearlman, Ann. Otol., etc., St Louis, *43*:739, 1934.
4. R. I. Williams, Trans. Pacif. Cst oto-ophthal. Soc., *36*:339, 1955. W. Bachmann and U. Legler, Acta otolaryng., Stockh., *73*:433, 1972.
5. A. Hilding, Arch. Otolaryng., Chicago, *15*:92, 1932. W. Messerklinger, Arch. klin. exp. Ohr.-, Nas.-, u. Kehlk-Heilk., *195*:138, 1969. R. Naessen, J. Laryng., *84*:1231, 1970.
6. V. E. Negus, Brit. med. J., *1*:367, 1956.
7. W. F. Harper, J. Anat., Lond., *83*:61, 1949, abstract. J. D. K. Dawes and M. M. L. Prichard, J. Anat., Lond., *87*:311, 1953. R. Tiedemann, Arch. Ohr.-, Nas.-, u. KehlkHeilk., *172*:257, 1958. D. Temesrekasi, Z. mikrosk. anat. Forsch., *80*:219, 1969.
8. J. N. Mackenzie, Johns Hopk. Hosp. Bull., 9:10, 1898. T. H. Holmes, H. Goodell, S. Wolf, and H. G. Wolff, *The Nose. An Experimental Study of Reactions within the Nose in Human Subjects during Varying Life Experiences,* Thomas, Springfield, Illinois, 1950.
9. R. Naessen, Acta otolaryng., Stockh., *70*:51, 1970.
10. C. G. Smith, J. comp. Neurol., *77*:589, 1942.
11. O. Larsell, Ann. Otol., etc., St Louis, *59*:414, 1950.
12. J. A. Harpman, Brit. med. J., *2*:497, 1951.
13. G. Weddell *et al.,* Brit. J. Surg., *33*:387, 1946.
14. J. M. Yoffey and C. K. Drinker, J. Anat., Lond., *74*:45, 1939.

15. W. M. Faber, Amer. J. Anat., *62*:121, 1937.
16. Z. W. Colson, Laryngoscope, St Louis, *58*:642, 1948. Proetz disagrees.
17. P. L. Blanton and N. L. Biggs, Amer. J. Anat., *124*:135, 1969. N. L. Biggs and P. L. Blanton, J. Biomech., *3*:255, 1970.
18. C. M. Jackson and C. E. Connor, Ann. Otol., etc., St Louis, *26*:585, 1917. See also G. H. Schumacher, H. J. Heyne, and R. Fanghänel, Anat. Anz., *130*:132 and 143, 1972.
19. S. Jovanovic, N. Lotric, and V. Radoiévitch, C. R. Ass. Anat., 47th reunion: 401, 1962.
20. O. E. Van Alyea, Arch. Otolaryng., Chicago, *24*:553, 1936. A summary of this author's work on the paranasal sinuses can be found in O. E. Van Alyea, *Nasal Sinuses. An Anatomic and Clinical Consideration,* Williams & Wilkins, Baltimore, 2nd ed., 1951.
21. E. Simon, Arch. Otolaryng., Chicago, *29*:640, 1939.
22. M. C. Myerson, Arch. Otolaryng., Chicago, *15*:80, 1932.
23. A. C. Hilding, Ann. Otol., etc., St Louis, *53*:35, 1944.
24. J. McMurray, Arch. Otolaryng., Chicago, *14*:581, 1931.
25. O. E. Van Alyea, Arch. Otolaryng., Chicago, *29*:881, 1939.
26. S. Jovanovic, Acta anat., *45*:133, 1961.
27. K. A. Kasper, Arch. Otolaryng., Chicago, *23*:322, 1936.
28. F. W. Dixon, Ann. Otol., etc., St Louis, *46*:687, 1937. O. E. Van Alyea, Arch. Otolaryng., Chicago, *34*:225, 1941. J. C. Haley, Med. Rec., S. Antonio, *42*:693, 1948. J. C. Peele, Laryngoscope, St Louis, *67*:208, 1957.
29. G. Hammer and C. Rådberg, Arch. Ohr.-, Nas.-, u. Kehlk-Heilk., *173*:278, 1958.
30. G. Hammer and C. Rådberg, Acta radiol., Stockh., *56*:401, 1961.

GENERAL READING

Proetz, A. W., *Essays on the Applied Physiology of the Nose,* Annals Publishing Co., St Louis, 2nd ed., 1953. An interesting functional account.

Schaeffer, J. P., *The Nose, Paranasal Sinuses, Nasolacrimal Passageways, and Olfactory Organ in Man,* Blakiston's Son & Co., Philadelphia, 1920. A classical text on the anatomy and development.

Terracol, J., and Ardouin, P., *Anatomie des fosses nasales et des cavités annexes,* Maloine, Paris, 1965. A detailed account of the nose and paranasal sinuses, including their development.

PHARYNX AND LARYNX 63

PHARYNX

The *pharynx*[1] is the part of the digestive system that is situated behind the nasal and oral cavities, and behind the larynx. It can conveniently be divided, therefore, into nasal, oral, and laryngeal parts: (1) nasopharynx, (2) oropharynx, and (3) laryngopharynx (fig. 63–1). The pharynx extends from the base of the skull down to the lower border of the cricoid cartilage (opposite C.V. 6), where it becomes continuous with the esophagus. The pharynx is a tube composed of muscular and fibrous layers, and it is lined by a mucous membrane.

The pharynx acts as a common channel for both deglutition and respiration, and the food and air pathways cross each other in the pharynx (fig. 63–2). In the anesthetized patient, the passage of air through the pharynx is facilitated by extension of the head.[2]

The pharynx is related above to the body of the sphenoid and the basilar part of the occipital bone; below it is continuous with the esophagus. In front, it opens into the nasal and oral cavities, and the larynx; behind, it is related to the prevertebral layer of fascia and the prevertebral mus-

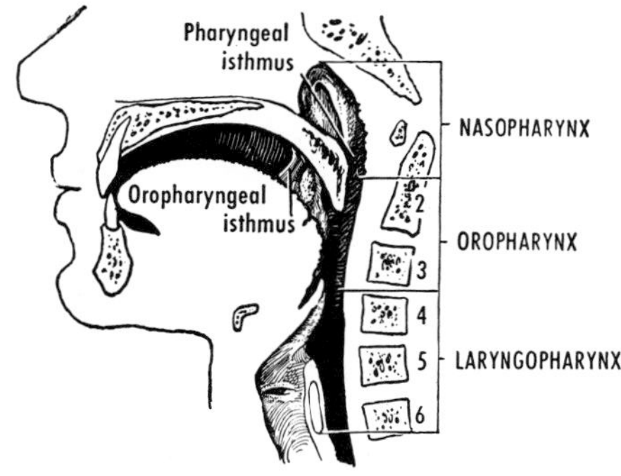

Figure 63–1 General arrangement of the major parts of the pharynx. Median section.

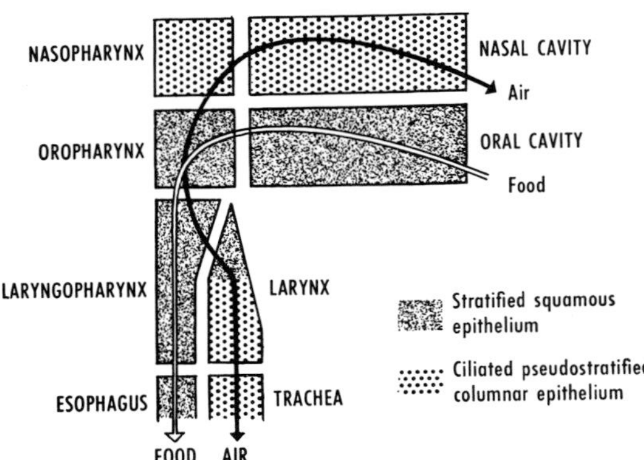

Figure 63–2 Scheme of respiratory and digestive cavities in the head and neck. Note that the pharynx acts as a common channel for both respiration and deglutition, and that the air and food pathways cross each other. In general, the purely air-carrying passages are lined with ciliated pseudostratified columnar epithelium, whereas the passages that transmit either food or food and air are lined with stratified squamous epithelium. Modified from Braus.

cles, and the upper six cervical vertebrae. Laterally, it is related to the styloid process and styloid muscles, the medial pterygoid muscle, the carotid sheath, and the thyroid gland, and it communicates with the auditory tube. Figures 63–3 and 63–4 show general views of the pharynx.

Lateral cervical cysts and fistulae are abnormal features sometimes found in the neck; their developmental origin is in considerable dispute.

SUBDIVISIONS OF PHARYNX

Nasopharynx

The nasopharynx, generally included as a part of the pharynx, may, in many respects, be regarded as the back portion of the nasal cavity.[3] More precisely, the anterior part of the nasopharynx resembles the nasal cavity, whereas the posterior part (beyond the tubal openings) is comparable to the oropharynx.[4] The nasal cavity and the nasopharynx are both functional components of the respiratory system. The nasopharynx communicates with the oropharynx through the *pharyngeal isthmus (nasopharyngeal hiatus)*, which is bounded by the soft palate, the palatopharyngeal arches, and the posterior wall of the pharynx. The isthmus is closed by muscular action during swallowing.[5] The choanae form the junction between the nasopharynx and the nasal cavity proper. Like the nasal cavity, the cavity of the nasopharynx is never obliterated, because the walls (other than the soft palate) are largely immobile.

Roof and Posterior Wall. The *fornix*, or roof, and the posterior wall of the nasopharynx form a continuous sloping surface that lies below the body of the sphenoid and the basilar part of the occipital bone (fig. 63–5).

PHARYNGEAL TONSIL. **A mass of lymphoid tissue, termed the pharyngeal tonsil (nasopharyngeal tonsil, more precisely), is embedded in the mucous membrane of the posterior wall of the nasopharynx. Enlarged nasopharyngeal tonsils are called "adenoids" and may cause respiratory obstruction,** leading to persistent mouth breathing and malgrowth of the face,[6] although this correlation has been denied. The tonsils grow during childhood but involute after puberty.

Like the nasopharynx as a whole, the nasopharyngeal tonsil is covered by pseudostratified columnar ciliated epithelium, which, however, may be replaced in the adult by stratified squamous epithelium. Its surface is characterized by a number of

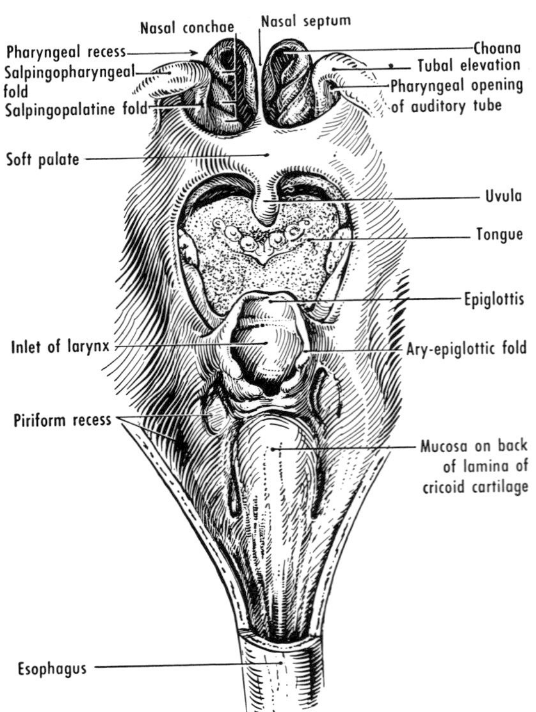

Figure 63–3 Posterior aspect of anterior wall of pharynx. Note the various cavities with which the pharynx communicates: nasal cavity, auditory tubes, oral cavity, larynx, and esophagus.

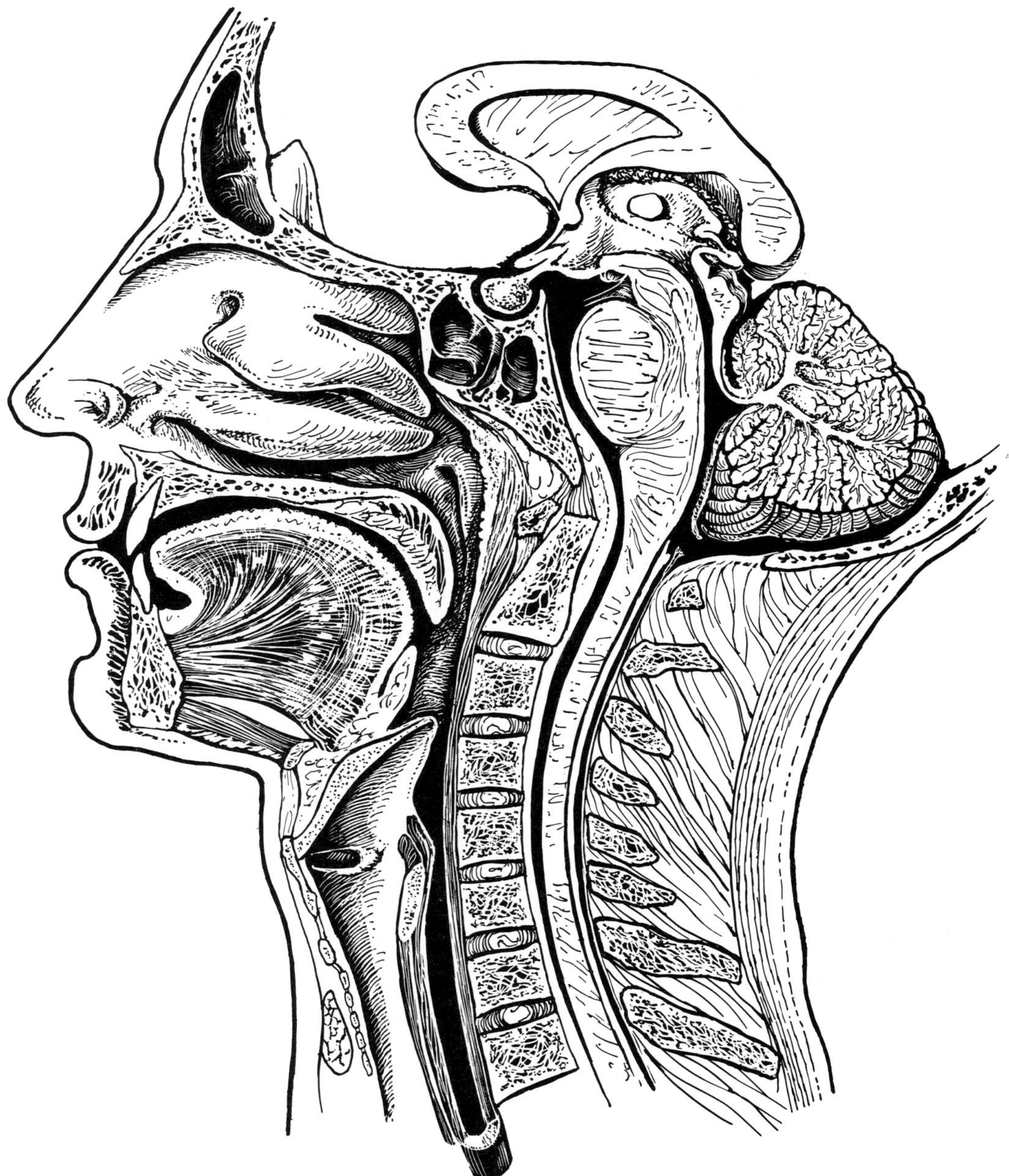

Figure 63–4 Sagittal section of the head and neck, almost in the median plane. A portion of the brain has been included. The various structures shown in this illustration have been given labels in other figures. Note the hypophysis, corpus callosum, septum pellucidum, pineal body, third ventricle, aqueduct, fourth ventricle, pons, cerebellum, and spinal cord; the seven cervical and first thoracic vertebrae; the frontal and sphenoidal air sinuses; the nasal conchae, the palate, and the opening of the auditory tube; the genioglossus and geniohyoid muscles; various structures bounding the nasal and oral cavities, the pharynx and the larynx; the trachea and the esophagus.

glandular ducts.[7] The *pharyngeal bursa* is a median pit intimately related to the nasopharyngeal tonsil.[8] The *pharyngeal hypophysis*,[9] a median structure located in the mucosa or the periosteum, resembles the adenohypophysis histologically and has a similar developmental origin (from the craniopharyngeal pouch in the embryo).

Lateral Wall. **Each lateral wall of the nasopharynx is marked by the *pharyngeal opening of the auditory tube*. The opening is located[10] approximately 1 to $1\frac{1}{2}$ cm: (1) below the roof of the pharynx, (2) in front of the posterior wall of the pharynx, (3) above**

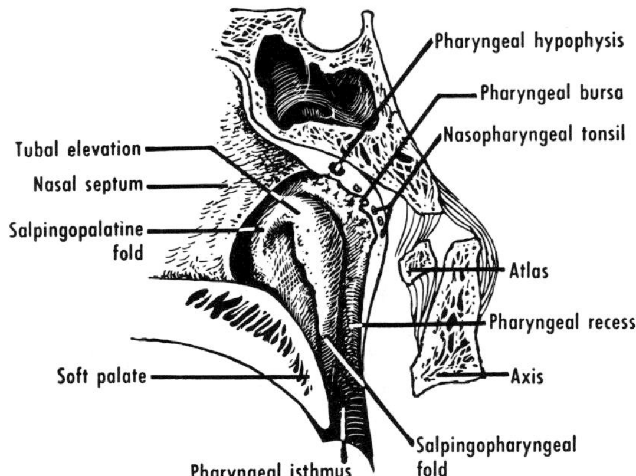

Figure 63–5 Medial view of the right lateral wall of the nasopharynx. For orientation, compare figure 63–4.

the level of the palate, and (4) behind the inferior nasal concha and the nasal septum. The auditory tube can be catheterized through a nostril. The opening is limited above and behind by the *tubal elevation (torus tubarius)*, which is produced by the cartilage of the tube. Folds of mucous membrane descend from the elevation to the palate *(salpingopalatine fold)* and the side wall of the pharynx *(salpingopharyngeal fold)*. Another fold (the *torus levatorius*), produced by the levator veli palatini, descends from the mouth of the tube to the soft palate. The levator has been described as "pouring out of the tube," but this appearance is seen only on elevation of the palate, whereas the pharyngeal opening of the tube appears as a vertical slit when at rest. **The part of the pharyngeal cavity behind the tubal elevation is termed the *pharyngeal recess*.** It extends backward and laterally between the longus capitis medially and the levator veli palatini laterally. Lymphoid tissue that is sometimes found in the mucous membrane of the recess is collectively termed the *tubal tonsil*.

AUDITORY TUBE. **The *auditory (pharyngotympanic) tube*[12] connects the nasopharynx to the tympanic cavity** (fig. 63–6). It equalizes the pressure of the external air and that contained in the tympanic cavity. **Infections may spread from the pharynx to the middle ear by way of the tube. The tube extends backward, laterally, and upward, and is about 3 to 4 cm in length. It consists of (1) a cartilaginous part, the anteromedial two-thirds, and (2) an osseous part, the posterolateral third.** The two parts meet at a narrowed and slightly angled region, termed the *isthmus*.

The auditory tube develops from the tubotympanic recess, which probably arises near the site of the first pharyngeal pouch in the embryo.

1. The *cartilaginous part* of the auditory tube is a diverticulum of the pharynx. It lies on the inferior aspect of the base of the skull, in a groove between the greater wing of the sphenoid and the petrous portion of the temporal bone (fig. 52–13, p. 563). The cartilage, elastic in type, is a fluted plate that is completed below by connective tissue. The mucous membrane of this part of the tube comprises mostly pseudostratified columnar ciliated epithelium. The tube is related laterally to the tensor veli palatini, the mandibular nerve, and the middle meningeal artery; medially, it is related to

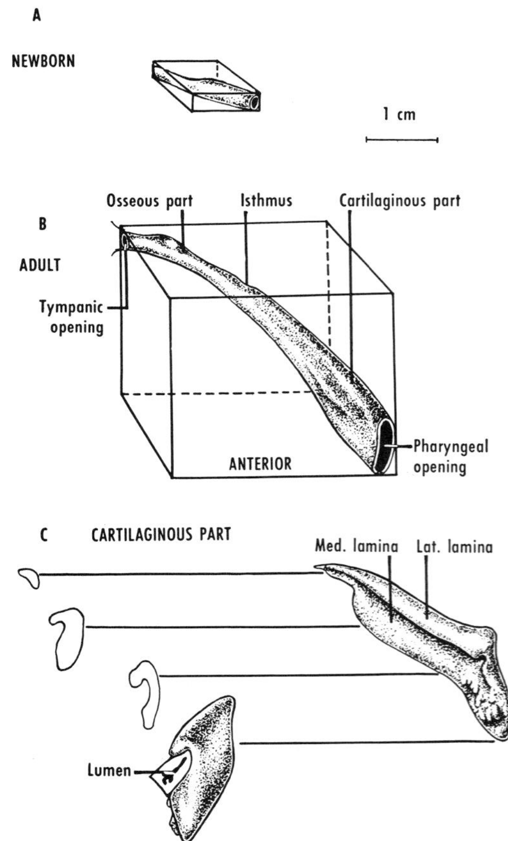

Figure 63–6 Anterior aspect of the right auditory tube. A, in the newborn. *B*, in the adult. *C*, the cartilaginous part of the tube, together with outlines of several cross-sections. Based on Graves and Edwards.

the levator veli palatini and the pharyngeal recess.

The cartilaginous part of the tube remains closed (perhaps owing to its elastic tissue[13]) **except on swallowing or yawning, during either of which its opening prevents excessive pressure in the middle ear.** The mechanism of opening (whether passive or muscular, and, if muscular, which muscle produces it, e.g., the tensor veli palatini) is in dispute. The tube may become blocked by swelling of its mucous membrane, as from a "cold in the head."

As one reaches a higher altitude, on ascending a mountain or in an airplane, the air becomes less dense (decreased atmospheric pressure), and the air within the tympanic cavity expands, pushing laterally on the tympanic membrane. In the absence of swallowing, the higher air pressure in the middle ear may force the auditory tube to open with a "click." On descending, the pressure changes are reversed and, in either event, pressure sensations are felt in the ears, and the hearing may become impaired temporarily. The pressures on each side of the membrane can be equalized readily by swallowing or yawning, procedures that open the auditory tube.

2. The *osseous part* of the auditory tube is a forward prolongation of the tympanic cavity. It occupies a semicanal in the petrous temporal and it may be regarded as a portion of the pneumatic area of the temporal bone.[14] It can be found on the under surface of the dried skull between the petrous and the downwardly turned edge of the tegmen tympani. This part of the tube is lined by a mucoperiosteum, which normally includes a non-ciliated, flattened or cuboidal epithelium. It is related superiorly to the semicanal for the tensor tympani, anterolaterally to the tympanic portion of the temporal bone, and posteromedially to the carotid canal.

Oropharynx

The oropharynx extends from the soft palate above to the superior border of the epiglottis below. It communicates in front with the oral cavity by the *faucial (oropharyngeal) isthmus*, which is bounded above by the soft palate, laterally by the palatoglossal arches, and below by the tongue (fig. 63–1). This region is characterized by a *lymphatic ring*, composed chiefly of the nasopharyngeal tonsil above, the palatine tonsils laterally, and the lingual tonsil below. It is often assumed that this tissue acts as a barrier to the spread of infections, but the function of the lymphatic ring is actually obscure.[15]

The mucous membrane of the epiglottis is reflected onto the base of the tongue (as the *median glosso-epiglottic fold*) and onto the lateral wall of the pharynx (as the *lateral glosso-epiglottic*, or the *pharyngo-epiglottic, fold*). **The space on each side of the median glosso-epiglottic fold is termed the *epiglottic vallecula*.**

Posteriorly, the oropharynx is related to the bodies of C.V. 2 and 3.

Each *lateral wall* of the orpharynx is characterized by the diverging *palatoglossal* and *palatopharyngeal arches* (often called the *anterior* and *posterior pillars of the fauces*, respectively). The arches are produced by two underlying muscles: the palatoglossus and the palatopharyngeus. The triangular recess between the two arches is termed the tonsillar fossa and lodges the palatine tonsil (fig. 63–1). The vague term *fauces* is used to include the oropharyngeal isthmus, the arches, the tonsillar fossa, and the tonsil.

Palatine Tonsils. A *tonsil* is a mass of lymphoid tissue containing reaction or germinal centers (secondary lymphatic follicles) and related to an epithelial surface in the pharynx. The functional significance of the tonsils is uncertain. **The palatine tonsils,[16] often referred to as merely as "the tonsils," are two masses of lymphoid tissue located one on each side of the oropharynx. Each is lodged in a *tonsillar fossa*, bounded by the palatoglossal and palatopharyngeal arches and the tongue.** The tonsil presents a free medial surface and a deep lateral surface.

The *medial surface* (fig. 63–7A) usually presents the *intratonsillar cleft*, commonly but inaccurately called the *supratonsillar fossa*. The medialward projection of the tonsil does not give a good indication of the size of the organ. Indeed it is difficult to gain an idea of the normal limits of variation in size because the tonsil is so frequently found hypertrophied as a result of inflammation. The medial surface also presents a number of pits that lead to blind mucosal tubules, called the *tonsillar crypts*[17] (fig. 63–7*B*). The crypts are clefts lined with stratified squamous epithelium, deep to which are lymphatic follicles.

The *lateral surface* of the tonsil is covered by a fibrous capsule, which is related laterally to a sheath of pharyngobasilar fascia, the paratonsillar vein, the superior constrictor muscle and/or the palatopharyngeus,[18] the medial pterygoid muscle, and the region of the angle of the mandible. The internal carotid artery is usually situated a few centimeters pos-

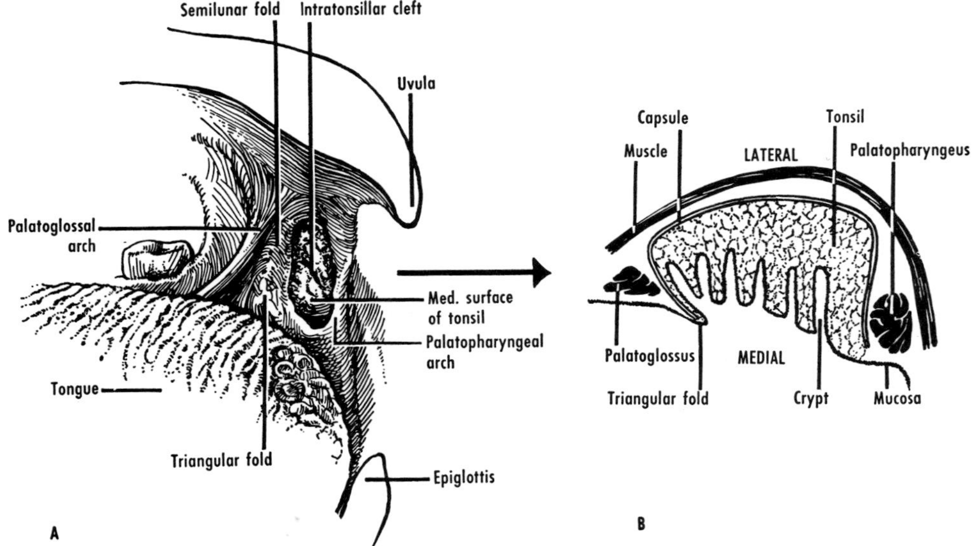

Figure 63–7 A, diagram of the right palatine tonsil and its surroundings. Medial aspect. *B*, horizontal section through the tonsil (at a greater magnification) in plane of arrow. Based on Fetterolf.

terolateral to the tonsil. The tonsils are not swallowed because they are firmly anchored by (1) a fibrous connection (suspensory ligament) between the front of the tonsillar capsule and the tongue, and (2) the partial insertion of the palatoglossus and the palatopharyngeus into the tonsillar capsule.

The tonsil is supplied by the external carotid artery, chiefly (if not entirely) by the tonsillar branch of the facial, which pierces the superior constrictor and enters the lower part of the lateral surface of the tonsil (hilus). Hemorrhage after tonsillectomy comes from the external palatine (paratonsillar) vein, a variable vessel that descends from the soft palate lateral to the tonsillar capsule, pierces the superior constrictor, and ends in the facial vein. Lymphatics drain into the upper deep cervical nodes, particularly to the jugulodigastric node. The tonsil is supplied by twigs from the glossopharyngeal nerve and from the pterygopalatine ganglion.

Shortly after puberty the tonsils begin to undergo involution.[19] A conspicuous reduction in the size of the tonsils occurs from the age of 30 years onward. The connective and lymphatic components disappear together, and bone, cartilage, and cysts may be found at any time.

Laryngopharynx

The laryngopharynx extends from the upper border of the epiglottis to the lower border of the cricoid cartilage, where it becomes continuous with the esophagus. Anteriorly, it presents the inlet of the larynx, and the back of the arytenoid and cricoid cartilages. Posteriorly, the laryngopharynx is related to the bodies of C.V. 4 to 6.

The *piriform recess*[20] or *fossa* is that part of the cavity of the laryngopharynx situated on each side of the inlet of the larynx (fig. 63–3). It lies between the thyrohyoid membrane and the thyroid cartilage laterally, and the ary-epiglottic fold and the arytenoid and cricoid cartilages medially. It is limited by the hyoid bone above and by the lower border of the cricoid cartilage below. The branches of the internal laryngeal nerve and the superior laryngeal vessels lie under cover of the mucous membrane of the piriform recess. **Foreign bodies may become lodged in the recess.**

STRUCTURE OF PHARYNX

The pharynx consists of four main coats, from within outward:

1. Mucous Membrane. The mucous membrane is continuous with that of the auditory tubes, and of the nasal, oral, and laryngeal cavities. The epithelium is pseudostratified columnar ciliated in the nasopharynx, stratified squamous in the oro- and laryngopharynx, and stratified columnar at the junction between these two types. Mixed glands are present.

2. Fibrous Coat. The fibrous coat is thick above, forming the *pharyngobasilar fascia (pharyngeal aponeurosis),* which is described as being attached to the base of the skull, the auditory tube, the posterior border of the medial pterygoid plate, the pterygomandibular ligament, the posterior end of the mylohyoid line of the mandible, the hyoid bone, and the thyroid and cricoid cartilages. The pharyngobasilar fascia serves to limit deformation of the nasopharynx and thereby helps to keep it patent. Posteriorly, the fibrous coat consists of a median raphe that is anchored above to the pharyngeal tubercle on the basilar part of the occipital bone.

3. Muscular Coat. The muscular coat comprises two layers of skeletal muscles (see below).

4. Fascial Coat. The fascial coat consists of the *buccopharyngeal fascia,* which covers the buccinator and the pharyngeal muscles, and blends above with the pharyngobasilar fascia.

MUSCLES OF PHARYNX

The wall of the pharynx is composed largely of two layers of muscles. The external, circular layer comprises three constrictors (fig. 63–8). The internal, chiefly longitudinal layer consists of two levators: the stylopharyngeus and the palatopharyngeus. **The constrictors have their fixed points in front, where they are attached to bones or cartilages, whereas they expand behind and overlap one another from below upward, and end in a median tendinous raphe.** The overlapping has been compared with that seen when three flower pots are placed one inside another (A. L. McGregor); the front wall of each is deficient, however. The muscular wall of the pharynx is invested by the buccopharyngeal fascia and lined by the pharyngobasilar fascia.

Inferior Constrictor. The inferior constrictor[21] arises from the arch of the

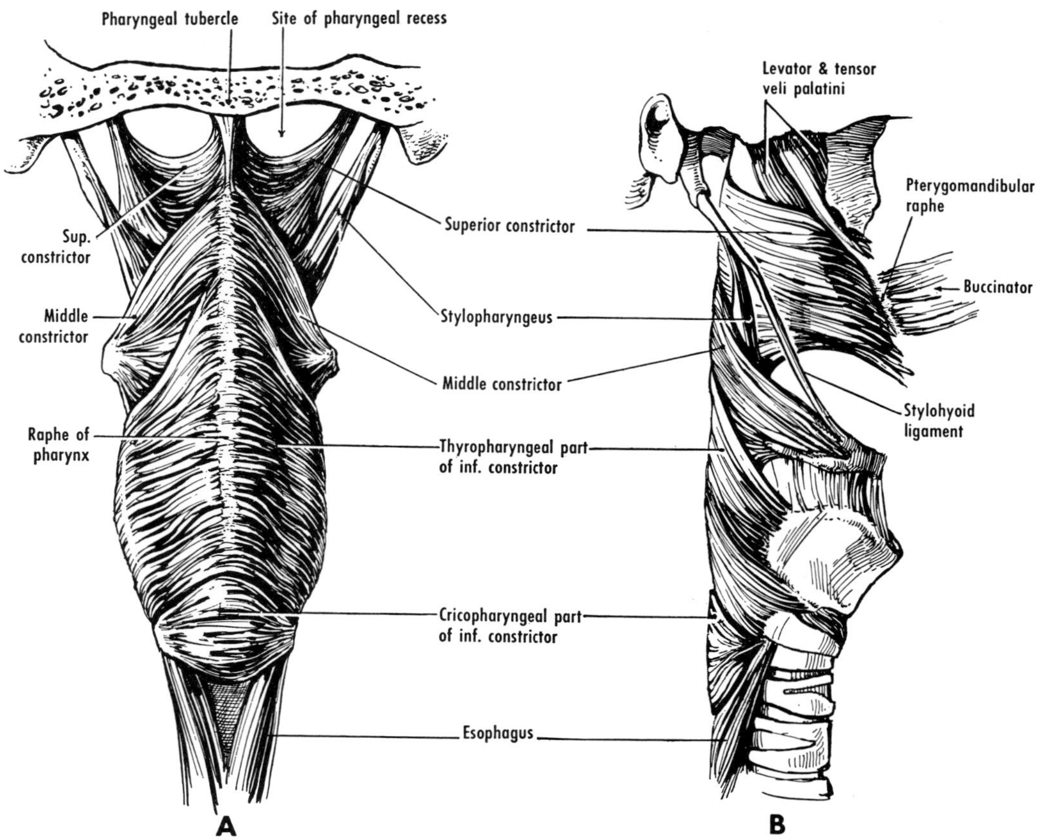

Figure 63–8 Muscles of the pharynx. *A,* posterior aspect. *B,* right lateral aspect.

cricoid cartilage *(cricopharyngeal part)* and from the inferior horn and the oblique line of the thyroid cartilage *(thyropharyngeal part)*. The separation into two parts, however, is frequently not clear. The cricopharyngeal fibers are horizontal in direction and are continuous with the circular fibers of the esophagus. They act as a sphincter[22] or "pinchcock" (C. Jackson) and, together with the uppermost circular fibers of the esophagus, prevent air from entering the esophagus. The cricopharyngeal sphincter relaxes suddenly during swallowing.[23] The thyropharyngeal fibers ascend obliquely, decussate posteriorly at the median raphe, and overlap the middle constrictor. They act in propulsion. The cricopharyngeal fibers shorten, whereas the thyropharyngeal fibers lengthen, the vocal folds of the larynx.[24] **A pharyngeal diverticulum may form posteriorly through the parts of the inferior constrictor (or sometimes below that muscle).**

The external and internal muscular layers of the upper portion of the esophagus consist of skeletal muscle that is attached (by the *crico-esophageal tendon*) to the back of the lamina of the cricoid cartilage.[25]

Middle Constrictor. The middle constrictor arises from the angle between the greater and lesser horns of the hyoid bone and from the stylohyoid ligament. Its fibers diverge backward and end in the median raphe. Its lowermost fibers descend under cover of the inferior constrictor; the highest fibers ascend and overlap the superior constrictor.

Superior Constrictor. The superior constrictor arises from (1) the side of the tongue and the mucous membrane of the mouth, (2) the mylohyoid line of the mandible, (3) the pterygomandibular raphe, and (4) the pterygoid hamulus but not from the posterior border of the medial pterygoid plate.[26] The fibers curve backward to end in the median raphe and in an aponeurosis that is attached to the pharyngeal tubercle on the basilar part of the occipital bone. A gap occurs between the upper border of the muscle and the base of the skull.

Palatopharyngeus. The palatopharyngeus occupies the palatopharyngeal fold. It arises from the posterior border of the bony palate and from the palatine aponeurosis. In the soft palate it is arranged in two strands, medial and lateral, separated by the levator veli palatini.[27] The strands unite and the entire muscle is inserted into the posterior border of the thyroid cartilage *(palatothyroideus)* and into the side of the pharynx and esophagus *(palatopharyngeus proper)*. The *salpingopharyngeus* extends from the cartilage of the auditory tube to the walls of the pharynx.

Stylopharyngeus. The stylopharyngeus arises from the medial aspect of the styloid process. It descends between the superior and middle constrictors and then passes under cover of the middle constrictor. It expands to end on the side of the pharynx and on the posterior border of the thyroid cartilage, becoming continuous with the palatopharyngeus.

Certain structures gain the palate or the pharynx in relation to the borders of the constrictors. Between superior constrictor and skull: levator veli palatini, auditory tube, ascending palatine artery. Between superior and middle constrictors: glossopharyngeal nerve and stylopharyngeus. Between middle and inferior constrictors: internal laryngeal nerve and superior laryngeal artery. Between inferior constrictor and esophagus: recurrent laryngeal nerve and inferior laryngeal artery.

Innervation of Muscles of Pharynx

The constrictors, the palatopharyngeus, and the salpingopharyngeus are supplied through the pharyngeal plexus from the pharyngeal branch of the vagus, which is believed to consist chiefly of fibers from the cranial part of the accessory nerve. The pharyngeal plexus is situated mainly on the middle constrictor. The inferior constrictor also receives branches from the external and recurrent laryngeal nerves. The stylopharyngeus is supplied by the glossopharyngeal nerve, which winds around the lateral side of the muscle.

Actions of Muscles of Pharynx

The constrictors constrict the wall of the pharynx upon its contents and they are active in swallowing.[28] The stylopharyngeus is frequently regarded as the chief elevator of the pharynx and the larynx but the levator veli palatini is also important.[29] The salpingopharyngeus has probably little effect on the auditory tube but assists in raising the walls of the pharynx during swallowing.[30] The chief action in which the muscles of the pharynx combine is swallowing.

DEGLUTITION

Deglutition,[31] or swallowing, is a complicated, neuromuscular act whereby food is transferred from the mouth, through the pharynx and esophagus, to the stomach. The word "bolus" is used for the mass of food (solid or even liquid) that is swallowed at one time. **Deglutition is commonly considered in three stages, occurring in (1) the mouth, (2) the pharynx, and (3) the esophagus,** respectively, and swallowing involves a consecutive activation of the walls of these three cavities. The pharyngeal stage, generally completed in less than a second, is the most rapid and most complex phase of deglutition. "A succession of coordinative efforts of the pharynx in swallowing" proceeds "in a consistent pattern when once reflexly initiated" and is "not dependent upon the mechanical action of the bolus."[31] Difficulty in swallowing is termed dysphagia.

Certain muscles contract concurrently to initiate the act.[32] By voluntary action the tongue passes the bolus almost vertically downward between the adducted palatopharyngeal folds. The bolus is prevented from entering the nasopharynx by contraction of the soft palate. As the bolus passes through the oropharynx, the walls of the pharynx are raised abruptly, and the hyoid bone and the larynx are elevated maximally. During deglutition the vestibule of the larynx is closed[33] but the epiglottis apparently adopts a variable position. (Conflicting reports have been published.[31]) The bolus is usually deviated laterally by the epiglottis and the ary-epiglottic folds into one or both lateral food channels (the piriform recesses of the laryngopharynx). Next, the laryngopharynx is elevated behind the obliterated vestibule of the larynx by the hyoid and laryngeal muscles. The sphincter formed by the cricopharyngeal part of the inferior constrictor, the oblique part of the cricothyroid,[34] and the uppermost circular fibers of the esophagus opens abruptly, and the bolus enters the esophagus.

The *pharyngeal ridge* is an elevation or bar on the posterior wall of the pharynx below the level of the soft palate. It is produced by transverse muscular fibers (the *palatopharyngeal sphincter*) that pull the posterior pharyngeal wall forward during swallowing but not normally during speech.[35] The fibers are derived from the superior constrictor[36] or from the palatopharyngeus.[37] The pharyngeal ridge can sometimes be seen in the living on elevation of the soft palate during effort or swallowing. It also marks the site of the change in the mucosa from the respiratory or nasal type (on the upper surface of the palate) to the pharyngeal type.

INNERVATION AND BLOOD SUPPLY OF PHARYNX

The motor, and most of the sensory, supply to the pharynx is by way of the *pharyngeal plexus*. This plexus, situated chiefly on the middle constrictor, is formed by the pharyngeal branches of the vagus and glossopharyngeal nerves, together with a more deeply placed[38] sympathetic branch from the superior cervical ganglion. The motor fibers in the plexus are from the 11th cranial nerve but are carried by the vagus and contribute to the supply of all the muscles of the pharynx and the soft palate except the stylopharyngeus (9th cranial nerve) and tensor veli palatini (5th cranial nerve). The sensory fibers in the plexus are from the glossopharyngeal[39] and they supply the greater portion of all three parts of the pharynx.

The pharynx is supplied chiefly by the ascending pharyngeal and inferior thyroid arteries, both derived from the external carotid. Venous plexuses are found beneath the mucous membrane[40] and on the back of the pharynx externally. The lymphatic vessels drain into the deep cervical nodes.

LARYNX

The ear, nose, and throat are frequently studied together from a clinical point of view under the general heading of otorhinolaryngology. The word throat is used for the parts of the neck in front of the vertebral column, especially the pharynx and the larynx. The specific study of the larynx is called laryngology.

The larynx is the organ that connects the lower part of the pharynx with the trachea. It serves (1) as a valve to guard the air passages, especially during swallowing, (2) for the maintenance of a patent airway, and (3) for vocalization.

The adult larynx is about 5 cm in length in the male, a little less in the female. The greater size in the male is due mostly to the increased growth of the larynx after puberty. The larynx is superficial anteriorly (fig. 63–9) and is related posteriorly to the laryngopharynx, the prevertebral layer of fascia and the prevertebral muscles, and the bodies of the cervical vertebrae (C.V. 3 to 6). Laterally, the larynx is related to the carotid sheath and its contents, the infrahyoid muscles, the sternomastoid, and the thyroid gland.

The larynx is elevated (particularly by

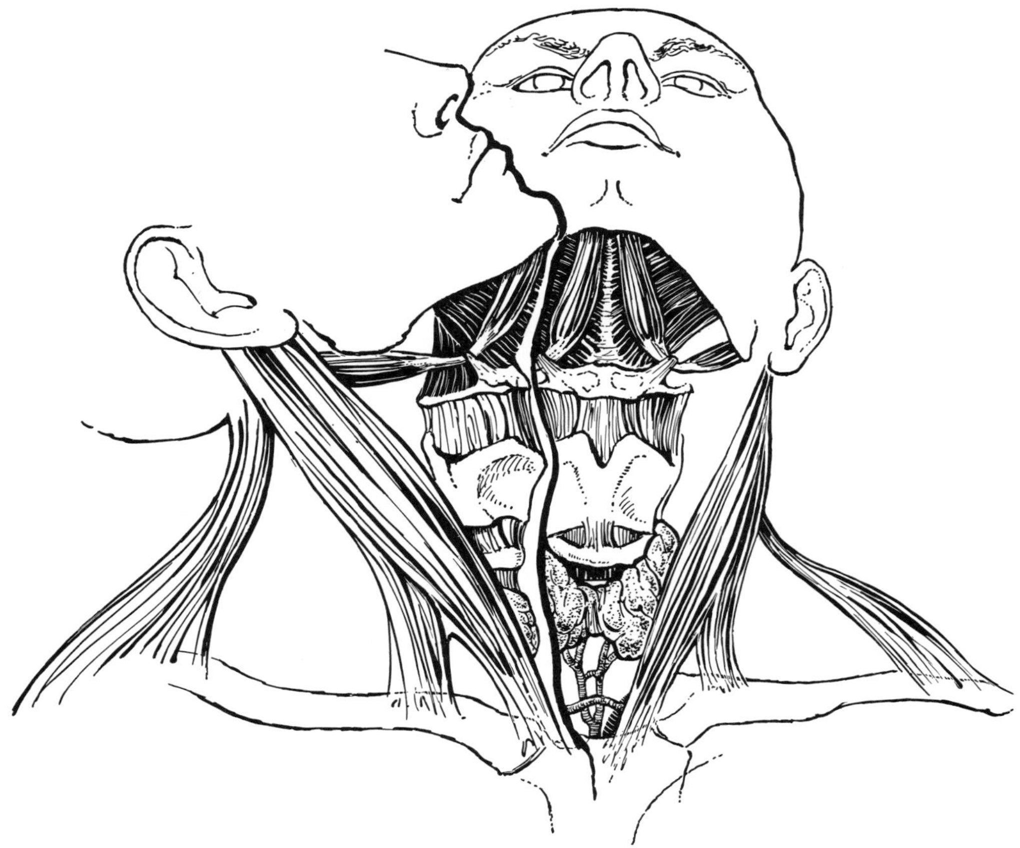

Figure 63–9 Drawing of the structures in or near **the anterior median line of the neck.** These include (1) the **symphysis menti,** (2) the **diaphragma oris** (mylohyoid muscles) crossed by the digastric muscles, (3) the **hyoid bone** (and median anastomoses between the suprahyoid branches of the right and left lingual arteries, and between the infrahyoid branches of the right and left superior thyroid arteries), (4) the **median thyrohyoid ligament** of the thyrohyoid membrane, (5) the **laryngeal prominence** formed by the thyroid cartilage, (6) the **cricothyroid ligament** (and the median anastomosis between the cricothyroid branches of the right and left superior thyroid arteries), (7) the **cricoid cartilage,** (8) the **cricotracheal ligament,** (9) the **trachea** and the **isthmus of the thyroid gland** (and the median anastomosis between the glandular branches of the right and left superior thyroid arteries), (10) the plexus formed by the **inferior thyroid veins,** (11) the **jugular arch** uniting the right and left anterior jugular veins, (12) occasionally small portions of the **brachiocephalic trunk** (and its inconstant branch, the arteria thyroidea ima), the **left brachiocephalic vein,** and the **thymus** (chiefly in childhood), and (13) the **jugular notch** of the manubrium sterni. The infrahyoid muscles have been removed for this illustration. The glottis (not seen) is approximately at the level of the midpoint of the anterior margin of the thyroid cartilage.

the palatopharyngeus) during extension of the head and during deglutition.

CARTILAGES OF LARYNX

The cartilages of the larynx are the thyroid, cricoid, epiglottic, arytenoid, corniculate, and cuneiform. The first three are single, the last three paired. The thyroid, cricoid, and arytenoid are composed of hyaline cartilage and may undergo calcification and/or endochondral ossification, thereby becoming visible radiographically (fig. 63–10). Ossification in the thyroid cartilage not uncommonly begins from about 20 years onward in the posterior border of each lamina; it is not possible to estimate age, however, from the extent of the ossification.[41] The other cartilages consist of elastic cartilage (as do also the apices and vocal processes of the arytenoids).

Thyroid Cartilage (fig. 63–11). The thyroid cartilage comprises two spring-like[42]

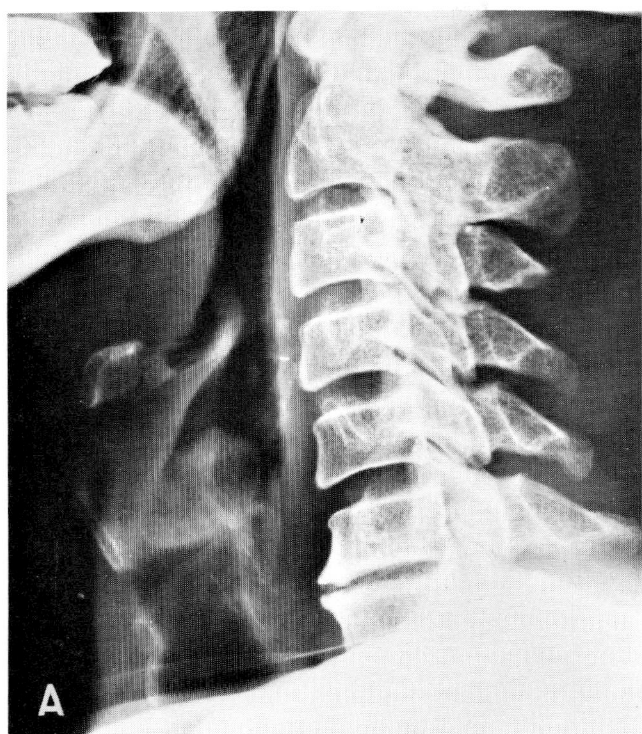

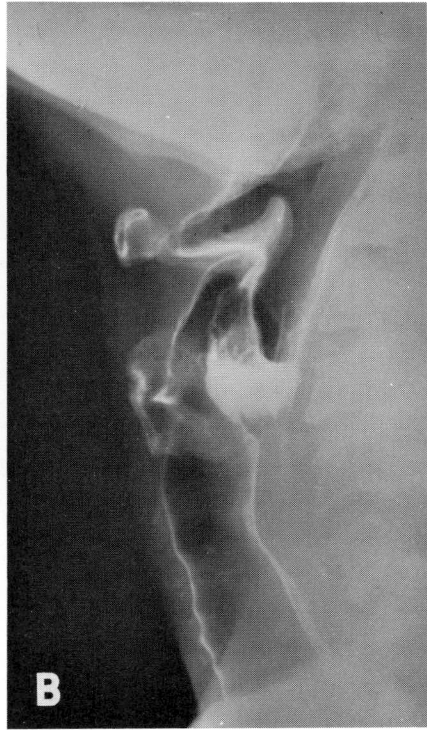

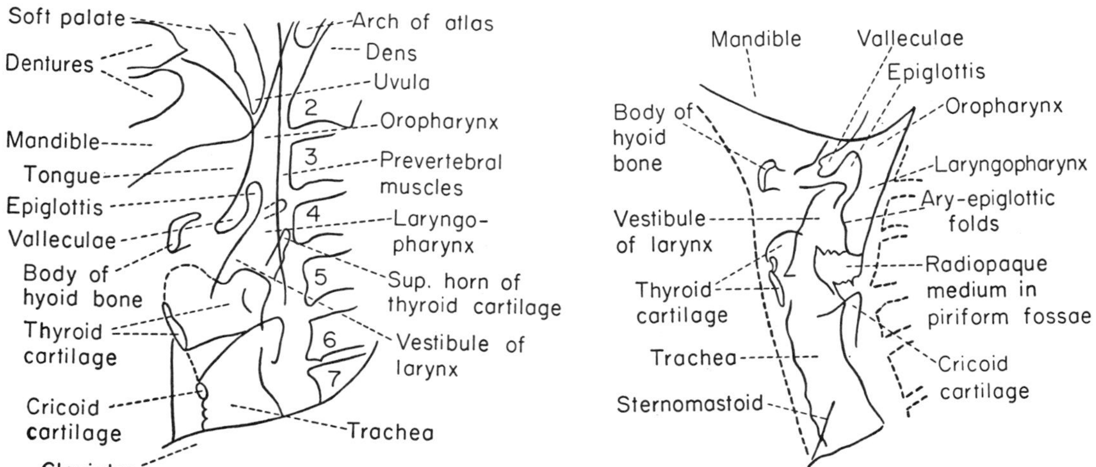

Figure 63–10 Larynx. Lateral radiograms showing the larynx *in vivo*, and key diagrams. In *B*, the mucous membrane of the pharynx, larynx, and trachea has been coated with a radio-opaque medium. *A*, courtesy of Sir Thomas Lodge, F.F.R., The Royal Hospital, Sheffield, England. *B*, from *Medical Radiography and Photography;* courtesy of Eastman Kodak Company, Rochester, New York.

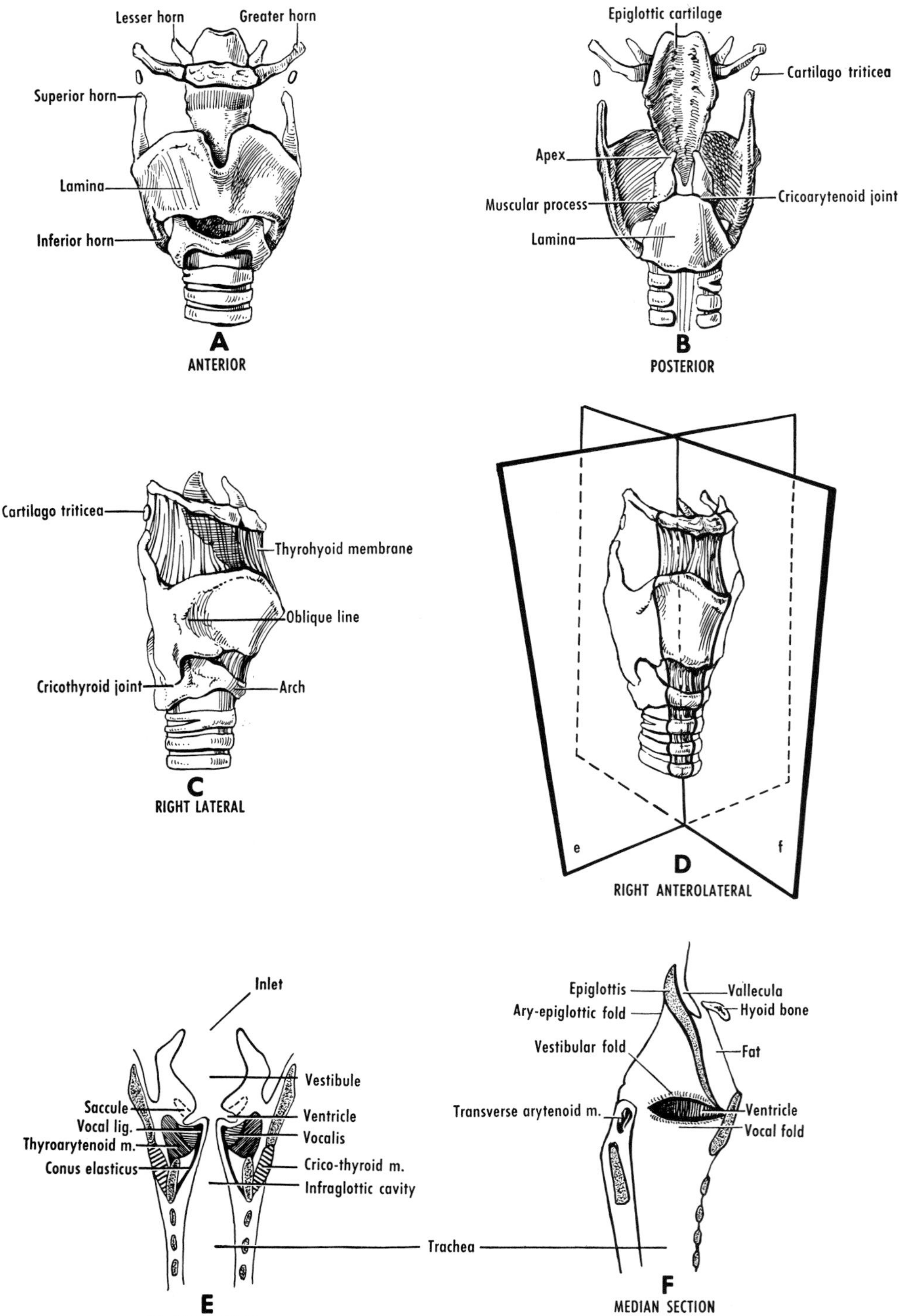

Figure 63–11 The larynx: (*A*) anterior, (*B*) posterior, and (*C*) right lateral aspects of the cartilages; (*D*) right anterolateral aspect, showing the planes of section of the next two diagrams, (*E*) coronal section, and (*F*) median section. Note the thyroid and cricoid cartilages in *A* to *F*, the hyoid bone and the epiglottic cartilage in *A* to *D* and in *F*, and the arytenoid cartilage in *B*. Based chiefly on von Lanz and Wachsmuth.

plates, termed *laminae,* which are fused in front but which diverge behind. The anterior borders of the laminae, fused below, diverge above (to form the *superior thyroid notch*). **Furthermore, the laminae produce a median elevation termed the *laryngeal prominence* ("Adam's apple"), which is palpable and is frequently visible in the living body.** The angle formed by the two laminae is approximately a right angle in the male and somewhat greater in the female. In the male, the thyroid angle is more acute, the laryngeal prominence more evident, the vocal folds longer, and the voice deeper in pitch. The posterior border of each lamina is prolonged upward and downward as cornua, or horns. The *superior horn* is anchored to the tip of the greater horn of the hyoid bone. The *inferior horn* presents a facet medially for articulation with the cricoid cartilage. The lateral surface of each lamina is crossed by an *oblique line,* to which are attached the inferior constrictor of the pharynx, the sternothyroid, and the thyrohyoid muscles.

Cricoid Cartilage (fig. 63–11). The cricoid cartilage is shaped like a signet ring. It comprises a posterior plate, termed its *lamina,* and a narrow, anterior part, or *arch.* On each side of the upper border of the lamina is a facet for articulation with the corresponding arytenoid cartilage. A depression on each side of the posterior surface of the lamina gives attachment to the posterior crico-arytenoid muscle and, in the median plane, a crest gives attachment to the esophagus (crico-esophageal tendon). **The lower border of the cricoid cartilage marks the end of the pharynx and larynx, and the commencement of the esophagus and trachea.** The lower border is connected to the first ring of the trachea by the cricotracheal ligament. Laterally, the cricoid cartilage presents a facet for articulation with the inferior horn of the thyroid cartilage. **The cricoid cartilage is at the level of C.V. 6, and its arch is palpable *in vivo*.**

Arytenoid Cartilages (fig. 63–11*B*). The arytenoid cartilages articulate with the upper border of the lamina of the cricoid cartilage. Each is shaped like a three-sided pyramid and presents an *apex* above and a *base* below. The apex curves backward and medially, and supports the corniculate cartilage. Two processes project from the base. The *vocal process* extends forward and gives attachment to the vocal ligament. The *muscular process* extends laterally and gives attachment to the thyro-arytenoid and the lateral and posterior crico-arytenoid muscles. The medial surface of the arytenoid is covered by the mucous membrane of the larynx. The posterior surface gives attachment to the transverse arytenoid muscle. The anterolateral surface gives attachment to the vocal and thyro-arytenoid muscles and to the vestibular ligament.

Corniculate Cartilages. These are a pair of nodules that sit on the apices of the arytenoid cartilages and are located in the ary-epiglottic folds of mucous membrane.

Cuneiform Cartilages. These are an inconstant pair of rods situated in the ary-epiglottic folds in front of the corniculate cartilages. (See figures 63–12*C* and 63–17, pp. 754, 761.)

Epiglottic Cartilage (fig. 63–11). The *epiglottis*[43] consists of the leaf-like epiglottic cartilage, largely covered by mucous membrane. **It is situated behind the root of the tongue and the body of the hyoid bone, and in front of the inlet of the larynx.** The epiglottic cartilage presents pits in which glands are lodged, and foramina through which nerves and blood vessels pass. The upper end of the cartilage is broad, whereas the lower end, the *stalk (petiolus),* is pointed and is attached to the back of the thyroid cartilage. The front of the epiglottis is separated by a pad of fat from the median thyrohyoid ligament. Taste buds are present in the posterior surface of the epiglottis and in the anterolateral surfaces of the arytenoids.[44] The lower part of the posterior surface of the epiglottis projects backward (and forms the *epiglottic tubercle*).

JOINTS OF LARYNX

Cricothyroid Joint. The cricothyroid joint on each side is a synovial joint between the side of the cricoid cartilage and the inferior horn of the thyroid cartilage. The main movement is rotation of the thyroid cartilage around a horizontal axis that passes through the joints of the two sides. Some gliding movements may also take place.

Crico-arytenoid Joint. The crico-arytenoid joint on each side is a synovial joint between the upper border of the lamina of the cricoid cartilage and the base of the arytenoid cartilage in the region of its mus-

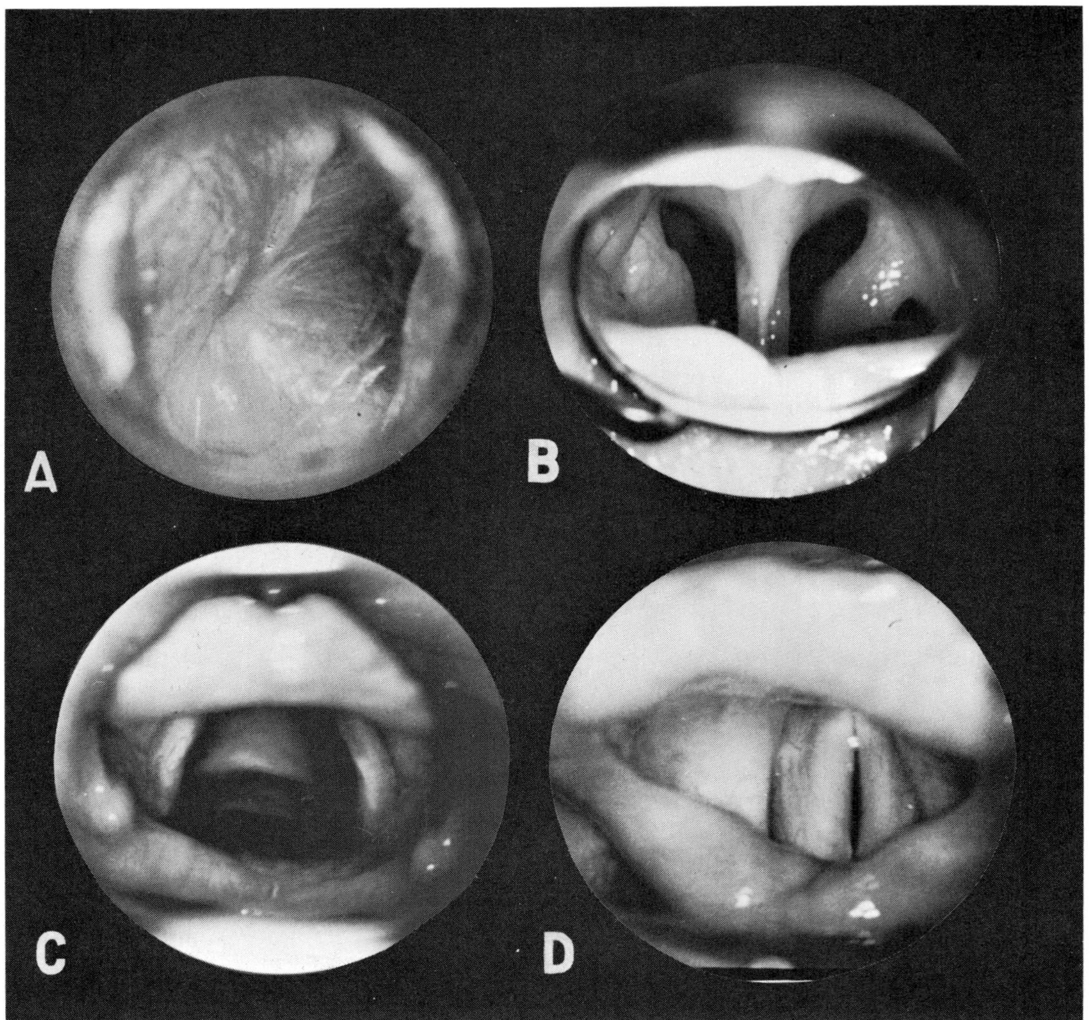

Figure 63–12 Ear, nose, and throat *in vivo*. *A*, the right tympanic membrane as seen through a speculum. Note the handle and lateral process of the malleus, and the cone of light. Compare figure 54–2. *B*, the nasopharynx and nasal cavities as seen in a mirror placed on the posterior pharyngeal wall. Note the posterior edge of the nasal septum, the inferior nasal conchae, and (on the right side of the illustration) the orifice of the auditory tube. Compare figure 62–5. *C*, the larynx on inspiration, as seen in a mirror placed on the posterior pharyngeal wall. Note the epiglottis, ary-epiglottic fold and cuneiform cartilage (on the left side of the illustration), vestibular and vocal folds, and trachea. Compare figure 63–17. *D*, the larynx on phonation, as seen in a mirror. Note the vestibular and vocal folds, and that the latter are now approximated. All photographs courtesy of Paul H. Holinger, M.D., Chicago, Illinois.

cular process. Strong ligamentous fibers anchor the arytenoid and prevent it, so to speak, from "falling into the larynx" (J. C. B. Grant). The arytenoid cartilages glide[45] and rotate[46] on the cricoid at these joints. As the vocal processes are carried laterally or medially, the rima glottidis is opened or closed, respectively.

LIGAMENTS OF LARYNX

Thyrohyoid Membrane. The thyrohyoid membrane connects the thyroid cartilage with the hyoid bone. It is attached to the *upper* border of the hyoid bone (fig. 63–11*F*), and a bursa intervenes between the membrane and the back of the body of the

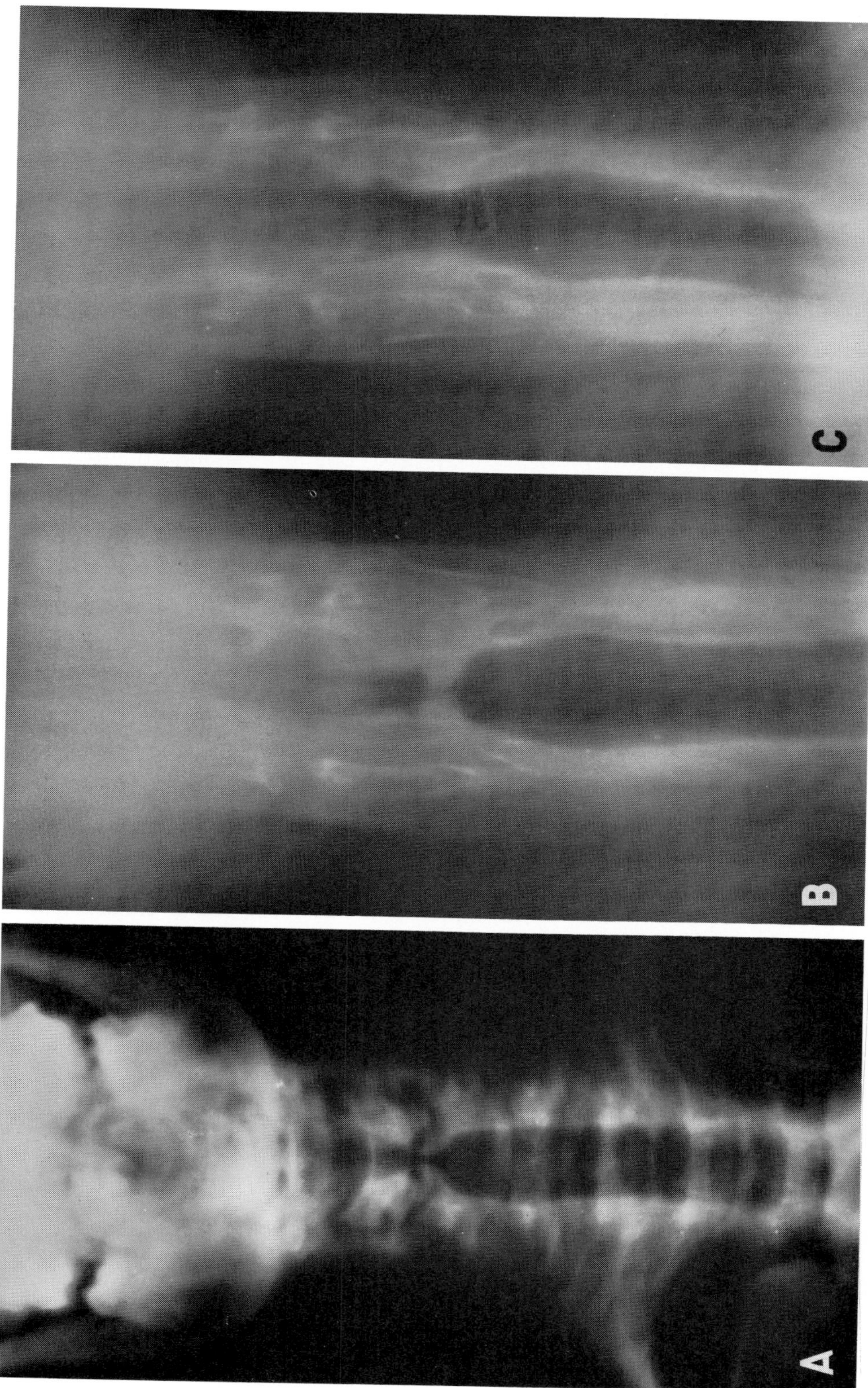

Figure 63–13 Anteroposterior tomograms of larynx *in vivo*. *A*, tomogram of a larynx during phonation. The cavities of the larynx and trachea (and also portions of the piriform recesses) appear radiolucent (black) because of the air that they contain. Note the vestibule, vestibular folds, ventricles, and vocal folds of the larynx, and compare figure 63–11*E*. *B* and *C*, tomograms of a larynx (*B*) during phonation and (*C*) at rest. Note the approximation of the vocal folds in *B*, and their separation in *C*. The hyoid bone and the thyroid cartilage can be identified as radio-opaque areas on each side. A, courtesy of Lt. A. V. Headley, Medical Department, U.S. Navy.

hyoid. The median part is thickened *(median thyrohyoid ligament).* The membrane is pierced on each side by the internal laryngeal nerve and the superior laryngeal vessels. The posterior margin of the membrane on each side *(lateral thyrohyoid ligament)* is thickened and connects the tip of the superior horn of the thyroid cartilage with the tip of the greater horn of the hyoid. A small nodule *(cartilago triticea)* may be found in the posterior margin of the membrane (fig. 63–11*C*).

Cricothyroid Ligament (fig. 63–9). The cricothyroid ligament connects the arch of the cricoid cartilage with the thyroid and with the vocal processes of the arytenoid cartilages. The terminology of this ligament or membrane varies from one author to another. The term *conus elasticus (cricovocal membrane)* (fig. 63–11*E*) is used for the elastic fibers that extend upward from the cricoid cartilage to the vocal ligaments (see below).

In acute respiratory obstruction, cricothyrotomy is preferable to tracheostomy (p. 690) for the non-surgeon.[47] A pillow is placed beneath the scapulae, the head is moderately extended, and the cricothyroid interval is palpated. The thyroid and cricoid cartilages are fixed between the thumb and the middle finger of the left hand, while the index finger is used as a marker for the cricothyroid ligament. The instrument of choice is either a "dagger-scissors" or a "dagger-dilator." The skin over the cricothyroid interval is pinched and cut transversely for about 1 cm, the cricothyroid ligament is then pierced transversely and dilated transversely, and an appropriate cannula is inserted. After cricothyrotomy, tracheostomy is performed within 24 to 48 hours if an opening is still required.

Vocal Ligament. The vocal ligament on each side extends from the thyroid cartilage in front to the vocal process of the arytenoid cartilage behind. It may be regarded as the upper border of the conus elasticus. It is composed of elastic fibers and is covered tightly by the vocal fold of mucous membrane (fig. 63–11*F*).

Vestibular (Ventricular) Ligament. The vestibular ligament on each side is an indefinite band that is situated above the vocal ligament. It extends from the thyroid cartilage in front to the anterolateral surface of the arytenoid cartilage behind. It is covered loosely by the vestibular fold of mucous membrane (fig. 63–11*F*).

Ligaments of Epiglottis. The epiglottis is attached to the hyoid bone *(hyo-epiglottic ligament),* the back of the tongue *(median glosso-epiglottic fold),* the side of the pharynx *(lateral glosso-epiglottic* or *pharyngo-epiglottic folds),* and the thyroid cartilage *(thyro-epiglottic ligament).* The depression on each side between the median and lateral glosso-epiglottic folds is termed the *epiglottic vallecula.* The elastic tissue connected with the side of the epiglottic cartilage, the upper border of which forms the basis of the ary-epiglottic fold, is termed the *quadrangular membrane.*

INLET OF LARYNX
(fig. 63–3)

The *aditus (inlet)* of the larynx leads from the laryngopharynx into the cavity of the larynx. It is set obliquely, facing largely backward. It is bounded anteriorly by the upper border of the epiglottis, on each side by the ary-epiglottic folds, and below and behind by an interarytenoid fold. Figure 63–12*C* shows a photograph of the inlet. The ary-epiglottic folds contain the ary-epiglottic muscles and the corniculate and cuneiform cartilages. The inlet is related laterally to the piriform recess of the laryngopharynx on each side (fig. 63–3). The ary-epiglottic folds provide lateral food channels which lead down the sides of the epiglottis, through the piriform recesses, and to the esophagus. Closure of the inlet protects the respiratory passages against the invasion of food and foreign bodies.

CAVITY OF LARYNX
(fig. 63–11*E*)

The cavity of the larynx is divided into three portions (vestibule, ventricles and the part between them, and infraglottic cavity) by two pairs of horizontal folds, the vestibular and the vocal (fig. 63–11*F*).

Vestibule

The vestibule of the larynx extends from the inlet to the vestibular folds. It is limited in front by the back of the epiglottis, on each side by the ary-epiglottic folds, and behind by an interarytenoid fold.

Ventricles

The ventricle *(sinus)*[48] of the larynx on each side extends from the ventricular fold above to the vocal fold below. Each ventricle resembles a canoe laid on its side, and the two ventricles communicate with each other by means of the median portion of the laryngeal cavity. The ventricles allow free movement of the vocal folds.

They are lined by stratified columnar epithelium.

Saccule. The saccule of the larynx *(appendix of the ventricle)* is a diverticulum that extends upward from the front of each ventricle. The secretion of the mixed glands of the saccule lubricates the vocal folds, and hence the saccule has been called the "oil can" of the vocal folds.[49] Lymphatic tissue is found in the wall of the saccule.[50] The saccules are highly variable in size[51] and may perforate the thyrohyoid membrane.

Vestibular (Ventricular) Folds. The two vestibular folds (fig. 63–11, *E* and *F*), or "false vocal cords," extend from the thyroid cartilage in front to the region of the cuneiform cartilages behind. Each consists of elastic tissue, together with fat, mucous glands, and muscle, and contains a ventricular ligament, which extends from the angle of the thyroid cartilage to the arytenoid cartilage above the vocal process. The *rima vestibuli* is the interval between the two folds. The vestibular folds are protective in function and do not normally affect voice production. The two folds probably meet during swallowing.

Glottis. **The glottis comprises the vocal folds and processes, together with the interval, the rima glottidis, between them.**

VOCAL FOLDS (fig. 63–11, *E* and *F*). **The two vocal folds, or "true vocal cords," are pearly white, mobile, musculomembranous shelves located below and medial to the ventricular folds. They extend from the angle of the thyroid cartilage in front to the vocal processes of the arytenoid cartilages behind.** Each contains a vocal ligament, which consists of elastic tissue derived from the conus elasticus. The vocalis, which is a part of the thyro-arytenoid muscle, forms the bulk of the vocal fold.

RIMA GLOTTIDIS. **The rima glottidis is the narrowest part of the laryngeal cavity** and can be seen through the wider rima vestibuli by laryngoscopy. The mucous membrane over each vocal ligament presents non-keratinizing, stratified squamous (and also pseudostratified columnar[52]) epithelium, is firmly bound down, and, because it is avascular, appears white. The vocal folds control the stream of air passing through the rima and hence are important in voice production.

The longer, anterior part of the rima glottidis (the *intermembranous part*) is located between the vocal folds, whereas its shorter, posterior portion (the *intercartilaginous part*) is situated between the arytenoid cartilages (fig. 63–15, *A* and *B*). The shape and size of the rima are altered by movements of the arytenoids (fig. 63–16). The rima is wider during inspiration and narrower during expiration. During quiet respiration, however, the lumen of the larynx remains wide open and the vocal folds are turned upward over the ventricles.[53] In phonation the vocal folds are turned down across the airway and are approximated. The rima is then merely a slit. The quiescent rima glottidis is nearly 2½ cm in length in the male, less than 2 cm in the female.

In surface anatomy, the rima glottidis is approximately on the level of the midpoint of the anterior margin of the thyroid cartilage.

Infraglottic Cavity

The infraglottic cavity is the lowest portion of the laryngeal cavity and extends from the rima glottidis above to the trachea below. It is limited by the cricothyroid ligament and the internal aspect of the cricoid cartilage. When the vocal folds are approximated, the infraglottic cavity is shaped like a dome, the roof of which is formed by the mucosa-covered conus elasticus.

CLOSURE OF LARYNX

There are three levels or tiers in the larynx[54] that can be closed by sphincteric muscles: (1) the inlet (ary-epiglottic sphincter), which is closed during deglutition and protects the respiratory passages against the invasion of food; (2) the vestibular folds, closure of which traps the air below, and makes possible an increase of intrathoracic pressure (as in coughing), or intra-abdominal pressure (as in micturition and defecation); and (3) the vocal folds, the approximation of which occurs in phonation.

MUCOUS MEMBRANE OF LARYNX

The mucosa of the larynx is continuous with that of the laryngopharynx above and with that of the trachea below. It is adherent over the back of the epiglottis and over the vocal ligaments but is loose elsewhere, so that abnormally it may become raised by submucous fluid, as in edema of the larynx. **The edema does not spread below**

the level of the vocal folds, being limited by the tight attachment of the mucosa to the vocal ligaments. The epithelium is stratified squamous in the upper part of the vestibule (including the ary-epiglottic folds) and over the vestibular[55] and vocal ligaments; elsewhere, including the ventricles, it is pseudostratified columnar ciliated. Mucous glands are numerous.

SENSORY INNERVATION OF LARYNX

The mucosa of the larynx receives its sensory innervation on each side chiefly from the internal laryngeal branch of the superior laryngeal nerve, which supplies the larynx as far down as the vocal folds. This nerve also contains secretory fibers to the mucous glands of the larynx, and probably proprioceptive fibers. The lower part of the larynx may receive sensory fibers from the recurrent laryngeal nerve. Sympathetic fibers reach the larynx through the recurrent and superior laryngeal nerves and along the arteries.

MUSCLES OF LARYNX

Extrinsic Muscles

The extrinsic muscles of the larynx are those that move the larynx as a whole. They may be classified as elevators and depressors. The elevators include the thyrohyoid, stylohyoid, mylohyoid, digastric, stylopharyngeus, and palatopharyngeus. The depressors include the omohyoid, sternohyoid, and sternothyroid.

Intrinsic Muscles

The chief intrinsic muscles of the larynx are the cricothyroid, posterior crico-arytenoid, lateral crico-arytenoid, transverse arytenoid, thyro-arytenoid, vocalis, and oblique arytenoid (fig. 63–14). All are paired, with the exception of the transverse arytenoid. Other muscular bundles, sometimes given special names, are of little practical importance. The muscles of the larynx are richly supplied with nerves and posses numerous muscle spindles.[56]

Although the laryngeal muscles are complicated, they may be classified into (1) a sphincteric or adductor group which closes the larynx, and (2) a pair of dilator or abductor muscles (the posterior crico-arytenoids) which open it.

Cricothyroid (fig. 63–14*A*). The most superficial of the laryngeal muscles, the cricothyroid receives an innervation different from that of the others. It arises from the lateral surface of the arch of the cricoid and is inserted fanwise into the lower border of the lamina of the thyroid cartilage (straight part of cricothyroid) and into the anterior border of the inferior horn (oblique part of cricothyroid). The cricothyroid is supplied by the external laryngeal nerve. The muscles of the two sides act on the cricothyroid joints and tilt the thyroid cartilage down and/or the cricoid up, thereby lengthening, tensing, and adducting the vocal folds. Other actions are also described.[57]

Posterior Crico-arytenoid (fig. 63–14*D*). The posterior crico-arytenoid arises from the back of the lamina of the cricoid and is inserted into the muscular process of the arytenoid. It draws the muscular process backward and thereby rotates the vocal process laterally, resulting in opening of the rima glottidis (fig. 63–15*A* and *B*). It is the only abductor of the vocal folds as a whole.

Lateral Crico-arytenoid (fig. 63–14*B*). The lateral crico-arytenoid extends from the arch of the cricoid to the muscular process of the arytenoid. It is frequently inseparable from the thyro-arytenoid mus-

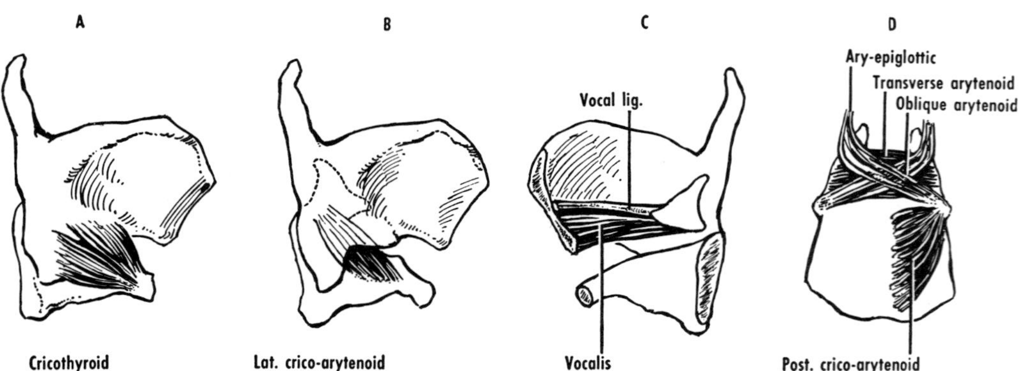

Figure 63–14 Intrinsic muscles of the larynx. *A* and *B*, right lateral aspect of the thyroid and cricoid cartilages. *C*, medial aspect of the right half of the thyroid and cricoid cartilages. *D*, posterior aspect of the arytenoid and cricoid cartilages.

cle. It draws the muscular process forward and thereby rotates the vocal process medially, resulting in closing of the rima glottidis, as in phonation.

Transverse Arytenoid (Interarytenoid) (fig. 63–14*D*). This muscle connects the medial surfaces of the two arytenoid cartilages. The fibers, therefore, approximate the arytenoids, thereby aiding in closing of the rima glottidis, as in phonation (fig. 63–16*B*).

Thyro-arytenoid[58] (fig. 63–11*E*). This is a variable muscle situated partly on the lateral aspect of the conus elasticus. It arises from the medial surface of the lamina of the thyroid cartilage and from the conus elasticus. It is inserted into the anterolateral surface and the muscular process of the arytenoid cartilage. Some of the fibers extend to the lateral border of the epiglottic cartilage. Its action is disputed.

Vocalis (Vocal Muscle) (figs. 63–14*C* and 63–11*E*). The vocalis lies medial to the thyro-arytenoid, with which it is fused. It arises from the angle between the laminae of the thyroid cartilage and is inserted into the vocal process of the arytenoid cartilage. It is not attached to the vocal ligament.[59] The precise arrangement of the fibers is complicated[60] and is in dispute. The muscle is perhaps responsible for local variations in the tension of the vocal fold during phonation and singing.

Oblique Arytenoid (fig. 63–14*D*). This muscle connects the muscular process of one arytenoid cartilage with the apex of the opposite cartilage. Some of the fibers are continued into the ary-epiglottic fold (as the *ary-epiglottic muscle*), and some of these may reach the epiglottis. The oblique arytenoid and the ary-epiglottic close the inlet of the larynx, as in deglutition.

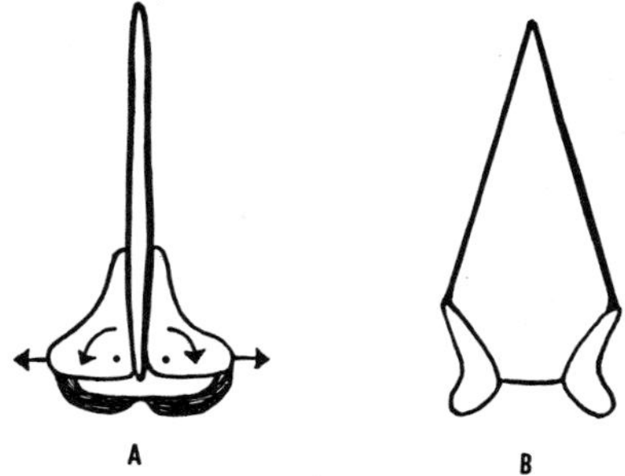

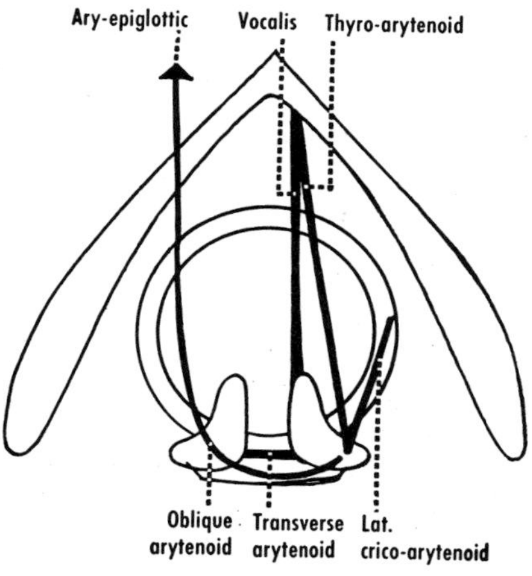

Figure 63–15 Simplified schemes of the intrinsic muscles of the larynx, as seen from above. *A* shows the posterior crico-arytenoid muscle (shaded). The arrows indicate (1) the lateral rotation of the arytenoid cartilages (around oblique axes), and (2) the associated lateral sliding of the arytenoid cartilages. *B* shows the resulting abduction of the vocal folds. *C* shows the attachments of the adductors.

Summary of Intrinsic Muscles

Three muscles arise from the cricoid cartilage: the cricothyroid, passing backward to the lamina and inferior horn of the thyroid cartilage; the lateral crico-arytenoid, extending backward to the muscular process; and the posterior crico-arytenoid, extending laterally to the muscular process (fig. 63–14). Two muscles, intimately related to each other, connect the thyroid and arytenoid cartilages: the thyro-arytenoid and the vocalis (fig. 63–15). Two muscles unite the arytenoid cartilages together: the transverse and oblique arytenoids.

MOTOR INNERVATION OF LARYNX

All the intrinsic muscles, with the exception of the cricothyroid, are supplied by the recurrent laryngeal nerve of the vagus.

The cricothyroid is supplied by the external laryngeal branch of the superior laryngeal nerve of the vagus. It is said to receive additional fibers, however, from the recurrent laryngeal nerve.[61] **The fibers to the various laryngeal muscles are believed to reach the vagus by way of the internal branch of the accessory nerve.** The transverse arytenoid receives additional fibers from the internal laryngeal branch of the superior laryngeal nerve.[62] Whether these fibers are motor[63] or proprioceptive[64] is disputed.

Injury to a recurrent laryngeal nerve may occur from a tumor or aortic aneurysm, or from trauma during thyroid surgery. The last-mentioned source of damage depends on "the variability rather than the vulnerability" of the nerve (Berlin).

Unilateral severance of a recurrent laryngeal nerve results in paralysis of all of the intrinsic muscles of the larynx, except the cricothyroid. The vocal fold moves mediad (usually to the paramedian position) owing to the action of the cricothyroid. It is sometimes stated that the abductor (posterior crico-arytenoid) is paralyzed first because the adductor fibers of the recurrent nerve are "more resistant," but considerable doubt has been cast on this. The voice is hoarse, although this effect may become masked by the unaffected fold crossing the median plane to meet the paralyzed one.

In bilateral paralysis of the recurrent laryngeal nerves, the vocal folds are usually paramedian or median (glottic chink), the voice is reduced to a hoarse whisper, and respiratory distress is apparent. Severance of both the recurrent and the external laryngeal nerve (the latter supplying the cricothyroid) leads to what is termed the intermediate or cadaveric position (intermediate between paramedian and abducted positions).[65]

GENERAL ACTIONS OF MUSCLES OF LARYNX

The intrinsic muscles of the larynx may be classified functionally as follows:

1. **External adductor and tensor: cricothyroid.**
2. **Internal adductors or sphincters: lateral crico-arytenoid, transverse arytenoid, thyro-arytenoid, vocalis, oblique arytenoid.**
3. **Internal abductor or dilator: posterior crico-arytenoid.**

The following is a simplified account of the actions of the muscles of the larynx.[66] The muscles of the larynx, including the thyro-arytenoid, are concerned with widening the chink of the glottis (abduction), as in respiration, or closing it (adduction), as in phonation (fig. 63–16). After closure of the glottis, the vocal folds can be tightened and lengthened. Abduction is carried out solely by the posterior crico-arytenoids, which, extending laterally from the back of the cricoid cartilage to the muscular processes, rotate the arytenoid cartilages laterally (fig. 63–15, *A* and *B*). Adduction is carried out by the lateral crico-arytenoids, which, extending backward from the arch of the cricoid cartilage to the muscular processes, rotate the arytenoid cartilages medially (fig. 63–15*C*). When the vocal folds are adducted, the gap that would otherwise occur posteriorly between the two arytenoid cartilages is closed by contraction of the transverse arytenoid. The adducted vocal folds are lengthened and placed under tension by the cricothyroid muscles.

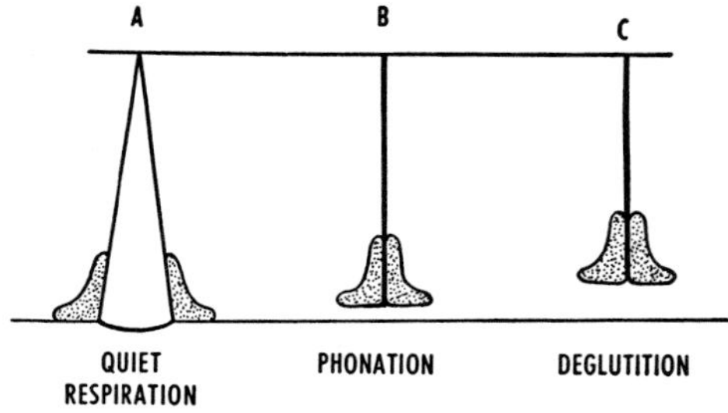

Figure 63–16 Three positions of the vocal folds. *A*, during quiet respiration, the folds are in gentle abduction. *B*, during phonation, the action of the adductors is opposed by the posterior crico-arytenoids. *C*, during deglutition, the action of the adductors is unopposed.

VOCALIZATION

Speech[67] involves (1) the expiration of air from the lungs by the diaphragm, and the abdominal and intercostal muscles; (2) the vibration of air (phonation) against the vocal folds, the tension and position of the folds being controlled by muscular action; and (3) resonance and articulation in the nasal, oral, and pharyngeal cavities, aided by the labial, lingual, and palatal muscles.

Tones produced by the larynx have the characteristic fundamental pitch of the sounds that emanate from the mouth. In other words, the larynx is a tone-producing organ that furnishes the raw sound which is modified into the voice by various resonating chambers above and below the larynx. The upper resonators (mouth, pharynx, and nose) are important organs of speech, as can be seen from the circumstance that, after removal of the larynx, the esophagus can serve as the source of sound for the production of speech, although control of pitch and volume is lacking.

The actions of the intrinsic muscles of the larynx are not entirely clear. It is generally believed, however, that the cricothyroid and the vocalis alter the length and tension of the vocal folds and are concerned in the production of various tones.

Cerebral development for speech is generally said to be linked with the development of hand preference, but actually "the left hemisphere is usually dominant for speech, regardless of handedness."[68] Right-handedness, present in about 90 to 95 per cent of the population of America and Britain, is claimed to be largely determined by heredity.[69] Other examples of lateral dominance are eye preference and foot preference.

Coughing and sneezing are respiratory reflexes in which the glottis is first closed and then opened suddenly, so that a burst of air is forced through the mouth or nose. Hiccup is an inspiratory reflex in which a staccato type of inspiration is produced by sudden contractions of the diaphragm, the glottis being partly or wholly closed. Laughing is produced by jerky expiration and is often accompanied by phonation ("ha, ha").

INNERVATION AND BLOOD SUPPLY OF LARYNX

The sensory and the motor innervation of the larynx have been noted (pp. 758 and 759). **In summary, the internal laryngeal nerve innervates the mucosa as far down as the vocal folds. The external laryngeal nerve sup-**

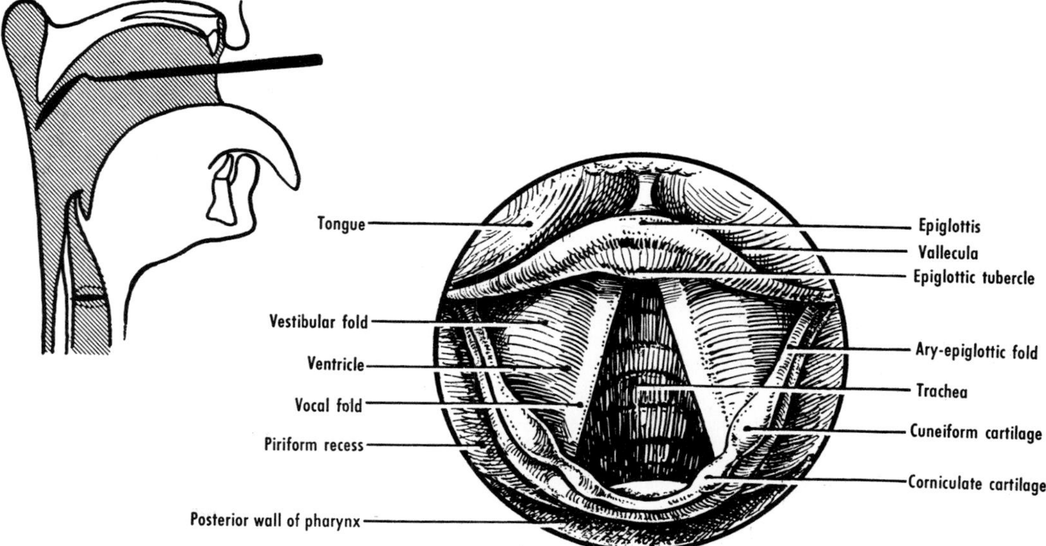

Figure 63–17 Indirect laryngoscopy. *A*, showing placement of mirror in the larynx. *B*, details of structures seen in the mirror during respiration. Compare figure 63–12*C*, p. 754. The cartilaginous rings of the upper part of the trachea can be seen through the open glottis.

plies the inferior constrictor of the pharynx and the cricothyroid. The recurrent laryngeal nerve supplies all of the laryngeal muscles except the cricothyroid, and also innervates the mucosa below the level of the vocal folds.

The larynx is supplied by the superior[70] and inferior laryngeal arteries, derived, respectively, from the superior and inferior thyroid arteries. They accompany, respectively, the internal and recurrent laryngeal nerves. Veins accompany the arteries, and lymphatics end in the deep cervical nodes.[71]

EXAMINATION OF LARYNX (LARYNGOSCOPY)

(fig. 63–12, *C* and *D*, p. 754)

The larynx can be examined *in vivo* by means of a mirror (indirect laryngoscopy) or a tubular instrument (direct laryngoscopy).

Indirect Laryngoscopy (fig. 63–17). The light from a lamp is reflected into the subject's mouth by a concave forehead mirror with a central aperture through which the examiner looks. An angled laryngeal mirror, warmed and oriented with its reflecting surface downward and forward, is placed against the anterior aspect of the uvula. The light illuminates the laryngeal mirror, in which an image of the larynx can be seen. The vocal folds become approximated (fig. 63–16) when the subject says "eh."

Direct Laryngoscopy. This procedure forms a part of peroral endoscopy, that is, the direct examination of the laryngopharynx, esophagus, and stomach, or of the larynx, trachea, and bronchi. Such examinations are performed by means of tubes provided with a source of illumination.

REFERENCES

1. J. F. Bosma and S. G. Fletcher, Ann. Otol., etc., St Louis, *70*:953, 1961 and *71*:134, 1962.
2. H. Ruben, N. Bentzen, and S. K. Saev, Lancet, *1*:849, 1960.
3. F. W. Jones, J. Anat., Lond., *74*:147, 1940.
4. K. Leela, R. Kanagasuntheram, and F. Y. Khoo, J. Anat., Lond., *117*:333, 1974.
5. J. S. Calnan, Brit. J. plast. Surg., *5*:286, 1955. R. Astley, J. Laryng., *72*:325, 1958.
6. R. D. Emslie, M. Massler, and J. D. Zwemer, J. Amer. dent. Ass., *44*:506, 1952.
7. L. B. Arey, Amer. J. Anat., *80*:203, 1947.
8. G. M. Dorrance, Arch. Otolaryng., Chicago, *13*:187, 1931.
9. R. H. Melchionna and R. A. Moore, Amer. J. Path., *14*:763, 1938. J. D. Boyd, J. Endocr., *14*:66, 1956. P. McGrath, J. Anat., Lond., *112*:185, 1972.
10. A. Mangiaracina, Arch. Otolaryng., Chicago, *35*:649, 1942.
11. F. Y. Khoo, R. Kanagasuntheram, and K. B. Chia, Arch. Otolaryng., Chicago, *86*:456, 1967.
12. C. S. Simkins, Arch. Otolaryng., Chicago, *38*:476, 1943. G. O. Graves and L. F. Edwards, Arch. Otolaryng., Chicago, *39*:359, 1944. J. Terracol, A. Corone, and Y. Guerrier, *La trompe d'Eustache*, Masson, Paris, 1949.
13. S. R. Guild, Ann. Otol., etc., St Louis, *64*:537, 1955.
14. A. Schwartzbart, Ann. Otol., etc., St Louis, *67*:241, 1958.
15. J. Hochfilzer, Laryngoscope, St Louis, *58*:712, 1948.
16. G. Fetterolf, Amer. J. med. Sci., *144*:37, 1912. D. Brown, J. Anat., Lond., *63*:82, 1928.
17. D. Kassay and A. Sandor, Arch. Otolaryng., Chicago, *75*:144, 1962.
18. T. W. Todd and R. H. Fowler, Amer. J. Anat., *40*:355, 1927.

19. G. Kelemen, Ann. Otol., etc., St Louis, *52*:419, 1943.
20. H. P. Schugt, Arch. Otolaryng., Chicago, *31*:626, 1940.
21. A. Birmingham, J. Anat., Lond., *33*:10, 1898. H. Brunner, Arch. Otolaryng., Chicago, *56*:616, 1952. Y. Guerrier and Ngo-Van-Hien, C. R. Ass. Anat., *39*:279, 1952.
22. O. V. Batson, Ann. Otol., etc., St Louis, *64*:47, 1955. See also C. Zaino *et al.*, *The Pharyngoesophageal Sphincter,* Thomas, Springfield, Illinois, 1970.
23. F. E. Fyke and C. F. Code, Gastroenterology, *29*:23, 1955. F. J. Ingelfinger, Physiol. Rev., *38*:533, 1958.
24. W. Zenker, Z. Anat. EntwGesch., *121*:550, 1960.
25. A. Birmingham, J. Anat., Lond., *33*:10, 1898. M. E. Sauer, Anat. Rec., *109*:691, 1951.
26. R. Locchi and J. M. de Castro, Rev. sudamer. Morf., *1*:44, 1943. M. C. Oldfield, Brit. J. Surg., *29*:197, 1941.
27. J. Whillis, J. Anat., Lond., *65*:92, 1930.
28. J. V. Basmajian and C. R. Dutta, Anat. Rec., *139*:561, 1961.
29. J. F. Bosma, Ann. Otol., etc., St Louis, *62*:51, 1953.
30. J. K. McMyn, J. Laryng., *55*:1, 1940.
31. V. E. Negus, Proc. R. Soc. Med., *36*:85, 1942, and J. Laryng., *58*:46, 1943. G. H. Ramsey *et al.*, Radiol., *64*:498, 1955. M. Atkinson *et al.*, J. clin. Invest., *36*:581, 1957. J. F. Bosma, Physiol. Rev., *37*:275, 1957. A. W. Hrycyshyn and J. V. Basmajian, Amer. J. Anat., *133*:333, 1972.
32. R. W. Doty and J. F. Bosma, J. Neurophysiol., *19*:44, 1956.
33. G. M. Ardran and F. H. Kemp, Brit. J. Radiol., *29*:205, 1956.
34. Y. Guerrier and Ngo-Van-Hien, C. R. Ass. Anat., *39*:279, 1952.
35. J. Calnan, Plast. reconstr. Surg., *13*:275, 1954.
36. J. Whillis, J. Anat., Lond., *65*:92, 1930.
37. F. W. Jones, J. Anat., Lond., *74*:147, 1940. R. H. Townshend, J. Laryng., *55*:154, 1940.
38. G. Laux and J.-B. Prioton, C. R. Ass. Anat., *43*:466, 1957.
39. O. V. Batson, Arch. Otolaryng., Chicago, *36*:212, 1942.
40. R. Kanagasuntheram, W. C. Wong, and H. L. Chan, J. Anat., Lond., *104*:361, 1969.
41. J. A. Keen and J. Wainwright, S. Afr. J. lab. clin. Med., *4*:83, 1958.
42. B. R. Fink, Anesthesiol., *39*:325, 1973, and *40*:58, 1974; Acta otolaryng., Stockh., *77*:295, 1974.
43. G. Leutert and W. Kreutz, Z. mikr.-anat. Forsch., *72*:96, 1964.
44. G. J. Romanes, Anat. Soc. G. B. and Ireland, July, 1953.
45. W. M. Maue and D. R. Dickson, Arch. Otolaryng., Chicago, *94*:432, 1971.
46. C. A. R. D. Snell, Koninl. nederl. Akad. Wetensch. Proc., *50*:1370, 1947. B. Sonesson, Z. Anat. EntwGesch., *121*:292, 1959. W. W. Sullivan, M. E. Sauer, and G. Corssen, Tex. Rep. Biol. Med., *18*:284, 1960.
47. D. S. Ruhe, G. V. Williams, and G. O. Proud, Trans. Amer. Acad. Ophthal. Otolaryng., *64*:182, 1960.
48. P. Lacoste, C. R. Ass. Anat., *47*:418, 1962.
49. A. O. Freedman, Arch. Otolaryng., Chicago, *28*:329, 1938.
50. B. F. Kingsbury, Amer. J. Anat., *72*:171, 1943.
51. E. N. Broyles, Ann. Otol., etc., St Louis, *68*:461, 1959.
52. J. A. Holliday, Laryngoscope, St Louis, *81*:1596, 1971.
53. G. M. Ardran, F. H. Kemp, and L. Manen, Brit. J. Radiol., *26*:497, 1953.
54. J. J. Pressman and G. Kelemen, Physiol. Rev., *35*:506, 1955.
55. E. S. Hopp, Laryngoscope, *65*:475, 1955.
56. M. F. Lucas Keene, J. Anat., Lond., *95*:25, 1961.
57. A. Mayet and K. Mündnich, Acta anat., *33*:273, 1958. V. E. Negus, Proc. R. Soc. Med., *40*:849, 1947.
58. E. M. Josephson, Arch. Otolaryng., Chicago, *6*:139, 1927.
59. A. Mayet, Acta anat., *24*:15, 1955. J. van den Berg and J. Moll, Z. Anat. EntwGesch., *118*:465, 1955. B. Schlosshauer and K.-H. Vosteen, Z. Anat. EntwGesch., *120*: 456, 1958. M. Gajo and A. Gellert, Anat. Anz., *126*: 59, 1970.
60. F. Wustrow, Z. Anat. EntwGesch., *116*:506, 1952. L. Gomez Oliveros, Bull. Ass. Anat., *49*:127, 1964.
61. A. Mayet, Anat. Anz., *103*:340, 1956.
62. T. F. M. Dilworth, J. Anat., Lond., *56*:48, 1921. D. D. Berlin and F. H. Lahey, Surg. Gynec. Obstet., *49*:102, 1929. M. Nordland, Surg. Gynec. Obstet., *51*:449, 1930.
63. P. H. Vogel, Amer. J. Anat., *90*:427, 1952.
64. A. F. Williams, J. Laryng., *65*:343, 1951.
65. V. E. Negus, Proc. R. Soc. Med., *40*:849, 1947. O. H. Meurman, Acta otolaryng., Stockh., *38*:460, 1950.
66. J. J. Pressman and G. Kelemen, Physiol. Rev., *35*:506, 1955.
67. M. C. Oldfield, Brit. J. Surg., *35*:173, 1947.
68. W. Penfield and L. Roberts, *Speech and Brain-Mechanisms,* Princeton University Press, 1959.
69. W. R. Brain, Lancet, 2:837, 1945, and *Speech Disorders,* Butterworth, London, 1961.
70. L. M. Speiden, G. Tucker, and R. Soulen, Canad. J. Otolaryng., *1*:219, 1972.
71. L. W. Welsh, Ann. Otol., etc., St Louis, *73*:569, 1964.

GENERAL READING

Tucker, G. F., *Human Larynx. Coronal Section Atlas,* Armed Forces Institute of Pathology, Washington, D.C., 1971.

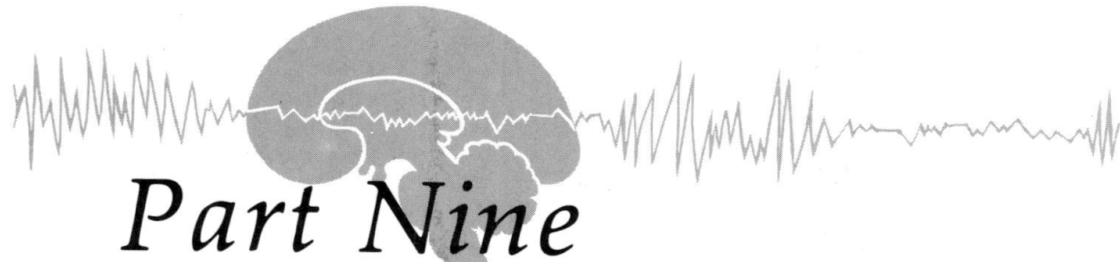

Part Nine

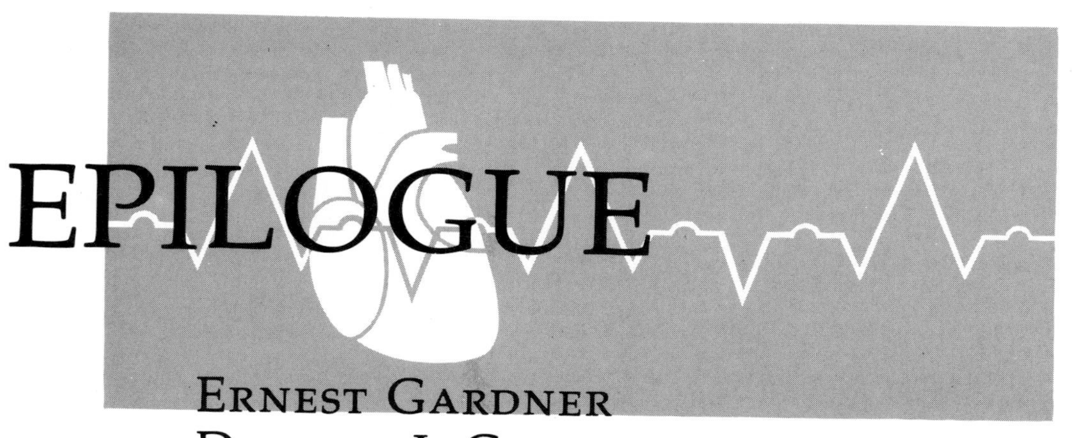

EPILOGUE

ERNEST GARDNER
DONALD J. GRAY
RONAN O'RAHILLY

SUMMARY OF BLOOD AND NERVE SUPPLY OF THE BODY

64

The purpose of this chapter is to outline the continuity of certain major arteries and nerves that traverse more than one region.

BLOOD SUPPLY

The blood supply of the limbs is summarized in figures 64–1[1] and 64–2.[2] Although the blood supply of the head and neck cannot be outlined briefly, important characteristics are shown in figures 53–21 (p. 607), 60–16 (p. 692), 60–18 (p. 697), 60–24 (p. 702), and 60–26 (p. 704). The major features of the great vessels of the trunk are shown in figures 29–7, 29–8, and 38–1 (pp. 290, 291, and 414).

NERVE SUPPLY

The following account summarizes the major peripheral nerves of the limbs, the cranial nerves, the cranial parasympathetic ganglia, and the sympathetic trunk. Brief descriptions of the motor and sensory defects resulting from section of each peripheral nerve and some of the cranial nerves are included.[3]

Upper Limb

Cervical and Brachial Plexuses. The ventral rami of the upper four cervical nerves unite to form the cervical plexus (figs 13–9 and 60–5, pp. 112, 682), whereas those of the lower four, together with the greater part of the ventral ramus of the first thoracic nerve, join to form the brachial plexus (figs. 13–8 and 13–9, pp. 111, 112).

The cervical plexus is located in front of the levator scapulae and the scalenus medius, under cover of the internal jugular vein and the sternocleidomastoid. Its branches supply the skin of the back of the head, the neck, and the shoulder, and certain muscles of the neck. In addition, the diaphragm is supplied by the phrenic nerve, which usually has two roots of origin: a principal root from C4 and an accessory root from C5 (by way of the brachial plexus). Its course in the thorax is described on page 332.

The brachial plexus is situated partly in the neck and partly in the axilla (cervicobrachial junction). It lies first between the scalenus anterior and the scalenus medius, and then in the posterior triangle of the neck. The plexus descends behind the concavity of the medial two-thirds of the clavicle, and accompanies the axillary artery into the axilla. The terminal branches of the plexus arise at the inferolateral border of the pectoralis minor.

The common, although not invariable, arrangement of branches is as follows. The ventral rami of the fifth and sixth cervical nerves unite to form the upper trunk, the ventral ramus of the seventh remains single as the middle trunk, and the eighth and first thoracic rami form the lower trunk. Each trunk then divides into an anterior and a posterior division (for the front and back of the limb, respectively). The anterior divi-

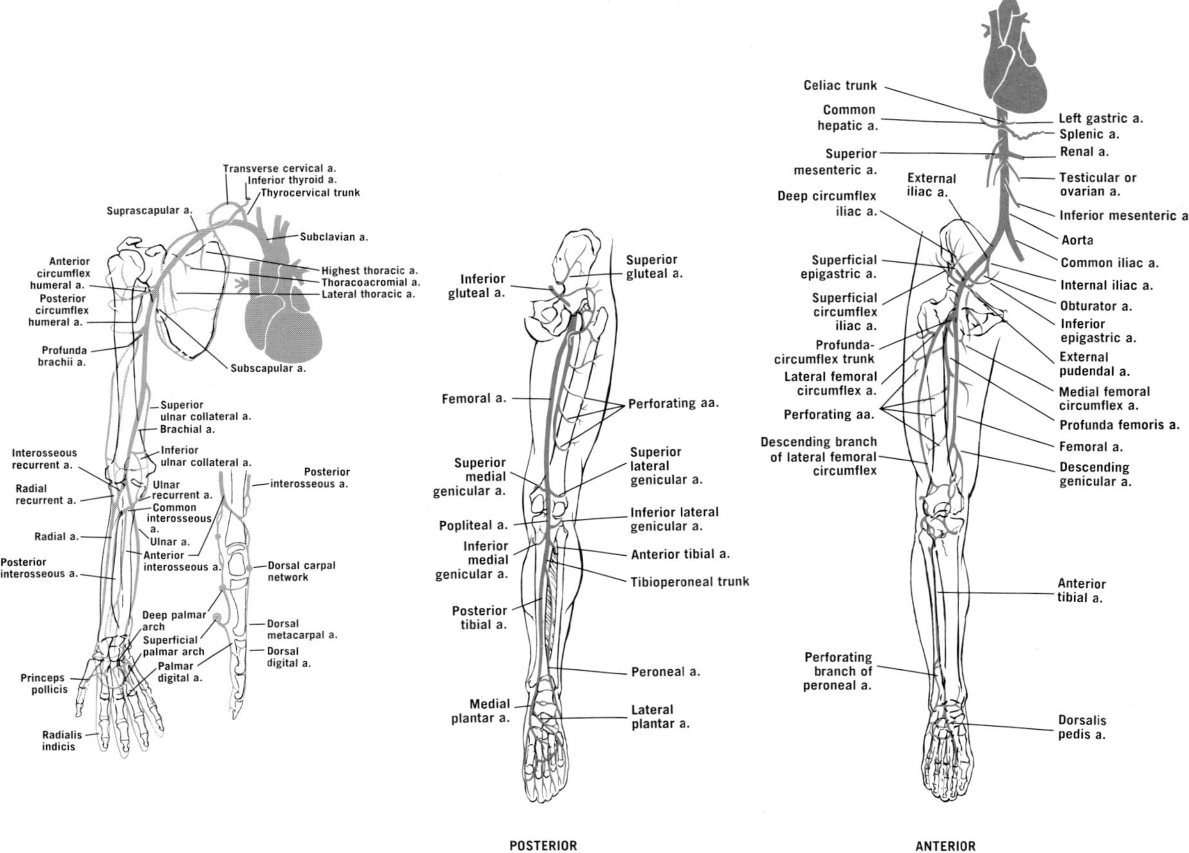

Figure 64–1 *Figure 64–2*

Figure 64–1 The arteries of the upper limb. The sagittal section (*lower right*) is based on Grant.

Figure 64–2 The arteries of the lower limb.

sions of the upper and middle trunks unite to form the lateral cord, the anterior division of the lower trunk forms the medial cord, and the three posterior divisions form the posterior cord. The terminal branches arise from the three cords.

The brachial plexus is thus composed successively of (1) ventral rami and trunks, which lie in the neck in relation to the subclavian artery (the lower trunk lies on the first rib behind the clavicle); (2) divisions, which lie behind the clavicle; and (3) cords and branches, which lie in the axilla in relation to the axillary artery. The branches of the brachial plexus (p. 111) include branches of the ventral rami, branches of the trunks, and branches of the cords. The important branches of the cords are the terminal branches, which are reviewed below.

Injuries to the brachial plexus are of great importance. Some may occur as a part of a neurovascular compression syndrome, in which weakness, pain, and sensory and vascular disorders of the upper limb follow abnormal compression, at the cervicobrachial junction, of the subclavian or axillary vessels, of the brachial plexus, or of both (p. 703). "Upper type" injuries, those to the fifth and sixth cervical nerves or to the upper trunk, are produced when the arm is pulled downward and the head is drawn away from the shoulder. After such an injury, the upper limb tends to lie in medial rotation, in a position referred to as "waiter's-tip hand." Upper type injuries occasionally occur during parturition (birth palsy, obstetric paralysis). "Lower type" injuries, those to the eighth cervical and first thoracic nerves or to the lower trunk, are produced when the arm is pulled upward. The short muscles of the hand are affected and a "claw hand," or *main en griffe*, results (p. 769).

Axillary Nerve (fig. 64–3). A terminal branch (C5, 6) of the posterior cord of the brachial plexus, it passes through the quadrilateral space, supplies the shoulder joint, teres minor, and deltoid, and gives off the upper lateral brachial cutaneous nerve.

SECTION OF NERVE. Sensation is lost in a small patch of skin over the deltoid.

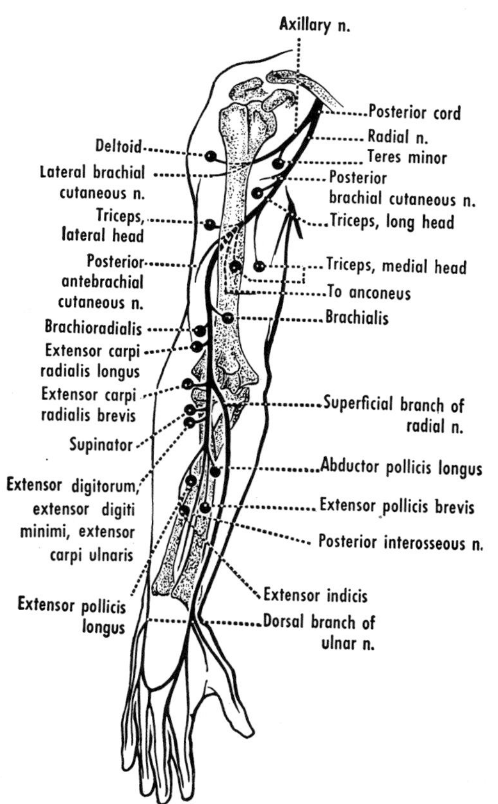

Figure 64–3 Schematic representation of the sequence of muscular branches of the axillary and radial nerves. The black circle in each instance represents a muscle; only the proximal part of the branch to the anconeus is shown. The sequences of muscular branches are those commonly encountered.[4]

The deltoid is paralyzed. The supraspinatus can still abduct the arm, but usually not to a horizontal level. The teres minor is paralyzed, and lateral rotation of the arm is weakened (but not lost, because the infraspinatus is intact).

Radial Nerve (fig. 64–3). A terminal branch (C5 to T1) of the posterior cord of the brachial plexus, it leaves the axilla, winds around the humerus, pierces the lateral intermuscular septum, descends between the brachialis and brachioradialis, and, at or below the lateral epicondyle, divides into superficial and deep branches. The radial nerve supplies the triceps, anconeus, brachioradialis, extensor carpi radialis longus (and often the brevis), gives off the posterior brachial and posterior antebrachial cutaneous nerves, and gives twigs to the brachialis and elbow joint. The deep branch, which frequently supplies the extensor carpi radialis brevis, pierces and supplies the supinator. Its continuation, the posterior interosseous nerve, supplies the other muscles of the back of the forearm and the joints of the hand. The superficial branch descends deep to the brachioradialis and emerges in the anatomical snuff-box, where it gives off its digital branches.

SECTION OF NERVE. Several important levels may be considered. If the lesion is in the axilla, all the muscles supplied by the radial nerve are paralyzed. Forearm extension is lost, flexion is weakened, and loss of wrist extension leads to a wristdrop (fig. 64–6). In addition, extension of the proximal phalanges is lost, adduction and abduction of the hand are weakened, and thumb movements are impaired. Sensory loss is small and unimportant because the overlap from adjacent nerves is considerable.

When the lesion is in the arm (e.g., in the radial groove), the motor loss is similar to that resulting from a lesion in the axilla, except that the triceps is unaffected, or only weakened.

If the lesion is in the cubital fossa (or involves the deep branch of the radial nerve at the neck of the radius), the muscles that extend the wrist are relatively unaffected. Hence, extension is lost at the metacarpophalangeal joints and thumb movements are impaired, but no wristdrop develops. Whether or not supination is weakened depends on the level of the lesion. If the lesion is in the cubital fossa and includes the superficial branch of the radial nerve, a patch of sensory loss may be found on the dorsum of the hand.

If the posterior interosseous nerve is sectioned after it leaves the supinator, only thumb movements are impaired (abduction, extension, opposition). The effect on the index finger of the loss of action of the extensor indicis may not be great.

Radial nerve fibers regenerate well, perhaps better than fibers of any other nerve.

Musculocutaneous Nerve (fig. 64–4). A terminal branch (C5 to 7) of the lateral cord, it pierces the coracobrachialis and descends between the biceps and brachialis. It supplies these three muscles, gives afferent twigs to the elbow joint, and continues as the lateral antebrachial cutaneous nerve.

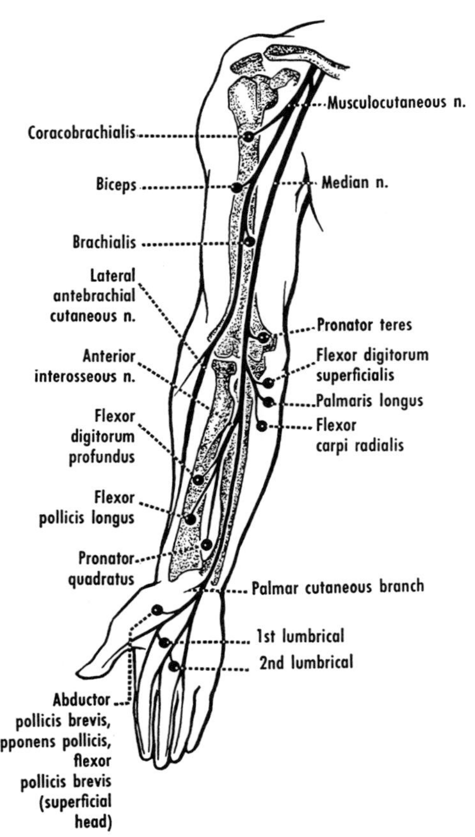

Figure 64–4 Schematic representation of the sequence of muscular and cutaneous branches of the musculocutaneous and median nerves. The black circle in each instance represents a muscle or muscle group. The sequences of muscular branches are those commonly encountered.[5]

SECTION OF NERVE. The nerve is not commonly involved alone. When it is, the main motor loss is severe weakness of forearm flexion (paralyzed brachialis and biceps) and weakness of supination (biceps). An unimportant loss of sensation occurs within the area of cutaneous distribution.

Median Nerve (fig. 64–4). Arising from medial and lateral cords (C5 to T1), it descends as a part of the neurovascular bundle of the arm, passes deep to the bicipital aponeurosis between the heads of the pronator teres, and descends on the deep surface of the flexor digitorum superficialis. It gives no branches in the arm, but it supplies all the muscles of the front of the forearm except the flexor carpi ulnaris and the medial part of the flexor digitorum profundus. It gives sensory twigs to the elbow joint and, more distally, a cutaneous branch to the palm. It enters the hand through the carpal canal, supplies the abductor pollicis brevis, opponens, superficial head of the flexor pollicis brevis, and the lateral two lumbricals, and then gives off its digital branches. The median nerve and its interosseous branch supply the radiocarpal and carpal joints; the digital branches supply the joints of the fingers. The motor fibers to the intrinsic muscles of the hand come mainly from the first thoracic segment of the spinal cord.

SECTION OF NERVE. Regardless of the level of injury, the important sensory loss is in the distribution of the digital branches. The anesthesia and the loss of muscular and joint proprioception impose a severe handicap on the proper use of the hand.

When the median nerve is sectioned above the elbow, elbow flexion may be interfered with only slightly, although pronation is lost. Flexion and abduction of the hand are impaired. Flexion of the interphalangeal joints is lost in the lateral two fingers, and impaired in the medial two fingers. (The ulnar nerve supplies the medial part of the profundus.) Movements of the thumb suffer severely, especially opposition. Opposition may not be lost completely, however, because more muscles than usual may be supplied by the ulnar nerve, and intact muscles may compensate in part for the paralyzed ones.

If the median nerve is sectioned at the wrist, only the intrinsic muscles of the

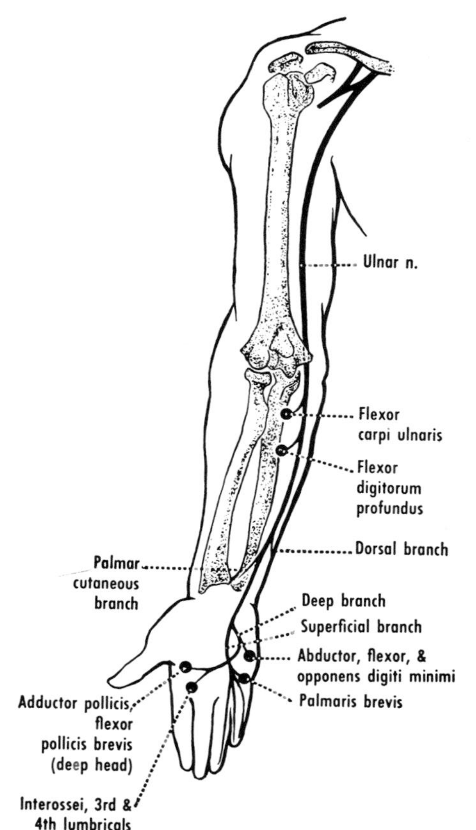

Figure 64–5 Schematic representation of the sequence of muscular and cutaneous branches of the ulnar nerve. The black circle in each instance represents a muscle or muscle group. The sequences of muscular branches are those commonly encountered.[6]

thumb (with defects as listed above) are affected. The sensory loss remains the same.

Painful disorders (e.g., causalgia) are common following median nerve lesions, especially if the injury is incomplete or crushing. Recovery after median nerve lesions is seldom as complete as that after radial nerve injuries.

Ulnar Nerve (fig. 64–5). A terminal branch (C7 to T1) of the medial cord (with a root from the lateral), it descends with the neurovascular bundle, pierces the medial intermuscular septum, and then descends behind the medial epicondyle between the two heads of the flexor carpi ulnaris. It gives a sensory twig to the elbow joint, supplies the flexor carpi ulnaris and the medial part of the flexor digitorum profundus, and descends in the forearm on the profundus. It gives a dorsal branch to the back of the hand, contributes a variable cutaneous branch to the palm, enters the palm lateral to the pisiform and medial to the hook of the hamate, gives off its digital branches, and supplies the hypothenar muscles, palmaris brevis, interossei, medial two lumbricals, adductor pollicis, and deep head of the flexor pollicis brevis. The ulnar nerve supplies the radiocarpal and carpal joints; the digital branches supply the joints of the fingers. The motor fibers to the intrinsic muscles of the hand arise from the first thoracic segment of the spinal cord.

SECTION OF NERVE. The anesthesia and the loss of muscular and articular proprioception are on the ulnar portion of the hand and on the front and back of the little and ring fingers. Recovery after ulnar nerve lesions is seldom complete.

If the ulnar nerve is sectioned above the elbow, adduction of the hand is impaired, and flexion of the distal interphalangeal joints of the medial two fingers is lost. (The proximal joints are controlled by the median nerve.) The main motor deficit is due to paralysis of the muscles of the hand. The interossei (and medial two lumbricals) are paralyzed, the fingers cannot be adducted or abducted, and adduction of the thumb is lost. The proximal phalanges cannot be flexed (especially those of the ring and little fingers), and they are therefore hyperextended by the unopposed long extensors. The middle and distal phalanges cannot be extended (especially those of the ring and little fingers), and those of the index and middle fingers are hyperflexed by the unopposed long flexors. The condition that results is known as clawhand, or *main en griffe* (fig. 64–6). The clawing is less marked in the medial two fingers because the profundus muscle to these fingers is paralyzed.

When the ulnar nerve is sectioned at the wrist (without involving tendons), the long flexors supplied by it are intact. Consequently, clawing is more marked.

If both the median and ulnar nerves are sectioned above the elbow, clawing is minimal because all the long flexors are paralyzed. As the nerves regenerate, the long flexors recover first. The intrinsic muscles of the hand are still paralyzed, and a clawhand results. It is thus characteristic of high, combined median and ulnar lesions that clawing is absent at first, but develops during recovery.

The disability from a combined lesion

is very great, regardless of the level, because of the important sensory losses and the paralysis of all the intrinsic muscles of the hand.

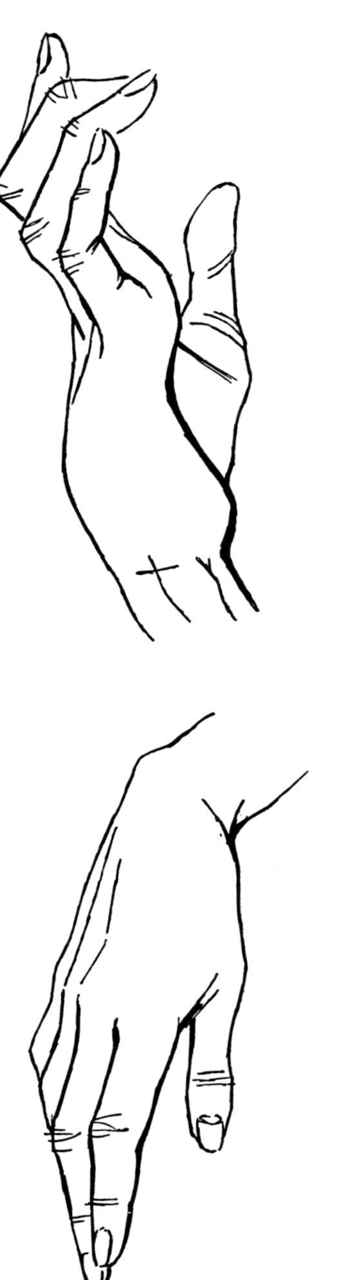

Figure 64–6 *Upper*, clawhand, drawn from a patient with an ulnar nerve lesion, and probably some median nerve damage, at the wrist. *Lower*, wristdrop, drawn from a patient with a radial nerve lesion. Both subjects were patients of Drs. J. L. Posch and R. D. Larsen, Detroit, Michigan.

When peripheral nerves are cut, the denervated skin becomes warmer and drier, owing to the loss of vasomotor fibers and the consequent vasodilatation and reduction of sweating. Thus, the determination of areas in which sweating is absent constitutes a test for sensory loss,[7] and is an important method of assessing injuries of the nerves to the hand.

Cutaneous and Dermatomal Distribution. When the ventral ramus of a spinal nerve enters a plexus and joins other rami, its component funiculi ultimately enter several of the nerves emerging from the plexus. Thus, as a general principle, any given spinal nerve contributes to several peripheral nerves and each peripheral nerve contains fibers derived from several spinal nerves (p. 33). The cutaneous nerve distribution to the upper limb is shown in figure 64–7, and the dermatomal distribution is shown in the figures inside the front cover.

Lower Limb

Lumbosacral Plexus. The ventral rami of the lumbar nerves enter the psoas major muscle and combine in a variable fashion to form the lumbar plexus (figs. 38–8, 38–9, and 41–4, pp. 426, 427, and 454). The second to fourth rami are usually described as forming the lumbar plexus proper. However, the lower part of the fourth ramus and all of the fifth enter the sacral plexus (the combined trunk is known as the lumbosacral trunk), and the two plexuses are commonly known as the lumbosacral plexus. The ventral ramus of the fourth lumbar is then the one ventral ramus that is common to both plexuses. In addition, the branches of the first lumbar nerve (p. 426) are usually described with the lumbar plexus. The lumbar plexus gives direct branches (L1 to L4) to the quadratus lumborum, psoas major, and psoas minor muscles. The major branches to the lower limb are reviewed below.

The ventral ramus of the fourth sacral nerve divides into an upper and a lower division; the upper division and the first three ventral rami combine with the lumbosacral nerve to form the sacral plexus, which lies in front of the piriformis muscle. The sacral plexus has 12 named branches. Five are distributed to the pelvis (p. 455). Of the remaining seven, which help to sup-

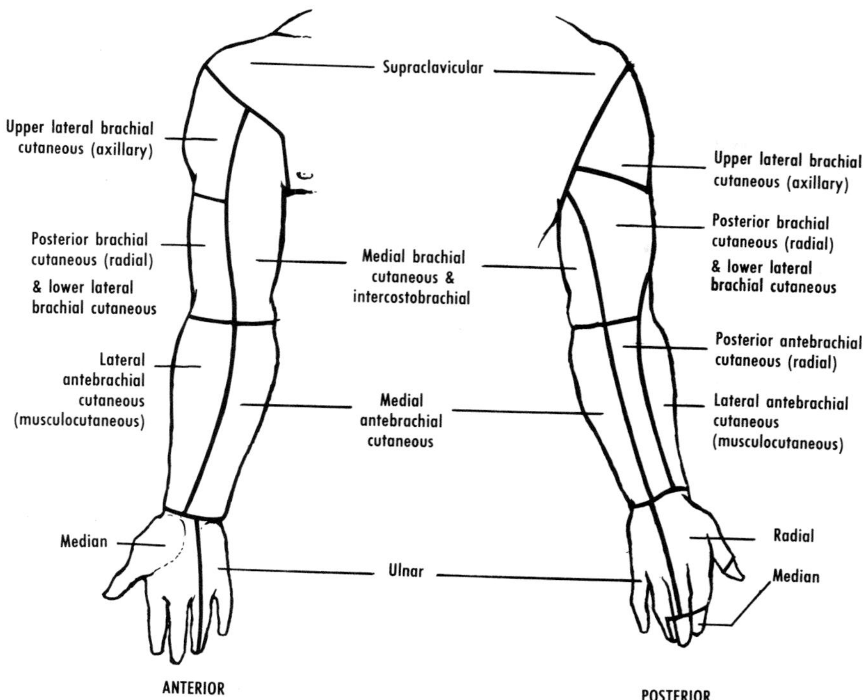

Figure 64–7 Approximate areas of cutaneous nerve distribution to the upper limb. These diagrams do not show overlap or variation. For example, the area in the arm supplied by the posterior brachial cutaneous nerve also receives fibers from the posterior antebrachial cutaneous nerve.

ply the buttock and lower limb, the gluteal and sciatic nerves are the most important and are summarized below.

Femoral Nerve (p. 218; fig. 64–8). A branch of the lumbar plexus (L2 to 4), it arises in the substance of the psoas major, descends in the groove between the psoas and iliacus, and enters the thigh behind the inguinal ligament, lateral to the femoral vessels. It supplies the skin of the front and medial aspect of the thigh, the anteromedial aspect of the leg, the iliacus, pectineus, sartorius, and quadriceps femoris muscles, and the hip and knee joints.

SECTION OF NERVE. Complete section of the femoral nerve above the level of origin of its branches is uncommon. When it does occur, sensation is impaired or lost on the anterior and medial aspects of the thigh, and on a small strip along the anteromedial aspect of the leg, from the knee to the ankle.

The major motor sign is paralysis of the quadriceps femoris. The leg cannot be extended, and flexion of the hip may be impaired. The patient can stand, and may be able to walk adequately on level ground, but finds it difficult to walk up or down stairs.

Femoral nerve injuries more commonly involve one or more of the branches of the nerve. The signs then vary according to which branch is injured.

Obturator Nerve (p. 215; fig. 64–8). A branch of the lumbar plexus (L3, 4), it arises in the substance of the psoas major. On leaving the medial aspect of the muscle it reaches the obturator foramen, where it divides into anterior and posterior branches, the continuations of which run in front of and behind the adductor brevis, respectively. The nerve and its branches supply the obturator externus, the adductor longus and brevis, a part of the adductor magnus, the gracilis, occasionally the pectineus, the skin on the medial side of the thigh (sometimes reaching the leg), and the hip and knee joints.

SECTION OF NERVE. The result is insignificant sensory loss on the medial side of the thigh. The chief motor disability is marked weakness of adduction. During walking, the unopposed abductors tend to swing the limb laterally.

Superior Gluteal Nerve (p. 208). A branch of the sacral plexus (L4, 5, S1), it passes through the greater sciatic foramen above the piriformis, accompanies the branches of the superior gluteal artery, and supplies the gluteus medius, gluteus minimus, tensor fasciae latae, and hip joint.

SECTION OF NERVE. There is no sensory loss. The characteristic motor loss is evidenced by the disabling gluteus medius limp. Flexion of the thigh is weakened, and medial rotation is severely impaired.

Inferior Gluteal Nerve (p. 209). A branch of the sacral plexus (L5, S1, 2), it

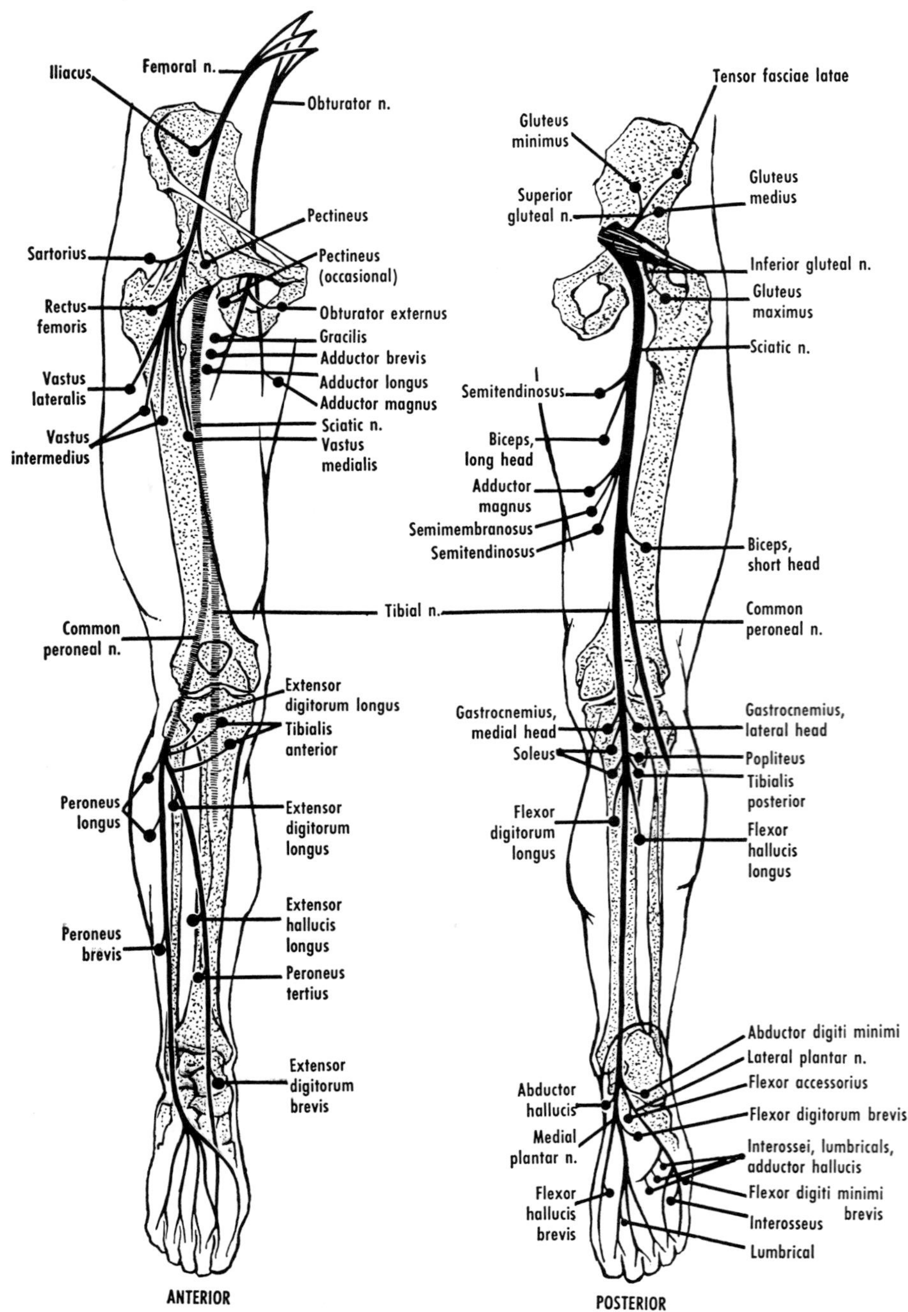

Figure 64–8 The sequences of branches of the sciatic nerve and of its tibial and common peroneal branches are those commonly encountered.[8] The sequences of branches of the femoral and the obturator nerves are based on Pitres and Testut.[9]

passes through the greater sciatic foramen below the piriformis and supplies the gluteus maximus.

SECTION OF NERVE. This nerve is unlikely to be injured without accompanying damage to the posterior femoral cutaneous nerve or to the sciatic nerve, or to both. There is no sensory loss. Owing to the paralysis of the gluteus maximus, extension of the thigh and trunk is impaired, but the defect may not be noticeable during ordinary walking.

Sciatic Nerve (pp. 209, 213; fig. 64–8). A branch of the sacral plexus (L4 to S3), it leaves the pelvis through the greater sciatic foramen, usually below the piriformis. It descends under cover of the gluteus maximus, between the greater trochanter and the ischial tuberosity, and enters the thigh behind the adductor magnus. It divides into tibial (L4 to S3) and common peroneal (L4 to S2) nerves in the lower third of the thigh and supplies the hamstrings and the extensor part of the adductor magnus.

The *tibial nerve* descends in the popliteal fossa and gives twigs to the triceps surae, plantaris, popliteus, tibialis posterior, and knee joint, and cutaneous branches to the calf. It then descends with the posterior tibial artery, supplies the soleus and deep calf muscles, gives cutaneous branches to the heel and sole and a twig to the ankle joint, and finally divides into medial and lateral plantar nerves. The medial plantar nerve supplies the abductor hallucis, the flexor digitorum brevis, and the tarsal joints, gives cutaneous branches to the sole, and then divides into plantar digital nerves that supply the flexor hallucis brevis, the first lumbrical, and the skin and joints of the medial four toes. The lateral plantar nerve supplies the quadratus plantae and abductor digiti minimi and gives cutaneous branches to the sole. Its terminal branches are (1) superficial, to the skin and joints of the fourth and fifth toes and the flexor digiti minimi brevis, and (2) deep, to the adductor hallucis, interossei, second to fourth lumbricals, and tarsal joints.

The *common peroneal nerve* descends in the popliteal fossa to the neck of the fibula. In the fossa, it gives twigs to the knee joint and cutaneous nerves to the calf. One of its terminal branches, the *deep peroneal nerve,* descends on the interosseous membrane with the anterior tibial artery. It supplies the tibialis anterior, extensor hallucis longus, extensor digitorum longus, peroneus tertius, and the ankle joint. Its terminal branches in the foot supply the tarsal joints, the extensor digitorum brevis, and, in part, the skin and joints of the first and second toes. The other terminal branch, the *superficial peroneal nerve,* descends in front of the fibula, supplies the peroneus longus and brevis (and sometimes the extensor digitorum brevis), and becomes cutaneous in the lower third of the leg. It descends superficial to the extensor retinacula and then divides into its terminal branches, which give cutaneous and articular twigs to all five toes.

SECTION OF SCIATIC NERVE. Complete section of the sciatic nerve is uncommon. In incomplete lesions, the peroneal component is nearly always the more severely damaged. If the injury is in the upper part of the thigh or in the gluteal region, the inferior gluteal or posterior femoral cutaneous nerve, or both, may be involved as well. In a complete sciatic section, much of the leg becomes useless. Extension at the hip is impaired, as is flexion at the knee, and all foot and ankle movements are lost. The loss of dorsiflexion at the ankle and of eversion of the foot results in a foot that hangs down in an equinovarus position.* This condition is also known as footdrop. Sensation is lost below the knee, except in the areas supplied by the saphenous and obturator nerves. The patient is able to stand, but his gait is peculiar because flexion at the hip is increased in order to clear the dropped foot off the ground.

If the sciatic nerve is completely sectioned in the midthigh, the nerves to the hamstrings are usually spared, and flexion at the knee is nearly normal. Extension at the hip is unimpaired.

Recovery from a sciatic lesion is slow and rarely complete.

SECTION OF TIBIAL NERVE. The important sensory loss is on the sole of the foot and on the plantar aspects of the toes. Depending on the level of the lesion, sensory deficits may result in the lower part of

---

*In talipes (clubfoot), the deformities are classified as equinus (foot plantar flexed), calcaneus (foot dorsiflexed), valgus (heel turned laterally), or varus (heel turned medially). Thus equinovarus denotes that the heel is elevated and turned medially. Similarly, calcaneovalgus signifies that the forefoot is raised and the heel turned laterally.

the leg. Sensation from the sole of the foot is important in posture and locomotion, and these functions are impaired. Trophic ulcers may occur.

The extent of motor loss depends upon the level of the lesion. If the lesion is in the popliteal fossa, all the calf muscles are paralyzed, as well as the intrinsic muscles of the foot (except the extensor digitorum brevis). Plantar flexion of the foot and toes is lost; the peroneus longus by itself cannot carry out normal plantar flexion. Inversion is impaired (loss of tibialis posterior). In walking, it is difficult to raise the heel off the ground, and the gait is shuffling. The atrophy of the small muscles of the foot leads to an increase in the concavity of the plantar arch (pes cavus). A calcaneovalgus position of the foot develops, owing to the unopposed action of the evertors and dorsiflexors.

If the tibial nerve is severed in the lower part of the leg, below the level of origin of muscular branches, only the intrinsic muscles of the foot are involved. Because the long flexors are intact, clawing of the digits may develop. The sensory loss is the main defect. In incomplete lesions of the tibial nerve, the small muscles of the foot are usually more affected than are other muscles.

SECTION OF COMMON PERONEAL NERVE. This nerve is more susceptible to injury than any other branch of the sciatic nerve.[10] In patients confined to bed, tension on the nerve because of the weight of blankets on upward-pointing toes has been known to cause footdrop, which may result also from the pressure of a plaster cast on the upper end of the fibula. Recovery is slow.

The sensory loss is on the dorsum of the foot and lateral aspect of the leg. Dorsiflexion and eversion of the foot are lost and footdrop results. (See Section of Sciatic Nerve.) The toes cannot be extended. When the foot is put on the ground at the beginning of the stance phase, it slaps the ground because the paralyzed dorsiflexors cannot decelerate plantar flexion properly.

SECTION OF DEEP PERONEAL NERVE. An unimportant patch of sensory loss occurs between the first and second toes. The dorsiflexors of the foot and the extensors of the toes are paralyzed, and footdrop and a steppage gait result. Inversion is somewhat impaired, and the peronei tend to evert the foot during dorsiflexion (swing phase). Pes valgus may develop. If the nerve is cut in the lower part of the leg, below the level of origin of muscular branches, only the extensor digitorum brevis is paralyzed. Extension of the big toe is impaired.

SECTION OF SUPERFICIAL PERONEAL NERVE. Sensation is impaired or lost on the dorsum of the foot and lateral aspect of the lower part of the leg. If the lesion is high in the leg, the peroneus longus and brevis are paralyzed. There is no footdrop, but eversion of the foot is lost. Consequently, the foot is inverted during dorsiflexion (swing phase). An equinovarus position may develop.

Cutaneous and Dermatomal Distribution. The cutaneous nerve distribution to the lower limb is shown in figure 64–9, and the dermatomal distribution is shown in the figures inside the front cover.

Cranial Nerves

The cranial nerves are numbered and named in table 64–1, which summarizes the key features of each. Additional features of the trigeminal and facial nerves, pertaining to their interrelationships and complex distribution, are outlined below. The ninth, tenth, and eleventh nerves likewise have complex functional and topographical relationships and, in the case of the tenth, an extensive distribution; these nerves are also outlined below.

Trigeminal Nerve. The trigeminal nerve is sensory from the face, anterior half of the scalp, teeth, mouth, nasal cavity, and paranasal sinuses, and motor to the muscles of mastication. It is attached to the side of the pons by a sensory and a motor root. The sensory root expands into the trigeminal ganglion, which gives rise to three large divisions: the ophthalmic, maxillary, and mandibular. The motor root, which also contains afferent fibers from the muscles of mastication, joins the mandibular division. The attachment of the trigeminal roots to the pons is in an area termed the cerebellopontine angle. In this vicinity, space-occupying lesions (e.g., tumors) generally involve several or all of the local nerves, namely, the trigeminal, facial, and vestibulocochlear, and sometimes the glossopharyngeal and vagus also.

The branches of the trigeminal nerve

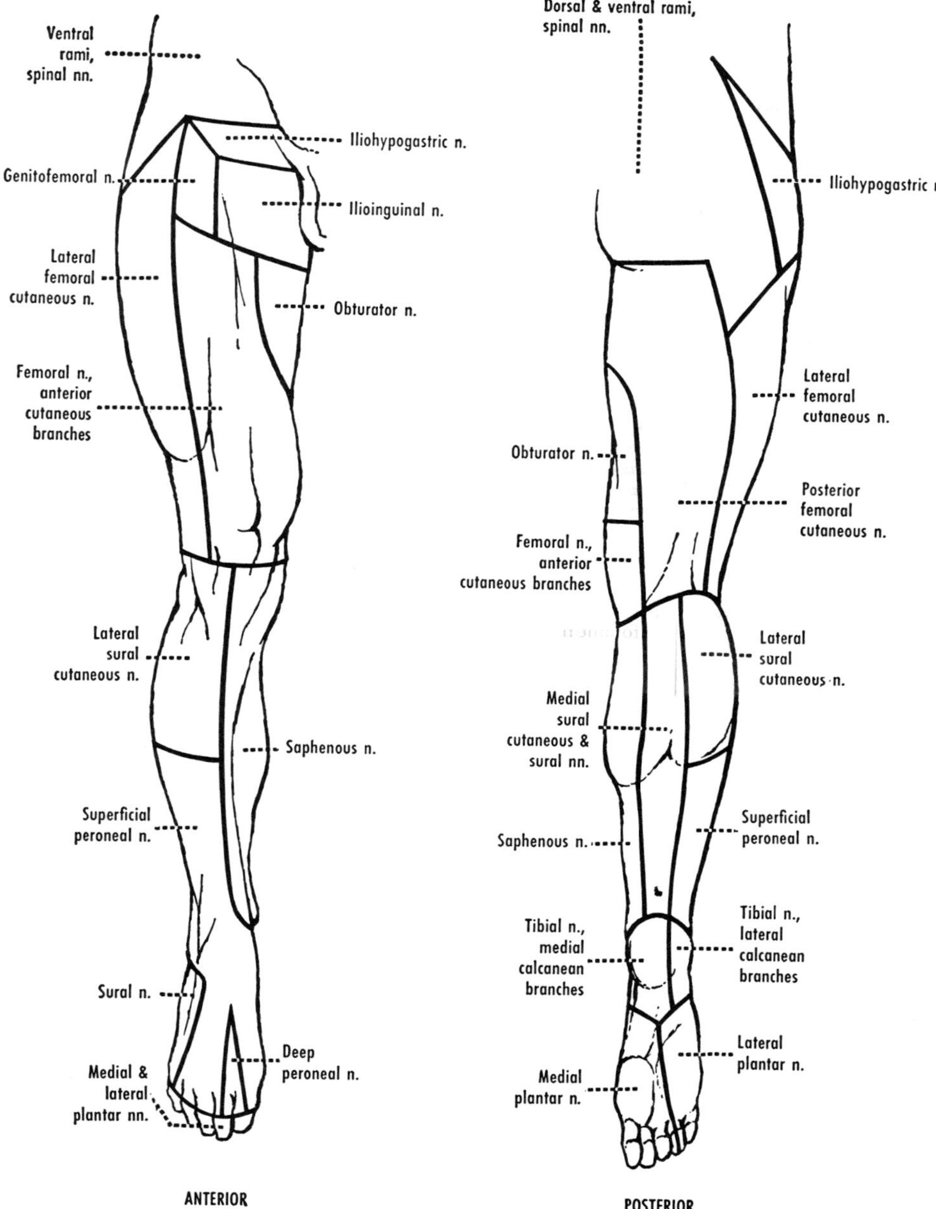

Figure 64–9 The distribution of cutaneous nerves to the lower limb. Neither overlap nor variation is shown.

are summarized in figure 64–10 and table 64–2. The ophthalmic nerve (first division, p. 630) runs forward in the dura of the lateral wall of the cavernous sinus and divides into lacrimal, frontal, and nasociliary nerves, which enter the orbit through the superior orbital fissure. The nasociliary nerve is the afferent limb of the corneal reflex (p. 632); the efferent limb is the facial nerve. The maxillary nerve (second division, p. 668) lies in the dura lateral to the cavernous sinus. It passes through the foramen rotundum and enters the pterygopalatine fossa. Then, as the infraorbital nerve, it gains the orbit through the inferior orbital fissure and ends on the face by emerging through the infraorbital foramen. The mandibular nerve (third division, p. 670), together with the motor root, passes through the foramen ovale to the infratem-

TABLE 64–1 Summary of Cranial Nerves

Nerve	*Attachment to Brain*	*Cranial Exit*	*Cells of Origin*	*Chief Components*	*Chief Functions*
1. Olfactory	Olfactory bulb	Cribriform plate	Nasal mucosa	Special visceral or somatic afferent	Smell
2. Optic	Optic chiasma	Optic canal	Retina (ganglion cells)	Special somatic afferent	Vision
3. Oculomotor	Midbrain, at medial border of cerebral peduncle	Superior orbital fissure	Midbrain	Somatic efferent	Movements of eyes
			Midbrain	General visceral efferent (parasympathetic)	Miosis and accommodation
4. Trochlear	Midbrain, below inferior colliculus	Superior orbital fissure	Midbrain	Somatic efferent	Movements of eyes
5. Trigeminal	Side of pons	Superior orbital fissure, foramen rotundum, and foramen ovale	Pons	Special visceral efferent	Chiefly movements of of mandible
			Trigeminal ganglion	General somatic afferent	Sensation in head
6. Abducent	Lower border of pons	Superior orbital fissure	Pons	Somatic efferent	Movements of eyes
7. Facial	Lower border of pons	Stylomastoid foramen	Pons	Special visceral efferent	Facial expression
			Pons	General viseral efferent (parasympathetic)	Secretion of tears and saliva
			Geniculate ganglion	Special visceral afferent	Taste
8. Vestibulocochlear	Lower border of pons	Does not leave skull	Vestibular ganglion	Special somatic afferent	Equilibration
			Spiral ganglion	Special somatic afferent	Hearing
9. Glossopharyngeal	Medulla, lateral to olive	Jugular foramen	Medulla (nucleus ambiguus)	Special visceral efferent	Elevation of pharynx
			Medulla (dorsal nucleus)	General visceral efferent (parasympathetic)	Secretion of saliva
			Inferior ganglion	General visceral afferent	Sensation in tongue and pharynx; visceral reflexes
			Inferior ganglion	Special visceral afferent	Taste
			Inferior ganglion	General somatic afferent	Sensation in external and middle ear
10. Vagus	Medulla, lateral to olive	Jugular foramen	Medulla (nucleus ambiguus)	Special visceral efferent	Movements of larynx
			Medulla (dorsal nucleus)	General visceral efferent (parasympathetic)	Movements and secretion of thoracic and abdominal viscera
			Inferior ganglion	General visceral afferent	Sensation in pharynx, larynx, and thoracic and abdominal viscera. Also visceral reflexes
			Inferior ganglion	Special visceral afferent	Taste
			Superior ganglion	General somatic afferent	Sensation in external ear

TABLE 64–1 Summary of Cranial Nerves *(Continued)*

Nerve	*Attachment to Brain*	*Cranial Exit*	*Cells of Origin*	*Chief Components*	*Chief Functions*
11. Accessory	Medulla, lateral to olive	Jugular foramen	Medulla (nucleus ambiguus)	Special visceral (?) efferent	Movements of pharynx and larynx
			Spinal cord (cervical)	Special visceral (?) efferent	Movements of head and shoulder
			Medulla (dorsal) nucleus)	General visceral efferent	Movements and secretion of thoracic and abdominal viscera
12. Hypoglossal	Medulla, between pyramid and olive	Hypoglossal canal	Medulla	Somatic efferent	Movements of tongue

poral fossa. As it does so, it is joined by the motor root and thereupon divides into branches that are classified into an anterior and a posterior division.

The jaw jerk is a reflex closure of the mouth when the muscles are quickly stretched by tapping the front of the mandible. The afferent impulses from the muscles are carried by the mandibular nerve and motor root to the mesencephalic nucleus of the trigeminal nerve. The efferent impulses from the motor nucleus of the trigeminal leave by the motor root and reach the muscles by the mandibular nerve.

Facial Nerve. The facial nerve consists of a larger part, which supplies the muscles of facial expression, and a smaller part, the nervus intermedius, which contains taste fibers from the anterior two-thirds of the tongue and secretomotor fibers for the lacrimal and salivary glands. The facial nerve has a complicated course in the temporal bone. Its two portions, which are attached to the lateral side of the brain stem at the junction of the pons and medulla (in the cerebellopontine angle), enter the internal acoustic meatus with the eighth nerve. The facial nerve then enters

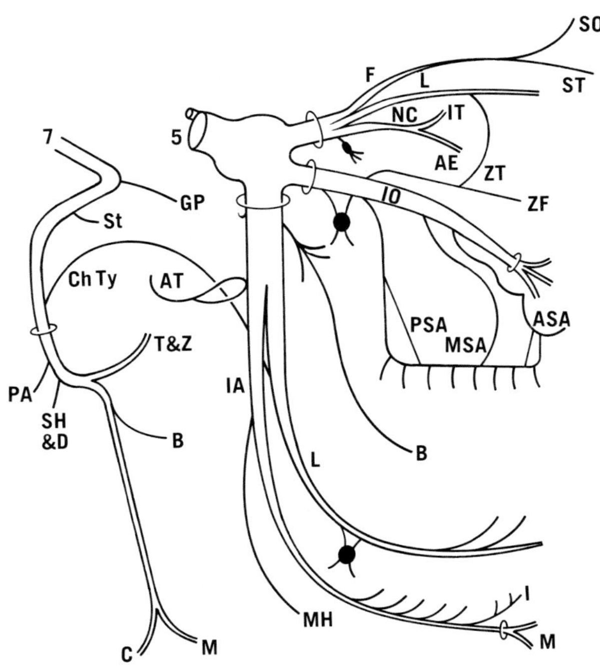

Figure 64–10 Schematic representation of the facial and trigeminal nerves and their branches. The branches to which the abbreviations refer are listed in tables 64–2 and 64–3.

TABLE 64–2 Branches of Trigeminal Nerve Shown in Figure 64–10

Abbreviation	*Nerve*
	Ophthalmic n.
F	Frontal n.
SO	Supraorbital n.
ST	Supratrochlear n.
L	Lacrimal n.
NC	Nasociliary n.
	Communicating branch to ciliary ganglion
IT	Infratrochlear n.
AE	Anterior ethmoidal n.
IO	*Maxillary and infraorbital nn.*
	Communicating branches to pterygopalatine ganglion
PSA	Posterior superior alveolar branches
Z	Zygomatic n.
ZT	Zygomaticotemporal branch
ZF	Zygomaticofacial branch
MSA	Middle superior alveolar branch
ASA	Anterior superior alveolar branch
	Mandibular n.
B	Buccal n.
AT	Auriculotemporal n.
IA	Inferior alveolar n.
MH	Mylohyoid n.
M	Mental n.
I	Incisive branch
L	Lingual n.
	Communicating branches to submandibular ganglion

the facial canal, expands into the geniculate ganglion, and turns sharply backward. It sweeps down behind the middle ear and emerges from the skull at the stylomastoid foramen. It finally enters the substance of the parotid gland, where it gives rise to its terminal branches for the facial muscles. The branches of the facial nerve are summarized in figure 64–10 and table 64–3.

Destruction of the facial nerve anywhere along its course from the brain stem to the parotid gland will cause a facial paralysis on the affected side. The lesion may be located by determining which disturbances, if any, are found in addition to the facial paralysis. If the lesion is between the brain stem and the geniculate ganglion, the signs are facial paralysis, loss of lacrimation in the corresponding eye, and loss of taste on the anterior two-thirds of the corresponding side of the tongue. Moreover, there may be an ipsilateral hyperacusis. (Sounds seem louder because the dampening effect of the stapedius is lost.) The eighth and fifth nerves may also be involved.

If the facial nerve is affected in the middle ear, similar signs are present, except that lacrimation may be normal. (The greater petrosal nerve, which carries secretomotor fibers for the lacrimal gland, leaves at the geniculate ganglion.) If the nerve is affected distal to the level of the chorda tympani, taste will also be spared.

In all cases of complete unilateral facial nerve paralysis, the corneal reflex on that side is lost.

Glossopharyngeal, Vagus, and Accessory Nerves. This complex of nerves is attached to the lateral side of the medulla oblongata and the upper portion of the cervical part of the spinal cord. Their motor fibers to skeletal muscle arise from the nucleus ambiguus in the medulla and the equivalent nuclear column in the spinal cord. The interrelationships and branches of these nerves are shown in figure 64–11. (See also fig. 60–18, p. 697.)

The *glossopharyngeal nerve* passes through the middle part of the jugular foramen, where it presents two ganglia, a superior and an inferior. It next courses between the internal jugular vein and the internal carotid artery, descends in front of the latter vessel, and then curves forward, passing between the superior and middle

TABLE 64–3 Branches of Facial Nerve Shown in Figure 64–10

Abbreviation	*Nerve*	
GP	Greater petrosal n.	In facial canal
St	N. to stapedius	
Ch Ty	Chorda tympani	
SH&D	Stylohyoid and digastric branches	Below base of skull
PA	Posterior auricular n.	
T&Z	Temporal and zygomatic branches	On face
B	Buccal branches	
M	Marginal branch of mandible	
C	Cervical branch	

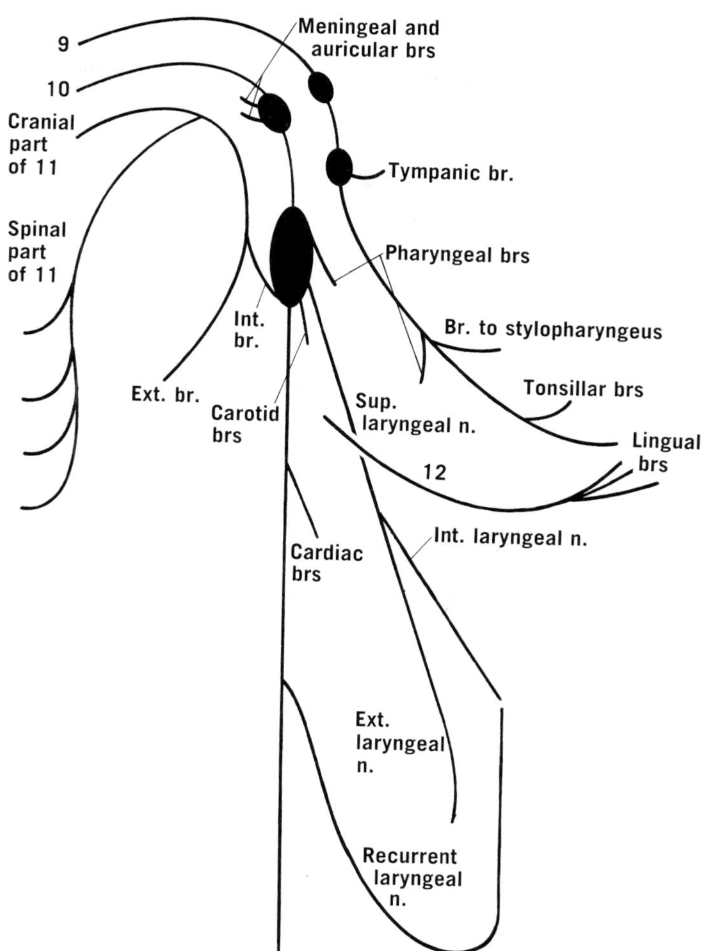

Figure 64–11 Schematic representation of the major branches of the glossopharyngeal, vagus, and accessory nerves in the head and neck.

constrictors of the pharynx. Its branches contain secretomotor fibers for the parotid gland, afferent fibers from the carotid sinus and carotid body, sensory fibers from the mucous membrane of the pharynx, tonsil, soft palate, and posterior part of the tongue, taste fibers from the posterior third of the tongue, and motor fibers to the stylopharyngeus. The functional components of the glossopharyngeal nerve are shown in figure 60–19 (p. 697). Irritative lesions of the glossopharyngeal nerve (glossopharyngeal neuralgia) cause pain on one side of the throat (especially in the region of the tonsil) and in the external acoustic meatus. Destructive lesions are characterized chiefly by ipsilateral loss of the gag reflex (loss of the sensory portion of the reflex arc) and ipsilateral loss of taste over the posterior third of the tongue.

The *vagus nerve* also passes through the middle of the jugular foramen, where it presents a superior and an inferior ganglion. Below the inferior ganglion, the vagus is joined by the internal branch of the accessory nerve. It then descends within the carotid sheath, enters the thorax (p. 333), contributes to the pulmonary plexuses, and next, with the vagus of the opposite side, forms the esophageal plexus. At the lower part of the esophagus, the plexus collects into an anterior and a posterior vagal trunk, which descend through the esophageal opening of the diaphragm to the anterior and posterior surfaces of the stomach, respectively (p. 422). Each trunk, however, contains fibers from both right and left vagus nerves. The general course of the vagus nerves is shown in figure 64–12. Specific branches are shown in figure 64–11

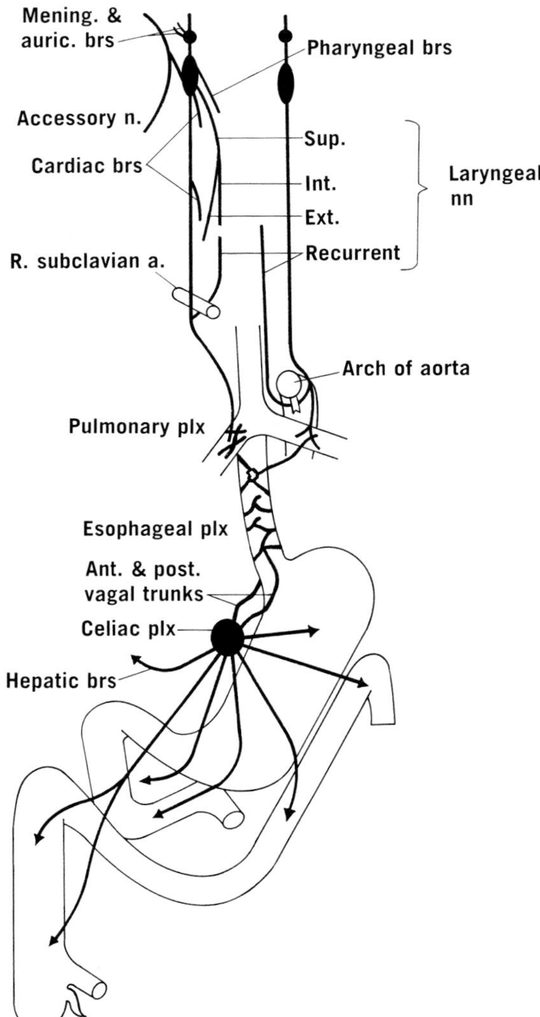

Figure 64–12 Schematic representation of the distribution of the vagus nerves.

and functional components in figures 31–8, 38–7, and 60–21, pp. 334, 425, and 698.

The vagal branches in the head and neck are (1) cardiac (parasympathetic) (p. 334), (2) sensory (from the meninges, the external acoustic meatus, and the mucous membrane of the larynx); and (3) taste (from the epiglottis and base of the tongue). The fibers of the accessory nerve that are distributed by the vagus are motor to the muscles of (1) the pharynx (except the stylopharyngeus), (2) the soft palate (except the tensor veli palatini), and (3) the larynx. The motor supply to the larynx is chiefly by the recurrent laryngeal nerves (all the muscles of the larynx except the cricothyroid). The right recurrent laryngeal arises in front of the right subclavian artery and winds around that vessel, whereas the left nerve arises on the left side of the arch of the aorta and winds around that vessel. The recurrent laryngeal nerves also supply the esophagus. The motor fibers to the striated muscle of the esophagus are probably of vagal rather than accessory origin.

Cardiac branches arise in the thorax, and contributions are given to the pulmonary and esophageal plexuses. In the abdomen, the anterior vagal trunk gives off hepatic branches and several gastric and celiac branches. The posterior vagal trunk likewise has a number of gastric and celiac branches. The vagal fibers that enter the celiac plexus course in the branches of the celiac and superior mesenteric plexuses to reach the stomach, pancreas, liver, small intestine, and the large intestine as far as the left colic flexure.

Section of a vagus nerve may have relatively little effect upon viscera because of the extensive intermingling of fibers from both vagus nerves. The chief effect, if the lesion is at the base of the skull, is ipsilateral paralysis of the laryngeal muscles. The gag reflex is also lost ipsilaterally, owing to the damage to the motor fibers. The uvula deviates to the normal side, especially when it is elevated as the patient says "ah." Swallowing is severely impaired.

Injury to a recurrent laryngeal nerve may occur from a tumor, from an aortic aneurysm, or from trauma during thyroid surgery. Unilateral severance results in paralysis of all the intrinsic muscles of the larynx on the same side, except the cricothyroid, and the vocal fold moves medially. In bilateral paralysis, the vocal folds are usually paramedian or median, the voice is reduced to a hoarse whisper, and respiratory distress is apparent.

The *accessory nerve* is formed by the union of a cranial and a spinal branch (fig. 64–11). Both portions traverse the jugular foramen, where they interchange some fibers. The cranial portion, or internal branch, joins the vagus at or just below the inferior ganglion and is distributed to the soft palate, pharynx, and larynx. The spinal portion, or external branch, runs backward, pierces the deep surface of the sternomastoid and supplies that muscle, and then crosses the posterior triangle of the neck obliquely to supply the trapezius.

It is important to note the close topographical relationship of the glossopharyngeal, vagus, and accessory nerves in and immediately below the jugular foramen, together with the hypoglossal nerve, which passes through the hypoglossal canal medial to the jugular foramen. A mass lesion in this region, such as a tumor, would involve all four nerves.

AUTONOMIC NERVOUS SYSTEM

Parasympathetic System

Preganglionic parasympathetic fibers issue from the brain stem by way of the third, seventh, ninth, tenth, and eleventh cranial nerves, and from the sacral portion of the spinal cord by way of the second and third or third and fourth ventral roots, or all three. The parasympathetic fibers of the accessory nerve are distributed by way of the vagus, and the ganglion cells are in or near the organ to be innervated. (This is true also of the ganglionic cells for the sacral parasympathetic fibers.)

The parasympathetic fibers in the third, seventh, and ninth cranial nerves end in the cranial sympathetic ganglia, namely, the ciliary, pterygopalatine, otic, and submandibular. The connections of these ganglia are shown in figure 64–13 and are summarized in table 64–4.

Sympathetic System

Sympathetic preganglionic fibers issue from the thoracic and upper lumbar levels

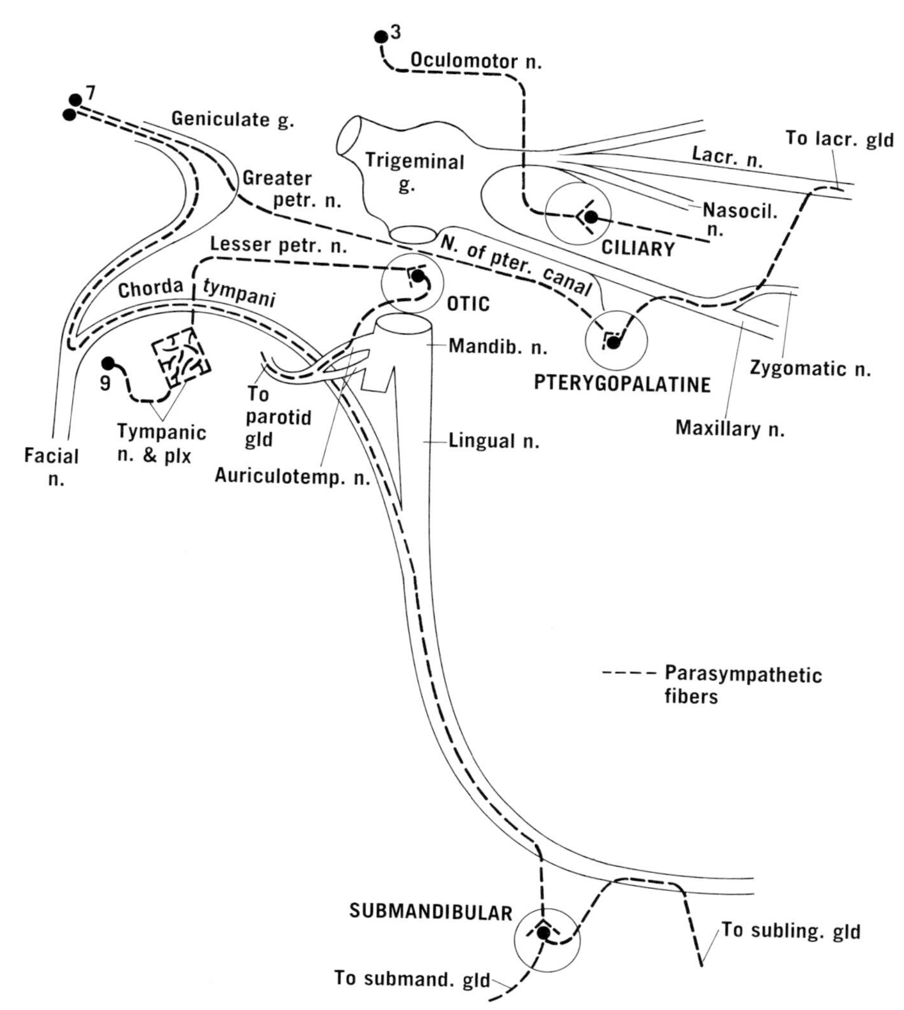

Figure 64–13 Schematic representation of the connections of the cranial parasympathetic ganglia. See also table 64–4.

TABLE 64–4 Parasympathetic Ganglia Associated with Cranial Nerves

Ganglion	*Page*	*Figure*	*Location*	*Parasympathetic Root*	*Sympathetic Root*	*Chief Distribution*
Ciliary	635	55–8	Lateral to optic n.	Oculomotor n.	Internal carotid plexus	Ciliary m. and sphincter pupillae Dilator pupillae and tarsal muscles
Pterygo-palatine	669	58–8	In pterygopalatine fossa	Greater petrosal (7) and n. of pterygoid canal	Internal carotid plexus	Lacrimal gland
Otic	672	58–10	Below foramen ovale	Lesser petrosal (9)	Plexus on middle meningeal a.	Parotid gland
Submandibular	676	59–4	On hyoglossus	Chorda tympani (7) by way of lingual n.	Plexus on facial a.	Submandibular and sublingual glands

of the spinal cord. These fibers travel in ventral roots and spinal nerves, and most of them reach the adjacent sympathetic trunk and ganglia by way of rami communicantes. Many of them synapse in the ganglia of the trunk, others continue through and reach ganglia of the prevertebral plexuses by way of splanchnic nerves, and still others synapse in *accessory* or *intermediate ganglia* (fig. 64–14). Of the postganglionic fibers arising in the trunk ganglia, some go directly to adjacent viscera and blood vessels. Most of the others return to spinal nerves and dorsal and ventral rami by way of rami communicantes. Some postganglionic fibers, however, reach the back and the proximal portions of the limbs by accompanying blood vessels.

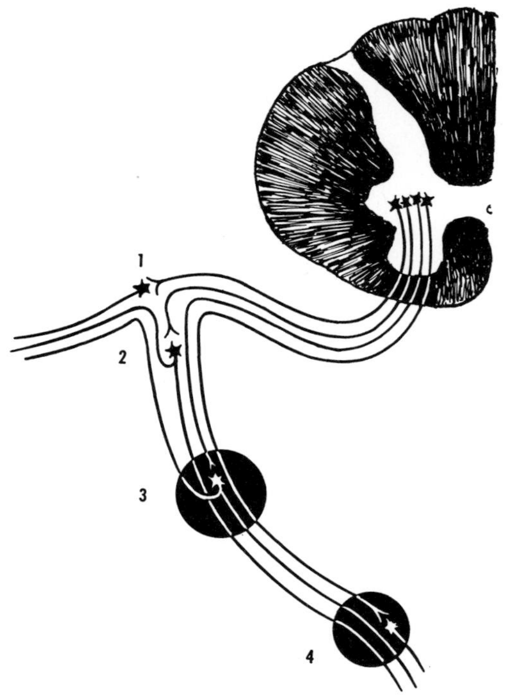

Figure 64–14 Locations of sympathetic ganglion cells. Ganglion cells may form accessory ganglia in (1) spinal nerves or ventral rami, or in (2) rami communicantes. Ganglion cells are also present in (3) ganglia of sympathetic trunk, and in (4) prevertebral ganglia, such as the splanchnic, celiac, or superior mesenteric.

The accessory ganglia occur chiefly in the cervical, lower thoracic, and upper lumbar regions. Lumbar or thoracolumbar sympathectomies may not be completely successful, because these accessory ganglia and their connections are usually not affected by the operation.

Sympathetic Trunk and Ganglia. The sympathetic trunks are long nerve strands, one on each side of the vertebral column; they extend from the base of the skull to the coccyx (fig. 64–15). Each usually presents 21 to 25 ganglia of varying sizes, but broader ranges have been recorded. The composition of the sympathetic trunk and the general nature of its branches are schematically represented in figure 64–16. This figure also shows that sensory fibers (probably chiefly pain) arising from the thoracic, abdominal, and some pelvic viscera pass through the sympathetic trunks. They reach spinal nerves by way of rami communicantes and enter the spinal cord by way of dorsal roots.

The cervical part of the sympathetic trunk consists of three or four ganglia connected by an intervening cord or cords. The superior cervical ganglion (p. 707) lies behind the internal carotid artery, in front of the longus capitis muscle, and extends from the level of the first to that of the second or third cervical vertebra. The middle cervical ganglion (p. 707) is quite variable and is often fused with either the superior or the vertebral ganglion. It usually lies just above the arch formed by the inferior thyroid artery, at the level of the sixth cervical vertebra. The vertebral ganglion (p. 707) usually lies in front of the vertebral artery, at about the level of the seventh cervical vertebra. The ansa subclavia loops in front of the subclavian artery and connects the cervicothoracic and the vertebral ganglia. The cervicothoracic (stellate) ganglion includes two parts: the inferior cervical and the first thoracic (occasionally the second and third also). These ganglia may be completely fused. The ganglionic mass lies usually at the level of the seventh cervical and first thoracic vertebrae, in front of the eighth cervical and first thoracic nerves, the seventh cervical transverse process, and the neck of the first rib, and behind the vertebral artery.

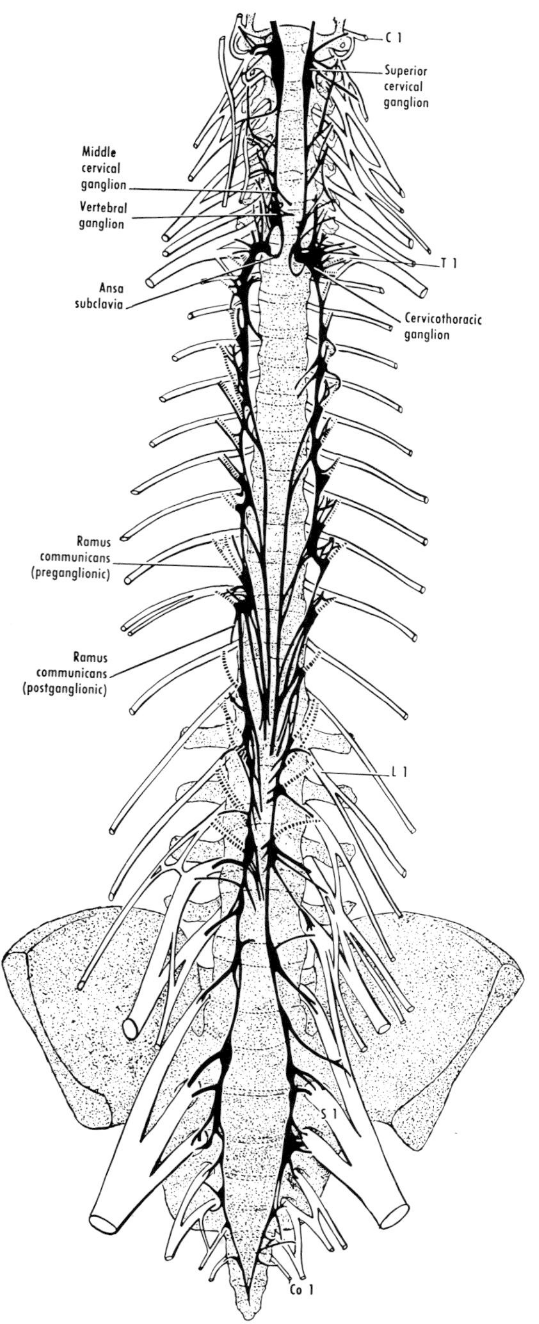

Figure 64–15 The sympathetic trunks. Preganglionic rami communicantes are shown as interrupted lines, postganglionic rami as solid black. Based on Pick and Sheehan.[11]

The sympathetic trunks enter the thorax from the neck, descend in front of the heads of the ribs and the posterior intercostal vessels and accompanying nerves, and enter the abdomen by piercing the crura of the diaphragm or by passing behind the medial arcuate ligaments. In the thorax, each trunk usually presents 11 or 12 separate ganglia of varying size (occasionally 10 or 13), including the cervicothoracic described above. Each ganglion has one to four rami communicantes.

In the abdomen and pelvis, the two sympathetic trunks descend on the vertebral column, adjacent to the psoas major muscles. The right trunk lies behind the inferior vena cava, the left one beside the aorta. The trunks continue into the pelvis, where they lie on the pelvic surface of the sacrum, medial to the upper three pelvic foramina and usually in front of the fourth. They end by uniting in front of the coccyx to form the ganglion impar.

In the lumbar region, the two trunks are seldom symmetrical and the ganglia are irregular in size, number, and position. Three to five ganglia are usually present, but there may be from two to six, and occasionally a trunk is an elongated ganglionic mass. The variations are such as to make identification of a specific ganglion very difficult. When the first lumbar ganglion is present, it lies between the crus of the diaphragm and the vertebral column and is often overlooked. Each lumbar ganglion sends two or more rami communicantes to two or more spinal nerves.

In the pelvis, the number of ganglia is

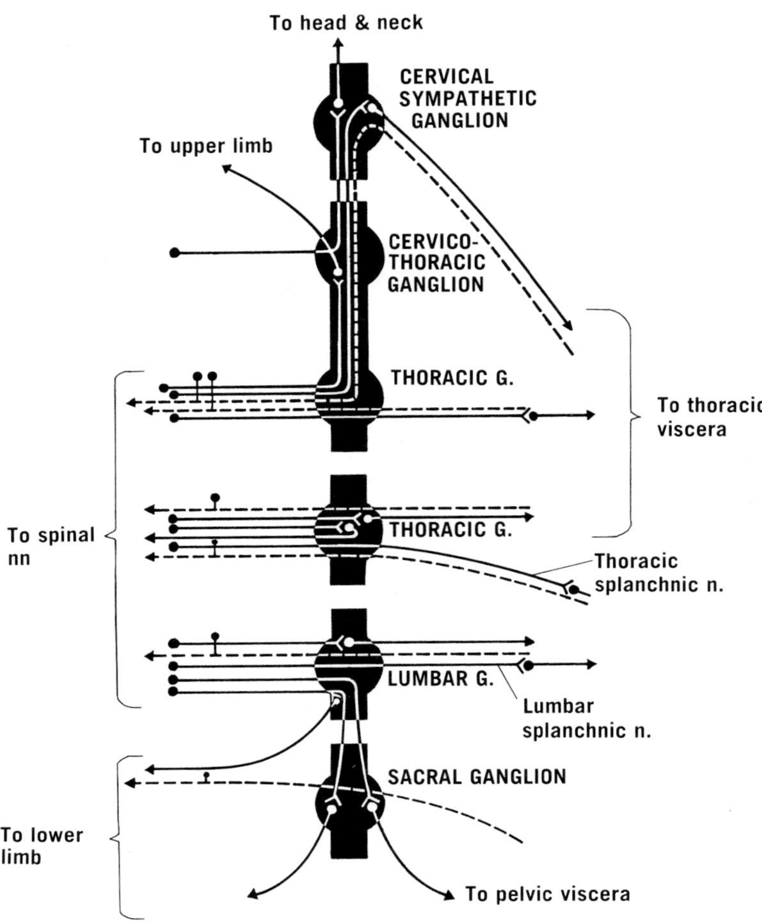

Figure 64–16 Functional components of the sympathetic trunk, with schematic representation of levels of origin of preganglionic fibers and sites of ganglion cells. Sensory fibers from viscera (mostly pain fibers) are shown as interrupted lines.

variable, but there are usually three or four. Each ganglion tends to be connected by rami communicantes with only one spinal nerve.

Prevertebral Ganglia. The prevertebral ganglia are found in the prevertebral plexuses, which are formed by branches of the vagus nerves and sympathetic trunks. Aside from a variable number of cardiac ganglia in the cardiac plexus, the chief prevertebral ganglia occur in the abdomen. They are named according to the plexus with which they are associated. The suprarenal medullae are comparable to sympathetic ganglia in that their cells are supplied by preganglionic sympathetic fibers.

Head and Neck. The preganglionic fibers for the head and neck arise from the first one or two segments of the thoracic part of the spinal cord and leave chiefly by way of the first thoracic ventral root. The preganglionic fibers enter the sympathetic trunk and synapse in cervical sympathetic ganglia. Loss of sympathetic supply to the head and neck results in Horner's syndrome (miosis, mild ptosis, mild enophthalmos, lack of sweating, and warm skin—all on the side of the damage) (p. 706). Some preganglionic fibers may leave by the eighth cervical ventral root. Fibers that control sweating in the face probably leave by way of the second thoracic ventral root. Section of this root (or rami communicantes) may result in a dry face on that side, but not necessarily in Horner's syndrome.

Upper Limb. Sympathetic fibers reach the upper limb chiefly by way of the brachial plexus. The preganglionic fibers arise from approximately the second to the ninth or tenth thoracic segment and reach the sympathetic trunk by way of the appropriate ventral roots and rami. They next ascend in the trunk and synapse in sympathetic ganglia (chiefly the cervicothoracic and vertebral). The postganglionic fibers then travel in rami communicantes to the ventral rami that form the brachial plexus. An intrathoracic ramus that connects the first and second thoracic nerves (and sometimes one between the second and third nerves) provides an additional route for sympathetic fibers to the brachial plexus. Finally, a certain number of postganglionic fibers (mostly from the cervicothoracic ganglion) accompany the subclavian artery and are distributed to vessels of the shoulder.

Thorax. Preganglionic fibers for the heart and coronary vessels arise from the upper four to six segments of the thoracic part of the spinal cord and synapse in the corresponding ganglia, as well as in cervical ganglia. Postganglionic fibers are distributed by way of cervical, cervicothoracic, and thoracic cardiac nerves.

Preganglionic fibers for the bronchial tree and pulmonary vessels arise from the upper four to six thoracic segments. Postganglionic fibers from the corresponding ganglia are distributed by direct branches to the pulmonary plexuses and in thoracic cardiac nerves.

Preganglionic fibers for the aorta arise from the upper thoracic segments, and those for the esophagus from the lower. Postganglionic fibers reach the aorta by direct branches, and the esophagus by direct branches and by way of the greater splanchnic nerves.

Preganglionic fibers for the thoracic and abdominal walls arise from all thoracic segments, and postganglionic fibers are distributed by the dorsal and ventral rami and meningeal branches of the spinal nerves.

Abdomen and Pelvis. Preganglionic fibers for the abdominal viscera arise from the lower half of the spinal cord and from the upper lumbar segments. They pass through the sympathetic trunk, course in splanchnic nerves, and synapse in prevertebral ganglia. Preganglionic fibers for the pelvic viscera arise in the upper lumbar segments and descend in the sympathetic trunks to lumbar and sacral ganglia. Postganglionic fibers are distributed by the hypogastric and associated plexuses.

Lower Limb. Preganglionic fibers arise from the lower thoracic and upper lumbar segments. They enter the sympathetic trunk and descend to lumbar and sacral ganglia, from which postganglionic fibers enter lumbar and sacral nerves. Some postganglionic fibers accompany the common iliac artery and its branches and supply vessels in the pelvis and upper portion of the thigh.

REFERENCES

1. For variations in arteries of the upper limb, see L. Dubreuil-Chambardel, *Variations des artères du membre supérieur,* Masson, Paris, 1926; B. Adachi, *Das Arterien System der Japaner,* Kaiserlich-Universität zu Kyoto, 1928; J. A. Keen, Amer. J. Anat., *108*:245, 1961. See also reference 22, page 121. The nerve supply of the arteries of the upper limb is described by J. Pick, Anat. Rec., *130*:103, 1958.
2. For variations in arteries of the lower limb, see B. Adachi, and J. A. Keen, cited above. A concise anatomical review of the nerve supply of the blood vessels of the lower limb has been published by G. M. Wyburn, Scot. med. J., *1*:201, 1956. For angiograms of collateral circulation with obstruction of the femoral artery, see R. F. Muller and M. M. Figley, Amer. J. Roentgenol., *77*:296, 1957.
3. Based particularly upon the following accounts: *Aids to the Investigation of Peripheral Nerve Injuries,* Med. res. Coun. (War) Memor. No. 7, London, revised 2nd ed., 1943; H. J. Seddon (ed.), *Peripheral Nerve Injuries,* Spec. Rep. Ser. med. Res. Coun., Lond., no. 282, 1954; W. Haymaker and B. Woodhall, *Peripheral Nerve Injuries,* Saunders, Philadelphia, 2nd ed., 1953; H. Seddon, *Surgical Disorders of the Peripheral Nerves,* Churchill Livingstone, Edinburgh, 1972.
4. S. Sunderland, J. comp. Neurol., *85*:93, 1946.
5. S. Sunderland and L. J. Ray, J. comp. Neurol., *85*:191, 1946.
6. S. Sunderland and E. S. R. Hughes, J. comp. Neurol., *85*:113, 1946.
7. E. Moberg, J. Bone Jt Surg., *40B*:454, 1958.
8. S. Sunderland and E. S. R. Hughes, J. comp. Neurol., *85*:205, 1946.
9. A. Pitres and L. Testut, *Les nerfs en schémas,* Doin, Paris, 1925.
10. S. Sunderland, Brit. J. Surg., *41*:300, 1953.
11. J. Pick and D. Sheehan, J. Anat., Lond., *80*:12, 1946.

GLOSSARY OF EPONYMOUS TERMS

Although eponyms should be avoided, they are in frequent use and a student often has occasion to seek the meaning of such terms. This glossary provides a list of anatomical eponyms, many of which are in common usage. The chief source has been J. Dobson, *Anatomical Eponyms*, Livingstone, Edinburgh, 2nd ed., 1962. Biographical notes concerning the workers commemorated in this glossary are given by Dobson, as are also citations of the publications in which the structures are described. Separate entries are used here to distinguish two or more workers who have the same surname (e.g., Meckel, Petit).

Adam's apple—laryngeal prominence
Addison's plane—transpyloric plane
Alberran's gland—the portion of the median lobe of the prostate immediately underlying the uvula of the urinary bladder
Albini's nodules—tiny nodules on the margins of the mitral and tricuspid valves
Alcock's canal—pudendal canal
Allen, fossa of—a fossa on the neck of the femur
Ammonis (Ammon), cornu—hippocampus
Arantius, bodies of (corpora arantii)—nodules of the aortic and pulmonary valves
  venous canal of (canalis arantii)—ductus venosus
Arnold's nerve—auricular branch of vagus
Auerbach's ganglia—ganglia in the myenteric plexus
  plexus—myenteric plexus

Ball's valves—anal valves
Bartholin's ducts—sublingual ducts that open into the submandibular duct
  glands—greater vestibular glands
Bauhin's glands—anterior lingual glands
  valve—ileocecal valve
Bell's muscle—the muscular strands from the ureteric orifices to the uvula, bounding the trigone of the urinary bladder
Bell's nerve—long thoracic nerve
Bellini's ducts—orifices of collecting tubules of kidney
  tubules—collecting tubules of the kidney
Bertin's columns—renal columns
  ligament—iliofemoral ligament
Bichat's ligament—lower part of dorsal sacroiliac ligament, sometimes known as the transverse iliac ligament
Bigelow's ligament—iliofemoral ligament
Billroth's cords—arrangement of red pulp in spleen
Blandin, glands of—anterior lingual glands
Botallo's duct—ductus arteriosus
  foramen—foramen ovale of the heart
  ligament—ligamentum arteriosum
Bourgery's ligament—oblique popliteal ligament
Bowman's capsule—glomerular capsule
  glands—serous glands in olfactory mucous membrane
  membrane—anterior limiting lamina of cornea
Breschet's bones—the suprasternal ossicles
Broca's convolution—inferior frontal gyrus of left cerebral hemisphere
Brödel's bloodless line—the line of division on the kidney, between the areas supplied by the anterior and posterior branches of the renal artery
Brodie's bursa—bursa of the semimembranosus tendon
Bruch's membrane—basal lamina of choroid
Brücke's muscle—meridional fibers of ciliary muscle
Brunn's cell nests—epithelial cell masses in the male urethra
Brunner's glands—duodenal glands
Buck's fascia—the deep fascia of the penis
Burns' ligament—falciform margin of the fascia lata at the saphenous opening
  space—fascial space above the jugular notch of the sternum

Calot, triangle of—cystohepatic triangle
Camper's fascia—superficial layer of the subcutaneous tissue (superficial fascia) of the abdomen
Chassaignac's space or bursa—the retromammary space between the deep layer of subcutaneous tissue and the pectoralis major
tubercle—carotid tubercle on the sixth cervical vertebra
Chopart's joint—transverse tarsal joint
Civinini, foramen of—pterygospinous foramen
Cleland's cutaneous ligaments—cutaneous ligaments of the digits
Cloquet's canal—hyaloid canal
gland—lymph node in the femoral ring
septum—femoral septum
Colles' fascia—membranous layer of the superficial perineal fascia
ligament—reflected inguinal ligament
Cooper's ligament—(1) pectineal ligament; (2) suspensory ligament of the breast
Corti, ganglion of—spiral ganglion
organ of—spiral organ
Cowper's glands—bulbourethral glands
Cruveilhier's nerve—vertebral nerve
plexus—plexus formed by dorsal rami of first three spinal nerves—"posterior cervical plexus"

Deaver's windows—fat-free portions of the mesentery framed by vascular arcades adjacent to the attached margin of the gut
Denonvillier's fascia—rectovesical septum
Descemet's membrane—posterior limiting lamina of cornea
Dorello's canal—foramen formed by attachment of petroclinoid ligament across notch at petrosphenoid junction; it contains the abducent nerve
Douglas, fold of—rectouterine fold
line of—the arcuate line of the posterior layer of the sheath of the rectus abdominis
pouch of—rectouterine pouch
Dupuytren's fascia—palmar fascia

von Ebner's glands—serous glands near vallate papillae
Edinger-Westphal nucleus—part of oculomotor nucleus in midbrain
Ellis' muscle—corrugator cutis ani muscle
Eustachian (Eustachi) tube—auditory tube (strictly, its cartilaginous part)

Fallopian (Falloppio) canal or **aqueduct**—facial canal
ligament or **arch**—inguinal ligament
tube—uterine tube
Ferrein's pyramids—medullary rays of the kidney
Flack's node—sinuatrial node
Flood's ligament—superior glenohumeral ligament
Folian (Folius) process (processus folii)—anterior process of malleus
Fontana, spaces of—spaces of iridocorneal angle
Frankenhauser's ganglion—uterovaginal plexus of nerves
Frankfort plane—orbitomeatal plane

Gärtner's duct—longitudinal duct of the epoöphoron
Galen, vein of—great cerebral vein
Gasserian (Gasser) ganglion—semilunar ganglion of trigeminal nerve
Gerdy's ligament—suspensory ligament of the axilla
Gerlach's tonsil—tubal tonsil
Gerota's capsule or **fascia**—renal fascia
Gimbernat's ligament—lacunar ligament
Giraldès, organ of—the paradidymis
Glaserian (Glaser) fissure—petrotympanic fissure
Glisson's capsule—fibrous capsule of the liver
Golgi apparatus or **complex**—a system of cytoplasmic membranous organelles or lipochondria
corpuscles—proprioceptive endings in tendons
-Mazzoni corpuscles—corpuscular nerve endings
Graafian (de Graaf) follicle—vesicular ovarian follicle
Gruber, ligament of—petroclinoid ligament
Grynfeltt's triangle—triangle bounded by the posterior border of the internal oblique, the anterior border of the quadratus lumborum, and above by the 12th rib
Guerin's valve or **valvule**—a fold of mucous membrane in the navicular fossa of the urethra
Guthrie's muscle—the sphincter urethrae

Haller's ductulus aberrans—diverticulum of the canal of the epididymis
layer—vascular lamina of choroid
rete—rete testis
Hannover, spaces of—zonular spaces
Harris' lines—transverse lines in long bones near the epiphysis, sometimes seen radiographically
Hartmann's critical point—the site on the large intestine where the lowest sigmoid artery anastomoses with the superior rectal artery
Hasner, valve of—lacrimal fold
Haversian (Havers) canals—spaces in compact bone
glands or **folds**—synovial pads or fringes of synovial membrane consisting largely of intra-articular fat
lamellae—bony layers surrounding a Haversian canal

system—a Haversian canal and its surrounding lamellae, the structural unit of bone (osteon)
Heister's valve—spiral folds of cystic duct
Henle's ligament—lateral expansion of the lateral edge of the rectus abdominis, together with transversalis fascia and transversus aponeurosis; forms the medial boundary of the femoral ring
loop—the looped portion of the renal tubule
spine—suprameatal spine
Herophili (Herophilus), torcular—confluence of sinuses
Hesselbach's fascia—cribriform fascia
ligament—interfoveolar ligament; a thickening of transversalis fascia (and possibly of the extraperitoneal tissue around the inferior epigastric vessels) extending from the inner edge of the deep inguinal ring, upward along the inferior epigastric vessels, toward the arcuate line
triangle—inguinal triangle
Heubner, artery of—a recurrent branch of the anterior cerebral artery
Hey's ligament—falciform margin of the fascia lata at the saphenous opening
Highmore, antrum of—maxillary sinus
body of—mediastinum testis
Hilton's line—the white line in the anal canal
His, bundle of—atrioventricular bundle
Horner's muscle—lacrimal part of orbicularis oculi
Houston's fold or **valve**—the middle one of three transverse rectal folds
Humphrey, ligament of—anterior meniscofemoral ligament
Hunter's canal—adductor canal
Huschke, foramen of—gap in developing tympanic ring
Hyrtl, porus of—pterygoalar foramen

Jackson's membrane—a peritoneal fold or adhesion between the cecum or ascending colon and the right abdominal wall
Jacobson's nerve—tympanic nerve from glossopharyngeal
organ—vomeronasal organ

Keith and Flack, node of—sinuatrial node
Kent's bundle—atrioventricular bundle
Kerckring's valves or **valvules**—circular folds of intestine
Kiesselbach's area—site in nose of junction between septal branches of superior labial and sphenopalatine arteries
Koch's node—sinuatrial node
Krause, glands of—accessory lacrimal glands near superior fornix of conjunctiva
Kupffer's cells—phagocytic stellate cells lining the sinusoids of the liver

Labbé, vein of—inferior anastomotic vein of brain
Langer's lines—cleavage lines of skin
Langerhans, islets of—pancreatic islets
Langley's ganglion—portions of submandibular ganglion in submandibular gland
Leydig's cells—interstitial cells of the testis
Lieberkühn's glands, crypts, or **follicles**—intestinal glands
Lieutaud's trigone—trigone of the urinary bladder
Lisfranc's joint—the tarsometatarsal joints
ligament—interosseous ligament between second metatarsal and medial cuneiform bone
tubercle—scalene tubercle on the first rib
Lister's tubercle—dorsal tubercle of radius
Listing's plane—equatorial plane of eye
Littre, glands of—urethral glands
Lockwood's ligament—sling for eyeball formed by muscular sheaths
Louis, angle of—sternal angle
Lower's tubercle—intervenous tubercle
Ludwig's ganglion—a ganglion associated with the cardiac plexus
Luschka, foramen of—lateral aperture of fourth ventricle
glomus or **glands of**—the glomus coccygeum
nerve of—(1) posterior ethmoidal nerve; (2) sometimes used to refer to the sinuvertebral nerve
tonsil of—nasopharyngeal tonsil

McBurney's point—the reputed site of maximal tenderness in appendicitis, "between 1½ and 2 inches from the right anterior superior iliac spine upon a line to the umbilicus"
Macewen's triangle—suprameatal triangle
Mackenrodt's ligament—lateral (transverse) cervical or cardinal ligament of the uterus
Magendie, foramen of—median aperture of fourth ventricle
Maier, sinus of—common channel into which lacrimal canaliculi open
Maissiat, bandelette of—iliotibial tract
Malpighian (Malpighi) canal—longitudinal duct of the epoöphoron
capsule—splenic capsule
corpuscles or **bodies**—splenic corpuscles
layer—germinative zone of epidermis
Marcille's triangle—triangle bounded by the medial margin of the psoas major, the lateral margin of the vertebral column, and below by the iliolumbar ligament; it contains the obturator nerve

Marshall's fold—fold of the left vena cava
vein—oblique vein of the left atrium
Mayo's vein—the prepyloric vein
Meckel's diverticulum—diverticulum ilei
Meckel's cave—cavum trigeminale
ganglion—pterygopalatine ganglion
Meibomian (Meibom) glands—tarsal glands
Meissner's corpuscles—specialized sensory nerve endings in skin
plexus—submucous plexus
Mercier's bar—interureteric ridge
Merkel's corpuscles or **discs**—one form of sensory nerve ending, found chiefly in skin
Moll, glands of—sudoriferous ciliary glands
Monro, foramen of—interventricular foramen of brain
Montgomery's tubercles or **glands**—enlarged sebaceous glands projecting from the surface of the areola of the nipple
Morgagni, columns of—anal columns
foramen of—(1) foramen caecum of tongue; (2) sternocostal triangle or foramen
hydatid of—appendix testis
lacunae of—urethral lacunae
sinus of—(1) interval between superior constrictor and base of skull; (2) ventricle of larynx
Müller's fibers— radial fibers in retina
muscle— (1) tarsal or palpebral muscle; (2) orbital muscle; (3) circular fibers of ciliary muscle

Nélaton's line—a projected line extending from the anterior superior iliac spine to the tuber of the ischium
Nissl bodies, granules, or **substance**—cytoplasmic chromidial substance of neurons (rough endoplasmic reticulum)
Nuck, canal of—patent processus vaginalis peritonei in the female
Nuhn, gland of—anterior lingual glands

O'Beirne's sphincter—circular muscle fibers at junction of sigmoid colon and rectum
Oddi, sphincter of—sphincteric muscle fibers around the termination of the bile duct

Pacchionian (Pacchioni) bodies—arachnoid granulations
Pacinian (Pacini) corpuscles or **bodies**—lamellated corpuscles
Passavant's ridge or **bar**—pharyngeal ridge
Pawlik's triangle—an area on the anterior wall of the vagina in contact with the base of the bladder and distinguished by the absence of vaginal rugae
Pecquet's cisterna—cisterna chyli
Petit's ligaments—uterosacral ligaments
Petit, spaces of—zonular spaces
Petit's triangle—a "triangle of lumbar hernia" between the crest of the ilium and the margins of the external oblique and latissimus dorsi muscles
Peyer's nodules—solitary lymphatic follicles
patches—aggregated lymphatic follicles in the ileum
Poupart's ligament—inguinal ligament
Prussak's space or **pouch**—part of epitympanic recess between flaccid part of tympanic membrane and neck of malleus
Purkinje fibers—cardiac muscle fibers of the conduction system, located beneath the endocardium

Ranvier, nodes of—interruptions of the myelin sheaths of nerve fibers
Reil, island of—insula of cerebral hemisphere
Reisseisen's muscle—the smooth muscle fibers of the smallest bronchi
Reissner's membrane—vestibular membrane
Remak's fibers—nonmyelinated nerve fibers
ganglion—autonomic ganglion
Retzius, cave of—retropubic (prevesical) space
veins of—retroperitoneal veins
Riolan's anastomosis—intermesenteric arterial communication between the superior and inferior mesenteric arteries
arc or **arcade**—(1) intermesenteric arterial communication between the superior and inferior mesenteric arteries; (2) the part of the marginal artery connecting the middle and left colic arteries; (3) the mesocolon; (4) the arch of the mesocolon
muscle—(1) ciliary bundle of palpebral part of orbicularis oculi; (2) cremaster muscle
Rivinus, ducts of—smaller ducts of sublingual gland
notch of—gap in tympanic ring
Robert, ligament of—posterior meniscofemoral ligament
Rolando, fissure of—central sulcus of cerebral hemisphere
Rosenmüller's fossa—pharyngeal recess
organ—epoöphoron
Rosenthal, vein of—basal vein of brain
Ruffini's bodies or **corpuscles**—specialized sensory nerve endings, found chiefly in deep tissues

Santorini, cartilage of—corniculate cartilage
caruncula of—orifice of the accessory pancreatic duct into the duodenum
duct of—accessory pancreatic duct

Sappey's plexus—plexus of lymphatics in the areolar area of the breast
  veins—venous plexus in the falciform ligament of the liver
Sattler's layer—vascular lamina of choroid
Scarpa's canals—lesser incisive canals
  fascia—membranous layer of the subcutaneous tissue of the abdomen
  ganglion—vestibular ganglion
  nerve—nasopalatine nerve
  triangle—femoral triangle
Schlemm's canal—sinus venosus sclerae
Schneiderian (Schneider) membrane—nasal mucous membrane
Schwalbe's pocket—depression between the tendinous arch of the levator ani and the lateral wall of the pelvis
  ring—anterior border ring of cornea
Schwann, sheath of—neurilemma
Sertoli's cells—sustentacular cells of the testis
Sharpey's fibers—connective tissue fibers penetrating bone from periosteum and tendon
Shenton's line—a continuous curved line seen radiographically and formed by the margin of the obturator foramen (superior ramus of the pubis) and the neck of the femur
Shrapnell's membrane—flaccid part of tympanic membrane
Sibson's fascia—suprapleural membrane
  muscle—scalenus minimus
Skene's tubules or **glands**—the paraurethral glands of the female
Spieghel's line—semilunar line of the muscles of the abdominal wall
Spigelian (Spieghel) lobe—caudate lobe of the liver
Stensen's canals—greater incisive canals
  duct—parotid duct
Stilling's canal—hyaloid canal
Stroud's pecten or **pectinated area**—the pecten of the anal canal
Sudeck's critical point—the site on the large intestine where the lowest sigmoid artery anastomoses with the superior rectal artery
Sylvius, aqueduct of—aqueduct in midbrain
  fissure of—lateral sulcus of cerebral hemisphere
  iter of—aqueduct in midbrain

Tawara, node of—atrioventricular node
Tenon's capsule—fascia bulbi
Thebesian (Thebesius) foramina (foramina thebesii)—openings of the venae cordis minimae
  valve—valve of the coronary sinus
  veins—venae cordis minimae
Toldt's fascia—fixation of fascial planes behind the body of the pancreas
Traube's space—the semilunar area on the chest wall over which the stomach is tympanitic on percussion
Treitz, fascia of—fascia behind the head of the pancreas
  muscle or **ligament of**—suspensory muscle of the duodenum
Treves, bloodless fold of—ileocecal fold
Trolard, vein of—superior anastomotic vein of brain

Valsalva, sinuses of—the sinuses of the aorta
Varolii (Varolius), pons—pons of brainstem
Vater, ampulla of—hepatopancreatic ampulla
  -Pacinian corpuscles—lamellated corpuscles
  tubercle of—greater duodenal papilla
Verga's ventricle (cavum vergae)—posterior extension of cavity of septum pellucidum
Vesalius, foramen of (foramen vesalii)—sphenoidal emissary foramen
  Os Vesalii or vesalianum—a separate tuberosity of the base of the fifth metatarsal bone
Vidian (Guido Guidi or **Vidus Vidius) nerve**—nerve of pterygoid canal
Vieussens, annulus of—ansa subclavia
Virchow-Robin spaces—perivascular spaces in brain and spinal cord

Waldeyer's organ—paradidymis
  ring—lymphatic ring of the pharynx
Weber's point—a point near the promontory of the sacrum that is the center of gravity of the body
Weitbrecht's fibers or **retinaculum**—retinacular fibers of the neck of the femur
  foramen ovale—a gap in the capsule of the shoulder joint between the glenohumeral ligaments
  ligament—oblique cord of the proximal radioulnar joint
Wharton's duct—submandibular duct
Wilkie's artery—supraduodenal artery
Willis, circle of—circulus arteriosus
Winslow, foramen of—epiploic foramen
  ligament of—oblique popliteal ligament
  pancreas of—uncinate process of the pancreas
Wirsung, duct of—pancreatic duct
Wolfring, glands of—accessory lacrimal glands
Wood's muscle—abductor ossis metatarsi quinti
Wormian (Worm) bones—sutural bones
Wrisberg, cartilage of—cuneiform cartilage
  ganglion of—cardiac ganglion
  ligament of—posterior meniscofemoral ligament
  nerve of—(1) medial brachial cutaneous nerve; (2) nervus intermedius of facial

Zeis, glands of—sebaceous ciliary glands
Zinn, annulus of—common tendinous ring of orbit
  zonule of—ciliary zonule
Zuckerkandl, bodies of—paired para-aortic bodies near the origin of the inferior mesenteric artery

INDEX

Items in this index are listed, for the most part, only under the nouns rather than under the descriptive adjectives; for example, the brachialis muscle will be found not under brachialis but under the heading Muscle. In this edition, also, subentries beginning with prepositions are listed under the prepositions; for example, under the heading Arch, the subheading "of aorta" will be found at "o" rather than at "a."

With few exceptions, eponyms have not been used in the text and are not listed in the index. They are given in the Glossary of Eponymous Terms (p. 787).

TABLE OF MEASUREMENTS*

ORGAN OR STRUCTURE	LINEAR MEASUREMENTS IN CM	WEIGHT IN GRAMS
Appendix, vermiform	9 (3–13)	
Bladder, gall	7–10 × 3	
Body, pineal	0.5–1 × 0.5 × 0.4	0.1–0.18
Brain		*Ca.* 1240 (*Ca.* 1000–1565) F. *Ca.* 1375 (*Ca.* 1100–1685) M.
Bronchus, left main **right main**	5 2½	
Canal, anal **inguinal**	3–4 4 (3–5)	
Colon	Ca. 150	
Column, vertebral	⅖ of body height	
Cord, spinal	Av. 45	
Duct, bile **cystic** **nasolacrimal** **parotid** **submandibular** **thoracic**	4–7½ 3–4 2 5 5 Av. 45	
Ductus deferens	45	
Duodenum	25–30	
Embryo at 8 weeks	3 C.R.	2–2.7
Epiphysis cerebri	0.5–1 × 0.5 × 0.4	0.1–0.18
Esophagus	25–30	
Eye	2.35 × 2.35 × 2.4	
Filum terminale	15–20	
Gland, parotid **pituitary** **submandibular** **suprarenal** **thyroid**	 2 × 1.5 × 0.5 6 × 3½ × 2	20–30 0.4–0.8 10–20 3–6 40 (20–70)
Heart		250 (198–279) F. 300 (256–390) M.
Hypophysis	2 × 1.5 × 0.5	0.4–0.8
Intestine, large **small**	*Ca.* 150 500–800	
Kidney	11–13 × 5–6 × 3–4	145 (120–175) F. 155 (115–220) M.
Liver	Variable	1020–2120 F. 1200–3020 M.
Lung, left **right**		375 (325–480) 450 (360–570)
Meatus, external acoustic **internal acoustic**	2½–3 1	
Membrane, tympanic	1	
Newborn	*Ca.* 33.6 C.R.	*Ca.* 3350 (2500–4000)
Orbit	3½ × 4	
Ovary	4 × 2½ × 1	*Ca.* 7
Pancreas	23 × 4½ × 4	110 (60–135)
Pharynx	12	
Placenta	15–20 × 3	*Ca.* 500
Prostate	3 × 3½ × 2	5–20
Rectum	12–15	
Spleen	Variable	55–400
Testis	4–5 × 2½ × 3	25 (20–27)
Thymus	Variable	20–40 (*aet.* 6–35)
Trachea **lumen in adult** **lumen in newborn**	9–15 1.2 0.1–0.7	
Tube, auditory **uterine**	3–4 10	
Ureter	25–30	
Urethra, female **male** { **prostatic** { **membranous** { **spongy**	4 20 3 1–2 15	
Uterus, multiparous **nulliparous**	9 × 6 × 3½ 8 × 4 × 2	110 (100–120) 35 (30–40)
Vagina, anterior wall **posterior wall**	7.5 9	
Vesicle, seminal	4.5 × 1.7 × 0.9	

*Adapted from various sources, including F. W. Sunderman and F. Boerner, *Normal Values in Clinical Medicine*, Saunders, Philadelphia, 1949, and J. Ludwig, *Current Methods of Autopsy Practice*, Saunders, Philadelphia, 1972. F.: female. M.: male. Difficulties in selecting normal organ weights necessitate that considerable caution be exercized before assuming that a figure found is abnormal.

MEDIAN TIMES OF APPEARANCE OF POSTNATAL OSSIFICATION CENTERS IN THE UPPER LIMB*

Bone	Postnatal Centers	50 Per Cent Appear	Fusion Complete Radiographically
Clavicle	Medial end	Adolescence	3rd decade
Scapula	Chief coracoid	*Ca.* birth	Adolescence or later
	Base of coracoid	Puberty	
	Acromion (2), medial border, inferior angle	Puberty or adolescence	
Humerus	Head	*Ca.* birth	20
	Greater tubercle	½ F. 1 M.	
	Lesser tubercle	3	
	Lateral epicondyle	9 F. 11 M.	13 F. 15 M.
	Capitulum, lat. part of trochlea	⅓	
	Medial part of trochlea	9 F. 10 M.	
	Medial epicondyle	3½ F. 6 M.	14 F. 16 M.
Radius	Proximal end	4 F. 5 M.	13 F. 16 M.
	Distal end	1	16 F. 18 M.
Ulna	Proximal end	8 F. 10 M.	13 F. 15 M.
	Distal end	5½ F. 7 M.	16 F. 18 M.
Capitate and hamate		⅓	
Triquetral		2 F. 2½ M.	
Lunate		2½ F. 4 M.	
Trapezium, trapezoid, scaphoid		4 F. 6 M.	
Pisiform		8 F. 10 M.	
Metacarpals	Heads (2–5) or base (1)	1–2½	14 F. 17 M.
Phalanges	Bases	1–3	14 F. 17 M.
Sesamoids		11 F. 13 M.	

* Ages, where different, are specified in years for female (F.) and male (M.).

(See back of this page for Table of Measurements.)